Dissection Experience

Customizable images enliven presentations and quizzes for lecture or lab.

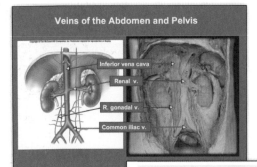

Veins of the Abdomen and Pelvis

Inferior vena cava
Renal v.
R. gonadal v.
Common iliac v.

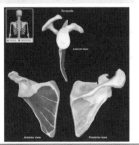

6. **Which muscle's origin is highlighted in red?**

a. Infraspinatus
b. *Subscapularis*
c. Teres minor
d. Teres major
e. Supraspinatus

Bonus: Which of the above muscles is not part of the rotator cuff?

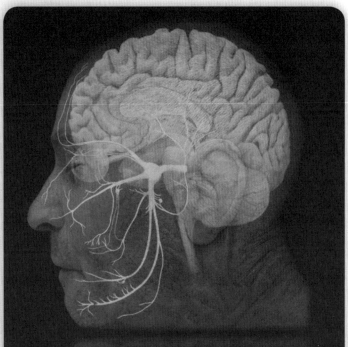

Layered cadaver dissections not available anywhere else!

www.aprevealed.com

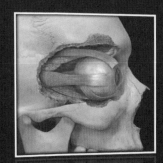
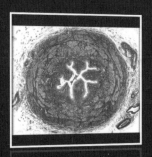
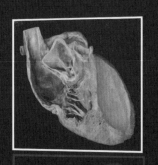
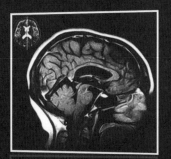

Portable cadavers can replace or enhance the laboratory experience.

Imaging
Correlate dissected anatomy with radiologic images like X-ray, MRI, and CT scans.

Self-Test
Gauge proficiency with quizzes and simulated practical exams.

Search
Type in a term to quickly locate any structure in the program.

McGraw Hill

Vander's
Human Physiology

ELEVENTH EDITION

Vander's Human Physiology

THE MECHANISMS OF BODY FUNCTION

Eric P. Widmaier
BOSTON UNIVERSITY

Hershel Raff
MEDICAL COLLEGE OF WISCONSIN
AURORA ST. LUKE'S MEDICAL CENTER

Kevin T. Strang
UNIVERSITY OF WISCONSIN–MADISON

McGraw-Hill
Higher Education

Boston   Burr Ridge, IL   Dubuque, IA   New York   San Francisco   St. Louis
Bangkok   Bogotá   Caracas   Kuala Lumpur   Lisbon   London   Madrid   Mexico City
Milan   Montreal   New Delhi   Santiago   Seoul   Singapore   Sydney   Taipei   Toronto

McGraw-Hill
Higher Education

VANDER'S HUMAN PHYSIOLOGY: THE MECHANISMS OF BODY FUNCTION, ELEVENTH EDITION

Published by McGraw-Hill, a business unit of The McGraw-Hill Companies, Inc., 1221 Avenue of the Americas, New York, NY 10020. Copyright © 2008 by The McGraw-Hill Companies, Inc. All rights reserved. Previous editions © 2006, 2004, 2001, 1998, 1994, 1990, 1985, 1980 and 1975. No part of this publication may be reproduced or distributed in any form or by any means, or stored in a database or retrieval system, without the prior written consent of The McGraw-Hill Companies, Inc., including, but not limited to, in any network or other electronic storage or transmission, or broadcast for distance learning.

Some ancillaries, including electronic and print components, may not be available to customers outside the United States.

⊕ This book is printed on recycled, acid-free paper containing 10% postconsumer waste.

2 3 4 5 6 7 8 9 0 DOW/DOW 0 9 8

ISBN 978–0–07–304962–5
MHID 0–07–304962–X

Publisher: *Michelle Watnick*
Executive Editor: *Colin H. Wheatley*
Developmental Editor: *Fran Schreiber*
Marketing Manager: *Lynn M. Breithaupt*
Senior Project Manager: *Jayne Klein*
Senior Production Supervisor: *Sherry L. Kane*
Senior Media Project Manager: *Tammy Juran*
Lead Media Producer: *John J. Theobald*
Senior Designer: *David W. Hash*
Cover/Interior Designer: *Christopher Reese*
Senior Photo Research Coordinator: *John C. Leland*
Photo Research: *LouAnn K. Wilson*
Supplement Producer: *Mary Jane Lampe*
Compositor: *Precision Graphics*
Typeface: *10/12 Galliard*
Printer: *R. R. Donnelley Willard, OH*
(USE) Cover Image: *Alveoli in the lung, the air sacs where the exchange of oxygen and carbon dioxide take place,* ©*Dr. David M. Phillips/Getty Images*

The credits section for this book begins on page 749 and is considered an extension of the copyright page.

Library of Congress Cataloging-in-Publication Data

Widmaier, Eric P.
  Vander's human physiology : the mechanisms of body function / Eric P. Widmaier, Hershel Raff, Kevin T. Strang. – 11th ed.
      p. cm.
  Includes bibliographical references and index.
  ISBN 978–0–07–304962–5 — ISBN 0–07–304962–X (hard copy : alk. paper)
  1. Human physiology—Textbooks. I. Raff, Hershel, 1953- II. Strang, Kevin T. III. Title.
  QP34.5.W47 2008
  612–dc22

                                                              2007028697

Dedication

To our wives

and children:

Maria, Ricky, and Carrie

Judy and Jonathan

Jake and Amy, and in

loving memory of LeeAnn

Meet the Authors

ERIC P. WIDMAIER received his Ph.D. in 1984 in Endocrinology from the *University of California at San Francisco*. His postdoctoral training was in endocrinology and physiology at the *Worcester Foundation for Experimental Biology*, and *The Salk Institute* in La Jolla, California. His research is focused on the control of body mass and metabolism in mammals, the mechanisms of hormone action, and the postnatal development of adrenal gland function. He is currently Professor of Biology at *Boston University*, where he teaches Systems Physiology and Comparative Physiology, and has been recognized with the Gitner Award for Distinguished Teaching by the College of Arts and Sciences, and the Metcalf Prize for Excellence in Teaching by Boston University. He is the author of numerous scientific and lay publications, including books about physiology for the general reader. He lives outside Boston with his wife, Maria, and children, Carrie and Ricky.

HERSHEL RAFF received his Ph.D. in Environmental Physiology from the *Johns Hopkins University* in 1981 and did postdoctoral training in Endocrinology at the *University of California at San Francisco*. He is now a Professor of Medicine (Endocrinology, Metabolism and Clinical Nutrition) and Physiology at the *Medical College of Wisconsin* and Director of the Endocrine Research Laboratory at *Aurora St. Luke's Medical Center*. At the *Medical College of Wisconsin*, he teaches systems physiology, neuroendocrinology, and endocrine pharmacology to medical and graduate students. He was an inaugural inductee into the Society of Teaching Scholars, and he has received the Beckman Basic Science Teaching Award from the Senior Class and the Outstanding Teacher Award from the Graduate Student Association. He is also an Adjunct Professor of Biomedical Sciences at *Marquette University*. He recently completed terms as Secretary-Treasurer of The Endocrine Society and as Associate Editor of Advances in Physiology Education. Dr. Raff's basic research focuses on the effects of low oxygen (hypoxia) at the organismal, cellular, and molecular levels. His clinical interest focuses on pituitary and adrenal diseases, with a special focus on Cushing's syndrome. His hobby is playing the piano and guitar. He resides outside Milwaukee with his wife, Judy, and son, Jonathan.

KEVIN T. STRANG received his Master's degree in Zoology (1988) and his Ph.D. in Physiology (1994) from the *University of Wisconsin at Madison*. His research area is cellular mechanisms of contractility modulation in cardiac muscle. He teaches a large undergraduate systems physiology course as well as first-year medical physiology in the *UW-Madison School of Medicine and Public Health*. He was elected to UW-Madison's Teaching Academy and serves on the steering commitee of the Institute for Cross-college Biology Education (ICBE). Teaching awards include the UW Medical Alumni Association's Distinguished Teaching Award for Basic Sciences, and the University of Wisconsin System's Underkofler/Alliant Energy Excellence in Teaching Award. Interested in teaching technology, Dr. Strang has created an interactive CD-ROM tutorial called "Anatomy of a Heart Attack," and has produced numerous animations for teaching physiology. He lives in Madison with his children, Jake and Amy.

Brief Contents

Table of Contents

CHAPTER 4

Movement of Molecules Across Cell Membranes   96

CHAPTER 5

Control of Cells by Chemical Messengers   120

CHAPTER 6

Neuronal Signaling and the Structure of the Nervous System   137

SECTION A   *Neural Tissue*

SECTION B   *Membrane Potentials*

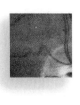

CHAPTER 7
Sensory Physiology 191

CHAPTER 8
Consciousness, the Brain, and Behavior 232

CHAPTER **9**

Muscle 254

CHAPTER **10**

Control of Body Movement 296

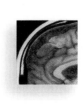

CHAPTER **11**

The Endocrine System 315

CHAPTER 12

Cardiovascular Physiology   359

CHAPTER **13**

Respiratory Physiology   442

CHAPTER

17

Reproduction   599

CHAPTER **18**

Defense Mechanisms of the Body 646

CHAPTER **19**

Medical Physiology: Integration Using Clinical Cases 683

Preface

From the Authors

We are very pleased to launch the 11th edition of *Vander's Human Physiology*. The current authors have attempted to maintain the highest standards of excellence, accuracy, and pedagogy developed by Arthur Vander, James Sherman, and Dorothy Luciano over the many years in which they educated countless thousands of students worldwide with this textbook. At the same time, we have been very attuned to the evolving needs of instructors and students in physiology, particularly those interested in a career in the health sciences. Thus, in addition to the usual updates of scientific material reflecting recent advances in physiology, this edition builds on the pedagogy that was expanded in the 10th edition. A new feature, called *Physiological Inquiries*, has been added to each chapter beginning with Chapter 4. These inquiries are associated with key figures throughout the chapters, encouraging students to stop and think about the broader implications of what they have just learned. In some cases, this may entail quantitative analyses, while in other cases it may involve understanding the material in an evolutionary context. We think that such exercises will further encourage students to think about what they are learning in new and more profound ways.

Similarly, a new chapter has been added to the end of the textbook (Chapter 19), called *Medical Physiology: Integration Using Clinical Cases*. Three case studies, adapted from real-life scenarios, are presented to the student in a way that requires the student to think critically and apply what has been learned throughout the semester to novel clinical situations. Along the way, students are asked to *Reflect and Review* the material as the case unfolds, providing them with a step-by-step interactive learning experience. We hope that users of the book will agree that the increased emphasis on pedagogy has enhanced the utility of the textbook as a learning tool.

We are as always deeply grateful for the many helpful insights, suggestions, and reviews from colleagues and students around the world. We remain indebted to Drs. Vander, Sherman, and Luciano for their trust and guidance, and to the wonderful staff at McGraw-Hill Higher Education for their support and professionalism.

New Case Study Chapter!

Guided Tour Through a Chapter

Colorized scanning
electron micrograph
(SEM) of freeze-
fractured muscle fibers.

The ability to use chemical energy
to produce force and movement
is present to a limited extent in
most cells, but in muscle cells it has
become dominant. Muscles generate
force and movements used to regulate
the internal environment, and they
also produce movements in the
external environment. In humans, the
ability to communicate, whether by
speech, writing, or artistic expression,
also depends on muscle contractions.
Indeed, it is only by controlling
muscle activity that the human mind
ultimately expresses itself.

Three types of muscle tissue can be
identified on the basis of structure,
contractile properties, and control
mechanisms—skeletal muscle, smooth
muscle, and cardiac muscle. Most
skeletal muscle, as the name implies, is
attached to bone, and its contraction is
responsible for supporting and moving
the skeleton. The contraction of
skeletal muscle is initiated by impulses
in the neurons to the muscle and is
usually under voluntary control.

Sheets of smooth muscle surround
various hollow organs and tubes.

Chapter Outline
Every chapter starts with an outline giving the reader a brief view of what is to be covered in that chapter.

Chapter Overview
Every chapter starts with a brief chapter overview.

ADDITIONAL CLINICAL EXAMPLES

A number of diseases can affect the contraction of skeletal muscle. Many of them are caused by defects in the parts of the nervous system that control contraction of the muscle fibers rather than by defects in the muscle fibers themselves. For example, *poliomyelitis* is a viral disease that destroys motor neurons, leading to the paralysis of skeletal muscle, and may result in death due to respiratory failure.

Muscle Cramps

Involuntary tetanic contraction of skeletal muscles produces *muscle cramps*. During cramping, action potentials fire at abnormally high rates, a much greater rate than occurs during maximal voluntary contraction. The specific cause of this high activity is uncertain, but it is probably related to electrolyte imbalances in the extracellular fluid surrounding both the muscle and nerve fibers. These imbalances may arise from overexercise or persistent dehydration, and they can directly induce action potentials in motor neurons and muscle fibers. Another theory is that chemical imbalances within the muscle stimulate sensory receptors in the muscle, and the motor neurons to the area are activated by reflex when those signals reach the spinal cord.

Hypocalcemic Tetany

Hypocalcemic tetany is the involuntary tetanic contraction of skeletal muscles that occurs when the extracellular calcium concentration falls to about 40 percent of its normal value. This may seem surprising, because we have seen that calcium is required for excitation-contraction coupling. However, recall that this calcium is sarcoplasmic reticulum calcium, not extracellular calcium. The effect of changes in extracellular calcium is exerted not on the sarcoplasmic reticulum calcium, but directly on the plasma membrane. Low extracellular calcium (**hypocalcemia**) increases the opening of sodium channels in excitable membranes, leading to membrane depolarization and the spontaneous firing of action potentials. This causes the increased muscle contractions, which are similar to muscular cramping. Chapter 11 discusses the mechanisms controlling the extracellular concentration of calcium ions.

Muscular Dystrophy

This disease is one of the most frequently encountered genetic diseases, affecting one in every 3500 males (but many fewer females) born in America. *Muscular dystrophy* is associated with the progressive degeneration of skeletal and cardiac muscle fibers, weakening the muscles and leading ultimately to death from respiratory or cardiac failure (Figure 9–31). The symptoms become evident at about 2 to 6 years of age, and most affected individuals do not survive far beyond the age of 20.
  The recessive gene responsible for a major form of muscular dystrophy (**Duchenne muscular dystrophy**) has

been identified on the X chromosome, and Duchenne muscular dystrophy is thus a sex-linked recessive disease. (As described in Chapter 17, girls have two X chromosomes and boys only one. Consequently, a girl with one abnormal X chromosome and one normal one will not generally develop the disease. This is why the disease is so much more common in boys.) This gene codes for a protein known as dystrophin, which is either present in a nonfunctional form or absent in patients with the disease. Dystrophin is a large protein that links cytoskeletal proteins to membrane glycoproteins. It resembles other known cytoskeletal proteins and may be involved in maintaining the structural integrity of the plasma membrane or of elements within the membrane, such as ion channels. In its absence, fibers subjected to repeated structural deformation during contraction are susceptible to membrane rupture and cell death. Preliminary attempts are being made to treat the disease by inserting the normal gene into dystrophic muscle cells.

Myasthenia Gravis

Myasthenia gravis is a collection of neuromuscular disorders characterized by muscle fatigue and weakness that progressively worsens as the muscle is used. It affects about one out of every 7500 Americans, occurring more often in women than men. The most common cause is the destruction of nicotinic ACh receptor proteins of the motor end plate, mediated by antibodies of a person's own immune system (see Chapter 18 for a description of autoimmune diseases). The release of ACh from the nerve terminals is normal, but the magnitude of the end-plate potential is markedly reduced because of the decreased availability of receptors. Even in normal muscle, the amount of ACh released with each action potential decreases with repetitive activity, and thus the magnitude of the resulting EPP falls. In normal muscle, however, the EPP remains well above the threshold necessary to initiate a muscle action potential. In contrast, after a few motor nerve impulses in a myasthenia gravis patient, the magnitude of the EPP falls below the threshold for initiating a muscle action potential.
  A number of approaches are currently used to treat the disease. One is to administer acetylcholinesterase inhibitors (e.g., *neostygmine*). This can partially compensate for the reduction in available ACh receptors by prolonging the time that acetylcholine is available at the synapse. Other therapies aim at blunting the immune response. Treatment with glucocorticoids is one way that immune function is suppressed (Chapter 11). Removal of the thymus gland (*thymectomy*) reduces the production of antibodies and reverses symptoms in about 50 percent of patients. *Plasmapheresis* is a treatment that involves removing the liquid fraction of blood (plasma), which contains the offending antibodies. A combination of these treatments has greatly reduced the mortality rate for myasthenia gravis.■

Additional Clinical Examples
The authors have drawn from their teaching experiences to provide students with real-life applications through clinical applications.

Summary Tables
Summary tables are used to bring together large amounts of information that may be scattered throughout the book or to summarize small or moderate amounts of information. The tables complement the accompanying figures to provide a rapid means of reviewing the most important material in the chapter.

Table 9–3	Characteristics of the Three Types of Skeletal Muscle Fibers		
	Slow-Oxidative Fibers (Type I)	**Fast-Oxidative-Glycolytic Fibers (Type IIa)**	**Fast-Glycolytic Fibers (Type IIb)***
Primary source of ATP production	Oxidative phosphorylation	Oxidative phosphorylation	Glycolysis
Mitochondria	Many	Many	Few
Capillaries	Many	Many	Few
Myoglobin content	High (red muscle)	High (red muscle)	Low (white muscle)
Glycolytic enzyme activity	Low	Intermediate	High
Glycogen content	Low	Intermediate	High
Rate of fatigue	Slow	Intermediate	Fast
Myosin-ATPase activity	Low	High	High
Contraction velocity	Slow	Fast	Fast
Fiber diameter	Small	Intermediate	Large
Motor unit size	Small	Intermediate	Large
Size of motor neuron innervating fiber	Small	Intermediate	Large

*Type IIb fibers are sometimes designated as type IIx in the human muscle physiology literature.

Physiological Inquiries–*NEW!*

You will now find critical-thinking and quantitative questions based on many figures found in most chapters. These questions are designed to help students become more engaged to learn a concept or process described in the art. These questions challenge a student to analyze the content of the figure.

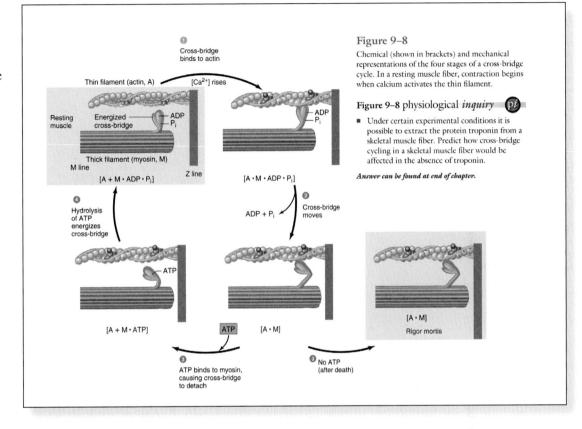

Figure 9–8

Chemical (shown in brackets) and mechanical representations of the four stages of a cross-bridge cycle. In a resting muscle fiber, contraction begins when calcium activates the thin filament.

Figure 9–8 physiological *inquiry*

- Under certain experimental conditions it is possible to extract the protein troponin from a skeletal muscle fiber. Predict how cross-bridge cycling in a skeletal muscle fiber would be affected in the absence of troponin.

Answer can be found at end of chapter.

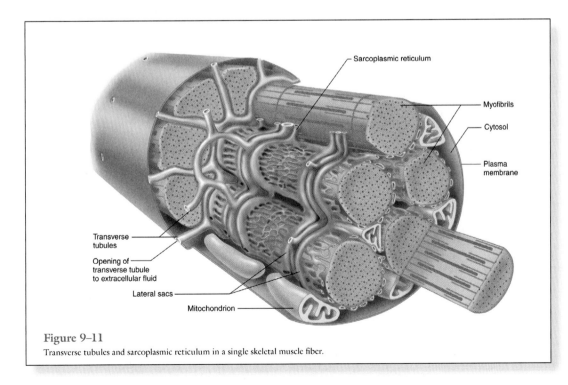

Figure 9–11

Transverse tubules and sarcoplasmic reticulum in a single skeletal muscle fiber.

Descriptive Art Style

A realistic three-dimensional perspective is included in many of the figures for greater clarity and understanding of concepts presented.

Guided Tour Through a Chapter

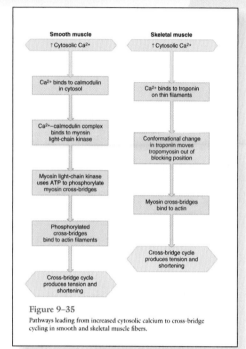

Figure 9–35

Pathways leading from increased cytosolic calcium to cross-bridge cycling in smooth and skeletal muscle fibers.

Flow Diagrams

Long a hallmark of this book, extensive use of flow diagrams is continued in this edition. They have been updated to assist in learning.

Key to Flow Diagrams

- The beginning boxes of the diagrams are color coded green.
- Other boxes are consistently color coded throughout the book.
- Structures are always shown in three-dimensional form.

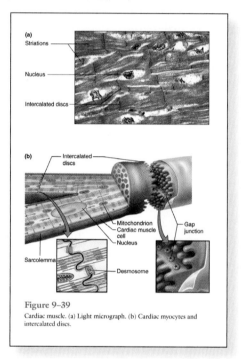

Figure 9–39

Cardiac muscle. (a) Light micrograph. (b) Cardiac myocytes and intercalated discs.

Multi-Level Perspective

Illustrations depicting complex structures or processes combine macroscopic and microscopic views to help students see the relationships between increasingly detailed drawings.

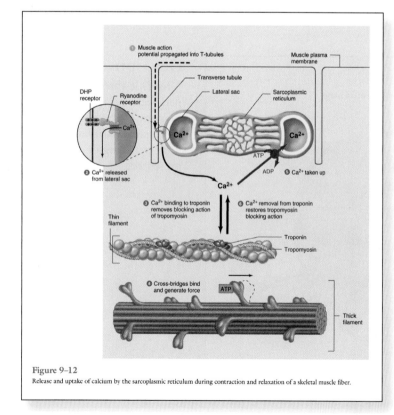

Figure 9–12

Release and uptake of calcium by the sarcoplasmic reticulum during contraction and relaxation of a skeletal muscle fiber.

Uniform Color-Coded Illustrations

Color coding is effectively used to promote learning. For example, there are specific colors for extracellular fluid, the intracellular fluid, muscle filaments, and transporter molecules.

End of Section

At the end of sections throughout the book you will find a summary, key terms, clinical terms, and review questions.

Table 9–6 Characteristics of Muscle Cells

Characteristic	Skeletal Muscle	Smooth Muscle: Single Unit	Smooth Muscle: Multiunit	Cardiac Muscle
Thick and thin filaments	Yes	Yes	Yes	Yes
Sarcomeres—banding pattern	Yes	No	No	Yes
Transverse tubules	Yes	No	No	Yes
Sarcoplasmic reticulum (SR)*	++++	+	+	++
Gap junctions between cells	No	Yes	Few	Yes
Source of activating calcium	SR	SR and extracellular	SR and extracellular	SR and extracellular
Site of calcium regulation	Troponin	Myosin	Myosin	Troponin
Speed of contraction	Fast-slow	Very slow	Very slow	Slow
Spontaneous production of action potentials by pacemakers	No	Yes	No	Yes in certain fibers, but most not spontaneously active
Tone (low levels of maintained tension in the absence of external stimuli)	No	Yes	No	No
Effect of nerve stimulation	Excitation	Excitation or inhibition	Excitation or inhibition	Excitation or inhibition
Physiological effects of hormones on excitability and contraction	No	Yes	Yes	Yes
Stretch of cell produces contraction	No	Yes	No	No

*Number of plus signs (+) indicates the relative amount of sarcoplasmic reticulum present in a given muscle type.

■ SECTION B SUMMARY

Structure of Smooth Muscle

I. Smooth muscle cells are spindle-shaped, lack striations, have a single nucleus, and are capable of cell division. They contain actin and myosin filaments and contract by a sliding-filament mechanism.

Smooth Muscle Contraction and Its Control

I. An increase in cytosolic calcium leads to the binding of calcium by calmodulin. The calcium-calmodulin complex then binds to myosin light-chain kinase, activating the enzyme, which uses ATP to phosphorylate smooth muscle myosin. Only phosphorylated myosin can bind to actin and undergo cross-bridge cycling.

II. Smooth muscle myosin has a low rate of ATP splitting, resulting in a much slower shortening velocity than in striated muscle. However, the tension produced per unit cross-sectional area is equivalent to that of skeletal muscle.

III. Two sources of the cytosolic calcium ions that initiate smooth muscle contraction are the sarcoplasmic reticulum and extracellular calcium. The opening of calcium channels in the smooth muscle plasma membrane and sarcoplasmic reticulum, mediated by a variety of factors, allows calcium ions to enter the cytosol.

IV. The increase in cytosolic calcium resulting from most stimuli does not activate all the cross-bridges. Therefore, smooth muscle tension can be increased by agents that increase the concentration of cytosolic calcium ions.

V. Table 9–5 summarizes the types of stimuli that can initiate smooth muscle contraction by opening or closing calcium channels in the plasma membrane or sarcoplasmic reticulum.

VI. Most, but not all, smooth muscle cells can generate action potentials in their plasma membrane upon membrane depolarization. The rising phase of the smooth muscle action

292 Chapter 9

potential is due to the influx of calcium ions into the cell through voltage-gated calcium channels.

VII. Some smooth muscles generate action potentials spontaneously, in the absence of any external input, because of pacemaker potentials in the plasma membrane that repeatedly depolarize the membrane to threshold. Slow waves are a pattern of spontaneous, periodic depolarization of the membrane potential seen in some smooth muscle pacemaker cells.

VIII. Smooth muscle cells do not have a specialized end-plate region. A number of smooth muscle cells may be influenced by neurotransmitters released from the varicosities on a single nerve ending, and a single smooth muscle cell may be influenced by neurotransmitters from more than one neuron. Neurotransmitters may have either excitatory or inhibitory effects on smooth muscle contraction.

IX. Smooth muscles can be classified broadly as single-unit or multiunit smooth muscle.

Cardiac Muscle

I. Cardiac muscle combines features of skeletal and smooth muscles. Like skeletal muscle, it is striated, is composed of myofibrils with repeating sarcomeres, has troponin in its thin filaments, has T-tubules that conduct action potentials, and has sarcoplasmic reticulum lateral sacs that store calcium. Like smooth muscle, cardiac muscle cells are small and single-nucleated, arranged in layers around hollow cavities, and connected by gap junctions.

II. Cardiac muscle excitation-contraction coupling involves entry of a small amount of calcium through L-type calcium channels, which triggers opening of ryanodine receptors that release a larger amount of calcium from the sarcoplasmic reticulum. Calcium activates the thin filament and cross-bridge cycling as in skeletal muscle.

III. Cardiac contractions and action potentials are prolonged, tetany does not occur, and both the strength and frequency of contraction are modulated by autonomic neurotransmitters and hormones.

IV. Table 9–6 summarizes and compares the features of skeletal, smooth, and cardiac muscles.

■ SECTION B KEY TERMS

dense body 284
intercalated disk 290
latch state 286
L-type calcium channel 290
multiunit smooth muscle 289
myosin light-chain kinase 285
myosin light-chain phosphatase 286
pacemaker potential 287
single-unit smooth muscle 289
slow waves 287
smooth muscle tone 287
varicosity 288

■ SECTION B REVIEW QUESTIONS

1. How does the organization of thick and thin filaments in smooth muscle fibers differ from that in striated muscle fibers?
2. Compare the mechanisms by which a rise in cytosolic calcium concentration initiates contractile activity in skeletal, smooth, and cardiac muscle cells.
3. What are the two sources of calcium that lead to the increase in cytosolic calcium that triggers contraction in smooth muscle?
4. What types of stimuli can trigger a rise in cytosolic calcium in smooth muscle cells?
5. What effect does a pacemaker potential have on a smooth muscle cell?
6. In what ways does the neural control of smooth muscle activity differ from that of skeletal muscle?
7. Describe how a stimulus may lead to the contraction of a smooth muscle cell without a change in the plasma membrane potential.
8. Describe the differences between single-unit and multiunit smooth muscles.
9. Compare and contrast the physiology of cardiac muscle with that of skeletal and smooth muscles.
10. Explain why cardiac muscle cannot undergo tetanic contractions.

Chapter 9 Test Questions

(Answers appear in Appendix A.)

1. Which is a false statement about skeletal muscle structure?
   a. A myofibril is composed of multiple muscle fibers.
   b. Most skeletal muscles attach to bones by connective-tissue tendons.
   c. Each end of a thick filament is surrounded by six thin filaments.
   d. A cross-bridge is a portion of the myosin molecule.
   e. Thin filaments contain actin, tropomyosin, and troponin.
2. Which is correct regarding a skeletal muscle sarcomere?
   a. M lines are found in the center of the I band.
   b. The I band is the space between one Z line and the next.
   c. The H zone is the region where thick and thin filaments overlap.
   d. Z lines are found in the center of the A band.
   e. The width of the A band is equal to the length of a thick filament.
3. When a skeletal muscle fiber undergoes a concentric isotonic contraction:
   a. M lines remain the same distance apart.
   b. Z lines move closer to the ends of the A bands.
   c. A bands become shorter.
   d. I bands become wider.
   e. M lines move closer to the end of the A band.
4. During excitation-contraction coupling in a skeletal muscle fiber
   a. the Ca^{2+}-ATPase pumps calcium into the T-tubule.
   b. action potentials propagate along the membrane of the sarcoplasmic reticulum.
   c. calcium floods the cytosol through the dihydropyridine (DHP) receptors.
   d. DHP receptors trigger the opening of lateral sac ryanodine receptor calcium channels.
   e. acetylcholine opens the DHP receptor channel.

Muscle 293

End of Chapter

At the end of the chapters you will find:

- Test Questions that are designed to test student comprehension of key concepts.
- Quantitative and Thought Questions that challenge the student to go beyond the memorization of facts, to solve problems and to encourage thinking about the meaning or broader significance of what has just been read.

completed.
   b. ADP is rapidly converted back to ATP by creatine phosphate.
   c. Glucose is metabolized in glycolysis, producing large quantities of ATP.
   d. The mitochondria immediately begin oxidative phosphorylation.
   e. Fatty acids are rapidly converted to ATP by oxidative glycolysis.
7. Which correctly characterizes a "fast-oxidative" type of skeletal muscle fiber?
   a. few mitochondria and high glycogen content
   b. low myosin ATPase rate and few surrounding capillaries
   c. low glycolytic enzyme activity and intermediate contraction velocity
   d. high myoglobin content and intermediate glycolytic enzyme activity
   e. small fiber diameter and fast onset of fatigue
8. Which is *false* regarding the structure of smooth muscle?
   a. The thin filament does not include the regulatory protein troponin.
   b. The thick and thin filaments are not organized in sarcomeres.
   c. Thick filaments are anchored to dense bodies instead of Z lines.
   d. The cells have a single nucleus.
   e. Single-unit smooth muscles have gap junctions connecting individual cells.
9. The role of myosin light-chain kinase in smooth muscle is to
   a. bind to calcium ions to initiate excitation-contraction coupling.
   b. phosphorylate cross-bridges, thus driving them to bind with the thin filament.
   c. split ATP to provide the energy for the power stroke of the cross-bridge cycle.
   d. dephosphorylate myosin light chains of the cross-bridge, thus relaxing the muscle.
   e. pump calcium from the cytosol back into the sarcoplasmic reticulum.
10. Single-unit smooth muscle differs from multiunit smooth muscle because
   a. single-unit muscle contraction speed is slow, while multiunit is fast.
   b. single-unit muscle has T-tubules, multiunit muscle does not.
   c. single-unit muscles are not innervated by autonomic nerves.
   d. single-unit muscle contracts when stretched, whereas multiunit muscle does not.
   e. single-unit muscle does not produce action potentials spontaneously, but multiunit muscle does.
11. Which of the following describes a similarity between cardiac and smooth muscle cells?
   a. An action potential always precedes contraction.
   b. The majority of the calcium that activates contraction comes from the extracellular fluid.
   c. Action potentials are generated by pacemaker potentials.
   d. An extensive system of T-tubules is present.
   e. Calcium release and contraction strength are graded.

Chapter 9 Quantitative and Thought Questions

(Answers appear in Appendix A.)

1. Which of the following corresponds to the state of myosin (M) under resting conditions and in rigor mortis? (a) M · ATP (b) M · ADP · P_i (c) A · M · ADP · P_i (d) A · M
2. If the transverse tubules of a skeletal muscle are disconnected from the plasma membrane, will action potentials trigger a contraction? Give reasons.
3. When a small load is attached to a skeletal muscle that is then tetanically stimulated, the muscle lifts the load in an isotonic contraction over a certain distance, but then stops shortening and enters a state of isometric contraction. With a heavier load, the distance shortened before entering an isometric contraction is shorter. Explain these shortening limits in terms of the length-tension relation of muscle.
4. What conditions will produce the maximum tension in a skeletal muscle fiber?
5. A skeletal muscle can often maintain a moderate level of active tension for long periods of time, even though many of its fibers become fatigued. Explain.
6. If the blood flow to a skeletal muscle were markedly decreased, which types of motor units would most rapidly undergo a severe reduction in their ability to produce ATP for muscle contraction? Why?
7. As a result of an automobile accident, 50 percent of the muscle fibers in the biceps muscle of a patient were destroyed. Ten months later, the biceps muscle was able to generate 80 percent of its original force. Describe the changes that took place in the damaged muscle that enabled it to recover.

294 Chapter 9

NEW TO THIS EDITION

- **Chapter 19** *Medical Physiology: Integration Using Clinical Cases.* Three case studies, adapted from real-life scenarios, are presented to the student in a way that requires the student to think critically and apply what has been learned throughout the semester to novel clinical situations.
- **Physiological Inquiries** These inquiries are associated with key figures throughout the chapters, encouraging students to stop and think about the broader implications of what they have just learned.

Chapter 1 Homeostasis: A Framework for Human Physiology

Introduction of the concept of dynamic constancy; *new figure* illustrating water distribution among body compartments; *new figure* illustrating changes in blood glucose levels during 24 hours, as an example of a homeostatic process.

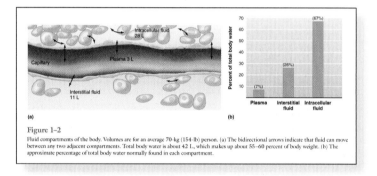

Figure 1–2
Fluid compartments of the body. Volumes are for an average 70-kg (154-lb) person. (a) The bidirectional arrows indicate that fluid can move between any two adjacent compartments. Total body water is about 42 L, which makes up about 55–60 percent of body weight. (b) The approximate percentage of total body water normally found in each compartment.

Chapter 2 Chemical Composition of the Body

Expanded discussion of free radicals and their relationship to aging; *trans* fatty acids.

Chapter 3 Cell Structure, Protein Function, and Metabolic Pathways

Expanded discussion of desmosomes, cadherins, and connexins; addition of vaults to section on cellular organelles.

Chapter 4 Movement of Molecules Across Cell Membranes

New description and illustrations of receptor-mediated endocytosis and potocytosis.

Chapter 5 Control of Cells by Chemical Messengers

Clarification and expanded description of types of receptors.

Chapter 6 Neuronal Signaling and the Structure of the Nervous System

New Additional Clinical Examples (nicotine; multiple sclerosis); updated, expanded discussion of mechanisms of neurotransmitter release; *new figure* elaborating autonomic neurotransmitters and their receptors.

Chapter 7 Sensory Physiology

Updated section on processing of visual signals, including on- and off-bipolar cell pathways fat-sensitive taste receptor; *new figure* demonstrating referred pain mechanism; *new figure* with updated explanation of photobleaching of cone pigments.

Chapter 8 Consciousness, The Brain, and Behavior

Expanded discussion of the physiological functions of sleep; ADHD and its treatment; anterograde amnesia.

Chapter 9 Muscle

New section on Cardiac Muscle; creatine supplements and their use in athletics.

Chapter 10 Control of Body Movement

Description of the use of deep-brain stimulation for Parkinson disease; *new figure* illustrating integration of spinal interneurons.

Chapter 11 The Endocrine System

Integration of endocrine control of Ca^{2+} balance into endocrine chapter; *new figure* illustrating the control of thyroid hormone secretion.

Chapter 12 Cardiovascular Physiology

New figure on autonomic neurotransmitters and receptors; *new figure* illustrating cellular mechanisms of sympathetic contractility; *new figure* illustrating the pathways of lymph flow; discussion of the use of statin drugs in treating hypercholesterolemia; cardiac cycle diagram keyed numerically to text discussion.

Chapter 13 Respiratory Physiology

Significant expansion of the control of respiration.

Chapter 14 The Kidneys and Regulation of Water and Inorganic Ions

Endocrine control of calcium homeostasis moved to Chapter 11; renal contribution to calcium homeostasis discussed in this chapter; expansion and update of section on glucose handling in the kidney; expansion and update of section on control of sodium transport in the proximal tubule; consolidation and clarification of section on countercurrent multiplier.

Chapter 15 The Digestion and Absorption of Food

New section on inflammatory bowel disease.

Chapter 16 Regulation of Organic Metabolism and Energy Balance

Expansion and update of sections on control of body weight.

Chapter 17 Reproduction

Discussion of different types of estrogen; update of section on meiosis; new summary of events from ovulation to implantation; *new figure* illustrating the control of gonadotropins in the luteal phase.

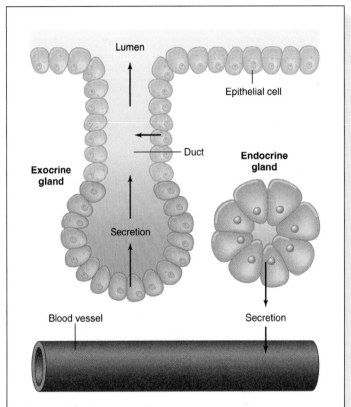

Figure 11–1

Exocrine gland secretions enter ducts, whereas hormones secreted by endocrine (ductless) glands diffuse directly into the blood.

Chapter 18 Defense Mechanisms of the Body

Updated information on clinical treatments for HIV infection.

Chapter 19 Medical Physiology: Integration Using Clinical Cases

New chapter describing the integrative, whole-body responses to homeostatic disturbances, including three complete Case Studies.

Teaching and Learning Supplements

McGraw-Hill offers various tools and technology products to support the eleventh edition of *Vander's Human Physiology*. Students can order supplemental study materials by contacting their campus bookstore. Instructors can obtain teaching aides by calling the McGraw-Hill Customer Service Department at 1-800-338-3987, by visiting our Human Physiology catalog at www.mhhe.com, or by contacting their local McGraw-Hill sales representative.

Laboratory Exercises in Human Physiology: A Clinical and Experimental Approach, Second Edition, by Lutterschmidt and Lutterschmidt

This laboratory manual contains 15 carefully selected laboratory exercises that coincide nicely with a typical human physiology course. The exercises will allow students to master fundamental principals in human physiology without overwhelming them with superfluous material. It also contains artwork from the *Vander's Human Physiology* text. Included in the laboratory manual is a copy of Ph.I.L.S. 3.0.

Physiology Interactive Lab Simulations (Ph.I.L.S.)

Ph.I.L.S. 3.0 contains 37 lab simulations that allow students to perform experiments without using expensive lab equipment or live animals. This easy-to-use software offers students the flexibility to change the parameters of every lab experiment, with no limit to the amount of times a student can repeat experiments or modify variables. This power to manipulate each experiment reinforces key physiology concepts by helping students to view outcomes, make predictions, and draw conclusions.

ARIS Course Management website (aris.mhhe.com)

McGraw-Hill's ARIS–Assessment, Review, and Instruction System–for *Vander's Human Physiology,* Eleventh Edition, is a complete electronic homework and course management system. Instructors can create and share course materials and assignments with colleagues with a few clicks of the mouse. Instructors can edit questions, import their own content, and create announcements and due dates for assignments. ARIS has automatic grading and reporting of easy-to-assign generated homework, quizzing, and testing. Once a student is registered in the course, all student activity within McGraw-Hill's ARIS website is automatically recorded and available to the instructor through a fully integrated grade book that can be downloaded to Excel. Instructors: To access ARIS, request registration information from your McGraw-Hill sales representative.

Text website (aris.mhhe.com)

The ARIS website that accompanies this text offers an extensive array of learning and teaching tools.

- Interactive Activities—Fun and exciting learning experiences await the student at the *Vander's Human Physiology* website. Chapters offer a series of interactive activities like art labeling, animations, vocabulary flashcards, and more!
- Practice Quizzes at the *Vander's Human Physiology* text website gauge student mastery of chapter content. Each chapter quiz is specially constructed to test student comprehension of key concepts. Immediate feedback to student responses explains why an answer is correct or incorrect.
- Tutorial Service is a free "homework hotline" which offers the students the opportunity to discuss text questions with our human physiology consultant.

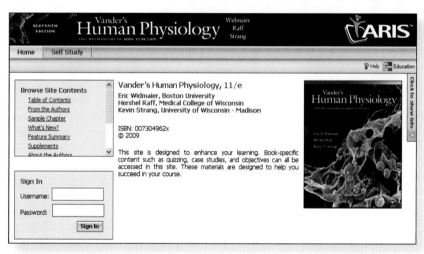

Presentation Center (aris.mhhe.com)

Build instructional materials wherever, whenever, and however they are needed!

Presentation Center is an online digital library containing assets such as photos, artwork, animations, and PowerPoints that can be used to create customized lectures, visually enhanced tests and quizzes, compelling course websites, or attractive printed support materials.

Access to your book, access to other McGraw-Hill books!

The Presentation Center library includes thousands of assets from many McGraw-Hill titles. This ever-growing resource gives instructors the power to utilize assets specific to an adopted textbook as well as content from all other books in the library.

Nothing could be easier!

Teaching and Learning Supplements

Anatomy & Physiology | Revealed®

This amazing multimedia tool is designed to help students learn and review human anatomy using cadaver specimens. Detailed cadaver photographs blended together with a state-of-the-art layering technique provide a uniquely interactive dissection experience. This easy-to-use program features the following sections:

- Dissection—Peel away layers of the human body to reveal structures beneath the surface. Structures can be pinned and labeled, just like in a real dissection lab. Each labeled structure is accompanied by detailed information and an audio pronunciation. Dissection images can be captured and saved.
- Animation—Compelling animations demonstrate muscle actions, clarify anatomical relationships, or explain difficult physiological concepts.
- Imaging—Labeled X-ray, MRI, and CT images familiarize students with the appearance of key anatomical structures as seen through different medical imaging techniques.
- Self-Test—Challenging exercises let students test their ability to identify anatomical structures in a timed practical exam format or traditional multiple choice. A results page provides analysis of test scores plus links back to all incorrectly identified structures for review.
- Anatomy Terms—This visual glossary of general terms includes directional and regional terms, as well as planes and terms of movement.

Test Bank

A computerized test bank that uses testing software to quickly create customized exams is available for this text. The user-friendly program allows instructors to search for questions by topic or format, edit existing questions or add new ones; and scramble questions for multiple versions of the same test. Word files of the test bank questions are provided for those instructors who prefer to work outside the test-generator software.

Instructor's Manual

The Instructor's Manual is available on the text website (aris.mhhe.com). It contains teaching/learning objectives, sample lecture outlines, and the answers to Review Questions for each chapter.

Course Delivery Systems

With help from our partners WebCT, Blackboard, Top-Class, eCollege, and other course management systems, professors can take complete control over their course content. Course cartridges containing text website content, online testing, and powerful student tracking features are readily available for use within these platforms.

eInstruction

This classroom performance system (CPS) utilizes wireless technology to bring interactivity into the classroom or lecture hall. Instructors and students receive immediate feedback through wireless response pads that are easy to use and engage students. eInstruction can assist instructors with:

- Taking attendance
- Administering quizzes and tests
- Creating a lecture with intermittent questions
- Using the CPS grade book to manage lectures and student comprehension
- Integrating interactivity into PowerPoint presentations.

MediaPhys CD-ROM

This interactive software tool offers detailed explanations, high quality illustrations and animations to provide students with a thorough introduction to the world of physiology—giving them a virtual tour of physiological processes. MediaPhys is filled with interactive activities and quizzes to help reinforce physiology concepts that are often difficult to understand.

Acknowledgments

The authors are deeply indebted to the following individuals for their contributions to the eleventh edition of *Vander's Human Physiology*. Their feedback on the tenth edition, their critique of the revised text, or their participation in a focus group provided invaluable assistance and greatly improved the final product. Any errors that may remain are solely the responsibility of the authors.

Thomas Adams
Michigan State University
Ateegh Al-Arabi
Johnson County Community College
Mark Alston
University of Tennessee Knoxville
Sharon Rohr Barnewall
Columbus State Community College
Steven Bassett
Southeast Community College
Erwin A. Bautista
UC Davis
Christina G. Benishin
University of Alberta
Ari Berkowitz
University of Oklahoma
Carol A. Britson
University of Mississippi
Brian P. Buggy
Aurora St. Luke's Medical Center
Phyllis Callahan
Miami University
Craig Canby
Des Moines University
James Cerletty
Medical College of Wisconsin
Edwin R. Chapman
University of Wisconsin–Madison
Pat Clark
IUPUI
Robert L. Conhaim
University of Wisconsin–Madison School of Medicine and Public Health
Nick Carvajal Cucuzzo
University of Florida
Gerald F. DiBona
University of Iowa
Jean-Pierre Dujardin
The Ohio State University
James S. Ferraro
Southern Illinois University
James W. Findling
Aurora St. Luke's Medical Center

John Fishback
Ozarks Technical Community College
Shawn W. Flanagan
The University of Iowa
Michael T. Griffin
Angelo State University
Jeffrey Grossman
University of Wisconsin–Madison School of Medicine and Public Health
Tara Haas
York University
Mary Ann Handel
The Jackson Laboratory
John P. Harley
Eastern Kentucky University and University of Kentucky
Janet L. Haynes
Long Island University
Stephen K. Henderson
California State University, Chico
Reinhold Hutz
University of Wisconsin–Milwaukee
Najma H. Javed
Ball State University
Leonard R. Johnson
University of Tennessee College of Medicine
Henry Kayongo-Male
South Dakota State University
Jack L. Keyes
Linfield College–Portland Campus
Leslie A. King
University of San Francisco
Loren W. Kline
University of Alberta
Penny Knoblich
Minnesota State University, Mankato
Ray Kumar
Northwestern State University
Stuart A. Levy
Aurora St. Luke's Medical Center
Mingyu Liang
Medical College of Wisconsin
Lenard R. Lichtenberger
University of Texas
William G. Loftin
Longview Community College
Andrew J. Lokuta
University of Wisconsin–Madison School of Medicine and Public Health
David S. Mallory
Marshall University

David L Mattson
Medical College of Wisconsin

Laurie Kelly McCorry
Massachusetts College of Pharmacy and Health Sciences

John S. McReynolds
University of Michigan

Katie Mechlin
Wright State University

Patricia A. Moberg
Community College of Rhode Island

Richard L. Moss
University of Wisconsin–Madison School of Medicine and Public Health

Judith A. Neubauer
UMDNJ-RW Johnson Med Sch

H. Clive Palfrey
University of Chicago

Frank L. Powell
University of California, San Diego

Elizabeth M. Rust
University of Michigan

Virginia K. Shea
University of North Carolina at Chapel Hill

Susan E.C. Simmons
Ivy Tech Community College, Bloomington, IN

Dexter F. Speck
University of Kentucky

Amanda Starnes
Emory University

William Stekiel
Medical College of Wisconsin

Darrell R. Stokes
Emory University

Rema G. Suniga
Ohio Northern University

Stephen Sup
Aurora St. Luke's Medical Center

Robert B. Tallitsch
University of Wisconsin–Madison School of Medicine and Public Health

Tom Tomasi
Missouri State University

Paula W. Trilling
Asheville-Buncombe Technical Community College

Gordon M. Wahler
Midwestern University

Curt Walker
Dixie State College

Joanne Westin
Case Western Reserve University

Meghan M. White
Saint Louis University

David F. Wilson
Miami University

Heather Wilson-Ashworth
Utah Valley State College

Carola Z. Wright
Mt. San Antonio College

Joan E. Zuckerman
Long Beach City College

The authors are indebted to the editors and staff at McGraw-Hill Higher Education who contributed to the development and publication of this text, particularly Developmental Editor Fran Schreiber, Executive Editor Colin Wheatley, Senior Project Manager Jayne Klein, and Publisher Michelle Watnick. We are also grateful for the excellent editing provided by copyeditor Sue Dillon. As always, we are thankful to Arthur Vander for his valuable input and counsel and to the many students and faculty who have provided us with critiques and suggestions for improvement.

Eric P. Widmaier

Hershel Raff

Kevin T. Strang

Maintenance of body temperature is an example of homeostasis.

Homeostasis: A Framework for Human Physiology

the purpose of this chapter is to provide an orientation to the subject of human physiology and the central role of homeostasis in the study of this science.

The Scope of Human Physiology

Stated most simply and broadly, **physiology** is the study of how living organisms work. As applied to human beings, its scope is extremely broad. At one end of the spectrum, it includes the study of individual molecules—for example, how a particular protein's shape and electrical properties allow it to function as a channel for ions to move into or out of a cell. At the other end, it is concerned with complex processes that depend on the integrated functions of many organs in the body—for example, how the heart, kidneys, and several glands all work together to cause the excretion of more sodium in the urine when a person has eaten salty food.

Physiologists are interested in function and integration—how parts of the body work together at various levels of organization and, most importantly, in the entire organism. Thus, even when physiologists study parts of organisms, all the way down to individual molecules, the intention is ultimately to apply the information they gain to the function of the whole body. As the nineteenth-century physiologist Claude Bernard put it: "After carrying out an analysis of phenomena, we must . . . always reconstruct our physiological synthesis, so as to see the *joint action* of all the parts we have isolated. . . ."

In this regard, a very important point must be made about the present and future status of physiology. It is easy for a student to gain the impression from a textbook that almost everything is known about the subject, but nothing could be farther from the truth for physiology. Many areas of function are still only poorly understood, such as how the workings of the brain produce conscious thought and memory.

Indeed, we can predict with certainty a continuing explosion of new physiological information and understanding. One of the major reasons is related to the recent landmark sequencing of the human genome. As the functions of all the proteins encoded by the genome are uncovered, their application to the functioning of the cells and organ systems discussed in this text will provide an ever-sharper view of how our bodies work. The integration of molecular biology with physiology has, in fact, led to the need for a new term to describe this growing area of research—**physiological genomics.** Nowadays, physiologists use the tools of molecular biology to ask not just *what* changes occur in the body in response to some external or internal stimulus, but *how* the changes are produced at the level of the gene.

Finally, in many areas of this text, we will relate physiology to medicine. Some disease states can be viewed as physiology "gone wrong," or **pathophysiology,** which makes an understanding of physiology essential for the study and practice of medicine. Indeed, many physiologists are actively engaged in research on the physiological bases of a wide range of diseases. In this text, we will give many examples of pathophysiology to illustrate the basic physiology that underlies the disease. A handy index of all the diseases and medical conditions discussed in this text appears in Appendix C. We begin our study of physiology by describing the organization of the structures of the human body.

How Is the Body Organized?

Cells: The Basic Units of Living Organisms

Before exploring how the human body works, it is necessary to understand the components of the body and their anatomical relationships to each other.

The simplest structural units into which a complex multicellular organism can be divided and still retain the functions characteristic of life are called **cells.** One of the unifying generalizations of biology is that certain fundamental activities are common to almost all cells and represent the minimal requirements for maintaining cell integrity and life. Thus, for example, a human liver cell and an amoeba are remarkably similar in terms of how they exchange materials with their immediate environments, obtain energy from organic nutrients, synthesize complex molecules, duplicate themselves, and detect and respond to signals in their immediate environments.

Each human organism begins as a single cell, a fertilized egg, which divides to create two cells, each of which divides in turn to result in four cells, and so on. If cell multiplication were the only event occurring, the end result would be a spherical mass of identical cells. During development, however, each cell becomes specialized for the performance of a particular function, such as producing force and movement or generating electric signals. The process of transforming an unspecialized cell into a specialized cell is known as **cell differentiation,** the study of which is one of the most exciting areas in biology today. Essentially all cells in a person have the same genes. How then is one unspecialized cell instructed to differentiate into a nerve cell, another into a muscle cell, and so on? What are the external chemical signals that constitute these "instructions," and how do they affect various cells differently? For the most part, the answers to these questions are only beginning to be understood.

In addition to differentiating, cells migrate to new locations during development and form selective adhesions with other cells to produce multicellular structures. In this manner, the cells of the body arrange themselves in various combinations to form a hierarchy of organized structures. Differentiated cells with similar properties aggregate to form **tissues,** such as nerve tissue or muscle tissue, which combine with other types of tissues to form **organs,** such as the heart, lungs, and kidneys. Organs, in turn, work together to form **organ systems,** such as the urinary system (**Figure 1–1**).

About 200 distinct kinds of cells can be identified in the body in terms of differences in structure and function. When cells are classified according to the broad types of function they perform, however, four major categories emerge: (1) muscle cells, (2) nerve cells, (3) epithelial cells, and (4) connective tissue cells. In each of these functional categories, several cell types perform variations of the specialized function. For example, there are three types of muscle cells—skeletal, cardiac, and smooth. These cells differ from each other in shape, in the mechanisms controlling their contractile activity, and in their location in the various organs of the body, but each of them is a muscle cell.

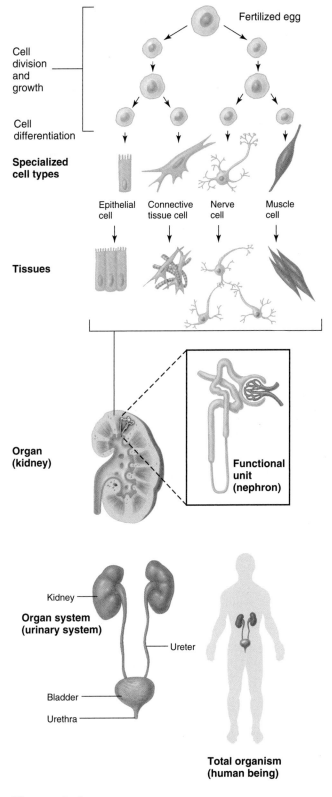

Cell division and growth

Cell differentiation

Specialized cell types

Fertilized egg

Epithelial cell | Connective tissue cell | Nerve cell | Muscle cell

Tissues

Organ (kidney)

Functional unit (nephron)

Organ system (urinary system)

Kidney

Ureter

Bladder

Urethra

Total organism (human being)

Figure 1–1

Levels of cellular organization. The nephron is not drawn to scale.

Muscle cells are specialized to generate the mechanical forces that produce movement. They may be attached through other structures to bones and produce movements of the limbs or trunk. They may be attached to skin, such as the muscles

producing facial expressions. They may also surround hollow cavities so that their contraction expels the contents of the cavity, as in the pumping of the heart. Muscle cells also surround many of the tubes in the body—blood vessels, for example—and their contraction changes the diameter of these tubes.

Nerve cells are specialized to initiate and conduct electrical signals, often over long distances. A signal may initiate new electrical signals in other nerve cells, or it may stimulate a gland cell to secrete substances or a muscle cell to contract. Thus, nerve cells provide a major means of controlling the activities of other cells. The incredible complexity of connections between nerve cells underlies such phenomena as consciousness and perception.

Epithelial cells are specialized for the selective secretion and absorption of ions and organic molecules, and for protection. They are located mainly at the surfaces that cover the body or individual organs, and they line the walls of various tubular and hollow structures within the body. Epithelial cells, which rest on an extracellular protein layer called the **basement membrane,** form the boundaries between compartments and function as selective barriers regulating the exchange of molecules. For example, the epithelial cells at the surface of the skin form a barrier that prevents most substances in the **external environment**—the environment surrounding the body—from entering the body through the skin. Epithelial cells are also found in glands that form from the invagination of epithelial surfaces.

Connective tissue cells, as their name implies, connect, anchor, and support the structures of the body. Some connective tissue cells are found in the loose meshwork of cells and fibers underlying most epithelial layers. Other types include adipose (fat-storing) cells, bone cells, red blood cells, and white blood cells.

Tissues

Most specialized cells are associated with other cells of a similar kind to form tissues. Corresponding to the four general categories of differentiated cells, there are four general classes of tissues: (1) **muscle tissue,** (2) **nerve tissue,** (3) **epithelial tissue,** and (4) **connective tissue.** The term *tissue* is used in different ways. It is formally defined as an aggregate of a single type of specialized cell. However, it is also commonly used to denote the general cellular fabric of any organ or structure—for example, kidney tissue or lung tissue, each of which in fact usually contains all four classes of tissue.

The immediate environment that surrounds each individual cell in the body is the extracellular fluid. Actually, this fluid is interspersed within a complex **extracellular matrix** consisting of a mixture of protein molecules and, in some cases, minerals, specific for any given tissue. The matrix serves two general functions: (1) It provides a scaffold for cellular attachments, and (2) it transmits information, in the form of chemical messengers, to the cells to help regulate their activity, migration, growth, and differentiation.

The proteins of the extracellular matrix consist of fibers—ropelike **collagen fibers** and rubberband-like **elastin fibers**—and a mixture of nonfibrous proteins that contain chains of complex sugars (carbohydrates). In some ways, the extracellular matrix is analogous to reinforced concrete. The fibers of the matrix, particularly collagen, which constitutes one-third of all bodily proteins, are like the reinforcing iron mesh or rods in the concrete. The carbohydrate-containing protein molecules are analogous to the surrounding cement. However, these latter molecules are not merely inert packing material, as in concrete, but function as adhesion/recognition molecules between cells. Thus, they are links in the communication between extracellular messenger molecules and cells.

Organs and Organ Systems

Organs are composed of the four kinds of tissues arranged in various proportions and patterns, such as sheets, tubes, layers, bundles, and strips. For example, the kidneys consist of (1) a series of small tubes, each composed of a single layer of epithelial cells; (2) blood vessels, whose walls contain varying quantities of smooth muscle and connective tissue; (3) exten-

sions from nerve cells that end near the muscle and epithelial cells; (4) a loose network of connective-tissue elements that are interspersed throughout the kidneys and include the protective capsule that surrounds the organ.

Many organs are organized into small, similar subunits often referred to as **functional units,** each performing the function of the organ. For example, the kidneys' functional units, called nephrons, contain the small tubes mentioned in the previous paragraph. The total production of urine by the kidneys is the sum of the amounts produced by the two million individual nephrons.

Finally we have the organ system, a collection of organs that together perform an overall function. For example, the kidneys, the urinary bladder, the tubes leading from the kidneys to the bladder, and the tube leading from the bladder to the exterior constitute the urinary system. **Table 1–1** lists the components and functions of the organ systems in the body.

Table 1–1	Organ Systems of the Body	
System	**Major Organs or Tissues**	**Primary Functions**
Circulatory	Heart, blood vessels, blood	Transport of blood throughout the body's tissues
Digestive	Mouth, salivary glands, pharynx, esophagus, stomach, large and small intestines, pancreas, liver, gallbladder	Digestion and absorption of nutrients and water; elimination of wastes
Endocrine	All glands or organs secreting hormones: Pancreas, testes, ovaries, hypothalamus, kidneys, pituitary, thyroid, parathyroid, adrenal, intestinal, thymus, heart, and pineal, and endocrine cells in other locations	Regulation and coordination of many activities in the body, including growth, metabolism, reproduction, blood pressure, electrolyte balance, and others
Immune	White blood cells, spleen, thymus (also see: Lymphatic system)	Defense against pathogens
Integumentary	Skin	Protection against injury and dehydration; defense against pathogens; regulation of body temperature
Lymphatic	Lymph vessels, lymph nodes	Collect extracellular fluid for return to circulation; participate in immune defenses
Musculoskeletal	Cartilage, bone, ligaments, tendons, joints, skeletal muscle	Support, protection, and movement of the body; production of blood cells
Nervous	Brain, spinal cord, peripheral nerves and ganglia, sense organs	Regulation and coordination of many activities in the body; detection of changes in the internal and external environments; states of consciousness; learning; cognition
Reproductive	Male: Testes, penis, and associated ducts and glands Female: Ovaries, fallopian tubes, uterus, vagina, mammary glands	Production of sperm; transfer of sperm to female Production of eggs; provision of a nutritive environment for the developing embryo and fetus; nutrition of the infant
Respiratory	Nose, pharynx, larynx, trachea, bronchi, lungs	Exchange of carbon dioxide and oxygen; regulation of hydrogen ion concentration
Urinary	Kidneys, ureters, bladder, urethra	Regulation of plasma composition through controlled excretion of salts, water, and organic wastes

To sum up, the human body can be viewed as a complex society of differentiated cells that combine structurally and functionally to carry out the functions essential to the survival of the entire organism. The individual cells constitute the basic units of this society, and almost all of these cells individually exhibit the fundamental activities common to all forms of life, such as metabolism and replication.

There is a paradox in this analysis, however. Why are the functions of the organ systems essential to the survival of the body when each cell seems capable of performing its own fundamental activities? As described in the next section, the resolution of this paradox is found in the isolation of most of the cells of the body from the external environment, and in the existence of a reasonably stable internal environment. The **internal environment** of the body refers to the fluids that surround cells and exist in the blood. These fluid compartments and one other—that which exists inside cells—are described next.

Body Fluid Compartments

Water is present within and around the cells of the body, and within all the blood vessels. Collectively, the fluid present in blood and in the spaces surrounding cells is called **extracellular fluid.** Of this, only about 20–25 percent is in the fluid portion of blood, the **plasma,** in which the various blood cells are suspended. The remaining 75–80 percent of the extracellular fluid, which lies around and between cells, is known as the **interstitial fluid.**

As the blood flows through the smallest of blood vessels in all parts of the body, the plasma exchanges oxygen, nutrients, wastes, and other metabolic products with the interstitial fluid. Because of these exchanges, concentrations of dissolved substances are virtually identical in the plasma and interstitial fluid, except for protein concentration. With this major exception—higher protein concentration in plasma than in interstitial fluid—the entire extracellular fluid may be considered to have a homogeneous composition. In contrast, the composition of the extracellular fluid is very different from that of the **intracellular fluid,** the fluid inside the cells. Maintaining differences in fluid composition across the cell membrane is an important way in which cells regulate their own activity. For example, intracellular fluid contains many different proteins that are important in regulating cellular events such as growth and metabolism. These proteins must be retained within the intracellular fluid, and are not required in the other fluid compartments.

In essence, the fluids in the body are enclosed in compartments. **Figure 1–2** summarizes the volumes of the body fluid compartments in terms of water, because water is by far the major component of the fluids. Water accounts for about 55–60 percent of normal body weight in an adult male, and slightly less in a female. (Females generally have more body fat than do males, and fat has a low water content.) Two-thirds of the water is intracellular fluid. The remaining one-third is extracellular. As described previously, 75–80 percent of this extracellular fluid is interstitial fluid, and 20–25 percent is plasma.

Compartmentalization is an important general principle in physiology. Compartmentalization is achieved by barriers between the compartments. The properties of the barriers determine which substances can move between compartments. These movements, in turn, account for the differences in composition of the different compartments. In the case of the body fluid compartments, plasma membranes that surround each cell separate the intracellular fluid from the extracellular fluid. Chapter 4 describes the properties of plasma membranes and how they account for the profound differences between intracellular and extracellular fluid. In contrast, the two components of extracellular fluid—the interstitial fluid and the blood plasma—are separated by the cellular wall of the smallest blood vessels, the capillaries. Chapter 12 discusses how this barrier normally keeps 75–80 percent of the extracellular fluid

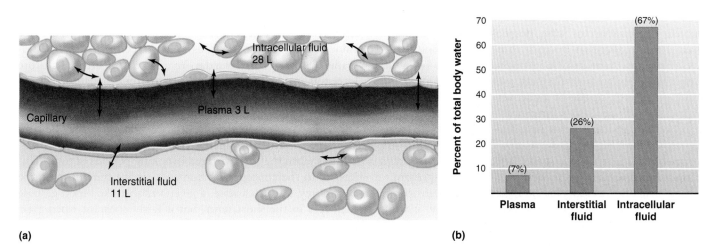

(a)　　　　　　　　　　　　　　　　　　　　**(b)**

Figure 1–2

Fluid compartments of the body. Volumes are for an average 70-kg (154-lb) person. (a) The bidirectional arrows indicate that fluid can move between any two adjacent compartments. Total body water is about 42 L, which makes up about 55–60 percent of body weight. (b) The approximate percentage of total body water normally found in each compartment.

in the interstitial compartment and restricts proteins mainly to the plasma. With this understanding of the structural organization of the body, and the way in which water is distributed throughout, we turn to a description of how a balance is achieved in the body's internal environment.

Homeostasis: A Defining Feature of Physiology

From the earliest days of physiology—at least as early as the time of Aristotle—physicians recognized that good health was somehow associated with a balance among the multiple life-sustaining forces ("humours") in the body. It would take millennia, however, for scientists to determine just what it was that was being balanced, and how this balance was achieved. The advent of modern tools of science, including the ordinary microscope, led to the discovery that the human body is composed of trillions of cells, each of which is packaged to permit movement of certain substances, but not others, across the cell membrane. Over the course of the nineteenth and twentieth centuries, it became clear that most cells are in contact with the interstitial fluid. The interstitial fluid, in turn, was found to be in a state of flux, with water and solutes, such as ions and gases, moving back and forth through it between the cell interiors and the blood in nearby capillaries (see Figure 1–2).

It was further determined by careful observation that most of the common physiological variables found in normal, healthy organisms—blood pressure, body temperature, and blood-borne factors such as oxygen, glucose, and sodium, for example—are maintained within a predictable range. This was true despite external environmental conditions that may be far from constant. Thus was born the idea, first put forth by the French physician and physiologist Claude Bernard, of a constant internal *milieu* that is a prerequisite for good health, a concept later refined by the American physiologist Walter Cannon, who coined the term *homeostasis*.

Originally, **homeostasis** was defined as a state of reasonably stable balance between physiological variables such as those just described. This simple definition cannot give one a complete appreciation of what homeostasis truly entails, however. There probably is no such thing as a physiological variable that is constant over long periods of time. In fact, some variables undergo fairly dramatic swings around an average value during the course of a day, yet are still considered in balance. That is because homeostasis is a *dynamic*, not a static, process. Consider swings in blood glucose levels over the course of a day (**Figure 1–3**). After a meal, blood glucose levels rise considerably. Clearly, such a large change from baseline cannot be considered stable or static. What is important, though, is that once blood glucose increases, compensatory mechanisms restore the glucose level toward the level it was at before the meal. These homeostatic compensatory mechanisms do not, however, overshoot to any significant degree in the opposite direction. That is, the blood glucose levels do not fall below the pre-meal level, or do so only moderately. In the case of glucose, the endocrine system is primarily responsible for this adjustment, but a wide variety of control systems may

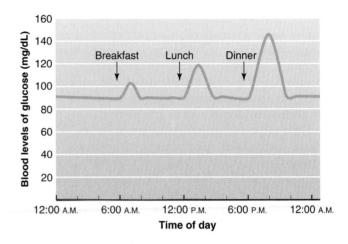

Figure 1–3

Changes in blood glucose levels during a typical 24-hour period. Note that glucose increases after each meal, more so after larger meals, and then returns to the pre-meal level in a short while. The profile shown here is that of a person who is homeostatic for blood glucose, even though levels of this sugar vary considerably throughout the day.

be initiated to regulate other processes. In later chapters, we will see how nearly every organ and tissue of the human body contributes to homeostasis, sometimes in multiple ways, and usually in concert with each other.

Thus, homeostasis does not imply that a given physiological function or variable is rigidly constant with respect to time, but that it fluctuates within a predictable and often narrow range. When disturbed up or down from the normal range, it is restored to normal.

What do we mean when we say that something varies within a normal range? This depends on just what we are monitoring. If the circulating arterial oxygen level of a healthy person breathing air at sea level is measured, it does not change much over the course of time, even if the person exercises. Such a system is said to be tightly controlled and to demonstrate very little variability or scatter around an average value. Blood glucose levels, as we have seen, may vary considerably over the course of a day. Yet, if the daily average glucose level was determined in the same person on many consecutive days, it would be much more predictable over days or even years than random, individual measurements of glucose over the course of a single day. In other words, there may be considerable variation in glucose values over short time periods, but less when they are averaged over long periods of time. This has led to the concept that homeostasis is a state of **dynamic constancy.** In such a state, a given variable like blood glucose may vary in the short term, but is fairly constant when averaged over the long term.

It is also important to realize that a person may be homeostatic for one variable, but not homeostatic for another. Homeostasis must be described differently, therefore, for each variable. For example, as long as the concentration of sodium in the blood remains within a few percent of its normal range, sodium homeostasis exists. However, a person in sodium homeostasis may suffer from other disturbances, such as abnormally high carbon dioxide levels in the blood resulting from lung disease, a condition that could be fatal. Just one nonhomeostatic variable, among the many that can be described, can have life-threatening consequences. Typically, though, if one system becomes dramatically out of balance, other systems in the body become nonhomeostatic as a consequence. In general, if all the major organ systems are operating in a homeostatic manner, a person is in good health. Certain kinds of disease, in fact, can be defined as the loss of homeostasis in one or more systems in the body. To elaborate on our earlier definition of physiology, therefore, when homeostasis is maintained, we refer to physiology; when it is not, we refer to pathophysiology.

General Characteristics of Homeostatic Control Systems

The activities of cells, tissues, and organs must be regulated and integrated with each other so that any change in the extracellular fluid initiates a reaction to correct the change. The compensating mechanisms that mediate such responses are performed by **homeostatic control systems.**

Consider an example of the regulation of body temperature. Our subject is a resting, lightly clad man in a room having a temperature of 20°C and moderate humidity. His internal body temperature is 37°C, and he is losing heat to the external environment because it is at a lower temperature. However, the chemical reactions occurring within the cells of his body are producing heat at a rate equal to the rate of heat loss. Under these conditions, the body undergoes no net gain or loss of heat, and the body temperature remains constant. The system is in a **steady state,** defined as a system in which a particular variable—temperature, in this case—is not changing, but energy—in this case, heat—must be added continuously to maintain a constant condition. Steady state differs from **equilibrium,** in which a particular variable is not changing but no input of energy is required to maintain the constancy. The steady-state temperature in our example is known as the **set point,** sometimes termed the operating point, of the thermoregulatory system.

This example illustrates a crucial generalization about homeostasis. Stability of an internal environmental variable is achieved by the balancing of inputs and outputs. In the previous example, the variable (body temperature) remains constant because metabolic heat production (input) equals heat loss from the body (output).

Now imagine that we lower the temperature of the room rapidly, say to 5°C, and keep it there. This immediately increases the loss of heat from our subject's warm skin, upsetting the balance between heat gain and loss. The body temperature therefore starts to fall. Very rapidly, however, a variety of homeostatic responses occur to limit the fall. **Figure 1–4** summarizes these responses. *The reader is urged to study Figure 1–4 and its legend carefully because the figure is typical of those used throughout the remainder of the book to*

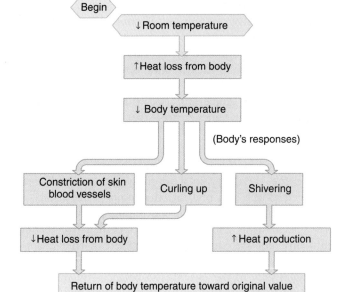

Figure 1–4

A homeostatic control system maintains body temperature when room temperature decreases. This flow diagram is typical of those used throughout this book to illustrate homeostatic systems, and several conventions should be noted. The "Begin" sign indicates where to start. The arrows next to each term within the boxes denote increases or decreases. The arrows connecting any two boxes in the figure denote cause and effect; that is, an arrow can be read as "causes" or "leads to." (For example, decreased room temperature "leads to" increased heat loss from the body.) In general, you should add the words "tends to" in thinking about these cause-and-effect relationships. For example, decreased room temperature tends to cause an increase in heat loss from the body, and curling up tends to cause a decrease in heat loss from the body. Qualifying the relationship in this way is necessary because variables like heat production and heat loss are under the influence of many factors, some of which oppose each other.

illustrate homeostatic systems, and the legend emphasizes several conventions common to such figures.

The first homeostatic response is that blood vessels to the skin become constricted (narrowed), reducing the amount of warm blood flowing through the skin. This reduces heat loss to the environment and helps maintain body temperature. At a room temperature of 5°C, however, blood vessel constriction cannot completely eliminate the extra heat loss from the skin. Like the person shown in the chapter opening photo, our subject curls up in order to reduce the surface area of the skin available for heat loss. This helps somewhat, but excessive heat loss still continues, and body temperature keeps falling, although at a slower rate. Clearly, then, if excessive heat loss (output) cannot be prevented, the only way of restoring the balance between heat input and output is to increase input, and this is precisely what occurs. Our subject begins to shiver, and the chemical reactions responsible for the skeletal muscular contractions that constitute shivering produce large quantities of heat.

Feedback Systems

The thermoregulatory system just described is an example of a **negative feedback** system, in which an increase or decrease in the variable being regulated brings about responses that tend to move the variable in the direction opposite ("negative" to) the direction of the original change. Thus, in our example, a decrease in body temperature led to responses that tended to increase the body temperature—that is, move it toward its original value.

Without negative feedback, oscillations like some of those described in this chapter would be much greater, and therefore the variability in a given system would increase. Negative feedback also prevents the compensatory responses to a loss of homeostasis from continuing unabated. Details of the mechanisms and characteristics of negative feedback within different systems will be addressed in later chapters. For now, it is important to recognize that negative feedback plays a vital part in the checks and balances on most physiological variables.

Negative feedback may occur at the organ, cellular, or molecular level. For instance, negative feedback regulates many enzymatic processes, as shown in schematic form in **Figure 1–5**. An enzyme is a protein that catalyzes chemical reactions (Chapter 3). In this example, the product formed from a substrate by an enzyme negatively feeds back to inhibit further action of the enzyme. This may occur by several processes, such as chemical modification of the enzyme by the product of the reaction. The production of energy within cells is a good example of a chemical process regulated by feedback. When a cell's energy stores are depleted, glucose molecules are enzymatically broken down to provide chemical energy that is stored in adenosine triphosphate (ATP). As ATP accumulates in the cell, it inhibits the activity of some of the enzymes involved in the breakdown of glucose. Thus, as ATP levels increase within a cell, further production of ATP slows down

due to negative feedback. Conversely, when ATP levels drop within a cell, negative feedback is removed, and more glucose is broken down so that more ATP can be produced.

As an aside, not all forms of feedback are negative. In some cases, **positive feedback** accelerates a process, leading to an "explosive" system. This is counter to the principle of homeostasis, because positive feedback has no obvious means of stopping. Not surprisingly, therefore, positive feedback is less common in nature than negative feedback. Nonetheless, there are examples in physiology where positive feedback is very important. One well-described example, which you will learn about in Chapter 17, is the process of parturition (birth). As the uterine muscles contract and a baby is forced through the birth canal during labor, signals are relayed via nerves to the brain. This initiates the secretion into the blood of a molecule called oxytocin, which is a potent stimulator of further uterine contractions. As the uterus contracts ever harder in response to oxytocin, more stretch occurs, and more signals are sent to the brain, resulting in yet more oxytocin secretion. This self-perpetuating cycle continues until finally the baby is born.

Resetting of Set Points

As we have seen, changes in the external environment can displace a variable from its set point. In addition, the set points for many regulated variables can be physiologically reset to a new value. A common example is fever, the increase in body temperature that occurs in response to infection and that is somewhat analogous to raising the setting of a home's thermostat. The homeostatic control systems regulating body temperature are still functioning during a fever, but they maintain the temperature at a higher value. This regulated rise in body temperature is adaptive for fighting the infection, because elevated temperature inhibits proliferation of some pathogens. In fact, this is why a fever is often preceded by chills and shivering. The set point for body temperature has been reset to a higher value, and the body responds by shivering to generate heat.

The fact that set points can be reset adaptively, as in the case of fever, raises important challenges for medicine, as another example illustrates. Plasma iron concentration decreases significantly during many infections. Until recently, it was assumed that this decrease was a symptom caused by the infectious organism and that it should be treated with iron supplements. In fact, just the opposite is true. The decrease in iron is brought about by the body's defense mechanisms and serves to deprive the infectious organisms of the iron they require to replicate. Several controlled studies have shown that iron replacement can make the illness much worse. Clearly it is crucial to distinguish between those deviations of homeostatically controlled variables that are truly part of a disease and those that, through resetting, are part of the body's defenses against the disease.

The examples of fever and plasma iron concentration may have left the impression that set points are reset only in response to external stimuli, such as the presence of bacteria, but this is not the case. Indeed, the set points for many regulated variables change on a rhythmical basis every day. For example, the set point for body temperature is higher during the day than at night.

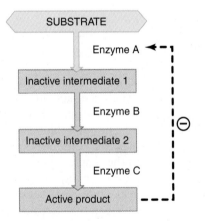

Figure 1–5

Hypothetical example of negative feedback (as denoted by the circled minus sign and dashed feedback line) occurring within a set of sequential chemical reactions. By inhibiting the activity of the first enzyme involved in the formation of a product, the product can regulate the rate of its own formation.

Although the resetting of a set point is adaptive in some cases, in others it simply reflects the clashing demands of different regulatory systems. This brings us to one more generalization. It is not possible for everything to be held constant by homeostatic control systems. In our example, body temperature was maintained despite large swings in ambient temperature, but only because the homeostatic control system brought about large changes in skin blood flow and skeletal muscle contraction. Moreover, because so many properties of the internal environment are closely interrelated, it is often possible to keep one property relatively constant only by moving others away from their usual set point. This is what we mean by "clashing demands."

The generalizations we have given about homeostatic control systems are summarized in **Table 1–2**. One additional point is that, as is illustrated by the regulation of body temperature, multiple systems often control a single parameter. The adaptive value of such redundancy is that it provides much greater fine-tuning and also permits regulation to occur even when one of the systems is not functioning properly because of disease.

Feedforward Regulation

Another type of regulatory process often used in conjunction with feedback systems is feedforward. Let us give an example of feedforward and then define it. The temperature-sensitive nerve cells that trigger negative feedback regulation of body temperature when it begins to fall are located inside the body. In addition, there are temperature-sensitive nerve cells in the skin, and these cells, in effect, monitor outside temperature. When outside temperature falls, as in our example, these nerve cells immediately detect the change and relay this information to the brain. The brain then sends out signals to the blood vessels and muscles, resulting in heat conservation and increased heat production. In this manner, compensatory thermoregulatory responses are activated *before* the colder outside temperature can cause the internal body temperature to fall. In another familiar example, the smell of food triggers nerve responses from smell receptors in the nose to the cells of the gastrointestinal system. This prepares the stomach and intestines for the process of digestion. Thus, the stomach begins to churn and produce acid even before we consume any food. Thus, **feedforward** regulation anticipates changes in regulated variables such as internal body temperature or energy availability, improves the speed of the body's homeostatic responses, and minimizes fluctuations in the level of the variable being regulated—that is, it reduces the amount of deviation from the set point.

In our examples, feedforward control utilizes a set of external or internal environmental detectors. It is likely, however, that many examples of feedforward control are the result of a different phenomenon—learning. The first times they occur, early in life, perturbations in the external environment probably cause relatively large changes in regulated internal environmental factors, and in responding to these changes the central nervous system learns to anticipate them and resist them more effectively. A familiar form of this is the increased heart rate that occurs in an athlete just before a competition begins.

Components of Homeostatic Control Systems

Reflexes

The thermoregulatory system we used as an example in the previous section, and many of the body's other homeostatic control systems, belong to the general category of stimulus-response sequences known as reflexes. Although in some reflexes we are aware of the stimulus and/or the response, many reflexes regulating the internal environment occur without our conscious awareness.

In the most narrow sense of the word, a **reflex** is a specific involuntary, unpremeditated, unlearned "built-in" response to a particular stimulus. Examples of such reflexes include pulling your hand away from a hot object or shutting your eyes as an object rapidly approaches your face. There are also many responses, however, that appear automatic and stereotyped but are actually the result of learning and practice. For example, an experienced driver performs many complicated acts in operating a car. To the driver these motions are, in large part, automatic, stereotyped, and unpremeditated, but they occur only because a great deal of conscious effort was spent learning them. We term such reflexes **learned,** or **acquired reflexes.** In general,

Table 1–2	Some Important Generalizations About Homeostatic Control Systems

1. Stability of an internal environmental variable is achieved by balancing inputs and outputs. It is not the absolute magnitudes of the inputs and outputs that matter, but the balance between them.

2. In negative feedback systems, a change in the variable being regulated brings about responses that tend to move the variable in the direction opposite the original change—that is, back toward the initial value (set point).

3. Homeostatic control systems cannot maintain complete constancy of any given feature of the internal environment. Therefore, any regulated variable will have a more-or-less narrow range of normal values depending on the external environmental conditions.

4. The set point of some variables regulated by homeostatic control systems can be reset—that is, physiologically raised or lowered.

5. It is not always possible for homeostatic control systems to maintain constancy in every variable in response to an environmental challenge. There is a hierarchy of importance, so that the constancy of certain variables may be altered markedly to maintain others within their normal range.

most reflexes, no matter how simple they may appear to be, are subject to alteration by learning.

The pathway mediating a reflex is known as the **reflex arc,** and its components are shown in **Figure 1–6**. A **stimulus** is defined as a detectable change in the internal or external environment, such as a change in temperature, plasma potassium concentration, or blood pressure. A **receptor** detects the environmental change. A stimulus acts upon a receptor to produce a signal that is relayed to an **integrating center.** The pathway the signal travels between the receptor and the integrating center is known as the **afferent pathway** (the general term *afferent* means "to carry to," in this case, to the integrating center).

An integrating center often receives signals from many receptors, some of which may respond to quite different types of stimuli. Thus, the output of an integrating center reflects the net effect of the total afferent input; that is, it represents an integration of numerous bits of information.

The output of an integrating center is sent to the last component of the system, whose change in activity constitutes the overall response of the system. This component is known as an **effector.** The information going from an integrating center to an effector is like a command directing the effector to alter its activity. The pathway along which this information travels is known as the **efferent pathway** (the general term *efferent* means "to carry away from," in this case, away from the integrating center).

Thus far we have described the reflex arc as the sequence of events linking a stimulus to a response. If the response produced by the effector causes a decrease in the magnitude of the stimulus that triggered the sequence of events, then the reflex leads to negative feedback and we have a typical homeostatic control system. Not all reflexes are associated with such feedback. For example, the smell of food stimulates the stomach to secrete molecules that are important for digestion, but these molecules do not eliminate the smell of food (the stimulus).

Figure 1–7 demonstrates the components of a negative feedback homeostatic reflex arc in the process of thermoregulation. The temperature receptors are the endings of

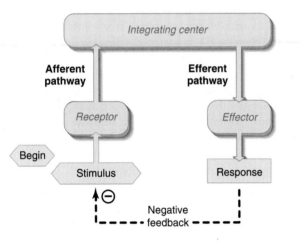

Figure 1–6

General components of a reflex arc that functions as a negative feedback control system. The response of the system has the effect of counteracting or eliminating the stimulus. This phenomenon of negative feedback is emphasized by the minus sign in the dashed feedback loop.

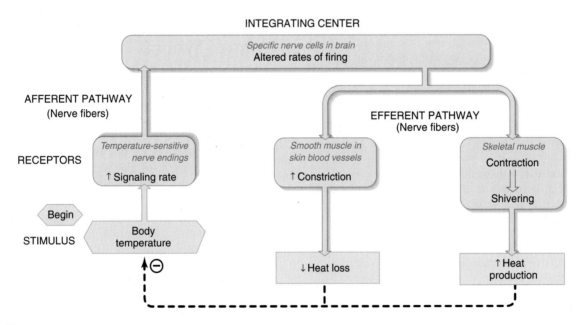

Figure 1–7

Reflex for minimizing the decrease in body temperature that occurs on exposure to a reduced external environmental temperature. This figure provides the internal components for the reflex shown in Figure 1–4. The dashed arrow and the ⊖ indicate the negative feedback nature of the reflex, denoting that the reflex responses cause the decreased body temperature to return toward normal. An additional flow-diagram convention is shown in this figure: Blue boxes always denote events that are occurring in anatomical structures (labeled in blue italic type in the upper portion of the box).

certain nerve cells in various parts of the body. They generate electrical signals in the nerve cells at a rate determined by the temperature. These electrical signals are conducted by the nerve fibers—the afferent pathway—to the brain, where the integrating center for temperature regulation is located. The integrating center, in turn, sends signals out along those nerve cells that cause skeletal muscles and the muscles in skin blood vessels to contract. The nerve fibers to the muscles are the efferent pathway, and the muscles are the effectors. The dashed arrow and the ⊖ indicate the negative feedback nature of the reflex.

Almost all body cells can act as effectors in homeostatic reflexes. There are, however, two specialized classes of tissues—muscle and gland—that are the major effectors of biological control systems. In the case of glands, for example, the effector may be a hormone secreted into the blood. A **hormone** is a type of chemical messenger secreted into the blood by cells of the endocrine system (see Table 1–1). Hormones may act on many different cells simultaneously because they circulate throughout the body.

Traditionally, the term *reflex* was restricted to situations in which the receptors, afferent pathway, integrating center, and efferent pathway were all parts of the nervous system, as in the thermoregulatory reflex. However, the principles are essentially the same when a blood-borne chemical messenger, rather than a nerve fiber, serves as the efferent pathway, or when a hormone-secreting gland (called an **endocrine gland**) serves as the integrating center. Thus, in the thermoregulation example, the integrating center in the brain not only sends signals by way of nerve fibers, as shown in Figure 1–7, but also causes the release of a hormone that travels via the blood to many cells, where it increases the amount of heat these cells produce. This hormone therefore also serves as an efferent pathway in thermoregulatory reflexes.

In our use of the term *reflex,* therefore, we include hormones as reflex components. Moreover, depending on the specific nature of the reflex, the integrating center may reside either in the nervous system or in an endocrine gland. In addition, an endocrine gland may act as both receptor and integrating center in a reflex. For example, the endocrine gland cells that secrete the hormone insulin, which lowers plasma glucose concentration, themselves detect increases in the plasma glucose concentration.

In conclusion, many reflexes function in a homeostatic manner to keep a physical or chemical variable of the body within its normal range. Any such system can be analyzed by answering the questions listed in **Table 1–3**.

Local Homeostatic Responses

In addition to reflexes, another group of biological responses, called **local homeostatic responses,** is of great importance for homeostasis. They are initiated by a change in the external or internal environment (that is, a stimulus), and they induce an alteration of cell activity with the net effect of counteracting the stimulus. Like a reflex, therefore, a local response is the result of a sequence of events proceeding from a stimulus. Unlike a reflex, however, the entire sequence occurs only in the area of the stimulus. For example, when cells of a tissue

Table 1–3	Questions to Be Asked About Any Homeostatic Response

1. What is the variable (for example, plasma potassium concentration, body temperature, blood pressure) that is maintained within a normal range in the face of changing conditions?

2. Where are the receptors that detect changes in the state of this variable?

3. Where is the integrating center to which these receptors send information and from which information is sent out to the effectors, and what is the nature of these afferent and efferent pathways?

4. What are the effectors, and how do they alter their activities so as to maintain the regulated variable near the set point of the system?

become very metabolically active, they secrete substances into the interstitial fluid that dilate local blood vessels. The resulting increased blood flow increases the rate at which nutrients and oxygen are delivered to that area. The significance of local responses is that they provide individual areas of the body with mechanisms for local self-regulation.

Intercellular Chemical Messengers

Essential to reflexes and local homeostatic responses, and therefore to homeostasis, is the ability of cells to communicate with one another. In this way, cells in the brain, for example, can be made aware of the status of activities of structures outside the brain, such as the heart, and help regulate those activities to meet new challenges. In the majority of cases, this communication between cells—intercellular communication—is performed by chemical messengers. There are three categories of such messengers: hormones, neurotransmitters, and paracrine/autocrine agents (**Figure 1–8**).

As noted earlier, a hormone functions as a chemical messenger that enables the hormone-secreting cell to communicate with cells acted upon by the hormone—its **target cells**—with the blood acting as the delivery system. Most nerve cells communicate with each other or with effector cells, such as muscles, by means of chemical messengers called **neurotransmitters.** Thus, one nerve cell alters the activity of another by releasing from its ending a neurotransmitter that diffuses through the extracellular fluid separating the two nerve cells and acts upon the second. Similarly, neurotransmitters released from nerve cells into the extracellular fluid in the immediate vicinity of effector cells constitute the controlling input to the effector cells. Neurotransmitters and their roles in nerve cell signaling will be covered in Chapter 6.

Chemical messengers participate not only in reflexes, but also in local responses. Chemical messengers involved in local communication between cells are known as **paracrine agents.** Paracrine agents are synthesized by cells and released,

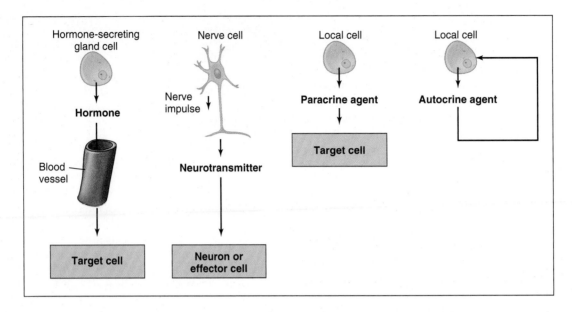

Figure 1–8

Categories of chemical messengers. With the exception of autocrine agents, all messengers act between cells—that is, *inter*cellularly.

once given the appropriate stimulus, into the extracellular fluid. They then diffuse to neighboring cells, some of which are their target cells. Note that, given this broad definition, neurotransmitters could be classified as a subgroup of paracrine agents, but by convention they are not. Paracrine agents are generally inactivated rapidly by locally existing enzymes so that they do not enter the bloodstream in large quantities.

There is one category of local chemical messengers that are not *inter*cellular messengers—that is, they do not communicate *between* cells. Rather, the chemical is secreted by a cell into the extracellular fluid and then acts upon the very cell that secreted it. Such messengers are termed **autocrine agents** (see Figure 1–8). Frequently a messenger may serve both paracrine and autocrine functions simultaneously—that is, molecules of the messenger released by a cell may act locally on adjacent cells as well as on the same cell that released the messenger.

One of the most exciting developments in physiology today is the identification of a growing number of paracrine/autocrine agents and the extremely diverse effects they exert. Their structures range from a simple gas such as nitric oxide to fatty acid derivatives such as the eicosanoids (Chapter 5), to peptides and amino acid derivatives. They tend to be secreted by multiple cell types in many tissues and organs. According to their structures and functions, they can be classified into families. For example, one such family constitutes the growth factors, encompassing more than 50 distinct molecules, each of which is highly effective in stimulating certain cells to divide and/or differentiate.

Stimuli for the release of paracrine/autocrine agents are also extremely varied. These include not only local chemical changes, such as in the concentration of oxygen, but neurotransmitters and hormones as well. In these two latter cases, the paracrine/autocrine agent often serves to oppose the effects the neurotransmitter or hormone induces locally. For example, the neurotransmitter norepinephrine strongly constricts blood vessels in the kidneys, but it simultaneously causes certain kidney cells to secrete paracrine agents that cause the same vessels to dilate. This provides a local negative feedback, in which the paracrine agents keep the action of norepinephrine from becoming too intense. This, then, is an example of homeostasis occurring at a highly localized level.

A point of great importance must be emphasized here to avoid later confusion. A nerve cell, endocrine gland cell, and other cell type may all secrete the same chemical messenger. Thus, a particular messenger may sometimes function as a neurotransmitter, as a hormone, or as a paracrine/autocrine agent. Norepinephrine, for example, is not only a neurotransmitter in the brain, it is also produced as a hormone by cells of the adrenal glands.

All types of intercellular communication described so far in this section involve secretion of a chemical messenger into the extracellular fluid. However, there are two important types of chemical communication between cells that do not require such secretion. In the first type, which occurs via gap junctions (Chapter 3), molecules move from one cell to an adjacent cell without ever entering the extracellular fluid. In the second type, the chemical messenger is not actually released from the cell producing it but rather is located in the plasma membrane of that cell. When the cell encounters another cell type capable of responding to the message, the two cells link up via the membrane-bound messenger. This type of signaling, sometimes termed "juxtacrine", is of particular importance in the growth and differentiation of tissues as well as in the functioning of cells that protect the body against microbes and other foreign agents (Chapter 18).

Processes Related to Homeostasis

Adaptation and Acclimatization

The term **adaptation** denotes a characteristic that favors survival in specific environments. Homeostatic control systems are inherited biological adaptations. An individual's ability to respond to a particular environmental stress is not fixed, however, but can be enhanced by prolonged exposure to that stress. This type of adaptation—the improved functioning of an already existing homeostatic system—is known as **acclimatization.**

Let us take sweating in response to heat exposure as an example and perform a simple experiment. On day 1 we expose a person for 30 min to a high temperature and ask her to do a standardized exercise test. Body temperature rises, and sweating begins after a certain period of time. The sweating provides a mechanism for increasing heat loss from the body and thus tends to minimize the rise in body temperature in a hot environment. The volume of sweat produced under these conditions is measured. Then, for a week, our subject enters the heat chamber for 1 or 2 h per day and exercises. On day 8, her body temperature and sweating rate are again measured during the same exercise test performed on day 1. The striking finding is that the subject begins to sweat sooner and much more profusely than she did on day 1. As a consequence, her body temperature does not rise to nearly the same degree. The subject has become acclimatized to the heat. She has undergone an adaptive change induced by repeated exposure to the heat and is now better able to respond to heat exposure.

The precise anatomical and physiological changes that bring about increased capacity to withstand change during acclimatization are highly varied. Typically, they involve an increase in the number, size, or sensitivity of one or more of the cell types in the homeostatic control system that mediate the basic response.

Acclimatizations are usually completely reversible. Thus, if the daily exposures to heat are discontinued, our subject's sweating rate will revert to the preacclimatized value within a relatively short time. If an acclimatization is induced very early in life, however, at a critical period for development of a structure or response, it is termed a **developmental acclimatization** and may be irreversible. For example, the barrel-shaped chests of natives of the Andes Mountains do not represent a genetic difference between them and their lowland compatriots. Rather, this is an irreversible acclimatization induced during the first few years of their lives by their exposure to the high-altitude, low-oxygen environment. The increase in chest size reflects the increase in lung size and function. The altered chest size remains even if the individual moves to a lowland environment later in life and stays there. Lowland persons who have suffered oxygen deprivation from heart or lung disease during their early years show precisely the same chest shape.

Biological Rhythms

As noted earlier, a striking characteristic of many body functions is the rhythmical changes they manifest. The most common type is the **circadian rhythm,** which cycles approximately once every 24 h. Waking and sleeping, body temperature, hormone concentrations in the blood, the excretion of ions into the urine, and many other functions undergo circadian variation (**Figure 1–9**).

What have biological rhythms to do with homeostasis? They add an anticipatory component to homeostatic control systems, in effect a feedforward system operating without detectors. The negative-feedback homeostatic responses we described earlier in this chapter are *corrective* responses. They are initiated *after* the steady state of the individual has been perturbed. In contrast, biological rhythms enable homeostatic mechanisms to be utilized immediately and automatically by activating them at times when a challenge is *likely* to occur but before it actually does occur. For example, there is a rhythm in the urinary excretion of potassium—excretion is high during the day and low at night. This makes sense because we ingest potassium in our food during the day, not at night when we are asleep. Therefore, the total amount of potassium in the body fluctuates less than if the rhythm did not exist.

A crucial point concerning most body rhythms is that they are internally driven. Environmental factors do not drive the rhythm but rather provide the timing cues important for **entrainment**, or setting of the actual hours of the rhythm. A classic experiment will clarify this distinction.

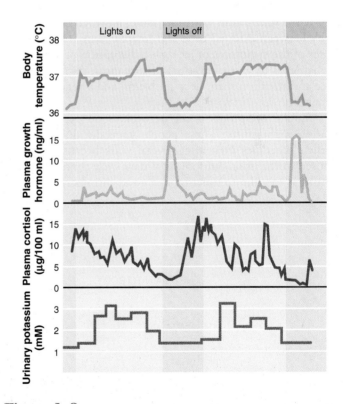

Figure 1–9

Circadian rhythms of several physiological variables in a human subject with room lights on (open bars at top) for 16 h and off (blue bars at top) for 8 h. Growth hormone and cortisol are hormones that regulate metabolism.

Adapted from Moore-Ede and Sulzman.

Subjects were put in experimental chambers that completely isolated them from their usual external environment, including knowledge of the time of day. For the first few days, they were exposed to a 24 h rest-activity cycle in which the room lights were turned on and off at the same time each day. Under these conditions, their sleep-wake cycles were 24 h long. Then, all environmental time cues were eliminated, and the subjects were allowed to control the lights themselves. Immediately, their sleep-wake patterns began to change. On average, bedtime began about 30 min later each day, and so did wake-up time. Thus a sleep-wake cycle persisted in the complete absence of environmental cues. Such a rhythm is called a **free-running rhythm.** In this case it was approximately 25 h rather than 24. This indicates that cues are required to entrain or set a circadian rhythm to 24 h.

The light-dark cycle is the most important environmental time cue in our lives, but not the only one. Others include external environmental temperature, meal timing, and many social cues. Thus, if several people were undergoing the experiment just described in isolation from each other, their free-running rhythms would be somewhat different, but if they were all in the same room, social cues would entrain all of them to the same rhythm.

Environmental time cues also function to **phase-shift** rhythms—in other words, to reset the internal clock. Thus if you jet west or east to a different time zone, your sleep-wake cycle and other circadian rhythms slowly shift to the new light-dark cycle. These shifts take time, however, and the disparity between external time and internal time is one of the causes of the symptoms of jet lag—a disruption of homeostasis that leads to gastrointestinal disturbances, decreased vigilance and attention span, sleep problems, and a general feeling of malaise.

Similar symptoms occur in workers on permanent or rotating night shifts. These people generally do not adapt to their schedules even after several years because they are exposed to the usual outdoor light-dark cycle (normal indoor lighting is too dim to function as a good entrainer). In recent experiments, night-shift workers were exposed to extremely bright indoor lighting while they worked and 8 h of total darkness during the day when they slept. This schedule produced total adaptation to night-shift work within 5 days.

What is the neural basis of body rhythms? In the part of the brain called the hypothalamus, a specific collection of nerve cells (the suprachiasmatic nucleus) functions as the principal **pacemaker,** or time clock, for circadian rhythms. How it keeps time independent of any external environmental cues is not fully understood, but it appears to involve the rhythmical turning on and off of critical genes in the pacemaker cells.

The pacemaker receives input from the eyes and many other parts of the nervous system, and these inputs mediate the entrainment effects exerted by the external environment. In turn, the pacemaker sends out neural signals to other parts of the brain, which then influence the various body systems, activating some and inhibiting others. One output of the pacemaker goes to the **pineal gland,** a gland within the brain that secretes the hormone **melatonin.** These neural signals from the pacemaker cause the pineal to secrete melatonin during darkness but not during daylight. It has been hypothesized, therefore, that melatonin may act as an important mediator to influence other organs either directly or by altering the activity of the parts of the brain that control these organs.

Balance in the Homeostasis of Chemical Substances in the Body

Many homeostatic systems regulate the balance between addition and removal of a chemical substance from the body. **Figure 1–10** is a generalized schema of the possible pathways involved in maintaining such balance. The **pool** occupies a position of central importance in the balance sheet. It is the body's readily available quantity of the substance and is often identical to the amount present in the extracellular fluid. The pool receives substances from and redistributes them to all the pathways.

The pathways on the left of Figure 1–10 are sources of net gain to the body. A substance may enter the body through

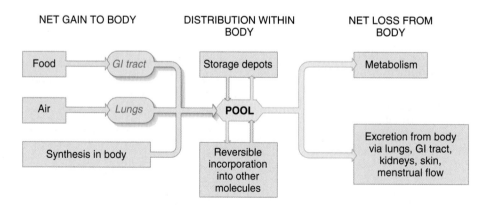

Figure 1–10

Balance diagram for a chemical substance.

the gastrointestinal (GI) tract or the lungs. Alternatively, a substance may be synthesized within the body from other materials.

The pathways on the right of the figure are causes of net loss from the body. A substance may be lost in the urine, feces, expired air, or menstrual fluid, as well as from the surface of the body as skin, hair, nails, sweat, or tears. The substance may also be chemically altered by enzymes and thus removed by metabolism.

The central portion of the figure illustrates the distribution of the substance within the body. The substance may be taken from the pool and accumulated in storage depots, such as the accumulation of fat in adipose tissue. Conversely, it may leave the storage depots to reenter the pool. Finally, the substance may be incorporated reversibly into some other molecular structure, such as fatty acids into plasma membranes. Incorporation is reversible because the substance is liberated again whenever the more complex structure is broken down. This pathway is distinguished from storage in that the incorporation of the substance into other molecules produces new molecules with specific functions.

Substances do not necessarily follow all pathways of this generalized schema. For example, minerals such as sodium cannot be synthesized, do not normally enter through the lungs, and cannot be removed by metabolism.

The orientation of Figure 1–10 illustrates two important generalizations concerning the balance concept: (1) During any period of time, total-body balance depends upon the relative rates of net gain and net loss to the body; and (2) the pool concentration depends not only upon the total amount of the substance in the body, but also upon exchanges of the substance *within* the body.

For any substance, three states of total-body balance are possible: (1) Loss exceeds gain, so that the total amount of the substance in the body is decreasing, and the person is in **negative balance;** (2) gain exceeds loss, so that the total amount of the substance in the body is increasing, and the person is in **positive balance;** and (3) gain equals loss, and the person is in **stable balance.**

Clearly a stable balance can be upset by a change in the amount being gained or lost in any single pathway in the schema. For example, increased sweating can cause severe negative water balance. Conversely, stable balance can be restored by homeostatic control of water intake and output.

Let us take sodium balance as another example. The control systems for sodium balance target the kidneys, and the systems operate by inducing the kidneys to excrete into the urine an amount of sodium approximately equal to the amount ingested daily. In this example, we assume for simplicity that all sodium loss from the body occurs via the urine (although some is also lost in perspiration). Now imagine a person with a daily intake and excretion of 7 g of sodium—not an unusual intake for most Americans—and a stable amount of sodium in her body (**Figure 1–11**). On day 2 of our experiment, the subject changes her diet so that her daily sodium consumption rises to 15 g—a large but still commonly observed intake—and remains there indefinitely. On this same day, the kidneys excrete into the urine somewhat more than 7 g of sodium,

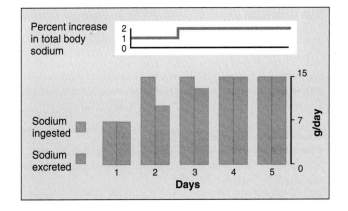

Figure 1–11

Effects of a continued change in the amount of sodium ingested on sodium excretion and total-body sodium balance. Stable sodium balance is reattained by day 4, but with some gain of total-body sodium.

but not all the ingested 15 g. The result is that some excess sodium is retained in the body that day—that is, the person is in positive sodium balance. The kidneys do somewhat better on day 3, but it is probably not until day 4 or 5 that they are excreting 15 g. From this time on, output from the body once again equals input, and sodium balance is once again stable. The delay of several days before stability is reached is quite typical for the kidneys' handling of sodium, but should not be assumed to apply to other homeostatic responses, most of which are much more rapid.

Although again in stable balance, the woman has perhaps 2 percent more sodium in her body than was the case when she was in stable balance ingesting 7 g. It is this 2 percent extra body sodium that constitutes the continuous error signal to the control systems, driving the kidneys to excrete 15 g/day rather than 7 g/day. Recall the generalization (Table 1–2, no. 3) that homeostatic control systems cannot maintain complete constancy of the internal environment *in the face of continued change in the perturbing event* because some change in the regulated variable (body sodium content, in our example) must persist to signal the need to maintain the compensating responses. An increase of 2 percent does not seem large, but it has been hypothesized that this small gain might facilitate the development of high blood pressure in some people.

In summary, homeostasis is a complex, dynamic process that regulates the adaptive responses of the body to changes in the external and internal environments. To work properly, homeostatic systems require a sensor to detect the environmental change, and a means to produce a compensatory response. Because compensatory responses require muscle activity, behavioral changes, or synthesis of chemical messengers such as hormones, homeostasis is only achieved by the expenditure of energy. The nutrients that provide this energy, and the cellular structures and chemical reactions that release the energy stored in the chemical bonds of the nutrients, are described in the following two chapters.

The Scope of Human Physiology

I. Physiology is the study of how living organisms work. Physiologists are interested in the regulation of body function.

II. Disease states are physiology "gone wrong" (pathophysiology).

How Is the Body Organized?

I. Cells are the simplest structural units into which a complex multicellular organism can be divided and still retain the functions characteristic of life.

II. Cell differentiation results in the formation of four categories of specialized cells:

a. Muscle cells generate the mechanical activities that produce force and movement.

b. Nerve cells initiate and conduct electrical signals.

c. Epithelial cells selectively secrete and absorb ions and organic molecules.

d. Connective tissue cells connect, anchor, and support the structures of the body.

III. Specialized cells associate with similar cells to form tissues: muscle tissue, nerve tissue, epithelial tissue, and connective tissue.

IV. Organs are composed of the four kinds of tissues arranged in various proportions and patterns. Many organs contain multiple small, similar functional units.

V. An organ system is a collection of organs that together perform an overall function.

Body Fluid Compartments

I. The body fluids are enclosed in compartments.

a. The extracellular fluid is composed of the interstitial fluid (the fluid between cells) and the blood plasma. Of the extracellular fluid, 75–80 percent is interstitial fluid, and 20–25 percent is plasma.

b. Interstitial fluid and plasma have essentially the same composition except that plasma contains a much higher concentration of protein.

c. Extracellular fluid differs markedly in composition from the fluid inside cells—the intracellular fluid.

d. Approximately one-third of body water is in the extracellular compartment, and two-thirds is intracellular.

II. The differing compositions of the compartments reflect the activities of the barriers separating them.

Homeostasis: A Defining Feature of Physiology

I. The body's internal environment is the extracellular fluid.

II. The function of organ systems is to maintain a stable internal environment—homeostasis.

III. Numerous variables within the body must be maintained homeostatically. When homeostasis is lost for one variable, it may trigger a series of changes in other variables.

General Characteristics of Homeostatic Control Systems

I. Homeostasis denotes the stable condition of the internal environment that results from the operation of compensatory homeostatic control systems.

a. In a negative feedback control system, a change in the variable being regulated brings about responses that tend to push the variable in the direction opposite to the original change. Negative feedback minimizes changes from the set point of the system, leading to stability.

b. Homeostatic control systems minimize changes in the internal environment but cannot maintain complete constancy.

c. Feedforward regulation anticipates changes in a regulated variable, improves the speed of the body's homeostatic responses, and minimizes fluctuations in the level of the variable being regulated.

Components of Homeostatic Control Systems

I. The components of a reflex arc are receptor, afferent pathway, integrating center, efferent pathway, and effector. The pathways may be neural or hormonal.

II. Local homeostatic responses are also stimulus-response sequences, but they occur only in the area of the stimulus, with neither nerves nor hormones involved.

Intercellular Chemical Messengers

I. Intercellular communication is essential to reflexes and local responses and is achieved by neurotransmitters, hormones, and paracrine or autocrine agents. Less common is intercellular communication through either gap junctions or cell-bound messengers.

Processes Related to Homeostasis

I. Acclimatization is an improved ability to respond to an environmental stress.

a. The improvement is induced by prolonged exposure to the stress with no change in genetic endowment.

b. If acclimatization occurs early in life, it may be irreversible and is known as a developmental acclimatization.

II. Biological rhythms provide a feedforward component to homeostatic control systems.

a. The rhythms are internally driven by brain pacemakers, but are entrained by environmental cues, such as light, which also serve to phase-shift (reset) the rhythms when necessary.

b. In the absence of cues, rhythms free run.

III. The balance of substances in the body is achieved by matching inputs and outputs. Total-body balance of a substance may be negative, positive, or stable.

■ KEY TERMS

acclimatization 13	endocrine gland 11
acquired reflex 9	entrainment 13
adaptation 13	epithelial cell 3
afferent pathway 10	epithelial tissue 3
autocrine agent 12	equilibrium 7
basement membrane 3	external environment 3
cell 2	extracellular fluid 5
cell differentiation 2	extracellular matrix 3
circadian rhythm 13	feedforward 9
collagen fiber 3	fiber 3
connective tissue 3	free-running rhythm 14
connective tissue cell 3	functional unit 4
developmental	homeostasis 6
acclimatization 13	homeostatic control system 7
dynamic constancy 6	hormone 11
effector 10	integrating center 10
efferent pathway 10	internal environment 5
elastin fiber 3	interstitial fluid 5

■ R E V I E W   Q U E S T I O N S

1. Describe the levels of cellular organization and state the four types of specialized cells and tissues.
2. List the organ systems of the body and give one-sentence descriptions of their functions.
3. Contrast the two categories of functions performed by every cell.
4. Name the two fluids that constitute the extracellular fluid. What are their relative proportions in the body, and how do they differ from each other in composition?
5. State the relative volumes of water in the body fluid compartments.
6. Describe five important generalizations about homeostatic control systems.
7. Contrast feedforward and negative feedback.
8. List the components of a reflex arc.
9. What is the basic difference between a local homeostatic response and a reflex?
10. List the general categories of intercellular messengers.
11. Describe the conditions under which acclimatization occurs. In what period of life might an acclimatization be irreversible? Are acclimatizations passed on to a person's offspring?
12. Under what conditions do circadian rhythms become free running?
13. How do phase shifts occur?
14. What is the most important environmental cue for entrainment of body rhythms?
15. Draw a figure illustrating the balance concept in homeostasis.
16. What are the three possible states of total-body balance of any chemical?

Chapter 1 Test Questions

(Answers appear in Appendix A.)

1. Which of the following is one of the four basic cell types in the body?
   a. respiratory
   b. epithelial
   c. endocrine
   d. integumentary
   e. immune

2. Which of the following is incorrect?
   a. Equilibrium requires a constant input of energy.
   b. Positive feedback is less common in nature than negative feedback.
   c. Homeostasis does not imply that a given variable is unchanging.
   d. Fever is an example of resetting a set point.
   e. Efferent pathways carry information away from the integrating center of a reflex arc.

3. In a reflex arc initiated by touching a hand to a hot stove, the effector will belong to which class of tissue?
   a. nerve
   b. connective
   c. muscle
   d. epithelial

4. In the absence of any environmental cues, a circadian rhythm is said to be
   a. entrained.
   b. in phase.
   c. free running.
   d. phase-shifted.
   e. no longer present.

5. Most of the water in the human body is found in
   a. the interstitial fluid compartment.
   b. the intracellular fluid compartment.
   c. the plasma compartment.
   d. the total extracellular fluid compartment.

Chapter 1 Quantitative and Thought Questions

(Answers appear in Appendix A.)

1. Eskimos have a remarkable ability to work in the cold without gloves and not suffer decreased skin blood flow. Does this prove that there is a genetic difference between Eskimos and other people with regard to this characteristic?

2. Explain how an imbalance in any given physiological variable might produce a change in one or more other variables.

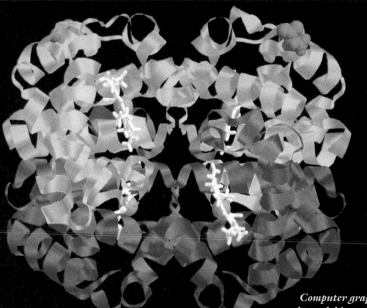

Computer graphic of hemoglobin molecule.

Chemical Composition of the Body

a toms and molecules are the chemical units of cell structure and function. In this chapter we describe the distinguishing characteristics of the major chemicals in the human body. The specific roles of these substances in physiology will be discussed in subsequent chapters. This chapter is, in essence, an expanded glossary of chemical terms and structures, and like a glossary, it should be consulted as needed.

Atoms

The units of matter that form all chemical substances are called **atoms.** Each type of atom—carbon, hydrogen, oxygen, and so on—is called a **chemical element.** A one- or two-letter symbol is used as a shorthand identification for each element. Although more than 100 elements exist in the universe, only 24 (**Table 2–1**) are known to be essential for the structure and function of the human body.

Table 2–1	Essential Chemical Elements in the Body
Element	**Symbol**
Major Elements: 99.3% of Total Atoms in the Body	
Hydrogen	H (63%)
Oxygen	O (26%)
Carbon	C (9%)
Nitrogen	N (1%)
Mineral Elements: 0.7% of Total Atoms in the Body	
Calcium	Ca
Phosphorus	P
Potassium	K (Latin *kalium*)
Sulfur	S
Sodium	Na (Latin *natrium*)
Chlorine	Cl
Magnesium	Mg
Trace Elements: Less than 0.01% of Total Atoms in the Body	
Iron	Fe (Latin *ferrum*)
Iodine	I
Copper	Cu (Latin *cuprum*)
Zinc	Zn
Manganese	Mn
Cobalt	Co
Chromium	Cr
Selenium	Se
Molybdenum	Mo
Fluorine	F
Tin	Sn (Latin *stannum*)
Silicon	Si
Vanadium	V

The chemical properties of atoms can be described in terms of three subatomic particles—**protons, neutrons, and electrons.** The protons and neutrons are confined to a very small volume at the center of an atom called the **atomic nucleus.** The electrons revolve in orbits at various distances from the nucleus. This miniature solar-system model of an atom is a gross oversimplification, but it is sufficient to provide a conceptual framework for understanding the chemical and physical interactions of atoms.

Each of the subatomic particles has a different electric charge. Protons have one unit of positive charge, electrons have one unit of negative charge, and neutrons are electrically neutral (**Table 2–2**). Because the protons are located in the atomic nucleus, the nucleus has a net positive charge equal to the number of protons it contains. The entire atom has no net electric charge, however, because the number of negatively charged electrons orbiting the nucleus equals the number of positively charged protons in the nucleus.

Atomic Number

Each chemical element contains a specific number of protons, and it is this number, known as the **atomic number,** that distinguishes one type of atom from another. For example, hydrogen, the simplest atom, has an atomic number of 1, corresponding to its single proton. As another example, calcium has an atomic number of 20, corresponding to its 20 protons. Because an atom is electrically neutral, the atomic number is also equal to the number of electrons in the atom.

Atomic Weight

Atoms have very little mass. A single hydrogen atom, for example, has a mass of only 1.67×10^{-24} g. The **atomic weight** scale indicates an atom's mass relative to the mass of other atoms. This scale is based upon assigning the carbon atom a mass of 12. On this scale, a hydrogen atom has an atomic weight of approximately 1, indicating that it has one-twelfth the mass of a carbon atom. A magnesium atom, with an atomic weight of 24, has twice the mass of a carbon atom.

Because the atomic weight scale is a *ratio* of atomic masses, it has no absolute units. The unit of atomic mass is known as a dalton. One dalton (d) equals one-twelfth the mass of a carbon atom. Thus, carbon has an atomic weight of 12, and a carbon atom has an atomic mass of 12 daltons.

Table 2–2	Characteristics of Major Subatomic Particles		
Particle	**Mass Relative to Electron**	**Electric Charge**	**Location in Atom**
Proton	1836	+1	Nucleus
Neutron	1839	0	Nucleus
Electron	1	–1	Orbiting the nucleus

Although the number of neutrons in the nucleus of an atom is often equal to the number of protons, many chemical elements can exist in multiple forms, called **isotopes,** which differ in the number of neutrons they contain. For example, the most abundant form of the carbon atom, ^{12}C, contains 6 protons and 6 neutrons, and thus has an atomic number of 6. Protons and neutrons are approximately equal in mass. Therefore, ^{12}C has an atomic weight of 12. The radioactive carbon isotope ^{14}C contains 6 protons and 8 neutrons, giving it an atomic number of 6 but an atomic weight of 14.

One **gram atomic mass** of a chemical element is the amount of the element, in grams, equal to the numerical value of its atomic weight. Thus, 12 g of carbon (assuming it is all ^{12}C) is 1 gram atomic mass of carbon. *One gram atomic mass of any element contains the same number of atoms.* For example, 1 g of hydrogen contains 6×10^{23} atoms, and 12 g of carbon, whose atoms have 12 times the mass of a hydrogen atom, also has 6×10^{23} atoms (the so-called Avogadro's number).

Atomic Composition of the Body

Just four of the body's essential elements (see Table 2–1)—hydrogen, oxygen, carbon, and nitrogen—account for over 99 percent of the atoms in the body.

The seven essential mineral elements are the most abundant substances dissolved in the extracellular and intracellular fluids. Most of the body's calcium and phosphorus atoms, however, make up the solid matrix of bone tissue.

The 13 essential **trace elements** are present in extremely small quantities, but they are nonetheless essential for normal growth and function. For example, iron plays a critical role in the blood's transport of oxygen.

Many other elements, in addition to the 24 listed in Table 2–1, may be detected in the body. These elements enter in the foods we eat and the air we breathe but are not essential for normal body function and may even interfere with normal body chemistry. For example, ingested arsenic has poisonous effects.

Molecules

Two or more atoms bonded together make up a **molecule.** For example, a molecule of water contains two hydrogen atoms and one oxygen atom, which can be represented as H_2O. The atomic composition of glucose, a sugar, is $C_6H_{12}O_6$, indicating that the molecule contains 6 carbon atoms, 12 hydrogen atoms, and 6 oxygen atoms. Such formulas, however, do not indicate how the atoms are linked together in the molecule.

Covalent Chemical Bonds

The atoms in molecules are held together by chemical bonds, which form when electrons transfer from one atom to another or when two atoms share electrons. The strongest chemical bond between two atoms, a **covalent bond,** forms when one electron in the outer electron orbit of each atom is shared between the two atoms (**Figure 2–1**). The atoms in most molecules found in the body are linked by covalent bonds.

The atoms of some elements can form more than one covalent bond and thus become linked simultaneously to two or more other atoms. Each type of atom forms a characteristic number of covalent bonds, which depends on the number of electrons in its outermost orbit. The number of chemical bonds formed by the four most abundant atoms in the body are hydrogen, one; oxygen, two; nitrogen, three; and carbon, four. When the structure of a molecule is diagrammed, each covalent bond is represented by a line indicating a pair of shared electrons. The covalent bonds of the four elements just mentioned can be represented as

$$H— \qquad —O— \qquad \overset{|}{—N—} \qquad \overset{|}{\underset{|}{—C—}}$$

A molecule of water, H_2O, can be diagrammed as

$$H—O—H$$

In some cases, two covalent bonds—a double bond—form between two atoms when they share two electrons from each atom. Carbon dioxide (CO_2) contains two double bonds:

$$O=C=O$$

Note that in this molecule the carbon atom still forms four covalent bonds and each oxygen atom only two.

Molecular Shape

When atoms are linked together, they form molecules with various shapes. Although we draw diagrammatic structures of molecules on flat sheets of paper, molecules are three-dimensional. When more than one covalent bond is formed with a given atom, the bonds are distributed around the atom in a pattern that may or may not be symmetrical (**Figure 2–2**).

Molecules are not rigid, inflexible structures. Within certain limits, the shape of a molecule can be changed without breaking the covalent bonds linking its atoms together. A covalent bond is like an axle around which the joined atoms can rotate. As illustrated in **Figure 2–3**, a sequence of six carbon atoms can assume a number of shapes by rotating around various covalent bonds. As we will see, the three-dimensional, flexible shape of molecules is one of the major factors governing molecular interactions.

Ions

A single atom is electrically neutral because it contains equal numbers of negative electrons and positive protons. If, however, an atom gains or loses one or more electrons, it acquires a net electric charge and becomes an **ion.** For example, when a sodium atom (Na), which has 11 electrons, loses one electron, it becomes a sodium ion (Na^+) with a net positive charge; it still has 11 protons, but it now has only 10 electrons. On the other hand, a chlorine atom (Cl), which has 17 electrons, can gain an electron and become a chloride ion (Cl^-) with a net negative charge—it now has 18 electrons but only 17 protons. Some atoms can gain or lose more than one electron to become ions with two or even three units of net electric charge (for example, the calcium ion Ca^{2+}).

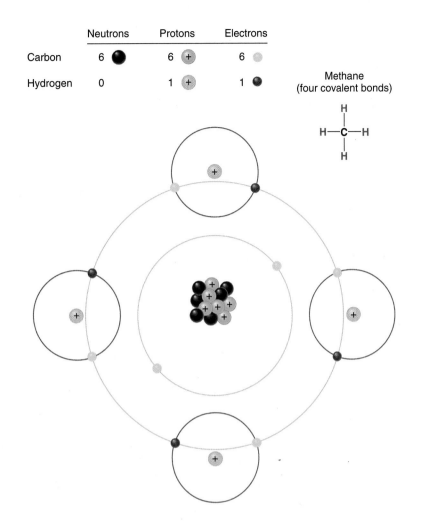

	Neutrons	Protons	Electrons
Carbon	6 ●	6 ⊕	6 ○
Hydrogen	0	1 ⊕	1 ●

Methane
(four covalent bonds)

$$H-\overset{\displaystyle H}{\underset{\displaystyle H}{C}}-H$$

Figure 2–1

Each of the four hydrogen atoms in a molecule of methane (CH_4) forms a covalent bond with the carbon atom by sharing its one electron with one of the electrons in carbon. Each shared pair of electrons—one electron from the carbon and one from a hydrogen atom—forms a covalent bond. The sizes of protons, neutrons, and electrons are not to scale.

Hydrogen atoms and most mineral and trace element atoms readily form ions. **Table 2–3** lists the ionic forms of some of these elements that are found in the body. Ions that have a net positive charge are called **cations,** while those that have a net negative charge are called **anions.** Because of their ability to conduct electricity when dissolved in water, the ionic forms of mineral elements are collectively referred to as **electrolytes.**

The process of ion formation, known as ionization, can occur in single atoms or in atoms that are covalently linked in molecules. Within molecules, two commonly encountered groups of atoms that undergo ionization are the **carboxyl group** (—COOH) and the **amino group** (—NH$_2$). The shorthand formula for only a portion of a molecule can be written as R—COOH or R—NH$_2$, where R signifies the remaining portion of the molecule. The carboxyl group ionizes when the oxygen linked to the hydrogen captures the hydrogen's only electron to form a carboxyl ion (R—COO$^-$), releasing a hydrogen ion (H$^+$):

$$R-COOH \rightleftharpoons R-COO^- + H^+$$

The amino group can bind a hydrogen ion to form an ionized amino group (R—NH$_3^+$):

$$R-NH_2 + H^+ \rightleftharpoons R-NH_3^+$$

The ionization of each of these groups can be reversed, as indicated by the double arrows; the ionized carboxyl group can combine with a hydrogen ion to form an un-ionized carboxyl group, and the ionized amino group can lose a hydrogen ion and become an un-ionized amino group.

Free Radicals

The electrons that revolve around the nucleus of an atom occupy regions known as orbitals, each of which can be occupied by one or more pairs of electrons, depending on the distance of the orbital from the nucleus. An atom is most stable when each orbital is occupied by its full complement of electrons. An atom containing a single (unpaired) electron in its outermost orbital is known as a **free radical,** as are molecules containing such atoms. Free radicals are unstable molecules that can react with other atoms, through the process known as oxidation. When a free radical oxidizes another atom, the free radical gains an electron and the other atom usually becomes a new free radical.

Free radicals are formed by the actions of certain enzymes in some cells, such as types of white blood cells

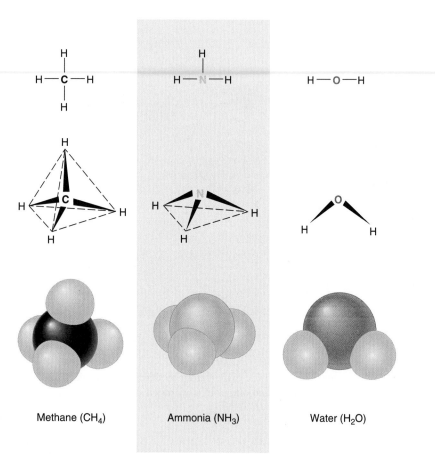

Methane (CH₄)   Ammonia (NH₃)   Water (H₂O)

Figure 2–2

Three different ways of representing the geometric configuration of covalent bonds around the carbon, nitrogen, and oxygen atoms bonded to hydrogen atoms.

Figure 2–3

Changes in molecular shape occur as portions of a molecule rotate around different carbon-to-carbon bonds, transforming this molecule's shape, for example, from a relatively straight chain (top) into a ring (bottom).

Table 2–3	Ionic Forms of Elements Most Frequently Encountered in the Body			
Chemical Atom	**Symbol**	**Ion**	**Chemical Symbol**	**Electrons Gained or Lost**
Hydrogen	H	Hydrogen ion	H^+	1 lost
Sodium	Na	Sodium ion	Na^+	1 lost
Potassium	K	Potassium ion	K^+	1 lost
Chlorine	Cl	Chloride ion	Cl^-	1 gained
Magnesium	Mg	Magnesium ion	Mg^{2+}	2 lost
Calcium	Ca	Calcium ion	Ca^{2+}	2 lost

that destroy pathogens. The free radicals are highly reactive, removing electrons from the outer orbits of molecules present in the pathogen cell membrane, for example. This mechanism begins the process whereby the pathogen is destroyed.

In addition, however, free radicals can be produced in the body following exposure to radiation or toxin ingestion. These free radicals can do considerable harm to the cells of the body. For example, oxidation due to long-term buildup of free radicals has been proposed as one cause of several different human diseases, notably eye, cardiovascular, and neural diseases associated with aging. Thus, it is important that free radicals be inactivated by molecules that can donate electrons to free radicals without becoming free radicals themselves. Examples of such protective molecules are the antioxidant vitamins C and E.

Free radicals are diagrammed with a dot next to the atomic symbol. Examples of biologically important free radicals are superoxide anion, $O_2 \cdot ^-$; hydroxyl radical, $OH \cdot$; and nitric oxide, $NO \cdot$. Note that a free radical configuration can occur in either an ionized (charged) or an un-ionized molecule.

Polar Molecules

As we have seen, when the electrons of two atoms interact, the two atoms may share the electrons equally, forming an electrically neutral covalent bond. Alternatively, one of the atoms may completely capture an electron from the other, forming two ions. Between these two extremes are bonds in which the electrons are not shared equally between the two atoms, but instead reside closer to one atom of the pair. This atom thus acquires a slight negative charge, while the other atom, having partly lost an electron, becomes slightly positive. Such bonds are known as **polar covalent bonds** (or, simply, polar bonds) because the atoms at each end of the bond have an opposite electric charge. For example, the bond between hydrogen and oxygen in a **hydroxyl group** (—OH) is a polar covalent bond in which the oxygen is slightly negative and the hydrogen slightly positive:

$$\overset{(\delta^-)\ (\delta^+)}{R—O—H}$$

(The δ^- and δ^+ symbols refer to atoms with a partial negative or positive charge, respectively. The R symbolizes the remainder of the molecule.) The electric charge associated with the ends of a polar bond is considerably less than the charge on a fully ionized atom. For example, the oxygen in the polarized hydroxyl group has only about 13 percent of the negative charge associated with the oxygen in an ionized carboxyl group, R—COO$^-$. Polar bonds do not have a *net* electric charge, as do ions, because they contain equal amounts of negative and positive charge.

Atoms of oxygen and nitrogen, which have a relatively strong attraction for electrons, form polar bonds with hydrogen atoms. In contrast, bonds between carbon and hydrogen atoms and between two carbon atoms are electrically neutral (**Table 2–4**).

Table 2–4	Examples of Nonpolar and Polar Bonds, and Ionized Chemical Groups
Nonpolar Bonds	$—\overset{\textstyle\vert}{\underset{\textstyle\vert}{C}}—H$ Carbon-hydrogen bond
	$—\overset{\textstyle\vert}{\underset{\textstyle\vert}{C}}—\overset{\textstyle\vert}{\underset{\textstyle\vert}{C}}—$ Carbon-carbon bond
Polar Bonds	$\overset{(\delta^-)\ (\delta^+)}{R—O—H}$ Hydroxyl group (R—OH)
	$\overset{(\delta^-)\ (\delta^+)}{R—S—H}$ Sulfhydryl group (R—SH)
	$\underset{R—N—R}{\overset{H(\delta^+)}{\vert(\delta^-)}}$ Nitrogen-hydrogen bond
Ionized Groups	$\underset{R—C—O^-}{\overset{O}{\|\|}}$ Carboxyl group (R—COO$^-$)
	$\underset{\underset{H}{\vert}}{\overset{\overset{H}{\vert^+}}{R—N—H}}$ Amino group (R—NH$_3^+$)
	$\underset{\underset{O^-}{\vert}}{\overset{\overset{O}{\|\|}}{R—O—P—O^-}}$ Phosphate group (R—PO$_4^{2-}$)

Different regions of a single molecule may contain nonpolar bonds, polar bonds, and ionized groups. Molecules containing significant numbers of polar bonds or ionized groups are known as **polar molecules,** whereas molecules composed predominantly of electrically neutral bonds are known as **nonpolar molecules.** As we will see, the physical characteristics of these two classes of molecules, especially their solubility in water, are quite different.

Hydrogen Bonds

The electrical attraction between the hydrogen atom in a polar bond in one molecule and an oxygen or nitrogen atom in a polar bond of another molecule forms a **hydrogen bond.** Such bonds may also form between atoms within the same molecule. Hydrogen bonds are represented in diagrams by dashed or dotted lines to distinguish them from covalent bonds (**Figure 2–4**). Hydrogen bonds are very weak, having only about 4 percent of the strength of the polar bonds between the hydrogen and oxygen atoms in a single molecule of water. Although hydrogen bonds are weak individually, when present in large numbers, they play an extremely important role in molecular interactions and in determining the shape of large

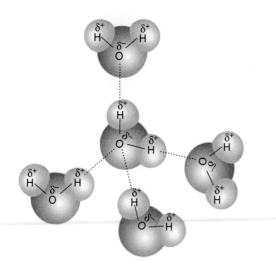

Figure 2–4

Five water molecules. Note that polarized covalent bonds link the hydrogen and oxygen atoms within each molecule and that hydrogen bonds occur between adjacent molecules. Hydrogen bonds are represented in diagrams by dashed or dotted lines, and covalent bonds by solid lines. The δ symbol means that a partial charge exists on that atom due to the unequal sharing of electrons between hydrogens and oxygen within a molecule.

molecules. This is of great importance for physiology, because the shape of large molecules often determines their functions. For example, the ability of a specific cell membrane "receptor" to recognize a large protein depends partly on the shape of both the protein and the receptor.

Water

Water is the most common molecule in the body. Out of every 100 molecules, 99 are water. The covalent bonds linking the two hydrogen atoms to the oxygen atom in a water molecule are polar. Therefore, the oxygen in water has a slight negative charge, and each hydrogen has a slight positive charge. The positively polarized regions near the hydrogen atoms of one water molecule are electrically attracted to the negatively polarized regions of the oxygen atoms in adjacent water molecules by hydrogen bonds (see Figure 2–4).

At body temperature, water exists as a liquid because the weak hydrogen bonds between water molecules are continuously forming and breaking. If the temperature is increased, the hydrogen bonds break more readily, and molecules of water escape into the gaseous state. However, if the temperature is lowered, hydrogen bonds break less frequently, so larger and larger clusters of water molecules form until at 0°C water freezes into a continuous crystalline matrix—ice.

Water molecules take part in many chemical reactions of the general type:

$$R_1—R_2 + H—O—H \rightleftharpoons R_1—OH + H—R_2$$

In this reaction, the covalent bond between R_1 and R_2 and the one between a hydrogen atom and oxygen in water are broken,

and the hydroxyl group and hydrogen atom are transferred to R_1 and R_2, respectively. Reactions of this type are known as hydrolytic reactions, or **hydrolysis.** Many large molecules in the body are broken down into smaller molecular units by hydrolysis, usually with the assistance of a class of molecules called enzymes. These reactions are usually reversible, a process known as **dehydration.** In dehydration, one net water molecule is removed to combine two small molecules into one larger one. Dehydration reactions are responsible for, among other things, building proteins and other polymers required by the body.

Other properties of water that are of importance in physiology include the colligative properties—those that depend on the number of dissolved substances, or **solutes,** in water. For example, water moves between fluid compartments by the process of osmosis, which you will learn about in detail in Chapter 4. In osmosis, water moves from regions of low solute concentrations to regions of high solute concentrations. The characteristics of solutes and solutions are described next.

Solutions

Substances dissolved in a liquid are known as solutes, and the liquid in which they are dissolved is the **solvent.** Solutes dissolve in a solvent to form a **solution.** Water is the most abundant solvent in the body, accounting for ≈60 percent of total body weight. A majority of the chemical reactions that occur in the body involve molecules that are dissolved in water, either in the intracellular or extracellular fluid. However, not all molecules dissolve in water.

Molecular Solubility

In order to dissolve in water, a substance must be electrically attracted to water molecules. For example, table salt (NaCl) is a solid crystalline substance because of the strong electrical attraction between positive sodium ions and negative chloride ions. This strong attraction between two oppositely charged ions is known as an **ionic bond.** When a crystal of sodium chloride is placed in water, the polar water molecules are attracted to the charged sodium and chloride ions (**Figure 2–5**). Clusters of water molecules surround the ions, allowing the sodium and chloride ions to separate from the salt crystal and enter the water—that is, to dissolve.

Molecules having a number of polar bonds and/or ionized groups will dissolve in water. Such molecules are said to be **hydrophilic,** or "water-loving." Thus, the presence of ionized groups (such as carboxyl and amino groups) or of polar groups (such as hydroxyl groups) in a molecule promotes solubility in water. In contrast, molecules composed predominantly of carbon and hydrogen are insoluble in water because their electrically neutral covalent bonds are not attracted to water molecules. These molecules are **hydrophobic,** or "water-fearing."

When hydrophobic molecules are mixed with water, two phases form, as occurs when oil is mixed with water. The strong attraction between polar molecules "squeezes" the nonpolar molecules out of the water phase. Such a separation is never 100 percent complete, however, so very small amounts of nonpolar solutes remain dissolved in the water phase.

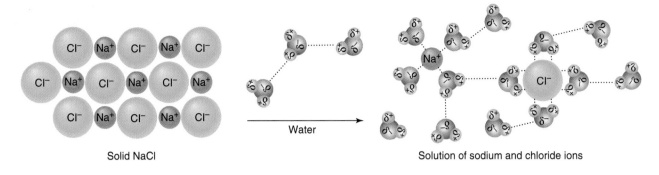

Figure 2–5

The ability of water to dissolve sodium chloride crystals depends upon the electrical attraction between the polar water molecules and the charged sodium and chloride ions.

Molecules that have a polar or ionized region at one end and a nonpolar region at the opposite end are called **amphipathic**—consisting of two parts. When mixed with water, amphipathic molecules form clusters, with their polar (hydrophilic) regions at the surface of the cluster where they are attracted to the surrounding water molecules. The nonpolar (hydrophobic) ends are oriented toward the interior of the cluster (**Figure 2–6**). Such an arrangement provides the maximal interaction between water molecules and the polar ends of the amphipathic molecules. Nonpolar molecules can dissolve in the central nonpolar regions of these clusters and thus exist in aqueous solutions in far higher amounts than would otherwise be possible based on their low solubility in water. As we will see, the orientation of amphipathic molecules plays an important role in cell membrane structure and in both the absorption of nonpolar molecules from the gastrointestinal tract and their transport in the blood.

Concentration

Solute **concentration** is defined as the amount of the solute present in a unit volume of solution. One measure of the amount of a substance is its mass expressed in grams. The unit of volume in the metric system is a liter (L). (One liter equals 1.06 quarts; see the conversion table at the back of the book for metric and English units.) Smaller units commonly used in physiology are the deciliter (dL, or 0.1 liter), the milliliter (ml, or 0.001 liter), and the microliter (μl, or 0.001 ml). The concentration of a solute in a solution can then be expressed as the number of grams of the substance present in one liter of solution (g/L).

A comparison of the concentrations of two different substances on the basis of the number of grams per liter of solution does not directly indicate how many molecules of each substance are present. For example, if the molecules of compound X are heavier than those of compound Y, 10 g of compound X will contain fewer molecules than 10 g of compound Y. Concentrations in units of grams per liter are most often used when the chemical structure of the solute is unknown. When the structure of a molecule is known, concentrations

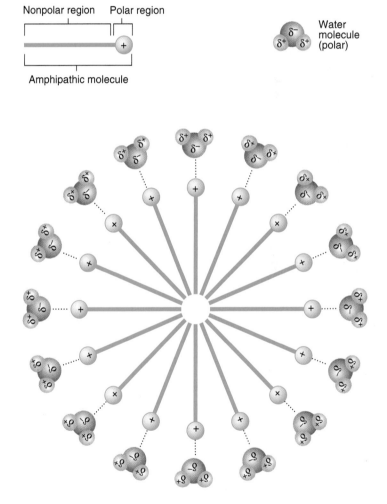

Figure 2–6

In water, amphipathic molecules aggregate into spherical clusters. Their polar regions form hydrogen bonds with water molecules at the surface of the cluster, while the nonpolar regions cluster together away from water.

are usually expressed as moles per liter, which provides a unit of concentration based upon the number of solute molecules in solution, as described next.

The **molecular weight** of a molecule is equal to the sum of the atomic weights of all the atoms in the molecule. For example, glucose ($C_6H_{12}O_6$) has a molecular weight of 180 $[(6 \times 12) + (12 \times 1) + (6 \times 16)] = 180$. One **mole** (abbreviated mol) of a compound is the amount of the compound in grams equal to its molecular weight. A solution containing 180 g of glucose (1 mol) in 1 L of solution is a 1 molar solution of glucose (1 mol/L). If 90 g of glucose were dissolved in 1 L of water, the solution would have a concentration of 0.5 mol/L. Just as 1 gram atomic mass of any element contains the same number of atoms, 1 mol (1 gram molecular mass) of any molecule will contain the same number of molecules—6×10^{23} (Avogadro's number). Thus, a 1 mol/L solution of glucose contains the same number of solute molecules per liter as a 1 mol/L solution of any other substance.

The concentrations of solutes dissolved in the body fluids are much less than 1 mol/L. Many have concentrations in the range of millimoles per liter (1 mmol/L = 0.001 mol/L), while others are present in even smaller concentrations—micromoles per liter (1 μmol/L = 0.000001 mol/L) or nanomoles per liter (1 nmol/L = 0.000000001 mol/L). By convention, the liter (L) term is sometimes dropped when referring to concentrations. Thus, a 1 mmol/L solution is often written as 1 mM (the capital "M" stands for "molar," and is defined as mol/L).

Hydrogen Ions and Acidity

As mentioned earlier, a hydrogen atom has a single proton in its nucleus orbited by a single electron. A hydrogen ion (H^+), formed by the loss of the electron, is thus a single free proton. Hydrogen ions form when the proton of a hydrogen atom in a molecule is released, leaving behind the hydrogen atom's electron. Molecules that release protons (hydrogen ions) in solution are called **acids,** for example:

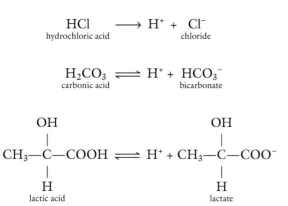

Conversely, any substance that can accept a hydrogen ion (proton) is termed a **base.** In the reactions shown, bicarbonate and lactate are bases because they can combine with hydrogen ions (note the two-way arrows in the two reactions). It is important to distinguish between the un-ionized acid and ionized base forms of these molecules. Also, note that by convention, separate terms are used for the acid forms, lactic acid

and carbonic acid, and the bases derived from the acids, lactate and bicarbonate. By combining with hydrogen ions, bases lower the hydrogen ion concentration of a solution.

When hydrochloric acid is dissolved in water, 100 percent of its atoms separate to form hydrogen and chloride ions, and these ions do not recombine in solution (note the one-way arrow in the preceding diagram). In the case of lactic acid, however, only a fraction of the lactic acid molecules in solution release hydrogen ions at any instant. Therefore, if a 1 mol/L solution of lactic acid is compared with a 1 mol/L solution of hydrochloric acid, the hydrogen ion concentration will be lower in the lactic acid solution than in the hydrochloric acid solution. Hydrochloric acid and other acids that are 100 percent ionized in solution are known as **strong acids,** whereas carbonic and lactic acids and other acids that do not completely ionize in solution are **weak acids.** The same principles apply to bases.

It is important to understand that the hydrogen ion concentration of a solution refers only to the hydrogen ions that are free in solution and not to those that may be bound, for example, to amino groups ($R—NH_3^+$). The **acidity** of a solution thus refers to the *free* (unbound) hydrogen ion concentration in the solution; the higher the hydrogen ion concentration, the greater the acidity. The hydrogen ion concentration is often expressed as the solution's **pH,** which is defined as the negative logarithm to the base 10 of the hydrogen ion concentration. The brackets around the symbol for the hydrogen ion in the following formula indicate concentration:

$$pH = -\log [H^+]$$

Thus, a solution with a hydrogen ion concentration of 10^{-7} mol/L has a pH of 7, whereas a more acidic solution with a higher H^+ concentration of 10^{-6} mol/L has a lower pH of 6. Note that as the acidity *increases,* the pH *decreases;* a change in pH from 7 to 6 represents a 10-fold increase in the hydrogen ion concentration.

Pure water, due to the ionization of some of the molecules into H^+ and OH^-, has a hydrogen ion concentration of 10^{-7} mol/L (pH = 7.0) and is termed a **neutral solution. Alkaline solutions** have a lower hydrogen ion concentration (a pH higher than 7.0), while those with a higher hydrogen ion concentration (a pH lower than 7.0) are **acidic solutions.** The extracellular fluid of the body has a hydrogen ion concentration of about 4×10^{-8} mol/L (pH = 7.4), with a homeostatic range of about pH 7.35 to 7.45, and is thus slightly alkaline. Most intracellular fluids have a slightly higher hydrogen ion concentration (pH 7.0 to 7.2) than extracellular fluids.

As we saw earlier, the ionization of carboxyl and amino groups involves the release and uptake, respectively, of hydrogen ions. These groups behave as weak acids and bases. Changes in the acidity of solutions containing molecules with carboxyl and amino groups alter the net electric charge on these molecules by shifting the ionization reaction to the right or left according to the general form:

$$R—COO^- + H^+ \rightleftharpoons R—COOH$$

For example, if the acidity of a solution containing lactate is increased by adding hydrochloric acid, the concentration of lactic acid will increase and that of lactate will decrease.

If the electric charge on a molecule is altered, its interaction with other molecules or with other regions within the same molecule changes, and thus its functional characteristics change. In the extracellular fluid, *hydrogen ion concentrations beyond the 10-fold pH range of 7.8 to 6.8 are incompatible with life if maintained for more than a brief period of time.* Even small changes in the hydrogen ion concentration can produce large changes in molecular interaction. For example, many enzymes in the body operate efficiently within very narrow ranges of pH. Should pH vary from the normal homeostatic range due to disease, these enzymes work at reduced levels, creating an even worse pathological situation.

This concludes our overview of atomic and molecular structure, water, and pH. We turn now to a description of the molecules essential for life in all living organisms, including humans. These are the carbon-based molecules required for forming the building blocks of cells, tissues, and organs; providing energy; and forming the genetic blueprints of all life.

Classes of Organic Molecules

Because most naturally occurring carbon-containing molecules are found in living organisms, the study of these compounds became known as organic chemistry. (Inorganic chemistry is the study of noncarbon-containing molecules.) However, the chemistry of living organisms, biochemistry, now forms only a portion of the broad field of organic chemistry.

One of the properties of the carbon atom that makes life possible is its ability to form four covalent bonds with other atoms, including with other carbon atoms. Because carbon atoms can also combine with hydrogen, oxygen, nitrogen, and sulfur atoms, a vast number of compounds can form from relatively few chemical elements. Some of these molecules are extremely large **(macromolecules),** composed of thousands of atoms. Such large molecules form when many smaller molecules, or subunits, link together. These large molecules are known as **polymers** (literally "many small parts"). The structure of macromolecules depends upon the structure of the subunits (monomers), the number of subunits bonded together, and the three-dimensional way in which the subunits are linked.

Most of the organic molecules in the body can be classified into one of four groups: carbohydrates, lipids, proteins, and nucleic acids (**Table 2–5**).

Carbohydrates

Although carbohydrates account for only about 1 percent of body weight, they play a central role in the chemical reactions that provide cells with energy. Carbohydrates are composed of carbon, hydrogen, and oxygen atoms in the proportions represented by the general formula $C_n(H_2O)_n$, where n is any whole number. It is from this formula that the class of molecules gets its name, **carbohydrate**—water-containing (hydrated) carbon atoms. Linked to most of the carbon atoms in a carbohydrate are a hydrogen atom and a hydroxyl group:

$$H—C—OH$$

Table 2–5	Major Categories of Organic Molecules in the Body		
Category	**Percent of Body Weight**	**Predominant Atoms**	**Subclass**	**Subunits**
Carbohydrates	1	C, H, O	Polysaccharides (and disaccharides)	Monosaccharides
Lipids	15	C, H	Triglycerides	3 fatty acids + glycerol
			Phospholipids	2 fatty acids + glycerol + phosphate + small charged nitrogen molecule
			Steroids	
Proteins	17	C, H, O, N	Peptides and polypeptides	Amino acids
Nucleic acids	2	C, H, O, N	DNA	Nucleotides containing the bases adenine, cytosine, guanine, thymine, the sugar deoxyribose, and phosphate
			RNA	Nucleotides containing the bases adenine, cytosine, guanine, uracil, the sugar ribose, and phosphate

It is the presence of numerous hydroxyl groups that makes carbohydrates readily soluble in water.

Most carbohydrates taste sweet, particularly the carbohydrates known as sugars. The simplest sugars are the monomers called **monosaccharides,** the most abundant of which is **glucose,** a six-carbon molecule ($C_6H_{12}O_6$). Glucose is often called "blood sugar" because it is the major monosaccharide found in the blood.

Two ways to represent the bonds between the atoms of a monosaccharide are illustrated in **Figure 2–7**. The first is the conventional way of drawing the structure of organic molecules, but the second gives a better representation of their three-dimensional shape. Five carbon atoms and an oxygen atom form a ring that lies in an essentially flat plane. The hydrogen and hydroxyl groups on each carbon lie above and below the plane of this ring. If one of the hydroxyl groups below the ring is shifted to a position above the ring, as shown in **Figure 2–8**, a different monosaccharide is produced.

Most monosaccharides in the body contain five or six carbon atoms and are called **pentoses** and **hexoses,** respectively. Larger carbohydrates can be formed by joining a number of monosaccharides together. Carbohydrates composed of two monosaccharides are known as **disaccharides. Sucrose,** or table sugar, is composed of two monosaccharides, glucose and fructose (**Figure 2–9**). The linking together of most monosaccharides involves a dehydration reaction in which a hydroxyl group is removed from one monosaccharide and a hydrogen atom is removed from the other, giving rise to a molecule of water and bonding the two sugars together through an oxygen atom. Conversely, hydrolysis of the disaccharide breaks this linkage by adding back the water and thus uncoupling the two monosaccharides. Other disaccharides frequently encountered are maltose (glucose-glucose), formed during the digestion of large carbohydrates in the intestinal tract, and lactose (glucose-galactose), present in milk.

When many monosaccharides are linked together to form polymers, the molecules are known as **polysaccharides.** Starch, found in plant cells, and **glycogen,** present in animal cells and often called "animal starch," are examples of polysaccharides (**Figure 2–10**). Both of these polysaccharides are composed of thousands of glucose molecules linked together in long chains, differing only in the degree of branching along the chain. Glycogen exists in the body as a reservoir of available fuel. Hydrolysis of glycogen, as occurs during periods of

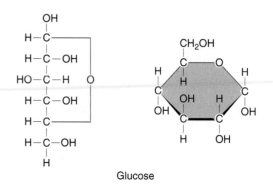

Glucose

Figure 2–7

Two ways of diagramming the structure of the monosaccharide glucose.

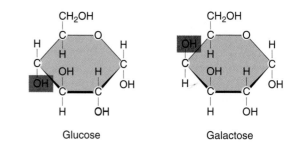

Glucose   Galactose

Figure 2–8

The structural difference between the monosaccharides glucose and galactose is based on whether the hydroxyl group at the position indicated lies below or above the plane of the ring.

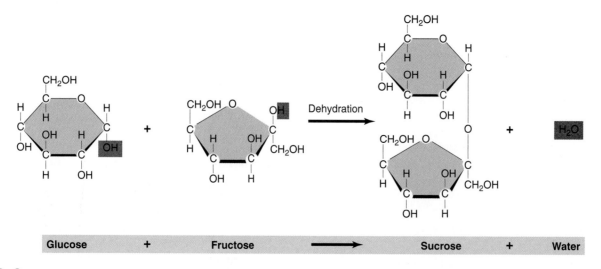

| Glucose | + | Fructose | ⟶ | Sucrose | + | Water |

Figure 2–9

Sucrose (table sugar) is a disaccharide formed when two monosaccharides, glucose and fructose, bond together through a dehydration reaction.

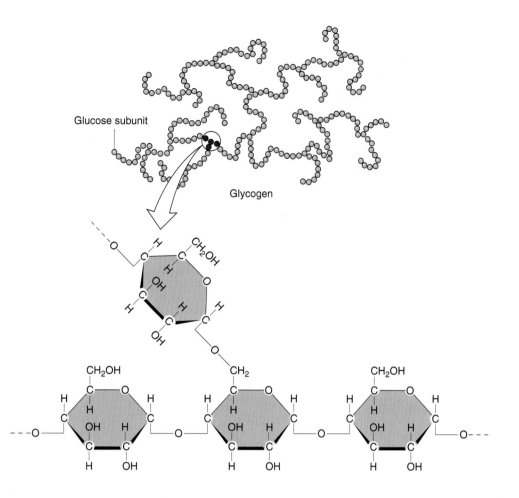

Figure 2–10

Many molecules of glucose joined end-to-end and at branch points form the branched-chain polysaccharide glycogen, shown here in diagrammatic form. The four red subunits in the glycogen molecule correspond to the four glucose subunits shown at the bottom.

fasting, leads to release of the glucose subunits into the blood, thereby preventing blood glucose from decreasing to dangerously low levels.

Lipids

Lipids are molecules composed predominantly (but not exclusively) of hydrogen and carbon atoms. These atoms are linked by neutral covalent bonds. Thus, lipids are nonpolar and have a very low solubility in water. Lipids, which account for about 40 percent of the organic matter in the average body (15 percent of the body weight), can be divided into four subclasses: fatty acids, triglycerides, phospholipids, and steroids. Like carbohydrates, lipids are important in physiology partly because they provide a valuable source of energy.

Fatty Acids

A **fatty acid** consists of a chain of carbon and hydrogen atoms with a carboxyl group at one end (**Figure 2–11**). Thus, fatty acids contain two oxygen atoms in addition to their complement of carbon and hydrogen. Fatty acids are synthesized in the body by the bonding together of two-carbon fragments, resulting most commonly in fatty acids of 16 or 18 carbon atoms. When all the carbons in a fatty acid are linked by single covalent bonds, the fatty acid is said to be a **saturated fatty acid,** because all the carbons are saturated with covalently bound H. Some fatty acids contain one or more double bonds, and these are known as **unsaturated fatty acids.** If one double bond is present, the fatty acid is **monounsaturated,** and if there is more than one double bond, **polyunsaturated (Figure 2–11a).**

Most naturally occurring unsaturated fatty acids exist in the *cis* position, with both hydrogens on the same side of the double-bonded carbons (see Figure 2–11). It is possible, however, to modify fatty acids during the processing of certain fatty foods, such that the hydrogens are on opposite sides of the double bond. These chemically altered fatty acids are known as **trans fatty acids.** The *trans* configuration imparts stability to the food for longer storage, and alters its flavor and consistency. However, *trans* fatty acids have recently been linked with a number of serious health conditions, including elevated blood levels of cholesterol.

Some fatty acids can be altered to produce a special class of molecules that regulate a number of cell functions. As Chapter 5 will describe in more detail, these modified fatty

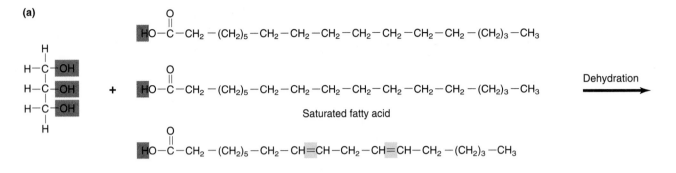

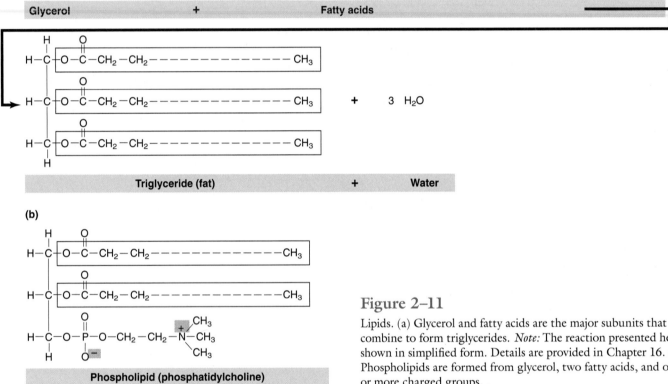

Figure 2–11

Lipids. (a) Glycerol and fatty acids are the major subunits that combine to form triglycerides. *Note:* The reaction presented here is shown in simplified form. Details are provided in Chapter 16. (b) Phospholipids are formed from glycerol, two fatty acids, and one or more charged groups.

acids—collectively termed eicosanoids—are derived from the 20-carbon, polyunsaturated fatty acid arachidonic acid.

Triglycerides

Triglycerides (also known as triacylglycerols) constitute the majority of the lipids in the body, and it is these molecules that are generally referred to simply as "fat." Triglycerides form when **glycerol,** a three-carbon alcohol, bonds to three fatty acids (see Figure 2–11a). Each of the three hydroxyl groups in glycerol is bonded to the carboxyl group of a fatty acid by a dehydration reaction.

The three fatty acids in a molecule of triglyceride need not be identical. Therefore, a variety of fats can be formed with fatty acids of different chain lengths and degrees of saturation. Animal fats generally contain a high proportion of saturated fatty acids, whereas vegetable fats contain more unsaturated fatty acids. Saturated fats tend to be solid at low temperatures.

Unsaturated fats, on the other hand, have a very low melting point, and thus they are liquids (oil) even at very low temperatures. Thus, heating a hamburger on the stove melts the saturated animal fats, leaving grease in the frying pan. When allowed to cool, however, the oily grease returns to its solid form.

Hydrolysis of triglycerides releases the fatty acids from glycerol and allows these products to then be metabolized to provide energy for cell functions. Thus, to store energy in the form of triglycerides and polysaccharides requires dehydration reactions, and both polymers break down to usable forms of fuel through hydrolysis.

Phospholipids

Phospholipids are similar in overall structure to triglycerides, with one important difference. The third hydroxyl group of glycerol, rather than being attached to a fatty acid, is linked to phosphate. In addition, a small polar or ionized nitrogen-containing molecule is usually attached to this phosphate (**Figure 2–11b**). These groups constitute a polar (hydrophilic) region at one end of the phospholipid, whereas the fatty acid chains provide a nonpolar (hydrophobic) region at the opposite end. Therefore, phospholipids are amphipathic. In water, they become organized into clusters, with their polar ends attracted to the water molecules. It is this property of phospholipids that permits them to form the lipid bilayers of plasma and intracellular membranes (Chapter 3).

Steroids

Steroids have a distinctly different structure from those of the other subclasses of lipid molecules. Four interconnected rings of carbon atoms form the skeleton of every steroid (**Figure 2–12**). A few hydroxyl groups, which are polar, may be attached to this ring structure, but they are not numerous enough to make a steroid water-soluble. Examples of steroids are cholesterol, cortisol from the adrenal glands, and female (estrogen) and male (testosterone) sex hormones secreted by the gonads.

Proteins

The term **protein** comes from the Greek *proteios* ("of the first rank"), which aptly describes their importance. Proteins account for about 50 percent of the organic material in the body (17 percent of the body weight), and they play critical roles in almost every physiological process. Proteins are composed of carbon, hydrogen, oxygen, nitrogen, and small amounts of other elements, notably sulfur. They are macromolecules, often containing thousands of atoms, and like most large molecules, they are formed when a large number of small subunits (monomers) bond together via dehydration reactions to create long chains.

Amino Acid Subunits

The subunits of proteins are **amino acids;** thus, proteins are polymers of amino acids. Every amino acid except one (proline) has an amino (—NH_2) and a carboxyl (—COOH) group bound to the terminal carbon in the molecule:

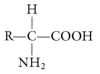

The third bond of this terminal carbon is bonded to a hydrogen and the fourth to the remainder of the molecule, which is known as the **amino acid side chain** (R in the formula). These side chains are relatively small, ranging from a single hydrogen to nine carbons with their associated hydrogens.

The proteins of all living organisms are composed of the same set of 20 different amino acids, corresponding to 20 different side chains. The side chains may be nonpolar (8 amino acids), polar (7 amino acids), or ionized (5 amino acids) (**Figure 2–13**). The human body can synthesize many amino acids, but several must be obtained in the diet; the latter are known as essential amino acids.

Polypeptides

Amino acids are joined together by linking the carboxyl group of one amino acid to the amino group of another. As in the formation of glycogen and triglycerides, a molecule of water is formed by dehydration (**Figure 2–14**). The bond formed between the amino and carboxyl group is called a **peptide bond.** Although peptide bonds are covalent, they can be broken

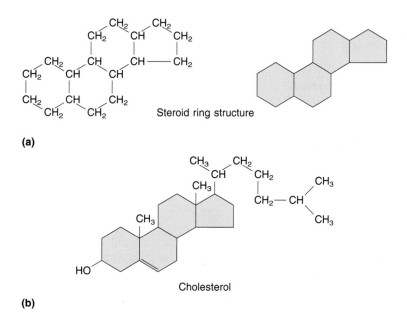

Figure 2–12

(a) Steroid ring structure, shown with all the carbon and hydrogen atoms in the rings and again without these atoms to emphasize the overall ring structure of this class of lipids. (b) Different steroids have different types and numbers of chemical groups attached at various locations on the steroid ring, as shown by the structure of cholesterol.

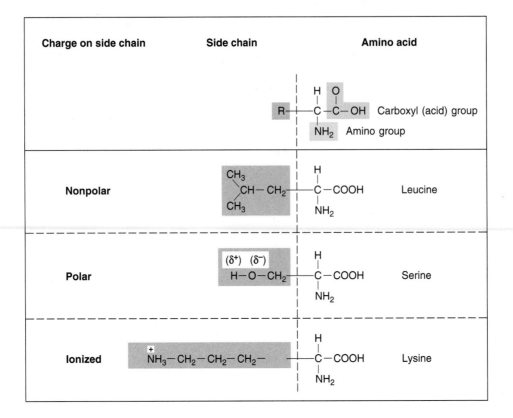

Figure 2–13
Representative structures of each class of amino acids found in proteins.

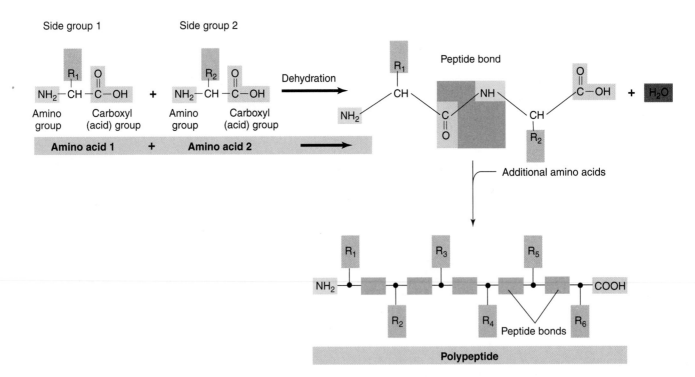

Figure 2–14
Linkage of amino acids by peptide bonds to form a polypeptide.

down by hydrolysis to yield individual amino acids, as happens in the stomach and intestines when we digest protein in our diet.

Note that when two amino acids are linked together, one end of the resulting molecule has a free amino group, and the other has a free carboxyl group. Additional amino acids can be linked by peptide bonds to these free ends. A sequence of amino acids linked by peptide bonds is known as a **polypeptide.** The peptide bonds form the backbone of the polypeptide, and the side chain of each amino acid sticks out from the chain. By convention, if the number of amino acids in a polypeptide is 50 or less, the molecule is known as a **peptide;** if the sequence is more than 50 amino acid units, it is known as a protein. The number 50 is somewhat arbitrary but is useful in distinguishing among large and small polypeptides. Small peptides have certain chemical properties that differ from proteins (e.g., peptides are generally soluble in acid, while proteins generally are not).

When one or more monosaccharides are covalently attached to the side chains of specific amino acids, the proteins are known as **glycoproteins.** These proteins are major components of connective tissue, and are also abundant in fluids like mucus, where they play a protective or lubricating role.

All proteins have multiple levels of structure that give each protein a unique shape. The shape of the protein determines its biological activity. In all cases, a protein's shape depends on its amino acid sequence, known as the primary structure of the protein.

Primary Protein Structure

Two variables determine the **primary structure** of a polypeptide: (1) the number of amino acids in the chain, and (2) the specific type of amino acid at each position along the chain (**Figure 2–15**). Each position along the chain can be occupied by any one of the 20 different amino acids. Consider the number of different peptides that can form that have a sequence of just three amino acids. Any one of the 20 different amino acids may occupy the first position in the sequence, any one of the 20 the second position, and any one of the 20 the third position, for a total of $20 \times 20 \times 20 = 20^3 = 8000$ possible sequences of three amino acids. If the peptide is six amino acids in length, $20^6 = 64,000,000$ possible combinations. Peptides that are only six amino acids long are still very small compared to proteins, which may have sequences of 1000 or more amino acids. Thus, with 20 different amino acids, an almost unlimited variety of polypeptides can be formed by altering both the amino acid sequence and the total number of amino acids in the chain. Only a fraction of these potential proteins is found in nature, however.

Secondary Protein Structure

A polypeptide is analogous to a string of beads, each bead representing one amino acid (see Figure 2–15). Moreover, because amino acids can rotate around bonds within a polypeptide chain, the chain is flexible and can bend into a number of shapes, just as a string of beads can be twisted into many configurations. The three-dimensional shape of a molecule is known as its **conformation** (**Figure 2–16**). The conformations of peptides and proteins play a major role in their functioning.

Figure 2–15

The position of each type of amino acid in a polypeptide chain and the total number of amino acids in the chain distinguish one polypeptide from another. The polypeptide illustrated contains 223 amino acids. Different amino acids are represented by different-colored circles. The bonds between various regions of the chain (red to red) represent covalent disulfide bonds between cysteine side chains.

Figure 2–16

Conformation (shape) of the protein molecule myoglobin. Each dot corresponds to the location of a single amino acid.

Adapted from Albert L. Lehninger.

Four major factors determine the conformation of a polypeptide chain once the amino acid sequence (primary structure) has been formed: (1) hydrogen bonds between portions of the chain or with surrounding water molecules; (2) ionic bonds between polar and ionized regions along the chain; (3) attraction between nonpolar (hydrophobic) regions; and (4) covalent bonds linking the side chains of two amino acids (**Figure 2–17**). (A fifth force, called **van der Waals forces,** causes a very weak attraction between two nonpolar atoms that are in very close proximity to each other.)

An example of the attractions between various regions along a polypeptide chain is the hydrogen bond that can occur between the hydrogen linked to the nitrogen atom in one peptide bond and the double-bonded oxygen atom in another peptide bond (**Figure 2–18**). Because peptide bonds occur at regular intervals along a polypeptide chain, the hydrogen bonds between them tend to force the chain into a coiled conformation known as an **alpha helix.** Hydrogen bonds can also form between peptide bonds when extended regions of a polypeptide chain run approximately parallel to each other, forming a relatively straight, extended region known as a **beta sheet** (**Figure 2–19**). However, for several reasons, a given region of a polypeptide chain may not assume either a helical or beta sheet conformation. For example, the sizes of the side chains and the ionic bonds between side chains with opposite charges can interfere with the repetitive hydrogen bonding required to produce these shapes. These irregular regions, known as loop conformations, occur in regions linking the more regular helical and beta sheet patterns (see Figure 2–19).

Beta sheets and alpha helices are regions of **secondary structure** of proteins. Secondary structure, therefore, is determined by primary structure. Secondary structure allows the protein to be defined in terms of "domains." For example, many helical domains are comprised primarily of hydrophobic amino acids. These regions tend to impart upon a protein the ability to anchor itself into a lipid bilayer, like that of a cell membrane.

Tertiary Protein Structure

Covalent bonds between certain side chains can also modify a protein's shape. For example, the side chain of the amino acid cysteine contains a sulfhydryl group (R—SH), which can react with a sulfhydryl group in another cysteine side chain to produce a **disulfide bond** (R—S—S—R) that joins the two amino acid side chains together (**Figure 2–20**). Disulfide bonds form covalent bonds between portions of a polypeptide chain, in contrast to the weaker and more easily broken

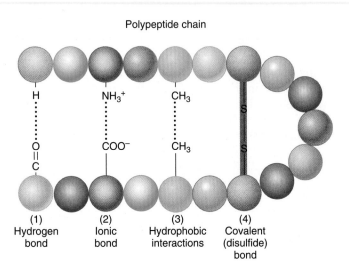

Figure 2–17

Factors that contribute to the folding of polypeptide chains and thus to their conformation are (1) hydrogen bonds between side chains or with surrounding water molecules, (2) ionic bonds between polar or ionized side chains, (3) hydrophobic attractive forces between nonpolar side chains, and (4) covalent bonds between side chains.

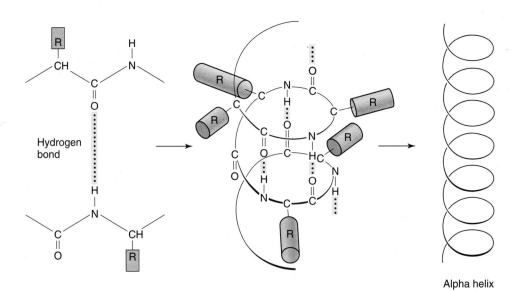

Figure 2–18

Hydrogen bonds between regularly spaced peptide bonds can produce a helical conformation in a polypeptide chain.

hydrogen and ionic bonds. **Table 2–6** provides a summary of the types of bonding forces that contribute to the conformation of polypeptide chains. These same bonds are also involved in other intermolecular interactions, which will be described in later chapters.

Thus, a protein can fold over on itself in a variety of ways, resulting in a three-dimensional shape **(tertiary structure)** characteristic of that protein (see Figure 2–16). This three-dimensional shape allows one protein to interact with other molecules, including other proteins.

Quaternary Protein Structure

When proteins are composed of more than one polypeptide chain, they are said to have **quaternary structure** and are known as **multimeric proteins** ("many parts"). The same

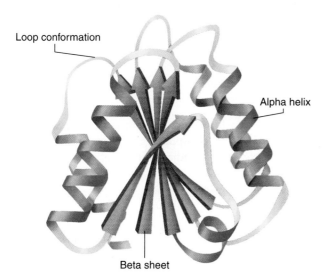

Figure 2–19

A ribbon diagram illustrating the pathway followed by the backbone of a single polypeptide chain. Helical regions (blue) are coiled, beta sheets (red) of parallel chains are shown as relatively straight arrows, and loop conformations (yellow) connect the various helical and beta sheet regions. Beginning at the end of the chain labeled "Beta sheet," a continuous chain of amino acids passes through various conformations.

factors that influence the conformation of a single polypeptide also determine the interactions between the polypeptides in a multimeric protein. Thus, the chains can be held together by interactions between various ionized, polar, and nonpolar side chains, as well as by disulfide covalent bonds between the chains.

The polypeptide chains in a multimeric protein may be identical or different. For example, hemoglobin, the protein that transports oxygen in the blood, is a multimeric protein with four polypeptide chains, two of one kind and two of another (**Figure 2–21**).

The primary structures (amino acid sequences) of a large number of proteins are known, but three-dimensional conformations have been determined for only a small number. Because of the multiple factors that can influence the folding of a polypeptide chain, it is not yet possible to predict accurately the conformation of a protein from its primary amino acid sequence. However, it should be clear that a change in the primary structure of a protein may alter its secondary, tertiary, and quaternary structures. Such an alteration in primary structure is called a **mutation.**

Even a single amino acid change resulting from a mutation may have devastating consequences, as occurs when a molecule of valine replaces a molecule of glutamic acid in the β chains of hemoglobin. The result of this change is a serious disease called *sickle cell anemia.* When red blood cells in a person with this disease are exposed to low oxygen levels, their hemoglobin precipitates. This contorts the red blood cells into a crescent shape, which makes the cells fragile and unable to function normally.

Nucleic Acids

Nucleic acids account for only 2 percent of the body's weight, yet these molecules are extremely important because they are responsible for the storage, expression, and transmission of genetic information. It is the expression of genetic information in the form of specific proteins that determines whether one is a human or a mouse, or whether a cell is a muscle cell or a nerve cell.

There are two classes of nucleic acids, **deoxyribonucleic acid (DNA)** and **ribonucleic acid (RNA).** DNA molecules store genetic information coded in the sequence of

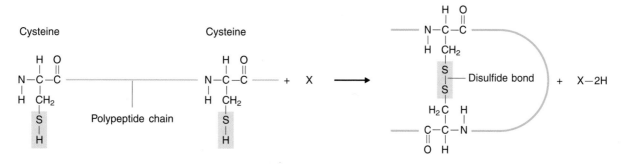

Figure 2–20

Formation of a disulfide bond between the side chains of two cysteine amino acids links two regions of the polypeptide together. The hydrogen atoms on the sulfhydryl groups of the cysteines are transferred to another molecule, X, during the formation of the disulfide bond.

Table 2–6	Bonding Forces Between Atoms and Molecules		
Bond	**Strength**	**Characteristics**	**Examples**
Hydrogen	Weak	Electrical attraction between polarized bonds, usually hydrogen and oxygen	Attractions between peptide bonds, forming the alpha helix structure of proteins, and between polar amino acid side chains, contributing to protein conformation; attractions between water molecules
Ionic	Strong	Electrical attraction between oppositely charged ionized groups	Attractions between ionized groups in amino acid side chains, contributing to protein conformation; attractions between ions in a salt
Hydrophobic interactions	Weak	Attraction between nonpolar molecules and groups when very close to each other	Attractions between nonpolar amino acids in proteins, contributing to protein conformation; attractions between lipid molecules
Covalent	Very strong	Shared electrons between atoms. Nonpolar covalent bonds share electrons equally, while in polar bonds the electrons reside closer to one atom in the pair	Most bonds linking atoms together to form molecules

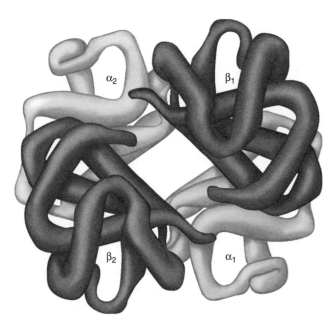

Figure 2–21

Hemoglobin, a multimeric protein composed of two identical alpha (α) chains or subunits and two identical beta (β) chains. (The iron-containing heme groups attached to each globin chain are not shown.)

their subunits, whereas RNA molecules are involved in decoding this information into instructions for linking together a specific sequence of amino acids to form a specific polypeptide chain.

Both types of nucleic acids are polymers and are therefore composed of linear sequences of repeating subunits. Each subunit, known as a **nucleotide,** has three components: a phosphate group, a sugar, and a ring of carbon and nitrogen atoms known as a base because it can accept hydrogen ions (**Figure 2–22**). The phosphate group of one nucleotide is linked to the sugar of the adjacent nucleotide to form a chain, with the bases sticking out from the side of the phosphate–sugar backbone (**Figure 2–23**).

DNA

The nucleotides in DNA contain the five-carbon sugar **deoxyribose** (hence the name "deoxyribonucleic acid"). Four different nucleotides are present in DNA, corresponding to the four different bases that can be bound to deoxyribose. These bases are divided into two classes: (1) the **purine** bases, **adenine** (A) and **guanine** (G), which have double rings of nitrogen and carbon atoms, and (2) the **pyrimidine** bases, **cytosine** (C) and **thymine** (T), which have only a single ring (see Figure 2–23).

A DNA molecule consists of not one but two chains of nucleotides coiled around each other in the form of a double helix (**Figure 2–24**). The two chains are held together by hydrogen bonds between a purine base on one chain and a pyrimidine base on the opposite chain. The ring structure of each base lies in a flat plane perpendicular to the phosphate–sugar backbone, like steps on a spiral staircase. This base pairing maintains a constant distance between the sugar–phosphate backbones of the two chains as they coil around each other.

Specificity is imposed on the base pairings by the location of the hydrogen-bonding groups in the four bases (**Figure 2–25**). Three hydrogen bonds form between the purine guanine and the pyrimidine cytosine (G–C pairing), while only two hydrogen bonds can form between the purine adenine and the pyrimidine thymine (A–T pairing). As a result, G is always paired with C, and A with T. It is this specificity that provides the mechanism for duplicating and transferring genetic information.

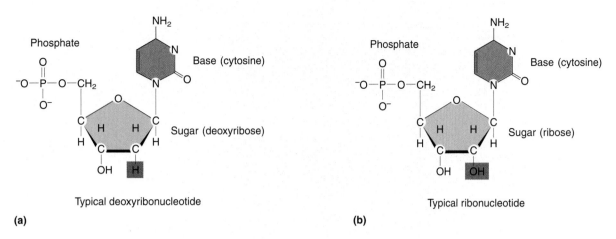

Figure 2–22

Nucleotide subunits of DNA and RNA. Nucleotides are composed of a sugar, a base, and a phosphate group. (a) Deoxyribonucleotides present in DNA contain the sugar deoxyribose. (b) The sugar in ribonucleotides, present in RNA, is ribose, which has an OH at a position in which deoxyribose has only a hydrogen atom.

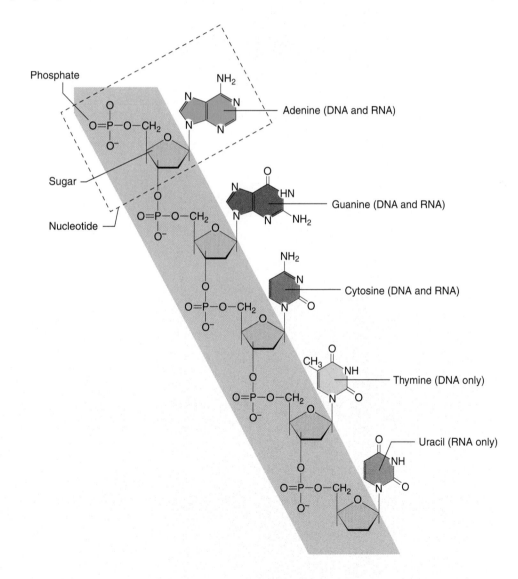

Figure 2–23

Phosphate-sugar bonds link nucleotides in sequence to form nucleic acids. Note that the pyrimidine base thymine is only found in DNA, and uracil is only present in RNA.

Figure 2–24

Base pairings between a purine and pyrimidine base link the two polynucleotide strands of the DNA double helix.

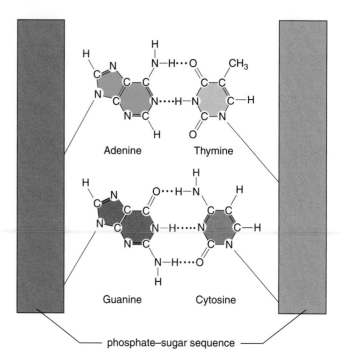

phosphate–sugar sequence

Figure 2–25

Hydrogen bonds between the nucleotide bases in DNA determine the specificity of base pairings: adenine with thymine and guanine with cytosine.

RNA

RNA molecules differ in only a few respects from DNA (**Table 2–7**): (1) RNA consists of a single (rather than a double) chain of nucleotides; (2) in RNA, the sugar in each nucleotide is **ribose** rather than deoxyribose; and (3) the pyrimidine base thymine in DNA is replaced in RNA by the pyrimidine base **uracil** (U) (see Figure 2–23), which can base-pair with the purine adenine (A–U pairing). The other three bases, adenine, guanine, and cytosine, are the same in both DNA and RNA. Because RNA contains only a single chain of nucleotides, portions of this chain can bend back upon themselves and undergo base pairing with nucleotides in the same chain or in other molecules of DNA or RNA.

ATP

The purine bases are important not only in DNA and RNA synthesis, but also in a molecule that serves as the molecular energy source for all cells.

The functioning of a cell depends upon its ability to extract and use the chemical energy in the organic molecules discussed in this chapter. For example, when, in the presence of oxygen, a cell breaks down glucose to carbon dioxide and water, energy is released. Some of this energy is in the form of heat, but a cell cannot use heat energy to perform its functions. The remainder of the energy is transferred to another important molecule that can in turn transfer it to yet another molecule or to energy-requiring processes. In all cells, from bacterial to human, **adenosine**

Table 2–7	Comparison of DNA and RNA Composition	
	DNA	**RNA**
Nucleotide sugar	Deoxyribose	Ribose
Nucleotide bases		
Purines	Adenine	Adenine
	Guanine	Guanine
Pyrimidines	Cytosine	Cytosine
	Thymine	Uracil
Number of chains	Two	One

triphosphate (ATP) (**Figure 2–26**) is the primary molecule that receives the transfer of energy from the breakdown of fuel molecules—carbohydrates, fats, and proteins.

Energy released from organic molecules is used to add phosphate groups to molecules of adenosine. This stored energy can then be released upon hydrolysis:

$$\text{ATP} + \text{H}_2\text{O} \longrightarrow \text{ADP} + \text{P}_i + \text{H}^+ + \text{energy}$$

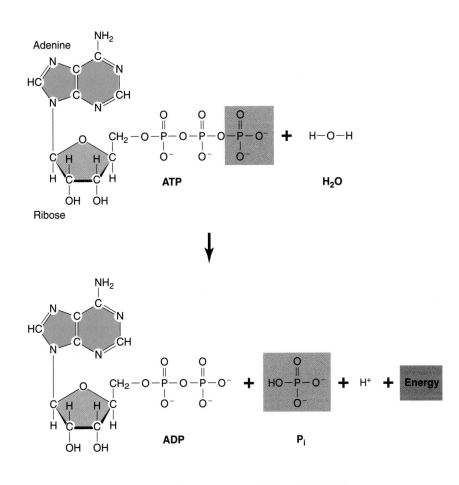

Figure 2–26

Chemical structure of ATP. Its breakdown to ADP and P_i is accompanied by the release of energy.

The products of the reaction are **adenosine diphosphate (ADP)**, inorganic phosphate (P_i) and H^+.

The energy derived from the hydrolysis of ATP is used by the cells for (1) the production of force and movement, as in muscle contraction; (2) active transport of molecules across membranes; and (3) synthesis of the organic molecules used in cell structures and functions.

We must emphasize that cells use ATP not to *store* energy but rather to *transfer* it. ATP is an energy-carrying molecule that transfers relatively small amounts of energy from fuel molecules to the cells for processes that require energy.

■ **SUMMARY**

Atoms

  I. Atoms are composed of three subatomic particles: positive protons and neutral neutrons, both located in the nucleus, and negative electrons revolving around the nucleus.

  II. The atomic number is the number of protons in an atom, and because atoms are electrically neutral, it is also the number of electrons.

  III. The atomic weight of an atom is the ratio of the atom's mass relative to that of a carbon-12 atom.

  IV. One gram atomic mass is the number of grams of an element equal to its atomic weight. One gram atomic mass of any element contains the same number of atoms—6×10^{23}.

Molecules

  I. Molecules are formed by linking atoms together.

  II. A covalent bond forms when two atoms share a pair of electrons. Each type of atom can form a characteristic number of covalent bonds: hydrogen forms one; oxygen, two; nitrogen, three; and carbon, four.

  III. Molecules have characteristic shapes that can be altered within limits by the rotation of their atoms around covalent bonds.

Ions

  I. When an atom gains or loses one or more electrons, it acquires a net electric charge and becomes an ion.

Free Radicals

  I. Free radicals are atoms or molecules that contain atoms having an unpaired electron in their outer electron orbital.

Chemical Composition of the Body

Polar Molecules

I. In polar covalent bonds, one atom attracts the bonding electrons more than the other atom of the pair.

II. The electrical attraction between hydrogen and an oxygen or nitrogen atom in a separate molecule or different region of the same molecule forms a hydrogen bond.

III. Water, a polar molecule, is attracted to other water molecules by hydrogen bonds.

Solutions

I. Substances dissolved in a liquid are solutes, and the liquid in which they are dissolved is the solvent. Water is the most abundant solvent in the body.

II. Substances that have polar or ionized groups dissolve in water by being electrically attracted to the polar water molecules.

III. In water, amphipathic molecules form clusters with the polar regions at the surface and the nonpolar regions in the interior of the cluster.

IV. The molecular weight of a molecule is the sum of the atomic weights of all its atoms. One mole of any substance is its molecular weight in grams and contains 6×10^{23} molecules.

V. Substances that release a hydrogen ion in solution are called acids. Those that accept a hydrogen ion are bases.

   a. The acidity of a solution is determined by its free hydrogen ion concentration; the greater the hydrogen ion concentration, the greater the acidity.

   b. The pH of a solution is the negative logarithm of the hydrogen ion concentration. As the acidity of a solution increases, the pH decreases. Acid solutions have a pH less than 7.0, whereas alkaline solutions have a pH greater than 7.0.

Classes of Organic Molecules

I. Carbohydrates are composed of carbon, hydrogen, and oxygen in the proportions $C_n(H_2O)_n$.

   a. The presence of the polar hydroxyl groups makes carbohydrates soluble in water.

   b. The most abundant monosaccharide in the body is glucose ($C_6H_{12}O_6$), which is stored in cells in the form of the polysaccharide glycogen.

II. Most lipids have many fewer polar and ionized groups than carbohydrates, a characteristic that makes them insoluble in water.

   a. Triglycerides (fats) form when fatty acids are bound to each of the three hydroxyl groups in glycerol.

   b. Phospholipids contain two fatty acids bound to two of the hydroxyl groups in glycerol, with the third hydroxyl bound to phosphate, which in turn is linked to a small charged or polar compound. The polar and ionized groups at one end of phospholipids make these molecules amphipathic.

   c. Steroids are composed of four interconnected rings, often containing a few hydroxyl and other groups.

   d. One fatty acid (arachidonic acid) can be converted to a class of signaling substances called eicosanoids.

III. Proteins, macromolecules composed primarily of carbon, hydrogen, oxygen, and nitrogen, are polymers of 20 different amino acids.

   a. Amino acids have an amino (—NH$_2$) and a carboxyl (—COOH) group bound to their terminal carbon atom.

   b. Amino acids are bound together by peptide bonds between the carboxyl group of one amino acid and the amino group of the next.

   c. The primary structure of a polypeptide chain is determined by (1) the number of amino acids in sequence, and (2) the type of amino acid at each position.

   d. Hydrogen bonds between peptide bonds along a polypeptide force much of the chain into an alpha helix (secondary structure).

   e. Covalent disulfide bonds can form between the sulfhydryl groups of cysteine side chains to hold regions of a polypeptide chain close to each other (tertiary structure).

   f. Multimeric proteins have multiple polypeptide chains (quaternary structure).

IV. Nucleic acids are responsible for the storage, expression, and transmission of genetic information.

   a. Deoxyribonucleic acid (DNA) stores genetic information.

   b. Ribonucleic acid (RNA) is involved in decoding the information in DNA into instructions for linking amino acids together to form proteins.

   c. Both types of nucleic acids are polymers of nucleotides, each containing a phosphate group, a sugar, and a base of carbon, hydrogen, oxygen, and nitrogen atoms.

   d. DNA contains the sugar deoxyribose and consists of two chains of nucleotides coiled around each other in a double helix. The chains are held together by hydrogen bonds between purine and pyrimidine bases in the two chains.

   e. Base pairings in DNA always occur between guanine and cytosine and between adenine and thymine.

   f. RNA consists of a single chain of nucleotides, containing the sugar ribose and three of the four bases found in DNA. The fourth base in RNA is the pyrimidine uracil rather than thymine. Uracil base-pairs with adenine.

   g. In all cells, energy from the catabolism of fuel molecules is transferred to ATP. Hydrolysis of ATP to ADP + P$_i$ then transfers this energy to power cell functions. ATP consists of the purine adenine coupled by high-energy bonds to three phosphate groups.

■ KEY TERMS

acid 26	dehydration 24
acidic solution 26	deoxyribonucleic acid
acidity 26	(DNA) 35
adenine 36	deoxyribose 36
adenosine diphosphate	disaccharide 28
(ADP) 39	disulfide bond 34
adenosine triphosphate	electrolyte 21
(ATP) 38	electron 19
alkaline solution 26	fatty acid 29
alpha helix 34	free radical 21
amino acid 31	glucose 28
amino acid side chain 31	glycerol 30
amino group 21	glycogen 28
amphipathic 25	glycoprotein 33
anion 21	gram atomic mass 20
atom 19	guanine 36
atomic nucleus 19	hexose 28
atomic number 19	hydrogen bond 23
atomic weight 19	hydrolysis 24
base 26	hydrophilic 24
beta sheet 34	hydrophobic 24
carbohydrate 27	hydroxyl group 23
carboxyl group 21	ion 20
cation 21	ionic bond 24
chemical element 19	isotope 20
concentration 25	lipid 29
conformation 33	macromolecule 27
covalent bond 20	mole 26
cytosine 36	molecular weight 26

■ CLINICAL TERMS

■ REVIEW QUESTIONS

1. Describe the electrical charge, mass, and location of the three major subatomic particles in an atom.
2. Which four kinds of atoms are most abundant in the body?
3. Describe the distinguishing characteristics of the three classes of essential chemical elements found in the body.
4. How many covalent bonds can be formed by atoms of carbon, nitrogen, oxygen, and hydrogen?
5. What property of molecules allows them to change their three-dimensional shape?
6. Describe how an ion is formed.
7. Draw the structures of an ionized carboxyl group and an ionized amino group.
8. Define a free radical.
9. Describe the polar characteristics of a water molecule.
10. What determines a molecule's solubility or lack of solubility in water?
11. Describe the organization of amphipathic molecules in water.
12. What is the molar concentration of 80 g of glucose dissolved in sufficient water to make 2 L of solution?
13. What distinguishes a weak acid from a strong acid?
14. What effect does increasing the pH of a solution have upon the ionization of a carboxyl group? An amino group?
15. Name the four classes of organic molecules in the body.
16. Describe the three subclasses of carbohydrate molecules.
17. To which subclass of carbohydrates do each of the following molecules belong: glucose, sucrose, and glycogen?
18. What properties are characteristic of lipids?
19. Describe the subclasses of lipids.
20. Describe the linkages between amino acids that form polypeptide chains.
21. What is the difference between a peptide and a protein?
22. What two factors determine the primary structure of a polypeptide chain?
23. Describe the types of interactions that determine the conformation of a polypeptide chain.
24. Describe the structure of DNA and RNA.
25. Describe the characteristics of base pairings between nucleotide bases.

Chapter 2 Test Questions

(Answers appear in Appendix A.)

1. A molecule that loses an electron to a free radical
   a. becomes more stable.
   b. becomes electrically neutral.
   c. becomes less reactive.
   d. is permanently destroyed.
   e. becomes a free radical itself.

2. Of the bonding forces between atoms and molecules, which are strongest?
   a. hydrogen bonds
   b. bonds between oppositely charged ionized groups
   c. bonds between nearby nonpolar groups
   d. covalent bonds
   e. bonds between polar groups

3. The process by which monomers of organic molecules are made into larger units
   a. requires hydrolysis.
   b. results in the generation of water molecules.
   c. is irreversible.
   d. occurs only with carbohydrates.
   e. results in the production of ATP.

4. Which of the following is not found in DNA?
   a. adenine
   b. uracil
   c. cytosine
   d. deoxyribose
   e. both b and d

5. Which of the following statements is incorrect about disulfide bonds?
   a. They form between two cysteine amino acids.
   b. They are noncovalent.
   c. They contribute to the tertiary structure of some proteins.
   d. They contribute to the quaternary structure of some proteins.
   e. They involve the loss of two hydrogen atoms.

chapter **3**

Cellular Structure, Proteins, and Metabolism

Color enhanced electron microscopic image of a liver cell.

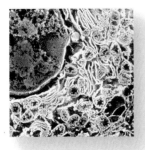

Cells are the structural and functional units of all living organisms. The word *cell* means "a small chamber." The human body is composed of trillions of cells, each a microscopic compartment (**Figure 3–1**). In this chapter, we briefly describe the structures found in most of the body's cells and list their functions.

Having learned the basic structures that comprise cells, we next turn our attention to how cellular proteins are produced, secreted, and degraded, and how proteins participate in the chemical reactions needed for all cells to survive.

Proteins are associated with practically every function living cells perform. One fact is crucial for an understanding of protein function, and thus the functioning of a living organism: Proteins have a unique shape or conformation that enables them to bind specific molecules on portions of their surfaces known as binding sites. Thus, this chapter includes a discussion of the properties of protein binding sites that apply to all proteins, as well as a description of how these properties are involved in one class of protein functions—the ability of enzymes to accelerate specific chemical reactions. We then apply this information to a description of the multitude of chemical reactions involved in metabolism.

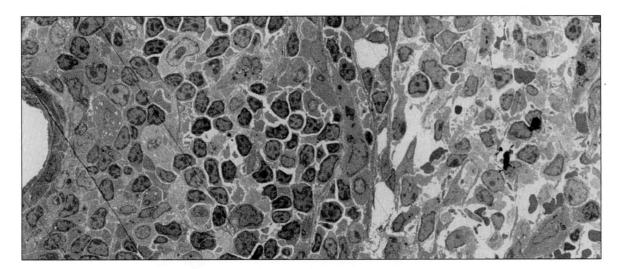

Figure 3–1

Cellular organization of tissues, as illustrated by a portion of spleen. Oval, clear spaces in the electron micrograph are blood vessels. Note the changes in overall appearance of this complex organ, as you move from left (closely compacted cells) to right (loosely arranged cells).

From Johannes A. G. Rhodin, *Histology, A Text & Atlas,* Oxford University Press, New York, 1974.

SECTION A   Cell Structure

Microscopic Observations of Cells

The smallest object that can be resolved with a microscope depends upon the wavelength of the radiation used to illuminate the specimen—the shorter the wavelength, the smaller the object that can be seen. While a light microscope can resolve objects as small as 0.2 μm in diameter, an electron microscope, which uses electron beams instead of light rays, can resolve structures as small as 0.002 μm. Typical sizes of cells and cellular components are illustrated in **Figure 3–2**.

Although living cells can be observed with a light microscope, this is not possible with an electron microscope. To form an image with an electron beam, most of the electrons must pass through the specimen, just as light passes through a specimen in a light microscope. However, electrons can penetrate only a short distance through matter; therefore, the observed specimen must be very thin. Cells to be observed with an electron microscope must be cut into sections on the order of 0.1 μm thick, which is about one-hundredth of the thickness of a typical cell.

Because electron micrographs, such as the one in **Figure 3–3**, are images of very thin sections of a cell, they can sometimes be misleading. Structures that appear as separate objects in the electron micrograph may actually be continuous

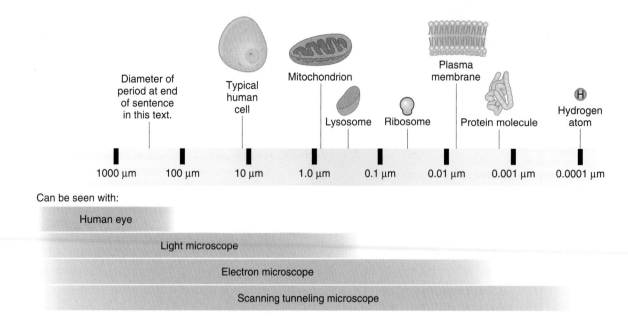

Figure 3–2

Typical sizes of cell structures, plotted on a logarithmic scale.

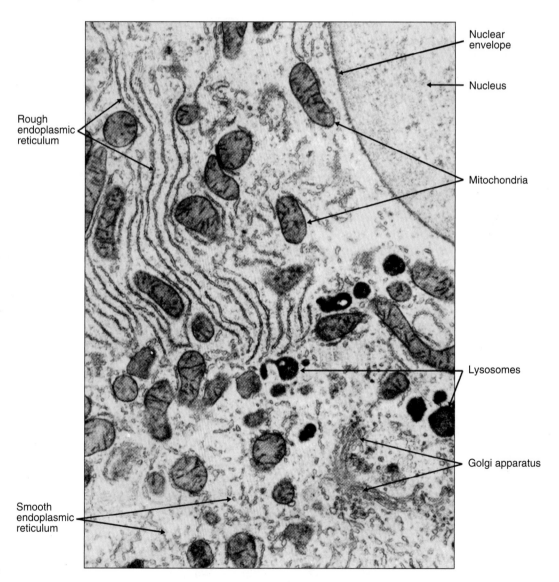

Figure 3–3

Electron micrograph of a thin section through a portion of a rat liver cell.

From K. R. Porter in T. W. Goodwin and O. Lindberg (eds.), *Biological Structure and Function*, vol. I, Academic Press, Inc., New York, 1961.

structures connected through a region lying outside the plane of the section. As an analogy, a thin section through a ball of string would appear to be a collection of separate lines and disconnected dots even though the piece of string was originally continuous.

Two classes of cells, **eukaryotic cells** and **prokaryotic cells,** can be distinguished by their structure. The cells of the human body, as well as those of other multicellular animals and plants, are eukaryotic (true-nucleus) cells. These cells contain a nuclear membrane surrounding the cell nucleus, and also contain numerous other membrane-bound structures. Prokaryotic cells, such as bacteria, lack these membranous structures. This chapter describes the structure of eukaryotic cells only.

Compare an electron micrograph of a section through a cell (see Figure 3–3) with a diagrammatic illustration of a typical human cell (**Figure 3–4**). What is immediately obvious from both figures is the extensive structure inside the cell. Cells are surrounded by a limiting barrier, the **plasma membrane,** which covers the cell surface. The cell interior is divided into a number of compartments surrounded by membranes. These membrane-bound compartments, along with some particles and filaments, are known as **cell organelles.** Each cell organelle performs specific functions that contribute to the cell's survival.

The interior of a cell is divided into two regions: (1) the **nucleus,** a spherical or oval structure usually near the center of the cell, and (2) the **cytoplasm,** the region outside the nucleus (**Figure 3–5**). The cytoplasm contains cell organelles and fluid surrounding the organelles, known as the **cytosol.** As described in Chapter 1, the term **intracellular fluid** refers to *all* the fluid inside a cell—in other words, cytosol plus the fluid inside all the organelles, including the nucleus. The chemical compositions of the fluids in these cell organelles may differ from that of the cytosol. The cytosol is by far the largest intracellular fluid compartment.

Membranes

Membranes form a major structural element in cells. Although membranes perform a variety of functions that are important in physiology (**Table 3–1**), their most universal role is to act as a selective barrier to the passage of molecules, allowing some molecules to cross while excluding others. The plasma membrane regulates the passage of substances into and out of the cell, whereas the membranes surrounding cell organelles allow the selective movement of substances between the organelles and the cytosol. One of the advantages of restricting the movements of molecules across membranes is confining the products of chemical reactions to specific cell organelles. The hindrance

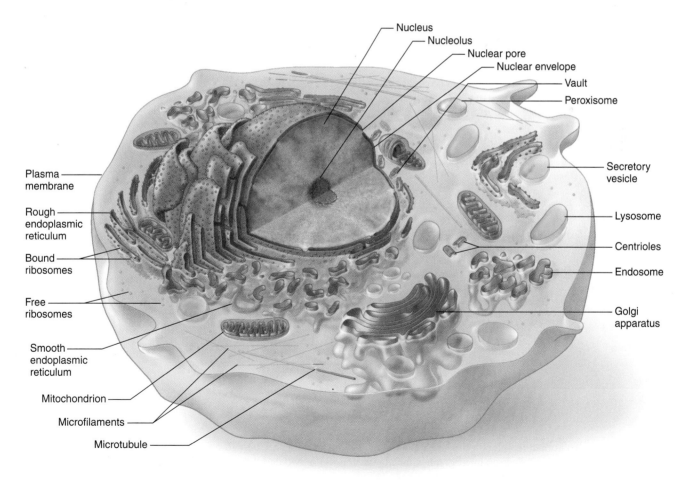

Figure 3–4

Structures found in most human cells. Not all structures are drawn to scale.

a membrane offers to the passage of substances can be altered to allow increased or decreased flow of molecules or ions across the membrane in response to various signals.

In addition to acting as a selective barrier, the plasma membrane plays an important role in detecting chemical signals from other cells and in anchoring cells to adjacent cells and to the extracellular matrix of connective-tissue proteins.

Membrane Structure

All membranes consist of a double layer of lipid molecules containing embedded proteins (**Figure 3–6**). The major membrane lipids are **phospholipids,** which are amphipathic molecules. One end has a charged or polar region, and the remainder of the molecule, which consists of two long fatty acid chains, is nonpolar. The phospholipids in cell membranes are organized into a bimolecular layer with the nonpolar fatty

acid chains in the middle. The polar regions of the phospholipids are oriented toward the surfaces of the membrane as a result of their attraction to the polar water molecules in the extracellular fluid and cytosol.

No chemical bonds link the phospholipids to each other or to the membrane proteins. Therefore, each molecule is free to move independently of the others. This results in considerable random lateral movement of both membrane lipids and proteins parallel to the surfaces of the bilayer. In addition, the long fatty acid chains can bend and wiggle back and forth. Thus, the lipid bilayer has the characteristics of a fluid, much like a thin layer of oil on a water surface, and this makes the membrane quite flexible. This flexibility, along with the fact that cells are filled with fluid, allows cells to undergo moderate changes in shape without disrupting their structural integrity. Like a piece of cloth, a membrane can be bent and folded but cannot be stretched without being torn.

The plasma membrane also contains about one molecule of cholesterol for each molecule of phospholipid, whereas intra-

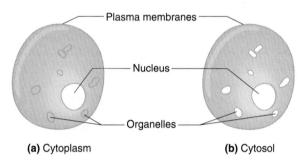

(a) Cytoplasm   **(b)** Cytosol

Figure 3–5

Comparison of cytoplasm and cytosol. (a) Cytoplasm (colored area) is the region of the cell outside the nucleus. (b) Cytosol (colored area) is the fluid portion of the cytoplasm outside the cell organelles.

Table 3–1	Functions of Cell Membranes
1. Regulate the passage of substances into and out of cells and between cell organelles and cytosol	
2. Detect chemical messengers arriving at the cell surface	
3. Link adjacent cells together by membrane junctions	
4. Anchor cells to the extracellular matrix	

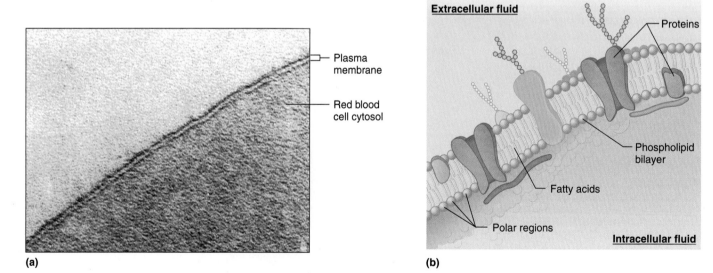

(a)   **(b)**

Figure 3–6

(a) Electron micrograph of a human red blood cell plasma membrane. Cell membranes are 6 to 10 nm thick, too thin to be seen without the aid of an electron microscope. In an electron micrograph, a membrane appears as two dark lines separated by a light interspace. The dark lines correspond to the polar regions of the proteins and lipids, whereas the light interspace corresponds to the nonpolar regions of these molecules. (b) Arrangement of the proteins and lipids in a membrane.

From J. D. Robertson in Michael Locke (ed.), *Cell Membranes in Development,* Academic Press, Inc., New York, 1964.

cellular membranes contain very little cholesterol. As described in Chapter 2, cholesterol is slightly amphipathic because of a single polar hydroxyl group (see Figure 2–12) on its nonpolar ring structure. Like the phospholipids, therefore, cholesterol is inserted into the lipid bilayer with its polar region at the bilayer surface and its nonpolar rings in the interior in association with the fatty acid chains. Cholesterol associates with certain classes of plasma membrane phospholipids and proteins, forming organized clusters that work together to pinch off portions of the plasma membrane to form vesicles that deliver their contents to various intracellular organelles, as Chapter 4 will describe.

There are two classes of membrane proteins: integral and peripheral. **Integral membrane proteins** are closely associated with the membrane lipids and cannot be extracted from the membrane without disrupting the lipid bilayer. Like the phospholipids, the integral proteins are amphipathic, having polar amino acid side chains in one region of the molecule and nonpolar side chains clustered together in a separate region. Because they are amphipathic, integral proteins are arranged in the membrane with the same orientation as amphipathic lipids—the polar regions are at the surfaces in association with polar water molecules, and the nonpolar regions are in the interior in association with nonpolar fatty acid chains (**Figure 3–7**). Like the membrane lipids, many of the integral proteins can move laterally in the plane of the membrane, but others are immobilized because they are linked to a network of peripheral proteins located primarily at the cytosolic surface of the membrane.

Most integral proteins span the entire membrane and are referred to as **transmembrane proteins.** Most of these transmembrane proteins cross the lipid bilayer several times (**Figure 3–8**). These proteins have polar regions connected by nonpolar segments that associate with the nonpolar regions of the lipids in the membrane interior. The polar regions of transmembrane proteins may extend far beyond the surfaces of the lipid bilayer. Some transmembrane proteins form channels through which ions or water can cross the membrane, whereas others are associated with the transmission of chemical signals across the membrane or the anchoring of extracellular and intracellular protein filaments to the plasma membrane.

Peripheral membrane proteins are not amphipathic and do not associate with the nonpolar regions of the lipids in the interior of the membrane. They are located at the membrane surface where they are bound to the polar regions of the integral membrane proteins (see Figure 3–7). Most of the peripheral proteins are on the cytosolic surface of the plasma membrane where they are associated with cytoskeletal elements that influence cell shape and motility.

The extracellular surface of the plasma membrane contains small amounts of carbohydrate covalently linked to some of the membrane lipids and proteins. These carbohydrates consist of short, branched chains of monosaccharides that extend from the cell surface into the extracellular fluid, where they form a layer known as the **glycocalyx.** These surface carbohydrates play important roles in enabling cells to identify and interact with each other.

The lipids in the outer half of the bilayer differ somewhat in kind and amount from those in the inner half, and, as we have seen, the proteins or portions of proteins on the outer surface differ from those on the inner surface. Many membrane

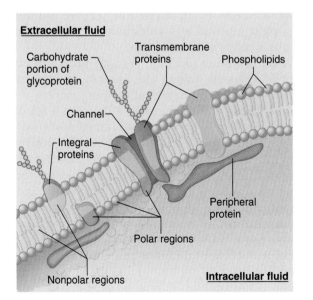

Figure 3–7
Arrangement of integral and peripheral membrane proteins in association with a bimolecular layer of phospholipids. Cholesterol molecules are omitted for the sake of clarity.

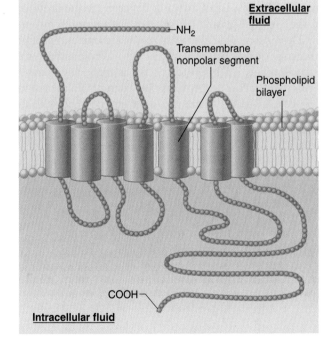

Figure 3–8
A typical transmembrane protein with multiple hydrophobic segments traversing the lipid bilayer. Each transmembrane segment is composed of nonpolar amino acids spiraled in an alpha helical conformation (shown as cylinders).

functions are related to these asymmetries in chemical composition between the two surfaces of a membrane.

All membranes have the general structure just described, which is known as the **fluid-mosaic model** because membrane proteins float in a sea of lipid (**Figure 3–9**). However, the proteins and, to a lesser extent, the lipids (the distribution of cholesterol, for example) in the plasma membrane differ from those in organelle membranes. Thus, the special functions of membranes, which depend primarily on the membrane proteins, may differ in the various membrane-bound organelles and in the plasma membranes of different types of cells.

Membrane Junctions

In addition to providing a barrier to the movements of molecules between the intracellular and extracellular fluids, plasma membranes are involved in the interactions between cells to form tissues. Most cells are packaged into tissues and are not free to move around the body. But even in tissues, there is usually a space between the plasma membranes of adjacent cells. This space, filled with extracellular (interstitial) fluid, provides a pathway for substances to pass between cells on their way to and from the blood.

The forces that organize cells into tissues and organs are poorly understood, but they depend, at least in part, on the ability of certain transmembrane proteins in the plasma membrane, known as **integrins,** to bind to specific proteins in the extracellular matrix and link them to membrane proteins on adjacent cells.

Many cells are physically joined at discrete locations along their membranes by specialized types of junctions, including desmosomes, tight junctions, and gap junctions. **Desmosomes** (**Figure 3–10a**) consist of a region between two adjacent cells where the apposed plasma membranes are separated by about 20 nm. Desmosomes are characterized by accumulations of protein known as dense plaques along the cytoplasmic surface of the plasma membrane. These proteins serve as anchoring points for cadherins. **Cadherins** are proteins that extend from the cell into the extracellular space, where they link up and bind with cad-

herins from an adjacent cell. In this way, two adjacent cells can be firmly attached to each other. The presence of numerous desmosomes between cells helps to provide the structural integrity of tissues in the body. In addition, other proteins such as keratin filaments anchor the cytoplasmic surface of desmosomes to interior structures of the cell. It is believed that this helps secure the desmosome in place and also provides structural support for the cell. Desmosomes hold adjacent cells firmly together in areas that are subject to considerable stretching, such as the skin. The specialized area of the membrane in the region of a desmosome is usually disk-shaped; these membrane junctions could be likened to rivets or spot-welds.

A second type of membrane junction, the **tight junction** (**Figure 3–10b**), forms when the extracellular surfaces of two adjacent plasma membranes join together so that no extracellular space remains between them. Unlike the desmosome, which is limited to a disk-shaped area of the membrane, the tight junction occurs in a band around the entire circumference of the cell.

Most epithelial cells are joined by tight junctions. For example, epithelial cells cover the inner surface of the intestinal tract, where they come in contact with the digestion products in the cavity (lumen) of the tract. During absorption, the products of digestion move across the epithelium and enter the blood. This movement could theoretically take place either through the extracellular space between the epithelial cells or through the epithelial cells themselves. For many substances, however, movement through the extracellular space is blocked by the tight junctions; this forces organic nutrients to pass through the cells, rather than between them. In this way, the selective barrier properties of the plasma membrane can control the types and amounts of substances absorbed. The ability of tight junctions to impede molecular movement between cells is not absolute. Ions and water can move through these junctions with varying degrees of ease in different epithelia. **Figure 3–10c** shows both a tight junction and a desmosome near the luminal border between two epithelial cells.

A third type of junction, the **gap junction,** consists of protein channels linking the cytosols of adjacent cells (**Figure 3–10d**). In the region of the gap junction, the two opposing plasma membranes come within 2 to 4 nm of each other, which allows specific proteins (called *connexins*) from the two membranes to join, forming small, protein-lined channels linking the two cells. The small diameter of these channels (about 1.5 nm) limits what can pass between the cytosols of the connected cells to small molecules and ions, such as sodium and potassium, and excludes the exchange of large proteins. A variety of cell types possess gap junctions, including the muscle cells of the heart, where they play a very important role in the transmission of electrical activity between the cells.

Cell Organelles

The contents of cells can be released by grinding a tissue against rotating glass or metal surfaces (homogenization) or using various chemical methods to break the plasma membrane. The cell organelles thus released can then be isolated by subjecting

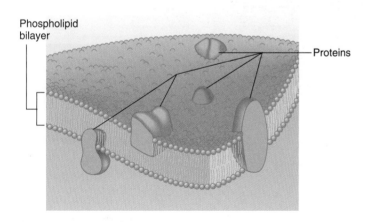

Phospholipid bilayer

Proteins

Figure 3–9

Fluid-mosaic model of cell membrane structure. Only proteins and phospholipids are shown; other membrane components are omitted for clarity. The proteins may move within the bilayer.

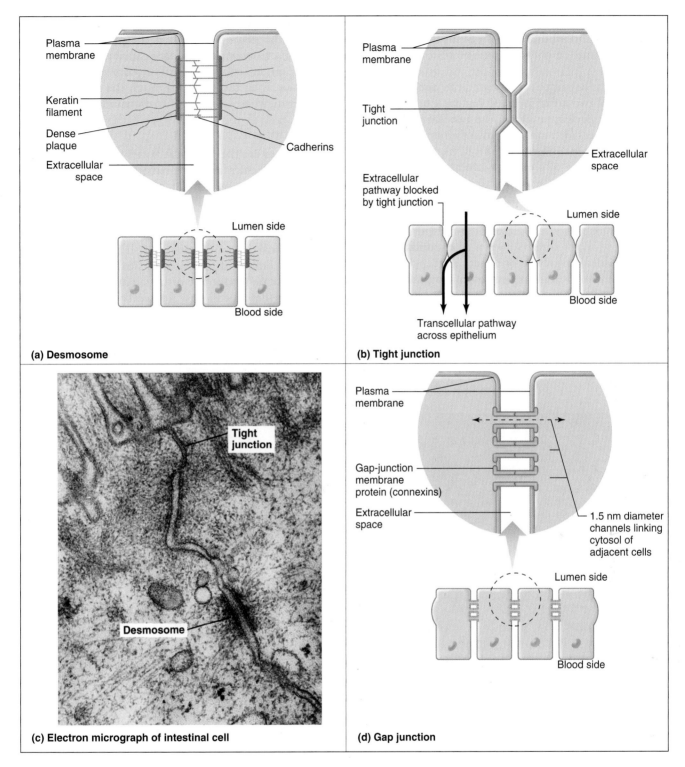

Figure 3–10

Three types of specialized membrane junctions: (a) desmosome; (b) tight junction; (c) electron micrograph of two intestinal epithelial cells joined by a tight junction near the luminal surface and a desmosome below the tight junction; and (d) gap junction.

Electron micrograph from M. Farquhar and G.E. Palade, *J. Cell. Biol.,* 17:375–412 (1963).

the homogenate to ultracentrifugation in which the mixture is spun at very high speeds. Cell organelles of different sizes and density settle out at various rates, so by controlling the speed and time of centrifugation, various fractions can be separated. We can then study these isolated cell organelles to learn their chemical composition and metabolic functions.

Nucleus

Almost all cells contain a single nucleus, the largest of the membrane-bound cell organelles. A few specialized cells, such as skeletal muscle cells, contain multiple nuclei, whereas mature red blood cells have none. The primary function of the nucleus is the storage and transmission of genetic information to the next generation of cells. This information, coded in molecules of DNA, is also used to synthesize the proteins that determine the structure and function of the cell, as described later in this chapter.

Surrounding the nucleus is a barrier, the **nuclear envelope,** composed of two membranes. At regular intervals along the surface of the nuclear envelope, the two membranes are joined to each other, forming the rims of circular openings known as **nuclear pores** (**Figure 3–11**). RNA molecules that determine the structure of proteins synthesized in the cytoplasm move between the nucleus and cytoplasm through these nuclear pores. Proteins that modulate the expression of various genes in DNA move into the nucleus through these pores.

Within the nucleus, DNA, in association with proteins, forms a fine network of threads known as **chromatin.** The threads are coiled to a greater or lesser degree, producing the variations in density seen in electron micrographs of the nucleus (see Figure 3–11). At the time of cell division, the chromatin threads become tightly condensed, forming rod-like bodies known as **chromosomes.**

The most prominent structure in the nucleus is the **nucleolus,** a densely staining filamentous region without a membrane. It is associated with specific regions of DNA that contain the genes for forming the particular type of RNA found in cytoplasmic organelles called ribosomes. This RNA and the protein components of ribosomes are assembled in the nucleolus, then transferred through the nuclear pores to the cytoplasm, where they form functional ribosomes.

Ribosomes

Ribosomes are the protein factories of a cell. On ribosomes, protein molecules are synthesized from amino acids, using genetic information carried by RNA messenger molecules from DNA in the nucleus. Ribosomes are large particles, about 20 nm in diameter, composed of about 70 to 80 proteins and several RNA molecules. As described in Section B, ribosomes consist of two subunits that are either floating free in the cytoplasm or combine during protein synthesis. In the latter case, the ribosomes bind to the organelle called rough endoplasmic reticulum (described next). A typical cell may contain as many as 10 million ribosomes.

The proteins synthesized on the free ribosomes are released into the cytosol, where they perform their varied functions. The proteins synthesized by ribosomes attached to

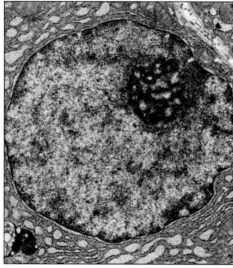

Nucleus

Structure: Largest organelle. Round or oval body located near the cell center. Surrounded by a nuclear envelope composed of two membranes. Envelope contains nuclear pores; messenger molecules pass between the nucleus and the cytoplasm through these pores. No membrane-bound organelles are present in the nucleus, which contains coiled strands of DNA known as chromatin. These condense to form chromosomes at the time of cell division.

Function: Stores and transmits genetic information in the form of DNA. Genetic information passes from the nucleus to the cytoplasm, where amino acids are assembled into proteins.

Nucleolus

Structure: Densely stained filamentous structure within the nucleus. Consists of proteins associated with DNA in regions where information concerning ribosomal proteins is being expressed.

Function: Site of ribosomal RNA synthesis. Assembles RNA and protein components of ribosomal subunits, which then move to the cytoplasm through nuclear pores.

Figure 3–11

Nucleus and nucleolus.

Electron micrograph courtesy of K. R. Porter.

the rough endoplasmic reticulum pass into the lumen of the reticulum and are then transferred to yet another organelle, the Golgi apparatus. They are ultimately secreted from the cell or distributed to other organelles.

Endoplasmic Reticulum

The most extensive cytoplasmic organelle is the network of membranes that forms the **endoplasmic reticulum** (**Figure 3–12**). These membranes enclose a space that is continuous throughout the network.

Two forms of endoplasmic reticulum can be distinguished: rough, or granular, and smooth, or agranular. The rough endoplasmic reticulum has ribosomes bound to its cytosolic surface, and it has a flattened-sac appearance. Rough endoplasmic reticulum is involved in packaging proteins that, after processing in the Golgi apparatus, are secreted by the cell or distributed to other cell organelles.

The smooth endoplasmic reticulum has no ribosomal particles on its surface and has a branched, tubular structure.

It is the site at which certain lipid molecules are synthesized, it plays a role in detoxification of certain hydrophobic molecules, and it also stores and releases calcium ions involved in controlling various cell activities.

Golgi Apparatus

The **Golgi apparatus** is a series of closely apposed, flattened membranous sacs that are slightly curved, forming a cup-shaped structure (**Figure 3–13**). Associated with this organelle, particularly near its concave surface, are a number of roughly spherical, membrane-enclosed vesicles.

Proteins arriving at the Golgi apparatus from the rough endoplasmic reticulum undergo a series of modifications as they pass from one Golgi compartment to the next. For example, carbohydrates are linked to proteins to form glycoproteins, and the length of the protein is often shortened by removing a terminal portion of the polypeptide chain. The Golgi apparatus sorts the modified proteins into discrete classes of transport vesicles that will travel to various cell organelles

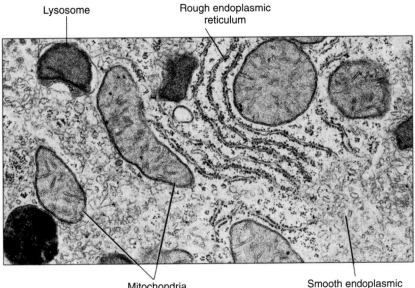

Lysosome

Rough endoplasmic reticulum

Mitochondria

Smooth endoplasmic reticulum

Rough endoplasmic reticulum

Structure: Extensive membranous network of flattened sacs. Encloses a space that is continuous throughout the organelle and with the space between the two nuclear-envelope membranes.
Has ribosomal particles attached to its cytosolic surface.

Function: Proteins synthesized on the attached ribosomes enter the lumen of the reticulum from which they are ultimately distributed to other organelles or secreted from the cell.

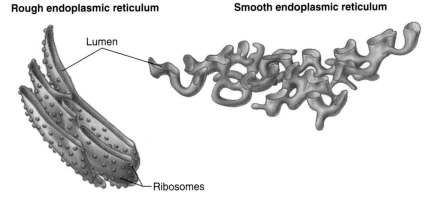

Rough endoplasmic reticulum

Smooth endoplasmic reticulum

Lumen

Ribosomes

Smooth endoplasmic reticulum

Structure: Highly branched tubular network that does not have attached ribosomes but may be continuous with the rough endoplasmic reticulum.

Function: Contains enzymes for fatty acid and steroid synthesis. Stores and releases calcium, which controls various cell activities.

Figure 3–12

Endoplasmic reticulum.

Electron micrograph from D. W. Fawcett, *The Cell, An Atlas of Fine Structure,* W. B. Saunders Company, Philadelphia, 1966.

Cellular Structure, Proteins, and Metabolism

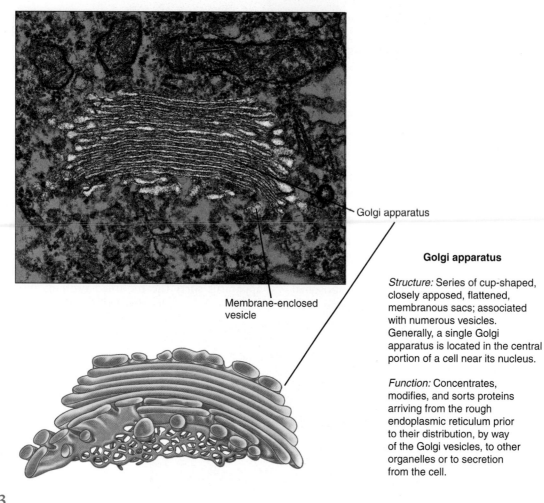

Golgi apparatus

Structure: Series of cup-shaped, closely apposed, flattened, membranous sacs; associated with numerous vesicles. Generally, a single Golgi apparatus is located in the central portion of a cell near its nucleus.

Function: Concentrates, modifies, and sorts proteins arriving from the rough endoplasmic reticulum prior to their distribution, by way of the Golgi vesicles, to other organelles or to secretion from the cell.

Golgi apparatus

Membrane-enclosed vesicle

Golgi apparatus

Figure 3–13

Golgi apparatus.

Electron micrograph from W. Bloom and D. W. Fawcett, *Textbook of Histology,* 9th ed. W. B. Saunders Company, Philadelphia, 1968.

or to the plasma membrane, where the protein contents of the vesicle are released to the outside of the cell. Vesicles containing proteins to be secreted from the cell are known as **secretory vesicles.** Such vesicles are found, for example, in certain endocrine gland cells, where protein hormones are released into the extracellular fluid to modify the activities of other cells.

Endosomes

A number of membrane-bound vesicular and tubular structures called **endosomes** lie between the plasma membrane and the Golgi apparatus. Certain types of vesicles that pinch off the plasma membrane travel to and fuse with endosomes. In turn, the endosome can pinch off vesicles that then move to other cell organelles or return to the plasma membrane. Like the Golgi apparatus, endosomes are involved in sorting, modifying, and directing vesicular traffic in cells.

Mitochondria

Mitochondria (singular, *mitochondrion*) participate in the chemical processes that transfer energy from the chemical bonds of nutrient molecules to newly created adenosine triphosphate (ATP) molecules, which are then made available to cells. Most of the ATP that cells use is formed in the mito-

chondria by a process called cellular respiration, which consumes oxygen and produces carbon dioxide, heat, and water.

Mitochondria are spherical or elongated, rodlike structures surrounded by an inner and an outer membrane (**Figure 3–14**). The outer membrane is smooth, whereas the inner membrane is folded into sheets or tubules known as **cristae,** which extend into the inner mitochondrial compartment, the **matrix.** Mitochondria are found throughout the cytoplasm. Large numbers of them, as many as 1000, are present in cells that utilize large amounts of energy, whereas less active cells contain fewer.

In addition to providing most of the energy needed to power physiological events such as muscle contraction, mitochondria also play a role in the synthesis of certain lipids, such as the hormones estrogen and testosterone (Chapter 11).

Lysosomes

Lysosomes are spherical or oval organelles surrounded by a single membrane (see Figure 3–4). A typical cell may contain several hundred lysosomes. The fluid within a lysosome is acidic and contains a variety of digestive enzymes. Lysosomes act as "cellular stomachs," breaking down bacteria and the debris from dead cells that have been engulfed by a cell. They

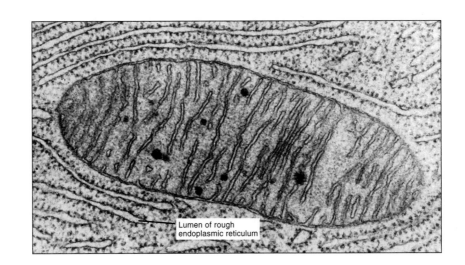

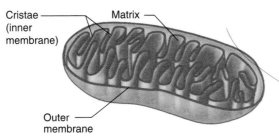

Lumen of rough endoplasmic reticulum

Cristae (inner membrane)   Matrix

Outer membrane

Mitochondrion

Structure: Rod- or oval-shaped body surrounded by two membranes. Inner membrane folds into matrix of the mitochondrion, forming cristae.

Function: Major site of ATP production, O_2 utilization, and CO_2 formation. Contains enzymes active in Krebs cycle and oxidative phosphorylation.

Figure 3–14

Mitochondrion.

Electron micrograph courtesy of K. R. Porter.

may also break down cell organelles that have been damaged and no longer function normally. They play an especially important role in the various cells that make up the defense systems of the body (Chapter 18).

Peroxisomes

Like lysosomes, **peroxisomes** are moderately dense oval bodies enclosed by a single membrane. Like mitochondria, peroxisomes consume molecular oxygen, although in much smaller amounts. This oxygen is not used in the transfer of energy to ATP, however. Instead it undergoes reactions that remove hydrogen from organic molecules including lipids, alcohol, and potentially toxic ingested substances. One of the reaction products is hydrogen peroxide, H_2O_2, thus the organelle's name. Hydrogen peroxide can be toxic to cells in high concentrations, but peroxisomes can also destroy hydrogen peroxide and thus prevent its toxic effects. Peroxisomes are also involved in the process by which fatty acids are broken down into 2-carbon fragments, which the cell can then use as a source for generating ATP.

Vaults

Vaults are recently discovered cytoplasmic structures composed of protein and a type of RNA called vault RNA (vRNA). These tiny structures have been described as barrel-shaped, but also as resembling the vaulted cathedrals found in many large buildings, thus their name. Although the functions of vaults are not certain, studies using electron microscopy and other methods have revealed that vaults tend to be associated with nuclear pores. This has led to the hypothesis that vaults are important for transport of molecules between the cytosol and the nucleus. In addition, at least one vault protein is believed to function in regulating a cell's sensitivity to certain drugs. For example, increased expression of this vault protein has been linked in some studies to drug resistance, including some drugs used in the treatment of cancer. If true, then vaults may someday provide a target for modulating the effectiveness of such drugs in human patients.

Cytoskeleton

In addition to the membrane-enclosed organelles, the cytoplasm of most cells contains a variety of protein filaments. This filamentous network is referred to as the cell's **cytoskeleton,** and, like the bony skeleton of the body, it is associated with processes that maintain and change cell shape and produce cell movements.

There are three classes of cytoskeletal filaments, based on their diameter and the types of protein they contain. In order of size, starting with the thinnest, they are (1) microfilaments, (2) intermediate filaments, and (3) microtubules (**Figure 3–15**). Microfilaments and microtubules can be assembled and disassembled rapidly, allowing a cell to alter these components of its cytoskeletal framework according to changing requirements. In contrast, intermediate filaments, once assembled, are less readily disassembled.

Microfilaments are composed of the contractile protein **actin,** and make up a major portion of the cytoskeleton in all cells. **Intermediate filaments** are most extensively developed

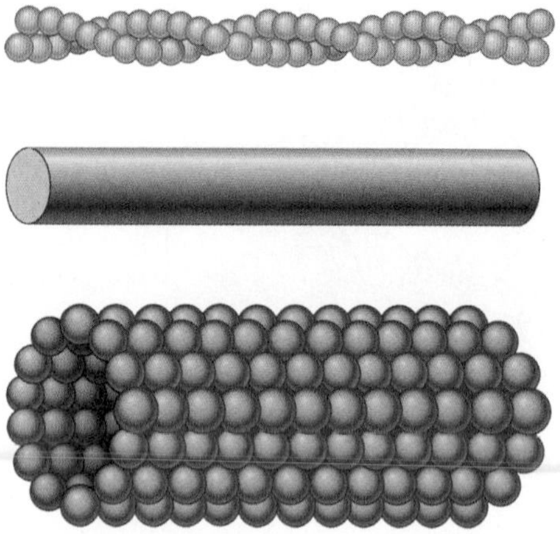

Cytoskeletal filaments	Diameter (nm)	Protein subunit
Microfilament	7	Actin
Intermediate filament	10	Several proteins
Microtubule	25	Tubulin

Figure 3–15

Cytoskeletal filaments associated with cell shape and motility.

in the regions of cells subject to mechanical stress (for example, in association with desmosomes).

Microtubules are hollow tubes about 25 nm in diameter, whose subunits are composed of the protein **tubulin.** They are the most rigid of the cytoskeletal filaments and are present in the long processes of nerve cells, where they provide the framework that maintains the processes' cylindrical shape. Microtubules also radiate from a region of the cell known as the **centrosome,** which surrounds two small, cylindrical bodies called **centrioles,** composed of nine sets of fused microtubules. The centrosome is a cloud of amorphous material that regulates the formation and elongation of microtubules. During cell division the centrosome generates the microtubular spindle fibers used in chromosome separation. Microtubules and microfilaments have also been implicated in the movements of organelles within the cytoplasm. These fibrous elements form tracks, and organelles are propelled along these tracks by contractile proteins attached to the surface of the organelles.

Cilia, the hairlike motile extensions on the surfaces of some epithelial cells, have a central core of microtubules organized in a pattern similar to that found in the centrioles. These microtubules, in combination with a contractile protein, produce movements of the cilia. In hollow organs lined with ciliated epithelium, the cilia wave back and forth, propelling the luminal contents along the surface of the epithelium.

■ **SECTION A SUMMARY**

Microscopic Observations of Cells

I. All living matter is composed of cells.
II. There are two types of cells: prokaryotic cells (bacteria) and eukaryotic cells (plant and animal cells).

Membranes

I. Every cell is surrounded by a plasma membrane.
II. Within each eukaryotic cell are numerous membrane-bound compartments, nonmembranous particles, and filaments, known collectively as cell organelles.

III. A cell is divided into two regions, the nucleus and the cytoplasm. The latter is composed of the cytosol and cell organelles other than the nucleus.
IV. The membranes that surround the cell and cell organelles regulate the movements of molecules and ions into and out of the cell and its compartments.
 a. Membranes consist of a bimolecular lipid layer, composed of phospholipids with embedded proteins.
 b. Integral membrane proteins are amphipathic proteins that often span the membrane, whereas peripheral membrane proteins are confined to the surfaces of the membrane.
V. Three types of membrane junctions link adjacent cells.
 a. Desmosomes link cells that are subject to considerable stretching.
 b. Tight junctions, found primarily in epithelial cells, limit the passage of molecules through the extracellular space between the cells.
 c. Gap junctions form channels between the cytosols of adjacent cells.

Cell Organelles

I. The nucleus transmits and expresses genetic information.
 a. Threads of chromatin, composed of DNA and protein, condense to form chromosomes when a cell divides.
 b. Ribosomal subunits are assembled in the nucleolus.
II. Ribosomes, composed of RNA and protein, are the sites of protein synthesis.
III. The endoplasmic reticulum is a network of flattened sacs and tubules in the cytoplasm.
 a. Rough endoplasmic reticulum has attached ribosomes and is primarily involved in the packaging of proteins to be secreted by the cell or distributed to other organelles.
 b. Smooth endoplasmic reticulum is tubular, lacks ribosomes, and is the site of lipid synthesis and calcium accumulation and release.
IV. The Golgi apparatus modifies and sorts the proteins that are synthesized on the rough or granular endoplasmic reticulum and packages them into secretory vesicles.
V. Endosomes are membrane-bound vesicles that fuse with vesicles derived from the plasma membrane and bud off vesicles that travel to other cell organelles.

VI. Mitochondria are the major cell sites that consume oxygen and produce carbon dioxide in chemical processes that transfer energy to ATP, which can then provide energy for cell functions.

VII. Lysosomes digest particulate matter that enters the cell.

VIII. Peroxisomes use oxygen to remove hydrogen from organic molecules and in the process form hydrogen peroxide.

IX. Vaults are cytoplasmic structures made of protein and RNA, and may be involved in cytoplasmic-nuclear transport.

X. The cytoplasm contains a network of three types of filaments that form the cytoskeleton: (1) microfilaments, (2) intermediate filaments, and (3) microtubules.

■ SECTION A KEY TERMS

actin 53	glycocalyx 47
cadherin 48	Golgi apparatus 51
cell organelle 45	integral membrane protein 47
centriole 54	integrin 48
centrosome 54	intermediate filament 53
chromatin 50	intracellular fluid 45
chromosome 50	lysosome 52
cilia 54	matrix 52
cristae 52	microfilament 53
cytoplasm 45	microtubule 54
cytoskeleton 53	mitochondrion 52
cytosol 45	nuclear envelope 50
desmosome 48	nuclear pore 50
endoplasmic reticulum 51	nucleolus 50
endosome 52	nucleus 45
eukaryotic cell 45	peripheral membrane
fluid-mosaic model 48	protein 47
gap junction 48	peroxisome 53

phospholipid 46	tight junction 48
plasma membrane 45	transmembrane protein 47
prokaryotic cell 45	tubulin 54
ribosome 50	vault 53
secretory vesicle 52	

■ SECTION A REVIEW QUESTIONS

1. Identify the location of cytoplasm, cytosol, and intracellular fluid within a cell.
2. Identify the classes of organic molecules found in cell membranes.
3. Describe the orientation of the phospholipid molecules in a membrane.
4. Which plasma membrane components are responsible for membrane fluidity?
5. Describe the location and characteristics of integral and peripheral membrane proteins.
6. Describe the structure and function of the three types of junctions found between cells.
7. What function does the nucleolus perform?
8. Describe the location and function of ribosomes.
9. Contrast the structure and functions of the rough and smooth endoplasmic reticulum.
10. What function does the Golgi apparatus perform?
11. What functions do endosomes perform?
12. Describe the structure and primary function of mitochondria.
13. What functions do lysosomes and peroxisomes perform?
14. List the three types of filaments associated with the cytoskeleton. Identify the structures in cells that are composed of microtubules.

SECTION B — Proteins

Genetic Code

The importance of proteins in physiology cannot be overstated. Proteins are involved in all physiological processes, from cell signaling to tissue remodeling to organ function. This section describes how cells synthesize, degrade, and, in some cases, secrete proteins. We begin with an overview of the genetic basis of protein synthesis.

As noted previously, the nucleus of cells contains DNA, which directs the synthesis of all proteins in the body. Molecules of DNA contain information, coded in the sequence of nucleotides, for protein synthesis. A sequence of DNA nucleotides containing the information that specifies the amino acid sequence of a single polypeptide chain is known as a **gene.** A gene is thus a unit of hereditary information. A single molecule of DNA contains many genes.

The total genetic information coded in the DNA of a typical cell in an organism is known as its **genome.** The human genome contains roughly 30,000 to 40,000 genes. Recently, scientists determined the nucleotide sequence of the entire human genome (approximately 3 billion nucleotides). This is

only a first step, however, because the function and regulation of most genes in the human genome remain unknown.

It is easy to misunderstand the relationship between genes, DNA molecules, and chromosomes. In all human cells other than the eggs or sperm, there are 46 separate DNA molecules in the cell nucleus, each molecule containing many genes. Each DNA molecule is packaged into a single chromosome composed of DNA and proteins, so there are 46 chromosomes in each cell. A chromosome contains not only its DNA molecule, but also a special class of proteins called **histones.** The cell's nucleus is a marvel of packaging. The very long DNA molecules, with lengths a thousand times greater than the diameter of the nucleus, fit into the nucleus by coiling around clusters of histones at frequent intervals to form complexes known as **nucleosomes.** There are about 25 million of these complexes on the chromosomes, resembling beads on a string.

Although DNA contains the information specifying the amino acid sequences in proteins, it does not itself participate directly in the assembly of protein molecules. Most of a cell's DNA is in the nucleus, whereas most protein synthesis occurs in the cytoplasm. The transfer of information from DNA to the site of protein synthesis is accomplished by RNA molecules, whose synthesis is governed by the information coded in DNA. Genetic information flows from DNA to RNA and then to protein (**Figure 3–16**). The process of transferring genetic information from DNA to RNA in the nucleus is known as **transcription.** The process that uses the coded information in RNA to assemble a protein in the cytoplasm is known as **translation.**

$$\text{DNA} \xrightarrow{\text{transcription}} \text{RNA} \xrightarrow{\text{translation}} \text{Protein}$$

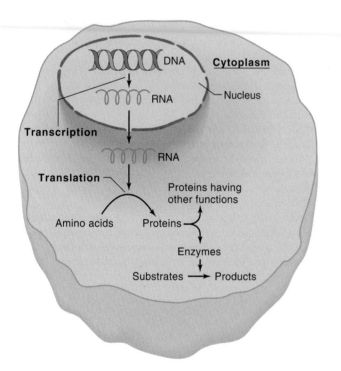

Figure 3–16

The expression of genetic information in a cell occurs through the *transcription* of coded information from DNA to RNA in the nucleus, followed by the *translation* of the RNA information into protein synthesis in the cytoplasm. The proteins then perform the functions that determine the characteristics of the cell.

As described in Chapter 2, a molecule of DNA consists of two chains of nucleotides coiled around each other to form a double helix. Each DNA nucleotide contains one of four bases—adenine (A), guanine (G), cytosine (C), or thymine (T)—and each of these bases is specifically paired by hydrogen bonds with a base on the opposite chain of the double helix. In this base pairing, A and T bond together and G and C bond together. Thus, both nucleotide chains contain a specifically ordered sequence of bases, with one chain complementary to the other. This specificity of base pairing forms the basis of the transfer of information from DNA to RNA and of the duplication of DNA during cell division.

The genetic language is similar in principle to a written language, which consists of a set of symbols, such as Λ, B, C, D, that form an alphabet. The letters are arranged in specific sequences to form words, and the words are arranged in linear sequences to form sentences. The genetic language contains only four letters, corresponding to the bases A, G, C, and T. The genetic words are three-base sequences that specify particular amino acids—that is, each word in the genetic language is only three letters long. This is termed a triplet code. The sequence of three-letter code words (triplets) along a gene in a single strand of DNA specifies the sequence of amino acids in a polypeptide chain (**Figure 3–17**). Thus, a gene is equivalent to a sentence, and the genetic information in the human genome is equivalent to a book containing 30,000 to 40,000 sentences. Using a single letter (A, T, C, G) to specify each of the four bases in the DNA nucleotides, it would require about 550,000 pages, each equivalent to this text page, to print the nucleotide sequence of the human genome.

The four bases in the DNA alphabet can be arranged in 64 different three-letter combinations to form 64 triplets ($4 \times 4 \times 4 = 64$). Thus, this code actually provides more than enough words to code for the 20 different amino acids that are found in proteins. This means that a given amino acid is usually specified by more than one triplet. For example, the four DNA triplets C—C—A, C—C—G, C—C—T, and C—C—C all specify the amino acid glycine. Only 61 of the 64 possible triplets are used to specify amino acids. The triplets that do not specify amino acids are known as **"stop" signals.** They perform the same function as a period at the end

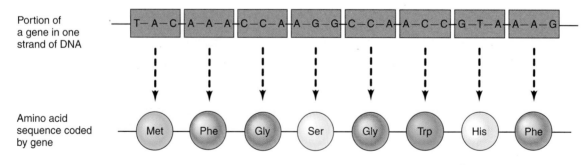

Figure 3–17

The sequence of three-letter code words in a gene determines the sequence of amino acids in a polypeptide chain. The names of the amino acids are abbreviated. Note that more than one three-letter code sequence can specify the same amino acid; for example, the amino acid phenylalanine (Phe) is coded by two triplet codes, A—A—A and A—A—G.

of a sentence—they indicate the end of a genetic message has been reached.

The genetic code is a universal language used by all living cells. For example, the triplets specifying the amino acid tryptophan are the same in the DNA of a bacterium, an amoeba, a plant, and a human being. Although the same triplets are used by all living cells, the messages they spell out—the sequences of triplets that code for a specific protein—vary from gene to gene in each organism. The universal nature of the genetic code supports the concept that all forms of life on earth evolved from a common ancestor.

Before we turn to the specific mechanisms by which the DNA code operates in protein synthesis, an important qualification is required. Although the information coded in genes is always first transcribed into RNA, there are several classes of RNA—including messenger RNA, ribosomal RNA, and transfer RNA. Only messenger RNA *directly* codes for the amino acid sequences of proteins, even though the other RNA classes participate in the overall process of protein synthesis. For this reason, the customary definition of a gene as the sequence of DNA nucleotides that specifies the amino acid sequence of a protein is true only for the vast majority of genes that are transcribed into messenger RNA.

Protein Synthesis

To repeat, the first step in using the genetic information in DNA to synthesize a protein is called transcription, and it involves the synthesis of an RNA molecule containing coded information that corresponds to the information in a single gene. The class of RNA molecules that specifies the amino acid sequence of a protein and carries this message from DNA to the site of protein synthesis in the cytoplasm is known as **messenger RNA (mRNA).**

Transcription: mRNA Synthesis

Recall from Chapter 2 that ribonucleic acids are single-chain polynucleotides whose nucleotides differ from DNA because they contain the sugar ribose (rather than deoxyribose) and the base uracil (rather than thymine). The other three bases—adenine, guanine, and cytosine—occur in both DNA and RNA. The pool of subunits used to synthesize mRNA are free (uncombined) ribonucleotide triphosphates: ATP, GTP, CTP, and UTP.

Recall also that the two polynucleotide chains in DNA are linked together by hydrogen bonds between specific pairs of bases: A–T and C–G. To initiate RNA synthesis, the two strands of the DNA double helix must separate so that the bases in the exposed DNA can pair with the bases in free ribonucleotide triphosphates (**Figure 3–18**). Free ribonucleotides containing U bases pair with the exposed A bases in DNA, and likewise, free ribonucleotides containing G, C, or A bases pair with the exposed DNA bases C, G, and T, respectively. Note that uracil, which is present in RNA but not DNA, pairs with the base adenine in DNA. In this way, the nucleotide sequence in one strand of DNA acts as a template that determines the sequence of nucleotides in mRNA.

The aligned ribonucleotides are joined together by the enzyme **RNA polymerase,** which hydrolyses the nucleotide triphosphates, releasing two of the terminal phosphate groups and joining the remaining phosphate in covalent linkage to the ribose of the adjacent nucleotide.

Because DNA consists of *two* strands of polynucleotides, both of which are exposed during transcription, it should theoretically be possible to form two individual RNA molecules, one from each strand. However, only one of the two potential RNAs is ever formed. Which of the two DNA strands is used as the **template strand** for RNA synthesis from a particular gene is determined by a specific sequence of DNA nucleotides called the **promoter,** which is located near the beginning of

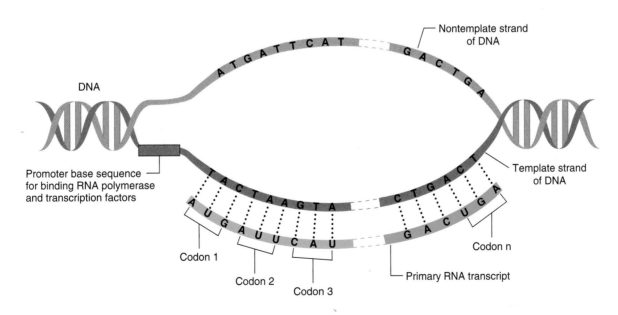

Figure 3–18

Transcription of a gene from the template strand of DNA to a primary mRNA transcript.

the gene on the strand to be transcribed (see Figure 3–18). It is to this promoter region that RNA polymerase binds and initiates transcription. Thus, for any given gene, only one DNA strand is transcribed, and that is the strand with the promoter region at the beginning of the gene.

Thus, the transcription of a gene begins when RNA polymerase binds to the promoter region of that gene. This initiates the separation of the two strands of DNA. RNA polymerase moves along the template strand, joining one ribonucleotide at a time (at a rate of about 30 nucleotides per second) to the growing RNA chain. Upon reaching a "stop" signal specifying the end of the gene, the RNA polymerase releases the newly formed RNA transcript, which is then translocated out of the nucleus where it binds to ribosomes in the cytoplasm.

In a given cell, typically only 10 to 20 percent of the genes present in DNA are transcribed into RNA. Genes are transcribed only when RNA polymerase can bind to their promoter sites. Cells use various mechanisms either to block or to make accessible the promoter region of a particular gene to RNA polymerase. Such regulation of gene transcription provides a means of controlling the synthesis of specific proteins and thereby the activities characteristic of a particular type of cell.

It must be emphasized that the base sequence in the RNA transcript is not identical to that in the template strand of DNA, because the RNA's formation depends on the pairing between *complementary,* not identical, bases (see Figure 3–18). A three-base sequence in RNA that specifies one amino acid is called a **codon.** Each codon is complementary to a three-base sequence in DNA. For example, the base sequence T—A—C in the template strand of DNA corresponds to the codon A—U—G in transcribed RNA.

Although the entire sequence of nucleotides in the template strand of a gene is transcribed into a complementary sequence of nucleotides known as the **primary RNA transcript,** only certain segments of most genes actually code for sequences of amino acids. These regions of the gene, known as **exons** (expression regions), are separated by noncoding sequences of nucleotides known as **introns** (intervening sequences). It is estimated that as much as 98.5 percent of human DNA is composed of intron sequences that do not contain protein-coding information. What role, if any, such large amounts of noncoding DNA may perform is unclear, although they have recently been postulated to exert some transcriptional regulation.

Before passing to the cytoplasm, a newly formed primary RNA transcript must undergo splicing (**Figure 3–19**) to remove the sequences that correspond to the DNA introns. This allows the formation of the continuous sequence of exons that will be translated into protein. Only after this splicing occurs is the RNA termed messenger RNA.

Splicing occurs in the nucleus and is performed by a complex of proteins and small nuclear RNAs known as a **spliceosome.** The spliceosome identifies specific nucleotide sequences at the beginning and end of each intron-derived segment in the primary RNA transcript, removes the segment, and splices the end of one exon-derived segment to the beginning of another to form mRNA with a continuous coding sequence. In some cases during the splicing process, the exon-derived segments from a single gene can be spliced together in different sequences, or some exon-derived segments can be deleted entirely. These processes result in the formation of different mRNA sequences from the same gene and give rise, in turn, to proteins with slightly different amino acid sequences.

Translation: Polypeptide Synthesis

After splicing, the mRNA moves through the pores in the nuclear envelope into the cytoplasm. Although the nuclear pores allow the diffusion of small molecules and ions between the nucleus and cytoplasm, they have specific energy-dependent

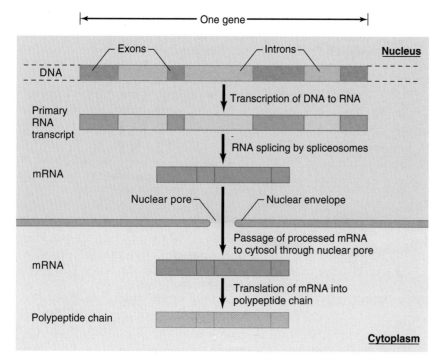

Figure 3–19

Spliceosomes remove the noncoding intron-derived segments from a primary RNA transcript and link the exon-derived segments together to form the mRNA molecule that passes through the nuclear pores to the cytosol. The lengths of the intron- and exon-derived segments represent the relative lengths of the base sequences in these regions.

mechanisms for the selective transport of large molecules such as proteins and RNA.

In the cytoplasm, mRNA binds to a ribosome, the cell organelle that contains the enzymes and other components required for the translation of mRNA into protein. Before describing this assembly process, we will examine the structure of a ribosome and the characteristics of two additional classes of RNA involved in protein synthesis.

Ribosomes and rRNA

A ribosome is a complex particle composed of about 70 to 80 different proteins in association with a class of RNA molecules known as **ribosomal RNA (rRNA).** The genes for rRNA are transcribed from DNA in a process similar to that for mRNA except that a different RNA polymerase is used. Ribosomal RNA transcription occurs in the region of the nucleus known as the nucleolus. Ribosomal proteins, like other proteins, are synthesized in the cytoplasm from the mRNAs specific for them. These proteins then move back through nuclear pores to the nucleolus, where they combine with newly synthesized rRNA to form two ribosomal subunits, one large and one small. These subunits are then individually transported to the cytoplasm, where they combine to form a functional ribosome during protein translation.

Transfer RNA

How do individual amino acids identify the appropriate codons in mRNA during the process of translation? By themselves, free amino acids do not have the ability to bind to the bases in mRNA codons. This process of identification involves the third major class of RNA, known as **transfer RNA (tRNA).** Transfer RNA molecules are the smallest (about 80 nucleotides long) of the major classes of RNA. The single chain of tRNA loops back upon itself, forming a structure resembling a cloverleaf with three loops (**Figure 3–20**).

Like mRNA and rRNA, tRNA molecules are synthesized in the nucleus by base-pairing with DNA nucleotides at specific tRNA genes; then they move to the cytoplasm. The key to tRNA's role in protein synthesis is its ability to combine with both a specific amino acid and a codon in ribosome-bound mRNA specific for that amino acid. This permits tRNA to act as the link between an amino acid and the mRNA codon for that amino acid.

A tRNA molecule is covalently linked to a specific amino acid by an enzyme known as aminoacyl-tRNA synthetase. There are 20 different aminoacyl-tRNA synthetases, each of which catalyzes the linkage of a specific amino acid to a specific type of tRNA. The next step is to link the tRNA, bearing its attached amino acid, to the mRNA codon for that amino acid. This is achieved by base-pairing between tRNA and mRNA. A three-nucleotide sequence at the end of one of the loops of tRNA can base-pair with a complementary codon in mRNA. This tRNA three-letter code sequence is appropriately termed an **anticodon.** Figure 3–20 illustrates the binding between mRNA and a tRNA specific for the amino acid tryptophan. Note that tryptophan is covalently linked to one end of tRNA and does not bind to either the anticodon region of tRNA or the codon region of mRNA.

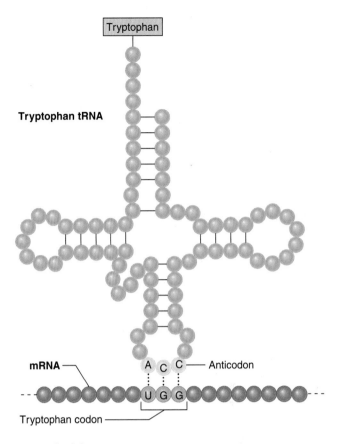

Figure 3–20

Base-pairing between the anticodon region of a tRNA molecule and the corresponding codon region of an mRNA molecule.

Protein Assembly

The process of assembling a polypeptide chain based on an mRNA message involves three stages—initiation, elongation, and termination. The initiation of synthesis occurs when a tRNA containing the amino acid methionine binds to the small ribosomal subunit. A number of proteins known as **initiation factors** are required to establish an initiation complex, which positions the methionine-containing tRNA opposite the mRNA codon that signals the start site at which assembly is to begin. The large ribosomal subunit then binds, enclosing the mRNA between the two subunits. This initiation phase is the slowest step in protein assembly, and factors that influence the activity of initiation factors can regulate the rate of protein synthesis.

Following the initiation process, the protein chain is elongated by the successive addition of amino acids (**Figure 3–21**). A ribosome has two binding sites for tRNA. Site 1 holds the tRNA linked to the portion of the protein chain that has been assembled up to this point, and site 2 holds the tRNA containing the next amino acid to be added to the chain. Ribosomal enzymes catalyze the linkage of the protein chain to the newly arrived amino acid. Following the formation of the peptide bond, the tRNA at site 1 is released from the ribosome, and the tRNA at site 2—now linked to the peptide chain—is transferred to site 1. The ribosome moves down one codon along the

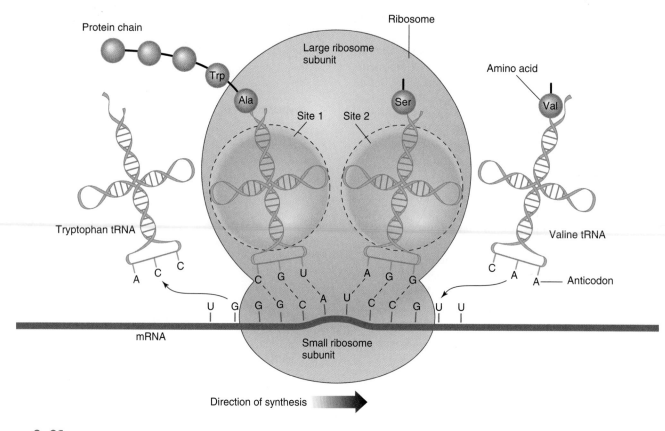

Figure 3–21

Sequence of events during protein synthesis by a ribosome.

mRNA, making room for the binding of the next amino acid–tRNA molecule. This process is repeated over and over as amino acids are added to the growing peptide chain, at an average rate of two to three per second. When the ribosome reaches a termination sequence in mRNA specifying the end of the protein, the link between the polypeptide chain and the last tRNA is broken, and the completed protein is released from the ribosome.

Messenger RNA molecules are not destroyed during protein synthesis, so they may be used to synthesize many more protein molecules. In fact, while one ribosome is moving along a particular strand of mRNA, a second ribosome may become attached to the start site on that same mRNA and begin the synthesis of a second identical protein molecule. Thus, a number of ribosomes, as many as 70, may be moving along a single strand of mRNA, each at a different stage of the translation process (**Figure 3–22**).

Molecules of mRNA do not, however, remain in the cytoplasm indefinitely. Eventually cytoplasmic enzymes break them down into nucleotides. Therefore, if a gene corresponding to a particular protein ceases to be transcribed into mRNA, the protein will no longer be formed after its cytoplasmic mRNA molecules have broken down.

Once a polypeptide chain has been assembled, it may undergo posttranslational modifications to its amino acid sequence. For example, the amino acid methionine that is used to identify the start site of the assembly process is cleaved from the end of most proteins. In some cases, other specific peptide

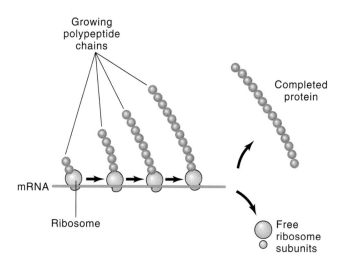

Figure 3–22

Several ribosomes can simultaneously move along a strand of mRNA, producing the same protein in different stages of assembly.

bonds within the polypeptide chain are broken, producing a number of smaller peptides, each of which may perform a different function. For example, as illustrated in **Figure 3–23**, five different proteins can be derived from the same mRNA as a result of posttranslational cleavage. The same initial polypep-

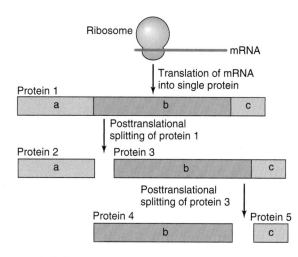

Figure 3–23

Posttranslational splitting of a protein can result in several smaller proteins, each of which may perform a different function. All these proteins are derived from the same gene.

tide may be split at different points in different cells depending on the specificity of the hydrolyzing enzymes present.

Carbohydrates and lipid derivatives are often covalently linked to particular amino acid side chains. These additions may protect the protein from rapid degradation by proteolytic enzymes or act as signals to direct the protein to those locations in the cell where it is to function. The addition of a fatty acid to a protein, for example, can lead the protein to anchor to a membrane as the nonpolar portion of the fatty acid inserts into the lipid bilayer.

The steps leading from DNA to a functional protein are summarized in **Table 3–2**.

Although 99 percent of eukaryotic DNA is located in the nucleus, a small amount is present in mitochondria. Mitochondrial DNA, like bacterial DNA, does not contain introns and is circular. These characteristics support the hypothesis that mitochondria arose during an early stage of evolution when an anaerobic cell ingested an aerobic bacterium that ultimately led to what we know today as mitochondria. Mitochondria also have the machinery, including ribosomes, for protein synthesis. However, the mitochondrial DNA contains the genes for only 13 mitochondrial proteins and a few of the rRNA and tRNA genes. Therefore, additional components are required for mitochondrial protein synthesis, and most of the mitochondrial proteins are coded by nuclear DNA genes. These components are synthesized in the cytoplasm and then transported into the mitochondria.

Regulation of Protein Synthesis

As noted earlier, in any given cell only a small fraction of the genes in the human genome are ever transcribed into mRNA and translated into proteins. Of this fraction, a small number of genes are continuously being transcribed into mRNA. The transcription of other genes, however, is regulated and can be turned on or off in response either to signals generated within the cell or to external signals the cell receives. In order for a gene to be tran-

| **Table 3–2** | Events Leading from DNA to Protein Synthesis |

Transcription

1. RNA polymerase binds to the promoter region of a gene and separates the two strands of the DNA double helix in the region of the gene to be transcribed.

2. Free ribonucleotide triphosphates base-pair with the deoxynucleotides in the template strand of DNA.

3. The ribonucleotides paired with this strand of DNA are linked by RNA polymerase to form a primary RNA transcript containing a sequence of bases complementary to the template strand of the DNA base sequence.

4. RNA splicing removes the intron-derived regions in the primary RNA transcript, which contain noncoding sequences, and splices together the exon-derived regions, which code for specific amino acids, producing a molecule of mRNA.

Translation

5. The mRNA passes from the nucleus to the cytoplasm, where one end of the mRNA binds to the small subunit of a ribosome.

6. Free amino acids are linked to their corresponding tRNAs by aminoacyl-tRNA synthetase.

7. The three-base anticodon in an amino acid–tRNA complex pairs with its corresponding codon in the region of the mRNA bound to the ribosome.

8. The amino acid on the tRNA is linked by a peptide bond to the end of the growing polypeptide chain.

9. The tRNA that has been freed of its amino acid is released from the ribosome.

10. The ribosome moves one codon step along mRNA.

11. Step 7 to 10 are repeated until a termination sequence is reached, and the completed protein is released from the ribosome.

12. In some cases, the protein undergoes posttranslational processing in which various chemical groups are attached to specific side chains and/or the protein is split into several smaller peptide chains.

scribed, RNA polymerase must be able to bind to the promoter region of the gene and be in an activated configuration.

Transcription of most genes is regulated by a class of proteins known as **transcription factors,** which act as gene switches, interacting in a variety of ways to activate or repress the initiation process that takes place at the promoter region of a particular gene. The influence of a transcription factor on transcription is not necessarily all or none, on or off; it may simply slow or speed up the initiation of the transcription process. The transcription factors, along with accessory proteins, form a **preinitiation complex** at the promoter that is needed to carry out the process of separating the DNA strands, removing any blocking nucleosomes in the region of the promoter, activating the bound RNA polymerase, and moving the complex along the template strand of DNA. Some transcription factors bind to regions of DNA that are far removed from the promoter region of the gene whose transcription they regulate. In this case, the DNA containing the bound transcription factor forms a loop that brings the transcription factor into contact with the promoter region, where it may then activate or repress transcription (**Figure 3–24**).

Many genes contain regulatory sites that a common transcription factor can influence; thus there does not need to be a different transcription factor for every gene. In addition, more than one transcription factor may interact to control the transcription of a given gene.

Because transcription factors are proteins, the activity of a particular transcription factor—that is, its ability to bind to DNA or to other regulatory proteins—can be turned on or off by allosteric or covalent modulation in response to signals a cell either receives or generates. Thus, specific genes can be regulated in response to specific signals.

To summarize, the rate of a protein's synthesis can be regulated at various points: (1) gene transcription into mRNA; (2) the initiation of protein assembly on a ribosome; and (3) mRNA degradation in the cytoplasm.

Mutation

Any alteration in the nucleotide sequence that spells out a genetic message in DNA is known as a **mutation.** Certain chemicals and various forms of ionizing radiation, such as x-rays, cosmic rays, and atomic radiation, can break the chemical bonds in DNA. This can result in the loss of segments of DNA or the incorporation of the wrong base when the broken bonds re-form. Environmental factors that increase the rate of mutation are known as **mutagens.**

Types of Mutations

The simplest type of mutation, known as a point mutation, occurs when a single base is replaced by a different one. For example, the base sequence C–G–T is the DNA triplet for the amino acid alanine. If guanine (G) is replaced by adenine (A), the sequence becomes C—A—T, which is the code for valine. If, however, cytosine (C) replaces thymine (T), the sequence becomes C—G—C, which is another code for alanine, and the

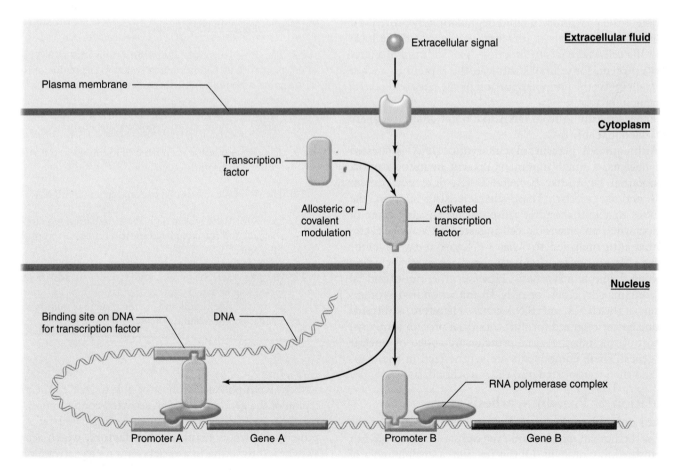

Figure 3–24

Transcription of gene B is modulated by the binding of an activated transcription factor directly to the promoter region. In contrast, transcription of gene A is modulated by the same transcription factor, which, in this case, binds to a region of DNA considerably distant from the promoter region.

amino acid sequence transcribed from the mutated gene would not be altered. On the other hand, if an amino acid code mutates to one of the termination triplets, the translation of the mRNA message will cease when this triplet is reached, resulting in the synthesis of a shortened, typically nonfunctional protein.

Assume that a mutation has altered a single triplet code in a gene, for example, alanine C—G—T changed to valine C—A—T, so that it now codes for a protein with one different amino acid. What effect does this mutation have upon the cell? The answer depends upon where in the gene the mutation has occurred. Although proteins are composed of many amino acids, the properties of a protein often depend upon a very small region of the total molecule, such as the binding site of an enzyme. If the mutation does not alter the conformation of the binding site, there may be little or no change in the protein's properties. On the other hand, if the mutation alters the binding site, a marked change in the protein's properties may occur.

What effects do mutations have upon the functioning of a cell? If a mutated, nonfunctional protein is part of a chemical reaction supplying most of a cell's chemical energy, the loss of the protein's function could lead to the death of the cell. In contrast, if the active protein were involved in the synthesis of a particular amino acid, and if the cell could also obtain that amino acid from the extracellular fluid, the cell function would not be impaired by the absence of the protein.

To generalize, a mutation may have any one of three effects upon a cell: (1) It may cause no noticeable change in cell function; (2) it may modify cell function, but still be compatible with cell growth and replication; (3) it may lead to cell death.

Mutations and Evolution

Mutations contribute to the evolution of organisms. Although most mutations result in either no change or an impairment of cell function, a very small number may alter the activity of a protein in such a way that it is more, rather than less, active, or they may introduce an entirely new type of protein activity into a cell. If an organism carrying such a mutant gene is able to perform some function more effectively than an organism lacking the mutant gene, the organism has a better chance of reproducing and passing on the mutant gene to its descendants. On the other hand, if the mutation produces an organism that functions less effectively than organisms lacking the mutation, the organism is less likely to reproduce and pass on the mutant gene. This is the principle of **natural selection.** Although any one mutation, if it is able to survive in the population, may cause only a very slight alteration in the properties of a cell, given enough time, a large number of small changes can accumulate to produce very large changes in the structure and function of an organism.

Protein Degradation

We have thus far emphasized protein synthesis, but the concentration of a particular protein in a cell at a particular time depends not only upon its rate of synthesis but also upon its rates of degradation and/or secretion.

Different proteins degrade at different rates. In part this depends on the structure of the protein, with some proteins having a higher affinity for certain proteolytic enzymes than others. A denatured (unfolded) protein is more readily digested than a protein with an intact conformation. Proteins can be targeted for degradation by the attachment of a small peptide, **ubiquitin,** to the protein. This peptide directs the protein to a protein complex known as a **proteasome,** which unfolds the protein and breaks it down into small peptides. Degradation is an important mechanism for confining the activity of a given protein to a precise window of time.

In summary, there are many steps in the path from a gene in DNA to a fully active protein that allow the rate of protein synthesis or the final active form of the protein to be altered (**Table 3–3**). By controlling these steps, extracellular or intracellular signals, as described in Chapter 5, can regulate the total amount of a specific protein in a cell.

Protein Secretion

Most proteins synthesized by a cell remain in the cell, providing structure and function for the cell's survival. Some proteins, however, are secreted into the extracellular fluid, where they act as signals to other cells or provide material for forming the extracellular matrix. Proteins are large, charged molecules that cannot diffuse through cell membranes. Thus, special mechanisms are required to insert them into or move them through membranes.

Proteins destined to be secreted from a cell or to become integral membrane proteins are recognized during the early stages of protein synthesis. For such proteins, the first 15 to 30 amino acids that emerge from the surface of the ribosome act as a recognition signal, known as the **signal sequence** or signal peptide.

The signal sequence binds to a complex of proteins known as a signal recognition particle, which temporarily inhibits further growth of the polypeptide chain on the ribosome. The signal recognition particle then binds to a specific membrane protein on the surface of the rough endoplasmic reticulum.

Table 3–3	Factors that Alter the Amount and Activity of Specific Cell Proteins
Process Altered	**Mechanism of Alteration**
1. Transcription of DNA	Activation or inhibition by transcription factors
2. Splicing of RNA	Activity of enzymes in spliceosome
3. mRNA degradation	Activity of RNAase
4. Translation of mRNA	Activity of initiating factors on ribosomes
5. Protein degradation	Activity of proteasomes
6. Allosteric and covalent modulation	Signal ligands, protein kinases, and phosphatases

This binding restarts the process of protein assembly, and the growing polypeptide chain is fed through a protein complex in the endoplasmic reticulum membrane into the lumen of the reticulum (**Figure 3–25**). Upon completion of protein assembly, proteins that are to be secreted end up in the lumen of the rough endoplasmic reticulum. Proteins that are destined to function as integral membrane proteins remain embedded in the reticulum membrane.

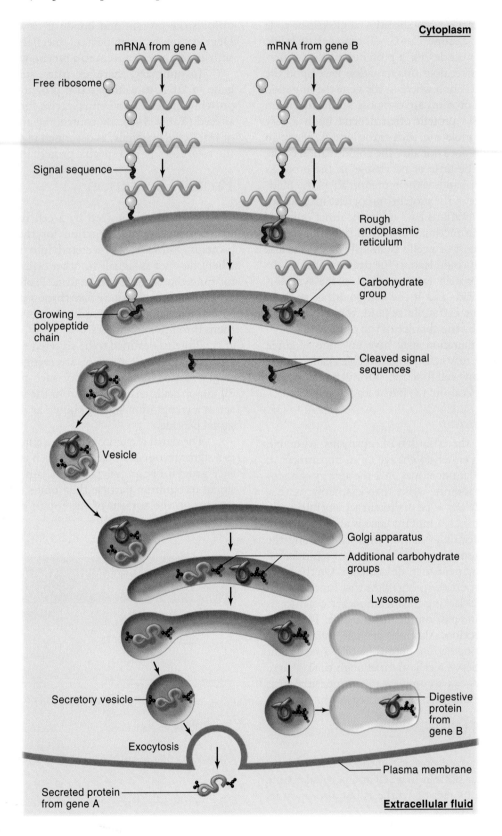

Figure 3–25

Pathway of proteins destined to be secreted by cells or transferred to lysosomes. An example of the latter might be a protein important in digestive functions in which a cell degrades other intracellular molecules.

Within the lumen of the endoplasmic reticulum, enzymes remove the signal sequence from most proteins, so this portion is not present in the final protein. In addition, carbohydrate groups are sometimes linked to various side chains in the proteins.

Following these modifications, portions of the reticulum membrane bud off, forming vesicles that contain the newly synthesized proteins. These vesicles migrate to the Golgi apparatus (see Figure 3–25) and fuse with the Golgi membranes.

Within the Golgi apparatus, the protein may undergo further modifications. For example, additional carbohydrate groups—important as recognition sites within the cell—may be added.

While in the Golgi apparatus, the many different proteins that have been funneled into this organelle are sorted out according to their final destinations. This sorting involves the binding of regions of a particular protein to specific proteins in the Golgi membrane that are destined to form vesicles targeted to a particular destination.

Following modification and sorting, the proteins are packaged into vesicles that bud off the surface of the Golgi membrane. Some of the vesicles travel to the plasma membrane, where they fuse with the membrane and release their contents to the extracellular fluid, a process known as exocytosis. Other vesicles may dock and fuse with lysosome membranes, delivering digestive enzymes to the interior of this organelle. Specific docking proteins on the surface of the membrane where the vesicle finally fuses recognize the specific proteins on the surface of the vesicle.

In contrast to this entire story, if a protein does not have a signal sequence, synthesis continues on a free ribosome until the completed protein is released into the cytosol. These proteins are not secreted but are destined to function within the cell. Many remain in the cytosol, where they function, for example, as enzymes in various metabolic pathways. Others are targeted to particular cell organelles. For example, ribosomal proteins are directed to the nucleus, where they combine with rRNA before returning to the cytosol as part of the ribosomal subunits. The specific location of a protein is determined by binding sites on the protein that bind to specific sites at the protein's destination. For example, in the case of the ribosomal proteins, they bind to sites on the nuclear pores that control access to the nucleus.

■ SECTION B SUMMARY

Genetic Code

I. Genetic information is coded in the nucleotide sequences of DNA molecules. A single gene contains either (a) the information that, via mRNA, determines the amino acid sequence in a specific protein, or (b) the information for forming rRNA, tRNA, or small nuclear RNAs, which assist in protein assembly.

II. Genetic information is transferred from DNA to mRNA in the nucleus (transcription); then mRNA passes to the cytoplasm, where its information is used to synthesize protein (translation).

III. The "words" in the DNA genetic code consist of a sequence of three nucleotide bases that specify a single amino acid. The sequence of three-letter codes along a gene determines the sequence of amino acids in a protein. More than one triplet can specify a given amino acid.

Protein Synthesis

I. Table 3–2 summarizes the steps leading from DNA to protein synthesis.

II. Transcription involves forming a primary RNA transcript by base-pairing with the template strand of DNA containing a single gene. Transcription also involves the removal of intron-derived segments by spliceosomes to form mRNA, which moves to the cytoplasm.

III. Translation of mRNA occurs on the ribosomes in the cytoplasm when the anticodons in tRNAs, linked to single amino acids, base-pair with the corresponding codons in mRNA.

IV. Protein transcription factors activate or repress the transcription of specific genes by binding to regions of DNA that interact with the promoter region of a gene.

V. Mutagens alter DNA molecules, resulting in the addition or deletion of nucleotides or segments of DNA. The result is an altered DNA sequence known as a mutation. A mutation may (1) cause no noticeable change in cell function, (2) modify cell function but still be compatible with cell growth and replication, or (3) lead to the death of the cell.

Protein Degradation

I. The concentration of a particular protein in a cell depends on: (1) the rate of the corresponding gene's transcription, (2) the rate of initiating protein assembly on a ribosome, (3) the rate at which mRNA is degraded, (4) the rate of protein digestion by enzymes associated with proteasomes, and (5) the rate of secretion, if any, of the protein from the cell.

Protein Secretion

I. Targeting of a protein for secretion depends on the signal sequence of amino acids that first emerge from a ribosome during protein synthesis.

■ SECTION B KEY TERMS

anticodon 59	primary RNA transcript 58
codon 58	promoter 57
exon 58	proteasome 63
gene 55	ribosomal RNA (rRNA) 59
genome 55	RNA polymerase 57
histone 55	signal sequence 63
initiation factor 59	spliceosome 58
intron 58	"stop" signal 56
messenger RNA (mRNA) 57	template strand 57
mutagen 62	transcription 56
mutation 62	transcription factor 61
natural selection 63	transfer RNA (tRNA) 59
nucleosome 55	translation 56
preinitiation complex 62	ubiquitin 63

■ SECTION B REVIEW QUESTIONS

1. Describe how the genetic code in DNA specifies the amino acid sequence in a protein.
2. List the four nucleotides found in mRNA.

3. Describe the main events in the transcription of genetic information from DNA into mRNA.
4. Explain the difference between an exon and an intron.
5. What is the function of a spliceosome?
6. Identify the site of ribosomal subunit assembly.
7. Describe the role of tRNA in protein assembly.
8. Describe the events of protein translation that occur on the surface of a ribosome.
9. Describe the effects of transcription factors on gene transcription.

10. List the factors that regulate the concentration of a protein in a cell.
11. What is the function of the signal sequence of a protein? How is it formed, and where is it located?
12. Describe the pathway that leads to the secretion of proteins from cells.
13. List the three general types of effects a mutation can have on a cell's function.

SECTION C

Protein-Binding Sites

Binding Site Characteristics

In the previous sections, we learned how the cellular machinery synthesizes and processes proteins. We now turn our attention to how proteins interact with each other and with other molecules.

The ability of various molecules and ions to bind to specific sites on the surface of a protein forms the basis for the wide variety of protein functions. A **ligand** is any molecule that is bound to the surface of a protein by one of the following forces: (1) electrical attractions between oppositely charged ionic or polarized groups on the ligand and the protein, or (2) weaker attractions due to hydrophobic forces between nonpolar regions on the two molecules. Note that this binding does not involve covalent bonds; in other words, binding is generally reversible. The region of a protein to which a ligand binds is known as a **binding site.** A protein may contain several binding sites, each specific for a particular ligand.

Chemical Specificity

The force of electrical attraction between oppositely charged regions on a protein and a ligand decreases markedly as the distance between them increases. The even weaker hydrophobic forces act only between nonpolar groups that are very close to each other. Therefore, for a ligand to bind to a protein, the ligand must be close to the protein surface. This proximity occurs when the shape of the ligand is complementary to the shape of the protein-binding site, so that the two fit together like pieces of a jigsaw puzzle (**Figure 3–26**).

The binding between a ligand and a protein may be so specific that a binding site can bind only one type of ligand and no other. Such selectivity allows a protein to identify (by binding) one particular molecule in a solution containing hundreds of different molecules. This ability of a protein binding site to bind specific ligands is known as **chemical specificity,** because the binding site determines the type of chemical that is bound.

In Chapter 2 we described how the conformation of a protein is determined by the location of the various amino acids along the polypeptide chain. Accordingly, proteins with different amino acid sequences have different shapes and, therefore, differently shaped binding sites, each with its own chemical specificity. As illustrated in **Figure 3–27**, the amino acids that interact with a ligand at a binding site need not be

adjacent to each other along the polypeptide chain, because the three-dimensional folding of the protein may bring various segments of the molecule into juxtaposition.

Although some binding sites have a chemical specificity that allows them to bind only one type of ligand, others are less specific and thus can bind a number of related ligands. For example, three different ligands can combine with the binding site of protein X in **Figure 3–28**, because a portion of each ligand is complementary to the shape of the binding site. In contrast, protein Y has a greater chemical specificity and can bind only one of the three ligands. It is the degree of specificity of proteins that determines, in part, the side effects

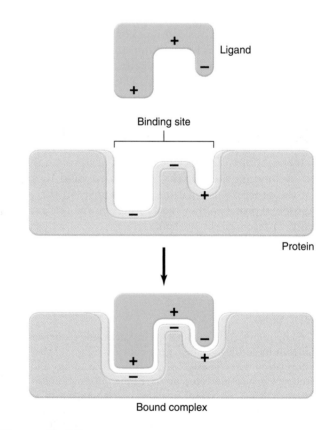

Figure 3–26

The complementary shapes of ligand and protein-binding site determine the chemical specificity of binding.

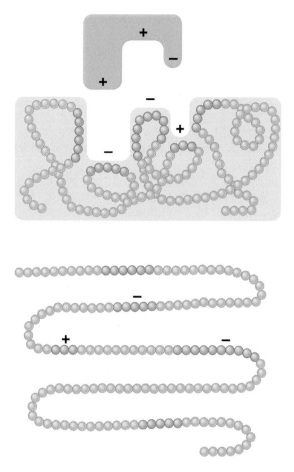

Figure 3–27

Amino acids that interact with the ligand at a binding site need not be at adjacent sites along the polypeptide chain, as indicated in this model showing the three-dimensional folding of a protein. The unfolded polypeptide chain appears at the bottom.

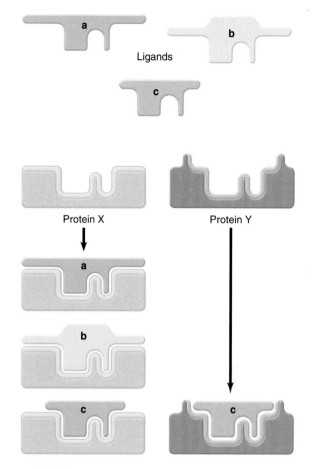

Figure 3–28

Protein X can bind all three ligands, which have similar chemical structures. Protein Y, because of the shape of its binding site, can bind only ligand c. Protein Y, therefore, has a greater chemical specificity than protein X.

of therapeutic drugs. For example, a drug (ligand) designed to treat high blood pressure may act by binding to certain proteins which, in turn, help restore pressure to normal. The same drug, however, may also bind to a lesser degree to other proteins, whose functions may be completely unrelated to blood pressure.

Affinity

The strength of ligand-protein binding is a property of the binding site known as **affinity.** The affinity of a binding site for a ligand determines how likely it is that a bound ligand will leave the protein surface and return to its unbound state. Binding sites that tightly bind a ligand are called high-affinity binding sites; those that weakly bind the ligand are low-affinity binding sites.

Affinity and chemical specificity are two distinct, although closely related, properties of binding sites. Chemical specificity, as we have seen, depends only on the shape of the binding site, whereas affinity depends on the strength of the attraction between the protein and the ligand. Thus, different proteins may be able to bind the same ligand—that is,

may have the same chemical specificity—but may have different affinities for that ligand. For example, a ligand may have a negatively charged ionized group that would bind strongly to a site containing a positively charged amino acid side chain, but would bind less strongly to a binding site having the same shape but no positive charge (**Figure 3–29**). In addition, the closer the surfaces of the ligand and binding site are to each other, the stronger the attractions. Thus, the more closely the ligand shape matches the binding site shape, the greater the affinity. In other words, shape can influence affinity as well as chemical specificity.

Saturation

An equilibrium is rapidly reached between unbound ligands in solution and their corresponding protein-binding sites. Thus, at any instant, some of the free ligands become bound to unoccupied binding sites, and some of the bound ligands are released back into solution. A single binding site is either occupied or unoccupied. The term **saturation** refers to the fraction of total binding sites that are occupied at any given time. When all the binding sites are occupied, the population of binding sites is 100 percent saturated. When half the available sites are occupied, the system is 50 percent saturated, and so on. A *single* binding site would also be 50 percent saturated if it were occupied by a ligand 50 percent of the time.

The percent saturation of a binding site depends upon two factors: (1) the concentration of unbound ligand in the solution, and (2) the affinity of the binding site for the ligand.

The greater the ligand concentration, the greater the probability of a ligand molecule encountering an unoccupied binding site and becoming bound. Thus, the percent saturation of binding sites increases with increasing ligand concentration until all the sites become occupied (**Figure 3–30**). Assuming that the ligand is a molecule that exerts a biological effect when it binds to a protein, the magnitude of the effect would also increase with increasing numbers of bound ligands until all the binding sites were occupied. Further increases in ligand concentration would produce no further effect because there would be no additional sites to be occupied. To generalize, a continuous increase in the magnitude of a chemical stimulus (ligand concentration) that exerts its effects by binding to proteins will produce an increased biological response until the point at which the protein-binding sites are 100 percent saturated.

The second factor determining the percent of binding site saturation is the affinity of the binding site. Collisions between molecules in a solution and a protein containing a bound ligand can dislodge a loosely bound ligand, just as tackling a football player may cause a fumble. If a binding site has a high affinity for a ligand, even a low ligand concentration will result in a high degree of saturation because, once bound to the site, the ligand is not easily dislodged. A low-affinity site, on the other hand, requires a higher concentration of ligand to achieve the same degree of saturation (**Figure 3–31**). One measure of binding site affinity is the ligand concentration necessary to produce 50 percent saturation; the lower the ligand concentration required to bind to half the binding sites, the greater the affinity of the binding site (see Figure 3–31).

Competition

As we have seen, more than one type of ligand can bind to certain binding sites (see Figure 3–28). In such cases, **competition** occurs between the ligands for the same binding site. In other words, the presence of multiple ligands able to bind

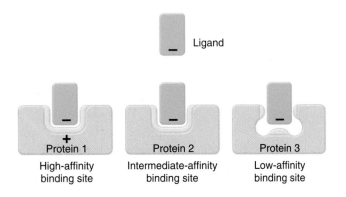

Figure 3–29

Three binding sites with the same chemical specificity but different affinities for a ligand.

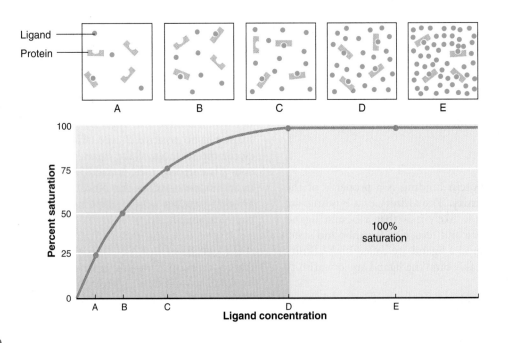

Figure 3–30

Increasing ligand concentration increases the number of binding sites occupied—that is, it increases the percent saturation. At 100 percent saturation, all the binding sites are occupied, and further increases in ligand concentration do not increase the number bound.

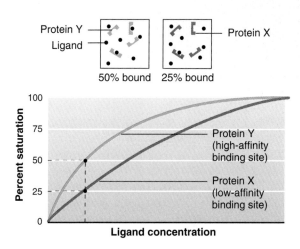

Protein Y ─── Ligand ─── Protein X

50% bound   25% bound

Figure 3–31

When two different proteins, X and Y, are able to bind the same ligand, the protein with the higher-affinity binding site (protein Y) has more bound sites at any given ligand concentration up to 100 percent saturation.

to the same binding site affects the percentage of binding sites occupied by any one ligand. If two competing ligands, A and B, are present, increasing the concentration of A will increase the amount of A that is bound, thereby decreasing the number of sites available to B and decreasing the amount of B that is bound.

As a result of competition, the biological effects of one ligand may be diminished by the presence of another. For example, many drugs produce their effects by competing with the body's natural ligands for binding sites. By occupying the binding sites, the drug decreases the amount of natural ligand that can be bound.

Regulation of Binding Site Characteristics

Because proteins are associated with practically everything that occurs in a cell, the mechanisms for controlling these functions center on the control of protein activity. There are two ways of controlling protein activity: (1) changing protein shape, which alters the binding of ligands, and (2) as described earlier in this chapter, regulating protein synthesis and degradation, which determines the types and amounts of proteins in a cell.

As described in Chapter 2, a protein's shape depends on electrical attractions between charged or polarized groups in various regions of the protein. Therefore, a change in the charge distribution along a protein or in the polarity of the molecules immediately surrounding it will alter its shape. The two mechanisms used by cells to selectively alter protein shape are known as allosteric modulation and covalent modulation. Before we describe these mechanisms, however, it should be emphasized that only certain key proteins are regulated by modulation. Most proteins are not subject to either of these types of modulation.

Allosteric Modulation

Whenever a ligand binds to a protein, the attracting forces between the ligand and the protein alter the protein's shape. For example, as a ligand approaches a binding site, these attracting forces can cause the surface of the binding site to bend into a shape that more closely approximates the shape of the ligand's surface.

Moreover, as the shape of a binding site changes, it produces changes in the shape of *other* regions of the protein, just as pulling on one end of a rope (the polypeptide chain) causes the other end of the rope to move. Therefore, when a protein contains *two* binding sites, the noncovalent binding of a ligand to one site can alter the shape of the second binding site and, therefore, the binding characteristics of that site. This is termed **allosteric** (other shape) **modulation** (**Figure 3–32a**), and such proteins are known as **allosteric proteins.**

One binding site on an allosteric protein, known as the **functional** (or active) **site,** carries out the protein's physiological function. The other binding site is the **regulatory site.** The ligand that binds to the regulatory site is known as a **modulator molecule,** because its binding allosterically modulates the shape, and thus the activity, of the functional site.

The regulatory site to which modulator molecules bind is the equivalent of a molecular switch that controls the functional site. In some allosteric proteins, the binding of the modulator molecule to the regulatory site turns on the functional site by changing its shape so that it can bind the functional ligand. In other cases, the binding of a modulator molecule turns off the functional site by preventing the functional site from binding its ligand. In still other cases, binding of the modulator molecule may decrease or increase the affinity of the functional site. For example, if the functional site is 50 percent saturated at a particular ligand concentration, the binding of a modulator molecule that increases the affinity of the functional site may increase its saturation to 75 percent. This concept will be especially important when we consider how gases are bound to a transport protein in the blood (Chapter 13).

To summarize, the activity of a protein can be increased without changing the concentration of either the protein or the functional ligand. By controlling the concentration of the modulator molecule, and thus the percent saturation of the regulatory site, the functional activity of an allosterically regulated protein can be increased or decreased.

We have spoken thus far only of interactions between regulatory and functional binding sites. There is, however, a way that functional sites can influence each other in certain proteins. These proteins are composed of more than one polypeptide chain held together by electrical attractions between the chains. There may be only one binding site, a functional binding site, on each chain. The binding of a functional ligand to one of the chains, however, can result in an alteration of the functional binding sites in the other chains. This happens because the change in shape of the chain that holds the bound ligand induces a change in the shape of the other chains. The interaction between the functional binding sites of a multimeric (more than one polypeptide chain) protein is known as **cooperativity.** It can result in a progressive increase in the affinity for ligand binding as more and more of the sites become occupied. Such a

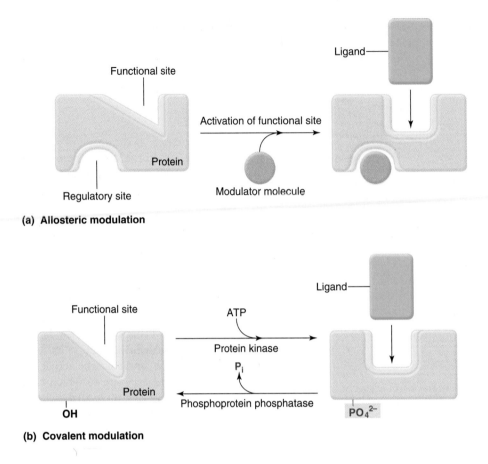

Functional site

Protein

Regulatory site

Activation of functional site

Modulator molecule

Ligand

Protein

(a) Allosteric modulation

Functional site

Protein

OH

ATP

Protein kinase

P_i

Phosphoprotein phosphatase

Ligand

PO_4^{2-}

(b) Covalent modulation

Figure 3–32

(a) Allosteric modulation and (b) covalent modulation of a protein's functional binding site.

situation occurs, for example, when oxygen binds to hemoglobin, a protein composed of four polypeptide chains, each containing one binding site for oxygen (Chapter 13).

Covalent Modulation

The second way to alter the shape and therefore the activity of a protein is by the covalent bonding of charged chemical groups to some of the protein's side chains. This is known as **covalent modulation.** In most cases, a phosphate group, which has a net negative charge, is covalently attached by a chemical reaction called **phosphorylation,** in which a phosphate group is transferred from one molecule to another. Phosphorylation of one of the side chains of certain amino acids in a protein introduces a negative charge into that region of the protein. This charge alters the distribution of electric forces in the protein and produces a change in protein conformation (**Figure 3–32b**). If the conformational change affects a binding site, it changes the binding site's properties. Although the mechanism is completely different, the effects produced by covalent modulation are similar to those of allosteric modulation—that is, a functional binding site may be turned on or off, or the affinity of the site for its ligand may be altered. Unlike allosteric modulation, which involves noncovalent binding of modulator molecules, covalent modulation requires chemical reactions in which covalent bonds are formed.

Most chemical reactions in the body are mediated by a special class of proteins known as enzymes, whose properties

will be discussed in Section D of this chapter. For now, suffice it to say that enzymes accelerate the rate at which reactant molecules, called substrates, are converted to different molecules called products. Two enzymes control a protein's activity by covalent modulation: One adds phosphate, and one removes it. Any enzyme that mediates protein phosphorylation is called a **protein kinase.** These enzymes catalyze the transfer of phosphate from a molecule of ATP to a hydroxyl group present on the side chain of certain amino acids:

$$\text{Protein} + \text{ATP} \xrightarrow{\text{protein kinase}} \text{Protein}-\text{PO}_4^{2-} + \text{ADP}$$

The protein and ATP are the substrates for protein kinase, and the phosphorylated protein and adenosine diphosphate (ADP) are the products of the reaction.

There is also a mechanism for removing the phosphate group and returning the protein to its original shape. This dephosphorylation is accomplished by a second class of enzymes known as **phosphoprotein phosphatases:**

$$\text{Protein}-\text{PO}_4^{2-} + \text{H}_2\text{O} \xrightarrow[\text{phosphatase}]{\text{phosphoprotein}} \text{Protein} + \text{HPO}_4^{2-}$$

The activity of the protein will depend on the relative activity of the kinase and phosphatase that controls the extent of the protein's phosphorylation. There are many protein kinases, each with specificities for different proteins, and several kinases may be present in the same cell. The chemical specificities of the phosphoprotein phosphatases are broader; a single enzyme can dephosphorylate many different phosphorylated proteins.

An important interaction between allosteric and covalent modulation results from the fact that protein kinases are themselves allosteric proteins whose activity can be controlled by modulator molecules. Thus, the process of covalent modulation is itself indirectly regulated by allosteric mechanisms. In addition, some allosteric proteins can also be modified by covalent modulation.

In Chapter 5 we will describe how cell activities can be regulated in response to signals that alter the concentrations of various modulator molecules. These modulator molecules, in turn, alter specific protein activities via allosteric and covalent modulations. **Table 3–4** summarizes the factors influencing protein function.

Table 3–4	Factors that Influence Protein Function

I. CHANGING PROTEIN SHAPE
   a. Allosteric modulation
   b. Covalent modulation
      i. Protein kinase activity
      ii. Phosphoprotein phosphatase activity

II. CHANGING PROTEIN CONCENTRATION
   a. Protein synthesis
   b. Protein degradation

■ **SECTION C SUMMARY**

Binding Site Characteristics
I. Ligands bind to proteins at sites with shapes complementary to the ligand shape.
II. Protein-binding sites have the properties of chemical specificity, affinity, saturation, and competition.

Regulation of Binding Site Characteristics
I. Protein function in a cell can be controlled by regulating either the shape of the protein or the amounts of protein synthesized and degraded.
II. The binding of a modulator molecule to the regulatory site on an allosteric protein alters the shape of the functional binding site, thereby altering its binding characteristics and the activity of the protein. The activity of allosteric proteins is regulated by varying the concentrations of their modulator molecules.
III. Protein kinase enzymes catalyze the addition of a phosphate group to the side chains of certain amino acids in a protein, changing the shape of the protein's functional binding site and thus altering the protein's activity by covalent modulation. A second enzyme is required to remove the phosphate group, returning the protein to its original state.

■ **SECTION C KEY TERMS**

affinity 67	ligand 66
allosteric modulation 69	modulator molecule 69
allosteric protein 69	phosphoprotein
binding site 66	phosphatase 70
chemical specificity 66	phosphorylation 70
competition 68	protein kinase 70
cooperativity 69	regulatory site 69
covalent modulation 70	saturation 67
functional site 69	

■ **SECTION C REVIEW QUESTIONS**

1. List the four characteristics of a protein-binding site.
2. List the types of forces that hold a ligand on a protein surface.
3. What characteristics of a binding site determine its chemical specificity?
4. Under what conditions can a single binding site have a chemical specificity for more than one type of ligand?
5. What characteristics of a binding site determine its affinity for a ligand?
6. What two factors determine the percent saturation of a binding site?
7. How is the activity of an allosteric protein modulated?
8. How does regulation of protein activity by covalent modulation differ from that by allosteric modulation?

SECTION D	Enzymes and Chemical Energy

Thus far, we have discussed the synthesis and regulation of proteins. In this section, we describe some of the major functions of proteins, specifically those that relate to facilitating chemical reactions.

Thousands of chemical reactions occur each instant throughout the body; this coordinated process of chemical change is termed **metabolism** (Greek, change). Metabolism involves the synthesis and breakdown of organic molecules required for cell structure and function and the release of chemical energy used for cell functions. The synthesis of organic molecules by cells is called **anabolism,** and their

breakdown, **catabolism.** For example, the synthesis of a triglyceride is an anabolic reaction, whereas the breakdown of a triglyceride to glycerol and fatty acids is a catabolic reaction.

The body's organic molecules undergo continuous transformation as some molecules are broken down while others of the same type are being synthesized. Molecularly, no person is the same at noon as at 8 o'clock in the morning because during even this short period, some of the body's structure has been broken down and replaced with newly synthesized molecules. In a healthy adult, the body's composition is in a steady state in which the anabolic and catabolic rates for the synthesis and breakdown of most molecules are equal.

Chemical Reactions

Chemical reactions involve (1) the breaking of chemical bonds in reactant molecules, followed by (2) the making of new chemical bonds to form the product molecules. Take, for example, a chemical reaction that occurs in the lungs, which permits the lungs to rid the body of carbon dioxide. In the reaction shown below, carbonic acid is transformed into carbon dioxide and water. Two of the chemical bonds in carbonic acid are broken, and the product molecules are formed by establishing two new bonds between different pairs of atoms:

$$H-O-C-O-H \longrightarrow O=C + H-O-H$$

broken   broken   formed   formed

$$\underset{\text{carbonic acid}}{H_2CO_3} \longrightarrow \underset{\text{carbon dioxide}}{CO_2} + \underset{\text{water}}{H_2O} + \text{Energy}$$

Because the energy contents of the reactants and products are usually different, and because energy can neither be created nor destroyed, energy must either be added or released during most chemical reactions. For example, the breakdown of carbonic acid into carbon dioxide and water releases energy because carbonic acid has a higher energy content than the sum of the energy contents of carbon dioxide and water.

The released energy takes the form of heat, the energy of increased molecular motion, which is measured in units of calories. One **calorie** (1 cal) is the amount of heat required to raise the temperature of 1 g of water 1° on the Celsius scale. Energies associated with most chemical reactions are several thousand calories per mole and are reported as **kilocalories** (1 kcal = 1000 cal).

Determinants of Reaction Rates

The rate of a chemical reaction (in other words, how many molecules of product form per unit of time) can be determined by measuring the change in the concentration of reactants or products per unit of time. The faster the product concentration increases or the reactant concentration decreases, the greater the rate of the reaction. Four factors (**Table 3–5**) influence the reaction rate: reactant concentration, activation energy, temperature, and the presence of a catalyst.

The lower the concentration of reactants, the slower the reaction simply because there are fewer molecules available to react. Conversely, the higher the concentration of reactants, the faster the reaction rate.

Given the same initial concentrations of reactants, however, not all reactions occur at the same rate. Each type of chemical reaction has its own characteristic rate, which depends upon what is called the activation energy for the reaction. In order for a chemical reaction to occur, reactant molecules must acquire enough energy—the **activation energy**—to enter an activated state in which chemical bonds can be broken and formed. The activation energy does not affect the

Table 3–5	Determinants of Chemical Reaction Rates
1. Reactant concentrations (higher concentrations: faster reaction rate)	
2. Activation energy (higher activation energy: slower reaction rate)	
3. Temperature (higher temperature: faster reaction rate)	
4. Catalyst (presence of catalyst: faster reaction rate)	

difference in energy content between the reactants and final products since the activation energy is released when the products are formed.

How do reactants acquire activation energy? In most of the metabolic reactions we will be considering, the reactants obtain activation energy when they collide with other molecules. If the activation energy required for a reaction is large, then the probability of a given reactant molecule acquiring this amount of energy will be small, and the reaction rate will be slow. Thus, the higher the activation energy required, the slower the rate of a chemical reaction.

Temperature is the third factor influencing reaction rates. The higher the temperature, the faster molecules move and the greater their impact when they collide. Therefore, one reason that increasing the temperature increases a reaction rate is that reactants have a better chance of acquiring sufficient activation energy from a collision. In addition, faster-moving molecules collide more often.

A **catalyst** is a substance that interacts with a reactant by altering the distribution of energy between the chemical bonds of the reactant, resulting in a decrease in the activation energy required to transform the reactant into product. Because less activation energy is required, a reaction will proceed at a faster rate in the presence of a catalyst. The chemical composition of a catalyst is not altered by the reaction, so *a single catalyst molecule can act over and over again to catalyze the conversion of many reactant molecules to products*. Furthermore, a catalyst does not alter the difference in the energy contents of the reactants and products.

Reversible and Irreversible Reactions

Every chemical reaction is, in theory, reversible. Reactants are converted to products (we will call this a "forward reaction"), and products are converted to reactants (a "reverse reaction"). The overall reaction is a **reversible reaction:**

$$\text{Reactants} \underset{\text{reverse}}{\overset{\text{forward}}{\rightleftharpoons}} \text{Products}$$

As a reaction progresses, the rate of the forward reaction will decrease as the concentration of reactants decreases. Simultaneously, the rate of the reverse reaction will increase as the concentration of the product molecules increases. Eventually the reaction will reach a state of **chemical equilibrium** in which the forward and reverse reaction rates are equal. At this point there will be no further change in the concentrations of reactants or products even though reactants will continue to be converted into products and products converted into reactants.

Consider our previous example in which carbonic acid breaks down into carbon dioxide and water. The products of this reaction, carbon dioxide and water, can also recombine to form carbonic acid. This occurs outside the lungs and is a means for safely transporting CO_2 in the blood in a nongaseous state.

$$CO_2 + H_2O + Energy \rightleftharpoons H_2CO_3$$

Carbonic acid has a greater energy content than the sum of the energies contained in carbon dioxide and water; therefore, energy must be added to the latter molecules in order to form carbonic acid. This energy is not activation energy but is an integral part of the energy balance. This energy can be obtained, along with the activation energy, through collisions with other molecules.

When chemical equilibrium has been reached, the concentration of products need not be equal to the concentration of reactants even though the forward and reverse reaction rates are equal. The ratio of product concentration to reactant concentration at equilibrium depends upon the amount of energy released (or added) during the reaction. The greater the energy released, the smaller the probability that the product molecules will be able to obtain this energy and undergo the reverse reaction to reform reactants. Therefore, in such a case, the ratio of product to reactant concentration at chemical equilibrium will be large. If there is no difference in the energy contents of reactants and products, their concentrations will be equal at equilibrium.

Thus, although all chemical reactions are reversible to some extent, reactions that release large quantities of energy are said to be **irreversible reactions** because almost all of the reactant molecules have been converted to product molecules when chemical equilibrium is reached. It must be emphasized that the energy released in a reaction determines the degree to which the reaction is reversible or irreversible. This energy is not the activation energy and it does not determine the reaction rate, which is governed by the four factors discussed earlier. The characteristics of reversible and irreversible reactions are summarized in **Table 3–6**.

Law of Mass Action

The concentrations of reactants and products play a very important role in determining not only the rates of the forward and reverse reactions, but also the direction in which the *net* reaction proceeds—whether reactants or products are accumulating at a given time.

Consider the following reversible reaction that has reached chemical equilibrium:

$$A + B \underset{\text{reverse}}{\overset{\text{forward}}{\rightleftharpoons}} C + D$$

$$\text{Reactants} \qquad \text{Products}$$

If at this point we increase the concentration of one of the reactants, the rate of the forward reaction will increase and lead to increased product formation. In contrast, increasing the concentration of one of the product molecules will drive the reaction in the reverse direction, increasing the formation of reactants. The direction in which the net reaction is proceeding can also be altered by *decreasing* the concentration of one of the participants. Thus, decreasing the concentration of one of the products drives the net reaction in the forward direction because it decreases the rate of the reverse reaction without changing the rate of the forward reaction.

These effects of reaction and product concentrations on the direction in which the net reaction proceeds are known as the **law of mass action.** Mass action is often a major determining factor controlling the direction in which metabolic pathways proceed because reactions in the body seldom come to chemical equilibrium. More typically, new reactant molecules are added and product molecules are simultaneously removed by other reactions.

Enzymes

Most of the chemical reactions in the body, if carried out in a test tube with only reactants and products present, would proceed at very low rates because they have high activation energies. To achieve the high reaction rates observed in living organisms, catalysts must lower the activation energies. These particular catalysts are called enzymes. Enzymes are protein molecules, so an **enzyme** can be defined as a protein catalyst. (Although some RNA molecules possess catalytic activity, the number of reactions they catalyze is very small, so we will restrict the term *enzyme* to protein catalysts.)

Table 3–6	Characteristics of Reversible and Irreversible Chemical Reactions
Reversible Reactions	$A + B \rightleftharpoons C + D$ + small amount of energy
	At chemical equilibrium, product concentrations are only slightly higher than reactant concentrations.
Irreversible Reactions	$E + F \longrightarrow G + H$ + large amount of energy
	At chemical equilibrium, almost all reactant molecules have been converted to product.

To function, an enzyme must come into contact with reactants, which are called **substrates** in the case of enzyme-mediated reactions. The substrate becomes bound to the enzyme, forming an enzyme-substrate complex, which then breaks down to release products and enzyme. The reaction between enzyme and substrate can be written:

$$\text{S} + \text{E} \rightleftharpoons \text{ES} \rightleftharpoons \text{P} + \text{E}$$

Substrate   Enzyme   Enzyme-substrate complex   Product   Enzyme

At the end of the reaction, the enzyme is free to undergo the same reaction with additional substrate molecules. The overall effect is to accelerate the conversion of substrate into product, with the enzyme acting as a catalyst. Note that an enzyme increases both the forward and reverse rates of a reaction and thus does not change the chemical equilibrium that is finally reached.

The interaction between substrate and enzyme has all the characteristics described previously for the binding of a ligand to a binding site on a protein—specificity, affinity, competition, and saturation. The region of the enzyme the substrate binds to is known as the enzyme's **active site** (a term equivalent to "binding site"). The shape of the enzyme in the region of the active site provides the basis for the enzyme's chemical specificity. Two models have been proposed to describe the interaction of an enzyme with its substrate(s). In one, the enzyme and substrate(s) fit together in a "lock-and-key" configuration. In another model, the substrate itself induces a shape change in the active site of the enzyme, which results in a highly specific binding interaction ("induced fit model") (**Figure 3–33**).

There are approximately 4000 different enzymes in a typical cell, each capable of catalyzing a different chemical reaction. Enzymes are generally named by adding the suffix *-ase* to the name of either the substrate or the type of reaction the enzyme catalyzes. For example, the reaction in which carbonic acid is broken down into carbon dioxide and water is catalyzed by the enzyme **carbonic anhydrase.**

The catalytic activity of an enzyme can be extremely large. For example, a single molecule of carbonic anhydrase can catalyze the conversion of about 100,000 substrate molecules to products in one second! The major characteristics of enzymes are listed in **Table 3–7.**

Cofactors

Many enzymes are inactive in the absence of small amounts of other substances known as **cofactors.** In some cases, the cofactor is a trace metal, such as magnesium, iron, zinc, or copper. Binding of one of the metals to an enzyme alters the enzyme's conformation so that it can interact with the substrate (this is a form of allosteric modulation). Because only a few enzyme molecules need be present to catalyze the conversion of large amounts of substrate to product, very small quantities of these trace metals are sufficient to maintain enzymatic activity.

In other cases, the cofactor is an organic molecule that directly participates as one of the substrates in the reaction, in which case the cofactor is termed a **coenzyme.** Enzymes that require coenzymes catalyze reactions in which a few atoms (for example, hydrogen, acetyl, or methyl groups) are either removed from or added to a substrate. For example:

$$\text{R—2 H} + \text{Coenzyme} \xrightarrow{\text{Enzyme}} \text{R} + \text{Coenzyme—2 H}$$

What distinguishes a coenzyme from an ordinary substrate is the fate of the coenzyme. In our example, the two hydrogen atoms that transfer to the coenzyme can then be transferred from the coenzyme to another substrate with the aid of a second enzyme. This second reaction converts the coenzyme back to its original form so that it becomes available to accept two more hydrogen atoms. A single coenzyme molecule can act over and over again to transfer molecular fragments from one reaction to another. Thus, as with metallic cofactors, only small quantities of coenzymes are necessary to maintain the enzymatic reactions in which they participate.

Coenzymes are derived from several members of a special class of nutrients known as **vitamins.** For example, the coenzymes **NAD⁺** (nicotinamide adenine dinucleotide) and **FAD** (flavine adenine dinucleotide) are derived from the B-vitamins niacin and riboflavin, respectively. As we will see, they play major roles in energy metabolism by transferring hydrogen from one substrate to another.

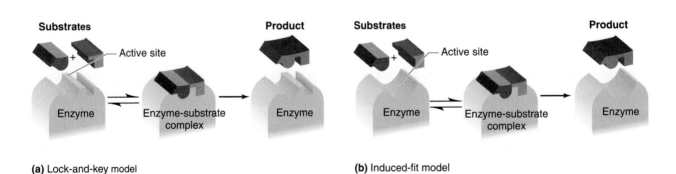

(a) Lock-and-key model   **(b)** Induced-fit model

Figure 3–33

Binding of substrate to the active site of an enzyme catalyzes the formation of products.

From M. S. Silberberg, *Chemistry:The Molecular Nature of Matter and Change* 3d ed., p. 701. The McGraw-Hill Companies, Inc., New York, NY, 2003.

Table 3–7	Characteristics of Enzymes

1. An enzyme undergoes no net chemical change as a consequence of the reaction it catalyzes.

2. The binding of substrate to an enzyme's active site has all the characteristics—chemical specificity, affinity, competition, and saturation—of a ligand binding to a protein.

3. An enzyme increases the rate of a chemical reaction but does not cause a reaction to occur that would not occur in its absence.

4. Some enzymes increase both the forward and reverse rates of a chemical reaction and thus do not change the chemical equilibrium finally reached. They only increase the rate at which equilibrium is achieved.

5. An enzyme lowers the activation energy of a reaction but does not alter the net amount of energy that is added to or released by the reactants in the course of the reaction.

Regulation of Enzyme-Mediated Reactions

The rate of an enzyme-mediated reaction depends on substrate concentration and on the concentration and activity (a term defined later in this section) of the enzyme that catalyzes the reaction. Body temperature is normally nearly constant, so changes in temperature do not directly alter the rates of metabolic reactions. Increases in body temperature can occur during a fever, however, and around muscle tissue during exercise, and such increases in temperature increase the rates of all metabolic reactions, including enzyme-catalyzed ones, in the affected tissues.

Substrate Concentration

Substrate concentration may be altered as a result of factors that alter the supply of a substrate from outside a cell. For example, there may be changes in its blood concentration due to changes in diet or the rate of substrate absorption from the intestinal tract. Intracellular substrate concentration can also be altered by cellular reactions that either utilize the substrate, and thus lower its concentration, or synthesize the substrate, and thereby increase its concentration.

The rate of an enzyme-mediated reaction increases as the substrate concentration increases, as illustrated in **Figure 3–34**, until it reaches a maximal rate, which remains constant despite further increases in substrate concentration. The maximal rate is reached when the enzyme becomes saturated with substrate—that is, when the active binding site of every enzyme molecule is occupied by a substrate molecule.

Enzyme Concentration

At any substrate concentration, including saturating concentrations, the rate of an enzyme-mediated reaction can be increased by increasing the enzyme concentration. In most metabolic reactions, the substrate concentration is much greater than the concentration of enzyme available to catalyze the reaction. Therefore, if the number of enzyme molecules is doubled, twice as many active sites will be available to bind substrate, and twice as many substrate molecules will be converted to product (**Figure 3–35**). Certain reactions proceed faster in some cells than in others because more enzyme molecules are present.

To change the concentration of an enzyme, either the rate of enzyme synthesis or the rate of enzyme breakdown must be altered. Because enzymes are proteins, this involves changing the rates of protein synthesis or breakdown.

Enzyme Activity

In addition to changing the rate of enzyme-mediated reactions by changing the *concentration* of either substrate or enzyme, the rate can be altered by changing **enzyme activity.** A change in enzyme activity occurs when either allosteric or covalent modulation alters the properties of the enzyme's active site. Such modulation alters the rate at which the binding site converts substrate to product, the affinity of the binding site for substrate, or both.

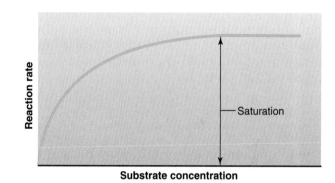

Figure 3–34

Rate of an enzyme-catalyzed reaction as a function of substrate concentration.

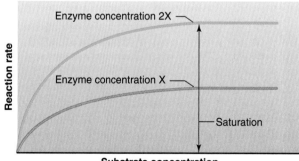

Figure 3–35

Rate of an enzyme-catalyzed reaction as a function of substrate concentration at two enzyme concentrations, X and 2X. Enzyme concentration 2X is twice the enzyme concentration of X, resulting in a reaction that proceeds twice as fast at any substrate concentration.

Figure 3–36 illustrates the effect of increasing the affinity of an enzyme's active site without changing the substrate or enzyme concentration. If the substrate concentration is less than the saturating concentration, the increased affinity of the enzyme's binding site results in an increased number of active sites bound to substrate, and thus an increase in the reaction rate.

The regulation of metabolism through the control of enzyme activity is an extremely complex process because, in many cases, more than one agent can alter the activity of an enzyme (**Figure 3–37**). The modulator molecules that allosterically alter enzyme activities may be product molecules of other cellular reactions. The result is that the overall rates of metabolism can adjust to meet various metabolic demands. In contrast, covalent modulation of enzyme activity is mediated by protein kinase enzymes that are themselves activated by various chemical signals the cell receives, for example, from a hormone.

Figure 3–38 summarizes the factors that regulate the rate of an enzyme-mediated reaction.

Multienzyme Reactions

The sequence of enzyme-mediated reactions leading to the formation of a particular product is known as a **metabolic pathway.** For example, the 19 reactions that convert glucose to carbon dioxide and water constitute the metabolic pathway for glucose catabolism. Each reaction produces only a small change in the structure of the substrate. By such a sequence of small steps, a complex chemical structure, such as glucose, can be transformed to the relatively simple molecular structures carbon dioxide and water.

Consider a metabolic pathway containing four enzymes (e_1, e_2, e_3, and e_4) and leading from an initial substrate A to the end product E, through a series of intermediates, B, C, and D:

$$A \xrightleftharpoons{e_1} B \xrightleftharpoons{e_2} C \xrightleftharpoons{e_3} D \xrightarrow{e_4} E$$

(The irreversibility of the last reaction is of no consequence for the moment.) By mass action, increasing the concentration of A will lead to an increase in the concentration of B (provided e_1 is not already saturated with substrate), and so on until eventually there is an increase in the concentration of the end product E.

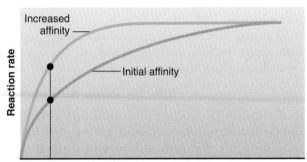

Figure 3–36

At a constant substrate concentration, increasing the affinity of an enzyme for its substrate by allosteric or covalent modulation increases the rate of the enzyme-mediated reaction. Note that increasing the enzyme's affinity does not increase the *maximal* rate of the enzyme-mediated reaction.

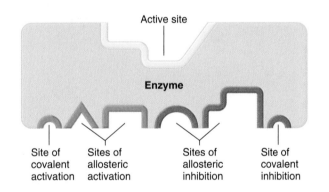

Figure 3–37

On a single enzyme, multiple sites can modulate enzyme activity, and therefore the reaction rate, by allosteric and covalent activation or inhibition.

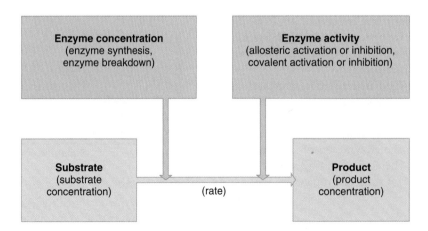

Figure 3–38

Factors that affect the rate of enzyme-mediated reactions.

Because different enzymes have different concentrations and activities, it would be extremely unlikely that the reaction rates of all these steps would be exactly the same. Thus, one step is likely to be slower than all the others. This step is known as the **rate-limiting reaction** in a metabolic pathway. None of the reactions that occur later in the sequence, including the formation of end product, can proceed more rapidly than the rate-limiting reaction because their substrates are supplied by the previous steps. By regulating the concentration or activity of the rate-limiting enzyme, the rate of flow through the whole pathway can be increased or decreased. Thus, it is not necessary to alter all the enzymes in a metabolic pathway to control the rate at which the end product is produced.

Rate-limiting enzymes are often the sites of allosteric or covalent regulation. For example, if enzyme e_2 is rate-limiting in the pathway just described, and if the end product E inhibits the activity of e_2, **end-product inhibition** occurs (**Figure 3–39**). As the concentration of the product increases, the inhibition of further product formation increases. Such inhibition, which is a form of negative feedback (Chapter 1), frequently occurs in synthetic pathways where the formation of end product is effectively shut down when it is not being utilized. This prevents unnecessary excessive accumulation of the end product.

Control of enzyme activity also can be critical for *reversing* a metabolic pathway. Consider the pathway we have been discussing, ignoring the presence of end-product inhibition of enzyme e_2. The pathway consists of three reversible reactions mediated by e_1, e_2, and e_3, followed by an irreversible reaction mediated by enzyme e_4. E can be converted into D, however, if the reaction is coupled to the simultaneous breakdown of a molecule that releases large quantities of energy. In other words, an irreversible step can be "reversed" by an alternative route, using a second enzyme and its substrate to provide the large amount of required energy. Two such high-energy irreversible reactions are indicated by bowed arrows to emphasize that two separate enzymes are involved in the two directions:

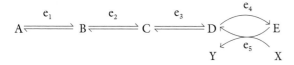

The direction of flow through the pathway can be regulated by controlling the concentration and/or activities of e_4

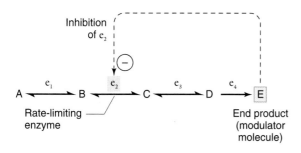

Figure 3–39

End-product inhibition of the rate-limiting enzyme in a metabolic pathway. The end product E becomes the modulator molecule that produces inhibition of enzyme e_2.

and e_5. If e_4 is activated and e_5 inhibited, the flow will proceed from A to E, whereas inhibition of e_4 and activation of e_5 will produce flow from E to A.

Another situation involving the differential control of several enzymes arises when there is a branch in a metabolic pathway. A single metabolite, C, may be the substrate for more than one enzyme, as illustrated by the pathway:

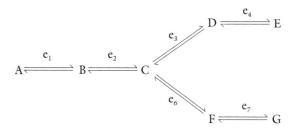

Altering the concentration and/or activities of e_3 and e_6 regulates the flow of metabolite C through the two branches of the pathway.

Considering the thousands of reactions that occur in the body and the permutations and combinations of possible control points, the overall result is staggering. The details of regulating the many metabolic pathways at the enzymatic level are beyond the scope of this book. In the remainder of this chapter, we consider only (1) the overall characteristics of the pathways by which cells obtain energy, and (2) the major pathways by which carbohydrates, fats, and proteins are broken down and synthesized.

■ **SECTION D SUMMARY**

In adults, the rates at which organic molecules are continuously synthesized (anabolism) and broken down (catabolism) are approximately equal.

Chemical Reactions

I. The difference in the energy content of reactants and products is the amount of energy (measured in calories) released or added during a reaction.

II. The energy released during a chemical reaction is either released as heat or transferred to other molecules.

III. The four factors that can alter the rate of a chemical reaction are listed in Table 3–5.

IV. The activation energy required to initiate the breaking of chemical bonds in a reaction is usually acquired through collisions between molecules.

V. Catalysts increase the rate of a reaction by lowering the activation energy.

VI. The characteristics of reversible and irreversible reactions are listed in Table 3–6.

VII. The net direction in which a reaction proceeds can be altered, according to the law of mass action, by increases or decreases in the concentrations of reactants or products.

Enzymes

I. Nearly all chemical reactions in the body are catalyzed by enzymes, the characteristics of which are summarized in Table 3–7.

II. Some enzymes require small concentrations of cofactors for activity.

a. The binding of trace metal cofactors maintains the conformation of the enzyme's binding site so that it is able to bind substrate.

b. Coenzymes, derived from vitamins, transfer small groups of atoms from one substrate to another. The coenzyme is regenerated in the course of these reactions and can do its work over and over again.

Regulation of Enzyme-Mediated Reactions

I. The rates of enzyme-mediated reactions can be altered by changes in temperature, substrate concentration, enzyme concentration, and enzyme activity. Enzyme activity is altered by allosteric or covalent modulation.

Multienzyme Reactions

I. The rate of product formation in a metabolic pathway can be controlled by allosteric or covalent modulation of the enzyme mediating the rate-limiting reaction in the pathway. The end product often acts as a modulator molecule, inhibiting the rate-limiting enzyme's activity.

II. An "irreversible" step in a metabolic pathway can be reversed by the use of two enzymes, one for the forward reaction and one for the reverse direction via another, energy-yielding reaction.

■ SECTION D KEY TERMS

■ SECTION D REVIEW QUESTIONS

1. How do molecules acquire the activation energy required for a chemical reaction?
2. List the four factors that influence the rate of a chemical reaction and state whether increasing the factor will increase or decrease the rate of the reaction.
3. What characteristics of a chemical reaction make it reversible or irreversible?
4. List five characteristics of enzymes.
5. What is the difference between a cofactor and a coenzyme?
6. From what class of nutrients are coenzymes derived?
7. Why are small concentrations of coenzymes sufficient to maintain enzyme activity?
8. List three ways to alter the rate of an enzyme-mediated reaction.
9. How can an "irreversible step" in a metabolic pathway be reversed?

SECTION E

Metabolic Pathways

Cellular Energy Transfer

Glycolysis

Glycolysis (from the Greek *glycos,* sugar, and *lysis,* breakdown) is a pathway that partially catabolizes carbohydrates, primarily glucose. It consists of 10 enzymatic reactions that convert a six-carbon molecule of glucose into two three-carbon molecules of **pyruvate,** the ionized form of pyruvic acid (**Figure 3–41**). The reactions produce a net gain of two molecules of ATP and four atoms of hydrogen, two transferred to NAD^+ and two released as hydrogen ions:

$$\text{Glucose} + 2\ \text{ADP} + 2\ P_i + 2\ NAD^+ \longrightarrow$$
$$2\ \text{Pyruvate} + 2\ \text{ATP} + 2\ \text{NADH} + 2\ H^+ + 2\ H_2O$$

These 10 reactions, *none of which utilizes molecular oxygen,* take place in the cytosol. Note (see Figure 3–41) that all the intermediates between glucose and the end product pyruvate contain one or more ionized phosphate groups. Plasma membranes are impermeable to such highly ionized molecules, and thus these molecules remain trapped within the cell.

Note that the early steps in glycolysis (reactions 1 and 3) each *use,* rather than produce, one molecule of ATP, to form phosphorylated intermediates. In addition, note that reaction 4 splits a six-carbon intermediate into two three-carbon molecules, and reaction 5 converts one of these

Enzymes are involved in many important physiological reactions that together promote a homeostatic state. In addition, enzymes are vital for the regulated production of cellular energy (ATP), which, in turn, is needed for such widespread events as muscle contraction, nerve cell function, and chemical signal transduction.

Cells use three distinct but linked metabolic pathways to transfer the energy released from the breakdown of fuel molecules to ATP. They are known as (1) glycolysis, (2) the Krebs cycle, and (3) oxidative phosphorylation (**Figure 3–40**). In the following section, we will describe the major characteristics of these three pathways, including the location of the pathway enzymes in a cell, the relative contribution of each pathway to ATP production, the sites of carbon dioxide formation and oxygen utilization, and the key molecules that enter and leave each pathway. Later, in Chapter 16, we will refer to these pathways when we describe the physiology of energy balance in the human body.

Several facts should be noted in Figure 3–40. First, glycolysis operates only on carbohydrates. Second, all the categories of nutrients—carbohydrates, fats, and proteins—contribute to ATP production via the Krebs cycle and oxidative phosphorylation. Third, mitochondria are the sites of the Krebs cycle and oxidative phosphorylation. Finally, one important generalization to keep in mind is that glycolysis can occur in either the presence or absence of oxygen, whereas both the Krebs cycle and oxidative phosphorylation require oxygen.

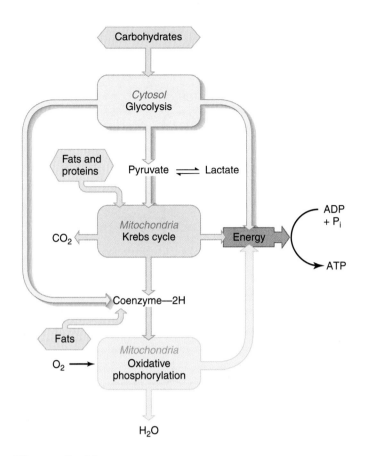

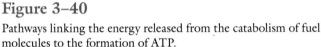

Figure 3–40

Pathways linking the energy released from the catabolism of fuel molecules to the formation of ATP.

three-carbon molecules into the other. Thus, at the end of reaction 5 we have two molecules of 3-phosphoglyceraldehyde derived from one molecule of glucose. Keep in mind, then, that from this point on, *two* molecules of each intermediate are involved.

The first formation of ATP in glycolysis occurs during reaction 7, when a phosphate group is transferred to ADP to form ATP. Since two intermediates exist at this point, reaction 7 produces two molecules of ATP, one from each intermediate. In this reaction, the mechanism of forming ATP is known as **substrate-level phosphorylation** because the phosphate group is transferred from a substrate molecule to ADP.

A similar substrate-level phosphorylation of ADP occurs during reaction 10, where again two molecules of ATP are formed. Thus, reactions 7 and 10 generate a total of four molecules of ATP for every molecule of glucose entering the pathway. There is a net gain, however, of only two molecules of ATP during glycolysis because two molecules of ATP are used in reactions 1 and 3.

The end product of glycolysis, pyruvate, can proceed in one of two directions, depending on the availability of molecular oxygen, which, as we stressed earlier, is *not* utilized in any of the glycolytic reactions themselves. If oxygen is present—that is, if **aerobic** conditions exist—pyruvate can enter the Krebs cycle and be broken down into carbon dioxide, as described in

the next section. In contrast, in the absence of oxygen (**anaerobic** conditions), pyruvate is converted to **lactate** (the ionized form of lactic acid) by a single enzyme-mediated reaction. In this reaction (**Figure 3–42**), two hydrogen atoms derived from $NADH^+ + H^+$ are transferred to each molecule of pyruvate to form lactate, and NAD^+ is regenerated. These hydrogens were originally transferred to NAD^+ during reaction 6 of glycolysis, so the coenzyme NAD^+ shuttles hydrogen between the two reactions during anaerobic glycolysis. The overall reaction for anaerobic glycolysis is:

$$\text{Glucose} + 2\ \text{ADP} + 2\ P_i \longrightarrow 2\ \text{Lactate} + 2\ \text{ATP} + 2\ H_2O$$

As stated in the previous paragraph, under aerobic conditions pyruvate is not converted to lactate but instead enters the Krebs cycle. Therefore, the mechanism just described for regenerating NAD^+ from $NADH^+ + H^+$ by forming lactate does not occur. The hydrogens of NADH are transferred to oxygen during oxidative phosphorylation, regenerating NAD^+ and producing H_2O, as described in detail in the discussion that follows.

In most cells, the amount of ATP glycolysis produces from one molecule of glucose is much smaller than the amount formed under aerobic conditions by the other two ATP-generating pathways—the Krebs cycle and oxidative phosphorylation. In special cases, however, glycolysis supplies most, or even all, of a cell's ATP. For example, erythrocytes contain the enzymes for glycolysis but have no mitochondria, which are required for the other pathways. All of their ATP production occurs, therefore, by glycolysis. Also, certain types of skeletal muscles contain considerable amounts of glycolytic enzymes but few mitochondria. During intense muscle activity, glycolysis provides most of the ATP in these cells and is associated with the production of large amounts of lactate. Despite these exceptions, most cells do not have sufficient concentrations of glycolytic enzymes or enough glucose to provide, by glycolysis alone, the high rates of ATP production necessary to meet their energy requirements.

Our discussion of glycolysis has focused upon glucose as the major carbohydrate entering the glycolytic pathway. However, other carbohydrates such as fructose, derived from the disaccharide sucrose (table sugar), and galactose, from the disaccharide lactose (milk sugar), can also be catabolized by glycolysis because these carbohydrates are converted into several of the intermediates that participate in the early portion of the glycolytic pathway. **Table 3–8** summarizes the major characteristics of glycolysis.

Krebs Cycle

The **Krebs cycle,** named in honor of Hans Krebs, who worked out the intermediate steps in this pathway (also known as the **citric acid cycle** or **tricarboxylic acid cycle**), is the second of the three pathways involved in fuel catabolism and ATP production. It utilizes molecular fragments formed during carbohydrate, protein, and fat breakdown, and it produces carbon dioxide, hydrogen atoms (half of which are bound to coenzymes), and small amounts of ATP. The enzymes for this pathway are located in the inner mitochondrial compartment, the matrix.

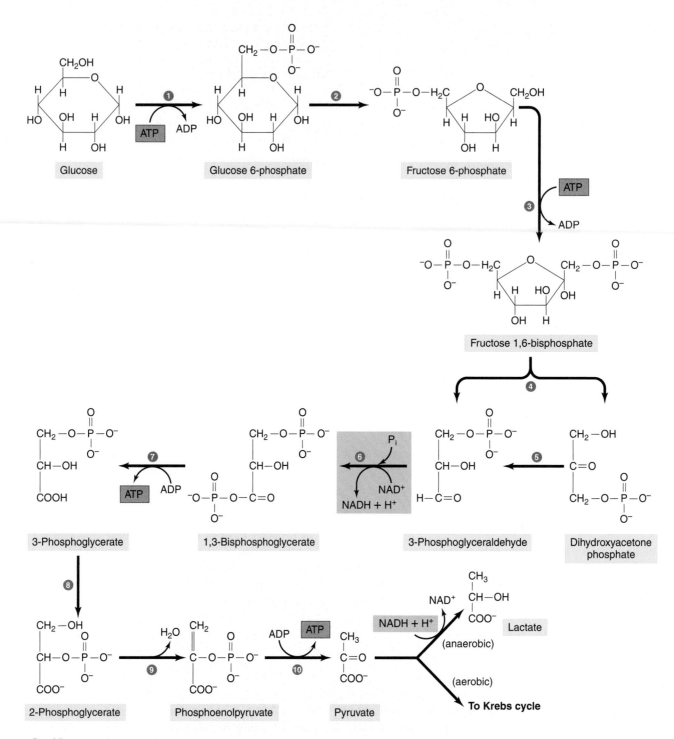

Figure 3–41

Glycolytic pathway. Under anaerobic conditions, every molecule of glucose that enters the pathway produces a net synthesis of two molecules of ATP. Note that at the pH existing in the body, the products produced by the various glycolytic steps exist in the ionized, anionic form (pyruvate, for example). They are actually produced as acids (pyruvic acid, for example) that then ionize.

The primary molecule entering at the beginning of the Krebs cycle is **acetyl coenzyme A (acetyl CoA):**

$$CH_3-\overset{\overset{\textstyle O}{\|}}{C}-S-CoA$$

Coenzyme A (CoA) is derived from the B vitamin pantothenic acid and functions primarily to transfer acetyl groups, which contain two carbons, from one molecule to another. These acetyl groups come either from pyruvate—the end product of aerobic glycolysis—or from the breakdown of fatty acids and some amino acids.

Pyruvate, upon entering mitochondria from the cytosol, is converted to acetyl CoA and CO_2 (**Figure 3–43**). Note that this reaction produces the first molecule of CO_2 formed thus far in the pathways of fuel catabolism, and that the reaction also transfers hydrogen atoms to NAD^+.

The Krebs cycle begins with the transfer of the acetyl group of acetyl CoA to the four-carbon molecule, oxaloacetate, to form the six-carbon molecule, citrate (**Figure 3–44**). At the third step in the cycle a molecule of CO_2 is produced, and again at the fourth step. Thus, two carbon atoms entered the cycle as part of the acetyl group attached to CoA, and two carbons (although not the same ones) have left in the form of CO_2. Note also that the oxygen that appears in the CO_2 is not derived from molecular oxygen, but from the carboxyl groups of Krebs cycle intermediates.

In the remainder of the cycle, the four-carbon molecule formed in reaction 4 is modified through a series of reactions to produce the four-carbon molecule oxaloacetate, which becomes available to accept another acetyl group and repeat the cycle.

Now we come to a crucial fact: In addition to producing carbon dioxide, intermediates in the Krebs cycle generate hydrogen atoms, most of which are transferred to the coenzymes NAD^+ and FAD to form NADH and $FADH_2$. This hydrogen transfer to NAD^+ occurs in each of steps 3, 4, and 8, and to FAD in reaction 6. These hydrogens will be transferred from the coenzymes, along with the free H^+, to oxygen in the next stage of fuel metabolism—oxidative phosphorylation. Because oxidative phosphorylation is necessary for regeneration of the hydrogen-free form of these coenzymes, *the Krebs cycle can operate only under aerobic conditions*. There is no pathway in the mitochondria that can remove the hydrogen from these coenzymes under anaerobic conditions.

So far we have said nothing of how the Krebs cycle contributes to the formation of ATP. In fact, the Krebs cycle *directly* produces only one high-energy nucleotide triphosphate. This

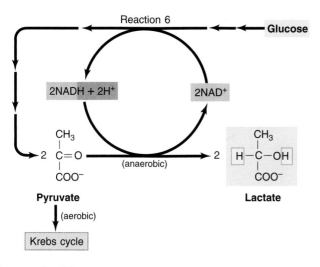

Figure 3–42

Under anaerobic conditions, the coenzyme NAD^+ utilized in the glycolytic reaction 6 (see Figure 3–41) is regenerated when it transfers its hydrogen atoms to pyruvate during the formation of lactate.

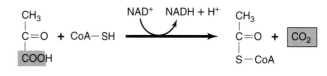

Figure 3–43

Formation of acetyl coenzyme A from pyruvic acid with the formation of a molecule of carbon dioxide.

Table 3–8	Characteristics of Glycolysis
Entering substrates	Glucose and other monosaccharides
Enzyme location	Cytosol
Net ATP production	2 ATP formed directly per molecule of glucose entering pathway can be produced in the absence of oxygen (anaerobically)
Coenzyme production	2 NADH + 2 H^+ formed under aerobic conditions
Final products	Pyruvate—under aerobic conditions
	Lactate—under anaerobic conditions
Net reaction	
Aerobic:	Glucose + 2 ADP + 2 P_i + 2 NAD^+ $\longrightarrow$ 2 pyruvate + 2 ATP + 2 NADH + 2 H^+ + 2 H_2O
Anaerobic:	Glucose + 2 ADP + 2 P_i $\longrightarrow$ 2 lactate + 2 ATP + 2 H_2O

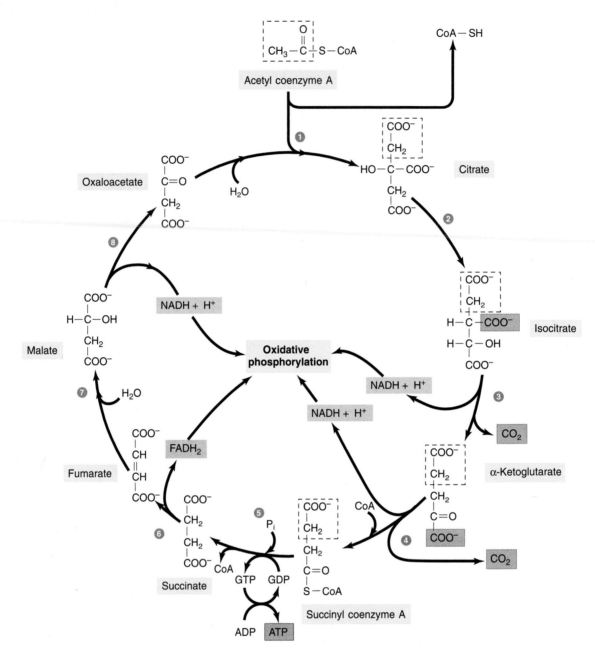

Figure 3–44

The Krebs cycle pathway. Note that the carbon atoms in the two molecules of CO_2 produced by a turn of the cycle are not the same two carbon atoms that entered the cycle as an acetyl group (identified by the dashed boxes in this figure).

occurs during reaction 5 in which inorganic phosphate is transferred to guanosine diphosphate (GDP) to form guanosine triphosphate (GTP). The hydrolysis of GTP, like that of ATP, can provide energy for some energy-requiring reactions. In addition, the energy in GTP can be transferred to ATP by the reaction

$$GTP + ADP \rightleftharpoons GDP + ATP$$

The formation of ATP from GTP is the only mechanism by which ATP is formed within the Krebs cycle. Why, then, is the Krebs cycle so important? Because the hydrogen atoms transferred to coenzymes during the cycle (plus the free hydrogen ions generated) are used in the next pathway, oxidative phosphorylation, to form large amounts of ATP.

The net result of the catabolism of one acetyl group from acetyl CoA by way of the Krebs cycle can be written:

$$Acetyl\ CoA + 3\ NAD^+ + FAD + GDP + P_i + 2\ H_2O \longrightarrow$$
$$2\ CO_2 + CoA + 3\ NADH + 3\ H^+ + FADH_2 + GTP$$

Table 3–9 summarizes the characteristics of the Krebs cycle reactions.

Oxidative Phosphorylation

Oxidative phosphorylation provides the third, and quantitatively most important, mechanism by which energy derived from fuel molecules can be transferred to ATP. The basic principle behind this pathway is simple: The energy transferred to

Chapter 3

Table 3–9	Characteristics of the Krebs Cycle
Entering substrate	Acetyl coenzyme A—acetyl groups derived from pyruvate, fatty acids, and amino acids
	Some intermediates derived from amino acids
Enzyme location	Inner compartment of mitochondria (the mitochondrial matrix)
ATP production	1 GTP formed directly, which can be converted into ATP
	Operates only under aerobic conditions even though molecular oxygen is not used directly in this pathway
Coenzyme production	3 NADH + 3 H$^+$ and 2 FADH$_2$
Final products	2 CO$_2$ for each molecule of acetyl coenzyme A entering pathway
	Some intermediates used to synthesize amino acids and other organic molecules required for special cell functions
Net reaction	Acetyl CoA + 3 NAD$^+$ + FAD + GDP + P$_i$ + 2 H$_2$O $\longrightarrow$ 2 CO$_2$ + CoA + 3 NADH + 3 H$^+$ + FADH$_2$ + GTP

ATP is derived from the energy released when hydrogen ions combine with molecular oxygen to form water. The hydrogen comes from the NADH + H$^+$ and FADH$_2$ coenzymes generated by the Krebs cycle, by the metabolism of fatty acids (see the discussion that follows), and, to a much lesser extent, during aerobic glycolysis. The net reaction is:

$$\tfrac{1}{2}O_2 + NADH + H^+ \longrightarrow H_2O + NAD^+ + Energy$$

Unlike the enzymes of the Krebs cycle, which are soluble enzymes in the mitochondrial matrix, the proteins that mediate oxidative phosphorylation are embedded in the inner mitochondrial membrane. The proteins for oxidative phosphorylation can be divided into two groups: (1) those that mediate the series of reactions that cause the transfer of hydrogen ions to molecular oxygen, and (2) those that couple the energy released by these reactions to the synthesis of ATP.

Most of the first group of proteins contain iron and copper cofactors, and are known as **cytochromes** (because in pure form they are brightly colored). Their structure resembles the red iron-containing hemoglobin molecule, which binds oxygen in red blood cells. The cytochromes form the components of the **electron transport chain,** in which two electrons from the hydrogen atoms are initially transferred either from NADH + H$^+$ or FADH$_2$ to one of the elements in this chain. These electrons are then successively transferred to other compounds in the chain, often to or from an iron or copper ion, until the electrons are finally transferred to molecular oxygen, which then combines with hydrogen ions (protons) to form water. These hydrogen ions, like the electrons, come from free hydrogen ions and the hydrogen-bearing coenzymes, having been released early in the transport chain when the electrons from the hydrogen atoms were transferred to the cytochromes.

Importantly, in addition to transferring the coenzyme hydrogens to water, this process regenerates the hydrogen-free form of the coenzymes, which then become available to accept two more hydrogens from intermediates in the Krebs cycle, glycolysis, or fatty acid pathway (as described in the discussion that follows). Thus, the electron transport chain provides the *aerobic* mechanism for regenerating the hydrogen-free form of the coenzymes, whereas, as described earlier, the *anaerobic* mechanism, which applies only to glycolysis, is coupled to the formation of lactate.

At each step along the electron transport chain, small amounts of energy are released. Because this energy is released in small steps, it can be coupled to the synthesis of several molecules of ATP in a controlled manner.

ATP is formed at three points along the electron transport chain. The mechanism by which this occurs is known as the **chemiosmotic hypothesis.** As electrons are transferred from one cytochrome to another along the electron transport chain, the energy released is used to move hydrogen ions (protons) from the matrix into the compartment between the inner and outer mitochondrial membranes (**Figure 3–45**), thus producing a source of potential energy in the form of a hydrogen-ion gradient across the membrane. At three points along the chain, a protein complex forms a channel in the inner mitochondrial membrane, allowing the hydrogen ions to flow back to the matrix side and, in the process, transfer energy to the formation of ATP from ADP and P$_i$. FADH$_2$ has a slightly lower chemical energy content than does NADH + H$^+$ and enters the electron transport chain at a point beyond the first site of ATP generation (see Figure 3–45). The process is not perfectly stoichiometric, however, and thus the transfer of electrons to oxygen produces approximately 2.5 and 1.5 molecules of ATP for each molecule of NADH + H$^+$ and FADH$_2$, respectively.

In summary, most ATP formed in the body is produced during oxidative phosphorylation as a result of processing hydrogen atoms that originated largely from the Krebs cycle during the breakdown of carbohydrates, fats, and proteins. The mitochondria, where the oxidative phosphorylation and the Krebs cycle reactions occur, are thus considered the powerhouses of the cell. In addition, most of the oxygen we breathe is consumed within these organelles, and most of the carbon dioxide we exhale is produced within them as well.

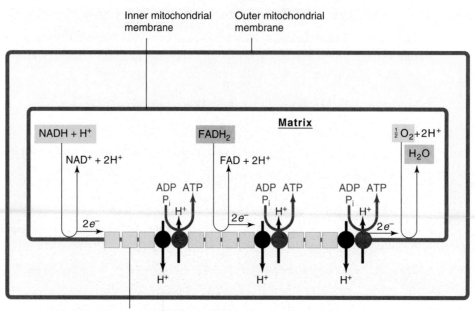

Figure 3–45

ATP is formed during oxidative phosphorylation by the flow of hydrogen ions across the inner mitochondrial membrane. A maximum of two or three molecules of ATP are produced per pair of electrons donated, depending on the point at which a particular coenzyme enters the electron transport chain.

Table 3–10 summarizes the key features of oxidative phosphorylation.

Reactive Oxygen Species

As we have just seen, the formation of ATP by oxidative phosphorylation involves the transfer of electrons and hydrogen to molecular oxygen. Several highly reactive transient oxygen derivatives can also be formed during this process—**hydrogen peroxide** and the free radicals **superoxide anion** and **hydroxyl radical:**

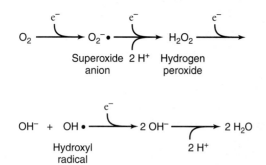

Although most of the electrons transferred along the electron transport chain go into the formation of water, small amounts can combine with oxygen to form reactive oxygen species. As described in Chapter 2, these species can react with and damage proteins, membrane phospholipids, and nucleic acids. Such damage has been implicated in the aging process and in inflammatory reactions to tissue injury. Some cells use these reactive molecules to kill invading bacteria.

Reactive oxygen molecules are also formed by the action of ionizing radiation on oxygen and by reactions of oxygen with heavy metals such as iron. Cells contain several enzymatic mechanisms for removing these reactive oxygen species and thus providing protection from their damaging effects.

Carbohydrate, Fat, and Protein Metabolism

Now that we have described the three pathways by which energy is transferred to ATP, let's consider how each of the three classes of energy-yielding nutrient molecules—carbohydrates, fats, and proteins—enters the ATP-generating pathways. We will also consider the synthesis of these fuel molecules and the pathways and restrictions governing their conversion from one class to another. These anabolic pathways are also used to synthesize molecules that have functions other than the storage and release of energy. For example, with the addition of a few enzymes, the pathway for fat synthesis is also used for synthesis of the phospholipids found in membranes.

The material presented in this section should serve as a foundation for understanding how the body copes with changes in fuel availability. The physiological mechanisms that regulate appetite, digestion, and absorption of food, transport of fuel sources in the blood and across cell membranes, and the body's responses to fasting and starvation are covered in Chapters 15 and 16.

Table 3–10 — Characteristics of Oxidative Phosphorylation

Entering substrates	Hydrogen atoms obtained from NADH + H$^+$ and FADH$_2$ formed (1) during glycolysis, (2) by the Krebs cycle during the breakdown of pyruvate and amino acids, and (3) during the breakdown of fatty acids Molecular oxygen
Enzyme location	Inner mitochondrial membrane
ATP production	3 ATP formed from each NADH + H$^+$ 2 ATP formed from each FADH$_2$
Final products	H$_2$O—one molecule for each pair of hydrogens entering pathway
Net reaction	$\frac{1}{2}$O$_2$ + NADH + H$^+$ + 3 ADP + 3 P$_i$ $\longrightarrow$ H$_2$O + NAD$^+$ + 3 ATP

Carbohydrate Metabolism

Carbohydrate Catabolism

In the previous sections, we described the major pathways of carbohydrate catabolism: the breakdown of glucose to pyruvate or lactate by way of the glycolytic pathway, and the metabolism of pyruvate to carbon dioxide and water by way of the Krebs cycle and oxidative phosphorylation.

The amount of energy released during the catabolism of glucose to carbon dioxide and water is 686 kcal/mol of glucose:

$$C_6H_{12}O_6 + 6\,O_2 \longrightarrow 6\,H_2O + 6\,CO_2 + 686\text{ kcal/mol}$$

About 40 percent of this energy is transferred to ATP. **Figure 3–46** summarizes the points at which ATP forms during glucose catabolism. A net gain of two ATP molecules occurs by substrate-level phosphorylation during glycolysis, and two more are formed during the Krebs cycle from GTP, one from each of the two molecules of pyruvate entering the cycle. The majority of ATP molecules glucose catabolism produces—34 ATP per molecule—form during oxidative phosphorylation from the hydrogens generated at various steps during glucose breakdown.

Because, in the absence of oxygen, only two molecules of ATP can form from the breakdown of glucose to lactate,

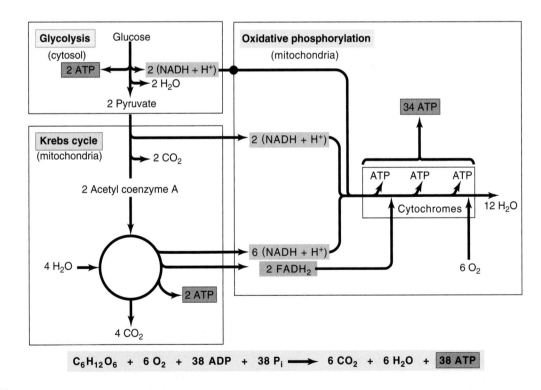

Figure 3–46

Pathways of glycolysis and aerobic glucose catabolism and their linkage to ATP formation.

the evolution of aerobic metabolic pathways greatly increases the amount of energy available to a cell from glucose catabolism. For example, if a muscle consumed 38 molecules of ATP during a contraction, this amount of ATP could be supplied by the breakdown of one molecule of glucose in the presence of oxygen or 19 molecules of glucose under anaerobic conditions.

It is important to note, however, that although only two molecules of ATP are formed per molecule of glucose under anaerobic conditions, large amounts of ATP can still be supplied by the glycolytic pathway if large amounts of glucose are broken down to lactate. This is not an efficient utilization of nutrients, but it does permit continued ATP production under anaerobic conditions, such as occur during intense exercise.

Glycogen Storage

A small amount of glucose can be stored in the body to provide a reserve supply for use when glucose is not being absorbed into the blood from the intestinal tract. Recall from Chapter 2 that it is stored as the polysaccharide **glycogen,** mostly in skeletal muscles and the liver.

Glycogen is synthesized from glucose by the pathway illustrated in **Figure 3–47**. The enzymes for both glycogen synthesis and glycogen breakdown are located in the cytosol. The first step in glycogen synthesis, the transfer of phosphate from a molecule of ATP to glucose, forming glucose 6-phosphate, is the same as the first step in glycolysis. Thus, glucose 6-phosphate can either be broken down to pyruvate or used to form glycogen.

As indicated in Figure 3–47, different enzymes are used to synthesize and break down glycogen. The existence of two pathways containing enzymes that are subject to both covalent and allosteric modulation provides a mechanism for regulating the flow between glucose and glycogen. When an excess of glucose is available to a liver or muscle cell, the enzymes in the glycogen synthesis pathway are activated, and the enzyme that breaks down glycogen is simultaneously inhibited. This combination leads to the net storage of glucose in the form of glycogen.

When less glucose is available, the reverse combination of enzyme stimulation and inhibition occurs, and net breakdown of glycogen to glucose 6-phosphate (known as **glycogenolysis**) ensues. Two paths are available to this glucose 6-phosphate: (1) In most cells, including skeletal muscle, it enters the glycolytic pathway where it is catabolized to provide the energy for ATP formation; (2) in liver (and kidney) cells, glucose 6-phosphate can be converted to free glucose by removal of the phosphate group, and the glucose is then able to pass out of the cell into the blood to fuel other cells.

Glucose Synthesis

In addition to being formed in the liver from the breakdown of glycogen, glucose can be synthesized in the liver and kidneys from intermediates derived from the catabolism of glycerol (a so-called sugar alcohol) and some amino acids. This process of generating new molecules of glucose from noncarbohydrate precursors is known as **gluconeogenesis.** The major substrate in gluconeogenesis is pyruvate, formed from lactate and from several amino acids during protein breakdown. In addition, glycerol derived from the hydrolysis of triglycerides can be converted into glucose via a pathway that does not involve pyruvate.

The pathway for gluconeogenesis in the liver and kidneys (**Figure 3–48**) makes use of many but not all of the enzymes used in glycolysis because most of these reactions are reversible. However, reactions 1, 3, and 10 (see Figure 3–41) are irreversible, and additional enzymes are required, therefore, to form glucose from pyruvate. Pyruvate is converted to phosphoenolpyruvate by a series of mitochondrial reactions in which CO_2 is added to pyruvate to form the four-carbon Krebs-cycle intermediate oxaloacetate. An additional series of reactions leads to the transfer of a four-carbon intermediate derived from oxaloacetate out of the mitochondria and its conversion to phosphoenolpyruvate in the cytosol. Phosphoenolpyruvate then reverses the steps of glycolysis back to the level of reaction 3, in which a different enzyme from that used in glycolysis is required to convert fructose 1,6-bisphosphate to fructose 6-phosphate. From this point on, the reactions are again reversible, leading to glucose 6-phosphate, which can be converted to glucose in the liver and kidneys or stored as glycogen. Because energy in the form of heat and ATP generation is released during the glycolytic breakdown of glucose to pyruvate, energy must be added to reverse this pathway. A total of six ATP are consumed in the reactions of gluconeogenesis per molecule of glucose formed.

Many of the same enzymes are used in glycolysis and gluconeogenesis, so the question arises: What controls the direction of the reactions in these pathways? What conditions determine whether glucose is broken down to pyruvate or whether pyruvate is converted into glucose? The answer lies in

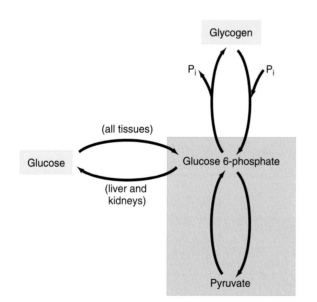

Figure 3–47

Pathways for glycogen synthesis and breakdown. Each bowed arrow indicates one or more irreversible reactions that requires different enzymes to catalyze the reaction in the forward and reverse directions.

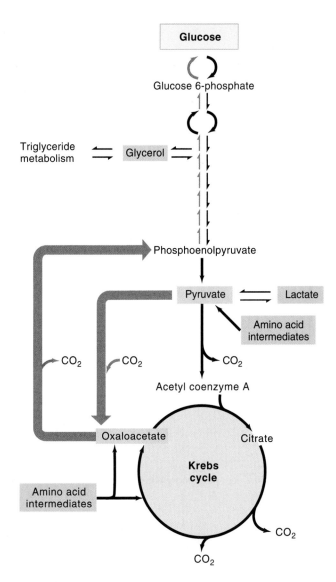

Figure 3–48

Gluconeogenic pathway by which pyruvate, lactate, glycerol, and various amino acid intermediates can be converted into glucose in the liver. Note the points at which each of these precursors, supplied by the blood, enters the pathway.

tions, approximately half the energy used by muscle, liver, and the kidneys is derived from the catabolism of fatty acids.

Although most cells store small amounts of fat, the preponderance of the body's fat is stored in specialized cells known as **adipocytes.** Almost the entire cytoplasm of each of these cells is filled with a single, large fat droplet. Clusters of adipocytes form **adipose tissue,** most of which is in deposits underlying the skin or surrounding internal organs. The function of adipocytes is to synthesize and store triglycerides during periods of food uptake and then, when food is not being absorbed from the intestinal tract, to release fatty acids and glycerol into the blood for uptake and use by other cells to provide the energy needed for ATP formation. The factors controlling fat storage and release from adipocytes during different physiological states will be described in Chapter 16. Here we will emphasize the pathway by which most cells catabolize fatty acids to provide the energy for ATP synthesis, and the pathway by which other fuel molecules are used to synthesize fatty acids.

Figure 3–49 shows the pathway for fatty acid catabolism, which is achieved by enzymes present in the mitochondrial matrix. The breakdown of a fatty acid is initiated by linking a molecule of coenzyme A to the carboxyl end of the fatty acid. This initial step is accompanied by the breakdown of ATP to AMP and two P_i.

The coenzyme-A derivative of the fatty acid then proceeds through a series of reactions, collectively known as **beta oxidation,** which split off a molecule of acetyl coenzyme A from the end of the fatty acid and transfer two pairs of hydrogen atoms to coenzymes (one pair to FAD and the other to NAD^+). The hydrogen atoms from the coenzymes then enter the oxidative phosphorylation pathway to form ATP.

When an acetyl coenzyme A is split from the end of a fatty acid, another coenzyme A is added (ATP is not required for this step), and the sequence is repeated. Each passage through this sequence shortens the fatty acid chain by two carbon atoms until all the carbon atoms have transferred to coenzyme A molecules. As we saw, these molecules then lead to production of CO_2 and ATP via the Krebs cycle and oxidative phosphorylation.

How much ATP is formed as a result of the total catabolism of a fatty acid? Most fatty acids in the body contain 14 to

the concentrations of glucose or pyruvate in a cell and in the control the enzymes exert in the irreversible steps in the pathway. This control is carried out via various hormones that alter the concentrations and activities of these key enzymes. For example, if blood sugar levels fall below normal, certain hormones are secreted into the blood and act on the liver. There, the hormones preferentially induce the expression of the gluconeogenic enzymes, thus favoring the formation of glucose.

Fat Metabolism

Fat Catabolism

Triglyceride (fat) consists of three fatty acids bound to glycerol (Chapter 2). Fat accounts for approximately 80 percent of the energy stored in the body (**Table 3–11**). Under resting condi-

Table 3–11	Fuel Content of a 70-kg Person			
	Total-Body Content, kg	Energy Content, kcal/g	Total-Body Energy Content kcal	%
Triglycerides	15.6	9	140,000	78
Proteins	9.5	4	38,000	21
Carbohydrates	0.5	4	2,000	1

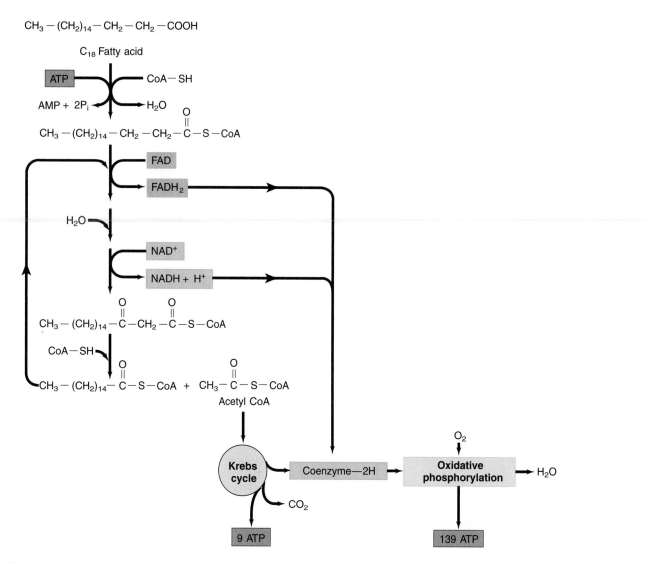

Figure 3–49

Pathway of fatty acid catabolism, which takes place in the mitochondria. The energy equivalent of two ATP is consumed at the start of the pathway.

22 carbons, 16 and 18 being most common. The catabolism of one 18-carbon saturated fatty acid yields 146 ATP molecules. In contrast, as we have seen, the catabolism of one glucose molecule yields a maximum of 38 ATP molecules. Thus, taking into account the difference in molecular weight of the fatty acid and glucose, the amount of ATP formed from the catabolism of a gram of fat is about $2\frac{1}{2}$ times greater than the amount of ATP produced by catabolizing 1 gram of carbohydrate. If an average person stored most of his or her energy as carbohydrate rather than fat, body weight would have to be approximately 30 percent greater in order to store the same amount of usable energy, and the person would consume more energy moving this extra weight around. Thus, a major step in fuel economy occurred when animals evolved the ability to store fuel as fat.

Fat Synthesis

The synthesis of fatty acids occurs by reactions that are almost the reverse of those that degrade them. However, the enzymes in the synthetic pathway are in the cytosol, whereas (as we have just seen) the enzymes catalyzing fatty acid breakdown are in the mitochondria. Fatty acid synthesis begins with cytoplasmic acetyl coenzyme A, which transfers its acetyl group to another molecule of acetyl coenzyme A to form a four-carbon chain. By repetition of this process, long-chain fatty acids are built up two carbons at a time. This accounts for the fact that all the fatty acids synthesized in the body contain an even number of carbon atoms.

Once the fatty acids are formed, triglycerides can be synthesized by linking fatty acids to each of the three hydroxyl groups in glycerol, more specifically, to a phosphorylated form of glycerol called α-**glycerol phosphate.** The synthesis of triglyceride is carried out by enzymes associated with the membranes of the smooth endoplasmic reticulum.

Compare the molecules produced by glucose catabolism with those required for synthesis of both fatty acids and α-glycerol phosphate. First, acetyl coenzyme A, the starting material for fatty acid synthesis, can be formed from pyruvate, the end product of glycolysis. Second, the other ingredients required for fatty acid synthesis—hydrogen-bound coenzymes and ATP—are produced during carbohydrate catabolism. Third, α-glycerol phosphate can be formed from a glucose intermediate. It should not be surprising, therefore, that much of the carbohydrate in food is converted into fat and stored in adipose tissue shortly after its absorption from the gastrointestinal tract.

It is very important to note that fatty acids or, more specifically, the acetyl coenzyme A derived from fatty acid breakdown, cannot be used to synthesize *new* molecules of glucose. We can see the reasons for this by examining the pathways for glucose synthesis (see Figure 3–48). First, because the reaction in which pyruvate is broken down to acetyl coenzyme A and carbon dioxide is irreversible, acetyl coenzyme A cannot be converted into pyruvate, a molecule that could lead to the production of glucose. Second, the equivalents of the two carbon atoms in acetyl coenzyme A are converted into two molecules of carbon dioxide during their passage through the Krebs cycle before reaching oxaloacetate, another takeoff point for glucose synthesis, and therefore they cannot be used to synthesize *net* amounts of oxaloacetate.

Thus, *glucose can readily be converted into fat, but the fatty acid portion of fat cannot be converted to glucose.*

Protein and Amino Acid Metabolism

In contrast to the complexities of protein synthesis, protein catabolism requires only a few enzymes, termed **proteases,** to break the peptide bonds between amino acids (a process called **proteolysis**). Some of these enzymes split off one amino acid at a time from the ends of the protein chain, whereas others break peptide bonds between specific amino acids within the chain, forming peptides rather than free amino acids.

Amino acids can be catabolized to provide energy for ATP synthesis, and they can also provide intermediates for the synthesis of a number of molecules other than proteins. Because there are 20 different amino acids, a large number of intermediates can be formed, and there are many pathways for processing them. A few basic types of reactions common to most of these pathways can provide an overview of amino acid catabolism.

Unlike most carbohydrates and fats, amino acids contain nitrogen atoms (in their amino groups) in addition to carbon, hydrogen, and oxygen atoms. Once the nitrogen-containing amino group is removed, the remainder of most amino acids can be metabolized to intermediates capable of entering either the glycolytic pathway or the Krebs cycle.

Figure 3–50 illustrates the two types of reactions by which the amino group is removed. In the first reaction, **oxidative deamination,** the amino group gives rise to a molecule of ammonia (NH_3) and is replaced by an oxygen atom derived from water to form a **keto acid,** a categorical name rather than the name of a specific molecule. The second means of removing an amino group is known as **transamination** and involves transfer of the amino group from an amino acid to a keto acid. Note that the keto acid to which the amino group is transferred becomes an amino acid. Cells can also use the nitrogen derived from amino groups to synthesize other important nitrogen-containing molecules, such as the purine and pyrimidine bases found in nucleic acids.

Figure 3–51 illustrates the oxidative deamination of the amino acid glutamic acid and the transamination of the amino acid alanine. Note that the keto acids formed are intermediates either in the Krebs cycle (α-ketoglutaric acid) or glycolytic pathway (pyruvic acid). Once formed, these keto acids can be metabolized to produce carbon dioxide and form ATP, or they can be used as intermediates in the synthetic pathway leading to the formation of glucose. As a third alternative, they can be used to synthesize fatty acids after their conversion to acetyl coenzyme A by way of pyruvic acid. Thus, amino acids can be used as a source of energy, and some can be converted into carbohydrate and fat.

The ammonia that oxidative deamination produces is highly toxic to cells if allowed to accumulate. Fortunately, it passes through cell membranes and enters the blood, which carries it to the liver. The liver contains enzymes that can combine two molecules of ammonia with carbon dioxide to form

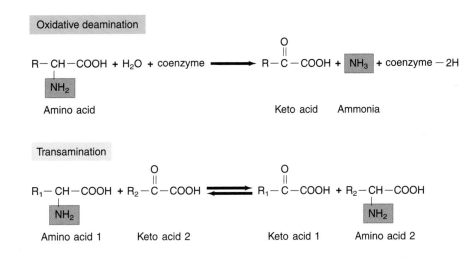

Figure 3–50

Oxidative deamination and transamination of amino acids.

urea. Thus, urea, which is relatively nontoxic, is the major nitrogenous waste product of protein catabolism. It enters the blood from the liver and is excreted by the kidneys into the urine.

Thus far, we have discussed mainly amino acid catabolism; now we turn to amino acid synthesis. The keto acids pyruvic acid and α-ketoglutaric acid can be derived from the breakdown of glucose; they can then be transaminated, as described previously, to form the amino acids glutamate and alanine. Thus glucose can be used to produce certain amino acids, provided other amino acids are available in the diet to supply amino groups for transamination. However, only 11 of the 20 amino acids can be formed by this process because 9 of the specific keto acids cannot be synthesized from other intermediates. We have to obtain the 9 amino acids corresponding to these keto acids from the food we eat, and they are thus known as **essential amino acids.**

Figure 3–52 provides a summary of the multiple routes by which the body handles amino acids. The amino acid pools, which consist of the body's total free amino acids, are derived from (1) ingested protein, which is degraded to amino acids during digestion in the intestinal tract, (2) the synthesis of nonessential amino acids from the keto acids derived from carbohydrates and fat, and (3) the continuous breakdown of body proteins. These pools are the source of amino acids for the resynthesis of body protein and a host of specialized amino acid derivatives, as well as for conversion to carbohydrate and fat. The body loses a very small quantity of amino acids and protein via the urine, skin, hair, fingernails, and, in women, the menstrual fluid. The major route for the loss of amino acids is not their excretion but rather their deamination, with the eventual excretion of the nitrogen atoms as urea in the urine. The terms **negative nitrogen balance** and **positive nitrogen balance** refer to whether there is a net loss or gain, respectively, of amino acids in the body over any period of time.

If any of the essential amino acids are missing from the diet, a negative nitrogen balance—that is, loss greater than gain—always results. The proteins that require a missing essential amino acid cannot be synthesized, and the other amino acids that would have been incorporated into these proteins are metabolized. This explains why a dietary requirement for protein cannot be specified without regard to the amino acid composition of that protein. Protein is graded in terms of how closely its relative proportions of essential amino acids approximate those in the average body protein. The highest quality proteins are found in animal products, whereas the quality of most plant proteins is lower. Nevertheless, it is quite possible to obtain adequate quantities of all essential amino acids from a mixture of plant proteins alone.

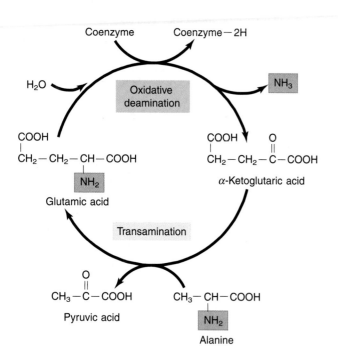

Figure 3–51

Oxidative deamination and transamination of the amino acids glutamic acid and alanine produce keto acids that can enter the carbohydrate pathways.

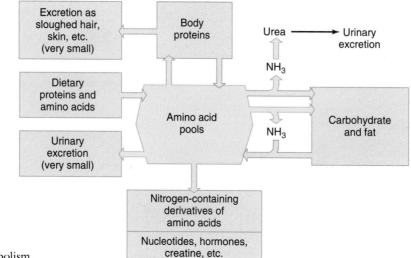

Figure 3–52

Pathways of amino acid metabolism.

Fuel Metabolism Summary

Having discussed the metabolism of the three major classes of organic molecules, we can now briefly review how each class is related to the others and to the process of synthesizing ATP. **Figure 3–53** illustrates the major pathways we have discussed and the relationships between the common intermediates. All three classes of molecules can enter the Krebs cycle through some intermediate, and thus all three can be used as a source of energy for the synthesis of ATP. Glucose can be converted into fat or into some amino acids by way of common intermediates such as pyruvate, oxaloacetate, and acetyl coenzyme A. Similarly, some amino acids can be converted into glucose and fat. Fatty acids cannot be converted into glucose because of the irreversibility of the reaction converting pyruvate to acetyl coenzyme A, but the glycerol portion of triglycerides can be converted into glucose. Fatty acids can be used to synthesize portions of the keto acids used to form some amino acids. Metabolism is thus a highly integrated process in which all classes of molecules can be used, if necessary, to provide energy, and in which each class of molecule can provide the raw materials required to synthesize most but not all members of other classes.

Essential Nutrients

About 50 substances required for normal or optimal body function cannot be synthesized by the body or are synthesized in amounts inadequate to keep pace with the rates at which they are broken down or excreted. Such substances are known as **essential nutrients** (**Table 3–12**). Because they are all removed from the body at some finite rate, they must be continually supplied in the foods we eat.

The term *essential nutrient* is reserved for substances that fulfill *two* criteria: (1) they must be essential for health, and (2) they must not be synthesized by the body in adequate amounts. Thus, glucose, although "essential" for normal metabolism, is not classified as an essential nutrient because the body normally can synthesize all it needs, from amino acids, for example. Furthermore, the quantity of an essential nutrient that must be present in the diet to maintain health is not a criterion for determining whether the substance is essential. Approximately 1500 g of water, 2 g of the amino acid methionine, but only about 1 mg of the vitamin thiamine are required per day.

Water is an essential nutrient because the body loses far more water in the urine and from the skin and respiratory tract than it can synthesize. (Recall that water forms as an end product of oxidative phosphorylation as well as from several other metabolic reactions.) Therefore, to maintain water balance, water intake is essential.

The mineral elements are examples of substances the body cannot synthesize or break down but that the body continually loses in the urine, feces, and various secretions. The major minerals must be supplied in fairly large amounts, whereas only small quantities of the trace elements are required.

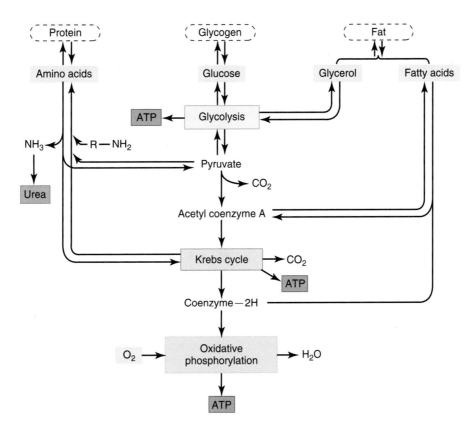

Figure 3–53

The relationships between the pathways for the metabolism of carbohydrate, fat, and protein.

Table 3–12 Essential Nutrients

Water

Mineral Elements

    7 major mineral elements (see Table 2–1)

    13 trace elements (see Table 2–1)

Essential Amino Acids

    Isoleucine

    Leucine

    Lysine

    Methionine

    Phenylalanine

    Threonine

    Tryptophan

    Tyrosine

    Valine

Essential Fatty Acids

    Linoleic

    Linolenic

Vitamins

    Water-soluble vitamins

        B_1: thiamine ⎫

        B_2: riboflavin ⎪

        B_6: pyridoxine ⎪

        B_{12}: cobalamine ⎬ Vitamin B complex

        Niacin ⎪

        Pantothenic acid ⎪

        Folic acid ⎪

        Biotin ⎭

        Lipoic acid

        Vitamin C

    Fat-soluble vitamins

        Vitamin A

        Vitamin D

        Vitamin E

        Vitamin K

Other Essential Nutrients

    Inositol

    Choline

    Carnitine

We have already noted that 9 of the 20 amino acids are essential. Two fatty acids, linoleic and linolenic acid, which contain a number of double bonds and serve important roles in chemical messenger systems, are also essential nutrients. Three additional essential nutrients—inositol, choline, and carnitine—have functions that will be described in later chap-

ters but do not fall into any common category other than being essential nutrients. Finally, the class of essential nutrients known as vitamins deserves special attention.

Vitamins

Vitamins are a group of 14 organic essential nutrients required in very small amounts in the diet. The exact chemical structures of the first vitamins to be discovered were unknown, and they were simply identified by letters of the alphabet. Vitamin B turned out to be composed of eight substances now known as the vitamin B complex. Plants and bacteria have the enzymes necessary for vitamin synthesis, and we get our vitamins by eating either plants or meat from animals that have eaten plants.

The vitamins as a class have no particular chemical structure in common, but they can be divided into the **water-soluble vitamins** and the **fat-soluble vitamins.** The water-soluble vitamins form portions of coenzymes such as NAD^+, FAD, and coenzyme A. The fat-soluble vitamins (A, D, E, and K) in general do not function as coenzymes. For example, vitamin A (retinol) is used to form the light-sensitive pigment in the eye, and lack of this vitamin leads to night blindness. The specific functions of each of the fat-soluble vitamins will be described in later chapters.

The catabolism of vitamins does not provide chemical energy, although some of them participate as coenzymes in chemical reactions that release energy from other molecules. Increasing the amount of a vitamin in the diet beyond a certain minimum does not necessarily increase the activity of those enzymes for which the vitamin functions as a coenzyme. Only very small quantities of coenzymes participate in the chemical reactions that require them, and increasing the concentration above this level does not increase the reaction rate.

The fate of large quantities of ingested vitamins varies depending upon whether the vitamin is water-soluble or fat-soluble. As the amount of water-soluble vitamins in the diet is increased, so is the amount excreted in the urine; thus the accumulation of these vitamins in the body is limited. On the other hand, fat-soluble vitamins can accumulate in the body because they are poorly excreted by the kidneys and because they dissolve in the fat stores in adipose tissue. The intake of very large quantities of fat-soluble vitamins can produce toxic effects.

■ SECTION E SUMMARY

Cellular Energy Transfer

I. The end products of glycolysis under aerobic conditions are ATP and pyruvate; the end products under anaerobic conditions are ATP and lactate.

  a. Carbohydrates are the only major fuel molecules that can enter the glycolytic pathway, and the enzymes that facilitate this pathway are located in the cytosol.

  b. During anaerobic glycolysis, hydrogen atoms are transferred to NAD^+, which then transfers them to pyruvate to form lactate, thus regenerating the original coenzyme molecule.

  c. The formation of ATP in glycolysis occurs by substrate-level phosphorylation, a process in which a phosphate group is transferred from a phosphorylated metabolic intermediate directly to ADP.

   d. During aerobic glycolysis, NADH + H⁺ transfers hydrogen atoms to the oxidative phosphorylation pathway.

II. The Krebs cycle catabolizes molecular fragments derived from fuel molecules and produces carbon dioxide, hydrogen atoms, and ATP. The enzymes that mediate the cycle are located in the mitochondrial matrix.

   a. Acetyl coenzyme A, the acetyl portion of which is derived from all three types of fuel molecules, is the major substrate entering the Krebs cycle. Amino acids can also enter at several places in the cycle by being converted to cycle intermediates.

   b. During one rotation of the Krebs cycle, two molecules of carbon dioxide are produced, and four pairs of hydrogen atoms are transferred to coenzymes. Substrate-level phosphorylation produces one molecule of GTP, which can be converted to ATP.

III. Oxidative phosphorylation forms ATP from ADP and P_i, using the energy released when molecular oxygen ultimately combines with hydrogen atoms to form water.

   a. The enzymes for oxidative phosphorylation are located on the inner membranes of mitochondria.

   b. Hydrogen atoms derived from glycolysis, the Krebs cycle, and the breakdown of fatty acids are delivered, most bound to coenzymes, to the electron transport chain. The electron transport chain then regenerates the hydrogen-free forms of the coenzymes NAD⁺ and FAD by transferring the hydrogens to molecular oxygen to form water.

   c. The reactions of the electron transport chain produce a hydrogen-ion gradient across the inner mitochondrial membrane. The flow of hydrogen ions back across the membrane provides the energy for ATP synthesis.

   d. Small amounts of reactive oxygen species, which can damage proteins, lipids, and nucleic acids, are formed during electron transport.

Carbohydrate, Fat, and Protein Metabolism

I. The aerobic catabolism of carbohydrates proceeds through the glycolytic pathway to pyruvate. Pyruvate enters the Krebs cycle and is broken down to carbon dioxide and to hydrogens, which are then transferred to coenzymes.

   a. About 40 percent of the chemical energy in glucose can be transferred to ATP under aerobic conditions; the rest is released as heat.

   b. Under aerobic conditions, 38 molecules of ATP can form from 1 molecule of glucose: 34 from oxidative phosphorylation, 2 from glycolysis, and 2 from the Krebs cycle.

   c. Under anaerobic conditions, 2 molecules of ATP can form from 1 molecule of glucose during glycolysis.

II. Carbohydrates are stored as glycogen, primarily in the liver and skeletal muscles.

   a. Different enzymes synthesize and break down glycogen. The control of these enzymes regulates the flow of glucose to and from glycogen.

   b. In most cells, glucose 6-phosphate is formed by glycogen breakdown and is catabolized to produce ATP. In liver and kidney cells, glucose can be derived from glycogen and released from the cells into the blood.

III. New glucose can be synthesized (gluconeogenesis) from some amino acids, lactate, and glycerol via the enzymes that catalyze reversible reactions in the glycolytic pathway. Fatty acids cannot be used to synthesize new glucose.

IV. Fat, stored primarily in adipose tissue, provides about 80 percent of the stored energy in the body.

   a. Fatty acids are broken down, two carbon atoms at a time, in the mitochondrial matrix by beta oxidation, to form acetyl coenzyme A and hydrogen atoms, which combine with coenzymes.

   b. The acetyl portion of acetyl coenzyme A is catabolized to carbon dioxide in the Krebs cycle, and the hydrogen atoms generated there, plus those generated during beta oxidation, enter the oxidative phosphorylation pathway to form ATP.

   c. The amount of ATP formed by the catabolism of 1 g of fat is about $2\frac{1}{2}$ times greater than the amount formed from 1 g of carbohydrate.

   d. Fatty acids are synthesized from acetyl coenzyme A by enzymes in the cytosol and are linked to α-glycerol phosphate, produced from carbohydrates, to form triglycerides by enzymes in the smooth endoplasmic reticulum.

V. Proteins are broken down to free amino acids by proteases.

   a. The removal of amino groups from amino acids leaves keto acids, which can either be catabolized via the Krebs cycle to provide energy for the synthesis of ATP or converted into glucose and fatty acids.

   b. Amino groups are removed by (1) oxidative deamination, which gives rise to ammonia, or by (2) transamination, in which the amino group is transferred to a keto acid to form a new amino acid.

   c. The ammonia formed from the oxidative deamination of amino acids is converted to urea by enzymes in the liver and then excreted in the urine by the kidneys.

VI. Some amino acids can be synthesized from keto acids derived from glucose, whereas others cannot be synthesized by the body and must be provided in the diet.

Essential Nutrients

I. Approximately 50 essential nutrients are necessary for health but cannot be synthesized in adequate amounts by the body and must therefore be provided in the diet.

II. A large intake of water-soluble vitamins leads to their rapid excretion in the urine, whereas large intakes of fat-soluble vitamins lead to their accumulation in adipose tissue and may produce toxic effects.

■ SECTION E KEY TERMS

α-glycerol phosphate 88	hydrogen peroxide 84
acetyl coenzyme A (acetyl CoA) 80	hydroxyl radical 84
	keto acid 89
adipocyte 87	Krebs cycle 79
adipose tissue 87	lactate 79
aerobic 79	negative nitrogen balance 90
anaerobic 79	oxidative deamination 89
beta oxidation 87	oxidative phosphorylation 82
chemiosmotic hypothesis 83	positive nitrogen balance 90
citric acid cycle 79	protease 89
cytochrome 83	proteolysis 89
electron transport chain 83	pyruvate 78
essential amino acid 90	substrate-level phosphorylation 79
essential nutrient 91	
fat-soluble vitamin 92	superoxide anion 84
gluconeogenesis 86	transamination 89
glycogen 86	tricarboxylic acid cycle 79
glycogenolysis 86	urea 90
glycolysis 78	water-soluble vitamin 92

1. What are the end products of glycolysis under aerobic and anaerobic conditions?
2. What are the major substrates entering the Krebs cycle, and what are the products formed?
3. Why does the Krebs cycle operate only under aerobic conditions even though it doesn't use molecular oxygen in any of its reactions?
4. Identify the molecules that enter the oxidative phosphorylation pathway and the products that form.
5. Where are the enzymes for the Krebs cycle located? The enzymes for oxidative phosphorylation? The enzymes for glycolysis?
6. How many molecules of ATP can form from the breakdown of one molecule of glucose under aerobic conditions? Under anaerobic conditions?
7. Describe the origin and effects of reactive oxygen molecules.
8. What molecules can be used to synthesize glucose?
9. Why can't fatty acids be used to synthesize glucose?
10. Describe the pathways used to catabolize fatty acids to carbon dioxide.
11. Why is it more efficient to store fuel as fat than as glycogen?
12. Describe the pathway by which glucose is converted into fat.
13. Describe the two processes by which amino groups are removed from amino acids.
14. What can keto acids be converted into?
15. What is the source of the nitrogen atoms in urea, and in what organ is urea synthesized?
16. Why is water considered an essential nutrient whereas glucose is not?
17. What is the consequence of ingesting large quantities of water-soluble vitamins? Fat-soluble vitamins?

Chapter 3 Test Questions

(Answers appear in Appendix A.)

1. Which cell structure contains the enzymes required for oxidative phosphorylation?
   a. mitochondria
   b. smooth endoplasmic reticulum
   c. rough endoplasmic reticulum
   d. endosomes
   e. peroxisomes

2. Which sequence regarding protein synthesis is correct?
   a. translation $\longrightarrow$ transcription $\longrightarrow$ mRNA synthesis
   b. transcription $\longrightarrow$ splicing of primary RNA transcript $\longrightarrow$ translocation of mRNA $\longrightarrow$ translation
   c. splicing of introns $\longrightarrow$ transcription $\longrightarrow$ mRNA synthesis translation
   d. transcription $\longrightarrow$ translation $\longrightarrow$ mRNA production
   e. tRNA enters nucleus $\longrightarrow$ transcription begins $\longrightarrow$ mRNA moves to cytoplasm $\longrightarrow$ protein synthesis begins

3. Which is *incorrect* regarding ligand:protein binding reactions?
   a. Allosteric modulation of the protein's binding site occurs directly at the binding site itself.
   b. Allosteric modulation can alter the affinity of the protein for the ligand.
   c. Phosphorylation of the protein is an example of covalent modulation.
   d. If two ligands can bind to the binding site of the protein, competition for binding will occur.
   e. Binding reactions are either electrical or hydrophobic in nature.

4. According to the law of mass action, in the following reaction:

$$CO_2 + H_2O \rightleftharpoons H_2CO_3$$

   a. Increasing the concentration of carbon dioxide will slow down the forward (left-to-right) reaction.
   b. Increasing the concentration of carbonic acid will accelerate the rate of the reverse (right-to-left) reaction.
   c. Increasing the concentration of carbon dioxide will speed up the reverse reaction.
   d. Decreasing the concentration of carbonic acid will slow down the forward reaction.
   e. No enzyme is required for either the forward or reverse reaction.

5. Which of the following can be converted to glucose by gluconeogenesis in the liver?
   a. fatty acid   d. ATP
   b. triglyceride   e. glycogen
   c. glycerol

6. Which of the following is true?
   a. Triglycerides have the least energy content per gram of the three major fuel sources in the body.
   b. Fat catabolism generates new triglycerides for storage in adipose tissue.
   c. By mass, the total body content of carbohydrates exceeds that of total triglycerides.
   d. Catabolism of fatty acids occurs in two-carbon steps.
   e. Triglycerides are the major lipids found in plasma membranes.

Chapter 3 Quantitative and Thought Questions

(Answers appear in Appendix A.)

1. A base sequence in a portion of one strand of DNA is
   A—G—T—G—C—A—A—G—T—C—T. Predict:
   a. the base sequence in the complementary strand of DNA.
   b. the base sequence in RNA transcribed from the sequence shown.

2. The triplet code in DNA for the amino acid histidine is
   G—T—A. Predict the mRNA codon for this amino acid and the tRNA anticodon.

3. If a protein contains 100 amino acids, how many nucleotides will be present in the gene that codes for this protein?

4. A variety of chemical messengers that normally regulate acid secretion in the stomach bind to proteins in the plasma membranes of the acid-secreting cells. Some of these binding reactions lead to increased acid secretion, others to decreased secretion. In what ways might a drug that causes decreased acid secretion be acting on these cells?

5. In one type of diabetes, the plasma concentration of the hormone insulin is normal, but the response of the cells that insulin usually binds to is markedly decreased. Suggest a reason for this in terms of the properties of protein binding sites.

6. The following graph shows the relation between the amount of acid secreted and the concentration of compound X, which stimulates acid secretion in the stomach by binding to a membrane protein. At a plasma concentration of 2 pM, compound X produces an acid secretion of 20 mmol/h.

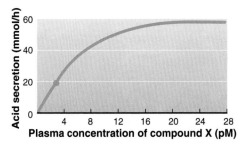

   a. Specify two ways in which acid secretion by compound X could be increased to 40 mmol/h.
   b. Why will increasing the concentration of compound X to 28 pM fail to produce more acid secretion than increasing the concentration of X to 18 pM?

7. In the following metabolic pathway, what is the rate of formation of the end product E if substrate A is present at a saturating concentration? The maximal rates (products formed per second) of the individual steps are indicated.

$$A \xrightarrow{30} B \xrightarrow{5} C \xrightarrow{20} D \xrightarrow{40} E$$

8. If the concentration of oxygen in the blood delivered to a muscle is increased, what effect will it have on the muscle's rate of ATP production?

9. During prolonged starvation, when glucose is not being absorbed from the gastrointestinal tract, what molecules can be used to synthesize new glucose?

10. Why do certain forms of liver disease produce an increase in the blood levels of ammonia?

Movement of Molecules Across Cell Membranes

Changes in red blood cell shape due to osmosis.

as we have seen, the contents of a cell are separated from the surrounding extracellular fluid by a thin layer of lipids and protein—the plasma membrane. In addition, membranes associated with mitochondria, endoplasmic reticulum, lysosomes, the Golgi apparatus, and the nucleus divide the intracellular fluid into several membrane-bound compartments. The movements of molecules and ions both between the various cell organelles and the cytosol, and between the cytosol and the extracellular fluid, depend on the properties of these membranes. The rates at which different substances move through membranes vary considerably and in some cases can be controlled—increased or decreased—in response to various signals. This chapter focuses upon the transport functions of membranes, with emphasis on the plasma membrane. There are several mechanisms by which substances pass through membranes, and we begin our discussion of these mechanisms with the process known as diffusion.

Diffusion

The molecules of any substance, be it solid, liquid, or gas, are in a continuous state of movement or vibration, and the warmer a substance is, the faster its molecules move. The average speed of this "thermal motion" also depends upon the mass of the molecule. At body temperature, a molecule of water moves at about 2500 km/h (1500 mi/h), whereas a molecule of glucose, which is 10 times heavier, moves at about 850 km/h. In solutions, such rapidly moving molecules cannot travel very far before colliding with other molecules, undergoing millions of collisions every second. Each collision alters the direction of the molecule's movement, so that the path of any one molecule becomes unpredictable. Because a molecule may at any instant be moving in any direction, such movement is random, with no preferred direction of movement.

The random thermal motion of molecules in a liquid or gas will eventually distribute them uniformly throughout a container. Thus, if we start with a solution in which a solute is more concentrated in one region than another (**Figure 4–1a**), random thermal motion will redistribute the solute from regions of higher concentration to regions of lower concentration until the solute reaches a uniform concentration throughout the solution (**Figure 4–1b**). This movement of molecules from one location to another solely as a result of their random thermal motion is known as **diffusion.**

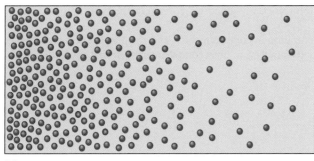

(a)

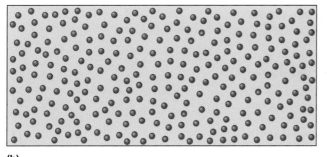

(b)

Figure 4–1

Diffusion. (a) Molecules initially concentrated in one region of a solution will, due to their random thermal motion, undergo a net diffusion from the region of higher concentration to the region of lower concentration. (b) With time, the molecules will become uniformly distributed throughout the solution.

Many processes in living organisms are closely associated with diffusion. For example, oxygen, nutrients, and other molecules enter and leave the smallest blood vessels (capillaries) by diffusion, and the movement of many substances across plasma membranes and organelle membranes occurs by diffusion.

Magnitude and Direction of Diffusion

Figure 4–2 illustrates the diffusion of glucose between two compartments of equal volume separated by a permeable barrier. Initially, glucose is present in compartment 1 at a concentration of 20 mmol/L, and there is no glucose in compartment 2. The random movements of the glucose molecules in compartment 1 carry some of them into compartment 2. The amount of material crossing a surface in a unit of time is known as a **flux.** This one-way flux of glucose from compartment 1 to compartment 2 depends on the concentration of glucose in compartment 1. If the number of molecules in a unit of volume is doubled, the flux of molecules across each surface of the unit will also be doubled because twice as many molecules will be moving in any direction at a given time.

After a short time, some of the glucose molecules that have entered compartment 2 will randomly move back into compartment 1 (see Figure 4–2, time B). The magnitude of the glucose flux from compartment 2 to compartment 1 depends upon the concentration of glucose in compartment 2 at any time.

The **net flux** of glucose between the two compartments at any instant is the difference between the two one-way fluxes. It is the net flux that determines the net gain of molecules in compartment 2 and the net loss from compartment 1.

Eventually the concentrations of glucose in the two compartments become equal at 10 mmol/L. The two one-way fluxes are then equal in magnitude but opposite in direction, and the net flux of glucose is zero (see Figure 4–2, time C). The system has now reached **diffusion equilibrium.** No further change in the glucose concentrations of the two compartments will occur because equal numbers of glucose molecules will continue to diffuse in both directions between the two compartments.

Several important properties of diffusion can be emphasized using this example. Three fluxes can be identified at any surface—the two one-way fluxes occurring in opposite directions from one compartment to the other, and the net flux, which is the difference between them (**Figure 4–3**). The net flux is the most important component in diffusion because it is the net amount of material transferred from one location to another. Although the movement of individual molecules is random, *the net flux always proceeds from regions of higher concentration to regions of lower concentration.* For this reason, we often say that substances move "downhill" by diffusion. The greater the difference in concentration between any two regions, the greater the magnitude of the net flux. Thus, the concentration difference determines both the direction and the magnitude of the net flux.

At any concentration difference, however, the magnitude of the net flux depends on several additional factors: (1) temperature—the higher the temperature, the greater the

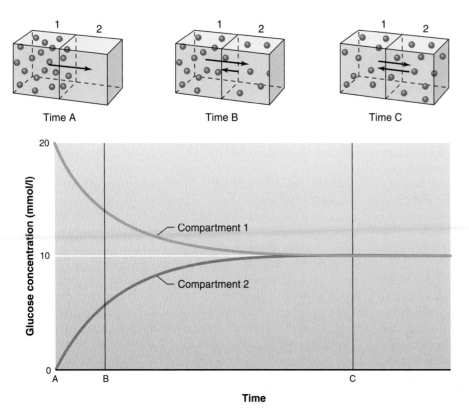

Time A   Time B   Time C

Figure 4–2 physiological *inquiry* (pi)

■ If at time C, additional glucose was added to compartment 1 such that its concentration was instantly increased to 15 mmol/L, what would the graph look like following time C? Draw the new graph on the figure and indicate the glucose concentrations in compartments 1 and 2 at diffusion equilibrium. (*Note:* It is not actually possible to instantly change the concentration of a substance in this way because it will immediately begin diffusing to the other compartment as it is added.)

Answer can be found at end of chapter.

Figure 4–2

Diffusion of glucose between two compartments of equal volume separated by a barrier permeable to glucose. Initially, time A, compartment 1 contains glucose at a concentration of 20 mmol/L, and no glucose is present in compartment 2. At time B, some glucose molecules have moved into compartment 2, and some of these are moving back into compartment 1. The length of the arrows represents the magnitudes of the one-way movements. At time C, diffusion equilibrium has been reached, the concentrations of glucose are equal in the two compartments (10 mmol/l), and the *net* movement is zero.

In the graph at the bottom of the figure, the green line represents the concentration in compartment 1, and the purple line represents the concentration in compartment 2. Note that at time C, glucose concentration is 10 mmol/L in both compartments. At that time, diffusion equilibrium has been reached.

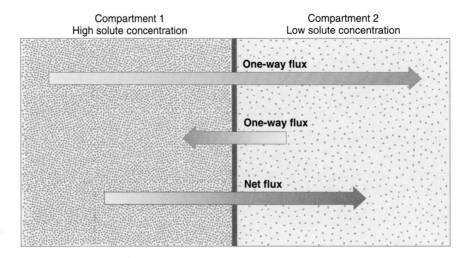

Figure 4–3

The two one-way fluxes occurring during the diffusion of solute across a boundary and the net flux, which is the difference between the two one-way fluxes. The net flux always occurs in the direction from higher to lower concentration. The length of the arrows indicates the magnitude of the flux.

speed of molecular movement and the greater the net flux; (2) mass of the molecule—large molecules such as proteins have a greater mass and lower speed than smaller molecules such as glucose, and thus have a smaller net flux; (3) surface area—the greater the surface area between two regions, the greater the space available for diffusion and thus the greater the net flux; and (4) the medium through which the molecules are moving—molecules diffuse more rapidly in air than in water because collisions are less frequent in a gas phase, and, as we will see, when a membrane is involved, its chemical composition influences diffusion rates.

Diffusion Rate Versus Distance

The distance over which molecules diffuse is an important factor in determining the rate at which they can reach a cell from the blood or move throughout the interior of a cell after crossing the plasma membrane. Although individual molecules travel at high speeds, the number of collisions they undergo prevents them from traveling very far in a straight line. Diffusion times increase in proportion to the *square* of the distance over which the molecules diffuse. For example, it takes glucose only a few seconds to reach diffusion equilibrium at a point 10 µm away from a source of glucose, but it would take over 11 years to reach the same concentration at a point 10 cm away from the source.

Thus, although diffusion equilibrium can be reached rapidly over distances of cellular dimensions, it takes a very long time when distances of a few centimeters or more are involved. For an organism as large as a human being, the diffusion of oxygen and nutrients from the body surface to tissues located only a few centimeters below the surface would be far too slow to provide adequate nourishment. Accordingly, the circulatory system provides the mechanism for rapidly moving materials over large distances using a pressure source (the heart). This process, known as bulk flow, is described in Chapter 12. Diffusion, on the other hand, provides movement over the short distance between the blood and tissue cells.

The rate at which diffusion can move molecules *within* a cell is one of the reasons cells must be small. A cell would not have to be very large before diffusion failed to provide sufficient nutrients to its central regions. For example, the center of a 20-µm diameter cell reaches diffusion equilibrium with extracellular oxygen in about 15 ms, but it would take 265 days to reach equilibrium at the center of a cell the size of a basketball.

Diffusion Through Membranes

The rate at which a substance diffuses across a plasma membrane can be measured by monitoring the rate at which its intracellular concentration approaches diffusion equilibrium with its concentration in the extracellular fluid. For simplicity's sake, let us assume that because the volume of extracellular fluid is large, its solute concentration will remain essentially constant as the substance diffuses into the small intracellular volume (**Figure 4–4**). As with all diffusion processes, the net flux, J, of material across the membrane is from the region of higher concentration (the extracellular solution in this case) to the region of lower concentration (the intracellular fluid).

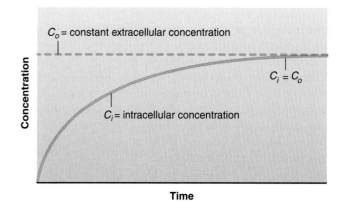

Figure 4–4

The increase in intracellular concentration as a solute diffuses from a constant extracellular concentration until diffusion equilibrium ($C_i = C_o$) is reached across the plasma membrane of a cell.

The magnitude of the net flux is directly proportional to the difference in concentration across the membrane ($C_o - C_i$), the surface area of the membrane A, and the membrane **permeability coefficient P:**

$$J = PA(C_o - C_i)$$

The numerical value of the permeability coefficient P is an experimentally determined number for a particular type of molecule at a given temperature, and it reflects the ease with which the molecule is able to move through a given membrane. In other words, the greater the permeability coefficient, the larger the net flux across the membrane for any given concentration difference and membrane surface area.

The rates at which molecules diffuse across membranes, as measured by their permeability coefficients, are a thousand to a million times slower than the diffusion rates of the same molecules through a water layer of equal thickness. Membranes, therefore, act as barriers that considerably slow the diffusion of molecules across their surfaces. The major factor limiting diffusion across a membrane is the hydrophobic interior of its lipid bilayer.

Diffusion Through the Lipid Bilayer

When the permeability coefficients of different organic molecules are examined in relation to their molecular structures, a correlation emerges. Whereas most polar molecules diffuse into cells very slowly or not at all, nonpolar molecules diffuse much more rapidly across plasma membranes—that is, they have large permeability constants. The reason is that nonpolar molecules can dissolve in the nonpolar regions of the membrane occupied by the fatty acid chains of the membrane phospholipids. In contrast, polar molecules have a much lower solubility in the membrane lipids. Increasing the lipid solubility of a substance by decreasing the number of polar or ionized groups it contains, will increase the number of molecules dissolved in the membrane lipids. This will increase the flux of the substance across the membrane. Oxygen, carbon dioxide, fatty acids, and steroid hormones are examples of nonpolar molecules that diffuse rapidly through the lipid portions of membranes. Most of the organic molecules that make up the intermediate stages of the various metabolic pathways (Chapter 3) are ionized or polar molecules, often containing an ionized phosphate group, and thus have a low solubility in the lipid bilayer. Most of these substances are retained within cells and organelles because they cannot diffuse across the lipid barrier of membranes.

Diffusion of Ions Through Protein Channels

Ions such as Na^+, K^+, Cl^-, and Ca^{2+} diffuse across plasma membranes at much faster rates than would be predicted from their very low solubility in membrane lipids. Moreover, different cells have quite different permeabilities to these ions, whereas nonpolar substances have similar permeabilities in different cells. The fact that artificial lipid bilayers containing no protein are practically impermeable to these ions indicates that the protein component of the membrane is responsible for these permeability differences.

As we have seen (Chapter 3), integral membrane proteins can span the lipid bilayer. Some of these proteins form **channels** that allow ions to diffuse across the membrane. A single protein may have a conformation similar to that of a doughnut, with the hole in the middle providing the channel for ion movement. More often, several proteins aggregate, each forming a subunit of the walls of a channel (**Figure 4–5**). The diameters of protein channels are very small, only slightly larger than those of the ions that pass through them. The small size of the channels prevents larger, polar, organic molecules from entering or leaving.

An important characteristic of ion channels is that they show a selectivity for the type of ion that can diffuse through them. This selectivity is based on the channel diameter, the charged and polar surfaces of the protein subunits that form the channel walls and electrically attract or repel the ions, and on the number of water molecules associated with the ions (so-called waters of hydration). For example, some channels (K^+ channels) allow only potassium ions to pass, while others are specific for sodium (Na^+ channels). For this reason, two membranes that have the same permeability to potassium because they have the same number of K^+ channels may have quite different permeabilities to sodium if they contain different numbers of Na^+ channels.

Role of Electrical Forces on Ion Movement

Thus far we have described the direction and magnitude of solute diffusion across a membrane in terms of the solute's concentration difference across the membrane, its solubility in the membrane lipids, the presence of membrane ion channels, and the area of the membrane. When describing the diffusion of ions, since they are charged, one additional factor must be considered: the presence of electrical forces acting upon the ions.

A separation of electrical charge exists across plasma membranes. This is known as a **membrane potential** (**Figure 4–6**), the origin of which will be described in Chapter 6 in the context of nerve cell function. The membrane potential provides an electrical force that influences the movement of ions across the membrane. Electrical charges of the same sign, both positive or both negative, repel each other, while opposite charges attract. For example, if the inside of a cell has a net negative charge with respect to the outside, as is true in most cells, there will be an electrical force attracting positive ions into the cell and repelling negative ions. Even if there were no difference in ion concentration across the membrane, there would still be a net movement of positive ions into and negative ions out of the cell because of the membrane potential. Thus, the direction and magnitude of ion fluxes across membranes depend on both the concentration difference *and* the electrical difference (the membrane potential). These two driving forces are collectively known as the **electrochemical gradient** across a membrane.

It is important to recognize that the two forces that make up the electrochemical gradient may oppose each other. Thus, the membrane potential may be driving potassium ions, for example, in one direction across the membrane, while the concentration difference for potassium is driving these ions in the opposite direction. The net movement of potassium in this case would be determined by the relative magnitudes of the two opposing forces—that is, by the electrochemical gradient across the membrane.

Regulation of Diffusion Through Ion Channels

Ion channels can exist in an open or closed state (**Figure 4–7**), and changes in a membrane's permeability to ions can occur rapidly as these channels open or close. The process of opening and closing ion channels is known as **channel gating,** like the opening and closing of a gate in a fence. A single ion channel may open and close many times each second, suggesting that the channel protein fluctuates between two (or more) conformations. Over an extended period of time, at any given electrochemical gradient, the total number of ions that pass through a channel depends on how often the channel opens and how long it stays open.

Three factors can alter the channel protein conformations, producing changes in how long or how often a channel opens. First, the binding of specific molecules to channel proteins may directly or indirectly produce either an allosteric or covalent change in the shape of the channel protein. Such channels are termed **ligand-gated channels,** and the ligands that influence them are often chemical messengers. Second, changes in the membrane potential can cause movement of the charged regions on a channel protein, altering its shape—these are **voltage-gated channels.** Third, physically deforming (stretching) the membrane may affect the conformation of some channel proteins—these are **mechanically-gated channels.**

A particular type of ion may pass through several different types of channels. For example, a membrane may contain ligand-gated K^+ channels, voltage-gated K^+ channels, and mechanically-gated K^+ channels. Moreover, the same membrane may have several types of voltage-gated K^+ channels, each responding to a different range of membrane voltage, or several types of ligand-gated K^+ channels, each responding to a different chemical messenger. The roles of these gated channels in cell communication and electrical activity will be discussed in Chapters 5 through 7.

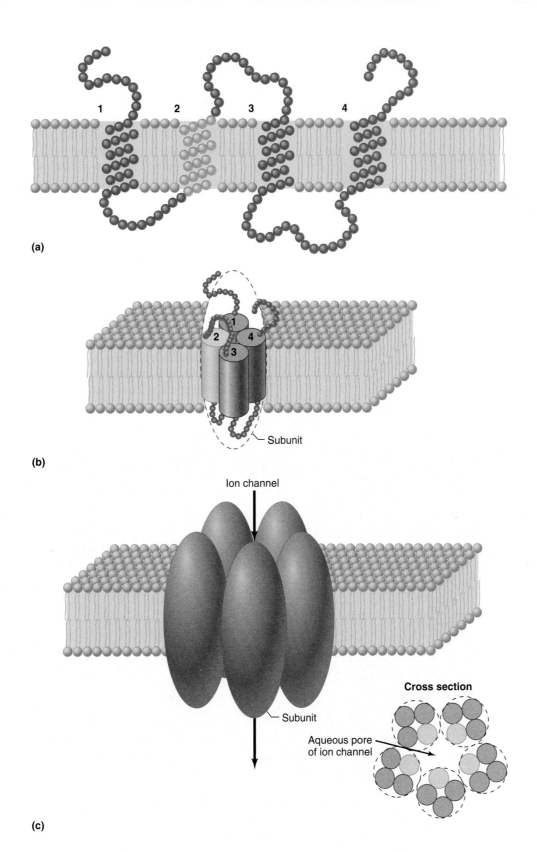

Figure 4–5

Model of an ion channel composed of five polypeptide subunits. (a) A channel subunit consisting of an integral membrane protein containing four transmembrane segments (1, 2, 3, and 4), each of which has an alpha helical configuration within the membrane. Although this model has only four transmembrane segments, some channel proteins have as many as 12. (b) The same subunit as in (a) shown in three dimensions within the membrane, with the four transmembrane helices aggregated together. (c) The ion channel consists of five of the subunits illustrated in (b), which form the sides of the channel. As shown in cross section, the helical transmembrane segment 2 (light purple) of each subunit forms each side of the channel opening. The presence of ionized amino acid side chains along this region determines the selectivity of the channel to ions. Although this model shows the five subunits as identical, many ion channels are formed from the aggregation of several different types of subunit polypeptides.

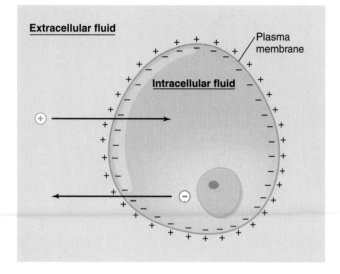

Figure 4–6

The separation of electrical charge across a plasma membrane (the membrane potential) provides the electrical force that drives positive ions into a cell and negative ions out.

and the nondiffusional movements of ions are mediated by integral membrane proteins known as **transporters** (or carriers). The movement of substances through a membrane by these **mediated-transport** systems depends on conformational changes in these transporters.

The transported solute must first bind to a specific site on a transporter, a site exposed to the solute on one surface of the membrane (**Figure 4–8**). A portion of the transporter then undergoes a change in shape, exposing this same binding site to the solution on the opposite side of the membrane. The dissociation of the substance from the transporter binding site completes the process of moving the material through the membrane. Using this mechanism, molecules can move in either direction, getting on the transporter on one side and off at the other. The diagram of the transporter in Figure 4–8 is only a model, because the specific conformational changes of any transport protein are still uncertain.

Many of the characteristics of transporters and ion channels are similar. Both involve membrane proteins and show chemical specificity. They do, however, differ in the number of molecules or ions crossing the membrane by way of these membrane proteins. Ion channels typically move several thousand times more ions per unit time than do transporters. In part, this is because a transporter must change its shape for each molecule transported across the membrane, while an open ion channel can support a continuous flow of ions without a change in conformation.

There are many types of transporters in membranes, each type having binding sites that are specific for a particular substance or a specific class of related substances. For example, although both amino acids and sugars undergo mediated transport, a protein that transports amino acids does not transport sugars, and vice versa. Just as with ion channels, the

Mediated-Transport Systems

Although diffusion through channels accounts for some of the transmembrane movement of ions, it does not account for all. Moreover, a number of other molecules, including amino acids and glucose, are able to cross membranes yet are too polar to diffuse through the lipid bilayer and too large to diffuse through ion channels. The passage of these molecules

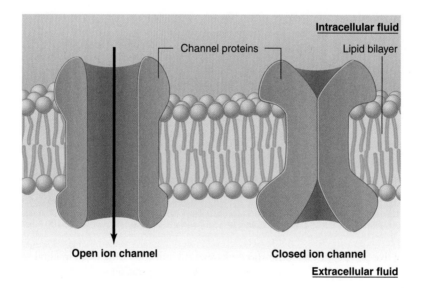

Figure 4–7

As a result of conformational changes in the proteins forming an ion channel, the channel may be open, allowing ions to diffuse across the membrane, or may be closed. The conformational change is grossly exaggerated for illustrative purposes. The actual conformational change is more likely to be just sufficient to allow or prevent an ion to fit through.

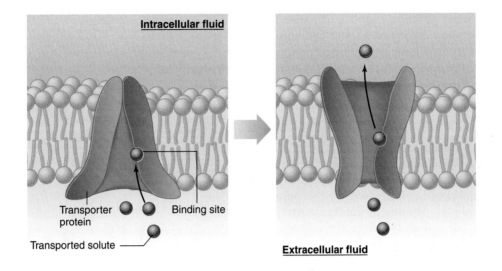

Figure 4–8

Model of mediated transport. A change in the conformation of the transporter exposes the transporter binding site first to one surface of the membrane then to the other, thereby transferring the bound solute from one side of the membrane to the other. This model shows net mediated transport from the extracellular fluid to the inside of the cell. In many cases, the net transport is in the opposite direction. The size of the conformational change is exaggerated for illustrative purposes in this and subsequent figures.

plasma membranes of different cells contain different types and numbers of transporters, and thus they exhibit differences in the types of substances transported and in their rates of transport.

Three factors determine the magnitude of solute flux through a mediated-transport system. The first of these is the extent to which the transporter binding sites are saturated, which depends on both the solute concentration and the affinity of the transporters for the solute. Second, the number of transporters in the membrane determines the flux at any level of saturation. The third factor is the rate at which the conformational change in the transport protein occurs. The flux through a mediated-transport system can be altered by changing any of these three factors.

For any transported solute there is a finite number of specific transporters in a given membrane at any particular moment. As with any binding site, as the concentration of the solute to be transported is increased, the number of occupied binding sites increases until the transporters become saturated—that is, until all the binding sites are occupied. When the transporter binding sites are saturated, the maximal flux across the membrane has been reached, and no further increase in solute flux will occur with increases in solute concentration. Contrast the solute flux resulting from mediated transport with the flux produced by diffusion through the lipid portion of a membrane (**Figure 4–9**). The flux due to diffusion increases in direct proportion to the increase in extracellular concentration, and there is no limit because diffusion does not involve binding to a fixed number of sites. (At very high ion concentrations, however, diffusion through ion channels may approach a limiting value because of the fixed number of channels available, just as there is an upper limit to the rate at which a crowd of people can pass through a single open doorway.)

When transporters are saturated, however, the maximal transport flux depends upon the rate at which the conformational changes in the transporters can transfer their binding sites from one surface to the other. This rate is much slower than the rate of ion diffusion through ion channels.

Thus far, we have described mediated transport as though all transporters had similar properties. In fact, two types of mediated transport exist—**facilitated diffusion** and

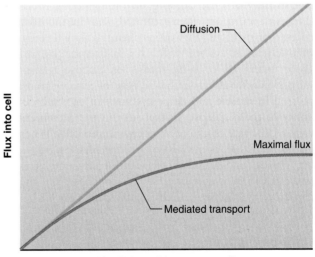

Figure 4–9

The flux of molecules diffusing into a cell across the lipid bilayer of a plasma membrane (green line) increases continuously in proportion to the extracellular concentration, whereas the flux of molecules through a mediated transport system (purple line) reaches a maximal value.

active transport. Facilitated diffusion uses a transporter to move solute "downhill" from a higher to a lower concentration across a membrane, as in Figure 4–8. Active transport uses a transporter coupled to an energy source to move solute "uphill" across a membrane—that is, *against* its electrochemical gradient.

Facilitated Diffusion

As in ordinary diffusion, in facilitated diffusion the net flux of a molecule across a membrane always proceeds from higher to lower concentration and continues until the concentrations on the two sides of the membrane become equal. At this point, equal numbers of molecules are binding to the transporter at the outer surface of the cell and moving into the cell as are binding at the inner surface and moving out. Neither diffusion nor facilitated diffusion is coupled to energy (ATP) derived from metabolism. Thus, they are incapable of moving solute from a lower to a higher concentration across a membrane.

Among the most important facilitated-diffusion systems in the body are those that move glucose across plasma membranes. Without such glucose transporters, cells would be virtually impermeable to glucose, a relatively large, polar molecule. It might be expected that as a result of facilitated diffusion the glucose concentration inside cells would become equal to the extracellular concentration. This does not occur in most cells, however, because glucose is metabolized to glucose 6-phosphate almost as quickly as it enters. Thus, the intracellular glucose concentration remains lower than the extracellular concentration, and there is a continuous net flux of glucose into cells.

Several distinct transporters are known to mediate the facilitated diffusion of glucose across cell membranes. Each transporter is coded by a different gene, and these genes are expressed in different types of cells. The transporters differ in the affinity of their binding sites for glucose, their maximal rates of transport when saturated, and the modulation of their transport activity by various chemical signals, such as the hormone insulin. As you will learn in Chapter 16, although glucose enters all cells by means of glucose transporters, insulin primarily affects only the type of glucose transporter expressed in muscle and adipose tissue. Insulin increases the number of these glucose transporters in the membrane and, therefore, the rate of glucose movement into cells. When insulin is not available, as in one type of the disease *diabetes mellitus,* muscle and adipose cells cannot efficiently transport glucose across their membranes. This contributes to the accumulation of glucose in the extracellular fluid that is a hallmark of the disease.

Active Transport

Active transport differs from facilitated diffusion in that it uses energy to move a substance *uphill* across a membrane—that is, against the substance's electrochemical gradient (**Figure 4–10**). As with facilitated diffusion, active transport requires a substance to bind to the transporter in the membrane. Because these transporters move the substance *uphill,* they are often referred to as pumps. As with facilitated-diffusion transporters, active-transport transporters exhibit specificity and

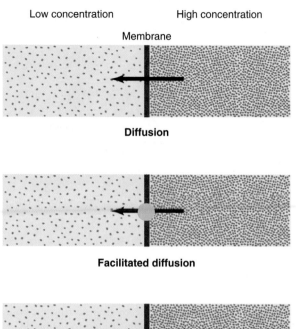

Low concentration   High concentration

Membrane

Diffusion

Facilitated diffusion

Active transport

Figure 4–10

Direction of net solute flux crossing a membrane by: diffusion (high to low concentration), facilitated diffusion (high to low concentration), and active transport (low to high concentration). The colored circles represent transporter molecules.

saturation—that is, the flux via the transporter is maximal when all transporter binding sites are occupied.

The net movement from lower to higher concentration and the maintenance of a higher steady-state concentration on one side of a membrane can be achieved only by the continuous input of energy into the active-transport process. Therefore, active transport must be coupled to the simultaneous flow of some energy source from a higher energy level to a lower energy level. Two means of coupling an energy flow to transporters are known: (1) the direct use of ATP in **primary active transport,** and (2) the use of an electrochemical gradient across a membrane to drive the process in **secondary active transport.**

Primary Active Transport

The hydrolysis of ATP by a transporter provides the energy for primary active transport. The transporter is actually an enzyme called an ATPase that catalyzes the breakdown of ATP and, in the process, phosphorylates itself. Phosphorylation of the transporter protein is a type of covalent modulation that changes the conformation of the transporter and the affinity of the transporter's solute binding site.

One of the best studied examples of primary active transport is the movement of sodium and potassium ions across plasma membranes by the Na$^+$/K$^+$-ATPase pump. This transporter, which is present in all cells, moves sodium ions from intracellular to extracellular fluid, and potassium ions in the opposite direction. In both cases, the movements of the ions are against their respective concentration gradients. **Figure 4–11** illustrates the sequence the Na$^+$/K$^+$-ATPase pump is believed to use to transport these two ions in opposite directions. (1) Initially the transporter, with an associated molecule of ATP, binds three sodium ions at high-affinity sites on the intracellular surface of the protein. Two binding sites also exist for K$^+$, but at this stage they are in a low-affinity state and thus do not bind intracellular K$^+$. (2) Binding of Na$^+$ results in activation of an inherent ATPase activity of the transporter protein, causing phosphorylation of the cytosolic surface of the transporter and releasing a molecule of ADP. (3) Phosphorylation results in a conformational change of the transporter, exposing the bound sodium ions to the extracellular fluid, and at the same time reducing the affinity of the binding sites for sodium. The sodium ions are released from their binding sites. (4) The new conformation of the transporter results in an increased affinity of the two binding sites for K$^+$, allowing two molecules of K$^+$ to bind to the transporter on the extracellular surface. (5) Binding of K$^+$ results in dephosphorylation of the transporter. This returns the transporter to its original conformation, resulting in reduced affinity of the K$^+$ binding sites and increased affinity of the Na$^+$ binding sites. K$^+$ is therefore released into the intracellular fluid, allowing new molecules of Na$^+$ (and ATP) to be bound at the intracellular surface.

The pumping activity of the Na$^+$/K$^+$-ATPase primary active transporter establishes and maintains the characteristic distribution of high intracellular potassium and low intracellular sodium relative to their respective extracellular concentrations (**Figure 4–12**). For each molecule of ATP hydrolyzed, this transporter moves three sodium ions out of a cell, and two potassium ions into a cell. This results in a net transfer of positive charge to the outside of the cell, and thus this transport process is not electrically neutral, a point that will be described in detail in Chapter 6 when we consider the electrical charge across plasma membranes of nerve cells.

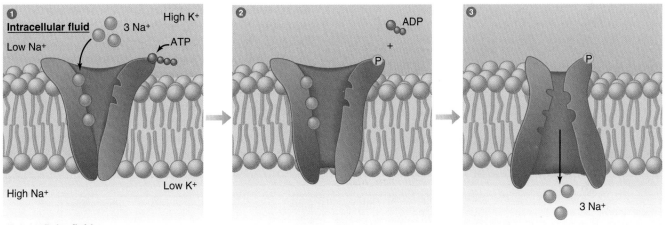

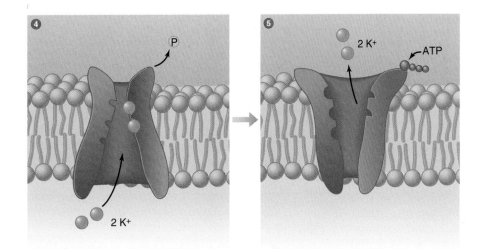

Figure 4–11

Active transport of Na$^+$ and K$^+$ mediated by the Na$^+$/K$^+$-ATPase pump. See text for the numbered sequence of events occurring during transport.

In addition to the Na⁺/K⁺-ATPase transporter, the major primary active-transport proteins found in most cells are (1) Ca^{2+}-ATPase; (2) H^+-ATPase; and (3) H^+/K^+-ATPase. Ca^{2+}-ATPase is found in the plasma membrane and several organelle membranes, including the membranes of the endoplasmic reticulum. In the plasma membrane, the direction of active calcium transport is from cytosol to extracellular fluid. In organelle membranes, it is from cytosol into the organelle lumen. Thus, active transport of Ca^{2+} out of the cytosol, via Ca^{2+}-ATPase, is one reason that the cytosol of most cells has a very low Ca^{2+} concentration, about 10^{-7} mol/L, compared with an extracellular Ca^{2+} concentration of 10^{-3} mol/L, 10,000 times greater.

H^+-ATPase is in the plasma membrane and several organelle membranes, including the inner mitochondrial and lysosomal membranes. In the plasma membrane, the H^+-ATPase moves hydrogen ions out of cells, and in this way helps maintain cellular pH.

H^+/K^+-ATPase is in the plasma membranes of the acid-secreting cells in the stomach and kidneys, where it pumps one hydrogen ion out of the cell and moves one potassium in for each molecule of ATP hydrolyzed.

Secondary Active Transport

Secondary active transport is distinguished from primary active transport by its use of an electrochemical gradient across a plasma membrane as its energy source, rather than phosphorylation of a transport molecule by ATP. In secondary active transport, the movement of an ion down its electrochemical gradient is coupled to the transport of another molecule, such as a nutrient like glucose or an amino acid.

Thus, transporters that mediate secondary active transport have two binding sites, one for an ion—typically but not always sodium—and another for the cotransported molecule. An example of such transport is shown in **Figure 4–13**. In this example, the electrochemical gradient for sodium is directed into the cell because of the higher concentration of sodium in the extracellular fluid and the excess negative charges inside the cell. The solute to be transported, however, must move *against* its concentration gradient, uphill into the cell. High-affinity binding sites for sodium exist on the extracellular surface of the transporter. Binding of sodium increases the affinity of the binding site for the transported solute. The transporter

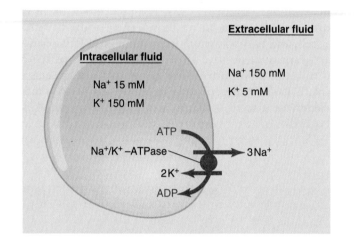

Figure 4–12

The primary active transport of sodium and potassium ions in opposite directions by the Na⁺/K⁺-ATPase in plasma membranes is responsible for the low sodium and high potassium intracellular concentrations. For each ATP hydrolyzed, three sodium ions move out of a cell, and two potassium ions move in.

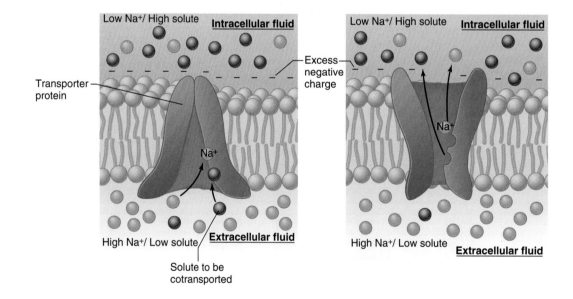

Figure 4–13

Secondary active transport model. In this example, the binding of a sodium ion to the transporter produces an allosteric increase in the affinity of the solute binding site at the extracellular surface of the membrane. Binding of Na⁺ and solute causes a conformational change in the transporter that exposes the binding sites to the intracellular fluid. Na⁺ diffuses down its electrochemical gradient into the cell, which returns the solute-binding site to a low-affinity state.

then undergoes a conformational change, which exposes the binding sites to the intracellular side of the membrane. When the transporter changes conformation, sodium moves into the intracellular fluid by simple diffusion down its electrochemical gradient. At the same time, the affinity of the solute binding site decreases, which releases the solute into the intracellular fluid. Once the transporter releases both molecules, the protein assumes its original conformation. The most important distinction, therefore, between primary and secondary active transport is that secondary active transport uses the stored energy of an electrochemical gradient to move both an ion and a second solute across a plasma membrane. The creation and maintenance of the electrochemical gradient, however, depends on the action of primary active transporters.

To summarize, the creation of a sodium concentration gradient across the plasma membrane by the primary active transport of sodium is a means of indirectly "storing" energy that can then be used to drive secondary active-transport pumps linked to sodium. Ultimately, however, the energy for secondary active transport is derived from metabolism in the form of the ATP that is used by the Na^+/K^+-ATPase to create the sodium concentration gradient. If the production of ATP were inhibited, the primary active transport of sodium would cease, and the cell would no longer be able to maintain a sodium concentration gradient across the membrane. This, in turn, would lead to a failure of the secondary active-transport systems that depend on the sodium gradient for their source of energy. Between 10 and 40 percent of the ATP a cell produces under resting conditions is used by the Na^+/K^+-ATPase to maintain the sodium gradient, which in turn drives a multitude of secondary active-transport systems.

As noted earlier, the net movement of sodium by a secondary active-transport protein is always from high extracellular concentration into the cell, where the concentration of sodium is lower. Thus, in secondary active transport, the movement of sodium is always *downhill*, while the net movement of the actively transported solute on the same transport protein is *uphill*, moving from lower to higher concentration. The movement of the actively transported solute can be either into the cell (in the same direction as sodium), in which case it is known as **cotransport,** or out of the cell (opposite the direction of sodium movement), which is called **countertransport** (**Figure 4–14**). The terms *symport* and *antiport* are also used to refer to the processes of cotransport and countertransport, respectively.

In summary, the distribution of substances between the intracellular and extracellular fluid is often unequal (**Table 4–1**) due to the presence in the plasma membrane of primary and secondary active transporters, ion channels, and the membrane

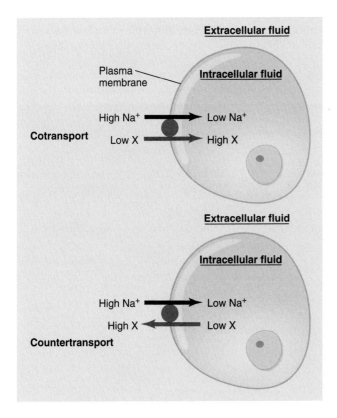

Figure 4–14

Cotransport and countertransport during secondary active transport driven by sodium. Sodium ions always move *down* their concentration gradient into a cell, and the transported solute always moves *up* its gradient. Both sodium and the transported solute X move in the same direction during cotransport, but in opposite directions during countertransport.

Table 4–1	Composition of Extracellular and Intracellular Fluids	
	Extracellular Concentration, mM	**Intracellular Concentration,* mM**
Na^+	140	12
K^+	5	150
Ca^{2+}	1	0.0001
Mg^{2+}	1.5	12
Cl^-	100	7
HCO_3^-	24	10
P_i	2	40
Amino acids	2	8
Glucose	5.6	1
ATP	0	4
Protein	0.2	4

*The intracellular concentrations differ slightly from one tissue to another, depending on the expression of plasma membrane ion channels and transporters. The intracellular concentrations shown above are typical of most cells. For Ca^{2+}, values represent free concentrations. Total calcium levels, including the portion sequestered by proteins or in organelles, approach 2.5 mM (extracellular) and 1.5 mM (intracellular).

potential. **Table 4–2** provides a summary of the major characteristics of the different pathways by which substances move through cell membranes, while **Figure 4–15** illustrates the variety of commonly encountered channels and transporters associated with the movement of substances across a typical plasma membrane.

Not included in Table 4–2 is the mechanism by which water moves across membranes. The special case whereby this polar molecule moves between body fluid compartments is covered next.

Osmosis

Water is a polar molecule that diffuses across the plasma membranes of most cells very rapidly. This process is facilitated by a family of membrane proteins known as **aquaporins** that form channels through which water can diffuse. The type and concentration of these water channels differ in different membranes. Consequently, some cells are more permeable to water than others. In some cells the number of aquaporin channels, and thus the permeability of the membrane to water, can be altered in response to various signals.

The net diffusion of water across a membrane is called **osmosis.** As with any diffusion process, there must be a concentration difference in order to produce a net flux. How can a difference in water concentration be established across a membrane?

The addition of a solute to water lowers the concentration of water in the solution compared to the concentration of pure water. For example, if a solute such as glucose is dissolved

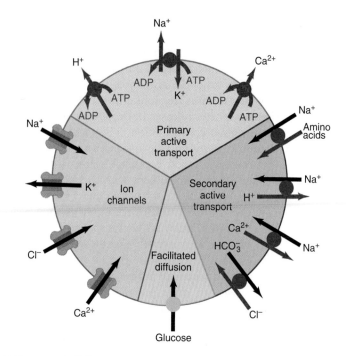

Figure 4–15

Movement of solutes across a typical plasma membrane involving membrane proteins. A specialized cell may contain additional transporters and channels not shown in this figure. Many of these membrane proteins can be modulated by various signals, leading to a controlled increase or decrease in specific solute fluxes across the membrane. The stoichiometry of cotransporters is not shown.

Table 4–2	Major Characteristics of Pathways by which Substances Cross Membranes				
	Diffusion			*Mediated Transport*	
	Through Lipid Bilayer	**Through Protein Channel**	**Facilitated Diffusion**	**Primary Active Transport**	**Secondary Active Transport**
Direction of net flux	High to low concentration	High to low concentration	High to low concentration	Low to high concentration	Low to high concentration
Equilibrium or steady state	$C_o = C_i$	$C_o = C_i$*	$C_o = C_i$	$C_o \neq C_i$	$C_o \neq C_i$
Use of integral membrane protein	No	Yes	Yes	Yes	Yes
Maximal flux at high concentration (saturation)	No	No	Yes	Yes	Yes
Chemical specificity	No	Yes	Yes	Yes	Yes
Use of energy and source	No	No	No	Yes: ATP	Yes: ion gradient (often Na^+)
Typical molecules using pathway	Nonpolar: O_2, CO_2, fatty acids	Ions: Na^+, K^+, Ca^{2+}	Polar: glucose	Ions: Na^+, K^+, Ca^{2+}, H^+	Polar: amino acids, glucose, some ions

*In the presence of a membrane potential, the intracellular and extracellular ion concentrations will not be equal at equilibrium.

in water, the concentration of water in the resulting solution is less than that of pure water. A given volume of a glucose solution contains fewer water molecules than an equal volume of pure water because each glucose molecule occupies space formerly occupied by a water molecule (**Figure 4–16**). In quantitative terms, a liter of pure water weighs about 1000 g, and the molecular weight of water is 18. Thus, the concentration of water in pure water is $1000/18 = 55.5$ M. The decrease in water concentration in a solution is approximately equal to the concentration of added solute. In other words, one solute molecule will displace one water molecule. The water concentration in a 1 M glucose solution is therefore approximately 54.5 M rather than 55.5 M. Just as adding water to a solution will dilute the solute, adding solute to a solution will "dilute" the water. *The greater the solute concentration, the lower the water concentration.*

It is essential to recognize that the degree to which the water concentration is decreased by the addition of solute depends upon the *number* of particles (molecules or ions) of solute in solution (the solute concentration) and not upon the *chemical nature* of the solute. For example, 1 mol of glucose in 1 L of solution decreases the water concentration to the same extent as does 1 mol of an amino acid, or 1 mol of urea, or 1 mol of any other molecule that exists as a single particle in solution. On the other hand, a molecule that ionizes in solution decreases the water concentration in proportion to the number of ions formed. For example, many simple salts dissociate nearly completely in water. For simplicity's sake, we will assume the dissociation is 100 percent at body temperature and at concentrations found in the blood. Therefore, 1 mol of sodium chloride in solution gives rise to 1 mol of sodium ions

and 1 mol of chloride ions, producing 2 mol of solute particles. This lowers the water concentration twice as much as 1 mol of glucose. By the same reasoning, if a 1 M $MgCl_2$ solution were to dissociate completely, it would lower the water concentration three times as much as would a 1 M glucose solution.

Because the water concentration in a solution depends upon the number of solute particles, it is useful to have a concentration term that refers to the total concentration of solute particles in a solution, regardless of their chemical composition. The total solute concentration of a solution is known as its **osmolarity.** One **osmol** is equal to 1 mol of solute particles. Thus, a 1 M solution of glucose has a concentration of 1 Osm (1 osmol per liter), whereas a 1 M solution of sodium chloride contains 2 osmol of solute per liter of solution. A liter of solution containing 1 mol of glucose and 1 mol of sodium chloride has an osmolarity of 3 Osm. A solution with an osmolarity of 3 Osm may contain 1 mol of glucose and 1 mol of sodium chloride, or 3 mol of glucose, or 1.5 mol of sodium chloride, or any other combination of solutes as long as the total solute concentration is equal to 3 Osm.

Although osmolarity refers to the concentration of solute particles, it also determines the water concentration in the solution because the higher the osmolarity, the lower the water concentration. The concentration of water in any two solutions having the same osmolarity is the same because the total number of solute particles per unit volume is the same.

Let us now apply these principles governing water concentration to osmosis of water across membranes. **Figure 4–17** shows two 1-L compartments separated by a membrane permeable to *both* solute and water. Initially the concentration of solute is 2 Osm in compartment 1 and 4 Osm in compartment 2. This difference in solute concentration means there is also a difference in water concentration across the membrane: 53.5 M in compartment 1 and 51.5 M in compartment 2. Therefore, a net diffusion of water from the higher concentration in 1 to the lower concentration in 2 will take place, and a net diffusion of solute in the opposite direction, from 2 to 1. When diffusion equilibrium is reached, the two compartments will have identical solute and water concentrations, 3 Osm and 52.5 M, respectively. One mol of water will have diffused from compartment 1 to compartment 2, and 1 mol of solute will have diffused from 2 to 1. Since 1 mol of solute has replaced 1 mol of water in compartment 1, and vice versa in compartment 2, no change in the volume occurs for either compartment.

If the membrane is now replaced by one *permeable to water but impermeable to solute* (**Figure 4–18**), the same *concentrations* of water and solute will be reached at equilibrium as before, but a change in the *volumes* of the compartments will also occur. Water will diffuse from 1 to 2, but there will be no solute diffusion in the opposite direction because the membrane is impermeable to solute. Water will continue to diffuse into compartment 2, therefore, until the water concentrations on the two sides become equal. The solute concentration in compartment 2 decreases as it is diluted by the incoming water, and the solute in compartment 1 becomes more concentrated as water moves out. When the water reaches diffusion equilibrium, the osmolarities of the

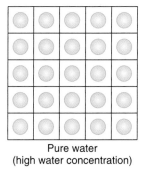
○ Water molecule

● Solute molecule

Pure water
(high water concentration)

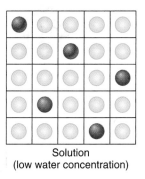
Solution
(low water concentration)

Figure 4–16

The addition of solute molecules to pure water lowers the water concentration in the solution.

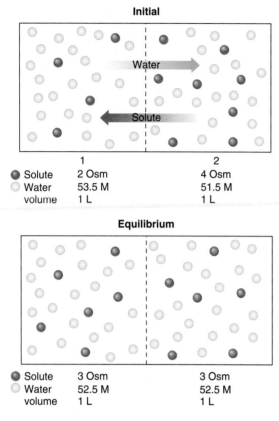

Figure 4–17

Between two compartments of equal volume, the net diffusion of water and solute across a membrane permeable to both leads to diffusion equilibrium of both, with no change in the volume of either compartment. (For clarity's sake, not all water molecules are shown in this figure or in Figure 4–18.)

compartments will be equal, and thus the solute concentrations must also be equal. To reach this state of equilibrium, enough water must pass from compartment 1 to 2 to increase the volume of compartment 2 by one-third and decrease the volume of compartment 1 by an equal amount. Note that it is the presence of a membrane impermeable to solute that leads to the volume changes associated with osmosis.

The two compartments in our example were treated as if they were infinitely expandable, so that the net transfer of water does not create a pressure difference across the membrane. In contrast, if the walls of compartment 2 in Figure 4–18 had only a limited capacity to expand, as occurs across plasma membranes, the movement of water into compartment 2 would raise the pressure in compartment 2, which would oppose further net water entry. Thus the movement of water into compartment 2 can be prevented by the application of pressure to compartment 2. This leads to an important definition: When a solution containing solutes is separated from pure water by a **semipermeable membrane** (a membrane permeable to water but not to solutes), the pressure that must be applied to the solution to prevent the net flow of water into it is termed the **osmotic pressure** of the solution. The greater the osmolarity of a solution, the greater its osmotic pressure. It is important

to recognize that the osmotic pressure of a solution does not push water molecules into the solution. Rather, it represents the amount of pressure that would have to be applied to the solution to *prevent* the net flow of water into the solution. Like osmolarity, the osmotic pressure of a solution is a measure of the solution's water concentration—the lower the water concentration, the higher the osmotic pressure.

Extracellular Osmolarity and Cell Volume

We can now apply the principles learned about osmosis to cells, which meet all the criteria necessary to produce an osmotic flow of water across a membrane. Both the intracellular and extracellular fluids contain water, and cells are surrounded by a membrane that is very permeable to water but impermeable to many substances **(nonpenetrating solutes).**

About 85 percent of the extracellular solute particles is sodium and chloride ions, which can diffuse into the cell through ion channels in the plasma membrane or enter the cell during secondary active transport. As we have seen, however, the plasma membrane contains Na^+/K^+-ATPase pumps that actively move sodium ions out of the cell. Thus, sodium moves into cells and is pumped back out, behaving as if it never

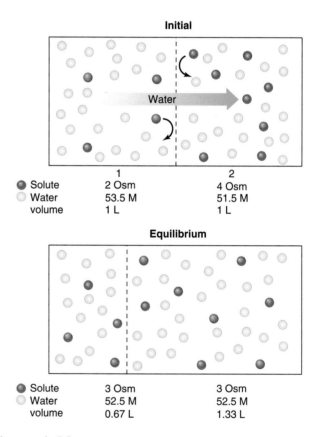

Figure 4–18

The movement of water across a membrane that is permeable to water but not to solute leads to an equilibrium state involving a change in the volumes of the two compartments. In this case, a net diffusion of water (0.33 L) occurs from compartment 1 to 2. (We will assume that the membrane in this example stretches as the volume of compartment 2 increases so that no significant change in compartment pressure occurs.)

entered in the first place; that is, extracellular sodium behaves as a nonpenetrating solute. Any chloride ions that enter cells are also removed as quickly as they enter, due to the electrical repulsion generated by the membrane potential and the action of secondary transporters. Like sodium, therefore, extracellular chloride ions behave as if they were nonpenetrating solutes.

Inside the cell, the major solute particles are potassium ions and a number of organic solutes. Most of the latter are large polar molecules unable to diffuse through the plasma membrane. Although potassium ions can diffuse out of a cell through potassium channels, they are actively transported back by the Na^+/K^+-ATPase pump. The net effect, as with extracellular sodium and chloride, is that potassium behaves as if it were a nonpenetrating solute, but in this case one confined to the intracellular fluid. Thus, sodium and chloride outside the cell and potassium and organic solutes inside the cell behave as nonpenetrating solutes on the two sides of the plasma membrane.

The osmolarity of the extracellular fluid is normally in the range of 285–300 mOsm (we will assume a value of 300 for the rest of this chapter). Because water can diffuse across plasma, water in the intracellular and extracellular fluids will come to diffusion equilibrium. At equilibrium, therefore, the osmolarities of the intracellular and extracellular fluids are the same—approximately 300 mOsm. Changes in extracellular osmolarity can cause cells, such as the red blood cells shown in the chapter opening photo, to shrink or swell as water molecules move across the plasma membrane.

If cells with an intracellular osmolarity of 300 mOsm are placed in a solution of nonpenetrating solutes having an osmolarity of 300 mOsm, they will neither swell nor shrink because the water concentrations in the intra- and extracellular fluid are the same, and the solutes cannot leave or enter.

Such solutions are said to be **isotonic** (**Figure 4–19**), meaning any solution that does not cause a change in cell size. Such solutions have the same concentration of *nonpenetrating* solutes as normal extracellular fluid. By contrast, **hypotonic** solutions have a nonpenetrating solute concentration lower than that found in cells, and therefore water moves by osmosis into the cells, causing them to swell. Similarly, solutions containing greater than 300 mOsm of nonpenetrating solutes (**hypertonic** solutions) cause cells to shrink as water diffuses out of the cell into the fluid with the lower water concentration. Note that the concentration of *nonpenetrating* solutes in a solution, not the total osmolarity, determines its tonicity—isotonic, hypotonic, or hypertonic. Penetrating solutes do not contribute to the tonicity of a solution.

Another set of terms—**isoosmotic, hypoosmotic,** and **hyperosmotic**—denotes the osmolarity of a solution relative to that of normal extracellular fluid without regard to whether the solute is penetrating or nonpenetrating. The two sets of terms are therefore not synonymous. For example, a 1-L solution containing 300 mOsmol of nonpenetrating NaCl and 100 mOsmol of urea, which can rapidly cross plasma membranes, would have a total osmolarity of 400 mOsm and would be hyperosmotic. It would, however, also be an isotonic solution, producing no change in the equilibrium volume of cells immersed in it. *Initially,* cells placed in this solution would shrink as water moved into the extracellular fluid. However, urea would quickly diffuse into the cells and reach the same concentration as the urea in the extracellular solution, and thus both the intracellular and extracellular solutions would soon reach the same osmolarity. Therefore, at equilibrium, there would be no difference in the water concentration across the membrane and thus no change in final cell volume, even though the extracellular fluid would remain hyperosmotic.

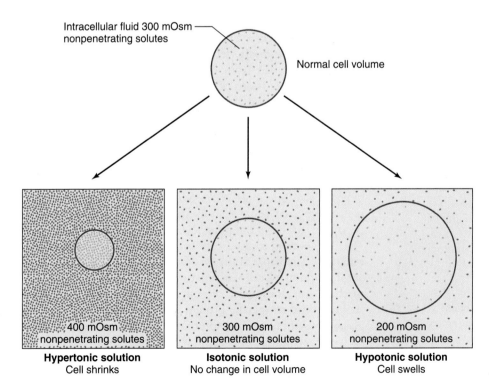

Figure 4–19

Changes in cell volume produced by hypertonic, isotonic, and hypotonic solutions.

Figure 4–19 physiological *inquiry* 🄿𝒾

- Blood volume must be restored in a person who has lost large amounts of blood due to serious injury. This is often accomplished by infusing isotonic NaCl solution into the blood. Why is this better than infusing an isoosmotic solution of a penetrating solute, such as urea?

Answer can be found at end of chapter.

As you will learn in Chapter 14, one of the major functions of the kidneys is to regulate the excretion of water in the urine so that the osmolarity of the extracellular fluid remains nearly constant in spite of variations in salt and water intake and loss, thereby preventing damage to cells from excessive swelling or shrinkage.

Table 4–3 provides a comparison of the various terms used to describe the osmolarity and tonicity of solutions.

Endocytosis and Exocytosis

In addition to diffusion and mediated transport, there is another pathway by which substances can enter or leave cells, one that does not require the molecules to pass through the structural matrix of the plasma membrane. When cells are observed under a microscope, regions of the plasma membrane can be seen to fold into the cell, forming small pockets that pinch off to produce intracellular, membrane-bound vesicles that enclose a small volume of extracellular fluid. This process is known as **endocytosis** (**Figure 4–20**). A similar

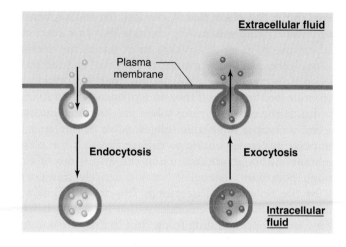

Figure 4–20

Endocytosis and exocytosis.

process in the reverse direction, **exocytosis,** occurs when membrane-bound vesicles in the cytoplasm fuse with the plasma membrane and release their contents to the outside of the cell (see Figure 4–20).

Endocytosis

There are three general types of endocytosis that may occur in a cell. These are fluid endocytosis, also known as **pinocytosis** ("cell drinking"), phagocytosis ("cell eating"), and receptor-mediated endocytosis (**Figure 4–21**).

In **fluid endocytosis,** an endocytotic vesicle encloses a small volume of extracellular fluid. This process is nonspecific because the vesicle simply engulfs the water in the extracellular fluid along with whatever solutes are present. These solutes may include ions, nutrients, or any other small extracellular molecule. Large macromolecules, other cells, and cell debris do not normally enter a cell via this process.

In **phagocytosis,** cells engulf bacteria or large particles such as cell debris from damaged tissues. In this form of endocytosis, extensions of the plasma membrane called pseudopodia fold around the surface of the particle, engulfing it entirely. The pseudopodia, with their engulfed contents, then fuse into large vesicles called **phagosomes** that are internalized into the cell. Phagosomes migrate to and fuse with lysosomes in the cytoplasm, and the contents of the phagosome are then destroyed by lysosomal enzymes and other molecules. While most cells undergo pinocytosis, only a few special types of cells, such as those of the immune system (Chapter 18), carry out phagocytosis.

In contrast to fluid endocytosis and phagocytosis, most cells have the capacity to *specifically* take up molecules that are important for cellular function or structure. In **receptor-mediated endocytosis,** certain molecules in the extracellular fluid bind to specific proteins on the outer surface of the plasma membrane. These proteins are called **receptors,** and each one recognizes one ligand with high affinity (see Chapter 3 for a discussion of ligand-protein interactions). In one form of receptor-mediated endocytosis, the receptor undergoes

Table 4–3	Terms Referring to the Osmolarity and Tonicity of Solutions*
Isotonic	A solution that does not cause a change in cell volume; one that contains 300 mOsmol/L of nonpenetrating solutes, regardless of the concentration of membrane-penetrating solutes present
Hypertonic	A solution that causes cells to shrink; one that contains greater than 300 mOsmol/L of nonpenetrating solutes, regardless of the concentration of membrane-penetrating solutes present
Hypotonic	A solution that causes cells to swell; one that contains less than 300 mOsmol/L of nonpenetrating solutes, regardless of the concentration of membrane-penetrating solutes present
Isoosmotic	A solution containing 300 mOsmol/L of solute, regardless of its composition of membrane-penetrating and nonpenetrating solutes
Hyperosmotic	A solution containing greater than 300 mOsmol/L of solutes, regardless of its composition of membrane-penetrating and nonpenetrating solutes
Hypoosmotic	A solution containing less than 300 mOsmol/L of solutes, regardless of its composition of membrane-penetrating and nonpenetrating solutes

*These terms are defined using an intracellular osmolarity of 300 mOsm, which is within the range for human cells but not an absolute fixed number.

a conformational change when it binds a ligand. Through a series of steps, a cytosolic protein called **clathrin** is recruited to the plasma membrane. Clathrin forms a cage-like structure that leads to the aggregation of ligand-bound receptors into a localized region of membrane, forming a depression, or **clathrin-coated pit,** which then invaginates and pinches off to form a clathrin-coated vesicle. By localizing ligand-receptor complexes to discrete patches of plasma membrane prior to endocytosis, cells may obtain concentrated amounts of ligands without having to engulf large amounts of extracellular fluid from many different sites along the membrane. Receptor-mediated endocytosis, therefore, leads to a selective concentration in the endocytotic vesicle of a specific ligand bound to one type of receptor.

Cholesterol is one example of a ligand that enters cells via clathrin-dependent, receptor-mediated endocytosis. Cholesterol is an important building block for plasma- and intracellular membranes, and most cells require a steady supply of this molecule. Cholesterol circulates in the blood, bound with pro-

teins in particles called lipoproteins. The protein components of lipoproteins are recognized by plasma membrane receptors. When the receptors bind the lipoproteins, endocytosis ensues and the cholesterol is delivered to the intracellular fluid. The rate at which this occurs can be regulated. For example, if a cell has sufficient supplies of cholesterol, the rate at which it replenishes its supply of lipoprotein receptors may decrease. Conversely, receptor production increases when cholesterol supplies are low. This is a type of negative feedback that acts to maintain the cholesterol content of the cell within a homeostatic range.

Once an endocytotic vesicle pinches off from the plasma membrane in receptor-mediated endocytosis, the clathrin coat is removed and clathrin proteins are recycled back to the membrane. The vesicles then have several possible fates, depending upon the cell type and the ligand that was engulfed. Some vesicles fuse with the membrane of an intracellular organelle, adding the contents of the vesicle to the lumen of that organelle. Other endocytotic vesicles pass through the cytoplasm and fuse with the plasma membrane on the opposite side of the cell, releasing their contents to the extracellular space. This provides a pathway for the transfer of large molecules, such as proteins, across the layers of cells that separate two fluid compartments in the body (for example, the blood and interstitial

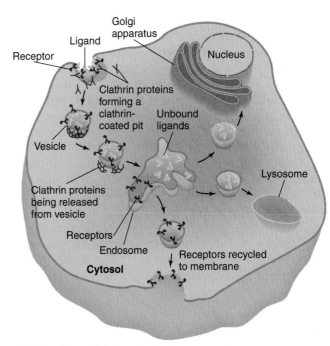

(c) Receptor-mediated endocytosis

Figure 4–21

Types of endocytosis. (a) In fluid endocytosis, solutes and water are nonspecifically brought into the cell from the extracellular fluid via endocytotic vesicles. (b) In phagocytosis, specialized cells form extensions of the plasma membrane called pseudopodia, which engulf bacteria or other large objects such as cell debris. The vesicles that form fuse with lysosomes, which contain enzymes and other molecules that destroy the vesicle contents. (c) In receptor-mediated endocytosis, a cell recognizes a specific extracellular ligand that binds to a plasma membrane receptor. The binding triggers endocytosis. In the example shown here, the ligand-receptor complexes are internalized via clathrin-coated vesicles, which merge with endosomes. Ligands may be routed to the Golgi apparatus for further processing, or to lysosomes. The receptors are typically recycled to the plasma membrane.

fluid). A similar process allows small amounts of macromolecules to move across the intestinal epithelium.

Most endocytotic vesicles fuse with a series of intracellular vesicles and tubular elements known as endosomes (Chapter 3), which lie between the plasma membrane and the Golgi apparatus. Like the Golgi apparatus, the endosomes perform a sorting function, distributing the contents of the vesicle and its membrane to various locations. Some of the contents of endocytotic vesicles are passed from the endosomes to the Golgi apparatus, where the ligands are modified and processed. Other vesicles fuse with lysosomes, organelles that contain digestive enzymes that break down large molecules such as proteins, polysaccharides, and nucleic acids. The fusion of endosomal vesicles with the lysosomal membrane exposes the contents of the vesicle to these digestive enzymes. Finally, in many cases, the receptors that were internalized with the vesicle get recycled back to the plasma membrane.

Another fate of endocytotic vesicles is seen in a special type of receptor-mediated endocytosis called potocytosis. **Potocytosis** is similar to other types of receptor-mediated endocytosis in that an extracellular ligand typically binds to a plasma membrane receptor, initiating formation of an intracellular vesicle. In potocytosis, however, the ligands appear to be primarily restricted to low-molecular-weight molecules such as certain vitamins, but have also been found to include the lipoprotein complexes just described. Potocytosis differs from clathrin-dependent, receptor-mediated endocytosis in the fate of the endocytotic vesicle. In potocytosis, tiny vesicles called **caveolae** (singular: caveolus, "little caves") pinch off from the plasma membrane and deliver their contents directly to the cell cytosol, rather than merging with lysosomes or other organelles. The small molecules within the caveolae may diffuse into the cytosol via channels, or be transported by carriers. Although their functions are still being actively investigated, caveolae have been implicated in a variety of important cellular functions, including cell signaling, transcellular transport, and cholesterol homeostasis.

Each episode of endocytosis removes a small portion of the membrane from the cell surface. In cells that have a great deal of endocytotic activity, more than 100 percent of the plasma membrane may be internalized in an hour, yet the membrane surface area remains constant. This is because the membrane is replaced at about the same rate by vesicle membrane that fuses with the plasma membrane during *exocytosis.* Some of the plasma membrane proteins taken into the cell during endocytosis are stored in the membranes of endosomes, and upon receiving the appropriate signal can be returned to fuse with the plasma membrane during exocytosis.

Exocytosis

Exocytosis performs two functions for cells: (1) It provides a way to replace portions of the plasma membrane that endocytosis has removed, and, in the process, to add new membrane components as well, and (2) it provides a route by which membrane-impermeable molecules (such as protein hormones) the cell synthesizes can be secreted into the extracellular fluid.

How does the cell package substances that are to be secreted by exocytosis into vesicles? Chapter 3 described the entry of newly formed proteins into the lumen of the endoplasmic reticulum and the protein's processing through the Golgi apparatus. From the Golgi apparatus, the proteins to be secreted travel to the plasma membrane in vesicles from which they can be released into the extracellular fluid by exocytosis. Very high concentrations of various organic molecules, such as neurotransmitters, can be held within vesicles by employing a combination of mediated transport across the vesicle membrane and binding of the transported substances to proteins within the vesicle.

The secretion of substances by exocytosis is triggered in most cells by stimuli that lead to an increase in cytosolic calcium concentration in the cell. As will be described in Chapters 5 and 6, these stimuli open calcium channels in the plasma membrane and/or the membranes of intracellular organelles. The resulting increase in cytosolic calcium concentration activates proteins required for the vesicle membrane to fuse with the plasma membrane and release the vesicle contents into the extracellular fluid. Material stored in secretory vesicles is available for rapid secretion in response to a stimulus, without delays that might occur if the material had to be synthesized after the stimulus arrived.

Epithelial Transport

Epithelial cells line hollow organs or tubes and regulate the absorption or secretion of substances across these surfaces. One surface of an epithelial cell generally faces a hollow or fluid-filled chamber, and the plasma membrane on this side is referred to as the **luminal membrane** (also known as the apical, or mucosal, membrane) of the epithelium. The plasma membrane on the opposite surface, which is usually adjacent to a network of blood vessels, is referred to as the **basolateral membrane** (also known as the serosal membrane).

There are two pathways by which a substance can cross a layer of epithelial cells: (1) by diffusion *between* the adjacent cells of the epithelium—the **paracellular pathway,** or (2) by movement *into* an epithelial cell across either the luminal or basolateral membrane, diffusion through the cytosol, and exit across the opposite membrane. This is termed the **transcellular pathway.**

Diffusion through the paracellular pathway is limited by the presence of tight junctions between adjacent cells, because these junctions form a seal around the luminal end of the epithelial cells (Chapter 3). Although small ions and water can diffuse to some degree through tight junctions, the amount of paracellular diffusion is limited by the tightness of the junctional seal and the relatively small area available for diffusion. The leakiness of the paracellular pathway varies in different types of epithelium, with some being very leaky and others very tight.

During transcellular transport, the movement of molecules through the plasma membranes of epithelial cells occurs via the pathways (diffusion and mediated transport) already described for movement across membranes. However, the transport and permeability characteristics of the luminal and basolateral membranes are not the same. These two membranes contain different ion channels and different transporters for mediated transport. As a result of these differences, substances can undergo a net movement from a low concentration on one side of an epithelium to a higher concentration on the other side, or in other words, can undergo active transport across the overall epithelial layer. Examples include the absorption of material from the gastrointestinal tract into the blood, the movement of substances between the kidney tubules and the blood during urine formation, and the secretion of salts and fluid by glands.

Figures 4–22 and 4–23 illustrate two examples of active transport across an epithelium. Sodium is actively transported across most epithelia from lumen to blood side in absorptive processes, and from blood side to lumen during secretion. In our example, the movement of sodium from the lumen into the epithelial cell occurs by diffusion through sodium channels in the luminal membrane (see Figure 4–22). Sodium diffuses into the cell because the intracellular concentration of sodium is kept low by the active transport of sodium back out of the cell across the basolateral membrane on the opposite side, where all of the Na$^+$/K$^+$-ATPase pumps are located. In

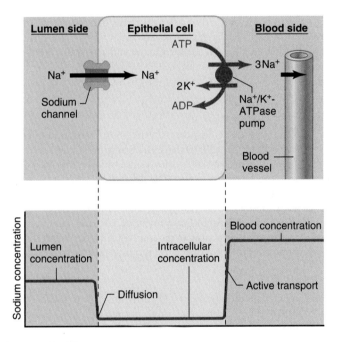

Figure 4–22

Active transport of sodium across an epithelial cell. The transepithelial transport of sodium always involves primary active transport out of the cell across one of the plasma membranes, typically via a Na$^+$/K$^+$-ATPase pump as shown here. The movement of sodium into the cell across the plasma membrane on the opposite side is always downhill. Sometimes, as in this example, it is by diffusion through sodium channels, whereas in other epithelia this downhill movement occurs through a secondary active transporter. Shown below the cell is the concentration profile of the transported solute across the epithelium.

Figure 4–22 physiological _inquiry_ _(pi)_

- What would happen in this situation if the cell's ATP supply was to decrease significantly?

Answer can be found at end of chapter.

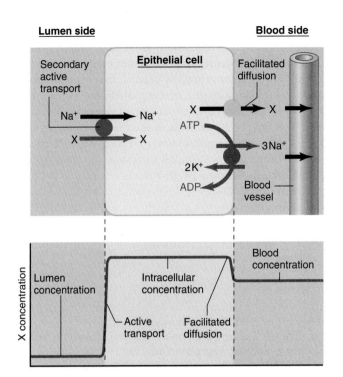

Figure 4–23

The transepithelial transport of most organic solutes (X) involves their movement into a cell through a secondary active transport driven by the downhill flow of sodium. The organic substance then moves out of the cell at the blood side down a concentration gradient by means of facilitated diffusion. Shown below the cell is the concentration profile of the transported solute across the epithelium.

other words, sodium moves downhill into the cell and then uphill out of it. The net result is that sodium can be moved from lower to higher concentration across the epithelium.

Figure 4–23 illustrates the active absorption of organic molecules across an epithelium. In this case, the entry of an organic molecule X across the luminal plasma membrane occurs via a secondary active transporter linked to the downhill movement of sodium into the cell. In the process, X moves from a lower concentration in the luminal fluid to a higher concentration in the cell. The substance exits across the basolateral membrane by facilitated diffusion, which moves the material from its higher concentration in the cell to a lower concentration in the extracellular fluid on the blood side. The concentration of the substance may be considerably higher on the blood side than in the lumen because the blood-side concentration can approach equilibrium with the high intracellular concentration created by the luminal membrane entry step.

Although water is not actively transported across cell membranes, net movement of water across an epithelium can occur by osmosis as a result of the active transport of solutes, especially sodium, across the epithelium. The active transport of sodium, as previously described, results in a decrease in the sodium concentration on one side of an epithelial layer (the luminal side in our example) and an increase on the other. These changes in solute concentration are accompanied by changes in the water concentration on the two sides because a change in solute concentration, as we have seen, produces a change in water concentration. The water concentration difference will cause water to move by osmosis from the low-sodium side to the high-sodium side of the epithelium (**Figure 4–24**). Thus, net movement of solute across an epithelium is accompanied by a flow of water in the same direction. If the epithelial cells are highly permeable to water, large net movements of water can occur with very small differences in osmolarity. As you will learn in Chapter 14, this is a major way in which epithelial cells of the kidney reabsorb water from the urine back into the blood.

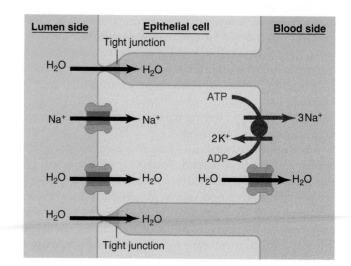

Figure 4–24

Net movements of water across an epithelium are dependent on net solute movements. The active transport of sodium across the cells, into the surrounding interstitial spaces, produces an elevated osmolarity in this region and a decreased osmolarity in the lumen. This leads to the osmotic flow of water across the epithelium in the same direction as the net solute movement. The water diffuses through water channels in the membrane and across the tight junctions between the epithelial cells.

■ SUMMARY

Diffusion

I. Diffusion is the movement of molecules from one location to another by random thermal motion.
  a. The net flux between two compartments always proceeds from higher to lower concentration.
  b. Diffusion equilibrium is reached when the concentrations of the diffusing substance in the two compartments become equal.

II. The magnitude of the net flux J across a membrane is directly proportional to the concentration difference across the membrane $C_o - C_i$, the surface area of the membrane A, and the membrane permeability coefficient P.

III. Nonpolar molecules diffuse through the lipid portions of membranes much more rapidly than do polar or ionized molecules because nonpolar molecules can dissolve in the lipids in the membrane.

IV. Ions diffuse across membranes by passing through ion channels formed by integral membrane proteins.
  a. The diffusion of ions across a membrane depends on both the concentration gradient and the membrane potential.
  b. The flux of ions across a membrane can be altered by opening or closing ion channels.

Mediated-Transport Systems

I. The mediated transport of molecules or ions across a membrane involves binding the transported solute to a transporter protein in the membrane. Changes in the conformation of the transporter move the binding site to the opposite side of the membrane, where the solute dissociates from the protein.
  a. The binding sites on transporters exhibit chemical specificity, affinity, and saturation.
  b. The magnitude of the flux through a mediated-transport system depends on the degree of transporter saturation, the number of transporters in the membrane, and the rate at which the conformational change in the transporter occurs.

II. Facilitated diffusion is a mediated-transport process that moves molecules from higher to lower concentration across a membrane by means of a transporter until the two concentrations become equal. Metabolic energy is not required for this process.

III. Active transport is a mediated-transport process that moves molecules against an electrochemical gradient across a membrane by means of a transporter and an input of energy.

  a. Primary active transport uses the phosphorylation of the transporter by ATP to drive the transport process.
  b. Secondary active transport uses the binding of ions (often sodium) to the transporter to drive the secondary transport process.
  c. In secondary active transport, the downhill flow of an ion is linked to the uphill movement of a second solute either in the same direction as the ion (cotransport) or in the opposite direction of the ion (countertransport).

Osmosis

I. Water crosses membranes by (1) diffusing through the lipid bilayer, and (2) diffusing through protein channels in the membrane.

II. Osmosis is the diffusion of water across a membrane from a region of higher water concentration to a region of lower water concentration. The osmolarity—total solute concentration in a solution—determines the water concentration: The higher the osmolarity of a solution, the lower the water concentration.

III. Osmosis across a membrane that is permeable to water but impermeable to solute leads to an increase in the volume of the compartment on the side that initially had the higher osmolarity, and a decrease in the volume on the side that initially had the lower osmolarity.

IV. Application of sufficient pressure to a solution will prevent the osmotic flow of water into the solution from a compartment of pure water. This pressure is called the osmotic pressure. The greater the osmolarity of a solution, the greater its osmotic pressure. Net water movement occurs from a region of lower osmotic pressure to one of higher osmotic pressure.

V. The osmolarity of the extracellular fluid is about 300 mOsm. Because water comes to diffusion equilibrium across cell membranes, the intracellular fluid has an osmolarity equal to that of the extracellular fluid.
  a. Na$^+$ and Cl$^-$ ions are the major effectively nonpenetrating solutes in the extracellular fluid; K$^+$ ions and various organic solutes are the major effectively nonpenetrating solutes in the intracellular fluid.
  b. Table 4–3 lists the terms used to describe the osmolarity and tonicity of solutions containing different compositions of penetrating and nonpenetrating solutes.

Endocytosis and Exocytosis

I. During endocytosis, regions of the plasma membrane invaginate and pinch off to form vesicles that enclose a small volume of extracellular material.
  a. The three classes of endocytosis are (1) fluid endocytosis, (2) phagocytosis, and (3) receptor-mediated endocytosis.
  b. Most endocytotic vesicles fuse with endosomes, which in turn transfer the vesicle contents to lysosomes for digestion by lysosomal enzymes.
  c. Potocytosis is a special type of receptor-mediated endocytosis in which vesicles called caveolae deliver their contents directly to the cytosol.
II. Exocytosis, which occurs when intracellular vesicles fuse with the plasma membrane, provides a means of adding components to the plasma membrane and a route by which membrane-impermeable molecules, such as proteins the cell synthesizes, can be released into the extracellular fluid.

Epithelial Transport

I. Molecules can cross an epithelial layer of cells by two pathways: (1) through the extracellular spaces between the cells—the paracellular pathway, and (2) through the cell, across both the luminal and basolateral membranes as well as the cell's cytoplasm—the transcellular pathway.
II. In epithelial cells, the permeability and transport characteristics of the luminal and basolateral plasma membranes differ, resulting in the ability of cells to actively transport a substance between the fluid on one side of the cell and the fluid on the opposite side.
III. The active transport of sodium through an epithelium increases the osmolarity on one side of the cell and decreases it on the other, causing water to move by osmosis in the same direction as the transported sodium.

■ KEY TERMS

active transport   104
aquaporin   108
basolateral membrane   114
caveolus   114
channel   100
channel gating   100
clathrin   113
clathrin-coated pit   113
cotransport   107
countertransport   107
diffusion   97
diffusion equilibrium   97
electrochemical gradient   100
endocytosis   112
exocytosis   112
facilitated diffusion   103
fluid endocytosis   112
flux   97
hyperosmotic   111
hypertonic   111
hypoosmotic   111
hypotonic   111
isoosmotic   111
isotonic   111
ligand-gated channel   100
luminal membrane   114
mechanically-gated channel   100
mediated transport   102
membrane potential   100
net flux   97
nonpenetrating solute   110
osmol   109
osmolarity   109
osmosis   108
osmotic pressure   110
paracellular pathway   114
permeability coefficient, P 99
phagocytosis   112
phagosome   112
pinocytosis   112
potocytosis   114
primary active transport   104
receptor   112
receptor-mediated endocytosis   112
secondary active transport   104
semipermeable membrane   110
transcellular pathway   114
transporter   102
voltage-gated channel   100

■ CLINICAL TERMS

diabetes mellitus   104

■ REVIEW QUESTIONS

1. What determines the direction in which net diffusion of a nonpolar molecule will occur?
2. In what ways can the net solute flux between two compartments separated by a permeable membrane be increased?
3. Why are membranes more permeable to nonpolar molecules than to most polar and ionized molecules?
4. Ions diffuse across cell membranes by what pathway?
5. When considering the diffusion of ions across a membrane, what driving force, in addition to the ion concentration gradient, must be considered?
6. Describe the mechanism by which a transporter of a mediated-transport system moves a solute from one side of a membrane to the other.
7. What determines the magnitude of flux across a membrane in a mediated-transport system?
8. What characteristics distinguish diffusion from facilitated diffusion?
9. What characteristics distinguish facilitated diffusion from active transport?
10. Describe the direction in which sodium ions and a solute transported by secondary active transport move during cotransport and countertransport.
11. How can the concentration of water in a solution be decreased?
12. If two solutions with different osmolarities are separated by a water-permeable membrane, why will a change occur in the volumes of the two compartments if the membrane is impermeable to the solutes, but no change in volume will occur if the membrane is permeable to solute?
13. Why do sodium and chloride ions in the extracellular fluid and potassium ions in the intracellular fluid behave as though they were nonpenetrating solutes?
14. What is the approximate osmolarity of the extracellular fluid? Of the intracellular fluid?
15. What change in cell volume will occur when a cell is placed in a hypotonic solution? In a hypertonic solution?
16. Under what conditions will a hyperosmotic solution be isotonic?
17. How do the mechanisms for actively transporting glucose and sodium across an epithelium differ?
18. By what mechanism does the active transport of sodium lead to the osmotic flow of water across an epithelium?

Chapter 4 Test Questions

(Answers appear in Appendix A.)

1. Which properties are characteristic of ion channels?
   a. They are usually lipids.
   b. They exist on one side of the plasma membrane, usually the intracellular side.
   c. They can open and close depending on the presence of any of three types of "gates."
   d. They permit movement of ions against concentration gradients.
   e. They mediate facilitated diffusion.

2. Which of the following does not directly or indirectly require an energy source?
   a. primary active transport
   b. operation of the sodium/potassium pump
   c. the mechanism used by cells to produce a calcium ion gradient across the plasma membrane
   d. facilitated transport of glucose across a plasma membrane
   e. secondary active transport

3. If a small amount of urea were added to an isoosmotic saline solution containing cells, what would be the result?
   a. The cells would shrink and remain that way.
   b. The cells would first shrink, but then be restored to normal volume after a brief period of time.
   c. The cells would swell and remain that way.
   d. The cells would first swell, but then be restored to normal volume after a brief period of time.
   e. The urea would have no effect, even transiently.

4. Which is (are) true of epithelial cells?
   a. They can only move uncharged molecules across their surfaces.
   b. They may have segregated functions on luminal and basolateral surfaces.
   c. They cannot form tight junctions.
   d. They depend upon the activity of Na^+/K^+-ATPase pumps for much of their transport functions.
   e. Both b and d are correct.

5. Which is *incorrect*?
   a. Diffusion of a solute through a membrane is considerably quicker than diffusion of the same solute through a water layer of equal thickness.
   b. A single ion, such as K^+, can diffuse through more than one type of channel.
   c. Lipid-soluble solutes diffuse more readily through the phospholipid bilayer of a plasma membrane than do water-soluble ones.
   d. The rate of facilitated diffusion of a solute is limited by the number of transporters in the membrane at any given time.
   e. A common example of cotransport is that of an ion and an organic molecule.

6. In considering diffusion of ions through an ion channel, which driving force(s) must be considered?
   a. the ion concentration gradient
   b. the electrical gradient
   c. osmosis
   d. facilitated diffusion
   e. both a and b

Chapter 4 Quantitative and Thought Questions

(Answers appear in Appendix A.)

1. In two cases (A and B), the concentrations of solute X in two 1-L compartments separated by a membrane through which X can diffuse are:

   Concentration of X, mM

Case	Compartment 1	Compartment 2
A	3	5
B	32	30

   a. In what direction will the net flux of *X* take place in case A and in case B?
   b. When diffusion equilibrium is reached, what will the concentration of solute in each compartment be in case A and in case B?
   c. Will A reach diffusion equilibrium faster, slower, or at the same rate as B?

2. When the extracellular concentration of the amino acid alanine is increased, the net flux of the amino acid leucine into a cell is decreased. How might this observation be explained?

3. If a transporter that mediates active transport of a substance has a lower affinity for the transported substance on the extracellular surface of the plasma membrane than on the intracellular surface, in what direction will there be a net transport of the substance across the membrane? (Assume that the rate of transporter conformational change is the same in both directions.)

4. Why will inhibition of ATP synthesis by a cell lead eventually to a decrease and, ultimately, cessation in secondary active transport?

5. Given the following solutions, which has the lowest water concentration? Which two have the same osmolarity?

Solution	*Concentration, mM*			
	Glucose	Urea	NaCl	$CaCl_2$
A	20	30	150	10
B	10	100	20	50
C	100	200	10	20
D	30	10	60	100

6. Assume that a membrane separating two compartments is permeable to urea but not permeable to NaCl. If compartment 1 contains 200 mmol/L of NaCl and 100 mmol/L of urea, and compartment 2 contains 100 mmol/L of NaCl and 300 mmol/L of urea, which compartment will have increased in volume when osmotic equilibrium is reached?

7. What will happen to cell volume if a cell is placed in each of the following solutions?

Concentration, mM

Solution	NaCl (nonpenetrating)	Urea (penetrating)
A	150	100
B	100	150
C	200	100
D	100	50

8. Characterize each of the solutions in question 7 as isotonic, hypotonic, hypertonic, isoosmotic, hypoosmotic, or hyperosmotic.

9. By what mechanism might an increase in intracellular sodium concentration lead to an increase in exocytosis?

Chapter 4 Answers to Physiological Inquires

Figure 4.2 As shown in the accompanying graph, there would be a net flux of glucose from compartment 1 to compartment 2, with diffusion equilibrium occurring at 12.5 mmol/L.

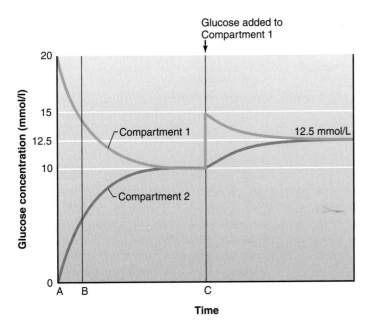

Figure 4.19 Because it is a nonpenetrating solute, infusion of isotonic NaCl restores blood volume without causing a redistribution of water between body fluid compartments due to osmosis. An isoosmotic solution of a penetrating solute, however, would only partially restore blood volume because some water would enter the intracellular fluid by osmosis as the solute enters cells. This could also result in damage to cells as their volume expands beyond normal.

Figure 4.22 Active transport of sodium across the basolateral (blood side) membrane would decrease, resulting in an increased intracellular concentration of Na^+. This would reduce the rate of Na^+ diffusion into the cell through the Na^+ channel on the lumen side because the diffusion gradient would be smaller.

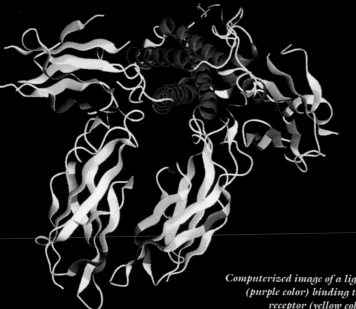

*Computerized image of a ligand
(purple color) binding to its
receptor (yellow color).*

chapter **5**

Control of Cells by Chemical Messengers

You learned in Chapter 1 how homeostatic control systems help maintain a normal balance of the body's internal environment. The operation of control systems requires that cells be able to communicate with each other, often over long distances. Much of this intercellular communication is mediated by chemical messengers. This chapter describes how these messengers interact with their target cells and how these interactions trigger intracellular chains of chemical events that lead to the cell's response. Throughout this chapter, the reader should carefully distinguish *inter*cellular (between cells) and *intra*cellular (within a cell) chemical messengers and communication. The material in this chapter will provide a foundation for understanding how the nervous, endocrine, and other organ systems work.

120

Receptors

In Chapter 1, you learned that several classes of chemical messengers can communicate a signal from one cell to another. These messengers include chemicals such as neurotransmitters, whose signals are mediated rapidly and over a short distance. Other messengers, such as hormones, communicate more slowly and over greater distances. Whatever the chemical messenger, however, the cell receiving the signal must have a way to detect the signal's presence. Once a cell detects a signal, a transduction mechanism is needed to convert that signal into a biologically meaningful response, such as the cell-division response to the delivery of growth-promoting signals.

The first step in the action of any intercellular chemical messenger is the binding of the messenger to specific target-cell proteins known as **receptors** or **receptor proteins.** In the general language of Chapter 3, a chemical messenger is a ligand, and the receptor protein has a binding site for that ligand. The binding of a messenger to a receptor protein initiates a sequence of events in the cell leading to the cell's response to that messenger, a process called signal transduction.

The term *receptor* can be the source of confusion because the same word is used to denote the "detectors" in a reflex arc, as Chapter 1 described. You should keep in mind that the term *receptor* has two distinct meanings, but the context in which the term is used makes the meaning clear.

What is the nature of the receptors with which intercellular chemical messengers combine? They are proteins or glycoproteins located either in the cell's plasma membrane or inside the cell, mainly in the nucleus. The plasma membrane is the much more common location, because a very large number of messengers are water-soluble and thus cannot diffuse across the lipid-rich plasma membrane. In contrast, the much smaller number of lipid-soluble messengers pass through membranes (mainly by diffusion but, in some cases, by mediated transport as well) to bind to their receptors located inside the cell.

Plasma membrane receptors are transmembrane proteins; that is, they span the entire membrane thickness. A typical plasma membrane receptor is illustrated in **Figure 5–1**. Like other transmembrane proteins, a plasma membrane receptor has hydrophobic segments within the membrane, one or more hydrophilic segments extending out from the membrane into the extracellular fluid, and other hydrophilic segments extending into the intracellular fluid. It is to the extracellular portions that the arriving chemical messenger binds. Also like other transmembrane proteins, certain receptors may be composed of two or more nonidentical subunits bound together.

The binding of a chemical messenger to its receptor protein initiates the events leading to the cell's response. The existence of receptor proteins explains a very important characteristic of intercellular communication—**specificity** (see **Table 5–1** for a glossary of terms concerning receptors). Although a given chemical messenger may come into contact with many different cells, it influences only certain cells and not others. This is because cells differ in the types of receptors they possess. Only certain cell types, often just one, possess the specific receptor protein

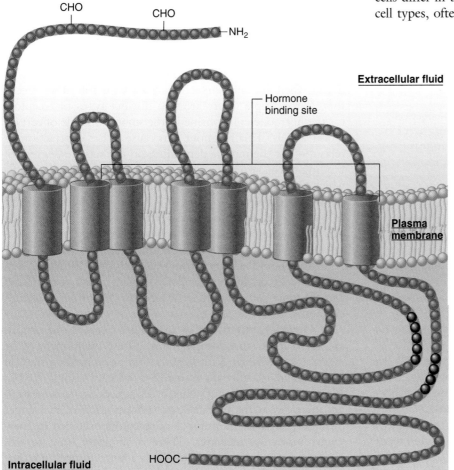

Figure 5–1

Structure of a receptor that binds the hormone epinephrine. The seven clusters of amino acids embedded in the phospholipid bilayer represent hydrophobic portions of the protein's alpha helix. Note that the binding site for the hormone includes several of the segments that extend into the extracellular fluid. The amino acids denoted by black circles represent sites at which intracellular substances can phosphorylate, and thereby regulate, the receptor.

Adapted from Dohlman et al.

Table 5–1	A Glossary of Terms Concerning Receptors
Receptor (receptor protein)	A specific protein in either the plasma membrane or the interior of a target cell that a chemical messenger combines with, thereby invoking a biologically relevant response in that cell.
Specificity	The ability of a receptor to bind only one type or a limited number of structurally related types of chemical messengers.
Saturation	The degree to which receptors are occupied by messengers. If all are occupied, the receptors are fully saturated; if half are occupied, the saturation is 50 percent, and so on.
Affinity	The strength with which a chemical messenger binds to its receptor.
Competition	The ability of different molecules very similar in structure to compete with each other to combine with the same receptor.
Antagonist	A molecule that competes for a receptor with a chemical messenger normally present in the body. The antagonist binds to the receptor but does not trigger the cell's response. Antihistamines are examples of antagonists.
Agonist	A chemical messenger that binds to a receptor and triggers the cell's response; often refers to a drug that mimics a normal messenger's action. Decongestants are examples of agonists.
Down-regulation	A decrease in the total number of target-cell receptors for a given messenger; may occur in response to chronic high extracellular concentration of the messenger.
Up-regulation	An increase in the total number of target-cell receptors for a given messenger; may occur in response to a chronic low extracellular concentration of the messenger.
Supersensitivity	The increased responsiveness of a target cell to a given messenger; may result from up-regulation of receptors.

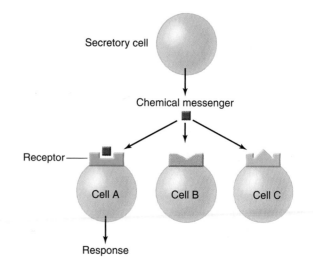

Figure 5–2

Specificity of receptors for chemical messengers. Only cell A has the appropriate receptor for this chemical messenger; therefore, it is the only one among the group that is a target cell for the messenger.

required to bind a given chemical messenger (**Figure 5–2**). In many cases, the receptor proteins for a group of messengers are structurally related. Thus, for example, scientists who study hormones refer to "superfamilies" of hormone receptors.

When different types of cells possess the same receptors for a particular messenger, the responses of the various cell types to that messenger may differ from each other. For example, the neurotransmitter norepinephrine causes the smooth muscle of

certain blood vessels to contract, but, via the same type of receptor, causes endocrine cells in the pancreas to secrete less insulin. In essence, then, the receptor functions as a molecular switch that elicits the cell's response when "switched on" by the messenger binding to it. Just as identical types of switches can be used to turn on a light or a radio, a single type of receptor can be used to produce quite different responses in different cell types.

Similar reasoning explains a more surprising phenomenon: A single cell may contain more than one different receptor type *for a single messenger*. When the messenger binds to one of these receptor types, it may produce a cellular response quite different from, indeed sometimes opposite to, that produced when the messenger combines with the other receptors. For example, there are two distinct types of receptors for the hormone epinephrine in the smooth muscle of certain blood vessels. This hormone can cause either contraction or relaxation of the muscle depending on the relative degrees of binding to the two different types of receptors. The degree to which the molecules of a particular messenger bind to different receptor types in a single cell is determined by the **affinity** of the different receptor types for the messenger. A receptor with high affinity will bind at lower concentrations of a messenger than will a receptor of low affinity.

You should not infer from these descriptions that a given cell has receptors for only one messenger. In fact, a single cell usually contains many different receptors for different chemical messengers.

Other characteristics of messenger-receptor interactions are **saturation** and **competition.** These phenomena were described in Chapter 3 for ligands binding to binding sites on proteins and are fully applicable here (and are summarized in **Figure 5–3**). In most systems, a cell's response to a messenger increases as the extracellular concentration of messenger increases, because the number of receptors occupied by messenger molecules increases. There is an upper limit to this responsiveness, however, because only a finite number of receptors are available, and they become saturated at some point.

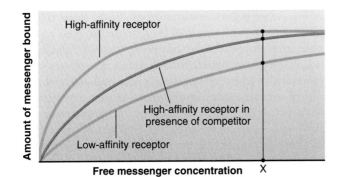

Figure 5–3

Characteristics of receptor binding to messengers. The receptors with high affinity will have more bound messenger at a given messenger concentration (e.g., concentration X). The presence of a competitor will reduce the amount of messenger bound, until at very high concentrations the receptors become saturated with messenger.

Competition is the ability of different messenger molecules that are very similar in structure to compete with each other for a receptor. Competition occurs physiologically with closely related messengers, and it also underlies the action of many drugs. If researchers or physicians wish to interfere with the action of a particular messenger, they can administer competing molecules that are similar enough to the endogenous messenger that they bind to the receptors for that messenger. However, the competing molecules fail to activate the receptor. This blocks the endogenous messenger from binding and yet does not trigger the cell's response. Such drugs are known as **antagonists** with regard to the usual chemical messenger. One example are the beta-blockers, used in the treatment of high blood pressure and other diseases. These drugs antagonize the ability of epinephrine and norepinephrine to bind to one of their receptors—the beta-adrenergic receptor. Because epinephrine and norepinephrine normally act to raise blood pressure (Chapter 12), beta-blockers tend to reduce blood pressure by acting as antagonists. Antihistamines are another example, and are useful in treating allergic symptoms brought on due to excess histamine secretion from cells known as mast cells (Chapter 18). Antihistamines are antagonists that block histamine from binding with cells and triggering an allergic response.

On the other hand, some drugs that bind to a particular receptor type do trigger the cell's response exactly as if the true (endogenous) chemical messenger had combined with the receptor. Such drugs, known as **agonists,** are used therapeutically to mimic the messenger's action. For example, the decongestant drugs phenylephrine, pseudoephedrine, and oxymetazoline mimic the action of epinephrine on a different class of receptors, called alpha-adrenergic receptors, in blood vessels. When alpha-adrenergic receptors are activated, the smooth muscles of blood vessels in the nose contract, resulting in vasoconstriction in the nasal passages and fewer sniffles.

Regulation of Receptors

Receptors are themselves subject to physiological regulation. The number of receptors a cell has, or the affinity of the receptors for their specific messenger, can be increased or decreased in certain systems. An important example of such regulation is the phenomenon of **down-regulation.** When a high extracellular concentration of a messenger is maintained for some time, the total number of the target cell's receptors for that messenger may decrease—that is, down-regulate. Down-regulation has the effect of reducing the target cells' responsiveness to frequent or intense stimulation by a messenger—that is, desensitizing them—and thus represents a local negative feedback mechanism.

Change in the opposite direction, called **up-regulation,** also occurs. Cells exposed for a prolonged period to very low concentrations of a messenger may come to have many more receptors for that messenger, thereby developing increased sensitivity (**supersensitivity**) to it. For example, when the nerves to a muscle are cut, the delivery of neurotransmitters from those nerves to the muscle is eliminated. Under these conditions, the muscle will contract in response to a much smaller amount of experimentally injected neurotransmitter than that to which a normal muscle responds. This happens because the receptors for the neurotransmitter have been up-regulated, resulting in supersensitivity.

Up-regulation and down-regulation are possible because there is a continuous degradation and synthesis of receptors. The main cause of down-regulation of plasma-membrane receptors is **internalization.** The binding of a messenger to its receptor can stimulate the internalization of the complex; that is, the messenger-receptor complex is taken into the cell by receptor-mediated endocytosis. This increases the rate of receptor degradation inside the cell. Thus, at high hormone concentrations, the number of plasma-membrane receptors of that type gradually decreases during down-regulation.

The opposite events also occur and contribute to up-regulation. The cell may contain stores of receptors in the membranes of intracellular vesicles. These are then inserted into the plasma membrane via exocytosis during up-regulation.

Another important mechanism of up-regulation and down-regulation is alteration of the expression of the genes that code for the receptors.

Signal Transduction Pathways

What are the sequences of events by which the binding of a chemical messenger (hormone, neurotransmitter, or paracrine/autocrine agent) to a receptor causes the cell to respond?

The combination of messenger with receptor causes a change in the conformation (three-dimensional shape) of the receptor. This event, known as **receptor activation,** is always the initial step leading to the cell's responses to the messenger. These responses can take the form of changes in: (1) the permeability, transport properties, or electrical state of the cell's plasma membrane; (2) the cell's metabolism; (3) the cell's secretory activity; (4) the cell's rate of proliferation and differentiation; or (5) the cell's contractile activity.

Despite the seeming variety of these five types of ultimate responses, there is a common denominator: They are all directly due to alterations of particular cell proteins. Let us examine a few examples of messenger-induced responses, all of which are described more fully in subsequent chapters. The neurotransmitter-induced generation of electrical signals in nerve cells reflects the altered conformation of membrane proteins (ion channels) through which ions can diffuse between extracellular and intracellular fluid. Similarly, changes in the rate of glucose secretion by the liver induced by the hormone epinephrine reflect the altered activity and concentration of enzymes in the metabolic pathways for glucose synthesis. Finally, muscle contraction induced by the neurotransmitter acetylcholine results from the altered conformation of contractile proteins.

Thus, receptor activation by a messenger is only the first step leading to the cell's ultimate response (contraction, secretion, and so on). The diverse sequences of events between receptor activation and cellular responses are termed **signal transduction pathways.** The "signal" is the receptor activation, and "transduction" denotes the process by which a stimulus is transformed into a response. The important question is: How does receptor activation influence the cell's internal proteins, which are usually critical for the response but may be located far from the receptor?

Signal transduction pathways differ between lipid-soluble and water-soluble messengers. As described earlier, the receptors for these two broad chemical classes of messenger are in different locations—the former inside the cell and the latter in the plasma membrane of the cell. The rest of this chapter elucidates the general principles of the signal transduction pathways that these two broad categories of receptors initiate.

Pathways Initiated by Lipid-Soluble Messengers

Lipid-soluble messengers generally act on cells by binding to intracellular receptor proteins. Lipid-soluble messengers include steroid hormones, the thyroid hormones, and the steroid derivative, 1,25-dihydroxy vitamin D. Structurally these hormones are all lipophilic, and their receptors constitute the **steroid-hormone receptor superfamily.** Although plasma-membrane receptors for a few of these messengers have been identified, most of the receptors in this superfamily are intracellular. When not bound to a messenger, the receptors are inactive. In a few cases, the inactive receptors are located in the cytosol and move into the nucleus after binding their hormone. Most of the inactive receptors in the steroid hormone superfamily, however, already reside in the cell nucleus, where they bind to and are activated by their respective ligands. Receptor activation leads to altered rates of gene transcription.

The messenger diffuses out of capillaries from plasma to the interstitial fluid. From there, the messenger diffuses across the cell's plasma membrane and nuclear membrane to enter the nucleus and bind to the receptor there **(Figure 5–4)**. The receptor, activated by the binding of hormone to it, then functions in the nucleus as a **transcription factor,** defined as any regulatory protein that directly influences gene transcription. The hormone-receptor complex binds to a specific sequence near a gene in DNA called a response element, an event that increases the rate of that gene's transcription into mRNA. The mRNA molecules move out of the nucleus to direct the synthesis, on ribosomes, of the protein the gene encodes. The result is an increase in the cellular concentration of the protein and/or its rate of secretion, and this accounts for the cell's ultimate response to the messenger. For example, if the protein encoded by the gene is an enzyme, the cell's response is an increase in the rate of the reaction catalyzed by that enzyme.

Two other points are important. First, more than one gene may be subject to control by a single receptor type. For example, the glucocorticoid hormone cortisol (Chapter 11) acts via one type of intracellular receptor to activate numerous genes involved in cellular metabolism and energy balance. Second, in some cases the transcription of a gene or genes may be *decreased* rather than increased by the activated receptor. Cortisol, for example, inhibits transcription of several genes whose protein products mediate inflammatory responses that occur following injury or infection (Chapter 18).

Pathways Initiated by Water-Soluble Messengers

Water-soluble messengers exert their actions on cells by binding to receptor proteins on the extracellular surface of the plasma membrane. Water-soluble messengers include most hormones, neurotransmitters, and paracrine/autocrine compounds. On the basis of the signal transduction pathways they initiate, plasma membrane receptors can be classified into the types listed in **Table 5–2** and illustrated in **Figure 5–5**.

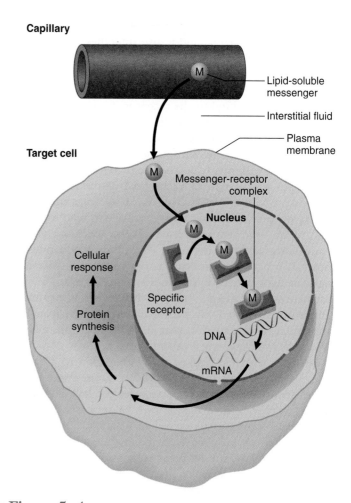

Figure 5–4

Mechanism of action of lipid-soluble messengers. This figure shows the receptor for these messengers in the nucleus. In some cases, the unbound receptor is in the cytosol rather than the nucleus, in which case the binding occurs there, and the messenger-receptor complex moves into the nucleus. For simplicity, a single messenger is shown binding to a single receptor. In many cases, however, two messenger/receptor complexes must bind together in order to activate a gene.

Table 5–2	Classification of Receptors Based on Their Locations and the Signal Transduction Pathways They Use

1. INTRACELLULAR RECEPTORS (Figure 5–4) (for lipid-soluble messengers) Function in the nucleus as transcription factors or suppressors to alter the rate of transcription of particular genes.

2. PLASMA MEMBRANE RECEPTORS (Figure 5–5) (for water-soluble messengers)

   a. Receptors that are ligand-gated ion channels.

   b. Receptors that themselves function as enzymes, such as receptor tyrosine kinases.

   c. Receptors that are bound to and activate cytoplasmic JAK kinases.

   d. G-protein-coupled receptors that activate G proteins, which in turn act upon effector proteins—either ion channels or enzymes—in the plasma membrane.

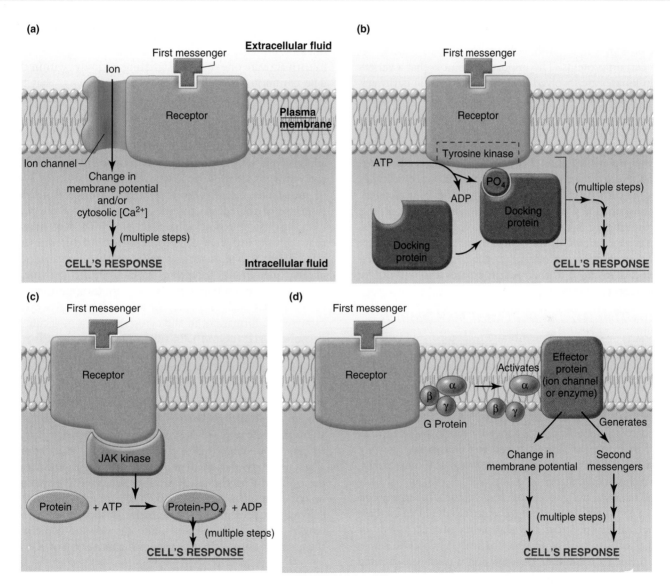

Figure 5–5

Mechanisms of action of water-soluble messengers (noted as "first messengers" in this and subsequent figures). (a) Signal transduction mechanism in which the receptor complex includes an ion channel. (b) Signal transduction mechanism in which the receptor itself functions as an enzyme, usually a tyrosine kinase. (c) Signal transduction mechanism in which the receptor activates a JAK kinase in the cytoplasm. (d) Signal transduction mechanism involving G proteins.

Figure 5–5 physiological *inquiry* pi

■ Many cells express more than one of the four types of receptors depicted in this figure. Why might this be?

Answer can be found at end of chapter.

Three notes on general terminology are essential for this discussion. First, the intercellular chemical messengers that reach the cell from the extracellular fluid and bind to their specific plasma membrane receptors are often referred to as **first messengers. Second messengers,** then, are substances that enter or are generated in the cytoplasm as a result of receptor activation by the first messenger. The second messengers diffuse throughout the cell to serve as chemical relays from the plasma membrane to the biochemical machinery inside the cell.

The third essential general term is **protein kinase.** As described in Chapter 3, protein kinase is the name for any enzyme that phosphorylates other proteins by transferring a phosphate group to them from ATP. Introduction of the phosphate group changes the conformation and/or activity of the phosphorylated protein, often itself an enzyme. There are many different protein kinases, and each type is able to phosphorylate only certain proteins. The important point is that a variety of protein kinases are involved in signal transduction pathways. These pathways may involve a series of reactions in which a particular inactive protein kinase is activated by phosphorylation and then catalyses the phosphorylation of another inactive protein kinase, and so on. At the ends of these sequences, the ultimate phosphorylation of key proteins, such as transporters, metabolic enzymes, ion channels, and contractile proteins, underlies the cell's biochemical response to the first messenger.

As described in Chapter 3, other enzymes do the reverse of protein kinases; that is, they dephosphorylate proteins. These enzymes, termed protein phosphatases, also participate in signal transduction pathways, but their roles are much less understood than those of the protein kinases and will not be described further in this chapter.

Receptors That Are Ligand-Gated Ion Channels

In the first type of plasma membrane receptor listed in Table 5–2, the protein that acts as the receptor is also an ion channel. Activation of the receptor by a first messenger (the ligand) results in a conformational change of the receptor such that it forms an open channel through the plasma membrane (**Figure 5–5a**). Because the opening of ion channels has been compared to the opening of a gate in a fence, these type of channels are known as **ligand-gated ion channels.** They are particularly prevalent in the plasma membranes of nerve cells, as you will learn in Chapter 6.

The opening of ligand-gated ion channels in response to binding of a first messenger to its receptor results in an increase in the net diffusion across the plasma membrane of one or more types of ions specific to that channel. As you will see in Chapter 6, such a change in ion diffusion is usually associated with a change in the electrical charge, or membrane potential, of a cell. This electrical signal is often the essential event in the cell's response to the messenger. In addition, when the channel is a calcium channel, its opening results in an increase, by diffusion, in cytosolic calcium concentration. Increasing cytosolic calcium is another essential event in the transduction pathway for many signaling systems.

Receptors That Function as Enzymes

The receptors in the second category of plasma membrane receptors listed in Table 5–2 have intrinsic enzyme activity.

With one major exception (discussed soon), the many receptors that possess intrinsic enzyme activity are all protein kinases (**Figure 5–5b**). Of these, the great majority specifically phosphorylate the portions of proteins that contain the amino acid tyrosine. Thus, these receptors are known as **receptor tyrosine kinases.**

The typical sequence of events for receptors with intrinsic tyrosine kinase activity is as follows. The binding of a specific messenger to the receptor changes the conformation of the receptor so that its enzymatic portion, located on the cytoplasmic side of the plasma membrane, is activated. This results in autophosphorylation of the receptor; that is, the receptor phosphorylates its own tyrosine groups. The newly created phosphotyrosines on the cytoplasmic portion of the receptor then serve as docking sites for cytoplasmic proteins. The bound docking proteins then bind and activate other proteins, which in turn activate one or more signaling pathways within the cell. The common denominator of these pathways is that they all involve activation of cytoplasmic proteins by phosphorylation.

The number of kinases that mediate these phosphorylations can be very large, and their names constitute a veritable alphabet soup—RAF, MEK, MAPKK, and many others. In all this complexity, it is easy to lose track of the point that the end result of all these pathways is the activation or synthesis of molecules, usually proteins, that ultimately mediate the response of the cell to the messenger. Most of the receptors with intrinsic tyrosine kinase activity bind first messengers that typically influence cell proliferation and differentiation.

There is one major exception to the generalization that plasma membrane receptors with inherent enzyme activity function as protein kinases. In this exception, the receptor functions both as a receptor and as a **guanylyl cyclase** to catalyse the formation, in the cytoplasm, of a molecule known as **cyclic GMP (cGMP).** In turn, cGMP functions as a second messenger to activate a protein kinase called **cGMP-dependent protein kinase.** This kinase phosphorylates specific proteins that then mediate the cell's response to the original messenger. As described in Chapter 7, receptors that function both as ligand-binding molecules and as guanylyl cyclases are present in high amounts in the retina of the vertebrate eye, where they are important for processing visual inputs. This signal transduction pathway is used by only a small number of messengers and should not be confused with the much more prevalent cAMP system to be described in a later section. Also, in certain cells, guanylyl cyclase enzymes are present in the cytoplasm. In these cases, a first messenger—nitric oxide—diffuses into the cell and combines with the guanylyl cyclase there to trigger the formation of cGMP.

Receptors That Interact with Cytoplasmic JAK Kinases

Recall that in the previous category, the receptor itself has intrinsic enzyme activity. In the next category of receptors (see Table 5–2 and **Figure 5–5c**), the enzymatic activity—again tyrosine kinase activity—resides not in the receptor but in a family of separate cytoplasmic kinases, termed **JAK kinases,** which are associated with the receptor. (The term *JAK* has several derivations, including "janus kinase.") In these cases, the receptor and its associated JAK kinase function as a unit. The binding of a first messenger to the receptor causes a conformational change in the receptor that leads to activation of the JAK kinase. Different receptors associate with different members of the JAK kinase family, and the different JAK kinases phosphorylate different target proteins, many of which act as transcription factors. The result of these pathways is the synthesis of new proteins, which mediate the cell's response to the first messenger. Signaling by cytokines—proteins secreted by cells of the immune system that play a critical role in immune defenses (Chapter 18)—occurs primarily via receptors linked to JAK kinases.

G-Protein-Coupled Receptors

The fourth category of plasma membrane receptors in Table 5–2 is by far the largest, including hundreds of distinct receptors (**Figure 5–5d**). Bound to the receptor is a protein complex located on the cytosolic surface of the plasma membrane and belonging to the family of heterotrimeric (containing three different subunits) proteins known as **G proteins.** The binding of a first messenger to the receptor changes the conformation of the receptor. This change increases the affinity of the alpha subunit of the G protein for GTP. When bound to GTP, the alpha subunit dissociates from the remaining two (beta and gamma) subunits of the trimeric G protein. This dissociation allows the activated alpha subunit to link up with still another plasma membrane protein, either an ion channel or an enzyme. These ion channels and enzymes are termed plasma membrane effector proteins because they mediate the next steps in the sequence of events leading to the cell's response.

In essence, then, a G protein serves as a switch to couple a receptor to an ion channel or to an enzyme in the plasma membrane. Thus, these receptors are known as **G-protein-coupled receptors.** The G protein may cause the ion channel to open, with a resulting change in electrical signals or,

in the case of calcium channels, changes in the cytosolic calcium concentration. Alternatively, the G protein may activate or inhibit the membrane enzyme with which it interacts. Such enzymes, when activated, cause the generation of second messengers inside the cell.

Once the alpha subunit of the G protein activates its effector protein, a GTP-ase activity inherent in the alpha subunit cleaves the GTP into GDP plus P_i. This cleavage renders the alpha subunit inactive, allowing it to recombine with its beta and gamma subunits. The beta and gamma subunits help anchor the alpha subunit in the membrane.

There are several subfamilies of plasma membrane G proteins, each with multiple distinct members, and a single receptor may be associated with more than one type of G protein. Moreover, some G proteins may couple to more than one type of plasma membrane effector protein. Thus, a first-messenger-activated receptor, via its G-protein couplings, can call into action a variety of plasma membrane effector proteins such as ion channels and enzymes. These molecules can, in turn, induce a variety of cellular events.

To illustrate some of the major points concerning G proteins, plasma membrane effector proteins, second messengers, and protein kinases, the next two sections describe the two most important effector protein enzymes regulated by G proteins—adenylyl cyclase and phospholipase C. In addition, the subsequent portions of the signal transduction pathways in which they participate are described.

Adenylyl Cyclase and Cyclic AMP

In this pathway (**Figure 5–6**), activation of the receptor by the binding of the first messenger (for example, the hormone epinephrine) allows the receptor to activate its associated G protein,

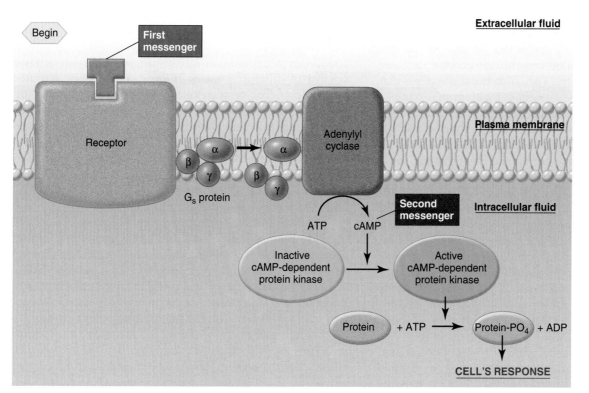

Figure 5–6

Cyclic AMP second-messenger system. Not shown in the figure is the existence of another regulatory protein, G_i, which certain receptors can react with to cause inhibition of adenylyl cyclase.

in this example known as G_s (the subscript s denotes "stimulatory"). This causes G_s to activate its effector protein, the membrane enzyme called **adenylyl cyclase** (also known as adenylate cyclase). The activated adenylyl cyclase, whose catalytic site is located on the cytosolic surface of the plasma membrane, catalyzes the conversion of cytosolic ATP molecules to cyclic 3′,5′-adenosine monophosphate, or **cyclic AMP (cAMP)** (**Figure 5–7**). Cyclic AMP then acts as a second messenger (see Figure 5–6). It diffuses throughout the cell to trigger the sequence of events leading to the cell's ultimate response to the first messenger. The action of cAMP eventually terminates when it is broken down to noncyclic AMP, a reaction catalyzed by the enzyme **phosphodiesterase** (see Figure 5–7). This enzyme is also subject to physiological control. Thus, the cellular concentration of cAMP can be changed either by altering the rate of its messenger-mediated generation or the rate of its phosphodiesterase-mediated breakdown. Caffeine and theophylline, the active ingredients of coffee and tea, are widely consumed stimulants that work partly by inhibiting phosphodiesterase activity, thus prolonging the actions of cAMP within a cell.

What does cAMP actually do inside the cell? It binds to and activates an enzyme known as **cAMP-dependent protein kinase,** also called protein kinase A (see Figure 5–6). Protein kinases phosphorylate other proteins—often enzymes—by transferring a phosphate group to them. The changes in the activity of proteins phosphorylated by cAMP-dependent protein kinase bring about the cell's response (secretion, contraction, and so on). Again, note that each of the various protein kinases that participate in the multiple signal transduction pathways described in this chapter has its own specific substrates.

In essence, then, the activation of adenylyl cyclase by a G protein initiates an "amplification cascade" of events that converts proteins in sequence from inactive to active forms. **Figure 5–8** illustrates the benefit of such a cascade. While it is active, a single enzyme molecule is capable of transforming into product not one but many substrate molecules, let us say 100. Therefore, one active molecule of adenylyl cyclase may catalyze the generation of 100 cAMP molecules. At each of the two subsequent enzyme-activation steps in our example, another 100-fold amplification occurs. Therefore, the end result is that a single molecule of the first messenger could, in this example, cause the generation of 1 million product molecules. This helps to explain how hormones and other messengers can be effective at extremely low extracellular concentrations. To take an actual example, one molecule of the hormone epinephrine can cause the liver to generate and release 10^8 molecules of glucose.

In addition, cAMP-activated protein kinase A can diffuse into the cell nucleus, where it can phosphorylate a protein that then binds to specific regulatory regions of certain genes. Such genes are said to be cAMP-responsive. Thus, the effects of cAMP can be rapid and independent of changes in gene activity, as in the example of epinephrine and glucose production, or slower and dependent upon the formation of new gene products.

How can cAMP's activation of a single molecule, cAMP-dependent protein kinase, be common to the great variety of biochemical sequences and cell responses initiated by cAMP-generating first messengers? The answer is that

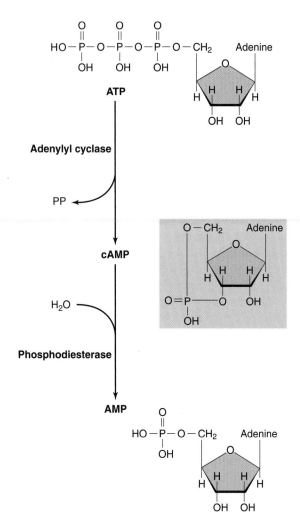

Figure 5–7

Structure of ATP, cAMP, and AMP, the last resulting from enzymatic inactivation of cAMP.

cAMP-dependent protein kinase can phosphorylate a large number of different proteins (**Figure 5–9**). Thus, activated cAMP-dependent protein kinase can exert multiple actions within a single cell and different actions in different cells. For example, epinephrine acts via the cAMP pathway on fat cells to stimulate the breakdown of triglyceride, a process that is mediated by one particular phosphorylated enzyme. In the liver, epinephrine acts via cAMP to stimulate both glycogenolysis and gluconeogenesis, processes that are mediated by phosphorylated enzymes that differ from those in fat cells.

Note that whereas phosphorylation mediated by cAMP-dependent protein kinase activates certain enzymes, it inhibits others. For example, the enzyme catalyzing the rate-limiting step in glycogen synthesis is inhibited by phosphorylation. This explains how epinephrine inhibits glycogen synthesis at the same time it stimulates glycogen breakdown by activating the enzyme that catalyzes the latter response.

Not mentioned thus far is the fact that receptors for some first messengers, upon activation by their messengers, *inhibit* adenylyl cyclase. This inhibition results in less, rather

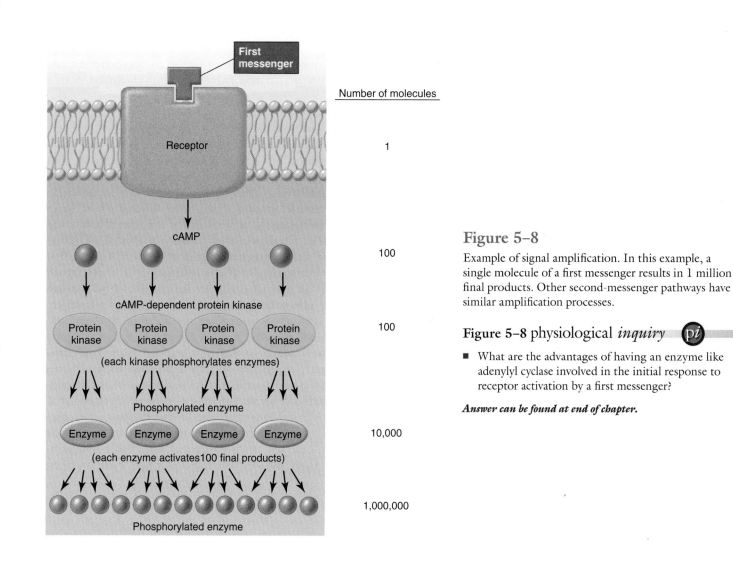

Number of molecules

1

100

100

10,000

1,000,000

Figure 5–8

Example of signal amplification. In this example, a single molecule of a first messenger results in 1 million final products. Other second-messenger pathways have similar amplification processes.

Figure 5–8 physiological *inquiry*

■ What are the advantages of having an enzyme like adenylyl cyclase involved in the initial response to receptor activation by a first messenger?

Answer can be found at end of chapter.

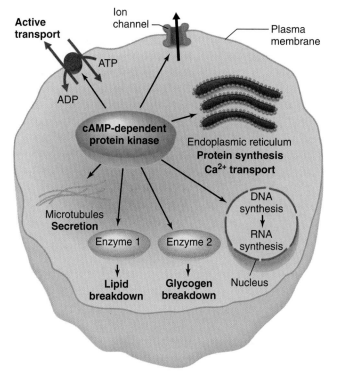

Figure 5–9

The variety of cellular responses induced by cAMP is due mainly to the fact that activated cAMP-dependent protein kinase can phosphorylate many different proteins, activating or inhibiting them. In this figure, the protein kinase is shown phosphorylating seven different proteins—a microtubular protein, an ATPase, an ion channel, a protein in the endoplasmic reticulum, a protein involved in DNA synthesis, and two enzymes.

Figure 5–9 physiological *inquiry*

■ Does a given protein kinase, such as cAMP-dependent protein kinase, phosphorylate the same proteins in all cells in which the kinase is present?

Answer can be found at end of chapter.

than more, generation of cAMP. This occurs because these receptors are associated with a different G protein known as G_i (the subscript i denotes "inhibitory"). Activation of G_i causes the inhibition of adenylyl cyclase. The result is to decrease the concentration of cAMP in the cell and thereby the phosphorylation of key proteins inside the cell.

Phospholipase C, Diacylglycerol, and Inositol Trisphosphate

In this system, a G protein called G_q gets activated by a receptor that has bound a first messenger. Activated G_q then activates a plasma membrane effector enzyme called **phospholipase C.** This enzyme catalyzes the breakdown of a plasma membrane phospholipid known as phosphatidylinositol bisphosphate, abbreviated PIP_2, to **diacylglycerol (DAG)** and **inositol trisphosphate (IP_3)** (**Figure 5–10**). Both DAG and IP_3 then function as second messengers but in very different ways.

DAG activates a class of protein kinases known collectively as **protein kinase C,** which then phosphorylate a large number of other proteins, leading to the cell's response.

IP_3, in contrast to DAG, does not exert its second messenger role by directly activating a protein kinase. Rather, IP_3, after entering the cytosol, binds to receptors located on the endoplasmic reticulum. These receptors are ligand-gated Ca^{2+} channels. When bonded to IP_3, the channels open. Because the concentration of calcium is much higher in the endoplasmic reticulum than in the cytosol, calcium diffuses out of this organelle into the cytosol, significantly increasing cytosolic calcium concentration. This increased calcium concentration then continues the sequence of events leading to the cell's response to the first messenger. We will pick up this thread in more detail in a later section. However, it is worth noting that one of the actions of Ca^{2+} is to help activate some forms of protein kinase C (which is how this kinase got its name—"C" for calcium).

Control of Ion Channels by G Proteins

A comparison of Figures 5–5d and 5–9 emphasizes one more important feature of G-protein function—its ability to both directly and indirectly gate ion channels. As shown in Figure 5–5d and described earlier, an ion channel can be the effector protein for a G protein. This situation is known as direct G-protein gating of plasma membrane ion channels because the G protein interacts directly with the channel. All the events occur in the plasma membrane and are independent of second messengers. Now look at Figure 5–9, and you will see that cAMP-dependent protein kinase can phosphorylate a plasma membrane ion channel, thereby causing it to open. As we have seen, the sequence of events leading to the activation of cAMP-dependent protein kinase proceeds through a G protein, so it should be clear that the opening of this channel is indirectly dependent on that G protein. To generalize, the *indirect G-protein gating* of ion channels utilizes a second-messenger pathway for the opening or closing of the channel. Not just cAMP-dependent protein kinase, but protein kinases involved in other signal transduction pathways can participate in reactions leading to such indirect gating. **Table 5–3** summarizes the three ways by which receptor activation by a first

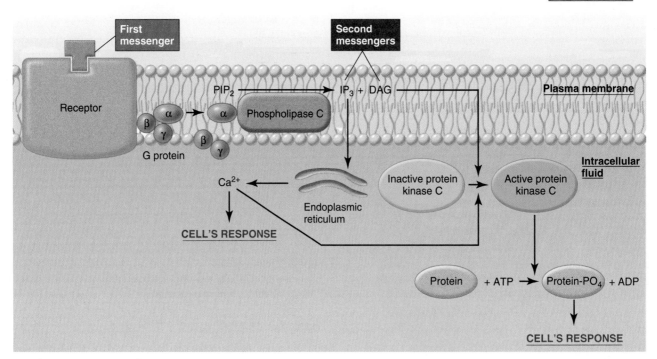

Figure 5–10

Mechanism by which an activated receptor stimulates the enzymatically-mediated breakdown of PIP_2 to yield IP_3 and DAG. IP_3 then causes the release of calcium ions from the endoplasmic reticulum, which together with DAG activate protein kinase C.

Table 5–3	Summary of Mechanisms by Which Receptor Activation Influences Ion Channels

1. The ion channel is part of the receptor.

2. A G protein directly gates the channel.

3. A G protein gates the channel indirectly via a second messenger.

Table 5–4	Calcium as a Second Messenger

Common Mechanisms By Which Stimulation of a Cell Leads to an Increase in Cytosolic Ca^{2+} Concentration:

1. Receptor activation

 a. Plasma-membrane calcium channels open in response to a first messenger; the receptor itself may contain the channel, or the receptor may activate a G protein that opens the channel via a second messenger.

 b. Calcium is released from the endoplasmic reticulum; this is mediated by second messengers, particularly IP_3 and calcium entering from the extracellular fluid.

 c. Active calcium transport out of the cell is inhibited by a second messenger.

2. Opening of voltage-gated calcium channels.

Major Mechanisms By Which an Increase in Cytosolic Ca^{2+} Concentration Induces the Cell's Responses:

1. Calcium binds to calmodulin. On binding calcium, the calmodulin changes shape, which allows it to activate or inhibit a large variety of enzymes and other proteins. Many of these enzymes are protein kinases.

2. Calcium combines with calcium-binding intermediary proteins other than calmodulin. These proteins then act in a manner analogous to calmodulin.

3. Calcium combines with and alters response proteins directly, without the intermediation of any specific calcium-binding protein.

messenger leads to opening or closing of ion channels, causing a change in membrane potential.

Calcium as a Second Messenger

The calcium ion (Ca^{2+}) functions as a second messenger in a great variety of cellular responses to stimuli, both chemical and electrical. The physiology of calcium as a second messenger requires an analysis of two broad questions: (1) How do stimuli cause the cytosolic calcium concentration to increase? (2) How does the increased calcium concentration elicit the cells' responses? Note that, for simplicity, our two questions are phrased in terms of an *increase* in cytosolic concentration. There are, in fact, first messengers that elicit a *decrease* in cytosolic calcium concentration and therefore a decrease in calcium's second-messenger effects. Now for the answer to the first question.

By means of active-transport systems in the plasma membrane and cell organelles, Ca^{2+} is maintained at an extremely low concentration in the cytosol. Consequently, there is always a large electrochemical gradient favoring diffusion of calcium into the cytosol via calcium channels in both the plasma membrane and the endoplasmic reticulum. A stimulus to the cell can alter this steady state by influencing the active-transport systems and/or the ion channels, resulting in a change in cytosolic calcium concentration.

The most common ways that receptor activation by a first messenger increases the cytosolic Ca^{2+} concentration have already been presented in this chapter and are summarized in the top part of **Table 5–4**.

The previous paragraph dealt with receptor-initiated sequences of events. This is a good place, however, to emphasize that there are calcium channels in the plasma membrane that are opened directly by an electrical stimulus to the membrane. Calcium can act as a second messenger, therefore, in response not only to chemical stimuli acting via receptors, but to electrical stimuli acting via voltage-gated calcium channels as well. Moreover, extracellular calcium entering the cell via these channels can, in certain cells, bind to calcium-sensitive channels in the endoplasmic reticulum and open them. In this manner, a small amount of extracellular calcium entering the cell can function as a second messenger to release a much larger amount of calcium from the endoplasmic reticulum. This phenomenon is called "calcium-induced calcium release." Thus, depending on the cell and the signal—first messenger or an electrical impulse—the major second messenger that releases calcium from the endoplasmic reticulum can be either IP_3 or calcium itself (see item 1b in the top of Table 5–4).

Now we turn to the question of how the increased cytosolic calcium concentration elicits the cells' responses (see bottom of Table 5–4). The common denominator of calcium's actions is its ability to bind to various cytosolic proteins, altering their conformation and thereby activating their function. One of the most important of these is a protein, found in virtually all cells, known as **calmodulin** (**Figure 5–11**). On binding with calcium, calmodulin changes shape, and this allows calcium-calmodulin to activate or inhibit a large variety of enzymes and other proteins, many of them protein kinases. Activation or inhibition of **calmodulin-dependent protein kinases** leads, via phosphorylation, to activation or inhibition of proteins involved in the cell's ultimate responses to the first messenger.

Calmodulin is not, however, the only intracellular protein influenced by calcium binding. For example, you will learn in Chapter 9 how calcium binds to a protein called troponin in certain types of muscle to initiate contraction.

Arachidonic Acid and Eicosanoids

The **eicosanoids** are a family of molecules produced from the polyunsaturated fatty acid **arachidonic acid,** which is present in plasma membrane phospholipids. The eicosanoids include the **cyclic endoperoxides,** the **prostaglandins,** the **thromboxanes,** and the **leukotrienes** (**Figure 5–12**). They

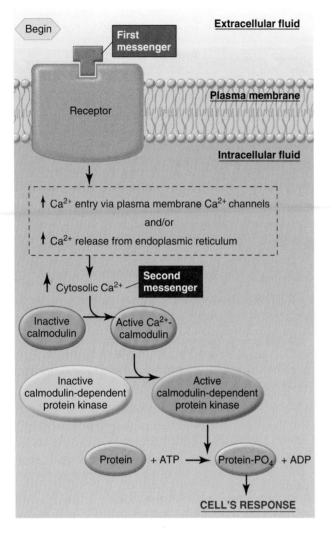

Figure 5–11

Calcium, calmodulin, and the calmodulin-dependent protein kinase system. (There are multiple calmodulin-dependent protein kinases.) Table 5–4 summarizes the mechanisms for increasing cytosolic calcium concentration.

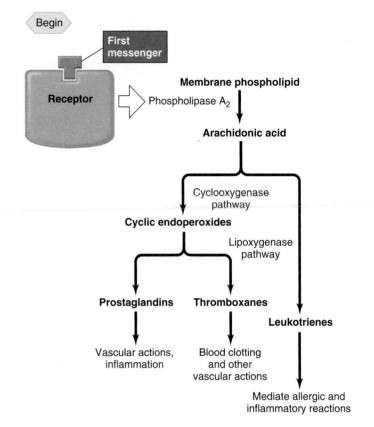

Figure 5–12

Pathways for eicosanoid synthesis and some of their major functions. Phospholipase A_2 is the one enzyme common to the formation of all the eicosanoids; it is the site at which stimuli act. Anti-inflammatory steroids inhibit phospholipase A_2. The step mediated by cyclooxygenase is inhibited by aspirin and other nonsteroidal anti-inflammatory drugs (NSAIDs). There are also drugs available that inhibit the lipoxygenase enzyme, thus blocking the formation of leukotrienes. These drugs may be helpful in controlling asthma, in which excess leukotrienes have been implicated in the allergic and inflammatory components of the disease.

are generated in many kinds of cells in response to an extracellular signal.

The synthesis of eicosanoids begins when an appropriate stimulus—hormone, neurotransmitter, paracrine agent, drug, or toxic agent—binds its receptor and activates an enzyme, **phospholipase A_2,** in the plasma membrane of the stimulated cell. As shown in Figure 5–12, this enzyme splits off arachidonic acid from the membrane phospholipids, and the arachidonic acid can then be metabolized by two pathways. One pathway is initiated by an enzyme called **cyclooxygenase (COX)** and leads ultimately to formation of the cyclic endoperoxides, prostaglandins, and thromboxanes. The other pathway is initiated by the enzyme **lipoxygenase** and leads to formation of the leukotrienes. Within both of these pathways, synthesis of the various specific eicosanoids is enzyme-mediated. Thus, beyond phospholipase A_2, the eicosanoid-pathway enzymes expressed in a particular cell determine which eicosanoids the cell synthesizes in response to a stimulus.

Each of the major eicosanoid subdivisions contains more than one member, as indicated by the use of the plural in referring to them (prostaglandins, for example). On the basis of structural differences, the different molecules within each subdivision are designated by a letter—for example, PGA and PGE for prostaglandins of the A and E types—which then may be further subdivided—for example, PGE_2.

Once they have been synthesized in response to a stimulus, the eicosanoids may in some cases act as intracellular messengers, but more often they are released immediately and act locally. Thus, the eicosanoids are usually categorized as paracrine and autocrine agents. After they act, they are quickly metabolized by local enzymes to inactive forms. The eicosanoids exert a wide array of effects, particularly on blood vessels and in inflammation. Many of these will be described in future chapters.

Because arachidonic acid transduces a signal from a messenger and its receptor into a cellular response (production and secretion of eicosanoids), it is sometimes considered a second messenger. Unlike the other second messengers discussed in this chapter, though, arachidonic acid also serves as a substrate to be converted into other products.

Finally, a word about drugs that influence the eicosanoid pathway, which are perhaps the most commonly used drugs in the world today. At the top of the list must come **aspirin,** which inhibits cyclooxygenase and, therefore, blocks the synthesis of the endoperoxides, prostaglandins, and thromboxanes. It and the new drugs that also block cyclooxygenase are collectively termed **nonsteroidal anti-inflammatory drugs (NSAIDs).** Their major uses are to reduce pain, fever, and inflammation. The term *nonsteroidal* distinguishes them from the adrenal steroids that are used in large doses as anti-inflammatory drugs; these steroids inhibit phospholipase A_2 and thus block the production of all eicosanoids.

Plasma Membrane Receptors and Gene Transcription

As described earlier in this chapter, the receptors for lipid-soluble messengers, once activated by hormone binding, act in the nucleus as transcription factors to increase or decrease the rate of gene transcription. We now emphasize that there are many other transcription factors inside cells and that the signal transduction pathways initiated by *plasma membrane* receptors often activate, by phosphorylation, these transcription factors. Thus, many first messengers that bind to plasma membrane receptors can also alter gene transcription via second messengers. For example, at least three of the proteins that cAMP-dependent protein kinase phosphorylates function as transcription factors.

Some of the genes influenced by transcription factors that are activated in response to first messengers are known collectively as **primary response genes,** or **PRGs** (also termed immediate-early genes). In many cases, especially those involving first messengers that influence the proliferation or differentiation of their target cells, the story does not stop with a PRG and the protein it encodes. In these cases, the protein the PRG encodes is itself a transcription factor for other genes (**Figure 5–13**). Thus, an initial transcription factor activated in the signal transduction pathway causes the synthesis of a different transcription factor, which in turn causes the synthesis of additional proteins, ones particularly important for the long-term biochemical events required for cellular proliferation and differentiation. A great deal of research is being done on the transcription factors PRGs encode because of their relevance to the abnormal growth and differentiation that is typical of cancer.

Cessation of Activity in Signal Transduction Pathways

Once initiated, signal transduction pathways are eventually shut off because chronic overstimulation of a cell can in some cases be detrimental. The key event is usually the cessation of receptor activation. Because organic second messengers are rapidly inactivated or broken down intracellularly (for example, cAMP by phosphodiesterase), and because calcium is continuously being pumped out of the cell or back into the endoplasmic reticulum, increases in the cytosolic concentrations of all these components are transient events. Such changes persist

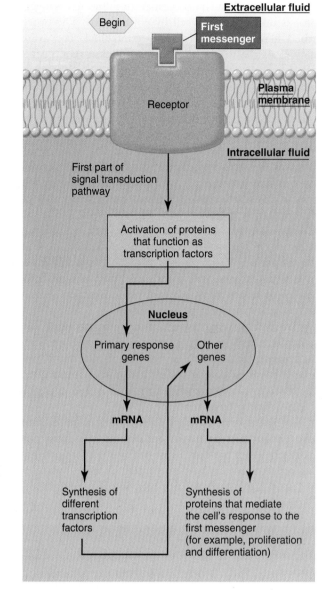

Figure 5–13

Role of multiple transcription factors and primary response genes (PRGs) in mediating protein synthesis in response to a first messenger binding to a plasma membrane receptor. The initial components of the signal transduction pathway are omitted for simplicity.

for only a brief time once the receptor is no longer being activated by a first messenger.

A major way that receptor activation ceases is by a decrease in the concentration of first messenger molecules in the region of the receptor. This occurs as enzymes in the vicinity metabolize the first messenger, as the first messenger is taken up by adjacent cells, or as it simply diffuses away.

In addition, receptors can be inactivated in at least three other ways: (1) the receptor becomes chemically altered (usually by phosphorylation), which may lower its affinity for a first messenger, and so the messenger is released; (2) phosphorylation of the receptor may prevent further G-protein binding to the receptor; and (3) plasma membrane receptors may be removed when the combination of first messenger and receptor is taken into the cell by endocytosis. The processes described here are physiologically controlled. For example, in many cases the inhibitory phosphorylation of a receptor is mediated by a protein kinase in the signal transduction pathway triggered by first-messenger binding to that very receptor. Thus, this receptor inactivation constitutes negative feedback.

This concludes our description of the basic principles of signal transduction pathways. It is essential to recognize that the pathways do not exist in isolation but may be active simultaneously in a single cell, undergoing complex interactions. This is possible because a single first messenger may trigger changes in the activity of more than one pathway and, much more importantly, because many different first messengers—often dozens—may simultaneously influence a cell. Moreover, a great deal of "cross-talk" can occur at one or more levels among the various signal transduction pathways. For example, active molecules generated in the cAMP pathway can alter the ability of receptors that, themselves, function as protein kinases to activate transcription factors.

Why should signal transduction pathways be so diverse and complex? The only way a cell can achieve controlled distinct effects in the face of the barrage of multiple first messengers, each often having more than one ultimate effect, is to have diverse pathways with branch points that may enhance one pathway and reduce another.

The biochemistry and physiology of plasma membrane signal transduction pathways are among the most rapidly expanding fields in biology. Most of this information, beyond the basic principles we have presented, exceeds the scope of this book. For example, the protein kinases we have identified are those that are closest in the various sequences to the original receptor activation. In fact, as noted earlier, there are often cascades of protein kinases in the remaining portions of the pathways. Moreover, a host of molecules other than protein kinases play "helper" roles.

Finally, for reference purposes, **Table 5–5** summarizes the biochemistry of the second messengers described in this chapter.

■ SUMMARY

Receptors

I. Receptors for chemical messengers are proteins or glycoproteins located either inside the cell or, much more commonly, in the plasma membrane. The binding of a messenger by a receptor manifests specificity, saturation, and competition.

II. Receptors are subject to physiological regulation by their own messengers. This includes down- and up-regulation.

III. Different cell types express different types of receptors; even a single cell may express multiple receptor types.

Signal Transduction Pathways

I. Binding a chemical messenger activates a receptor, and this initiates one or more signal transduction pathways leading to the cell's response.

II. Lipid-soluble messengers bind to receptors inside the target cell. The activated receptor acts in the nucleus as a transcription factor to alter the rate of transcription of specific genes, resulting in a change in the concentration or secretion of the proteins the genes encode.

Table 5–5	Reference Table of Important Second Messengers	
Substance	**Source**	**Effects**
Arachidonic acid	Converted into eicosanoids by cytoplasmic enzymes	Eicosanoids exert paracrine and autocrine effects, such as smooth muscle relaxation
Calcium	Enters cell through plasma membrane ion channels or is released from endoplasmic reticulum	Activates calmodulin and other calcium-binding proteins; calcium-calmodulin activates calmodulin-dependent protein kinases. Also activates protein kinase C
Cyclic AMP (cAMP)	A G protein activates plasma membrane adenylyl cyclase, which catalyzes the formation of cAMP from ATP	Activates cAMP-dependent protein kinase (protein kinase A)
Cyclic GMP (cGMP)	Generated from guanosine triphosphate in a reaction catalyzed by a plasma membrane receptor with guanylyl cyclase activity	Activates cGMP-dependent protein kinase (protein kinase G)
Diacylglycerol (DAG)	A G protein activates plasma membrane phospholipase C, which catalyzes the generation of DAG and IP_3 from plasma membrane phosphatidylinositol bisphosphate (PIP_2)	Activates protein kinase C
Inositol trisphosphate (IP_3)	See DAG above	Releases calcium from endoplasmic reticulum

III. Water-soluble messengers bind to receptors on the plasma membrane. The pathways induced by activation of the receptor often involve second messengers and protein kinases.
   a. The receptor may be a ligand-gated ion channel. The channel opens, resulting in an electrical signal in the membrane and, when calcium channels are involved, an increase in the cytosolic calcium concentration.
   b. The receptor may itself be an enzyme. With one exception, the enzyme activity is that of a protein kinase, usually a tyrosine kinase. The exception is the receptor that functions as a guanylyl cyclase to generate cyclic GMP.
   c. The receptor may activate a cytosolic JAK kinase associated with it.
   d. The receptor may interact with an associated plasma membrane G protein, which in turn interacts with plasma membrane effector proteins—ion channels or enzymes.
   e. Very commonly, the receptor may stimulate, via a Gs protein, or inhibit, via a Gi protein, the membrane effector enzyme adenylyl cyclase, which catalyzes the conversion of cytosolic ATP to cyclic AMP. Cyclic AMP acts as a second messenger to activate intracellular cAMP-dependent protein kinase, which phosphorylates proteins that mediate the cell's ultimate responses to the first messenger.
   f. The receptor may activate, via a G protein, the plasma membrane enzyme phospholipase C, which catalyzes the formation of diacylglycerol (DAG) and inositol trisphosphate (IP_3). DAG activates protein kinase C, and IP_3 acts as a second messenger to release calcium from the endoplasmic reticulum.

IV. The receptor, via a G protein, may directly open or close (gate) an adjacent ion channel. This differs from indirect G-protein gating of channels, in which a second messenger acts upon the channel.

V. The calcium ion is one of the most widespread second messengers.
   a. An activated receptor can increase cytosolic calcium concentration by causing certain calcium channels in the plasma membrane and/or endoplasmic reticulum to open. Voltage-gated calcium channels can also influence cytosolic calcium concentration.
   b. Calcium binds to one of several intracellular proteins, most often calmodulin. Calcium-activated calmodulin activates or inhibits many proteins, including calmodulin-dependent protein kinases.

VI. Arachidonic acid is released from phospholipids in the plasma membrane to act as a unique type of second messenger. Eicosanoids are derived from the fatty acid arachidonic acid. They exert widespread intra- and extracellular effects on cell activity.

VII. The signal transduction pathways triggered by activated plasma membrane receptors may influence genetic expression by activating transcription factors. In some cases, the primary response genes (PRGs) influenced by these transcription factors code for still other transcription factors. This is particularly true in pathways initiated by first messengers that stimulate their target cell's proliferation or differentiation.

VIII. Cessation of receptor activity occurs when the first messenger molecule concentration decreases or when the receptor is chemically altered or internalized, in the case of plasma membrane receptors.

■ REVIEW QUESTIONS

1. What is the chemical nature of receptors? Where are they located?
2. Explain why different types of cells may respond differently to the same chemical messenger.
3. Describe how the metabolism of receptors can lead to down-regulation or up-regulation.
4. What is the first step in the action of a messenger on a cell?
5. Describe the signal transduction pathway that lipid-soluble messengers use.
6. Classify plasma membrane receptors according to the signal transduction pathways they initiate.
7. What is the result of opening a membrane ion channel?
8. Contrast receptors that have intrinsic enzyme activity with those associated with cytoplasmic JAK kinases.
9. Describe the role of plasma membrane G proteins.
10. Draw a diagram describing the adenylyl cyclase-cAMP system.
11. Draw a diagram illustrating the phospholipase C/DAG/IP_3 system.
12. What are the two general mechanisms by which first messengers elicit an increase in cytosolic calcium concentration? What are the sources of the calcium in each mechanism?
13. How does the calcium-calmodulin system function?
14. Describe the manner in which activated plasma membrane receptors influence gene expression.

Chapter 5 Test Questions

(Answers appear in Appendix A.)

Match the receptor feature (a–e) with the best choice (1–3; you can use an answer more than once):

  a. affinity
  b. saturation
  c. competition
  d. down-regulation
  e. specificity

 1. defines the situation when all receptor binding sites are occupied by a messenger

 2. defines the strength of receptor binding to a messenger

 3. reflects the fact that a receptor normally binds only to a single messenger

 4. Which of the following intracellular or plasma-membrane proteins require Ca^{2+} for full activity?
  a. calmodulin
  b. Janus kinase (JAK)
  c. protein kinase A
  d. guanylyl cyclase
  e. all of the above

 5. Which is correct?
  a. Protein kinase A phosphorylates tyrosine residues.
  b. Protein kinase C is activated by cAMP.
  c. The subunit of G_s proteins that activates adenylyl cyclase is the β-subunit.
  d. Lipid-soluble messengers typically act on receptors in the cell cytosol or nucleus.
  e. The binding site of a typical plasma-membrane receptor for its messenger is located on the cytosolic surface of the receptor.

 6. Inhibition of which enzyme(s) would inhibit the conversion of arachidonic acid to leukotrienes?
  a. cyclooxygenase
  b. lipoxygenase
  c. phospholipase A_2
  d. adenylyl cyclase
  e. both b and c

Chapter 5 Quantitative and Thought Questions

(Answers appear in Appendix A.)

 1. Patient A is given a drug that blocks the synthesis of all eicosanoids, whereas patient B is given a drug that blocks the synthesis of leukotrienes but none of the other eicosanoids. What are the enzymes these drugs most likely block?

 2. Certain nerves to the heart release the neurotransmitter norepinephrine. If these nerves are removed in experimental animals, the heart becomes extremely sensitive to the administration of a drug that is an agonist of norepinephrine. Explain why, in terms of receptor physiology.

 3. A particular hormone is known to elicit, completely by way of the cyclic AMP system, six different responses in its target cell. A drug is found that eliminates one of these responses but not the other five. Which of the following, if any, could the drug be blocking: the hormone's receptors, G_s protein, adenylyl cyclase, or cyclic AMP?

 4. If a drug were found that blocked all calcium channels directly linked to G proteins, would this eliminate the role of calcium as a second messenger? Why or why not?

 5. Explain why the effects of a first messenger do not immediately cease upon removal of the messenger.

Chapter 5 Answers to Physiological Inquiries

Figure 5–5 Expressing more than one type of receptor allows a cell to respond to more than one type of first messenger. For example, one first messenger might activate a particular biochemical pathway in a cell by activating one type of receptor and signaling pathway. By contrast, another first messenger acting on a different receptor and activating a different signaling pathway, might inhibit the same biochemical process. In this way, the biochemical process can be tightly regulated.

Figure 5–8 Enzymes can generate large amounts of product without being consumed. This is an extremely efficient way to generate a second messenger like cAMP. Enzymes have many other advantages (Table 3–7) including the ability to have their activities fine-tuned by other inputs (Figures 3–35 to 3–38).

This enables the cell to adjust its response to a first messenger depending on the other conditions present.

Figure 5–9 Not necessarily. In some cases, a kinase may phosphorylate the same protein in many different types of cells. However, many cells also express certain cell-specific proteins that are not found in all tissues, and some of these proteins may be substrates for cAMP-dependent protein kinase. Thus, the proteins that are phosphorylated by a given kinase depend upon the cell type, which makes the cellular response tissue-specific. As an example, in the kidneys cAMP-dependent protein kinase phosphorylates proteins that insert water channels in cell membranes and thereby reduce urine volume, while in heart muscle the same kinase phosphorylates calcium channels that increase the strength of muscle contraction.

chapter 6

Neuronal Signaling and the Structure of the Nervous System

Micrograph of stem cells differentiating into neurons (red) and astrocytes (green).

i n order to coordinate the functions of the cells of the human body, two control systems exist. One, the endocrine system (covered in Chapter 11), is a collection of blood-borne messengers that work relatively slowly. The other is the nervous system, a rapid control system that is the focus of this chapter. Together they regulate most internal functions and organize and control the activities we know collectively as human behavior. These activities include not only such easily observed acts as smiling and walking, but also experiences such as feeling angry, being motivated, having an idea, or remembering a long-past event. Such experiences, which we attribute to the "mind," relate to the integrated activities of nerve cells in as yet unidentified ways.

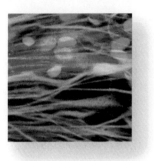

The nervous system is composed of trillions of cells distributed in a network throughout the brain, spinal cord, and periphery. These cells communicate with each other by electrical and chemical signals. They maintain homeostasis by coordinating the functions of internal organs, as well as mediating sensation, controlling movements, and encoding the fabulous complexity that is the human mind. In this chapter, we discuss the structure of individual nerve cells, the chemical and electrical mechanisms underlying nerve cell function, and the basic organization and major divisions of the nervous system.

SECTION A

Neural Tissue

The various structures of the nervous system are intimately interconnected, but for convenience we divide them into two parts: (1) the **central nervous system (CNS),** composed of the brain and spinal cord, and (2) the **peripheral nervous system,** consisting of the nerves that connect the brain or spinal cord with the body's muscles, glands, and sense organs.

The basic unit of the nervous system is the individual nerve cell, or **neuron.** Neurons operate by generating electrical signals that move from one part of the cell to another part of the same cell or to neighboring cells. In most neurons, the electrical signal causes the release of chemical messengers—**neurotransmitters**—to communicate with other cells. Most neurons serve as **integrators** because their output reflects the balance of inputs they receive from thousands or hundreds of thousands of other neurons that impinge upon them.

Structure and Maintenance of Neurons

Neurons occur in a wide variety of sizes and shapes, but all share features that allow cell-to-cell communication. Long extensions, or **processes,** connect neurons to each other and perform the neuron's input and output functions. As shown in **Figure 6–1,** most neurons contain a cell body and two types of processes—dendrites and axons.

As in other types of cells, a neuron's **cell body** (or **soma**) contains the nucleus and ribosomes and thus has the genetic information and machinery necessary for protein synthesis. The **dendrites** are a series of highly branched outgrowths of the cell body. They and the cell body receive most of the inputs from other neurons, with the dendrites taking a more important role in this regard than the cell body. The branching dendrites increase the cell's surface area—some neurons may have as many as 400,000 dendrites! Thus, dendrites increase a cell's capacity to receive signals from many other neurons.

The **axon,** sometimes also called a **nerve fiber,** is a long process that extends from the cell body and carries output to its target cells. Axons range in length from a few microns to over a meter. The region where the axon connects to the cell body is known as the **initial segment** (or **axon hillock**). The initial segment is the "trigger zone" where, in most neurons, the electrical signals are generated. These signals then propagate away from the cell body along the axon or, sometimes, back along the dendrites. The axon may have branches, called

collaterals. Near their ends, both the axon and its collaterals undergo further branching (see Figure 6–1). The greater the degree of branching of the axon and axon collaterals, the greater the cell's sphere of influence.

Each branch ends in an **axon terminal,** which is responsible for releasing neurotransmitters from the axon. These chemical messengers diffuse across an extracellular gap to the cell opposite the terminal. Alternatively, some neurons release their chemical messengers from a series of bulging areas along the axon known as **varicosities.**

The axons of many neurons are covered by **myelin** (**Figure 6–2**), which usually consists of 20 to 200 layers of highly modified plasma membrane wrapped around the axon by a nearby supporting cell. In the brain and spinal cord, these myelin-forming cells are the **oligodendrocytes.** Each oligodendrocyte may branch to form myelin on as many as 40

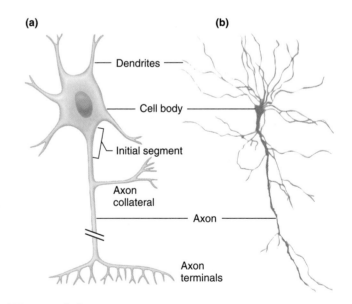

Figure 6–1

(a) Diagrammatic representation of a neuron. The break in the axon indicates that axons may extend for long distances; in fact, they may be 5000 to 10,000 times longer than the cell body is wide. This neuron is a common type, but there are a wide variety of neuronal morphologies, one of which has no axon. (b) A neuron as observed through a microscope. The axon terminals cannot be seen at this magnification.

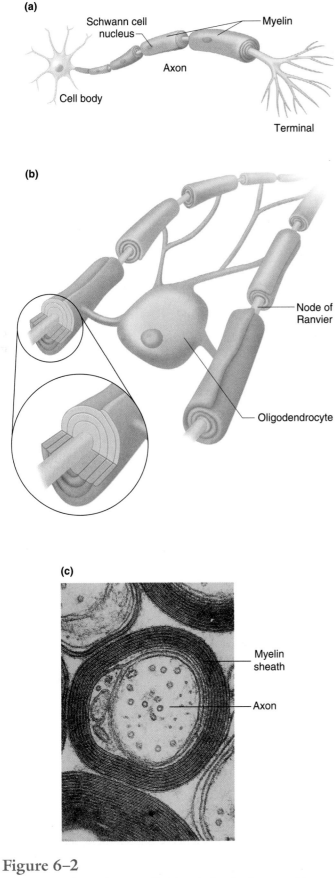

(a)

Schwann cell
nucleus

Myelin

Axon

Cell body

Terminal

(b)

Node of
Ranvier

Oligodendrocyte

(c)

Myelin
sheath

Axon

Figure 6–2

Myelin formed by Schwann cells (a) and oligodendrocytes (b) on axons. Electron micrograph of transverse sections of myelinated axons in brain (c).

axons. In the peripheral nervous system, cells called **Schwann cells** form individual myelin sheaths at regular intervals along the axons. The spaces between adjacent sections of myelin where the axon's plasma membrane is exposed to extracellular fluid are the **nodes of Ranvier.** The myelin sheath speeds up conduction of the electrical signals along the axon and conserves energy.

To maintain the structure and function of the cell axon, various organelles and other materials must move as far as one meter between the cell body and the axon terminals. This movement, termed **axonal transport,** depends on a scaffolding of microtubule "rails" running the length of the axon (Chapter 3) and specialized types of "motor proteins" known as **kinesins** and **dyneins** (**Figure 6–3**). At one end, these double-headed motor proteins bind to their cellular cargo, while the other end uses energy derived from the hydrolysis of ATP to "walk" along the microtubules. Kinesin transport mainly occurs from the cell body toward the axon terminals (**anterograde**), and is important in moving nutrient molecules, enzymes, mitochondria, neurotransmitter-filled vesicles, and other organelles. Dynein movement is in the other direction (**retrograde**), carrying recycled membrane vesicles, growth factors, and other chemical signals that can affect the neuron's morphology, biochemistry, and connectivity. Retrograde transport is also the route by which some harmful agents invade the central nervous system, including tetanus toxin and the herpes simplex, rabies, and polio viruses.

Functional Classes of Neurons

Neurons can be divided into three functional classes: afferent neurons, efferent neurons, and interneurons (**Figure 6–4**). **Afferent neurons** convey information from the tissues and organs of the body *into* the central nervous system. **Efferent neurons** convey information from the central nervous system *out* to effector cells like muscle, gland, or other nerve cells. **Interneurons** connect neurons *within* the central nervous system. As a rough estimate, for each afferent neuron entering the central nervous system, there are 10 efferent neurons and 200,000 interneurons. Thus, the great majority of neurons are interneurons.

At their peripheral ends (the ends farthest from the central nervous system), afferent neurons have **sensory receptors,** which respond to various physical or chemical changes in their environment by generating electrical signals in the neuron. The receptor region may be a specialized portion of the plasma membrane or a separate cell closely associated with the neuron ending. (Recall from Chapter 5 that the term *receptor* has two distinct meanings, the one defined here and the other referring to the specific proteins a chemical messenger combines with to exert its effects on a target cell.) Afferent neurons propagate electrical signals from their receptors into the brain or spinal cord.

Afferent neurons are unusual because they have only a single process, usually considered an axon. Shortly after leaving the cell body, the axon divides. One branch, the peripheral process, begins at the receptors. The other branch, the central process, enters the central nervous system to form

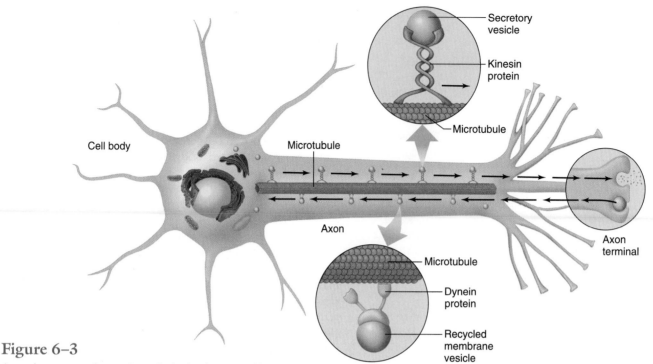

Figure 6–3

Axonal transport along microtubules by dynein and kinesin.

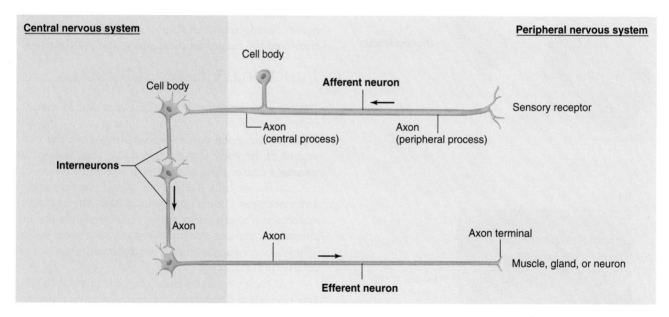

Figure 6–4

Three classes of neurons. The arrows indicate the direction of transmission of neural activity. Afferent neurons in the peripheral nervous system generally receive input at sensory receptors. Efferent components of the peripheral nervous system may terminate on muscle, gland, or neuron effectors. Both afferent and efferent components may consist of two neurons, not one as shown here.

junctions with other neurons. Note in Figure 6–4 that for afferent neurons both the cell body and the long axon are outside the central nervous system, and only a part of the central process enters the brain or spinal cord.

Generally, the cell bodies and dendrites of efferent neurons are within the central nervous system, and the axons extend out to the periphery (a notable exception is the enteric nervous system of the gastrointestinal tract, described in

Chapter 15). Groups of afferent and efferent neurons form the **nerves** of the peripheral nervous system. Note that a nerve fiber is a single axon, and a nerve is a bundle of axons (fibers) bound together by connective tissue.

Interneurons lie entirely within the central nervous system. They account for over 99 percent of all neurons and have a wide range of physiological properties, shapes, and functions. The number of interneurons interposed between specific afferent and efferent neurons varies according to the complexity of the action they control. The knee-jerk reflex elicited by tapping below the kneecap requires no interneurons—the afferent neurons interact directly with efferent neurons. In contrast, when you hear a song or smell a certain perfume that evokes memories of someone you once knew, millions of interneurons may be involved.

Table 6–1 summarizes the characteristics of the three functional classes of neurons.

The anatomically specialized junction between two neurons where one neuron alters the electrical and chemical activity of another is called a **synapse.** At most synapses, the signal is transmitted from one neuron to another by neurotransmitters, a term that also includes the chemicals efferent neurons use to communicate with effector cells (e.g., a muscle cell). The neurotransmitters released from one neuron alter the receiving neuron by binding with specific protein receptors on the membrane of the receiving neuron. (Once again, do not confuse this use of the term *receptor* with the sensory receptors at the peripheral ends of afferent neurons.)

Most synapses occur between an axon terminal of one neuron and a dendrite or the cell body of a second neuron. Sometimes, however, synapses occur between two dendrites or between a dendrite and a cell body or between an axon terminal and a second axon terminal. A neuron that conducts a signal toward a synapse is called a **presynaptic neuron,** whereas a neuron conducting signals away from a synapse is a **postsynaptic neuron. Figure 6–5** shows how, in a multineuronal pathway, a single neuron can be postsynaptic to one cell and presynaptic to another. A postsynaptic neuron may have thousands of synaptic junctions on the surface of its dendrites and cell body, so that signals from many presynaptic neurons can affect it.

Table 6–1	Characteristics of Three Classes of Neurons

I. AFFERENT NEURONS
   A. Transmit information into the central nervous system from receptors at their peripheral endings
   B. Cell body and the long peripheral process of the axon are in the peripheral nervous system; only the short central process of the axon enters the central nervous system
   C. Most have no dendrites (do not receive inputs from other neurons)

II. EFFERENT NEURONS
   A. Transmit information out of the central nervous system to effector cells, particularly muscles, glands, or other neurons
   B. Cell body, dendrites, and a small segment of the axon are in the central nervous system; most of the axon is in the peripheral nervous system

III. INTERNEURONS
   A. Function as integrators and signal changers
   B. Integrate groups of afferent and efferent neurons into reflex circuits
   C. Lie entirely within the central nervous system
   D. Account for 99 percent of all neurons

Glial Cells

Neurons account for only about 10 percent of the cells in the central nervous system. The remainder are **glial cells,** also called neuroglia (*glia* = glue). However, because neurons branch more extensively than glia do, neurons occupy about 50 percent of the volume of the brain and spinal cord.

Glial cells surround the soma, axon, and dendrites of neurons and provide them with physical and metabolic support (**Figure 6–6**). As noted earlier, one type of glial cell, the oligodendrocyte, forms the myelin covering of CNS axons.

A second type of glial cell, the **astrocyte,** helps regulate the composition of the extracellular fluid in the central nervous system by removing potassium ions and neurotransmitters around synapses. Another important function of astrocytes is to stimulate the formation of tight junctions between the cells that make up the walls of capillaries found in the central nervous system. This forms the **blood-brain barrier,** which prevents toxins and other substances from entering the brain. Astrocytes also sustain the neurons metabolically—for example, by providing glucose and removing ammonia. In developing embryos, astrocytes guide neurons as they migrate to their ultimate destination, and they stimulate neuronal growth by secreting growth factors. In addition, astrocytes have many neuron-like characteristics. For example, they have ion channels, receptors for certain neurotransmitters and the

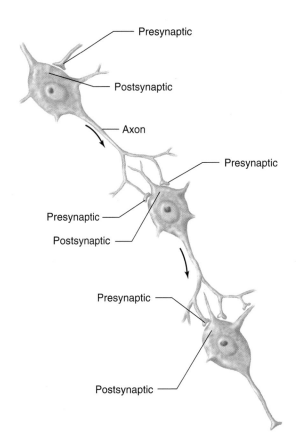

Figure 6–5

A neuron postsynaptic to one cell can be presynaptic to another. Arrows indicate direction of neural transmission.

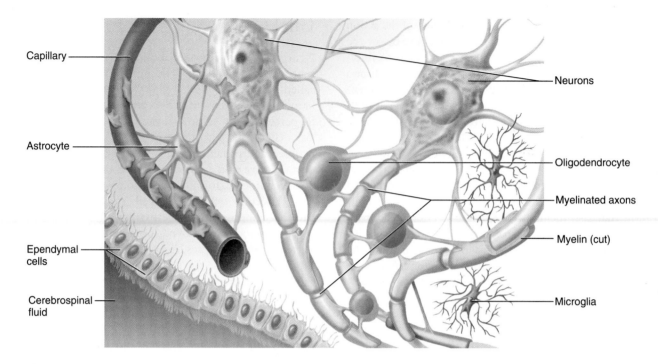

Figure 6–6

Glial cells of the central nervous system.

enzymes for processing them, and the capability of generating weak electrical responses. Thus, in addition to all their other roles, it is speculated that astrocytes may take part in information signaling in the brain.

A third type of glial cell, the **microglia,** is a specialized, macrophage-like cell (Chapter 18) that performs immune functions in the central nervous system. Lastly, **ependymal cells** line the fluid-filled cavities within the brain and spinal cord and regulate the production and flow of cerebrospinal fluid, which will be described later.

Schwann cells, the glial cells of the peripheral nervous system, have most of the properties of the central nervous system glia. As mentioned earlier, Schwann cells produce the myelin sheath of peripheral nerve fibers.

Neural Growth and Regeneration

The elaborate networks of nerve cell processes that characterize the nervous system are remarkably similar in all human beings and depend upon the outgrowth of specific axons to specific targets.

Development of the nervous system in the embryo begins with a series of divisions of undifferentiated precursor cells **(stem cells)** that can develop into neurons or glia. After the last cell division, each neuronal daughter cell differentiates, migrates to its final location, and sends out processes that will become its axon and dendrites. A specialized enlargement, the **growth cone,** forms the tip of each extending axon and is involved in finding the correct route and final target for the process.

As the axon grows, it is guided along the surfaces of other cells, most commonly glial cells. Which route the axon follows depends largely on attracting, supporting, deflecting, or inhibiting influences exerted by several types of molecules.

Some of these molecules, such as cell adhesion molecules, reside on the membranes of the glia and embryonic neurons. Others are soluble **neurotrophic factors** (growth factors for neural tissue) in the extracellular fluid surrounding the growth cone or its distant target.

Once the target of the advancing growth cone is reached, synapses form. The synapses are active, however, before their final maturation. This early activity, in part, determines their final function. During these early stages of neural development, which occur during all trimesters of pregnancy and into infancy, alcohol and other drugs, radiation, malnutrition, and viruses can exert effects that cause permanent damage to the developing fetal nervous system.

A surprising aspect of development of the nervous system occurs after growth and projection of the axons. Many of the newly formed neurons and synapses degenerate. In fact, as many as 50 to 70 percent of neurons undergo a programmed self-destruction called **apoptosis** in the developing central nervous system! Exactly why this seemingly wasteful process occurs is unknown, although neuroscientists speculate that this refines or fine-tunes connectivity in the nervous system.

The basic shapes and locations of existing neurons in the mature central nervous system do not change. The creation and removal of synaptic contacts begun during fetal development continue, however, though at a slow pace throughout life as part of normal growth, learning, and aging. Division of neuron precursor stem cells is largely complete before birth.

If axons are severed, they can repair themselves and restore significant function provided that the damage occurs outside the central nervous system and does not affect the neuron's cell body. After such an injury, the axon segment that is separated from the cell body degenerates. The part of the axon still attached to the cell body then gives rise to a growth cone, which grows out to the effector organ so that function is sometimes restored. Return of function following a peripheral nerve injury is delayed because axon regrowth proceeds at a rate of only 1 mm per day. So, for example, if afferent neurons from your thumb were damaged by an injury in the area of your shoulder, it might take two years for sensation in your thumb to be restored.

In humans, spinal injuries typically crush rather than cut the tissue, leaving the axons intact. In this case, a primary problem is self-destruction (apoptosis) of the nearby oligodendrocytes. When these cells die and their associated axons lose their myelin coat, the axons cannot transmit information effectively. Severed axons within the central nervous system may sprout, but no significant regeneration of the axon occurs across the damaged site, and there are no well-documented reports of significant return of function. Either some basic difference of central nervous system neurons or some property of their environment, such as inhibitory factors associated with nearby glia, prevents their functional regeneration.

Researchers are trying a variety of ways to provide an environment that will support axonal regeneration in the central nervous system. They are creating tubes to support regrowth of the severed axons, redirecting the axons to regions of the spinal cord that lack growth-inhibiting factors, preventing apoptosis of the oligodendrocytes so myelin can be maintained, and supplying neurotrophic factors that support recovery of the damaged tissue.

Medical researchers are also attempting to restore function to damaged or diseased brains by implanting progenitor stem cells that will develop into new neurons and replace missing neurotransmitters or neurotrophic factors. Alternatively, pieces of fetal brain or tissues from the patient that produce the needed neurotransmitters or growth factors have been implanted. For example, in patients with **Parkinson's disease,** a degenerative nervous system disease resulting in progressive loss of movement, the implantation of tissue from posterior portions of a fetal brain into the affected area has been somewhat successful in restoring motor function. (Ethical concerns have rendered the future of this technique uncertain, however.)

■ SECTION A SUMMARY

Structure and Maintenance of Neurons

I. The nervous system is divided into two parts. The central nervous system (CNS) comprises the brain and spinal cord, and the peripheral nervous system consists of nerves extending from the CNS.
II. The basic unit of the nervous system is the nerve cell, or neuron.
III. The cell body and dendrites receive information from other neurons.
IV. The axon (nerve fiber), which may be covered with sections of myelin separated by nodes of Ranvier, transmits information to other neurons or effector cells.

Functional Classes of Neurons

I. Neurons are classified in three ways:
   a. *Afferent neurons* transmit information into the CNS from receptors at their peripheral endings.
   b. *Efferent neurons* transmit information out of the CNS to effector cells.

   c. *Interneurons* lie entirely within the CNS and form circuits with other interneurons or connect afferent and efferent neurons.
II. Neurotransmitters, which are released by a presynaptic neuron and combine with protein receptors on a postsynaptic neuron, transmit information across a synapse.

Glial Cells

I. The CNS also contains glial cells, which help regulate the extracellular fluid composition, sustain the neurons metabolically, form myelin and the blood-brain barrier, serve as guides for developing neurons, provide immune functions, and regulate cerebrospinal fluid.

Neural Growth and Regeneration

I. Neurons develop from stem cells, migrate to their final locations, and send out processes to their target cells.
II. Cell division to form new neurons is markedly slowed after birth.
III. After degeneration of a severed axon, damaged peripheral neurons may regrow the axon to their target organ. In the CNS, there is some regeneration of neurons, but it is not yet known how significant this is for function.

■ SECTION A KEY TERMS

afferent neuron 139	kinesin 139
anterograde 139	microglia 142
apoptosis 142	myelin 138
astrocyte 141	nerve 140
axon 138	nerve fiber 138
axon hillock 138	neuron 138
axon terminal 138	neurotransmitter 138
axonal transport 139	neurotrophic factor 142
blood-brain barrier 141	node of Ranvier 139
cell body 138	oligodendrocyte 138
central nervous system	peripheral nervous system 138
(CNS) 138	postsynaptic neuron 141
collateral 138	presynaptic neuron 141
dendrite 138	process 138
dynein 139	retrograde 139
efferent neuron 139	Schwann cell 139
ependymal cell 142	sensory receptor 139
glial cell 141	soma 138
growth cone 142	stem cell 142
initial segment 138	synapse 141
integrator 138	varicosity 138
interneuron 139	

■ SECTION A CLINICAL TERMS

Parkinson's disease   143

■ SECTION A REVIEW QUESTIONS

1. Describe the direction of information flow through a neuron in response to input from another neuron. What is the relationship between the presynaptic neuron and the postsynaptic neuron?
2. Contrast the two uses of the word *receptor.*
3. Where are afferent neurons, efferent neurons, and interneurons located in the nervous system? Are there places where all three could be found?

Basic Principles of Electricity

As discussed in Chapter 4, the predominant solutes in the extracellular fluid are sodium and chloride ions. The intracellular fluid contains high concentrations of potassium ions and ionized nondiffusible molecules, particularly proteins with negatively charged side chains and phosphate compounds. Electrical phenomena resulting from the distribution of these charged particles occur at the cell's plasma membrane and play a significant role in signal integration and cell-to-cell communication, the two major functions of the neuron.

Charges of the same type repel each other—positive charge repels positive charge, and negative charge repels negative charge. In contrast, oppositely charged substances attract each other and will move toward each other if not separated by some barrier (**Figure 6–7**).

Separated electrical charges of opposite sign have the potential to do work if they are allowed to come together. This potential is called an **electrical potential** or, because it is determined by the difference in the amount of charge between two points, a **potential difference.** The electrical potential difference is often referred to simply as the **potential.** The units of electrical potential are volts. The total charge that can be separated in most biological systems is very small, so the potential differences are small and are measured in millivolts (1 mV = 0.001 V).

The movement of electrical charge is called a **current.** The electrical potential between charges tends to make them flow, producing a current. If the charges are opposite, the current brings them toward each other; if the charges are alike, the current increases the separation between them. The amount of charge that moves—in other words, the current—depends on the potential difference between the charges and on the nature of the material or structure through which they are moving. The hindrance to electrical charge movement is known as **resistance.** If resistance is high, the current flow will be low. The effect of voltage V and resistance R on current I is expressed in **Ohm's law:**

$$I = \frac{V}{R}$$

Materials that have a high electrical resistance reduce current flow and are known as insulators. Materials that have a low resistance allow rapid current flow and are called conductors.

Water that contains dissolved ions is a relatively good conductor of electricity because the ions can carry the current. As we have seen, the intracellular and extracellular fluids contain many ions and can therefore carry current. Lipids, however, contain very few charged groups and cannot carry current. Therefore, the lipid layers of the plasma membrane are regions of high electrical resistance separating the intracellular fluid and the extracellular fluid, two low-resistance water compartments.

The Resting Membrane Potential

All cells under resting conditions have a potential difference across their plasma membranes, with the inside of the cell negatively charged with respect to the outside (**Figure 6–8a**). This potential is the **resting membrane potential.**

By convention, extracellular fluid is assigned a voltage of zero, and the polarity (positive or negative) of the membrane potential is stated in terms of the sign of the excess charge on the inside of the cell. For example, if the intracellular fluid has an excess of negative charge and the potential difference across the membrane has a magnitude of 70 mV, we say that the membrane potential is –70 mV (inside relative to outside).

The magnitude of the resting membrane potential varies from about –5 to –100 mV, depending upon the type of cell. In neurons, it is generally in the range of –40 to –90 mV (**Figure 6–8b**). The resting membrane potential holds steady unless changes in electrical current alter the potential.

The resting membrane potential exists because of a tiny excess of negative ions inside the cell and an excess of positive ions outside. The excess negative charges inside are electrically attracted to the excess positive charges outside the cell, and vice versa. Thus, the excess charges (ions) collect in a thin shell tight against the inner and outer surfaces of the plasma membrane (**Figure 6–9**), whereas the bulk of the intracellular and extracellular fluids remain neutral. Unlike the diagrammatic representation in Figure 6–9, the number

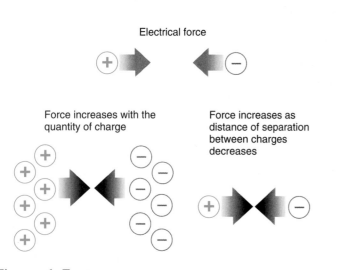

Electrical force

Force increases with the quantity of charge

Force increases as distance of separation between charges decreases

Figure 6–7

The electrical force of attraction between positive and negative charges increases with the quantity of charge and with decreasing distance between charges.

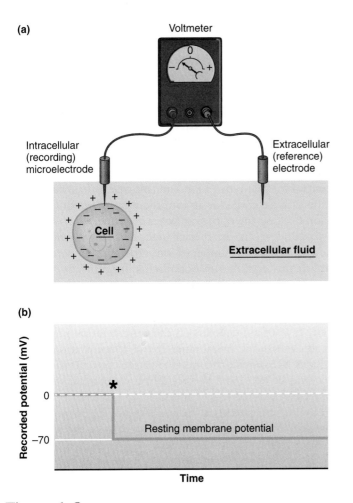

(a) Voltmeter

Intracellular
(recording)
microelectrode

Extracellular
(reference)
electrode

Cell

Extracellular fluid

(b)

Recorded potential (mV)

0

−70

Resting membrane potential

Time

Figure 6–8

(a) Apparatus for measuring membrane potentials. The voltmeter records the difference between the intracellular and extracellular electrodes. (b) The potential difference across a plasma membrane as measured by an intracellular microelectrode. The asterisk indicates the moment the electrode entered the cell.

of positive and negative charges that have to be separated across a membrane to account for the potential is actually an infinitesimal fraction of the total number of charges in the two compartments.

Table 6–2 lists the concentrations of sodium, potassium, and chloride ions in the extracellular fluid and in the intracellular fluid of a representative nerve cell. Each of these ions has a 10- to 30-fold difference in concentration between the inside and the outside of the cell. Although this table appears to contradict our earlier assertion that the bulk of the intra- and extracellular fluids are electrically neutral, there are many other ions, including Mg^{2+}, Ca^{2+}, H^+, HCO_3^-, HPO_4^{2-}, SO_4^{2-}, amino acids, and proteins, in both fluid compartments. Of the diffusable ions, sodium, potassium, and chloride ions are present in the highest concentrations, and the membrane permeability to each is independently determined. Sodium and potassium generally play the most important roles in generating the resting membrane potential, but in some cells chloride is also a factor. Note that the sodium and chloride

concentrations are lower inside the cell than outside, and that the potassium concentration is greater inside the cell. The concentration differences for sodium and potassium are established by the action of the sodium-potassium pump (Na^+/K^+-ATPase, Chapter 4) that pumps sodium out of the cell and potassium into it. The reason for the chloride distribution varies among cell types, as will be described later.

The magnitude of the resting membrane potential depends mainly on two factors: (1) differences in specific ion concentrations in the intracellular and extracellular fluids, and (2) differences in membrane permeabilities to the different ions, which reflect the number of open channels for the different ions in the plasma membrane.

To understand how concentration differences for sodium and potassium create membrane potentials, first consider what happens when the membrane is permeable (has open channels)

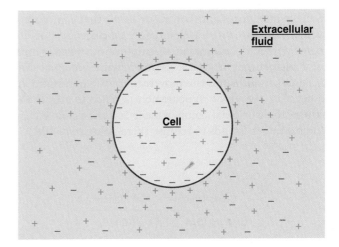

Extracellular fluid

Cell

Figure 6–9

The excess positive charges outside the cell and the excess negative charges inside collect in a tight shell against the plasma membrane. In reality, these excess charges are only an extremely small fraction of the total number of ions inside and outside the cell.

Table 6–2	Distribution of Major Mobile Ions Across the Plasma Membrane of a Typical Nerve Cell	
	Concentration, mmol/L	
Ion	**Extracellular**	**Intracellular**
Na^+	145	15
Cl^-	100	7*
K^+	5	150

A more accurate measure of electrical driving force can be obtained using mEq/L, which factors in ion valence. Since all the ions in this table have a valence of 1, the mEq/L is the same as the mmol/L concentration.

*Intracellular chloride concentration varies significantly between neurons due to differences in expression of membrane transporters and channels.

to only one ion (**Figure 6–10**). In this hypothetical situation, it is assumed that the membrane contains potassium channels but no sodium or chloride channels. Initially, compartment 1 contains 0.15 M NaCl, compartment 2 contains 0.15 M KCl, and no ion movement occurs because the channels are closed (**Figure 6–10a**). There is no potential difference across the membrane because the two compartments contain equal numbers of positive and negative ions. The positive ions are different—sodium versus potassium, but the *total* numbers of positive ions in the two compartments are the same, and each positive ion balances a chloride ion.

However, if these potassium channels are opened, potassium will diffuse down its concentration gradient from compartment 2 into compartment 1 (**Figure 6–10b**). Sodium ions will not be able to move across the membrane. After a few potassium ions have moved into compartment 1, that compartment will have an excess of positive charge, leaving behind an excess of negative charge in compartment 2 (**Figure 6–10c**). Thus, a potential difference has been created across the membrane.

This introduces another major factor that can cause net movement of ions across a membrane: an electrical potential. As compartment 1 becomes increasingly positive and compartment 2 increasingly negative, the membrane potential difference begins to influence the movement of the potassium ions. The negative charge of compartment 2 tends to attract them back into their original compartment and the positive charge of compartment 1 tends to repulse them (**Figure 6–10d**).

As long as the flux or movement of ions due to the potassium concentration gradient is greater than the flux due to the membrane potential, net movement of potassium will occur from compartment 2 to compartment 1 (see Figure 6–10d), and the membrane potential will progressively increase. However, eventually the membrane potential will become negative enough to produce a flux equal but opposite to the flux produced by the concentration gradient (**Figure 6–10e**). The membrane potential at which these two fluxes become equal in magnitude but opposite in direction is called the **equilibrium potential** for that type of ion—in this case, potassium. At the equilibrium potential for an ion, there is no *net* movement of the ion because the opposing fluxes are equal, and the potential will undergo no further change. It is worth emphasizing once again that the number of ions crossing the membrane to establish this equilibrium potential is insignificant compared to the number originally present in compartment 2, so there is no measurable change in the potassium concentration.

The magnitude of the equilibrium potential (in mV) for any type of ion depends on the concentration gradient for that ion across the membrane. If the concentrations on the two sides were equal, the flux due to the concentration gradient would be zero, and the equilibrium potential would also be zero. The larger the concentration gradient, the larger the equilibrium potential because a larger, electrically driven movement of ions will be required to balance the movement due to the concentration difference.

Now consider the situation when the membrane separating the two compartments is replaced with one that contains only sodium channels. A parallel situation will occur (**Figure 6–11**). Na^+ ions will initially move from compartment 1 to compartment 2. When compartment 2 is positive with respect to compartment 1, the difference in electrical charge across the membrane will begin to drive Na^+ ions from compartment 2 back to compartment 1, and eventually net movement of sodium will cease. Again, at the equilibrium potential, the movement of ions due to the concentration gradient is equal but opposite to the movement due to the electrical gradient.

Thus, the equilibrium potential for one ion species can be different in magnitude *and* direction from those for other ion species, depending on the concentration gradients between the intracellular and extracellular compartments for each ion. If the concentration gradient for any ion is known, the equilibrium potential for that ion can be calculated by means of the Nernst equation.

The **Nernst equation** describes the equilibrium potential for any ion species—that is, the electrical potential necessary to balance a given ionic concentration gradient across a membrane so that the net flux of the ion is zero. The Nernst equation is

$$E_{\text{ion}} = \frac{61}{Z} \log\left(\frac{C_{\text{o}}}{C_{\text{i}}}\right)$$

where

E_{ion} = equilibrium potential for a particular ion, in mV
C_{i} = intracellular concentration of the ion
C_{o} = extracellular concentration of the ion
Z = the valence of the ion

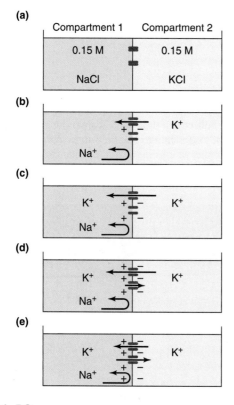

Figure 6–10

Generation of a potential across a membrane due to diffusion of K^+ through potassium channels (red). Arrows represent ion movements. See the text for a complete explanation of the steps a–e.

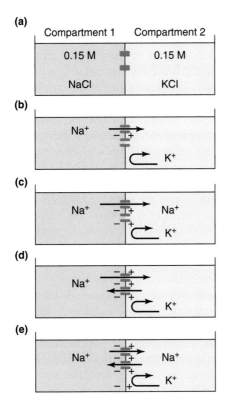

Figure 6–11

Generation of a potential across a membrane due to diffusion of Na⁺ through sodium channels (blue). Arrows represent ion movements. See the text for a fuller explanation.

61 is a constant value that takes into account the universal gas constant, the temperature (37°C), and the Faraday electrical constant.

Using the concentration gradients from Table 6–2, the equilibrium potentials for sodium (E_{Na}) and potassium (E_K) are

$$E_{Na} = \frac{61}{+1} \log\left(\frac{145}{15}\right) = +60 \text{ mV}$$

$$E_K = \frac{61}{+1} \log\left(\frac{5}{150}\right) = -90 \text{ mV}$$

Thus, at these typical concentrations, sodium flux through open channels will tend to bring the membrane potential toward +60 mV, while potassium flux will bring it toward –90 mV. If the concentration gradients change, the equilibrium potentials will change.

When channels for more than one ion species are open in the membrane at the same time, the permeabilities and concentration gradients for all the ions must be considered when accounting for the membrane potential. For a given concentration gradient, the greater the membrane permeability to an ion species, the greater the contribution that ion species will make to the membrane potential. Given the concentration gradients and relative membrane permeabilities (P_{ion}) for sodium, potassium, and chloride, the potential of a membrane

(V_m) can be calculated using the **Goldman-Hodgkin-Katz (GHK) equation:**

$$V_m = 61 \log \frac{P_K[K_o] + P_{Na}[Na_o] + P_{Cl}[Cl_i]}{P_K[K_i] + P_{Na}[Na_i] + P_{Cl}[Cl_o]}$$

The GHK equation is essentially an expanded version of the Nernst equation that takes into account individual ion permeabilities. In fact, setting the permeabilities of any two ions to zero gives the equilibrium potential for the remaining ion. Note that the chloride concentrations are reversed as compared to sodium and potassium (the inside concentration is in the numerator and the outside in the denominator) because chloride is an anion, and its movement has the opposite effect on the membrane potential. In an actual nerve cell at rest, there are many more open potassium channels than sodium channels; chloride permeability generally falls in between. Typical values for relative permeabilities are: $P_K = 1$, $P_{Na} = 0.04$, and $P_{Cl} = 0.45$. Inserting those values (along with the concentrations in Table 6–2) into the GHK equation allows us to calculate the resting membrane potential taking all of these ions into account:

$$V_m = 61 \log \frac{(1)(5) + (.04)(145) + (.45)(7)}{(1)(150) + (.04)(15) + (.45)(100)} = -70 \text{ mV}$$

The contributions of sodium and potassium to the overall membrane potential are a function of their concentration gradients and relative permeabilities. The concentration gradients determine their equilibrium potentials, and the relative permeability determines how strongly the resting membrane potential is influenced toward those potentials. Potassium has by far the highest permeability, which explains why a typical neuron's resting membrane potential is much closer to the equilibrium potential for potassium than for sodium (**Figure 6–12**). Based on its permeability, you might think that chloride would also have a strong influence on the resting membrane potential. This turns out not to be the case, for reasons that we will return to shortly.

In other words, the resting potential is generated across the plasma membrane largely because of the movement of potassium out of the cell down its concentration gradient through open or so-called **leak potassium channels,** so that the inside of the cell becomes negative with respect to the outside. Even though potassium flux has more impact on the resting membrane potential than does sodium flux, the resting membrane potential is not *equal* to the potassium equilibrium potential, because a small number of sodium channels are open in the resting state. Some sodium ions continually move into the cell, canceling the effect of an equivalent number of potassium ions simultaneously moving out. Thus, ion channels allow net movement of sodium into the cell and potassium out of the cell.

Over time, the concentration of intracellular sodium and potassium ions does not change, however, because the Na⁺/K⁺-ATPase pump maintains the sodium and potassium concentrations at stable levels. In a resting cell, the number of ions the pump moves equals the number of ions that move in

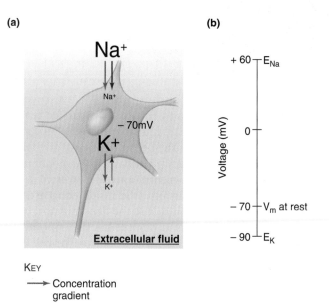

(a)

Na⁺

−70mV

K+

Extracellular fluid

(b)

+ 60 — E_Na

0

− 70 — V_m at rest

− 90 — E_K

Voltage (mV)

KEY

→ Concentration gradient

→ Electrical gradient

Figure 6–12

Forces influencing sodium and potassium ions at the resting membrane potential. (a) At a resting membrane potential of –70 mV both the concentration and electrical gradients favor inward movement of sodium, while the potassium concentration and electrical gradients are in opposite directions. (b) The greater permeability and movement of potassium maintains the resting membrane potential at a value near E_K.

Figure 6–12 physiological *inquiry* (*pi*)

- Would lowering a neuron's intracellular $[K^+]$ by 1 mM have the same effect on resting membrane potential as raising the extracellular fluid $[K^+]$ by 1 mM?

Answer can be found at end of chapter.

the opposite direction through membrane channels down their concentration and/or electrical gradients (described collectively in Chapter 4 as the *electrochemical gradient*). As long as the concentration gradients remain stable and the ion permeabilities of the plasma membrane do not change, the electrical potential across the resting membrane will also remain constant.

Thus far, we have described the membrane potential as due purely and directly to the passive movement of ions down their electrochemical gradients, with the concentration gradients maintained by membrane pumps. However, the Na⁺/K⁺-ATPase pump not only maintains the concentration gradients for these ions, but also helps to establish the membrane potential more directly. The Na⁺/K⁺-ATPase pumps actually move three sodium ions out of the cell for every two potassium ions that they bring in. This unequal transport of positive ions makes the inside of the cell more negative than it would be from ion diffusion alone. When a pump moves net charge across the membrane and contributes directly to the membrane potential, it is known as an **electrogenic pump.**

In most cells, the electrogenic contribution to the membrane potential is quite small. It must be reemphasized,

however, that even though the electrogenic contribution of the Na⁺/K⁺-ATPase pump is small, the pump always makes an essential *indirect* contribution to the membrane potential because it maintains the concentration gradients down which the ions diffuse to produce most of the charge separation that makes up the potential.

Figure 6–13 summarizes in three conceptual steps how a resting membrane potential develops. First, the action of the Na⁺/K⁺-ATPase pump sets up the concentration gradients for sodium and potassium (**Figure 6–13a**). These concentration gradients determine the equilibrium potentials for the two ions—that is, the value to which each ion would bring the membrane potential if it were the only permeating ion. Simultaneously, the pump has a small electrogenic effect on the membrane due to the fact that three sodium ions are pumped out for every two potassium ions pumped in. The next step shows that initially there is a greater flux of potassium out of the cell than sodium into the cell (**Figure 6–13b**). This is because in a resting membrane there are a greater number of open potassium channels than there are sodium channels. Because there is greater net efflux than influx of positive ions during this step, a significant negative membrane potential develops, with the value approaching that of the potassium equilibrium potential. In the steady-state resting neuron, the flux of ions across the membrane reaches a dynamic balance (**Figure 6–13c**). Because the membrane potential is not equal to the equilibrium potential for either ion, there is a small but steady leak of sodium into the cell and potassium out of the cell. The concentration gradients do not dissipate over time, however, because ion movement by the Na⁺/K⁺-ATPase pump exactly balances the rate at which the ions leak through open channels.

Now let's return to the behavior of chloride ions in this system. The plasma membranes of many cells also have chloride channels but do not contain chloride-ion pumps. Therefore, in these cells chloride concentrations simply shift until the equilibrium potential for chloride is equal to the resting membrane potential. In other words, the negative membrane potential determined by sodium and potassium moves chloride out of the cell, and the chloride concentration inside the cell becomes lower than that outside. This concentration gradient produces a diffusion of chloride back into the cell that exactly opposes the movement out because of the electrical potential.

In contrast, some cells have a nonelectrogenic active transport system that moves chloride out of the cell, generating a strong concentration gradient. In these cells, the membrane potential is not at the chloride equilibrium potential, and net chloride diffusion into the cell contributes to the excess negative charge inside the cell; that is, net chloride diffusion makes the membrane potential more negative than it would otherwise be.

We noted earlier that most of the negative charge in neurons is accounted for not by chloride ions but by negatively charged organic molecules, such as proteins and phosphate compounds. Unlike chloride, however, these molecules do not readily cross the plasma membrane. Instead they remain inside the cell, where their charge contributes to the total negative charge within the cell.

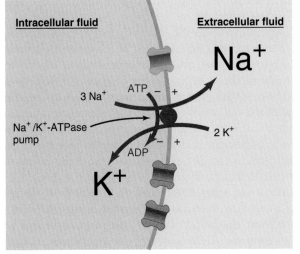

(a)

Intracellular fluid | Extracellular fluid

Na⁺

3 Na⁺ ATP − +

Na⁺/K⁺-ATPase
pump

2 K⁺

ADP − +

K⁺

(b)

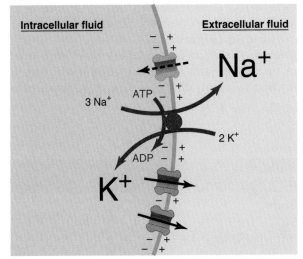

Intracellular fluid | Extracellular fluid

Na⁺

3 Na⁺ ATP

2 K⁺

ADP

K⁺

(c)

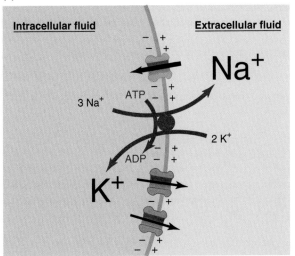

Intracellular fluid | Extracellular fluid

Na⁺

3 Na⁺ ATP

2 K⁺

ADP

K⁺

Figure 6–13

Summary of steps establishing the resting membrane potential.
(a) Na⁺/K⁺-ATPase pump establishes concentration gradients and
generates a small negative potential. (b) Greater net movement of
potassium than sodium makes the membrane potential more negative
on the inside. (c) At a steady negative resting membrane potential,
ion fluxes through the channels and pump balance each other.

Graded Potentials and Action Potentials

Transient changes in the membrane potential from its resting level produce electrical signals. Such changes are the most important way that nerve cells process and transmit information. These signals occur in two forms: graded potentials and action potentials. Graded potentials are important in signaling over short distances, whereas action potentials are the long-distance signals of nerve and muscle membranes.

The terms *depolarize, repolarize,* and *hyperpolarize* are used to describe the direction of changes in the membrane potential relative to the resting potential (**Figure 6–14**). The resting membrane potential, at −70 mV, is polarized. "Polarized" simply means that the outside and inside of a cell have a different net charge. The membrane is **depolarized** when its potential becomes less negative (closer to zero) than the resting level. **Overshoot** refers to a reversal of the membrane potential polarity—that is, when the inside of a cell becomes positive relative to the outside. When a membrane potential that has been depolarized returns toward the resting value, it is **repolarizing**. The membrane is **hyperpolarized** when the potential is more negative than the resting level.

The changes in membrane potential that the neuron uses as signals occur because of changes in the permeability of the cell membrane to ions. Recall from Chapter 4 that some channels in the membrane are gated; that is, opened or closed by mechanical, electrical, or chemical stimuli. When a neuron receives a chemical signal from a neighboring neuron, for instance, some channels will open, allowing greater ionic current across the membrane. The greater movement of ions down their concentration gradient alters the membrane potential so that it is either depolarized or hyperpolarized relative to the resting state. We will see that particular characteristics of these gated channels play a role in determining the nature of the electrical signal generated.

Graded Potentials

Graded potentials are changes in membrane potential that are confined to a relatively small region of the plasma membrane. They are usually produced when some specific change

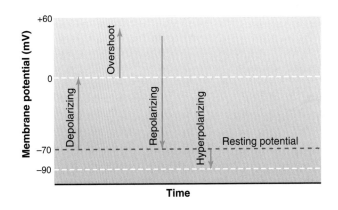

Figure 6–14

Depolarizing, repolarizing, hyperpolarizing, and overshoot changes in membrane potential relative to the resting potential.

in the cell's environment acts on a specialized region of the membrane. They are called graded potentials simply because the magnitude of the potential change can vary (is "graded"). Graded potentials are given various names related to the location of the potential or the function they perform; for instance, receptor potential, synaptic potential, and pacemaker potential (**Table 6–3**).

Whenever a graded potential occurs, charge flows between the place of origin of this potential and adjacent regions of the plasma membrane, which are still at the resting potential. In **Figure 6–15**, a small region of a membrane has been depolarized by transient application of a chemical signal, briefly opening membrane channels and producing a potential less negative than that of adjacent areas. Inside the cell, positive charge (positive ions) will flow through the intracellular fluid away from the depolarized region and toward the more negative, resting regions of the membrane. Simultaneously, outside the cell, positive charge will flow from the more positive region of the resting membrane toward the less positive regions the depolarization just created. Note that this local current moves positive charges toward the depolarization site along the outside of the membrane and away from the depolarization site along the inside of the membrane. Thus, it produces a decrease in the amount of charge separation (i.e., depolarization) in the membrane sites adjacent to the originally depolarized region, and the signal moves along the membrane.

Depending upon the initiating event, graded potentials can occur in either a depolarizing or a hyperpolarizing direction (**Figure 6–16a**), and their magnitude is related to the magnitude of the initiating event (**Figures 6–15b, 6–16b**). In addition to the movement of ions on the inside and the outside of the cell, charge is lost across the membrane because the membrane is permeable to ions through open membrane channels. The result is that the change in membrane potential decreases as the distance increases from the initial site of the potential change (Figure 6–15b, **Figure 6–16c**). Current flows much like water flows through a leaky hose, decreasing just as water flow decreases the farther along the leaky hose you are from the faucet. In fact, plasma membranes are so leaky to ions that these currents die out almost completely within a few millimeters of their point of origin. Because of this, local current is **decremental;** that is, the flow of charge decreases as the distance from the site of origin of the graded potential increases (**Figure 6–17**).

Because the electrical signal decreases with distance, graded potentials (and the local current they generate) can function as signals only over very short distances (a few

Table 6–3	A Miniglossary of Terms Describing the Membrane Potential
Potential or potential difference	The voltage difference between two points
Membrane potential or transmembrane potential	The voltage difference between the inside and outside of a cell
Equilibrium potential	The voltage difference across a membrane that produces a flux of a given ion species that is equal but opposite to the flux due to the concentration gradient of that same ion species
Resting membrane potential or resting potential	The steady transmembrane potential of a cell that is not producing an electric signal
Graded potential	A potential change of variable amplitude and duration that is conducted decrementally; it has no threshold or refractory period
Action potential	A brief all-or-none depolarization of the membrane, reversing polarity in neurons; it has a threshold and refractory period and is conducted without decrement
Synaptic potential	A graded potential change produced in the postsynaptic neuron in response to the release of a neurotransmitter by a presynaptic terminal; it may be depolarizing (an excitatory postsynaptic potential or EPSP) or hyperpolarizing (an inhibitory postsynaptic potential or IPSP)
Receptor potential	A graded potential produced at the peripheral endings of afferent neurons (or in separate receptor cells) in response to a stimulus
Pacemaker potential	A spontaneously occurring graded potential change that occurs in certain specialized cells
Threshold potential	The membrane potential at which an action potential is initiated

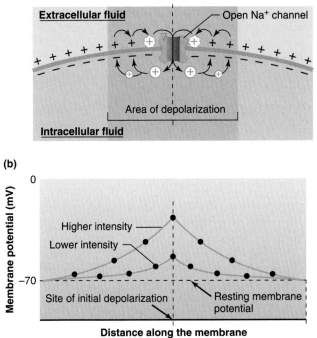

(a)

Extracellular fluid — Open Na⁺ channel

Area of depolarization

Intracellular fluid

(b)

Membrane potential (mV)

Higher intensity
Lower intensity

Site of initial depolarization
Resting membrane potential

Distance along the membrane

Figure 6–15

Depolarizing graded potentials can be produced when transient application of a chemical stimulus opens ion channels at a specific location. These channels close relatively quickly when the signal molecules dissociate and diffuse away. (a) Local current through ion channels depolarizes adjacent regions. (b) Different stimulus intensities result in different degrees of depolarization, and regions of the membrane more distant from a given stimulus are depolarized less.

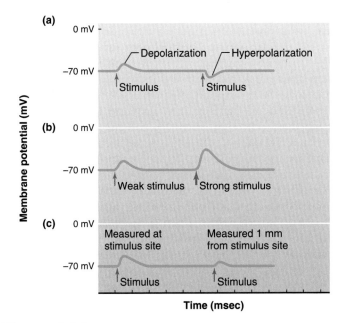

(a)

0 mV
Depolarization — Hyperpolarization
−70 mV
Stimulus   Stimulus

(b)

0 mV
−70 mV
Weak stimulus   Strong stimulus

(c)

0 mV
Measured at stimulus site   Measured 1 mm from stimulus site
−70 mV
Stimulus   Stimulus

Time (msec)

Figure 6–16

Graded potentials can be recorded under experimental conditions in which the stimulus strength can vary. Such experiments show that graded potentials (a) can be depolarizing or hyperpolarizing, (b) can vary in size, (c) are conducted decrementally. The resting membrane potential is −70 mV.

millimeters). However, if additional stimuli occur before the graded potential has died away, these can be added to the depolarization from the first stimulus. This process, termed **summation,** is particularly important for sensation, as Chapter 7 will discuss. Graded potentials are the only means of communication some neurons use, while in other cells they play very important roles in the initiation of signaling over longer distances, as described next.

Action Potentials

Action potentials are very different from graded potentials. They are large alterations in the membrane potential; the membrane potential may change 100 mV, from −70 to +30 mV, and then repolarize to its resting potential. Action potentials are generally very rapid (as brief as 1–4 milliseconds) and may repeat at frequencies of several hundred per second. Nerve and muscle cells as well as some endocrine, immune, and reproductive cells have plasma membranes capable of producing action potentials. These membranes are called **excitable membranes,** and their ability to generate action potentials is known as **excitability.** Whereas all cells are capable of conducting graded potentials, only excitable membranes can conduct action potentials. The propagation of action potentials down the axon is the mechanism the nervous system uses to communicate over long distances.

What properties of ion channels allow them to generate these large, rapid changes in membrane potential, and how are action potentials propagated along an excitable membrane? These questions are addressed in the following sections.

Voltage-Gated Ion Channels

As described in Chapter 4, there are many types of ion channels, and several different mechanisms that regulate the opening of the different types. **Ligand-gated channels** open in response to the binding of signaling molecules (as shown in Figure 6–15), and **mechanically gated channels** open in response to physical deformation (stretching) of the plasma membranes. While these types of channels often serve as the initial stimulus for an action

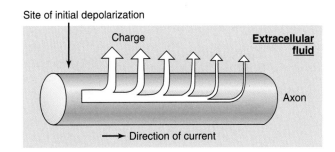

Site of initial depolarization

Charge

Extracellular fluid

Axon

Direction of current

Figure 6–17

Leakage of charge (predominantly potassium ions) across the plasma membrane reduces the local current at sites farther along the membrane from the site of initial depolarization.

potential, it is **voltage-gated channels** that give a membrane the ability to undergo action potentials. There are dozens of different types of voltage-gated ion channels, varying by which ion they conduct (e.g., sodium, potassium, calcium, or chloride) and in how they behave as the membrane voltage changes. For now, we will focus on the particular types of sodium and potassium channels that mediate most neuronal action potentials.

Figure 6–18 summarizes the relevant characteristics of these channels. Sodium and potassium channels are similar in having sequences of charged amino acid residues in their structure that make the channels reversibly change shape in response to changes in membrane voltage. When the membrane is at negative potentials (for example, at the resting membrane potential) both types of channels tend to close, whereas membrane depolarization tends to open them. Two key differences, however, allow these channels to play different roles in the production of action potentials. First, sodium channels are much faster to respond to changes in membrane voltage. When an area of a membrane is suddenly depolarized, local sodium channels open well before the potassium channels do, and if the membrane is then repolarized to negative voltages, the potassium channels are slower to close. The second key difference is that sodium channels have an extra feature in their cytosolic region, known as an **inactivation gate.** This structure, sometimes visualized as a "ball-and-chain," limits the flux of sodium ions by blocking the channel shortly after depolarization opens it. When the membrane repolarizes, the channel closes, forcing the inactivation gate back out of the pore and allowing the channel to return to the closed

state with no sodium flux occurring. Integrating these channel properties with the basic principles governing membrane potentials, we can now explain how action potentials occur.

Action Potential Mechanism

In our previous coverage of resting membrane potential and graded potentials, we saw that the membrane potential depends upon the concentration gradients and membrane permeabilities of different ions, particularly sodium and potassium. This is true of the action potential as well. During an action potential, transient changes in membrane permeability allow sodium and potassium ions to move down their concentration gradients. **Figure 6–19** illustrates the steps that occur during an action potential.

In step 1 of the figure, the resting membrane potential is close to the potassium equilibrium potential because there are more open potassium channels than sodium channels. Note that these leak channels are distinct from the voltage-gated channels just described. An action potential begins with a depolarizing stimulus; for example, when a neurotransmitter binds to a specific ion channel and allows sodium to enter the cell (review Figure 6–15). This initial depolarization stimulates the opening of some voltage-gated sodium channels, and further entry of sodium through those channels adds to the local membrane depolarization. When the membrane reaches a critical **threshold potential** (step 2), depolarization becomes a **positive feedback** loop. Sodium entry causes depolarization, which opens more voltage-gated sodium channels, which causes more depolarization, and so on. This process is represented as a large upstroke of the membrane potential (step 3), and it overshoots so that the membrane actually becomes positive on the inside and negative on the outside. In this phase, the membrane approaches, but does not quite reach, the sodium equilibrium potential (+60 mV).

As the membrane potential approaches its peak value (step 4), the sodium permeability abruptly declines as inactivation gates break the cycle of positive feedback by blocking the open sodium channels. Meanwhile, the depolarized state of the membrane has begun to open the relatively sluggish voltage-gated potassium channels, and the resulting elevated potassium flux out of the cell rapidly repolarizes the membrane toward its resting value (step 5). The return of the membrane to a negative potential causes voltage-gated sodium channels to go from their inactivated state back to the closed state (without opening, as described earlier), and potassium channels to also return to the closed state. Because voltage-gated potassium channels close relatively slowly, immediately after an action potential there is a period when potassium permeability remains above resting levels, and the membrane is transiently hyperpolarized toward the potassium equilibrium potential (step 6). This portion of the action potential is known as the **after-hyperpolarization.** Once the voltage-gated potassium channels finally close, however, the resting membrane potential is restored (step 7). Thus, while voltage-gated sodium channels operate in a positive feedback mode at the beginning of an action potential, voltage-gated potassium channels bring the action potential to an end and induce their own closing through a **negative feedback** process (**Figure 6–20**).

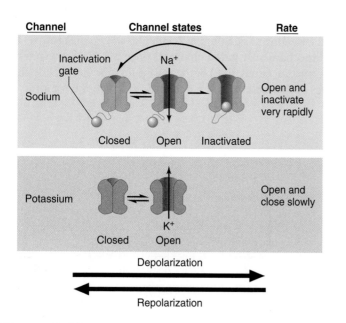

Figure 6–18

Behavior of voltage-gated sodium and potassium channels. Depolarization of the membrane causes sodium channels to rapidly open, then undergo inactivation followed by the opening of potassium channels. When the membrane repolarizes to negative voltages, both channels return to the closed state.

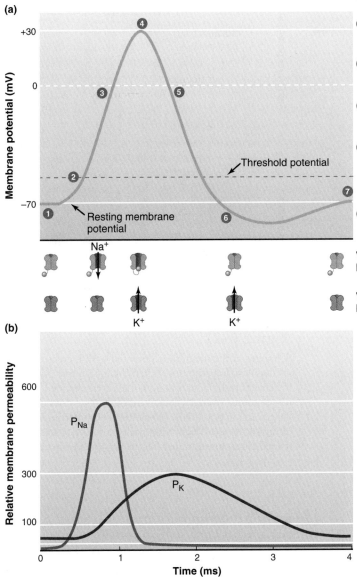

(a)

(b)

① Steady resting membrane potential is near E_K, $P_K > P_{Na}$, due to leak K^+ channels.

② Local membrane is brought to threshold voltage by a depolarizing stimulus.

③ Current through opening voltage-gated Na^+ channels rapidly depolarizes the membrane, causing more Na^+ channels to open.

④ Inactivation of Na^+ channels and delayed opening of voltage-gated K^+ channels halts membrane depolarization.

⑤ Outward current through open voltage-gated K^+ channels repolarizes the membrane back to a negative potential.

⑥ Persistent current through slowly closing voltage-gated K^+ channels hyperpolarizes membrane toward E_K; Na^+ channels return from inactivated state to closed state(without opening).

⑦ Closure of voltage-gated K^+ channels returns the membrane potential to its resting value.

Voltage-gated Na^+ channel

Voltage-gated K^+ channel

Figure 6–19

The changes in (a) membrane potential (mV) and (b) relative membrane permeability (P) to sodium and potassium ions during an action potential. Steps 1–7 are described in more detail in the text.

Figure 6–19 physiological *inquiry* (p*i*)

■ If extracellular [Na^+] is elevated (and you ignore any effects of a change in osmolarity), how would the resting potential and action potential of a neuron change?

Answer can be found at end of chapter.

You might think that large movements of ions across the membrane are required to produce such large changes in membrane potential. Actually, the number of ions that cross the membrane during an action potential is extremely small compared to the total number of ions in the cell, producing only infinitesimal changes in the intracellular ion concentrations. Yet, if this tiny number of additional ions crossing the membrane with repeated action potentials were not eventually moved back across the membrane, the concentration gradients of sodium and potassium would gradually dissipate, and action potentials could no longer be generated. As might be expected, cellular accumulation of sodium and loss of potassium are prevented by the continuous action of the membrane Na^+/K^+-ATPase pumps.

As explained previously, not all membrane depolarizations in excitable cells trigger the positive feedback relationship that leads to an action potential. Action potentials occur only when the initial stimulus plus the current through the sodium channels it opens are sufficient to elevate the membrane potential beyond the threshold potential. Stimuli that are just strong enough to depolarize the membrane to this level are **threshold stimuli** (**Figure 6–21**). The threshold of most excitable membranes is about 15 mV less negative than the resting membrane potential. Thus, if the resting potential of a neuron is –70 mV, the threshold potential may be –55 mV. At depolarizations less than threshold, the positive feedback cycle cannot get started. In such cases, the membrane will return to its resting level as soon as the stimulus is removed, and no action potential will be generated. These weak depolarizations are **subthreshold potentials,** and the stimuli that cause them are **subthreshold stimuli.**

Stimuli of *more than* threshold magnitude elicit action potentials, but as can be seen in Figure 6–21, the action potentials resulting from such stimuli have exactly the same amplitude

(a)

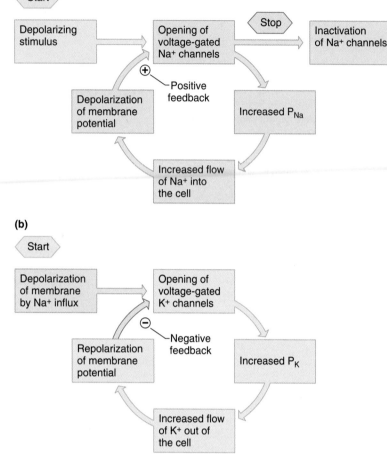

(b)

Figure 6–20

Feedback control in voltage-gated ion channels. (a) Sodium channels exert positive feedback on membrane potential. (b) Potassium channels exert negative feedback.

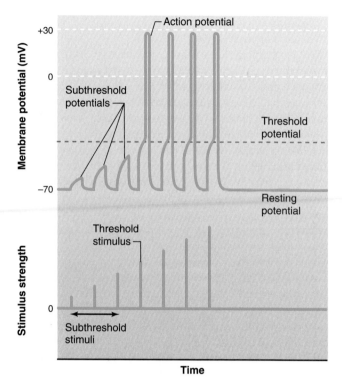

Figure 6–21

Changes in the membrane potential with increasing strength of depolarizing stimulus. When the membrane potential reaches threshold, action potentials are generated. Increasing the stimulus strength above threshold level does not cause larger action potentials. (The afterhyperpolarization has been omitted from this figure for clarity, and the absolute value of threshold is not indicated because it varies from cell to cell.)

as those caused by threshold stimuli. This is because once threshold is reached, membrane events are no longer dependent upon stimulus strength. Rather, the depolarization generates an action potential because the positive feedback cycle is operating. Action potentials either occur maximally or they do not occur at all. Another way of saying this is that action potentials are **all-or-none.**

The firing of a gun is a mechanical analogy that shows the principle of all-or-none behavior. The magnitude of the explosion and the velocity at which the bullet leaves the gun do not depend on how hard the trigger is squeezed. Either the trigger is pulled hard enough to fire the gun, or it is not; the gun cannot be fired halfway.

Because of its all-or-none nature, a single action potential cannot convey information about the magnitude of the stimulus that initiated it. How then do you distinguish between a loud noise and a whisper, a light touch and a pinch? This information, as we will discuss later, depends upon the number and patterns of action potentials transmitted per unit of time (i.e., their frequency) and not upon their magnitude.

The generation of action potentials is prevented by *local anesthetics* such as procaine (*Novocaine®*) and lidocaine

(*Xylocaine®*) because these drugs block voltage-gated sodium channels, preventing them from opening in response to depolarization. Without action potentials, graded signals generated in the periphery—in response to injury, for example—cannot reach the brain and give rise to the sensation of pain.

Some animals produce toxins that work by interfering with nerve conduction in the same way that local anesthetics do. For example, the ovary of the puffer fish produces an extremely potent toxin, *tetrodotoxin,* that binds to voltage-gated sodium channels and prevents the sodium component of the action potential. In Japan, chefs who prepare this delicacy are specially trained to completely remove the ovaries before serving the puffer fish dish called fugu. Individuals who eat improperly prepare fugu may die, even if they ingest only a tiny quantity of tetrodotoxin.

Refractory Periods

During the action potential, a second stimulus, no matter how strong, will not produce a second action potential. That region of the membrane is then said to be in its **absolute refractory period.** This occurs during the period when the voltage-gated sodium channels are either already open or have proceeded to the inactivated state during the first action potential. The inactivation gate that has blocked these channels must be removed by repolarizing the membrane and closing the pore before the channels can reopen to the second stimulus.

Following the absolute refractory period, there is an interval during which a second action potential can be produced, but only if the stimulus strength is considerably greater than usual. This is the **relative refractory period,** which can last 1 to 15 ms or longer and coincides roughly with the period of afterhyperpolarization. During the relative refractory period, some but not all of the voltage-gated sodium channels have returned to a resting state, and some of the potassium channels that repolarized the membrane are still open. From this relative refractory state it is possible for a new stimulus to depolarize the membrane above the threshold potential, but only if the stimulus is large in magnitude or outlasts the relative refractory period.

The refractory periods limit the number of action potentials an excitable membrane can produce in a given period of time. Most nerve cells respond at frequencies of up to 100 action potentials per second, and some may produce much higher frequencies for brief periods. Refractory periods contribute to the separation of these action potentials so that individual electrical signals pass down the axon. The refractory periods also are key in determining the direction of action potential propagation, as we will discuss in the following section.

Action Potential Propagation

The action potential can only travel the length of a neuron if each point along the membrane is depolarized to its threshold potential as the action potential moves down the axon (**Figure 6–22**). As with graded potentials (refer back to Figure 6–15a), the membrane is depolarized at each point along the way with respect to the adjacent portions of the membrane, which are still at the resting membrane potential. The difference between the potentials causes ions to flow, and this local current depolarizes the adjacent membrane where it causes the voltage-gated sodium channels located there to open. The current entering during an action potential is sufficient to easily depolarize the adjacent membrane to the threshold potential.

The new action potential produces local currents of its own that depolarize the region adjacent to it (**Figure 6–22b**), producing yet another action potential at the next site, and so on, to cause **action potential propagation** along the length of the membrane. Thus, there is a sequential opening and closing of sodium and potassium channels along the membrane. It is like lighting a trail of gunpowder—the action potential doesn't move, but it "sets off" a new action potential in the region of the axon just ahead of it. Because each action potential depends on the positive feedback cycle of a new group of sodium channels where the action potential is occurring, the action potential arriving at the end of the membrane is virtually identical in form to the initial one. Thus, action potentials are not conducted decrementally as are graded potentials.

Because a membrane area that has just undergone an action potential is refractory and cannot immediately undergo another, the only direction of action potential propagation is away from a region of membrane that has recently been active.

If the membrane through which the action potential must travel is not refractory, excitable membranes can conduct action potentials in either direction, with the direction of propagation determined by the stimulus location. For example, the action potentials in skeletal muscle cells are initiated near the middle of the cells and propagate toward the two ends. In most nerve cells, however, action potentials are initiated at one end of the cell and propagate toward the other end as shown in Figure 6–22. The propagation ceases when the action potential reaches the end of an axon.

The velocity with which an action potential propagates along a membrane depends upon fiber diameter and whether or not the fiber is myelinated. The larger the fiber diameter, the faster the action potential propagates. This is because a large fiber offers less resistance to local current; more ions will flow in a given time, bringing adjacent regions of the membrane to threshold faster.

Myelin is an insulator that makes it more difficult for charge to flow between intracellular and extracellular fluid compartments. Because there is less "leakage" of charge across the myelin, a local current can spread farther along an axon. Moreover, the concentration of voltage-gated sodium channels in the myelinated region of axons is low. Therefore, action potentials occur only at the nodes of Ranvier, where the myelin coating is interrupted and the concentration of voltage-gated sodium channels is high (**Figure 6–23**). Thus, action potentials jump from one node to the next as they propagate along a myelinated fiber, and for this reason such propagation is called **saltatory conduction** (Latin, *saltare*, to leap).

Propagation via saltatory conduction is faster than propagation in nonmyelinated fibers of the same axon diameter because less charge leaks out through the myelin-covered sections of the membrane. More charge arrives at the node adjacent to the active node, and an action potential is generated there sooner than if the myelin were not present. Moreover, because ions cross the membrane only at the nodes of Ranvier, the membrane pumps need to restore fewer ions. Myelinated axons are therefore metabolically more efficient than unmyelinated ones. Thus, myelin adds speed, reduces metabolic cost, and saves room in the nervous system because the axons can be thinner. The loss of myelin at one or several places in the nervous system occurs in the disease *multiple sclerosis.* This slows or blocks the propagation of impulses, which results in poor coordination, lack of sensation, and partial paralysis (see Additional Clinical Examples at the end of this section).

Conduction velocities range from about 0.5 m/s (1 mi/h) for small-diameter, unmyelinated fibers to about 100 m/s (225 mi/h) for large-diameter, myelinated fibers. At 0.5 m/s, an action potential would travel the distance from the toe to the brain of an average-sized person in about 4 s; at a velocity of 100 m/s, it takes about 0.02 s. Perhaps you've dropped a heavy object on your toe and noticed that an immediate, sharp pain (carried by large-diameter, myelinated neurons) occurs well before the onset of a dull, throbbing ache (transmitted along small, unmyelinated neurons).

Generation of Action Potentials

In our description of action potentials thus far, we have spoken of "stimuli" as the initiators of action potentials. These stimuli bring the membrane to the threshold potential, and voltage-gated sodium channels trigger the all-or-none action

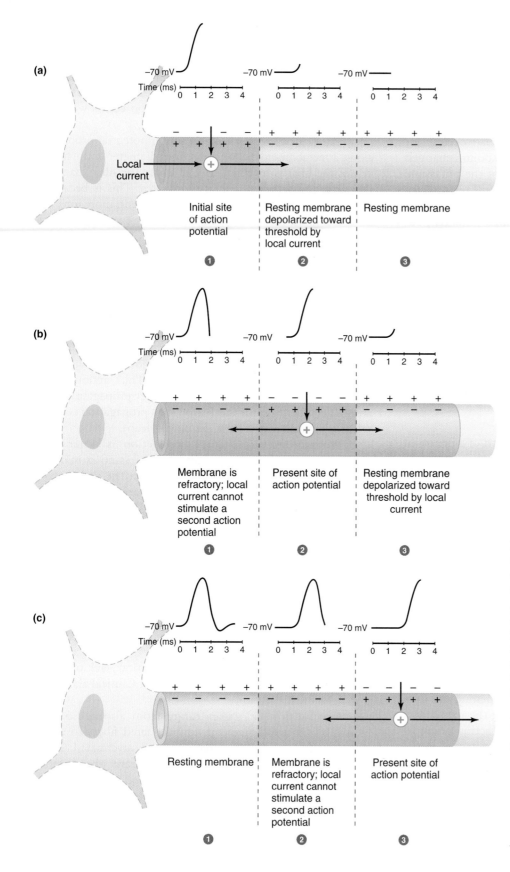

(a)

Local current

Initial site of action potential ①

Resting membrane depolarized toward threshold by local current ②

Resting membrane ③

(b)

Membrane is refractory; local current cannot stimulate a second action potential ①

Present site of action potential ②

Resting membrane depolarized toward threshold by local current ③

(c)

Resting membrane ①

Membrane is refractory; local current cannot stimulate a second action potential ②

Present site of action potential ③

Figure 6–22

One-way propagation of an action potential. For simplicity, potentials are shown only on the upper membrane, local currents are shown only on the inside of the membrane, and repolarizing currents are not shown. (a) Action potential is initiated in region 1, and local currents depolarize region 2. (b) Action potential in region 2 generates local currents; region 3 is depolarized toward threshold, but region 1 is refractory. (c) Action potential in region 3 generates local currents, but region 2 is refractory.

Figure 6–22 physiological *inquiry* ⓟ*i*

■ Striking the ulnar nerve in your elbow against a hard surface (sometimes called "hitting your funny bone") initiates action potentials in the middle of sensory and motor axons traveling in that nerve. In which direction will those action potentials propagate?

Answer can be found at end of chapter.

potential. How is the threshold potential attained, and how do various types of neurons actually generate action potentials?

In afferent neurons, the initial depolarization to threshold is achieved by a graded potential—here called a **receptor potential,** which is generated in the sensory receptors at the peripheral ends of the neurons. These are the ends farthest from the central nervous system where the nervous system receives information. In all other neurons, the depolarization to threshold is due either to a graded potential generated by synaptic input to the neuron, known as a **synaptic potential,**

Chapter 6

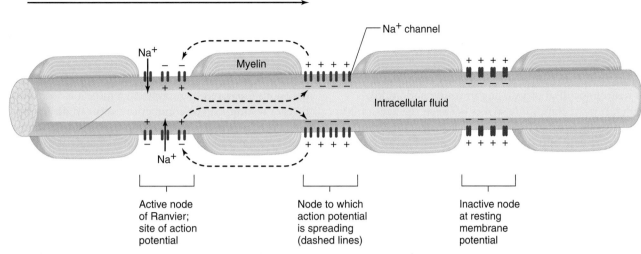

Figure 6–23

Myelinization and saltatory conduction of action potentials. Potassium channels are not depicted.

or to a spontaneous change in the neuron's membrane potential, known as a **pacemaker potential.** The next section will address the production of synaptic potentials, and Chapter 7 will discuss the production of receptor potentials.

Triggering of action potentials by pacemaker potentials is an inherent property of certain neurons (and other excitable cells, including certain smooth-muscle and cardiac-muscle cells). In these cells, the activity of different types of ion channels in the plasma membrane causes a graded depolarization of the membrane—the pacemaker potential. If threshold is reached, an action potential occurs; the membrane then repolarizes and

again begins to depolarize. There is no stable, resting membrane potential in such cells because of the continuous change in membrane permeability. The rate at which the membrane depolarizes to threshold determines the action potential frequency. Pacemaker potentials are implicated in many rhythmical behaviors, such as breathing, the heartbeat, and movements within the walls of the stomach and intestines.

Because of the effects of graded changes in membrane potential on action potential generation, a review of graded and action potentials is recommended. The differences between graded potentials and action potentials are listed in **Table 6–4**.

Table 6–4	Differences between Graded Potentials and Action Potentials
Graded Potential	**Action Potential**
Amplitude varies with size of the initiating event.	All-or-none. Once membrane is depolarized to threshold, amplitude is independent of the size of the initiating event.
Can be summed.	Cannot be summed.
Has no threshold.	Has a threshold that is usually about 15 mV depolarized relative to the resting potential.
Has no refractory period.	Has a refractory period.
Is conducted decrementally; that is, amplitude decreases with distance.	Is conducted without decrement; the depolarization is amplified to a constant value at each point along the membrane.
Duration varies with initiating conditions.	Duration is constant for a given cell type under constant conditions.
Can be a depolarization or a hyperpolarization.	Is only a depolarization.
Initiated by environmental stimulus (receptor), by neurotransmitter (synapse), or spontaneously.	Initiated by a graded potential.
Mechanism depends on ligand-gated channels or other chemical or physical changes.	Mechanism depends on voltage-gated channels.

Multiple Sclerosis

Multiple sclerosis (MS) ranks second to only trauma as a cause of neurologic disability arising in young and middle-aged adults. It most commonly strikes between the ages of 20 and 50, and twice as often in females as in males. It currently affects approximately 400,000 Americans and as many as 3 million people worldwide. MS is an autoimmune condition in which the myelin sheaths surrounding axons in the central nervous system are attacked and destroyed by antibodies and cells of the immune system. The loss of insulating myelin sheaths results in increased leak of potassium through voltage-gated channels. This results in hyperpolarization and failure of action potential conduction of neurons in the brain and spinal cord. Depending upon the location of the affected neurons, symptoms can include muscle weakness, fatigue, decreased motor coordination, slurred speech, blurred or hazy vision, bladder dysfunction, pain or other sensory disturbances, and cognitive dysfunction. In many patients, the symptoms are markedly worsened when body temperature is elevated; for example, by exercise, a hot shower, or hot weather.

The severity and rate of progression of MS varies enormously among individuals, ranging from isolated, episodic attacks with complete recovery in between, to steadily progressing neurological disability. In the latter case, MS can ultimately be fatal as brainstem centers responsible for respiratory and cardiovascular function are destroyed.

Because of this variability, diagnosing MS can be difficult. A person having several of these symptoms on two or more occasions separated by more than a month is a candidate for further testing. Nerve-conduction tests can detect slowed or failed action potential conduction in the motor, sensory, and visual systems, and cerebrospinal fluid analysis can reveal the presence of an abnormal immune reaction against myelin. The most definitive evidence, however, is usually the visualization by magnetic resonance imaging (MRI) of multiple, scarred (sclerotic) areas within the brain and spinal cord, from which this disease derives its name.

The cause of multiple sclerosis is not known, but it appears to result from a combination of genetic and environmental factors. It tends to run in families and is more common among Caucasians than in other racial groups. The participation of environmental triggers is suggested by occasional clusters of disease outbreaks, and also by the observation that the prevalence of MS in people of Japanese descent rises significantly when they move to the United States. Among the suspects for the environmental trigger is infection early in life with a virus, such as those that cause measles, herpes, chicken pox, or influenza. There is presently no cure for multiple sclerosis, but certain drugs that suppress the immune response have been proven to reduce the severity and slow the progression of the disease. A variety of effective drugs and therapies are now available to help MS patients cope with their specific symptoms. ■

■ **SECTION B SUMMARY**

Basic Principles of Electricity

I. Separated electrical charges create the potential to do work, as occurs when charged particles produce an electrical current as they flow down a potential gradient. The lipid barrier of the plasma membrane is a high-resistance insulator that keeps charged ions separated, while ionic current flows readily in the aqueous intracellular and extracellular fluids.

The Resting Membrane Potential

I. Membrane potentials are generated mainly by the diffusion of ions and are determined by (a) the ionic concentration differences across the membrane, and (b) the membrane's relative permeability to different ions.
   a. Plasma membrane Na^+/K^+-ATPase pumps maintain low intracellular sodium concentration and high intracellular potassium concentration.
   b. In almost all resting cells, the plasma membrane is much more permeable to potassium than to sodium, so the membrane potential is close to the potassium equilibrium potential—that is, the inside is negative relative to the outside.
   c. The Na^+/K^+-ATPase pumps directly contribute a small component of the potential because they are electrogenic.

Graded Potentials and Action Potentials

I. Neurons signal information by graded potentials and action potentials (APs).
II. Graded potentials are local potentials whose magnitude can vary and that die out within 1 or 2 mm of their site of origin.
III. An AP is a rapid change in the membrane potential during which the membrane rapidly depolarizes and repolarizes. At the peak, the potential reverses and the membrane becomes positive inside. APs provide long-distance transmission of information through the nervous system.
   a. APs occur in excitable membranes because these membranes contain many voltage-gated sodium channels. These channels open as the membrane depolarizes, causing a positive feedback opening of more voltage-gated sodium channels and moving the membrane potential toward the sodium equilibrium potential.
   b. The AP ends as the sodium channels inactivate and potassium channels open, restoring resting conditions.
   c. Depolarization of excitable membranes triggers an AP only when the membrane potential exceeds a threshold potential.
   d. Regardless of the size of the stimulus, if the membrane reaches threshold, the APs generated are all the same size.
   e. A membrane is refractory for a brief time following an AP.
   f. APs are propagated without any change in size from one site to another along a membrane.

g. In myelinated nerve fibers, APs manifest saltatory conduction.

h. APs can be triggered by depolarizing graded potentials in sensory neurons, at synapses, or in some cells by pacemaker potentials.

Additional Clinical Examples

I. Multiple sclerosis is an autoimmune disease in which antibodies attack the myelin sheaths surrounding neurons. The resulting failure of action potential propagation causes neurological disability that can vary in severity and rate of progression. The causes of MS are currently unknown and there is no cure, but drugs that suppress immune system function slow the progression of the disease.

■ SECTION B KEY TERMS

absolute refractory period 154	negative feedback 152
action potential 151	Nernst equation 146
action potential	Ohm's law 144
propagation 155	overshoot 149
after-hyperpolarization 152	pacemaker potential 157
all-or-none 154	positive feedback 152
current 144	potential 144
decremental 150	potential difference 144
depolarized 149	receptor potential 156
electrical potential 144	relative refractory period 155
electrogenic pump 148	repolarizing 149
equilibrium potential 146	resistance 144
excitability 151	resting membrane potential 144
excitable membrane 151	saltatory conduction 155
Goldman-Hodgkin-Katz	subthreshold potential 153
(GHK) equation 147	subthreshold stimulus 153
graded potential 149	summation 151
hyperpolarized 149	synaptic potential 156
inactivation gate 152	threshold potential 152
leak potassium channels 147	threshold stimulus 153
ligand-gated channels 151	voltage-gated channels 152
mechanically gated	
channels 151	

■ SECTION B CLINICAL TERMS

local anesthetics 154	tetrodotoxin 154
multiple sclerosis 155	Xylocaine® 154
Novocaine® 154	

■ SECTION B REVIEW QUESTIONS

1. Describe how negative and positive charges interact.
2. Contrast the abilities of intracellular and extracellular fluids and membrane lipids to conduct electrical current.
3. Draw a simple cell; indicate where the concentrations of Na^+, K^+, and Cl^- are high and low and the electrical potential difference across the membrane when the cell is at rest.
4. Explain the conditions that give rise to the resting membrane potential. What effect does membrane permeability have on this potential? What role do Na^+/K^+-ATPase membrane pumps play in the membrane potential? Is this role direct or indirect?
5. Which two factors involving ion diffusion determine the magnitude of the resting membrane potential?
6. Explain why the resting membrane potential is not equal to the potassium equilibrium potential.
7. Draw a graded potential and an action potential on a graph of membrane potential versus time. Indicate zero membrane potential, resting membrane potential, and threshold potential; indicate when the membrane is depolarized, repolarizing, and hyperpolarized.
8. List the differences between graded potentials and action potentials.
9. Describe how ion movement generates the action potential.
10. What determines the activity of the voltage-gated sodium channel?
11. Explain threshold and the relative and absolute refractory periods in terms of the ionic basis of the action potential.
12. Describe the propagation of an action potential. Contrast this event in myelinated and unmyelinated axons.
13. List three ways in which action potentials can be initiated in neurons.

SECTION C
Synapses

As defined earlier, a synapse is an anatomically specialized junction between two neurons, at which the electrical activity in a presynaptic neuron influences the electrical activity of a postsynaptic neuron. Anatomically, synapses include parts of the presynaptic and postsynaptic neurons and the extracellular space between these two cells. According to the latest estimate, there are approximately 10^{14} (100 trillion!) synapses in the CNS.

Activity at synapses can increase or decrease the likelihood that the postsynaptic neuron will fire action potentials by producing a brief, graded potential in the postsynaptic membrane. The membrane potential of a postsynaptic neuron is brought closer to threshold (i.e., depolarized) at an **excitatory synapse,** and it is either driven farther from threshold (i.e., hyperpolarized) or stabilized at its resting potential at an **inhibitory synapse.**

Hundreds or thousands of synapses from many different presynaptic cells can affect a single postsynaptic cell **(convergence),** and a single presynaptic cell can send branches to affect many other postsynaptic cells **(divergence, Figure 6–24).**

Convergence allows information from many sources to influence a cell's activity; divergence allows one information source to affect multiple pathways.

The level of excitability of a postsynaptic cell at any moment (i.e., how close its membrane potential is to threshold) depends on the number of synapses active at any one time and the number that are excitatory or inhibitory. If the membrane of the postsynaptic neuron reaches threshold, it will generate action potentials that are propagated along its axon to the terminal branches, which in turn influence the excitability of other cells.

Functional Anatomy of Synapses

There are two types of synapses: **electrical** and **chemical.** At electrical synapses, the plasma membranes of the pre- and postsynaptic cells are joined by gap junctions (Chapter 3). These allow the local currents resulting from arriving action potentials

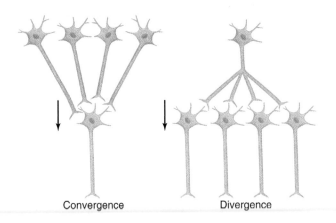

Figure 6–24

Convergence of neural input from many neurons onto a single neuron, and divergence of output from a single neuron onto many others. Arrows indicate the direction of transmission of neural activity.

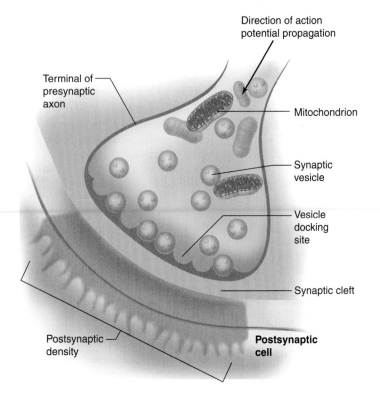

Figure 6–25

Diagram of a chemical synapse. Some vesicles are docked at the presynaptic membrane, ready for release. The postsynaptic membrane is distinguished microscopically by the postsynaptic density, which contains proteins associated with the receptors.

to flow directly across the junction through the connecting channels from one neuron to the other. This depolarizes the membrane of the second neuron to threshold, continuing the propagation of the action potential. Communication between cells via electrical synapses is extremely rapid. Although numerous in cardiac and smooth muscles, electrical synapses are not commonly found in the mammalian nervous system.

Figure 6–25 shows the basic structure of a typical chemical synapse. The axon of the presynaptic neuron ends in a slight swelling, the axon terminal, which holds the **synaptic vesicles** that contain the neurotransmitter. The postsynaptic membrane adjacent to the axon terminal has a high density of intrinsic and extrinsic membrane proteins that make up a specialized area called the **postsynaptic density.** Note that in actuality the size and shape of the pre- and postsynaptic elements can vary greatly (**Figure 6–26**). A 10- to 20-nm extracellular space, the **synaptic cleft,** separates the pre- and postsynaptic neurons and prevents *direct* propagation of the current from the presynaptic neuron to the postsynaptic cell. Instead, signals are transmitted across the synaptic cleft by means of a chemical messenger—a neurotransmitter—released from the presynaptic axon terminal. Sometimes more than one neurotransmitter may be simultaneously released from an axon, in which case the additional neurotransmitter is called a **cotransmitter.** These neurotransmitters have different receptors on the postsynaptic cell.

In general, the neurotransmitter is stored on the presynaptic side of the synaptic cleft, whereas receptors for the neurotransmitters are on the postsynaptic side. Therefore, most chemical synapses operate in only one direction. One-way conduction across synapses causes action potentials to transmit along a given multineuronal pathway in one direction.

Mechanisms of Neurotransmitter Release

As indicated in **Figure 6–27**, neurotransmitter is stored in small vesicles with lipid bilayer membranes. Prior to activation, many vesicles are docked on the presynaptic membrane

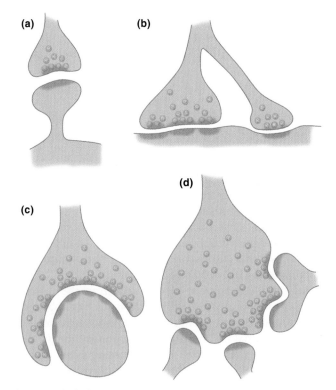

Figure 6–26

Synapses appear in many forms, as demonstrated here in views (a) to (d). The presynaptic terminal contains synaptic vesicles.

Redrawn from Walmsley et al.

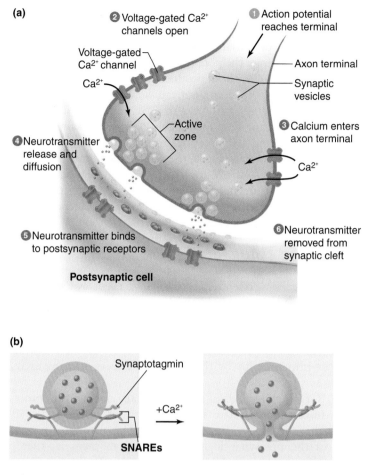

(a)

② Voltage-gated Ca²⁺ channels open

Voltage-gated Ca²⁺ channel

Ca²⁺

① Action potential reaches terminal

Axon terminal

Synaptic vesicles

Active zone

④ Neurotransmitter release and diffusion

③ Calcium enters axon terminal

Ca²⁺

⑤ Neurotransmitter binds to postsynaptic receptors

⑥ Neurotransmitter removed from synaptic cleft

Postsynaptic cell

(b)

Synaptotagmin

+Ca²⁺

SNAREs

Figure 6–27

(a) Neurotransmitter storage and release at the synapse and binding to the postsynaptic receptor. Voltage-gated calcium channels in the terminal open in response to an action potential, triggering release of neurotransmitter. (b) Magnified view showing details of neurotransmitter release. Calcium entering the terminal binds to synaptotagmin, stimulating SNARE proteins to induce membrane fusion and neurotransmitter release. *Note:* The SNARE complex is known to involve more proteins than are shown here. (SNARE = Soluble N-ethylmaleimide-sensitive fusion protein attachment protein receptor)

at release regions known as **active zones,** while others are dispersed within the terminal. Neurotransmitter release is initiated when an action potential reaches the terminal of the presynaptic membrane. A key feature of neuron terminals is that in addition to the sodium and potassium channels found elsewhere in the neuron, they also possess voltage-gated calcium channels. Depolarization during the action potential opens these calcium channels, and because the electrochemical gradient favors calcium influx, calcium flows into the axon terminal.

Calcium activates processes that lead to the fusion of docked vesicles with the synaptic terminal membrane (**Figure 6–27b**). Prior to the arrival of an action potential, vesicles are loosely docked in the active zones by the interaction of a group of proteins, some of which are anchored in the vesicle membrane and others found in the membrane of the terminal.

These proteins are collectively known as **SNAREs** (soluble N-ethylmaleimide-sensitive fusion protein attachment protein receptors). Calcium entering during depolarization binds to a separate family of proteins associated with the vesicle, **synaptotagmins,** triggering a conformational change in the SNARE complex that leads to membrane fusion and neurotransmitter release.

Activation of the Postsynaptic Cell

Once neurotransmitters are released from the presynaptic axon terminal, they diffuse across the cleft. A fraction of these molecules bind to receptors on the plasma membrane of the postsynaptic cell. The activated receptors themselves may be ion channels, which designates them as **ionotropic receptors.** Alternatively, the receptors may act indirectly on separate ion channels through a G protein and/or a second messenger, a type referred to as **metabotropic receptors.** In either case, the result of the binding of neurotransmitter to receptor is the opening or closing of specific ion channels in the postsynaptic plasma membrane, which eventually leads to functional changes in that neuron. These channels belong, therefore, to the class of ligand-gated channels controlled by receptors, as discussed in Chapter 5, and are distinct from voltage-gated channels.

Because of the sequence of events involved, there is a very brief **synaptic delay**—about 0.2 msec—between the arrival of an action potential at a presynaptic terminal and the membrane potential changes in the postsynaptic cell.

Neurotransmitter binding to the receptor is a transient event. As with any binding site, the bound ligand—in this case, the neurotransmitter—is in equilibrium with the unbound form. Thus, if the concentration of unbound neurotransmitter in the synaptic cleft decreases, the number of occupied receptors will decrease. The ion channels in the postsynaptic membrane return to their resting state when the neurotransmitter is no longer bound. Unbound neurotransmitters are removed from the synaptic cleft when they (1) are actively transported back into the presynaptic axon terminal (in a process called **reuptake**) or, in some cases, into nearby glial cells; (2) diffuse away from the receptor site; or (3) are enzymatically transformed into inactive substances, some of which are transported back into the axon terminal for reuse.

The two kinds of chemical synapses—excitatory and inhibitory—are differentiated by the effects of the neurotransmitter on the postsynaptic cell. Whether the effect is excitatory or inhibitory depends on the type of signal transduction mechanism brought into operation when the neurotransmitter binds to a receptor and on the type of channel the receptor influences.

Excitatory Chemical Synapses

At an excitatory synapse, the postsynaptic response to the neurotransmitter is a depolarization, bringing the membrane potential closer to threshold. The usual effect of the activated receptor on the postsynaptic membrane at such synapses is to open nonselective channels that are permeable to sodium, potassium, and other small, positively charged ions. These ions then are free to move according to the electrical and chemical gradients across the membrane.

There are both electrical and concentration gradients driving sodium into the cell, while for potassium, the electrical gradient opposes the concentration gradient (review Figure 6–12). Opening channels that are permeable to all small positively charged ions therefore results in the simultaneous movement of a relatively small number of potassium ions out of the cell and a larger number of sodium ions into the cell. Thus, the *net* movement of positive ions is into the postsynaptic cell, causing a slight depolarization. This potential change is called an **excitatory postsynaptic potential** (**EPSP, Figure 6–28**).

The EPSP is a graded potential that spreads decrementally away from the synapse by local current. Its only function is to bring the membrane potential of the postsynaptic neuron closer to threshold.

Inhibitory Chemical Synapses

At inhibitory synapses, the potential change in the postsynaptic neuron is generally a hyperpolarizing graded potential called an **inhibitory postsynaptic potential** (**IPSP, Figure 6–29**). Alternatively, there may be no IPSP but rather *stabilization* of the membrane potential at its existing value. In either case, activation of an inhibitory synapse lessens the likelihood that the postsynaptic cell will depolarize to threshold and generate an action potential.

At an inhibitory synapse, the activated receptors on the postsynaptic membrane open chloride or potassium channels; sodium permeability is not affected. In those cells that actively regulate intracellular chloride concentrations via active transport out of the cell, the chloride equilibrium potential is more negative than the resting potential. Therefore, as chloride channels open, chloride enters the cell, producing a hyperpolarization—that is, an IPSP. In cells that do not actively transport chloride, the equilibrium potential for chloride is equal to the resting membrane potential. Therefore, a rise in chloride ion permeability does not change the membrane potential but is able to increase chloride's influence on the membrane potential. This makes it more difficult for other ion types to change the potential and stabilizes the membrane at the resting level without producing a hyperpolarization.

Increased potassium permeability, when it occurs in the postsynaptic cell, also produces an IPSP. Earlier we noted that if a cell membrane were permeable only to potassium ions, the resting membrane potential would equal the potassium equilibrium potential; that is, the resting membrane potential would be about –90 mV instead of –70 mV. Thus, with an increased potassium permeability, more potassium ions leave the cell and the membrane moves closer to the potassium equilibrium potential, causing a hyperpolarization.

Synaptic Integration

In most neurons, one excitatory synaptic event by itself is not enough to reach threshold in the postsynaptic neuron. For example, a single EPSP may be only 0.5 mV, whereas changes of about 15 mV are necessary to depolarize the neuron's membrane to threshold. This being the case, an action potential can be initiated only by the combined effects of many excitatory synapses.

Of the thousands of synapses on any one neuron, probably hundreds are active simultaneously or close enough in time that the effects can add together. The membrane potential of the postsynaptic neuron at any moment is, therefore, the result of all the synaptic activity affecting it at that moment. A depolarization of the membrane toward threshold occurs when excitatory synaptic input predominates, and either a hyperpolarization or stabilization occurs when inhibitory input predominates (**Figure 6–30**).

A simple experiment can demonstrate how EPSPs and IPSPs interact, as shown in **Figure 6–31**. Assume there are

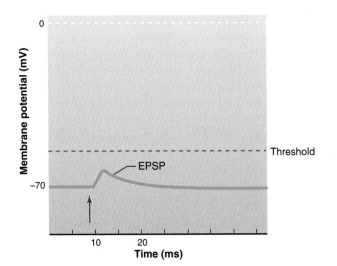

Figure 6–28

Excitatory postsynaptic potential (EPSP). Stimulation of the presynaptic neuron is marked by the arrow. Drawn larger than normal; typical EPSP = 0.5 mV.

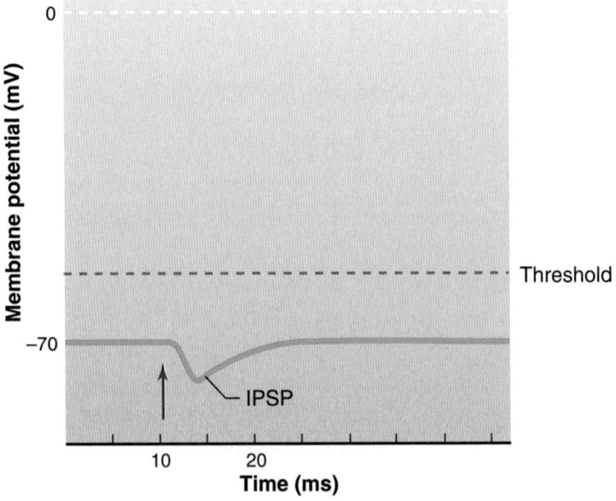

Figure 6–29

Inhibitory postsynaptic potential (IPSP). Stimulation of the presynaptic neuron is marked by the arrow. (This hyperpolarization is drawn larger than a typical IPSP.)

three synaptic inputs to the postsynaptic cell: The synapses from axons A and B are excitatory, and the synapse from axon C is inhibitory. There are laboratory stimulators on axons A, B, and C so that each can be activated individually. An electrode is placed in the cell body of the postsynaptic neuron that will record the membrane potential. In part 1 of the experiment, we will test the interaction of two EPSPs by stimulating axon A and then, after a short time, stimulating it again. Part 1 of Figure 6–31 shows that no interaction occurs between the two EPSPs. The reason is that the change in membrane potential associated with an EPSP is fairly short-lived. Within a few milliseconds (by the time we stimulate axon A for the second time), the postsynaptic cell has returned to its resting condition.

In part 2, we stimulate axon A for the second time before the first EPSP has died away; the second synaptic potential adds to the previous one and creates a greater depolarization than from one input alone. This is called **temporal summation** because the input signals arrive from the same presynaptic cell at different *times*. The potentials summate because there are a greater number of open ion channels and, therefore, a greater flow of positive ions into the cell. In part 3 of Figure 6–31, axon B is stimulated alone to determine its response, and then axons A and B are stimulated simultaneously. The two EPSPs that result also summate in the postsynaptic neuron; this is called **spatial summation** because the two inputs occurred at different *locations* on the same cell. The interaction of multiple EPSPs through spatial and temporal summation can increase the inward flow of positive ions and bring the postsynaptic membrane to threshold so that action potentials are initiated (see part 4 of Figure 6–31).

So far we have tested only the patterns of interaction of excitatory synapses. Because EPSPs and IPSPs are due to oppositely directed local currents, they tend to cancel each other, and there is little or no net change in membrane potential when both A and C are stimulated (see Figure 6–31, part 5). Inhibitory potentials can also show spatial and temporal summation.

Depending on the postsynaptic membrane's resistance and on the amount of charge moving through the ligand-gated channels, the synaptic potential will spread to a greater or lesser degree across the plasma membrane of the cell. The membrane of a large area of the cell becomes slightly depolarized during activation of an excitatory synapse and slightly hyperpolarized or stabilized during activation of an inhibitory synapse, although these graded potentials will decrease with distance from the synaptic junction (**Figure 6–32**). Inputs from more than one synapse can result in summation of the synaptic potentials, which may then trigger an action potential.

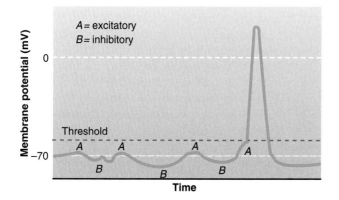

Figure 6–30

Intracellular recording from a postsynaptic cell during times of (A) excitatory synaptic activity when the cell is depolarized, and (B) inhibitory synaptic activity when the membrane hyperpolarizes.

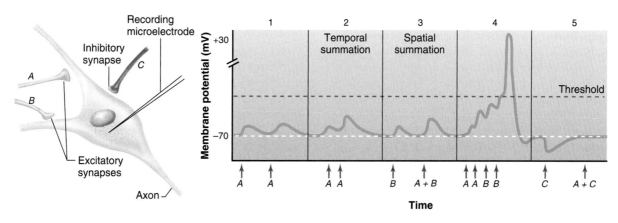

Figure 6–31

Interaction of EPSPs and IPSPs at the postsynaptic neuron. Presynaptic neurons (A–C) were stimulated at times indicated by the arrows, and the resulting membrane potential was recorded in the postsynaptic cell by a recording microelectrode.

Figure 6–31 physiological *inquiry* (pi)

■ If this postsynaptic neuron had no active chloride pumps and the synapse from neuron C opened chloride channels, how would the traces in panel 5 be different from what is shown?

Answer can be found at end of chapter.

Neuronal Signaling and the Structure of the Nervous System

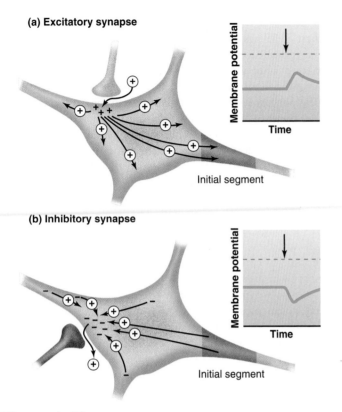

(a) Excitatory synapse

Membrane potential

Time

Initial segment

(b) Inhibitory synapse

Membrane potential

Time

Initial segment

Figure 6–32

Comparison of excitatory and inhibitory synapses, showing current direction through the postsynaptic cell following synaptic activation. (a) Current through the postsynaptic cell is away from the excitatory synapse, and may depolarize the initial segment. (b) Current through the postsynaptic cell is toward the inhibitory synapse and may hyperpolarize the initial segment. The arrow on the chart indicates moment of stimulus.

In the previous examples, we referred to the threshold of the postsynaptic neuron as though it were the same for all parts of the cell. However, different parts of the neuron have different thresholds. In general, the initial segment has a lower threshold (i.e., much closer to the resting potential) than the membrane of the cell body and dendrites. This is due to a higher density of voltage-gated sodium channels in this area of the membrane. Therefore, the initial segment is extremely responsive to small changes in the membrane potential that occur in response to synaptic potentials on the cell body and dendrites. The initial segment reaches threshold whenever enough EPSPs summate. The resulting action potential is then propagated from this point down the axon.

The fact that the initial segment usually has the lowest threshold explains why the locations of individual synapses on the postsynaptic cell are important. A synapse located near the initial segment will produce a greater voltage change in the initial segment than will a synapse on the outermost branch of a dendrite because it will expose the initial segment to a larger local current. In some neurons, however, signals from dendrites distant from the initial segment may be boosted by the presence of some voltage-gated sodium channels in parts of those dendrites.

Postsynaptic potentials last much longer than action potentials. In the event that cumulative EPSPs cause the initial segment to still be depolarized to threshold after an action potential has been fired and the refractory period is over, a second action potential will occur. In fact, as long as the membrane is depolarized to threshold, action potentials will continue to arise. Neuronal responses almost always occur in bursts of action potentials rather than as single, isolated events.

Synaptic Strength

Individual synaptic events—whether excitatory or inhibitory—have been presented as though their effects are constant and reproducible. Actually, enormous variability occurs in the postsynaptic potentials that follow a presynaptic input. The effectiveness or strength of a given synapse is influenced by both presynaptic and postsynaptic mechanisms.

A presynaptic terminal does not release a constant amount of neurotransmitter every time it is activated. One reason for this variation involves calcium concentration. Calcium that has entered the terminal during previous action potentials is pumped out of the cell or (temporarily) into intracellular organelles. If calcium removal does not keep up with entry, as can occur during high-frequency stimulation, calcium concentration in the terminal, and consequently the amount of neurotransmitter released upon subsequent stimulation, will be greater than usual. The greater the amount of neurotransmitter released, the greater the number of ion channels opened in the postsynaptic membrane, and the larger the amplitude of the EPSP or IPSP in the postsynaptic cell.

The neurotransmitter output of some presynaptic terminals is also altered by activation of membrane receptors on the terminals themselves. Activation of these presynaptic receptors influences calcium influx into the terminal and thus the number of neurotransmitter vesicles that release neurotransmitter into the synaptic cleft. These presynaptic receptors may be associated with a second synaptic ending known as an **axo-axonic synapse,** in which an axon terminal of one neuron ends on an axon terminal of another. For example, in **Figure 6–33** the neurotransmitter released by A binds with receptors on B, resulting in a change in the amount of neurotransmitter released from B in response to action potentials. Thus, neuron A has no direct effect on neuron C, but it has an important influence on the ability of B to influence C. Neuron A is thus exerting a presynaptic effect on the synapse between B and C. Depending upon the type of presynaptic receptors activated by the neurotransmitter from neuron A, the presynaptic effect may decrease the amount of neurotransmitter released from B **(presynaptic inhibition)** or increase it **(presynaptic facilitation).**

Axo-axonic synapses such as A in Figure 6–33 can alter the calcium concentration in axon terminal B or even affect neurotransmitter synthesis there. If the calcium concentration increases, the number of vesicles releasing neurotransmitter from B increases. Decreased calcium reduces the number of vesicles releasing transmitter. Axo-axonic synapses are important because they selectively control one specific input to the postsynaptic neuron C. This type of synapse is particularly common in the modulation of sensory input.

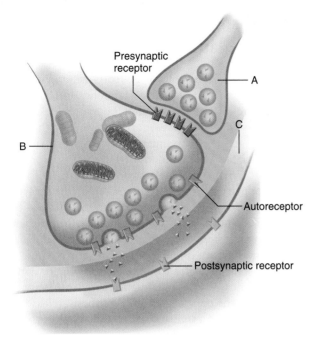

Figure 6–33

A presynaptic (axo-axonic) synapse between axon terminal A and axon terminal B. C is the final postsynaptic cell body.

Some receptors on the presynaptic terminal are not associated with axo-axonic synapses. Instead, they are activated by neurotransmitters or other chemical messengers released by nearby neurons or glia or even by the axon terminal itself. In the last case, the receptors are called **autoreceptors** (Figure 6–33) and provide an important feedback mechanism the neuron can use to regulate its own neurotransmitter output. In most cases, the released neurotransmitter acts on autoreceptors to decrease its own release, thereby providing negative feedback control.

Postsynaptic mechanisms for varying synaptic strength also exist. For example, as described in Chapter 5, many types and subtypes of receptors exist for each kind of neurotransmitter. The different receptor types operate by different signal transduction mechanisms and have different—sometimes even opposite—effects on the postsynaptic mechanisms they influence. Moreover, a given signal transduction mechanism may be regulated by multiple neurotransmitters, and the various second-messenger systems affecting a channel may interact with each other.

Recall, too, from Chapter 5 that the number of receptors is not constant, varying with up- and down-regulation, for example. Also, the ability of a given receptor to respond to its neurotransmitter can change. Thus, in some systems, a receptor responds once and then temporarily fails to respond despite the continued presence of the receptor's neurotransmitter, a phenomenon known as **receptor desensitization.**

Imagine the complexity when a cotransmitter (or several cotransmitters) is released with the neurotransmitter to act upon postsynaptic receptors and maybe upon presynaptic receptors as well! Clearly, the possible variations in transmission

are great at even a single synapse, and this provides a mechanism by which synaptic strength can be altered in response to changing conditions, a characteristic known as *plasticity*.

Modification of Synaptic Transmission by Drugs and Disease

The great majority of drugs that act on the nervous system do so by altering synaptic mechanisms and thus synaptic strength. Drugs act by interfering with or stimulating normal processes in the neuron involved in neurotransmitter synthesis, storage, and release, and in receptor activation. The synaptic mechanisms labeled in **Figure 6–34** are important to synaptic function and are vulnerable to the effects of drugs.

The long-term effects of drugs are sometimes difficult to predict because the imbalances the initial drug action produces are soon counteracted by feedback mechanisms that normally regulate the processes. For example, if a drug interferes with the action of a neurotransmitter by inhibiting the rate-limiting enzyme in its synthetic pathway, the neurons may respond by increasing the rate of precursor transport into the axon terminals to maximize the use of any available enzyme.

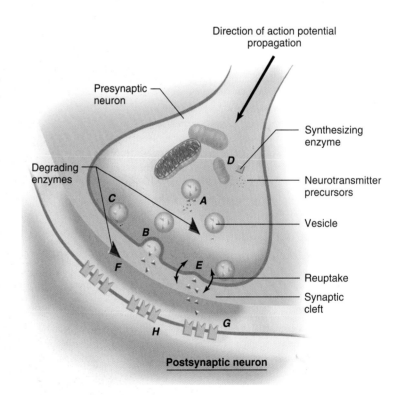

Figure 6–34

Possible actions of drugs on a synapse. A drug might: (A) Increase leakage of neurotransmitter from vesicle to cytoplasm, exposing it to enzyme breakdown, (B) increase transmitter release into cleft, (C) block transmitter release, (D) inhibit transmitter synthesis, (E) block transmitter reuptake, (F) block cleft enzymes that metabolize transmitter, (G) bind to receptor on postsynaptic membrane to block (antagonist) or mimic (agonist) transmitter action, and (H) inhibit or stimulate second-messenger activity within postsynaptic cell.

Recall from Chapter 5 that drugs that bind to a receptor and produce a response similar to the normal activation of that receptor are called **agonists,** and drugs that bind to the receptor but are unable to activate it are **antagonists.** By occupying the receptors, antagonists prevent binding of the normal neurotransmitter at the synapse. Specific agonists and antagonists can affect receptors on both pre- and postsynaptic membranes.

Diseases can also affect synaptic mechanisms. For example, the toxin that causes the neurological disorder tetanus is produced by the bacillus *Clostridium tetani* **(tetanus toxin).** It is a protease that destroys SNARE proteins in the presynaptic terminal so that fusion of vesicles with the membrane is prevented, inhibiting neurotransmitter release. Because tetanus toxin specifically affects inhibitory neurons, tetanus is characterized by an increase in muscle contraction. The toxin of the *Clostridium botulinum* bacilli, which causes **botulism,** also affects neurotransmitter release from synaptic vesicles by interfering with actions of SNARE proteins. However, because it targets the excitatory synapses that activate muscles, botulism is characterized by muscle paralysis.

Table 6–5 summarizes the factors that determine synaptic strength.

Table 6–5	Factors that Determine Synaptic Strength

I. PRESYNAPTIC FACTORS
  A. *Availability of neurotransmitter*
    1. Availability of precursor molecules
    2. Amount (or activity) of the rate-limiting enzyme in the pathway for neurotransmitter synthesis
  B. *Axon terminal membrane potential*
  C. *Axon terminal calcium*
  D. *Activation of membrane receptors on presynaptic terminal*
    1. Axo-axonic synapses
    2. Autoreceptors
    3. Other receptors
  E. *Certain drugs and diseases, which act via the above mechanisms A–D*

II. POSTSYNAPTIC FACTORS
  A. *Immediate past history of electrical state of postsynaptic membrane (e.g., excitation or inhibition from temporal or spatial summation)*
  B. *Effects of other neurotransmitters or neuromodulators acting on postsynaptic neuron*
  C. *Up- or down-regulation and desensitization of receptors*
  D. *Certain drugs and diseases*

III. GENERAL FACTORS
  A. *Area of synaptic contact*
  B. *Enzymatic destruction of neurotransmitter*
  C. *Geometry of diffusion path*
  D. *Neurotransmitter reuptake*

Neurotransmitters and Neuromodulators

We have emphasized the role of neurotransmitters in eliciting EPSPs and IPSPs. However, certain chemical messengers elicit complex responses that cannot be simply described as EPSPs or IPSPs. The word *modulation* is used for these complex responses, and the messengers that cause them are called **neuromodulators.** The distinctions between neuromodulators and neurotransmitters are not always clear. In fact, certain neuromodulators are often synthesized by the presynaptic cell and co-released with the neurotransmitter. To add to the complexity, many hormones, paracrine agents, and messengers that the immune system uses serve as neuromodulators.

Neuromodulators often modify the postsynaptic cell's response to specific neurotransmitters, amplifying or dampening the effectiveness of ongoing synaptic activity. Alternatively, they may change the presynaptic cell's synthesis, release, reuptake, or metabolism of a transmitter. In other words, they alter the effectiveness of the synapse.

In general, the receptors for neurotransmitters influence ion channels that directly affect excitation or inhibition of the postsynaptic cell. These mechanisms operate within milliseconds. Receptors for neuromodulators, on the other hand, more often bring about changes in metabolic processes in neurons, often via G proteins coupled to second-messenger systems. Such changes, which can occur over minutes, hours, or even days, include alterations in enzyme activity or, through influences on DNA transcription, in protein synthesis. Thus, neurotransmitters are involved in rapid communication, whereas neuromodulators tend to be associated with slower events such as learning, development, motivational states, or even some sensory or motor activities.

The number of substances known to act as neurotransmitters or neuromodulators is large and still growing. **Table 6–6** provides a framework for categorizing that list. A huge amount of information has accumulated concerning the synthesis, metabolism, and mechanisms of action of these messengers—material well beyond the scope of this book. The following sections will therefore present only some basic generalizations about some of the neurotransmitters that are deemed most important. For simplicity's sake, we use the term *neurotransmitter* in a general sense, realizing that sometimes the messenger may more appropriately be described as a neuromodulator. A note on terminology should also be included here: Neurons are often referred to as *-ergic;* the missing prefix is the type of neurotransmitter the neuron releases. For example, *dopaminergic* applies to neurons that release the neurotransmitter dopamine.

Acetylcholine

Acetylcholine (ACh) is a major neurotransmitter in the peripheral nervous system at the neuromuscular junction (Chapter 9) and in the brain. Neurons that release ACh are called **cholinergic** neurons. The cell bodies of the brain's cholinergic neurons are concentrated in relatively few areas, but their axons are widely distributed.

Table 6-6	Classes of Some of the Chemicals Known or Presumed to Be Neurotransmitters or Neuromodulators

1. Acetylcholine (ACh)

2. Biogenic amines

   Catecholamines
       Dopamine (DA)
       Norepinephrine (NE)
       Epinephrine (Epi)
   Serotonin (5-hydroxytryptamine, 5-HT)
   Histamine

3. Amino acids

   Excitatory amino acids; for example, glutamate
   Inhibitory amino acids; for example, gamma-
       aminobutyric acid (GABA) and glycine

4. Neuropeptides

   For example, endogenous opioids, oxytocin, tachykinins

5. Miscellaneous

   Gases; for example, nitric oxide
   Purines; for example, adenosine and ATP

Acetylcholine (ACh) is synthesized from choline and acetyl coenzyme A in the cytoplasm of synaptic terminals, and stored in synaptic vesicles. After it is released and activates receptors on the postsynaptic membrane, the concentration of ACh at the postsynaptic membrane decreases (thereby stopping receptor activation) due to the action of the enzyme **acetylcholinesterase.** This enzyme is located on the pre- and postsynaptic membranes and rapidly destroys ACh, releasing choline and acetate. The choline is then transported back into the presynaptic axon terminals where it is reused in the synthesis of new ACh. The ACh concentration at the receptors is also reduced by simple diffusion away from the synapse and eventual breakdown of the molecule by an enzyme in the blood. Some chemical weapons, such as the nerve gas **Sarin,** inhibit acetylcholinesterase, causing a buildup of ACh in the synaptic cleft. This results in overstimulation of postsynaptic ACh receptors, initially causing uncontrolled muscle contractions, but ultimately leading to receptor desensitization and paralysis.

There are two types of ACh receptors, and they are distinguished by their responsiveness to two different drugs. Recall that although a receptor is considered specific for a given ligand, such as ACh, most receptors will recognize natural or synthetic compounds that exhibit some degree of chemical similarity to that ligand. Some ACh receptors respond not only to acetylcholine but to the drug nicotine, and have therefore come to be known as **nicotinic receptors.** The nicotinic receptor is an excellent example of a receptor that contains an ion channel (i.e., a ligand-gated channel); in this case the channel is permeable to both sodium and potassium ions.

Nicotinic receptors are present at the neuromuscular junction and, as Chapter 9 will explain, several nicotinic receptor antagonists are toxins that induce paralysis. Nicotinic receptors in the brain are important in cognitive functions and behavior. Their presence on presynaptic terminals in reward pathways of the brain suggest an explanation for why tobacco use is addictive. As another example, one cholinergic system that employs nicotinic receptors plays a major role in attention, learning, and memory by reinforcing the ability to detect and respond to meaningful stimuli.

Neurons associated with the ACh system degenerate in people with **Alzheimer's disease,** a brain disease that is usually age-related and is the most common cause of declining intellectual function in late life. Alzheimer's disease affects 10 to 15 percent of people over age 65, and 50 percent of people over age 85. Because of the degeneration of cholinergic neurons, this disease is associated with a decreased amount of ACh in certain areas of the brain and even the loss of the postsynaptic neurons that would have responded to it. These defects and those in other neurotransmitter systems that are affected in this disease are related to the declining language and perceptual abilities, confusion, and memory loss that characterize Alzheimer's victims. The exact causes of this degeneration are unknown.

The other type of cholinergic receptor is stimulated not only by acetylcholine but by the mushroom poison muscarine; therefore, these are called **muscarinic receptors.** These receptors couple with G proteins, which then alter the activity of a number of different enzymes and ion channels. They are prevalent at cholinergic synapses in the brain and at junctions of neurons that innervate many glands and organs, notably the heart. **Atropine** is an antagonist of muscarinic receptors with many clinical uses, such as in eye drops that dilate the pupils for an eye exam.

Biogenic Amines

The **biogenic amines** are small, charged molecules that are synthesized from amino acids and contain an amino group ($R-NH_2$). The most common biogenic amines are dopamine, norepinephrine, serotonin, and histamine. Epinephrine, another biogenic amine, is not a common neurotransmitter in the central nervous system but is the major *hormone* the adrenal medulla secretes. Norepinephrine is an important neurotransmitter in both the central and peripheral components of the nervous system.

Catecholamines

Dopamine, norepinephrine (NE), and **epinephrine** all contain a catechol ring (a six-carbon ring with two adjacent hydroxyl groups) and an amine group; thus they are called **catecholamines.** The catecholamines are formed from the amino acid tyrosine and share the same two initial steps in their synthetic pathway (**Figure 6–35**). Synthesis of catecholamines begins with the uptake of tyrosine by the axon terminals and its conversion to another precursor, L-dihydroxy-phenylalanine (**L-dopa**) by the rate-limiting enzyme in the pathway, tyrosine hydroxylase. Depending on the enzymes present in the terminals, any one of the three catecholamines may ultimately be

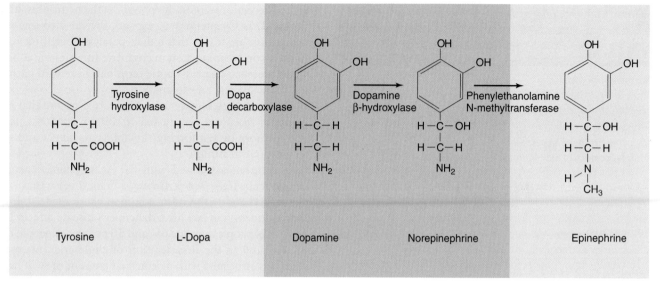

Figure 6–35

Catecholamine biosynthetic pathway. Tyrosine hydroxylase is the rate-limiting enzyme, but which neurotransmitter is ultimately released from a neuron depends on which of the other three enzymes are present in that cell. The colored screen indicates the more common CNS catecholamine neurotransmitters.

released. Autoreceptors on the presynaptic terminals strongly modulate synthesis and release of the catecholamines.

After activation of the receptors on the postsynaptic cell, the catecholamine concentration in the synaptic cleft declines, mainly because a membrane transporter protein actively transports the catecholamine back into the axon terminal. The catecholamine neurotransmitters are also broken down in both the extracellular fluid and the axon terminal by enzymes such as **monoamine oxidase (MAO).** Drugs known as MAO inhibitors increase the amount of norepinephrine and dopamine in a synapse by slowing their metabolic degradation. They are used in the treatment of mood disorders such as some types of depression.

Within the central nervous system, the cell bodies of the catecholamine-releasing neurons lie in parts of the brain called the brainstem and hypothalamus. Although these neurons are relatively few in number, their axons branch greatly and may go to virtually all parts of the brain and spinal cord. These neurotransmitters play essential roles in states of consciousness, mood, motivation, directed attention, movement, blood-pressure regulation, and hormone release, all functions that later chapters will cover in more detail.

The British word for epinephrine is "adrenaline." However, nerve fibers that release either epinephrine or norepinephrine are referred to as **adrenergic** fibers. Norepinephrine-releasing fibers are also sometimes called **noradrenergic.**

There are two major classes of receptors for norepinephrine and epinephrine: **alpha-adrenergic receptors** and **beta-adrenergic receptors** (also called alpha-adrenoceptors and beta-adrenoceptors). All catecholamine receptors are metabotropic, and thus use second messengers to transfer a signal from the surface of the cell to the cytoplasm. Beta-adrenoceptors act via stimulatory G proteins to increase cAMP in the postsynaptic

cell. There are three subclasses of beta-receptors, β_1, β_2, and β_3, which function in different ways in different tissues (see Table 6–11). Alpha-adrenoceptors exist in two subclasses, α_1 and α_2. They act presynaptically to inhibit norepinephrine release (α_2) or postsynaptically to either stimulate or inhibit activity at different types of potassium channels (α_1). The subclasses of alpha- and beta-receptors are distinguished by the drugs that influence them and their second-messenger systems.

Serotonin

While not a catecholamine, **serotonin** (5-hydroxy-tryptamine, or 5-HT) is an important biogenic amine. It is produced from tryptophan, an essential amino acid. Its effects generally have a slow onset, indicating that it works as a neuromodulator. Serotonin-releasing neurons innervate virtually every structure in the brain and spinal cord and operate via at least 16 different receptor types.

In general, serotonin has an excitatory effect on pathways that are involved in the control of muscles, and an inhibitory effect on pathways that mediate sensations. The activity of serotonergic neurons is lowest or absent during sleep and highest during states of alert wakefulness. In addition to their contributions to motor activity and sleep, serotonergic pathways also function in the regulation of food intake, reproductive behavior, and emotional states such as mood and anxiety.

Serotonin reuptake blockers such as **paroxetine (Paxil®)** are thought to aid in the treatment of depression by inactivating the 5-HT transporter and increasing the synaptic concentration of the neurotransmitter. Interestingly, such drugs are often associated with decreased appetite but paradoxically cause weight gain due to disruption of enzymatic pathways that regulate fuel metabolism. This is one example of how the use of reuptake inhibitors for a specific neurotransmitter—one with widespread actions—can cause unwanted side effects. The drug lysergic acid diethylamide (***LSD***) is thought to block serotonin receptors in the brain, thereby preventing normal serotonergic neurotransmission. However, it is not clear how this action produces the intense visual hallucinations that are produced by ingestion of this drug.

Serotonin is also present in many nonneural cells (e.g., blood platelets and certain cells of the immune system and digestive tract). In fact, the brain contains only 1 to 2 percent of the body's serotonin.

Amino Acid Neurotransmitters

In addition to the neurotransmitters that are synthesized from amino acids, several amino acids themselves function as neurotransmitters. Although the amino acid neurotransmitters chemically fit the category of biogenic amines, neurophysiologists traditionally put them into a category of their own. The amino acid neurotransmitters are by far the most prevalent neurotransmitters in the central nervous system, and they affect virtually all neurons there.

Glutamate

There are a number of **excitatory amino acids, aspartate** being one example, but the most common neurotransmitter at excitatory synapses in the CNS is the amino acid **glutamate.** As with other neurotransmitter systems, pharmacological manipu-

lation of the receptors for glutamate has permitted identification of specific receptor subtypes by their ability to bind natural and synthetic ligands. Although metabotropic glutamate receptors do exist, the vast majority are ionotropic, with two important subtypes being found in postsynaptic membranes. They are designated as **AMPA receptors** (identified by their binding to α-amino-3 hydroxy-5 methyl-4 isoxazole proprionic acid) and **NMDA receptors** (which bind *N*-methyl-*D*-aspartate).

Cooperative activity of AMPA and NMDA receptors has been implicated in a phenomenon called **long-term potentiation (LTP).** This mechanism couples frequent activity across a synapse with lasting changes in the strength of signaling across that synapse, and is thus thought to be a cellular process underlying learning and memory. **Figure 6–36** outlines the mechanism in stepwise fashion. When a presynaptic neuron fires action potentials (step 1), glutamate is released from presynaptic terminals (step 2) and binds to both AMPA and NMDA receptors on postsynaptic membranes (step 3). AMPA receptors function just like the excitatory postsynaptic receptors discussed earlier—when glutamate binds, the channel becomes permeable to both sodium and potassium, but the larger entry of sodium creates a depolarizing EPSP of the postsynaptic cell (step 4). By contrast, NMDA-receptor channels also mediate a substantial calcium flux, but opening them requires more than just glutamate binding. A magnesium ion blocks NMDA channels when the membrane voltage is near the negative resting potential, and to

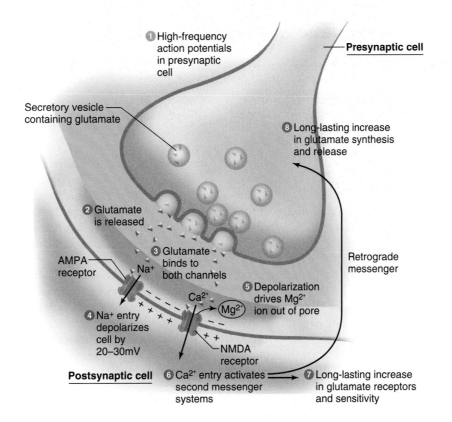

Figure 6–36

Long-term potentiation at glutamatergic synapses. Episodes of intense firing across a synapse result in structural and chemical changes that amplify the strength of synaptic signaling during subsequent activation. See text for description of each step. Note that both AMPA and NMDA receptors are nonspecific cation channels, but the main current through AMPA channels is sodium whereas NMDA channels allow significant calcium current.

drive it out of the way the membrane must be significantly depolarized by the current through AMPA channels (step 5). This explains why it requires a high frequency of presynaptic action potentials to complete the long-term potentiation mechanism: At low frequencies there is insufficient temporal summation of AMPA-receptor EPSPs to provide the 20–30 mV of depolarization needed to move the magnesium ion, and so the NMDA receptors do not open. When the depolarization is sufficient, however, NMDA receptors do open, allowing calcium to enter the postsynaptic cell (step 6). Calcium then activates a second-messenger cascade in the postsynaptic cell that includes persistent activation of two different protein kinases, and which increases the sensitivity of the postsynaptic neuron to glutamate (step 7). This second-messenger system can also activate long-term enhancement of presynaptic glutamate release via a retrograde messenger not yet identified (step 8). Each subsequent action potential arriving along this presynaptic cell will cause a greater depolarization of the postsynaptic membrane. Thus, repeatedly and intensely activating a particular pattern of synaptic firing (as you might when studying for an exam) causes chemical and structural changes that facilitate future activity along those same pathways (as might occur when recalling what you learned).

NMDA receptors have also been implicated in mediating **excitotoxicity.** This is a phenomenon in which the injury or death of some brain cells (due, for example, to blocked or ruptured blood vessels) rapidly spreads to adjacent regions. When glutamate-containing cells die and their membranes rupture, the flood of glutamate excessively stimulates AMPA and NMDA receptors on nearby neurons. The excessive stimulation of those neurons causes the accumulation of toxic levels of intracellular calcium, which in turn kills those neurons and causes *them* to rupture, and the wave of damage progressively spreads. Recent experiments and clinical trials suggest that administering NMDA receptor antagonists may help minimize the spread of cell death following injuries to the brain.

GABA

GABA (gamma-aminobutyric acid) is the major inhibitory neurotransmitter in the brain. Although it is not one of the 20 amino acids used to build proteins, it is classified with the amino acid neurotransmitters because it is a modified form of glutamate. With few exceptions, GABA neurons in the brain are small interneurons that dampen activity within neural circuits. Postsynaptically, GABA may bind to ionotropic or metabotropic receptors. The ionotropic receptor increases chloride flux into the cell, resulting in hyperpolarization of the postsynaptic membrane. In addition to the GABA binding site, this receptor has several additional binding sites for other compounds, including steroids, barbiturates, ethanol, and benzodiazepines. Benzodiazepine drugs such as *Xanax*® and *Valium*® reduce anxiety, guard against seizures, and induce sleep, by increasing chloride flux through the GABA receptor.

Glycine

Glycine is the major neurotransmitter released from inhibitory interneurons in the spinal cord and brainstem. It binds to ionotropic receptors on postsynaptic cells that allow chloride to enter, thus hyperpolarizing or stabilizing the resting membrane potential. Normal function of glycinergic neurons is essential for maintaining a balance of excitatory and inhibitory activity in spinal cord integrating centers that regulate skeletal muscle contraction. This becomes apparent in cases of poisoning with the neurotoxin **strychnine,** an antagonist of glycine receptors. Victims experience hyperexcitability throughout the nervous system, which leads to convulsions, spastic contraction of skeletal muscles, and ultimately death due to impairment of the muscles of respiration.

Neuropeptides

The **neuropeptides** are composed of two or more amino acids linked together by peptide bonds. Some 85 neuropeptides have been identified, but their physiological roles are often unknown. It seems that evolution has selected the same chemical messengers for use in widely differing circumstances, and many of the neuropeptides had been previously identified in nonneural tissue where they function as hormones or paracrine agents. They generally retain the name they were given when first discovered in the nonneural tissue.

The neuropeptides are formed differently from other neurotransmitters, which are synthesized in the axon terminals by very few enzyme-mediated steps. The neuropeptides, in contrast, are derived from large precursor proteins, which in themselves have little, if any, inherent biological activity. The synthesis of these precursors, directed by mRNA, occurs on ribosomes, which exist only in the cell body and large dendrites of the neuron, often a considerable distance from axon terminals or varicosities where the peptides are released.

In the cell body, the precursor protein is packaged into vesicles, which are then moved by axonal transport into the terminals or varicosities, where the protein is cleaved by specific peptidases. Many of the precursor proteins contain multiple peptides, which may be different or be copies of one peptide. Neurons that release one or more of the peptide neurotransmitters are collectively called **peptidergic.** In many cases, neuropeptides are cosecreted with another type of neurotransmitter and act as neuromodulators.

The amount of peptide released from vesicles at synapses is significantly lower than the amount of nonpeptidergic neurotransmitters such as catecholamines. In addition, neuropeptides can diffuse away from the synapse and affect other neurons at some distance, in which case they are referred to as neuromodulators. The actions of these neuromodulators are longer-lasting (on the order of several hundred milliseconds) than when peptides or other molecules act as neurotransmitters. After release, peptides can interact with either ionotropic or metabotropic receptors. They are eventually broken down by peptidases located in the neuronal membrane.

Endogenous opioids, a group of neuropeptides that includes **beta-endorphin,** the **dynorphins,** and the **enkephalins**—have attracted much interest because their receptors are the sites of action of opiate drugs such as *morphine* and *codeine.* The opiate drugs are powerful *analgesics* (that is, they relieve pain without loss of consciousness), and the endogenous opioids undoubtedly play a role in regulating pain. The opioids have been implicated in the runner's "second wind," when the athlete feels a boost of energy and a decrease in pain and effort, and in the general feeling of well-being experienced after a bout of strenuous exercise, the so-called "runner's high." There is also evidence that the opioids play a role in eating and drinking behavior, in regulation of the cardiovascular system, and in mood and emotion.

Substance P, another of the neuropeptides, is a transmitter released by afferent neurons that relay sensory information into the central nervous system. It is known to be involved in pain sensation.

Miscellaneous

Surprisingly, at least two gases—**nitric oxide** and **carbon monoxide**—serve as neurotransmitters. Gases are not released from presynaptic vesicles, nor do they bind to postsynaptic plasma membrane receptors. They simply diffuse from their sites of origin in one cell into the intracellular fluid of nearby cells. These gases serve as messengers between some neurons and between neurons and effector cells. Both are produced by cytosolic enzymes and bind to and activate guanylyl cyclase in the recipient cell. This increases the concentration of the second-messenger cyclic GMP in that cell.

Nitric oxide plays a role in a bewildering array of neurally mediated events—learning, development, drug tolerance, penile erection, and sensory and motor modulation, to name a few. Paradoxically, it is also implicated in neural damage that results, for example, from the stoppage of blood flow to the brain or from a head injury. In later chapters we will see that nitric oxide is produced not only in the central and peripheral nervous systems, but also by a variety of non-neural cells, and it plays an important paracrine role in the cardiovascular and immune systems, among others.

Other nontraditional neurotransmitters include the purines, **ATP** and **adenosine,** which are considered excitatory neuromodulators. ATP is present in all pre-synaptic vesicles and is coreleased with one or more of the classical neurotransmitters in response to calcium influx into the terminal. Adenosine is derived from ATP via extracellular enzymatic activity. Both presynaptic and postsynaptic receptors have been described for adenosine, though the exact roles these substances play in neurotransmission is unknown.

Neuroeffector Communication

Thus far we have described the effects of neurotransmitters released at synapses. Many neurons of the peripheral nervous system end, however, not at synapses on other neurons but at neuroeffector junctions on muscle and gland cells. The neurotransmitters released by these efferent neurons' terminals or varicosities provide the link by which electrical activity of the nervous system regulates effector cell activity.

The events that occur at neuroeffector junctions are similar to those at a synapse. The neurotransmitter is released from the efferent neuron upon the arrival of an action potential at the neuron's axon terminals or varicosities. The neurotransmitter then diffuses to the surface of the effector cell, where it binds to receptors on that cell's plasma membrane. The receptors may be directly under the axon terminal or varicosity, or they may be some distance away so that the diffusion path the neurotransmitter follows is long. The receptors on the effector cell may be either ionotropic or metabotropic. The response (altered muscle contraction or glandular secretion) of the effector cell will be described in later chapters. As we will see in the next section, the major neurotransmitters released at neuroeffector junctions are acetylcholine and norepinephrine.

ADDITIONAL CLINICAL EXAMPLES

Ethanol: A Pharmacological Hand Grenade

After caffeine, the ethanol found in alcoholic beverages is the second most widely used drug in the world. Although the psychological and behavioral effects of ethanol ingestion have been known about (and sought after) for much of recorded human history, only recently are the physiological mechanisms of its complex effects and side effects being investigated. Experiments done in the 1890s showed that the lipid solubility of different alcohols varied with the number of carbon molecules in their structure, which in turn correlated with their anesthetizing/intoxicating strength. This led to the hypothesis that ethanol simply dissolved and disrupted the lipid bilayers of neurons, causing generalized malfunction of the brain. It has more recently become clear that while other alcohols do effectively dissolve the plasma membrane (often with irreversible and lethal consequences), the two-carbon ethanol molecule has specific pharmacological effects on a wide variety of cellular proteins.

Among ethanol's targets are proteins involved in synaptic transmission throughout the brain. Effects on dopaminergic and endogenous opioid signaling result in short-term mood elevation or euphoria, and may also explain the long-term addictive effects some people experience. Ethanol has a global depressant effect on the activity of the brain and brainstem, arising from the fact that it strongly inhibits glutamate signaling (the brain's main excitatory neurotransmitter) while stimulating GABA signaling (the brain's main inhibitory neurotransmitter). Thus, as a person's blood alcohol content rises, there is a progressive reduction in overall mental processing capability, and side-effects begin to emerge such as reduced sensory perception (hearing in particular), motor incoordination, impaired judgment, memory loss, and unconsciousness. Very high doses of ethanol are sometimes fatal, due to suppression of brainstem centers responsible for regulating the cardiovascular and respiratory systems.∎

■ SECTION C SUMMARY

I. An excitatory synapse brings the membrane of the postsynaptic cell closer to threshold. An inhibitory synapse hyperpolarizes the postsynaptic cell or stabilizes it at its resting level.

II. Whether a postsynaptic cell fires action potentials depends on the number of synapses that are active and whether they are excitatory or inhibitory.

III. Neurotransmitters are chemical messengers that pass from one neuron to another and modify the electrical or metabolic function of the recipient cell.

Functional Anatomy of Synapses

I. A neurotransmitter, which is stored in synaptic vesicles in the presynaptic axon terminal, carries the signal from a pre- to a postsynaptic neuron.

Mechanisms of Neurotransmitter Release

I. Depolarization of the axon terminal raises the calcium concentration within the terminal, which causes the release of neurotransmitter into the synaptic cleft.

II. The neurotransmitter diffuses across the synaptic cleft and binds to receptors on the postsynaptic cell; the activated receptors usually open ion channels.

Activation of the Postsynaptic Cell

I. At an excitatory synapse, the electrical response in the postsynaptic cell is called an excitatory postsynaptic potential (EPSP). At an inhibitory synapse, it is an inhibitory postsynaptic potential (IPSP).

II. Usually at an excitatory synapse, channels in the postsynaptic cell that are permeable to sodium, potassium, and other small positive ions open; at inhibitory synapses, channels to chloride and/or potassium open.

Synaptic Integration

I. The postsynaptic cell's membrane potential is the result of temporal and spatial summation of the EPSPs and IPSPs at the many active excitatory and inhibitory synapses on the cell.

II. Action potentials are generally initiated by the temporal and spatial summation of many EPSPs.

Synaptic Strength

I. Synaptic effects are influenced by pre- and postsynaptic events, drugs, and diseases (Table 6–5).

Neurotransmitters and Neuromodulators

I. In general, neurotransmitters cause EPSPs and IPSPs, and neuromodulators cause, via second messengers, more complex metabolic effects in the postsynaptic cell.

II. The actions of neurotransmitters are usually faster than those of neuromodulators.

III. A substance can act as a neurotransmitter at one type of receptor and as a neuromodulator at another.

IV. The major classes of known or suspected neurotransmitters and neuromodulators are listed in Table 6–6.

Neuroeffector Communication

I. The junction between a neuron and an effector cell is called a neuroeffector junction.

II. The events at a neuroeffector junction (release of neurotransmitter into an extracellular space, diffusion of neurotransmitter to the effector cell, and binding with a receptor on the effector cell) are similar to those at a synapse.

Additional Clinical Examples

I. Ethanol alters brain function by targeting proteins involved in synaptic transmission throughout the brain. By inhibiting glutamate and enhancing GABA signaling, it has a global depressant effect on the nervous system. Its effects on dopaminergic and endogenous opioid signaling result in euphoria, mood elevation, and occasionally addiction. High doses are fatal due to suppression of cardiovascular and respiratory centers in the brainstem.

■ SECTION C KEY TERMS

acetylcholine (ACh) 166	glutamate 169
acetylcholinesterase 167	glycine 170
active zones 161	inhibitory postsynaptic potential
adenosine 171	(IPSP) 162
adrenergic 168	inhibitory synapse 159
agonist 166	ionotropic receptor 161
alpha-adrenergic receptor 168	L-dopa 167
AMPA receptor 169	long-term potentiation
antagonist 166	(LTP) 169
aspartate 169	metabotropic receptor 161
ATP 171	monoamine oxidase
autoreceptor 165	(MAO) 168
axo-axonic synapse 164	muscarinic receptor 167
beta-adrenergic receptor 168	neuromodulator 166
beta-endorphin 170	neuropeptide 170
biogenic amine 167	nicotinic receptor 167
carbon monoxide 171	nitric oxide 171
catecholamine 167	NMDA receptor 169
chemical synapse 159	noradrenergic 168
cholinergic 166	norepinephrine (NE) 167
convergence 159	peptidergic 170
cotransmitter 160	postsynaptic density 160
divergence 159	presynaptic facilitation 164
dopamine 167	presynaptic inhibition 164
dynorphin 170	receptor desensitization 165
electrical synapse 159	reuptake 161
endogenous opioid 170	serotonin 168
enkephalin 170	SNARE proteins 161
epinephrine 167	spatial summation 163
excitatory amino acid 169	strychnine 170
excitatory postsynaptic potential	substance P 171
(EPSP) 162	synaptic cleft 160
excitatory synapse 159	synaptic delay 161
excitotoxicity 170	synaptic vesicle 160
GABA (gamma-aminobutyric	synaptotagmin 161
acid) 170	temporal summation 163

■ SECTION C CLINICAL TERMS

Alzheimer's disease 167	morphine 170
analgesics 170	paroxetine (Paxil®) 168
atropine 167	Sarin 167
botulism 166	tetanus toxin 166
codeine 170	Valium® 170
LSD 169	Xanax® 170

1. Contrast the postsynaptic mechanisms of excitatory and inhibitory synapses.
2. Explain how synapses allow neurons to act as integrators; include the concepts of facilitation, temporal and spatial summation, and convergence in your explanation.
3. List at least eight ways in which the effectiveness of synapses may be altered.
4. Discuss differences between neurotransmitters and neuromodulators.
5. Discuss the relationship among dopamine, norepinephrine, and epinephrine.

SECTION D   Structure of the Nervous System

We now survey the anatomy and broad functions of the major structures of the central and peripheral nervous systems. **Figure 6–37** provides a conceptual overview of the organization of the nervous system for you to refer to as we discuss the various subdivisions in this section and in later chapters.

First, we must deal with some potentially confusing terminology. Recall that a long extension from a single neuron is called an axon or a nerve fiber and that the term *nerve* refers to a group of many axons that are traveling together to and from the same general location in the peripheral nervous system. There are no nerves in the central nervous system. Rather, a group of axons traveling together in the central nervous system is called a **pathway,** a **tract,** or, when it links the right and left halves of the central nervous system, a **commissure.**

Information can pass through the central nervous system down *long neural pathways.* In these pathways, neurons with long axons carry information directly between the brain and spinal cord or between large regions of the brain. In addition, information can travel through multisynaptic pathways made up of many neurons and many synaptic connections. Because synapses are the sites where new information can be integrated into neural messages, there are many opportunities for neural processing along the multisynaptic pathways. The long pathways, on the other hand, consist of chains of only a

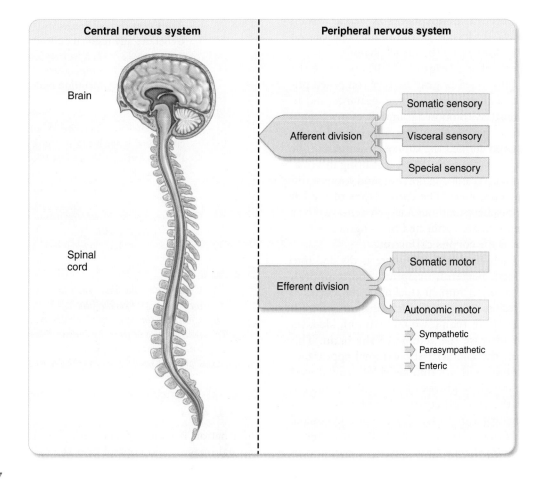

Figure 6–37

Overview of the structural and functional organization of the nervous system.

few neurons connected in sequence. Because the long pathways contain few synapses, there are fewer opportunities for alteration in the information they transmit.

The cell bodies of neurons with similar functions are often clustered together. Groups of neuron cell bodies in the peripheral nervous system are called **ganglia** (singular, *ganglion*). In the central nervous system, they are called **nuclei** (singular, *nucleus*), not to be confused with cell nuclei.

Central Nervous System: Brain

During development, the central nervous system forms from a long tube. As the anterior part of the tube, which becomes the brain, folds during its continuing formation, four different regions become apparent. These regions become the four subdivisions of the brain: the **cerebrum, diencephalon, brainstem,** and **cerebellum** (**Figure 6–38**). The cerebrum and diencephalon together constitute the **forebrain.** The brainstem consists of the **midbrain, pons,** and **medulla oblongata.** The brain also contains four interconnected cavities, the **cerebral ventricles,** which are filled with fluid.

Overviews of the brain subdivisions are included here and in **Table 6–7**, but details of their functions are given more fully in Chapters 7, 8, and 10.

Forebrain

The larger component of the forebrain, the cerebrum, consists of the right and left **cerebral hemispheres** as well as certain other structures on the underside of the brain. The central core of the forebrain is formed by the diencephalon.

The cerebral hemispheres (**Figure 6–39**) consist of the **cerebral cortex,** an outer shell of **gray matter** composed primarily of cell bodies that give the area a gray appearance, and an inner layer of **white matter,** composed primarily of myelinated fiber tracts. This in turn overlies cell clusters, which are also gray matter and are collectively termed the **subcortical nuclei.** The fiber tracts consist of the many nerve fibers that bring information into the cerebrum, carry information out, and connect different areas within a hemisphere. The cortex layers of the left and right cerebral hemispheres, although largely separated by a deep longitudinal division, are connected by a massive bundle of nerve fibers known as the **corpus callosum.**

The cortex of each cerebral hemisphere is divided into four **lobes:** the **frontal, parietal, occipital,** and **temporal.** Although it averages only 3 mm in thickness, the cortex is highly folded. This results in an area containing cortical neurons that is four times larger than it would be if unfolded, yet does not appreciably increase the volume of the brain. This folding also results in the characteristic external appearance of the human cerebrum, with its sinuous ridges called **gyri** (singular, *gyrus*) separated by grooves called **sulci** (singular, *sulcus*). The cells of the cerebral cortex are organized in six layers. The cortical neurons are of two basic types: pyramidal cells (named for the shape of their cell bodies) and nonpyramidal cells. The **pyramidal cells** form the major output cells of the cortex, sending their axons to other parts of the cortex and to other parts of the central nervous system.

The cerebral cortex is the most complex integrating area of the nervous system. In the cerebral cortex, basic afferent

Table 6–7	Summary of Functions of the Major Parts of the Brain

I. FOREBRAIN
  A. *Cerebral hemispheres*
    1. Contain the cerebral cortex, which participates in perception (Chapter 7), the generation of skilled movements (Chapter 10), reasoning, learning, and memory (Chapter 8)
    2. Contain subcortical nuclei, including those that participate in coordination of skeletal muscle activity (Chapter 10)
    3. Contain interconnecting fiber pathways
  B. *Thalamus*
    1. Acts as a synaptic relay station for sensory pathways on their way to the cerebral cortex (Chapter 7)
    2. Participates in control of skeletal muscle coordination (Chapter 10)
    3. Plays a key role in awareness (Chapter 8)
  C. *Hypothalamus*
    1. Regulates anterior pituitary gland function (Chapter 11)
    2. Regulates water balance (Chapter 14)
    3. Participates in regulation of autonomic nervous system (Chapters 6 and 16)
    4. Regulates eating and drinking behavior (Chapter 16)
    5. Regulates reproductive system (Chapters 11 and 17)
    6. Reinforces certain behaviors (Chapter 8)
    7. Generates and regulates circadian rhythms (Chapters 1, 7, 11, and 16)
    8. Regulates body temperature (Chapter 16)
    9. Participates in generation of emotional behavior (Chapter 8)
  D. *Limbic system*
    1. Participates in generation of emotions and emotional behavior (Chapter 8)
    2. Plays essential role in most kinds of learning (Chapter 8)

II. CEREBELLUM
  A. *Coordinates movements, including those for posture and balance (Chapter 10)*
  B. *Participates in some forms of learning (Chapter 8)*

III. BRAINSTEM
  A. *Contains all the fibers passing between the spinal cord, forebrain, and cerebellum*
  B. *Contains the reticular formation and its various integrating centers, including those for cardiovascular and respiratory activity (Chapters 12 and 13)*
  C. *Contains nuclei for cranial nerves III through XII*

information is collected and processed into meaningful perceptual images, and control over the systems that govern the movement of the skeletal muscles is refined. Nerve fibers enter the cortex predominantly from the diencephalon, specifically from a region known as the thalamus as well as from other regions of the cortex and areas of the brainstem. Some of the

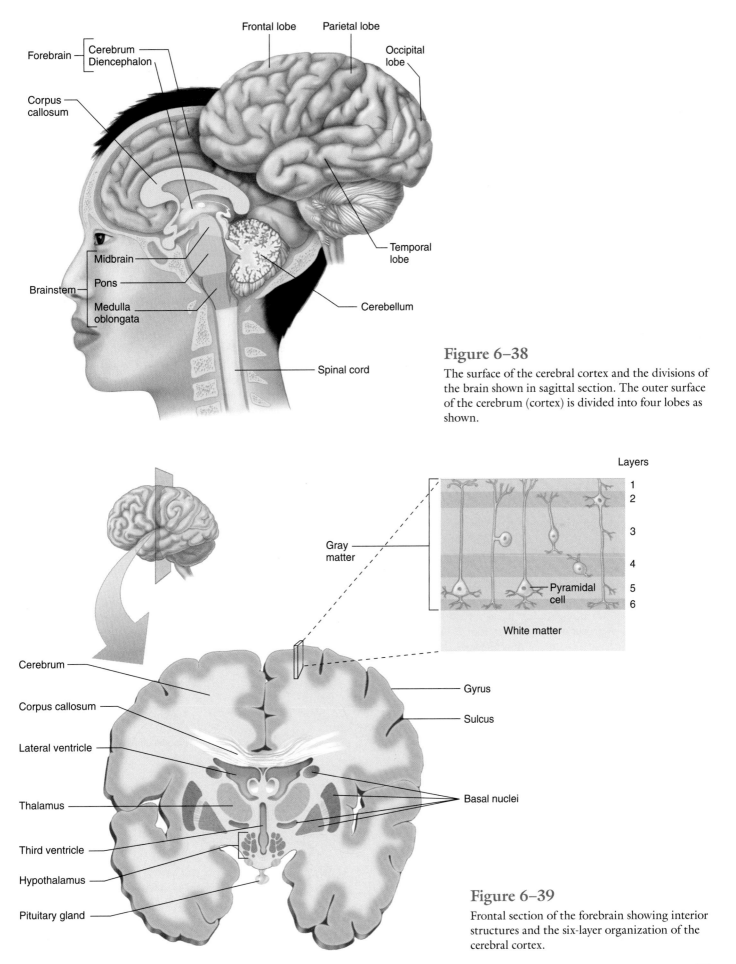

Figure 6–38

The surface of the cerebral cortex and the divisions of the brain shown in sagittal section. The outer surface of the cerebrum (cortex) is divided into four lobes as shown.

Figure 6–39

Frontal section of the forebrain showing interior structures and the six-layer organization of the cerebral cortex.

input fibers convey information about specific events in the environment, whereas others control levels of cortical excitability, determine states of arousal, and direct attention to specific stimuli.

The subcortical nuclei are heterogeneous groups of gray matter that lie deep within the cerebral hemispheres. Predominant among them are the **basal nuclei** (also known as **basal ganglia**), which play an important role in controlling movement and posture and in more complex aspects of behavior.

The diencephalon, which is divided in two by the narrow third cerebral ventricle, is the second component of the forebrain. It contains two major parts: the thalamus and the hypothalamus. The **thalamus** is a collection of several large nuclei that serve as synaptic relay stations and important integrating centers for most inputs to the cortex. It also plays a key role in general arousal and focused attention.

The **hypothalamus** lies below the thalamus and is on the undersurface of the brain. Although it is a tiny region that accounts for less than 1 percent of the brain's weight, it contains different cell groups and pathways that form the master command center for neural and endocrine coordination. Indeed, the hypothalamus is the single most important control area for homeostatic regulation of the internal environment. Behaviors having to do with preservation of the individual (for example, eating and drinking) and preservation of the species (reproduction) are among the many functions of the hypothalamus. The hypothalamus lies directly above and modulates the function of the **pituitary gland,** an important endocrine structure, which is attached to the hypothalamus by a stalk (Chapter 11).

Thus far we have described discrete anatomical areas of the forebrain. Some of these forebrain areas, consisting of both gray and white matter, are also classified together in a functional system called the **limbic system.** This interconnected group of brain structures includes portions of frontal-lobe cortex, temporal lobe, thalamus, and hypothalamus, as well as the fiber pathways that connect them (**Figure 6–40**). Besides being connected with each other, the parts of the limbic system connect with many other parts of the central nervous system. Structures within the limbic system are associated with learning, emotional experience and behavior, and a wide variety of visceral and endocrine functions. In fact, the hypothalamus coordinates much of the output of the limbic system into behavioral and endocrine responses.

Cerebellum

The cerebellum consists of an outer layer of cells, the cerebellar cortex (don't confuse this with the cerebral cortex), and several deeper cell clusters. Although the cerebellum does not initiate voluntary movements, it is an important center for coordinating movements and for controlling posture and balance. To carry out these functions, the cerebellum receives information from the muscles and joints, skin, eyes and ears,

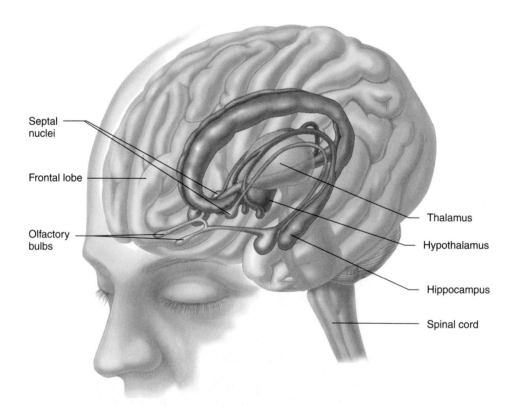

Septal nuclei

Frontal lobe

Olfactory bulbs

Thalamus

Hypothalamus

Hippocampus

Spinal cord

Figure 6–40

Structures of the limbic system (violet) and their anatomic relation to the hypothalamus (purple) are shown in this partially transparent view of the brain.

viscera, and the parts of the brain involved in control of movement. Although the cerebellum's function is almost exclusively motor, it is implicated in some forms of learning.

Brainstem

All the nerve fibers that relay signals between the forebrain, cerebellum, and spinal cord pass through the brainstem. Running through the core of the brainstem and consisting of loosely arranged neuron cell bodies intermingled with bundles of axons is the **reticular formation,** the one part of the brain absolutely essential for life. It receives and integrates input from all regions of the central nervous system and processes a great deal of neural information. The reticular formation is involved in motor functions, cardiovascular and respiratory control, and the mechanisms that regulate sleep and wakefulness and that focus attention. Most of the biogenic amine neurotransmitters are released from the axons of cells in the reticular formation and, because of the far-reaching projections of these cells, these neurotransmitters affect all levels of the nervous system.

Some reticular formation neurons send axons for considerable distances up or down the brainstem and beyond, to most regions of the brain and spinal cord. This pattern explains the very large scope of influence that the reticular formation has over other parts of the central nervous system and explains the widespread effects of the biogenic amines.

The pathways that convey information from the reticular formation to the upper portions of the brain stimulate arousal and wakefulness. They also direct attention to specific events by selectively stimulating neurons in some areas of the brain while inhibiting others. The fibers that descend from the reticular formation to the spinal cord influence activity in both efferent and afferent neurons. Considerable interaction takes place between the reticular pathways that go up to the forebrain, down to the spinal cord, and to the cerebellum. For example, all three components function in controlling muscle activity.

The reticular formation encompasses a large portion of the brainstem, and many areas within the reticular formation serve distinct functions. For example, some reticular formation neurons are clustered together, forming brainstem nuclei and integrating centers. These include the cardiovascular, respiratory, swallowing, and vomiting centers, all of which we will discuss in later chapters. The reticular formation also has nuclei important in eye-movement control and the reflex orientation of the body in space.

In addition, the brainstem contains nuclei involved in processing information for 10 of the 12 pairs of **cranial nerves.** These are the peripheral nerves that connect directly with the brain and innervate the muscles, glands, and sensory receptors of the head, as well as many organs in the thoracic and abdominal cavities.

Central Nervous System: Spinal Cord

The spinal cord lies within the bony vertebral column (**Figure 6–41**). It is a slender cylinder of soft tissue about as big around as the little finger. The central butterfly-shaped area (in cross section) of gray matter is composed of interneurons, the cell

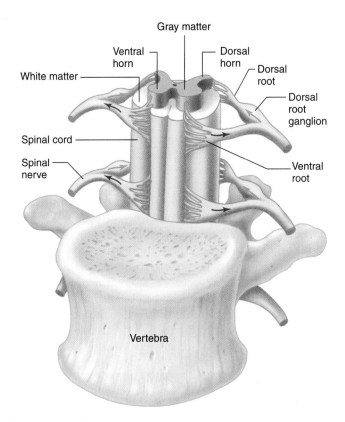

Figure 6–41

Section of the spinal cord, ventral view. The arrows indicate the direction of transmission of neural activity.

bodies and dendrites of efferent neurons, the entering axons of afferent neurons, and glial cells. The regions of gray matter projecting toward the back of the body are called the **dorsal horns,** whereas those oriented toward the front are the **ventral horns.**

The gray matter is surrounded by white matter, which consists of groups of myelinated axons. These groups of fiber tracts run longitudinally through the cord, some descending to relay information *from* the brain to the spinal cord, others ascending to transmit information *to* the brain. Pathways also transmit information between different levels of the spinal cord.

Groups of afferent fibers that enter the spinal cord from the peripheral nerves enter on the dorsal side of the cord via the **dorsal roots** (see Figure 6–41). Small bumps on the dorsal roots, the **dorsal root ganglia,** contain the cell bodies of these afferent neurons. The axons of efferent neurons leave the spinal cord on the ventral side via the **ventral roots.** A short distance from the cord, the dorsal and ventral roots from the same level combine to form a **spinal nerve,** one on each side of the spinal cord.

Peripheral Nervous System

Neurons in the peripheral nervous system transmit signals between the central nervous system and receptors and effectors in all other parts of the body. As noted earlier, the axons are grouped into bundles called nerves. The peripheral nervous system has 43 pairs of nerves: 12 pairs of cranial nerves and 31 pairs that connect with the spinal cord as the spinal nerves. **Table 6–8** lists the cranial nerves and summarizes the information they transmit. The 31 pairs of spinal nerves are designated by the vertebral levels from which they exit: cervical, thoracic, lumbar, sacral, and coccygeal (**Figure 6–42**). The eight pairs of cervical nerves control the muscles and glands and receive sensory input from the neck, shoulders, arms, and hands. The 12 pairs of thoracic nerves are associated with the chest and upper abdomen. The five pairs of lumbar nerves are associated with the lower abdomen, hips, and legs, and the five pairs of sacral nerves are associated with the genitals and lower digestive tract. (A single pair of coccygeal nerves associated with the tailbone brings the total to 31 pairs.)

These peripheral nerves can contain nerve fibers that are the axons of efferent neurons, afferent neurons, or both. Therefore, fibers in a nerve may be classified as belonging to the **efferent** or the **afferent division** of the peripheral nervous system (refer back to Figure 6–37). All the spinal nerves contain both afferent and efferent fibers, whereas some of the cranial nerves (the optic nerves from the eyes, for example) contain only afferent fibers.

As noted earlier, afferent neurons convey information from sensory receptors at their peripheral endings to the central nervous system. The long part of their axon is outside the central nervous system and is part of the peripheral nervous system. Afferent neurons are sometimes called primary afferents or first-order neurons because they are the first cells entering the central nervous system in the synaptically linked chains of neurons that handle incoming information.

Efferent neurons carry signals out from the central nervous system to muscles or glands. The efferent division of the peripheral nervous system is more complicated than the afferent, being subdivided into a **somatic nervous system** and an **autonomic nervous system.** These terms are somewhat misleading because they suggest the presence of additional nervous systems distinct from the central and peripheral systems. Keep in mind that these terms together make up the efferent division of the peripheral nervous system.

Table 6–8	The Cranial Nerves	
Name	**Fibers**	**Comments**
I. Olfactory	Afferent	Carries input from receptors in olfactory (smell) neuroepithelium. Not a true nerve.
II. Optic	Afferent	Carries input from receptors in eye. Not a true nerve.
III. Oculomotor	Efferent	Innervates skeletal muscles that move eyeball up, down, and medially and raise upper eyelid; innervates smooth muscles that constrict pupil and alter lens shape for near and far vision.
	Afferent	Transmits information from receptors in muscles.
IV. Trochlear	Efferent	Innervates skeletal muscles that move eyeball downward and laterally.
	Afferent	Transmits information from receptors in muscles.
V. Trigeminal	Efferent	Innervates skeletal chewing muscles.
	Afferent	Transmits information from receptors in skin; skeletal muscles of face, nose, and mouth; and teeth sockets.
VI. Abducens	Efferent	Innervates skeletal muscles that move eyeball laterally.
	Afferent	Transmits information from receptors in muscles.
VII. Facial	Efferent	Innervates skeletal muscles of facial expression and swallowing; innervates nose, palate, and lacrimal and salivary glands.
	Afferent	Transmits information from taste buds in front of tongue and mouth.
VIII. Vestibulocochlear	Afferent	Transmits information from receptors in ear.
IX. Glossopharyngeal	Efferent	Innervates skeletal muscles involved in swallowing and parotid salivary gland.
	Afferent	Transmits information from taste buds at back of tongue and receptors in auditory-tube skin.
X. Vagus	Efferent	Innervates skeletal muscles of pharynx and larynx and smooth muscle and glands of thorax and abdomen.
	Afferent	Transmits information from receptors in thorax and abdomen.
XI. Accessory	Efferent	Innervates neck skeletal muscles.
XII. Hypoglossal	Efferent	Innervates skeletal muscles of tongue.

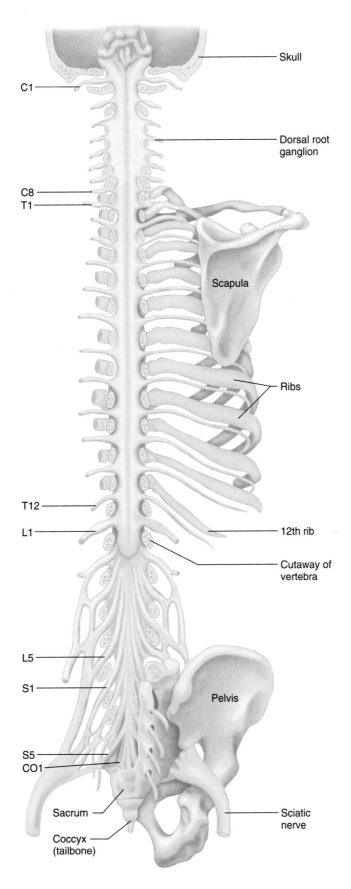

Skull

C1

Dorsal root
ganglion

C8
T1

Scapula

Ribs

T12

L1

12th rib

Cutaway of
vertebra

L5

S1

Pelvis

S5
CO1

Sacrum

Sciatic
nerve

Coccyx
(tailbone)

Figure 6–42

Dorsal view of the spinal cord and spinal nerves. Parts of the skull and vertebrae have been cut away; the ventral roots of the spinal nerves are not visible. In general, the eight cervical nerves (C) control the muscles and glands and receive sensory input from the neck, shoulders, arms, and hands. The 12 thoracic nerves (T) are associated with the shoulders, chest, and upper abdomen. The five lumbar nerves (L) are associated with the lower abdomen, hips and legs, and the five sacral nerves (S) are associated with the genitals and lower digestive tract.

Redrawn from FUNDAMENTAL NEUROANATOMY by Walle J. H. Nauta and Michael Fiertag. Copyright © 1986 by W. H. Freeman and Company. Reprinted by permission.

The simplest distinction between the somatic and autonomic systems is that the neurons of the somatic division innervate skeletal muscle, whereas the autonomic neurons innervate smooth and cardiac muscle, glands, and neurons in the gastrointestinal tract. Other differences are listed in **Table 6–9**.

The somatic portion of the efferent division of the peripheral nervous system is made up of all the nerve fibers going from the central nervous system to skeletal muscle cells. The cell bodies of these neurons are located in groups in the brainstem or the ventral horn of the spinal cord. Their large-diameter, myelinated axons leave the central nervous system and pass without any synapses to skeletal muscle cells. The neurotransmitter these neurons release is acetylcholine. Because activity in the somatic neurons leads to contraction of the innervated skeletal muscle cells, these neurons are called **motor neurons.** Excitation of motor neurons leads only to the *contraction* of skeletal muscle cells; there are no somatic neurons that inhibit skeletal muscles. Muscle relaxation involves the inhibition of the motor neurons in the spinal cord.

Autonomic Nervous System

The efferent innervation of tissues other than skeletal muscle is by way of the autonomic nervous system. A special case occurs in the gastrointestinal tract, where autonomic neurons innervate a nerve network in the wall of the intestinal tract. Chapter 15 will describe this network, termed the **enteric nervous system,** in more detail.

In contrast to the somatic nervous system, the autonomic nervous system is made up of two neurons in series that connect the central nervous system and the effector cells (**Figure 6–43**). The first neuron has its cell body in the central nervous system. The synapse between the two neurons is outside the central nervous system in a cell cluster called an **autonomic ganglion.** The neurons passing between the central nervous system and the ganglia are called **preganglionic neurons;** those passing between the ganglia and the effector cells are **postganglionic neurons.**

Anatomical and physiological differences within the autonomic nervous system are the basis for its further subdivision into **sympathetic** and **parasympathetic** divisions (review Figure 6–37). The neurons of the sympathetic and parasympathetic divisions leave the central nervous system at different levels—the sympathetic fibers from the thoracic (chest) and lumbar regions of the spinal cord, and the parasympathetic fibers from the brainstem and the sacral portion of the spinal

Table 6–9	Peripheral Nervous System: Somatic and Autonomic Divisions
Somatic	
1. Consists of a single neuron between central nervous system and skeletal muscle cells	
2. Innervates skeletal muscle	
3. Can lead only to muscle excitation	
Autonomic	
1. Has two-neuron chain (connected by a synapse) between central nervous system and effector organ	
2. Innervates smooth and cardiac muscle, glands, and GI neurons	
3. Can be either excitatory or inhibitory	

cord (**Figure 6–44**). Therefore, the sympathetic division is also called the thoracolumbar division, and the parasympathetic is called the craniosacral division.

The two divisions also differ in the location of ganglia. Most of the sympathetic ganglia lie close to the spinal cord and form the two chains of ganglia—one on each side of the cord—known as the **sympathetic trunks** (see Figure 6–44). Other sympathetic ganglia, called collateral ganglia—the celiac, superior mesenteric, and inferior mesenteric ganglia—are in the abdominal cavity, closer to the innervated organ (see Figure 6–44). In contrast, the parasympathetic ganglia lie within, or very close to, the organs that the postganglionic neurons innervate.

Preganglionic sympathetic neurons leave the spinal cord only between the first thoracic and second lumbar segments, whereas sympathetic *trunks* extend the entire length of the cord, from the cervical levels high in the neck down to the sacral levels. The ganglia in the extra lengths of sympathetic trunks receive preganglionic neurons from the thoracolumbar regions because some of the preganglionic neurons, once in the sympathetic trunks, turn to travel upward or downward for several segments before forming

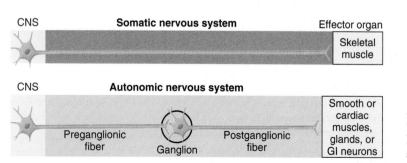

Figure 6–43

Efferent division of the peripheral nervous system, including an overall plan of the somatic and autonomic nervous systems.

Chapter 6

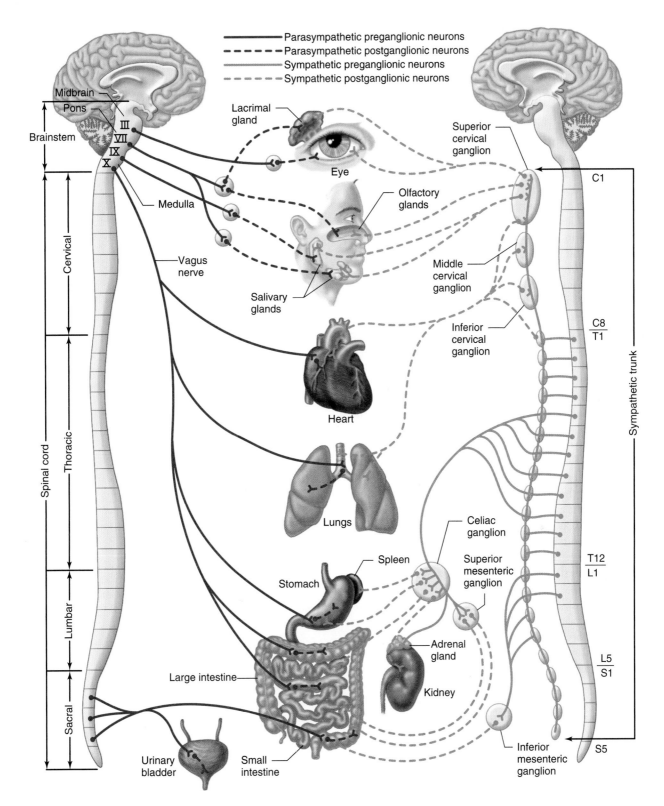

Figure 6–44

The parasympathetic (at left) and sympathetic (at right) divisions of the autonomic nervous system. Although single nerves are shown exiting the brainstem and spinal cord, all represent paired (left and right) nerves. Only one sympathetic trunk is indicated, although there are two, one on each side of the spinal cord. The celiac, superior mesenteric, and inferior mesenteric ganglia are collateral ganglia. Not shown are the fibers passing to the liver, blood vessels, genitalia, and skin glands.

synapses with postganglionic neurons (**Figure 6–45**, numbers 1 and 4). Other possible paths the sympathetic fibers might take are shown in Figure 6–45, numbers 2, 3, and 5.

Due in part to differences in their anatomy, the overall activation pattern within the sympathetic and parasympathetic systems tends to be different. The close anatomical association of the sympathetic ganglia and the marked divergence of presynaptic sympathetic neurons make that division tend to respond as a single unit. Although small segments are occasionally activated independently, it is thus more typical for increased sympathetic activity to occur body-wide when circumstances warrant activation. The parasympathetic system, in contrast, exhibits less divergence, and thus it tends to activate specific organs in a pattern finely tailored to each given physiological situation.

In both the sympathetic and parasympathetic divisions, acetylcholine is the neurotransmitter released between pre- and postganglionic neurons in autonomic ganglia (**Figure 6–46**). In the parasympathetic division, acetylcholine is also the neurotransmitter between the postganglionic neuron and the effector cell. In the sympathetic division, norepinephrine is usually the transmitter between the postganglionic neuron and the effector cell. We say "usually" because a few sympathetic postganglionic endings release acetylcholine (e.g., sympathetic pathways that regulate sweating). At many autonomic synapses, one or more cotransmitters are stored and released with the major neurotransmitter. These include ATP, dopamine, and several of the neuropeptides, all of which seem to play a relatively small role.

In addition to the classical autonomic neurotransmitters just described, there is a widespread network of postganglionic neurons recognized as nonadrenergic and noncholinergic. These neurons use nitric oxide and other neurotransmitters to mediate some forms of blood vessel dilation and to regulate various gastrointestinal, respiratory, urinary, and reproductive functions.

Many of the drugs that stimulate or inhibit various components of the autonomic nervous system affect receptors for acetylcholine and norepinephrine. Recall that there are several types of receptors for each neurotransmitter. The great majority of acetylcholine receptors in the autonomic ganglia are nicotinic receptors. In contrast, the acetylcholine receptors on smooth muscle, cardiac muscle, and gland cells are muscarinic receptors. The cholinergic receptors on skeletal muscle fibers, innervated by the *somatic* motor neurons, not autonomic neurons, are nicotinic receptors (**Table 6–10**).

One set of postganglionic neurons in the sympathetic division never develops axons. Instead, they form an endocrine gland, the **adrenal medulla** (see Figure 6–46). Upon activation by preganglionic sympathetic axons, cells of the adrenal medulla release a mixture of about 80 percent epinephrine and 20 percent norepinephrine into the blood (plus small amounts of other substances, including dopamine, ATP, and neuropeptides). These catecholamines, properly called hormones rather than neurotransmitters in this circumstance, are transported via the blood to effector cells having receptors sensitive to them. The receptors may be the same adrenergic receptors that

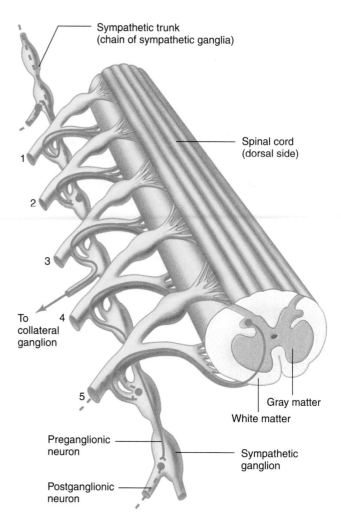

Figure 6–45

Relationship between a sympathetic trunk and spinal nerves (1 through 5) with the various courses that preganglionic sympathetic neurons (solid lines) take through the sympathetic trunk. Dashed lines represent postganglionic neurons. A mirror image of this exists on the opposite side of the spinal cord.

are located near the release sites of sympathetic postganglionic neurons and are normally activated by the norepinephrine released from these neurons. In other cases, the receptors may be located in places that are not near the neurons and are therefore activated only by the circulating epinephrine or norepinephrine. The overall effect of these catecholamines is slightly different due to the fact that some adrenergic receptor subtypes have a higher affinity for epinephrine (e.g., β_2), whereas others have a higher affinity for norepinephrine (e.g., α_1).

Table 6–11 is a reference list of the effects of autonomic nervous system activity, which will be described in later chapters. Note that the heart and many glands and smooth muscles are innervated by both sympathetic and parasympathetic fibers; that is, they receive **dual innervation.** Whatever effect one division has on the effector cells, the other division usually has the opposite effect. (Several exceptions to this rule are indicated in Table 6–11.) Moreover, the two divisions are

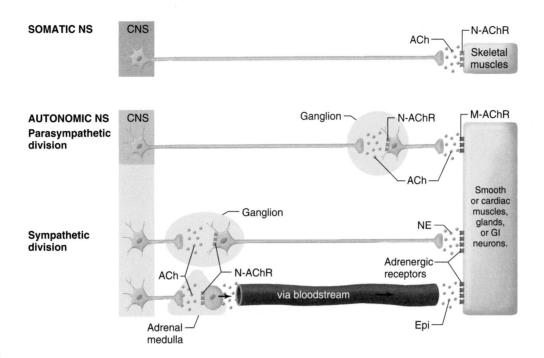

Figure 6–46

Transmitters used in the various components of the peripheral efferent nervous system. Notice that the first neuron exiting the central nervous system—whether in the somatic or the autonomic nervous system—releases acetylcholine. In a very few cases, postganglionic sympathetic neurons may release a transmitter other than norepinephrine. (ACh, acetylcholine; NE, norepinephrine; Epi, epinephrine; N-AChR, nicotinic acetylcholine receptor; M-AChR, muscarinic acetylcholine receptor.)

Figure 6–46 physiological *inquiry* pi

■ How would the effects differ between a drug that blocks muscarinic acetylcholine receptors versus one that blocks nicotinic acetylcholine receptors?

Answer can be found at end of chapter.

Table 6–10	Locations of Receptors for Acetylcholine, Norepinephrine, and Epinephrine

I. Receptors for acetylcholine
  a. Nicotinic receptors
     On postganglionic neurons in the autonomic ganglia
     At neuromuscular junctions of skeletal muscle
     On some central nervous system neurons
  b. Muscarinic receptors
     On smooth muscle
     On cardiac muscle
     On gland cells
     On some central nervous system neurons
     On some neurons of autonomic ganglia (although the great majority of receptors at this site are nicotinic)

II. Receptors for norepinephrine and epinephrine
     On smooth muscle
     On cardiac muscle
     On gland cells
     On some central nervous system neurons

usually activated reciprocally; that is, as the activity of one division increases, the activity of the other decreases. Think of this like a person driving a car with one foot on the brake and the other on the accelerator. Either depressing the brake (parasympathetic) or relaxing the accelerator (sympathetic) will slow the car. Dual innervation by neurons that cause opposite responses provides a very fine degree of control over the effector organ.

A useful generalization is that the sympathetic system increases its activity under conditions of physical or psychological stress. Indeed, a generalized activation of the sympathetic system is called the **fight-or-flight response,** describing the situation of an animal forced to either challenge an attacker or run from it. All resources for physical exertion are activated: heart rate and blood pressure increase; blood flow increases to the skeletal muscles, heart, and brain; the liver releases glucose; and the pupils dilate. Simultaneously, the activity of the gastrointestinal tract and blood flow to the skin are inhibited by sympathetic firing. In contrast, when the parasympathetic system is activated, a person is in a **rest-or-digest** state in which homeostatic functions are predominant.

The two divisions of the autonomic nervous system rarely operate independently, and autonomic responses generally represent the regulated interplay of both divisions. Autonomic

Table 6–11 Some Effects of Autonomic Nervous System Activity

Effector Organ	Receptor Type*	Sympathetic Nervous System Effect	Parasympathetic Nervous System Effect†
Eyes			
Iris muscle	α_1	Contracts radial muscle (widens pupil)	Contracts sphincter muscle (makes pupil smaller)
Ciliary muscle	β_2	Relaxes (flattens lens for far vision)	Contracts (allows lens to become more convex for near vision)
Heart			
SA node	β_1	Increases heart rate	Decreases heart rate
Atria	β_1, β_2	Increases contractility	Decreases contractility
AV node	β_1, β_2	Increases conduction velocity	Decreases conduction velocity
Ventricles	β_1, β_2	Increases contractility	Decreases contractility slightly
Arterioles			
Coronary	α_1, α_2	Constricts	—‡
	β_2	Dilates	
Skin	α_1, α_2	Constricts	—
Skeletal muscle	α_1	Constricts	—
	β_2	Dilates	
Abdominal viscera	α_1	Constricts	—
Kidneys	α_1	Constricts	—
Salivary glands	α_1, α_2	Constricts	Dilates
Veins	α_1, α_2	Constricts	—
	β_2	Dilates	
Lungs			
Bronchial muscle	β_2	Relaxes	Contracts
Salivary glands	α_1	Stimulates watery secretion	Stimulates watery secretion
	β	Stimulates enzyme secretion	
Stomach			
Motility, tone	$\alpha_1, \alpha_2, \beta_2$	Decreases	Increases
Sphincters	α_1	Contracts	Relaxes
Secretion	(?)	Inhibits (?)	Stimulates
Intestine			
Motility	$\alpha_1, \alpha_2, \beta_1, \beta_2$	Decreases	Increases
Sphincters	α_1	Contracts (usually)	Relaxes (usually)
Secretion	α_2	Inhibits	Stimulates
Gallbladder	β_2	Relaxes	Contracts
Liver	α_1, β_2	Glycogenolysis and gluconeogenesis	—
Pancreas			
Exocrine glands	α	Inhibits secretion	Stimulates secretion
Endocrine glands	α_2	Inhibits secretion	—
	β_2	Stimulates secretion	
Fat cells	α_2, β_3	Increases fat breakdown	—
Kidneys	β_1	Increases renin secretion	—
Urinary bladder			
Bladder wall	β_2	Relaxes	Contracts
Sphincter	α_1	Contracts	Relaxes
Uterus	α_1	Contracts in pregnancy	Variable
	β_2	Relaxes	
Reproductive tract (male)	α_1	Ejaculation	Erection
Skin			
Muscles causing hair erection	α_1	Contracts	—
Sweat glands	α_1	Secretion from hands, feet, and armpits	—
	AChR	Generalized abundant, dilute secretion	—

Table adapted from "Goodman and Gilman's The Pharmacological Basis of Therapeutics," Laurence L. Brunton, John S. Lazo, and Keither L. Parker, eds., 11th ed., McGraw-Hill, New York, 2006.

*Note that many effector organs contain both alpha-adrenergic and beta-adrenergic receptors. Activation of these receptors may produce either the same or opposing effects. For simplicity, except for the arterioles and a few other cases, only the dominant sympathetic effect is given when the two receptors oppose each other.

†These effects are all mediated by muscarinic receptors.

‡A dash means these cells are not innervated by this branch of the autonomic nervous system or that these nerves do not play a significant physiological role.

responses usually occur without conscious control or awareness, as though they were indeed autonomous (in fact, the autonomic nervous system has been called the "involuntary" nervous system). However, it is wrong to assume that this is always the case, for some visceral or glandular responses can be learned and thus, to an extent, voluntarily controlled.

Blood Supply, Blood-Brain Barrier, and Cerebrospinal Fluid

As mentioned earlier, the brain lies within the skull, and the spinal cord lies within the vertebral column. Between the soft neural tissues and the bones that house them are three types of membranous coverings called **meninges:** the thick **dura mater** next to the bone, the **arachnoid mater** in the middle, and the thin **pia mater** next to the nervous tissue (**Figure 6–47**). The **subarachnoid space** between the arachnoid and pia is filled with **cerebrospinal fluid (CSF).** The meninges and their specialized parts protect and support the central nervous system, and they circulate and absorb the cerebrospinal fluid. *Meningitis* is an infection of the meninges that originates in the CSF of the subarachnoid space and that results in increased intracranial pressure and, in some cases, seizures and loss of consciousness.

As described previously, CSF is produced by ependymal cells, which make up a specialized epithelial structure called the **choroid plexus.** The black arrows in Figure 6–47 show the flow of CSF. It circulates through the interconnected ventricular system to the brainstem, where it passes through small openings out to a space between the meninges on the surface of the brain and spinal cord. Aided by circulatory, respiratory, and

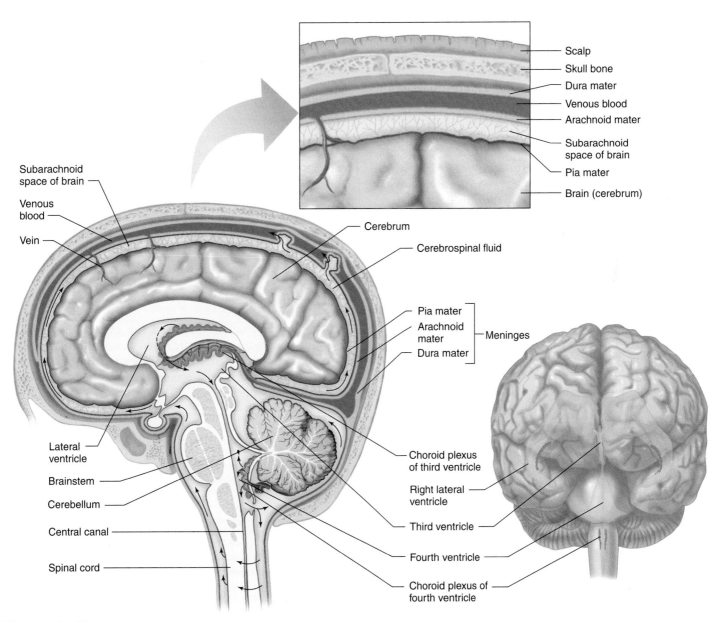

Figure 6–47

The four interconnected ventricles of the brain. The lateral ventricles form the first two. The choroid plexus forms the cerebrospinal fluid (CSF), which flows out of the ventricular system at the brainstem (arrows).

Neuronal Signaling and the Structure of the Nervous System

postural pressure changes, the fluid ultimately flows to the top of the outer surface of the brain, where most of it enters the bloodstream through one-way valves in large veins. Thus, the central nervous system literally floats in a cushion of cerebrospinal fluid. Because the brain and spinal cord are soft, delicate tissues, they are somewhat protected by this shock-absorbing fluid from sudden and jarring movements. If the outflow is obstructed, cerebrospinal fluid accumulates, causing *hydrocephalus* ("water on the brain"). In severe, untreated cases, the resulting elevation of pressure in the ventricles causes compression of the brain's blood vessels, which may lead to inadequate blood flow to the neurons, neuronal damage, and mental retardation.

Under normal conditions, glucose is the only substrate metabolized by the brain to supply its energy requirements, and most of the energy from the oxidative breakdown of glucose is transferred to ATP. The brain's glycogen stores are negligible, so it depends upon a continuous blood supply of glucose and oxygen. In fact, the most common form of brain damage is caused by a decreased blood supply to a region of the brain. When neurons in the region are without a blood supply and deprived of nutrients and oxygen for even a few minutes, they cease to function and die. This neuronal death, when it results from vascular disease, is called a *stroke.*

Although the adult brain makes up only 2 percent of the body weight, it receives 12 to 15 percent of the total blood supply, which supports its high oxygen utilization. If the blood flow to a region of the brain is reduced to 10 to 25 percent of its normal level, energy-dependent membrane ion pumps begin to fail, membrane ion gradients decrease, extracellular potassium concentration increases, and membranes depolarize.

The exchange of substances between blood and extracellular fluid in the central nervous system is different from the more-or-less unrestricted diffusion of nonprotein substances from blood to extracellular fluid in the other organs of the body. A complex group of **blood-brain barrier** mechanisms closely control both the kinds of substances that enter the extracellular fluid of the brain and the rates at which they enter. These mechanisms minimize the ability of many harmful substances to reach the neurons, but they also reduce the access of the immune system to the brain.

The blood-brain barrier, which comprises the cells that line the smallest blood vessels in the brain, has both anatomical structures, such as tight junctions, and physiological transport systems that handle different classes of substances in different ways. Substances that dissolve readily in the lipid components of the plasma membranes enter the brain quickly. Therefore, the extracellular fluid of the brain and spinal cord is a product of, but chemically different from, the blood.

The blood-brain barrier accounts for some drug actions, too, as we can see from the following scenario. Morphine differs chemically from heroin only slightly: morphine has two hydroxyl groups, whereas heroin has two acetyl groups ($—COCH_3$). This small difference renders heroin more lipid-soluble and able to cross the blood-brain barrier more readily than morphine. As soon as heroin enters the brain, however, enzymes remove the acetyl groups from heroin and change it to morphine. The morphine, less soluble in lipid, is then effectively trapped in the brain, where it may have prolonged effects. Other drugs that have rapid effects in the central nervous system because of their high lipid solubility are barbiturates, nicotine, caffeine, and alcohol.

Many substances that do not dissolve readily in lipids, such as glucose and other important substrates of brain metabolism, nonetheless enter the brain quite rapidly by combining with membrane transport proteins in the cells that line the smallest brain blood vessels. Similar transport systems also move substances out of the brain and into the blood, preventing the buildup of molecules that could interfere with brain function.

A barrier is also present between the blood in the capillaries of the choroid plexuses and the cerebrospinal fluid, and cerebrospinal fluid is a selective secretion. For example, potassium and calcium concentrations are slightly lower in cerebrospinal fluid than in plasma, whereas the sodium and chloride concentrations are slightly higher. The choroid plexuses also trap toxic heavy metals such as lead, thus affording a degree of protection to the brain.

The cerebrospinal fluid and the extracellular fluid of the central nervous system are, over time, in diffusion equilibrium. Thus, the restrictive, selective barrier mechanisms in the capillaries and choroid plexuses regulate the extracellular environment of the neurons of the brain and spinal cord.

ADDITIONAL CLINICAL EXAMPLES

Nicotine

Nicotine is the world's third most widely used drug, behind caffeine and alcohol. *Nicotine* is a plant alkaloid compound that constitutes 1 to 2 percent of smoking tobacco. It is also contained in treatments for smoking cessation such as nasal sprays, chewing gums, and transdermal patches. Its hydrophobic structure allows rapid absorption through lung capillaries, mucous membranes, skin, and the blood-brain barrier. Nicotine binds tightly, and lends its name to a type of neurotransmitter receptor distributed widely in the nervous and muscular systems—nicotinic acetylcholine receptors (N-AChRs). These receptors mediate (1) end-plate potentials at neuromuscular junctions; (2) excitatory postsynaptic potentials within ganglia of the autonomic nervous system; (3) the release of catecholamines from the adrenal medulla; and (4) presynaptic facilitation of excitatory neurotransmitter release at widespread synapses in the brain, including the release of dopamine in the brain's principal "reward" pathway.

Nicotine's physiological effects are a complex result of stimulation and desensitization of N-AChRs at these diverse synapses. For example, at low doses nicotine activates autonomic ganglia and stimulates the release of catecholamines from the adrenal medulla. The sympathetic components of these pathways dominate control of the cardiovascular system under these conditions, and so heart rate and blood pressure increase. Persistent high blood pressure and increased work on the heart are part of the reason that chronic nicotine use contributes to cardiovascular disease. In the gastrointestinal system, parasympathetic effects tend to dominate, leading to activation of intestinal smooth muscle motor activity. Brainstem control centers that regulate gastrointestinal functions are also extremely sensitive to nicotine, and vomiting or diarrhea can sometimes occur in individuals who ingest high nicotine doses or in individuals who have not been previously exposed to nicotine. At higher doses of nicotine, the N-AChRs in these autonomic pathways tend to desensitize and thus there is a depression of all autonomic responses. At all doses of nicotine, the neuromuscular junction receptors desensitize so rapidly that the predominant effect on the musculature is relaxation.

Perhaps the most significant effect of nicotine is its stimulation of excitatory neurotransmitter release in the central nervous system, particularly the release of dopamine in the reward center pathways of the brain. These pathways mediate pleasurable sensations associated with behaviors that increase the survival of individuals and species, such as feeding and sexual activity. Beginning just 7 seconds after inhaling tobacco smoke, nicotine produces an overall cognitive stimulation and euphoric sensation that strongly reinforces the desire to smoke. Because of desensitization of the N-AChRs, however, these effects wear off and the user needs more nicotine to regain the pleasurable sensation. With chronic use, a **tolerance** to nicotine develops such that it takes progressively higher concentrations to achieve a given effect. Overt **addiction** occurs when a smoker requires continuous nicotine reinforcement just to feel "normal." Nicotine carries a higher risk of addiction than do most commonly used legal and illegal drugs, such as alcohol, cocaine, and heroin. About one-third of those who use nicotine become addicted, and a variety of unpleasant withdrawal symptoms occur if nicotine use is stopped, including: irritability, impatience, hostility, anxiety, depressed mood, difficulty concentrating, decreased heart rate, increased appetite and weight gain. The addictive power of nicotine most likely accounts for why more than 20 percent of adult Americans continue to smoke tobacco despite the well-publicized fact that almost half a million people die each year from its effects. ■

■ SECTION D SUMMARY

Central Nervous System: Brain

I. The brain is divided into six regions: cerebrum, diencephalon, midbrain, pons, medulla oblongata, and cerebellum.
II. The cerebrum, made up of right and left cerebral hemispheres, and the diencephalon together form the forebrain. The cerebral cortex forms the outer shell of the cerebrum and is divided into the parietal, frontal, occipital, and temporal lobes.
III. The diencephalon contains the thalamus and hypothalamus.
IV. The limbic system is a set of deep forebrain structures associated with learning and emotion.
V. The cerebellum plays a role in posture, movement, and some kinds of memory.
VI. The midbrain, pons, and medulla oblongata form the brainstem, which contains the reticular formation.

Central Nervous System: Spinal Cord

I. The spinal cord is divided into two areas: central gray matter, which contains nerve cell bodies and dendrites; and white matter, which surrounds the gray matter and contains myelinated axons organized into ascending or descending tracts.
II. The axons of the afferent and efferent neurons form the spinal nerves.

Peripheral Nervous System

I. The peripheral nervous system consists of 43 paired nerves—12 pairs of cranial nerves and 31 pairs of spinal nerves. Most nerves contain the axons of both afferent and efferent neurons.
II. The efferent division of the peripheral nervous system is divided into somatic and autonomic parts. The somatic fibers innervate skeletal muscle cells and release the neurotransmitter acetylcholine.

Autonomic Nervous System

I. The autonomic nervous system innervates cardiac and smooth muscle, glands, and gastrointestinal tract neurons. Each autonomic pathway consists of a preganglionic neuron with its cell body in the CNS and a postganglionic neuron with its cell body in an autonomic ganglion outside the CNS.
II. The autonomic nervous system is divided into sympathetic and parasympathetic components. The preganglionic neurons in both the sympathetic and parasympathetic divisions release acetylcholine; the postganglionic parasympathetic neurons release mainly acetylcholine; and the postganglionic sympathetics release mainly norepinephrine.
III. The adrenal medulla is a hormone-secreting part of the sympathetic nervous system and secretes mainly epinephrine.
IV. Many effector organs that the autonomic nervous system innervates receive dual innervation from the sympathetic and parasympathetic division of the autonomic nervous system.

Blood Supply, Blood-Brain Barrier, and Cerebrospinal Fluid

I. Inside the skull and vertebral column, the brain and spinal cord are enclosed in and protected by the meninges.
II. Brain tissue depends on a continuous supply of glucose and oxygen for metabolism.

III. The brain ventricles and the space within the meninges are filled with cerebrospinal fluid, which is formed in the ventricles.

IV. The blood-brain barrier closely regulates the chemical composition of the extracellular fluid of the CNS.

Additional Clinical Examples

I. Nicotine, a plant alkaloid found in tobacco products, stimulates acetylcholine receptors found at neuromuscular junctions and in widespread parts of the nervous system. It causes a complex pattern of stimulation and desensitization of those receptors, and effects include muscle relaxation, blood pressure elevation, and stimulation of brain reward pathways. The latter effect explains why nicotine has a higher addiction risk than most commonly used drugs or alcohol.

■ SECTION D KEY TERMS

adrenal medulla 182
afferent division of the peripheral nervous system 178
arachnoid mater 185
autonomic ganglion 180
autonomic nervous system 178
basal ganglia 176
basal nuclei 176
blood-brain barrier 186
brainstem 174
cerebellum 174
cerebral cortex 174
cerebral hemisphere 174
cerebral ventricle 174
cerebrospinal fluid (CSF) 185
cerebrum 174
choroid plexus 185
commissure 173
corpus callosum 174
cranial nerve 177
diencephalon 174
dorsal horn 177
dorsal root 177
dorsal root ganglia 177
dual innervation 182
dura mater 185
efferent division of the peripheral nervous system 178

enteric nervous system 180
fight-or-flight response 183
forebrain 174
frontal lobe 174
ganglion 174
gray matter 174
gyrus 174
hypothalamus 176
limbic system 176
medulla oblongata 174
meninges 185
midbrain 174
motor neuron 180
nucleus 174
occipital lobe 174
parasympathetic division of the autonomic nervous system 180
parietal lobe 174
pathway 173
pia mater 185
pituitary gland 176
pons 174
postganglionic neuron 180
preganglionic neuron 180
pyramidal cell 174
rest-or-digest 183
reticular formation 177

somatic nervous system 178
spinal nerve 177
subarachnoid space 185
subcortical nucleus 174
sulcus 174
sympathetic division of the autonomic nervous system 180

sympathetic trunk 180
temporal lobe 174
thalamus 176
tract 173
ventral horn 177
ventral root 177
white matter 174

■ SECTION D CLINICAL TERMS

addiction 187
hydrocephalus 186
meningitis 185

nicotine 186
stroke 186
tolerance 187

■ SECTION D REVIEW QUESTIONS

1. Make an organizational chart showing the central nervous system, peripheral nervous system, brain, spinal cord, spinal nerves, cranial nerves, forebrain, brainstem, cerebrum, diencephalon, midbrain, pons, medulla oblongata, and cerebellum.

2. Draw a cross section of the spinal cord showing the gray and white matter, dorsal and ventral roots, dorsal root ganglion, and spinal nerve. Indicate the general locations of pathways.

3. List two functions of the thalamus.

4. List the functions of the hypothalamus, and discuss how they relate to homeostatic control.

5. Make a peripheral nervous system chart indicating the relationships among afferent and efferent divisions, somatic and autonomic nervous systems, and sympathetic and parasympathetic divisions.

6. Contrast the somatic and autonomic divisions of the efferent nervous system; mention at least three characteristics of each.

7. Name the neurotransmitter released at each synapse or neuroeffector junction in the somatic and autonomic systems.

8. Contrast the sympathetic and parasympathetic components of the autonomic nervous system; mention at least four characteristics of each.

9. Explain how the adrenal medulla can affect receptors on various effector organs despite the fact that its cells have no axons.

10. The chemical composition of the CNS extracellular fluid is different from that of blood. Explain how this difference is achieved.

Chapter 6 Test Questions

(Answers appear in Appendix A.)

1. Which best describes an afferent neuron?
   a. The cell body is in the central nervous system and the peripheral axon terminal is in the skin.
   b. The cell body is in the dorsal root ganglion and the central axon terminal is in the spinal cord.
   c. The cell body is in the ventral horn of the spinal cord and the axon ends on skeletal muscle.
   d. The dendrites are in the peripheral nervous system and the axon terminal is in the dorsal root.
   e. All parts of the cell are within the central nervous system.

2. Which incorrectly pairs a glial cell type with an associated function?
   a. astrocytes; formation of the blood-brain barrier
   b. microglia; performance of immune function in the central nervous system
   c. oligodendrocytes; formation of myelin sheaths on axons in the peripheral nervous system
   d. ependymal cells; regulation of production of cerebrospinal fluid
   e. astrocytes; removal of potassium ions and neurotransmitters from the brain's extracellular fluid

3. If the extracellular chloride concentration is 110 mmol/L, and a particular neuron maintains an intracellular chloride concentration of 4 mmol/L, at what membrane potential would chloride be closest to electrochemical equilibrium in that cell?
   a. +80 mV
   b. +60 mV
   c. 0 mV
   d. −86 mV
   e. −100 mV

4. Consider the five experiments below, in which the concentration gradient for sodium was varied. In which case(s) would sodium tend to leak out of the cell if the membrane potential was experimentally held at +42 mV?

Experiment	Extracellular Na$^+$ (mmol/L)	Intracellular Na$^+$ (mmol/L)
A	50	15
B	60	15
C	70	15
D	80	15
E	90	15

   a. A only
   b. B only
   c. C only
   d. A, B, and C
   e. D and E

5. Which is a true statement about the resting membrane potential in a typical neuron?
   a. The membrane potential is closer to the sodium equilibrium potential than to the potassium equilibrium potential.
   b. The chloride permeability is higher than that for sodium or potassium.
   c. The membrane potential is at the equilibrium potential for potassium.
   d. There is no ion movement at the steady resting membrane potential.
   e. Ion movement by the Na$^+$/K$^+$-ATPase pump is equal and opposite to the leak of ions through sodium and potassium channels.

6. If a ligand-gated channel permeable to both sodium and potassium was briefly opened at a specific location on the membrane of a typical resting neuron, what would result?
   a. Local currents on the inside of the membrane would flow away from that region.
   b. Local currents on the outside of the membrane would flow away from that region.
   c. Local currents would travel without decrement all along the cell's length.
   d. A brief local hyperpolarization of the membrane would result.
   e. Fluxes of sodium and potassium would be equal, so no local currents would flow.

7. Which ion channel state correctly describes the phase of the action potential it is associated with?
   a. Voltage-gated sodium channels are inactivated in a resting neuronal membrane.
   b. Open voltage-gated potassium channels cause the depolarizing upstroke of the action potential.
   c. Open voltage-gated potassium channels cause afterhyperpolarization.
   d. The sizable leak through voltage-gated potassium channels determines the value of the resting membrane potential.
   e. Opening of voltage-gated chloride channels is the main factor causing rapid repolarization of the membrane at the end of an action potential.

8. Two neurons, A and B, synapse onto a third neuron, C. If neurotransmitter from A opens ligand-gated channels permeable to sodium and potassium, while neurotransmitter from B opens ligand-gated chloride channels, which of the following statements is true?
   a. An action potential in neuron A causes a depolarizing EPSP in neuron B.
   b. An action potential in neuron B causes a depolarizing EPSP in neuron C.
   c. Simultaneous action potentials in A and B will cause hyperpolarization of neuron C.
   d. Simultaneous action potentials in A and B will cause less depolarization of neuron C than if only neuron A fired an action potential.
   e. An action potential in neuron B will bring neuron C closer to its action potential threshold than would an action potential in neuron A.

9. Which correctly associates a neurotransmitter with one of its characteristics?
    a. Dopamine is a catecholamine synthesized from the amino acid tyrosine.
    b. Glutamate is released by most inhibitory interneurons in the spinal cord.
    c. Serotonin is an endogenous opioid associated with "runner's high."
    d. GABA is the neurotransmitter that mediates long-term potentiation.
    e. Neuropeptides are synthesized in the axon terminals of the neurons that release them.

10. Which of these synapses does not have acetylcholine as its primary neurotransmitter?
    a. synapse of a postganglionic parasympathetic neuron onto a heart cell
    b. synapse of a postganglionic sympathetic neuron onto a smooth muscle cell
    c. synapse of a preganglionic sympathetic neuron onto a postganglionic neuron
    d. synapse of a somatic efferent neuron onto a skeletal muscle cell
    e. synapse of a preganglionic sympathetic neuron onto adrenal medullary cells

Chapter 6 Quantitative and Thought Questions

(Answers appear in Appendix A.)

1. Neurons are treated with a drug that instantly and permanently stops the Na^+/K^+-ATPase pumps. Assume for this question that the pumps are not electrogenic. What happens to the resting membrane potential immediately and over time?

2. Extracellular potassium concentration in a person is increased with no change in intracellular potassium concentration. What happens to the resting potential and the action potential?

3. A person has received a severe blow to the head but appears to be all right. Over the next weeks, however, he develops loss of appetite, thirst, and sexual capacity, but no loss in sensory or motor function. What part of the brain do you think may have been damaged?

4. A person is taking a drug that causes, among other things, dryness of the mouth and speeding of the heart rate but no impairment of the ability to use the skeletal muscles. What type of receptor does this drug probably block? (Table 6–11 will help you answer this.)

5. Some cells are treated with a drug that blocks chloride channels, and the membrane potential of these cells becomes slightly depolarized (less negative). From these facts, predict whether the plasma membrane of these cells actively transports chloride and, if so, in what direction.

6. If the enzyme acetylcholinesterase were blocked with a drug, what malfunctions would occur in the heart and skeletal muscle?

7. The compound tetraethylammonium (TEA) blocks the voltage-gated changes in potassium permeability that occur during an action potential. After administration of TEA, what changes would you expect in the action potential? In the afterhyperpolarization?

Chapter 6 Answers to Physiological Inquiries

Figure 6–12 No. Changing the ECF [K^+] has a greater effect on E_K (and thus the resting membrane potential). This is because the ratio of external to internal potassium is changed more when ECF levels go from 5 to 6 mM (a 20 percent increase) than when ICF levels are lowered from 150 to 149 mM (a 0.7 percent decrease). You can confirm this with the Nernst equation: Inserting typical values, when [K_o] = 5 mM and [K_i] = 150 mM, the calculated value of EK = –90.1 mV. If you change [K_i] to 149 mM, the calculated value of E_K = –89.9 mV, which is not very different. By comparison, changing [K_o] to 6 mM causes a greater change, with the resulting E_K = –85.3 mV.

Figure 6–19 The value of the resting potential would change very little because the permeability of resting membranes to sodium is very low. However, during an action potential, the membrane voltage would rise more steeply and reach a more positive value due to the larger electrochemical gradient for Na^+ entry through open voltage-gated channels.

Figure 6–22 In all of the neurons, action potentials will propagate in both directions from the elbow—up the arm toward the spinal cord and down the arm toward the hand. Action potentials traveling upward along afferent pathways will continue through synapses into the CNS to be perceived as pain, tingling, vibration, and other sensations of the lower arm. In contrast, action potential signals traveling backward up motor axons will die out once they reach the cell bodies because synapses found there are "one way" in the opposite direction.

Figure 6–31 When neuron C alone fired there would be no change from the resting membrane potential because increased chloride conductance would effectively clamp the membrane potential at that voltage. This is because in a cell with no chloride pumping, the chloride equilibrium potential and resting membrane potential have the same value. However, if A and C simultaneously fired action potentials, there would be a depolarization about half as large as that produced when A alone fired an action potential.

Figure 6–46 The muscarinic receptor blocker would only inhibit parasympathetic pathways, where acetylcholine released from postganglionic neurons binds to muscarinic receptors on target organs. This would reduce the ability to stimulate "rest-or-digest" processes while leaving the sympathetic "fight-or-flight" response intact. On the other hand, a nicotinic acetylcholine receptor blocker would inhibit all autonomic control of target organs because those receptors are found at the ganglion in both parasympathetic and sympathetic pathways.

Sensory Physiology

Image of the retina showing its blood vessels converging on the optic disc.

t he neural mechanisms that process afferent sensory information bring about awareness of our internal and external world. Such information is communicated to the CNS from the skin, muscles, and viscera as well as from the visual, auditory, vestibular, and chemical sensory systems. In this chapter you will learn how input to the brain and spinal cord from sensory neurons is crucial for our interactions with the world around us and for the body's maintenance of homeostasis.

A **sensory system** is a part of the nervous system that consists of sensory receptor cells that receive stimuli from the external or internal environment, the neural pathways that conduct information from the receptors to the brain or spinal cord, and those parts of the brain that deal primarily with processing the information.

Information that a sensory system processes may or may not lead to conscious awareness of the stimulus. For example, while you would immediately notice a change when leaving an air-conditioned house on a hot summer day, your blood pressure can fluctuate significantly without your awareness. Regardless of whether the information reaches consciousness, it is called **sensory information.** If the information does reach consciousness, it can also be called a **sensation.** A person's understanding of the sensation's meaning is called **perception.** For example, feeling pain is a sensation, but awareness that a tooth hurts is a perception. Sensations and perceptions occur after the CNS modifies or processes sensory information. This processing can accentuate, dampen, or otherwise filter sensory afferent information. At present we have little understanding of the final processing stages that cause patterns of action potentials to become sensations or perceptions.

The initial step of sensory processing is the transformation of stimulus energy first into graded potentials—the receptor potentials—and then into action potentials in afferent neurons. The pattern of action potentials in particular neurons is a code that provides information about the world even though, as is frequently the case with symbols, the action potentials differ vastly from what they represent. Intuitively, it might seem that sensory systems operate like familiar electrical equipment, but this is true only up to a point. As an example, compare telephone transmission with our auditory (hearing) sensory system. The telephone changes sound waves into electrical impulses, which are then transmitted along wires to the receiver. Thus far the analogy holds. (Of course, the mechanisms by which electrical currents and action potentials are transmitted are quite different, but this does not affect our argument.) The telephone receiver then changes the coded electrical impulses back into sound waves. Here is the crucial difference, for our brain does not physically translate the code into sound. Instead, the coded information itself or some correlate of it is what we *perceive* as sound.

Sensory Receptors

Information about the external world and about the body's internal environment exists in different forms—pressure, temperature, light, odorants, sound waves, chemical concentration, and so on. **Sensory receptors** at the peripheral ends of afferent neurons change this information into graded potentials that can initiate action potentials, which travel into the central nervous system. The receptors are either specialized endings of afferent neurons (**Figure 7–1a**) or separate cells that signal the afferent neurons by releasing chemical messengers (**Figure 7–1b**).

To avoid confusion, recall from Chapter 5 that the term *receptor* has two completely different meanings. One meaning is that of "sensory receptor," as just defined. The second usage is for the individual proteins in the plasma membrane or inside the cell that bind specific chemical messengers, triggering an intracellular signal transduction pathway that culminates in the cell's response. The potential confusion between these two meanings is magnified by the fact that the stimuli for some sensory receptors (e.g., those involved in taste and smell) are chemicals that bind to receptor proteins in the plasma membrane of the sensory receptor. If you are in doubt as to which meaning is intended, add the adjective "sensory" or "protein" to see which makes sense in the context.

To repeat, regardless of the original form of the signal that activates sensory receptors, the information must be translated into the language of graded potentials or action potentials. The energy or chemical that impinges upon and activates a sensory receptor is known as a **stimulus.** The process by which a stimulus—a photon of light, say, or the mechanical stretch of a tissue—is transformed into an electrical response is known as **sensory transduction.**

There are many types of sensory receptors, each of which responds much more readily to one form of stimulus than to others. The type of stimulus to which a particular receptor responds in normal functioning is known as its **adequate stimulus.** In addition, within the general stimulus type that serves as a receptor's adequate stimulus, a particular receptor may respond best (i.e., at lowest threshold) to only a very narrow range of stimulus energies. For example, different individual receptors in the eye respond best to light (the adequate stimulus) in different wavelengths.

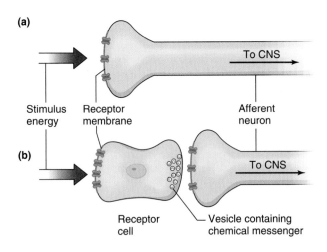

Figure 7–1

Schematic diagram of two types of sensory receptors. The sensitive membrane region that responds to a stimulus is either (a) an ending of an afferent neuron or (b) on a separate cell adjacent to an afferent neuron. Ion channels (shown in purple) on the receptor membrane alter ion flux and initiate stimulus transduction.

Most sensory receptors are exquisitely sensitive to their specific stimulus. For example, some olfactory receptors respond to as few as three or four odor molecules in the inspired air, and visual receptors can respond to a single photon, the smallest quantity of light.

Virtually all sensory receptors, however, can be activated by different types of stimuli if the intensity is sufficiently high. For example, the receptors of the eye normally respond to light, but they can be activated by an intense mechanical stimulus, like a poke in the eye. Note, however, that the sensation of *light* is still experienced in response to a poke in the eye. Regardless of how the receptor is stimulated, any given receptor gives rise to only one sensation.

There are several general classes of receptors that are characterized by the type of stimulus to which they are sensitive. As the name indicates, **mechanoreceptors** respond to mechanical stimuli, such as pressure or stretch, and are responsible for many types of sensory information, including touch, blood pressure, and muscle tension. These stimuli alter the permeability of ion channels on the receptor membrane, changing the membrane potential. **Thermoreceptors** detect both sensations of cold and warmth, and **photoreceptors** respond to particular light wavelengths. **Chemoreceptors** respond to the binding of particular chemicals to the receptor membrane. This type of receptor provides the senses of smell and taste and detects blood pH and oxygen concentration. **Nociceptors** are specialized nerve endings that respond to a number of different painful stimuli, such as heat or tissue damage.

The Receptor Potential

The transduction process in all sensory receptors involves the opening or closing of ion channels that receive information about the internal and external world, either directly or through the second-messenger system. The ion channels are present in a specialized region of the receptor membrane located at the distal tip of the cell's single axon or on associated specialized sensory cells (Figure 7–1). The gating of these ion channels allows a change in the ion fluxes across the receptor membrane, which in turn produces a change in the membrane potential. This change is a graded potential called a **receptor potential.** The different mechanisms that affect ion channels in the various types of sensory receptors are described throughout this chapter.

The specialized receptor membrane region where the initial ion channel changes occur does not generate action potentials. Instead, local current flows a short distance along the axon to a region where the membrane has voltage-gated ion channels and can generate action potentials. In myelinated afferent neurons, this region is usually at the first node of Ranvier. The receptor potential, like the synaptic potential discussed in Chapter 6, is a graded response to different stimulus intensities (**Figure 7–2**) and diminishes as it travels along the membrane.

If the receptor membrane is on a separate cell, the receptor potential there alters the release of neurotransmitter from that cell. The neurotransmitter diffuses across the extracellular cleft between the receptor cell and the afferent neuron and binds to receptor proteins on the afferent neuron. Thus, this junction is a synapse. The combination of neurotransmitter with its binding sites generates a graded potential in the afferent neuron analogous to either an excitatory postsynaptic potential or, in some cases, an inhibitory postsynaptic potential.

As is true of all graded potentials, the magnitude of a receptor potential (or a graded potential in the axon adjacent to the receptor cell) decreases with distance from its origin.

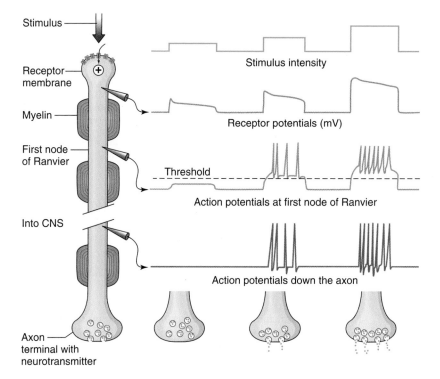

Figure 7–2

Stimulation of an afferent neuron with a receptor ending. Electrodes measure graded potentials and action potentials at various points in response to different stimulus intensities. Action potentials arise at the first node of Ranvier in response to a suprathreshold stimulus, and the action potential frequency and neurotransmitter release increase as the stimulus and receptor potential become larger.

Figure 7–2 physiological *inquiry*

■ How would this afferent pathway be affected by a drug that blocks voltage-gated calcium channels?

Answer can be found at end of chapter.

However, if the amount of depolarization at the first node in the afferent neuron is large enough to bring the membrane there to threshold, action potentials are initiated, which then propagate along the nerve fiber (see Figure 7–2). The only function of the graded potential is to trigger action potentials. (See Figure 6–16 to review the properties of graded potentials.)

As long as the receptor potential keeps the afferent neuron depolarized to a level at or above threshold, action potentials continue to fire and propagate along the afferent neuron. Moreover, an increase in the graded potential magnitude causes an increase in the action potential frequency in the afferent neuron (up to the limit imposed by the neuron's refractory period) and an increase in neurotransmitter release at the afferent neuron's central axon terminal (see Figure 7–2). Although the graded potential magnitude determines action potential *frequency*, it does not determine action potential *magnitude*. The action potential is "all-or-none," meaning that its magnitude is independent of the strength of the initiating stimulus.

Factors that control the magnitude of the receptor potential include stimulus strength, rate of change of stimulus strength, temporal summation of successive receptor potentials (see Figure 6–31), and a process called **adaptation.** This last process is a decrease in receptor sensitivity, which results in a decrease in action potential frequency in an afferent neuron despite a stimulus of constant strength (**Figure 7–3**). Degrees of adaptation vary widely among different types of sensory receptors.

Primary Sensory Coding

Converting stimulus energy into a signal that conveys the relevant sensory information to the central nervous system is termed **coding.** Important characteristics of a stimulus include the type of energy it represents, its intensity, and the location of the body it affects. Coding begins at the receptive neurons in the peripheral nervous system.

A single afferent neuron with all its receptor endings makes up a **sensory unit.** In a few cases, the afferent neuron has a single receptor, but generally the peripheral end of an afferent neuron divides into many fine branches, each terminating with a receptor.

The area of the body that, when stimulated, leads to activity in a particular afferent neuron is called the **receptive field** for that neuron (**Figure 7–4**). Receptive fields of neighboring afferent neurons usually overlap so that stimulation of a single point activates several sensory units. Thus, activation at a single sensory unit almost never occurs. As we will see, the degree of overlap varies in different parts of the body.

Stimulus Type

Another term for stimulus type (heat, cold, sound, or pressure, for example) is stimulus **modality.** Modalities can be divided into submodalities: Cold and warm are submodalities of temperature, whereas salt, sweet, bitter, and sour are submodalities of taste. The type of sensory receptor a stimulus activates plays the primary role in coding the stimulus modality.

As mentioned earlier, a given receptor type is particularly sensitive to one stimulus modality—the adequate stimulus—because of the signal transduction mechanisms and ion channels incorporated in the receptor's plasma membrane. For example, receptors for vision contain pigment molecules whose shape is transformed by light. These receptors also have intracellular mechanisms that cause changes in the pigment molecules to alter the activity of membrane ion channels and generate a receptor potential. In contrast, receptors in the skin do not have light-sensitive pigment molecules, so they cannot respond to light.

All the receptors of a single afferent neuron are preferentially sensitive to the same type of stimulus; for example, they are all sensitive to cold or all to pressure. Adjacent sensory units, however, may be sensitive to different types of stimuli. Because the receptive fields for different modalities overlap, a single stimulus, such as an ice cube on the skin, can simultaneously give rise to the sensations of touch and temperature.

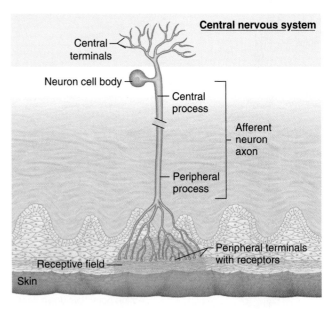

Figure 7–4

A sensory unit including the location of sensory receptors, the processes reaching peripherally and centrally from the cell body, and the terminals in the CNS. Also shown is the receptive field of this neuron.

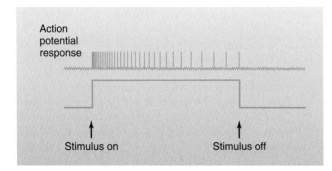

Figure 7–3

Action potentials in a single afferent nerve fiber showing adaptation to a stimulus of constant strength. Note that the frequency of action potentials decreases even before the stimulus is turned off.

Stimulus Intensity

How do we distinguish a strong stimulus from a weak one when the information about both stimuli is relayed by action potentials that are all the same size? The frequency of action potentials in a single receptor is one way, because increased stimulus strength means a larger receptor potential and more frequent action potential firing (review Figure 7–2).

As the strength of a local stimulus increases, receptors on adjacent branches of an afferent neuron are activated, resulting in a summation of their local currents. **Figure 7–5** shows a record of an experiment in which increased stimulus intensity to the receptors of a sensory unit is reflected in increased action potential frequency in its afferent neuron.

In addition to increasing the firing frequency in a single afferent neuron, stronger stimuli usually affect a larger area and activate similar receptors on the endings of *other* afferent neurons. For example, when you touch a surface lightly with a finger, the area of skin in contact with the surface is small, and only the receptors in that skin area are stimulated. Pressing down firmly increases the area of skin stimulated. This "calling in" of receptors on additional afferent neurons is known as **recruitment.**

Stimulus Location

A third type of information to be signaled is the location of the stimulus—in other words, where the stimulus is being applied. It should be noted that in vision, hearing, and smell, stimulus location is interpreted as arising from the site from which the stimulus originated rather than the place on our body where the stimulus was actually applied. For example, we interpret the sight and sound of a barking dog as occurring in that furry thing on the other side of the fence rather than in a specific region of our eyes and ears. We will have more to say about this later; we deal here with the senses in which the stimulus is located to a site on the body.

Stimulus location is coded by the site of a stimulated receptor, as well as by the fact that action potentials from each receptor travel along unique pathways to a specific region of the CNS associated only with that particular modality and body location. These distinct anatomical pathways are sometimes referred to as **labeled lines.** The precision, or **acuity,** with which we can locate and discern one stimulus from an adjacent one depends upon the amount of convergence of neuronal input in the specific ascending pathways: The greater the convergence, the less the acuity. Other factors affecting acuity are the size of the receptive field covered by a single sensory unit (**Figure 7–6a**), the density of sensory units, and the amount of overlap in nearby receptive fields. For example, it is easy to discriminate between two adjacent stimuli (two-point discrimination) applied to the skin on your lips, where the sensory units are small and numerous, but it is harder to do so on the back, where the relatively few sensory units are large and widely spaced (**Figure 7–6b**). Locating sensations from internal organs is less precise than from the skin because

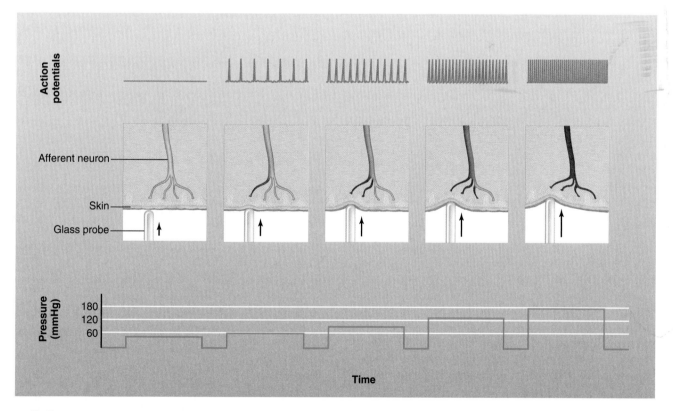

Figure 7–5

Action potentials from an afferent fiber leading from the pressure receptors of a single sensory unit increase in frequency as branches of the afferent neuron are stimulated by pressures of increasing magnitude.

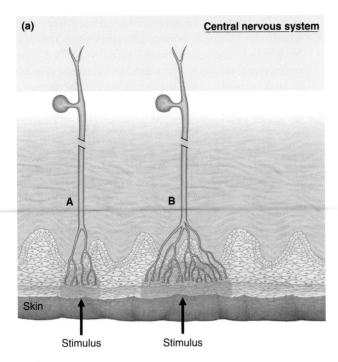

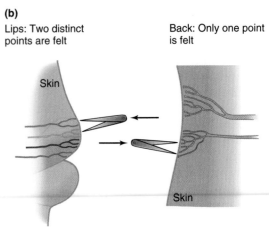

Figure 7–6

The influence of sensory unit size and density on acuity. (a) The information from neuron A indicates the stimulus location more precisely than does that from neuron B because A's receptive field is smaller. (b) Two-point discrimination is finer on the lips than on the back, due to the lips' numerous sensory units with small receptive fields.

Figure 7–6b physiological *inquiry* 🅟🅘

- Make a prediction about the relative size of the brain region devoted to processing lip sensations versus that for the brain region that processes sensations from your back.

Answer can be found at end of chapter.

there are fewer afferent neurons in the internal organs and each has a larger receptive field.

It is fairly easy to see why a stimulus to a neuron that has a small receptive field can be located more precisely than a stimulus to a neuron with a large receptive field (see Figure 7–6). However, more subtle mechanisms also exist that allow us to localize distinct stimuli within the receptive field of a single neuron. In some cases, receptive field overlap aids stimulus localization even though, intuitively, overlap would seem to "muddy" the image. In the next few paragraphs we will examine how this works.

An afferent neuron responds most vigorously to stimuli applied at the center of its receptive field because the receptor density—that is, the number of receptors in a given area—is greatest there. The response decreases as the stimulus is moved toward the receptive field periphery. Thus, a stimulus activates more receptors and generates more action potentials if it occurs at the center of the receptive field (point A in **Figure 7–7**). The firing frequency of the afferent neuron is also related to stimulus strength, however. Thus, a high frequency of impulses in the single afferent nerve fiber of Figure 7–7 could mean either that a moderately intense stimulus was

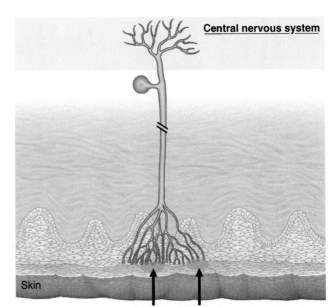

Figure 7–7

Two stimulus points, A and B, in the receptive field of a single afferent neuron. The density of nerve endings around area A is greater than around B, so the frequency of action potentials in response to a stimulus in area A will be greater than the response to a similar stimulus in B.

applied to the center at A or that a strong stimulus was applied to the periphery at B. Thus, neither the intensity nor the location of the stimulus can be detected precisely with a single afferent neuron.

Since the receptor endings of different afferent neurons overlap, however, a stimulus will trigger activity in more than one sensory unit. In **Figure 7–8**, neurons A and C, stimulated near the edges of their receptive fields where the receptor density is low, fire action potentials less frequently than does neuron B, stimulated at the center of its receptive field. A high action potential frequency in neuron B occurring simultaneously with lower frequencies in A and C provides the brain with a more accurate localization of the stimulus near the center of neuron B's receptive field. Once this location is known, the brain can use the firing frequency of neuron B to determine stimulus intensity.

Lateral Inhibition

The phenomenon of **lateral inhibition** is the most important mechanism enabling the localization of a stimulus site. In lateral inhibition, information from afferent neurons whose receptors are at the edge of a stimulus is strongly inhibited compared to information from the stimulus's center. **Figure 7–9** shows one neuronal arrangement that accomplishes lateral inhibition. The afferent neuron in the center (B) has a higher initial firing frequency than the neurons on either side (A and C). The number of action potentials transmitted in the lateral pathways is further decreased by inhibitory inputs from inhibitory interneurons stimulated by the central neuron. While the lateral afferent neurons (A and C) also exert inhibition on the central pathway, their lower initial firing frequency has less of an effect. Thus, lateral inhibition enhances the *contrast* between the center and periphery of a stimulated region, thereby increasing the brain's ability to localize a sensory input. Lateral inhibition can occur at different levels in the sensory pathways but typically happens at an early stage.

Lateral inhibition can be demonstrated by pressing the tip of a pencil against your finger. With your eyes closed, you can localize the pencil point precisely, even though the region

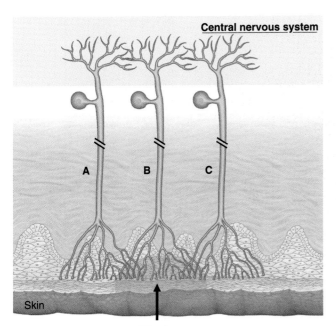

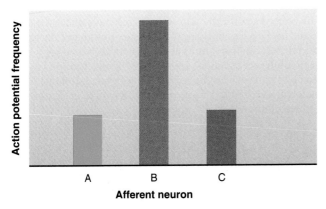

Figure 7–8

A stimulus point falls within the overlapping receptive fields of three afferent neurons. Note the difference in receptor response (i.e., the action potential frequency in the three neurons) due to the difference in receptor distribution under the stimulus (fewer receptor endings for A and C than for B).

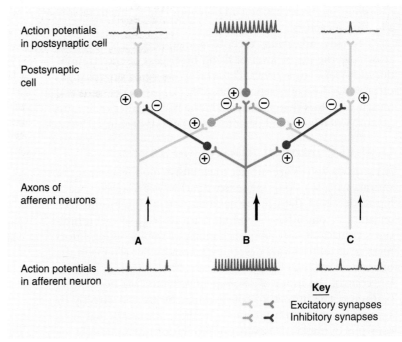

Figure 7–9

Afferent pathways showing lateral inhibition. Three sensory units have overlapping receptive fields. Because the central fiber B at the beginning of the pathway (bottom of figure) is firing at the highest frequency, it inhibits the lateral neurons (via inhibitory interneurons) to a greater extent than the lateral neurons inhibit the central pathway.

around the pencil tip is also indented, activating mechanoreceptors within this region (**Figure 7–10**). Exact localization is possible because lateral inhibition removes the information from the peripheral regions.

Lateral inhibition is utilized to the greatest degree in the pathways providing the most accurate localization. For example, skin hair movements, which we can locate quite well, activate pathways that have significant lateral inhibition, but temperature and pain, which we can locate relatively poorly, activate pathways that use lateral inhibition to a lesser degree. Lateral inhibition is essential for retinal processing, where it enhances visual acuity.

Stimulus Duration

Receptors differ in the way they respond to a constantly maintained stimulus. The action potential frequency at the beginning of the stimulus generally indicates the stimulus strength, but after this initial response, the frequency differs widely in different types of receptors. As **Figure 7–11** shows, some receptors respond very rapidly at the stimulus onset, but, after their initial burst of activity, fire only very slowly or stop firing altogether during the remainder of the stimulus. These are the **rapidly adapting receptors.** The rapid adaptation of these receptors codes for a restricted response in time to a stimulus, and they are important in signaling rapid change (e.g., vibrating or moving stimuli). Some receptors adapt so rapidly that they fire only a single action potential at the onset of a stimulus—a so-called "on response"—while others respond at the beginning of the stimulus and again at its removal—so-called "on-off responses." The rapid fading of the sensation of clothes pressing on your skin is due to rapidly adapting receptors.

Slowly adapting receptors maintain their response at or near the initial level of firing regardless of the stimulus duration (see Figure 7–11). These receptors signal slow changes or prolonged events, such as those that occur in the joint and muscle receptors that participate in the maintenance of upright posture when you stand or sit for long periods of time.

Central Control of Afferent Information

All sensory signals are subject to extensive modification at the various synapses along the sensory pathways before they reach higher levels of the central nervous system. Inhibition from collaterals from other ascending neurons (e.g., lateral inhibition) reduces or even abolishes much of the incoming information, as do pathways descending from higher centers in the brain. The reticular formation and cerebral cortex, in particular, control the input of afferent information via descending pathways. The inhibitory controls may be exerted directly by synapses on the axon terminals of the primary afferent neurons (an example of presynaptic inhibition) or indirectly via interneurons that affect other neurons in the sensory pathways (**Figure 7–12**).

In some cases (e.g., in the pain pathways), the afferent input is continuously inhibited to some degree. This provides the flexibility of either removing the inhibition, so as to allow a greater degree of signal transmission, or increasing the inhibition, so as to block the signal more completely.

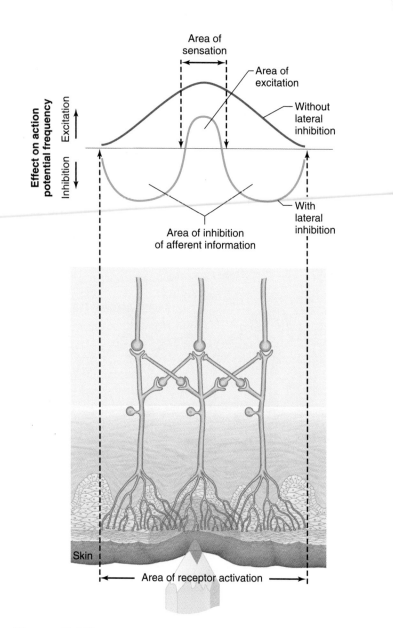

Figure 7–10

A pencil tip pressed against the skin activates receptors under the pencil tip and in the adjacent tissue. The sensory unit under the tip inhibits additional stimulated units at the edge of the stimulated area. Lateral inhibition produces a central area of excitation surrounded by an area where the afferent information is inhibited. The sensation is localized to a more restricted region than that in which all three units are actually stimulated.

Neural Pathways in Sensory Systems

The afferent sensory neurons form the first link in a chain consisting of three or more neurons connected end to end by synapses. A bundle of parallel, three-neuron chains together form a **sensory pathway.** The chains in a given pathway run parallel to each other in the central nervous system and, with one exception, carry information to the part of the cerebral cortex responsible for conscious recognition of the information. Sensory pathways are also called **ascending pathways** because they go "up" to the brain.

Ascending Pathways

The central processes of the afferent neurons enter the brain or spinal cord and synapse upon interneurons there. The central processes may diverge to terminate on several, or many, interneurons (**Figure 7–13a**) or converge so that the processes of many afferent neurons terminate upon a single interneuron (**Figure 7–13b**). The interneurons upon which the afferent neurons synapse are termed second-order neurons, and these in turn synapse with third-order neurons, and so on, until the information (coded action potentials) reaches the cerebral cortex.

Most sensory pathways convey information about only a single type of sensory information. Thus, one pathway conveys information only from mechanoreceptors, whereas another is influenced by information only from thermoreceptors. This allows the brain to distinguish the different types of sensory information even though all of it is being transmitted by essentially the same signal, the action potential. The ascending pathways in the spinal cord and brain that carry information about single types of stimuli are known as the **specific ascending pathways.** The specific ascending pathways pass to the brainstem and thalamus, and the final neurons in the pathways go from there to specific sensory areas of the cerebral cortex (**Figure 7–14**). (The olfactory pathways send some branches to terminate in parts of the limbic system rather than to the thal-

amus.) For the most part, the specific pathways cross to the side of the central nervous system that is opposite to the location of their sensory receptors. Thus, information from receptors on the right side of the body is transmitted to the left cerebral hemisphere, and vice versa.

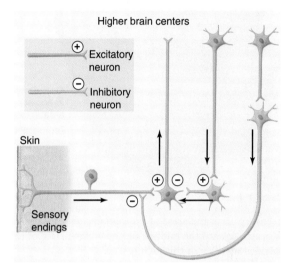

Figure 7–12

Descending pathways may influence sensory information by directly inhibiting the central terminals of the afferent neuron (an example of presynaptic inhibition) or via an interneuron that affects the ascending pathway by inhibitory synapses. Arrows indicate the direction of action potential transmission.

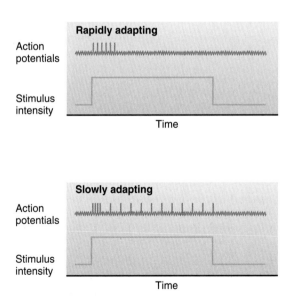

Figure 7–11

Rapidly and slowly adapting receptors. The top line in each graph indicates the action potential firing of the afferent nerve fiber from the receptor, and the bottom line, application of the stimulus. Some rapidly adapting receptors also generate a brief burst of action potentials when a stimulus ceases—an "off" response (not shown here).

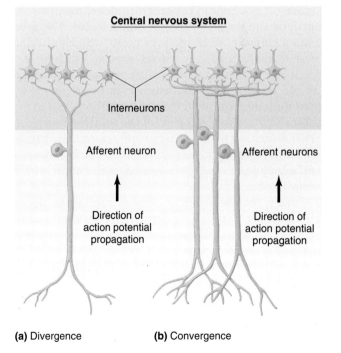

(a) Divergence   **(b)** Convergence

Figure 7–13

(a) Divergence of an afferent neuron on to many interneurons.
(b) Convergence of input from several afferent neurons onto single interneurons.

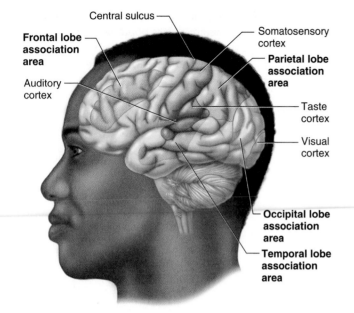

Figure 7–14

Primary sensory areas and areas of association cortex. The olfactory cortex is located toward the midline on the undersurface of the frontal lobes (not visible in this picture).

The specific ascending pathways that transmit information from **somatic receptors**—that is, the receptors in the framework or outer walls of the body, including skin, skeletal muscle, tendons, and joints—go to the **somatosensory cortex.** This is a strip of cortex that lies in the parietal lobe of the brain just posterior to the **central sulcus,** which separates the parietal and frontal lobes (see Figure 7–14). The specific ascending pathways from the eyes go to a different primary cortical receiving area, the **visual cortex,** which is in the occipital lobe. The specific ascending pathways from the ears go to the **auditory cortex,** which is in the temporal lobe. Specific ascending pathways from the taste buds pass to a cortical area adjacent to the region of the somatosensory cortex where information from the face is processed. The pathways serving olfaction project the portions of the limbic system and the **olfactory cortex,** which is located on the undersurface of the frontal lobes.

Finally, the processing of afferent information does not end in the primary cortical receiving areas, but continues from these areas to association areas in the cerebral cortex where complex integration occurs.

In contrast to the specific ascending pathways, neurons in the **nonspecific ascending pathways** are activated by sensory units of several different types (**Figure 7–15**) and therefore signal general information. In other words, they indicate that *something* is happening, without specifying just what or where. A given ascending neuron in a nonspecific ascending pathway may respond, for example, to input from several afferent neurons, each activated by a different stimulus, such as maintained skin pressure, heating, and cooling. Such pathway neurons are called **polymodal neurons.** The nonspecific

ascending pathways, as well as collaterals from the specific ascending pathways, end in the brainstem reticular formation and regions of the thalamus and cerebral cortex that are not highly discriminative, but are important in controlling alertness and arousal.

Association Cortex and Perceptual Processing

The **cortical association areas** presented in Figure 7–14 are brain areas that lie outside the primary cortical sensory or motor areas but are adjacent to them. The association areas are not considered part of the sensory pathways, but they play a role in the progressively more complex analysis of incoming information.

Although neurons in the earlier stages of the sensory pathways are necessary for perception, information from the primary sensory cortical areas undergoes further processing after it is relayed to a cortical association area. The region of association cortex closest to the primary sensory cortical area processes the information in fairly simple ways and serves basic sensory-related functions. Regions farther from the primary sensory areas process the information in more complicated ways. These include, for example, greater contributions from areas of the brain serving arousal, attention, memory, and language. Some of the neurons in these latter regions also integrate input concerning two or more types of sensory stimuli. Thus, an association area neuron receiving input from both the visual cortex and the "neck" region of the somatosensory cortex might integrate visual information with sensory information about head position. In this way, for example, a viewer understands a tree is vertical even if his or her head is tipped sideways.

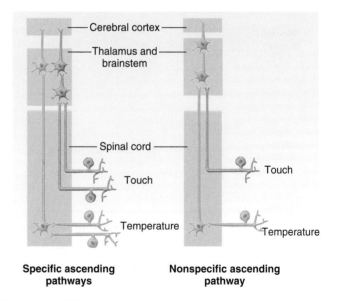

Figure 7–15

Diagrammatic representation of two specific ascending sensory pathways and a nonspecific ascending sensory pathway.

Axons from neurons of the parietal and temporal lobes go to association areas in the frontal lobes that are part of the limbic system. Through these connections, sensory information can be invested with emotional and motivational significance.

Further perceptual processing involves not only arousal, attention, learning, memory, language, and emotions, but also comparison of the information presented via one type of sensation with that presented through another. For example, we may hear a growling dog, but our perception of the event and our emotional response vary markedly, depending upon whether our visual system detects the sound source to be an angry animal or a tape recording.

Factors That Affect Perception

We put great trust in our sensory-perceptual processes despite the inevitable modifications we know the nervous system makes. Factors known to affect our perceptions of the real world include:

1. Sensory receptor mechanisms (e.g., adaptation) and processing of the information along afferent pathways can influence afferent information.

2. Factors such as emotions, personality, experience, and social background can influence perceptions so that two people can be exposed to the same stimuli and yet perceive them differently.

3. Not all information entering the central nervous system gives rise to conscious sensation. Actually, this is a very good thing because many unwanted signals are generated by the extreme sensitivity of our sensory receptors. For example, under ideal conditions, the rods of the eye can detect the flame of a candle 17 miles away. The hair cells of the ear can detect vibrations of an amplitude much lower than those caused by blood flow through the ears' blood vessels and can even detect molecules in random motion bumping against the ear drum. It is possible to detect one action potential generated by a certain type of mechanoreceptor. Although these receptors are capable of giving rise to sensations, much of their information is canceled out by receptor or central mechanisms to be discussed later. In other afferent pathways, information is not canceled out—it simply does not feed into parts of the brain that give rise to a conscious sensation. To use an example cited earlier, stretch receptors in the walls of some of the largest blood vessels monitor blood pressure as part of reflex regulation of this pressure, but people have no conscious awareness of their blood pressure.

4. We lack suitable receptors for many energy forms. For example, we cannot directly detect ionizing radiation and radio or television waves.

5. Damaged neural networks may give faulty perceptions as in the bizarre phenomenon known as **phantom limb,** in which a limb lost by accident or amputation is experienced as though it were still in place. The missing limb is perceived to be the site of tingling, touch, pressure, warmth, itch, wetness, pain, and even fatigue. It seems that the sensory neural networks in the central nervous system that are normally triggered by receptor activation are, instead, activated independent of peripheral input. The activated neural networks continue to generate the usual sensations, which the brain perceives as arising from the missing receptors.

6. Some drugs alter perceptions. In fact, the most dramatic examples of a clear difference between the real world and our perceptual world can be found in drug-induced hallucinations.

In summary, for perception to occur, there can be no separation of the three processes involved—transducing stimulus energy into action potentials by the receptor, transmitting data through the nervous system, and interpreting data. Sensory information is processed at each synapse along the afferent pathways and at many levels of the central nervous system, with the more complex stages receiving input only after the more elementary systems have processed the information. This hierarchical processing of afferent information along individual pathways is an important organizational principle of sensory systems. As we will see, a second important principle is that information is processed by *parallel* pathways, each of which handles a limited aspect of the neural signals generated by the sensory transducers. A third principle is that information at each stage along the pathway is modified by "top-down" influences serving the emotions, attention, memory, and language. Every synapse along the afferent pathway adds an element of organization and contributes to the sensory experience so that what we perceive is not a simple—or even an absolutely accurate—image of the stimulus that originally activated our receptors.

We conclude our general introduction to sensory system pathways and coding with a summary of the general principles of the organization of the sensory systems (**Table 7–1**).

Table 7–1	Principles of Sensory System Organization

1. Specific sensory receptor types are sensitive to certain modalities and submodalities.

2. A specific sensory pathway codes for a particular modality or submodality.

3. The specific ascending pathways are crossed so that sensory information is generally processed by the side of the brain opposite the stimulated side of the body.

4. In addition to other synaptic relay points, most specific ascending pathways synapse in the thalamus on their way to the cortex.

5. Information is organized such that initial cortical processing of the various modalities occurs in different parts of the brain.

6. Specific ascending pathways are subject to descending controls.

I. Sensory processing begins with the transformation of stimulus energy into graded potentials and then into action potentials in nerve fibers.

II. Information carried in a sensory system may or may not lead to a conscious awareness of the stimulus.

Sensory Receptors

I. Receptors translate information from the external and internal environments into graded potentials, which then generate action potentials.

   a. Receptors may be either specialized endings of afferent neurons or separate cells at the ends of the neurons.

   b. Receptors respond best to one form of stimulus energy, but they may respond to other energy forms if the stimulus intensity is abnormally high.

   c. Regardless of how a specific receptor is stimulated, activation of that receptor can only lead to perception of one type of sensation. Not all receptor activations lead, however, to conscious sensations.

II. The transduction process in all sensory receptors involves—either directly or indirectly—the opening or closing of ion channels in the receptor. Ions then flow across the membrane, causing a receptor potential.

   a. Receptor potential magnitude and action potential frequency increase as stimulus strength increases.

   b. Receptor potential magnitude varies with stimulus strength, rate of change of stimulus application, temporal summation of successive receptor potentials, and adaptation.

Primary Sensory Coding

I. The type of stimulus perceived is determined in part by the type of receptor activated. All receptors of a given sensory unit respond to the same stimulus modality.

II. Stimulus intensity is coded by the rate of firing of individual sensory units and by the number of sensory units activated.

III. Perception of the stimulus location depends on the size of the receptive field covered by a single sensory unit and on the overlap of nearby receptive fields. Lateral inhibition is a means by which ascending pathways increase sensory acuity.

IV. Stimulus duration is coded by slowly adapting receptors.

V. Information coming into the nervous system is subject to control by both ascending and descending pathways.

Neural Pathways in Sensory Systems

I. A single afferent neuron with all its receptor endings is a sensory unit.

   a. Afferent neurons, which usually have more than one receptor of the same type, are the first neurons in sensory pathways.

   b. The area of the body that, when stimulated, causes activity in a sensory unit or other neuron in the ascending pathway of that unit is called the receptive field for that neuron.

II. Neurons in the specific ascending pathways convey information about only a single type of stimulus to specific primary receiving areas of the cerebral cortex.

III. Nonspecific ascending pathways convey information from more than one type of sensory unit to the brainstem reticular formation and regions of the thalamus that are not part of the specific ascending pathways.

Association Cortex and Perceptual Processing

I. Information from the primary sensory cortical areas is elaborated after it is relayed to a cortical association area.

   a. The primary sensory cortical area and the region of association cortex closest to it process the information in fairly simple ways and serve basic sensory-related functions.

   b. Regions of association cortex farther from the primary sensory areas process the sensory information in more complicated ways.

   c. Processing in the association cortex includes input from areas of the brain serving other sensory modalities, arousal, attention, memory, language, and emotions.

acuity 195	polymodal neuron 200
adaptation 194	rapidly adapting receptor 198
adequate stimulus 192	receptive field 194
ascending pathway 199	receptor potential 193
auditory cortex 200	recruitment 195
central sulcus 200	sensation 192
chemoreceptor 193	sensory information 192
coding 194	sensory pathway 198
cortical association area 200	sensory receptor 192
labeled lines 195	sensory system 192
lateral inhibition 197	sensory transduction 192
mechanoreceptor 193	sensory unit 194
modality 194	slowly adapting receptor 198
nociceptor 193	somatic receptor 200
nonspecific ascending pathway 200	somatosensory cortex 200
olfactory cortex 200	specific ascending pathway 199
perception 192	stimulus 192
photoreceptor 193	thermoreceptor 193
	visual cortex 200

phantom limb 201

1. Distinguish between a sensation and a perception.
2. Define the term *adequate stimulus*.
3. Describe the general process of transduction in a receptor that is a cell separate from the afferent neuron. Include in your description the following terms: specificity, stimulus, receptor potential, neurotransmitter, graded potential, and action potential.
4. List several ways in which the magnitude of a receptor potential can vary.
5. Differentiate between the function of rapidly adapting and slowly adapting receptors.
6. Describe the relationship between sensory information processing in the primary cortical sensory areas and in the cortical association areas.
7. List several ways in which sensory information can be distorted.
8. How does the nervous system distinguish between stimuli of different types?
9. How does the nervous system code information about stimulus intensity?
10. Describe the general mechanism of lateral inhibition and explain its importance in sensory processing.
11. Make a diagram showing how a specific ascending pathway relays information from peripheral receptors to the cerebral cortex.

Specific Sensory Systems

Somatic Sensation

Sensation from the skin, muscles, bones, tendons, and joints, or **somatic sensation,** is initiated by a variety of sensory receptors collectively called somatic receptors (**Figure 7–16**). Some of these receptors respond to mechanical stimulation of the skin, hairs, and underlying tissues, whereas others respond to temperature or chemical changes. Activation of somatic receptors gives rise to the sensations of touch, pressure, awareness of the position of the body parts and their movement, temperature, and pain. The receptors for visceral sensations, which arise in certain organs of the thoracic and abdominal cavities, are the same types as the receptors that give rise to somatic sensations. Some organs, such as the liver, have no sensory receptors at all. Each sensation is associated with a specific receptor type. In other words, distinct receptors exist for heat, cold, touch, pressure, limb position or movement, and pain.

Touch and Pressure

Stimulation of a variety of mechanoreceptors in the skin (see Figure 7–16) leads to a wide range of touch and pressure experiences—hair bending, deep pressure, vibrations, and superficial touch, for example. These mechanoreceptors are highly specialized nerve endings encapsulated in elaborate cellular structures.

The details of the mechanoreceptors vary, but generally the nerve endings are linked to networks of collagen fibers within a capsule. These networks transmit the mechanical tension in the capsule to ion channels in the nerve endings and activate them.

The skin mechanoreceptors adapt at different rates. About half of them adapt rapidly (i.e., they fire only when the stimulus is changing), and the others adapt slowly. Activation of rapidly adapting receptors gives rise to the sensations of touch, movement, and vibration, whereas slowly adapting receptors give rise to the sensation of pressure.

In both categories, some receptors have small, well-defined receptive fields and can provide precise information about the contours of objects indenting the skin. As might be expected, these receptors are concentrated at the fingertips. In contrast, other receptors have large receptive fields with obscure boundaries, sometimes covering a whole finger or a large part of the palm. These receptors are not involved in detailed spatial discrimination but signal information about skin stretch and joint movement.

Sense of Posture and Movement

The senses of posture and movement are complex. The major receptors responsible for these senses are the muscle-spindle stretch receptors. These mechanoreceptors occur in skeletal

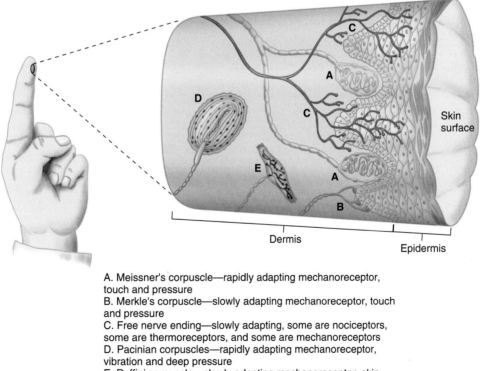

A. Meissner's corpuscle—rapidly adapting mechanoreceptor, touch and pressure
B. Merkle's corpuscle—slowly adapting mechanoreceptor, touch and pressure
C. Free nerve ending—slowly adapting, some are nociceptors, some are thermoreceptors, and some are mechanoreceptors
D. Pacinian corpuscles—rapidly adapting mechanoreceptor, vibration and deep pressure
E. Ruffini corpuscle—slowly adapting mechanoreceptor, skin stretch

Figure 7–16

Skin receptors. Some nerve fibers have free endings not related to any apparent receptor structure. Thicker, myelinated axons, on the other hand, end in receptors that have a complex structure. (Not drawn to scale; for example, Pacinian corpuscles are actually four to five times larger than Meissner's corpuscles.)

muscles and respond both to the absolute magnitude of muscle stretch and to the rate at which the stretch occurs (to be described in Chapter 10). Vision and the vestibular organs (the sense organs of balance) also support the senses of posture and movement. Mechanoreceptors in the joints, tendons, ligaments, and skin also play a role. The term **kinesthesia** refers to the sense of movement at a joint.

Temperature

Recent experiments have identified mechanisms by which two types of thermoreceptors respond to temperature in the skin, one group detecting cold and the other warmth. Cold-sensing receptors have nonselective cation channels that open in response to temperatures below body temperature. These channels are active over a broad range of temperatures ranging from 35°C down to near 0°C, with an influx of sodium depolarizing the associated afferent neurons. The plant compound **menthol** activates these same channels, explaining the perception of coolness experienced when it is applied to the skin. At temperatures from 30°C up to about 50°C, warmth-sensing thermoreceptors are activated. Nonselective cation channels found in those neurons depolarize the cell in response to warm temperatures. Interestingly, **capsaicin** (a chemical found in chili peppers) and

ethanol also activate these channels, explaining the burning sensation caused by eating some spicy foods or drinking a shot of whiskey. Extremes of temperature that cause tissue damage activate pain receptors, which are described next.

Pain

A stimulus that causes or is on the verge of causing tissue damage usually elicits a sensation of pain. Receptors for such stimuli are known as **nociceptors.** They respond to intense mechanical deformation, excessive heat, and many chemicals. Examples of the latter include neuropeptide transmitters, bradykinin, histamine, cytokines, and prostaglandins, several of which are released by damaged cells. Some of these chemicals are secreted by cells of the immune system (described in Chapter 18) that have moved into the injured area. These substances act by binding to specific ligand-gated ion channels on the nociceptor plasma membrane. In contrast to mechanoreceptors, nociceptors are free nerve endings without any form of specialization.

The primary afferents having nociceptor endings synapse on ascending neurons after entering the central nervous system (**Figure 7–17a**). Glutamate and the neuropeptide, substance P, are among the neurotransmitters released at these synapses.

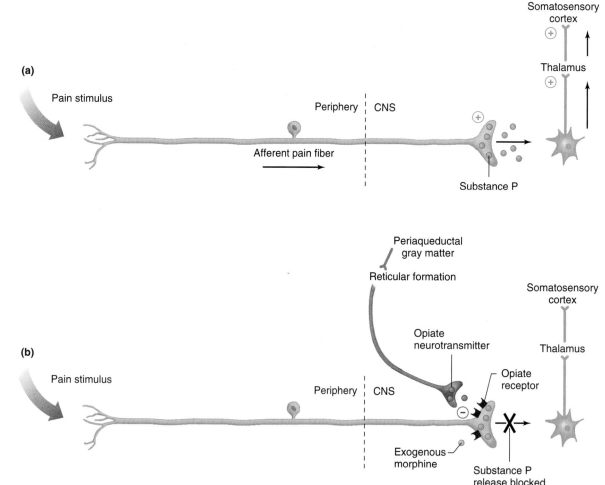

Figure 7–17

Cellular pathways of pain transmission and modulation. (a) Painful stimulation releases the neuropeptide substance P from afferent fibers in the spinal cord. (b) Substance P release is blocked by a descending analgesic system using axo-axonic synapses on the afferent neuron. Details of this system not shown include descending neurons releasing norepinephrine and serotonin onto spinal interneurons that, in turn, release opiate neurotransmitters. Morphine inhibits pain in a similar manner.

Chapter 7

When incoming nociceptive afferents activate interneurons, it may lead to the phenomenon of **referred pain,** in which the sensation of pain is experienced at a site other than the injured or diseased tissue. For example, during a heart attack, a person often experiences pain in the left arm. Referred pain occurs because both visceral and somatic afferents often converge on the same neurons in the spinal cord (**Figure 7–18a**).

Excitation of the somatic afferent fibers is the more usual source of afferent discharge, so we "refer" the location of receptor activation to the somatic source even though, in the case of visceral pain, the perception is incorrect. **Figure 7–18b** shows the typical distribution of referred pain from visceral organs.

Pain differs significantly from the other somatosensory modalities. After transduction of the first noxious stimuli into

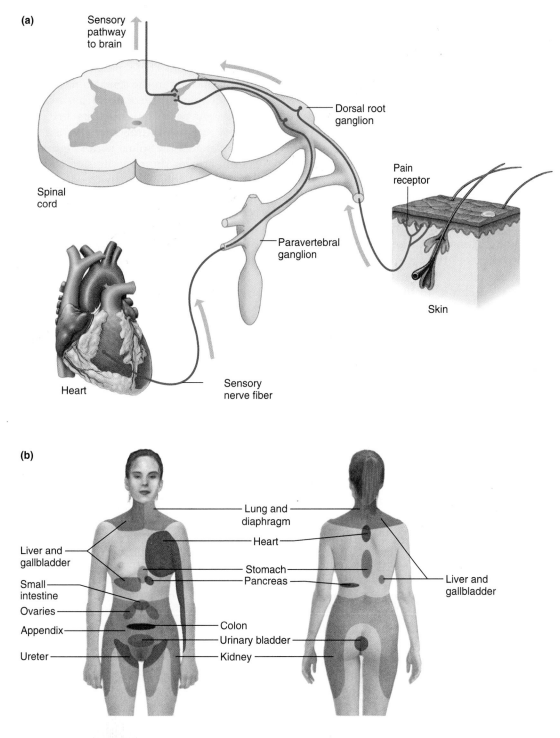

Figure 7–18

Referred pain. (a) Convergence of visceral and somatic afferent neurons onto ascending pathways. (b) Regions of the body surface where we typically perceive referred pain from visceral organs.

action potentials in the afferent neuron, a series of changes occur in components of the pain pathway—including the ion channels in the nociceptors themselves—that alter the way these components respond to subsequent stimuli. Both increased and decreased sensitivity to painful stimuli can occur. When these changes result in an increased sensitivity to painful stimuli, known as *hyperalgesia,* the pain can last for hours after the original stimulus is over. Thus, the pain experienced in response to stimuli occurring even a short time after the original stimulus (and the reactions to that pain) can be more intense than the initial pain. Moreover, probably more than any other type of sensation, pain can be altered by past experiences, suggestion, emotions (particularly anxiety), and the simultaneous activation of other sensory modalities. Thus, the level of pain experienced is not solely a physical property of the stimulus.

Analgesia is the selective suppression of pain without effects on consciousness or other sensations. Electrical stimulation of specific areas of the central nervous system can produce a profound reduction in pain, a phenomenon called *stimulation-produced analgesia,* by inhibiting pain pathways. This occurs because descending pathways that originate in these brain areas selectively inhibit the transmission of information originating in nociceptors (**Figure 7–17b**). The descending axons end at lower brainstem and spinal levels on interneurons in the pain pathways as well as on the synaptic terminals of the afferent nociceptor neurons themselves. Some of the neurons in these inhibitory pathways release morphine-like endogenous opioids (Chapter 6). These opioids inhibit the propagation of input through the higher levels of the pain system. Thus, infusion of morphine can provide relief in many cases of intractable pain by binding to and activating opioid receptors at the level of entry of the active nociceptor fibers. This is separate from morphine's effect on the brain.

The body's endogenous-opioid systems also mediate other phenomena known to relieve pain. In recent clinical studies, 55 to 85 percent of patients experienced pain relief when treated with *acupuncture,* an ancient Chinese therapy involving the insertion of needles into specific locations on the skin. This success rate was similar to that seen when patients were treated with morphine (70 percent). In studies comparing morphine versus a *placebo* (injections of sugar that patients *thought* was the drug), 35 percent of those receiving the placebo experienced pain relief. Acupuncture is thought to activate afferent neurons leading to spinal cord and midbrain centers that release endogenous opioids and other neurotransmitters implicated in pain relief. It seems likely that pathways descending from the cortex activate those same regions to exert the placebo effect. Thus, exploiting the body's built-in analgesia mechanisms can be an effective means of controlling pain.

Also of use to lessening pain is *transcutaneous electric nerve stimulation (TENS),* in which the painful site itself or the nerves leading from it are stimulated by electrodes placed on the surface of the skin. TENS works because the stimulation of nonpain, low-threshold afferent fibers (e.g., the fibers from touch receptors) leads to the inhibition of neurons in the pain pathways. You perform a low-tech version of this phenomenon when you vigorously rub your scalp at the site of a painful bump on the head.

Neural Pathways of the Somatosensory System

After entering the central nervous system, the afferent nerve fibers from the somatic receptors synapse on neurons that form the specific ascending pathways going primarily to the somatosensory cortex via the brainstem and thalamus. They also synapse on interneurons that give rise to the nonspecific ascending pathways. There are two major somatosensory pathways (there are different pathways for sensory input from the face). These pathways are organized differently from each other in the spinal cord and brain (**Figure 7–19**). The ascending **anterolateral pathway,** also called the spinothalamic pathway, makes its first synapse between the sensory receptor neuron and a second neuron located in the gray matter of the spinal cord (**Figure 7–19a**). This second neuron crosses the opposite side and projects through the anterolateral column of the cord to the thalamus, where it synapses on cortically projecting neurons. The anterolateral pathway processes pain and temperature information. The second major pathway for somatic sensation is the **dorsal column pathway** (**Figure 7–19b**). This, too, is named for the section of white matter (the dorsal columns) through which the sensory receptor neurons project to the brainstem, where the first synapse occurs. As in the anterolateral pathway, the second synapse is in the thalamus, from which projections are sent to the somatosensory cortex.

Note that the pathways cross from the side where the afferent neurons enter the central nervous system to the opposite side either in the spinal cord (anterolateral system) or in the brainstem (dorsal column system). Thus, sensory pathways from somatic receptors on the left side of the body terminate in the somatosensory cortex of the right cerebral hemisphere.

In the somatosensory cortex, the endings of the axons of the specific somatic pathways are grouped according to the peripheral location of the receptors that give input to the pathways (**Figure 7–20**). The parts of the body that are most densely innervated—fingers, thumb, and lips—are represented by the largest areas of the somatosensory cortex. There are qualifications, however, to this seemingly precise picture. There is considerable overlap of the body part representations, and the sizes of the areas can change with sensory experience. The phantom limb phenomenon described in the first section of this chapter provides a good example of the dynamic nature of the somatosensory cortex. Studies of amputees have shown that cortical areas formerly responsible for a missing arm and hand are commonly "re-wired" to respond to sensory inputs originating in the face (note the proximity of the cortical regions representing these areas in Figure 7–20). As the somatosensory cortex undergoes this reorganization, a touch on a person's cheek is often perceived as a touch on his or her missing arm.

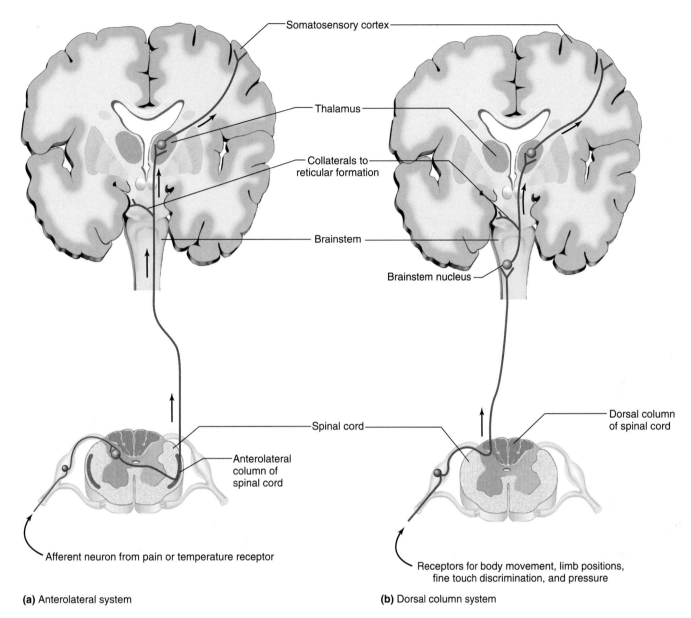

Somatosensory cortex

Thalamus

Collaterals to
reticular formation

Brainstem

Brainstem nucleus

Spinal cord

Dorsal column
of spinal cord

Anterolateral
column of
spinal cord

Afferent neuron from pain or temperature receptor

Receptors for body movement, limb positions,
fine touch discrimination, and pressure

(a) Anterolateral system

(b) Dorsal column system

Figure 7–19

(a) The anterolateral system. (b) The dorsal column system. Information carried over collaterals to the reticular formation in (a) and (b) contribute to alertness and arousal mechanisms.

Figure 7–19 physiological *inquiry* (p*i*)

■ If an accident severed the left half of a person's spinal cord at the mid-thoracic level but left the right side intact, what pattern of sensory deficits would occur?

Answer can be found at end of chapter.

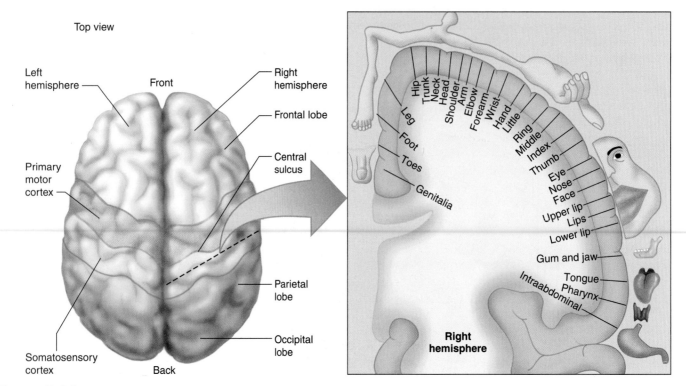

Top view

Left hemisphere — | Front | — Right hemisphere

— Frontal lobe

Primary motor cortex —

— Central sulcus

— Parietal lobe

Somatosensory cortex —

— Occipital lobe

Back

Right hemisphere

Hip, Trunk, Neck, Head, Shoulder, Arm, Elbow, Forearm, Wrist, Hand, Little, Ring, Middle, Index, Thumb, Leg, Foot, Toes, Genitalia, Eye, Nose, Face, Upper lip, Lips, Lower lip, Gum and jaw, Tongue, Pharynx, Intraabdominal

Figure 7–20

The location of pathway terminations for different parts of the body in somatosensory cortex, although there is actually much overlap between the cortical regions. The left half of the body is represented on the right hemisphere of the brain, and the right half of the body is represented on the left hemisphere, which is not shown here.

(a)

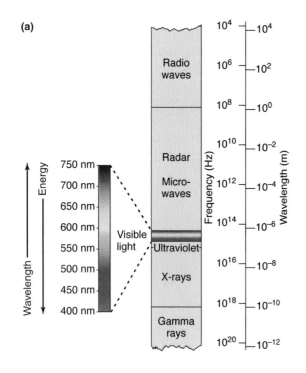

Energy / Wavelength

750 nm, 700 nm, 650 nm, 600 nm, 550 nm, 500 nm, 450 nm, 400 nm

Visible light

Radio waves, Radar, Micro-waves, Ultraviolet, X-rays, Gamma rays

Frequency (Hz): 10^4, 10^6, 10^8, 10^{10}, 10^{12}, 10^{14}, 10^{16}, 10^{18}, 10^{20}

Wavelength (m): 10^4, 10^2, 10^0, 10^{-2}, 10^{-4}, 10^{-6}, 10^{-8}, 10^{-10}, 10^{-12}

(b)

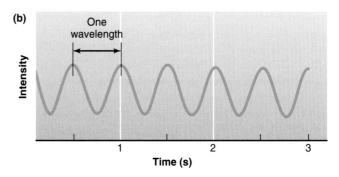

One wavelength

Intensity

Time (s): 1, 2, 3

Figure 7–21

The electromagnetic spectrum. (a) Visible light ranges in wavelength from 400 to 750 nm (1nm = 1 billionth of a meter). (b) Wavelength is the inverse of frequency. The frequency of this wave is 2 Hz (cycles/s).

Light

The receptors of the eye are sensitive only to that tiny portion of the vast spectrum of electromagnetic radiation that we call visible light (**Figure 7–21a**). Radiant energy is described in terms of wavelengths and frequencies. The **wavelength** is the distance between two successive wave peaks of the electromagnetic radiation (**Figure 7–21b**). Wavelengths vary from several kilometers at the long-wave radio end of the spectrum to trillionths of a meter at the gamma-ray end. The **frequency** (in hertz, the number of cycles per second) of the radiation wave varies inversely with wavelength. Those wavelengths capable of

Vision

The eyes are composed of an optical portion, which focuses the visual image on the receptor cells, and a neural component, which transforms the visual image into a pattern of graded and action potentials.

stimulating the receptors of the eye—the **visible spectrum**—are between about 400 and 750 nm. Different wavelengths of light within this band are perceived as different colors.

Overview of Eye Anatomy

The eye is a three-layered, fluid-filled ball, divided into two chambers (**Figure 7–22**). The **sclera** forms a white capsule around the eye, except at its anterior surface where it is specialized into the clear **cornea.** The tough, fibrous sclera serves as the insertion point for external muscles that move the eyeballs within their sockets. The underlying **choroid** layer is darkly pigmented to absorb light rays at the back of the eyeball, while in the front the choroid layer is specialized into the **iris** (the structure associated with eye color), the **ciliary muscle,** and the **zonular fibers.** Circular and radial smooth muscle fibers of the iris determine the diameter of the **pupil,** the anterior opening that allows light into the eye. Activity of the ciliary muscle and the resulting tension on the zonular fibers determines the shape of the crystalline **lens** just behind the iris. The **retina** is an extension of the brain lining the inner, posterior surface of the eye, containing numerous types of neurons as well as the eye's sensory cells, called **photoreceptors.** Features of the retina that can be viewed through the pupil with an **ophthalmoscope** include: (1) the **fovea centralis,** a region specialized to deliver the highest visual acuity; (2) the **optic disc,** where neurons carrying information from the photoreceptors exit the eye as the **optic nerve;** and (3) numerous blood vessels lying on the inner surface of the retina. The anterior chamber

of the eye, between the iris and the cornea, is filled with a clear fluid called **aqueous humor.** The posterior chamber of the eye, between the lens and the retina, is filled with a viscous, jellylike substance known as **vitreous humor.**

The Optics of Vision

A ray of light can be represented by a line drawn in the direction in which the wave is traveling. Light waves diverge in all directions from every point of a visible object. When a light wave crosses from air into a denser medium like glass or water, the wave changes direction at an angle that depends on the density of the medium and the angle at which it strikes the surface (**Figure 7–23a**). This bending of light waves, called **refraction,** is the mechanism allowing us to focus an accurate image of an object onto the retina.

When light waves diverging from a point on an object pass from air into the curved surfaces of the cornea and lens of the eye, they are refracted inward, converging back into a point on the retina (**Figure 7–23b**). The cornea plays a larger quantitative role than the lens in focusing light waves because the waves are refracted more in passing from air into the cornea than they are when passing into and out of the lens. Objects in the center of the field of view are focused onto the fovea centralis, with the image formed upside down and reversed right to left relative to the original source.

Light waves from objects close to the eye strike the cornea at greater angles and must be refracted more in order to reconverge on the retina. Although, as previously noted, the

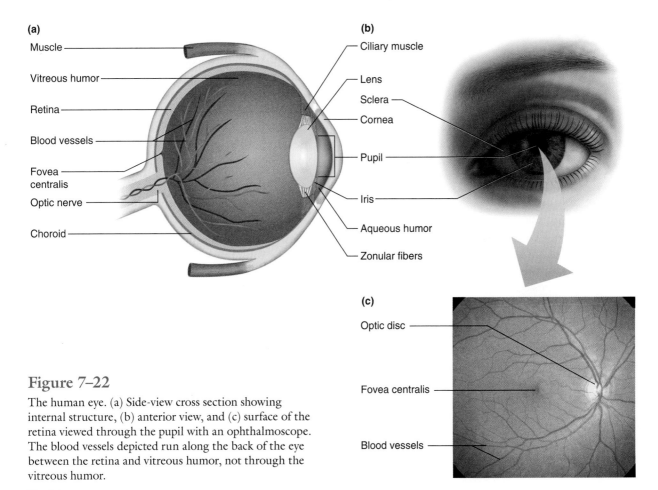

Figure 7–22

The human eye. (a) Side-view cross section showing internal structure, (b) anterior view, and (c) surface of the retina viewed through the pupil with an ophthalmoscope. The blood vessels depicted run along the back of the eye between the retina and vitreous humor, not through the vitreous humor.

(a)
Muscle
Vitreous humor
Retina
Blood vessels
Fovea centralis
Optic nerve
Choroid

(b)
Ciliary muscle
Lens
Sclera
Cornea
Pupil
Iris
Aqueous humor
Zonular fibers

(c)
Optic disc
Fovea centralis
Blood vessels

(a)

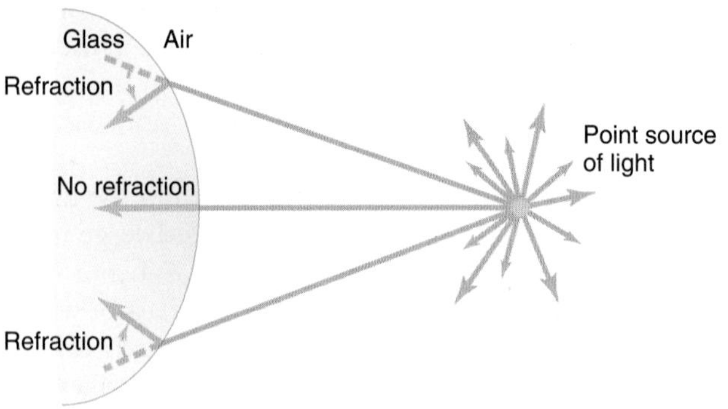

Glass · Air

Refraction

No refraction

Point source of light

Refraction

(b)

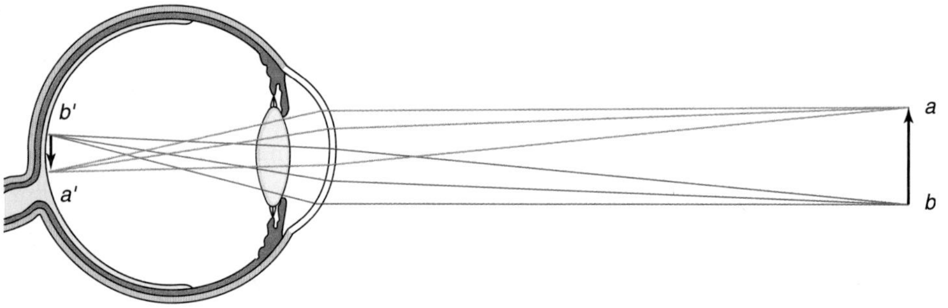

b'

a'

a

b

Figure 7–23

Focusing point sources of light. (a) When diverging light rays enter a dense medium at an angle to its convex surface, refraction bends them inward. (b) Refraction of light by the lens system of the eye. For simplicity, we show light refraction only at the surface of the cornea, where the greatest refraction occurs. Refraction also occurs in the lens and at other sites in the eye. Incoming light from a (above) and b (below) is bent in opposite directions, resulting in b′ being above a′ on the retina.

cornea performs the greater part quantitatively of focusing the visual image on the retina, all *adjustments* for distance are made by changes in lens shape. Such changes are part of the process known as **accommodation.**

The shape of the lens is controlled by the ciliary muscle and the tension it applies to the zonular fibers, which attach the ciliary muscle to the lens (**Figure 7–24**). The ciliary muscle, which is stimulated by parasympathetic nerves, is circular, like a sphincter, so that it draws nearer to the central lens as it contracts. As the muscle contracts, it lessens the tension on the zonular fibers. Conversely, when the ciliary muscle relaxes, the diameter of the ring of muscle increases and the tension on the zonular fibers also increases. Therefore, the shape of the lens is altered by contraction and relaxation of the ciliary muscle. To focus on distant objects, the ciliary muscle relaxes and the zonular fibers pull the lens into a flattened, oval shape. Contraction of the ciliary muscles focuses the eye on near objects, by releasing the tension on the zonular fibers, which allows the natural elasticity of the lens to return it to a more spherical shape (**Figure 7–25**). The shape of the lens determines to what degree the light waves are refracted and how they project onto the retina. Constriction of the pupil also occurs when the ciliary muscle contracts, which helps sharpen the image further.

As people age, the lens tends to lose elasticity, reducing its ability to assume a spherical shape. The result is a progressive decline in the ability to accommodate for near vision. This condition, known as **presbyopia,** is a normal part of the aging process and is the reason that people around 45 years of age may have to begin wearing reading glasses or bifocals for close work.

The cells that make up most of the lens lose their internal membranous organelles early in life and are thus transparent, but they lack the ability to replicate. The only lens cells that retain the capacity to divide are on the lens surface, and as new cells form, older cells come to lie deeper within the lens. With increasing age, the central part of the lens becomes denser and stiffer and acquires a coloration that progresses from yellow to black.

The changes in lens color that occur with aging are responsible for **cataract,** an opacity (clouding) of the lens that is one of the most common eye disorders. Early changes in lens color do not interfere with vision, but vision is impaired as the process slowly continues. The opaque lens can be removed surgically. With the aid of an implanted artificial lens or compensating corrective lenses, effective vision can be restored, although the ability to accommodate is lost.

Cornea and lens shape and eyeball length determine the point where light rays converge. Defects in vision occur if the eyeball is too long in relation to the focusing power of the lens (**Figure 7–26a**). In this case, the images of faraway objects focus at a point in front of the retina. This **nearsighted,** or **myopic,** eye is unable to see distant objects clearly. Near objects are clear to a person with this condition, but without the normal rounding of the lens that occurs via accommodation. In contrast, if the eye is too short for the lens, images of near objects are focused behind the retina (**Figure 7–26b**). This eye is *farsighted,* or *hyperopic,* and while a person with this condition has poor near vision, distant objects can be seen if the accommodation reflex is activated to increase the curvature of the lens. The use of corrective lenses for near- and farsighted vision is shown in Figure 7–26.

Defects in vision also occur when the lens or cornea does not have a smoothly spherical surface, a condition known as *astigmatism.* Corrective lenses can usually compensate for these surface imperfections.

The size and shape of a person's eye over time depends in part on the volume of the aqueous humor and vitreous humor. These two fluids are colorless and permit the transmission of light from the front of the eye to the retina. The aqueous humor is constantly formed by special vascular tissue that overlies the ciliary muscle and drains away through a canal in front of the iris at the edge of the cornea. In some instances, the aqueous humor forms faster than it is removed, which results in increased pressure within the eye. *Glaucoma,* the leading cause of irreversible blindness, is a disease in which retinal cells are damaged as a result of increased pressure within the eye.

Just as the aperture of a camera can be varied to alter the amount of light that enters, the iris regulates the diameter of the pupil. The color of the iris is of no importance as long

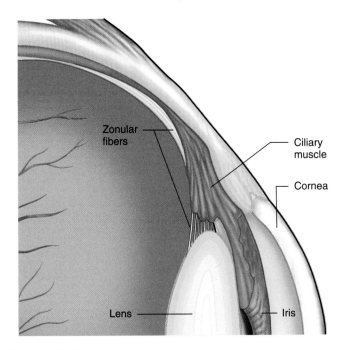

Figure 7–24

Ciliary muscle, zonular fibers, and lens of the eye.

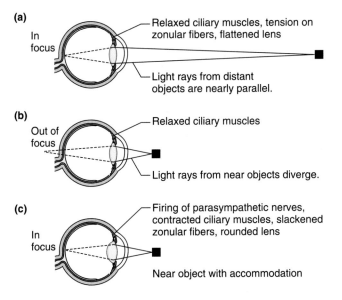

Figure 7–25

Accommodation for near vision. (a) Light rays from distant objects are more parallel, and they focus onto the retina when the lens is less curved. (b) Diverging light rays from near objects do not focus on the retina when the ciliary muscles are relaxed. (c) Accommodation increases the curvature of the lens, focusing the image of near objects onto the retina.

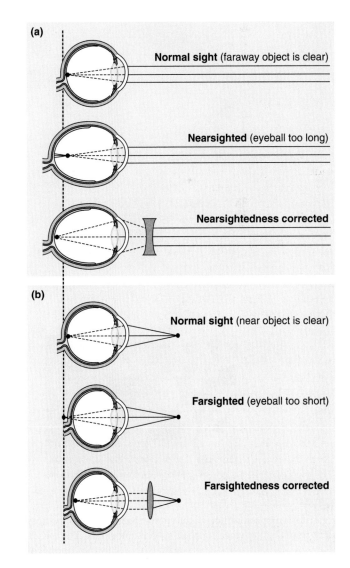

Figure 7–26

Correction of vision defects. (a) Nearsightedness. (b) Farsightedness.

as the tissue is sufficiently opaque to prevent the passage of light. The iris is composed of two rings of smooth muscle that are innervated by autonomic nerves. Stimulation of sympathetic nerves to the iris enlarges the pupil by causing radially arranged muscle fibers to contract. Stimulation of parasympathetic fibers to the iris makes the pupil smaller by causing the sphincter muscle fibers, which circle around the pupil, to contract.

These neurally induced changes occur in response to light-sensitive reflexes. Bright light causes a decrease in the diameter of the pupil, which reduces the amount of light entering the eye and restricts the light to the central part of the lens for more accurate vision. The constriction of the pupil also protects the retina from damage induced by very bright light, such as direct rays from the sun. Conversely, the pupil enlarges in dim light, when maximal illumination is needed. Changes also occur as a result of emotion or pain. Abnormal or absent response of the pupil to changes in light can indicate damage to the midbrain from trauma or tumors, or can also be a telltale sign when a person is under the influence of narcotics like heroin.

Photoreceptor Cells and Phototransduction

Figure 7–27 shows a detailed view of the retina. The photoreceptor cells have a tip, or **outer segment,** composed of stacked layers of membrane, called **discs.** The discs hold the chemical substances that respond to light. The photoreceptors also have an **inner segment** that contains the nucleus, mitochondria,

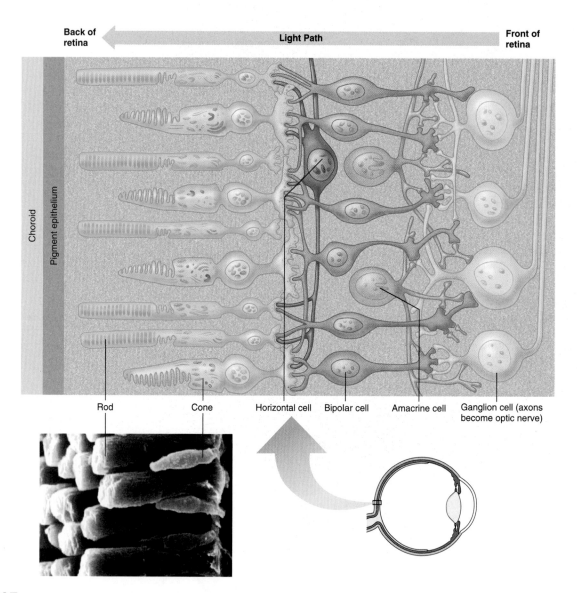

Figure 7–27

Organization of the retina. Light enters through the cornea, passes through the aqueous humor, pupil, vitreous humor, and the front surface of the retina before reaching the photoreceptor cells. The membranes that contain the light-sensitive proteins form discrete discs in the rods but are continuous with the plasma membrane in the cones, which accounts for the comblike appearance of these latter cells. Horizontal and amacrine cells, depicted here in purple and orange, provide lateral inhibition between neurons of the retina. At the lower left is a scanning electron micrograph of rods and cones.

Redrawn from Dowling and Boycott.

and other organelles, and the synaptic terminal that connects the photoreceptor to the next neurons in the retina. The two types of photoreceptors are called **rods** and **cones** because of the shapes of their light-sensitive outer segments. In cones, the light-sensitive discs are formed from in-foldings of the surface plasma membrane, whereas in rods, the disc membranes are intracellular structures. Note that the light-sensitive portions of the photoreceptor cells face *away* from the incoming light, and the light must pass through all the cell layers of the retina before reaching and stimulating the photoreceptors. Two pigmented layers, the choroid and the **pigment epithelium** of the back of the retina, absorb light that has bypassed the photoreceptors. This prevents its reflection and scattering back through the rods and cones, which would cause the visual image to blur. The rods are extremely sensitive and respond to very low levels of illumination, whereas the cones are considerably less sensitive and respond only when the light is bright.

The photoreceptors contain molecules called **photopigments,** which absorb light. There are four unique photopigments in the retina, one found in rods **(rhodopsin),** and one in each of three different types of cones. Photopigments contain membrane-bound proteins called **opsins,** which surround and bind a **chromophore** molecule. The chromophore in all types of photopigments is **retinal,** a derivative of vita-min A. This is the part of the photopigment that is light-sensitive. The opsin differs in each of the four photopigments. Each type of opsin binds to the chromophore in a different way. Because of this, each of the four photopigments absorbs light most effectively at a specific part of the visible spectrum. For example, one photopigment absorbs light most effectively at long wavelengths (sometimes designated as "red" cones), whereas another absorbs short wavelengths ("blue" cones).

The membranous discs of the outer segment are arranged in parallel to the surface of the retina (**Figure 7–28**). This layered arrangement maximizes the membrane surface area. In fact, each photoreceptor may contain over a billion molecules of photopigment, providing an extremely effective trap for light.

The photoreceptor is unique because it is the only type of sensory cell that is depolarized when it is at rest (i.e., in the dark), and *hyperpolarized* in response to its adequate stimulus (see Figure 7–28). In the absence of light, action of the membrane-bound enzyme **guanylyl cyclase** converts GTP into a high intracellular concentration of the second messenger molecule, cyclic GMP (cGMP). The cGMP maintains the ligand-gated cation channels in the outer segment membrane in the open state, and a persistent influx of sodium and calcium results. Thus, in the dark, cGMP concentrations are high, and the photoreceptor cell is maintained in a relatively depolarized state.

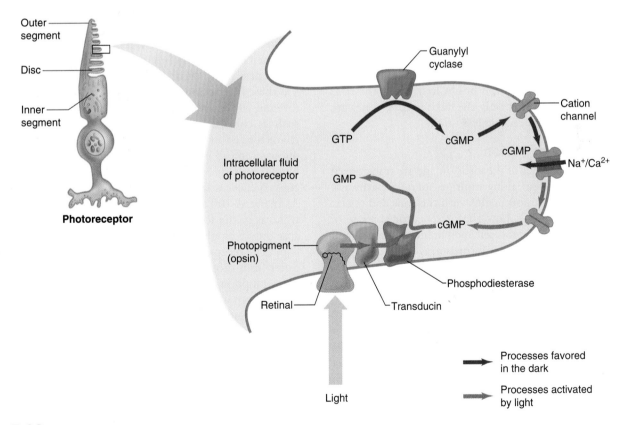

Figure 7–28

Phototransduction in a cone cell. Retinal is the chromophore in the photopigment. Stimulation of the cGMP phosphodiesterase in the cytoplasm of the disc produces the decrease in cGMP that closes the cation channels. For simplicity, the proteins are shown widely spaced in the membrane. In fact, all of these proteins are densely interspersed within the cone disc membrane. Phototransduction in rods is basically identical, except the membraneous discs are contained completely within the cell's cystol (see Figure 7–27), and the cGMP-gated ion channels are in the surface membrane rather than the disc membranes.

When light shines on the photoreceptors, a cascade of events leads to hyperpolarization of the photoreceptor cell membrane. Retinal molecules in the disc membrane assume a new conformation induced by the absorption of energy from photons. This stimulates an interaction between associated opsins and another protein that is a member of the G-protein family (see Chapter 5) called **transducin.** Transducin activates the enzyme **phosphodiesterase,** which rapidly degrades cGMP. The decrease in cytoplasmic cGMP concentration allows the cation channels to close, and the loss of depolarizing current allows the membrane potential to hyperpolarize. After its activation by light, retinal changes back to its resting shape by way of several mechanisms that do not depend on light, but are enzyme-mediated.

If you move from a place of bright sunlight into a darkened room, a temporary "blindness" takes place until the photoreceptors can undergo **dark adaptation.** In the low levels of illumination of the darkened room, vision can only be supplied by the rods, which have greater sensitivity than the cones. During the exposure to bright light, however, the rods' rhodopsin has been completely activated, making the rods insensitive to light. Rhodopsin cannot respond fully again until it is restored to its resting state, a process requiring several minutes. Dark adaptation occurs, in part, as enzymes regenerate the initial form of rhodopsin, which can respond to light. Vitamin A is necessary for good night vision because it is required for the synthesis of the retinal portion of the rhodopsin.

In contrast to dark adaptation, **light adaptation** occurs when you step from a dark place into a bright one. Initially, the eye is extremely sensitive to light, and the visual image is too bright and has poor contrast. However, the rhodopsin is soon used up ("bleached" by the bright light), and the rods become unresponsive so that only the less-sensitive cones are operating and the image becomes less bright.

Neural Pathways of Vision

The distinct characteristics of the visual image are transmitted through the visual system along multiple, parallel pathways. The neural pathway of vision begins with the rods and cones. We just described in detail how the presence or absence of light influences photoreceptor cell membrane potential, and now we will consider how this information is encoded, transmitted to the brain, and processed.

Light signals are converted into action potentials through the interaction of photoreceptors with **bipolar cells** and **ganglion cells.** Photoreceptor and bipolar cells only undergo graded responses; ganglion cells are the first cells in the pathway where action potentials can be initiated. Photoreceptors interact with bipolar and ganglion cells in two distinct ways, designated as "ON-pathways" and "OFF-pathways." In both types, photoreceptors are depolarized in the absence of light, causing the neurotransmitter glutamate to be released onto bipolar cells. Light striking either pathway hyperpolarizes the photoreceptors, resulting in a decrease in glutamate release onto bipolar cells. The key difference of the two pathways lies in the type of glutamate receptors found on the bipolar cells, causing them to respond exactly the opposite in the presence and absence of light.

Glutamate released onto ON-pathway bipolar cells binds to metabotropic receptors that cause enzymatic breakdown of cGMP, which hyperpolarizes the bipolar cells by a mechanism similar to that occurring when light strikes a photoreceptor cell. When the bipolar cells are hyperpolarized, they are prevented from releasing excitatory neurotransmitter onto their associated ganglion cells. Thus, in the absence of light, ganglion cells of the ON-pathway are not stimulated to fire action potentials. These processes reverse, however, when light strikes the photoreceptors: glutamate release from photoreceptors declines, ON-bipolar cells depolarize, excitatory neurotransmitter is released, the ganglion cells are depolarized, and action potentials propagate to the brain.

OFF-pathway bipolar cells have ionotropic glutamate receptors that are nonselective cation channels, that depolarize the bipolar cells when glutamate binds. Depolarization of these bipolar cells stimulates them to release excitatory neurotransmitter onto their associated ganglion cells, stimulating them to fire action potentials. Thus, the OFF-pathway generates action potentials in the absence of light, and reversal of these processes inhibits action potentials when light does strike the photoreceptors. The co-existence of these ON and OFF pathways in each region of the retina greatly improves image resolution by increasing the brain's ability to perceive contrast at edges or borders.

Stimulation of ganglion cells is actually more complex than just described—a significant amount of signal processing occurs within the retina before action potentials actually travel to the brain. Photoreceptors, bipolar cells, and ganglion cells are interconnected by **horizontal cells** and **amacrine cells,** which pass information between adjacent areas of the retina. Via these interactions, the ganglion cells are made to respond differentially to the various characteristics of visual images, such as color, intensity, form, and movement. The retina is characterized by its very great amount of convergence; many photoreceptors can synapse on each bipolar cell, and many bipolar cells synapse on a single ganglion cell. The amount of convergence varies by photoreceptor type and retinal region. As many as 100 rod cells converge onto a single bipolar cell, whereas in the fovea region only one or a few cone cells synapse onto a bipolar cell.

The receptive fields in the retina have characteristics that differ from those in the somatosensory system. If you were to shine pinpoints of light onto the retina and at the same time record from a ganglion cell, you would see that the receptive field for that cell is round. Furthermore, the response of the ganglion cell would be either depolarization or hyperpolarization, depending on the location of the stimulus within that single field. Because of different inputs from bipolar cells to the ganglion cell, each receptive field has an inner core that responds differently than the area surrounding the core. There can be "ON center/OFF surround" or "OFF center/ON surround" cells, so named because the responses are either depolarization (ON) or hyperpolarization (OFF) in the two areas of the field. The usefulness of this organization is that the existence of a clear edge between the "ON" and "OFF" areas of the receptive field increases the contrast between the area that is receiving light and the area around it, increasing visual acuity. Thus, a great deal of information processing takes place at this early stage of the sensory pathway.

The axons of the ganglion cells form the output from the retina—the optic nerve, which is cranial nerve II (**Figure 7–29**). The two optic nerves meet at the base of the brain to form the **optic chiasm,** where some of the fibers cross and travel within **optic tracts** to the opposite side of the brain, providing both cerebral hemispheres with input from each eye.

Parallel processing of information continues all the way to and within the cerebral cortex to the highest stages of visual neural networks. Cells in this pathway respond to electrical signals that are generated initially by the photoreceptors' response to light. Optic nerve fibers project to several structures in the brain, the largest number passing to the thalamus (specifically to the lateral geniculate nucleus of the thalamus, see Figure 7–29), where the information from the different ganglion cell types is kept distinct. In addition to the input from the retina, many neurons of the lateral geniculate nucleus also receive input from the brainstem reticular formation and input relayed back from the visual cortex. These nonretinal inputs can control the transmission of information from the retina to the visual cortex and may be involved in our ability to shift attention between vision and the other sensory modalities.

The lateral geniculate nucleus sends action potentials to the visual cortex, the primary visual area of the cerebral cortex (see Figures 7–14 and 7–29). Different aspects of visual information are carried in parallel pathways and are processed simultaneously in a number of independent ways in different parts of the cerebral cortex before they are reintegrated to produce the conscious sensation of sight and the perceptions associated with it. The cells of the visual pathways are organized to handle information about line, contrast, movement, and color. They do not, however, form a picture in the brain. Rather, they form a spatial and temporal pattern of electrical activity.

We mentioned that a substantial number of fibers of the visual pathway project to regions of the brain other than the visual cortex. For example, visual information is transmitted to the **suprachiasmatic nucleus,** which lies just above the optic chiasm and functions as part of the "biological clock." Information about cycles of light intensity is used to entrain this neuronal clock to a 24-hour day. Other visual information passes to the brainstem and cerebellum, where it is used in the coordination of eye and head movements, fixation of gaze, and change in pupil size.

Color Vision

The colors we perceive are related to the wavelengths of light that the pigments in the objects of our visual world reflect, absorb, or transmit. For example, an object appears red because it absorbs shorter wavelengths, which would be perceived as blue, while it reflects the longer wavelengths, perceived as red, to excite the photopigment of the retina most sensitive to red. Light perceived as white is a mixture of all wavelengths, and black is the absence of all light.

Color vision begins with activation of the photopigments in the cone photoreceptor cells. Human retinas have three kinds of cones—one responding optimally at long wavelengths ("red" cones), one at medium wavelengths ("green" cones),

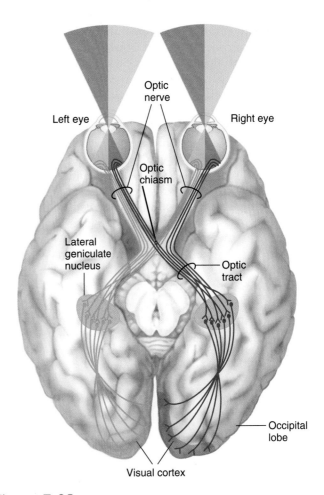

Figure 7–29

Visual pathways viewed in cross section from above.

Figure 7–29 physiological *inquiry* (pi)

- Three patients have suffered destruction of different portions of their visual pathway. Patient 1 has lost the right optic tract, patient 2 has lost the nerve fibers that cross at the optic chiasm, and patient 3 has lost the left occipital lobe. Draw a picture of what each person would perceive through each eye when looking at a white wall.

Answer can be found at end of chapter.

and the other stimulated best at short wavelengths ("blue" cones). Although each type of cone is excited most effectively by light of one particular wavelength, there is actually a range of wavelengths within which a response will occur. Thus, for any given wavelength, the three cone types are excited to different degrees (**Figure 7–30**). For example, in response to light of 531-nm wavelengths, the green cones respond maximally, the red cones less, and the blue not at all. Our sensation of the shade of green at this wavelength depends upon the relative outputs of these three types of cone cells and the comparison made by higher-order cells in the visual system.

The pathways for color vision follow those that Figure 7–29 describes. Ganglion cells of one type respond to a broad band of wavelengths. In other words, they receive input from all three types of cones, and they signal not specific color but general brightness. Ganglion cells of a second type code specific colors. These latter cells are also called **opponent color cells** because they have an excitatory input from one type of cone receptor and an inhibitory input from another. For example, the cell in **Figure 7–31** increases its rate of firing when viewing a blue light but decreases it when a red light replaces the blue. The cell gives a weak response when stimulated with a white light because the light contains both blue and red wavelengths. Other more complicated patterns also exist. The output from these cells is recorded by multiple—

and as yet unclear—strategies in visual centers of the brain. Our ability to discriminate color also depends on the *intensity* of light striking the retina. In brightly lit conditions, the differential response of the cones allows for good color vision. In dim light, however, only the highly sensitive rods are able to respond. Though rods are activated over a range of wavelengths that overlap with those that activate the cones (see Figure 7–30), there is no mechanism for distinguishing between frequencies. Thus, objects that appear vividly colored in bright daylight are perceived in shades of gray at night.

Eye Movement

The fovea centralis of the retina (see Figure 7–22) is specialized in several ways to provide the highest visual acuity. It is comprised of densely packed cones with minimal convergence through the bipolar and ganglion cell layers. In addition, light rays are scattered less on the way to the fovea than in other retinal regions, because the interneuron layers and the blood vessels are displaced to the edges, forming a shallow pit. To focus the most important point in the visual image (the fixation point) on the fovea and keep it there, the eyeball must be able to move. Six skeletal muscles attached to the outside of each eyeball (identified in **Figure 7–32**) control its movement. These muscles perform two basic movements, fast and slow.

The fast movements, called **saccades,** are small, jerking movements that rapidly bring the eye from one fixation point to another to allow a search of the visual field. In addition, saccades move the visual image over the receptors, thereby preventing adaptation. Saccades also occur during certain periods of sleep when dreaming occurs, though these movements are not thought to be involved in "watching" the visual imagery of dreams.

Slow eye movements are involved both in tracking visual objects as they move through the visual field and during com-

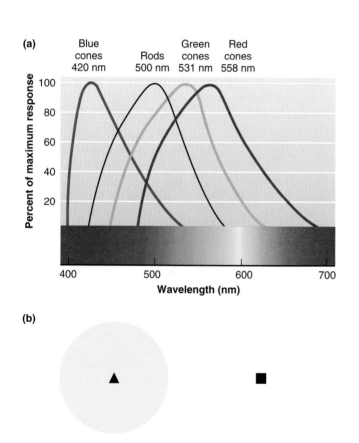

Figure 7–30

The sensitivities of the photopigments in the normal human retina.
(a) The frequency of action potentials in the optic nerve is directly related to a photopigment's absorption of light. Under bright lighting conditions, the three types of cones respond over different frequency ranges. In dim light, only the rods respond.
(b) Demonstration of cone cell fatigue and after-image. Hold very still and stare at the triangle inside the yellow circle for 30 seconds. Then, shift your gaze to the square and wait for the image to appear around it.

Figure 7–30b physiological *inquiry* *pi*

■ What color was the image you saw while you stared at the square? Why did you perceive that particular color?

Answer can be found at end of chapter.

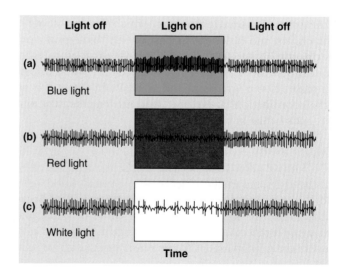

Figure 7–31

Response of a single opponent color ganglion cell to blue, red, and white lights.

Redrawn from Hubel and Wiesel.

pensation for movements of the head. The control centers for these compensating movements obtain their information about head movement from the vestibular system, which we will describe shortly. Control systems for the other slow movements of the eyes require the continuous feedback of visual information about the moving object.

Hearing

The sense of hearing is based on the physics of sound and the physiology of the external, middle, and inner ear, the nerves to the brain, and the brain parts involved in processing acoustic information.

Sound

Sound energy is transmitted through a gaseous, liquid, or solid medium by setting up a vibration of the medium's molecules, air being the most common medium. When there are no molecules, as in a vacuum, there can be no sound. Anything capable of disturbing molecules—for example, vibrating objects—can serve as a sound source. **Figure 7–33** demonstrates the basic mechanism of sound production using a tuning fork as an example. When struck, the tuning fork vibrates, creating disturbances of air molecules that make up the sound wave. The sound wave consists of zones of compression, in which the molecules are close together and the pressure is increased, alternating with zones of rarefaction, where the molecules are farther apart and the pressure is lower (**Figure 7–33a–d**). As the air molecules bump against each other, the zones of compression and rarefaction ripple outward, and the sound wave is transmitted over distance.

A sound wave measured over time (**Figure 7–33e**) consists of rapidly alternating pressures that vary continuously from a high during compression of molecules, to a low during rarefaction, and back again. The difference between the pressure of molecules in zones of compression and rarefaction determines the wave's amplitude, which is related to the loudness of the sound; the greater the amplitude, the louder the sound. The frequency of vibration of the sound source (i.e., the number of zones of compression or rarefaction in a given time) determines the pitch we hear; the faster the vibration, the higher the pitch. The sounds heard most keenly by human ears are those from sources vibrating at frequencies between 1000 and 4000 Hz (hertz, or cycles per second), but the entire range of frequencies audible to human beings extends from 20 to 20,000 Hz. Sound waves with sequences of pitches are generally perceived as musical. The addition of other frequencies, called overtones, to the basic sound wave gives the sound its characteristic quality, or timbre.

We can distinguish about 400,000 different sounds. For example, we can distinguish the note A played on a piano from the same note on a violin. We can also selectively *not* hear sounds, tuning out the background noise of a party to concentrate on a single voice.

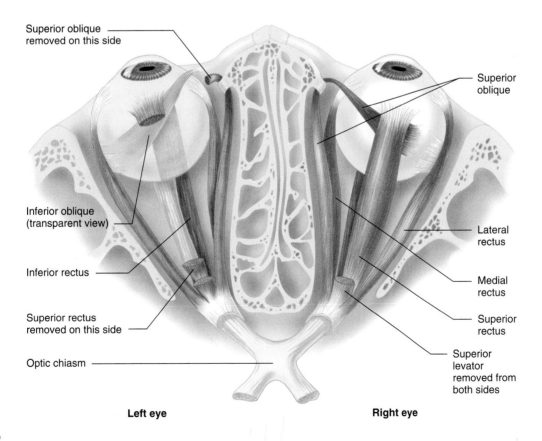

Superior oblique removed on this side

Superior oblique

Inferior oblique (transparent view)

Inferior rectus

Superior rectus removed on this side

Optic chiasm

Lateral rectus

Medial rectus

Superior rectus

Superior levator removed from both sides

Left eye

Right eye

Figure 7–32
A superior view of the muscles that move the eyes to direct the gaze and provide convergence.

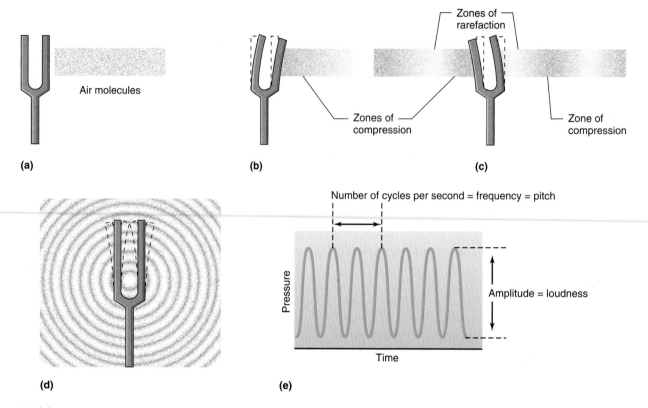

Figure 7–33
Formation of sound waves from a vibrating tuning fork.

Sound Transmission in the Ear

The first step in hearing is the entrance of sound waves into the **external auditory canal** (**Figure 7–34**). The shapes of the outer ear (the pinna, or auricle) and the external auditory canal help to amplify and direct the sound. The sound waves reverberate from the sides and end of the external auditory canal, filling it with the continuous vibrations of pressure waves.

The **tympanic membrane** (eardrum) is stretched across the end of the external auditory canal, and as air molecules push against the membrane, they cause it to vibrate at the same frequency as the sound wave. Under higher pressure during a zone of compression, the tympanic membrane bows inward. The distance the membrane moves, although always very small, is a function of the force with which the air molecules hit it and is related to the sound pressure and therefore its loudness. During the subsequent zone of rarefaction, the membrane returns to its original position. The exquisitely sensitive tympanic membrane responds to all the varying pressures of the sound waves, vibrating slowly in response to low-frequency sounds and rapidly in response to high-frequency ones.

The tympanic membrane separates the external auditory canal from the **middle ear,** an air-filled cavity in the temporal bone of the skull. The pressures in the external auditory canal and middle ear cavity are normally equal to atmospheric pressure. The middle ear cavity is exposed to atmospheric pressure through the **eustachian tube,** which connects the middle ear to the pharynx. The slitlike ending of this tube in the pharynx is normally closed, but muscle movements open the tube during yawning, swallowing, or sneezing. A difference in pressure can be produced with sudden changes in altitude (as in an ascending or descending elevator or airplane). When the pressure outside the ear and in the ear canal changes, the pressure in the middle ear initially remains constant because the eustachian tube is closed. This pressure difference can stretch the tympanic membrane and cause pain. This problem is relieved by voluntarily yawning or swallowing, which opens the eustachian tube and allows the pressure in the middle ear to equilibrate with the new atmospheric pressure.

The second step in hearing is the transmission of sound energy from the tympanic membrane through the middle ear cavity to the **inner ear.** The inner ear, called the **cochlea,** is a spiral-shaped passage in the temporal bone filled with a fluid called **endolymph.** The temporal bone also houses other passages, including the semicircular canals, which contain the sensory organs for balance and movement. These passages are connected to the cochlea but will be discussed later.

Because liquid is more difficult to move than air, the sound pressure transmitted to the inner ear must be amplified. This is achieved by a movable chain of three small bones, the **malleus, incus,** and **stapes** (**Figure 7–35**). These bones act as a piston and couple the vibrations of the tympanic membrane to the **oval window,** a membrane-covered opening separating the middle and inner ears. The total force of a sound wave applied to the tympanic membrane is transferred to the oval window, but because the oval window is much

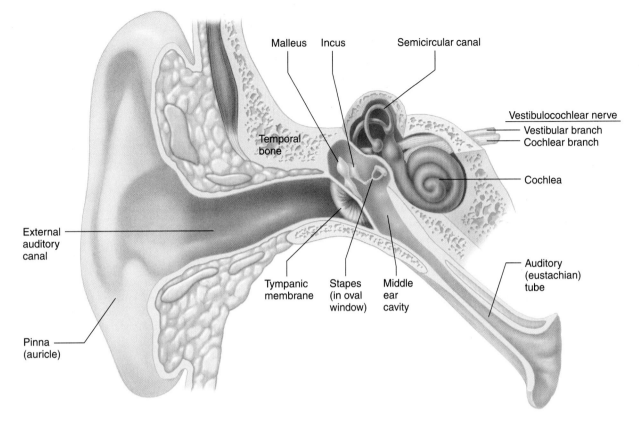

Figure 7–34

The human ear. In this and the following two figures, violet indicates the outer ear, green the middle ear, and blue the inner ear. The malleus, incus, and stapes are bones and components of the middle ear compartment. Actually, the auditory tube is closed except during pharynx movements such as swallowing or yawning.

smaller than the tympanic membrane, the force per unit area (i.e., the pressure) is increased 15 to 20 times. Additional advantage is gained through the lever action of the middle-ear bones. The amount of energy transmitted to the inner ear can be lessened by the contraction of two small skeletal muscles in the middle ear. The **tensor tympani** attaches to the malleus and tympanic membrane, and its contraction dampens their movement. The **stapedius** attaches to the stapes and similarly controls its mobility. These muscles contract reflexively to protect the delicate receptor apparatus of the inner ear from continuous, loud sounds. They cannot, however, protect against sudden, intermittent loud sounds, which is why it is crucial for people to wear ear protection in environments like a gun firing range. These muscles also contract reflexively when you vocalize to reduce the loudness of your own voice, and optimize hearing over certain frequency ranges.

The entire system described thus far involves the transmission of sound energy into the cochlea, where the receptor cells are located. The cochlea is almost completely divided lengthwise by a fluid-filled membranous tube, the **cochlear duct,** which follows the cochlear spiral (see Figure 7–35) and contains the sensory receptors of the auditory system. On either side of the cochlear duct are endolymph-filled compartments: the **scala vestibuli,** which is on the side of the cochlear

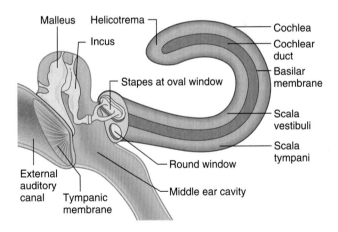

Figure 7–35

Relationship between the middle ear bones and the cochlea. Movement of the stapes against the membrane covering the oval window sets up pressure waves in the fluid-filled scala vestibuli. These waves cause vibration of the cochlear duct and the basilar membrane. Some of the pressure is transmitted around the helicotrema directly into the scala tympani. The cochlea is shown uncoiled for clarity.

Redrawn from Kandel and Schwartz.

duct that begins at the oval window; and the **scala tympani,** which is below the cochlear duct and ends in a second membrane-covered opening to the middle ear, the **round window.** The scala vestibuli and scala tympani meet at the end of the cochlear duct at the **helicotrema** (see Figure 7–35).

Sound waves in the ear canal cause in-and-out movement of the tympanic membrane, which moves the chain of middle-ear bones against the membrane covering the oval window, causing it to bow into the scala vestibuli and back out (**Figure 7–36**). This movement creates waves of pressure in the scala vestibuli. The wall of the scala vestibuli is largely bone, and there are only two paths by which the pressure waves can dissipate. One path is to the helicotrema, where the waves pass around the end of the cochlear duct into the scala tympani. However, most of the pressure is transmitted from the scala vestibuli across the cochlear duct. Pressure changes in the scala tympani are relieved by movements of the membrane within the round window.

The side of the cochlear duct nearest to the scala tympani is formed by the **basilar membrane** (**Figure 7–37**), upon which sits the **organ of Corti,** which contains the ear's

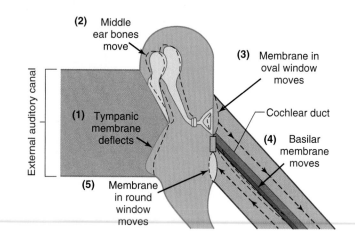

Figure 7–36

Transmission of sound vibrations through the middle and inner ear. Sound waves coming through the external auditory canal move the tympanic membrane (1), which sets off a sequence of events, moving the bones of the middle ear (2), the membrane in the oval window (3), the basilar membrane (4), and the round window membrane (5).

Redrawn from Davis and Silverman.

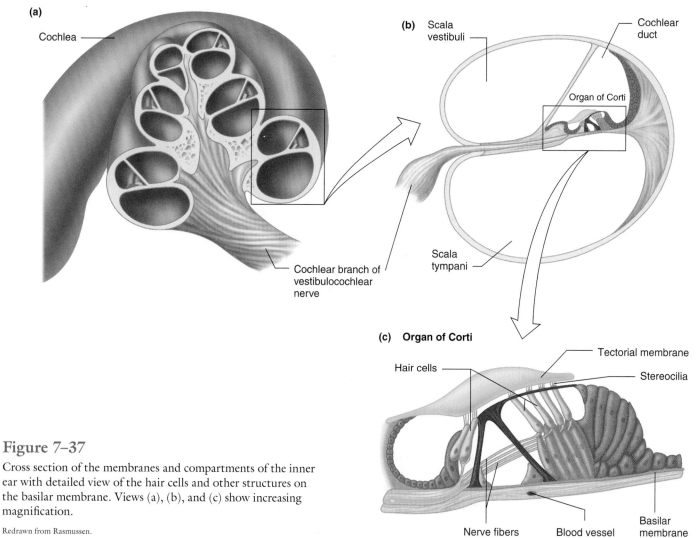

Figure 7–37

Cross section of the membranes and compartments of the inner ear with detailed view of the hair cells and other structures on the basilar membrane. Views (a), (b), and (c) show increasing magnification.

Redrawn from Rasmussen.

sensitive receptor cells. Pressure differences across the cochlear duct cause the basilar membrane to vibrate.

The region of maximal displacement of the vibrating basilar membrane varies with the frequency of the sound source. The properties of the membrane nearest the middle ear are such that this region vibrates most easily—that is, undergoes the greatest movement—in response to high-frequency (high-pitched) tones. As the frequency of the sound is lowered, vibration waves travel farther along the membrane toward the helicotrema. Progressively more distant regions of the basilar membrane vibrate maximally in response to progressively lower tones.

Hair Cells of the Organ of Corti

The receptor cells of the organ of Corti are called **hair cells.** These cells are mechanoreceptors that have hairlike **stereocilia** protruding from one end (**Figure 7–37c**). The hair cells transform the pressure waves in the cochlea into receptor potentials. Movements of the basilar membrane stimulate the hair cells because they are attached to the membrane.

The stereocilia of the hair cells are in contact with the **tectorial membrane** (see Figure 7–37c), which overlies the organ of Corti. As pressure waves displace the basilar membrane, the hair cells move in relation to the stationary tectorial membrane, and, consequently, the stereocilia bend (**Figure 7–38**). When the stereocilia are bent toward the tallest member of a bundle, fibrous connections called **tip links** pull open mechanically-gated cation channels, and the resulting charge influx depolarizes the membrane. This opens voltage-gated calcium channels near the base of the cell, which triggers neurotransmitter release. Unlike other extracellular fluids, the endolymph surrounding the stereocilia has a high concentration of K^+, so an influx of K^+ (rather than Na^+) is what depolarizes the hair cell. Bending the hair cells in the opposite direction slackens the tip links, closing the channels and allowing the cell to rapidly repolarize. Thus, as sound waves vibrate the basilar membrane, the stereocilia are bent back and forth, the membrane potential of the hair cells rapidly oscillates, and bursts of neurotransmitter are released onto afferent neurons.

The neurotransmitter released from hair cells is glutamate, which binds to and activates protein-binding sites on the terminals of 10 or so afferent neurons. This causes the generation of action potentials in the neurons, the axons of which join to form the cochlear branch of the **vestibulocochlear nerve** (cranial nerve VIII). The greater the energy (loudness) of the sound wave, the greater the frequency of action potentials generated in the afferent nerve fibers. Because of its position on the basilar membrane, each hair cell responds to a limited range of sound frequencies, with one particular frequency stimulating it most strongly.

In addition to the protective reflexes involving the tensor tympani and stapedius muscles, efferent nerve fibers from the brainstem regulate the activity of certain hair cells and dampen their response, which also protects them. Despite these protective mechanisms, the hair cells are easily damaged or even destroyed by exposure to high-intensity noises such as amplified rock concerts, engines of jet planes, and construction equipment. Lesser noise levels also cause damage if exposure is chronic.

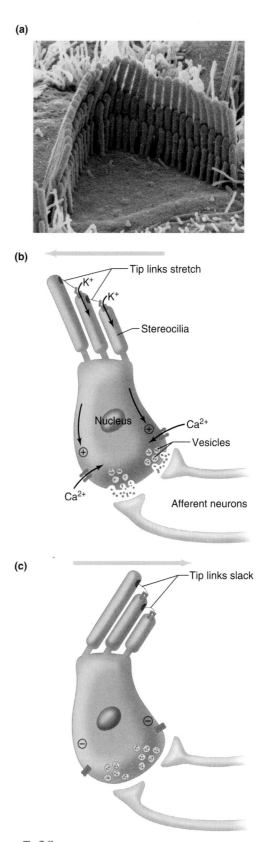

Figure 7–38

Mechanism for neurotransmitter release in the hair cell of the auditory system. (a) Scanning electron micrograph shows the size gradation in a bundle of stereocilia at the top of a single hair cell (tectorial membrane removed). (b) Bending stereocilia in one direction depolarizes the cell and stimulates neurotransmitter release, while bending in the opposite direction (c) repolarizes the cell and stops the release.

Neural Pathways in Hearing

Cochlear nerve fibers enter the brainstem and synapse with interneurons there. Fibers from both ears often converge on the same neuron. Many of these interneurons are influenced by the different arrival times and intensities of the input from the two ears. The different arrival times of low-frequency sounds and the different intensities of high-frequency sounds are used to determine the direction of the sound source. If, for example, a sound is louder in the right ear or arrives sooner at the right ear than at the left, we assume that the sound source is on the right. The shape of the outer ear (the pinna; see Figure 7–34) and movements of the head are also important in localizing the sound source.

From the brainstem, the information is transmitted via a multineuron pathway to the thalamus and on to the auditory cortex in the temporal lobe (see Figure 7–14). The neurons responding to different pitches (frequencies) are arranged along the auditory cortex in an orderly manner much as signals from different regions of the body are represented at different sites in the somatosensory cortex. Different areas of the auditory system are further specialized; some neurons respond best to complex sounds such as those used in verbal communication. Others signal the location, movement, duration, or loudness of a sound.

Electronic devices can help compensate for damage to the intricate middle ear, cochlea, or neural structures. **Hearing aids** amplify incoming sounds, which then pass via the ear canal to the same cochlear mechanisms used for normal sound. When substantial damage has occurred, however, and hearing aids cannot correct the deafness, electronic devices known as **cochlear implants** may restore functional hearing. In response to sound, cochlear implants directly stimulate the cochlear nerve with tiny electrical currents so that sound signals are transmitted directly to the auditory pathways, bypassing the cochlea.

Vestibular System

Hair cells are also found in the **vestibular apparatus** of the inner ear. This connected series of fluid-filled, membranous tubes also connects with the cochlear duct (**Figure 7–39**). The hair cells detect changes in the motion and position of the head. The vestibular apparatus consists of three membranous **semicircular canals** and two saclike swellings, the **utricle** and **saccule,** all of which lie in tunnels in the temporal bone on each side of the head. The bony tunnels of the inner ear that house the vestibular apparatus and cochlea have such a complicated shape that they are sometimes called the **labyrinth.**

The Semicircular Canals

The semicircular canals detect angular acceleration during *rotation* of the head along three perpendicular axes. The three axes of the semicircular canals are those activated while nodding the head up and down as in signifying "yes," shaking the head from side to side as in signifying "no," and tipping the head so the ear touches the shoulder (**Figure 7–40**).

Receptor cells of the semicircular canals, like those of the organ of Corti, contain hairlike stereocilia. These stereocilia are closely ensheathed by a gelatinous mass, the **cupula,**

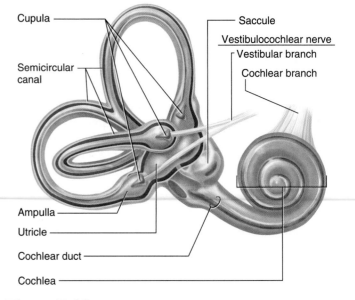

Figure 7–39

A tunnel in the temporal bone contains a fluid-filled membranous duct system. The semicircular canals, utricle, and saccule make up the vestibular apparatus. This system is connected to the cochlear duct. The purple structure within the ampullae are the cupulae, which contain the hair (receptor) cells.

Redrawn from Hudspeth.

Figure 7–40

Orientation of the semicircular canals within the labyrinth. Each plane of orientation is perpendicular to the others. Together they allow detection of movements in all directions.

which extends across the lumen of each semicircular canal at the **ampulla,** a slight bulge in the wall of each duct (**Figure 7–41**). Whenever the head moves, the semicircular canal within its bony enclosure and the attached bodies of the hair cells all move with it. The fluid filling the duct, however, is not attached to the

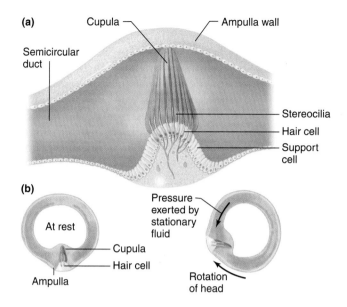

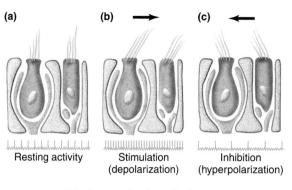

Discharge rate of vestibular nerve

Figure 7–41

(a) Organization of a cupula and ampulla. (b) Relation of the cupula to the ampulla when the head is at rest and when it is accelerating.

Figure 7–42

The relationship between the position of hairs in the ampulla and action potential firing in afferent neurons. (a) Resting activity. (b) Movement of hairs in one direction increases the action potential frequency in the afferent nerve activated by the hair cell. (c) Movement in the opposite direction decreases the rate relative to the resting state.

skull, and because of inertia, tends to retain its original position (i.e., to be "left behind"). Thus, the moving ampulla is pushed against the stationary fluid, which causes bending of the stereocilia and alteration in the rate of release of a chemical transmitter from the hair cells. This transmitter activates the nerve terminals synapsing with the hair cells.

The speed and magnitude of rotational head movements determine the direction in which the stereocilia are bent and the hair cells stimulated. Movement of these mechanoreceptors causes depolarization of the hair cell and neurotransmitter release as in cochlear hair cells. Neurotransmitter is released from the hair cells at rest, and the release increases or decreases from this resting rate according to the direction in which the hairs are bent. Each hair cell receptor has one direction of maximum neurotransmitter release; when its stereocilia are bent in this direction, the receptor cell depolarizes. When the stereocilia are bent in the opposite direction, the cell hyperpolarizes (**Figure 7–42**). The frequency of action potentials in the afferent nerve fibers that synapse with the hair cells is related to both the amount of force bending the stereocilia on the receptor cells and to the direction in which this force is applied.

When the head continuously turns at a steady velocity, the duct fluid begins to move at the same rate as the rest of the head, and the stereocilia slowly return to their resting position. For this reason, the hair cells are stimulated only during *changes* in the rate of rotation (i.e., during acceleration or deceleration) of the head.

The Utricle and Saccule

The utricle and saccule (see Figure 7–39) provide information about *linear*—up and down, back and forth—acceleration and changes in head position relative to the forces of gravity. Here, too, the receptor cells are mechanoreceptors sensitive to the dis-

placement of projecting hairs. The hair cells in the utricle point nearly straight up when you stand, and they respond when you tip your head away from the horizontal, or to linear accelerations in the horizontal plane. In the saccule, hair cells project at right angles to those in the utricle, and they respond when you move from a lying to a standing position, or to vertical accelerations like those produced when you jump on a trampoline. The utricle and saccule are slightly more complex than the ampullae.

The stereocilia projecting from the hair cells are covered by a gelatinous substance in which tiny stones, or **otoliths,** are embedded. The otoliths, which are calcium carbonate crystals, make the gelatinous substance heavier than the surrounding fluid. In response to a change in position, the gelatinous otolithic material moves according to the forces of gravity and pulls against the hair cells so that the stereocilia on the hair cells bend and the receptor cells are stimulated. **Figure 7–43** demonstrates how otolith organs are stimulated by a change in head position.

Vestibular Information and Pathways

Vestibular information is used in three ways. One is to control the eye muscles so that, in spite of changes in head position, the eyes can remain fixed on the same point. *Nystagmus* is a large, jerky, back-and-forth movement of the eyes that can occur in response to unusual vestibular input in normal people; it can also be a pathological sign. Nystagmus is noticeable when a person spins in a swiveling chair for about 20 seconds, then abruptly stops the chair. For a short time after the motion ceases, the fluid in the semicircular canals continues to spin and the person's eyes will involuntarily move as though attempting to track objects spinning past the field of view.

The second use of vestibular information is in reflex mechanisms for maintaining upright posture and balance. The vestibular apparatus plays a role in the support of the head during movement, orientation of the head in space, and reflexes accompanying locomotion. Very few postural reflexes,

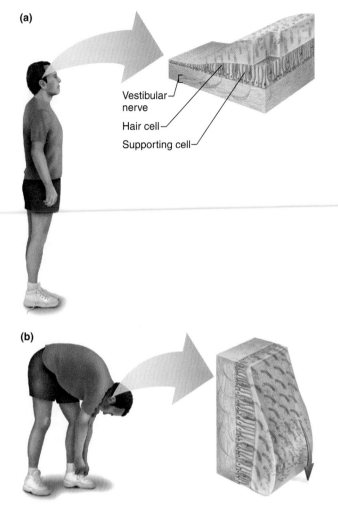

(a)

Vestibular nerve

Hair cell

Supporting cell

(b)

Figure 7–43

Effect of head position on otolith organ of the utricle. (a) Upright position, hair cells are not bent. (b) Gravity bends the hair cells when the head tilts forward.

however, depend exclusively on input from the vestibular system despite the fact that the vestibular organs are sometimes called the sense organs of balance.

The third use of vestibular information is in providing conscious awareness of the position and acceleration of the body, perception of the space surrounding the body, and memory of spatial information.

Information about hair cell stimulation is relayed from the vestibular apparatus to nuclei within the brainstem via the vestibular branch of the vestibulocochlear nerve. It is transmitted via a multineuronal pathway through the thalamus to a system of vestibular centers in the parietal lobe. Descending projections are also sent from the brainstem nuclei to the spinal cord to influence postural reflexes. Vestibular information is integrated with information from the joints, tendons, and skin, leading to the sense of posture **(proprioception)** and movement.

Unexpected inputs from the vestibular system and other sensory systems can induce *vertigo,* defined as an illusion of movement—usually spinning—accompanied by feelings of

nausea and lightheadedness. Such disruptions can result from strokes, irritation of the labyrinths by infection, loose particles of calcium carbonate in the semicircular canals, or inhibition of vestibular inputs caused by excess alcohol consumption. A similar phenomenon can occur when there is a mismatch in information from the various sensory systems. For example, many amusement parks feature widescreen virtual thrill rides in which your eyes take you on a dizzying helicopter ride, while your vestibular system signals that you are not moving at all. *Motion sickness* also involves the vestibular system, occurring when you experience unfamiliar patterns of linear and rotational acceleration and adaptation to them has not yet occurred.

Chemical Senses

Recall that receptors sensitive to specific chemicals are **chemoreceptors.** Some of these respond to chemical changes in the internal environment; two examples are the oxygen and hydrogen ion receptors in certain large blood vessels (Chapter 13). Others respond to external chemical changes. In this category are the receptors for taste and smell, which affect a person's appetite, saliva flow, gastric secretions, and avoidance of harmful substances.

Taste

The specialized sense organs for taste (also called **gustation**) are the 10,000 or so **taste buds** found in the mouth and throat, the vast majority on the upper surface and sides of the tongue. Taste buds are small groups of cells arranged like orange slices around a hollow pore, and are found in the walls of visible structures called **lingual papillae** (**Figure 7–44**). Some of the cells serve mainly as support cells, but others are specialized epithelial cells that act as receptors for various chemicals in the food we eat. Small, hairlike projections increase the surface area of taste receptor cells, and contain integral membrane proteins that transduce the presence of a given chemical into a receptor potential. At the bottom of taste buds are **basal cells,** which divide and differentiate to continually replace taste receptor cells damaged in the harsh environment of the mouth. To enter the pores of the taste buds and come into contact with taste-receptor cells, food molecules must be dissolved in liquid—either ingested or provided by secretions of the salivary glands. Try placing sugar or salt on your tongue after thoroughly drying it; no taste sensation occurs.

Many different chemicals can generate the sensation of taste by differentially activating a few basic types of taste receptors. Taste modalities generally fall into five different categories according to the receptor type most strongly activated; sweet, sour, salty, bitter, and **umami** (pronounced "oo-MOM-mee"). This latter category is named after the Japanese word for "delicious." This taste is associated with the taste of glutamate and similar amino acids, and is sometimes described as conveying the sense of flavorfulness. Glutamate (or monosodium glutamate, MSG) is a common additive used to enhance the flavor of foods in traditional Asian cuisine. In addition to these known taste receptors, there are likely others yet to be discovered. For example, recent experiments suggest that a fatty acid transport protein first identified in the lingual papillae of rodents may soon be added to the list. Research has shown that blocking these transporters inhibits the preference for the taste of foods with high lipid content, and reduces the production of fat-digesting enzymes by the gastrointestinal system. If confirmed in humans, this fatty acid transporter could become the sixth member of the taste receptor family, and might explain our tendency to over-indulge on high-calorie, high-fat foods.

Each group of tastes has a distinct signal transduction mechanism. Salt taste is detected by a simple mechanism in which ingested sodium enters channels in the receptor cell membrane, depolarizing the cell and stimulating the production of action potentials in the associated sensory neuron. Sour taste is stimulated by foods with high acid content, such as lemons, which contain citric acid. Hydrogen ions block potassium channels in the sour receptors, and the loss of the hyperpolarizing potassium leak current depolarizes the receptor cell. Sweet receptors have integral membrane proteins that bind natural sugars like glucose, as well as artificial sweetener molecules like saccharin and aspartame. Binding of sugars to these receptors activates a G-protein-coupled second-messenger pathway (Chapter 5) that ultimately blocks potassium channels and thus generates a depolarizing receptor potential. Bitter flavor is associated with many poisonous substances, especially plant alkaloids like strychnine and arsenic. There is an obvious evolutionary advantage in recognizing a wide variety of poisonous substances, and thus there are many varieties of bitter receptors. All of those types, however, generate receptor potentials via G-protein-mediated second-messenger pathways and ultimately evoke the negative sensation of bitter flavor. Umami receptor cells also depolarize via a G-protein-coupled receptor mechanism.

Each afferent neuron synapses with more than one receptor cell, and the taste system is organized into independent coded pathways into the central nervous system.

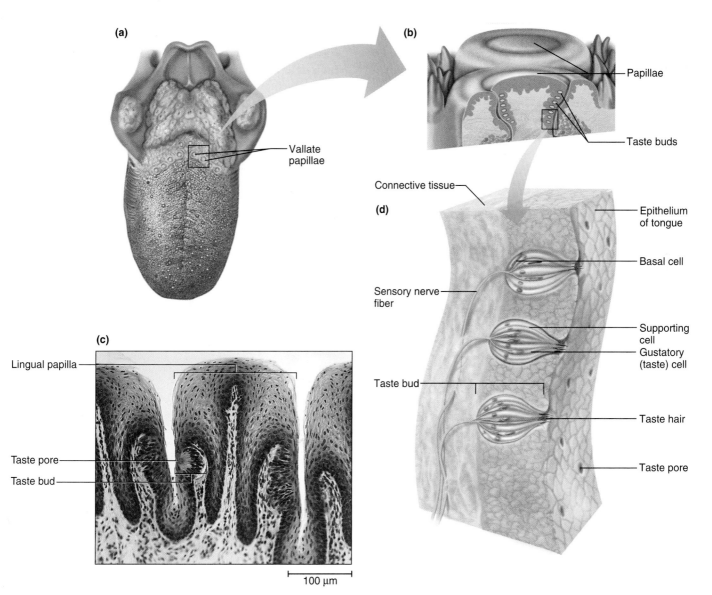

Figure 7–44

Taste receptors. (a) Top view of the tongue showing several types of lingual papillae. (b) and (c) Cross section of one type of papilla with taste buds. (d) Pores in the sides of papillae open into taste buds, which are composed of support cells, taste receptor cells, and basal cells.

Single receptor cells, however, respond in varying degrees to substances that fall into more than one taste category. This property is analogous to the overlapping sensitivities of photoreceptors to different wavelengths. Awareness of the specific taste of a substance depends also upon the pattern of firing in other types of sensory neurons. For example, sensations of pain (hot spices), texture, and temperature contribute to taste.

The pathways for taste in the central nervous system project to the parietal cortex, near the "mouth" region of the somatosensory cortex (see Figure 7–14).

Smell

Eighty percent of the flavor of food is actually contributed by the sense of smell, or **olfaction.** This is illustrated by the common experience of finding that food lacks taste when a head cold blocks your nasal passages. The odor of a substance is directly related to its chemical structure. We can recognize and identify thousands of different odors with great accuracy. Thus, neural circuits that deal with olfaction must encode information about different chemical structures, store (learn) the different code patterns that represent the different structures, and at a later time recognize a particular neural code to identify the odor.

The olfactory receptor neurons, the first cells in the pathways that give rise to the sense of smell, lie in a small patch of epithelium called the **olfactory epithelium,** in the upper part of the nasal cavity (**Figure 7–45a**). Olfactory receptor neurons live for only about two months, so they are constantly being replaced by new cells produced from stem cells in the olfactory epithelium. The mature cells are specialized afferent neurons that have a single, enlarged dendrite that extends to the surface of the epithelium. Several long, nonmotile cilia extend from the tip of the dendrite and lie along the sur-

face of the olfactory epithelium (**Figure 7–45b**) where they are bathed in mucus. The cilia contain the receptor proteins (binding sites) for olfactory stimuli. The axons of the neurons form the olfactory nerve, which is cranial nerve I.

For us to detect an odorous substance (an **odorant**), molecules of the substance must first diffuse into the air and pass into the nose to the region of the olfactory epithelium. Once there, they dissolve in the mucus that covers the epithelium and then bind to specific odorant receptors on the cilia. Proteins in the mucus may interact with the odorant molecules, transport them to the receptors, and facilitate their binding to the receptors. Stimulated odorant receptors activate a G-protein-mediated pathway that increases cAMP, which in turn opens nonselective cation channels and depolarizes the cell.

Although there are many thousands of olfactory receptor cells, each contains only one of the 1000 or so different plasma membrane odorant receptor types. In turn, each of these types responds only to a specific chemically related group of odorant molecules. Each odorant has characteristic chemical groups that distinguish it from other odorants, and each of these groups activates a different plasma membrane odorant receptor type. Thus, the identity of a particular odorant is determined by the activation of a precise combination of plasma membrane receptors, each of which is contained in a distinct group of olfactory receptor cells.

The axons of the olfactory receptor cells synapse in the brain structures known as **olfactory bulbs,** which lie on the undersurface of the frontal lobes. Axons from olfactory receptor cells that share a common receptor specificity synapse together on certain olfactory bulb neurons, thereby maintaining the specificity of the original stimulus. In other words, specific odorant receptor cells activate only certain olfactory

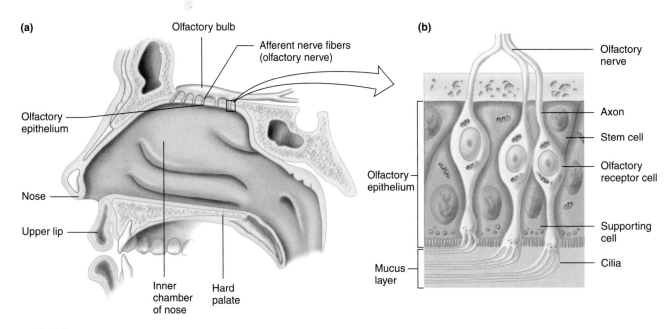

(a) Olfactory bulb
Afferent nerve fibers (olfactory nerve)
Olfactory epithelium
Nose
Upper lip
Inner chamber of nose
Hard palate

(b) Olfactory nerve
Axon
Stem cell
Olfactory receptor cell
Olfactory epithelium
Supporting cell
Cilia
Mucus layer

Figure 7–45

(a) Location and (b) enlargement of a portion of the olfactory epithelium showing the structure of the olfactory receptor cells. In addition to these cells, the olfactory epithelium contains stem cells, which give rise to new receptors and supporting cells.

bulb neurons, allowing the brain to determine which receptors have been stimulated. The codes used to transmit olfactory information probably use both spatial (which neurons are firing?) and temporal (what is the frequency of the action potential responses?) components.

Information passes from the olfactory bulbs to the olfactory cortex and other parts of the limbic system. This region is intimately associated with emotional, food-getting, and sexual behaviors and may play a role in processing olfactory information relevant to these activities. Some of the areas of the olfactory cortex then send projections to other regions of the frontal cortex. Different odors elicit different patterns of electrical activity in several cortical areas, allowing humans to discriminate between some 10,000 different odorants even though they have only 1000 or so different olfactory receptor types.

Olfactory discrimination varies with attentiveness, hunger (sensitivity is greater in hungry subjects), gender (women in general have keener olfactory sensitivities than men), smoking (decreased sensitivity has been repeatedly associated with smoking), age (the ability to identify odors decreases with age, and a large percentage of elderly persons cannot detect odors at all), and state of the olfactory mucosa (as we have mentioned, the sense of smell decreases when the mucosa is congested, as in a head cold).

ADDITIONAL CLINICAL EXAMPLES

Hearing and Balance: Losing Both at Once

A patient comes to the doctor with complaints of dizziness and ringing in one of his ears. He tells the doctor that these symptoms come and go and have been present for about a year. Recently, these attacks have increased in severity and have been accompanied with loss of balance and vomiting. The symptoms usually last for four to five hours, during which time he lies on the bed with his affected ear uppermost. Based on these symptoms and after many tests to rule out disorders of the cerebellum, the doctor diagnoses *Ménière's disease.* This is a disease of the auditory and vestibular systems that results from increased pressure in the inner ear, including the cochlea and the semicircular canals. The disease is believed to be caused by deficits in absorption and renewal of the endolymph and the distension of the structures in these areas. As a result, both hearing and vestibular disruptions occur, and complete or partial hearing loss may be permanent in the affected ear. Ménière's disease may sometimes resolve itself without treatment after months or years, but in many cases it is permanent.

Color Blindness

At high light intensities, as in daylight vision, most people—92 percent of the male population and over 99 percent of the female population—have normal color vision. However, there are several types of defects in color vision that result from mutations in the cone pigments. The most common form of *color blindness,* red-green color blindness, is present predominantly in men, affecting 1 out of 12. Color blindness in women is extremely rare (1 out of 200). Men with red-green color blindness either lack the red or the green cone pigments entirely or have them in an abnormal form. Because of this, the discrimination between shades of these colors is poor.

Color blindness results from a recessive mutation in one or more genes encoding the cone pigments. Genes encoding the red and green cone pigments are located very close to each other on the X chromosome, while the gene encoding the blue chromophore is located on chromosome 7. Because of this close association of the red and green genes on the X chromosome, there is a greater likelihood that crossover will occur during meiosis, thus eliminating or changing the spectral characteristics of the red and green pigments produced. This, in part, accounts for the fact that red-green defects are not always complete, and that some color-blind individuals under some conditions can distinguish shades of red or green. In males, the presence of only a single X chromosome means that a single recessive allele from the mother will result in color blindness, even though the mother herself may have normal color vision due to having one normal X chromosome. It also means that 50 percent of the male offspring of that mother will be expected to be color blind. Individuals who have red-green color blindness will not be able to see the number in **Figure 7–46.** ∎

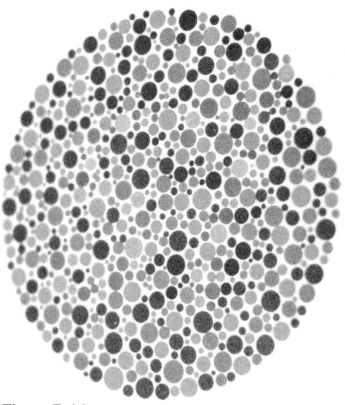

Figure 7–46

Image used for testing red-green color vision. With normal color vision the number 57 is visible, while no number is apparent to those with a red-green defect.

Somatic Sensation

I. A variety of receptors sensitive to one (or a few) stimulus types provide sensory function of the skin and underlying tissues.

II. Information about somatic sensation enters both specific and nonspecific ascending pathways. The specific pathways cross to the opposite side of the brain.

III. The somatic sensations include touch, pressure, the senses of posture and movement, temperature, and pain.

 a. Rapidly adapting mechanoreceptors of the skin give rise to sensations such as vibration, touch, and movement, whereas slowly adapting ones give rise to the sensation of pressure.

 b. Skin receptors with small receptive fields are involved in fine spatial discrimination, whereas receptors with larger receptive fields signal less spatially precise touch-pressure sensations.

 c. A major receptor type responsible for the senses of posture and kinesthesia is the muscle spindle stretch receptor.

 d. Cold receptors are sensitive to decreasing temperature; warmth receptors signal information about increasing temperature.

 e. Tissue damage and immune cells release chemical agents that stimulate specific receptors that give rise to the sensation of pain.

 f. Stimulation-produced analgesia, transcutaneous nerve stimulation (TENS), and acupuncture control pain by blocking transmission in the pain pathways.

Vision

I. Light is defined by its wavelength or frequency.

II. The light that falls on the retina is focused by the cornea and lens.

 a. Lens shape changes (accommodation) to permit viewing near or distant objects so that they are focused on the retina.

 b. Stiffening of the lens with aging interferes with accommodation. Cataracts decrease the amount of light transmitted through the lens.

 c. An eyeball too long or too short relative to the focusing power of the lens causes nearsighted or farsighted vision, respectively.

III. The photopigments of the rods and cones are made up of a protein component (opsin) and a chromophore (retinal).

 a. The rods and each of the three cone types have different opsins, which make each of the four receptor types sensitive to different ranges of light wavelengths.

 b. When light falls upon the chromophore, the photic energy causes the chromophore to change shape, which triggers a cascade of events leading to hyperpolarization of the photoreceptors and decreased neurotransmitter release from them. When exposed to darkness, the rods and cones are depolarized and therefore release more neurotransmitter than in light.

IV. The rods and cones synapse on bipolar cells, which synapse on ganglion cells.

 a. Ganglion cell axons form the optic nerves, which lead into the brain.

 b. The optic nerve fibers from the medial half of each retina cross to the opposite side of the brain in the optic chiasm. The fibers from the optic nerves terminate in the lateral geniculate nuclei of the thalamus, which send fibers to the visual cortex.

 c. Visual information is also relayed to areas of the brain dealing with biological rhythms.

V. Coding in the visual system occurs along parallel pathways, in which different aspects of visual information, such as color, form, movement, and depth, are kept separate from each other.

VI. The colors we perceive are related to the wavelength of light. Different wavelengths excite one of the three cone photopigments most strongly.

 a. Certain ganglion cells are excited by input from one type of cone cell and inhibited by input from a different cone type.

 b. Our sensation of color depends on the output of the various opponent color cells and the processing of this output by brain areas involved in color vision.

VII. Six skeletal muscles control eye movement to scan the visual field for objects of interest, keep the fixation point focused on the fovea centralis despite movements of the object or the head, and prevent adaptation of the photoreceptors.

Hearing

I. Sound energy is transmitted by movements of pressure waves.

 a. Sound wave frequency determines pitch.

 b. Sound wave amplitude determines loudness.

II. The sequence of sound transmission is as follows:

 a. Sound waves enter the external auditory canal and press against the tympanic membrane, causing it to vibrate.

 b. The vibrating membrane causes movement of the three small middle-ear bones; the stapes vibrates against the oval window membrane.

 c. Movements of the oval window membrane set up pressure waves in the fluid-filled scala vestibuli, which cause vibrations in the cochlear duct wall, setting up pressure waves in the fluid there.

 d. These pressure waves cause vibrations in the basilar membrane, which is located on one side of the cochlear duct.

 e. As this membrane vibrates, the hair cells of the organ of Corti move in relation to the tectorial membrane.

 f. Movement of the hair cells' stereocilia stimulates the hair cells to release glutamate, which activates receptors on the peripheral ends of the afferent nerve fibers.

III. Separate parts of the basilar membrane vibrate maximally in response to one particular sound frequency; high frequency is detected near the oval window and low frequency toward the far end of the cochlear duct.

Vestibular System

I. A vestibular apparatus lies in the temporal bone on each side of the head and consists of three semicircular canals, a utricle, and a saccule.

II. The semicircular canals detect angular acceleration during rotation of the head, which causes bending of the stereocilia on their hair cells.

III. Otoliths in the gelatinous substance of the utricle and saccule move in response to changes in linear acceleration and the position of the head relative to gravity, and stimulate the stereocilia on the hair cells.

Chemical Senses

I. The receptors for taste lie in taste buds throughout the mouth, principally on the tongue. Different types of taste receptors operate by different mechanisms.

II. Olfactory receptors, which are part of the afferent olfactory neurons, lie in the upper nasal cavity.

Chapter 7

a. Odorant molecules, once dissolved in the mucus that bathes the olfactory receptors, bind to specific receptors (protein binding sites). Each olfactory receptor cell has one or at most a few of the 1000 different receptor types.

b. Olfactory pathways go to the limbic system.

Additional Clinical Examples

I. Increased fluid in the cochlea and vestibular apparatus results in dizziness and loss of function.

II. Color blindness is due to abnormalities of the cone pigments resulting from genetic mutation.

■ SECTION B KEY TERMS

accommodation 210
amacrine cell 214
ampulla 222
anterolateral pathway 206
aqueous humor 209
basal cell 224
basilar membrane 220
bipolar cell 214
capsaicin 204
chemoreceptor 224
choroid 209
chromophore 213
ciliary muscle 209
cochlea 218
cochlear duct 219
cone 213
cornea 209
cupula 222
dark adaptation 214
disc 212
dorsal column pathway 206
endolymph 218
eustachian tube 218
external auditory canal 218
fovea centralis 209
frequency 208
ganglion cell 214
guanylyl cyclase 213
gustation 224
hair cell 221
helicotrema 220
horizontal cell 214
incus 218
inner ear 218
inner segment 212
iris 209
kinesthesia 204
labyrinth 222

lens 209
light adaptation 214
lingual papillae 224
malleus 218
menthol 204
middle ear 218
nociceptors 204
odorant 226
olfaction 226
olfactory bulb 226
olfactory epithelium 226
opponent color cell 216
opsin 213
optic chiasm 215
optic disc 209
optic nerve 209
optic tracts 215
organ of Corti 220
otolith 223
outer segment 212
oval window 218
phosphodiesterase 214
photopigment 213
photoreceptor 209
pigment epithelium 213
proprioception 224
pupil 209
refraction 209
retina 209
retinal 213
rhodopsin 213
rod 213
round window 220
saccade 216
saccule 222
scala tympani 220
scala vestibuli 219
sclera 209

semicircular canal 222
somatic sensation 203
stapedius 219
stapes 218
stereocilia 221
suprachiasmatic nucleus 215
taste bud 224
tectorial membrane 221
tensor tympani 219
tip link 221

transducin 214
tympanic membrane 218
umami 224
utricle 222
vestibular apparatus 222
vestibulocochlear nerve 221
visible spectrum 209
vitreous humor 209
wavelength 208
zonular fiber 209

■ SECTION B CLINICAL TERMS

acupuncture 206
analgesia 206
astigmatism 211
cataract 210
cochlear implant 222
color blindness 227
farsighted 211
glaucoma 211
hearing aid 222
hyperalgesia 206
hyperopic 211
Ménière's disease 227
motion sickness 224

myopic 210
nearsighted 210
nystagmus 223
ophthalmoscope 209
placebo 206
presbyopia 210
referred pain 205
stimulation-produced analgesia 206
transcutaneous electric nerve stimulation (TENS) 206
vertigo 224

■ SECTION B REVIEW QUESTIONS

1. Describe the similarities between pain and the other somatic sensations. Describe the differences.
2. List the structures through which light must pass before it reaches the photopigment in the rods and cones.
3. Describe the events that take place during accommodation for near vision.
4. What changes take place in neurotransmitter release from the rods or cones when they are exposed to light?
5. Beginning with the ganglion cells of the retina, describe the visual pathway.
6. List the sequence of events that occurs between the entry of a sound wave into the external auditory canal and the firing of action potentials in the cochlear nerve.
7. Describe the anatomical relationship between the cochlea and the cochlear duct.
8. What is the relationship between head movement and cupula movement in a semicircular canal?
9. What causes the release of neurotransmitter from the utricle and saccule receptor cells?
10. In what ways are the sensory systems for taste and olfaction similar? In what ways are they different?

Chapter 7 Test Questions

(Answers appear in Appendix A.)

1. Choose the TRUE statement:
   a. The modality of energy a given sensory receptor responds to in normal functioning is known as the "adequate stimulus" for that receptor.
   b. Receptor potentials are "all-or-none"; that is, they have the same magnitude regardless of the strength of the stimulus.
   c. When the frequency of action potentials along sensory neurons is constant as long as a stimulus continues, it is called "adaptation."
   d. When sensory units have large receptive fields, the acuity of perception is greater.
   e. The "modality" refers to the intensity of a given stimulus.

2. Using a single intracellular recording electrode, in what part of a sensory neuron could you simultaneously record both receptor potentials and action potentials?
   a. in the cell body
   b. at the node of Ranvier nearest the peripheral end
   c. at the receptor membrane where the stimulus occurs
   d. at the central axon terminals within the CNS
   e. there is no single point where both can be measured

3. Which best describes "lateral inhibition" in sensory processing?
   a. Presynaptic axo-axonal synapses reduce neurotransmitter release at excitatory synapses.
   b. When a stimulus is maintained for a long time, action potentials from sensory receptors decrease in frequency with time.
   c. Descending inputs from the brainstem inhibit afferent pain pathways in the spinal cord.
   d. Inhibitory interneurons decrease action potentials from receptors at the periphery of a stimulated region.
   e. Receptor potentials increase in magnitude with the strength of a stimulus.

4. What region of the brain contains the primary visual cortex?
   a. the occipital lobe
   b. the frontal lobe
   c. the temporal lobe
   d. the somatosensory cortex
   e. the parietal lobe association area

5. Which type of receptor does NOT encode a somatic sensation?
   a. muscle spindle stretch receptor
   b. nociceptor
   c. Pacinian corpuscle
   d. thermoreceptor
   e. cochlear hair cell

6. Which best describes the vision of a person with uncorrected nearsightedness?
   a. The eyeball is too long; far objects focus on the retina when the ciliary muscle contracts.
   b. The eyeball is too long; near objects focus on the retina when the ciliary muscle is relaxed.
   c. The eyeball is too long; near objects cannot be focused on the retina.
   d. The eyeball is too short; far objects cannot be focused on the retina.
   e. The eyeball is too short; near objects focus on the retina when the ciliary muscle is relaxed.

7. If a patient suffers a stroke that destroys the optic tract on the right side of the brain, which of the following visual defects will result?
   a. Complete blindness will result.
   b. There will be no vision in the left eye, but vision will be normal in the right eye.
   c. The patient will not perceive images of objects striking the left half of the retina in the left eye.
   d. The patient will not perceive images of objects striking the right half of the retina in the right eye.
   e. Neither eye will perceive objects in the right side of the patient's field of view.

8. Which correctly describes a step in auditory signal transduction?
   a. Displacement of the basilar membrane with respect to the tectorial membrane stimulates stereocilia on the hair cells.
   b. Pressure waves on the oval window cause vibrations of the malleus, which are transferred via the stapes to the round window.
   c. Movement of the stapes causes oscillations in the tympanic membrane, which is in contact with the endolymph.
   d. Oscillations of the stapes against the oval window set up pressure waves in the semicircular canals.
   e. The malleus, incus, and stapes are found in the inner ear, within the cochlea.

9. A standing subject looking over her left shoulder suddenly rotates her head to look over her right shoulder. How does the vestibular system detect this motion?
   a. The utricle goes from a vertical to a horizontal position, and otoliths stimulate stereocilia.
   b. Stretch receptors in neck muscles send action potentials to the vestibular apparatus, which relays them to the brain.
   c. Fluid within the semicircular canals remains stationary, bending the cupula and stereocilia as the head rotates.
   d. The movement causes endolymph in the cochlea to rotate from right to left, stimulating inner hair cells.
   e. Counter-rotation of the aqueous humor activates a nystagmus response.

10. Which category of taste receptor cells does MSG (monosodium glutamate) most strongly stimulate?
    a. salty
    b. bitter
    c. sweet
    d. umami
    e. sour

Chapter 7 Quantitative and Thought Questions

(Answers appear in Appendix A.)

1. Describe several mechanisms by which pain could theoretically be controlled medically or surgically.
2. At what two sites would central nervous system injuries interfere with the perception that heat is being applied to the right side of the body? At what single site would a central nervous system injury interfere with the perception that heat is being applied to either side of the body?
3. What would vision be like after a drug has destroyed all the cones in the retina?
4. Damage to what parts of the cerebral cortex could explain the following behaviors? (a) A person walks into a chair placed in her path. (b) The person does not walk into the chair, but she does not know what the chair can be used for.

Chapter 7 Answers to Physiological Inquiries

Figure 7–2 Receptor potentials would not be affected because they are not mediated by voltage-gated channels. Action potential propagation to the central nervous system would also be normal because it depends only on voltage-gated sodium and potassium channels. The drug would inhibit neurotransmitter release from the central axon terminal, however, because vesicle exocytosis requires calcium entry through voltage-gated channels.

Figure 7–6b Although the skin area of your lips is much smaller than that of your back, the much larger number of sensory neurons originating in your lips requires a larger processing area within the somatosensory cortex of your brain. See Figure 7–20 for a diagrammatic representation of cortical areas involved in sensory processing.

Figure 7–19 Sensation of all body parts above the level of the injury would be normal. Below the level of the injury, however, there would be a mixed pattern of sensory loss. Fine touch, pressure, and body position sensation would be lost from the left side of the body because that information ascends in the spinal cord on the side that it enters without crossing the midline until it reaches the brainstem. Pain and temperature sensation would be lost from the right side of the body below the injury because those pathways cross immediately upon entry and ascend in the opposite side of the spinal cord.

Figure 7–29

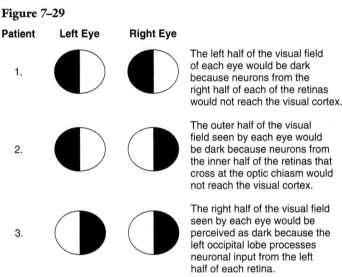

The left half of the visual field of each eye would be dark because neurons from the right half of each of the retinas would not reach the visual cortex.

The outer half of the visual field seen by each eye would be dark because neurons from the inner half of the retinas that cross at the optic chiasm would not reach the visual cortex.

The right half of the visual field seen by each eye would be perceived as dark because the left occipital lobe processes neuronal input from the left half of each retina.

Figure 7–30b Most people who stare at the yellow background, perceive an after-image of a blue circle around the square. This is because prolonged staring at the color yellow activates most of the available retinal in the photopigments of both red and green cones (see Figure 7–30a), effectively fatiguing them into a state of reduced sensitivity. When you shift your gaze to the white background (white light contains all wavelengths of light), only the blue cones are available to respond, so you perceive a blue circle until the red and green cones recover.

Brain function is monitored by
an electroencephalogram (EEG).

Consciousness, the Brain, and Behavior

i n this chapter, we discuss how the nervous system determines our states of consciousness and the ways in which the brain produces complex behaviors. Although recent advances in electrophysiology and brain-imaging techniques are yielding fascinating insights, there is still much that we don't know about these topics. If you can imagine that, for any given neuron, there may be as many as 200,000 other neurons connecting to it through synapses, you can begin to appreciate the complexity of the systems that control even the simplest behavior.

States of Consciousness

The term *consciousness* includes two distinct concepts: **states of consciousness** and **conscious experiences.** The first concept refers to whether a person is awake, asleep, drowsy, and so on. The second refers to experiences a person is aware of—thoughts, feelings, perceptions, ideas, dreams, reasoning—during any of the states of consciousness.

A person's state of consciousness is defined in two ways: (1) by behavior, covering the spectrum from maximum attentiveness to coma, and (2) by the pattern of brain activity that can be recorded electrically. This record, known as the **electroencephalogram (EEG),** portrays the electrical potential difference between different points on the surface of the scalp. The EEG is such an important tool in identifying the different states of consciousness that we begin with it.

Electroencephalogram

Neural activity is manifested by the electrical signals known as graded potentials and action potentials (Chapter 6). It is possible to record the electrical activity in the brain's neurons, particularly those in the cortex near the surface of the brain, from the outside of the head. Electrodes, which are wires attached to the head by a salty paste that conducts electricity, pick up electrical signals generated in the brain and transmit them to a machine that records them as the EEG.

While we often think of electrical activity in neurons in terms of action potentials, action potentials do not usually contribute directly to the EEG. Rather, EEG patterns are largely due to graded potentials, in this case summed postsynaptic potentials (Chapter 6) in the many hundreds of thousands of brain neurons that underlie the recording electrodes. The majority of the electrical signal recorded in the EEG originates in the pyramidal cells of the cortex. The processes of these large cells lie perpendicular to the brain's surface, and the EEG records postsynaptic potentials in their dendrites.

EEG patterns are complex waveforms with large variations in both amplitude and frequency (**Figure 8–1**). (The properties of a wave are summarized in Figure 7–21.) The wave's amplitude, measured in microvolts (μV), indicates how much electrical activity of a similar type is going on beneath the recording electrodes at any given time. A high amplitude indicates that many neurons are being activated simultaneously. In other words, it indicates the degree of synchronous firing of whichever neurons are generating the synaptic activity. On the other hand, a low amplitude indicates that these neurons are less activated or are firing asynchronously. The amplitude may range from 0.5 to 100 μV. Note that EEG amplitudes are about 1000 times smaller than the amplitude of an action potential

The wave's frequency indicates how often the wave cycles from its maximal amplitude to its minimal amplitude and back. The frequency is measured in hertz (Hz, or cycles per second) and may vary from 1 to 40 Hz or higher. Four distinct frequency ranges are characteristic of EEG patterns. In general, lower EEG frequencies indicate less responsive states, such as sleep, whereas higher frequencies indicate increased alertness. As we will see, one stage of sleep is an exception to this general relationship.

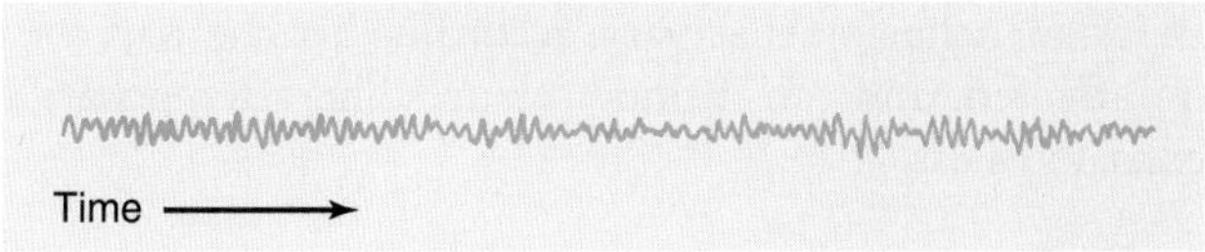

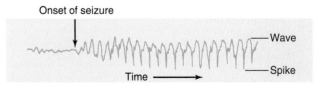

Figure 8–1

EEG patterns are wavelike. This represents a typical EEG recorded from the parietal or occipital lobe of an awake, relaxed person over approximately four seconds. EEG wave amplitudes are generally 20–100 μV, with a duration of about 50 msec.

The cause of the wavelike nature, or rhythmicity, of the EEG is not certain, nor is it known exactly where in the brain it originates. Current thinking is that clusters of neurons in the thalamus play a critical role; they provide a fluctuating output through nerve fibers leading from the thalamus to the cortex. This output, in turn, causes a rhythmic pattern of synaptic activity in the pyramidal neurons of the cortex. As noted previously, the cortical synaptic activity—not the activity of the deep thalamic structures—comprises most of a recorded EEG signal.

The synchronicity of the cortical synaptic activity (in other words, the amplitude of the EEG) does reflect the degree of synchronous firing of the thalamic neuronal clusters that are generating the EEG. These clusters receive input from other brain areas involved in controlling the conscious state. The purpose these oscillations serve in brain electrical activity is unknown. Theories range from the "idling hypothesis," which says that it is easier to get brain activity up and running from an "idle" as opposed to a "cold start," to the "epiphenomenon hypothesis," which says that the oscillations are simply the by-product of neuronal activity and have no functional significance at all.

The EEG is a useful clinical tool because wave patterns are abnormal over diseased or damaged brain areas (e.g., because of tumors, blood clots, hemorrhage, regions of dead tissue, and high or low blood sugar). Moreover, a shift from a less synchronized pattern of electrical activity (low-amplitude EEG) to a highly synchronized pattern can be a prelude to the electrical storm that signifies an epileptic seizure. *Epilepsy* is a common (occurs in about 1 percent of the population) neurological disease that appears in mild, intermediate, and severe forms. It is associated with abnormal synchronized discharges of cerebral neurons. These discharges are reflected in the EEG as recurrent waves having distinctive high amplitudes (up to 1000 μV) and individual spikes or combinations of spikes and waves (**Figure 8–2**). Epilepsy is also associated with stereotyped

Figure 8–2

Spike-and-wave pattern in the EEG of a patient during an epileptic seizure.

changes in behavior that vary according to the part of the brain affected, and can include seizures and a temporary loss of consciousness.

The Waking State

Behaviorally, the waking state is far from homogeneous, reflecting the wide variety of things you might be doing at any given moment. The most prominent EEG wave pattern of an awake, relaxed adult whose eyes are closed is a slow oscillation of 8 to 13 Hz known as the **alpha rhythm** (**Figure 8–3a**). The alpha rhythm is recorded best over the parietal and occipital lobes and is associated with decreased levels of attention. When alpha rhythms are generated, subjects commonly report that they feel relaxed and happy. However, people who normally experience more alpha rhythm than usual have not been shown to be psychologically different from those with less.

When people are attentive to an external stimulus or are thinking hard about something, the alpha rhythm is replaced by lower-amplitude, high-frequency (>13 Hz) oscillations, the **beta rhythm** (**Figure 8–3b**). This transformation, known as the **EEG arousal,** is associated with the act of paying attention to a stimulus rather than with the act of perception itself. For example, if people open their eyes in a completely dark room and try to see, EEG arousal occurs even though they perceive no visual input. With decreasing attention to repeated stimuli, the EEG pattern reverts to the alpha rhythm.

Sleep

The EEG pattern changes profoundly in sleep. As a person becomes increasingly drowsy, a decrease occurs in alpha-wave amplitude and frequency. When sleep actually occurs, the EEG shifts toward slower-frequency, higher-amplitude wave patterns known as the **theta rhythm** (4–8 Hz) and the **delta rhythm** (slower than 4 Hz) (**Figure 8–4**). Changes in posture, ease of arousal, threshold for sensory stimuli, and motor output accompany these EEG changes.

There are two phases of sleep whose names depend on whether or not the eyes move behind the closed eyelids:

NREM (nonrapid eye movement) and **REM** (rapid eye movement) sleep. The EEG waves during NREM sleep are of high amplitude and slow frequency, so NREM sleep is also referred to as slow-wave sleep. The initial phase of sleep—NREM sleep—is itself divided into four stages. Each successive stage is characterized by an EEG pattern with a slower frequency and higher amplitude than the preceding one (**Figure 8–4a**).

Sleep begins with the progression from stage 1 to stage 4, which normally takes 30 to 45 min. The process then reverses itself; the EEG ultimately resuming the low-voltage, high-frequency, asynchronous pattern characteristic of the alert, awake state (**Figure 8–4b**). Instead of the person waking, however, the behavioral characteristics of sleep continue at this time, but this sleep also includes rapid eye movement (REM sleep).

REM sleep is also called **paradoxical sleep** because the sleeper is difficult to arouse despite having an EEG characteristic of the alert, awake state. In fact, brain O_2 consumption is higher during REM sleep than during the NREM or awake states! When awakened during REM sleep, subjects generally report that they have been dreaming. This is true even in people who generally do not remember dreaming when they awaken on their own.

If uninterrupted, sleep continues in this cyclical fashion, tending to move from stages 1, 2, and 3, to 4 then back up from 4 to 3, 2, and 1, where NREM sleep is punctuated by an episode of REM sleep. Continuous recordings of adults show that the average total night's sleep comprises four or five such

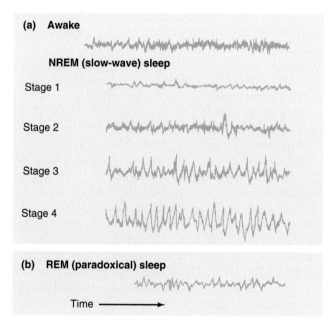

Figure 8–4

The EEG record of a person (a) passing from the awake state to deep sleep (stage 4) and (b) during REM sleep. The high-amplitude waves of the late stages of slow-wave sleep (theta and delta rhythms) demonstrate the synchronous activity pattern in cortical neurons. The desynchronous pattern during REM sleep is similar to that observed in awake individuals.

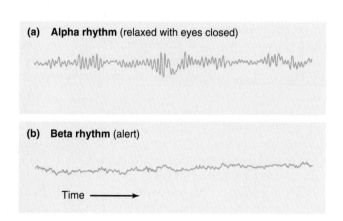

Figure 8–3

The (a) alpha and (b) beta rhythms shown on an EEG.

cycles, each lasting 90 to 100 min (**Figure 8–5**). In young adults, REM sleep constitutes 20 to 25 percent of the total sleeping time; this fraction tends to decline progressively with aging. The time spent in REM sleep increases toward the end of an undisturbed night. Initially, as one moves from drowsiness to stage 1 sleep, there is a considerable tension in the postural muscles, but the muscles become progressively more relaxed as NREM sleep progresses. Sleepers awakened during NREM sleep rarely report dreaming.

With several exceptions, skeletal muscle tension, already decreased during NREM sleep, is markedly inhibited during REM sleep. Exceptions occur in the eye muscles, which cause the rapid bursts of sweeping eye movements (see Figure 8–5), and the motor neurons to the muscles of respiration. In one form of the disease known as *sleep apnea,* however, stimulation of the respiratory muscles temporarily ceases, sometimes hundreds of times during a night. The resulting decreases in oxygen levels repeatedly awaken the apnea sufferer, who is deprived of slow-wave and REM sleep. As a result, this disease is associated with excessive—and sometimes dangerous—sleepiness during the day (refer to Chapter 13 for a more complete discussion of sleep apnea).

During the sleep cycle, many changes occur throughout the body, in addition to altered muscle tension. During NREM sleep, for example, there are pulsatile releases of growth hormone and the gonadotropic hormones from the anterior pituitary (Chapter 11), as well as decreases in blood pressure, heart rate, and respiratory rate. REM sleep is associated with an increase and irregularity in blood pressure, heart rate, and respiratory rate. Moreover, twitches of the facial muscles or limb muscles may occur (despite the generalized lack of skeletal muscle tone), as may erection of the penis and engorgement of the clitoris.

Although we spend about one-third of our lives sleeping, the functions of sleep are not completely understood. Many lines of research, however, suggest that it is a fundamental necessity of a complex nervous system. Sleep, or a sleep-like state, is a characteristic found throughout the animal kingdom, from organisms as simple as fruit flies to the most complex primates. Studies of sleep deprivation in humans and other animals suggest that it is a homeostatic requirement, similar to the need for food and water. Lack of sleep impairs the immune system, causes cognitive and memory deficits, and ultimately leads to psychosis and even death. Much of the sleep research on humans has focused on the importance of sleep for learning and memory formation. EEG studies show that during sleep the brain experiences reactivation of neural pathways stimulated during the prior awake state, and that subjects deprived of sleep show less effective memory retention. Based on these and other findings, many scientists believe that part of the restorative value of sleep lies in facilitating chemical and structural changes responsible for dampening the overall activity in the brain's neural networks while conserving and strengthening synapses in pathways associated with information that is important to learn and remember.

Table 8–1 summarizes the sleep states.

Neural Substrates of States of Consciousness

Periods of sleep and wakefulness alternate about once a day; that is, they manifest a circadian rhythm consisting on average of 8 h sleep and 16 h awake. Within the sleep portion of this circadian cycle, NREM sleep and REM sleep alternate, as we have seen. As we shift from the waking state through NREM sleep to REM sleep, attention shifts to internally generated stimuli (dreams) so that we are largely insensitive to external stimuli. Memory decreases (dreams are generally forgotten much faster than events we experience while awake), and postural muscles lose tone so our heads nod and we slump in our chairs. The tight rules for determining what is real become relaxed, allowing the bizarre happenings of our dreams.

What physiological processes drive these cyclic changes in states of consciousness? Nuclei in both the brainstem and hypothalamus are involved.

Recall from Chapter 6 that neurons of certain brainstem nuclei give rise to axons that diverge to affect wide areas of

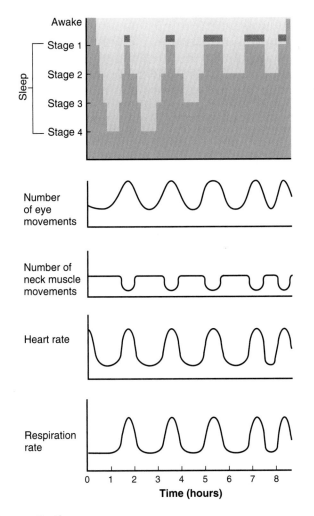

Figure 8–5

Schematic representations of the EEG and other physiological changes associated with the stages of sleep. Purple bars indicate periods of REM sleep. During periods of REM sleep, muscle movements decrease, while ocular movements, heart rate, and respiration rate increase.

Table 8–1 Sleep-Wakefulness Stages

Stage	Behavior	EEG (See Figures 8–3 and 8–4)
Alert wakefulness	Awake, alert with eyes open.	Beta rhythm (faster than 13 Hz).
Relaxed wakefulness	Awake, relaxed with eyes closed.	Mainly alpha rhythm (8–13 Hz) over the parietal and occipital lobes. Changes to beta rhythm in response to internal or external stimuli.
Relaxed drowsiness	Fatigued, tired, or bored; eyelids may narrow and close; head may start to droop; momentary lapses of attention and alertness. Sleepy but not asleep.	Decrease in alpha-wave amplitude and frequency.
NREM (slow-wave) sleep		
Stage 1	Light sleep; easily aroused by moderate stimuli or even by neck muscle jerks triggered by muscle stretch receptors as head nods; continuous lack of awareness.	Alpha waves reduced in frequency, amplitude, and percentage of time present; gaps in alpha rhythm filled with theta (4–8 Hz) and delta (slower than 4 Hz) activity.
Stage 2	Further lack of sensitivity to activation and arousal.	Alpha waves replaced by random waves of greater amplitude.
Stages 3 and 4	Deep sleep; in stage 4, activation and arousal occur only with vigorous stimulation.	Much theta and delta activity, predominant delta in stage 4.
REM (paradoxical) sleep	Deepest sleep; greatest relaxation and difficulty of arousal; begins 50–90 min after sleep onset, episodes repeated every 60–90 min, each episode lasting about 10 min; dreaming occurs, rapid eye movements behind closed eyelids; marked increase in brain O_2 consumption.	EEG resembles that of alert awake state.

the brain in a highly specific manner, forming a fiber system known as the reticular formation, also sometimes called the **reticular activating system (RAS).** This system is, in fact, composed of several separate divisions, distinguished by their anatomical distribution and neurotransmitters. The divisions originate in different nuclei within the brainstem, and some of them send fibers to those areas of the thalamus that influence the EEG. Components of the RAS that release norepinephrine, serotonin, or acetylcholine—functioning in this instance more as neuromodulators—are most involved in controlling the various states of consciousness.

One hypothesis about how sleep-wake cycles are generated proposes that alternating reciprocal activity of different types of neurons in the RAS causes shifts from one state to the other. In this model (**Figure 8–6**), the waking state and REM sleep are at opposite ends of a spectrum: During waking, the aminergic neurons (those that release norepinephrine or serotonin) dominate, and during REM sleep the cholinergic neurons are dominant. NREM sleep, according to this model, is intermediate between the two extremes.

The aminergic neurons, which are active during the waking state, facilitate a state of arousal, enhancing both attention to perceptions of the outer world and the motor activity that characterizes awake behavior. These neurons also inhibit certain of the cholinergic brainstem neurons. As the aminergic neurons stop firing, the cholinergic neurons, released from inhibition, increase their activity.

There are two additional areas in the forebrain (**Figure 8–7**) that are involved in the control of sleep-wake cycles. The preoptic area of the hypothalamus promotes slow-wave sleep by inhibitory GABAergic inputs to the thalamocortical neurons and to the midbrain reticular formation. It also inhibits activity within a center in the posterior hypothalamus that stimulates wakefulness. These latter neurons use histamine as a neurotransmitter and they also project to the RAS. The drowsiness that occurs in people using antihistamines may be a result of blocking the histaminergic transmission originating in the posterior hypothalamus.

Finally, the basic rhythm of the sleep-wake cycle is influenced by the biological clock function of the suprachiasmatic nucleus. This nucleus of the hypothalamus regulates the timing of sleep and awake periods relative to periods of light and darkness, that is, the circadian rhythm (Chapter 1) of the states of consciousness. The nucleus stimulates the pineal gland's production of melatonin (Chapters 1 and 11). Although melatonin has been used as a "natural" substance

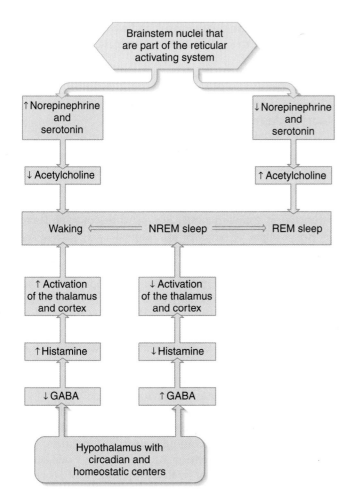

Figure 8–6

A model showing how alternating activity within neurons in the brainstem and hypothalamus may influence the differing states of consciousness. The changes in aminergic and cholinergic influence are discussed in the text.

Figure 8–6 physiological *inquiry*

- Explain why some drugs prescribed to treat allergic reactions cause drowsiness as a side effect.

Answer can be found at end of chapter.

for treating insomnia and jet lag, it has not yet proven effective as a sleeping pill. It has, however, been shown to induce lower body temperature, a key event in falling asleep.

In addition to these neurotransmitters, over 30 other chemical substances that affect sleep have been found in blood, urine, cerebrospinal fluid, and brain tissue. For example, interleukin 1, one of the cytokines in a family of intercellular messengers having an important role in the body's immune defense system (Chapter 18) fluctuates in parallel with normal sleep-wake cycles. However, none of these substances has been confirmed as a physiologically important factor in inducing sleep.

Coma and Brain Death

The term *coma* describes a severe decrease in mental function due to structural, physiological, or metabolic impairment of the brain. A person in a coma exhibits a sustained loss of the capacity for arousal even in response to vigorous stimulation. There is no outward behavioral expression of any mental function, the eyes are closed, and sleep-wake cycles disappear. Coma can result from extensive damage to the cerebral cortex, damage to the brainstem arousal mechanisms, interruptions of the connections between the brainstem and cortical areas, metabolic dysfunctions, brain infections, or an overdose of certain drugs, such as sedatives, sleeping pills, narcotics, or ethanol.

Patients in an irreversible coma often enter a ***persistent vegetative state*** in which sleep-wake cycles are present even though the patient is unaware of his surroundings. Individuals in a persistent vegetative state may smile, or cry, or seem to react to elements of their environment. However, there is no evidence that they can comprehend these behaviors.

But a coma—even an irreversible coma—is not equivalent to death. We are left, then, with the question: When is a person actually dead? This question often has urgent medical, legal, and social consequences. For example, with the need for viable tissues for organ transplantation it becomes imperative to know just when a person is "dead" so that the organs can be removed as soon after death as possible.

Brain death is widely accepted by doctors and lawyers as the criterion for death, despite the viability of other organs. Brain death occurs when the brain no longer functions and has no possibility of functioning again.

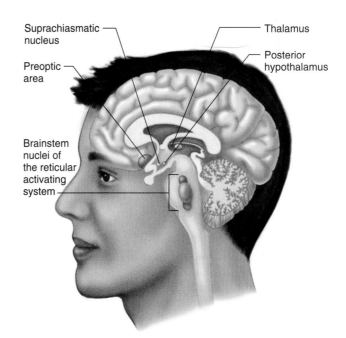

Figure 8–7

Brain structures involved in the sleep-wakefulness cycles. Cells within the thalamus are influenced by output from other areas and generate rhythmic firing patterns that produce the EEG.

The problem now becomes practical. How do we know when a person (e.g., someone in a coma) is brain dead? There is general agreement that the criteria listed in **Table 8–2**, if met, denote brain death. Notice that the cause of a coma must be known because comas due to drug poisoning are usually reversible. Also, the criteria specify that there be no evidence of functioning neural tissues above the spinal cord because fragments of spinal reflexes may remain for several hours or longer after the brain is dead. The criterion for lack of spontaneous respiration (apnea) can be difficult to check; if the patient is on a respirator, it is of course inadvisable to remove him or her for the 10-min test because of the danger of further brain damage due to lack of oxygen. Therefore, apnea is diagnosed if there is no spontaneous attempt to "fight" the respirator; that is, if the patient's reflexes do not drive respiration at a rate or depth different from those of the respirator.

Conscious Experiences

Having run the gamut of the states of consciousness from the awake, alert state to coma and brain death, we deal now with the conscious experiences during the awake state.

Conscious experiences are those things we are aware of—either internal, such as an idea, or external, such as an object or event. The most obvious aspect of this phenomenon is sensory awareness, but we are also aware of inner states such as fatigue, thirst, and happiness. We are aware of the passing of time, of what we are presently thinking about, and of consciously recalling a fact learned in the past. We are aware of

Table 8–2	Criteria for Brain Death

I. The nature and duration of the coma must be known.
   a. Known structural damage to brain or irreversible systemic metabolic disease.
   b. No chance of drug intoxication, especially from paralyzing or sedative drugs.
   c. No sign of brain function for 6 h in cases of known structural cause and when no drug or alcohol is involved; otherwise, 12–24 h without signs of brain function plus a negative drug screen.

II. Cerebral and brainstem function are absent.
   a. No response to painful stimuli other than spinal cord reflexes.
   b. Pupils unresponsive to light.
   c. No eye movement in response to ice-water stimulation of the vestibular reflex.
   d. Apnea (no spontaneous breathing) for 10 min.
   e. Systemic circulation may be intact.
   f. Purely spinal reflexes may be retained.

III. Supplementary (optional) criteria:
   a. Flat EEG (wave amplitudes less than 2 (μV).
   b. Responses absent in vital brainstem structures.
   c. Greatly reduced cerebral circulation.

reasoning and exerting self-control, and we are aware of directing our attention to specific events. Not least, we are aware of "self."

Basic to the concept of conscious experience is the question of attention.

Selective Attention

The term **selective attention** means avoiding the distraction of irrelevant stimuli while seeking out and focusing on stimuli that are momentarily important. Both voluntary and reflex mechanisms affect selective attention. An example of voluntary control of directed attention familiar to students is ignoring distracting events in a busy library while studying there.

An example of selective attention occurs when a novel stimulus is presented to a relaxed subject showing an alpha EEG pattern. This causes the EEG to shift to the beta rhythm. If the stimulus has meaning for the individual, behavioral changes also occur. The person stops what he or she is doing and looks around, listening intently and turning to face toward the stimulus source, a behavior called the **orienting response.** If the person is concentrating hard and is not distracted by the novel stimulus, the orienting response does not occur. It is also possible to focus attention on a particular stimulus without making any behavioral response.

For attention to be directed only toward stimuli that are meaningful, the nervous system must have the means to evaluate the importance of incoming sensory information. Thus, even before we focus attention on an object in our sensory world and become aware of it, a certain amount of processing has already occurred. This so-called **preattentive processing** directs our attention toward the part of the sensory world that is of particular interest and prepares the brain's perceptual processes for it.

If a stimulus is repeated but is found irrelevant, the behavioral response to the stimulus progressively decreases, a process known as **habituation.** For example, when a loud bell is sounded for the first time, it may evoke an orienting response because the person might be frightened by or curious about the novel stimulus. After several ringings, however, the individual makes progressively less response and eventually may ignore the bell altogether. An extraneous stimulus of another type or the same stimulus at a different intensity can restore the orienting response.

Habituation involves a depression of synaptic transmission in the involved pathway, possibly related to a prolonged inactivation of calcium channels in presynaptic axon terminals. Such inactivation results in a decreased calcium influx during depolarization and, therefore, a decrease in the amount of neurotransmitter released by a terminal in response to action potentials.

Neural Mechanisms for Selective Attention

Directing our attention to an object involves several distinct neurological processes. First, our attention must be disengaged from its present focus. Then, attention must be moved to the new focus. Attention must then be engaged at the new focus. Finally, there must be an increased level of arousal that produces prolonged attention to the new focus.

One of the areas that plays an important role in orienting and selective attention is in the brainstem, where the interaction of various sensory modalities in single cells can be detected experimentally. The receptive fields of the different modalities overlap. For example, a visual and auditory input from the same location in space will significantly enhance the firing rates of certain of these so-called multisensory cells, whereas the same type of stimuli originating at different places will have little effect on or may even inhibit their response. Thus, weak clues add together to enhance each other's significance so we pay attention to the event, whereas we might ignore an isolated small clue.

The locus ceruleus, a nucleus in the brainstem pons, which projects to the parietal cortex and to many other parts of the central nervous system, is also implicated in selective attention. The system of fibers leading from the locus ceruleus helps determine which brain area is to gain temporary predominance in the ongoing stream of the conscious experience. Norepinephrine, the transmitter these neurons release, acts as a neuromodulator to enhance the signals transmitted by certain sensory inputs so the difference increases between them and weaker signals. Thus, neurons of the locus ceruleus improve information processing during selective attention.

There are also multisensory neurons in association areas of the cerebral cortex (Chapter 6). Whereas the brainstem neurons are concerned with the orienting movements associated with paying attention to a specific stimulus, the cortical multisensory neurons are more involved in the perception of the stimulus. Neuroscientists are only beginning to understand how the various areas of the attentional system interact.

Some insights into neural mechanisms of selective attention are being gained from the study of individuals diagnosed with *attention deficit hyperactivity disorder (ADHD)*. This condition typically begins early in childhood and is the most common neurobehavioral problem in school-aged children (3 to 5 percent). ADHD is characterized by abnormal difficulty in maintaining selective attention, and/or impulsiveness and hyperactivity. Investigation has yet to reveal clear environmental causes, but there is good evidence for a genetic basis because ADHD tends to run in families. Functional imaging studies of the brains of children with ADHD have indicated dysfunction of brain regions in which catecholamine signaling is prominent, including the basal nuclei and prefrontal cortex. In support of this, the most effective medication used to treat ADHD is *methylphenidate* (Ritalin®), a drug that increases synaptic concentrations of dopamine and norepinephrine.

Neural Mechanisms of Conscious Experiences

All conscious experiences are popularly attributed to the workings of the "mind," a word that conjures up the image of a nonneural "me," a phantom interposed between afferent and efferent impulses. The implication is that the mind is something more than neural activity. Most neuroscientists agree, however, that the mind represents a summation of neural activity in the brain at any given moment and does not require anything more than the brain. Physiologists, however, have only a beginning understanding of the brain mechanisms that give rise to mind or to conscious experiences.

We will describe in this section how some neuroscientists have speculated about this problem. The thinking begins with the assumption that conscious experience requires neural processes—either graded potentials or action potentials—somewhere in the brain. At any moment, certain of these processes correlate with conscious awareness, and others do not. A key question here is: What is different about the processes we are aware of?

A further assumption is that the neural activity that corresponds to a conscious experience resides not in a single anatomical cluster of "consciousness neurons," but rather in a set of neurons that are temporarily functioning together in a specific way. Because we can become aware of many different things, we further assume that this grouping of neurons can vary—shifting, for example, among parts of the brain that deal with visual or auditory stimuli, memories or new ideas, emotions or language.

Consider the perception of a visual object. As we discussed in Chapter 7, different aspects of something we see are processed by different areas of the visual cortex—the object's color by one part, its motion by another, its location in the visual field by another, and its shape by still another—but we see *one* object. Not only do we perceive it; we may also know its name and function. Moreover, as we see an object, we can sometimes also hear or smell it, which requires participation of brain areas other than the visual cortex.

The simultaneous participation of different groups of neurons in a conscious experience can also be inferred for the olfactory system. Repugnant or alluring odors evoke different reactions, although they are both processed in the olfactory pathway. Neurons involved in emotion are also clearly involved in this type of perception.

Neurons from the various parts of the brain that simultaneously process different aspects of the information related to the object we see are said to form a "temporary set" of neurons. It is suggested that the synchronous activity of the neurons in the temporary set leads to conscious awareness of the object we are seeing.

As we become aware of still other events—perhaps a memory related to this object—the set of neurons involved in the synchronous activity shifts, and a different temporary set forms. In other words, it is suggested that specific relevant neurons in many areas of the brain function together to form the unified activity that corresponds to awareness.

What parts of the brain might be involved in such a temporary neuronal set? Clearly the cerebral cortex is involved. Removal of specific areas of the cortex abolishes awareness of only specific types of consciousness. For example, in a syndrome called *sensory neglect,* damage to association areas of the parietal cortex causes the injured person to neglect parts of the body or parts of the visual field as though they do not exist. Stroke patients with parietal lobe damage often do not acknowledge the presence of a paralyzed part of their body or will only be able to describe some but not all elements in a visual field. **Figure 8–8** shows an example of sensory neglect as evidenced in drawings made by a patient with parietal lobe damage on the right side of the brain. Patients such as these are completely unaware of the left-hand parts of the visual

Model   Patient's copy

Figure 8–8

Unilateral visual neglect in a patient with right parietal lobe damage. Although patients such as these are not impaired visually, they do not perceive part of their visual world. The drawings on the right were copied by the patient from the drawings on the left.

image. Subcortical areas such as the thalamus and basal ganglia may also be directly involved in conscious experience, but it seems that the hippocampus and cerebellum are not.

Saying that we can use one set of neurons and then shift to a new set at a later time may be the same as saying we can focus attention on—that is, bring into conscious awareness—one object or event and then shift our focus of attention to another object or event at a later time. Thus, the mechanisms of conscious awareness and attention must be intimately related.

Motivation and Emotion

Motivation is a factor in most, if not all, behaviors, while emotions accompany many of our conscious experiences. Motivated behaviors such as sexual behaviors play a part in controlling much of our day-to-day behavior, while emotions may help us to achieve the goals we set for ourselves as well as express our feelings.

Motivation

Those processes responsible for the goal-directed quality of behavior are the **motivations,** or "drives" for that behavior. Motivation can lead to hormonal, autonomic, and behavioral

responses. **Primary motivated behavior** is behavior related directly to homeostasis—that is, the maintenance of a relatively stable internal environment, such as getting something to drink when you are thirsty. In such homeostatic goal-directed behavior, specific body "needs" are satisfied. Thus, in our example, the perception of need results from a drop in body water concentration, and the correlate of need satisfaction is the return of body water concentration to normal. We will discuss the neurophysiological integration of much homeostatic goal-directed behavior later (thirst and drinking, Chapter 14; food intake and temperature regulation, Chapter 16).

In many kinds of behavior, however, the relation between the behavior and the primary goal is indirect. For example, the selection of a particular flavor of soft drink has little if any apparent relation to homeostasis. The motivation in this case is secondary. Much of human behavior fits into this latter category and is influenced by habit, learning, intellect, and emotions—factors that can be lumped together under the term "incentives." Often, it is difficult to distinguish between primary and secondary goals. For instance, you may find certain foods more appealing if you are deficient in some of the elements in that food. Sometimes the primary homeostatic goals and secondary goals conflict, as, for example, during a religious fast.

The concepts of reward and punishment are inseparable from motivation. Rewards are things that organisms work for or things that make the behavior that leads to them occur more often—in other words, positive reinforcers. Punishments are the opposite.

The neural system subserving reward and punishment is part of the reticular activating system, which you will recall arises in the brainstem and comprises several components. The component involved in motivation is known as the **mesolimbic dopamine pathway**—*meso-* because it arises in the midbrain (mesencephalon) area of the brainstem; *limbic* because it goes to areas of the limbic system, such as the prefrontal cortex, the nucleus accumbens, and the under-surface of the frontal lobe (**Figure 8–9**); and *dopamine* because its fibers release the neurotransmitter dopamine. The mesolimbic dopamine pathway is implicated in evaluating the availability of incentives and reinforcers (asking, "Is it worth it?" for example) and translating the evaluation into action.

Much of the available information concerning the neural substrates of motivation has been obtained by studying behavioral responses of animals to rewarding or punishing stimuli. One way in which this can be done is by using the technique of **brain self-stimulation.** In this technique, an unanesthetized experimental animal regulates the rate at which electric stimuli are delivered through electrodes implanted in discrete brain areas. The small electrical charges given to the brain cause the local neurons to depolarize, thus mimicking what might happen if these neurons were to fire spontaneously. The experimental animal is placed in a box containing a lever it can press (**Figure 8–10**). If no stimulus is delivered to the brain when the bar is pressed, the animal usually presses it occasionally at random.

If, in contrast, a stimulus is delivered to the brain as a result of a bar press, a different behavior occurs, depending on the location of the electrodes. If the animal increases the bar-

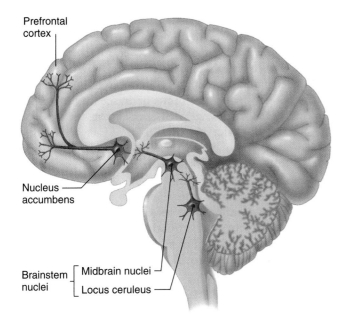

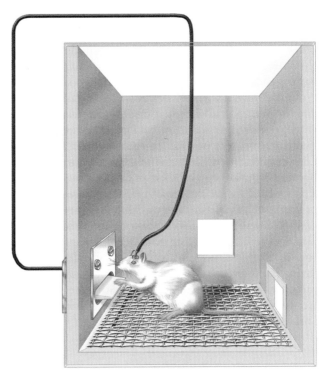

Figure 8–9

Schematic drawing of the mesolimbic dopamine pathway. Various psychoactive substances are thought to work in these areas to enhance brain reward.

Figure 8–10

Apparatus for self-stimulation experiments. Rats do not appear to be bothered by the implanted electrode. In fact, they work hard to get the electrical stimulation.

Adapted from Olds.

pressing rate above the level of random presses, the electric stimulus is by definition rewarding. If the animal decreases the press rate below the random level, the stimulus is punishing.

Thus, the rate of bar pressing with the electrode in different brain areas is taken to be a measure of the effectiveness of the reward or punishment. Different pressing rates are found for different brain regions.

Scientists expected the hypothalamus to play a role in motivation because the neural centers for the regulation of eating, drinking, temperature control, and sexual behavior are there (Chapter 6). Indeed, they found that brain self-stimulation of the lateral regions of the hypothalamus serves as a positive reward. Animals with electrodes in these areas have been known to bar press to stimulate their brains 2000 times per hour continuously for 24 h until they drop from exhaustion! In fact, electrical stimulation of the lateral hypothalamus is more rewarding than external rewards. Hungry rats, for example, often ignore available food for the sake of electrically stimulating their brains at that location.

Although the rewarding sites—particularly those for primary motivated behavior—are more densely packed in the lateral hypothalamus than anywhere else in the brain, self-stimulation can occur in a large number of brain areas. Thus, while the hypothalamus coordinates the sequenced hormonal, autonomic, and behavioral responses of motivated behaviors, motivated behaviors based on learning involve additional integrative centers, including the cortex and limbic system. Motivated behaviors also use integrating centers in the brainstem and spinal cord—in other words, all levels of the nervous system can be involved.

Chemical Mediators

Dopamine is a major neurotransmitter in the pathway that mediates the brain reward systems and motivation. For this reason, drugs that increase synaptic activity in the dopamine pathways increase self-stimulation rates—that is, they provide positive reinforcement. Amphetamines are an example of such a drug because they increase the presynaptic release of dopamine. Conversely, drugs such as chlorpromazine, an antipsychotic agent that blocks dopamine receptors and lowers activity in the catecholamine pathways, are negatively reinforcing. The catecholamines are also, as we will see, implicated in the pathways involved in learning. This is not unexpected because rewards and punishments are believed to constitute incentives for learning.

Emotion

Emotion can be considered in terms of a relation between an individual and the environment based on the individual's evaluation of the environment (is it pleasant or hostile, for example), disposition toward the environment (am I happy and attracted to the environment or fearful of it?), and the actual physical response to it. While analyzing the physiological bases of emotion, it is helpful to distinguish: (1) the anatomical sites where the emotional value of a stimulus is determined; (2) the hormonal, autonomic, and outward expressions and displays of response to the stimulus (so-called **emotional behavior**); and (3) the conscious experience, or **inner emotions,** such as feelings of fear, love, anger, joy, anxiety, hope, and so on.

Emotional behavior can be studied more easily than the anatomical systems or inner emotions because it includes responses that can be measured externally (in terms of behavior). For example, stimulation of certain regions of the lateral hypothalamus causes an experimental animal to arch its back, puff out the fur on its tail, hiss, snarl, bare its claws and teeth, flatten its ears, and strike. Simultaneously, its heart rate, blood pressure, respiration, salivation, and concentrations of plasma epinephrine and fatty acids all increase. Clearly, this behavior typifies that of an enraged or threatened animal. Moreover, the animal's behavior can be changed from savage to docile and back again simply by activating different areas of the limbic system, such as parts of the amygdala (**Figure 8–11**).

Emotional behavior includes such complex behaviors as the passionate defense of a political ideology and such simple actions as laughing, sweating, crying, or blushing. Emotional behavior is achieved by the autonomic and somatic nervous systems under the influence of integrating centers such as those we just mentioned, and provides an outward sign that the brain's "emotion systems" are activated.

The cerebral cortex plays a major role in directing many of the motor responses during emotional behavior (e.g., to approach or avoid a situation). Moreover, forebrain structures, including the cerebral cortex, account for the modulation, direction, understanding, or even inhibition of emotional behaviors.

Although limbic areas of the brain seem to handle inner emotions, there is no single "emotional system." The amygdala, a cluster of nuclei deep in the tip of each temporal lobe (see Figure 8–11), and the region of association cortex on the lower surface of the frontal lobe, however, are central to most emotional states (**Figure 8–12**). The amygdala interacts with other parts of the brain via extensive reciprocal connections that can influence emotions about external stimuli, decision making, memory, attention, homeostatic processes, and behavioral responses. For example, it sends output to the hypothalamus, which is central to autonomic and hormonal homeostatic processes.

The limbic areas have been stimulated in awake human beings undergoing neurosurgery. These patients reported vague feelings of fear or anxiety during periods of stimulation to certain areas. Stimulation of other areas induced pleasurable sensations, which the subjects found hard to define precisely. The cerebral cortex, however, elaborates the conscious experience of inner emotion.

Altered States of Consciousness

States of consciousness may be different from the commonly experienced ones like wakefulness and drowsiness. Other, more bizarre sensations, such as those occurring with hypnosis, mind-altering drugs, and certain diseases, are referred to as *altered states of consciousness.* These altered states are also characteristic of psychiatric illnesses.

Schizophrenia

One of the diseases that induces altered states of consciousness is *schizophrenia,* a disease in which information is not properly regulated in the brain. The amazingly diverse symptoms

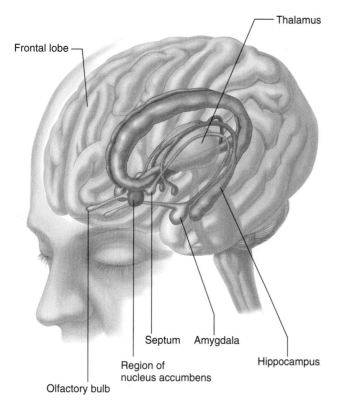

Figure 8–11

Brain structures involved in emotion, motivation, and the affective disorders. The limbic system is shaded purple.

From: *Brain, Mind, and Behavior* by Floyd E. Bloom and Arlyne Lazerson. Copyright © 1985, 1988 by Educational Broadcasting Corporation. Reprinted by permission of W. H. Freeman and Company.

Figure 8–11 physiological *inquiry* p*i*

■ What may have favored the evolution of emotions?

Answer can be found at end of chapter.

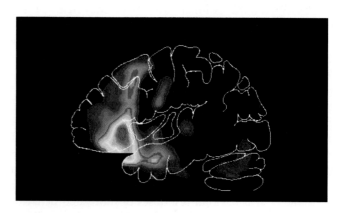

Figure 8–12

Computer image showing areas of neural activity during a sad thought.

Marcus E. Raichle, MD, Washington University School of Medicine.

of schizophrenia include hallucinations, especially "hearing" voices, and delusions, such as the belief that one has been chosen for a special mission or is being persecuted by others. Schizophrenics become withdrawn, are emotionally unresponsive, and experience inappropriate moods. They may also experience abnormal motor behavior, which can include total immobilization (*catatonia*). The symptoms vary from person to person.

The causes of schizophrenia remain unclear. Recent studies suggest that the disease reflects a developmental disorder in which neurons migrate or mature abnormally during brain formation. The abnormality may be due to a genetic predisposition or multiple environmental factors such as viral infections and malnutrition during fetal life or early childhood. The brain abnormalities involve diverse neural circuits and neurotransmitter systems that regulate basic cognitive processes. A widely accepted explanation for schizophrenia suggests that certain dopamine pathways are overactive. This hypothesis is supported by the fact that amphetamine-like drugs, which enhance dopamine signaling, make the symptoms worse, and by the fact that the most therapeutically beneficial drugs used in treating schizophrenia block dopamine receptors.

Schizophrenia affects approximately one in every 100 people and typically appears in the late teens or early twenties just as brain development nears completion. Currently there is no prevention or cure for the disease, although drugs can often control the symptoms. In a small number of cases, there has been complete recovery.

The Mood Disorders: Depressions and Bipolar Disorders

The term **mood** refers to a pervasive and sustained inner emotion that affects a person's perception of the world. In addition to being part of the conscious experience of the person, others can observe it. In healthy people, moods can be normal, elated, or depressed, and people generally feel that they have some degree of control over their moods. That sense of control is lost, however, in the *mood disorders,* which include depressive disorders and bipolar disorders. Along with schizophrenia, the mood disorders represent the major psychiatric illnesses today.

In the *depressive disorders (depression),* the prominent features are a pervasive sadness; a loss of energy, interest, or pleasure; anxiety; irritability; disturbed sleep; and thoughts of death or suicide. Depression can occur on its own, independent of any other illness, or it can arise secondary to other medical disorders. It is associated with decreased neuronal activity and metabolism in the anterior part of the limbic system and nearby prefrontal cortex. These same brain regions show abnormalities, albeit inconsistent ones, in bipolar disorders.

The term *bipolar disorders* describes swings between mania and depression. Episodes of *mania* are characterized by an abnormally and persistently elated mood, sometimes with euphoria (that is, an exaggerated sense of well-being), racing thoughts, excessive energy, overconfidence, and irritability.

Although the major biogenic amine neurotransmitters (norepinephrine, dopamine, and serotonin) and acetylcholine have all been implicated, the causes of the mood disorders are unknown.

Current treatment of the mood disorders emphasizes drugs and psychotherapy. The classical anti-depressant drugs are of three types. The *tricyclic antidepressant drugs* such as Elavil®, Norpramin®, and Sinequan® interfere with serotonin and/or norepinephrine reuptake by presynaptic endings. The *monoamine oxidase inhibitors* interfere with the enzyme responsible for the breakdown of these same two neurotransmitters. A third class of antidepressant drugs, the *serotonin-specific reuptake inhibitors (SSRIs),* are the most widely used antidepressant drugs and include Prozac®, Paxil®, and Zoloft®. As their name—SSRI—suggests, these drugs selectively inhibit serotonin reuptake by presynaptic terminals. In all three classes, the result is an increased concentration of serotonin and (except for the third class) norepinephrine in the extracellular fluid at synapses.

The biochemical effects of antidepressant medications occur immediately but the beneficial antidepressant effects appear only after several weeks of treatment. Thus, the known biochemical effect must be only an early step in a complex sequence that leads to a therapeutic effect of these drugs. Consistent with the long latency of the antidepressant effect is the recent evidence that the ultimate mechanism of these drugs may involve the growth of new neurons in the hippocampus. Chronic stress is a known trigger of depression in some people, and it has also been shown to inhibit neurogenesis in animals. In addition, careful measurements of the hippocampus in chronically depressed patients show that it tends to be smaller than in matched, nondepressed individuals. Finally, while antidepressant drugs normally have measurable effects on behavior in animal models of depression, it was recently shown that those effects disappear completely when steps are taken to prevent neurogenesis.

A major drug used in treating patients with bipolar disorder is the chemical element lithium, sometimes given in combination with anticonvulsant drugs. It is highly specific, normalizing both the manic and depressing moods and slowing down thinking and motor behavior without causing sedation. In addition, it decreases the severity of the swings between mania and depression that occur in the bipolar disorders. In some cases, lithium is even effective in depression not associated with mania. Lithium may interfere with the formation of signaling molecules of the inositol phosphate family, thereby decreasing the postsynaptic neurons' response to neurotransmitters that utilize this signal transduction pathway (Chapter 5).

Psychotherapy of various kinds can also be helpful in the treatment of depression. An alternative treatment when drug therapy and psychotherapy are not effective is *electroconvulsive therapy (ECT).* As the name suggests, pulses of electric current are used to activate a large number of neurons in the brain simultaneously, thereby inducing a convulsion, or seizure. The patient is under anesthesia and prepared with a muscle relaxant to minimize the effects of the convulsion on the musculoskeletal system. A series of ECT treatments alters neurotransmitter function by causing changes in the sensitivity of certain serotonin and adrenergic postsynaptic receptors. A modern, milder therapy that stimulates the brain with electromagnets is described at the end of this chapter. Another nondrug therapy used for the type of annual depression known as *seasonal affective depressive disorder (SADD)* is *phototherapy,* which exposes the patient to bright light for several hours per day during the winter months. Although light is thought to relieve depression by suppressing melatonin secretion from the pineal gland, as yet there is little evidence to support this claim.

Psychoactive Substances, Dependence, and Tolerance

In the previous sections, we mentioned several drugs used to combat altered states of consciousness. Psychoactive substances are also used as "recreational" drugs in a deliberate attempt to elevate mood and produce unusual states of consciousness ranging from meditational states to hallucinations. Virtually all the psychoactive substances exert their actions either directly or indirectly by altering neurotransmitter-receptor interactions in the biogenic amine—particularly dopamine—pathways. For example, the primary effect of cocaine comes from its ability to block the reuptake of dopamine into the presynaptic axon terminal. As mentioned in Chapter 6, psychoactive substances are often chemically similar to neurotransmitters such as dopamine, serotonin (**Figure 8–13**), and norepinephrine, and they interact with the receptors activated by these transmitters.

Dependence

Substance dependence, the term now preferred to *addiction,* has two facets that may occur either together or independently: (1) a *psychological dependence* that is experienced as a craving for a substance and an inability to stop using the substance at will; and (2) a *physical dependence* that requires one to take the substance to avoid *withdrawal,* which is the spectrum of unpleasant physiological symptoms that occurs with cessation of substance use. Substance dependence is diagnosed if three or more of the characteristics listed in **Table 8–3** occur within a 12-month period. **Table 8–4** lists the dependence-producing potential of various drugs.

Several neuronal systems are involved in substance dependence, but most psychoactive substances act on the mesolimbic dopamine pathway (Figure 8–9). In addition to the actions of this system mentioned earlier in the context of motivation and emotion, the mesolimbic dopamine pathway allows a person to experience pleasure in response to pleasurable events or in response to certain substances. Although the major neu-

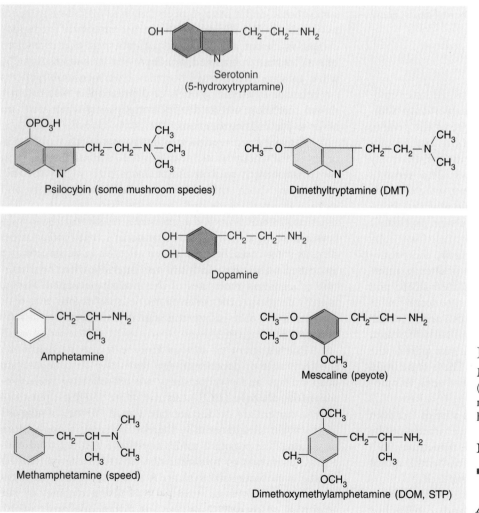

Figure 8–13

Molecular similarities between neurotransmitters (orange) and some substances that elevate mood. At high doses, these substances can cause hallucinations.

Figure 8–13 physiological *inquiry*

- How would you expect dimethyltryptamine (DMT) to affect sleeping behavior?

Answer can be found at end of chapter.

Table 8–3	Diagnostic Criteria for Substance Dependence

Substance dependence is indicated when three or more of the following occur within a 12-month period.

1. Tolerance, as indicated by
   a. a need for increasing amounts of the substance to achieve the desired effect, or
   b. decreasing effects when continuing to use the same amount of the substance.

2. Withdrawal, as indicated by
   a. appearance of the characteristic withdrawal symptoms upon terminating use of the substance, or
   b. use of the substance (or one closely related to it) to relieve or avoid withdrawal symptoms.

3. Use of the substance in larger amounts or for longer periods of time than intended.

4. Persistent desire for the substance; unsuccessful attempts to cut down or control use of the substance.

5. A great deal of time is spent in activities necessary to obtain the substance, use it, or recover from its effects.

6. Occupational, social, or recreational activities are given up or reduced because of substance use.

7. Use of the substance is continued despite knowledge that one has a physical or psychological problem that the substance is likely to exacerbate.

Adapted from DSM-IV, Diagnostic and Statistical Manual of Mental Disorders, 4th ed. American Psychiatric Association, Arlington, VA, 2000.

Table 8–4	Potential of Various Substances to Cause Dependence

If 100 people regularly use a substance, how many will become dependent on it?	
Nicotine	33
Heroin	25
Cocaine	16
Alcohol	15
Amphetamines	11
Marijuana	9

rotransmitter implicated in substance dependence is dopamine, other neurotransmitters, including GABA, enkephalin, serotonin, and glutamate, are also involved.

Tolerance

Tolerance to a substance occurs when increasing doses of the substance are required to achieve effects that initially occurred in response to a smaller dose. That is, it takes more of the substance to do the same job. Moreover, tolerance can develop to another substance as a result of taking the initial substance, a phenomenon called **cross-tolerance.** Cross-tolerance may develop if the physiological actions of the two substances are similar. Tolerance and cross-tolerance can occur with many classes of substances, not just psychoactive substances.

Tolerance may develop because the presence of the substance stimulates the synthesis of the enzymes that degrade it. As substance concentrations increase, so do the concentrations of these enzymes. Thus, more of the substance must be administered to produce the same plasma concentrations of the substance and, hence, the same initial effect.

Alternatively, tolerance can develop as a result of changes in the number and/or sensitivity of receptors that respond to

the substance, the amount or activity of enzymes involved in neurotransmitter synthesis, the reuptake transport molecules, or the signal transduction pathways in the postsynaptic cell.

Learning and Memory

Learning is the acquisition and storage of information as a consequence of experience. It is measured by an increase in the likelihood of a particular behavioral response to a stimulus. Generally, rewards or punishments are crucial ingredients of learning, as are contact with and manipulation of the environment. **Memory** is the relatively permanent storage form of learned information, although, as we will see, it is not a single, unitary phenomenon. Rather, the brain processes, stores, and retrieves information in different ways to suit different needs.

Memory

The term **memory encoding** defines the neural processes that change an experience into the memory of that experience—in other words, the physiological events that lead to memory formation. This section addresses three questions: Are there different kinds of memories, where do they occur in the brain, and what happens physiologically to make them occur?

New scientific facts about memory are being generated at a tremendous pace, and there is as yet no unifying theory as to how memory is encoded, stored, and retrieved. However, memory can be viewed in two broad categories: **Declarative memory** is the retention and recall of conscious experiences that can therefore be put into words (declared). One example is the memory of having perceived an object or event and, therefore, recognizing it as familiar and maybe even knowing the specific time and place when the memory originated. A second example would be one's general knowledge of the world, such as names and facts. The hippocampus, amygdala, and diencephalon—all parts of the limbic system—are required for the formation of declarative memories.

The second broad category of memory, **procedural memory,** can be defined as the memory of how to do things. In other words, it is the memory for skilled behaviors independent of any conscious understanding, as for example, riding a bicycle. Individuals can suffer severe deficits in declarative memory but have intact procedural memory. One case study describes a pianist who learned a new piece to accompany a singer at a concert but had no recollection the following morning of having performed the composition. He could remember how to play the music but could not remember having done so. The category of procedural memory also includes learned emotional responses, such as fear of spiders, and the classic example of Pavlov's dog, which learned to salivate at the sound of a bell after the dog had previously associated the bell with food. The primary areas of the brain involved in procedural memory are regions of sensorimotor cortex, the basal nuclei, and the cerebellum.

Another way to classify memory is in terms of duration—does it last for a long or only a short time? **Working memory,** also known as **short-term memory,** registers and retains incoming information for a short time—a matter of seconds to minutes—after its input. In other words, it is the memory that we use when we keep information consciously "in mind." For example, you might look up a number in the phone book and remember it only long enough to walk across the room and dial it. Working memory makes possible a temporary impression of one's present environment in a readily accessible form and is an essential ingredient of many forms of higher mental activity. Short-term memories may be converted into **long-term memories,** which may be stored for days to years and recalled at a later time. The process by which short-term memories become long-term memories is called **consolidation.**

Focusing attention is essential for many memory-based skills. The longer the span of attention in working memory, the better the chess player, the greater the ability to reason, and the better a student is at understanding complicated sentences and drawing inferences from texts. In fact, there is a strong correlation between working memory and standard measures of intelligence. Conversely, the specific memory deficit that occurs in the early stages of **Alzheimer's disease,** a condition marked by dementia and serious memory losses, may be in this attention-focusing component of working memory.

The Neural Basis of Learning and Memory

The neural mechanism and parts of the brain involved vary for different types of memory. Short-term encoding and long-term memory storage occur in different brain areas for both declarative and procedural memories (**Figure 8–14**).

But what is happening during memory formation on a cellular level? Conditions such as coma, deep anesthesia, electroconvulsive shock, and insufficient blood supply to the brain, all of which interfere with the electrical activity of the brain, also interfere with working memory. Thus, it is assumed that working memory requires ongoing graded or action potentials. Working memory is interrupted when a person becomes unconscious from a blow on the head, and memories are abolished for all that happened for a variable period of time before

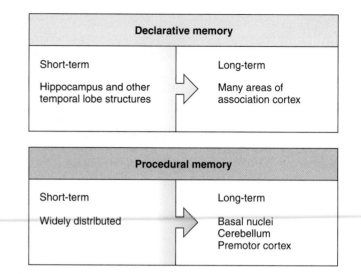

Figure 8–14

Brain areas involved in encoding and storage of declarative and procedural memories.

the blow, a condition called *retrograde amnesia.* (*Amnesia* is defined as the loss of memory.) Working memory is also susceptible to external interference, such as an attempt to learn conflicting information. On the other hand, long-term memory can survive deep anesthesia, trauma, or electroconvulsive shock, all of which disrupt the normal patterns of neural conduction in the brain. Thus, working memory requires electrical activity in the neurons.

Another type of amnesia is referred to as *anterograde amnesia.* It results from damage to the limbic system and associated structures, including the hippocampus, thalamus, and hypothalamus. Patients with this condition lose their ability to consolidate short-term declarative memories into long-term memories. While they can remember stored information and events that occurred before their brain injury, they can only retain anything that happens from that point forward in time as long as it exists in working memory. One particularly striking case report describes a woman whose husband died of a heart attack shortly after she suffered a stroke that permanently damaged her limbic system. At regular intervals during her recovery and thereafter, she would cheerfully inquire whether her husband would be coming for a visit, only to repeatedly re-experience the emotional shock and grief when told of his fate.

The problem of exactly how memories are stored in the brain is still unsolved, but some of the pieces of the puzzle are falling into place. One model for memory is **long-term potentiation (LTP),** in which certain synapses undergo a long-lasting increase in their effectiveness when they are heavily used. Review Figure 6–36, which details how this occurs at glutamatergic synapses. An analogous process, **long-term depression (LTD),** *decreases* the effectiveness of synaptic contacts between neurons. The mechanism of this suppression of activity appears to be mainly via changes in the channels in the postsynaptic membrane.

It is generally accepted that long-term memory formation involves processes that alter gene expression and result in the synthesis of new proteins. This is achieved by a cascade of second messengers that activate genes in the cell's DNA. These new proteins may be involved in the increased number of synapses that have been demonstrated after long-term memory formation. They may also be involved in structural changes in individual synapses (e.g., by an increase in the number of receptors on the postsynaptic membrane). This ability of neural tissue to change because of activation is known as **plasticity.**

Additional Facts Concerning Learning and Memory

Certain types of learning depend not only on factors such as attention, motivation, and various neurotransmitters, but also on certain hormones. For example, the hormones epinephrine, ACTH, and vasopressin affect the retention of learned experiences. These hormones are normally released in stressful or even mildly stimulating experiences, suggesting that the hormonal consequences of our experiences affect our memories of them.

Two of the opioid peptides, enkephalin and endorphin, interfere with learning and memory, particularly when the lesson involves a painful stimulus. They may inhibit learning simply because they decrease the emotional (fear, anxiety) component of the painful experience associated with the learning situation, thereby decreasing the motivation necessary for learning to occur.

Memories can be encoded very rapidly, sometimes after just one trial, and they can be retained over extended periods. Information can be retrieved from memory stores after long periods of disuse, and the common notion that memory, like muscle, atrophies with lack of use is not always true. Also, unlike working memory, memory storage apparently has an unlimited capacity because people's memories never seem so full that they cannot learn something new. Although we have mentioned specific areas of the brain that are active in learning, we want to stress that memory traces are laid down in specific neural systems throughout the brain, and different types of memory tasks utilize different systems. In even a simple memory task, such as trying to recall a certain word from a previously seen word list, different specific parts of the brain are activated in sequence. It is as though several small "processors" are linked together in a memory system for specific memory tasks.

Table 8–5 summarizes some general principles about learning and memory.

Cerebral Dominance and Language

The two cerebral hemispheres appear to be nearly symmetrical, but each has anatomical, chemical, and functional specializations. We have already mentioned that the left hemisphere deals with the somatosensory (Chapter 7) and motor (Chapter 10) functions of the right side of the body, and vice versa. In addition, in 90 percent of the population the left hemisphere is specialized to produce language—the conceptualization of

Table 8–5	General Principles of Learning and Memory

1. There are multiple memory systems in the brain.

2. Working memory requires changes in existing neural circuits, whereas long-term memory requires new protein synthesis and growth.

3. These changes may involve multiple cellular mechanisms within single neurons.

4. Second-messenger systems appear to play a role in mediating cellular changes.

5. Changes in the properties of membrane channels are often correlated with learning and memory.

Adapted from John M. Beggs et al. "Learning and Memory: Basic Mechanisms," in Michael J. Zigmond, Floyd E. Bloom, Story C. Landis, James L. Roberts, and Larry R. Squire, eds., *Fundamental Neuroscience,* Academic Press, San Diego, CA, 1999.

what one wants to say or write, the neural control of the act of speaking or writing, and recent verbal memory. This is even true of the sign language some deaf people use.

Language is a complex code that includes the acts of listening, seeing, reading, and speaking. The major centers for language function are in the left hemisphere in the temporal, parietal, and frontal cortex next to the **sylvian fissure,** which separates the temporal lobe from the frontal and parietal lobes (**Figure 8–15**). Each of the various regions deals with a separate aspect of language. For example, distinct areas are specialized for hearing, seeing, speaking, and generating words (**Figure 8–16**). There are even distinct brain networks for different categories of things, such as "animals" and "tools." The cerebellum is important in speaking and writing, which involve coordinated muscle contractions. Males and females typically use different brain areas for language processing, probably reflecting different strategies (**Figure 8–17**).

Much of our knowledge about how language is produced has been obtained from patients who have suffered brain damage and, as a result, have one or more defects in language, known as *aphasias.* For example, in most people, damage to the left cerebral hemisphere, but not to the right, interferes with the capacity for language manipulation, and damage to different areas of the left cerebral hemisphere affects language use differently.

Damage to the temporal region known as **Wernicke's area** (see Figure 8–15) generally results in aphasias that are more closely related to *comprehension*—the individuals have difficulty understanding spoken or written language even though their hearing and vision are unimpaired. Although they may have fluent speech, their speech is incomprehensible. In contrast, damage to **Broca's area,** the language area in the frontal cortex responsible for the articulation of speech, can cause *expressive* aphasias—the individuals have difficulty carrying out the coordinated respiratory and oral movements necessary for

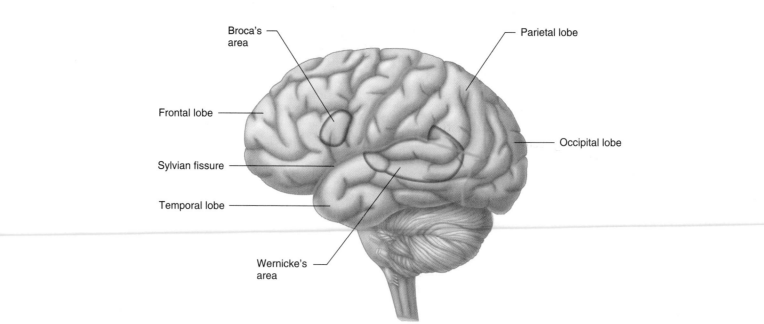

Figure 8–15

Areas found clinically to be involved in the comprehension (Wernicke's area) and motor (Broca's area) aspects of language. Blue lines indicate divisions of the cortex into frontal, parietal, temporal, and occipital lobes.

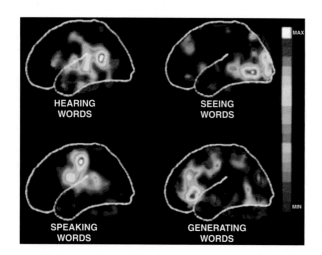

Figure 8–16

PET scans reveal areas of increased blood flow in specific parts of the temporal, occipital, parietal, and frontal lobes during various language-based activities.

Courtesy of Dr. Marcus E. Raichle.

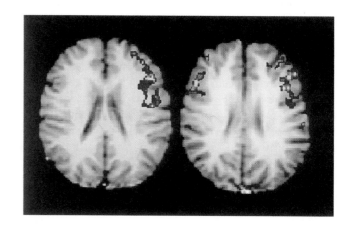

Figure 8–17

Images of the active areas of the brain in a male (left) and a female (right) during a language task. Note that both sides of the woman's brain are used in processing language, but the man's brain is more compartmentalized.

Shaywitz et al., 1995 NMR Research/Yale Medical School.

language even though they can move their lips and tongues. They understand spoken language and know what they want to say but have trouble forming words and putting them into grammatical order.

The potential for the development of language-specific mechanisms in the left hemisphere is present at birth, but the assignment of language functions to specific brain areas is fairly flexible in the early years of life. Thus, for example, damage to the language areas of the left hemisphere during infancy or early childhood causes temporary, minor language impairment until the right hemisphere can take over. However, similar damage acquired during adulthood typically causes permanent, devastating language deficits. By puberty, the brain's ability to transfer language functions to the right hemisphere is less successful, and often language skills are lost permanently.

Differences between the two hemispheres are usually masked by the integration that occurs via the corpus callosum and other pathways that connect the two sides of the brain. However, the separate functions of the left and right hemispheres have been uncovered by studying patients in which the two hemispheres have been separated surgically for treatment of severe epilepsy. These so-called **split brain** patients participated in studies in which they were asked to hold and identify an object such as a ball in their left or right hand behind a barrier that prevented them from seeing the object. Subjects who held the ball in their right hand, were able to say that it was a ball, but persons who held the ball in their left hand were unable to name it. Because the processing of sensory information occurs on the side of the brain opposite to the sensation, this result demonstrated conclusively that the left hemisphere contains a language center that is not present in the right hemisphere. These striking results provide the strongest evidence that at least some of the functions of the two hemispheres are distinct.

Although language skills emerge spontaneously in all normal children in all societies, there is a critical period during childhood when exposure to language is necessary for these skills to develop, just as the ability to see depends upon effective visual input early in life. The critical period is thought to end at puberty or earlier. The dramatic change at puberty in the possibility of learning language, or the ease of learning a second language, occurs as the brain attains its structural, biochemical, and functional maturity at that time.

As part of these basic language skills, the left hemisphere contains the "rules" for general grammatical principles; thus, it is much more skilled at changing verb tenses and constructing possessives than is the right hemisphere. It is also dominant for language usages that occur in sequences over time, such as speaking the first part of a word before the last. In addition, the left hemisphere seems to constantly form theories about how the world works, to find relationships between events, and to assess where one stands in relation to the world. The left hemisphere has been called "the interpreter."

The right hemisphere, on the other hand, more typically handles sensory information in basic ways, such as the perception of faces and other three-dimensional objects.

Memories are handled differently in the two hemispheres, too. Verbal memories are more apt to be associated with the left hemisphere, and nonverbal memories (e.g., visual patterns or nonverbal memories that convey emotions) with the right. Even the emotional responses of the two hemispheres seem to be different; for example, the left hemisphere has more ability to understand the emotional states of oneself or others. When electroconvulsive therapy is administered in the treatment of depression, however, better effects are often obtained when the electrodes are placed over the right hemisphere. The two sides of the brain also differ in their sensitivity to psychoactive drugs.

Conclusion

Mental tasks are the result of synchronized activity in vast neuronal networks made up of many functional regions of the cerebral cortex, subcortical nuclei, and brainstem. Also important are the pathways that reciprocally connect these sites and orchestrate their performance during specific tasks. In fact, the nervous system is so abundantly interconnected that it is difficult to know where any particular subsystem begins or ends.

Material in this chapter has often been highly qualified on a molecular level, largely because the information simply is not known. For example, the scientific literature is full of statements such as, "The study of the neural basis of language is in flux," and "The understanding of learning is one of nature's most closely guarded secrets." We also caution the reader that listing brain sites that show increased activity during substance dependence, sleep, language, or learning does little to explain how or why these phenomena occur. For the topics discussed in this chapter, the answer to the question "How?" which makes up the stuff of physiology, is often simply, "We don't yet know." However, this leaves us with the exciting prospect of future discoveries on the new frontier of the brain.

ADDITIONAL CLINICAL EXAMPLES

Limbic System Dysfunction

The case of H.M. illustrates that formation of declarative memories requires limbic structures of the temporal lobe. In 1953, a young man known as H. M. underwent bilateral removal of the amygdala and large parts of the hippocampus as a treatment for persistent, untreatable epilepsy. Although his epileptic condition improved after this surgery, he was now afflicted with anterograde amnesia. He still had a normal IQ and a normal working memory. He could retain information for minutes as long as he was not distracted; however, he could not form long-term memories. If he was introduced to someone on one day, on the next day he did not recall having previously met that person. Nor could he remember any events that occurred after his surgery, although his memory for events prior to the surgery was intact. Interestingly, H. M. had normal procedural memory, and could learn new puzzles and motor tasks as readily as normal individuals. This case was the first to draw attention to the critical importance of the temporal lobe in the formation of long-term declarative memories and to suggest that structures in this region are necessary for the conversion of short-term into long-term memories. Additional cases now demonstrate that the hippocampus is the primary structure involved in this process. Because H. M. retained memories from before the surgery, his case showed that the hippocampus is not involved in the *storage* of declarative memories.

The case of S. M. illustrates that the amygdala processes fearful emotions. The patient in this study suffered from a rare disease (**Urbach-Wiethe** disease) in which the anterior and medial portions of the temporal lobe atrophied, essentially destroying the amygdala bilaterally. Intelligence and memory formation remained intact. However, this individual now lacked the ability to express fear in appropriate situations and could not recognize fearful expressions in other people. Therefore, in humans the amygdala is important for at least one emotion—fear.

Repetitive Transcranial Magnetic Stimulation

Although shocking the brain with electroconvulsive therapy is certainly effective in treating depression and other mental disorders, the use of this therapy has been limited due to the requirement for general anesthesia and the potential for side effects. ***Repetitive transcranial magnetic stimulation (rTMS)*** shows great promise not only as a gentler alternative to shock therapy, but also as a powerful research tool for exploring the human brain. (Do not confuse this technology with the practice of placing ordinary magnets on the skin to treat various ailments, which has no scientifically documented basis.)

In rTMS, circular or figure-eight shaped metallic coils are placed against the skull overlying specific brain regions, and then brief, powerful electrical currents are applied at frequencies between 1 and 25 pulses per second. The resulting magnetic field induces current to flow through cortical neuronal networks directly beneath the coil. The immediate effect is similar to shock therapy: Neural activity is transiently disordered or sometimes silenced in that brain region. However, no anesthesia is required and no pain, convulsion, or memory loss occurs. Depending on the frequency and treatment regimen applied, the lasting effects of rTMS can cause either an increase or a decrease in the overall activity of the targeted area.

This technology provides a powerful, noninvasive research tool for studying brain function in normal, healthy people. Functional mapping of the human cortex is an example of how rTMS is being used in basic research. For example, placing the coil just above a person's left ear will cause the muscles of the right thumb to twitch. Such functional associations have historically been demonstrable only in subjects with their brains exposed during surgical procedures. Researchers are currently using rTMS to investigate cortical pathways involved in vision, language, learning, and drug reactions.

Electrical stimulation through rTMS also shows great promise as a therapeutic tool for the treatment of some mental disorders. In recent clinical trials, two to four weeks of daily rTMS stimulation of the left prefrontal cortex resulted in marked improvement of patients with major depression who hadn't responded to medication. Medical scientists are hopeful that refinements in technique will also lead to breakthroughs in the treatment of obsessive-compulsive disorder, mania, schizophrenia, and other psychiatric illnesses.

Head Trauma and Conscious State

Knowing what we do about consciousness and cognitive function, it is not surprising that damage to the cerebral cortex might result in confused and disoriented behaviors and even epilepsy-like seizures and coma. ***Concussion*** is the usually brief loss of consciousness that occurs after injury due to a blow to the head, a fall, or some other form of blunt trauma. Pupillary constriction in response to light is normal. When the patient awakens, he or she is usually confused for several hours, exhibiting amnesia for events that occurred immediately before and after the injury. In contrast, ***intracranial hemorrhage*** results from damage to vasculature in and around the brain and can be associated with skull fracture, violent shaking, and sudden accelerative forces such as those that would occur during an automobile accident. Blood may collect between the skull and the dura mater (an ***epidural hematoma***), or between the arachnoid mater and the surrounding meninges or within the brain (***subdural hematoma***). Intracranial hemorrhage often occurs without loss of consciousness, and symptoms such as headache, motor dysfunction, and loss of pupillary reflexes may not occur until several hours or days afterward. It is treated when necessary by drainage of the blood from the affected area. One reason why it is important to watch the behavior of a person with concussion is to recognize whether the initial trauma has resulted in an intracranial hemorrhage. ■

■ SUMMARY

States of Consciousness

I. The electroencephalogram provides one means of defining the states of consciousness.
   a. Electrical currents in the cerebral cortex due predominantly to summed postsynaptic potentials are recorded as the EEG.
   b. Slower EEG waves correlate with less responsive behaviors.
   c. Rhythm generators in the thalamus are probably responsible for the wavelike nature of the EEG.
   d. EEGs are used to diagnose brain disease and damage.
II. Alpha rhythms and, during EEG arousal, beta rhythms characterize the EEG of an awake person.
III. NREM sleep progresses from stage 1 (faster, lower-amplitude waves) through stage 4 (slower, higher-amplitude waves) and then back again, followed by an episode of REM sleep. There are generally four or five of these cycles per night.

IV. Aminergic and cholinergic brainstem centers, via their projections forward to the cerebrum as components of the reticular activating system, interact with the thalamus to regulate the sleep-wake cycles. Hypothalamic nuclei also play a role.

Conscious Experiences

I. Brain structures involved in selective attention determine which brain areas gain temporary predominance in the ongoing stream of conscious experience.
II. Conscious experiences may occur because a set of neurons temporarily function together, with the neurons that comprise the set changing as the focus of attention changes.

Motivation and Emotion

I. Behaviors that satisfy homeostatic needs are primary motivated behaviors. Behavior not related to homeostasis is a result of secondary motivation.

a. Repetition of a behavior indicates it is rewarding, and avoidance of a behavior indicates it is punishing.

b. The mesolimbic dopamine pathway, which goes to prefrontal cortex and parts of the limbic system, mediates emotion and motivation.

c. Dopamine is the primary neurotransmitter in the brain pathway that mediates motivation and reward.

II. Three aspects of emotion—anatomical and physiological bases for emotion, emotional behavior, and inner emotions—can be distinguished. The limbic system integrates inner emotions and behavior.

Altered States of Consciousness

I. Hyperactivity in a brain dopaminergic system is implicated in schizophrenia.

II. The mood disorders are caused, at least in part, by disturbances in transmission at brain synapses mediated by dopamine.

III. Many psychoactive drugs, which are often chemically related to neurotransmitters, result in substance dependence, withdrawal, and tolerance. The mesolimbic dopamine pathway is implicated in substance abuse.

Learning and Memory

I. The brain processes, stores, and retrieves information in different ways to suit different needs.

II. Memory encoding involves cellular or molecular changes specific to different memories.

III. Declarative memories are involved in remembering facts and events. Procedural memories are memories of how to do things.

IV. Short-term memories are converted into long-term memories by a process known as consolidation.

V. Prefrontal cortex and limbic regions of the temporal lobe are important brain areas for some forms of memory.

VI. Formation of long-term memory probably involves changes in second-messenger systems and protein synthesis.

Cerebral Dominance and Language

I. The two cerebral hemispheres differ anatomically, chemically, and functionally. In 90 percent of the population, the left hemisphere is superior at producing language and in performing other tasks that require rapid changes over time.

II. The development of language functions occurs in a critical period that closes at puberty.

III. After damage to the dominant hemisphere, the opposite hemisphere can acquire some language function—the younger the patient, the greater the transfer of function.

Conclusion

I. Many brain areas are involved in the performance of even simple mental tasks.

II. Each function is localized to a specific brain area, but, because many units are involved, widely distributed brain areas take part in mental tasks.

III. Little is known about how consciousness and behavior are actually formed in the brain.

Additional Clinical Examples

I. Structures in the temporal lobe are necessary for memory formation and for the expression and recognition of fear.

II. Repetitive transcranial magnetic stimulation uses electromagnetic pulses to noninvasively stimulate regions of the cortex. It is under investigation for treating depression and other disorders.

III. Head trauma can result in severe neurological dysfunction, and diagnosis of the type of brain injury depends in part on whether the patient loses consciousness immediately after the accident.

■ KEY TERMS

alpha rhythm 234
beta rhythm 234
brain self-stimulation 240
Broca's area 247
conscious experience 233
consolidation 246
declarative memory 245
delta rhythm 234
EEG arousal 234
electroencephalogram (EEG) 233
emotional behavior 241
habituation 238
inner emotion 241
learning 245
long-term depression (LTD) 246
long-term memory 246
long-term potentiation (LTP) 246
memory 245
memory encoding 245
mesolimbic dopamine pathway 240
mood 243
motivation 240
NREM sleep 234
orienting response 238
paradoxical sleep 234
plasticity 247
preattentive processing 238
primary motivated behavior 240
procedural memory 246
REM sleep 234
reticular activating system (RAS) 236
selective attention 238
short-term memory 246
states of consciousness 233
sylvian fissure 247
theta rhythm 234
Wernicke's area 247
working memory 246

■ CLINICAL TERMS

altered states of consciousness 242
Alzheimer's disease 246
amnesia 246
anterograde amnesia 246
aphasia 247
attention deficit hyperactivity disorder (ADHD) 239
bipolar disorder 243
brain death 237
catatonia 243
coma 237
concussion 250
cross-tolerance 245
depressive disorder (depression) 243
electroconvulsive therapy (ECT) 243
epidural hematoma 250
epilepsy 233
intracranial hemorrhage 250
mania 243
methylphenidate 239
monoamine oxidase inhibitor 243
mood disorder 243
persistent vegetative state 237
phototherapy 244
physical dependence 244
psychological dependence 244
repetitive transcranial magnetic stimulation (rTMS) 250
retrograde amnesia 246
schizophrenia 242
seasonal affective depressive disorder (SADD) 244
sensory neglect 239
serotonin-specific reuptake inhibitors (SSRI) 243
sleep apnea 235
split brain 249
subdural hematoma 250
substance dependence 244
tolerance 245
tricyclic antidepressant drug 243
Urback-Wiethe disease 249
withdrawal 244

1. State the two criteria used to define one's state of consciousness.
2. What type of neural activity is recorded as the EEG?
3. Draw EEG records that show alpha and beta rhythms, the stages of NREM sleep, and REM sleep. Indicate the characteristic wave frequencies of each.
4. Distinguish NREM sleep from REM sleep.
5. Briefly describe a neural mechanism that determines the states of consciousness.
6. Name the criteria used to distinguish brain death from coma.
7. Describe the orienting response as a form of directed attention.
8. Distinguish primary from secondary motivated behavior.
9. Explain how rewards and punishments are anatomically related to emotions.
10. Explain what brain self-stimulation can tell about emotions and rewards and punishments.
11. Name the primary neurotransmitter that mediates the brain reward systems.
12. Distinguish inner emotions from emotional behavior. Name the brain areas involved in each.
13. Describe the role of the limbic system in emotions.
14. Name the major neurotransmitters involved in schizophrenia and the mood disorders.
15. Describe a mechanism that could explain tolerance and withdrawal.
16. Distinguish the types of memory.

Chapter 8 Test Questions

(Answers appear in Appendix A.)

Match the following states of consciousness with the electroencephalogram patterns described below. (Use each once.)
   a. relaxed, awake, eyes closed
   b. stage 4 nonrapid eye movement (NREM) sleep
   c. rapid eye movement (REM) sleep
   d. epileptic seizure

1. Very high-amplitude, recurrent waves, associated with sharp spikes

2. Low-amplitude, high-frequency waves, similar to the attentive awake state

3. Irregular, slow frequency, high-amplitude, "alpha" rhythm

4. Regular, very slow frequency, very high-amplitude "delta" rhythm

5. Which correctly associates neurotransmitter activity with a state of consciousness?
   a. low GABA, high histamine: REM sleep
   b. low acetylcholine, high serotonin: REM sleep
   c. high acetylcholine, low norepinephrine: waking state
   d. high GABA, low histamine: REM sleep
   e. high GABA, low histamine: NREM sleep

6. Which best describes "habituation?"
   a. seeking out and focusing on momentarily important stimuli
   b. decreased behavioral response to a persistent irrelevant stimulus
   c. halting current activity and orienting toward a novel stimulus
   d. evaluation of the importance of sensory stimuli that occurs prior to focusing attention
   e. strengthening of synapses that are repeatedly stimulated during learning

7. The mesolimbic dopamine pathway is most closely associated with
   a. shifting between states of consciousness.
   b. emotional behavior.
   c. motivation and reward behaviors.
   d. perception of fear.
   e. primary visual perception.

8. Antidepressant medications most commonly target what neurotransmitter?
   a. acetylcholine
   b. dopamine
   c. histamine
   d. serotonin
   e. glutamate

9. Which is a true statement about memory?
   a. Consolidation converts short-term memories into long-term memories.
   b. Working memory stores information for years, perhaps indefinitely.
   c. In retrograde amnesia, the ability to form new memories is lost.
   d. The cerebellum is an important site of storage for declarative memory.
   e. Destruction of the hippocampus erases all previously stored memories.

10. Broca's area
   a. is in the parietal association cortex, and is responsible for language comprehension.
   b. is in the right frontal lobe, and is responsible for memory formation.
   c. is in the left frontal lobe, and is responsible for articulation of speech.
   d. is in the occipital lobe, and is responsible for interpreting body language.
   e. is part of the limbic system, and is responsible for the perception of fear.

Chapter 8 Quantitative and Thought Questions

(Answers appear in Appendix A.)

1. Explain why patients given drugs to treat Parkinson's disease (Chapters 6 and 10) sometimes develop symptoms similar to those of schizophrenia.

2. Explain how clinical observations of individuals with various aphasias help physiologists understand the neural basis of language.

Chapter 8 Answers to Physiological Inquiries

Figure 8–6 Among the drugs used to treat allergic reactions are antihistamines. They are prescribed because of their ability to block histamine's contributions to the inflammatory response, which include vasodilation and leakiness of small blood vessels (see Table 18–12). Because a decrease in histamine is associated with the induction of NREM sleep, drowsiness is a common side effect of antihistamines. Fortunately, antihistamines have been developed that do not cross the blood-brain barrier, and thus do not have this side effect, (e.g., loratadine [Claritin®, Alavert®]).

Figure 8–11 There are many ways in which emotions could potentially contribute to survival and reproduction. The perception of fear aids survival by stimulating avoidance or caution in potentially dangerous situations, like coming into contact with potentially venomous spiders or snakes, or walking near the edge of a high cliff. Our tendency to be disgusted by the smell of rotting food and fecal matter may have evolved as a protection against infection by potentially harmful bacteria or pathogens. Anger and rage could contribute to both survival and reproduction by facilitating our ability to fight for mates or territory, or for self-defense. Emotions like happiness and love may have evolved to encourage the safety of kinships and pair-bonding with mates.

Figure 8–13 An increase in serotonin concentrations is associated with the waking state (refer back to Figure 8–6), so sleep is inhibited by DMT and other drugs that simulate serotonin action. For this same reason, sleeplessness is also a common side effect of antidepressant medications discussed earlier in the text (e.g., serotonin-specific reuptake inhibitors) because they increase serotonin levels in the brain.

Colorized scanning electron micrograph (SEM) of freeze-fractured muscle fibers.

the ability to use chemical energy to produce force and movement is present to a limited extent in most cells, but in muscle cells it has become dominant. Muscles generate force and movements used to regulate the internal environment, and they also produce movements in the external environment. In humans, the ability to communicate, whether by speech, writing, or artistic expression, also depends on muscle contractions. Indeed, it is only by controlling muscle activity that the human mind ultimately expresses itself.

Three types of muscle tissue can be identified on the basis of structure, contractile properties, and control mechanisms—skeletal muscle, smooth muscle, and cardiac muscle. Most skeletal muscle, as the name implies, is attached to bone, and its contraction is responsible for supporting and moving the skeleton. The contraction of skeletal muscle is initiated by impulses in the neurons to the muscle and is usually under voluntary control.

Sheets of smooth muscle surround various hollow organs and tubes,

including the stomach, intestines, urinary bladder, uterus, blood vessels, and airways in the lungs. Contraction of the smooth muscle surrounding hollow organs may propel the luminal contents through the organ, or it may regulate internal flow by changing the tube diameter. In addition, small bundles of smooth muscle cells are attached to the hairs of the skin and iris of the eye. The autonomic nervous system, hormones, autocrine/paracrine agents, and other local chemical signals control smooth muscle contraction. Some smooth muscles contract autonomously, however, even in the absence of such signals. In contrast to skeletal muscle, smooth muscle is not normally under voluntary control.

Cardiac muscle is the muscle of the heart. Its contraction propels blood through the circulatory system. Like smooth muscle, it is regulated by the autonomic nervous system, hormones, and autocrine/paracrine agents, and it can undergo spontaneous contractions.

Although there are significant differences among these three types of muscle, the force-generating mechanism is similar in all of them. This chapter will describe skeletal muscle first, followed by smooth and cardiac muscle. Cardiac muscle, which combines some of the properties of both skeletal and smooth muscle, will be described in more depth in Chapter 12 in association with its role in the circulatory system.

SECTION A — Skeletal Muscle

Structure

The most striking feature seen when viewing **skeletal muscle** through a microscope is a distinct series of alternating light and dark bands perpendicular to the long axis. Because **cardiac muscle** shares this characteristic striped pattern, these two types are both referred to as **striated muscle.** The third basic muscle type, **smooth muscle,** derives its name from the fact that it lacks this striated appearance. **Figure 9–1** compares the appearance of skeletal muscle cells to cardiac and smooth muscle cells.

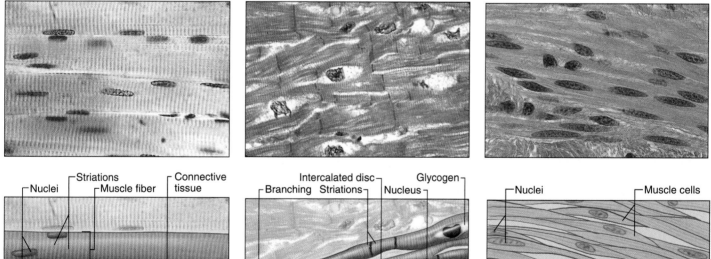

(a) Skeletal muscle **(b)** Cardiac muscle **(c)** Smooth muscle

Figure 9–1

Comparison of skeletal muscle (a) to cardiac (b) and smooth (c) muscle. Both skeletal and cardiac muscle have a striated appearance. Cardiac and smooth muscle cells tend to have a single nucleus, while skeletal muscle fibers are multinucleated.

A single skeletal muscle cell is known as a **muscle fiber.** Each muscle fiber is formed during development by the fusion of a number of undifferentiated, mononucleated cells, known as **myoblasts,** into a single cylindrical, multinucleated cell. Skeletal muscle differentiation is completed around the time of birth, and these differentiated fibers continue to increase in size from infancy to adulthood, but no new fibers are formed from myoblasts. Adult skeletal muscle fibers have diameters between 10 and 100 μm and lengths that may extend up to 20 cm.

If skeletal muscle fibers are destroyed after birth as a result of injury, they cannot be replaced by the division of other existing muscle fibers. New fibers can be formed, however, from undifferentiated cells known as **satellite cells,** which are located adjacent to the muscle fibers and undergo differentiation similar to that followed by embryonic myoblasts. This capacity for forming new skeletal muscle fibers is considerable but will generally not restore a severely damaged muscle to full strength. Much of the compensation for a loss of muscle tissue occurs through an increase in the size **(hypertrophy)** of the remaining muscle fibers.

The term **muscle** refers to a number of muscle fibers bound together by connective tissue (**Figure 9–2**). The relationship between a single muscle fiber and a muscle is analogous to that between a single neuron and a nerve, which is composed of the axons of many neurons. Skeletal muscles are usually attached to bones by bundles of collagen fibers known as **tendons.**

In some muscles, the individual fibers extend the entire length of the muscle, but in most, the fibers are shorter, often oriented at an angle to the longitudinal axis of the muscle. The transmission of force from muscle to bone is like a number of people pulling on a rope, each person corresponding to a single muscle fiber and the rope corresponding to the connective tissue and tendons.

Some tendons are very long, with the site where the tendon attaches to the bone far removed from the end of the muscle.

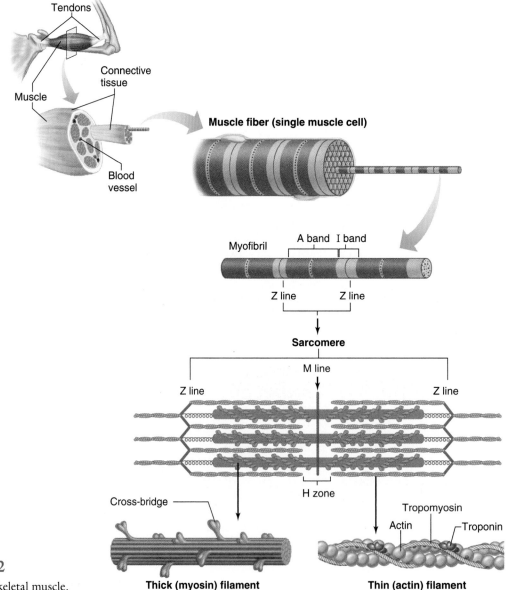

Figure 9–2

Structure of skeletal muscle.

For example, some of the muscles that move the fingers are in the forearm (wiggle your fingers and feel the movement of the muscles in your lower arm). These muscles are connected to the fingers by long tendons.

The striated pattern in skeletal (and cardiac) muscle results from the arrangement of numerous thick and thin filaments in the cytoplasm into **myofibrils,** approximately cylindrical bundles 1 to 2 μm in diameter (see Figure 9–2). Most of the cytoplasm of a fiber is filled with myofibrils, each extending from one end of the fiber to the other and linked to the tendons at the ends of the fiber.

The thick and thin filaments in each myofibril (see Figure 9–2 and **Figure 9–3**) are arranged in a repeating pattern along the length of the myofibril. One unit of this repeating pattern is known as a **sarcomere** (Greek, *sarco,* muscle; *mer,* part). The **thick filaments** are composed almost entirely of the protein **myosin.** The **thin filaments** (which are about half the diameter of the thick filaments) are principally composed of the protein **actin,** as well as two other proteins— **troponin** and **tropomyosin**—that play important roles in regulating contraction.

The thick filaments are located in the middle of each sarcomere, where their orderly, parallel arrangement produces a wide, dark band known as the **A band** (see Figures 9–2 and 9–3). Each sarcomere contains two sets of thin filaments, one at each end. One end of each thin filament is anchored to a network of interconnecting proteins known as the **Z line,** whereas the other end overlaps a portion of the thick filaments. Two successive Z lines define the limits of one sarcomere. Thus, thin filaments from two adjacent sarcomeres are anchored to the two sides of each Z line. (The term "line" refers to the appearance of these structures in two dimensions. Because myofibrils are cylindrical, it is more realistic to think of them as "Z disks.")

A light band known as the **I band** (see Figures 9–2 and 9–3) lies between the ends of the A bands of two adjacent sarcomeres and contains those portions of the thin filaments that do not overlap the thick filaments. The I band is bisected by the Z line.

Two additional bands are present in the A-band region of each sarcomere (see Figure 9–3). The **H zone** is a narrow, light band in the center of the A band. It corresponds to the space between the opposing ends of the two sets of thin filaments in each sarcomere. A narrow, dark band in the center of the H zone, known as the **M line** (also technically a disk), corresponds to proteins that link together the central region of adjacent thick filaments. In addition, filaments composed of the elastic protein **titin** extend from the Z line to the M line and are linked to both the M-line proteins and the thick filaments. Both the M-line linkage between thick filaments and the titin filaments act to maintain the alignment of thick filaments in the middle of each sarcomere.

A cross section through the A bands (**Figure 9–4**), shows the regular arrangement of overlapping thick and thin filaments. Each thick filament is surrounded by a hexagonal array of six thin filaments, and each thin filament is surrounded by a triangular arrangement of three thick filaments. Altogether there are twice as many thin as thick filaments in the region of filament overlap.

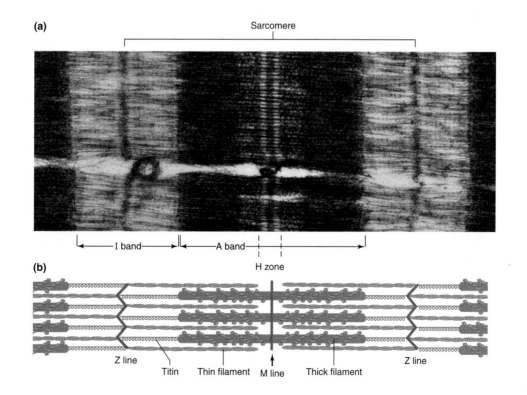

Figure 9–3

(a) High magnification of a sarcomere within myofibrils. (b) Arrangement of the thick and thin filaments in the sarcomere shown in (a).

(a)

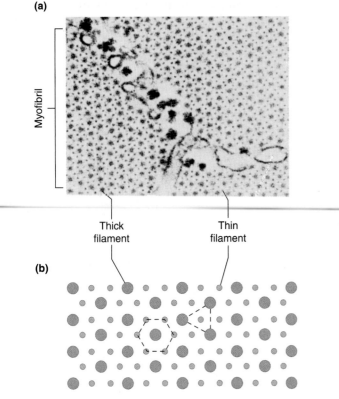

Myofibril

Thick filament

Thin filament

(b)

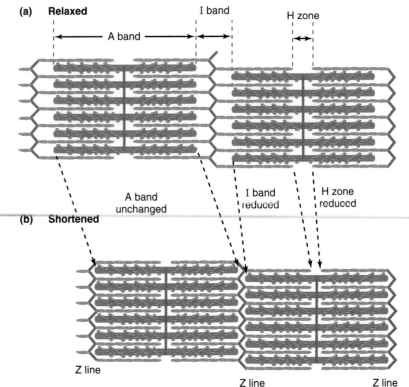

(a) Relaxed   I band   H zone

A band

(b) Shortened

A band unchanged   I band reduced   H zone reduced

Z line

Z line   Z line

Figure 9–4

(a) Electron micrograph of a cross section through three myofibrils in a single skeletal muscle fiber. (b) Hexagonal arrangements of the thick and thin filaments in the overlap region in a single myofibril. Six thin filaments surround each thick filament, and three thick filaments surround each thin filament. Titin filaments and cross-bridges are not shown.

From H. E. Huxley, *J. Mol. Biol.,* 37:507–520 (1968).

Figure 9–4 physiological *inquiry* ⓟ*i*

■ Draw a cross-section diagram like the one in part b for a slice taken: (1) in the H zone, (2) in the I band, (3) at the M line, and (4) at the Z line (ignore titin).

Answer can be found at end of chapter.

The space between overlapping thick and thin filaments is bridged by projections known as **cross-bridges.** These are portions of myosin molecules that extend from the surface of the thick filaments toward the thin filaments (see Figure 9–2). During muscle contraction, the cross-bridges make contact with the thin filaments and exert force on them.

Molecular Mechanisms of Skeletal Muscle Contraction

The term **contraction,** as used in muscle physiology, does not necessarily mean "shortening." It simply refers to activation of the force-generating sites within muscle fibers—the cross-bridges. For example, holding a dumbbell at a constant position requires

Figure 9–5

The sliding of thick filaments past overlapping thin filaments shortens the sarcomere with no change in thick or thin filament length. The I band and H zone are reduced.

muscle contraction, but not muscle shortening. Following contraction, the mechanisms that generate force are turned off, and tension declines, allowing **relaxation** of the muscle fiber.

Sliding-Filament Mechanism

When force generation produces shortening of a skeletal muscle fiber, the overlapping thick and thin filaments in each sarcomere move past each other, propelled by movements of the cross-bridges. During this shortening of the sarcomeres, there is no change in the lengths of either the thick or thin filaments (**Figure 9–5**). This is known as the **sliding-filament mechanism** of muscle contraction.

During shortening, each myosin cross-bridge attached to a thin filament actin molecule moves in an arc much like an oar on a boat. This swiveling motion of many cross-bridges forces the thin filaments attached to successive Z lines to move toward the center of the sarcomere, thereby shortening the sarcomere (**Figure 9–6**). One stroke of a cross-bridge produces only a very small movement of a thin filament relative to a thick filament. As long as a muscle fiber remains activated, however, each cross-bridge repeats its swiveling motion many times, resulting in large displacements of the filaments. It is worth noting that a common pattern of muscle shortening involves one end of the muscle remaining at a fixed position while the other end shortens toward it. In this case, as filaments slide and each sarcomere shortens internally, the center of each sarcomere also slides toward the fixed end of

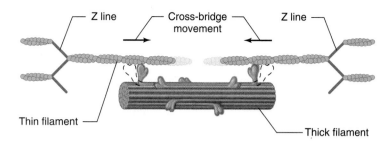

Figure 9–6

Cross-bridges in the thick filaments bind to actin in the thin filaments and undergo a conformational change that propels the thin filaments toward the center of a sarcomere. (Only a few of the approximately 200 cross-bridges in each thick filament are shown.)

the muscle (this is depicted in Figure 9–5). Thus, the ability of a muscle fiber to generate force and movement depends on the interaction of the contractile proteins actin and myosin.

An actin molecule is a globular protein composed of a single polypeptide that polymerizes with other actins to form two intertwined, helical chains (**Figure 9–7**). These chains make up the core of a thin filament. Each actin molecule contains a binding site for myosin. The myosin molecule, on the other hand, is composed of two large polypeptide **heavy chains** and four smaller **light chains.** These polypeptides combine to form a molecule that consists of two globular heads (containing heavy and light chains) and a long tail formed by the two intertwined heavy chains (see Figure 9–7). The tail of each myosin molecule lies along the axis of the thick filament, and the two globular heads extend out to the sides, forming the cross-bridges. Each globular head contains two binding sites, one for actin and one for ATP. The ATP binding site also serves as an enzyme—an ATPase that hydrolyzes the bound ATP, harnessing its energy for contraction.

The myosin molecules in the two ends of each thick filament are oriented in opposite directions, so that their tail ends are directed toward the center of the filament. Because of this arrangement, the power strokes of the cross-bridges move the attached thin filaments at the two ends of the sarcomere toward the center during shortening (see Figure 9–6).

The sequence of events that occurs between the time a cross-bridge binds to a thin filament, moves, and then is set to repeat the process is known as a **cross-bridge cycle.** Each cycle consists of four steps: (1) attachment of the cross-bridge to a thin filament, (2) movement of the cross-bridge, producing tension in the thin filament, (3) detachment of the cross-bridge from the thin filament, and (4) energizing the cross-bridge so it can again attach to a thin filament and repeat the cycle. Each cross-bridge undergoes its own cycle of movement independently of the other cross-bridges. At any instant during contraction, only a portion of the cross-bridges are attached to the thin filaments, producing tension, while others are in a detached portion of their cycle.

Figure 9–8 illustrates the chemical and physical events during the four steps of the cross-bridge cycle. In a resting muscle fiber, the cytoplasmic calcium concentration is low, and the myosin cross-bridges (M) cannot bind to actin (A). The cross-bridges, however, are in an energized state produced by splitting ATP, and the hydrolysis products (ADP and inorganic phosphate) are still bound to myosin. This energy storage in myosin is analogous to the storage of potential energy in a stretched spring.

Cross-bridge cycling is initiated when calcium enters the cytoplasm (by a mechanism that will be described shortly). The cycle begins with the binding of an energized myosin cross-bridge to a thin filament actin molecule (step 1):

$$\text{Step 1} \qquad A + M \cdot ADP \cdot P_i \underset{\substack{\text{actin} \\ \text{binding}}}{\longrightarrow} A \cdot M \cdot ADP \cdot P_i$$

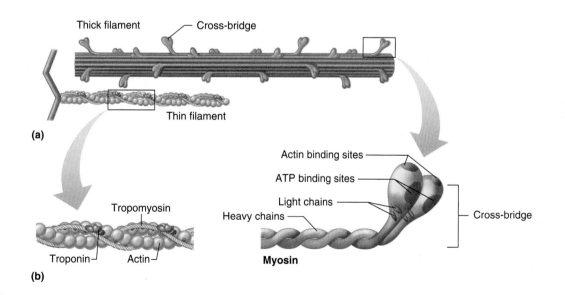

Figure 9–7

(a) The heavy chains of myosin molecules form the core of a thick filament. The myosin molecules are oriented in opposite directions in either half of a thick filament. (b) Structure of thin filament and myosin molecule. Cross-bridge binding sites on actin are covered by tropomyosin. The two globular heads of each myosin molecule extend from the sides of a thick filament, forming a cross-bridge.

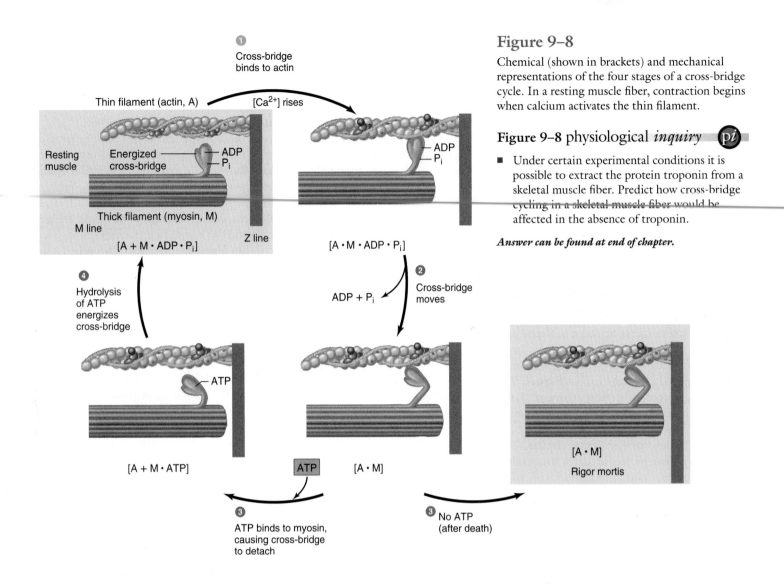

①
Cross-bridge
binds to actin

Thin filament (actin, A)　　　[Ca²⁺] rises

Resting
muscle　　Energized —— ADP
　　　　　cross-bridge —— Pᵢ

Thick filament (myosin, M)

M line　　　　　　　　　　Z line

$[A + M \cdot ADP \cdot P_i]$

$[A \cdot M \cdot ADP \cdot P_i]$

④
Hydrolysis
of ATP
energizes
cross-bridge

②
Cross-bridge
moves

$ADP + P_i$

—— ATP

$[A + M \cdot ATP]$　　　ATP　　$[A \cdot M]$

$[A \cdot M]$
Rigor mortis

③
ATP binds to myosin,
causing cross-bridge
to detach

③ No ATP
(after death)

Figure 9–8

Chemical (shown in brackets) and mechanical
representations of the four stages of a cross-bridge
cycle. In a resting muscle fiber, contraction begins
when calcium activates the thin filament.

Figure 9–8 physiological *inquiry*　**p***i*

- Under certain experimental conditions it is
  possible to extract the protein troponin from a
  skeletal muscle fiber. Predict how cross-bridge
  cycling in a skeletal muscle fiber would be
  affected in the absence of troponin.

Answer can be found at end of chapter.

The binding of energized myosin to actin triggers the release
of the strained conformation of the energized bridge, which produces the movement of the bound cross-bridge (sometimes called
the **power stroke**) and the release of P_i and ADP (step 2):

Step 2　　$A \cdot M \cdot ADP \cdot P_i \longrightarrow A \cdot M + ADP + P_i$
　　　　　　　　　cross-bridge
　　　　　　　　　movement

This sequence of energy storage and release by myosin is analogous to the operation of a mousetrap: Energy is stored in the
trap by cocking the spring (ATP hydrolysis) and released after
springing the trap (binding to actin).

During the cross-bridge movement, myosin is bound
very firmly to actin, and this linkage must be broken to allow
the cross-bridge to be re-energized and repeat the cycle. The
binding of a new molecule of ATP to myosin breaks the link
between actin and myosin (step 3):

Step 3　　　$A \cdot M + ATP \longrightarrow A + M \cdot ATP$
　　　　　　　　　cross-bridge
　　　　　　　　dissociation from actin

The dissociation of actin and myosin by ATP is an example of
allosteric regulation of protein activity. The binding of ATP at

one site on myosin decreases myosin's affinity for actin bound
at another site. Note that ATP is not split in this step; that is,
it is not acting as an energy source, but only as an allosteric
modulator of the myosin head that weakens the binding of
myosin to actin.

Following the dissociation of actin and myosin, the ATP
bound to myosin is split (step 4), thereby re-forming the energized state of myosin and returning the cross-bridge to its pre-
power-stroke position.

Step 4　　　$A + M \cdot ATP \longrightarrow A + M \cdot ADP \cdot P_i$
　　　　　　　　　ATP hydrolysis

Note that the hydrolysis of ATP (step 4) and the movement of
the cross-bridge (step 2) are not simultaneous events. If calcium is still present at this time, the cross-bridge can reattach
to a new actin molecule in the thin filament and the cross-
bridge cycle repeats. (In the event that the muscle is generating force without actually shortening, the cross-bridge will
reattach to the same actin molecule as in the previous cycle.)

Thus, ATP performs two distinct roles in the cross-
bridge cycle: (1) The energy released from ATP *hydrolysis* ultimately provides the energy for cross-bridge movement, and

(2) ATP *binding* (not hydrolysis) to myosin breaks the link formed between actin and myosin during the cycle, allowing the cycle to repeat.

The importance of ATP in dissociating actin and myosin during step 3 of a cross-bridge cycle is illustrated by **rigor mortis,** the gradual stiffening of skeletal muscles that begins several hours after death and reaches a maximum after about 12 hours. The ATP concentration in cells, including muscle cells, declines after death because the nutrients and oxygen the metabolic pathways require to form ATP are no longer supplied by the circulation. In the absence of ATP, the breakage of the link between actin and myosin does not occur (see Figure 9–8). The thick and thin filaments remain bound to each other by immobilized cross-bridges, producing a rigid condition in which the thick and thin filaments cannot be pulled past each other. The stiffness of rigor mortis disappears about 48 to 60 hours after death as the muscle tissue decomposes.

Roles of Troponin, Tropomyosin, and Calcium in Contraction

How does the presence of calcium in the cytoplasm regulate the cross-bridge cycle? The answer requires a closer look at the thin filament proteins, troponin and tropomyosin (**Figure 9–9**).

Tropomyosin is a rod-shaped molecule composed of two intertwined polypeptides with a length approximately equal to that of seven actin molecules. Chains of tropomyosin molecules are arranged end to end along the actin thin filament. These tropomyosin molecules partially cover the myosin-binding site on each actin molecule, thereby preventing the cross-bridges from making contact with actin. Each tropomyosin molecule is held in this blocking position by the smaller globular protein, troponin. Troponin, which interacts with both actin and tropomyosin, is composed of three subunits designated by the letters I (inhibitory), T (tropomyosin-binding) and C (calcium-binding). One molecule of troponin binds to each molecule of tropomyosin and regulates the access to myosin-binding sites on the seven actin molecules in contact with tropomyosin. This is the status of a resting muscle fiber; troponin and tropomyosin cooperatively block the interaction of cross-bridges with the thin filament.

What enables cross-bridges to bind to actin and begin cycling? For this to occur, tropomyosin molecules must move away from their blocking positions on actin. This happens when calcium binds to specific binding sites on the calcium-binding subunit of troponin. The binding of calcium produces a change in the shape of troponin, which relaxes its inhibitory grip and allows tropomyosin to move away from the myosin-binding site on each actin molecule. Conversely, the removal of calcium from troponin reverses the process, turning off contractile activity.

Thus, cytosolic calcium-ion concentration determines the number of troponin sites occupied by calcium, which in turn determines the number of actin sites available for cross-bridge binding. Electrical events in the muscle plasma membrane, which we will now discuss, control changes in cytosolic calcium concentration.

Excitation-Contraction Coupling

Excitation-contraction coupling refers to the sequence of events by which an action potential in the plasma membrane of a muscle fiber leads to the cross-bridge activity just described. The skeletal muscle plasma membrane is an excitable membrane capable of generating and propagating action potentials by mechanisms similar to those described for nerve cells (Chapter 6). An action potential in a skeletal muscle fiber lasts 1 to 2 ms and is completed before any signs of mechanical activity begin (**Figure 9–10**). Once begun, the mechanical activity following an action potential may last 100 ms or more. The electrical activity in the plasma membrane does not directly act upon the contractile proteins, but instead produces a state of increased cytosolic calcium concentration, which continues to activate the contractile apparatus long after the electrical activity in the membrane has ceased.

In a resting muscle fiber, the concentration of free, ionized calcium in the cytosol surrounding the thick and thin filaments is very low, only about 10^{-7} mol/L. At this low calcium concentration, very few of the calcium-binding sites on troponin are occupied, and thus cross-bridge activity is blocked by tropomyosin. Following an action potential, there is a rapid increase in cytosolic calcium concentration, and calcium binds to troponin, removing the blocking effect of tropomyosin and allowing cross-bridge cycling. The source of the increased cytosolic calcium is the **sarcoplasmic reticulum** within the muscle fiber.

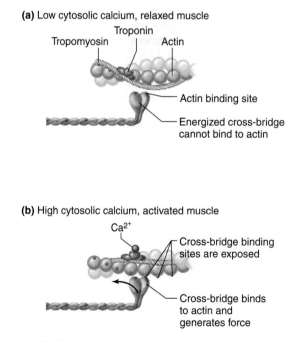

(a) Low cytosolic calcium, relaxed muscle

Tropomyosin — Troponin — Actin

Actin binding site

Energized cross-bridge cannot bind to actin

(b) High cytosolic calcium, activated muscle

Ca²⁺

Cross-bridge binding sites are exposed

Cross-bridge binds to actin and generates force

Figure 9–9

Activation of cross-bridge cycling by calcium. (a) Without calcium ions bound, troponin holds tropomyosin over cross-bridge binding sites on actin. (b) When calcium binds to troponin, tropomyosin is allowed to move away from cross-bridge binding sites on actin, and cross-bridges can bind to actin.

Sarcoplasmic Reticulum

The sarcoplasmic reticulum in muscle is homologous to the endoplasmic reticulum found in most cells. This structure forms a series of sleeve-like segments around each myofibril (**Figure 9–11**). At the end of each segment are two enlarged regions, known as **lateral sacs,** that are connected to each other by a series of smaller tubular elements. Calcium stored in the lateral sacs is released following membrane excitation.

A separate tubular structure, the **transverse tubule (T-tubule),** lies directly between, and is intimately associated with, the lateral sacs of adjacent segments of the sarcoplasmic reticulum. The T-tubules and lateral sacs surround the myofibrils at the region of the sarcomeres where the A bands and I bands meet. The lumen of the T-tubule is continuous with the extracellular fluid surrounding the muscle fiber. The membrane of the T-tubule, like the plasma membrane, is able to propagate action potentials. Once initiated in the plasma membrane, an action potential is rapidly conducted over the surface of the fiber and into its interior by way of the T-tubules.

A specialized mechanism couples T-tubule action potentials with calcium release from the sarcoplasmic reticulum (**Figure 9–12**, step 2). The T-tubules are in intimate contact with the lateral sacs of the sarcoplasmic reticulum, connected by structures known as **junctional feet** or **"foot processes."** This junction involves two integral membrane proteins, one in the T-tubule membrane, and the other in the membrane of the sarcoplasmic reticulum. The T-tubule protein is a modified voltage-sensitive calcium channel known as the **dihydropyridine (DHP) receptor** (so named because it binds the class of drugs called dihydropyridines). The main role of the DHP receptor, however, is not to conduct calcium, but rather to act as a voltage sensor. The protein embedded in the sarcoplasmic reticulum membrane is known as the **ryanodine receptor** (because it binds to the plant alkaloid ryanodine). This large molecule not only includes the foot process, but also forms a

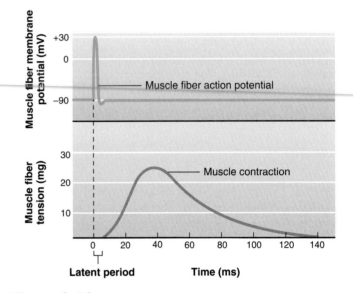

Figure 9–10

Time relationship between a skeletal muscle fiber action potential and the resulting contraction and relaxation of the muscle fiber.

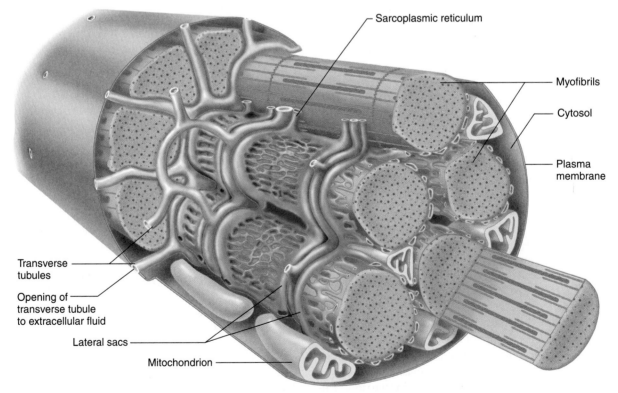

Figure 9–11

Transverse tubules and sarcoplasmic reticulum in a single skeletal muscle fiber.

calcium channel. During a T-tubule action potential, charged amino acid residues within the DHP receptor protein induce a conformational change, which acts via the foot process to open the ryanodine receptor channel. Calcium is thus released from the lateral sacs of the sarcoplasmic reticulum into the cytosol, activating cross-bridge cycling. The rise in cytosolic calcium in response to a single action potential is normally enough to briefly saturate all troponin binding sites on the thin filaments.

A contraction is terminated by removal of calcium from troponin, which is achieved by lowering the calcium concentration in the cytosol back to its prerelease level. The membranes of the sarcoplasmic reticulum contain primary active-transport proteins—Ca^{2+}-ATPases—that pump calcium ions from the cytosol back into the lumen of the reticulum. As we just saw, calcium is released from the reticulum when an action potential begins in the T-tubule, but the pumping of the released calcium back into the reticulum requires a much longer time. Therefore, the cytosolic calcium concentration remains elevated, and the contraction continues for some time after a single action potential.

To reiterate, just as contraction results from the release of calcium ions stored in the sarcoplasmic reticulum, so contraction ends and relaxation begins as calcium is pumped back into the reticulum (see Figure 9–12). ATP is required to provide the energy for the calcium pump—the third major role of ATP in muscle contraction (**Table 9–1**).

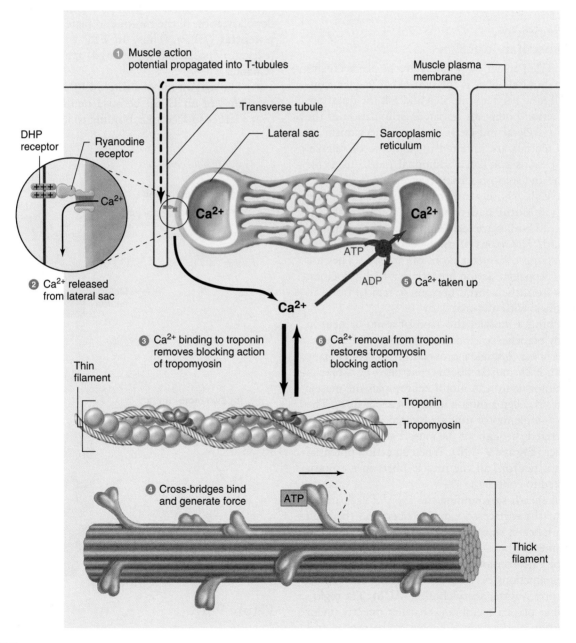

Figure 9–12

Release and uptake of calcium by the sarcoplasmic reticulum during contraction and relaxation of a skeletal muscle fiber.

Table 9–1	Functions of ATP in Skeletal Muscle Contraction

1. Hydrolysis of ATP by myosin energizes the cross-bridges, providing the energy for force generation.

2. Binding of ATP to myosin dissociates cross-bridges bound to actin, allowing the bridges to repeat their cycle of activity.

3. Hydrolysis of ATP by the Ca^{2+}-ATPase in the sarcoplasmic reticulum provides the energy for the active transport of calcium ions into the reticulum, lowering cytosolic calcium to prerelease levels, ending the contraction, and allowing the muscle fiber to relax.

Membrane Excitation: The Neuromuscular Junction

We have just seen that an action potential in the plasma membrane of a skeletal muscle fiber is the signal that triggers contraction. We will now back up one step and ask the question: How are these action potentials initiated? Stimulation of the nerve fibers to a skeletal muscle is the only mechanism by which action potentials are initiated in this type of muscle. In subsequent sections you'll see additional mechanisms for activating cardiac and smooth muscle contraction.

The nerve cells whose axons innervate skeletal muscle fibers are known as **motor neurons** (or somatic efferent neurons), and their cell bodies are located in either the brainstem or the spinal cord. The axons of motor neurons are myelinated and are the largest-diameter axons in the body. They are therefore able to propagate action potentials at high velocities, allowing signals from the central nervous system to travel to skeletal muscle fibers with minimal delay.

Upon reaching a muscle, the axon of a motor neuron divides into many branches, each branch forming a single junction with a muscle fiber. A single motor neuron innervates many muscle fibers, but each muscle fiber is controlled by a branch from only one motor neuron. A motor neuron plus the muscle fibers it innervates is called a **motor unit** (**Figure 9–13a**). The muscle fibers in a single motor unit are located in one muscle, but they are scattered throughout the muscle and are not adjacent to each other (**Figure 9–13b**). When an action potential occurs in a motor neuron, all the muscle fibers in its motor unit are stimulated to contract.

The myelin sheath surrounding the axon of each motor neuron ends near the surface of a muscle fiber, and the axon divides into a number of short processes that lie embedded in grooves on the muscle fiber surface (**Figure 9–14a**). The axon terminals of a motor neuron contain vesicles similar to the vesicles found at synaptic junctions between two neurons. The vesicles contain the neurotransmitter **acetylcholine (ACh).** The region of the muscle fiber plasma membrane that lies directly under the terminal portion of the axon is known as the **motor end plate.** The junction of an axon terminal with the motor end plate is known as a **neuromuscular junction** (**Figure 9–14b**).

Figure 9–15 shows the events occurring at the neuromuscular junction. When an action potential in a motor neuron arrives at the axon terminal, it depolarizes the plasma membrane, opening voltage-sensitive calcium channels and allowing calcium ions to diffuse into the axon terminal from the extracellular fluid. This calcium binds to proteins that enable the membranes of acetylcholine-containing vesicles to fuse with the neuronal plasma membrane, thereby releasing acetylcholine into the extracellular cleft separating the axon terminal and the motor end plate.

ACh diffuses from the axon terminal to the motor end plate where it binds to ionotropic receptors (of the nicotinic type; see Chapter 6). The binding of ACh opens an ion channel in each receptor protein; both sodium and potassium ions can pass through these channels. Because of the differences in electrochemical gradients across the plasma membrane (Chapter 6), more sodium moves in than potassium out, producing a local depolarization of the motor end plate known as an **end-plate potential (EPP).** Thus, an EPP is analogous to an EPSP (excitatory postsynaptic potential) at a neuron-neuron synapse (Chapter 6).

The magnitude of a single EPP is, however, much larger than that of an EPSP because neurotransmitter is released over a larger surface area, binding to many more receptors and

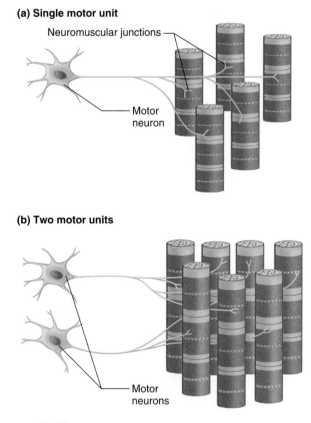

(a) Single motor unit

Neuromuscular junctions

Motor neuron

(b) Two motor units

Motor neurons

Figure 9–13

(a) Single motor unit consisting of one motor neuron and the muscle fibers it innervates. (b) Two motor units and their intermingled fibers in a muscle.

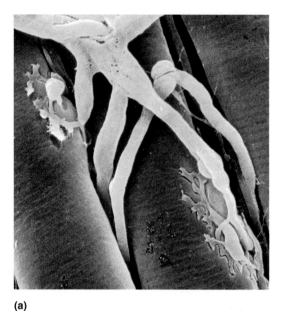

(a)

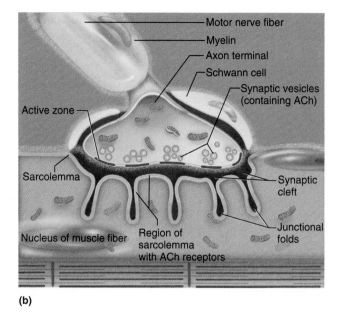

Motor nerve fiber
Myelin
Axon terminal
Schwann cell
Synaptic vesicles
(containing ACh)
Active zone
Sarcolemma
Synaptic
cleft
Junctional
folds
Nucleus of muscle fiber
Region of
sarcolemma
with ACh receptors

(b)

Figure 9–14

The neuromuscular junction. (a) Scanning electron micrograph showing branching of motor axons, with terminals embedded in grooves in the muscle fiber's surface. (b) Structure of a neuromuscular junction.

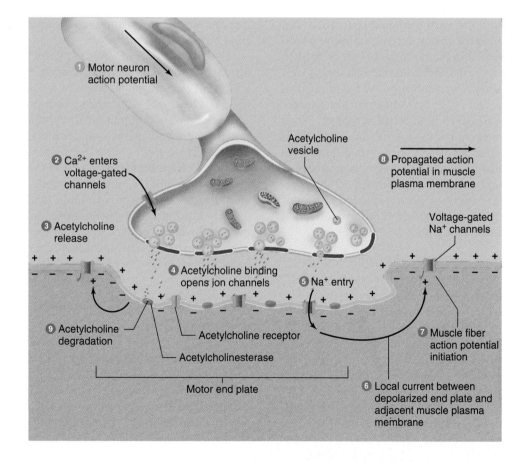

1 Motor neuron action potential

2 Ca^{2+} enters voltage-gated channels

3 Acetylcholine release

Acetylcholine vesicle

8 Propagated action potential in muscle plasma membrane

Voltage-gated Na^+ channels

4 Acetylcholine binding opens ion channels

5 Na^+ entry

9 Acetylcholine degradation

Acetylcholine receptor

Acetylcholinesterase

Motor end plate

7 Muscle fiber action potential initiation

6 Local current between depolarized end plate and adjacent muscle plasma membrane

Figure 9–15

Events at the neuromuscular junction that lead to an action potential in the muscle fiber plasma membrane. Although potassium also exits the muscle cell when ACh receptors are open, sodium entry and depolarization dominates, as shown here.

opening many more ion channels. For this reason, one EPP is normally more than sufficient to depolarize the muscle plasma membrane adjacent to the end-plate membrane to its threshold potential, initiating an action potential. This action potential is then propagated over the surface of the muscle fiber by the same mechanism described in Chapter 6 for the propagation of action potentials along unmyelinated axon membranes. Most neuromuscular junctions are located near the middle of a muscle fiber, and newly generated muscle action potentials propagate from this region in both directions toward the ends of the fiber and throughout the T-tubule network.

To repeat, every action potential in a motor neuron normally produces an action potential in each muscle fiber in its motor unit. This is quite different from synaptic junctions between neurons, where multiple EPSPs must occur in order for threshold to be reached and an action potential elicited in the postsynaptic membrane.

There is a second difference between interneuronal synapses and neuromuscular junctions. As we saw in Chapter 6, IPSPs (inhibitory postsynaptic potentials) are produced at some synaptic junctions. They hyperpolarize or stabilize the postsynaptic membrane and decrease the probability of its firing an action potential. In contrast, inhibitory potentials do not occur in human skeletal muscle; *all neuromuscular junctions are excitatory.*

In addition to receptors for ACh, the synaptic junction contains the enzyme **acetylcholinesterase,** which breaks down ACh, just as it does at ACh-mediated synapses in the nervous system. Choline is then transported back into the axon terminals, where it is reused in the synthesis of new ACh. ACh bound to receptors is in equilibrium with free ACh in the cleft between the nerve and muscle membranes. As the concentration of free ACh falls because of its breakdown by acetylcholinesterase, less ACh is available to bind to the receptors. When the receptors no longer contain bound ACh, the ion channels in the end plate close. The depolarized end plate returns to its resting potential and can respond to the subsequent arrival of ACh released by another neuron action potential.

Table 9–2 summarizes the sequence of events that lead from an action potential in a motor neuron to the contraction and relaxation of a skeletal muscle fiber.

Disruption of Neuromuscular Signaling

There are many ways by which disease or drugs can modify events at the neuromuscular junction. For example, the deadly South American arrowhead poison *curare* binds strongly to nicotinic ACh receptors, but it does not open their ion channels and acetylcholinesterase does not destroy it. When a receptor is occupied by curare, ACh cannot bind to the receptor. Therefore, although the motor nerves still conduct normal action potentials and release ACh, there is no resulting EPP in the motor end plate and no contraction. Because the skeletal muscles responsible for breathing, like all skeletal muscles, depend upon neuromuscular transmission to initiate their contraction, curare poisoning can cause death by asphyxiation. Drugs similar to curare (*succinylcholine* is one example) are used in small amounts to prevent muscular contractions during certain types of surgical procedures when it is neces-

sary to immobilize the surgical field. The use of such paralytic agents also reduces the required dose of general anesthetic, allowing patients to recover faster and with fewer complications. Patients are artificially ventilated to maintain respiration until the drug has been removed from the system.

Neuromuscular transmission can also be blocked by inhibiting acetylcholinesterase. Some organophosphates, which are the main ingredients in certain pesticides and "nerve gases" (the latter developed for chemical warfare), inhibit this enzyme. In the presence of such agents, ACh is released normally upon the arrival of an action potential at the axon terminal and binds to the end-plate receptors. The ACh is not destroyed, however, because the acetylcholinesterase is inhibited. The ion channels in the end plate therefore remain open, producing a maintained depolarization of the end plate and the muscle plasma membrane adjacent to the end plate. A skeletal muscle membrane maintained in a depolarized state cannot generate action potentials because the voltage-gated sodium channels in the membrane become inactivated, which requires repolarization to reverse. After prolonged exposure to ACh, the receptors of the motor end plate become insensitive to it, preventing any further depolarization. Thus, the muscle does not contract in response to subsequent nerve stimulation, and the result is skeletal muscle paralysis and death from asphyxiation. Note that nerve gases also cause ACh to build up at muscarinic synapses, for example, where parasympathetic neurons inhibit cardiac pacemaker cells (Chapter 12). Thus, the antidote for nerve gas exposure must include the muscarinic receptor antagonist *atropine.*

A third group of substances, including the toxin produced by the bacterium *Clostridium botulinum,* blocks the release of acetylcholine from nerve terminals. Botulinum toxin is an enzyme that breaks down proteins of the SNARE complex that are required for the binding and fusion of ACh vesicles with the plasma membrane of the axon terminal (review Figure 6–27). This toxin, which produces the food poisoning called *botulism,* is one of the most potent poisons known because of the very small amount necessary to produce an effect. Application of botulinum toxin is increasingly being used for clinical and cosmetic procedures, including the inhibition of overactive extraocular muscles, prevention of excessive sweat gland activity, treatment of migraine headaches, and reduction of aging-related skin wrinkles.

Mechanics of Single-Fiber Contraction

The force exerted on an object by a contracting muscle is known as muscle **tension,** and the force exerted on the muscle by an object (usually its weight) is the **load.** Muscle tension and load are opposing forces. Whether a fiber shortens depends on the relative magnitudes of the tension and the load. For muscle fibers to shorten, and thereby move a load, muscle tension must be greater than the opposing load.

When a muscle develops tension but does not shorten (or lengthen), the contraction is said to be **isometric** (constant length). Such contractions occur when the muscle supports a

Table 9–2   Sequence of Events Between a Motor Neuron Action Potential and Skeletal Muscle Fiber Contraction

1. Action potential is initiated and propagates to motor neuron axon terminals.

2. Calcium enters axon terminals through voltage-gated calcium channels.

3. Calcium entry triggers release of ACh from axon terminals.

4. ACh diffuses from axon terminals to motor end plate in muscle fiber.

5. ACh binds to nicotinic receptors on motor end plate, increasing their permeability to Na^+ and K^+.

6. More Na^+ moves into the fiber at the motor end plate than K^+ moves out, depolarizing the membrane and producing the end plate potential (EPP).

7. Local currents depolarize the adjacent muscle cell plasma membrane to its threshold potential, generating an action potential that propagates over the muscle fiber surface and into the fiber along the T-tubules.

8. Action potential in T-tubules induces DHP receptors to pull open ryanodine receptor channels, allowing release of Ca^{2+} from lateral sacs of sarcoplasmic reticulum.

9. Ca^{2+} binds to troponin on the thin filaments, causing tropomyosin to move away from its blocking position, thereby uncovering cross-bridge binding sites on actin.

10. Energized myosin cross-bridges on the thick filaments bind to actin:
$$A + M \cdot ADP \cdot P_i \rightarrow A \cdot M \cdot ADP \cdot P_i$$

11. Cross-bridge binding triggers release of ATP hydrolysis products from myosin, producing an angular movement of each cross-bridge:
$$A \cdot M \cdot ADP \cdot P_i \rightarrow A \cdot M + ADP + P_i$$

12. ATP binds to myosin, breaking linkage between actin and myosin and thereby allowing cross-bridges to dissociate from actin:
$$A \cdot M + ATP \rightarrow A + M \cdot ATP$$

13. ATP bound to myosin is split, energizing the myosin cross-bridge:
$$M \cdot ATP \rightarrow M \cdot ADP \cdot P_i$$

14. Cross-bridges repeat steps 10 to 13, producing movement (sliding) of thin filaments past thick filaments. Cycles of cross-bridge movement continue as long as Ca^{2+} remains bound to troponin.

15. Cytosolic Ca^{2+} concentration decreases as Ca^{2+}-ATPase actively transports Ca^{2+} into sarcoplasmic reticulum.

16. Removal of Ca^{2+} from troponin restores blocking action of tropomyosin, the cross-bridge cycle ceases, and the muscle fiber relaxes.

load in a constant position or attempts to move an otherwise supported load that is greater than the tension developed by the muscle. A contraction in which the muscle changes length, while the load on the muscle remains constant, is **isotonic** (constant tension).

Depending on the relative magnitudes of muscle tension and the opposing load, isotonic contractions can be associated with either shortening or lengthening of a muscle. When tension exceeds the load, shortening occurs and it is referred to as **concentric contraction.** On the other hand, if an unsupported load is greater than the tension generated by cross-bridges, the result is a **lengthening contraction (eccentric contraction).** In this situation, the load pulls the muscle to a longer length in spite of the opposing force produced by the cross-bridges. Such lengthening contractions occur when an object being supported

by muscle contraction is lowered, as when the knee extensors in your thighs are used to lower you to a seat from a standing position. It must be emphasized that in these situations the lengthening of muscle fibers is not an active process produced by the contractile proteins, but a consequence of the external forces being applied to the muscle. In the absence of external lengthening forces, a fiber will only *shorten* when stimulated; it will never lengthen. All three types of contractions—isometric, concentric, eccentric—occur in the natural course of everyday activities.

During each type of contraction, the cross-bridges repeatedly go through the four steps of the cross-bridge cycle illustrated in Figure 9–8. During step 2 of a concentric isotonic contraction, the cross-bridges bound to actin rotate through their power stroke, causing shortening of the sarcomeres. In contrast, during an isometric contraction, the bound cross-bridges do exert a force on the thin filaments, but they are unable to move it. Rather than the filaments sliding; the rotation during the power stroke is absorbed within the structure of the cross-bridge in this circumstance. If isometric contraction is prolonged, cycling cross-bridges repeatedly re-bind to the same actin molecule. During a lengthening contraction, the load pulls the cross-bridges in step 2 backward toward the Z lines while they are still bound to actin and exerting force. The events of steps 1, 3, and 4 are the same in all three types of contractions. Thus, the chemical changes in the contractile proteins during each type of contraction are the same. The end result (shortening, no length change, or lengthening) is determined by the magnitude of the load on the muscle.

Contraction terminology applies to both single fibers and whole muscles. In this section, we describe the mechanics of single-fiber contractions. Later we will discuss the factors controlling the mechanics of whole-muscle contraction.

Twitch Contractions

The mechanical response of a muscle fiber to a single action potential is known as a **twitch. Figure 9–16a** shows the main features of an isometric twitch. Following the action potential, there is an interval of a few milliseconds, known as the

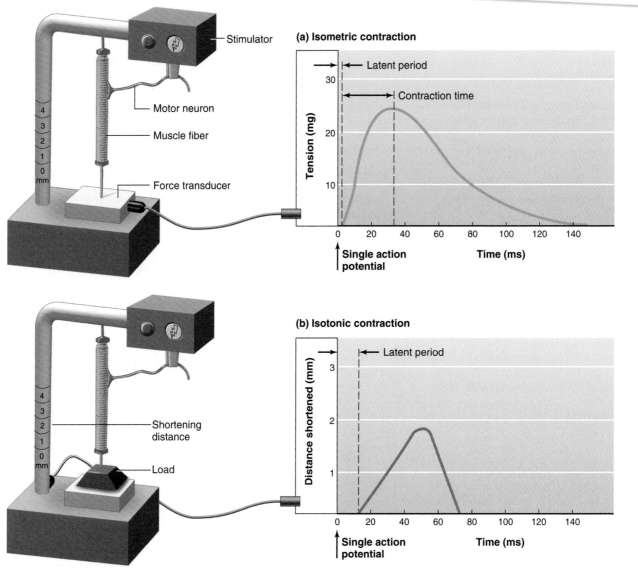

Figure 9–16

(a) Measurement of tension during a single isometric twitch of a skeletal muscle fiber. (b) Measurement of shortening during a single isotonic twitch of a skeletal muscle fiber.

Figure 9–16 physiological *inquiry* (pi)

- Assuming that the same muscle fiber is used in these two experiments, estimate the magnitude of the load (in mg) being lifted in the isotonic experiment.

Answer can be found at end of chapter.

latent period, before the tension in the muscle fiber begins to increase. During this latent period, the processes associated with excitation-contraction coupling are occurring. The time interval from the beginning of tension development at the end of the latent period to the peak tension is the **contraction time.** Not all skeletal muscle fibers have the same twitch contraction time. Some fast fibers have contraction times as short as 10 ms, whereas slower fibers may take 100 ms or longer. The total duration of a contraction depends in part on the time that cytosolic calcium remains elevated so that cross-bridges can continue to cycle. This is closely related to the Ca^{2+}-ATPase activity in the sarcoplasmic reticulum; activity is greater in fast-twitch fibers and less in slow-twitch fibers. Twitch duration also depends on how long it takes for cross-bridges to complete their cycle and detach after the removal of calcium from the cytosol.

Comparing isotonic and isometric twitches in the same muscle fiber, you can see from **Figure 9–16b** that the latent period in an isotonic twitch is longer than that in an isometric contraction. However, the duration of the mechanical event—shortening—is briefer in an isotonic twitch than the duration of force generation in an isometric twitch. The reason for these differences is most easily explained by referring to the measuring devices shown in Figure 9–16. In the isometric experiment, twitch tension begins to rise as soon as the first cross-bridge attaches, so the latent period is due only to the excitation-contraction coupling delay. By contrast, in the isotonic twitch experiment, the latent period includes both the time for excitation-contraction coupling and the extra time it takes to accumulate enough attached cross-bridges to lift the load off of the platform. Similarly, at the end of the twitch, the isotonic load comes back to rest on the platform well before all of the cross-bridges have detached in the isometric experiment.

Moreover, the characteristics of an isotonic twitch depend upon the magnitude of the load being lifted (**Figure 9–17**).

At heavier loads: (1) the latent period is longer, (2) the velocity of shortening (distance shortened per unit of time) is slower, (3) the duration of the twitch is shorter, and (4) the distance shortened is less.

A closer look at the sequence of events in an isotonic twitch explains this load-dependent behavior. As just explained, shortening does not begin until enough cross-bridges have attached and the muscle tension just exceeds the load on the fiber. Thus, before shortening, there is a period of *isometric* contraction during which the tension increases. The heavier the load, the longer it takes for the tension to increase to the value of the load, when shortening will begin. If the load on a fiber is increased, eventually a load is reached that the fiber is unable to lift, the velocity and distance of shortening decrease to zero, and the contraction will become completely isometric.

Load-Velocity Relation

It is a common experience that light objects can be moved faster than heavy objects. The isotonic twitch experiments illustrated in Figure 9–17 demonstrate that this phenomenon arises in part at the level of individual muscle fibers. When the initial shortening velocity (slope) of a series of isotonic twitches is plotted as a function of the load on a single fiber, the result is a hyperbolic curve (**Figure 9–18**). The shortening velocity is maximal when there is no load and is zero when the load is equal to the maximal isometric tension. At loads greater than the maximal isometric tension, the fiber will *lengthen* at a velocity that increases with load.

The shortening velocity is determined by the rate at which individual cross-bridges undergo their cyclical activity. Because

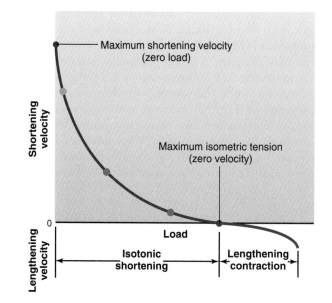

Figure 9–18
Velocity of skeletal muscle fiber shortening and lengthening as a function of load. Note that the force on the cross-bridges during a lengthening contraction is greater than the maximum isometric tension. The center three points correspond to the rate of shortening (slope) of the curves in Figure 9–17.

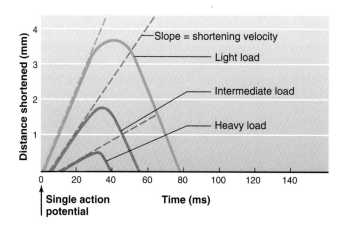

Figure 9–17
Isotonic twitches with different loads. The distance shortened, velocity of shortening, and duration of shortening all decrease with increased load, whereas the time from stimulation to the beginning of shortening increases with increasing load.

one ATP splits during each cross-bridge cycle, the rate of ATP splitting determines the shortening velocity. Increasing the load on a cross-bridge slows its forward movement during the power stroke. This reduces the overall rate of ATP hydrolysis, and thus decreases the velocity of shortening.

Frequency-Tension Relation

Because a single action potential in a skeletal muscle fiber lasts only 1 to 2 ms but the twitch may last for 100 ms, it is possible for a second action potential to be initiated during the period of mechanical activity. **Figure 9–19** illustrates the tension generated during isometric contractions of a muscle fiber in response to multiple stimuli. The isometric twitch following the first stimulus, S_1, lasts 150 ms. The second stimulus, S_2, applied to the muscle fiber 200 ms after S_1, when the fiber has completely relaxed, causes a second identical twitch. When a stimulus is applied before a fiber has completely relaxed from a twitch, it induces a contractile response with a peak tension greater than that produced in a single twitch (S_3 and S_4). If the interval between stimuli is reduced further, the resulting peak tension is even greater (S_5 and S_6). Indeed, the mechanical response to S_6 is a smooth continuation of the mechanical response already induced by S_5.

The increase in muscle tension from successive action potentials occurring during the phase of mechanical activity is known as **summation.** Do not confuse this with the summation of neuronal postsynaptic potentials described in Chapter 6. Postsynaptic potential summation involves additive voltage effects on the membrane whereas here we are observing the effect of additional attached cross-bridges. A maintained contraction in response to repetitive stimulation is known as a **tetanus** (tetanic contraction). At low stimulation frequencies, the tension may oscillate as the muscle fiber partially relaxes between stimuli, producing an **unfused tetanus.** A **fused tetanus,** with no oscillations, is produced at higher stimulation frequencies (**Figure 9–20**).

As the frequency of action potentials increases, the level of tension increases by summation until a maximal fused tetanic tension is reached, beyond which tension no longer increases even with further increases in stimulation frequency. This maximal tetanic tension is about three to five times greater than the isometric twitch tension. Different muscle fibers have different contraction times, so the stimulus frequency that will produce a maximal tetanic tension differs from fiber to fiber.

Why is tetanic tension so much greater than twitch tension? We can explain summation of tension in part by considering the relative timing of calcium availability and cross-bridge binding. The isometric tension produced by a muscle fiber at any instant depends mainly on the total number of cross-bridges bound to actin and undergoing the power stroke of the cross-bridge cycle. Recall that a single action potential in a skeletal muscle fiber briefly releases enough calcium to saturate troponin, and all the myosin-binding sites on the thin filaments are therefore *initially* available. But the binding of energized cross-bridges to these sites (step 1 of the cross-bridge cycle) takes time, while the calcium released into the cytosol begins to be pumped back into the sarcoplasmic reticulum almost immediately. Thus, after a single action potential, the calcium concentration begins to fall and the troponin/tropomyosin complex reblocks many binding sites before cross-bridges have had time to attach to them.

In contrast, during a tetanic contraction, the successive action potentials each release calcium from the sarcoplasmic reticulum before all the calcium from the previous action potential has been pumped back into the reticulum. This results in a persistent elevation of cytosolic calcium concentration, which prevents a decline in the number of available binding sites on the thin filaments. Under these conditions, more binding sites remain available, and many more cross-bridges become bound to the thin filaments.

Other causes of the lower tension seen in a single twitch are elastic structures, such as muscle tendons and the protein titin, which delay the transmission of cross-bridge force to the ends of a fiber. Because a single twitch is so brief, cross-bridge activity is already declining before force has been fully transmitted through these structures. This is less of a factor during tetanic stimulation because of the much longer duration of cross-bridge activity and force generation.

Length-Tension Relation

The spring-like characteristic of the protein titin (see Figure 9–3), which is attached to the Z line at one end and the thick filaments at the other, is responsible for most of the *passive* elastic properties of relaxed muscles. With increased stretch, the passive tension in a relaxed fiber increases, not from active cross-bridge movements but from elongation of the titin filaments. If the stretched fiber is released, it will return to an equilibrium length, much like what occurs when releasing a

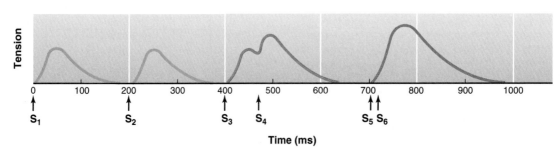

Figure 9–19

Summation of isometric contractions produced by shortening the time between stimuli.

Chapter 9

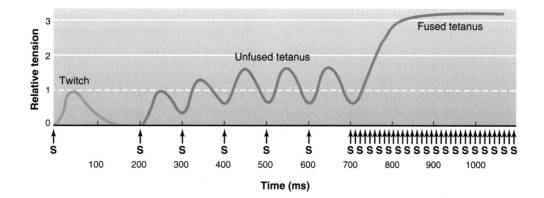

Figure 9–20

Isometric contractions produced by multiple stimuli (S) at 10 stimuli per second (unfused tetanus) and 100 stimuli per second (fused tetanus), as compared with a single twitch.

Figure 9–20 physiological *inquiry* (pi)

- If the twitch contraction time is 35 ms and twitch duration is 150 ms, estimate the range of stimulation frequencies (stimuli per second) over which unfused tetanic contractions will occur.

Answer can be found at end of chapter.

stretched rubber band. By a different mechnism, the amount of *active* tension a muscle fiber develops during contraction can also be altered by changing the length of the fiber. If you stretch a muscle fiber to various lengths and tetanically stimulate it at each length, the magnitude of the active tension will vary with length, as **Figure 9–21** shows. The length at which the fiber develops the greatest isometric active tension is termed the **optimal length, L_0.**

When a muscle fiber length is 60 percent of L_0, the fiber develops no tension when stimulated. As the length increases from this point, the isometric tension at each length is increased up to a maximum at L_0. Further lengthening leads to a *drop* in tension. At lengths of 175 percent of L_0 or beyond, the fiber develops no tension when stimulated.

When skeletal muscles are relaxed, the lengths of most fibers are near L_0 and thus near the optimal lengths for force generation. The length of a relaxed fiber can be altered by the load on the muscle or the contraction of other muscles that stretch the relaxed fibers, but the extent to which the relaxed length will change is limited by the muscle's attachments to bones. It rarely exceeds a 30 percent change from L_0 and is often much less. Over this range of lengths, the ability to develop tension never falls below about half of the tension that can be developed at L_0 (see Figure 9–21).

We can partially explain the relationship between fiber length and the fiber's capacity to develop active tension during contraction in terms of the sliding-filament mechanism. Stretching a relaxed muscle fiber pulls the thin filaments past the thick filaments, changing the amount of overlap between them. Stretching a fiber to 175 percent of L_0 pulls the filaments apart to the point where there is no overlap. At this point, there can be no cross-bridge binding to actin and no development of tension. As the fiber shortens toward L_0, more and more filament overlap occurs, and the tension developed upon stimulation increases in proportion to the increased number of cross-bridges in the overlap region. Filament overlap is greatest

at L_0, allowing the maximal number of cross-bridges to bind to the thin filaments, thereby producing maximal tension.

The tension decline at lengths less than L_0 is the result of several factors. For example, (1) the overlapping sets of thin filaments from opposite ends of the sarcomere may interfere with the cross-bridges' ability to bind and exert force, and (2) at very short lengths, the Z lines collide with the ends of the relatively rigid thick filaments, creating an internal resistance to sarcomere shortening.

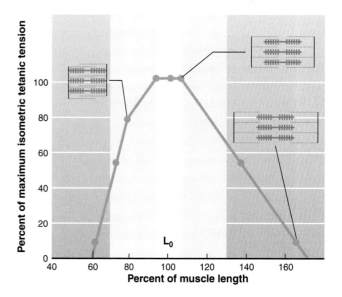

Figure 9–21

Variation in active isometric tetanic tension with muscle fiber length. Each point is the peak tension measured when the fiber was held at the indicated length and a fused, tetanic stimulus was applied. The blue band represents the range of length changes that can normally occur in the body.

Skeletal Muscle Energy Metabolism

As we have seen, ATP performs three functions directly related to muscle fiber contraction and relaxation (see Table 9–1). In no other cell type does the rate of ATP breakdown increase so much from one moment to the next as in a skeletal muscle fiber when it goes from rest to a state of contractile activity. The ATP breakdown may change 20- to several hundred-fold depending on the type of muscle fiber. The small supply of preformed ATP that exists at the start of contractile activity would only support a few twitches. If a fiber is to sustain contractile activity, metabolism must produce molecules of ATP as rapidly as they break down during the contractile process.

There are three ways a muscle fiber can form ATP (**Figure 9–22**): (1) phosphorylation of ADP by **creatine phosphate,** (2) oxidative phosphorylation of ADP in the mitochondria, and (3) phosphorylation of ADP by the glycolytic pathway in the cytosol.

Phosphorylation of ADP by creatine phosphate (CP) provides a very rapid means of forming ATP at the onset of contractile activity. When the chemical bond between creatine (C) and phosphate is broken, the amount of energy released is about the same as that released when the terminal phosphate bond in ATP is broken. This energy, along with the phosphate group, can be transferred to ADP to form ATP in a reversible reaction catalyzed by creatine kinase:

$$CP + ADP \underset{}{\overset{\text{creatine kinase}}{\rightleftharpoons}} C + ATP$$

Although creatine phosphate is a high-energy molecule, its energy cannot be released by myosin to drive cross-bridge activity. During periods of rest, muscle fibers build up a concentration of creatine phosphate approximately five times that of ATP. At the beginning of contraction, when the ATP concentration begins to fall and that of ADP to rise, owing to the increased rate of ATP breakdown by myosin, mass action favors the formation of ATP from creatine phosphate. This energy transfer is so rapid that the concentration of ATP in a muscle fiber changes very little at the start of contraction, whereas the concentration of creatine phosphate falls rapidly.

Although the formation of ATP from creatine phosphate is very rapid, requiring only a single enzymatic reaction, the amount of ATP that this process can form is limited by the initial concentration of creatine phosphate in the cell. (Many athletes in sports that require rapid power output consume creatine supplements in hopes of increasing the pool of immediately available ATP in their muscles.) If contractile activity is to continue for more than a few seconds, however, the muscle must be able to form ATP from the other two sources listed previously. The use of creatine phosphate at the start of contractile activity provides the few seconds necessary for the slower, multienzyme pathways of oxidative phosphorylation and glycolysis to increase their rates of ATP formation to levels that match the rates of ATP breakdown.

At moderate levels of muscular activity, most of the ATP used for muscle contraction is formed by oxidative phosphorylation, and during the first 5 to 10 min of such exercise, breakdown of muscle glycogen to glucose provides the major fuel contributing to oxidative phosphorylation. For the next 30 min or so, blood-borne fuels become dominant, blood glucose and fatty acids contributing approximately equally; beyond this period, fatty acids become progressively more important, and the muscle's glucose utilization decreases.

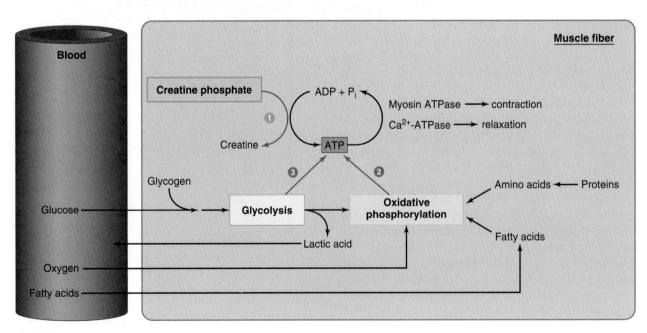

Figure 9–22

The three sources of ATP production during muscle contraction: (1) creatine phosphate, (2) oxidative phosphorylation, and (3) glycolysis.

If the intensity of exercise exceeds about 70 percent of the maximal rate of ATP breakdown, however, glycolysis contributes an increasingly significant fraction of the total ATP generated by the muscle. The glycolytic pathway, although producing only small quantities of ATP from each molecule of glucose metabolized, can produce ATP quite rapidly when enough enzymes and substrate are available, and it can do so in the absence of oxygen (under anaerobic conditions). The glucose for glycolysis can be obtained from two sources: the blood or the stores of glycogen within the contracting muscle fibers. As the intensity of muscle activity increases, a greater fraction of the total ATP production is formed by anaerobic glycolysis. This is associated with a corresponding increase in the production of lactic acid.

At the end of muscle activity, creatine phosphate and glycogen levels in the muscle have decreased. To return a muscle fiber to its original state, therefore, these energy-storing compounds must be replaced. Both processes require energy, and so a muscle continues to consume increased amounts of oxygen for some time after it has ceased to contract. In addition, extra oxygen is required to metabolize accumulated lactic acid and return the blood and interstitial fluid oxygen concentrations to pre-exercise values. These processes are evidenced by the fact that you continue to breathe deeply and rapidly for a period of time immediately following intense exercise. This elevated oxygen consumption following exercise repays the **oxygen debt**—that is, the increased production of ATP by oxidative phosphorylation following exercise is used to restore the energy reserves in the form of creatine phosphate and glycogen.

Muscle Fatigue

When a skeletal muscle fiber is repeatedly stimulated, the tension the fiber develops eventually decreases even though the stimulation continues (**Figure 9–23**). This decline in muscle tension as a result of previous contractile activity is known as **muscle fatigue.** Additional characteristics of fatigued muscle

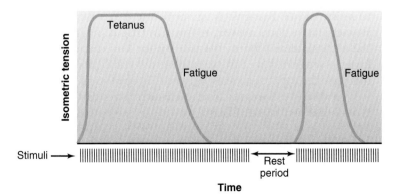

Figure 9–23

Muscle fatigue during a maintained isometric tetanus and recovery following a period of rest.

are a decreased shortening velocity and a slower rate of relaxation. The onset of fatigue and its rate of development depend on the type of skeletal muscle fiber that is active, the intensity and duration of contractile activity, and the degree of an individual's fitness.

If a muscle is allowed to rest after the onset of fatigue, it can recover its ability to contract upon restimulation (see Figure 9–23). The rate of recovery depends upon the duration and intensity of the previous activity. Some muscle fibers fatigue rapidly if continuously stimulated but also recover rapidly after a brief rest. This type of fatigue (high-frequency fatigue) accompanies high-intensity, short-duration exercise, such as weight lifting. In contrast, low-frequency fatigue develops more slowly with low-intensity, long-duration exercise, such as long-distance running, which includes cyclical periods of contraction and relaxation. This type of fatigue requires much longer periods of rest, often up to 24 hours, before the muscle achieves complete recovery.

It might seem logical that depletion of energy in the form of ATP would account for fatigue, but the ATP concentration in fatigued muscle is only slightly lower than in a resting muscle, and not low enough to impair cross-bridge cycling. If contractile activity were to continue without fatigue, the ATP concentration could decrease to the point that the cross-bridges would become linked in a rigor configuration, which is very damaging to muscle fibers. Thus, muscle fatigue may have evolved as a mechanism for preventing the onset of rigor.

Many factors can contribute to the fatigue of skeletal muscle. Fatigue from high-intensity, short-duration exercise is thought to involve at least three different mechanisms:

1. **Conduction Failure.** The muscle action potential can fail to be conducted into the fiber along the T-tubules, which halts the release of calcium from the sarcoplasmic reticulum. This conduction failure results from the buildup of potassium ions in the small volume of the T-tubule during the repolarization of repetitive action potentials. Elevated external potassium concentration leads to a persistent depolarization of the membrane potential, and eventually causes a failure to produce action potentials in the T-tubular membrane (due to inactivation of sodium channels). Recovery is rapid with rest as the accumulated potassium diffuses out of the tubule or is pumped back into the cell, restoring excitability.

2. **Lactic Acid Buildup.** Elevated hydrogen ion concentration alters protein conformation and activity. Thus, the acidification of muscle by lactic acid may alter a number of muscle proteins, including actin and myosin, as well as the proteins involved in calcium release. The function of the Ca^{2+}-ATPase pumps of the sarcoplasmic reticulum is also affected, which may in part explain the impaired relaxation of fatigued muscle. Recent single-fiber experiments performed at body temperature suggest that high acidity does not directly hinder contractile proteins, so effects on calcium handling may predominate.

3. **Inhibition of Cross-Bridge Cycling.** The buildup of ADP and P_i within muscle fibers during intense activity may directly inhibit cross-bridge cycling (in particular step 2) by mass action. Slowing the rate of this step delays cross-bridge detachment from actin, and thus slows the overall rate of cross-bridge cycling. These changes contribute to the reduced shortening velocity and impaired relaxation observed in muscle fatigue resulting from high-intensity exercise.

With low-intensity, long-duration exercise, a number of processes have been implicated in fatigue, but no single process can completely account for it. The three factors just listed may play minor roles in this type of exercise as well, but it appears that depletion of fuel substrates may be more important. Although ATP depletion is not a cause of fatigue, the decrease in muscle glycogen, which supplies much of the fuel for contraction, correlates closely with fatigue onset. In addition, low blood glucose (hypoglycemia) and dehydration have been demonstrated to increase fatigue. Thus, a certain level of carbohydrate metabolism appears necessary to prevent fatigue during low-intensity exercise, but the mechanism of this requirement is unknown.

Another type of fatigue quite different from muscle fatigue occurs when the appropriate regions of the cerebral cortex fail to send excitatory signals to the motor neurons. This is called **central command fatigue,** and it may cause a person to stop exercising even though the muscles are not fatigued. An athlete's performance depends not only on the physical state of the appropriate muscles but also upon the "will to win"—that is, the ability to initiate central commands to muscles during a period of increasingly distressful sensations.

Types of Skeletal Muscle Fibers

Skeletal muscle fibers do not all have the same mechanical and metabolic characteristics. Different types of fibers can be identified on the basis of (1) their maximal velocities of shortening—fast or slow—and (2) the major pathway they use to form ATP—oxidative or glycolytic.

Fast and slow fibers contain forms of myosin that differ in the maximal rates at which they split ATP. This, in turn, determines the maximal rate of cross-bridge cycling and thus the maximal shortening velocity. Fibers containing myosin with high ATPase activity are classified as **fast fibers,** and are also sometimes referred to as type II fibers. Several subtypes of fast myosin can be distinguished based on small variations in their structure. By contrast, fibers containing myosin with lower ATPase activity are called **slow fibers,** or type I fibers. Although the rate of cross-bridge cycling is about four times faster in fast fibers than in slow fibers, the force produced by both types of cross-bridges is about the same.

The second means of classifying skeletal muscle fibers is according to the type of enzymatic machinery available for synthesizing ATP. Some fibers contain numerous mitochondria and thus have a high capacity for oxidative phosphoryla-

tion. These fibers are classified as **oxidative fibers.** Most of the ATP such fibers produce is dependent upon blood flow to deliver oxygen and fuel molecules to the muscle. Not surprisingly, therefore, these fibers are surrounded by many small blood vessels. They also contain large amounts of an oxygen-binding protein known as **myoglobin,** which increases the rate of oxygen diffusion within the fiber and provides a small store of oxygen. The large amounts of myoglobin present in oxidative fibers give the fibers a dark red color, and thus oxidative fibers are often referred to as **red muscle fibers.**

In contrast, **glycolytic fibers** have few mitochondria but possess a high concentration of glycolytic enzymes and a large store of glycogen. Corresponding to their limited use of oxygen, these fibers are surrounded by relatively few blood vessels and contain little myoglobin. The lack of myoglobin is responsible for the pale color of glycolytic fibers and their designation as **white muscle fibers.**

On the basis of these two characteristics, three types of skeletal muscle fibers can be distinguished:

1. **Slow-oxidative fibers** (Type I) combine low myosin-ATPase activity with high oxidative capacity.
2. **Fast-oxidative-glycolytic fibers** (Type IIa) combine high myosin-ATPase activity with high oxidative capacity and intermediate glycolytic capacity.
3. **Fast-glycolytic fibers** (Type IIb) combine high myosin-ATPase activity with high glycolytic capacity.

Note that the fourth theoretical possibility—slow-glycolytic fibers—is not found.

In addition to these biochemical differences, there are also size differences. Glycolytic fibers generally have much larger diameters than oxidative fibers (**Figure 9–24**). This fact has significance for tension development. The number of thick and thin filaments per unit of cross-sectional area is about the same in all types of skeletal muscle fibers. Therefore, the larger the diameter of a muscle fiber, the greater the total number of thick and thin filaments acting in parallel to produce force, and the greater the maximum tension (greater strength) it can develop. Accordingly, the average glycolytic fiber, with its larger diameter, develops more tension when it contracts than does an average oxidative fiber.

These three types of fibers also differ in their capacity to resist fatigue. Fast-glycolytic fibers fatigue rapidly, whereas slow-oxidative fibers are very resistant to fatigue, which allows them to maintain contractile activity for long periods with little loss of tension. Fast-oxidative-glycolytic fibers have an intermediate capacity to resist fatigue (**Figure 9–25**).

Table 9–3 summarizes the characteristics of the three types of skeletal muscle fibers.

Whole-Muscle Contraction

As described earlier, whole muscles are made up of many muscle fibers organized into motor units. All the muscle fibers in a single motor unit are of the same fiber type. Thus, you can apply the fiber designation to the motor unit and refer to slow-oxidative-glycolytic motor units, fast-oxidative motor units, and fast-glycolytic motor units.

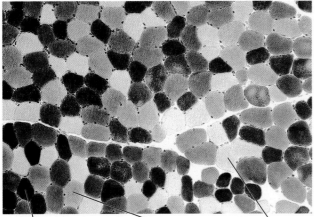

Slow-oxidative fiber   Fast-oxidative-glycolytic fiber   Fast-glycolytic fiber

Figure 9–24

Muscle fiber types in normal human muscle, prepared using ATPase stain. Darkest fibers are slow-oxidative type; lighter-colored fibers are fast-oxidative-glycolytic and fast-glycolytic fibers.

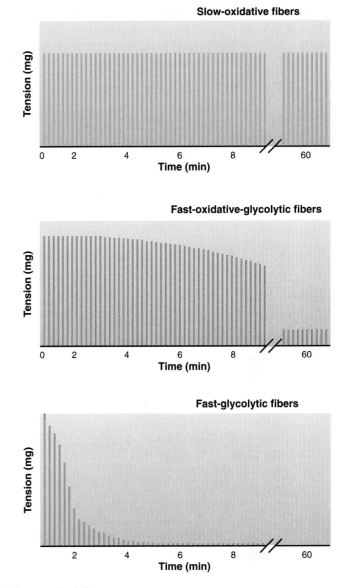

Figure 9–25

The rate of fatigue development in the three fiber types. Each vertical line is the contractile response to a brief tetanic stimulus and relaxation. The contractile responses occurring between about 9 min and 60 min are not shown on the figure.

Most muscles are composed of all three motor unit types interspersed with each other (**Figure 9–26**). No muscle has only a single fiber type. Depending on the proportions of the fiber types present, muscles can differ considerably in their maximal contraction speed, strength, and fatigability. For example, the muscles of the back, which must be able to maintain their activity for long periods of time without fatigue while supporting an upright posture, contain large numbers of slow-oxidative and fast-oxidative-glycolytic fibers. In contrast, the muscles in the arms, which may be called upon to produce large amounts of tension over a short time period, as when lifting a heavy object, have a greater proportion of fast-glycolytic fibers.

We will next use the characteristics of single fibers to describe whole-muscle contraction and its control.

Control of Muscle Tension

The total tension a muscle can develop depends upon two factors: (1) the amount of tension developed by each fiber, and (2) the number of fibers contracting at any time. By controlling these two factors, the nervous system controls whole-muscle tension as well as shortening velocity. The conditions that determine the amount of tension developed in a single fiber have been discussed previously and are summarized in **Table 9–4**.

The number of fibers contracting at any time depends on: (1) the number of fibers in each motor unit (motor unit size), and (2) the number of active motor units.

Motor unit size varies considerably from one muscle to another. The muscles in the hand and eye, which produce very delicate movements, contain small motor units. For example, one motor neuron innervates only about 13 fibers in an eye muscle. In contrast, in the more coarsely controlled muscles of the legs, each motor unit is large, containing hundreds and, in some cases, several thousand fibers. When a muscle is composed of small motor units, the total tension the muscle pro-

duces can be increased in small steps by activating additional motor units. If the motor units are large, large increases in tension will occur as each additional motor unit is activated. Thus, finer control of muscle tension is possible in muscles with small motor units.

The force a single fiber produces, as we have seen earlier, depends in part on the fiber diameter—the greater the diameter, the greater the force. We have also noted that fast-glycolytic fibers have the largest diameters. Thus, a motor unit composed of 100 fast-glycolytic fibers produces more force than a motor unit composed of 100 slow-oxidative fibers. In addition, fast-glycolytic motor units tend to have more muscle fibers. For both of these reasons, activating a fast-glycolytic

Table 9–3   Characteristics of the Three Types of Skeletal Muscle Fibers

	Slow-Oxidative Fibers (Type I)	Fast-Oxidative-Glycolytic Fibers (Type IIa)	Fast-Glycolytic Fibers (Type IIb)*
Primary source of ATP production	Oxidative phosphorylation	Oxidative phosphorylation	Glycolysis
Mitochondria	Many	Many	Few
Capillaries	Many	Many	Few
Myoglobin content	High (red muscle)	High (red muscle)	Low (white muscle)
Glycolytic enzyme activity	Low	Intermediate	High
Glycogen content	Low	Intermediate	High
Rate of fatigue	Slow	Intermediate	Fast
Myosin-ATPase activity	Low	High	High
Contraction velocity	Slow	Fast	Fast
Fiber diameter	Small	Intermediate	Large
Motor unit size	Small	Intermediate	Large
Size of motor neuron innervating fiber	Small	Intermediate	Large

*Type IIb fibers are sometimes designated as type IIx in the human muscle physiology literature.

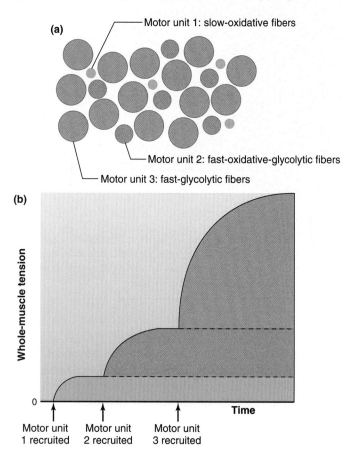

(a) Motor unit 1: slow-oxidative fibers

Motor unit 2: fast-oxidative-glycolytic fibers

Motor unit 3: fast-glycolytic fibers

(b)

Whole-muscle tension

0

Time

Motor unit 1 recruited   Motor unit 2 recruited   Motor unit 3 recruited

Figure 9–26

(a) Diagram of a cross section through a muscle composed of three types of motor units. (b) Tetanic muscle tension resulting from the successive recruitment of the three types of motor units. Note that motor unit 3, composed of fast-glycolytic fibers, produces the greatest rise in tension because it is composed of the largest-diameter fibers and contains the largest number of fibers per motor unit.

Table 9–4   Factors Determining Muscle Tension

1. Tension developed by each fiber
   a. Action potential frequency (frequency-tension relation)
   b. Fiber length (length-tension relation)
   c. Fiber diameter
   d. Fatigue

2. Number of active fibers
   a. Number of fibers per motor unit
   b. Number of active motor units

motor unit will produce more force than activating a slow-oxidative motor unit.

The process of increasing the number of motor units that are active in a muscle at any given time is called **recruitment.** It is achieved by activating excitatory synaptic inputs to more motor neurons. The greater the number of active motor neurons, the more motor units recruited, and the greater the muscle tension.

Motor neuron size plays an important role in the recruitment of motor units. The size of a motor neuron refers to the diameter of the nerve cell body, which usually correlates with the diameter of its axon. Given the same number of sodium ions entering a cell at a single excitatory synapse in a large and in a small motor neuron, the small neuron will undergo a greater depolarization because these ions will be distributed over a smaller membrane surface area. Accordingly, given the same level of synaptic input, the smallest neurons will be recruited first—that is, they will begin to generate action potentials first. The larger neurons will be recruited only as the level of synaptic input increases. Because the smallest motor neurons innervate the slow-oxidative motor units (see Table 9–3), these motor units are recruited first, followed by fast-oxidative-glycolytic motor units, and finally, during very strong contractions, by fast-glycolytic motor units (see Figure 9–26).

Thus, during moderate-strength contractions, such as those that occur in most endurance types of exercise, relatively few fast-glycolytic motor units are recruited, and most of the activity occurs in the more fatigue-resistant oxidative fibers. The large fast-glycolytic motor units, which fatigue rapidly, begin to be recruited when the intensity of contraction exceeds about 40 percent of the maximal tension the muscle can produce.

In summary, the neural control of whole-muscle tension involves (1) the frequency of action potentials in individual motor units (to vary the tension generated by the fibers in that unit) and (2) the recruitment of motor units (to vary the number of active fibers). Most motor neuron activity occurs in bursts of action potentials, which produce tetanic contractions of individual motor units rather than single twitches. Recall that the tension of a single fiber increases only three- to fivefold when going from a twitch to a maximal tetanic contraction. Therefore, varying the frequency of action potentials in the neurons supplying them provides a way to make only three- to fivefold adjustments in the tension of the recruited motor units. The force a whole muscle exerts can be varied over a much wider range than this, from very delicate movements to extremely powerful contractions, by recruiting motor units. Thus, recruitment provides the primary means of varying tension in a whole muscle. Recruitment is controlled by the central commands from the motor centers in the brain to the various motor neurons (Chapter 10).

Control of Shortening Velocity

As we saw earlier, the velocity at which a single muscle fiber shortens is determined by (1) the load on the fiber and (2) whether the fiber is a fast or slow fiber. Translated to a whole muscle, these characteristics become (1) the load on the whole muscle and (2) the types of motor units in the muscle. For the whole muscle, however, recruitment becomes a third very important factor, one that explains how the shortening velocity can be varied from very fast to very slow even though the load on the muscle remains constant. Consider, for the sake of illustration, a muscle composed of only two motor units of the same size and fiber type. One motor unit by itself will lift a 4-g load more slowly than a 2-g load because the shortening velocity decreases with increasing load. When both units are active and a 4-g load is lifted, each motor unit bears only half the load, and its fibers will shorten as if it were lifting only a 2-g load. In other words, the muscle will lift the 4-g load at a higher velocity when both motor units are active. Recruitment of motor units thus leads to increases in both force and velocity.

Muscle Adaptation to Exercise

The regularity with which a muscle is used, as well as the duration and intensity of its activity, affect the properties of the muscle. If the neurons to a skeletal muscle are destroyed or the neuromuscular junctions become nonfunctional, the denervated muscle fibers will become progressively smaller in diameter, and the amount of contractile proteins they contain will decrease. This condition is known as ***denervation atrophy.*** A muscle can also atrophy with its nerve supply intact if the muscle is not used for a long period of time, as when a broken arm or leg is immobilized in a cast. This condition is known as ***disuse atrophy.***

In contrast to the decrease in muscle mass that results from a lack of neural stimulation, increased amounts of contractile activity—in other words, exercise—can produce an increase in the size (hypertrophy) of muscle fibers as well as changes in their capacity for ATP production.

Exercise that is of relatively low intensity but long duration (popularly called "aerobic exercise"), such as running or swimming, produces increases in the number of mitochondria in the fibers that are recruited in this type of activity. In addition, the number of capillaries around these fibers also increases. All these changes lead to an increase in the capacity for endurance activity with a minimum of fatigue. (Surprisingly, fiber diameter decreases slightly, and thus there is a small decrease in the maximal strength of muscles as a result of endurance training.) As we will see in later chapters, endurance exercise produces changes not only in the skeletal muscles but also in the respiratory and circulatory systems, changes that improve the delivery of oxygen and fuel molecules to the muscle.

In contrast, short-duration, high-intensity exercise (popularly called "strength training"), such as weight lifting, affects primarily the fast-glycolytic fibers, which are recruited during strong contractions. These fibers undergo an increase in fiber diameter (hypertrophy) due to the increased synthesis of actin and myosin filaments, which form more myofibrils. In addition, glycolytic activity is increased by increasing the synthesis of glycolytic enzymes. The result of such high-intensity exercise is an increase in the strength of the muscle and the bulging muscles of a conditioned weight lifter. Such muscles, although very powerful, have little capacity for endurance, and they fatigue rapidly.

Exercise produces limited change in the types of myosin enzymes the fibers form and thus little change in the proportions of fast and slow fibers in a muscle. As described previously, however, exercise does change the rates at which metabolic enzymes are synthesized, leading to changes in the proportion of oxidative and glycolytic fibers within a muscle. With endurance training, there is a decrease in the number of fast-glycolytic fibers and an increase in the number of fast-oxidative-glycolytic fibers as the oxidative capacity of the fibers increases. The reverse occurs with strength training as fast-oxidative-glycolytic fibers convert to fast-glycolytic fibers.

The signals responsible for all these changes in muscle with different types of activity are unknown. They are related to the frequency and intensity of the contractile activity in the muscle fibers and thus to the pattern of action potentials produced in the muscle over an extended period of time.

Because different types of exercise produce quite different changes in the strength and endurance capacity of a muscle, an individual performing regular exercise to improve muscle performance must choose a type of exercise compatible with the type of activity he or she ultimately wishes to perform. Thus, lifting weights will not improve the endurance of a long-distance runner, and jogging will not produce the increased strength a weight lifter desires. Most exercises, however, produce some effect on both strength and endurance.

These changes in muscle in response to repeated periods of exercise occur slowly over a period of weeks. If regular exercise ceases, the changes in the muscle that occurred as a result of the exercise will slowly revert to their unexercised state.

The maximum force a muscle generates decreases by 30 to 40 percent between the ages of 30 and 80. This decrease in tension-generating capacity is due primarily to a decrease in average fiber diameter. Some of the change is simply the result of diminishing physical activity and can be prevented by regular exercise. The ability of a muscle to adapt to exercise, however, decreases with age: The same intensity and duration of exercise in an older individual will not produce the same amount of change as in a younger person.

This effect of aging, however, is only partial, and there is no question that even in the elderly, exercise can produce significant adaptation. Aerobic training has received major attention because of its effect on the cardiovascular system (Chapter 12). Strength training to even a modest degree, however, can partially prevent the loss of muscle tissue that occurs with aging. Moreover, it helps maintain stronger bones and joints.

Extensive exercise by an individual whose muscles have not been used in performing that particular type of exercise leads to muscle soreness the next day. This soreness is the result of a mild inflammation in the muscle, which occurs whenever tissues are damaged (Chapter 18). The most severe inflammation results from lengthening contractions, indicating that the lengthening of a muscle fiber by an external force produces greater muscle damage than does either shortening or isometric contraction. Thus, exercising by gradually lowering weights will produce greater muscle soreness than an equivalent amount of weight lifting. This explains a phenomenon well-known to athletic trainers: the shortening contractions of leg muscles used to run *up* flights of stairs result in far less soreness than the lengthening contractions used for running *down*.

The effects of anabolic steroids on skeletal muscle growth and strength are described in Chapter 17.

Lever Action of Muscles and Bones

A contracting muscle exerts a force on bones through its connecting tendons. When the force is great enough, the bone moves as the muscle shortens. A contracting muscle exerts only a pulling force, so that as the muscle shortens, the bones it is attached to are pulled toward each other. **Flexion** refers to the *bending* of a limb at a joint, whereas **extension** is the *straightening* of a limb (**Figure 9–27**). These opposing motions require at least two muscles, one to cause flexion and the other extension. Groups of muscles that produce oppositely directed movements at a joint are known as **antagonists.** For example, from Figure 9–27 we can see that contraction of the biceps causes flexion of the arm at the elbow, whereas contraction of the antagonistic muscle, the triceps, causes the arm to extend. Both muscles exert only a pulling force upon the forearm when they contract.

Sets of antagonistic muscles are required not only for flexion-extension, but also for side-to-side movements or rota-

tion of a limb. The contraction of some muscles leads to two types of limb movement, depending on the contractile state of other muscles acting on the same limb. For example, contraction of the gastrocnemius muscle in the calf causes a flexion of the leg at the knee, as in walking (**Figure 9–28**). However, contraction of the gastrocnemius muscle with the simultaneous contraction of the quadriceps femoris (which causes extension of the lower leg) prevents the knee joint from bending, leaving only the ankle joint capable of moving. The foot is extended, and the body rises on tiptoe.

The muscles, bones, and joints in the body are arranged in lever systems. The basic principle of a lever is illustrated by the flexion of the arm by the biceps muscle (**Figure 9–29**), which exerts an upward pulling tension on the forearm about 5 cm away from the elbow joint. In this example, a 10-kg weight held in the hand exerts a downward load of 10 kg about 35 cm from the elbow. A law of physics tells us that the forearm is in mechanical equilibrium when the product of the downward load (10 kg) and its distance from the elbow (35 cm) is equal to the product of the iso-

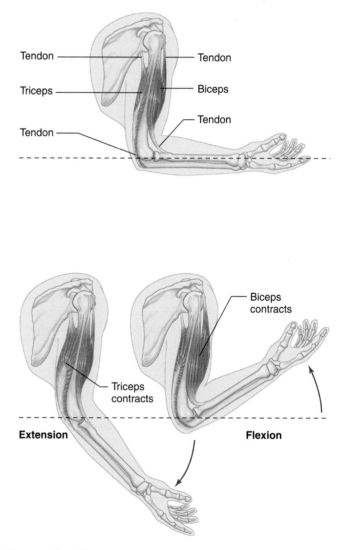

Figure 9–27

Antagonistic muscles for flexion and extension of the forearm.

metric tension exerted by the muscle (X), and its distance from the elbow (5 cm); that is, $10 \times 35 = 5 \times X$. Thus $X = 70$ kg. The important point is that this system is working at a mechanical disadvantage because the tension exerted by the muscle (70 kg) is considerably greater than the load (10 kg) it is supporting.

However, the mechanical disadvantage that most muscle lever systems operate under is offset by increased maneuverability. As illustrated in **Figure 9–30**, when the biceps shortens 1 cm, the hand moves through a distance of 7 cm. Because the muscle shortens 1 cm in the same amount of time that the hand moves 7 cm, the velocity at which the hand moves is seven times greater than the rate of muscle shortening. The

lever system amplifies the velocity of muscle shortening so that short, relatively slow movements of the muscle produce faster movements of the hand. Thus, a pitcher can throw a baseball at 90 to 100 mph even though his muscles shorten at only a small fraction of this velocity.

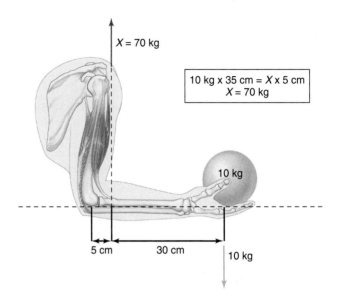

Figure 9–29

Mechanical equilibrium of forces acting on the forearm while supporting a 10-kg load.

Figure 9–29 physiological *inquiry* (p*i*)

■ Describe what would happen if this weight was mounted on a rod that moved it 10 cm farther away from the elbow and the tension generated by the muscle was increased to 85 kg.

Answer can be found at end of chapter.

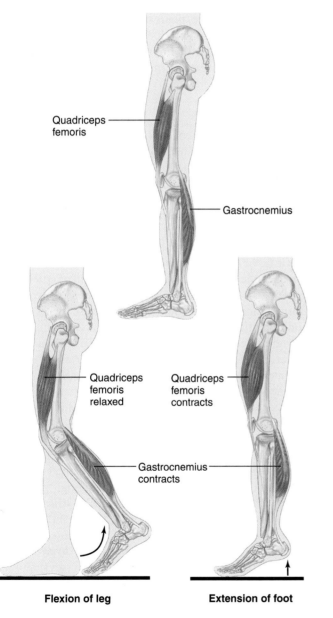

Flexion of leg   **Extension of foot**

Figure 9–28

Contraction of the gastrocnemius muscle in the calf can lead either to flexion of the leg, if the quadriceps femoris muscle is relaxed, or to extension of the foot, if the quadriceps is contracting, preventing the knee joint from bending.

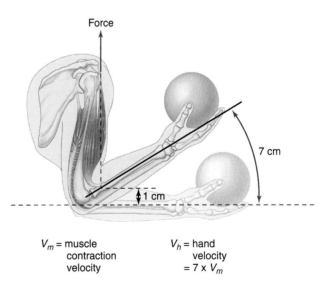

V_m = muscle contraction velocity

V_h = hand velocity = 7 × V_m

Figure 9–30

The lever system of the arm amplifies the velocity of the biceps muscle, producing a greater velocity of the hand. The range of movement is also amplified (1 cm of shortening by the muscle produces 7 cm of movement by the hand).

A number of diseases can affect the contraction of skeletal muscle. Many of them are caused by defects in the parts of the nervous system that control contraction of the muscle fibers rather than by defects in the muscle fibers themselves. For example, *poliomyelitis* is a viral disease that destroys motor neurons, leading to the paralysis of skeletal muscle, and may result in death due to respiratory failure.

Muscle Cramps

Involuntary tetanic contraction of skeletal muscles produces *muscle cramps.* During cramping, action potentials fire at abnormally high rates, a much greater rate than occurs during maximal voluntary contraction. The specific cause of this high activity is uncertain, but it is probably related to electrolyte imbalances in the extracellular fluid surrounding both the muscle and nerve fibers. These imbalances may arise from overexercise or persistent dehydration, and they can directly induce action potentials in motor neurons and muscle fibers. Another theory is that chemical imbalances within the muscle stimulate sensory receptors in the muscle, and the motor neurons to the area are activated by reflex when those signals reach the spinal cord.

Hypocalcemic Tetany

Hypocalcemic tetany is the involuntary tetanic contraction of skeletal muscles that occurs when the extracellular calcium concentration falls to about 40 percent of its normal value. This may seem surprising, because we have seen that calcium is required for excitation-contraction coupling. However, recall that this calcium is sarcoplasmic reticulum calcium, not extracellular calcium. The effect of changes in extracellular calcium is exerted not on the sarcoplasmic reticulum calcium, but directly on the plasma membrane. Low extracellular calcium (**hypocalcemia**) increases the opening of sodium channels in excitable membranes, leading to membrane depolarization and the spontaneous firing of action potentials. This causes the increased muscle contractions, which are similar to muscular cramping. Chapter 11 discusses the mechanisms controlling the extracellular concentration of calcium ions.

Muscular Dystrophy

This disease is one of the most frequently encountered genetic diseases, affecting one in every 3500 males (but many fewer females) born in America. *Muscular dystrophy* is associated with the progressive degeneration of skeletal and cardiac muscle fibers, weakening the muscles and leading ultimately to death from respiratory or cardiac failure (**Figure 9–31**). The symptoms become evident at about 2 to 6 years of age, and most affected individuals do not survive far beyond the age of 20.

The recessive gene responsible for a major form of muscular dystrophy (***Duchenne muscular dystrophy***) has been identified on the X chromosome, and Duchenne muscular dystrophy is thus a sex-linked recessive disease. (As described in Chapter 17, girls have two X chromosomes and boys only one. Consequently, a girl with one abnormal X chromosome and one normal one will not generally develop the disease. This is why the disease is so much more common in boys.) This gene codes for a protein known as dystrophin, which is either present in a nonfunctional form or absent in patients with the disease. Dystrophin is a large protein that links cytoskeletal proteins to membrane glycoproteins. It resembles other known cytoskeletal proteins and may be involved in maintaining the structural integrity of the plasma membrane or of elements within the membrane, such as ion channels. In its absence, fibers subjected to repeated structural deformation during contraction are susceptible to membrane rupture and cell death. Preliminary attempts are being made to treat the disease by inserting the normal gene into dystrophic muscle cells.

Myasthenia Gravis

Myasthenia gravis is a collection of neuromuscular disorders characterized by muscle fatigue and weakness that progressively worsens as the muscle is used. It affects about one out of every 7500 Americans, occurring more often in women than men. The most common cause is the destruction of nicotinic ACh receptor proteins of the motor end plate, mediated by antibodies of a person's own immune system (see Chapter 18 for a description of autoimmune diseases). The release of ACh from the nerve terminals is normal, but the magnitude of the end-plate potential is markedly reduced because of the decreased availability of receptors. Even in normal muscle, the amount of ACh released with each action potential decreases with repetitive activity, and thus the magnitude of the resulting EPP falls. In normal muscle, however, the EPP remains well above the threshold necessary to initiate a muscle action potential. In contrast, after a few motor nerve impulses in a myasthenia gravis patient, the magnitude of the EPP falls below the threshold for initiating a muscle action potential.

A number of approaches are currently used to treat the disease. One is to administer acetylcholinesterase inhibitors (e.g., *neostygmine*). This can partially compensate for the reduction in available ACh receptors by prolonging the time that acetylcholine is available at the synapse. Other therapies aim at blunting the immune response. Treatment with glucocorticoids is one way that immune function is suppressed (Chapter 11). Removal of the thymus gland (*thymectomy*) reduces the production of antibodies and reverses symptoms in about 50 percent of patients. *Plasmapheresis* is a treatment that involves removing the liquid fraction of blood (plasma), which contains the offending antibodies. A combination of these treatments has greatly reduced the mortality rate for myasthenia gravis. ■

Figure 9–31

Boy with Duchenne muscular dystrophy. Muscles of the hip girdle and trunk are the first to weaken, requiring patients to use their arms to "climb up" the legs in order to go from lying to standing.

■ SECTION A SUMMARY

I. There are three types of muscle—skeletal, smooth, and cardiac. Skeletal muscle is attached to bones and moves and supports the skeleton. Smooth muscle surrounds hollow cavities and tubes. Cardiac muscle is the muscle of the heart.

Structure

I. Skeletal muscles, composed of cylindrical muscle fibers (cells), are linked to bones by tendons at each end of the muscle.

II. Skeletal muscle fibers have a repeating, striated pattern of light and dark bands due to the arrangement of the thick and thin filaments within the myofibrils.

III. Actin-containing thin filaments are anchored to the Z lines at each end of a sarcomere. Their free ends partially overlap the myosin-containing thick filaments in the A band at the center of the sarcomere.

Molecular Mechanisms of Skeletal Muscle Contraction

I. When a skeletal muscle fiber actively shortens, the thin filaments are propelled toward the center of their sarcomere by movements of the myosin cross-bridges that bind to actin.

   a. The two globular heads of each cross-bridge contain a binding site for actin and an enzymatic site that splits ATP.

   b. The four steps occurring during each cross-bridge cycle are summarized in Figure 9–8. The cross-bridges undergo repeated cycles during a contraction, each cycle producing only a small increment of movement.

   c. The three functions of ATP in muscle contraction are summarized in Table 9–1.

II. In a resting muscle, tropomyosin molecules that are in contact with the actin subunits of the thin filaments block the attachment of cross-bridges to actin.

III. Contraction is initiated by an increase in cytosolic calcium concentration. The calcium ions bind to troponin, producing a change in its shape that is transmitted via tropomyosin to uncover the binding sites on actin, allowing the cross-bridges to bind to the thin filaments.

   a. The rise in cytosolic calcium concentration is triggered by an action potential in the plasma membrane. The action potential is propagated into the interior of the fiber along the transverse tubules to the region of the sarcoplasmic reticulum, where dihydropyridine receptors sense the voltage change and pull open ryanodine receptors, releasing calcium ions from the reticulum.

   b. Relaxation of a contracting muscle fiber occurs as a result of the active transport of cytosolic calcium ions back into the sarcoplasmic reticulum.

IV. Branches of a motor neuron axon form neuromuscular junctions with the muscle fibers in its motor unit. Each muscle fiber is innervated by a branch from only one motor neuron.

   a. Acetylcholine released by an action potential in a motor neuron binds to receptors on the motor end plate of the muscle membrane, opening ion channels that allow the passage of sodium and potassium ions, which depolarize the end-plate membrane.

   b. A single action potential in a motor neuron is sufficient to produce an action potential in a skeletal muscle fiber.

   c. Figure 9–15 summarizes events at the neuromuscular junction.

V. Table 9–2 summarizes the events leading to the contraction of a skeletal muscle fiber.

Mechanics of Single-Fiber Contraction

I. Contraction refers to the turning on of the cross-bridge cycle. Whether there is an accompanying change in muscle length depends upon the external forces acting on the muscle.

II. Three types of contractions can occur following activation of a muscle fiber: (1) an isometric contraction in which the muscle generates tension but does not change length; (2) an isotonic contraction in which the muscle shortens (concentric), moving a load; and (3) a lengthening (eccentric) contraction in which the external load on the muscle causes the muscle to lengthen during the period of contractile activity.

III. Increasing the frequency of action potentials in a muscle fiber increases the mechanical response (tension or shortening) up to the level of maximal tetanic tension.

IV. Maximum isometric tetanic tension is produced at the optimal sarcomere length L_0. Stretching a fiber beyond its optimal length or decreasing the fiber length below L_0 decreases the tension generated.

V. The velocity of muscle fiber shortening decreases with increases in load. Maximum velocity occurs at zero load.

Skeletal Muscle Energy Metabolism

I. Muscle fibers form ATP by the transfer of phosphate from creatine phosphate to ADP, by oxidative phosphorylation of ADP in mitochondria, and by substrate-level phosphorylation of ADP in the glycolytic pathway.

II. At the beginning of exercise, muscle glycogen is the major fuel consumed. As the exercise proceeds, glucose and fatty acids from the blood provide most of the fuel, and fatty acids become progressively more important during prolonged exercise. When the intensity of exercise exceeds about 70 percent of maximum, glycolysis begins to contribute an increasing fraction of the total ATP generated.

III. A variety of factors cause muscle fatigue, including internal changes in acidity, cross-bridge inhibition, glycogen depletion, and excitation-contraction coupling failure, but not a lack of ATP.

Types of Skeletal Muscle Fibers

I. Three types of skeletal muscle fibers can be distinguished by their maximal shortening velocities and the predominate pathway they use to form ATP: slow-oxidative, fast-oxidative-glycolytic, and fast-glycolytic fibers.

   a. Differences in maximal shortening velocities are due to different myosin enzymes with high or low ATPase activities, giving rise to fast and slow fibers.

   b. Fast-glycolytic fibers have a larger average diameter than oxidative fibers and therefore produce greater tension, but they also fatigue more rapidly.

II. All the muscle fibers in a single motor unit belong to the same fiber type, and most muscles contain all three types.

III. Table 9–3 summarizes the characteristics of the three types of skeletal muscle fibers.

Whole-Muscle Contraction

I. The tension produced by whole-muscle contraction depends on the amount of tension each fiber develops and the number of active fibers in the muscle (Table 9–4).

II. Muscles that produce delicate movements have a small number of fibers per motor unit, whereas large powerful muscles have much larger motor units.

III. Fast-glycolytic motor units not only have large-diameter fibers but also tend to have large numbers of fibers per motor unit.

IV. Increases in muscle tension are controlled primarily by increasing the number of active motor units in a muscle, a process known as recruitment. Slow-oxidative motor units are recruited first during weak contractions, then fast-oxidative-glycolytic motor units, and finally fast-glycolytic motor units during very strong contractions.

V. Increasing motor-unit recruitment increases the velocity at which a muscle will move a given load.

VI. Exercise can alter a muscle's strength and susceptibility to fatigue.

   a. Long-duration, low-intensity exercise increases a fiber's capacity for oxidative ATP production by increasing the number of mitochondria and blood vessels in the muscle, resulting in increased endurance.

   b. Short-duration, high-intensity exercise increases fiber diameter as a result of increased synthesis of actin and myosin, resulting in increased strength.

VII. Movement around a joint requires two antagonistic groups of muscles: one flexes the limb at the joint and the other extends the limb.

VIII. The lever system of muscles and bones requires muscle tension far greater than the load in order to sustain a load in an isometric contraction, but the lever system produces a shortening velocity at the end of the lever arm that is greater than the muscle-shortening velocity.

Additional Clinical Examples

I. Muscle cramps are involuntary tetanic contractions occurring when overexercise and dehydration cause electrolyte imbalances in the fluid surrounding muscle and nerve fibers.

II. When extracellular calcium falls below normal, sodium channels of nerve and muscle open spontaneously, which causes the excessive muscle contractions of hypocalcemic tetany.

III. Muscular dystrophies, the most commonly occurring genetic disorders, result from defects of a muscle membrane-stabilizing protein known as dystrophin. Muscles of individuals with this condition progressively degenerate with use.

IV. Myasthenia gravis is an autoimmune disorder in which destruction of ACh receptors of the motor end plate causes progressive loss of the ability to activate skeletal muscles.

■ SECTION A KEY TERMS

A band 257	end-plate potential (EPP) 264
acetylcholine (ACh) 264	excitation-contraction
acetylcholinesterase 266	coupling 261
actin 257	extension 278
antagonist 278	fast fiber 274
cardiac muscle 255	fast-glycolytic fiber 274
central command fatigue 274	fast-oxidative-glycolytic
concentric contraction 267	fiber 274
contraction 258	flexion 278
contraction time 269	foot process 262
creatine phosphate 272	fused tetanus 270
cross-bridge 258	glycolytic fiber 274
cross-bridge cycle 259	H zone 257
dihydropyridine (DHP)	heavy chains 259
receptor 262	hypertrophy 256
eccentric contraction 267	hypocalcemia 280

■ SECTION A CLINICAL TERMS

■ SECTION A REVIEW QUESTIONS

1. List the three types of muscle cells and their locations.
2. Diagram the arrangement of thick and thin filaments in a striated muscle sarcomere, and label the major bands that give rise to the striated pattern.
3. Describe the organization of myosin, actin, tropomyosin, and troponin molecules in the thick and thin filaments.
4. Describe the four steps of one cross-bridge cycle.
5. Describe the physical state of a muscle fiber in rigor mortis and the conditions that produce this state.
6. What three events in skeletal muscle contraction and relaxation depend on ATP?
7. What prevents cross-bridges from attaching to sites on the thin filaments in a resting skeletal muscle?
8. Describe the role and source of calcium ions in initiating contraction in skeletal muscle.
9. Describe the location, structure, and function of the sarcoplasmic reticulum in skeletal muscle fibers.
10. Describe the structure and function of the transverse tubules.
11. Describe the events that result in the relaxation of skeletal muscle fibers.
12. Define a motor unit and describe its structure.
13. Describe the sequence of events by which an action potential in a motor neuron produces an action potential in the plasma membrane of a skeletal muscle fiber.
14. What is an end-plate potential, and what ions produce it?
15. Compare and contrast the transmission of electrical activity at a neuromuscular junction with that at a synapse.
16. Describe isometric, concentric, and eccentric contractions.
17. What factors determine the duration of an isotonic twitch in skeletal muscle? An isometric twitch?
18. What effect does increasing the frequency of action potentials in a skeletal muscle fiber have upon the force of contraction? Explain the mechanism responsible for this effect.
19. Describe the length-tension relationship in skeletal muscle fibers.
20. Describe the effect of increasing the load on a skeletal muscle fiber on the velocity of shortening.
21. What is the function of creatine phosphate in skeletal muscle contraction?
22. What fuel molecules are metabolized to produce ATP during skeletal muscle activity?
23. List the factors responsible for skeletal muscle fatigue.
24. What component of skeletal muscle fibers accounts for the differences in the fibers' maximal shortening velocities?
25. Summarize the characteristics of the three types of skeletal muscle fibers.
26. Upon what three factors does the amount of tension developed by a whole skeletal muscle depend?
27. Describe the process of motor-unit recruitment in controlling (a) whole-muscle tension and (b) velocity of whole-muscle shortening.
28. During increases in the force of skeletal muscle contraction, what is the order of recruitment of the different types of motor units?
29. What happens to skeletal muscle fibers when the motor neuron to the muscle is destroyed?
30. Describe the changes that occur in skeletal muscles following a period of (a) long-duration, low-intensity exercise training; and (b) short-duration, high-intensity exercise training.
31. How are skeletal muscles arranged around joints so that a limb can push or pull?
32. What are the advantages and disadvantages of the muscle-bone-joint lever system?

Smooth and Cardiac Muscle

Having described in depth the mechanisms underlying skeletal muscle function, we now turn our attention to the other muscle types, beginning with smooth muscle. Two characteristics are common to all smooth muscles: they lack the cross-striated banding pattern found in skeletal and cardiac fibers (which makes them "smooth"), and the nerves to them are derived from the autonomic division of the nervous system rather than the somatic division. Thus, smooth muscle is not normally under direct voluntary control.

Smooth muscle, like skeletal muscle, uses cross-bridge movements between actin and myosin filaments to generate force, and calcium ions to control cross-bridge activity. However, the organization of the contractile filaments and the process of excitation-contraction coupling are quite different in smooth muscle. Furthermore, there is considerable diversity among smooth muscles with respect to the excitation-contraction coupling mechanism.

Structure of Smooth Muscle

Each smooth muscle cell is spindle-shaped, with a diameter between 2 and 10 µm, and length ranging from 50 to 400 µm. They are much smaller than skeletal muscle fibers, which are 10 to 100 µm wide and can be tens of centimeters long (see Figure 9–1). Skeletal fibers are sometimes large enough to run the entire length of the muscles they are found in, whereas many individual smooth muscle cells are generally interconnected to form sheetlike layers of cells (**Figure 9–32**). Skeletal muscle fibers are multinucleate cells that are unable to divide once they have differentiated; smooth muscle cells have a single nucleus and have the capacity to divide throughout the life of an individual. A variety of paracrine agents can stimulate smooth muscle cells to divide, often in response to tissue injury.

Just like skeletal muscle fibers, smooth muscle cells have thick myosin-containing filaments and thin actin-containing filaments. Although tropomyosin is present in the thin filaments, the regulatory protein troponin is absent. The thin filaments are anchored either to the plasma membrane or to cytoplasmic structures known as **dense bodies,** which are functionally similar to the Z lines in skeletal muscle fibers. Note in **Figure 9–33** that the filaments are oriented slightly diagonally to the long axis of the cell. When the fiber shortens, the regions of the plasma membrane between the points where actin is attached to the membrane balloon out. The thick and thin filaments are not organized into myofibrils, as in striated muscles, and there is no regular alignment of these filaments into sarcomeres, which accounts for the absence of a banding pattern. Nevertheless, smooth muscle contraction occurs by a sliding-filament mechanism.

The concentration of myosin in smooth muscle is only about one-third of that in striated muscle, whereas the actin content can be twice as great. In spite of these differences, the maximal tension per unit of cross-sectional area devel-

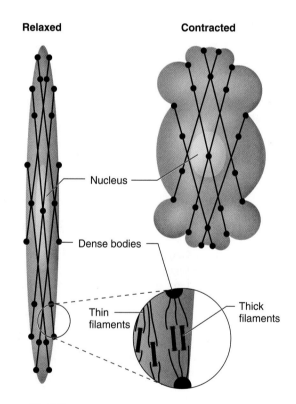

Relaxed   **Contracted**

Nucleus

Dense bodies

Thin filaments

Thick filaments

Figure 9–33

Thick and thin filaments in smooth muscle are arranged in slightly diagonal chains that are anchored to the plasma membrane or to dense bodies within the cytoplasm. When activated, the thick and thin filaments slide past each other, causing the smooth muscle fiber to shorten and thicken.

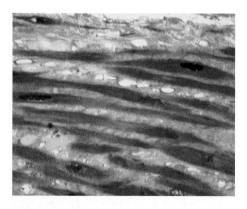

Figure 9–32

Photomicrograph of a sheet of smooth muscle cells. Note the spindle shape, single nucleus, and lack of striations.

oped by smooth muscles is similar to that developed by skeletal muscle.

The isometric tension produced by smooth muscle fibers varies with fiber length in a manner qualitatively similar to that observed in skeletal muscle—tension development is highest at intermediate lengths and lower at shorter or longer lengths. However, in smooth muscle significant force is generated over a relatively broad range of muscle lengths compared to that of skeletal muscle. This property is highly adaptive because most smooth muscles surround hollow structures and organs that undergo changes in volume with accompanying changes in the lengths of the smooth muscle fibers in their walls. Even with relatively large increases in volume, as during the accumulation of large amounts of urine in the bladder, the smooth muscle fibers in the wall retain some ability to develop tension, whereas such distortion might stretch skeletal muscle fibers beyond the point of thick- and thin-filament overlap.

Smooth Muscle Contraction and Its Control

Changes in cytosolic calcium concentration control the contractile activity in smooth muscle fibers, as in striated muscle. However, there are significant differences in the way calcium activates cross-bridge cycling and in the mechanisms by which stimulation leads to alterations in calcium concentration.

Cross-Bridge Activation

Because smooth muscle lacks the calcium-binding protein troponin, tropomyosin is never held in a position that blocks cross-bridge access to actin. Thus, the thin filament cannot act as the switch that regulates cross-bridge cycling. *Instead, cross-bridge cycling in smooth muscle is controlled by a calcium-regulated enzyme that phosphorylates myosin.* Only the phosphorylated form of smooth muscle myosin can bind to actin and undergo cross-bridge cycling.

The following sequence of events occurs after a rise in cytosolic calcium in a smooth muscle fiber (**Figure 9–34**): (1) Calcium binds to calmodulin, a calcium-binding protein that is present in the cytosol of most cells (Chapter 5) and whose structure is related to that of troponin. (2) The calcium-calmodulin complex binds to another cytosolic protein, **myosin light-chain kinase,** thereby activating the enzyme. (3) Active myosin light-chain kinase then uses ATP to phosphorylate myosin light chains in the globular head of myosin. (4) Phosphorylation of myosin drives the cross-bridge away from the thick filament backbone, allowing it to bind to actin. (5) Cross-bridges go through repeated cycles of force generation as long as myosin light chains are phosphorylated. A key difference here is that calcium-mediated changes in the thick filaments turn on cross-bridge activity in smooth muscle, whereas in striated muscle, calcium mediates changes in the thin filaments.

The smooth muscle form of myosin has a very low rate of ATPase activity, on the order of 10 to 100 times less than that

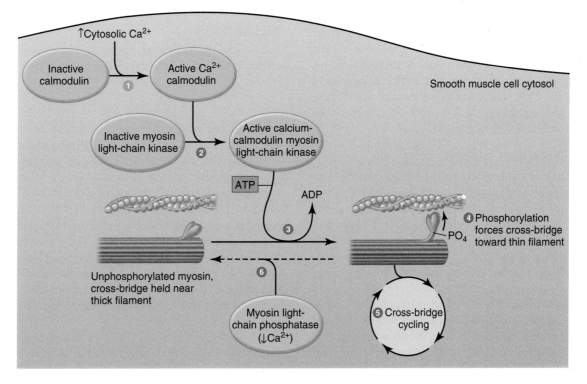

Figure 9–34

Activation of smooth muscle contraction by calcium. See text for description of the numbered steps.

of skeletal muscle myosin. Because the rate of ATP splitting determines the rate of cross-bridge cycling and, thus, shortening velocity, smooth muscle shortening is much slower than that of skeletal muscle. Due to this slow rate of energy usage, smooth muscle does not undergo fatigue during prolonged periods of activity. Note the distinction between the two roles of ATP in smooth muscle: splitting one ATP to transfer a phosphate onto myosin light chain *(phosphorylation)* starts a cross-bridge cycling, after which one ATP per cycle is split *(hydrolysis)* to provide the energy for force generation.

To relax a contracted smooth muscle, myosin must be dephosphorylated because dephosphorylated myosin is unable to bind to actin. This dephosphorylation is mediated by the enzyme **myosin light-chain phosphatase,** which is continuously active in smooth muscle during periods of rest and contraction (step 6 in Figure 9–34). When cytosolic calcium rises, the rate of myosin phosphorylation by the activated kinase exceeds the rate of dephosphorylation by the phosphatase, and the amount of phosphorylated myosin in the cell increases, producing a rise in tension. When the cytosolic calcium concentration decreases, the rate of phosphorylation falls below that of dephosphorylation, and the amount of phosphorylated myosin decreases, producing relaxation.

In some smooth muscles, when stimulation is persistent and the cytosolic calcium concentration remains elevated, the rate of ATP splitting by the cross-bridges declines even though isometric tension is maintained. This condition, known as the **latch state,** occurs when a phosphorylated cross-bridge becomes dephosphorylated while still attached to actin. In this circumstance it can maintain tension in an almost rigorlike state without movement. Dissociation of these dephosphorylated cross-bridges from actin by the binding of ATP does occur, but at a much slower rate than dissociation of phosphorylated bridges. The net result is the ability to maintain tension for long periods of time with a very low rate of ATP consumption. A good example of the usefulness of this mechanism is seen in sphincter muscles of the gastrointestinal tract, where smooth muscle must maintain contraction for prolonged periods. **Figure 9–35** compares the activation of smooth and skeletal muscles.

Sources of Cytosolic Calcium

Two sources of calcium contribute to the rise in cytosolic calcium that initiates smooth muscle contraction: (1) the sarcoplasmic reticulum and (2) extracellular calcium entering the cell through plasma-membrane calcium channels. The amount of calcium each of these two sources contributes differs among various smooth muscles, some being more dependent on extracellular calcium than the stores in the sarcoplasmic reticulum, and vice versa.

First we'll examine the role of the sarcoplasmic reticulum. The total quantity of this organelle in smooth muscle is smaller than in skeletal muscle, and it is not arranged in any specific pattern in relation to the thick and thin filaments. Moreover, there are no T-tubules connected to the plasma membrane in smooth muscle. The small cell diameter and the slow rate of contraction do not require such a rapid mechanism for getting an excitatory signal into the muscle

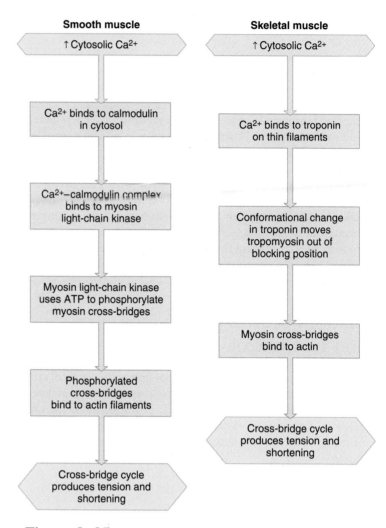

Figure 9–35
Pathways leading from increased cytosolic calcium to cross-bridge cycling in smooth and skeletal muscle fibers.

cell. Portions of the sarcoplasmic reticulum are located near the plasma membrane, however, forming associations similar to the relationship between T-tubules and the lateral sacs in skeletal muscle. Action potentials in the plasma membrane can be coupled to the release of sarcoplasmic reticulum calcium at these sites. In some types of smooth muscles, action potentials are not necessary for calcium release. Instead, second messengers released from the plasma membrane, or generated in the cytosol in response to the binding of extracellular chemical messengers to plasma-membrane receptors, can trigger the release of calcium from the more centrally located sarcoplasmic reticulum (review Figure 5–10 for a specific example).

What about extracellular calcium in excitation-contraction coupling? There are voltage-sensitive calcium channels in the plasma membranes of smooth muscle cells, as well as calcium channels controlled by extracellular chemical messengers. The calcium concentration in the extracellular fluid is 10,000 times greater than in the cytosol, thus the opening of calcium channels in the plasma membrane results in an increased flow of calcium into the cell. Because of the small cell size, the entering calcium does not have far to diffuse to reach binding sites within the cell.

Removal of calcium from the cytosol to bring about relaxation is achieved by the active transport of calcium back into the sarcoplasmic reticulum as well as out of the cell across the plasma membrane. The rate of calcium removal in smooth muscle is much slower than in skeletal muscle, with the result that a single twitch lasts several seconds in smooth muscle compared to a fraction of a second in skeletal muscle.

The degree of activation also differs between muscle types. In skeletal muscle a single action potential releases sufficient calcium to saturate all troponin sites on the thin filaments, whereas only a portion of the cross-bridges are activated in a smooth muscle fiber in response to most stimuli. Therefore, the tension generated by a smooth muscle cell can be *graded* by varying cytosolic calcium concentration. The greater the increase in calcium concentration, the greater the number of cross-bridges activated, and the greater the tension.

In some smooth muscles, the cytosolic calcium concentration is sufficient to maintain a low level of basal cross-bridge activity in the absence of external stimuli. This activity is known as **smooth muscle tone.** Factors that alter the cytosolic calcium concentration also vary the intensity of smooth muscle tone.

As in our description of skeletal muscle, we have approached the question of excitation-contraction coupling in smooth muscle by first describing the coupling (the changes in cytosolic calcium). Now we must back up a step and ask what constitutes the excitation that elicits these changes in calcium concentration.

Membrane Activation

In contrast to skeletal muscle, in which membrane activation is dependent on a single input—the somatic neurons to the muscle—many inputs to a smooth muscle plasma membrane can alter the contractile activity of the muscle (**Table 9–5**).

Table 9–5	Inputs Influencing Smooth Muscle Contractile Activity

1. Spontaneous electrical activity in the plasma membrane of the muscle cell

2. Neurotransmitters released by autonomic neurons

3. Hormones

4. Locally induced changes in the chemical composition (paracrine agents, acidity, oxygen, osmolarity, and ion concentrations) of the extracellular fluid surrounding the cell

5. Stretch

Some of these increase contraction, while others inhibit it. Moreover, at any one time, the smooth muscle plasma membrane may be receiving multiple inputs, with the contractile state of the muscle dependent on the relative intensity of the various inhibitory and excitatory stimuli. All these inputs influence contractile activity by altering cytosolic calcium concentration as described in the previous section.

Some smooth muscles contract in response to membrane depolarization, whereas others can contract in the absence of any membrane potential change. Interestingly, in smooth muscles in which action potentials occur, calcium ions, rather than sodium ions, carry a positive charge into the cell during the rising phase of the action potential—that is, depolarization of the membrane opens voltage-gated calcium channels, producing calcium-mediated rather than sodium-mediated action potentials.

Another important point applies to electrical activity and cytosolic calcium concentration in smooth muscle. Unlike the situation in skeletal muscle, in smooth muscle cytosolic calcium concentration can be increased (or decreased) by graded depolarizations (or hyperpolarizations) in membrane potential, which increase or decrease the number of open calcium channels.

Spontaneous Electrical Activity

Some types of smooth muscle cells generate action potentials spontaneously in the absence of any neural or hormonal input. The plasma membranes of such cells do not maintain a constant resting potential. Instead, they gradually depolarize until they reach the threshold potential and produce an action potential. Following repolarization, the membrane again begins to depolarize (**Figure 9–36a**), so that a sequence of action potentials occurs, producing a rhythmic state of contractile activity. The membrane potential change occurring during the spontaneous depolarization to threshold is known as a **pacemaker potential.**

Other smooth muscle pacemaker cells have a slightly different pattern of activity. The membrane potential drifts up and down due to regular variation in ion flux across the membrane. These periodic fluctuations are called **slow waves** (**Figure 9–36b**). When an excitatory input is superimposed, as when food enters a segment of the gastrointestinal tract, slow waves are depolarized above threshold, and action potentials lead to smooth muscle contraction.

Pacemaker cells are found throughout the gastrointestinal tract, and thus gut smooth muscle tends to contract rhythmically even in the absence of neural input. Some cardiac muscle fibers and a few neurons in the central nervous system also have pacemaker potentials and can spontaneously generate action potentials in the absence of external stimuli.

Nerves and Hormones

The contractile activity of smooth muscles is influenced by neurotransmitters released by autonomic nerve endings. Unlike skeletal muscle fibers, smooth muscle cells do not have a specialized motor end-plate region. As the axon of a postganglionic autonomic neuron enters the region of smooth muscle

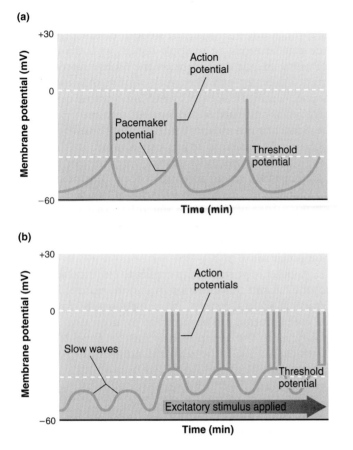

(a)

(b)

Figure 9–36

Generation of action potentials in smooth muscle fibers.
(a) Some smooth muscle cells have pacemaker potentials that drift to threshold at regular intervals. (b) Pacemaker cells with a slow wave pattern drift periodically toward threshold; excitatory stimuli can depolarize the cell to reach threshold and fire action potentials.

cells, it divides into many branches, each branch containing a series of swollen regions known as **varicosities** (**Figure 9–37**). Each varicosity contains many vesicles filled with neurotransmitter, some of which are released when an action potential passes the varicosity. Varicosities from a single axon may be located along several muscle cells, and a single muscle cell may be located near varicosities belonging to postganglionic fibers of both sympathetic and parasympathetic neurons. Therefore, a number of smooth muscle cells are influenced by the neurotransmitters released by a single nerve fiber, and a single smooth muscle cell may be influenced by neurotransmitters from more than one neuron.

Whereas some neurotransmitters enhance contractile activity, others decrease contractile activity. Thus, in contrast to skeletal muscle, which receives only excitatory input from its motor neurons, smooth muscle tension can be either increased or decreased by neural activity.

Moreover, a given neurotransmitter may produce opposite effects in different smooth muscle tissues. For example, norepinephrine, the neurotransmitter released from most postganglionic sympathetic neurons, enhances contraction of most vascular smooth muscle by acting on alpha-adrenergic receptors. By contrast, the same neurotransmitter produces relaxation of airway (bronchiolar) smooth muscle by acting on beta-2 adrenergic receptors. Thus, the type of response (excitatory or inhibitory) depends not on the chemical messenger *per se*, but on the receptors the chemical messenger binds to in the membrane and on the intracellular signaling mechanisms those receptors activate.

In addition to receptors for neurotransmitters, smooth muscle plasma membranes contain receptors for a variety of hormones. Binding of a hormone to its receptor may lead to either increased or decreased contractile activity.

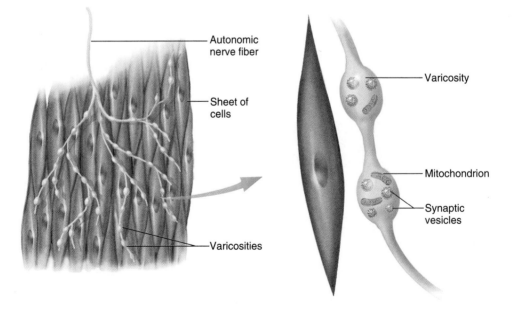

Figure 9–37

Innervation of smooth muscle by a postganglionic autonomic neuron. Neurotransmitter, released from varicosities along the branched axon, diffuses to receptors on muscle cell plasma membranes. Both sympathetic and parasympathetic neurons follow this pattern, often overlapping in their distribution.

Chapter 9

Although most changes in smooth muscle contractile activity induced by chemical messengers are accompanied by a change in membrane potential, this is not always the case. Second messengers—for example, inositol trisphosphate—can cause the release of calcium from the sarcoplasmic reticulum, producing a contraction without a change in membrane potential (review Figure 5–10).

Local Factors

Local factors, including paracrine agents, acidity, oxygen concentration, osmolarity, and the ion composition of the extracellular fluid, can also alter smooth muscle tension. Responses to local factors provide a means for altering smooth muscle contraction in response to changes in the muscle's immediate internal environment, which can lead to regulation that is independent of long-distance signals from nerves and hormones.

Many of these local factors induce smooth muscle relaxation. Nitric oxide (NO) is one of the most commonly encountered paracrine agents that produces smooth muscle relaxation. NO is released from some nerve terminals as well as from a variety of epithelial and endothelial cells. Because of the short life span of this reactive molecule, it acts as a paracrine agent, influencing only those cells that are very near its release site.

Some smooth muscles can also respond by contracting when they are stretched. Stretching opens mechanosensitive ion channels, leading to membrane depolarization. The resulting contraction opposes the forces acting to stretch the muscle.

It is important to remember that seldom is a single agent acting on a smooth muscle. Instead, the state of contractile activity at any moment depends on the net magnitude of the signals promoting contraction versus those promoting relaxation.

Types of Smooth Muscle

The great diversity of the factors that can influence the contractile activity of smooth muscles from various organs has made it difficult to classify smooth muscle fibers. Many smooth muscles can be placed, however, into one of two groups, based on the electrical characteristics of their plasma membrane: **single-unit smooth muscles** and **multiunit smooth muscles.**

Single-Unit Smooth Muscle

The muscle cells in a single-unit smooth muscle undergo synchronous activity, both electrical and mechanical; that is, the whole muscle responds to stimulation as a single unit. This occurs because each muscle cell is linked to adjacent fibers by gap junctions, which allow action potentials occurring in one cell to propagate to other cells by local currents. Therefore, electrical activity occurring anywhere within a group of single-unit smooth muscle cells can be conducted to all the other connected cells (**Figure 9–38**).

Some of the cells in a single-unit muscle are pacemaker cells that spontaneously generate action potentials. These action potentials are conducted by way of gap junctions to the rest of the cells, most of which are not capable of pacemaker activity.

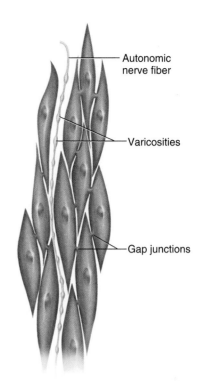

Figure 9–38

Innervation of a single-unit smooth muscle is often restricted to only a few cells in the muscle. Electrical activity is conducted from cell to cell throughout the muscle by way of the gap junctions between the cells.

Nerves, hormones, and local factors can alter the contractile activity of single-unit smooth muscles using the variety of mechanisms described previously for smooth muscles in general. The extent to which these muscles are innervated varies considerably in different organs. The nerve terminals are often restricted to the regions of the muscle that contain pacemaker cells. The activity of the entire muscle can be controlled by regulating the frequency of the pacemaker cells' action potentials.

One additional characteristic of single-unit smooth muscles is that a contractile response can often be induced by stretching the muscle. In several hollow organs—the stomach, for example—stretching the smooth muscles in the walls of the organ as a result of increases in the volume of material in the lumen initiates a contractile response.

The smooth muscles of the intestinal tract, uterus, and small-diameter blood vessels are examples of single-unit smooth muscles.

Multiunit Smooth Muscle

Multiunit smooth muscles have no or few gap junctions. Each cell responds independently, and the muscle behaves as multiple units. Multiunit smooth muscles are richly innervated by branches of the autonomic nervous system. The contractile response of the whole muscle depends on the number of muscle cells that are activated and on the frequency of nerve stimulation. Although stimulation of the nerve fibers to the muscle leads to some degree of depolarization and a contractile response, action potentials do not occur in most multiunit smooth muscles. Circulating hormones can increase or decrease contractile activity in multiunit smooth muscle, but stretching does not induce contraction in this type of muscle. The smooth muscles in the large airways to the lungs, in large arteries, and attached to the hairs in the skin are multiunit smooth muscles.

It must be emphasized that most smooth muscles do not show all the characteristics of either single-unit or multiunit smooth muscles. These two prototypes represent the two extremes in smooth muscle characteristics, with many smooth muscles having overlapping characteristics.

Cardiac Muscle

The third general type of muscle, cardiac muscle, is found only in the heart. Although many details about cardiac muscle will be discussed in the context of the cardiovascular system in Chapter 12, a brief explanation of its function and how it compares to skeletal and smooth muscle is presented here.

Cellular Structure of Cardiac Muscle

Cardiac muscle combines properties of both skeletal and smooth muscle. Like skeletal muscle, it has a striated appearance due to regularly repeating sarcomeres composed of myosin-containing thick filaments interdigitating with thin filaments that contain actin. Troponin and tropomyosin are also present in the thin filament, and they have the same functions as in skeletal muscle. Cellular membranes include a T-tubule system and associated calcium-loaded sarcoplasmic reticulum. The mechanism by which these membranes interact to release calcium is different than in skeletal muscle, however, as will be discussed shortly.

Like smooth muscle cells, individual cardiac muscle cells are relatively small (100 µm long and 20 µm in diameter) and generally contain a single nucleus. Adjacent cells are joined end-to-end at structures called **intercalated disks,** within which are desmosomes that hold the cells together and to which the myofibrils are attached (**Figure 9–39**). Also found within the intercalated disks are gap junctions similar to those found in single-unit smooth muscle. Cardiac muscle cells also are arranged in layers and surround hollow cavities, in this case the blood-filled chambers of the heart. When muscle in the walls of cardiac chambers contracts, it acts like a squeezing fist and exerts pressure on the blood inside.

Excitation-Contraction Coupling in Cardiac Muscle

As in skeletal muscle, contraction of cardiac muscle cells occurs in response to a membrane action potential that propagates through the T-tubules, but the mechanisms linking that excitation to the generation of force exhibit features of both skeletal and smooth muscles (**Figure 9–40**). Depolarization during cardiac muscle cell action potentials is in part due to an influx of Ca^{2+} ions through voltage-gated channels. These cal-

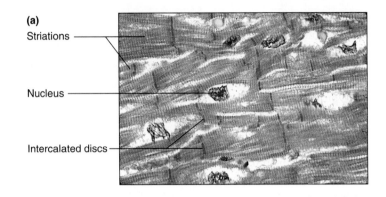

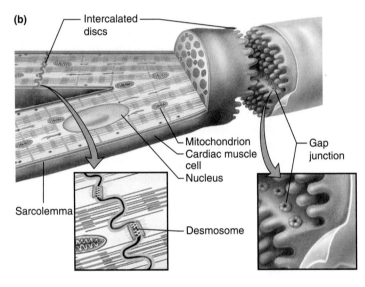

Figure 9–39

Cardiac muscle. (a) Light micrograph. (b) Cardiac myocytes and intercalated discs.

cium channels are known as **L-type calcium channels,** and are modified versions of the DHP receptors that act as the voltage-sensor in skeletal muscle cell excitation-contraction coupling. Not only does this entering calcium participate in depolarization of the plasma membrane and cause a small rise in cytosolic calcium concentration, but it also serves as a trigger for the release of a much larger amount of calcium from the sarcoplasmic reticulum. This occurs because ryanodine receptors in the cardiac sarcoplasmic reticulum lateral sacs are calcium channels, but rather than being opened directly by voltage as in skeletal muscle, they are opened by the binding of trigger calcium in the cytosol. Once cytosolic calcium is elevated, thin filament activation, cross-bridge cycling, and force generated occur by the same basic mechanisms described for skeletal muscle (review Figures 9–8 and 9–9).

Thus, even though most of the calcium initiating cardiac muscle contraction comes from the sarcoplasm reticulum, the process—unlike that in skeletal muscle—is dependent on the movement of extracellular calcium into the cytosol. Contraction ends when the cytosolic calcium concentration is

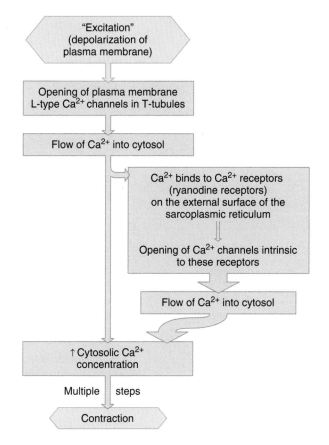

Figure 9–40

Excitation-contraction coupling in cardiac muscle.

restored to its original extremely low resting value by primary active Ca^{2+}-ATPase pumps and Na^+/Ca^{2+} countertransporters in the sarcoplasmic reticulum and sarcolemma. The amount of calcium returned to the extracellular fluid and into the sarcoplasmic reticulum exactly matches the amounts that entered the cytosol during excitation. During a single twitch contraction of cardiac muscle in a person at rest, the amount of calcium entering the cytosol is only sufficient to expose about 30 percent of the cross-bridge attachment sites on the thin filament. As Chapter 12 will describe, however, hormones and neurotransmitters of the autonomic nervous system modulate the amount of calcium released during excitation-contraction coupling, enabling the strength of cardiac muscle contractions to be varied. Cardiac muscle contractions are thus graded in a manner similar to that of smooth muscle contractions.

The "L" in L-type calcium channels stands for "long-lasting current," and this property of cardiac calcium channels underlies an important feature of this muscle type—cardiac muscle cannot undergo tetanic contractions. Unlike skeletal muscle, in which the membrane action potential is extremely brief (1–2 ms) and force generation lasts much longer (20–100 ms), in cardiac muscle the action potential and twitch are both prolonged due to the long-lasting calcium current (**Figure 9–41**). Because the plasma

membrane remains refractory to additional stimuli as long as it is depolarized (review Figure 6–22), it isn't possible to initiate multiple cardiac action potentials during the time frame of a single twitch. This is critical for the heart's function as an oscillating pump, because it must alternate between being relaxed and filling with blood, and contracting to eject blood.

A final question to consider is, "What initiates action potentials in cardiac muscle?" Certain specialized cardiac muscle cells exhibit pacemaker potentials that generate action potentials spontaneously, similar to the mechanism for smooth muscle described in Figure 9–36a. Because cardiac cells are linked via gap junctions, when an action potential is initiated by a pacemaker cell it propagates rapidly throughout the entire heart. A single heartbeat corresponds to the initiation and conduction of a single action potential. In addition to the modulation of calcium release and the strength of contraction, Chapter 12 will also discuss how hormones and autonomic neurotransmitters modify the frequency of cardiac pacemaker cell depolarization, and thus vary the heart rate.

Table 9–6 summarizes and compares the properties of the different types of muscle.

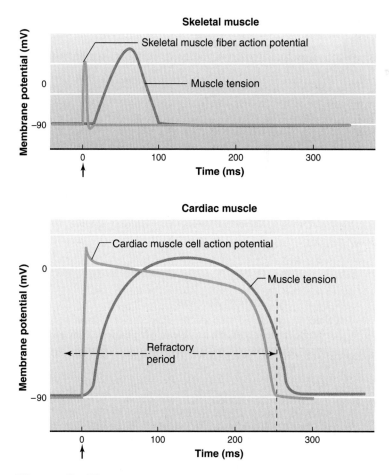

Figure 9–41

Timing of action potentials and twitch tension in skeletal and cardiac muscles. Muscle tension not drawn to scale.

Table 9–6 Characteristics of Muscle Cells

Characteristic	Skeletal Muscle	Smooth Muscle Single Unit	Smooth Muscle Multiunit	Cardiac Muscle
Thick and thin filaments	Yes	Yes	Yes	Yes
Sarcomeres—banding pattern	Yes	No	No	Yes
Transverse tubules	Yes	No	No	Yes
Sarcoplasmic reticulum (SR)*	++++	+	+	++
Gap junctions between cells	No	Yes	Few	Yes
Source of activating calcium	SR	SR and extracellular	SR and extracellular	SR and extracellular
Site of calcium regulation	Troponin	Myosin	Myosin	Troponin
Speed of contraction	Fast-slow	Very slow	Very slow	Slow
Spontaneous production of action potentials by pacemakers	No	Yes	No	Yes in certain fibers, but most not spontaneously active
Tone (low levels of maintained tension in the absence of external stimuli)	No	Yes	No	No
Effect of nerve stimulation	Excitation	Excitation or inhibition	Excitation or inhibition	Excitation or inhibition
Physiological effects of hormones on excitability and contraction	No	Yes	Yes	Yes
Stretch of cell produces contraction	No	Yes	No	No

*Number of plus signs (+) indicates the relative amount of sarcoplasmic reticulum present in a given muscle type.

■ **SECTION B SUMMARY**

Structure of Smooth Muscle

I. Smooth muscle cells are spindle-shaped, lack striations, have a single nucleus, and are capable of cell division. They contain actin and myosin filaments and contract by a sliding-filament mechanism.

Smooth Muscle Contraction and Its Control

I. An increase in cytosolic calcium leads to the binding of calcium by calmodulin. The calcium-calmodulin complex then binds to myosin light-chain kinase, activating the enzyme, which uses ATP to phosphorylate smooth muscle myosin. Only phosphorylated myosin can bind to actin and undergo cross-bridge cycling.

II. Smooth muscle myosin has a low rate of ATP splitting, resulting in a much slower shortening velocity than in striated muscle. However, the tension produced per unit cross-sectional area is equivalent to that of skeletal muscle.

III. Two sources of the cytosolic calcium ions that initiate smooth muscle contraction are the sarcoplasmic reticulum and extracellular calcium. The opening of calcium channels in the smooth muscle plasma membrane and sarcoplasmic reticulum, mediated by a variety of factors, allows calcium ions to enter the cytosol.

IV. The increase in cytosolic calcium resulting from most stimuli does not activate all the cross-bridges. Therefore, smooth muscle tension can be increased by agents that increase the concentration of cytosolic calcium ions.

V. Table 9–5 summarizes the types of stimuli that can initiate smooth muscle contraction by opening or closing calcium channels in the plasma membrane or sarcoplasmic reticulum.

VI. Most, but not all, smooth muscle cells can generate action potentials in their plasma membrane upon membrane depolarization. The rising phase of the smooth muscle action potential is due to the influx of calcium ions into the cell through voltage-gated calcium channels.

VII. Some smooth muscles generate action potentials spontaneously, in the absence of any external input, because of pacemaker potentials in the plasma membrane that repeatedly depolarize the membrane to threshold. Slow waves are a pattern of spontaneous, periodic depolarization of the membrane potential seen in some smooth muscle pacemaker cells.

VIII. Smooth muscle cells do not have a specialized end-plate region. A number of smooth muscle cells may be influenced by neurotransmitters released from the varicosities on a single nerve ending, and a single smooth muscle cell may be influenced by neurotransmitters from more than one neuron. Neurotransmitters may have either excitatory or inhibitory effects on smooth muscle contraction.

IX. Smooth muscles can be classified broadly as single-unit or multiunit smooth muscle.

Cardiac Muscle

I. Cardiac muscle combines features of skeletal and smooth muscles. Like skeletal muscle, it is striated, is composed of myofibrils with repeating sarcomeres, has troponin in its thin filaments, has T-tubules that conduct action potentials, and has sarcoplasmic reticulum lateral sacs that store calcium. Like smooth muscle, cardiac muscle cells are small and single-nucleated, arranged in layers around hollow cavities, and connected by gap junctions.

II. Cardiac muscle excitation-contraction coupling involves entry of a small amount of calcium through L-type calcium channels, which triggers opening of ryanodine receptors that release a larger amount of calcium from the sarcoplasmic reticulum. Calcium activates the thin filament and cross-bridge cycling as in skeletal muscle.

III. Cardiac contractions and action potentials are prolonged, tetany does not occur, and both the strength and frequency of contraction are modulated by autonomic neurotransmitters and hormones.

IV. Table 9–6 summarizes and compares the features of skeletal, smooth, and cardiac muscles.

■ SECTION B KEY TERMS

dense body 284
intercalated disk 290
latch state 286
L-type calcium channel 290
multiunit smooth muscle 289
myosin light-chain kinase 285

myosin light-chain
  phosphatase 286
pacemaker potential 287
single-unit smooth muscle 289
slow waves 287
smooth muscle tone 287
varicosity 288

■ SECTION B REVIEW QUESTIONS

1. How does the organization of thick and thin filaments in smooth muscle fibers differ from that in striated muscle fibers?
2. Compare the mechanisms by which a rise in cytosolic calcium concentration initiates contractile activity in skeletal, smooth, and cardiac muscle cells.
3. What are the two sources of calcium that lead to the increase in cytosolic calcium that triggers contraction in smooth muscle?
4. What types of stimuli can trigger a rise in cytosolic calcium in smooth muscle cells?
5. What effect does a pacemaker potential have on a smooth muscle cell?
6. In what ways does the neural control of smooth muscle activity differ from that of skeletal muscle?
7. Describe how a stimulus may lead to the contraction of a smooth muscle cell without a change in the plasma membrane potential.
8. Describe the differences between single-unit and multiunit smooth muscles.
9. Compare and contrast the physiology of cardiac muscle with that of skeletal and smooth muscles.
10. Explain why cardiac muscle cannot undergo tetanic contractions.

Chapter 9 Test Questions

(Answers appear in Appendix A.)

1. Which is a false statement about skeletal muscle structure?
   a. A myofibril is composed of multiple muscle fibers.
   b. Most skeletal muscles attach to bones by connective-tissue tendons.
   c. Each end of a thick filament is surrounded by six thin filaments.
   d. A cross-bridge is a portion of the myosin molecule.
   e. Thin filaments contain actin, tropomyosin, and troponin.

2. Which is correct regarding a skeletal muscle sarcomere?
   a. M lines are found in the center of the I band.
   b. The I band is the space between one Z line and the next.
   c. The H zone is the region where thick and thin filaments overlap.
   d. Z lines are found in the center of the A band.
   e. The width of the A band is equal to the length of a thick filament.

3. When a skeletal muscle fiber undergoes a concentric isotonic contraction:
   a. M lines remain the same distance apart.
   b. Z lines move closer to the ends of the A bands.
   c. A bands become shorter.
   d. I bands become wider.
   e. M lines move closer to the end of the A band.

4. During excitation-contraction coupling in a skeletal muscle fiber
   a. the Ca^{2+}-ATPase pumps calcium into the T-tubule.
   b. action potentials propagate along the membrane of the sarcoplasmic reticulum.
   c. calcium floods the cytosol through the dihydropyridine (DHP) receptors.
   d. DHP receptors trigger the opening of lateral sac ryanodine receptor calcium channels.
   e. acetylcholine opens the DHP receptor channel.

5. Why is the latent period longer during an isotonic twitch of a skeletal muscle fiber than it is during an isometric twitch?
   a. Excitation-contraction coupling is slower during an isotonic twitch.
   b. Action potentials propagate more slowly when the fiber is shortening, so extra time is required to activate the entire fiber.
   c. In addition to the time for excitation-contraction coupling, it takes extra time for enough cross-bridges to attach to make the tension in the muscle fiber greater than the load.
   d. Fatigue sets in much more quickly during isotonic contractions, and when muscles are fatigued the cross-bridges move much more slowly.
   e. The latent period is longer because isotonic twitches only occur in slow (Type I) muscle fibers.

6. What prevents a drop in muscle fiber ATP concentration during the first few seconds of intense contraction?
   a. Because cross-bridges are pre-energized, ATP is not needed until several cross-bridge cycles have been completed.
   b. ADP is rapidly converted back to ATP by creatine phosphate.
   c. Glucose is metabolized in glycolysis, producing large quantities of ATP.
   d. The mitochondria immediately begin oxidative phosphorylation.
   e. Fatty acids are rapidly converted to ATP by oxidative glycolysis.

7. Which correctly characterizes a "fast-oxidative" type of skeletal muscle fiber?
   a. few mitochondria and high glycogen content
   b. low myosin ATPase rate and few surrounding capillaries
   c. low glycolytic enzyme activity and intermediate contraction velocity
   d. high myoglobin content and intermediate glycolytic enzyme activity
   e. small fiber diameter and fast onset of fatigue

8. Which is *false* regarding the structure of smooth muscle?
   a. The thin filament does not include the regulatory protein troponin.
   b. The thick and thin filaments are not organized in sarcomeres.
   c. Thick filaments are anchored to dense bodies instead of Z lines.
   d. The cells have a single nucleus.
   e. Single-unit smooth muscles have gap junctions connecting individual cells.

9. The role of myosin light-chain kinase in smooth muscle is to
   a. bind to calcium ions to initiate excitation-contraction coupling.
   b. phosphorylate cross-bridges, thus driving them to bind with the thin filament.
   c. split ATP to provide the energy for the power stroke of the cross-bridge cycle.
   d. dephosphorylate myosin light chains of the cross-bridge, thus relaxing the muscle.
   e. pump calcium from the cytosol back into the sarcoplasmic reticulum.

10. Single-unit smooth muscle differs from multiunit smooth muscle because
    a. single-unit muscle contraction speed is slow, while multiunit is fast.
    b. single-unit muscle has T-tubules, multiunit muscle does not.
    c. single-unit muscles are not innervated by autonomic nerves.
    d. single-unit muscle contracts when stretched, whereas multiunit muscle does not.
    e. single-unit muscle does not produce action potentials spontaneously, but multiunit muscle does.

11. Which of the following describes a similarity between cardiac and smooth muscle cells?
    a. An action potential always precedes contraction.
    b. The majority of the calcium that activates contraction comes from the extracellular fluid.
    c. Action potentials are generated by pacemaker potentials.
    d. An extensive system of T-tubules is present.
    e. Calcium release and contraction strength are graded.

Chapter 9 Quantitative and Thought Questions

(Answers appear in Appendix A.)

1. Which of the following corresponds to the state of myosin (M) under resting conditions and in rigor mortis? (a) $M \cdot ATP$ (b) $M \cdot ADP \cdot P_i$ (c) $A \cdot M \cdot ADP \cdot P_i$ (d) $A \cdot M$

2. If the transverse tubules of a skeletal muscle are disconnected from the plasma membrane, will action potentials trigger a contraction? Give reasons.

3. When a small load is attached to a skeletal muscle that is then tetanically stimulated, the muscle lifts the load in an isotonic contraction over a certain distance, but then stops shortening and enters a state of isometric contraction. With a heavier load, the distance shortened before entering an isometric contraction is shorter. Explain these shortening limits in terms of the length-tension relation of muscle.

4. What conditions will produce the maximum tension in a skeletal muscle fiber?

5. A skeletal muscle can often maintain a moderate level of active tension for long periods of time, even though many of its fibers become fatigued. Explain.

6. If the blood flow to a skeletal muscle were markedly decreased, which types of motor units would most rapidly undergo a severe reduction in their ability to produce ATP for muscle contraction? Why?

7. As a result of an automobile accident, 50 percent of the muscle fibers in the biceps muscle of a patient were destroyed. Ten months later, the biceps muscle was able to generate 80 percent of its original force. Describe the changes that took place in the damaged muscle that enabled it to recover.

8. In the laboratory, if an isolated skeletal muscle is placed in a solution that contains no calcium ions, will the muscle contract when it is stimulated (1) directly by depolarizing its membrane, or (2) by stimulating the nerve to the muscle? What would happen if it were a smooth muscle?

9. The following experiments were performed on a single-unit smooth muscle in the gastrointestinal tract.
   a. Stimulating the parasympathetic nerves to the muscle produced a contraction.
   b. Applying a drug that blocks the voltage-sensitive sodium channels in most plasma membranes led to a failure to contract upon stimulating the parasympathetic nerves.

c. Applying a drug that binds to muscarinic receptors (Chapter 6), and hence blocks the action of ACh at these receptors, did not prevent the muscle from contracting when the parasympathetic nerve was stimulated.

From these observations, what might you conclude about the mechanism by which parasympathetic nerve stimulation produces a contraction of the smooth muscle?

10. Some endocrine tumors secrete a hormone that leads to elevation of extracellular fluid calcium concentrations. How might this affect cardiac muscle?

Chapter 9 Answers to Physiological Inquiries

Figure 9–4

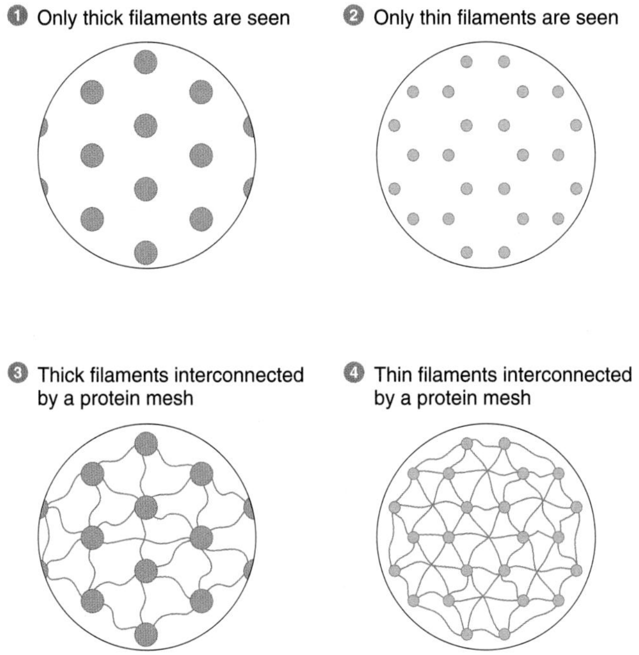

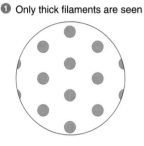

1 Only thick filaments are seen

2 Only thin filaments are seen

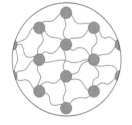

3 Thick filaments interconnected by a protein mesh

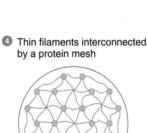

4 Thin filaments interconnected by a protein mesh

Figure 9–8   As long as ATP is available, cross-bridges would cycle continuously regardless of whether calcium was present.

Figure 9–16   The weight in the isotonic experiment is approximately 14 mg. This can be estimated by determining the time at which the isotonic load begins to move on the lower graph (approximately 12 ms), then using the upper graph to assess the amount of tension generated by the fiber at that point in time.

Figure 9–20   Unfused tetanic contractions will occur between 6.7 and 28.6 stimuli per second. In order for an unfused tetanus to occur, the interval between stimuli must be less than 150 ms but greater than 35 ms. (If the interval was greater than 150 ms, twitches would not summate, and if less than 35 ms, a fused tetanus would occur.) To calculate the corresponding frequencies:
1 stimulus/150 ms × 1000 ms/sec = 6.7 stimuli/sec
1 stimulus/35 ms × 1000 ms/sec = 28.6 stimuli/sec

Figure 9–29   The force acting upward on the forearm (85 × 5 = 425) would be less than the downward-acting force (10 × 45 = 450), so the muscle would undergo a lengthening (eccentric) contraction and the weight would move toward the ground.

Control of Body Movement

Carrying out a coordinated movement is a complicated process involving nerves, muscles, and bones. Consider the events associated with reaching out and grasping an object. The fingers are first extended (straightened) to reach around the object, and then flexed (bent) to grasp it. The degree of extension will depend upon the size of the object (Is it a golf ball or a soccer ball?), and the force of flexion will depend upon its weight and consistency (a bowling ball or a balloon?). Simultaneously, the wrist, elbow, and shoulder are extended, and the trunk is inclined toward the object. The shoulder, elbow, and wrist are stabilized to support first the weight of the arm and hand and then the added weight of the object. Through all this, the body maintains upright posture and balance despite its continuously shifting position.

The building blocks for these movements—as for all movements—are **motor units,** each comprising one motor neuron together with all the skeletal muscle fibers that this neuron innervates (Chapter 9). The motor neurons are the final common pathway out of the central nervous system because all neural influences on skeletal muscle converge on the motor neurons and can only affect skeletal muscle through them.

All the motor neurons that supply a given muscle make up the **motor neuron pool** for the muscle. The cell bodies of the pool for a given muscle are close to each other either in the ventral horn of the spinal cord or in the brainstem.

Within the brainstem or spinal cord, the axons of many neurons synapse on a motor neuron to control its activity. Although no single input to a motor neuron is essential for contraction of the muscle fibers it innervates, a balanced input from all sources is necessary to provide the precision and speed of normally coordinated actions. For example, if inhibitory synaptic input to a given motor neuron is decreased, the still-normal excitatory input to that neuron will be unopposed and the motor neuron firing will increase, leading to excessive contraction. This is exactly what happens in the disease *tetanus.*

It is important to realize that movements—even simple movements such as flexing a finger—are rarely achieved by just one muscle. All body movements are achieved by activation, in a precise sequence, of many motor units in various muscles.

This chapter deals with the interrelated neural inputs that converge upon motor neurons to control their activity. We first present a general model of how the motor system functions and then describe each component of the model in detail. Keep in mind throughout this chapter that many of the contractions that skeletal muscles execute, particularly the muscles involved in postural support, are isometric (Chapter 9). Even though the muscle is active during these contractions, no movement occurs. In the following discussions, the general term *muscle movement* includes these isometric contractions. In addition, remember that all information in the nervous system is transmitted in the form of graded potentials or action potentials.

Motor Control Hierarchy

Throughout the nervous system, the neurons involved in controlling skeletal muscles can be thought of as being organized in a hierarchical fashion, with each level of the hierarchy having a certain task in motor control (**Figure 10–1**). To begin a movement, a general intention such as "pick up sweater" or "write signature" or "answer telephone" is generated at the highest level of the motor control hierarchy. These higher centers include many regions of the brain, including those involved in memory, emotions, and motivation. Very little is known, however, as to exactly where intentions for movement form in the brain.

Information is relayed from these higher-center "command" neurons to parts of the brain that make up the middle level of the motor control hierarchy. The middle-level structures specify the individual postures and movements needed to carry out the intended action. In our example of picking up a sweater, structures of the middle hierarchical level coordinate the commands that tilt the body and extend the arm and hand toward the sweater and shift the body's weight to maintain balance. The middle-level hierarchical structures are located in parts of the cerebral cortex as well as in the cerebellum, subcortical nuclei, and brainstem (see Figure 10–1 and **Figure 10–2a and b**). These structures have extensive interconnections, as the arrows in Figure 10–1 indicate.

As the neurons in the middle level of the hierarchy receive input from the command neurons, they simultaneously receive afferent information from receptors in the muscles, tendons, joints, and skin, as well as from the vestibular apparatus and eyes. These afferent signals relay information to the middle-level neurons about the starting positions of the body parts that are "commanded" to move. They also relay information about the nature of the space just outside the body in which a movement will take place. Neurons of the middle level of the hierarchy integrate all of this afferent information with the signals from the command neurons to create a **motor program**—defined as the pattern of neural activity required to properly perform the desired movement. Impairment of a person's sensory pathways also results in slow and uncoordinated voluntary movement.

The information determined by the motor program is transmitted via **descending pathways** to the local level of the

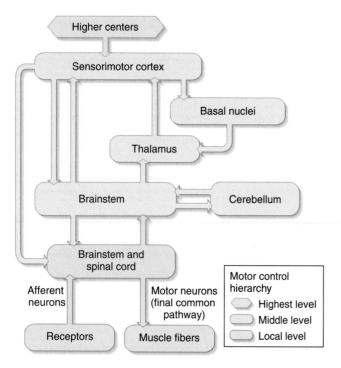

Figure 10–1

The conceptual hierarchical organization of the neural systems controlling body movement. Motor neurons control all the skeletal muscles of the body. Sensorimotor cortex includes those parts of the cerebral cortex that act together to control skeletal muscle activity. The middle level of the hierarchy also receives input from the vestibular apparatus and eyes (not shown in the figure).

motor control hierarchy. There the motor neurons to the muscles exit the brainstem or spinal cord. The local level of the hierarchy includes the afferent neurons, motor neurons, and the interneurons whose function is related to them. Local-level neurons determine exactly which motor neurons will be activated to achieve the desired action and when this will happen. Note in Figure 10–1 that the descending pathways to the local level arise only in the sensorimotor cortex and brainstem. Other brain areas, notably the basal nuclei (also referred to as the basal ganglia), thalamus, and cerebellum, exert their effects on the local level only indirectly, via the descending pathways from the cerebral cortex and brainstem.

The motor programs are continuously adjusted during the course of most movements. As the initial motor program begins and the action gets underway, brain regions at the middle level of the hierarchy continue to receive a constant stream of updated afferent information about the movements taking place. Afferent information about the position of the body and its parts in space is called **proprioception.** Say, for example, that the sweater you are picking up is wet and heavier than you expected so that the initially determined strength of muscle contraction is not sufficient to lift it. Any discrepancies between the intended and actual movements are detected, program corrections are determined, and the corrections are relayed to the local level of the hierarchy and the motor neurons. Reflex circuits acting entirely at the local level are also important in refining ongoing movements. Thus, many proprioceptive inputs are processed, and influence ongoing movements, without ever reaching the level of conscious perception.

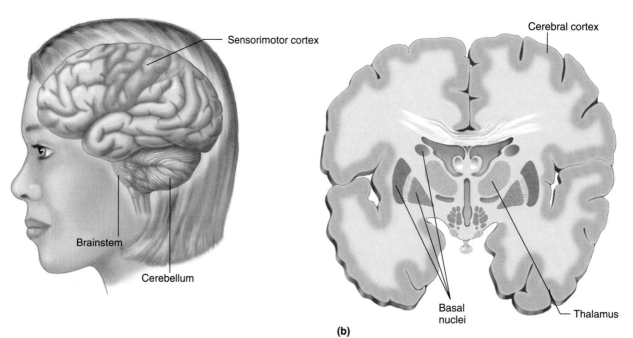

(a) **(b)**

Figure 10–2

(a) Side view of the brain showing three of the five components of the middle level of the motor control hierarchy. Figure 10–10 shows details of the sensorimotor cortex. (b) Cross section of the brain showing the thalamus and basal nuclei.

If a complex movement is repeated often, learning takes place and the movement becomes skilled. Then, the initial information from the middle hierarchical level is more accurate, and fewer corrections need to be made. Movements performed at high speed without concern for fine control are made solely according to the initial motor program.

Table 10–1 summarizes the structures and functions of the motor control hierarchy.

Voluntary and Involuntary Actions

Given such a highly interconnected and complicated neuroanatomical basis for the motor system, it is difficult to use the phrase **voluntary movement** with any real precision. We will use it, however, to refer to actions that have the following characteristics: (1) the movement is accompanied by a conscious awareness of what we are doing and why we are doing it, and (2) our attention is directed toward the action or its purpose.

The term *involuntary,* on the other hand, describes actions that do not have these characteristics. *Unconscious, automatic,* and *reflex* often serve as synonyms for *involuntary,* although in the motor system the term *reflex* has a more precise meaning.

Despite our attempts to distinguish between voluntary and involuntary actions, almost all motor behavior involves both components, and we cannot easily make a distinction between the two. For example, even such a highly conscious act as walking involves many reflexive components, as the pattern of contraction of leg muscles is subconsciously varied to adapt to obstacles or uneven terrain.

Thus, most motor behavior is neither purely voluntary nor purely involuntary, but falls somewhere between these two. Moreover, actions shift along this continuum according to the frequency with which they are performed. When a person first learns to drive a car with a manual transmission, for example, shifting gears requires a great deal of conscious attention. With practice, those same actions become automatic. On the other hand, reflex behaviors, which are all the way at the involuntary end of the spectrum, can with special effort sometimes be voluntarily modified or even prevented.

We now turn to an analysis of the individual components of the motor control system. We will begin with local control mechanisms because their activity serves as a base upon which the descending pathways exert their influence. Keep in mind throughout these descriptions that motor neurons always form the final common pathway to the muscles.

Local Control of Motor Neurons

The local control systems are the relay points for instructions to the motor neurons from centers higher in the motor control hierarchy. In addition, the local control systems play a major role in adjusting motor unit activity to unexpected obstacles to movement and to painful stimuli in the surrounding environment.

To carry out these adjustments, the local control systems use information carried by afferent fibers from sensory receptors in the muscles, tendons, joints, and skin of the body parts to be moved. As noted earlier, the afferent fibers also transmit information to higher levels of the hierarchy.

Interneurons

Most of the synaptic input to motor neurons from the descending pathways and afferent neurons does not go directly to motor neurons, but rather to interneurons that synapse with the motor neurons. Interneurons comprise 90 percent of spinal cord neurons, and they are of several types. Some are near the motor neuron they synapse upon and thus are called local interneurons. Others have processes that extend up or down short distances in the spinal cord and brainstem, or even throughout much of the length of the central nervous system. The interneurons with longer processes are important for integrating complex movements such as stepping forward with your left foot as you throw a baseball with your right arm.

The interneurons are important elements of the local level of the motor control hierarchy, integrating inputs not only from higher centers and peripheral receptors but from other interneurons as well (**Figure 10–3**). They are crucial in determining which muscles are activated and when. This is especially important in coordinating repetitive, rhythmic activities like walking or running, for which spinal cord interneurons encode pattern generator circuits responsible for activating and inhibiting limb movements in an alternating sequence. Moreover, interneurons can act as "switches" that enable a movement to be turned on or off under the command of higher motor centers. For example, if you pick up a

Table 10–1	Conceptual Motor Control Hierarchy for Voluntary Movements

I. Higher centers
  a. Function: form complex plans according to individual's intention and communicates with the middle level via command neurons.
  b. Structures: areas involved with memory and emotions, supplementary motor area, and association cortex. All these structures receive and correlate input from many other brain structures.

II. The middle level
  a. Function: converts plans received from the highest level to a number of smaller motor programs that determine the pattern of neural activation required to perform the movement. These programs are broken down into subprograms that determine the movements of individual joints. The programs and subprograms are transmitted through descending pathways to the local control level.
  b. Structures: sensorimotor cortex, cerebellum, parts of basal nuclei, some brainstem nuclei.

III. The local level
  a. Function: specifies tension of particular muscles and angle of specific joints at specific times necessary to carry out the programs and subprograms transmitted from the middle control levels.
  b. Structures: levels of brainstem or spinal cord from which motor neurons exit.

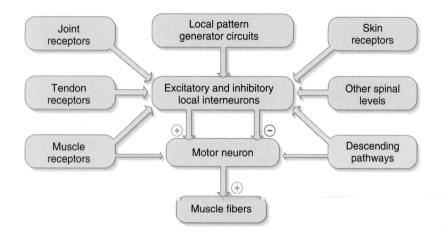

Figure 10–3
Converging inputs to local interneurons that control motor neuron activity. Plus signs indicate excitatory synapses and minus signs an inhibitory synapse. Neurons in addition to those shown may synapse directly onto motor neurons.

Figure 10–3 physiological *inquiry*

- Many spinal cord interneurons release the neurotransmitter glycine, which opens chloride channels on postsynaptic cell membranes. Given that the plant-derived chemical strychnine blocks glycine receptors, predict the symptoms of strychnine poisoning.

Answer can be found at end of chapter.

hot plate, a local reflex arc will be initiated by pain receptors in the skin of your hands, normally causing you to drop the plate. If it contains your dinner, however, descending commands can inhibit the local activity, and you can hold onto the plate until you can put it down safely.

Local Afferent Input
As just noted, afferent fibers usually impinge upon the local interneurons (in one case that will be discussed shortly, they synapse directly on motor neurons). The afferent fibers bring information from sensory receptors located in three places: (1) in the skeletal muscles controlled by the motor neurons, (2) in other nearby muscles, such as those with antagonistic actions, and (3) in the tendons, joints, and skin of body parts affected by the action of the muscle.

These receptors monitor the length and tension of the muscles, movement of the joints, and the effect of movements on the overlying skin. In other words, the movements themselves give rise to afferent input that, in turn, influences how the movement proceeds. As we will see next, their input sometimes provides negative feedback control over the muscles, and also contributes to the conscious awareness of limb and body position.

Length-Monitoring Systems
Stretch receptors embedded within muscles monitor muscle length and the rate of change in muscle length. These receptors consist of peripheral endings of afferent nerve fibers wrapped around modified muscle fibers, several of which are enclosed in a connective-tissue capsule and are collectively called a **muscle spindle** (**Figure 10–4**). The modified muscle fibers within the spindle are known as **intrafusal fibers.** The skeletal muscle fibers that form the bulk of the muscle and generate its force and movement are the **extrafusal fibers.**

Within a given spindle are two kinds of stretch receptors. One responds best to how much the muscle has been stretched **(nuclear chain fibers),** whereas the other responds to both the magnitude of the stretch and the speed with which it occurs **(nuclear bag fibers).** Although the two kinds of stretch receptors are separate entities, we will refer to them collectively as the **muscle-spindle stretch receptors.**

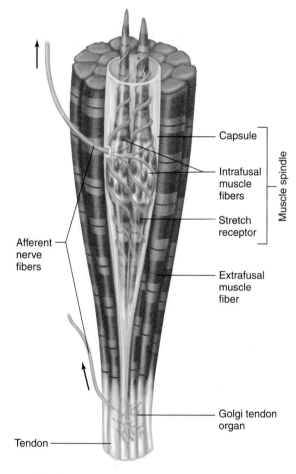

Figure 10–4
A muscle spindle and Golgi tendon organ. The muscle spindle is exaggerated in size compared to the extrafusal muscle fibers. The Golgi tendon organ will be discussed later in the chapter.

Adapted from Elias, Pauly, and Burns.

The muscle spindles are parallel to the extrafusal fibers. Thus, an external force stretching the muscle also pulls on the intrafusal fibers, stretching them and activating their receptor endings (**Figure 10–5a**). The more or the faster the muscle is

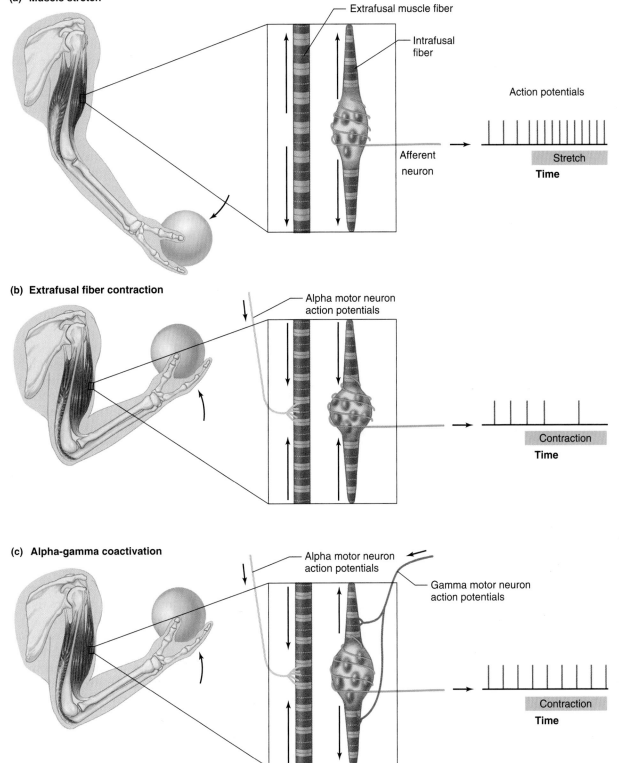

Figure 10–5

(a) Passive stretch of the muscle by an external load activates the spindle stretch receptors and causes an increased rate of action potentials in the afferent nerve. (b) Contraction of the extrafusal fibers removes tension on the stretch receptors and lowers the rate of action potential firing. (c) Simultaneous activation of alpha and gamma motor neurons results in maintained stretch of the central region of intrafusal fibers. Afferent information about muscle length continues to reach the central nervous system.

stretched, the greater the rate of receptor firing. In contrast, contraction of the extrafusal fibers and the resultant shortening of the muscle remove tension on the spindle and slow the rate of firing in the stretch receptor (**Figure 10–5b**).

When the afferent fibers from the muscle spindle enter the central nervous system, they divide into branches that take different paths. In **Figure 10–6**, path A makes excitatory synapses directly onto motor neurons that go back to the muscle that was stretched, thereby completing a reflex arc known as the **stretch reflex.**

This reflex is probably most familiar in the form of the **knee jerk,** part of a routine medical examination. The exam-

iner taps the patellar tendon (see Figure 10–6), which passes over the knee and connects extensor muscles in the thigh to the tibia in the lower leg. As the tendon is pushed in by tapping, the thigh muscles it is attached to are stretched, and all the stretch receptors within these muscles are activated. This stimulates a burst of action potentials in the afferent nerve fibers from the stretch receptors, and these action potentials activate excitatory synapses on the motor neurons that control these same muscles. The motor units are stimulated, the thigh muscles contract, and the patient's lower leg extends to give the knee jerk. The proper performance of the knee jerk tells the physician that the afferent fibers, the balance of synaptic input to the motor neurons, the motor neurons, the neuromuscular junctions, and the muscles themselves are functioning normally.

Because the afferent nerve fibers in the stretched muscle synapse directly on the motor neurons to that muscle without any interneurons, this portion of the stretch reflex is called **monosynaptic.** Stretch reflexes have the only known monosynaptic reflex arcs. All other reflex arcs are **polysynaptic;** they have at least one interneuron, and usually many, between the afferent and efferent neurons.

In path B of Figure 10–6, the branches of the afferent nerve fibers from stretch receptors end on inhibitory interneurons. When activated, these inhibit the motor neurons controlling antagonistic muscles whose contraction would interfere with the reflex response. In the knee jerk, for example, neurons to muscles that flex the knee are inhibited. This component of the stretch reflex is polysynaptic. The activation of neurons to one muscle with the simultaneous inhibition of neurons to its antagonistic muscle is called **reciprocal innervation.** This is characteristic of many movements, not just the stretch reflex.

Path C in Figure 10–6 activates motor neurons of **synergistic muscles**—that is, muscles whose contraction assists the intended motion. In the example of the knee jerk reflex, this would include other muscles that extend the leg.

In path D of Figure 10–6, the axon of the afferent neuron continues to the brainstem and synapses there with interneurons that form the next link in the pathway that conveys information about the muscle length to areas of the brain dealing with motor control. This information is especially important during slow, controlled movements such as the performance of an unfamiliar action. Ascending paths also provide information that contributes to the conscious perception of the position of a limb.

Alpha-Gamma Coactivation

As indicated in Figure 10–5b, stretch on the intrafusal fibers decreases when the muscle shortens. In this example, slackening of the spindle stretch receptors greatly reduces the action potentials along the afferent neuron, and there can be no indication of any further changes in muscle length the entire time the muscle is shortening. A mechanism exists to prevent this loss of information: the two ends of each intrafusal

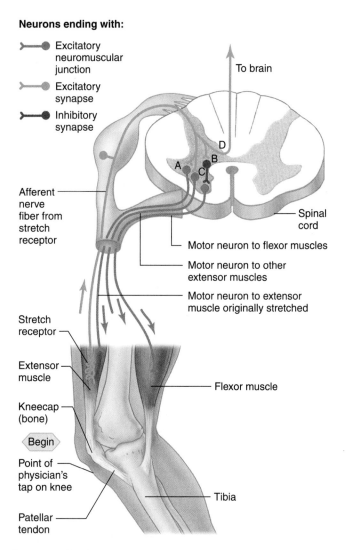

Neurons ending with:

- Excitatory neuromuscular junction
- Excitatory synapse
- Inhibitory synapse

To brain

D
B
A C

Afferent nerve fiber from stretch receptor

Spinal cord

Motor neuron to flexor muscles

Motor neuron to other extensor muscles

Motor neuron to extensor muscle originally stretched

Stretch receptor

Extensor muscle

Flexor muscle

Kneecap (bone)

Begin

Point of physician's tap on knee

Tibia

Patellar tendon

Figure 10–6

Neural pathways involved in the knee jerk reflex. Tapping the patellar tendon stretches the extensor muscle, causing (paths A and C) compensatory contraction of this and other extensor muscles, (path B) relaxation of flexor muscles, and (path D) information about muscle length to go to the brain. Arrows indicate direction of action potential propagation.

muscle fiber are stimulated to contract during the shortening of the extrafusal fibers. This maintains tension in the central region of the intrafusal fiber, where the stretch receptors are located (see Figure 10–5c). It is important to recognize that the intrafusal fibers are not large enough or strong enough to shorten a whole muscle and move joints; their sole job is to maintain tension on the spindle stretch receptors.

The intrafusal fibers contract in response to activation by motor neurons, but the motor neurons supplying them are not the same ones that activate the extrafusal muscle fibers. The motor neurons controlling the extrafusal muscle fibers are larger and are classified as **alpha motor neurons,** whereas the smaller motor neurons whose axons innervate the intrafusal fibers are known as **gamma motor neurons** (see Figure 10–5c).

The cell bodies of alpha and gamma motor neurons to a given muscle lie close together. Both types are activated by interneurons in their immediate vicinity and sometimes directly by neurons of the descending pathways. In fact, during many voluntary and involuntary movements, they are **coactivated**—that is, excited at almost the same time. Coactivation ensures that information about muscle length will be continuously available to provide for adjustment during ongoing actions and to plan and program future movements.

Tension-Monitoring Systems

Any given set of inputs to a given set of motor neurons can lead to various degrees of tension in the muscles they innervate. The tension depends on muscle length, the load on the muscles, and the degree of muscle fatigue. Therefore, feedback is necessary to inform the motor control systems of the tension actually achieved.

Some of this feedback is provided by vision (you can see whether you are lifting or lowering an object) as well as by afferent input from skin, muscle, and joint receptors. An additional receptor type specifically monitors how much tension the contracting motor units are exerting (or is being imposed on the muscle by external forces if the muscle is being stretched).

The receptors employed in this tension-monitoring system are the **Golgi tendon organs,** which are located in the tendons near their junction with the muscle (see Figure 10–4). Endings of afferent nerve fibers wrap around collagen bundles in the tendon, bundles that are slightly bowed in the resting state. When the muscle is stretched or the attached extrafusal muscle fibers contract, tension is exerted on the tendon. This tension straightens the collagen bundles and distorts the receptor endings, activating them. The tendon is typically stretched much more by an active contraction of the muscle than when the whole muscle is passively stretched (**Figure 10–7**). Thus, the Golgi tendon organs discharge in response to the tension generated by the contracting muscle and initiate action potentials that are transmitted to the central nervous system.

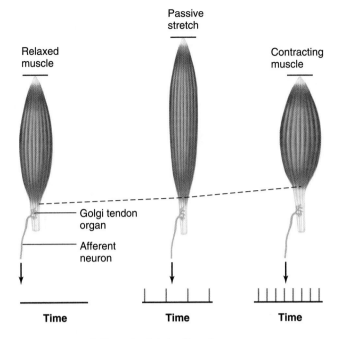

Action potentials in afferent neurons

Figure 10–7

Activation of Golgi tendon organs. Compared to when a muscle is contracting, passive stretch of the relaxed muscle produces less stretch of the tendon and fewer action potentials from the Golgi tendon organ.

Branches of the afferent neuron from the Golgi tendon organ cause widespread inhibition of the contracting muscle and its synergists via interneurons (A in **Figure 10–8**). They also stimulate the motor neurons of the antagonistic muscles (B in Figure 10–8). Note that this reciprocal innervation is the opposite of that produced by the muscle-spindle afferents. This difference reflects the different functional roles of the two systems: The muscle spindle provides local homeostatic control of muscle *length*, while the Golgi tendon organ provides local homeostatic control of muscle *tension*. In addition, the activity of afferent fibers from these two receptors supplies the higher-level motor control systems with information about muscle length and tension, which can be used to modify an ongoing motor program.

The Withdrawal Reflex

In addition to the afferent information from the spindle stretch receptors and Golgi tendon organs of the activated muscle, other input is transmitted to the local motor control systems. For example, painful stimulation of the skin, as occurs from stepping on a tack, activates the flexor muscles and inhibits the extensor muscles of the **ipsilateral** (on the same side of the body) leg. The resulting action moves the affected limb away from the harmful stimulus, and is thus known as a **withdrawal reflex**

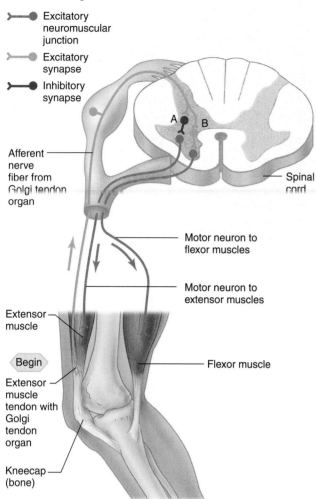

Neurons ending with:

- Excitatory neuromuscular junction
- Excitatory synapse
- Inhibitory synapse

Afferent nerve fiber from Golgi tendon organ

Spinal cord

A B

Motor neuron to flexor muscles

Motor neuron to extensor muscles

Extensor muscle

Begin

Extensor muscle tendon with Golgi tendon organ

Flexor muscle

Kneecap (bone)

Figure 10–8

Neural pathways underlying the Golgi tendon organ component of the local control system. In this diagram, contraction of the extensor muscles causes tension in the Golgi tendon organ and increases the rate of action potential firing in the afferent nerve fiber. By way of interneurons, this increased activity results in (path A) inhibition of the motor neuron of the extensor muscle and its synergists and (path B) excitation of the flexor muscles' motor neurons. Arrows indicate the direction of action potential propagation.

Figure 10–8 physiological *inquiry*

- Explain how the Golgi tendon organ protects against excessive force exertion that might tear a muscle or tendon.

Answer can be found at end of chapter.

(**Figure 10–9**). The same stimulus causes just the opposite response in the **contralateral** leg (on the opposite side of the body from the stimulus). Motor neurons to the extensors are activated while the flexor muscle motor neurons are inhibited. This **crossed-extensor reflex** enables the contralateral leg to support the body's weight as the injured foot is lifted by flexion (see Figure 10–9). This concludes our discussion of the local level of motor control.

The Brain Motor Centers and the Descending Pathways They Control

We now turn our attention to the motor centers in the brain and the descending pathways that direct the local control system (review Figure 10–1).

Cerebral Cortex

The cerebral cortex plays a critical role in both the planning and ongoing control of voluntary movements, functioning in both the highest and middle levels of the motor control hierarchy. The term **sensorimotor cortex** is used to include all those parts of the cerebral cortex that act together to control muscle movement. A large number of nerve fibers that give rise to descending pathways for motor control come from two areas of sensorimotor cortex on the posterior part of the frontal lobe: the **primary motor cortex** (sometimes called simply the **motor cortex**) and the **premotor area** (**Figure 10–10**). The neurons of the motor cortex that control muscle groups in various parts of the body are arranged anatomically into a **somatotopic map,** as shown in **Figure 10–11.**

Other areas of sensorimotor cortex include the **supplementary motor cortex,** which lies mostly on the surface on the frontal lobe where the cortex folds down between the two hemispheres (see Figure 10–10b), the **somatosensory cortex,** and parts of the **parietal-lobe association cortex** (see Figures 10–10a and b).

Although these areas are anatomically and functionally distinct, they are heavily interconnected, and individual muscles or movements are represented at multiple sites. Thus, the cortical neurons that control movement form a neural network, meaning that many neurons participate in each single movement. In addition, any one neuron may function in more than one movement. The neural networks can be distributed across multiple sites in parietal and frontal cortex, including the sites named in the preceding two paragraphs. The interaction of the neurons within the networks is flexible so that the neurons are capable of responding differently under different circumstances. This adaptability enhances the possibility of integrating incoming neural signals from diverse sources and the final coordination of many parts into a smooth, purposeful movement. It probably also accounts for the remarkable variety of ways in which we can approach a goal. For example, you can comb your hair with the right hand or the left, starting at the back of your head or the front. This same adaptability also accounts for some of the learning that occurs in all aspects of motor behavior.

We have described the various areas of sensorimotor cortex as giving rise, either directly or indirectly, to pathways descending to the motor neurons. However, these areas are not the prime initiators of movement, and other brain areas are certainly involved. We currently don't know exactly where or how intentional movements are initiated.

Association areas of the cerebral cortex also play a role in motor control. For example, neurons of parietal association cortex are important in the visual control of reaching and

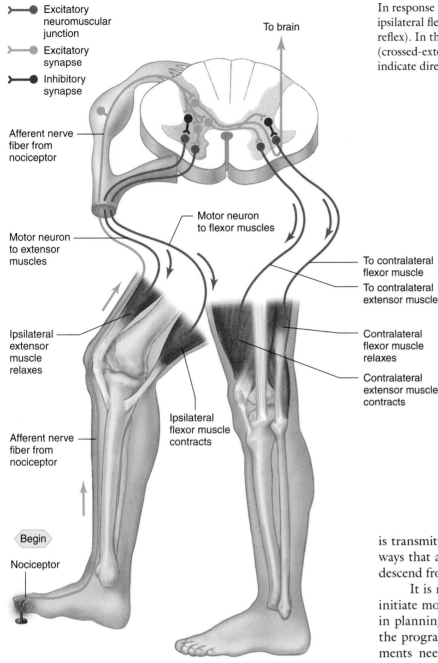

Neurons ending with:

⊶● Excitatory neuromuscular junction

⊶● Excitatory synapse

⊶● Inhibitory synapse

To brain

Afferent nerve fiber from nociceptor

Motor neuron to flexor muscles

Motor neuron to extensor muscles

To contralateral flexor muscle

To contralateral extensor muscle

Ipsilateral extensor muscle relaxes

Contralateral flexor muscle relaxes

Contralateral extensor muscle contracts

Afferent nerve fiber from nociceptor

Ipsilateral flexor muscle contracts

Begin

Nociceptor

Figure 10–9

In response to pain detected by nociceptors (Chapter 7), the ipsilateral flexor muscle's motor neuron is stimulated (withdrawal reflex). In the case illustrated, the opposite limb is extended (crossed-extensor reflex) to support the body's weight. Arrows indicate direction of action potential propagation.

grasping. These neurons play an important role in matching motor signals concerning the pattern of hand action with signals from the visual system concerning the three-dimensional features of the objects to be grasped.

During activation of the cortical areas involved in motor control, subcortical mechanisms also become active. It is to these areas of the motor control system that we now turn.

Subcortical and Brainstem Nuclei

Numerous highly interconnected structures lie in the brainstem and within the cerebrum beneath the cortex, where they interact with the cortex to control movements. Their influence is transmitted indirectly to the motor neurons both by pathways that ascend to the cerebral cortex and by pathways that descend from some of the brainstem nuclei.

It is not known to what extent, if any, these structures initiate movements, but they definitely play a prominent role in planning and monitoring them. Their role is to establish the programs that determine the specific sequence of movements needed to accomplish a desired action. Subcortical and brainstem nuclei are also important in learning skilled movements.

Prominent among the subcortical nuclei are the paired **basal nuclei** (see Figure 10–2b), which consist of a closely related group of separate nuclei. (These structures are often referred to as **basal ganglia,** but their presence within the central nervous system makes the term *nuclei* more technically correct.) They form a link in some of the looping parallel circuits through which activity in the motor system is transmitted from a specific region of sensorimotor cortex to the basal nuclei, from there to the thalamus, and then back to the cortical area where the circuit started (review Figure 10–1). Some of these circuits facilitate movements and others suppress them. The importance of the basal nuclei is particularly apparent in certain disease states, as we discuss next.

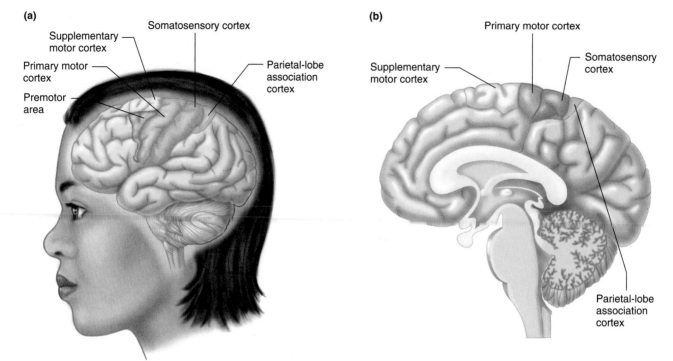

Figure 10–10

(a) The major motor areas of cerebral cortex. (b) Midline view of the right side of the brain showing the supplementary motor cortex, which lies in the part of the cerebral cortex that is folded down between the two cerebral hemispheres. Other cortical motor areas also extend onto this area. The premotor, supplementary motor, primary motor, somatosensory, and parietal-lobe association cortexes together make up the sensorimotor cortex.

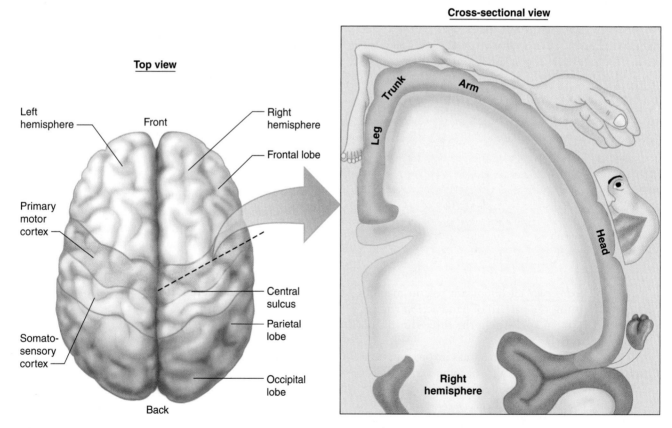

Figure 10–11

Somatotopic map of major body areas in primary motor cortex. Within the broad areas, no one area exclusively controls the movement of a single body region, and there is much overlap and duplication of cortical representation. Relative sizes of body structures are proportional to the number of neurons dedicated to their motor control. Only the right motor cortex, which principally controls muscles on the left side of the body, is shown.

Parkinson's Disease

In *Parkinson's disease,* the input to the basal nuclei is diminished, the interplay of the facilitory and inhibitory circuits is unbalanced, and activation of the motor cortex (via the basal nuclei-thalamus limb of the circuit just mentioned) is reduced. Clinically, Parkinson's disease is characterized by a reduced amount of movement *(akinesia),* slow movements *(bradykinesia),* muscular rigidity, and a tremor at rest. Other motor and nonmotor abnormalities may also be present. For example, a common set of symptoms includes a change in facial expression resulting in a masklike, unemotional appearance, a shuffling gait with loss of arm swing, and a stooped and unstable posture.

Although the symptoms of Parkinson's disease reflect inadequate functioning of the basal nuclei, a major part of the initial defect arises in neurons of the **substantia nigra** ("black substance"), a brainstem nucleus that gets its name from the dark pigment in its cells. These neurons normally project to the basal nuclei, where they release dopamine from their axon terminals. The substantia nigra neurons degenerate in Parkinson's disease, and the amount of dopamine they deliver to the basal nuclei is reduced. This decreases the subsequent activation of the sensorimotor cortex.

The drugs used to treat Parkinson's disease are all designed to restore dopamine activity in the basal nuclei, and fall into three main categories: (1) agonists of dopamine receptors, (2) inhibitors of the enzymes that metabolize dopamine at synapses, and (3) precursors of dopamine itself. The most widely prescribed drug is *Levodopa (L-dopa),* which falls into the third category. L-dopa enters the bloodstream, crosses the blood-brain barrier, and is converted to dopamine (dopamine itself is not used as medication because it cannot cross the blood-brain barrier, and it has many systemic side effects). The newly formed dopamine activates receptors in the basal nuclei and improves the symptoms of the disease. Other therapies include the lesioning (destruction) of overactive areas of the basal nuclei, or deep-brain stimulation of underactive areas. The latter is accomplished with surgically implanted electrodes connected to an electric pulse generator resembling a cardiac artificial pacemaker (Chapter 12). Still highly controversial is the transplantation of cells derived from various sources into the basal nuclei, including fetal cells, cells that have been genetically engineered, or cells taken from dopamine-secreting tissues in the patient's own body. In most cases, the injected cells are neurons, but more recently, undifferentiated embryonic stem cells have shown the ability to begin producing dopamine when injected into damaged areas of the brain. Regardless of their source, the implanted cells have shown some promise in reversing the symptoms of Parkinson's disease.

Cerebellum

The cerebellum is behind the brainstem (see Figure 10–2a). It influences posture and movement indirectly by means of input to brainstem nuclei and (by way of the thalamus) to regions of the sensorimotor cortex that give rise to pathways that descend to the motor neurons. The cerebellum receives information both from the sensorimotor cortex (relayed via brainstem nuclei) and from the vestibular system, eyes, ears, skin, muscles, joints, and tendons—that is, from some of the very receptors that movement affects.

One role of the cerebellum in motor functioning is to provide timing signals to the cerebral cortex and spinal cord for precise execution of the different phases of a motor program, in particular the timing of the agonist/antagonist components of a movement. It also helps coordinate movements that involve several joints and stores the memories of these movements so they are easily achieved the next time they are tried.

The cerebellum also participates in planning movements—integrating information about the nature of an intended movement with information about the surrounding space. The cerebellum then provides this as a feedforward signal to the brain areas responsible for refining the motor program. Moreover, during the course of the movement, the cerebellum compares information about what the muscles *should* be doing with information about what they actually *are* doing. If a discrepancy develops between the intended movement and the actual one, the cerebellum sends an error signal to the motor cortex and subcortical centers to correct the ongoing program.

The role of the cerebellum in programming movements can best be appreciated when seeing its absence in individuals with *cerebellar disease.* They typically cannot perform limb or eye movements smoothly, but move with a tremor—a so-called *intention tremor* that increases as a movement nears its final destination. (Note that this differs from patients with Parkinson's disease, who have a tremor while at rest.) People with cerebellar disease also cannot start or stop movements quickly or easily, and they cannot combine the movements of several joints into a single smooth, coordinated motion. The role of the cerebellum in the precision and timing of movements can be appreciated when you consider the complex tasks it helps us accomplish. For example, a tennis player sees a ball fly over the net, anticipates its flight path, runs along an intersecting path, and swings the racquet through an arc that will intercept the ball with the speed and force required to return it to the other side of the court. People with cerebellar damage cannot achieve this level of coordinated, precise, learned movement.

Unstable posture and awkward gait are two other symptoms characteristic of cerebellar disease. For example, people with cerebellar damage walk with their feet wide apart, and they have such difficulty maintaining balance that their gait appears drunken. Visual input helps compensate for some of the loss of motor coordination—patients can stand on one foot with eyes open, but not closed. A final symptom involves difficulty in learning new motor skills. Individuals with cerebellar disease find it hard to modify movements in response to new situations. Unlike damage to areas of sensorimotor cortex, cerebellar damage does not cause paralysis or weakness.

Finally, there is growing evidence that, in addition to its motor functions, the human cerebellum may also participate in higher cognitive activity. Though making up only one-tenth of the brain's volume, the cerebellum contains at least half of the total neurons. Anatomists have determined that cerebellar neurons project to regions of cortex beyond the sensorimotor areas, including those involved in memory and attention. Patients with cerebellar disease have been observed to display a flatness of emotion, and a reduction of cerebellar neurons has been suggested to underlie the abnormal sense of self-awareness in some cases of ***autism.***

Descending Pathways

The influence exerted by the various brain regions on posture and movement occurs via descending pathways to the motor neurons and the interneurons that affect them. The pathways are of two types: the **corticospinal pathways,** which, as their name implies, originate in the cerebral cortex, and a second group we will refer to as the **brainstem pathways,** which originate in the brainstem.

Fibers from both types of descending pathways end at synapses on alpha and gamma motor neurons or on interneurons that affect them. Sometimes these are the same interneurons that function in local reflex arcs, thereby ensuring that the descending signals are fully integrated with local information before the activity of the motor neurons is altered. In other cases, the interneurons are part of neural networks involved in posture or locomotion. The ultimate effect of the descending pathways on the alpha motor neurons may be excitatory or inhibitory.

Importantly, some of the descending fibers affect *afferent* systems. They do this via (1) presynaptic synapses on the terminals of afferent neurons as these fibers enter the central nervous system, or (2) synapses on interneurons in the ascending pathways. The overall effect of this descending input to afferent systems is to regulate their influence on either the local or brain motor control areas, thereby altering the importance of a particular bit of afferent information or sharpening its focus. This descending (motor) control over ascending (sensory) information provides another example to show that there is no real functional separation between the motor and sensory systems.

Corticospinal Pathway

The nerve fibers of the corticospinal pathways have their cell bodies in the sensorimotor cortex and terminate in the spinal cord. The corticospinal pathways are also called the **pyramidal tracts** or **pyramidal system** because of their triangular shape as they pass along the ventral surface of the medulla oblongata. In the medulla oblongata near the junction of the spinal cord and brainstem, most of the corticospinal fibers cross, or **decussate,** the spinal cord to descend on the opposite side (**Figure 10–12**). Thus, the skeletal muscles on the left side of the body are controlled largely by neurons in the right half of the brain, and vice versa.

As the corticospinal fibers descend through the brain from the cerebral cortex, they are accompanied by fibers of the **corticobulbar pathway** ("bulbar" means "pertaining to the brainstem"), a pathway that begins in the sensorimotor cortex and ends in the brainstem. The corticobulbar fibers control, directly or indirectly via interneurons, the motor neurons that innervate muscles of the eye, face, tongue, and throat. These fibers provide the main source of control for voluntary move-

ment of the muscles of the head and neck, whereas the corticospinal fibers serve this function for the muscles of the rest of the body. For convenience, we will include the corticobulbar pathway in the general term *corticospinal pathways.*

Convergence and divergence are hallmarks of the corticospinal pathway. For example, a great number of different neuronal sources converge on neurons of the sensorimotor cortex, which is not surprising when you consider the many factors that can affect motor behavior. As for the descending pathways, neurons from wide areas of the sensorimotor cortex converge onto single motor neurons at the local level so that multiple brain areas usually control single muscles. Also,

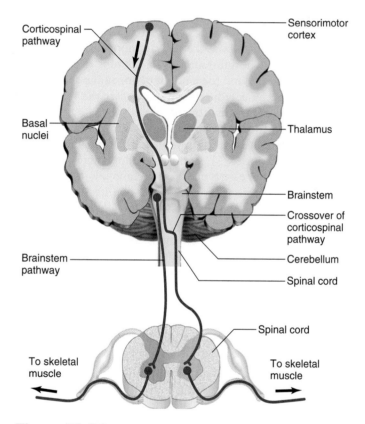

Figure 10–12

The corticospinal and brainstem pathways. Most of the corticospinal fibers cross in the brainstem to descend in the opposite side of the spinal cord, but the brainstem pathways are mostly uncrossed. The descending neurons are shown synapsing directly onto motor neurons in the spinal cord, but they commonly synapse onto local interneurons.

Adapted from Gardner.

Figure 10–12 physiological *inquiry*

- If a blood clot blocked a cerebral blood vessel supplying a small region of the right cerebral cortex, just in front of the central sulcus in the deep groove between the hemispheres, what symptoms might result?

Answer can be found at end of chapter.

axons of single corticospinal neurons diverge markedly to synapse with a number of different motor neuron populations at various levels of the spinal cord, thereby ensuring that the motor cortex can coordinate many different components of a movement.

This apparent "blurriness" of control is surprising when you think of the delicacy with which you can move a fingertip, because it is the corticospinal pathways that control rapid, fine movements of the distal extremities, such as those you make when you manipulate an object with your fingers. After damage occurs to the corticospinal pathways, all movements are slower and weaker, individual finger movements are absent, and it is difficult to release a grip.

Brainstem Pathways

Axons from neurons in the brainstem also form pathways that descend into the spinal cord to influence motor neurons. These pathways are sometimes referred to as the **extrapyramidal system,** or indirect pathways, to distinguish them from the corticospinal (pyramidal) pathways. However, no general term is widely accepted for these pathways, and for convenience we will refer to them collectively as the brainstem pathways.

Axons of some of the brainstem pathways cross from their side of origin in the brainstem to affect muscles on the opposite side of the body, but most remain uncrossed (see Figure 10–12). In the spinal cord, the fibers of the brainstem pathways descend as distinct clusters, named according to their sites of origin. For example, the vestibulospinal pathway descends to the spinal cord from the vestibular nuclei in the brainstem, whereas the reticulospinal pathway descends from neurons in the brainstem reticular formation.

The brainstem pathways are especially important in controlling muscles of the trunk for upright posture, balance, and walking.

Concluding Comments on the Descending Pathways

As stated previously, the corticospinal neurons generally have their greatest influence over motor neurons that control muscles involved in fine, isolated movements, particularly those of the fingers and hands. The brainstem descending pathways, in contrast, are involved more with coordination of the large muscle groups used in the maintenance of upright posture, in locomotion, and in head and body movements when turning toward a specific stimulus.

There is, however, much interaction between the descending pathways. For example, some fibers of the corticospinal pathway end on interneurons that play important roles in posture, whereas fibers of the brainstem descending pathways sometimes end directly on the alpha motor neurons to control discrete muscle movements. Because of this redundancy, one system may compensate for loss of function resulting from damage to the other system, although the compensation is generally not complete.

The distinctions between the corticospinal and brainstem descending pathways are not clear-cut. All movements, whether automatic or voluntary, require the continuous coordinated interaction of both types of pathways.

Muscle Tone

Even when a skeletal muscle is relaxed, there is a slight and uniform resistance when it is stretched by an external force. This resistance is known as **muscle tone,** and it can be an important diagnostic tool for clinicians assessing a patient's neuromuscular function.

Muscle tone is due both to the passive elastic properties of the muscles and joints and to the degree of ongoing alpha motor neuron activity. When a person is deeply relaxed, the alpha motor neuron activity probably makes little contribution to the resistance to stretch. As the person becomes increasingly alert, however, more activation of the alpha motor neurons occurs and muscle tone increases.

Abnormal Muscle Tone

Abnormally high muscle tone, called *hypertonia,* accompanies a number of disease processes and is seen very clearly when a joint is moved passively at high speeds. The increased resistance is due to a greater-than-normal level of alpha motor neuron activity, which keeps a muscle contracted despite the person's attempt to relax it. Hypertonia usually occurs with disorders of the descending pathways that normally inhibit the motor neurons.

Clinically, the descending pathways and neurons of the motor cortex are often referred to as the **upper motor neurons** (a confusing misnomer because they are not really motor neurons at all). Abnormalities due to their dysfunction are classed, therefore, as *upper motor neuron disorders.* Thus, hypertonia indicates an upper motor neuron disorder. In this clinical classification, the alpha motor neurons—the true motor neurons—are termed **lower motor neurons.**

Spasticity is a form of hypertonia in which the muscles do not develop increased tone until they are stretched a bit, and after a brief increase in tone, the contraction subsides for a short time. The period of "give" occurring after a time of resistance is called the *clasp-knife phenomenon.* (When an examiner bends the limb of a patient with this condition, it is like folding a pocket knife—at first, the spring resists the bending motion, but once bending begins, it closes easily.) Spasticity may be accompanied by increased responses of motor reflexes such as the knee jerk, and by decreased coordination and strength of voluntary actions. *Rigidity* is a form of hypertonia in which the increased muscle contraction is continual and the resistance to passive stretch is constant (as occurs in the disease tetanus, which is described in detail at the end of this section). Two other forms of hypertonia that can occur suddenly in individual or multiple muscles are *spasms,* which are brief contractions, and *cramps,* which are prolonged and painful.

Hypotonia is a condition of abnormally low muscle tone, accompanied by weakness, atrophy (a decrease in muscle bulk), and decreased or absent reflex responses. Dexterity and coordination are generally preserved unless profound weakness is present. While hypotonia may develop after cerebellar disease, it more frequently accompanies disorders of the alpha motor neurons (lower motor neurons), neuromuscular junctions, or the muscles themselves. The term *flaccid,* which means "weak" or "soft," is often used to describe hypotonic muscles.

Maintenance of Upright Posture and Balance

The skeleton supporting the body is a system of long bones and a many-jointed spine that cannot stand erect against the forces of gravity without the support provided through coordinated muscle activity. The muscles that maintain upright posture—that is, support the body's weight against gravity—are controlled by the brain and by reflex mechanisms "wired into" the neural networks of the brainstem and spinal cord. Many of the reflex pathways previously introduced (e.g., the stretch and crossed-extensor reflexes) are active in posture control.

Added to the problem of maintaining upright posture is that of maintaining balance. A human being is a very tall structure balanced on a relatively small base, with the center of gravity quite high, just above the pelvis. For stability, the center of gravity must be kept within the base of support the feet provide (**Figure 10–13**). Once the center of gravity has moved beyond this base, the body will fall unless one foot is shifted to broaden the base of support. Yet, people can operate under conditions of unstable equilibrium because complex interacting **postural reflexes** maintain their balance.

The afferent pathways of the postural reflexes come from three sources: the eyes, the vestibular apparatus, and the receptors involved in proprioception (joint, muscle, and touch receptors, for example). The efferent pathways are the alpha motor neurons to the skeletal muscles, and the integrating centers are neuron networks in the brainstem and spinal cord.

In addition to these integrating centers, there are centers in the brain that form an internal representation of the body's geometry, its support conditions, and its orientation with respect to vertical. This internal representation serves two purposes: (1) it provides a reference frame for the perception of the body's position and orientation in space and for planning actions, and (2) it contributes to stability via the motor controls involved in maintaining upright posture.

There are many familiar examples of using reflexes to maintain upright posture; one is the crossed-extensor reflex.

As one leg is flexed and lifted off the ground, the other is extended more strongly to support the weight of the body, and the positions of various parts of the body are shifted to move the center of gravity over the single, weight-bearing leg. This shift in the center of gravity, as **Figure 10–14** demonstrates, is an important component in the stepping mechanism of locomotion.

It is clear that afferent input from several sources is necessary for optimal postural adjustments, yet interfering with any one of these inputs alone does not cause a person to topple over. Blind people maintain their balance quite well with only a slight loss of precision, and people whose vestibular mechanisms have been destroyed have very little disability in everyday life as long as their visual system and somatic receptors are functioning.

The conclusion to be drawn from such examples is that the postural control mechanisms are not only effective and flexible, they are also highly adaptable.

Walking

Walking requires the coordination of hundreds of muscles, each activated to a precise degree at a precise time. We initiate walking by allowing the body to fall forward to an unstable position and then moving one leg forward to provide support. When the extensor muscles are activated on the supported side of the body to bear the body's weight, the contralateral extensors are inhibited by reciprocal innervation to allow the nonsupporting limb to flex and swing forward. The cyclical, alternating movements of walking are brought about largely by networks of interneurons in the spinal cord at the local level. The interneuron networks coordinate the output of the various motor neuron pools that control the appropriate muscles of the arms, shoulders, trunk, hips, legs, and feet.

The network neurons rely on both plasma-membrane spontaneous pacemaker properties and patterned synaptic activity to establish their rhythms. At the same time, however, the networks are remarkably adaptable, and a single network can generate many different patterns of neural activity, depending upon its inputs. These inputs come from other local interneurons, afferent fibers, and descending pathways.

These complex spinal cord neural networks can even produce the rhythmical movement of limbs in the absence of

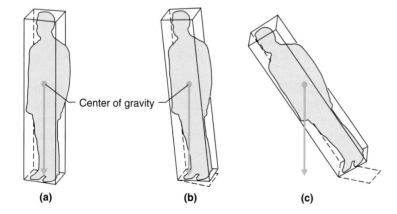

(a)   (b)   (c)

Figure 10–13

The center of gravity is the point in an object at which, if a string were attached and pulled up, all the downward force due to gravity would be exactly balanced. (a) The center of gravity must remain within the upward vertical projections of the object's base (the tall box outlined in the drawing) if stability is to be maintained. (b) Stable conditions: The box tilts a bit, but the center of gravity remains within the base area, so the box returns to its upright position. (c) Unstable conditions: The box tilts so far that its center of gravity is not above any part of the object's base—the dashed rectangle on the floor—and the object will fall.

Chapter 10

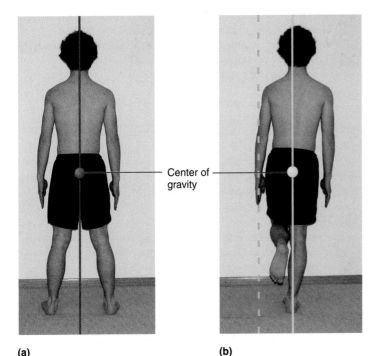

(a)

(b)

Center of gravity

Figure 10–14

Postural changes with stepping. (a) Normal standing posture. The center of gravity falls directly between the two feet. (b) As the left foot is raised, the whole body leans to the right so that the center of gravity shifts over the right foot.

command inputs from descending pathways. This was demonstrated in classical experiments involving animals with their cerebrums surgically separated from their spinal cords just above the brainstem. Though sensory perception and voluntary movement were completely absent, when suspended in a position that brought the limbs into contact with a treadmill, normal walking and running actions were initiated by afferent inputs arising from contact with the moving surface. This demonstrates that afferent inputs and spinal cord neural networks contribute substantially to the coordination of locomotion.

Under normal conditions, neural activation occurs in the cerebral cortex, cerebellum, and brainstem as well as in the spinal cord during locomotion. Moreover, middle and higher levels of the motor control hierarchy are necessary for postural control, voluntary override commands (like breaking stride to jump over a puddle), and adaptations to the environment (like walking across a stream on unevenly spaced stepping stones). The fact that damage to even small areas can cause marked disturbances in gait attests to the ultimate importance of the sensorimotor cortex.

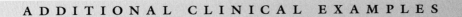

ADDITIONAL CLINICAL EXAMPLES

A number of disorders of motor control were discussed earlier in this chapter (Parkinson's disease, cerebellar disease, upper motor neuron disorders). We now return to a disease that was alluded to in the introduction to this chapter, and which demonstrates the importance of the balance between excitatory and inhibitory circuits in controlling motor function.

Tetanus

Tetanus is a neurological disorder that results from a decrease of the inhibitory input to alpha motor neurons. It occurs when spores of the soil bacterium **Clostridium tetani** invade a poorly oxygenated wound. Proliferation of the bacterium under anaerobic conditions induces it to secrete a neurotoxin that specifically targets inhibitory interneurons in the brainstem and spinal cord. Blockage of neurotransmitter release from these interneurons allows the normal excitatory inputs to dominate control of the alpha motor neurons, and

the result is high-frequency action potential firing that causes increased muscle tone and spasms.

Because the toxin attacks interneurons by traveling backward along the axons of alpha motor neurons, muscles with short motor neurons are affected first. Muscles of the head fall into this category, in particular those that move the jaw. The jaw rigidly clamps shut, because the muscles that close it are much stronger than those that open it. Appearance of this symptom early in the disease process explains the common name of this condition, **lockjaw.**

Treatment for tetanus includes: (1) administering antibiotics to kill the bacteria, (2) injecting antibodies that bind the toxin (tetanus immune globulin, or TIG), (3) providing drugs to relax and/or paralyze spastic muscles, and (4) mechanically ventilating the lungs, to maintain air flow despite spastic or paralyzed respiratory muscles. With prompt treatment, approximately 90 percent of patients with tetanus make a complete recovery within a few months.■

I. Skeletal muscles are controlled by their motor neurons. All the motor neurons that control a given muscle form a motor neuron pool.

Motor Control Hierarchy

I. The neural systems that control body movements can be conceptualized as being arranged in a motor control hierarchy.
   a. The highest level determines the general intention of an action.
   b. The middle level establishes a motor program and specifies the postures and movements needed to carry out the intended action, taking into account sensory information that indicates the body's position.
   c. The local level ultimately determines which motor neurons will be activated.
   d. As the movement progresses, information about what the muscles are doing feeds back to the motor control centers, which make program corrections.
   e. Almost all actions have conscious and unconscious components.

Local Control of Motor Neurons

I. Most direct input to motor neurons comes from local interneurons, which themselves receive input from peripheral receptors, descending pathways, and other interneurons.
II. Muscle spindle stretch receptors monitor muscle length and the velocity of changes in length.
   a. Activation of these receptors initiates the stretch reflex, which inhibits motor neurons of ipsilateral antagonists and activates those of the stretched muscle and its synergists. This provides negative feedback control of muscle length.
   b. Tension on the stretch receptors is maintained during muscle contraction by gamma efferent activation to the spindle muscle fibers.
   c. Alpha and gamma motor neurons are generally coactivated.
III. Golgi tendon organs monitor muscle tension. Through interneurons, they activate inhibitory synapses on motor neurons of the contracting muscle and excitatory synapses on motor neurons of ipsilateral antagonists. This provides negative feedback control of muscle tension.
IV. The withdrawal reflex excites the ipsilateral flexor muscles and inhibits the ipsilateral extensors. The crossed-extensor reflex excites the contralateral extensor muscles during excitation of the ipsilateral flexors.

The Brain Motor Centers and the Descending Pathways They Control

I. Neurons in the motor cortex are anatomically arranged in a somatotopic map.
II. Different areas of sensorimotor cortex have different functions, but much overlap in activity.
III. The basal nuclei form a link in a circuit that originates in and returns to sensorimotor cortex. These subcortical nuclei facilitate some motor behaviors and inhibit others.
IV. The cerebellum coordinates posture and movement and plays a role in motor learning.
V. The corticospinal pathways pass directly from the sensorimotor cortex to motor neurons in the spinal cord (or brainstem, in

the case of the corticobulbar pathways) or, more commonly, to interneurons near the motor neurons.
   a. In general, neurons on one side of the brain control muscles on the other side of the body.
   b. Corticospinal pathways control predominately fine, precise movements.
   c. Some corticospinal fibers affect the transmission of information in afferent pathways.
VI. Other descending pathways arise in the brainstem and are involved mainly in the coordination of large groups of muscles used in posture and locomotion.
VII. There is some duplication of function between the two descending pathways.

Muscle Tone

I. Hypertonia, as seen in spasticity and rigidity, usually occurs with disorders of the descending pathways.
II. Hypotonia can be seen with cerebellar disease or, more commonly, with disease of the alpha motor neurons or muscle.

Maintenance of Upright Posture and Balance

I. Maintenance of posture and balance depends upon inputs from the eyes, vestibular apparatus, and somatic proprioceptors.
II. To maintain balance, the body's center of gravity must be maintained over the body's base.
III. The crossed-extensor reflex is a postural reflex.

Walking

I. The activity of interneuron networks in the spinal cord brings about the cyclical, alternating movements of locomotion.
II. These pattern generators are controlled by corticospinal and brainstem descending pathways and affected by feedback and motor programs.

Additional Clinical Examples

I. *Tetanus* is a disease of muscle rigidity caused when a bacterial toxin blocks the release of an inhibitory neurotransmitter.

■ **K E Y   T E R M S**

alpha motor neuron 303	motor cortex 304
basal ganglia 305	motor neuron pool 297
basal nuclei 305	motor program 297
brainstem pathway 308	motor unit 297
coactivated 303	muscle spindle 300
contralateral 304	muscle-spindle stretch
corticobulbar pathway 308	receptor 300
corticospinal pathway 308	muscle tone 309
crossed-extensor reflex 304	nuclear bag fiber 300
decussation 308	nuclear chain fiber 300
descending pathway 297	parietal-lobe association
extrafusal fiber 300	cortex 304
extrapyramidal system 309	polysynaptic 302
gamma motor neuron 303	postural reflex 310
Golgi tendon organ 303	premotor area 304
intrafusal fiber 300	primary motor cortex 304
ipsilateral 303	proprioception 298
knee jerk 302	pyramidal system 308
lower motor neurons 309	pyramidal tract 308
monosynaptic 302	reciprocal innervation 302

■ CLINICAL TERMS

■ REVIEW QUESTIONS

1. Describe motor control in terms of the conceptual motor control hierarchy. Use the following terms: highest, middle, and local levels; motor program; descending pathways; motor neuron.
2. List the characteristics of voluntary actions.

3. Picking up a book, for example, has both voluntary and involuntary components. List the components of this action and indicate whether each is voluntary or involuntary.
4. List the inputs that can converge on the interneurons active in local motor control.
5. Draw a muscle spindle within a muscle, labeling the spindle, intrafusal and extrafusal muscle fibers, stretch receptors, afferent fibers, and alpha and gamma efferent fibers.
6. Describe the components of the knee jerk reflex (stimulus, receptor, afferent pathway, integrating center, efferent pathway, effector, and response).
7. Describe the major function of alpha-gamma coactivation.
8. Distinguish among the following areas of the cerebral cortex: sensorimotor, primary motor, premotor, and supplementary motor.
9. Contrast the two major types of descending motor pathways in terms of structure and function.
10. Describe the roles that the basal nuclei and cerebellum play in motor control.
11. Explain how hypertonia might result from disease of the descending pathways.
12. Explain how hypotonia might result from lower motor neuron disease.
13. Explain the role the crossed-extensor reflex plays in postural stability.
14. Explain the role of the interneuronal networks in walking, incorporating in your discussion the following terms: interneuron, reciprocal innervation, synergist, antagonist, and feedback.

Chapter 10 Test Questions

(Answers appear in Appendix A.)

1. Which is a correct statement regarding the hierarchical organization of motor control?
   a. Skeletal muscle contraction can only be initiated by neurons in the cerebral cortex.
   b. The basal nuclei participate in the creation of a motor program that specifies the pattern of neural activity required for a voluntary movement.
   c. Neurons in the cerebellum have long axons that synapse directly on alpha motor neurons in the ventral horn of the spinal cord.
   d. The cell bodies of alpha motor neurons are found in the primary motor region of the cerebral cortex.
   e. Neurons with cell bodies in the basal nuclei can either form excitatory or inhibitory synapses onto skeletal muscle cells.

2. In the stretch reflex:
   a. Golgi tendon organs activate contraction in extrafusal muscle fibers connected to that tendon.
   b. Lengthening of muscle-spindle receptors in a muscle leads to contraction in an antagonist muscle.
   c. Action potentials from muscle-spindle receptors in a muscle form monosynaptic excitatory synapses on motor neurons to extrafusal fibers within the same muscles.
   d. Slackening of intrafusal fibers within a muscle activates gamma motor neurons that form excitatory synapses with extrafusal fibers within that same muscle.

   e. Afferent neurons to the sensorimotor cortex stimulate the agonist muscle to contract and the antagonist muscle to be inhibited.

3. Which would result in reflex contraction of the extensor muscles of the right leg?
   a. stepping on a tack with the left foot
   b. stretching the flexor muscles in the right leg
   c. dropping a hammer on the right big toe
   d. action potentials from Golgi tendon organs in extensors of the right leg
   e. action potentials from muscle-spindle receptors in flexors of the right leg

4. If implanted electrodes were used to stimulate action potentials in gamma motor neurons to flexors of the left arm, which would be the most likely result?
   a. inhibition of the flexors of the left arm
   b. a decrease in action potentials from muscle-spindle receptors in the left arm
   c. a decrease in action potentials from Golgi tendon organs in the left arm
   d. an increase in action potentials along alpha motor neurons to flexors in the left arm
   e. contraction of flexor muscles in the right arm

5. Where is the primary motor cortex found?
   a. in the cerebellum
   b. in the occipital lobe of the cerebrum
   c. between the somatosensory cortex and the premotor area of the cerebrum
   d. in the ventral horn of the spinal cord
   e. just posterior to the parietal lobe assocation cortex

True or False

6. Neurons in the primary motor cortex of the right cerebral hemisphere mainly control muscles on the left side of the body.

7. Patients with upper motor neuron disorders generally have reduced muscle tone and flaccid paralysis.

8. Neurons descending in the corticospinal pathway control mainly trunk musculature and postural reflexes, whereas neurons of the brainstem pathways control fine motor movements of the distal extremities.

9. In patients with Parkinson's disease, an excess of dopamine from neurons of the substantia nigra causes intention tremors when the person performs voluntary movements.

10. The disease tetanus results when a bacterial toxin blocks the release of inhibitory neurotransmitter.

Chapter 10 Quantitative and Thought Questions

(Answers appear in Appendix A.)

1. What changes would occur in the knee jerk reflex after destruction of the gamma motor neurons?

2. What changes would occur in the knee jerk reflex after destruction of the alpha motor neurons?

3. Draw a cross section of the spinal cord and a portion of the thigh (similar to Figure 10–6) and "wire up" and activate the neurons so the leg becomes a stiff pillar; that is, so the knee does not bend.

4. We have said that hypertonia is usually considered a sign of disease of the descending motor pathways. How might it also result from abnormal function of the alpha motor neurons?

5. What neurotransmitters/receptors might be effective targets for drugs used to prevent the muscle spasms characteristic of the disease tetanus?

Chapter 10 Answers to Physiological Inquiries

Figure 10–3 Recall that when chloride channels are opened, a neuron is inhibited from depolarizing to threshold (see Figure 6–29 and accompanying text). Thus, the neurons of the spinal cord that release glycine are inhibitory interneurons. By specifically blocking glycine receptors, strychnine shifts the balance of inputs to motor neurons in favor of excitatory interneurons, resulting in excessive excitation. Poisoning victims experience excessive and uncontrollable muscle contractions body-wide, and when the respiratory muscles are affected, asphyxiation can occur. These symptoms are similar to those observed in the disease state tetanus, which is described at the end of this chapter.

Figure 10–8 Tendons are stretched more by actively contracting muscles than when muscles are passively stretched (see Figure 10–7). Thus, during very intense contractions that have the potential to cause injury, Golgi tendon organs are strongly activated. The resulting high-frequency action potentials arriving in the spinal cord stimulate interneurons that inhibit motor neurons to the muscle associated with that tendon, thus reducing the force and protecting the muscle.

Figure 10–12 When a region of the brain is deprived of oxygen and nutrients for even a short time, it often results in a stroke—neuronal cell death (see Chapter 6, section D). Because the right primary motor cortex was damaged in this case, the patient would have impaired motor function on the left side of the body. Given the midline location of the lesion, the leg would be most affected (see Figure 10–11).

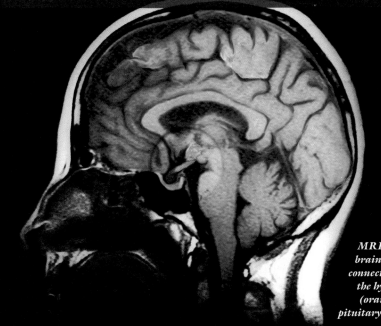

MRI of a human brain showing the connection between the hypothalamus (orange) and the pituitary gland (red)

chapter

11

The Endocrine System

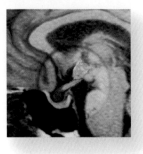

the endocrine system, along with the nervous system, is one of the body's two major communication systems. Communication within the nervous system is rapid, whereas the signals sent by the endocrine system may have much longer delays and last for much greater lengths of time. The **endocrine system** consists of all those glands, called **endocrine glands,** that secrete hormones. **Hormones** are chemical messengers that enter the blood, which carries them from endocrine glands to the cells upon which they act. The cells a particular hormone influences are the target cells for that hormone.

Endocrine glands are distinguished from another type of gland in the body, called exocrine glands. Endocrine glands release their secretory products into the blood; exocrine glands secrete their products into a duct, from where the secretions either exit the body (as in sweat) or enter the lumen of another organ, such as the intestines (**Figure 11–1**).

Table 11–1 summarizes most of the endocrine glands, the hormones they secrete, and the major functions the hormones control. The endocrine system differs from most of the other organ systems of the body in that the various glands are not anatomically connected; however, they do form a system in the functional sense. The reader may be puzzled to see some organs—the heart, for instance—that clearly have other functions yet are listed as part of the endocrine system. The explanation is that, in addition to the cells that carry out the organ's other functions, the organ also contains cells that secrete hormones. This illustrates the fact that organs are made up of different types of cells.

Note also in Table 11–1 that the hypothalamus, a part of the brain, is considered part of the endocrine system. This is because the chemical messengers released by certain neuron terminals in both the hypothalamus and its extension, the posterior pituitary, do not function as neurotransmitters affecting adjacent cells, but rather enter the blood as neurohormones. The blood then carries them to their sites of action.

Table 11–1 demonstrates that there are a large number of endocrine glands and hormones. This chapter is not meant to be all-inclusive. Some of the hormones listed in Table 11–1 are best considered in the context of the control systems in which they participate. For example, the pancreatic hormones (insulin and glucagon) are described in Chapter 16, on organic metabolism.

One phenomenon evident from Table 11–1 is that a single gland may secrete multiple hormones. The usual pattern in such cases is that a single cell type secretes only one hormone, so that multiple hormone secretion reflects the presence of different types of endocrine cells in the same gland. In a few cases, however, a single cell may secrete more than one hormone.

Finally, a chemical messenger secreted by an endocrine gland cell may also be secreted by other cell types and serves in these other locations as a neurotransmitter or paracrine/autocrine agent. For example, prolactin is secreted not only by the anterior pituitary but by cells of the immune system, where it is thought to exert paracrine/autocrine functions. Somatostatin, a hormone produced by the hypothalamus, is also secreted by cells of the stomach, where it has local paracrine actions.

The aims of this chapter are to present (1) the general principles of endocrinology—that is, a structural and functional analysis of hormones in general that transcends individual glands; and (2) an analysis of several of the most important hormonal systems. The reader is advised to review the material in Chapter 5 before continuing.

Hormone Structures and Synthesis

Hormones fall into three major chemical classes: (1) amines, (2) peptides and proteins, and (3) steroids.

Amine Hormones

The **amine hormones** are all derivatives of the amino acid tyrosine. They include the **thyroid hormones,** the catecholamines **epinephrine** and **norepinephrine** (produced by the adrenal medulla), and **dopamine** (produced by the hypothalamus). The structure and synthesis of the iodine-containing thyroid hormones will be described in detail in Section C of this chapter. For now, their structures are shown in **Figure 11–2**. Chapter 6 described the steps of catecholamine synthesis; the structures of the amine hormones are reproduced here in Figure 11–2.

There are two adrenal glands, one on top of each kidney. Each **adrenal gland** is composed of an inner **adrenal medulla,** which secretes amine hormones, and a surrounding **adrenal cortex,** which secretes steroid hormones. As Chapter 6 described, the adrenal medulla is really a modified sympathetic ganglion whose cell bodies do not have axons. Instead, they release their secretions into the blood, thereby fulfilling a criterion for an endocrine gland.

The adrenal medulla secretes mainly two amine hormones, epinephrine and norepinephrine. Recall from Chapter 6 that these molecules constitute, with dopamine, the chemical family of catecholamines. In humans, the adrenal medulla secretes approximately four times more epinephrine than norepinephrine. This is because the adrenal medulla expresses high amounts of an enzyme called phenyl-N-methyltransferase (PNMT), which catalyzes the step that converts norepinephrine to epinephrine. Epinephrine and norepinephrine exert actions similar to those of the sympathetic nerves, which, because they do not express PNMT, make only norepinephrine. These actions are described in various chapters and summarized in Section B.

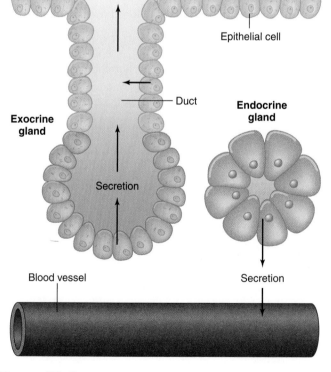

Figure 11–1

Exocrine gland secretions enter ducts, whereas hormones secreted by endocrine (ductless) glands diffuse directly into the blood.

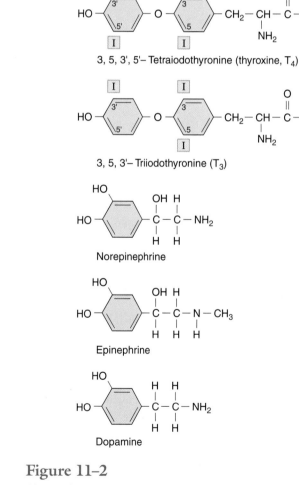

Figure 11–2

Chemical structures of the amine hormones: thyroxine, triiodothyronine, norepinephrine, epinephrine, and dopamine. The two thyroid hormones differ by only one iodine atom, a difference noted in the abbreviations T_3 and T_4. The position of the carbon atoms in the two rings of T_3 and T_4 are numbered.

Table 11–1 Summary of the Hormones

Site Produced (Endocrine Gland)	Hormone	Major Function* Is Control Of:
Adipose tissue cells	Leptin, several others	Appetite; metabolic rate; reproduction
Adrenal:		
Adrenal cortex	Cortisol	Organic metabolism; response to stress; immune system; development
	Androgens	Sex drive in women; adrenarche
	Aldosterone	Sodium and potassium excretion by kidneys
Adrenal medulla	Epinephrine and norepinephrine	Organic metabolism; cardiovascular function; response to stress
Gastrointestinal tract	Gastrin	Gastrointestinal tract motility and acid secretion
	Ghrelin	Appetite
	Secretin	Exocrine and endocrine secretions from pancreas
	Cholecystokinin (CCK)[†]	Secretion of bile from gallbladder
	Glucose-dependent insulinotropic peptide (GIP)	Insulin secretion
	Motilin	Gastrointestinal tract motility
Gonads:		
Ovaries: female	Estrogen (Estradiol in humans)	Reproductive system; secondary sex characteristics; growth and development; development of ovarian follicles
	Progesterone	Endometrium and pregnancy
	Inhibin	Follicle-stimulating hormone (FSH) secretion
	Relaxin	Relaxation of cervix and pubic ligaments
Testes: male	Androgen (Testosterone and Dihydrotestosterone)	Reproductive system; secondary sex characteristics; growth and development; sex drive; gamete development
	Inhibin	FSH secretion
	Müllerian-inhibiting substance (MIS)	Regression of Müllerian ducts
Heart	Atrial natriuretic peptide (ANP, atriopeptin)	Sodium excretion by kidneys; blood pressure
Hypothalamus	Hypophysiotropic hormones:	Secretion of hormones by the anterior pituitary
	Corticotropin-releasing hormone (CRH)	Secretion of adrenocorticotropic hormone (ACTH)
	Thyrotropin-releasing hormone (TRH)	Secretion of thyroid-stimulating hormone (TSH)
	Growth hormone-releasing hormone (GHRH)	Secretion of growth hormone (GH)
	Somatostatin (SS)	Secretion of growth hormone
	Gonadotropin-releasing hormone (GnRH)	Secretion of luteinizing hormone (LH) and follicle-stimulating hormone (FSH)
	Dopamine (DA)	Secretion of prolactin (PRL)
Kidneys	Erythropoietin (EPO)	Erythrocyte production in bone marrow
	1,25-dihydroxyvitamin D	Calcium absorption in GI tract
Leukocytes, macrophages, endothelial cells, and fibroblasts	Cytokines[‡] (these include the interleukins, colony-stimulating factors, interferons, tumor necrosis factors)	Immune defenses; immune cell growth and secretory processes
Liver and other cells	Insulin-like growth factor-1 (IGF-1)	Cell division and growth of bone and other tissues
Pancreas	Insulin	Plasma glucose, amino acids and fatty acids
	Glucagon	Plasma glucose
	Somatostatin (SS)	Secretion of insulin and glucagon
Parathyroids	Parathyroid hormone (PTH, parathormone)	Plasma calcium and phosphate; synthesis of 1,25-dihydroxyvitamin D

(continued)

Table 11–1 Summary of the Hormones *(continued)*

Site Produced (Endocrine Gland)	Hormone	Major Function* Is Control Of:
Pituitary glands:		
Anterior pituitary	Growth hormone (somatotropin)	Growth, mainly via local production of IGF-1; protein, carbohydrate, and lipid metabolism
	Thyroid-stimulating hormone (thyrotropin)	Thyroid gland activity and growth
	Adrenocorticotropic hormone (corticotropin)	Adrenal cortex activity and growth
	Prolactin	Milk production in breast
	Gonadotropic hormones:	
	Follicle-stimulating hormone	
	Males	Gamete production
	Females	Ovarian follicle growth
	Luteinizing hormone:	
	Males	Testicular production of testosterone
	Females	Ovarian production of estradiol; ovulation
	β-lipotropin and β-endorphin	Possibly fat mobilization and analgesia during stress
Posterior pituitary§	Oxytocin	Milk let-down; uterine motility
	Vasopressin (antidiuretic hormone, ADH)	Blood pressure; water excretion by the kidneys
Placenta	Human chorionic gonadotropin (hCG)	Secretion of progesterone and estrogen by corpus luteum
	Estrogens	See Gonads: ovaries
	Progesterone	See Gonads: ovaries
	Human placental lactogen (hPL)	Breast development; organic metabolism
Thymus	Thymopoietin	T-lymphocyte function
Thyroid	Thyroxine (T_4) and triiodothyronine (T_3)	Metabolic rate; growth; brain development and function
	Calcitonin	Plasma calcium in some vertebrates (role unclear in humans)
Multiple cell types	Growth factors‡ (e.g., epidermal growth factor)	Growth and proliferation of specific cell types
Other (produced in blood)	Angiotensin II	Blood pressure; production of aldosterone from adrenal cortex

*This table does not list all functions of all hormones.

†The names and abbreviations in parentheses are synonyms.

‡Some classifications include the cytokines under the category of growth factors.

§The posterior pituitary stores and secretes these hormones; they are made in the hypothalamus.

The other catecholamine hormone, dopamine, is synthesized by neurons in the hypothalamus. Dopamine is released into a special circulatory system called a portal system (see Section B), where it acts to regulate the activity of certain cells in the pituitary gland.

Peptide and Protein Hormones

Most hormones are either peptides or proteins. They range in size from small peptides having only three amino acids to small proteins (some of which are glycoproteins). For convenience, we will refer to all these hormones as **peptide hormones.**

In many cases, peptide hormones are initially synthesized on the ribosomes of endocrine cells as larger proteins known as preprohormones, which are then cleaved to **prohormones** by proteolytic enzymes in the rough endoplasmic reticulum (**Figure 11–3**). The prohormone is then packaged into secretory vesicles by the Golgi apparatus. In this process, the prohormone is cleaved to yield the active hormone and other

peptide chains found in the prohormone. Therefore, when the cell is stimulated to release the contents of the secretory vesicles by exocytosis, the other peptides are secreted along with the hormone. In certain cases they, too, may exert hormonal effects. In other words, instead of just one peptide hormone, the cell may be secreting multiple peptide hormones that differ in their effects on target cells.

As mentioned in Chapters 5 and 6, many peptides serve as both neurotransmitters (or neuromodulators) and as hormones. For example, some of the hormones secreted by the endocrine glands in the gastrointestinal tract are also produced by neurons in the brain, where they may function as neuromodulators.

Steroid Hormones

The third family of hormones is the steroids, the lipids whose ringlike structure was described in Chapter 2. **Steroid hormones** are primarily produced by the adrenal cortex and the

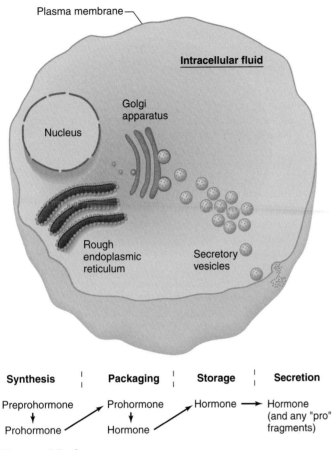

Figure 11–3 shows "Plasma membrane", "Intracellular fluid", "Golgi apparatus", "Nucleus", "Rough endoplasmic reticulum", "Secretory vesicles".

Synthesis	Packaging	Storage	Secretion
Preprohormone ↓ Prohormone	Prohormone ↓ Hormone	Hormone	Hormone (and any "pro" fragments)

Figure 11–3

Typical synthesis and secretion of peptide hormones. Secretion of stored secretory vesicles occurs by the process of exocytosis.

Figure 11–3 physiological *inquiry* (pi)

■ What is the advantage of packaging peptide hormones in secretory vesicles?

Answer can be found at end of chapter.

gonads (testes and ovaries) as well as by the placenta during pregnancy. **Figure 11–4** shows some examples of steroid hormones. In addition, the hormone 1,25-dihydroxyvitamin D, the active form of vitamin D, is a steroid.

The general process of steroid hormone synthesis is illustrated in **Figure 11–5.** In both the gonads and the adrenal cortex, the cells are stimulated by the binding of a pituitary gland hormone to its receptor (step 1 in Figure 11–5). These receptors are linked to G-proteins, which activate adenylyl cyclase and thus cAMP production (step 2). The subsequent activation of protein kinase A results in phosphorylation of numerous cytosolic and membrane proteins (step 3), which facilitate the subsequent steps in the process.

All of the steroid hormones are derived from cholesterol. Cells producing steroid hormones synthesize some of their own cholesterol for this purpose from molecules of acetate. Typically, however, most of the cholesterol enters the cells in the form of low-density lipoproteins (LDLs). Receptors on the cell surface bind to circulating LDLs; when this occurs, the LDL/receptor complex is internalized into the cell cytosol by endocytosis. Once inside the cell, the cholesterol is stored in an esterified form in large, non-membrane-bound lipid droplets. When the cell is stimulated, free cholesterol is released from the lipid droplet by the action of the enzyme **cholesterol esterase,** which is activated by protein kinase A (step 4). Carrier proteins then transport the free cholesterol to the mitochondria (step 5). Once at the mitochondria, the cholesterol must be transported across the outer membrane to the inner mitochondrial membrane, which contains the enzymes required to process cholesterol into steroid hormones. These enzymes, called **cytochrome P450s,** are a large family of related proteins that can attach hydroxyl groups to carbon atoms and that participate in cleaving carbon-carbon bonds. In the case of cholesterol and steroid hormones, the carbon atoms may be within the ring structure or in branches arising from the rings. As the P450 enzymes modify cholesterol,

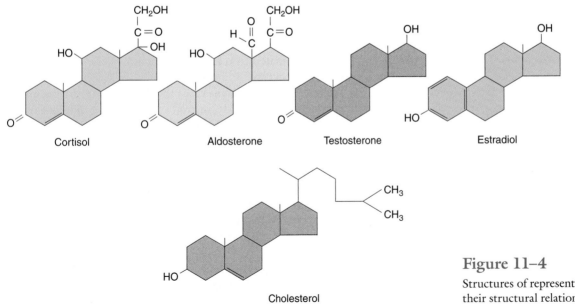

Cortisol   Aldosterone   Testosterone   Estradiol

Cholesterol

Figure 11–4

Structures of representative steroid hormones and their structural relationship to cholesterol.

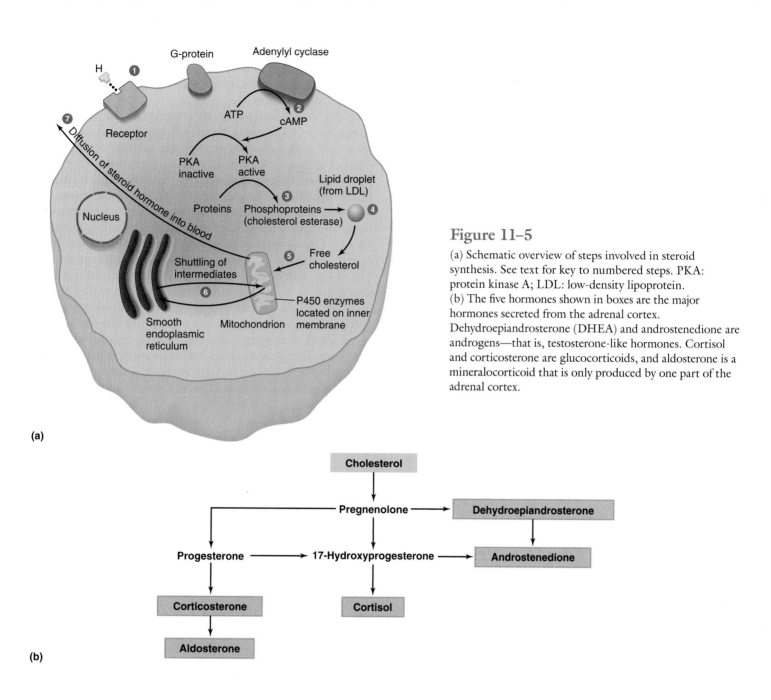

Figure 11–5

(a) Schematic overview of steps involved in steroid synthesis. See text for key to numbered steps. PKA: protein kinase A; LDL: low-density lipoprotein.
(b) The five hormones shown in boxes are the major hormones secreted from the adrenal cortex. Dehydroepiandrosterone (DHEA) and androstenedione are androgens—that is, testosterone-like hormones. Cortisol and corticosterone are glucocorticoids, and aldosterone is a mineralocorticoid that is only produced by one part of the adrenal cortex.

intermediates form, which must be further modified before the final steroid hormone is produced. These modifications, which also include a dehydrogenation reaction mediated by an enzyme different from the P450s, occur in both the mitochondria and in the smooth endoplasmic reticulum. Thus, the intermediates are shuttled back and forth between the two organelles (step 6). The final product depends upon the cell type and the types and amounts of the enzymes it expresses. Cells in the ovary, for example, express large amounts of the enzyme needed to convert testosterone to estradiol, whereas cells in the testes do not express significant amounts of this enzyme and thus make primarily testosterone.

Once formed, the lipophilic steroid hormones are not stored in the cytosol, but instead diffuse out through the lipid bilayer of the plasma membrane into the interstitial fluid and from there into the circulation (step 7). Because of their lipid nature, steroid hormones are not highly soluble in blood, and thus they are largely transported in plasma bound to carrier proteins such as albumin.

The next sections describe the pathways the adrenal cortex and gonads follow for steroid synthesis. Those for the placenta are somewhat unusual and are briefly discussed in Chapter 17.

Hormones of the Adrenal Cortex

The five major hormones secreted by the adrenal cortex are aldosterone, cortisol, corticosterone, dehydroepiandrosterone (DHEA), and androstenedione (**Figure 11–5b**). **Aldosterone** is known as a **mineralocorticoid** because its effects are on salt (mineral) balance, mainly on the kidneys' handling of sodium, potassium, and hydrogen ions. Its actions are described in detail in Chapter 14. Briefly, production of aldosterone is under the control of a circulating hormone called **angiotensin II,** which acts on plasma membrane receptors in the adrenal cortex to activate the inositol trisphosphate second messenger pathway. Note that this is different from the more common cAMP-mediated mechanism by which steroid hormones are produced, as previously described. Once synthesized, aldosterone enters the circulation and acts on cells of the kidneys to stimulate sodium and water retention, and potassium and hydrogen excretion in the urine.

Cortisol and corticosterone are called **glucocorticoids** because they have important effects on the metabolism of glucose and other organic nutrients. Cortisol is the predominant glucocorticoid in humans, and so we will not deal with corticosterone in future discussions. In addition to its effects on organic metabolism, cortisol exerts many other effects, including facilitation of the body's responses to stress and regulation of the immune system (Section D).

Dehydroepiandrosterone (DHEA) and androstenedione belong to the class of hormones known as **androgens,** which also includes the major male sex hormone, **testosterone,** produced by the testes. All androgens have actions similar to those of testosterone. Because the adrenal androgens are much less potent than testosterone, they are of little physiological significance in the adult male. They do, however, play roles in the adult female, and in both sexes in the fetus and at puberty, as described in Chapter 17.

The adrenal cortex is not a homogeneous gland but is composed of three distinct layers (**Figure 11–6**). The outer layer—the zona glomerulosa—possesses very high concentrations of the enzymes required to convert corticosterone to aldosterone, but lacks the enzymes required for the formation of cortisol and androgens. Thus, this layer synthesizes and secretes aldosterone but not the other major adrenal cortical hormones. In contrast, the zona fasciculata and zona reticularis have just the opposite enzyme profile. They, therefore, secrete no aldosterone but much cortisol and androgen.

In certain disease states, the adrenal cortex may secrete decreased or increased amounts of various steroids. For example, an absence of the enzymes for the formation of cortisol by the adrenal cortex can result in the shunting of the cortisol precursors into the androgen pathway. (Look at Figure 11–5b to imagine how this might happen.) In a woman, one result of the large increase in androgen secretion would be *masculinization.*

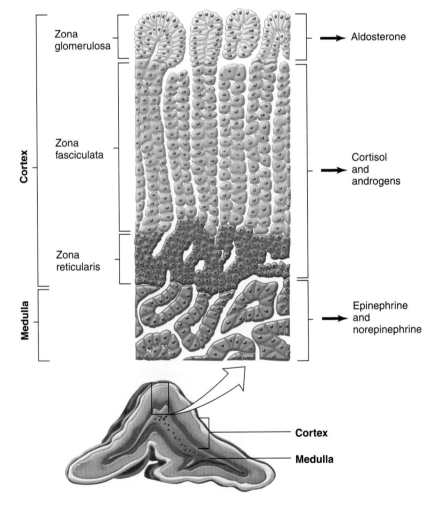

Figure 11–6

Section through an adrenal gland showing both the medulla and the zones of the cortex, as well as the hormones they secrete.

Hormones of the Gonads

Compared to the adrenal cortex, the gonads have very different concentrations of key enzymes in their steroid pathways. Endocrine cells in both the testes and the ovaries lack the enzymes needed to produce aldosterone and cortisol. They possess high concentrations of enzymes in the androgen pathways leading to androstenedione, as in the adrenal cortex. In addition, the endocrine cells in the testes contain a high concentration of the enzyme that converts androstenedione to testosterone, which is therefore the major androgen secreted by the testes (**Figure 11–7**). The ovarian endocrine cells synthesize the female sex hormones, which are collectively known as **estrogens** (estrone and estradiol). **Estradiol** is the predominant estrogen present during a woman's lifetime. The ovarian endocrine cells have a high concentration of the enzyme aromatase, which transforms androgens to estrogens (see Figure 11–7). Thus, estradiol, rather than testosterone, is the major steroid hormone secreted by the ovaries.

Very small amounts of testosterone do leak out of ovarian endocrine cells, however, and very small amounts of estradiol are produced from testosterone in the testes. Moreover, following their release into the blood by the gonads and the adrenal cortex, steroid hormones may undergo further conversion in the blood or organs. For example, testosterone is converted to estradiol in some of its target cells. Thus, the major male and female sex hormones—testosterone and estradiol, respectively—are not unique to males and females. The relative concentrations of the hormones, however, are quite different in the two sexes.

Finally, endocrine cells of the corpus luteum, an ovarian structure that arises following ovulation, secrete another major steroid hormone, **progesterone.** This steroid is critically important for uterine maturation during the menstrual cycle, and for maintaining a pregnancy (Chapter 17).

Hormone Transport in the Blood

Most peptide and all catecholamine hormones are water-soluble. Therefore, with the exception of a few peptides, these hormones are transported simply dissolved in plasma (**Table 11–2**). In contrast, the poorly soluble steroid hormones and thyroid hormones circulate in the blood largely bound to plasma proteins.

Even though the steroid and thyroid hormones exist in plasma mainly bound to large proteins, small concentrations of these hormones do exist dissolved in the plasma. The dissolved, or free, hormone is in equilibrium with the bound hormone:

Free hormone + Binding protein $\rightleftharpoons$

Hormone-protein complex

The total hormone concentration in plasma is the sum of the free and bound hormones. It is important to realize, however, that only the *free* hormone can diffuse out of capillaries and encounter its target cells. Therefore, the concentration of the free hormone is what is biologically important rather than the concentration of the total hormone, most of which

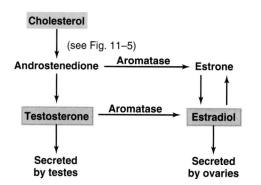

Figure 11–7

Gonadal production of steroids. Only the ovaries have high concentrations of the enzyme (aromatase) required to produce the estrogens estrone and estradiol.

Table 11–2	Categories of Hormones			
Chemical Class	**Major Form in Plasma**	**Location of Receptors**	**Most Common Signaling Mechanisms***	**Rate of Excretion/ Metabolism**
Peptides and catecholamines	Free (unbound)	Plasma membrane	1. Second messengers (e.g., cAMP, Ca^{2+}, IP_3) 2. Enzyme activation by receptor (e.g., JAK) 3. Intrinsic enzymatic activity of receptor (e.g., tyrosine autophosphorylation)	Fast (minutes)
Steroids and thyroid hormone	Protein-bound	Intracellular	Intracellular receptors directly alter gene transcription	Slow (hours to days)

*The diverse mechanisms of action of hormones were discussed in detail in Chapter 5.

is bound. As we will see next, the degree of protein binding also influences the rate of metabolism and the excretion of the hormone.

Hormone Metabolism and Excretion

Once a hormone has acted on a target tissue, the concentration of the hormone in the blood must be returned to normal. This is necessary to prevent excessive, possibly harmful effects from the prolonged exposure of target cells to hormones. A hormone's concentration in the plasma depends upon (1) its rate of secretion by the endocrine gland, and (2) its rate of removal from the blood. Removal, or "clearance," of the hormone occurs either by excretion or by metabolic transformation. The liver and the kidneys are the major organs that excrete or metabolize hormones.

The liver and kidneys, however, are not the only routes for eliminating hormones. Sometimes the hormone is metabolized by the cells upon which it acts. Very importantly, in the case of peptide hormones, endocytosis of hormone-receptor complexes on plasma membranes enables cells to remove the hormones rapidly from their surfaces and catabolize them intracellularly. The receptors are then often recycled to the plasma membrane.

In addition, enzymes in the blood and tissues rapidly break down catecholamine and peptide hormones. These hormones therefore tend to remain in the bloodstream for only brief periods—minutes to an hour. In contrast, because protein-bound hormones are protected from excretion or metabolism by enzymes, removal of the circulating steroid and thyroid hormones generally takes longer, often several hours. Thyroid hormone remains in the plasma for days due to very high binding to plasma proteins.

In some cases, metabolism of a hormone *activates* the hormone rather than inactivates it. In other words, the secreted hormone may be relatively inactive until metabolism transforms it. One example is testosterone, which is converted either to estradiol or dihydrotestosterone in certain of its target cells. These molecules, rather than testosterone itself, then bind to receptors inside the target cell and elicit the cell's response.

Figure 11–8 summarizes the fates of hormones after their secretion.

Mechanisms of Hormone Action

Hormone Receptors

Before continuing, you may want to review the mechanisms of ligand:receptor interactions and the signaling pathways activated by their interactions discussed in Chapter 5. The following material is presented in the specific context of hormones and their receptors.

Because hormones are transported in the blood, they can reach virtually all tissues. Yet the response to a hormone is highly specific, involving only the target cells for that hormone. The ability to respond depends upon the presence of specific receptors for those hormones on or in the target cells.

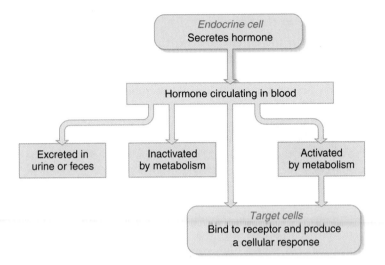

Figure 11–8

Possible fates and actions of a hormone following its secretion by an endocrine cell. Not all paths apply to all hormones.

As emphasized in Chapter 5, the response of a target cell to a chemical messenger is the final event in a sequence that begins when the messenger binds to specific cell receptors. As that chapter described, the receptors for peptide hormones and catecholamines are proteins located in the plasma membranes of the target cells. In contrast, the receptors for steroid hormones and the thyroid hormones are proteins located mainly *inside* the target cells.

Hormones can influence the ability of target cells to respond by regulating hormone receptors. Again, Chapter 5 described basic concepts of receptor modulation such as up-regulation and down-regulation. In the context of hormones, **up-regulation** is an increase in the number of a hormone's receptors, often resulting from a prolonged exposure to a low concentration of the hormone. This has the effect of increasing target-cell responsiveness to the hormone. **Down-regulation** is a decrease in receptor number, often from exposure to high concentrations of the hormone. This temporarily decreases target-cell responsiveness to the hormone, thus preventing overstimulation.

In some cases, hormones can down-regulate or up-regulate not only their own receptors but the receptors for other hormones as well. If one hormone induces a loss of a second hormone's receptors, the result will be a reduction of the second hormone's effectiveness. On the other hand, a hormone may induce an increase in the number of receptors for a second hormone. In this case, the effectiveness of the second hormone is increased. This latter phenomenon, in some cases, underlies the important hormone-hormone interaction known as permissiveness. In general terms, **permissiveness** means that hormone A must be present for the full strength of hormone B's effect. A low concentration of hormone A is usually all that is needed for this permissive effect, which may be due to A's ability to up-regulate B's receptors. For example, epinephrine causes a large release of fatty acids from adipose tissue, but only in the presence of permissive amounts of thyroid hormones (**Figure 11–9**). One reason is that thyroid hormones stimulate the synthesis of beta-adrenergic receptors for epinephrine in adipose tissue; thus, the tissue becomes much more sensitive to epinephrine. It should be noted, however, that receptor up-regulation does not explain all cases of permissiveness. Often, the explanation is not known or is due to changes in the signaling pathway that mediates the actions of a given hormone.

Events Elicited by Hormone-Receptor Binding

The events initiated when a hormone binds to its receptor—that is, the mechanisms by which the hormone elicits a cellular response—are one or more of the signal transduction pathways that apply to all chemical messengers, as described in Chapter 5. In other words, there is nothing unique about the mechanisms that hormones initiate as compared to those used by neurotransmitters and paracrine/autocrine agents, and so we will only briefly review them at this point (see Table 11–2).

Effects of Peptide Hormones and Catecholamines

As stated previously, the receptors for peptide hormones and the catecholamine hormones are located on the outer surface of the target cell's plasma membrane. This location is impor-

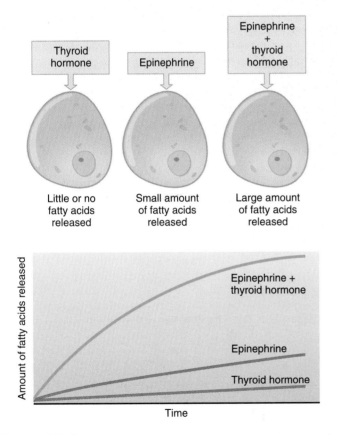

Little or no fatty acids released | Small amount of fatty acids released | Large amount of fatty acids released

Figure 11–9

The ability of thyroid hormone to "permit" epinephrine-induced release of fatty acids from adipose tissue cells. Thyroid hormone exerts this effect by causing an increased number of epinephrine receptors on the cell. Thyroid hormone by itself stimulates only a small amount of fatty acid release.

tant because these hormones are too large and hydrophilic to diffuse through the plasma membrane. When activated by hormone binding, the receptors trigger one or more of the signal transduction pathways described for plasma membrane receptors in Chapter 5. That is, the activated receptors directly influence: (1) enzyme activity that is part of the receptor; (2) activity of cytoplasmic janus kinases associated with the receptor; or (3) G proteins coupled in the plasma membrane to effector proteins—ion channels and enzymes—that generate second messengers such as cAMP and Ca^{2+}. The opening or closing of ion channels changes the electrical potential across the membrane. When a calcium channel is involved, the cytosolic concentration of this important ionic second messenger changes. The changes in enzyme activity are usually very rapid (e.g., due to phosphorylation) and produce changes in the activity of various cellular proteins. In some cases, the signal transduction pathways also lead to activation or inhibition of particular genes, causing a change in the synthesis rate of the proteins coded for by these genes. Thus, peptide hormones and catecholamines may exert both rapid (non-genomic) and delayed (gene transcription) actions on the same target cell.

Effects of Steroid and Thyroid Hormones

Structurally, the steroid hormones and the thyroid hormones are all lipophilic, and their receptors, which are intracellular, constitute part of the steroid hormone receptor superfamily. As described in Chapter 5, the binding of hormone to one of these receptors leads to the activation (or in some cases, inhibition) of the transcription of particular genes, causing a change in the synthesis rate of the proteins coded for by those genes. The ultimate result of changes in the concentrations of these proteins is an enhancement or inhibition of particular processes the cell carries out, or a change in the cell's rate of protein secretion.

Surprisingly, in addition to having intracellular receptors, some target cells also have plasma membrane receptors for certain of the steroid hormones, notably progesterone and estradiol. In such cases, the signal transduction pathways initiated by the plasma membrane receptors elicit rapid, nongenomic cell responses, whereas the intracellular receptors mediate a delayed response, requiring new protein synthesis. The physiological significance of the membrane receptors in humans is still under investigation, but it is clear from animal studies that these receptors are functional in other vertebrates.

Pharmacological Effects of Hormones

The administration of very large quantities of a hormone for medical purposes may have effects on an individual that are not usually seen in a healthy person. These **pharmacological effects** can also occur in diseases involving the secretion of excessive amounts of hormones. Pharmacological effects are of great importance in medicine because hormones are often used in large doses as therapeutic agents. Perhaps the most common example is that of very potent synthetic forms of cortisol, such as prednisone, which is administered to suppress allergic and inflammatory reactions. In such situations, a host of unwanted effects may be observed (as described in Additional Clinical Examples at the end of Section D).

Inputs that Control Hormone Secretion

Many hormones are secreted in short bursts, with little or no release occurring between bursts. Therefore, plasma concentrations of hormones may fluctuate rapidly over brief time periods. Hormones may also undergo 24-hour cyclical variations in their secretory rates, with different circadian patterns for different hormones. Some are clearly linked to sleep; for example, the secretion of growth hormone increases during the early period of sleep and decreases during the rest of the night and day. The mechanisms underlying these cycles are traceable to cyclical variations in the activity of neural pathways involved in the hormone's release.

Hormone secretion is mainly under the control of three types of inputs to endocrine cells (**Figure 11–10**): (1) changes in the plasma concentrations of mineral ions or organic nutrients; (2) neurotransmitters released from neurons impinging on the endocrine cell; and (3) another hormone (or, in some cases, a paracrine/autocrine agent) acting on the endocrine cell. There is actually a fourth type of input—chemical and physical factors in the lumen of the gastrointestinal tract—but it applies mainly to the hormones secreted by the gastrointestinal tract and will be described in Chapter 15.

Before we look more closely at each category, we must stress that more than one input may influence hormone secretion. For example, insulin secretion is controlled by the extracellular concentrations of glucose and other nutrients, by both sympathetic and parasympathetic neurons to the insulin-secreting endocrine cells, and by several hormones acting on these cells. Thus, endocrine cells, like neurons, may be subject to multiple, simultaneous, often opposing inputs, and the resulting output—the rate of hormone secretion—reflects the integration of all these inputs.

The term *secretion* applied to a hormone denotes its release by exocytosis from the cell. In some cases, hormones such as steroid hormones are not secreted *per se*, but instead diffuse through the cell's plasma membrane into the extracellular space. Secretion or release by diffusion is sometimes accompanied by increased synthesis of the hormone. For simplicity in this chapter and the rest of the book, we will generally not distinguish between these possibilities when we refer to stimulation or inhibition of hormone "secretion."

Control by Plasma Concentrations of Mineral Ions or Organic Nutrients

There are multiple hormones whose secretion is directly controlled, at least in part, by the plasma concentrations of specific mineral ions or organic nutrients. In each case, a major function of the hormone is to regulate, through negative feedback, the plasma concentration of the ion or nutrient controlling its secretion. For example, insulin secretion is stimulated by an increase in plasma glucose concentration. Insulin, in turn, acts on skeletal muscle and adipose tissue to promote facilitated diffusion of glucose across the plasma membranes into the cytosol. The effect of insulin, therefore, is to decrease the plasma glucose concentration (**Figure 11–11**). Another example is the regulation of Ca^{2+} homeostasis by parathyroid hormone (PTH), as described in detail in Section F. This hormone is produced by cells of the parathyroid glands, which, as their name implies, are located in close proximity to the thyroid gland. When plasma Ca^{2+} concentration decreases, PTH secretion is directly stimulated. PTH exerts several actions, in coordination with other hormones, that restore plasma Ca^{2+} to normal.

Control by Neurons

As stated earlier, the adrenal medulla is a modified sympathetic ganglion and thus is stimulated by sympathetic preganglionic fibers. In addition to controlling the adrenal medulla, the autonomic nervous system influences other endocrine glands (**Figure 11–12**). Both parasympathetic and sympathetic inputs to these other glands may occur, some inhibitory and some stimulatory. Examples are the secretion of insulin and the gastrointestinal hormones, which are stimulated by neurons of the parasympathetic nervous system and inhibited by sympathetic neurons.

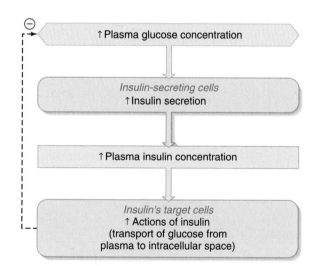

Figure 11–11

Example of how the direct control of hormone secretion by the plasma concentration of a substance, in this case by an organic nutrient, results in negative-feedback control of the substance's plasma concentration. In other cases, the regulated plasma substance may be a mineral, such as Ca^{2+}.

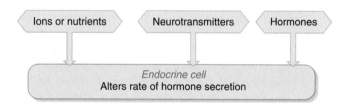

Figure 11–10

Inputs that act directly on endocrine gland cells to stimulate or inhibit hormone secretion.

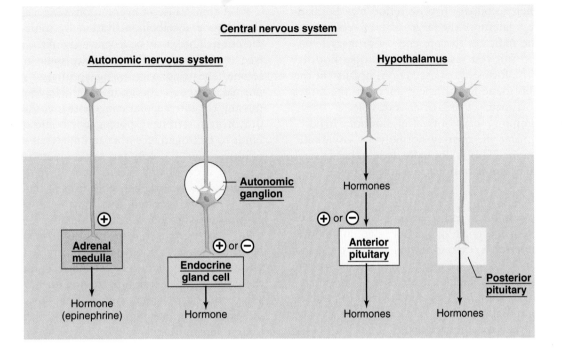

Figure 11–12

Pathways by which the nervous system influences hormone secretion. The autonomic nervous system controls hormone secretion by the adrenal medulla and many other endocrine glands. Certain neurons in the hypothalamus, some of which terminate in the posterior pituitary, secrete hormones. The secretion of hypothalamic hormones from the posterior pituitary and the effects of other hypothalamic hormones on the anterior pituitary are described later in this chapter. The ⊕ and ⊖ symbols indicate stimulatory and inhibitory actions, respectively.

Thus far, our discussion of neural control of hormone release has been limited to the role of the autonomic nervous system (see Figure 11–12). However, one large group of hormones—those secreted by the hypothalamus and its extension, the posterior pituitary—are under the direct control not of autonomic neurons, but of neurons in the brain itself (see Figure 11–12). This category will be described in detail in Section B.

Control by Other Hormones

In many cases, the secretion of a particular hormone is directly controlled by the blood concentration of another hormone. There are often sequences in which the only function of the first hormones in the sequence is to stimulate the secretion of the next. A hormone that stimulates the secretion of another hormone is often referred to as a **tropic hormone.** The tropic hormones usually stimulate not only secretion but also the growth of the stimulated gland. When referring to growth-promoting actions, the term *trophic* is often used. These types of hormonal sequences are covered in detail in Section B.

In addition to stimulatory actions, some hormones such as those in a multihormone sequence inhibit secretion of other hormones.

Types of Endocrine Disorders

Because there is such a wide variety of hormones and endocrine glands, disorders within the endocrine system may vary considerably in terms of symptoms. For example, endocrine disease may manifest as an imbalance in metabolism, leading to weight gain or loss; as a failure to grow or develop normally in early life; as an abnormally high or low blood pressure; as a loss of reproductive fertility; or as mental and emotional changes, to name a few. Despite these varied symptoms, which depend upon the particular hormone affected, essentially all endocrine diseases can be categorized in one of four ways. These include (1) too little hormone (*hyposecretion*); (2) too much hormone (*hypersecretion*); (3) reduced responsiveness of the target cells (*hyporesponsiveness*) to hormone; and (4) increased responsiveness of the target cells (*hyperresponsiveness*).

Hyposecretion

An endocrine gland may be secreting too little hormone because the gland cannot function normally, a condition termed *primary hyposecretion.* Examples of primary hyposecretion include (1) partial destruction of a gland such as the adrenal cortex, leading to decreased cortisol secretion, (2) an enzyme deficiency resulting in decreased synthesis of the hormone, and (3) dietary deficiency of iodine, leading to decreased secretion of thyroid hormones. There are many other causes such as infections and exposure to toxic chemicals, all having the common denominator of damaging the endocrine gland.

The other major cause of hyposecretion is *secondary hyposecretion.* In this case, the endocrine gland is not damaged initially, but is receiving too little of its tropic hormones.

One way to differentiate the two major causes of hyposecretion is to administer the tropic hormone. If the target gland responds, then secondary hyposecretion may be diagnosed. By contrast, a failure of the target gland to respond to the tropic hormone indicates the presence of primary hyposecretion. However, this test result sometimes does not distinguish between the forms of hyposecretion, because in the absence of tropic hormone for sufficient duration, the target gland atrophies. Fortunately, this is often reversible.

The most common means of treating hormone hyposecretion is to administer the missing hormone or a synthetic analog of the hormone. This is normally done either by oral (pill) or nasal (spray) administration, or by injection.

Hypersecretion

A hormone can also undergo either **primary hypersecretion** (the gland is secreting too much of the hormone on its own) or **secondary hypersecretion** (excessive stimulation of the gland by its tropic hormone). One of the most common causes of primary or secondary hypersecretion is the presence of a hormone-secreting endocrine-cell tumor. These tumors tend to produce their hormones continually at a high rate, even in the absence of stimulation (i.e., they are autonomous).

For the diagnosis of primary versus secondary hypersecretion, the concentrations of the hormone and, if relevant, its tropic hormone are measured. If both concentrations are elevated, then the hypersecretion is secondary. If the hypersecretion is primary, there will be a decreased concentration of the tropic hormone because of negative feedback from the high concentration of the hypersecreted hormone.

When an endocrine tumor causes hypersecretion, the tumor can often be removed surgically or destroyed with radiation if it is confined to a small area. In many cases, drugs that inhibit a hormone's synthesis can block hypersecretion. Alternatively, the situation can be treated with drugs that do not alter the hormone's secretion, but instead block the hormone's actions on its target cells.

Hyporesponsiveness and Hyperresponsiveness

In some cases, a component of the endocrine system may not be functioning normally, even though there is nothing wrong with hormone secretion. The problem is that the target cells do not respond normally to the hormone, a condition termed hyporesponsiveness. An important example of a disease resulting from hyporesponsiveness is the most common form of **diabetes mellitus** (called type 2 diabetes mellitus), in which the target cells of the hormone insulin are hyporesponsive to this hormone.

One cause of hyporesponsiveness is deficiency of or abnormal receptors for the hormone. For example, some individuals who are genetically male have a defect manifested by the absence of receptors for androgens. Their target cells are unable to bind androgens, and the result is lack of development of certain male characteristics, as though the hormones were not being produced (see Chapter 17 for additional details).

In a second type of hyporesponsiveness, the receptors for a hormone may be normal, but some event occurring after the hormone binds to receptors may be defective. For example, the activated receptor might be unable to stimulate formation of cyclic AMP or another component of the signaling pathway for that hormone.

A third cause of hyporesponsiveness applies to hormones that require metabolic activation by some other tissue after secretion. There may be a deficiency of the enzymes that catalyze the activation. For example, some men secrete testosterone (the major circulating androgen) normally and have normal receptors for androgens. However, these men are missing the intracellular enzyme that converts testosterone to dihydrotestosterone, a potent metabolite of testosterone that binds to androgen receptors and mediates some of the actions of testosterone on secondary sex characteristics.

In situations characterized by hyporesponsiveness to a hormone, the plasma concentration of the hormone in question is usually elevated, but the response of target cells to administered hormone is diminished.

Finally, hyperresponsiveness to a hormone can also occur and cause problems. For example, thyroid hormone causes an up-regulation of certain receptors for epinephrine; therefore, hypersecretion of thyroid hormone causes, in turn, a hyperresponsiveness to epinephrine. One result of this is the increased heart rate typical of people with elevated levels of thyroid hormones.

■ **SECTION A SUMMARY**

I. The endocrine system is one of the body's two major communications systems. It consists of all the glands that secrete hormones, which are chemical messengers the blood carries from the endocrine glands to target cells elsewhere in the body.

Hormone Structures and Synthesis

I. The amine hormones are the iodine-containing thyroid hormones and the catecholamines secreted by the adrenal medulla and the hypothalamus.

II. The majority of hormones are peptides, many of which are synthesized as larger molecules, which are then cleaved into active fragments.

III. Steroid hormones are produced from cholesterol by the adrenal cortex and the gonads, and by the placenta during pregnancy.

   a. The predominant steroid hormones produced by the adrenal cortex are the mineralocorticoid aldosterone, the glucocorticoid cortisol, and two androgens, DHEA and androstenedione.

   b. The ovaries produce mainly estradiol and progesterone, and the testes mainly testosterone.

Hormone Transport in the Blood

I. Peptide hormones and catecholamines circulate dissolved in the plasma but steroid and thyroid hormones circulate mainly bound to plasma proteins.

Hormone Metabolism and Excretion

I. The liver and kidneys are the major organs that remove hormones from the plasma by metabolizing or excreting them.

II. The peptide hormones and catecholamines are rapidly removed from the blood, whereas the steroid and thyroid hormones are removed more slowly, in part because they circulate bound to plasma proteins.

III. After their secretion, some hormones are metabolized to more active molecules in their target cells or other organs.

Mechanisms of Hormone Action

I. The great majority of receptors for steroid and thyroid hormones are inside the target cells; those for the peptide hormones and catecholamines are on the plasma membrane.

II. Hormones can cause up-regulation and down-regulation of their own receptors and those of other hormones. The induction of one hormone's receptors by another hormone increases the first hormone's effectiveness and may be essential to permit the first hormone to exert its effects.

III. Receptors activated by peptide hormones and catecholamines utilize one or more of the signal transduction pathways available to plasma-membrane receptors; the result is altered membrane potential or protein activity in the cell.

IV. Intracellular receptors activated by steroid and thyroid hormones function as transcription factors, combining with DNA in the nucleus and inducing the transcription of DNA into mRNA; the result is increased synthesis of particular proteins.

V. In pharmacological doses, hormones can have effects not seen under ordinary circumstances.

Inputs that Control Hormone Secretion

I. The secretion of a hormone may be controlled by the plasma concentration of an ion or nutrient that the hormone regulates, by neural input to the endocrine cells, and by one or more hormones.

II. The autonomic nervous system is the neural input controlling many hormones. Neuron endings from the sympathetic and parasympathetic nervous systems terminate directly on cells within some endocrine glands, thereby regulating hormone secretion.

Types of Endocrine Disorders

I. Endocrine disorders may be classified as hyposecretion, hypersecretion, and target-cell hypo- or hyperresponsiveness.
   a. Primary disorders are those in which the defect is in the cells that secrete the hormone.
   b. Secondary disorders are those in which there is too much or too little tropic hormone.
   c. Hyporesponsiveness is due to an alteration in the receptors for the hormone, to disordered postreceptor events, or to failure of normal metabolic activation of the hormone in cases requiring such activation.

II. These disorders can be distinguished by measurements of the hormone and any tropic hormones under both basal conditions and during experimental stimulation of the hormone's secretion.

■ SECTION A REVIEW QUESTIONS

1. What are the three general chemical classes of hormones?
2. Which catecholamine is secreted in the largest amount by the adrenal medulla, and why?
3. What are the major hormones produced by the adrenal cortex? By the testes? By the ovaries?
4. Which classes of hormones are carried in the blood mainly as unbound, dissolved hormone? Mainly bound to plasma proteins?
5. Do protein-bound hormones diffuse out of capillaries?
6. Which organs are the major sites of hormone excretion and metabolic transformation?
7. How do the rates of metabolism and excretion differ for the various classes of hormones?
8. List some metabolic transformations that prohormones and some hormones must undergo before they become biologically active.
9. Contrast the locations of receptors for the various classes of hormones.
10. How do hormones influence the concentrations of their own receptors and those of other hormones? How does this explain permissiveness in hormone action?
11. Describe the sequence of events when peptide or catecholamine hormones bind to their receptors.
12. Describe the sequence of events when steroid or thyroid hormones bind to their receptors.
13. What are the direct inputs to endocrine glands controlling hormone secretion?
14. How does control of hormone secretion by plasma mineral ions and nutrients achieve negative feedback control of these substances?
15. What roles does the autonomic nervous system play in controlling hormone secretion?
16. What groups of hormone-secreting cells receive input from neurons located in the brain rather than in the autonomic nervous system?
17. How would you distinguish between primary and secondary hyposecretion of a hormone? Between hyposecretion and hyporesponsiveness?

The Hypothalamus and Pituitary Gland

Control Systems Involving the Hypothalamus and Pituitary

The **pituitary gland,** or hypophysis, lies in a pocket (called the sella turcica) of the sphenoid bone at the base of the brain (**Figure 11–13**), just below the **hypothalamus.** The pituitary is connected to the hypothalamus by the **infundibulum,** a stalk containing nerve fibers and small blood vessels. In human beings, the pituitary gland is composed of two adjacent lobes—the **anterior pituitary** (toward the front of the head; also called the adenohypophysis) and the **posterior pituitary** (toward the back of the head; also called the neurohypophysis). The anterior pituitary arises embryologically from an invagination of the pharynx called Rathke's pouch, whereas the posterior pituitary is actually an extension of the neural components of the hypothalamus. In many mammalian species a well-developed intermediate lobe is

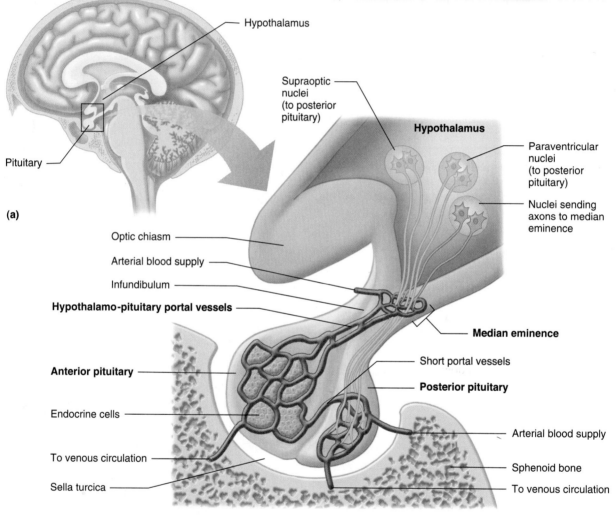

(a)

(b)

Figure 11–13

(a) Relation of the pituitary gland to the brain and hypothalamus. (b) Neural and vascular connections between the hypothalamus and pituitary. Hypothalamic neurons from the paraventricular and supraoptic nuclei run down the infundibulum to end in the posterior pituitary, whereas others (shown for simplicity as a single nucleus, but in reality several nuclei, including some cells from the paraventricular nuclei) end in the median eminence. Almost the entire blood supply to the anterior pituitary comes via the hypothalamo-pituitary portal vessels, which originate in the median eminence. (The short portal vessels, which originate in the posterior pituitary, carry only a small fraction of the blood leaving the posterior pituitary and supply only a small fraction of the blood received by the anterior pituitary.)

Figure 11–13 physiological *inquiry* (p*i*)

■ Why does it take only minute quantities of hypophysiotropic hormones to regulate anterior pituitary gland hormone secretion?

Answer can be found at end of chapter.

Chapter 11

found between the anterior and posterior portions of the pituitary, but this is not the case in humans. The intermediate lobe in such animals produces a hormone called melanocyte-stimulating hormone that controls coat color.

The axons of two well-defined clusters of hypothalamic neurons (the supraoptic and paraventricular nuclei) pass down the infundibulum and end within the posterior pituitary in close proximity to capillaries (the smallest of blood vessels) (**Figure 11–13b**). Thus, these neurons do not form a synapse with other neurons. Instead, their terminals end directly on capillaries.

In contrast to the neural connections between the hypothalamus and posterior pituitary, there are no important neural connections between the hypothalamus and anterior pituitary. There is, however, an unusual blood vessel connection (see Figure 11–13b). The capillaries at the junction of the hypothalamus and infundibulum (the **median eminence**) recombine to form the **hypothalamo-pituitary portal vessels.** The term *portal* denotes blood vessels that connect two capillary beds. The hypothalamo-pituitary portal vessels pass down the stalk connecting the hypothalamus and pituitary and enter the anterior pituitary, where they drain into a second capillary bed, the anterior pituitary capillaries. Thus, the hypothalamo-pituitary portal vessels offer a local route for blood flow directly from the hypothalamus to the cells of the anterior pituitary. This offers the advantage of a rapid response and minimizes the amount of hypothalamic hormone that must be synthesized to reach an effective blood concentration (because the hormone is not diluted into the general circulation of the body).

Posterior Pituitary Hormones

We emphasized that the posterior pituitary is really a neural extension of the hypothalamus (see Figure 11–13). The hormones are not synthesized in the posterior pituitary itself but in the hypothalamus, specifically in the cell bodies of the supraoptic and paraventricular nuclei, whose axons pass down the infundibulum and end in the posterior pituitary. Enclosed in small vesicles, the hormone moves down the axons to accumulate at the axon terminals in the posterior pituitary. Stimuli such as neurotransmitters generate action potentials in the neurons; these action potentials propagate to the axon terminals and trigger the release of the stored hormone by exocytosis. The hormone then enters capillaries in the posterior pituitary to be carried away by the blood returning to the heart. In this way, the brain can receive stimuli and respond as if it were an endocrine organ. By releasing its hormones into the general circulation, the posterior pituitary can modify the function of distant organs.

The two posterior pituitary hormones are **oxytocin** and **vasopressin.** Oxytocin stimulates contraction of smooth muscle cells in the breasts, which results in milk ejection during lactation, and stimulates contraction of uterine smooth muscle during labor. Although oxytocin is also present in males, its functions in males are uncertain. Vasopressin acts on smooth muscle cells around blood vessels to cause muscle contraction, which constricts blood vessels and increases blood pressure. Vasopressin also acts within the kidneys to decrease water excretion in the urine, thus retaining fluid in the body and helping to maintain blood volume. Because of its kidney function, vasopressin is also known as **antidiuretic hormone,** or **ADH** (a loss of excess water in the urine is known as a *diuresis*).

Anterior Pituitary Hormones and the Hypothalamus

Hypothalamic neurons different from those that produce the hormones released from the posterior pituitary, secrete hormones that control the secretion of all the anterior pituitary hormones. For simplicity's sake, Figure 11–13 depicts these neurons as arising from a single nucleus, but in fact several hypothalamic nuclei send axons whose terminals end in the median eminence. The hypothalamic hormones that regulate anterior pituitary function are collectively termed **hypophysiotropic hormones** (recall that another name for the pituitary is hypophysis); they are also commonly called hypothalamic releasing or inhibiting hormones. "Hypophysiotropic hormones" denotes only those hormones from the *hypothalamus* that influence the anterior pituitary. We will see later that nonhypothalamic hormones can also influence the anterior pituitary, but they are not categorized as hypophysiotropic hormones.

With one exception (dopamine), each of the hypophysiotropic hormones is the first in a *three*-hormone sequence: (1) A hypophysiotropic hormone controls the secretion of (2) an anterior pituitary hormone, which controls the secretion of (3) a hormone from some other endocrine gland (**Figure 11–14**). This last hormone then acts on its target cells. The adaptive value of such sequences is that they permit a variety of types of important hormonal feedback. They also allow amplification of a response of a small number of hypothalamic neurons into a large peripheral hormonal signal. We begin our description of these sequences in the middle—that is, with the anterior pituitary hormones—because the names and actions of the hypophysiotropic hormones are mostly based on the names of the anterior pituitary hormones.

Overview of Anterior Pituitary Hormones

As shown in Table 11–1, the anterior pituitary secretes at least eight hormones, but only six have well-established functions. All peptides, these six "classical" hormones are **follicle-stimulating hormone (FSH), luteinizing hormone (LH), growth hormone** (GH, also known as somatotropin), **thyroid-stimulating hormone** (TSH, also known as thyrotropin), **prolactin,** and **adrenocorticotropic hormone (ACTH,** also known as corticotropin). Each of the last four is secreted by a distinct cell type in the anterior pituitary, whereas FSH and LH, collectively termed **gonadotropic hormones** (or gonadotropins) because they stimulate the gonads, are usually secreted by the same cells.

What about the other two peptides—**beta-lipotropin** and **beta-endorphin**—secreted by the anterior pituitary? Their physiological roles, if any, in humans are unclear. In animal studies, however, beta-endorphin has been shown to have potent pain-killing effects, and beta-lipotropin can mobilize fats in the circulation to provide extra fuel. Both of these functions may contribute to an animal's ability to cope with stressful challenges.

Figure 11–15 summarizes the target organs and major functions of the six classical anterior pituitary hormones. Note that the only major function of two of the six is to stimulate their target cells to secrete other hormones (and to maintain the growth and function of these cells). Thyroid-stimulating hormone induces the thyroid to secrete thyroxine and triiodothyronine. Adrenocorticotropic hormone, meaning "hormone that stimulates the adrenal cortex," stimulates that gland to secrete cortisol.

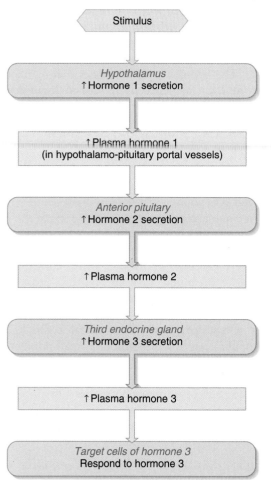

Figure 11–14

Typical sequential pattern by which a hypophysiotropic hormone (hormone 1 from the hypothalamus) controls the secretion of an anterior pituitary hormone (hormone 2), which in turn controls the secretion of a hormone by a third endocrine gland (hormone 3). The hypothalamo-pituitary portal vessels are illustrated in Figure 11–13.

Three other anterior pituitary hormones also stimulate the secretion of another hormone but have an additional function as well. Follicle-stimulating hormone and luteinizing hormone stimulate the gonads to secrete the sex hormones—estradiol and progesterone from the ovaries, or testosterone from the testes—but in addition, they regulate the growth and development of ova and sperm. The actions of FSH and LH are described in detail in Chapter 17. Growth hormone stimulates the liver to secrete a growth-promoting peptide hormone known as **insulin-like growth factor 1 (IGF-1),** and in addition, exerts direct effects on metabolism (Section E in this chapter).

Prolactin is unique among the six classical anterior pituitary hormones in that its major function is not to exert control over the secretion of a hormone by another endocrine gland. Its most important action is to stimulate development of the mammary glands and milk production by direct effects upon the breasts. During lactation, prolactin exerts a secondary action to inhibit gonadotropin secretion, thus decreasing fertility when a woman is breast-feeding. In the male, prolactin may facilitate several components of reproductive function, although its precise roles are uncertain.

Hypophysiotropic Hormones

As stated previously, secretion of the anterior pituitary hormones is largely regulated by hormones produced by the hypothalamus and collectively called hypophysiotropic hormones. These hormones are secreted by neurons that originate in discrete nuclei of the hypothalamus and terminate in the median eminence around the capillaries that are the origins of the hypothalamo-pituitary portal vessels. The generation of action potentials in these neurons causes them to secrete their hormones, much as action potentials cause other neurons to release neurotransmitters by exocytosis. Hypothalamic hormones, however, enter the capillaries and are carried by the hypothalamo-pituitary portal vessels to the anterior pituitary (**Figure 11–16**). There, they act upon the various anterior pituitary cells to control their hormone secretions.

Thus, these hypothalamic neurons secrete hormones in a manner identical to that described previously for the hypothalamic neurons whose axons end in the posterior pituitary. In both cases the hormones are synthesized in

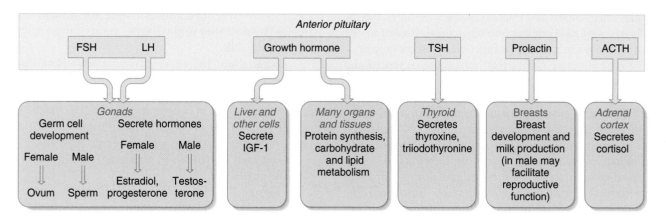

Figure 11–15

Targets and major functions of the six classical anterior pituitary hormones.

hypothalamic neurons, pass down axons to the neuron terminals, and are released in response to action potentials in the neurons. Two crucial differences, however, distinguish the two systems: (1) The axons of the hypothalamic neurons that secrete the posterior pituitary hormones leave the hypothalamus and end in the posterior pituitary, whereas those that secrete the hypophysiotropic hormones remain in the hypothalamus, ending in the median eminence. (2) Most of the posterior pituitary capillaries into which the posterior pituitary hormones are secreted immediately drain into the main bloodstream, which carries the hormones to the heart for distribution to the entire body. In contrast, the hypophysiotropic hormones enter capillaries in the median eminence of the hypothalamus that do not directly join the main bloodstream, but empty into the hypothalamo-pituitary portal vessels, which carry them to the anterior pituitary. Once the hypophysiotropic hormones leave the capillaries in the anterior pituitary, they bathe the pituitary cells. If a cell has an appropriate receptor for a given hypophysiotropic hormone, that cell will respond by increasing or decreasing the secretion of its pituitary hormone. If a pituitary hormone is secreted, it will diffuse into the same capillaries that delivered the hypophysiotropic hormone. These capillaries then drain into veins, which enter the general blood circulation, where the pituitary hormones can come into contact with their target cells.

There are multiple, discrete hypophysiotropic hormones, each secreted by a particular group of hypothalamic neurons and influencing the release of one or, in at least one case, two of the anterior pituitary hormones. For simplicity, **Figure 11–17**

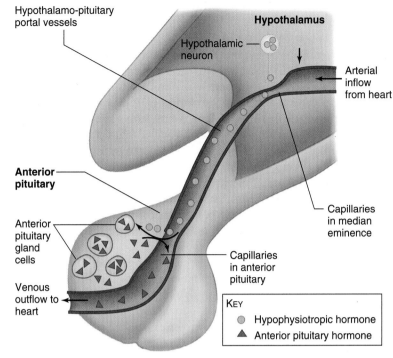

Figure 11–16

Hormone secretion by the anterior pituitary is controlled by hypophysiotropic hormones released by hypothalamic neurons and reaching the anterior pituitary by way of the hypothalamo-pituitary portal vessels.

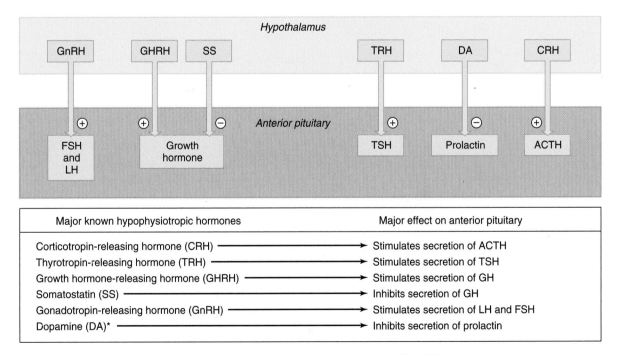

Major known hypophysiotropic hormones	Major effect on anterior pituitary
Corticotropin-releasing hormone (CRH)	Stimulates secretion of ACTH
Thyrotropin-releasing hormone (TRH)	Stimulates secretion of TSH
Growth hormone-releasing hormone (GHRH)	Stimulates secretion of GH
Somatostatin (SS)	Inhibits secretion of GH
Gonadotropin-releasing hormone (GnRH)	Stimulates secretion of LH and FSH
Dopamine (DA)*	Inhibits secretion of prolactin

*Dopamine is a catecholamine; all the other hypophysiotropic hormones are peptides. Evidence exists for PRL-releasing hormones, but they have not been unequivocally identified. One possibility is that TRH serves this role in addition to its actions on TSH.

Figure 11–17

The effects of definitely established hypophysiotropic hormones on the anterior pituitary. The hypophysiotropic hormones reach the anterior pituitary via the hypothalamo-pituitary portal vessels. The ⊕ and ⊖ symbols indicate stimulatory and inhibitory actions, respectively.

and the text of this chapter summarize only those hypophysiotropic hormones that have clearly documented physiological roles in humans.

Several of the hypophysiotropic hormones are named for the anterior pituitary hormone whose secretion they control. Thus, secretion of ACTH (corticotropin) is stimulated by **corticotropin-releasing hormone (CRH),** secretion of growth hormone is stimulated by **growth hormone-releasing hormone (GHRH),** secretion of thyroid-stimulating hormone (thyrotropin) is stimulated by **thyrotropin-releasing hormone (TRH),** and secretion of both luteinizing hormone and follicle-stimulating hormone (the gonadotropins) is stimulated by **gonadotropin-releasing hormone (GnRH).**

Note, however, in Figure 11–17, that two of the hypophysiotropic hormones do not *stimulate* the release of an anterior pituitary hormone but rather *inhibit* its release. One of them, **somatostatin (SS),** inhibits the secretion of growth hormone. The other, **dopamine (DA),** inhibits the secretion of prolactin.

As Figure 11–17 shows, growth hormone is controlled by *two* hypophysiotropic hormones—somatostatin, which inhibits its release, and growth hormone–releasing hormone, which stimulates it. The rate of growth hormone secretion depends, therefore, upon the relative amounts of the opposing hormones released by the hypothalamic neurons, as well as upon the relative sensitivities of the anterior pituitary to them. Such dual controls may also exist for the other anterior pituitary hormones, but the importance of such control, if it exists, is uncertain.

Figure 11–18 summarizes the information presented in Figures 11–15 and 11–17 to illustrate the full sequence of hypothalamic control of endocrine function.

Given that the hypophysiotropic hormones control anterior pituitary function, we must now ask: What controls secretion of the hypophysiotropic hormones? Some of the neurons that secrete hypophysiotropic hormones may possess spontaneous activity, but the firing of most of them requires neural and hormonal input.

Neural Control of Hypophysiotropic Hormones

Neurons of the hypothalamus receive stimulatory and inhibitory synaptic input from virtually all areas of the central nervous system, and specific neural pathways influence the secretion of the individual hypophysiotropic hormones. A large number of neurotransmitters, such as the catechol amines and serotonin, are released at the synapses on the hormone-secreting hypothalamic neurons. Not surprisingly, therefore, drugs that influence these neurotransmitters can alter the secretion of the hypophysiotropic hormones.

Figure 11–19 illustrates one example of the role of neural input to the hypothalamus. Corticotropin-releasing hormone (CRH) from the hypothalamus stimulates the anterior pituitary to secrete ACTH, which in turn stimulates the adrenal cortex to secrete cortisol. A wide variety of sensory stimuli resulting from physical or emotional stress act via neural pathways to the hypothalamus to increase CRH secretion and, therefore, ACTH and cortisol secretion. Even in the absence of stressful stimuli, however, cortisol secretion varies in a regular manner during a 24-hour period because neural rhythms within the central nervous system also impinge upon the hypothalamic neurons that secrete CRH.

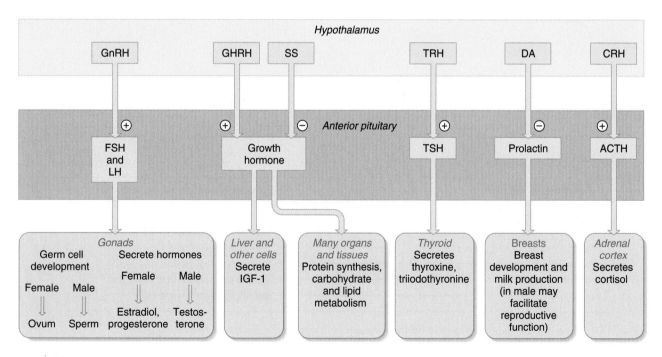

Figure 11–18

A combination of Figures 11–15 and 11–17 summarizes the hypothalamic–anterior pituitary system. The ⊕ and ⊖ symbols indicate stimulatory and inhibitory actions, respectively.

Hormonal Feedback Control of the Hypothalamus and Anterior Pituitary

A prominent feature of each of the hormonal sequences initiated by a hypophysiotropic hormone is negative feedback exerted upon the hypothalamo-pituitary system by one or more of the hormones in its sequence. For example, in the CRH-ACTH-cortisol sequence (see Figure 11–19), the final hormone, cortisol, acts upon the hypothalamus to reduce synthesis and secretion of CRH. In addition, cortisol acts directly on the anterior pituitary to reduce the response of the ACTH-secreting cells to CRH. Thus, by a double-barreled action, cortisol exerts a negative feedback control over its own secretion.

Such a system is effective in dampening hormonal responses—that is, in limiting the extremes of hormone secretory rates. For example, when a painful stimulus elicits increased secretion, in turn, of CRH, ACTH, and cortisol, the resulting elevation in plasma cortisol concentration feeds back to inhibit the hypothalamus and anterior pituitary. Therefore, cortisol secretion does not rise as much as it would without negative feedback.

The situation described for cortisol, in which the hormone secreted by the third endocrine gland in a sequence exerts a negative feedback effect over the anterior pituitary and/or hypothalamus, is known as a **long-loop negative feedback** (**Figure 11–20**). This type of feedback exists for each of the three-hormone sequences initiated by a hypophysiotropic hormone.

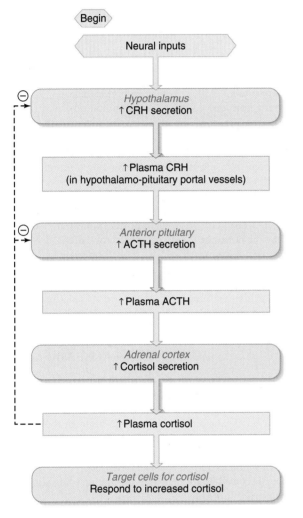

Figure 11–19

CRH-ACTH-cortisol sequence. Neural inputs include those related to stressful stimuli and nonstress inputs like circadian rhythms. Cortisol exerts a negative feedback (⊖ symbols) control over the system by acting on (1) the hypothalamus to inhibit CRH synthesis and secretion and (2) the anterior pituitary to inhibit ACTH production.

Figure 11–19 physiological *inquiry* (pi)

- What hormonal changes would be expected if a patient developed a benign tumor of the left adrenal cortex that secreted large amounts of cortisol in the absence of external stimulation? What would happen to the right adrenal gland?

Answer can be found at end of chapter.

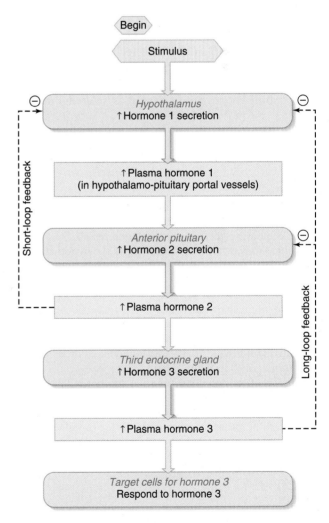

Figure 11–20

Short-loop and long-loop feedbacks. Long-loop feedback is exerted on the hypothalamus and/or anterior pituitary by the third hormone in the sequence. Short-loop feedback is exerted by the anterior pituitary hormone on the hypothalamus.

Long-loop feedback does not exist for prolactin because this is one anterior pituitary hormone that does not have major control over another endocrine gland—that is, it does not participate in a three-hormone sequence. Nonetheless, there is negative feedback in the prolactin system, for this hormone itself acts upon the hypothalamus to *stimulate* the secretion of dopamine, which then *inhibits* the secretion of prolactin. The influence of an anterior pituitary hormone on the hypothalamus is known as a **short-loop negative feedback** (see Figure 11–20). Like prolactin, several other anterior pituitary hormones, including growth hormone, also exert such feedback on the hypothalamus.

The Role of "Nonsequence" Hormones on the Hypothalamus and Anterior Pituitary

Keep in mind that there are many stimulatory and inhibitory hormonal influences on the hypothalamus and/or anterior pituitary other than those that fit the feedback patterns just described. In other words, a hormone that is not itself in a particular hormonal sequence may nevertheless exert important influences on the secretion of the hypophysiotropic or anterior pituitary hormones in that sequence. For example, estradiol markedly enhances the secretion of prolactin by the anterior pituitary, even though estradiol secretion is not normally controlled by prolactin. Thus, the sequences we have been describing should not be viewed as isolated units.

■ SECTION B SUMMARY

Control Systems Involving the Hypothalamus and Pituitary

I. The pituitary gland, comprising the anterior pituitary and the posterior pituitary, is connected to the hypothalamus by a stalk containing nerve axons and blood vessels.

II. Specific axons, whose cell bodies are in the hypothalamus, terminate in the posterior pituitary and release oxytocin and vasopressin.

III. The anterior pituitary secretes growth hormone (GH), thyroid-stimulating hormone (TSH), adrenocorticotropic hormone (ACTH), prolactin, and two gonadotropic hormones—follicle-stimulating hormone (FSH) and luteinizing hormone (LH). The functions of these hormones are summarized in Figure 11–15.

IV. Secretion of the anterior pituitary hormones is controlled mainly by hypophysiotropic hormones secreted into capillaries in the median eminence of the hypothalamus and reaching the anterior pituitary via the portal vessels connecting the hypothalamus and anterior pituitary. The actions of the hypophysiotropic hormones on the anterior pituitary are summarized in Figure 11–17.

V. The secretion of each hypophysiotropic hormone is controlled by neuronal and hormonal input to the hypothalamic neurons producing it.

   a. In each of the three-hormone sequences beginning with a hypophysiotropic hormone, the third hormone exerts negative feedback effects on the secretion of the hypothalamic and/or anterior pituitary hormone.

   b. The anterior pituitary hormone may exert a short-loop negative feedback inhibition of the hypothalamic releasing hormone(s) controlling it.

   c. Hormones not in a particular sequence can also influence secretion of the hypothalamic and/or anterior pituitary hormones in that sequence.

■ SECTION B KEY TERMS

adrenocorticotropic hormone (ACTH) 331
anterior pituitary 330
antidiuretic hormone (ADH) 331
beta-endorphin 331
beta-lipotropin 331
corticotropin-releasing hormone (CRH) 334
dopamine (DA) 334
follicle-stimulating hormone (FSH) 331
gonadotropic hormone 331
gonadotropin-releasing hormone (GnRH) 334
growth hormone (GH) 331
growth hormone–releasing hormone (GHRH) 334
hypophysiotropic hormone 331

hypothalamo-pituitary portal vessel 331
hypothalamus 330
infundibulum 330
insulin-like growth factor 1 (IGF-1) 332
long-loop negative feedback 335
luteinizing hormone (LH) 331
median eminence 331
oxytocin 331
pituitary gland 330
posterior pituitary 330
prolactin 331
short-loop negative feedback 336
somatostatin (SS) 334
thyroid-stimulating hormone (TSH) 331
thyrotropin-releasing hormone (TRH) 334
vasopressin 331

■ SECTION B REVIEW QUESTIONS

1. Describe the anatomical relationships between the hypothalamus and the pituitary.

2. Name the two posterior pituitary hormones and describe their site of synthesis and mechanism of release.

3. List all six well-established anterior pituitary hormones and their major functions.

4. List the major hypophysiotropic hormones and the hormone whose release each controls.

5. What kinds of inputs control secretion of the hypophysiotropic hormones?

6. Diagram the CRH-ACTH-cortisol system.

7. What is the difference between long-loop and short-loop negative feedback in the hypothalamo-anterior pituitary system?

Synthesis of Thyroid Hormones

Thyroid hormones exert widespread and diverse effects throughout the body. The actions of TH are so important— and the consequences of imbalances in TH concentrations so severe—that it is worth examining thyroid gland function in additional detail.

The thyroid gland produces two **iodine**-containing molecules of physiological importance, **thyroxine** (called T_4 because it contains four iodines) and **triiodothyronine** (T_3, three iodines; review Figure 11–2). T_4 generally is converted into T_3 by enzymes known as deiodinases in target cells. We will therefore consider T_3 to be the major thyroid hormone, even though T_4 is the major secretory product of the thyroid and total T_4 concentrations are higher in the blood.

The thyroid gland is a bilobed structure that sits within the neck straddling the trachea (**Figure 11–21a**). It first becomes functional early in fetal life. By adulthood, the thyroid gland weighs about 25 grams.

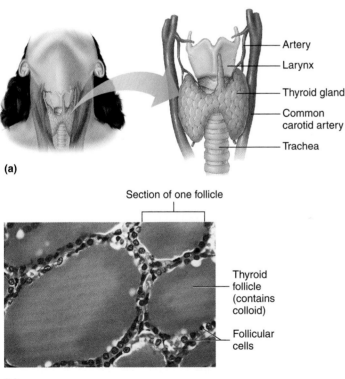

(a)

Artery
Larynx
Thyroid gland
Common carotid artery
Trachea

Section of one follicle

Thyroid follicle (contains colloid)

Follicular cells

(b)

Figure 11–21

Location of the bilobed thyroid gland (a), and a cross section through several adjoining follicles filled with colloid (b).

Within the thyroid gland are numerous **follicles,** each composed of an enclosed sphere of highly specialized cells surrounding a core containing a protein-rich material called the **colloid** (**Figure 11–21b**). The follicular cells participate in almost all phases of thyroid hormone synthesis and secretion. Synthesis begins when circulating iodide is cotransported with sodium ions across the follicular cell plasma membrane (step 1 in **Figure 11–22**). Once inside the follicular cell, the bulky iodide ion cannot diffuse back into the interstitial fluid; this is known as **iodide trapping.** The sodium is eventually pumped back out of the cell by Na^+/K^+-ATPases.

The trapped, negatively charged iodide ions diffuse down their electrical and concentration gradients to the lumenal border of the follicular cells (step 2). The colloid of the follicles contains large amounts of a protein called **thyroglobulin.** The iodide that diffuses to the colloid is rapidly oxidized at the lumenal surface of the follicular cells to iodine free radicals; these are then attached to the phenolic rings of tyrosine molecules within the amino acid structure of thyroglobulin (step 3). Thyroglobulin itself is synthesized by the follicular cells and secreted by exocytosis into the follicle lumen. The enzyme responsible for oxidizing iodides and attaching them to tyrosines on thyroglobulin in the colloid is called **thyroid peroxidase,** and it, too, is synthesized by follicular cells. Iodines may be added to either of two positions on a given tyrosine within thyroglobulin. A tyrosine with one iodine attached is called **monoiodotyrosine (MIT);** if two iodines are attached, the product is **diiodotyrosine (DIT).** The precise mechanism of what happens next is still somewhat unclear. The phenolic ring of a molecule of MIT or DIT is removed from the remainder of its tyrosine and coupled to another DIT on the thyroglobulin molecule (step 4). This reaction may also be mediated by thyroid peroxidase. If two DIT molecules are coupled, the result is thyroxine (T_4). If one MIT and one DIT are coupled, the result is T_3.

Finally, when thyroid hormone is needed in the blood, extensions of the colloid-facing membranes of follicular cells engulf portions of the colloid (with its iodinated thyroglobulin) by endocytosis (step 5). The thyroglobulin, with its coupled MITs and DITs, is brought into contact with lysosomes in the cell interior (step 6). Proteolysis of thyroglobulin releases T_3 and T_4, which then diffuse out of the follicular cell into the interstitial fluid and from there to the blood (step 7).

Control of Thyroid Function

Essentially all of the actions of the follicular cells are stimulated by TSH, which, as we have seen, is stimulated by TRH. The basic control mechanism of TSH production is the negative feedback action of TH on the anterior pituitary

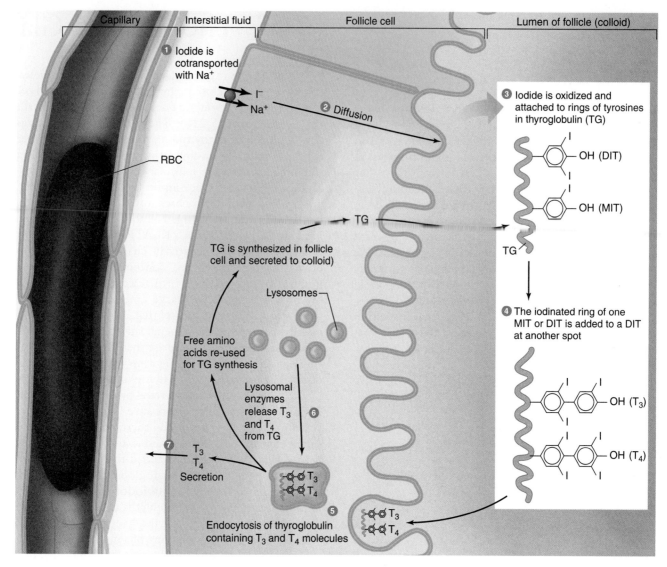

Figure 11–22

Steps involved in T_3 and T_4 formation. Steps are keyed to the text.

and, to a lesser extent, the hypothalamus (**Figure 11–23**). TSH is also a trophic hormone, as stated earlier. TSH not only stimulates T_3 and T_4 production, it also increases protein synthesis in follicular cells, increases DNA replication and cell division, and increases the amount of rough endoplasmic reticulum and other cellular machinery required by follicular cells for protein synthesis. Thus, if a thyroid cell is exposed to higher TSH levels than normal, it will undergo **hypertrophy;** that is, it will increase in size. An enlarged thyroid gland from any cause is called a *goiter.* There are several ways in which goiters can occur, in addition to increased exposure of the thyroid gland to TSH, as will be described at the end of this section and in Chapter 19.

Actions of Thyroid Hormones

TH receptors are present in the nuclei of most of the cells of the body, unlike receptors for many other hormones, whose distribution is more limited. Thus, the actions of T_3 and T_4 are widespread and affect many organs and tissues. The receptors are located in the nucleus and can bind both T_3 and T_4, but have a much higher affinity for T_3. Because most of the T_4 that enters cells is deiodinated to T_3, most receptor binding sites are generally occupied by T_3. As described in Chapter 5, T_3 and T_4 act by inducing gene transcription and protein synthesis.

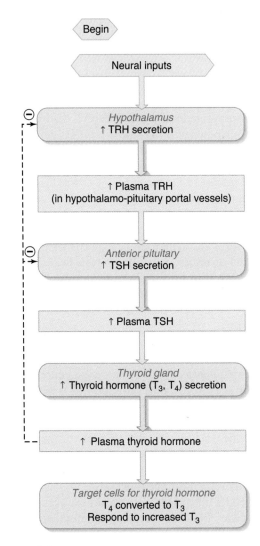

Begin

Neural inputs

⊖ *Hypothalamus*
↑ TRH secretion

↑ Plasma TRH
(in hypothalamo-pituitary portal vessels)

⊖ *Anterior pituitary*
↑ TSH secretion

↑ Plasma TSH

Thyroid gland
↑ Thyroid hormone (T₃, T₄) secretion

↑ Plasma thyroid hormone

Target cells for thyroid hormone
T₄ converted to T₃
Respond to increased T₃

Figure 11–23

TRH-TSH-thyroid hormone sequence. T_3 and T_4 inhibit secretion of TSH and TRH by negative feedback, indicated by the ⊖ symbol.

Metabolic Actions

Thyroid hormones have several effects on carbohydrate and lipid metabolism, although not to the extent as other hormones such as insulin. Nonetheless, TH stimulates carbohydrate absorption from the small intestine and increases fatty acid release from adipocytes. These actions provide energy to maintain metabolic rate at a high level, and are consistent with one of the major actions of TH, which is to stimulate the activity of Na^+/K^+-ATPases throughout the body. Recall from Chapter 3 that ATP negatively feeds back on glycolytic enzymes within cells to maintain its cellular concentration. A decrease in cellular stores of ATP therefore triggers an increase in glycolysis, which results in the burning of additional fuel to restore ATP levels. One of the by-products of this process is heat. Thus, as ATP is consumed by Na^+/K^+-ATPases at a high rate due to TH activation, the cellular stores of ATP must be maintained by increased metabolism of fuels. This **calorigenic** action of TH represents a very significant fraction of the total heat produced each day in a typical person.

Permissive Actions

Many of the actions of TH are attributable to its permissive effects on catecholamines. TH up-regulates beta-adrenergic receptors in many tissues, notably the heart and nervous system. Thus, it should not be surprising that the symptoms of excess thyroid hormone concentration closely resemble some of the symptoms of excess epinephrine and norepinephrine (sympathetic nervous system activity). That is because the increased thyroid hormone potentiates the actions of the catecholamines, even though the latter are within normal levels. Because of this potentiating effect, people with hyperthyroidism are often treated with drugs that block beta-adrenergic receptors to alleviate the anxiety, nervousness, and "racing heart" associated with excessive sympathetic activity.

Growth and Development

As described in Section E of this chapter, TH is needed for normal production of growth hormone. Therefore, in the absence of TH, growth in children is decreased. In addition, though, TH is among the most important developmental hormones for the nervous system. During fetal life, TH exerts many effects on central nervous system development, including the formation of nerve terminals and the production of synapses, the growth of dendrites and dendritic extensions (called "spines"), and the formation of myelin. Absence of TH during fetal life results in a poorly developed nervous system and a form of mental retardation called *cretinism.* The most common cause of cretinism around the world (but rare in the United States) is dietary iodine deficiency in the mother. Without iodine in her diet, little iodine is made available to the fetus. Thus, even though the fetal thyroid may be normal, it cannot manufacture sufficient TH. If the condition is discovered and corrected with iodine and TH administration shortly after birth, mental and physical retardation can be prevented. Some evidence suggests, however, that completely normal mental function may not be restored. Furthermore, if the treatment is not initiated in the early neonatal period, cretinism cannot be reversed. Despite the availability of iodized salt products in many countries, cretinism is still a common disorder in some parts of the world, particularly in mountainous regions where snow and rainwater leach iodine out of the soil.

The effects of TH on nervous system function are not limited to fetal and neonatal life. For example, TH is needed for proper nerve/muscle reflexes and for normal cognition in adults.

Because thyroid diseases are among the most common of all endocrine diseases, it is worth describing in detail the causes and symptoms of hyposecretion of TH (hypothyroidism) and hypersecretion of TH (hyperthyroidism).

Hypothyroidism and Hyperthyroidism

Any condition characterized by plasma levels of TH that are chronically below normal is known as *hypothyroidism.* Most cases of hypothyroidism—about 95 percent—are primary defects resulting from damage to or loss of functional thyroid tissue, or to inadequate iodine consumption.

One form of hypothyroidism that exists around the world is caused by iodine deficiency. In such cases, the synthesis of TH is compromised, leading to a decrease in the plasma levels of those hormones. This, in turn, releases negative feedback on the hypothalamus and pituitary, and TRH levels become chronically elevated in the portal circulation of the anterior pituitary. Plasma TSH concentration is also elevated due to the increased TRH. The resulting overstimulation of the thyroid can produce goiters that can achieve astounding sizes if untreated (**Figure 11–24**). This form of hypothyroidism is reversible if iodine is added to the diet. It is extremely rare in the United States because of the widespread use of iodized salt, in which one in every 10,000 molecules of NaCl is replaced with NaI.

The most common cause of hypothyroidism in the U.S. is due to autoimmune destruction of the thyroid

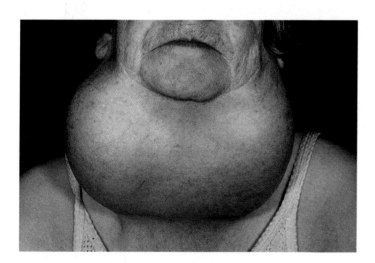

Figure 11–24
Goiter at an advanced stage.

gland (*autoimmune thyroiditis*). One such situation is *Hashimoto's disease,* in which cells of the immune system called T-cells attack and destroy thyroid tissue. Like many other autoimmune diseases, Hashimoto's disease is more common in women and can slowly progress with age. As thyroid hormones begin to decrease, TSH levels increase due to the decreased negative feedback, and a goiter can develop. The usual treatment for autoimmune thyroiditis is daily replacement with pharmaceutical preparations of T_4. This restores the thyroid hormone levels in the blood and causes the TSH levels to decrease to normal.

The signs and symptoms of hypothyroidism in adults may be mild or severe, depending on the degree of hormone deficiency. These include an increased sensitivity to cold (*cold intolerance*) and a tendency toward weight gain. Both of these symptoms are related to the decreased calorigenic actions normally produced by thyroid hormone. Many of the other symptoms appear to be diffuse and nonspecific, such as fatigue, changes in skin tone, hair, appetite, gastrointestinal function, and mental function (decreased concentration). The basis of the last effect is uncertain, but it has been postulated to be related in part to a decrease in cerebral blood flow, which in turn may be secondary to a decrease in the pumping activity of the heart. Recall that thyroid hormones are permissive for the actions of epinephrine and norepinephrine, and that these two catecholamines are responsible in part for maintaining normal cardiac function.

In severe, untreated hypothyroidism, certain hydrophilic polymers called glycosaminoglycans accumulate in the interstitial space in scattered regions of the body. Normally, thyroid hormones act to prevent overexpression of these extracellular compounds that are secreted by connective tissue cells. In the absence of TH, therefore, these hydrophilic molecules accumulate, and water tends to be trapped with them. This combination causes a characteristic puffiness of the face and other regions that is known as *myxedema.*

As in the case of hypothyroidism, there are a variety of ways in which *hyperthyroidism,* or *thyrotoxicosis,* can develop. Among these are hormone-secreting tumors of the thyroid (very rare), but the most common form of hyperthyroidism is an autoimmune disease called *Graves' disease.* Like Hashimoto's disease, Graves' disease is more common in women and may have a gradual onset, typically beginning between the ages of 20 and 40. The plasma of patients with Graves' disease contains antibodies that recognize the TSH receptor. Unlike other autoimmune diseases in which a receptor protein is attacked (e.g., myasthenia gravis, Chapter 9), the anti-TSH receptor antibodies actually stimulate the receptor.

Thus, circulating antibodies are produced that bind to and activate the TSH receptor. The thyroid gland responds to these antibodies as if TSH itself were binding to the receptor. This response causes hypertrophy of the thyroid and increased TH production, because there is no way for thyroid hormones to turn off antibody production by negative feedback. T_3 and T_4 do, however, feed back on the hypothalamus and pituitary gland, as usual. Thus, Graves' disease is characterized by elevated thyroid hormone levels despite suppressed TSH. The hypertrophy produced by the constant stimulation of the gland results in a goiter. This is an important point: The presence of a goiter in and of itself does not distinguish between hypothyroidism and hyperthyroidism, because both types of disease can result in thyroid hypertrophy.

The signs and symptoms of thyrotoxicosis can be predicted in part from the previous discussion about hypothyroidism. Hyperthyroid patients tend to have ***heat intolerance,*** weight loss, increased appetite, and often show signs of increased sympathetic nervous system activity (anxiety, tremors, jumpiness, increased heart rate).

Hyperthyroidism can be very serious, particularly because of its effects on the cardiovascular system (largely secondary to its permissive actions on catecholamines). It may be treated with drugs that inhibit thyroid hormone synthesis, by surgical removal of the thyroid gland, or by destroying a portion of the thyroid using radioactive iodine. In the last case, the radioactive iodine is ingested. Because the thyroid is the chief region of iodine uptake in the body, most of the radioactivity appears within the gland, where its high-energy gamma rays partly destroy the tissue. Following treatment, patients may need to take thyroid replacement pills each day to maintain their TSH and T_3/T_4 levels, unless sufficient functional thyroid tissue remains after radiotherapy.■

■ SECTION C SUMMARY

Synthesis of Thyroid Hormones

I. T_3 and T_4 are synthesized by sequential iodinations of thyroglobulin in the thyroid follicle lumen, or colloid. Iodinated tyrosines on thyroglobulin are coupled to produce either T_3 or T_4.

II. The enzyme responsible for T_3/T_4 synthesis is thyroid peroxidase.

Control of Thyroid Function

I. All of the synthetic steps involved in T_3/T_4 synthesis are stimulated by TSH. TSH also stimulates uptake of iodide, where it is trapped in the follicle.

II. TSH causes growth (hypertrophy) of thyroid tissue. Excessive exposure of the thyroid to TSH can cause goiter.

Actions of Thyroid Hormones

I. T_3/T_4 increase the metabolic rate, and thus promote consumption of calories (calorigenic effect). This results in heat production.

II. The actions of the sympathetic nervous system are potentiated by T_3/T_4. This is called the permissive action of T_3/T_4.

III. Thyroid hormones are essential for normal growth and development—particularly of the nervous system—during fetal life and childhood.

Additional Clinical Examples

I. Hypothyroidism most commonly results from autoimmune destruction of all or part of the thyroid. It is characterized by weight gain, fatigue, cold intolerance, and changes in skin tone and mentation. It may also result in goiter.

II. Hyperthyroidism is also typically the result of an autoimmune disorder. It is characterized by weight loss, heat intolerance, irritability and anxiety, and often goiter.

■ SECTION C KEY TERMS

calorigenic 339	iodine 337
colloid 337	monoiodotyrosine (MIT) 337
diiodotyrosine (DIT) 337	thyroglobulin 337
follicle 337	thyroid peroxidase 337
hypertrophy 338	thyroxine (T_4) 337
iodide trapping 337	triiodothyronine (T_3) 337

■ SECTION C CLINICAL TERMS

autoimmune thyroiditis 340	heat intolerance 341
cold intolerance 340	hyperthyroidism 340
cretinism 339	hypothyroidism 340
goiter 338	myxedema 340
Graves' disease 340	thyrotoxicosis 340
Hashimoto's disease 340	

■ SECTION C REVIEW QUESTIONS

1. Describe the steps leading to T_3 and T_4 production, beginning with the transport of iodide into the thyroid follicular cell.
2. What are the major actions of TSH on thyroid function and growth?
3. What is the major way in which the TRH/TSH/TH pathway is regulated?
4. Explain why the symptoms of hyperthyroidism may be confused with a disorder of the autonomic nervous system.
5. Explain how both hypothyroidism and hyperthyroidism can result in the appearance of a goiter.

The Endocrine Response to Stress

Much of this book is concerned with the body's response to **stress** in its broadest meaning as an environmental change that must be adapted to if health and life are to be maintained. Thus, any change in external temperature, water intake, or other factor sets into motion mechanisms designed to prevent a significant change in some physiological variable. In this section, the basic endocrine response to stress is described in the sense of real or potential threats to homeostasis. These threats comprise an immense number of situations, including physical trauma, prolonged exposure to cold, prolonged heavy exercise, infection, shock, decreased oxygen supply, sleep deprivation, pain, fright, and other emotional stresses.

It is obvious that the response to cold exposure must be very different from that to infection or fright, but in one respect the response to all these situations is the same: Invariably, the adrenal cortex's secretion of the glucocorticoid hormone cortisol is increased. Activity of the sympathetic nervous system, including release of the hormone epinephrine from the adrenal medulla, also usually increases in response to stress.

The increased cortisol secretion during stress is mediated mainly by the hypothalamo-anterior pituitary system described earlier. As previously illustrated in Figure 11–19, neural input to the hypothalamus from portions of the nervous system responding to a particular stress induces secretion of CRH. This hormone is carried by the hypothalamo-pituitary portal vessels to the anterior pituitary, where it stimulates ACTH secretion. ACTH in turn circulates to the adrenal cortex and stimulates cortisol release.

The secretion of ACTH, and therefore of cortisol, is stimulated by several hormones in addition to hypothalamic CRH. These include vasopressin, which usually increases in response to stress. A most interesting recent finding is that some of the cytokines (secretions from cells that comprise the immune system; Chapter 18) also stimulate ACTH secretion both directly and by stimulating the secretion of CRH. These cytokines provide a means for eliciting an endocrine stress response when the immune system is stimulated. The possible significance of this relationship for immune function is described next and in Chapter 18.

Physiological Functions of Cortisol

Although the effects of cortisol are most dramatically illustrated during the response to stress, cortisol exerts many important actions even in nonstress situations. For example, cortisol has permissive actions on the reactivity to epinephrine and norepinephrine of muscle cells that surround blood vessels. Thus, basal levels of cortisol are needed to maintain normal blood pressure. Likewise, basal levels of cortisol are required to maintain the cellular concentrations of certain enzymes involved in metabolic homeostasis. These enzymes are located primarily in the liver, and they act to increase hepatic glucose production between meals, thus preventing plasma glucose levels from decreasing below normal.

Two important systemic actions of cortisol are its anti-inflammatory and anti-immune functions. The mechanisms by which cortisol inhibits immune system function are numerous and complex. Cortisol inhibits the production of both leukotrienes and prostaglandins, both of which are involved in inflammation. Cortisol also stabilizes lysosomal membranes in damaged cells, preventing the release of their proteolytic contents. In addition, cortisol reduces capillary permeability in injured areas (thus, reducing fluid leakage to the interstitium), and it suppresses the growth and function of certain key immune cells. Thus, cortisol may serve as a "brake" on the immune system, which would tend to overreact to minor infections in the absence of cortisol. Indeed, in diseases in which cortisol is greatly reduced, an increased incidence of autoimmune disease is sometimes noticed. Such diseases are characterized by a person's own immune system launching an attack against part of the body.

During fetal and neonatal life, cortisol is also an extremely important developmental hormone. It has been implicated in the proper differentiation of numerous tissues and glands, including various parts of the brain, the adrenal medulla, the intestine and, notably, the lungs. In the latter case, cortisol is very important for the production of surfactant, which reduces surface tension in the lungs (see Chapter 13).

Thus, although it is common to define the actions of cortisol in the context of the stress response, it is worth remembering that the maintenance of a homeostatic situation in the absence of external stresses is also a critical function of cortisol.

Functions of Cortisol in Stress

Table 11–3 summarizes the major effects of increased cortisol during stress. The effects on organic metabolism are to mobilize fuels—to increase the plasma concentrations of amino acids, glucose, glycerol, and free fatty acids. These effects are ideally suited to meet a stressful situation. First, an animal faced with a potential threat is usually forced to forego eating, making these metabolic changes essential for survival during fasting. Second, the amino acids liberated by catabolism of body protein not only provide a source of glucose, via hepatic gluconeogenesis, but also constitute a potential source of amino acids for tissue repair should injury occur.

A few of the medically important implications of these cortisol-induced effects on organic metabolism are as follows: (1) any patient who is ill or is subjected to surgery catabolizes considerable quantities of body protein; (2) a diabetic who suffers an infection requires more insulin than usual; and (3) a child subjected to severe stress of any kind may manifest retarded growth.

Cortisol has important effects during stress other than those on organic metabolism. It increases the ability of vascular smooth muscle to contract in response to norepinephrine. Therefore, a patient with insufficient cortisol faced with even a moderate stress, which usually releases unknown vasodilators,

Table 11–3	Effects of Increased Plasma Cortisol Concentration During Stress

1. Effects on organic metabolism
   a. Stimulation of protein catabolism in bone, lymph, muscle, and elsewhere
   b. Stimulation of liver uptake of amino acids and their conversion to glucose (gluconeogenesis)
   c. Maintenance of plasma glucose levels
   d. Stimulation of triglyceride catabolism in adipose tissue, with release of glycerol and fatty acids into the blood

2. Enhanced vascular reactivity (increased ability to maintain vasoconstriction in response to norepinephrine and other stimuli)

3. Unidentified protective effects against the damaging influences of stress

4. Inhibition of inflammation and specific immune responses

5. Inhibition of nonessential functions (e.g., reproduction and growth)

may develop hypotension, due primarily to a marked decrease in total peripheral resistance caused by the vasodilators.

As item 3 in Table 11–3 notes, we still do not know the other reasons, in addition to the effect on vascular smooth muscle, why increased cortisol is so important for the body's optimal response to stress—that is, for its ability to resist the damaging influences of stress. What is clear is that a person exposed to severe stress can die, usually of circulatory failure, if his or her plasma cortisol concentration is too low.

Effect 4 in Table 11–3 stems originally from the fact that administration of large amounts of cortisol or its synthetic analogs profoundly reduces the inflammatory response to injury or infection. Because of this effect, cortisol is a valuable tool in the treatment of *allergy, arthritis,* other inflammatory diseases, and graft rejection. These anti-inflammatory and anti-immune effects have generally been classified among the various *pharmacological* effects of cortisol because it was assumed they could be achieved only by very large doses. It is now clear, however, that such effects also occur, albeit to a lesser degree, at the plasma concentrations achieved during stress. Thus, the increased plasma cortisol typical of infection or trauma exerts a dampening effect on the body's immune responses, protecting against possible damage from excessive inflammation. This effect explains the significance of the fact, mentioned earlier, that several cytokines (immune cell secretions) stimulate the secretion of ACTH and thereby cortisol. Such stimulation is part of a negative feedback system in which the increased cortisol then partially blocks the inflammatory processes in which the cytokines participate. Moreover, cortisol normally dampens the fever an infection causes.

In summary, stress is a broadly defined situation in which there exists a real or potential threat to homeostasis. In such a scenario, it is important to maintain blood pressure, to provide extra fuel sources in the blood, and to temporarily

shut down nonessential functions. Cortisol is the most important hormone that carries out these activities. Thus, cortisol enhances vascular reactivity, catabolizes protein and fat to provide energy, and inhibits growth and reproduction. The price the body pays during stress is that cortisol is strongly catabolic. Thus, cells of the immune system, bone, muscles, skin, and numerous other tissues undergo catabolism to provide substrates for gluconeogenesis. In the short term, this is not of any major consequence. Chronic exposure to stress, however, can in fact lead to severe decreases in bone density, immune function, and reproductive fertility.

Other Hormones Released During Stress

Other hormones that are usually released during many kinds of stress are aldosterone, vasopressin (ADH), growth hormone, glucagon, and beta-endorphin (which is coreleased from the anterior pituitary with ACTH). Insulin secretion usually decreases. Vasopressin and aldosterone act to retain water and sodium within the body, an important adaptation in the face of potential losses by hemorrhage or sweating. Vasopressin also stimulates the secretion of ACTH. The overall effects of the changes in growth hormone, glucagon, and insulin are, like those of cortisol and epinephrine, to mobilize energy stores. The role, if any, of beta-endorphin in stress may be related to its painkilling effects.

In addition, the sympathetic nervous system plays a key role in the stress response. Activation of the sympathetic nervous system during stress is often termed the **fight-or-flight response.** A list of the major effects of increased sympathetic activity, including secretion of epinephrine from the adrenal medulla, almost constitutes a guide to how to meet emergencies in which physical activity may be required and bodily damage may occur (**Table 11–4**).

Table 11–4	Actions of the Sympathetic Nervous System, Including Epinephrine Secreted by the Adrenal Medulla, During Stress

1. Increased hepatic and muscle glycogenolysis (provides a quick source of glucose)

2. Increased breakdown of adipose tissue triglyceride (provides a supply of glycerol for gluconeogenesis and of fatty acids for oxidation)

3. Decreased fatigue of skeletal muscle

4. Increased cardiac function (e.g., increased heart rate)

5. Diversion of blood from viscera to skeletal muscles by means of vasoconstriction in the former beds and vasodilation in the latter

6. Increased lung ventilation by stimulating brain breathing centers and dilating airways

This list of hormones whose secretion rates are altered by stress is by no means complete. It is likely that the secretion of almost every known hormone may be influenced by stress. For example, prolactin is increased, although the adaptive significance of this change is unclear. By contrast, the pituitary gonadotropins and the sex steroids are decreased. As noted previously, reproduction is not an essential function when life is in danger.

Psychological Stress and Disease

Throughout this section we have emphasized the adaptive value of the body's various responses to stress. It is now clear, however, that psychological stress, particularly if chronic, can have deleterious effects on the body, constituting important links in mind-body interactions. For example, it is very likely that the increased plasma cortisol associated with psychological stress can decrease the activity of the immune system enough to reduce the body's resistance to infection and, perhaps, cancer. It can also worsen the symptoms of diabetes because of its effects on blood glucose, and it may possibly cause an increase in the death rate of certain neurons.

Similarly, it is possible that prolonged and repeated activation of the sympathetic nervous system by psychological stress may enhance the development of certain diseases, particularly *atherosclerosis* (the accumulation of plaques in arteries) and *hypertension* (high blood pressure). For example, it is easy to imagine that the increased blood lipid concentration and cardiac work could contribute to the former disease. Thus, the body's adaptive stress responses, if excessive or inappropriate, may play a causal role in the development of diseases.

ADDITIONAL CLINICAL EXAMPLES

Adrenal Insufficiency and Cushing's Syndrome

Cortisol is one of the few hormones absolutely essential for life. The complete absence of cortisol leads to the body's inability to maintain homeostasis, particularly when confronted with a stress such as infection, which is usually fatal within days without cortisol. The general term for any situation in which plasma levels of cortisol are chronically lower than normal is *adrenal insufficiency.* Patients with adrenal insufficiency suffer from a diffuse array of symptoms, depending on the severity and cause of the disease. These patients typically report weakness, lethargy, and loss of appetite. Examination may reveal low blood pressure (in part because cortisol is needed to permit the full extent of the cardiovascular actions of epinephrine) and low blood sugar, especially after fasting (because of the loss of the normal metabolic actions of cortisol).

The causes of adrenal insufficiency are several. *Primary adrenal insufficiency* is due to a loss of adrenal cortical function, as may occur, for example, when infectious diseases such as *tuberculosis* infiltrate the adrenal glands and destroy them. The adrenals can also (rarely) be destroyed by invasive tumors. Most commonly, however, the syndrome is due to autoimmune attack, in which the immune system mistakenly recognizes some component of a person's own adrenal cells as "foreign." The resultant immune reaction causes inflammation and eventually the destruction of many of the cells of the adrenal glands. Because of this, all of the zones of the adrenal cortex are affected. Thus, not only cortisol but also aldosterone levels are decreased below normal in primary adrenal insufficiency. This decrease in aldosterone concentration creates the additional problem of an imbalance in sodium, potassium, and water in the blood because aldosterone is a key regulator of those variables. The loss of salt and water balance may lead to *hypotension* (low blood pressure). Primary adrenal insufficiency from any of these causes is also known as *Addison's disease,* after the nineteenth-century physician who first discovered the syndrome. Like many other autoimmune diseases (described in detail in Chapter 18), Addison's disease progresses slowly, and its symptoms at first glance are nonspecific and generalized. In fact, it may be misdiagnosed as *chronic fatigue syndrome,* or even a psychological disorder, because some patients with primary adrenal insufficiency may exhibit anxiety or emotional problems. The diagnosis is made by measuring plasma concentrations of cortisol. In primary adrenal insufficiency, cortisol levels are well below normal, whereas ACTH levels are greatly increased due to the loss of the negative feedback actions of cortisol. Treatment of this disease requires daily oral administration of glucocorticoids and mineralocorticoids. In addition, the patient must carefully monitor his or her diet to ensure an adequate consumption of carbohydrates and controlled potassium and sodium intake.

Adrenal insufficiency can also be due to a deficiency of ACTH—*secondary adrenal insufficiency*—which may arise from pituitary disease. Its symptoms are often less dramatic than primary adrenal insufficiency, because aldosterone secretion, which doesn't rely on ACTH, is maintained by other mechanisms.

Adrenal insufficiency can be life-threatening if not treated aggressively. The flip side of this disorder, *excess* glucocorticoids, is usually not as immediately dangerous but can also be very severe. In *Cushing's Syndrome,* there is excess cortisol in the blood even in the nonstressed individual. The cause may be a primary defect (e.g., a cortisol-secreting tumor of the adrenal) or may be secondary (usually due to an ACTH-secreting tumor of the pituitary gland). In the latter case, the condition is known as *Cushing's disease,* which accounts for most cases of Cushing's Syndrome. The increased blood levels of cortisol tend to promote uncontrolled catabolism of bone, muscle, skin, and other organs. Thus, bone strength diminishes and can even lead to **osteoporosis** (loss of bone mass); muscles weaken, and the skin becomes thinned and easily bruised. The increased catabolism may produce such a large quantity of precursors for hepatic gluconeogenesis that blood sugar increases to levels observed in diabetes. Thus, a person with Cushing's Syndrome may show some of the same symptoms as a person with diabetes. Equally troubling is the possibility of **immunosuppression,** brought about by the anti-immune actions of cortisol. Cushing's Syndrome is often associated with loss of fat mass from the extremities, and with redistribution of the fat in the trunk, face, and the back of the neck. Combined with an increased appetite, often triggered by high concentrations of cortisol, this results in obesity and a characteristic facial appearance in many patients (**Figure 11–25**). A further problem associated with Cushing's Syndrome is the possibility of developing hypertension (high blood pressure). This is due not to increased aldosterone production, but instead to the pharmacological effects of cortisol, including cortisol's ability to potentiate the effects of epinephrine and norepinephrine on the heart and blood vessels. At high enough concentrations, cortisol exerts aldosterone-like actions on the kidney, resulting in salt and water retention, which contributes to hypertension.

Treatment of Cushing's Syndrome depends on the cause. In Cushing's disease, for example, surgical removal of the pituitary tumor, if possible, is the best alternative. Unfortunately, these tumors may regrow over a number of years, necessitating a second pituitary surgery or even adrenalectomy. Unilateral adrenalectomy is the appropriate treatment for an adrenal tumor making too much cortisol.

Of importance is the fact that glucocorticoids are often used therapeutically to treat inflammation, lung disease, and other disorders. If glucocorticoids are administered at a high enough dosage for long periods, the side effect of such treatment can be Cushing's Syndrome.■

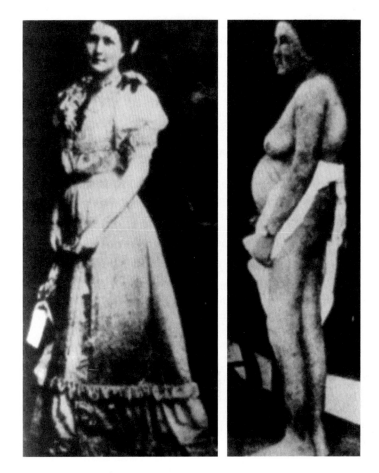

Figure 11–25

A young patient from the original series of Harvey Cushing. Left: before onset of disease. Right: after development of Cushing's Syndrome.

■ **SECTION D SUMMARY**

Physiological Functions of Cortisol

I. Cortisol is released from the adrenal cortex upon stimulation with ACTH. ACTH, in turn, is stimulated by the release of corticotropin-releasing hormone (CRH) from the hypothalamus.

II. The physiological functions of cortisol are to maintain the responsiveness of target cells to epinephrine and norepinephrine, to provide a "check" on the immune system, to participate in energy homeostasis, and to promote normal differentiation of tissues during fetal life.

Functions of Cortisol in Stress

I. The stimulus that activates the CRH/ACTH/cortisol endocrine pathway is stress, which encompasses a wide array of sensory and physical inputs that disrupt, or potentially disrupt, homeostasis.

II. In response to stress, the usual physiological functions of cortisol are enhanced as cortisol levels in the plasma increase. Thus, gluconeogenesis, lipolysis, and inhibition of insulin actions increase. This results in increased blood levels of fuel sources (glucose, fatty acids) needed to cope with stressful situations.

III. Cortisol levels at high concentrations also inhibit "nonessential" processes, such as reproduction, during stressful situations, and inhibit immune function.

Other Hormones Released During Stress

I. In addition to CRH, ACTH, and cortisol, several other hormones are released during stress. Beta-endorphin is coreleased with ACTH and may act to reduce pain. Vasopressin stimulates ACTH secretion and also acts on the kidney to increase water retention. Other hormones that are increased in the blood by stress are aldosterone, growth hormone, and glucagon. Insulin secretion, by contrast, decreases during stress.

II. Epinephrine is secreted from the adrenal medulla during stress, in response to stimulation from the sympathetic nervous system. The norepinephrine from sympathetic neuron terminals, combined with the circulating epinephrine, prepare the body for stress in several ways. These include increased heart rate and heart pumping strength, increased ventilation, increased shunting of blood to skeletal muscle, and increased generation of fuel sources that are released into the blood.

Psychological Stress and Disease

I. Chronic psychological stress can negatively affect health by contributing to the progression or development of several diseases, among them hypertension and atherosclerosis. Of greater concern is that chronic psychological stress may cause immunosuppression, and this in turn has been postulated to contribute to the advent of certain types of cancer.

Additional Clinical Examples

I. Adrenal insufficiency may result from adrenal destruction (primary adrenal insufficiency, or Addison's disease) or from hyposecretion of ACTH (secondary adrenal insufficiency).

II. Adrenal insufficiency is associated with decreased ability to maintain blood pressure (due to loss of aldosterone) and blood sugar. It may be fatal if untreated.

III. Cushing's Syndrome is the result of chronically elevated plasma cortisol concentration. When the cause of the increased cortisol is secondary to an ACTH-secreting pituitary tumor, the condition is known as Cushing's disease.

IV. Cushing's Syndrome is associated with hypertension, high blood sugar, redistribution of body fat, obesity, and muscle and bone weakness. If untreated, it can also lead to immunosuppression.

■ **SECTION D KEY TERMS**

fight-or-flight response 343   stress 342

■ **SECTION D CLINICAL TERMS**

Addison's disease 344
adrenal insufficiency 344
allergy 343
arthritis 343
atherosclerosis 344
chronic fatigue syndrome 344
Cushing's disease 344
Cushing's Syndrome 344
hypertension 344

hypotension 344
immunosuppression 345
osteoporosis 345
primary adrenal
   insufficiency 344
secondary adrenal
   insufficiency 344
tuberculosis 344

■ **SECTION D REVIEW QUESTIONS**

1. Diagram the CRH-ACTH-cortisol pathway.
2. List the physiological functions of cortisol.
3. Define stress, and list the functions of cortisol during stress.
4. List the major effects of activation of the sympathetic nervous system during stress.
5. Contrast the symptoms of adrenal insufficiency and Cushing's Syndrome.

SECTION E — Endocrine Control of Growth

One of the major functions of the endocrine system is to control growth. At least a dozen hormones directly or indirectly (e.g., hypophysiotropic hormones) play important roles in controlling growth. This complex process is influenced by genetics, endocrine function, and a variety of environmental factors, including nutrition. The growth process involves cell division and net protein synthesis throughout the body, but a person's height is determined specifically by bone growth, particularly of the vertebral column and legs. We first provide an overview of bone and the growth process before describing the roles of hormones in determining growth rates.

Bone Growth

Bone is a living tissue consisting of a protein (collagen) matrix upon which calcium salts, particularly calcium phosphates, are deposited. A growing long bone is divided, for descriptive purposes, into the ends, or **epiphyses,** and the remainder, the **shaft.** The portion of each epiphysis in contact with the shaft is a plate of actively proliferating cartilage, the **epiphyseal growth plate** (**Figure 11–26**). **Osteoblasts,** the bone-forming cells at the shaft edge of the epiphyseal growth plate, convert the carti-

laginous tissue at this edge to bone while cells called **chondrocytes** simultaneously lay down new cartilage in the interior of the plate. In this manner, the epiphyseal growth plate remains intact (indeed, actually widens) and is gradually pushed away from the center of the bony shaft as the shaft lengthens.

Linear growth of the shaft can continue as long as the epiphyseal growth plates exist, but ceases when the growth plates are converted to bone as a result of hormonal influences at puberty. This is known as **epiphyseal closure** and occurs at different times in different bones. Thus, a person's **bone age** can be determined by x-raying the bones and determining which ones have undergone epiphyseal closure.

As shown in **Figure 11–27**, children manifest two periods of rapid increase in height, the first during the first two years of life and the second during puberty. Note that increase in height is not necessarily correlated with the rates of growth of specific organs.

The pubertal growth spurt lasts several years in both sexes, but growth during this period is greater in boys. In addition, boys grow more before puberty because they begin puberty approximately two years later than girls. These factors account for the differences in average height between men and women.

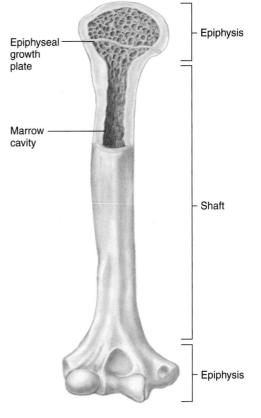

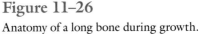

Figure 11–26
Anatomy of a long bone during growth.

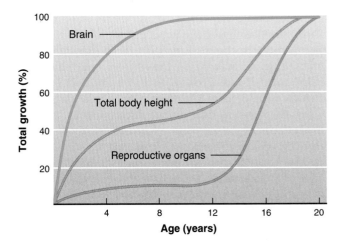

Figure 11–27
Relative growth in brain, total body height (a measure of long-bone and vertebral growth), and reproductive organs. Note that brain growth is nearly complete by age 5, whereas maximal height (maximal bone lengthening) and reproductive-organ size are not reached until the late teens.

Environmental Factors Influencing Growth

Adequate nutrition and freedom from disease are the primary environmental factors influencing growth. Lack of sufficient amounts of amino acids, fatty acids, vitamins, or minerals interferes with growth. Total protein and nutrients to provide energy must also be adequate.

The growth-inhibiting effects of malnutrition can be seen at any time of development but are most profound when they occur early in life. Thus, maternal malnutrition may cause growth retardation in the fetus. Because low birth weight is strongly associated with increased infant mortality, prenatal malnutrition causes increased numbers of prenatal and early postnatal deaths. Moreover, irreversible stunting of brain development may be caused by prenatal malnutrition. During infancy and childhood, too, malnutrition can interfere with both intellectual development and total body growth.

Following a temporary period of stunted growth due to malnutrition or illness, and given proper nutrition and recovery from illness, a child can manifest a remarkable growth spurt **(catch-up growth)** that brings the child up to the normal height expected for his or her age. The mechanisms that account for this accelerated growth are unknown.

Hormonal Influences on Growth

The hormones most important to human growth are growth hormone, insulin-like growth factors I and II, thyroid hormones, insulin, testosterone, and estrogens, all of which exert widespread effects. In addition to all these hormones, a large group of peptide **growth factors** exert effects, most of them acting as paracrine and autocrine agents to stimulate differentiation and/or cell division of certain cell types. The general term for a chemical that stimulates cell division is a **mitogen.**

There are also peptide **growth-inhibiting factors** that modulate growth by inhibiting cell division in specific tissues. The growth factors and growth-inhibiting factors are usually produced by multiple cell types rather than by discrete endocrine glands.

The various hormones and growth factors do not all stimulate growth at the same periods of life. For example, fetal growth is largely independent of growth hormone, the thyroid hormones, and the sex steroids, all of which stimulate growth during childhood and adolescence.

Growth Hormone and Insulin-Like Growth Factors

Growth hormone, secreted by the anterior pituitary gland, has little or no effect on fetal growth, but is the most important hormone for postnatal growth. Its major growth-promoting effect is (indirect, as we will see) stimulation of cell division in its many target tissues. Thus, growth hormone promotes bone lengthening by stimulating maturation and cell division of the chondrocytes in the epiphyseal plates, thereby continuously widening the plates and providing more cartilaginous material for bone formation.

Importantly, growth hormone exerts most of its cell division-stimulating (mitogenic) effect not *directly* on cells but *indirectly* through the mediation of a mitogen whose synthesis and release are induced by growth hormone. This mitogen is called **insulin-like growth factor 1 (IGF-1)** (previously known as somatomedin C). Despite its similarity to insulin, this messenger has its own unique effects distinct from those of insulin. Under the influence of growth hormone, IGF-1 is secreted by the liver, enters the blood, and functions as a hormone. In addition, growth hormone stimulates many other types of cells, including bone, to secrete IGF-1, and at these sites IGF-1 functions as an autocrine or paracrine agent. The relative importance of IGF-1 as a hormone versus autocrine/paracrine agent in any given organ or tissue remains controversial.

Current concepts of how growth hormone and IGF-1 interact on the epiphyseal plates of bone are as follows: (1) growth hormone stimulates the chondrocyte precursor cells (prechondrocytes) and/or young differentiating chondrocytes in the epiphyseal plates to differentiate into chondrocytes; (2) during this differentiation, the cells begin both to secrete IGF-1 and to become responsive to IGF-1; (3) the IGF-1 then acts as an autocrine or paracrine agent (probably along with blood-borne IGF-1) to stimulate the differentiating chondrocytes to undergo cell division.

The importance of IGF-1 in mediating the major growth-promoting effect of growth hormone is illustrated by the fact that *dwarfism* (abnormally short stature) can be caused not only by decreased growth hormone secretion, but also by decreased production of IGF-1 or failure of the tissues to respond to IGF-1. For example, one uncommon form of short stature (called *growth hormone insensitivity syndrome,* or *Laron Dwarfism*), is due to a genetic mutation that causes the growth hormone receptor to fail to respond to growth hormone. The result is failure to produce IGF-1 in response to growth hormone.

The secretion and activity of IGF-1 can be influenced by the nutritional status of the individual and by many hormones other than growth hormone. For example, malnutrition during childhood inhibits the production of IGF-1 even though plasma growth hormone concentration is elevated and should be stimulating IGF-1 secretion.

In addition to its specific growth-promoting effect on cell division via IGF-1, growth hormone directly stimulates protein synthesis in various tissues and organs, particularly muscle. It does this by increasing cells' amino acid uptake and both the synthesis and activity of ribosomes. All of these events are essential for protein synthesis. This anabolic effect on protein metabolism facilitates the ability of tissues and organs to enlarge. **Table 11–5** summarizes the multiple effects of growth hormone.

Figure 11–28 shows the control of growth hormone secretion. Briefly, the control system begins with two of the

Table 11–5	Major Effects of Growth Hormone

1. Promotes growth: Induces precursor cells in bone and other tissues to differentiate and secrete insulin-like growth factor 1 (IGF-1), which stimulates cell division. Also stimulates liver to secrete IGF-1.

2. Stimulates protein synthesis, predominantly in muscle.

3. Anti-insulin effects (particularly at high concentrations):
   a. Renders adipocytes more responsive to stimuli that induce breakdown of triglycerides, releasing fatty acids into the blood.
   b. Stimulates gluconeogenesis.
   c. Reduces the ability of insulin to stimulate glucose uptake by adipose and muscle cells, resulting in higher blood glucose levels.

hypophysiotropic hormones secreted by the hypothalamus. Growth hormone secretion is stimulated by growth hormone-releasing hormone (GHRH) and inhibited by somatostatin. As a result of changes in these two signals, which are usually 180 degrees out of phase with each other (i.e., one is high when the other is low), growth hormone secretion occurs in episodic bursts and manifests a striking daily rhythm. During most of the day, little or no growth hormone is secreted, although bursts may be elicited by certain stimuli, including stress, hypoglycemia, and exercise. In contrast, 1 to 2 hours after a person falls asleep, one or more larger, prolonged bursts of secretion may occur. The negative feedback controls that growth hormone and IGF-1 exert on the hypothalamus and anterior pituitary are summarized in Figure 11–28.

In addition to the hypothalamic controls, a variety of hormones—notably the sex hormones, insulin, and the thyroid hormones, as described in the paragraphs that follow—influence the secretion of growth hormone. The net result of all these inputs is that the secretion rate of growth hormone is highest during adolescence (the period of most rapid growth), next highest in children, and lowest in adults. The decreased growth hormone secretion associated with aging is responsible, in part, for the decrease in lean-body and bone mass, the expansion of adipose tissue, and the thinning of the skin that occur as people age.

The ready availability of human growth hormone produced by recombinant-DNA technology has greatly facilitated the treatment of children with short stature due to growth hormone deficiency. Controversial at present is the administration of growth hormone to short children who do not have growth hormone deficiency, to athletes in an attempt to increase muscle mass, and to normal elderly persons to reverse growth hormone–related aging changes. It should be clear from Table 11–5 that

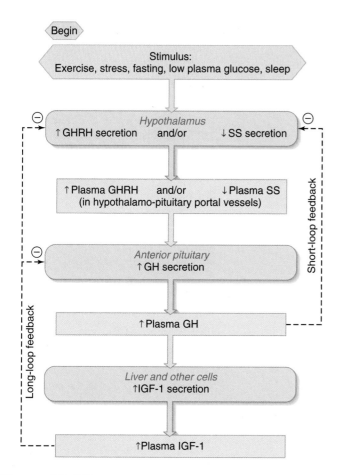

Begin

Stimulus:
Exercise, stress, fasting, low plasma glucose, sleep

⊖ *Hypothalamus*
↑ GHRH secretion   and/or   ↓ SS secretion   ⊖

↑ Plasma GHRH   and/or   ↓ Plasma SS
(in hypothalamo-pituitary portal vessels)

⊖ *Anterior pituitary*
↑ GH secretion

↑ Plasma GH

Liver and other cells
↑IGF-1 secretion

↑Plasma IGF-1

Short-loop feedback

Long-loop feedback

Figure 11–28

Hormonal pathways controlling the secretion of growth hormone (GH) and insulin-like growth factor I (IGF-1). At the hypothalamus, the minus sign ⊖ denotes that the input inhibits the secretion of growth hormone–releasing hormone (GHRH) and/or stimulates the release of somatostatin (SS). Which of these mechanisms predominates in humans is currently unknown. Not shown in the figure is that several hormones not in the sequence (e.g., the thyroid hormones) influence growth hormone secretion via effects on the hypothalamus and/or anterior pituitary.

administration of GH to an otherwise healthy individual (such as an athlete) can lead to serious side-effects. Abuse of GH in such situations can lead to symptoms similar to those of diabetes mellitus, as well as numerous other problems. The consequences of chronically elevated growth hormone levels are dramatically illustrated in the disease called acromegaly (see Additional Clinical Examples at the end of this section).

As noted, growth hormone plays little if any role in the growth of the embryo and fetus. One would suppose, therefore, that this would also be true for IGF-1, but this is not the case. IGF-1 is required for normal fetal total-body growth and, spe-

cifically, for normal maturation of the fetal nervous system. The stimulus for IGF-1 secretion during prenatal life is unknown.

Finally, it should be noted that there is another messenger—**insulin-like growth factor 2 (IGF-2)**—that is closely related to IGF-1. IGF-2, the secretion of which is *independent* of growth hormone, is also a crucial mitogen during the prenatal period. It continues to be secreted throughout life, but its postnatal function is not known.

Thyroid Hormones

Thyroid hormones are essential for normal growth because they are required for both the synthesis and the growth-promoting effects of growth hormone. Thus, infants and children with hypothyroidism manifest retarded growth due to slowed bone growth.

Insulin

The major actions of insulin are described in Chapter 16. Insulin is an anabolic hormone that promotes the entry of glucose and amino acids from the extracellular fluid into cells. Insulin stimulates storage of fat and inhibits protein degradation. Thus, it is not surprising that adequate amounts of insulin are necessary for normal growth. Its inhibitory effect on protein degradation is particularly important with regard to growth.

In addition to this general anabolic effect, however, insulin exerts direct growth-promoting effects on cell differentiation and cell division during fetal life, and possibly during childhood.

Sex Hormones

As Chapter 17 will explain, sex hormone secretion (testosterone in the male and estrogens in the female) begins to increase between the ages of 8 and 10 and reaches a plateau over the next 5 to 10 years. A normal pubertal growth spurt, which reflects growth of the long bones and vertebrae, requires this increased production of the sex hormones. The major growth-promoting effect of the sex hormones is to stimulate the secretion of growth hormone and IGF-1.

Unlike growth hormone, however, the sex hormones not only stimulate bone growth, but ultimately *stop* it by inducing epiphyseal closure. The dual effects of the sex hormones explain the pattern seen in adolescence—rapid lengthening of the bones culminating in complete cessation of growth for life.

In addition to these dual effects on bone, testosterone, but not estrogen, exerts a direct anabolic effect on protein synthesis in many nonreproductive organs and tissues of the body. This accounts, at least in part, for the increased muscle mass of men in comparison to women.

This effect of testosterone is also why athletes sometimes use androgens called ***anabolic steroids*** in an attempt to increase muscle mass and strength. These steroids include testosterone, synthetic androgens, and the hormones dehydroepiandrosterone (DHEA) and androstenedione. However, these steroids have multiple potential toxic side effects, such as liver damage, increased risk of prostate cancer, and infertility. Moreover, in females, they can produce masculinization.

Cortisol

Cortisol, the major hormone the adrenal cortex secretes in response to stress, can have potent *antigrowth* effects under certain conditions. When present in high concentration, it inhibits DNA synthesis and stimulates protein catabolism in many organs, and it inhibits bone growth. Moreover, it breaks down bone and inhibits the secretion of growth hormone. For all these reasons, in children, the elevation in plasma cortisol that accompanies infections and other stresses is, at least in part, responsible for the retarded growth that occurs with chronic illness. Furthermore, the administration of pharmacological glucocorticoid therapy for asthma or other disorders may temporarily decrease linear growth in children.

This completes our survey of the major hormones that affect growth. **Table 11–6** summarizes their actions.

Table 11–6	Major Hormones Influencing Growth
Hormone	**Principal Actions**
Growth hormone	Major stimulus of postnatal growth: Induces precursor cells to differentiate and secrete insulin-like growth factor I (IGF-1), which stimulates cell division
	Stimulates liver to secrete IGF-1
	Stimulates protein synthesis
Insulin	Stimulates fetal growth
	Stimulates postnatal growth by stimulating secretion of IGF-1
	Stimulates protein synthesis
Thyroid hormones	Permissive for growth hormone's secretion and actions
	Permissive for development of the central nervous system
Testosterone	Stimulates growth at puberty, in large part by stimulating the secretion of growth hormone
	Causes eventual epiphyseal closure
	Stimulates protein synthesis in male
Estrogen	Stimulates the secretion of growth hormone at puberty
	Causes eventual epiphyseal closure
Cortisol	Inhibits growth
	Stimulates protein catabolism

ADDITIONAL CLINICAL EXAMPLES

Acromegaly and Gigantism

Acromegaly and *gigantism* arise when chronic, excess amounts of growth hormone are secreted into the blood. In almost all cases, acromegaly and gigantism are caused by benign (noncancerous) tumors of the anterior pituitary gland that secrete growth hormone at very high rates. These tumors are typically very slow growing, and, if they occur after puberty, it may be decades before a person realizes there is something seriously wrong.

If the tumor arises before puberty, when the epiphyseal growth plates are still open, the individual will develop gigantism ("pituitary giant") and grow to extraordinary heights (**Figure 11–29**). Some pituitary giants have reached heights over 8 feet! If the tumor arises after puberty, when linear growth is no longer possible, the condition is known as acromegaly. Such people will be of normal height but will manifest many other symptoms that also occur in pituitary giants.

Even when linear growth is no longer possible (after puberty), very high plasma levels of GH and IGF-1 result in the thickening of many bones in the body, most noticeably in the hands, feet, and head. The jaw, particularly, enlarges to give the characteristic facial appearance called *prognathism* that is associated with acromegaly. In addition, many internal organs such as the heart also become enlarged, and this can interfere with their ability to function normally.

All adults continue to make and secrete GH even after growth ceases. That is because GH has metabolic actions in addition to its effects on growth. The major actions of GH in metabolism are to increase blood sugar levels, increase blood fatty acid levels, and decrease target cell sensitivity to insulin. Not surprisingly, therefore, one of the stimuli that increases GH levels in the normal adult is a lowering of blood sugar or fatty acids. The secretion of GH during these metabolic crises, however, is transient; once plasma glucose

or fatty acid concentrations are restored to normal, GH levels decrease to baseline. In acromegaly, however, GH levels are always elevated. Thus, acromegaly is associated with elevated plasma levels of glucose and fatty acids, similar to the levels in diabetes. As in Cushing's Syndrome (Section D), therefore, the presence of chronically elevated levels of GH may result in diabetic-like symptoms.

Treatment of gigantism and acromegaly usually requires surgical removal of the pituitary tumor. The residual normal pituitary tissue is then sufficient to maintain baseline GH levels. If this treatment is impossible or not successful, treatment with long-acting analogs of somatostatin is sometimes necessary. (Recall that somatostatin is the hypothalamic hormone that normally inhibits GH secretion.)■

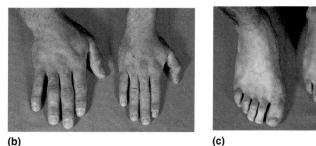

(a)   (b)   (c)

Figure 11–29

Gigantism and acromegaly in one individual of a pair of identical twins. Note the increased height and facial bone thickening (a), as well as the bone thickening of hands (b) and feet (c).

Figure 11–29 physiological *inquiry* (pi)

■ Did the individual in (a) develop excess growth hormone production before or after puberty?

Answer can be found at end of chapter.

Bone Growth

I. A bone lengthens as osteoblasts at the shaft edge of the epiphyseal growth plates convert cartilage to bone while new cartilage is being laid down in the plates.

II. Growth ceases when the plates are completely converted to bone.

Environmental Factors Influencing Growth

I. The major environmental factors influencing growth are nutrition and disease.

II. Maternal malnutrition during pregnancy may produce irreversible growth stunting and mental deficiency in offspring.

Hormonal Influences on Growth

I. Growth hormone is the major stimulus of postnatal growth.

a. It stimulates the release of IGF-1 from the liver and many other cells, and IGF-1 then acts locally (and perhaps also as a hormone) to stimulate cell division.

b. Growth hormone also acts directly on cells to stimulate protein synthesis.

c. Growth hormone secretion is highest during adolescence.

II. Because thyroid hormones are required for growth hormone synthesis and the growth-promoting effects of this hormone, they are essential for normal growth during childhood and adolescence. They are also permissive for brain development during infancy.

III. Insulin stimulates growth mainly during fetal life.

IV. Mainly by stimulating growth hormone secretion, testosterone and estrogen promote bone growth during adolescence, but these hormones also cause epiphyseal closure. Testosterone also stimulates protein synthesis.

V. High concentrations of cortisol inhibit growth and stimulate protein catabolism.

Additional Clinical Examples

I. Gigantism occurs when excess growth hormone is secreted during the prepubertal period. If excess growth hormone secretion occurs after linear growth has stopped, the condition—known as acromegaly—results in bone thickening and metabolic derangements.

SECTION F Endocrine Control of Ca^{2+} Homeostasis

Many of the hormones of the body control functions that, while important, are not necessarily vital for survival, such as growth. By contrast, some hormones control functions so vital that the absence of the hormone would be catastrophic, even life-threatening. One such function is Ca^{2+} homeostasis.

Extracellular calcium concentration normally remains within a narrow range. Large deviations in either direction can cause problems. For example, a low plasma calcium concentration increases the excitability of nerve and muscle plasma membranes. A high plasma calcium concentration causes cardiac arrhythmias as well as depressed neuromuscular excitability via its effects on membrane potential.

Effector Sites for Calcium Homeostasis

Calcium homeostasis depends on an interplay among bone, the kidneys, and the gastrointestinal tract. The activities of the gastrointestinal tract and kidneys determine the net intake and output of calcium for the entire body and, thereby, the overall state of calcium balance. In contrast, interchanges of calcium between extracellular fluid and bone do not alter total-body balance, but instead change the *distribution* of calcium within the body. We begin, therefore, with a discussion of the cellular and mineral composition of bone.

Bone

Approximately 99 percent of total-body calcium is contained in bone. Therefore, the flux of calcium into and out of bone is paramount in controlling plasma calcium concentration.

Bone is a special connective tissue made up of several cell types surrounded by a collagen matrix, called **osteoid,** upon which are deposited minerals, particularly the crystals of calcium and phosphate known as **hydroxyapatite.** In some instances, bones have central marrow cavities where blood cells form. Approximately one-third of a bone, by weight, is osteoid, and two-thirds is mineral (the bone cells contribute negligible weight).

The three types of bone cells involved in bone formation and breakdown are osteoblasts, osteocytes, and osteoclasts (**Figure 11–30**). As described in Section E, osteoblasts are

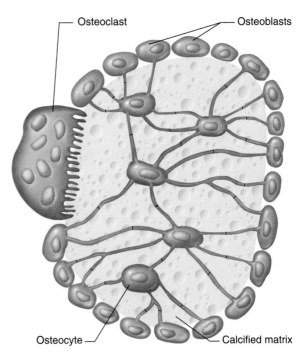

Figure 11–30

Cross section through a small portion of bone. The light tan area is mineralized osteoid. The osteocytes have long processes that extend through small canals and connect with each other and to osteoblasts via tight junctions.

Adapted from Goodman.

the bone-forming cells. They secrete collagen to form a surrounding matrix, which then becomes calcified **(mineralization).** Once surrounded by calcified matrix, the osteoblasts are called **osteocytes.** The osteocytes have long cytoplasmic processes that extend throughout the bone and form tight junctions with other osteocytes. **Osteoclasts** are large, multinucleated cells that break down (resorb) previously formed bone by secreting hydrogen ions, which dissolve the crystals, and hydrolytic enzymes, which digest the osteoid.

Throughout life, bone is constantly remodeled by the osteoblasts and osteoclasts working together. Osteoclasts resorb old bone, and then osteoblasts move into the area and lay down new matrix, which becomes mineralized. This process depends in part on the stresses that gravity and muscle tension impose on the bones, stimulating osteoblastic activity. Many hormones, as summarized in **Table 11–7**, and a variety of autocrine/paracrine growth factors produced locally in the bone also play a role. Of the hormones listed, only parathyroid hormone is controlled primarily by plasma calcium concentration. Nonetheless, changes in the other listed hormones have important influences on bone mass and plasma calcium concentration.

Kidneys

As you will learn in Chapter 14, the kidneys filter the blood in order to eliminate soluble wastes. In the process, cells in the tubules that comprise the functional units of the kidneys recapture most of the necessary solutes that got filtered, to minimize their loss in the urine. Thus, the urinary excretion of calcium is the difference between the amount filtered and the amount recaptured (reabsorbed). The control of calcium excretion is exerted mainly on reabsorption. Reabsorption decreases when plasma calcium concentration increases, and increases when plasma calcium decreases.

The hormonal controllers of calcium also regulate phosphate balance. Phosphate, too, is subject to a combination of filtration and reabsorption, with the latter hormonally controlled.

Gastrointestinal Tract

The absorption of such solutes as sodium and potassium from the gastrointestinal tract normally approximates 100 percent. In contrast, a considerable amount of ingested calcium is not absorbed from the intestine and simply leaves the body along with the feces. Moreover, the active transport system that achieves calcium absorption is under hormonal control. Thus, large regulated increases or decreases can occur in the amount of calcium absorbed from the diet. Hormonal control of this absorptive process is the major means for regulating total-body calcium balance, as we see next.

Hormonal Controls

The two major hormones that regulate plasma calcium concentration are parathyroid hormone and 1,25-dihydroxyvitamin D. A third hormone, calcitonin, plays a limited role, if any.

Parathyroid Hormone

Bone, kidneys, and the gastrointestinal tract are subject, directly or indirectly, to control by a protein hormone called **parathyroid hormone (PTH),** produced by the **parathyroid glands.** These endocrine glands are in the neck, embedded in the posterior surface of the thyroid gland, but are distinct from it (**Figure 11–31**). Parathyroid hormone production is controlled by the extracellular calcium concentration acting directly on the secretory cells via a plasma membrane calcium receptor. *Decreased* plasma calcium concentration *stimulates* parathyroid hormone secretion, and an increased plasma calcium concentration does just the opposite.

Table 11–7	Summary of Major Hormonal Influences on Bone Mass

Hormones that favor bone formation and increased bone mass

Insulin
Growth hormone
Insulin-like growth factor I (IGF-1)
Estrogen
Testosterone
Calcitonin

Hormones that favor increased bone resorption and decreased bone mass

Parathyroid hormone (chronic elevations)
Cortisol
Thyroid hormones (T_4 and T_3)

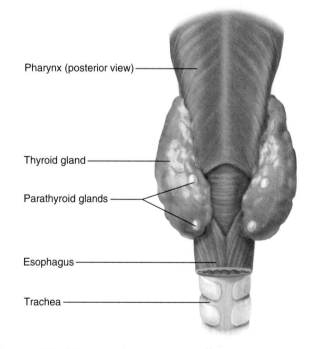

Pharynx (posterior view)

Thyroid gland

Parathyroid glands

Esophagus

Trachea

Figure 11–31

The parathyroid glands. There are usually four parathyroid glands embedded in the posterior surface of the thyroid gland.

Parathyroid hormone exerts multiple actions that increase extracellular calcium concentration, thus compensating for the decreased concentration that originally stimulated secretion of this hormone (**Figure 11–32**):

1. It directly increases the resorption of bone by osteoclasts, which causes calcium (and phosphate) to move from bone into extracellular fluid.
2. It directly stimulates the formation of another hormone, called 1,25-dihydroxyvitamin D, which then increases intestinal absorption of calcium (and phosphate). Thus, the effect of parathyroid hormone on the intestinal tract is indirect.
3. It directly increases calcium reabsorption in the kidneys, thus decreasing urinary calcium excretion.
4. It directly *reduces* the reabsorption of phosphate in the kidneys, thus raising its urinary excretion. This keeps plasma phosphate from increasing at a time when parathyroid hormone causes an increased release of both calcium and phosphate from bone, and increased 1,25-dihydroxyvitamin D is increasing both calcium and phosphate absorption in the intestine.

1,25-Dihydroxyvitamin D

The term **vitamin D** denotes a group of closely related compounds. **Vitamin D₃ (cholecalciferol)** is formed by the action of ultraviolet radiation (from sunlight, usually) on a cholesterol derivative (7-dehydrocholesterol) in skin. **Vitamin D₂ (ergocalciferol)** is derived from plants. Both can be found in vitamin pills and enriched foods and are collectively called vitamin D.

Because of clothing and decreased outdoor living, people are often dependent upon dietary vitamin D. For this reason, it was originally classified as a vitamin. Regardless of source, vitamin D is metabolized by the addition of hydroxyl groups, first in the liver by the enzyme 25-hydroxylase and then in certain kidney cells by 1-hydroxylase (**Figure 11–33**). The end result of these changes is **1,25-dihydroxyvitamin D**

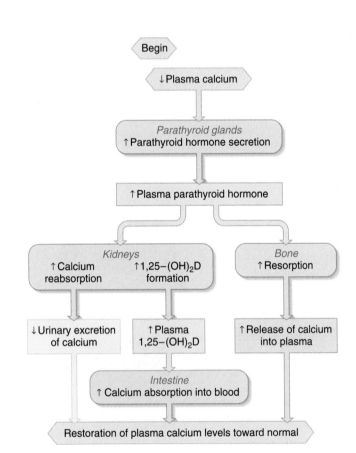

Figure 11–32

Mechanisms that allow parathyroid hormone to reverse a reduction in plasma calcium concentration toward normal. See Figure 11–33 for a more complete description of 1,25-$(OH)_2$D (1,25-dihydroxyvitamin D). Parathyroid hormone and 1,25-$(OH)_2$D are also involved in the control of phosphate levels.

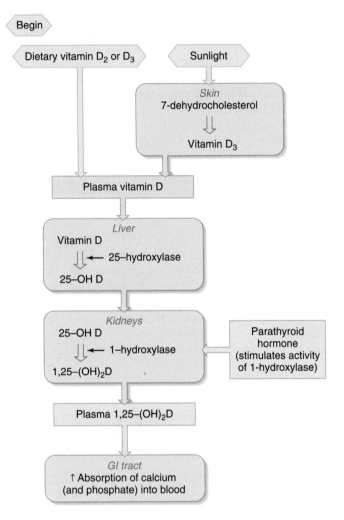

Figure 11–33

Metabolism of vitamin D to the active form, 1,25-$(OH)_2$D.

Figure 11–33 physiological *inquiry*

- Sarcoidosis is a disease that affects a variety of organs (usually the lungs). It is characterized by the development of nodules of inflamed tissue known as granulomas. These granulomas can express significant 1-hydroxylase activity that is not controlled by parathyroid hormone. What will happen to plasma calcium and parathyroid hormone levels under these circumstances?

Answer can be found at end of chapter.

(abbreviated **1,25-(OH)₂D,** also called calcitriol), the active form of vitamin D. It should be clear from this description that 1,25-(OH)₂D is synthesized in the body, so it is a hormone.

The major action of 1,25-(OH)₂D is to stimulate the intestinal absorption of calcium. Thus, the major event in vitamin D deficiency is decreased intestinal calcium absorption, resulting in decreased plasma calcium.

The blood concentration of 1,25-(OH)₂D is subject to physiological control. The major control point is the second hydroxylation step that occurs in the kidney (1-hydroxylase), which is stimulated by parathyroid hormone. Because a low plasma calcium concentration stimulates the secretion of parathyroid hormone, the production of 1,25-(OH)₂D is increased. Both hormones work together to restore plasma calcium to normal.

Calcitonin

Calcitonin is a peptide hormone secreted by cells (called parafollicular cells) that are within the thyroid gland but are distinct from the thyroid follicles. Calcitonin decreases plasma calcium concentration, mainly by inhibiting osteoclasts, thereby reducing bone resorption. Its secretion is stimulated by an increased plasma calcium concentration, just the opposite of the stimulus for parathyroid hormone secretion. Unlike parathyroid hormone and 1,25-(OH)₂D, however, calcitonin plays no role in the normal day-to-day regulation of plasma calcium regulation in humans. It may be a factor in decreasing bone resorption when plasma calcium is very high.

Metabolic Bone Diseases

Various diseases reflect abnormalities in the metabolism of bone. **Rickets** (in children) and **osteomalacia** (in adults) are conditions in which mineralization of bone matrix is deficient, causing the bones to be soft and easily fractured. In addition, a child suffering from rickets is typically severely bowlegged due to weight-bearing on the developing leg bones. A major cause of rickets and osteomalacia is deficiency of vitamin D.

In contrast to these diseases, in **osteoporosis** both matrix and minerals are lost as a result of an imbalance between bone resorption and bone formation. The resulting decrease in bone mass and strength leads to an increased incidence of fractures. Osteoporosis can occur in people who are immobilized ("disuse osteoporosis"), in people who have an excessive plasma concentration of a hormone that favors bone resorption, and in people who have a deficient plasma concentration of a hormone that favors bone formation (see Table 11–7). It is most commonly seen, however, with aging. Everyone loses bone as he or she ages, but osteoporosis is more common in elderly women than in men for several reasons: women have a smaller bone mass to begin with, and the loss that occurs with aging occurs more rapidly, particularly after menopause removes the bone-promoting influence of estrogen.

Prevention is the focus of attention for osteoporosis. Treatment of postmenopausal women with estrogen or its synthetic analogs is very effective in reducing the rate of bone loss. A regular weight-bearing exercise program, such as brisk walking and stair-climbing, is also helpful. Adequate dietary calcium (1000 to 1500 mg/day) and vitamin D intake throughout life are important to build up and maintain bone mass. Several agents also provide effective therapy once osteoporosis is established. Most prominent is a group of drugs, called **bisphosphonates,** that interfere with the resorption of bone by osteoclasts. Other antiresorptive agents include calcitonin and **selective estrogen receptor modulators (SERMs),** which, as their name implies, act by interacting with estrogen receptors, thus compensating for the low estrogen after menopause.

ADDITIONAL CLINICAL EXAMPLES

Hyper- and Hypocalcemia

There are a variety of pathophysiological disorders that lead to abnormally high (**hypercalcemia**) or low (**hypocalcemia**) calcium levels in the blood.

A common cause of hypercalcemia is **primary hyperparathyroidism.** This is usually caused by a benign tumor (adenoma) of one of the four parathyroid glands. These tumors are composed of abnormal cells that are not adequately suppressed by extracellular calcium. As a result, the adenoma secretes parathyroid hormone (PTH) in excess, leading to an increase in calcium resorption from bone, increased kidney reabsorption of calcium and the increased production of 1,25-(OH)₂D from the kidney. This results in an increase in calcium absorption in the GI tract. Primary hyperparathyroidism is most effectively treated by surgical removal of the parathyroid tumor.

Certain types of cancer can lead to **humoral hypercalcemia of malignancy.** The cause of the hypercalcemia is often the release of a molecule that is chemically similar to PTH, called **PTH-related peptide (PTHrp),** and that has effects similar to those of PTH. This chemical is produced by certain types of cancerous cells (for example, breast cancer cells). However, authentic PTH release from the normal parathyroid glands is decreased due to the hypercalcemia caused by PTHrp released from the cancer cells. The most effective treatment of humoral hypercalcemia of malignancy is to treat the cancer that is releasing PTHrp. In addition, drugs (like bisphosphonates) that decrease bone resorption can also provide effective treatment.

Finally, excessive ingestion of vitamin D can lead to hypercalcemia despite the fact that PTH levels will be very low.

Regardless of the cause, hypercalcemia causes significant symptoms primarily from its effects on excitable tissues.

Among these symptoms are tiredness and lethargy with muscle weakness, as well as nausea and vomiting (due to effects on the GI tract).

Hypocalcemia can result from a loss of parathyroid gland function (*primary hypoparathyroidism*); one cause of this is the removal of parathyroid glands that rarely occurs when a person with thyroid disease has his or her thyroid gland surgically removed. Because PTH is low, $1,25\text{-}(OH)_2D$ production from the kidney is also decreased. Decreases in both hormones lead to decreases in bone resorption, kidney calcium reabsorption, and GI calcium absorption. Resistance to the effects of PTH in target tissue can also lead to the symptoms of hypoparathyroidism, even though in such cases PTH levels in the blood tend to be elevated. This condition is called *pseudohypoparathyroidism.*

Another interesting hypocalcemic state is *secondary hyperparathyroidism.* Failure to absorb vitamin D from the gastrointestinal tract, or decreased kidney $1,25\text{-}(OH)_2D$ production, which can occur in kidney disease, can lead to secondary hyperparathyroidism. The decreased plasma calcium that results from decreased GI absorption of calcium results in stimulation of the parathyroid glands. While the increased parathyroid hormone does act to restore plasma calcium toward normal, it does so at the expense of significant loss of calcium from bone and the acceleration of metabolic bone disease.

The symptoms of hypocalcemia are also due to its effects on excitable tissue. It increases the excitability of nerves and muscles, which can lead to CNS effects (seizures), muscle spasms (*hypocalcemic tetany*), and nerve excitability. Long-term treatment of hypoparathyroidism involves giving calcium and $1,25\text{-}(OH)_2D$ (calcitriol). Poor absorption is treated with supplemental dietary calcium and high doses of dietary vitamin D. If high doses of dietary vitamin D do not overcome the problem, injections of vitamin D can be administered.■

■ **SECTION F SUMMARY**

Effector Sites for Calcium Homeostasis

I. The effector sites for the regulation of plasma calcium concentration are bone, the gastrointestinal tract, and the kidneys.
II. Approximately 99 percent of total-body calcium is contained in bone as minerals on a collagen matrix. Bone is constantly remodeled as a result of the interaction of osteoblasts and osteoclasts, a process that determines bone mass and provides a means for raising or lowering plasma calcium concentration.
III. Calcium is actively absorbed by the gastrointestinal tract, and this process is under hormonal control.
IV. The amount of calcium excreted in the urine is the difference between the amount filtered and the amount reabsorbed, the latter process being under hormonal control.

Hormonal Controls

I. Parathyroid hormone (PTH) increases plasma calcium concentration by influencing all of the effector sites.
   a. It stimulates kidney reabsorption of calcium, bone resorption with release of calcium into the blood, and formation of the hormone 1,25-dihydroxyvitamin D, which stimulates calcium absorption by the intestine.
   b. It also inhibits the reabsorption of phosphate in the kidneys, leading to increased excretion of phosphate in the urine.
II. Vitamin D is formed in the skin or ingested and then undergoes hydroxylations in the liver and kidneys. The kidneys express the enzyme that catalyzes the production of the active form, 1,25-dihydroxyvitamin D. This process is greatly stimulated by parathyroid hormone.

Metabolic Bone Disease

I. Osteomalacia (adults) and rickets (children) is a disease in which the mineralization of bone is deficient—usually due to inadequate vitamin D intake, absorption, or activation.

II. Osteoporosis is a loss of bone density (loss of matrix and minerals).
   a. Bone resorption exceeds formation.
   b. It is most common in postmenopausal (estrogen-deficient) women.
   c. It can be prevented by exercise, adequate calcium and vitamin D intake, and medications (such as bisphosphonates).

Additional Clinical Examples

I. Hypercalcemia (chronically elevated plasma calcium levels) can occur from several causes.
   a. Primary hyperparathyroidism is usually caused by a benign adenoma, which produces too much PTH. Increased PTH causes hypercalcemia by increasing bone resorption of calcium, increasing kidney reabsorption of calcium, and increasing kidney production of $1,25\text{-}(OH)_2D$, which increases calcium absorption in the intestines.
   b. Humoral hypercalcemia of malignancy is often due to the production of PTH-related peptide (PTHrp) from cancer cells. PTHrp acts like PTH.
   c. Excessive vitamin D intake may also result in hypercalcemia.
II. Hypocalcemia (chronically decreased plasma calcium levels) can also be traced to several causes.
   a. Low PTH levels from primary hypoparathyroidism (loss of parathyroid function) lead to hypocalcemia by decreasing bone resorption of calcium, decreasing urinary reabsorption of calcium, and decreasing renal production of $1,25\text{-}(OH)_2D$.
   b. Pseudohypoparathyroidism is caused by target organ resistance to the action of PTH.
   c. Secondary hyperparathyroidism is caused by vitamin D deficiency due to inadequate intake, absorption, or activation in the kidney (e.g. in kidney disease).

■ SECTION F REVIEW QUESTIONS

1. Describe bone remodeling.
2. Describe the handling of calcium by the kidneys and gastrointestinal tract.
3. What controls the secretion of parathyroid hormone, and what are this hormone's four major effects?
4. Describe the formation and action of 1,25-$(OH)_2$D. How does parathyroid hormone influence the production of this hormone?

Chapter 11 Test Questions

(Answers appear in Appendix A.)

1–5: Match the hormone with the function or feature (choices a–e):

Hormone:

1. vasopressin
2. ACTH
3. oxytocin
4. prolactin
5. luteinizing hormone

Function:

a. tropic for the adrenal cortex
b. is controlled by an amine-derived hormone of the hypothalamus
c. antidiuresis
d. stimulation of testosterone production
e. stimulation of uterine contractions during labor

6. In the figure below, which hormone (A or B) binds to receptor X with higher affinity?

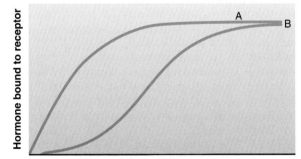

Concentration of free hormone

(y-axis: **Hormone bound to receptor**)

7. Which is NOT a symptom of Cushing's disease?
a. high blood pressure
b. bone loss
c. suppressed immune function
d. goiter
e. hyperglycemia (increased blood glucose)

8. Tremors, nervousness, and increased heart rate can all be symptoms of
a. increased activation of the sympathetic nervous system.
b. excessive secretion of epinephrine from the adrenal medulla.
c. hyperthyroidism.
d. hypothyroidism.
e. Answers a, b, and c are all correct.

9. Which of the following could theoretically result in dwarfism?
a. pituitary tumor making excess thyroid-stimulating hormone
b. inactivating mutations of IGF-1 receptors
c. delayed onset of puberty
d. decreased hypothalamic levels of somatostatin
e. normal plasma GH, but decreased feedback of growth hormone on GHRH

10. Choose the correct statement.
a. During times of stress, cortisol acts as an anabolic hormone in muscle and adipose tissue.
b. A deficiency of thyroid hormones would result in increased cellular concentrations of Na^+/K^+-ATPase pumps in target tissues.
c. The posterior pituitary is connected to the hypothalamus by long portal vessels.
d. The major way in which cholesterol is delivered to steroidogenic glands is by diffusion of free cholesterol through the plasma membrane.
e. A lack of iodide in the diet will have no significant effect on circulating thyroid hormones for at least several weeks.

11. Low plasma calcium intake causes
a. a PTH-mediated increase in 25-OH D.
b. a decrease in renal 1-hydroxylase activity.
c. a decrease in the urinary excretion of calcium.
d. a decrease in bone resorption.
e. an increase in vitamin D release from the skin.

12. Which of the following is *not* consistent with primary hyperparathyroidism?
    a. hypercalcemia
    b. elevated plasma 1,25-$(OH)_2$D
    c. phosphaturia
    d. a decrease in calcium resorption from bone
    e. an increase in calcium reabsorption in the kidney

True/False

13. T_4 is the chief circulating form of thyroid hormone, but is less active than T_3.

14. Acromegaly is usually associated with hypoglycemia and hypotension.

15. Thyroid hormones and cortisol are both permissive for epinephrine's actions.

Chapter 11 Quantitative and Thought Questions

(Answers appear in Appendix A.)

1. In an experimental animal, the sympathetic preganglionic fibers to the adrenal medulla are cut. What happens to the plasma concentration of epinephrine at rest and during stress?

2. During pregnancy there is an increase in the liver's production and, consequently, the plasma concentration of the major plasma binding protein for the thyroid hormones (TH). This causes a sequence of events involving feedback that results in an increase in the plasma concentration of TH, but no evidence of hyperthyroidism. Describe the sequence of events.

3. A child shows the following symptoms: deficient growth; failure to show sexual development; decreased ability to respond to stress. What is the most likely cause of all these symptoms?

4. If all the neural connections between the hypothalamus and pituitary were severed, the secretion of which pituitary hormones would be affected? Which pituitary hormones would not be affected?

5. Typically, an antibody to a peptide combines with the peptide and renders it nonfunctional. If an animal were given an antibody to somatostatin, the secretion of which anterior pituitary hormone would change and in what direction?

6. A drug that blocks the action of norepinephrine is injected directly into the hypothalamus of an experimental animal, and the secretion rates of several anterior pituitary hormones are observed to change. How is this possible, given the fact that norepinephrine is not a hypophysiotropic hormone?

7. A person is receiving very large doses of a cortisol-like drug to treat her arthritis. What happens to her secretion of cortisol?

8. A person with symptoms of hypothyroidism (for example, sluggishness and intolerance to cold) is found to have abnormally low plasma concentrations of T_4, T_3, and TSH. After an injection of TRH, the plasma concentrations of all three hormones increase. Where is the site of the defect leading to the hypothyroidism?

9. A full-term newborn infant is abnormally small. Is this most likely due to deficient growth hormone, deficient thyroid hormones, or deficient nutrition during fetal life?

10. Why might the administration of androgens to stimulate growth in a small 12-year-old male turn out to be counterproductive?

Chapter 11 Answers to Physiological Inquiries

Figure 11–3 By storing large amounts of hormone in an endocrine cell, the plasma level of the hormone can be increased within seconds when the cell is stimulated. Such rapid responses may be critical for an appropriate response to a challenge to homeostasis. Packaging peptides in this way also prevents intracellular degradation.

Figure 11–13 Because the amount of blood into which the hypophysiotropic hormones are secreted is far less than would be the case if they were secreted into the general circulation of the body, the absolute amount of hormone required to achieve a given concentration is much less. This means that the cells of the hypothalamus need only synthesize a tiny amount of hypophysiotropic hormone to reach concentrations in the portal blood vessels that are physiologically active (i.e., can activate receptors on pituitary cells). This allows for the tight control of the anterior pituitary gland by a very small number of discrete neurons within the hypothalamus.

Figure 11–19 Plasma cortisol levels would increase. This would result in decreased ACTH levels in the systemic blood, and

CRH levels in the portal vein blood, due to increased negative feedback at the pituitary and hypothalamus, respectively. The right adrenal gland would shrink in size (atrophy) as a consequence of the reduced ACTH levels (decreased "trophic" stimulation of the adrenal cortex).

Figure 11–29 His elevated height compared to the height of his identical twin indicates that growth hormone had been elevated in his blood prior to puberty, resulting in gigantism. Once growth ceased, his features enlarged due to continued excess GH, resulting in acromegaly. This patient is referred to as an "acromegaloid giant."

Figure 11–33 The 1-hydroxylase activity will stimulate the conversion of 25-OH D to 1,25-$(OH)_2$D in the granulomas themselves; the 1,25-$(OH)_2$D will then diffuse out of the granuloma cells and enter the plasma, leading to increased calcium absorption in the gastrointestinal tract. This will increase plasma calcium, which in turn will suppress parathyroid hormone production and, consequently, plasma parathyroid hormone levels will decrease. This is a form of secondary hypoparathyroidism.

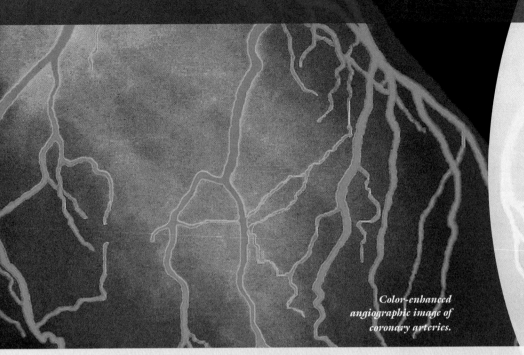

chapter

12

Cardiovascular Physiology

Color-enhanced angiographic image of coronary arteries.

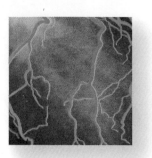

beyond a distance of a few cell diameters, diffusion—the random movement of substances from a region of higher concentration to one of lower concentration—is too slow to meet the metabolic requirements of cells. In large, multicellular organisms, therefore, some mechanism other than diffusion is needed to transport molecules rapidly over the long distances between internal cells and the body's surface, and between the various specialized tissues and organs. This purpose is achieved by the **circulatory system,** which includes a pump (the **heart**), a set of interconnected tubes (**blood vessels,** or **vascular system**), and a mixture of extracellular fluid and cells that fills the tubes (the **blood**). This body-wide transport system is often termed the **cardiovascular system.**

SECTION A Overall Design of the Circulatory System

System Overview

The three principal components that make up the circulatory system are the heart, blood vessels, and the blood itself. Each will be discussed in greater detail in subsequent sections, but we will begin with an overview of the anatomical design of the system and a discussion of some of the physical factors that determine its function.

Blood is composed of **formed elements** (cells and cell fragments) suspended in a liquid called **plasma.** Dissolved in the plasma are a large number of proteins, nutrients, metabolic wastes, and other molecules being transported between organ systems. The cells are the **erythrocytes** (red blood cells) and the **leukocytes** (white blood cells), and the cell fragments are the **platelets.** More than 99 percent of blood cells are erythrocytes, which carry oxygen. The leukocytes protect against infection and cancer, and the platelets function in blood clotting. The constant motion of the blood keeps all the cells dispersed throughout the plasma.

The **hematocrit** is defined as the percentage of blood volume that is erythrocytes. It is measured by centrifuging (spinning at high speed) a sample of blood. The erythrocytes are forced to the bottom of the centrifuge tube, the plasma remains on top, and the leukocytes and platelets form a very thin layer between them (**Figure 12–1**). The normal hematocrit is approximately 45 percent in men and 42 percent in women.

The volume of blood in a 70-kg (154-pound) person is approximately 5.5 L. If we take the hematocrit to be 45 percent, then:

$$\text{Erythrocyte volume} = 0.45 \times 5.5\ \text{L} = 2.5\ \text{L}$$

Because the volume occupied by leukocytes and platelets is normally negligible, the plasma volume equals the difference between blood volume and erythrocyte volume; therefore, in our 70-kg person:

$$\text{Plasma volume} = 5.5\ \text{L} - 2.5\ \text{L} = 3.0\ \text{L}$$

The rapid flow of blood throughout the body is produced by pressures created by the pumping action of the heart.

This type of flow is known as **bulk flow** because all constituents of the blood move in one direction together. The extraordinary degree of branching of blood vessels ensures that almost all cells in the body are within a few cell diameters of at least one of the smallest branches, the capillaries. Nutrients and metabolic end products move between capillary blood and the interstitial fluid by diffusion. Movements

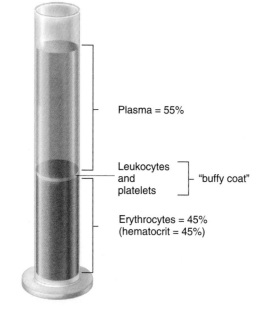

Plasma = 55%

Leukocytes
and
platelets

"buffy coat"

Erythrocytes = 45%
(hematocrit = 45%)

Figure 12–1

Measurement of the hematocrit by centrifugation. The values shown are typical for a healthy male. Due to the presence of a thin layer of leukocytes and platelets between the plasma and red cells, the value for plasma is actually very slightly less than 55 percent.

Figure 12–1 physiological *inquiry*

■ Estimate the hematocrit of a person with a plasma volume of 3 L and total blood volume of 4.5 L.

Answer can be found at end of chapter.

between the interstitial fluid and the cell interior are accomplished by both diffusion and mediated transport across the plasma membrane.

At any given moment, only about 5 percent of the total circulating blood is actually in the capillaries. Yet it is this 5 percent that is performing the ultimate functions of the entire cardiovascular system: the supplying of nutrients and the removal of metabolic end products and other cell secretions. All other components of the system serve the overall function of getting adequate blood flow through the capillaries.

As British physiologist William Harvey reported in 1628, the cardiovascular system forms a closed loop, so that blood pumped out of the heart through one set of vessels returns to the heart by a different set. There are actually two circuits (**Figure 12–2**), both originating and terminating in the heart, which is divided longitudinally into two functional halves. Each half of the heart contains two chambers: an upper chamber—the **atrium**—and a lower chamber—the **ventricle.** The atrium on each side empties into the ventricle on that side, but there is no direct blood flow between the two atria or the two ventricles in the adult heart.

The **pulmonary circulation** includes blood pumped from the right ventricle through the lungs and then to the left atrium. It is then pumped through the **systemic circulation** from the left ventricle through all the organs and tissues of the body except the lungs, and then to the right atrium. In both circuits, the vessels carrying blood away from the heart are called **arteries,** and those carrying blood from body organs and tissues back toward the heart are called **veins.**

In the systemic circuit, blood leaves the left ventricle via a single large artery, the **aorta** (see Figure 12–2). The arteries of the systemic circulation branch off the aorta, dividing into progressively smaller vessels. The smallest arteries branch into **arterioles,** which branch into a huge number (estimated at 10 billion) of very small vessels, the **capillaries,** which unite to form larger-diameter vessels, the **venules.** The arterioles, capillaries, and venules are collectively termed the **microcirculation.**

The venules in the systemic circulation then unite to form larger vessels, the veins. The veins from the various peripheral organs and tissues unite to produce two large veins, the **inferior vena cava,** which collects blood from below the heart, and the **superior vena cava,** which collects blood from above the heart (for simplicity, these are depicted as a single vessel in Figure 12–2). These two veins return the blood to the right atrium.

The pulmonary circulation is composed of a similar circuit. Blood leaves the right ventricle via a single large artery, the **pulmonary trunk,** which divides into the two **pulmonary arteries,** one supplying the right lung and the other the left. In the lungs, the arteries continue to branch, ultimately forming capillaries that unite into venules and then veins. The blood leaves the lungs via four **pulmonary veins,** which empty into the left atrium.

As blood flows through the lung capillaries, it picks up oxygen supplied to the lungs by breathing. Therefore, the blood in the pulmonary veins, left side of the heart, and systemic

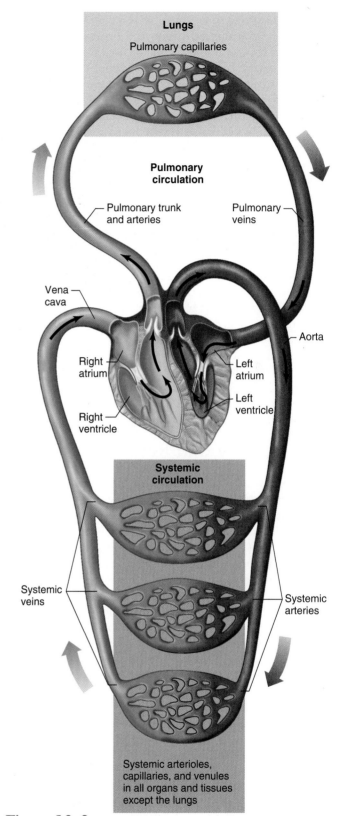

Figure 12–2

The systemic and pulmonary circulations. As depicted by the color change from blue to red, blood becomes fully oxygenated (red) as it flows through the lungs and then loses some oxygen (red to blue) as it flows through the other organs and tissues. For simplicity, the arteries and veins leaving and entering the heart are depicted as single vessels; in reality, this is true for the arteries but there are multiple pulmonary veins and two venae cavae (see Figure 12–6).

arteries has a high oxygen content. As this blood flows through the capillaries of peripheral tissues and organs, some of this oxygen leaves the blood to enter and be used by cells, resulting in the lower oxygen content of systemic venous blood.

As shown in Figure 12–2, blood can pass from the systemic veins to the systemic arteries only by first being pumped through the lungs. Thus the blood returning from the body's peripheral organs and tissues via the systemic veins is oxygenated before it is pumped back to them.

Note that the lungs receive all the blood pumped by the right side of the heart, whereas the branching of the systemic arteries results in a parallel pattern so that each of the peripheral organs and tissues receives only a fraction of the blood pumped by the left ventricle (see the three capillary beds shown in Figure 12–2). This arrangement guarantees that all systemic tissues receive freshly oxygenated blood, and allows for independent variation in blood flow through different tissues as their metabolic activities change. For reference, the typical distribution of the blood pumped by the left ventricle in an adult at rest is given in **Figure 12–3**.

Finally, there are some exceptions to the usual anatomical pattern described in this section for the systemic circulation, notably the liver and the pituitary gland. In those organs, blood passes through two capillary beds, arranged in series, before returning to the heart. As described in Chapter 11, this pattern is known as a **portal system.**

Pressure, Flow, and Resistance

An important feature of the cardiovascular system is the relationship among blood pressure, blood flow, and the resistance to blood flow. As applied to blood, these factors are collectively referred to as **hemodynamics.** In all parts of the system, blood flow (F) is always from a region of higher pressure to one of lower pressure. The pressure exerted by any fluid is called a **hydrostatic pressure,** but this is usually shortened simply to "pressure" in descriptions of the cardiovascular system, and it denotes the force exerted by the blood. This force is generated in the blood by the contraction of the heart, and its magnitude varies throughout the system for reasons later sections will describe. The units for the rate of flow are volume per unit time, usually liters per minute (L/min). The units for the pressure difference (ΔP) driving the flow are millimeters of mercury (mmHg) because historically blood pressure was measured by determining how high the blood pressure could drive a column of mercury. It is not the absolute pressure at any point in the cardiovascular system that determines flow rate, but the difference in pressure between the relevant points (**Figure 12–4**).

Knowing only the pressure difference between two points will not tell you the flow rate, however. For this, you also need to know the **resistance (R)** to flow—that is, how difficult it is for blood to flow between two points at any given pressure difference. Resistance is the measure of the friction that impedes flow. The basic equation relating these variables is:

$$F = \Delta P/R \qquad (12\text{--}1)$$

In words, flow rate is directly proportional to the pressure difference between two points and inversely proportional to the resistance. This equation applies not only to the cardiovascular system, but to any system in which liquid or air moves by bulk flow (e.g., in the urinary and respiratory systems).

Organ	Flow at rest ml/min
Brain	650 (13%)
Heart	215 (4%)
Skeletal muscle	1030 (20%)
Skin	430 (9%)
Kidneys	950 (20%)
Abdominal organs	1200 (24%)
Other	525 (10%)
Total	5000 (100%)

Figure 12–3

Distribution of systemic blood flow to the various organs and tissues of the body at rest. Figure 12–61 shows blood flow changes during exercise.

Adapted from Chapman and Mitchell.

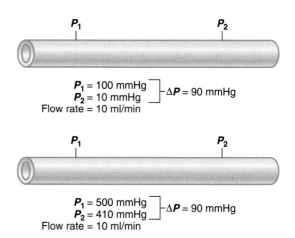

Figure 12–4

Flow between two points within a tube is proportional to the pressure difference between the points. The flows in these two identical tubes are the same (10 mL/min) because the pressure *differences* are the same.

Resistance cannot be measured directly, but it can be calculated from the directly measured F and ΔP. For example, in Figure 12–4 the resistances in both tubes can be calculated to be 90 mmHg ÷ 10 ml/min = 9 mmHg/ml per minute.

This example illustrates how resistance can be calculated, but what is it that actually determines the resistance? The distinction between how a thing is calculated or measured and its determinants may seem confusing, but consider the following: By standing on a scale you *measure* your weight, but your weight is not *determined* by the scale but rather by how much you eat and exercise, and so on. One determinant of resistance is the fluid property known as **viscosity,** which is a function of the friction between molecules of a flowing fluid; the greater the friction, the greater the viscosity. The other determinants of resistance are the length and radius of the tube through which the fluid is flowing, because these characteristics affect the surface area inside the tube and thus determine the amount of friction between the fluid and the wall of the tube. The following equation defines the contributions of these three determinants:

$$R = \frac{8L\eta}{\pi r^4} \qquad (12\text{–}2)$$

where η = fluid viscosity
L = length of the tube
r = inside radius of the tube
8/π = a mathematical constant

In other words, resistance is directly proportional to both the fluid viscosity and the vessel's length, and inversely proportional to the fourth power of the vessel's radius.

Blood viscosity is not fixed but increases as hematocrit increases. Changes in hematocrit, therefore, can have significant effects on resistance to flow in certain situations. In extreme dehydration, for example, the reduction in body water leads to a relative increase in hematocrit and thus, in the viscosity of the blood. Under most physiological conditions, however, the hematocrit and, thus, the viscosity of blood is relatively constant and does not play a role in controlling resistance.

Similarly, because the lengths of the blood vessels remain constant in the body, length is also not a factor in the control of resistance along these vessels. In contrast, the radii of the blood vessels do not remain constant, and so vessel radius—the $1/r^4$ term in our equation—is the most important determinant of changes in resistance along the blood vessels. **Figure 12–5** demonstrates how radius influences the frictional resistance and thus the flow of fluids through a tube. Decreasing the radius of a tube twofold increases its resistance sixteenfold. If ΔP is held constant in this example, flow through the tube decreases 16-fold because $F = \Delta P/R$.

Note that the equation relating pressure, flow, and resistance applies not only to flow through blood vessels, but also to the flows into and out of the various chambers of the heart. These flows occur through valves, and the resistance a valvular opening offers determines the flow through the valve at any given pressure difference across it.

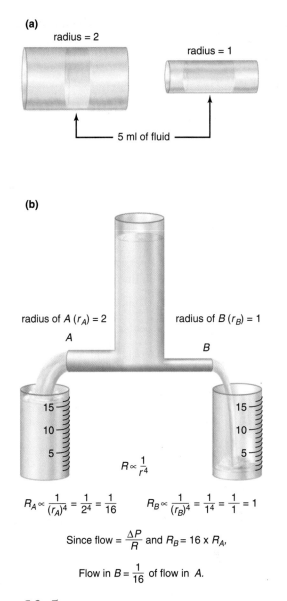

(a)

radius = 2

radius = 1

5 ml of fluid

(b)

radius of A (r_A) = 2

A

radius of B (r_B) = 1

B

$R \propto \dfrac{1}{r^4}$

$R_A \propto \dfrac{1}{(r_A)^4} = \dfrac{1}{2^4} = \dfrac{1}{16}$ $R_B \propto \dfrac{1}{(r_B)^4} = \dfrac{1}{1^4} = \dfrac{1}{1} = 1$

Since flow = $\dfrac{\Delta P}{R}$ and $R_B = 16 \times R_A$,

Flow in B = $\dfrac{1}{16}$ of flow in A.

Figure 12–5

Effect of tube radius (r) on resistance (R) and flow. (a) A given volume of fluid is exposed to far more surface friction against the walls of a smaller tube. (b) Given the same pressure gradient, flow through a tube decreases 16-fold when the radius is halved.

Figure 12–5 physiological *inquiry*

- If outlet B in Figure 12–5b had two individual outlet tubes, each with a radius of 1, would the flow be equal to side A?

Answer can be found at end of chapter.

As you read on, remember that *the ultimate function of the cardiovascular system is to ensure adequate blood flow through the capillaries of various organs.* Refer to the summary in **Table 12–1** as you read the description of each component to focus on how they contribute to this goal.

Table 12–1	The Cardiovascular System
Component	**Function**
Heart	
Atria	Chambers through which blood flows from veins to ventricles. Atrial contraction adds to ventricular filling but is not essential for it.
Ventricles	Chambers whose contractions produce the pressures that drive blood through the pulmonary and systemic vascular systems and back to the heart.
Vascular system	
Arteries	Low-resistance tubes conducting blood to the various organs with little loss in pressure. They also act as pressure reservoirs for maintaining blood flow during ventricular relaxation.
Arterioles	Major sites of resistance to flow; responsible for the pattern of blood flow distribution to the various organs; participate in the regulation of arterial blood pressure.
Capillaries	Major sites of nutrient, metabolic end product, and fluid exchange between blood and tissues.
Venules	Sites of nutrient, metabolic end product, and fluid exchange between blood and tissues.
Veins	Low-resistance conduits for blood flow back to the heart. Their capacity for blood is adjusted to facilitate this flow.
Blood	
Plasma	Liquid portion of blood that contains dissolved nutrients, ions, wastes, gases, and other substances. Its composition equilibrates with that of the interstitial fluid at the capillaries.
Cells	Includes erythrocytes that function mainly in gas transport, leukocytes that function in immune defenses, and platelets (cell fragments) for blood clotting.

■ SECTION A SUMMARY

System Overview

I. The key components of the circulatory system are the heart, blood vessels, and blood.

II. The cardiovascular system consists of two circuits: the pulmonary circulation—from the right ventricle to the lungs and then to the left atrium, and the systemic circulation—from the left ventricle to all peripheral organs and tissues and then to the right atrium.

III. Arteries carry blood away from the heart, and veins carry blood toward the heart.

  a. In the systemic circuit, the large artery leaving the left side of the heart is the aorta, and the large veins emptying into the right side of the heart are the superior vena cava and inferior vena cava. The analogous vessels in the pulmonary circulation are the pulmonary trunk and the four pulmonary veins.

  b. The microcirculation consists of the vessels between arteries and veins: the arterioles, capillaries, and venules.

Pressure, Flow, and Resistance

I. Flow between two points in the cardiovascular system is directly proportional to the pressure difference between those points and inversely proportional to the resistance: $F = \Delta P / R$.

II. Resistance is directly proportional to the viscosity of a fluid and to the length of the tube. It is inversely proportional to the fourth power of the tube's radius, which is the major variable controlling changes in resistance.

■ SECTION A KEY TERMS

aorta 361	leukocytes 360
arteriole 361	microcirculation 361
artery 361	plasma 360
atrium 361	platelet 360
blood 360	portal system 362
blood vessels 360	pulmonary artery 361
bulk flow 360	pulmonary circulation 361
capillary 361	pulmonary trunk 361
cardiovascular system 360	pulmonary vein 361
circulatory system 360	resistance (R) 362
erythrocytes 360	superior vena cava 361
formed elements 360	systemic circulation 361
heart 360	vascular system 360
hematocrit 360	vein 361
hemodynamics 362	ventricle 361
hydrostatic pressure 362	venule 361
inferior vena cava 361	viscosity 363

■ SECTION A REVIEW QUESTIONS

1. What is the oxygen status of arterial and venous blood in the systemic versus the pulmonary circulation?
2. State the formula relating flow, pressure difference, and resistance.
3. What are the three determinants of resistance?
4. Which determinant of resistance is varied physiologically to alter blood flow?
5. How does variation in hematocrit influence the hemodynamics of blood flow?

Anatomy

The heart is a muscular organ enclosed in a fibrous sac, the **pericardium,** and located in the chest (**Figure 12–6**). The inner layer of the pericardium is closely affixed to the heart and is called the **epicardium.** The extremely narrow space between the outer wall of the pericardium and the epicardium is filled with a watery fluid that serves as a lubricant as the heart moves within the sac.

The walls of the heart, the **myocardium,** are composed primarily of cardiac muscle cells. The inner surface of the cardiac chambers, as well as the inner wall of all blood vessels, is lined by a thin layer of cells known as **endothelial cells,** or **endothelium.**

As noted earlier, the human heart is divided into right and left halves, each consisting of an atrium and a ventricle. The two ventricles are separated by a muscular wall, the **interventricular septum.** Located between the atrium and ventricle in each half of the heart are the **atrioventricular (AV) valves,** which permit blood to flow from atrium to ventricle but not from ventricle to atrium. The right AV valve is called the **tricuspid valve** because it has three fibrous flaps, or

cusps (**Figure 12–7**). The left AV valve has two flaps and is thus called the **bicuspid valve.** (Its resemblance to a bishop's headgear has earned the left AV valve another commonly used name, **mitral valve.**)

The opening and closing of the AV valves is a passive process resulting from pressure differences across the valves. When the blood pressure in an atrium is greater than in the ventricle, the valve is pushed open and blood flows from atrium to ventricle. In contrast, when a contracting ventricle achieves an internal pressure greater than that in its connected atrium, the AV valve between them is forced closed. Therefore, blood does not normally move back into the atria, but is forced into the pulmonary trunk from the right ventricle and into the aorta from the left ventricle.

To prevent the AV valves from being pushed up into the atria (a condition called *prolapse*), the valves are fastened to muscular projections **(papillary muscles)** of the ventricular walls by fibrous strands **(chordae tendineae).** The papillary muscles do not open or close the valves. They act only to limit the valves' movements and prevent them from being everted.

The openings of the right ventricle into the pulmonary trunk and of the left ventricle into the aorta also contain valves,

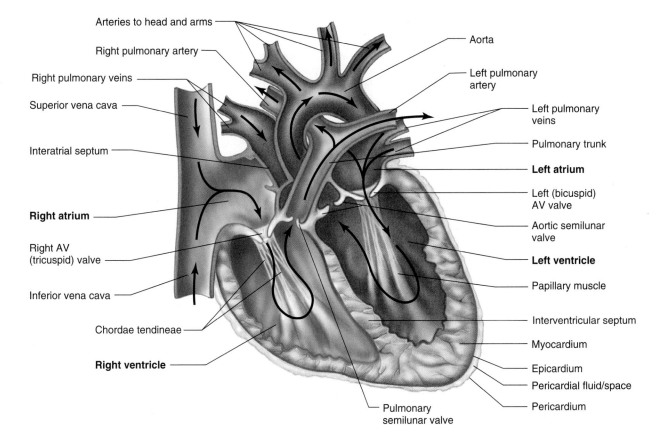

Figure 12–6

Diagrammatic section of the heart. The arrows indicate the direction of blood flow.

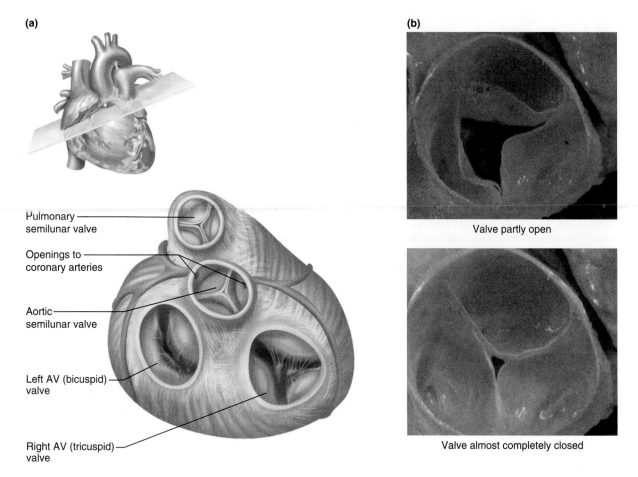

(a)

Pulmonary
semilunar valve

Openings to
coronary arteries

Aortic
semilunar valve

Left AV (bicuspid)
valve

Right AV (tricuspid)
valve

(b)

Valve partly open

Valve almost completely closed

Figure 12–7

Valves of the heart. (a) Superior view of the heart with the atria removed. (b) Photographs of the pulmonary valve from the pulmonary trunk looking down into the right ventricle. In the first photo, the valve is in the process of opening as blood flows through it from the right ventricle into the pulmonary trunk. In the second photo, the valve is in the process of closing, the cusps being forced together when the pressure of the blood in the pulmonary trunk is greater than the pressure in the right ventricle.

From R. Carola, J. P. Harley, and C. R. Noback, *Human Anatomy and Physiology,* McGraw-Hill, New York, 1990 (photos by Dr. Wallace McAlpine).

the **pulmonary** and **aortic valves,** respectively (see Figures 12–6 and 12–7). These valves are also referred to as the semilunar valves, due to the half-moon shape of the cusps. These valves permit blood to flow into the arteries during ventricular contraction but prevent blood from moving in the opposite direction during ventricular relaxation. Like the AV valves, they act in a purely passive manner. Whether they are open or closed depends upon the pressure differences across them. Vessels supplying the heart with oxygenated blood originate from behind the cusps of the aortic valve.

Another important point concerning the heart valves is that, when open, they offer very little resistance to flow. Consequently, very small pressure differences across them suffice to produce large flows. In disease states, however, a valve may become narrowed so that it offers a high resistance to flow even when open. In such a state, the contracting cardiac chamber must produce an unusually high pressure to cause flow across the valve.

There are no valves at the entrances of the superior and inferior venae cavae (plural of vena cava) into the right atrium,

and of the pulmonary veins into the left atrium. However, atrial contraction pumps very little blood back into the veins because atrial contraction constricts their sites of entry into the atria, greatly increasing the resistance to backflow. (Actually, a little blood is ejected back into the veins, and this accounts for the venous pulse that can often be seen in the neck veins when the atria are contracting.)

Figure 12–8 summarizes the path of blood flow through the entire cardiovascular system.

Cardiac Muscle

The cardiac muscle cells of the myocardium are arranged in layers that are tightly bound together and completely encircle the blood-filled chambers. When the walls of a chamber contract, they come together like a squeezing fist and exert pressure on the blood they enclose.

Like smooth and skeletal muscle, cardiac muscle is an excitable tissue that converts chemical energy in the form of ATP into force generation. Action potentials propagate along cell membranes, calcium enters the cytosol, and the cycling of

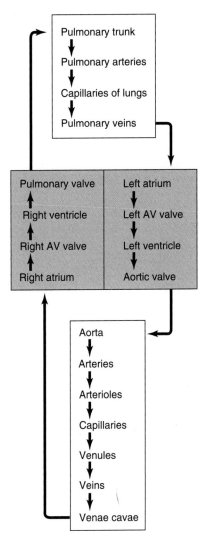

Figure 12–8

Path of blood flow through the entire cardiovascular system. All the structures within the colored box are located in the heart.

Figure 12–8 physiological *inquiry*

- How would this diagram be different if it was drawn for a systemic portal vessel?

Answer can be found at end of chapter.

force-generating cross-bridges is activated. Details of the cellular structure and function of cardiac muscle are discussed in Chapter 9.

Approximately 1 percent of cardiac cells do not function in contraction, but have specialized features that are essential for normal heart excitation. These cells constitute a network known as the **conducting system** of the heart and are in contact with the cardiac muscle cells via gap junctions. The conducting system initiates the heartbeat and helps spread the impulse rapidly throughout the heart.

One final point about cardiac muscle is that certain cells in the atria secrete the family of peptide hormones collectively called atrial natriuretic peptide, described in Chapter 14.

Innervation

The heart receives a rich supply of sympathetic and parasympathetic nerve fibers, the latter contained in the vagus nerves (**Figure 12–9**). The sympathetic postganglionic fibers, which innervate the entire heart, release primarily norepinephrine, whereas the parasympathetics terminate mainly on cells found in the atria and release primarily acetylcholine. The receptors for norepinephrine on cardiac muscle are mainly beta-adrenergic. The hormone epinephrine, from the adrenal medulla, combines with the same receptors as norepinephrine and exerts the same actions on the heart. The receptors for acetylcholine are of the muscarinic type.

Blood Supply

The blood being pumped through the heart chambers does not exchange nutrients and metabolic end products with the myocardial cells. They, like the cells of all other organs, receive their blood supply via arteries that branch from the aorta. The arteries supplying the myocardium are the **coronary arteries,** and the blood flowing through them is the **coronary blood flow.** The coronary arteries exit from the very first part of the aorta (see Figure 12–7a) and lead to a branching network of small arteries, arterioles, capillaries, venules, and veins similar to those in other organs. Most of the cardiac veins drain into a single large vein, the coronary sinus, which empties into the right atrium.

Heartbeat Coordination

The heart is a dual pump in that the left and right sides of the heart pump blood separately, but simultaneously, into the systemic and pulmonary vessels. Efficient pumping of blood requires that the atria contract first, followed almost immediately by the ventricles. Contraction of cardiac muscle, like that of skeletal muscle and many smooth muscles, is triggered by depolarization of the plasma membrane. Gap junctions interconnect myocardial cells and allow action potentials to spread from one cell to another. Thus, the initial excitation of one cardiac cell eventually results in the excitation of all

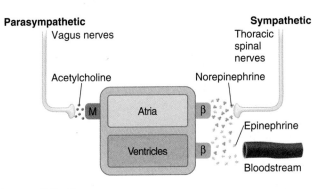

Figure 12–9

Autonomic innervation of heart. Neurons shown represent postganglionic neurons in the pathways. M = muscarinic-type ACh receptor; β = beta-adrenergic receptor.

cardiac cells. This initial depolarization normally arises in a small group of conducting-system cells called the **sinoatrial (SA) node,** located in the right atrium near the entrance of the superior vena cava (**Figure 12–10**). The action potential then spreads from the SA node throughout the atria and then into and throughout the ventricles. This pattern raises two questions: (1) what is the path of spread of excitation, and (2) what causes the SA node to "fire"? We'll deal initially with the first question and then return to the second question in the next section.

Sequence of Excitation

The SA node is the normal pacemaker for the entire heart. Its depolarization normally generates the action potential that leads to depolarization of all other cardiac muscle cells, and so its discharge rate determines the **heart rate,** the number of times the heart contracts per minute.

The action potential initiated in the SA node spreads throughout the myocardium, passing from cell to cell by way of gap junctions. Depolarization first spreads through the muscle cells of the atria, with conduction rapid enough that the right and left atria contract at essentially the same time.

The spread of the action potential to the ventricles involves a more complicated conducting system (see Figure 12–10 and **Figure 12–11**). The link between atrial depolarization and ventricular depolarization is a portion of the conducting system called the **atrioventricular (AV) node,** located at the base of

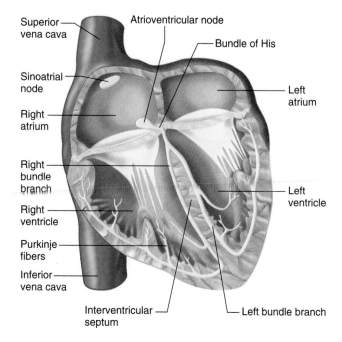

Figure 12–10

Conducting system of the heart.

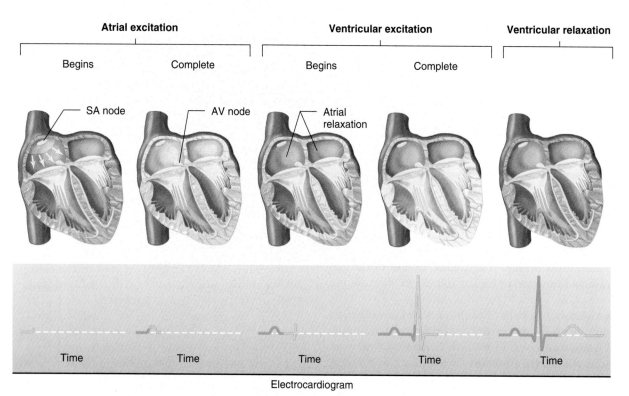

Figure 12–11

Sequence of cardiac excitation. The yellow color denotes areas that are depolarized. Impulse spreads from right atrium to left atrium via the atrial muscle cells where the atria share a wall. The electrocardiogram monitors the spread of the signal.

Adapted from Rushmer.

Chapter 12

the right atrium. The action potential spreading through the muscle cells of the right atrium causes depolarization of the AV node. This node has a particularly important characteristic: *The propagation of action potentials through the AV node is relatively slow* (requiring approximately 0.1 s). This delay allows atrial contraction to be completed before ventricular excitation occurs.

After leaving the AV node, the action potential propagates down the wall—the interventricular septum—between the two ventricles. This pathway has conducting-system fibers called the **bundle of His** (or atrioventricular bundle) after its discoverer (pronounced *Hiss*). The AV node and the bundle of His constitute the only electrical connection between the atria and the ventricles. Except for this pathway, the atria are completely separated from the ventricles by a layer of nonconducting connective tissue.

Within the interventricular septum, the bundle of His divides into **right and left bundle branches,** which eventually leave the septum to enter the walls of both ventricles. These fibers in turn make contact with **Purkinje fibers,** large conducting cells that rapidly distribute the impulse throughout much of the ventricles. Finally, the Purkinje fibers make contact with ventricular myocardial cells, which spread the impulse through the rest of the ventricles.

The rapid conduction along the Purkinje fibers and the diffuse distribution of these fibers cause depolarization of all right and left ventricular cells more or less simultaneously and ensures a single coordinated contraction. Actually, though, depolarization and contraction do begin slightly earlier in the bottom (apex) of the ventricles and then spread upward. The result is an efficient contraction that moves blood toward the exit valves, like squeezing a tube of toothpaste from the bottom up.

Cardiac Action Potentials and Excitation of the SA Node

The mechanism by which action potentials are conducted along the membranes of heart cells is basically similar to that seen in other excitable tissues like neurons and skeletal muscle cells. However, different types of heart cells express unique combinations of ion channels that produce different action potential shapes. This specializes them for particular roles in the spread of excitation through the heart.

Figure 12–12a illustrates a typical ventricular myocardial cell action potential. The plasma membrane permeability changes that underlie it are shown in **Figure 12–12b**. As in skeletal muscle cells and neurons, the resting membrane is much more permeable to potassium than to sodium. Therefore, the resting membrane potential is much closer to the potassium equilibrium potential (–90 mV) than to the sodium equilibrium potential (+60 mV). Similarly, the depolarizing phase of the action potential is due mainly to the opening of voltage-gated sodium channels. Sodium entry depolarizes the cell and sustains the opening of more sodium channels in positive feedback fashion. At almost the same time, the permeability to potassium decreases as potassium channels that were leaking at rest close, and this also contributes to the membrane depolarization.

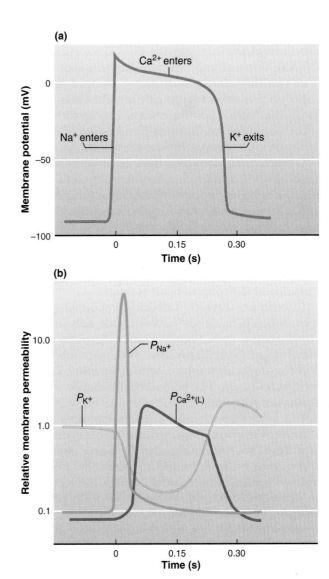

Figure 12–12

(a) Membrane potential recording from a ventricular muscle cell. Labels indicate key ionic movements in each phase. (b) Simultaneously measured permeabilities (P) to potassium, sodium, and calcium during the action potential of (a). Several subtypes of potassium channels contribute to P_{K^+}.

Also like in skeletal muscle cells and neurons, the increased sodium permeability is very transient because the sodium channels inactivate quickly. However, unlike these other excitable tissues, in cardiac muscle the reduction in sodium permeability is not accompanied by membrane repolarization. The membrane remains depolarized at a plateau of about 0 mV (see Figure 12–12a). The reasons for this continued depolarization are (1) potassium permeability stays below the resting value (i.e., the potassium channels mentioned previously remain closed), and (2) a marked increase occurs in the membrane permeability to calcium. The second reason is the more important of the two, and the explanation for it follows.

In myocardial cells, the original membrane depolarization causes voltage-gated calcium channels in the plasma membrane to open, which results in a flow of calcium ions down their electrochemical gradient into the cell. These channels open much more slowly than do sodium channels, and, because of the fact that they remain open for a prolonged period, they are often referred to as **L-type calcium channels** (L = long-lasting). The flow of positive calcium ions into the cell just balances the flow of positive potassium ions out of the cell and keeps the membrane depolarized at the plateau value.

Ultimately, repolarization does occur when the calcium channels slowly inactivate, and another subtype of potassium channels opens. These potassium channels are just like the ones described in neurons and skeletal muscle; they open in response to depolarization (after a delay), and close once the potassium current has repolarized the membrane to negative values.

The action potentials of atrial muscle cells are similar in shape to those just described for ventricular cells, although the duration of their plateau phase is shorter.

In contrast, there are extremely important differences between action potentials of cardiac muscle cells and those in the conducting system. **Figure 12–13a** illustrates the action potential of a cell from the SA node. Note that the SA node cell does not have a steady resting potential, but instead undergoes a slow depolarization. This gradual depolarization is known as a **pacemaker potential;** it brings the membrane potential to threshold, at which point an action potential occurs.

Three ion channel mechanisms, which are shown in **Figure 12–13b**, contribute to the pacemaker potential. The first is a progressive reduction in potassium permeability. Potassium channels, which opened during the repolarization phase of the previous action potential, gradually close due to the membrane's return to negative potentials. Second, pacemaker cells have a unique set of channels that, unlike most voltage-gated channels, open when the membrane potential is at *negative* values. These nonspecific cation channels conduct mainly an inward, depolarizing sodium current, and because of their unusual gating behavior have been termed "funny," or **F-type sodium channels.** (Do not confuse this type of sodium channel with the one that causes the action potential upstroke in neurons, skeletal muscle, and cardiac muscle cells.) The third pacemaker channel is a type of calcium channel that opens only briefly but contributes an inward calcium current and an important final depolarizing boost to the pacemaker potential. These channels are called **T-type calcium channels** (T = transient).

Once the pacemaker mechanisms have brought the nodal cell to threshold, an action potential occurs. The depolarizing phase is caused by calcium influx through the L-type calcium channels. After a delay, the opening of potassium channels repolarizes the membrane. The return to negative potentials activates the pacemaker mechanisms once again, and the cycle continues.

Thus, the pacemaker potential provides the SA node with **automaticity,** the capacity for spontaneous, rhythmical self-excitation. The slope of the pacemaker potential—that is, how quickly the membrane potential changes per unit time—determines how quickly threshold is reached and the next action potential is elicited. The inherent rate of the SA node—the rate exhibited in the total absence of any neural or hormonal input to the node—is approximately 100 depolar-

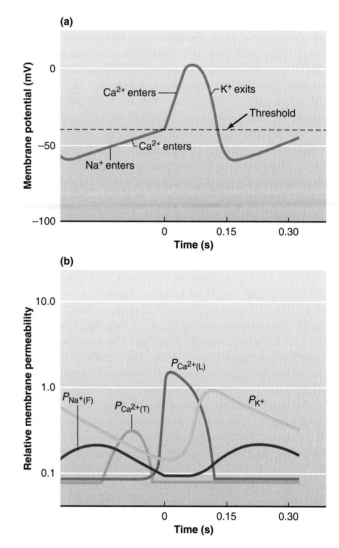

Figure 12–13

(a) Membrane potential recording from a cardiac nodal cell. Labels indicate key ionic movements in each phase. (b) Simultaneously measured permeabilities through four different ion channels during the action potential of (a).

Figure 12–13 physiological *inquiry*

■ Conducting cells of the ventricles contain all of the ion channel types found in both cardiac muscle cells and node cells. Draw a graph showing a Purkinje cell action potential.

Answer can be found at end of chapter.

izations per minute. (We will discuss later why a person's resting heart rate is typically slower than that.)

Several other portions of the conducting system are capable of generating pacemaker potentials, but the inherent rate of these other areas is slower than that of the SA node, so they normally are driven to threshold by action potentials from the SA node and do not manifest their own rhythm. However, they can do so under certain circumstances and are then called *ectopic pacemakers,* an example of which is given in the next paragraph.

Recall that excitation travels from the SA node to both ventricles only through the AV node; therefore, drug- or disease-induced malfunction of the AV node may reduce or completely eliminate the transmission of action potentials from the atria to the ventricles. If this occurs, autorhythmic cells in the bundle of His and Purkinje network, no longer driven by the SA node, begin to initiate excitation at their own inherent rate and become the pacemaker for the ventricles. Their rate is quite slow, generally 25 to 40 beats/min, and it is completely out of synchrony with the atrial contractions, which continue at the normal, higher rate of the SA node. Under such conditions, the atria are ineffective because they are often contracting against closed AV valves. Fortunately, atrial pumping is relatively unimportant for cardiac function except during strenuous exercise.

The current treatment for all severe *AV conduction disorders,* as well as for many other abnormal rhythms, is permanent surgical implantation of an *artificial pacemaker* that electrically stimulates the ventricular cells at a normal rate.

The Electrocardiogram

The **electrocardiogram** (**ECG** or EKG—the *k* is from the German *elektrokardiogramm*) is primarily a tool for evaluating the electrical events within the heart. The action potentials of cardiac muscle cells can be viewed as batteries that cause charge to move throughout the body fluids. These moving charges, or currents, are caused by all the action potentials occurring simultaneously in many individual myocardial cells and can be detected by recording electrodes at the surface of the skin. **Figure 12–14a** illustrates a typical normal ECG recorded as the potential difference between the right and left wrists. (Review Figure 12–11 for an illustration of how this waveform corresponds in time with the spread of an action potential through the heart.) The first deflection, the **P wave,** corresponds to current flows during atrial depolarization. The second deflection, the **QRS complex,** occurring approximately 0.15 s later, is the result of ventricular depolarization. It is a complex deflection because the paths taken by the wave of depolarization through the thick ventricular walls differ from instant to instant, and the currents generated in the body fluids change direction accordingly. Regardless of its form (for example, the Q and/or S portions may be absent), the deflection is still called a QRS complex. The final deflection, the **T wave,** is the result of ventricular repolarization. Atrial repolarization is usually not evident on the ECG because it occurs at the same time as the QRS complex.

A typical clinical ECG makes use of multiple combinations of recording locations on the limbs and chest (called **ECG leads**) so as to obtain as much information as possible concerning different areas of the heart. The shapes and sizes of the P wave, QRS complex, and T wave vary with the electrode locations. For reference, see **Figure 12–15** and **Table 12–2,** which describe the placement of electrodes for the different ECG leads.

To reiterate, the ECG is not a direct record of the changes in membrane potential across individual cardiac muscle cells. Instead, it is a measure of the currents generated in the extra-

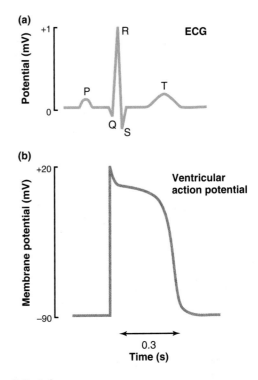

Figure 12–14

(a) Typical electrocardiogram recorded from electrodes placed on the wrists. P represents atrial depolarization; QRS, ventricular depolarization; T, ventricular repolarization. (b) Ventricular action potential recorded from a single ventricular muscle cell. Note the correspondence of the QRS complex with depolarization and the correspondence of the T wave with repolarization.

cellular fluid by the changes occurring simultaneously in many cardiac cells. To emphasize this point, **Figure 12–14b** shows the simultaneously occurring changes in membrane potential in a single ventricular cell.

Because many myocardial defects alter normal impulse propagation, and thereby the shapes and timing of the waves, the ECG is a powerful tool for diagnosing certain types of heart disease. **Figure 12–16** gives one example. However, note that the ECG provides information concerning only the electrical activity of the heart. Thus, if something is wrong with the heart's mechanical activity, but this defect does not give rise to altered electrical activity, the ECG will not be of diagnostic value.

Excitation-Contraction Coupling

The mechanisms linking cardiac muscle cell action potentials to contraction were described in detail in the chapter on muscle physiology (Chapter 9; review Figure 9–40). The small amount of extracellular calcium entering through L-type calcium channels during the plateau of the action potential triggers the release of a larger quantity of calcium from the ryanodine receptors in the sarcoplasmic reticulum membrane. Calcium activation of the thin filament and cross-bridge cycling then lead to generation of force, just as in skeletal muscle (review

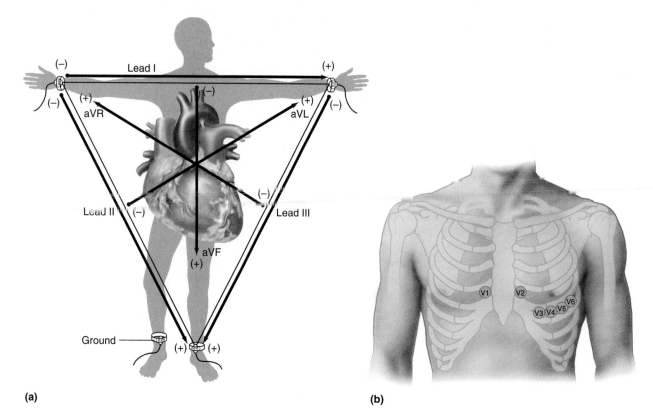

(a)   **(b)**

Figure 12–15

Placement of electrodes in electrocardiography. Each of the twelve leads uses a different combination of reference (negative pole) and recording (positive pole) electrodes, thus providing different angles for "viewing" the electrical activity of the heart. (a) The standard limb leads (I, II, and III) form a triangle between electrodes on the wrists and left leg (the right leg is a ground electrode). Augmented leads bisect the angles of the triangle by combining two electrodes as reference. (For example, for lead aVL the right wrist and foot are combined as the negative pole, thus creating a reference point along the line between them, pointing toward the recording electrode on the left wrist.) (b) The precordial leads (V_1–V_6) are recording electrodes placed on the chest as shown, with the limb leads combined into a reference point at the center of the heart.

Table 12–2	Electrocardiography Leads	
Name of Lead	**Electrode Placement**	
Standard Limb Leads	*Reference (−) Electrode*	*Recording (+) Electrode*
Lead I	Right arm	Left arm
Lead II	Right arm	Left leg
Lead III	Left arm	Left leg
Augmented Limb Leads		
aVR	Left arm and left leg	Right arm
aVL	Right arm and left leg	Left arm
aVF	Right arm and left arm	Left leg
Precordial (Chest) Leads		
V1	Combined limb leads	4th intercostal space, right of sternum
V2	" " "	4th intercostal space, left of sternum
V3	" " "	5th intercostal space, left of sternum
V4	" " "	5th intercostal space, centered on clavicle
V5	" " "	5th intercostal space, left of V4
V6	" " "	5th intercostal space, under left arm

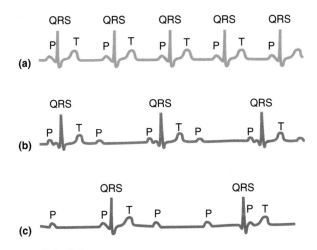

Figure 12–16

Electrocardiograms from a healthy person and from two people suffering from atrioventricular block. (a) A normal ECG. (b) Partial block. Damage to the AV node permits only every-other atrial impulse to be transmitted to the ventricles. Note that every second P wave is not followed by a QRS and T. (c) Complete block. There is no synchrony between atrial and ventricular electrical activities, and the ventricles are being driven by a very slow pacemaker cell in the bundle of His.

Figure 12–16 physiological *inquiry*

- Some people have a potentially lethal defect of ventricular muscle, in which the current through voltage-gated K⁺ channels is delayed and reduced. How could this defect be detected on their ECG recordings?

Answer can be found at end of chapter.

Figures 9–8 and 9–9). Contraction ends when calcium is returned to the sarcoplasmic reticulum and extracellular fluid by Ca^{2+}-ATPase pumps and Na^+/Ca^{2+} countertransporters.

The amount that cytosolic calcium concentration increases during excitation is a major determinant of the strength of cardiac muscle contraction. You may recall that in skeletal muscle, a single action potential releases sufficient calcium to fully saturate the troponin sites that activate contraction. By contrast, the amount of calcium released from the sarcoplasmic reticulum in cardiac muscle during a resting heartbeat is not usually sufficient to saturate all troponin sites. Therefore, the number of active cross-bridges, and thus the strength of contraction, can be increased if more calcium is released from the sarcoplasmic reticulum. The mechanisms that vary cytosolic calcium concentration will be discussed later.

Refractory Period of the Heart

Ventricular muscle, unlike skeletal muscle, is incapable of any significant degree of summation of contractions, and this is a very good thing. Imagine that cardiac muscle were able to undergo a prolonged tetanic contraction. During this period, no ventricular filling could occur because filling can occur

only when the ventricular muscle is relaxed, and the heart would therefore cease to function as a pump.

The inability of the heart to generate tetanic contractions is the result of the long absolute **refractory period** of cardiac muscle, defined as the period during and following an action potential when an excitable membrane cannot be re-excited. As in the case of neurons and skeletal muscle fibers, the main mechanism is the inactivation of sodium channels. The absolute refractory period of skeletal muscle is much shorter (1 to 2 ms) than the duration of contraction (20 to 100 ms), and a second contraction can therefore be elicited before the first is over (summation of contractions). In contrast, because of the prolonged, depolarized plateau in the cardiac muscle action potential, the absolute refractory period of cardiac muscle lasts almost as long as the contraction (250 ms), and the muscle cannot be re-excited in time to produce summation (**Figure 12–17**; also review Figure 9–41).

Mechanical Events of the Cardiac Cycle

The orderly process of depolarization described in the previous sections triggers a recurring **cardiac cycle** of atrial and ventricular contractions and relaxations (**Figure 12–18**). First we will present an overview of the cycle, naming the phases and key events. A closer look at the cycle will follow, with a discussion of the pressure and volume changes that cause the events.

The cycle is divided into two major phases, both named for events in the ventricles: the period of ventricular contraction and blood ejection called **systole,** and the alternating period of ventricular relaxation and blood filling, **diastole.** For a typical heart rate of 72 beats/min, each cardiac cycle lasts approximately 0.8 s, with 0.3 s in systole and 0.5 s in diastole.

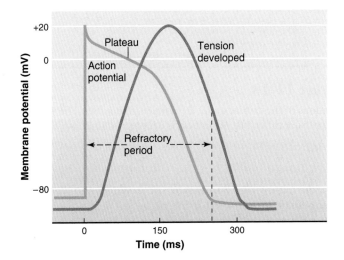

Figure 12–17

Relationship between membrane potential changes and contraction in a ventricular muscle cell. The refractory period lasts almost as long as the contraction. Tension scale not shown.

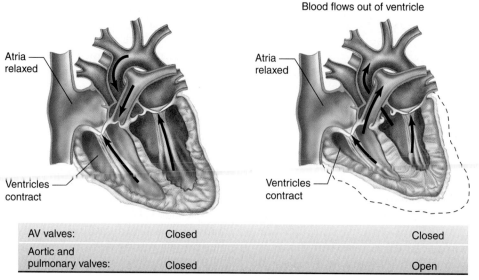

(a) Systole

Isovolumetric ventricular contraction

Atria relaxed

Ventricles contract

Ventricular ejection
Blood flows out of ventricle

Atria relaxed

Ventricles contract

AV valves:	Closed	Closed
Aortic and pulmonary valves:	Closed	Open

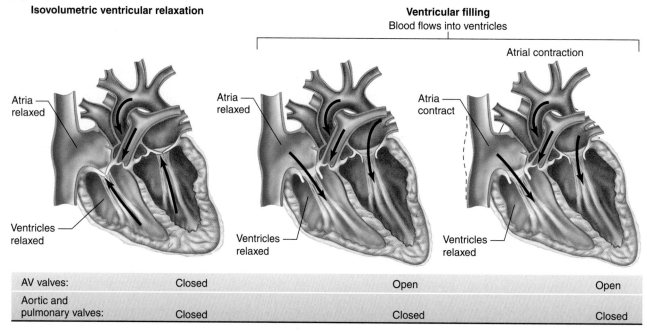

(b) Diastole

Isovolumetric ventricular relaxation

Atria relaxed

Ventricles relaxed

Ventricular filling
Blood flows into ventricles

Atrial contraction

Atria relaxed

Ventricles relaxed

Atria contract

Ventricles relaxed

AV valves:	Closed	Open	Open
Aortic and pulmonary valves:	Closed	Closed	Closed

Figure 12–18

Divisions of the cardiac cycle: (a) systole; (b) diastole. The phases of the cycle are identical in both halves of the heart. The direction in which the pressure difference *favors* flow is denoted by an arrow; note, however, that flow will not actually occur if a valve prevents it.

As Figure 12–18 illustrates, both systole and diastole can be subdivided into two discrete periods. During the first part of systole, the ventricles are contracting but all valves in the heart are closed, and so no blood can be ejected. This period is termed **isovolumetric ventricular contraction** because the ventricular volume is constant. The ventricular walls are developing tension and squeezing on the blood they enclose, raising the ventricular blood pressure. However, because the volume of blood in the ventricles is constant and because blood, like

water, is essentially incompressible, the ventricular muscle fibers cannot shorten. Thus, isovolumetric ventricular contraction is analogous to an isometric skeletal muscle contraction: The muscle develops tension, but it does not shorten.

Once the rising pressure in the ventricles exceeds that in the aorta and pulmonary trunk, the aortic and pulmonary valves open, and the **ventricular ejection** period of systole occurs. Blood is forced into the aorta and pulmonary trunk as the contracting ventricular muscle fibers shorten. The volume of blood ejected from each ventricle during systole is called the **stroke volume (SV).**

During the first part of diastole, the ventricles begin to relax, and the aortic and pulmonary valves close. (Physiologists and clinical cardiologists do not all agree on the dividing line between systole and diastole; as presented here, the dividing line is the point at which ventricular contraction stops and the pulmonary and aortic valves close.) At this time the AV valves are also closed; thus, no blood is entering or leaving the ventricles. Ventricular volume is not changing, therefore, and this period is called **isovolumetric ventricular relaxation.** Note, then, that the only times during the cardiac cycle that all valves are closed are the periods of isovolumetric ventricular contraction and relaxation.

Next, the AV valves open, and **ventricular filling** occurs as blood flows in from the atria. Atrial contraction occurs at the end of diastole, after most of the ventricular filling has taken place. This is an important point: The ventricle receives blood throughout most of diastole, not just when the atrium contracts. Indeed, in a person at rest, approximately 80 percent of ventricular filling occurs before atrial contraction.

This completes the basic orientation. We can now analyze, using **Figure 12–19**, the pressure and volume changes that occur in the left atrium, left ventricle, and aorta during the cardiac cycle. Events on the right side of the heart are identical except for the absolute pressures.

Mid-Diastole to Late Diastole

Our analysis of events in the left atrium and ventricle and the aorta begins at the far left of Figure 12–19 with the events of mid- to late diastole. The highlighted numbers that follow correspond to the numbered events shown in that figure.

(1) The left atrium and ventricle are both relaxed, but atrial pressure is very slightly higher than ventricular pressure.

(2) The AV valve is forced open by this pressure difference, and blood entering the atrium from the pulmonary veins continues on into the ventricle.

To reemphasize a point made earlier: All the valves of the heart offer very little resistance when they are open, so only very small pressure differences across them are required to produce relatively large flows.

(3) Note that at this time—indeed, throughout all of diastole—the aortic valve is closed because the aortic pressure is higher than the ventricular pressure.

(4) Throughout diastole, the aortic pressure is slowly falling because blood is moving out of the arteries and through the vascular system.

(5) In contrast, ventricular pressure is rising slightly because blood is entering the relaxed ventricle from the atrium, thereby expanding the ventricular volume.

(6) Near the end of diastole, the SA node discharges, as signified by the P wave of the ECG.

(7) Contraction of the atrium causes a rise in atrial pressure.

(8) The elevated atrial pressure forces a small additional volume of blood into the ventricle, sometimes referred to as the "atrial kick."

(9) This brings us to the end of ventricular diastole, so the amount of blood in the ventricle at this time is called the **end-diastolic volume (EDV).**

Systole

Thus far, the ventricle has been relaxed as it fills with blood. But immediately following the atrial contraction, the ventricles begin to contract.

(10) From the AV node, the wave of depolarization passes into and throughout the ventricular tissue—as signified by the QRS complex of the ECG—and this triggers ventricular contraction.

(11) As the ventricle contracts, ventricular pressure rises very rapidly, and almost immediately this pressure exceeds the atrial pressure.

(12) This change in pressure gradient forces the AV valve to close, thus preventing the backflow of blood into the atrium.

(13) Because the aortic pressure still exceeds the ventricular pressure at this time, the aortic valve remains closed, and the ventricle cannot empty despite its contraction. For a brief time, then, all valves are closed during this phase of isovolumetric ventricular contraction.

(14) This brief phase ends when the rapidly rising ventricular pressure exceeds aortic pressure.

(15) The pressure gradient now forces the aortic valve to open, and ventricular ejection begins.

(16) The ventricular volume curve shows that ejection is rapid at first and then tapers off.

(17) The amount of blood remaining after ejection is called the **end-systolic volume (ESV).**

Note that the ventricle does not empty completely. The amount of blood that does exit during each cycle is the difference between what it contained at the end of diastole and what remains at the end of systole. Thus:

$$\text{Stroke volume} = \text{End-diastolic volume} - \text{End-systolic volume}$$
$$\text{SV} \qquad\qquad \text{EDV} \qquad\qquad \text{ESV}$$

As Figure 12–19 shows, typical values for an adult at rest are stroke volume = 70 ml, end-diastolic volume = 135 ml, and end-systolic volume = 65 ml.

(18) As blood flows into the aorta, the aortic pressure rises along with the ventricular pressure. Throughout ejection, only very small pressure differences exist between ventricle and aorta because the aortic valve opening offers little resistance to flow.

(19) Note that peak ventricular and aortic pressures are reached before the end of ventricular ejection; that is, these pressures start to fall during the last part of systole despite continued ventricular contraction. This is because the strength of ventricular contraction diminishes during the last part of systole.

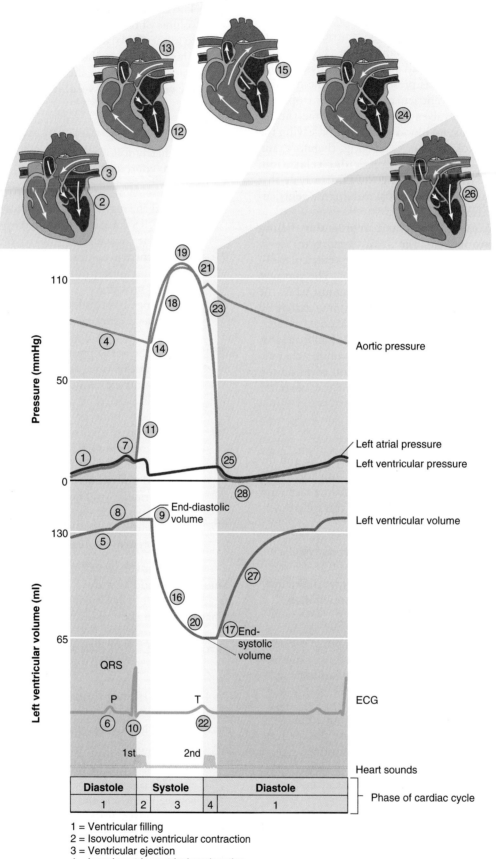

Figure 12–19

Summary of events in the left atrium, left ventricle, and aorta during the cardiac cycle (sometimes called the "Wiggers" diagram). See text for a description of the numbered steps.

20. This force reduction is evidenced by the reduced rate of blood ejection during the last part of systole.

21. The volume in the aorta, and therefore the pressure, falls as the rate of blood ejection from the ventricles becomes slower than the rate at which blood drains out of the arteries into the tissues.

Early Diastole

The phase of diastole begins as the ventricular muscle relaxes, and ejection comes to an end.

22. Recall that the T wave of ECG corresponds to the end of the plateau phase of ventricular action potentials—that is, to the onset of ventricular repolarization.

23. As the ventricles relax, the ventricular pressure falls below aortic pressure, which remains significantly elevated due to the volume of blood that just entered.

24. This change in the pressure gradient forces the aortic valve to close, and the AV valve also remains closed because the ventricular pressure is still higher than atrial pressure. For a brief time, then, all valves are again closed during this phase of isovolumetric ventricular relaxation.

25. This phase ends as the rapidly decreasing ventricular pressure falls below atrial pressure.

26. This change in pressure gradient results in the opening of the AV valve.

27. Venous blood that had accumulated in the atrium since the AV valve closed, flows rapidly into the ventricles.

28. The rate of blood flow is enhanced during this initial filling phase by a rapid drop of ventricular pressure. This occurs because the ventricle's previous contraction compressed the elastic elements of the chamber in such a way that the ventricle actually tends to recoil outward once systole is over. This expansion, in turn, lowers ventricular pressure more rapidly than would otherwise occur and may even create a negative (subatmospheric) pressure. Thus, some energy is stored within the myocardium during contraction, and its release during the subsequent relaxation aids filling.

The fact that most of ventricular filling is completed during early diastole is of great importance. It ensures that filling is not seriously impaired during periods when the heart is beating very rapidly, and the duration of diastole and therefore total filling time are reduced. However, when rates of approximately 200 beats/min or more are reached, filling time becomes inadequate, and the volume of blood pumped during each beat decreases. The clinical significance of this will be described in Section E.

Early ventricular filling also explains why the conduction defects that eliminate the atria as effective pumps do not seriously impair ventricular filling, at least in otherwise normal individuals at rest. This is true, for example, of *atrial fibrillation*, a state in which the cells of the atria contract in a completely uncoordinated manner and so fail to serve as effective pumps. Thus, the atria may be conveniently viewed as little more than continuations of the large veins.

Pulmonary Circulation Pressures

The pressure changes in the right ventricle and pulmonary arteries (**Figure 12–20**) are qualitatively similar to those just described for the left ventricle and aorta. There are striking quantitative differences, however. Typical pulmonary artery systolic and diastolic pressures are 25 and 10 mmHg, respectively, compared to systemic arterial pressures of 120 and 80 mmHg. Thus, the pulmonary circulation is a low-pressure system, for reasons to be described later. This difference is clearly reflected in the ventricular architecture—the right ventricular wall is much thinner than the left. Despite its lower pressure during contraction, however, the stroke volumes of the two ventricles are identical.

Heart Sounds

Two **heart sounds** resulting from cardiac contraction are normally heard through a stethoscope placed on the chest wall. The first sound, a soft low-pitched *lub*, is associated with closure of the AV valves; the second sound, a louder *dup*, is associated with closure of the pulmonary and aortic valves. Note in Figure 12–19 that the *lub* marks the onset of systole while the *dup* occurs at the onset of diastole. These sounds, which result from vibrations caused by the closing valves, are perfectly normal, but other sounds, known as **heart murmurs,** can be a sign of heart disease.

Murmurs can be produced by heart defects that cause blood flow to be turbulent. Normally, blood flow through valves and vessels is **laminar**—that is, it flows in smooth concentric layers (**Figure 12–21**). Turbulent flow can be caused by blood flowing rapidly in the usual direction through an abnormally narrowed valve **(stenosis),** by blood flowing backward through a damaged, leaky valve **(insufficiency),** or by blood flowing between the two atria or two ventricles through a small hole in the wall separating them (called a **septal defect**).

The exact timing and location of the murmur provide the physician with a powerful diagnostic clue. For example, a

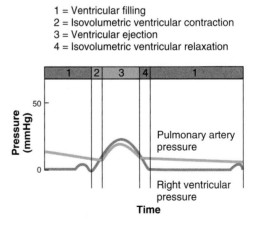

1 = Ventricular filling
2 = Isovolumetric ventricular contraction
3 = Ventricular ejection
4 = Isovolumetric ventricular relaxation

Figure 12–20

Pressures in the right ventricle and pulmonary artery during the cardiac cycle. Note that the absolute pressures are lower than in the left ventricle and aorta.

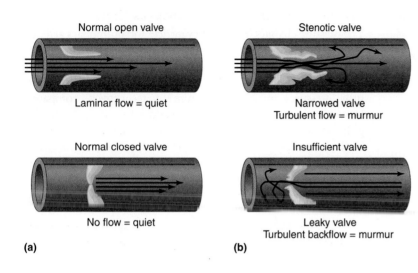

Normal open valve

Laminar flow = quiet

Normal closed valve

No flow = quiet

(a)

Stenotic valve

Narrowed valve
Turbulent flow = murmur

Insufficient valve

Leaky valve
Turbulent backflow = murmur

(b)

Figure 12–21

Heart valve defects causing turbulent blood flow and murmurs.
(a) Normal valves allow smooth, laminar flow of blood in the
forward direction when open and prevent backward flow of
blood when closed. No sound is heard in either state.
(b) Stenotic valves cause rapid, turbulent forward flow of
blood, making a high-pitched, whistling murmur. Valve
insufficiency results in turbulent backward flow when the valve
should be closed, causing a low-pitched gurgling murmur.

Figure 12–21 physiological *inquiry*

- What valve defect(s) would be indicated by the following
  sequence of heart sounds?

  "lub-whistle-dup-gurgle"

  Answer can be found at end of chapter.

murmur heard throughout systole suggests a stenotic pulmonary or aortic valve, an insufficient AV valve, or a hole in the interventricular septum. In contrast, a murmur heard during diastole suggests a stenotic AV valve or an insufficient pulmonary or aortic valve.

The Cardiac Output

The volume of blood each ventricle pumps, usually expressed in liters per minute, is called the **cardiac output (CO).** Cardiac output is also the volume of blood flowing through either the systemic or the pulmonary circuit per minute.

The cardiac output is determined by multiplying the heart rate (HR)—the number of beats per minute—and the stroke volume (SV)—the blood volume ejected by each ventricle with each beat:

$$CO = HR \times SV$$

Thus, if each ventricle has a rate of 72 beats/min and ejects 70 ml of blood with each beat, the cardiac output is:

$$CO = 72 \text{ beats/min} \times 0.07 \text{ L/beat} = 5.0 \text{ L/min}$$

These values are within the normal range for a resting, average-sized adult. Coincidentally, total blood volume is also approximately 5 L, so essentially all the blood is pumped around the circuit once each minute. During periods of strenuous exercise in well-trained athletes, the cardiac output may reach 35 L/min; the entire blood volume is pumped around the circuit seven times a minute! Even sedentary, untrained individuals can reach cardiac outputs of 20–25 L/min during exercise.

The following description of the factors that alter the two determinants of cardiac output—heart rate and stroke volume—applies in all respects to both the right and left sides of the heart because stroke volume and heart rate are the same for both under steady-state conditions. Note that heart rate and stroke volume do not always change in the same direction. For example, stroke volume decreases following blood loss, whereas heart rate increases. These changes produce opposing effects on cardiac output.

Control of Heart Rate

Rhythmical beating of the heart at a rate of approximately 100 beats/min will occur in the complete absence of any nervous or hormonal influences on the SA node. This is the inherent autonomous discharge rate of the SA node. The heart rate may be lower or higher than this, however, because the SA node is normally under the constant influence of nerves and hormones.

A large number of parasympathetic and sympathetic postganglionic fibers end on the SA node. Activity in the parasympathetic (vagus) nerves causes the heart rate to decrease, whereas activity in the sympathetic nerves causes an increase. In the resting state, there is considerably more parasympathetic activity to the heart than sympathetic, so the normal resting heart rate of about 70 beats/min is well below the inherent rate of 100 beats/min.

Figure 12–22 illustrates how sympathetic and parasympathetic activity influence SA node function. Sympathetic

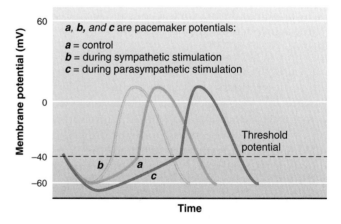

Figure 12–22

Effects of sympathetic and parasympathetic nerve stimulation on the slope of the pacemaker potential of an SA-nodal cell. Note that parasympathetic stimulation not only reduces the slope of the pacemaker potential but also causes the membrane potential to be more negative before the pacemaker potential begins.

Adapted from Hoffman and Cranefield.

stimulation increases the slope of the pacemaker potential by increasing the F-type channel permeability. Because the main current through these channels is sodium entering the cell, faster depolarization results. This causes the SA node cells to reach threshold more rapidly and the heart rate to increase. Stimulation of the parasympathetics has the opposite effect—the slope of the pacemaker potential decreases due to a reduction in the inward current. Threshold is thus reached more slowly, and heart rate decreases. Parasympathetic stimulation also hyperpolarizes the plasma membranes of SA node cells by increasing their permeability to potassium. The pacemaker potential thus starts from a more negative value (closer to the potassium equilibrium potential) and has a reduced slope.

Factors other than the cardiac nerves can also alter heart rate. Epinephrine, the main hormone liberated from the adrenal medulla, speeds the heart by acting on the same beta-adrenergic receptors in the SA node as norepinephrine released from neurons. The heart rate is also sensitive to changes in body temperature, plasma electrolyte concentrations, hormones other than epinephrine, and a metabolite—adenosine—produced by myocardial cells. These factors are normally of lesser importance, however, than the cardiac nerves. **Figure 12–23** summarizes the major determinants of heart rate.

As stated in the previous section on innervation, sympathetic and parasympathetic neurons innervate not only the SA node but other parts of the conducting system as well. Sympathetic stimulation increases conduction velocity through the entire cardiac conducting system, whereas parasympathetic stimulation decreases the rate of spread of excitation through the atria and the AV node.

Control of Stroke Volume

The second variable that determines cardiac output is stroke volume—the volume of blood each ventricle ejects during each contraction. Recall that the ventricles do not completely empty themselves during contraction. Therefore, a more forceful contraction can produce an increase in stroke volume by causing greater emptying. Changes in the stroke volume can be produced by a variety of factors, but three are dominant under most physiological and pathophysiological conditions: (1) changes in the end-diastolic volume (the volume of blood in the ventricles just before contraction, sometimes referred to as the **preload**); (2) changes in the magnitude of sympathetic nervous system input to the ventricles; and (3) changes in **afterload** (i.e., the arterial pressures against which the ventricles pump).

Relationship Between Ventricular End-Diastolic Volume and Stroke Volume: The Frank-Starling Mechanism

The mechanical properties of cardiac muscle form the basis for an inherent mechanism for altering stroke volume: The ventricle contracts more forcefully during systole when it has been filled to a greater degree during diastole. In other words, all other factors being equal, the stroke volume increases as the end-diastolic volume increases. This is illustrated graphically as a **ventricular function curve** (**Figure 12–24**). This relationship between stroke volume and end-diastolic volume is known as the **Frank-Starling mechanism** (also called Starling's law of the heart) in recognition of the two physiologists who identified it.

What accounts for the Frank-Starling mechanism? Basically it is simply a length-tension relationship, as described for skeletal muscle in Chapter 9, because end-diastolic volume is a major determinant of how stretched the ventricular sarcomeres are just before contraction. Thus, the greater the end-diastolic volume, the greater the stretch, and the more forceful the contraction. However, a comparison of Figure 12–24 with Figure 9–21 reveals an important difference between the length-tension relationship in skeletal and cardiac muscle. The normal point for cardiac muscle in a resting individual is not at its optimal length for contraction, as it is for most resting skeletal muscles, but is on the rising phase of the curve. For this reason, greater filling causes additional stretching of the cardiac muscle fibers and increases the force of contraction.

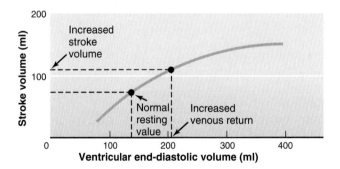

Figure 12–24

A ventricular function curve, which expresses the relationship between end-diastolic ventricular volume and stroke volume (the Frank-Starling mechanism). The horizontal axis could have been labeled "sarcomere length," and the vertical "contractile force." In other words, this is a length-tension curve, analogous to that for skeletal muscle (see Figure 9–21). At very high volumes, force (and thus stroke volume) declines as in skeletal muscle (not shown).

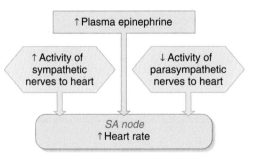

Figure 12–23

Major factors influencing heart rate. All effects are exerted upon the SA node. The figure shows how heart rate is increased; reversal of all the arrows in the boxes would illustrate how heart rate is decreased.

The significance of the Frank-Starling mechanism is as follows: At any given heart rate, an increase in the **venous return**—the flow of blood from the veins into the heart—automatically forces an increase in cardiac output by increasing end-diastolic volume and thus stroke volume. One important function of this relationship is maintaining the equality of right and left cardiac outputs. Should the right side of the heart, for example, suddenly begin to pump more blood than the left, the increased blood flow to the left ventricle would automatically produce an increase in left ventricular output. This ensures that blood will not accumulate in the pulmonary system.

Sympathetic Regulation

Sympathetic nerves are distributed to the entire myocardium. The sympathetic neurotransmitter norepinephrine acts on beta-adrenergic receptors to increase ventricular **contractility,** defined as the strength of contraction *at any given end-diastolic volume*. Plasma epinephrine acting on these receptors also increases myocardial contractility. Thus, the increased force of contraction and stroke volume resulting from sympathetic nerve stimulation or epinephrine is independent of a change in end-diastolic ventricular volume.

Note that a change in contraction force due to increased end-diastolic volume (the Frank-Starling mechanism) does not reflect increased contractility. Increased contractility is specifically defined as an increased contraction force at *any* given end-diastolic volume.

The relationship between the Frank-Starling mechanism and the cardiac sympathetic nerves is illustrated in

Figure 12–25. The green ventricular function curve is the same as that shown in Figure 12–24. The orange ventricular function curve was obtained for the same heart during sympathetic nerve stimulation. The Frank-Starling mechanism still applies, but during nerve stimulation, the stroke volume is greater at any given end-diastolic volume. In other words, the increased contractility leads to a more complete ejection of the end-diastolic ventricular volume.

One way to quantify contractility is through the **ejection fraction (EF),** defined as the ratio of stroke volume (SV) to end-diastolic volume (EDV):

$$EF = SV/EDV$$

Expressed as a percentage, the ejection fraction normally averages between 50 and 75 percent under resting conditions. Increased contractility causes an increased ejection fraction.

Not only does increased sympathetic nerve activity to the myocardium cause a more powerful contraction, it also causes both the contraction and relaxation of the ventricles to occur more quickly (**Figure 12–26**). These latter effects are quite important because, as described earlier, increased sympathetic activity to the heart also increases heart rate. As heart rate increases, the time available for diastolic filling decreases, but the quicker contraction and relaxation induced simultaneously by the sympathetic neurons partially compensate for this problem by permitting a larger fraction of the cardiac cycle to be available for filling.

Cellular mechanisms involved in sympathetic regulation of myocardial contractility are shown in **Figure 12–27**. Adrenergic receptors activate a G-protein-coupled cascade that includes the production of cAMP and activation of a protein kinase. A number of proteins involved in excitation-contraction

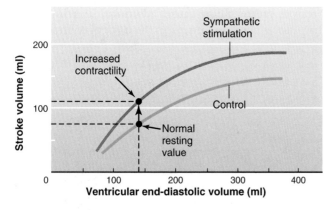

Figure 12–25

Effects on stroke volume of stimulating the sympathetic nerves to the heart. Stroke volume is increased at any given end-diastolic volume; that is, the sympathetic stimulation increases ventricular contractility.

Figure 12–25 physiological *inquiry* (*pi*)

■ From this figure, estimate the ejection fraction and end-systolic volumes under control and sympathetic stimulation conditions at an end-diastolic volume of 140 mL.

Answer can be found at end of chapter.

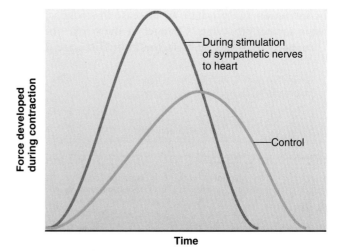

Figure 12–26

Effects of sympathetic stimulation on ventricular contraction and relaxation. Note that both the rate of force development and the rate of relaxation increase, as does the maximum force developed. All these changes reflect an increased contractility.

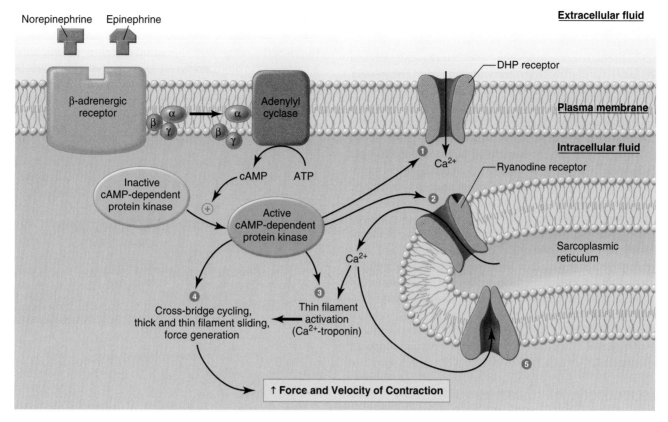

Figure 12–27

Mechanisms of sympathetic effects on cardiac muscle cell contractility. In some of the pathways, the kinase phosphorylates proteins that are not shown.

coupling are phosphorylated by the kinase, which alters their activity. These include:

1. DHP receptors (L-type calcium channels) in the plasma membrane;
2. the ryanodine receptor and associated proteins in the sarcoplasmic reticulum membrane;
3. thin filament proteins; in particular, troponin;
4. thick filament proteins associated with the cross-bridges; and
5. proteins involved in pumping calcium back into the sarcoplasmic reticulum.

Due to these alterations, cytosolic calcium concentration rises more quickly and reaches a greater value during excitation, calcium returns to its pre-excitation value more quickly following excitation, and the rate of cross-bridge activation and cycling are accelerated. The net result is the stronger, faster contraction observed during sympathetic activation of the heart.

There is little parasympathetic innervation of the ventricles, so the parasympathetic system normally has only a negligible direct effect on ventricular contractility.

Table 12–3 summarizes the effects of the autonomic nerves on cardiac function.

	Table 12–3	**Effects of Autonomic Nerves on the Heart**
Area Affected	**Sympathetic Nerves (norepinephrine on β-adrenergic receptors)**	**Parasympathetic Nerves (ACh on muscarinic receptors)**
SA node	Increased heart rate	Decreased heart rate
AV node	Increased conduction rate	Decreased conduction rate
Atrial muscle	Increased contractility	Decreased contractility
Ventricular muscle	Increased contractility	No significant effect

Afterload

An increased arterial pressure tends to reduce stroke volume. This is because, in analogy to the situation in skeletal muscle, the arterial pressure constitutes the "load" (technically termed the *afterload*) for contracting ventricular muscle. The greater this load, the less the contracting muscle fibers can shorten at a given contractility. This factor will not be dealt with further, because in the normal heart, several inherent adjustments minimize the overall influence of arterial pressure on stroke volume. However, in the sections on high blood pressure and heart failure, we will see that alterations in vascular resistance and long-term elevations of arterial pressure can weaken the heart and thereby influence stroke volume.

In summary, the two most important physiologic controllers of stroke volume are the Frank-Starling mechanism, which depends on changes in end-diastolic volume, and ventricular contractility, which is influenced by cardiac sympathetic nerves and circulating epinephrine. The contribution of each of these two mechanisms in specific physiological situations is described in later sections.

Figure 12–28 integrates the factors that determine stroke volume and heart rate into a summary of the control of cardiac output.

Measurement of Cardiac Function

Human cardiac output can be measured by a variety of methods. Moreover, two- and three-dimensional images of the heart can be obtained throughout the entire cardiac cycle. For example, in *echocardiography,* ultrasonic waves are beamed at the heart, and returning echoes are electronically plotted by computer to produce continuous images of the heart. This technique can detect the abnormal functioning of cardiac valves or contractions of the cardiac walls, and can also be used to measure ejection fraction.

Echocardiography is a noninvasive technique because everything used remains external to the body. Other visualization techniques are invasive. One, *cardiac angiography,* requires the temporary threading of a thin, flexible tube called a catheter through an artery or vein into the heart. A dye is then injected through the catheter during high-speed x-ray videography. This technique is useful not only for evaluating cardiac function, but also for identifying narrowed coronary arteries.

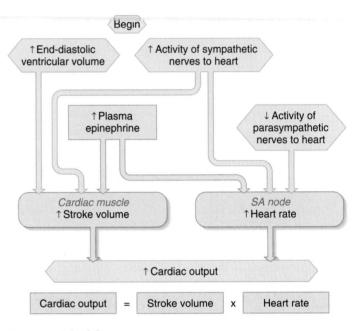

Figure 12–28

Major factors determining cardiac output. Reversal of all arrows in the boxes would illustrate how cardiac output is decreased.

Hypertrophic Cardiomyopathy

Hypertrophic cardiomyopathy is one of the most common genetically transmitted cardiac diseases, occurring in one out of 500 people. As the name implies, it is characterized by an increase in thickness of the heart muscle, in particular the interventricular septum and the wall of the left ventricle. In conjunction with wall thickening, there is a disruption of the orderly array of myocytes and conducting cells within the walls. The thickening of the septum interferes with the ejection of blood through the aortic valve, particularly during exercise, which can prevent cardiac output from rising sufficiently to meet tissue needs. The heart itself is commonly a victim of this reduction in blood flow, and one symptom that can be an early warning sign is the associated chest pain **(angina pectoris).** Moreover, disruption of the conduction

pathway can lead to dangerous, sometimes fatal arrhythmias. Many people with this disease have no symptoms, though, and thus it can go undetected until it has progressed to an advanced stage. For these reasons, hypertrophic cardiomyopathy is the leading cause of sudden cardiac death in young athletes. If it progresses without treatment, it can lead to *heart failure* (further discussed in Section E). Although the mechanisms by which this disease process develops are not completely understood, all of the genetic mutations that have been found to cause it involve proteins of the contractile system, including troponin, tropomyosin, and myosin. Depending on the severity of the condition when it is discovered, treatments include administering drugs that prevent arrhythmias, surgical repair of the septum and valve, or even heart transplantation.■

Anatomy

I. The atrioventricular (AV) valves prevent flow from the ventricles back into the atria.

II. The pulmonary and aortic valves prevent backflow from the pulmonary trunk into the right ventricle and from the aorta into the left ventricle, respectively.

III. Cardiac muscle cells are joined by gap junctions that permit the conduction of action potentials from cell to cell.

IV. The myocardium also contains specialized cells that constitute the conducting system of the heart, initiating the cardiac action potentials and speeding their spread through the heart.

Heartbeat Coordination

I. Action potentials must be initiated in cardiac cells for contraction to occur.

   a. The rapid depolarization of the action potential in atrial and ventricular muscle cells is due mainly to a positive-feedback increase in sodium permeability.

   b. Following the initial rapid depolarization, the cardiac muscle cell membrane remains depolarized (the plateau phase) for almost the entire duration of the contraction because of prolonged entry of calcium into the cell through plasma-membrane L-type calcium channels.

II. The SA node generates the action potential that leads to depolarization of all other cardiac cells.

   a. The SA node manifests a pacemaker potential, which brings its membrane potential to threshold and initiates an action potential.

   b. The impulse spreads from the SA node throughout both atria and to the AV node, where a small delay occurs. The impulse then passes into the bundle of His, right and left bundle branches, Purkinje fibers, and ventricular muscle cells.

III. Calcium, mainly released from the sarcoplasmic reticulum (SR), functions as the excitation-contraction coupler in cardiac muscle, as in skeletal muscle, by combining with troponin.

   a. The major signal for calcium release from the SR is extracellular calcium entering through voltage-gated L-type calcium channels in the T-tubular membrane during the action potential.

   b. This "trigger" calcium opens ryanodine receptor calcium channels in the sarcoplasmic reticulum membrane.

   c. The amount of calcium released does not usually saturate all troponin binding sites, so the number of active cross-bridges can increase if cytosolic calcium increases still further.

IV. Cardiac muscle cannot undergo summation of contractions because it has a very long refractory period.

Mechanical Events of the Cardiac Cycle

I. The cardiac cycle is divided into systole (ventricular contraction) and diastole (ventricular relaxation).

   a. At the onset of systole, ventricular pressure rapidly exceeds atrial pressure, and the AV valves close. The aortic and pulmonary valves are not yet open, however, so no ejection occurs during this isovolumetric ventricular contraction.

   b. When ventricular pressures exceed aortic and pulmonary trunk pressures, the aortic and pulmonary valves open, and the ventricles eject the blood.

   c. When the ventricles relax at the beginning of diastole, the ventricular pressures fall significantly below those in the aorta and pulmonary trunk, and the aortic and pulmonary valves close. Because the AV valves are also still closed, no change in ventricular volume occurs during this isovolumetric ventricular relaxation.

   d. When ventricular pressures fall below the pressures in the right and left atria, the AV valves open, and the ventricular filling phase of diastole begins.

   e. Filling occurs very rapidly at first so that atrial contraction, which occurs at the very end of diastole, usually adds only a small amount of additional blood to the ventricles.

II. The amount of blood in the ventricles just before systole is the end-diastolic volume. The volume remaining after ejection is the end-systolic volume, and the volume ejected is the stroke volume.

III. Pressure changes in the systemic and pulmonary circulations have similar patterns, but the pulmonary pressures are much lower.

IV. The first heart sound is due to the closing of the AV valves, and the second to the closing of the aortic and pulmonary valves.

V. Murmurs can result from narrowed or leaky valves, as well as from holes in the interventricular septum.

The Cardiac Output

I. The cardiac output is the volume of blood each ventricle pumps and equals the product of heart rate and stroke volume.

   a. Heart rate is increased by stimulation of the sympathetic nerves to the heart and by epinephrine; it is decreased by stimulation of the parasympathetic nerves to the heart.

   b. Stroke volume is increased mainly by an increase in end-diastolic volume (the Frank-Starling mechanism) and by an increase in contractility due to sympathetic-nerve stimulation or to epinephrine. Afterload can also play a significant role in certain situations.

Measurement of Cardiac Function

I. Methods of measuring cardiac function include echocardiography, for assessing wall and valve funcion, and cardiac angiography, for determining coronary blood flow.

Additional Clinical Examples

I. Hypertrophic cardiomyopathy is a disease caused by a genetic mutation in genes coding for cardiac contractile proteins. It can lead to heart failure, as well as sudden death caused by arrhythmia.

afterload 379	diastole 373
aortic valve 366	ECG lead 371
atrioventricular (AV) node 368	ejection fraction (EF) 380
atrioventricular (AV) valve 365	electrocardiogram (ECG) 371
automaticity 370	end-diastolic volume
bicuspid valve 365	(EDV) 375
bundle of His 369	endothelial cell 365
cardiac cycle 373	endothelium 365
cardiac output (CO) 378	end-systolic volume (ESV) 375
chordae tendineae 365	epicardium 365
conducting system 367	Frank-Starling mechanism 379
contractility 380	F-type sodium channel 370
coronary artery 367	heart rate 368
coronary blood flow 367	heart sounds 377
cusp 365	interventricular septum 365

■ SECTION B CLINICAL TERMS

■ SECTION B REVIEW QUESTIONS

1. List the structures through which blood passes from the systemic veins to the systemic arteries.
2. Contrast and compare the structure of cardiac muscle with that of skeletal and smooth muscle.
3. Describe the autonomic innervation of the heart, including the types of receptors involved.
4. Draw a ventricular muscle cell action potential. Describe the changes in membrane permeability that underlie the potential changes.
5. Contrast action potentials in ventricular muscle cells with SA-node action potentials. What is the pacemaker potential due to, and what is its inherent rate? By what mechanism does the SA node function as the pacemaker for the entire heart?
6. Describe the spread of excitation from the SA node through the rest of the heart.
7. Draw and label a normal ECG. Relate the P, QRS, and T waves to the atrial and ventricular action potentials.
8. Describe the sequence of events leading to excitation-contraction coupling in cardiac muscle.
9. What prevents the heart from undergoing summation of contractions?
10. Draw a diagram of the pressure changes in the left atrium, left ventricle, and aorta throughout the cardiac cycle. Show when the valves open and close, when the heart sounds occur, and the pattern of ventricular ejection.
11. Contrast the pressures in the right ventricle and pulmonary trunk with those in the left ventricle and aorta.
12. What causes heart murmurs in diastole? In systole?
13. Write the formula relating cardiac output, heart rate, and stroke volume; give normal values for a resting adult.
14. Describe the effects of the sympathetic and parasympathetic nerves on heart rate. Which is dominant at rest?
15. What are the major factors influencing force of contraction?
16. Draw a ventricular function curve illustrating the Frank-Starling mechanism.
17. Describe the effects of the sympathetic nerves on cardiac muscle during contraction and relaxation.
18. Draw a pair of curves relating end-diastolic volume and stroke volume during different levels of sympathetic stimulation.
19. Summarize the effects of the autonomic nerves on the heart.
20. Draw a flow diagram summarizing the factors determining cardiac output.

SECTION C — The Vascular System

The functional and structural characteristics of the blood vessels change with successive branching. Yet the entire cardiovascular system, from the heart to the smallest capillary, has one structural component in common: a smooth, single-celled layer of endothelial cells, or endothelium, which lines the inner (blood-contacting) surface of the vessels. Capillaries consist only of endothelium, whereas all other vessels have additional layers of connective tissue and smooth muscle. Endothelial cells have a large number of active functions. These are summarized for reference in **Table 12–4** and are described in relevant sections of this chapter and others.

We have previously described the pressures in the aorta and pulmonary arteries during the cardiac cycle. **Figure 12–29** illustrates the pressure changes that occur along the rest of the systemic and pulmonary vascular systems. Text sections dealing with the individual vascular segments will describe the reasons for these changes in pressure. For the moment, note only that by the time the blood has completed its journey back to the atrium in each circuit, virtually all the pressure originally generated by the ventricular contraction has dissipated. The reason pressure at any point in the vascular system is lower than that at an earlier point is that the blood vessels offer resistance to the flow from one point to the next.

Arteries

The aorta and other systemic arteries have thick walls containing large quantities of elastic tissue. Although they also have smooth muscle, arteries can be viewed most conveniently as elastic tubes. Because the arteries have large radii, they serve as low-resistance tubes conducting blood to the various organs. Their second major function, related to their elasticity, is to act as a "pressure reservoir" for maintaining blood flow through the tissues during diastole, as we will describe next.

Table 12–4	Functions of Endothelial Cells

1. Serve as a physical lining that blood cells do not normally adhere to in heart and blood vessels.

2. Serve as a permeability barrier for the exchange of nutrients, metabolic end products, and fluid between plasma and interstitial fluid; regulate transport of macromolecules and other substances.

3. Secrete paracrine agents that act on adjacent vascular smooth muscle cells; including vasodilators—prostacyclin and nitric oxide (endothelium-derived relaxing factor, EDRF)—and vasoconstrictors—notably endothelin-1.

4. Mediate angiogenesis (new capillary growth).

5. Play a central role in vascular remodeling by detecting signals and releasing paracrine agents that act on adjacent cells in the blood vessel wall.

6. Contribute to the formation and maintenance of extracellular matrix.

7. Produce growth factors in response to damage.

8. Secrete substances that regulate platelet clumping, clotting, and anticlotting.

9. Synthesize active hormones from inactive precursors (Chapter 14).

10. Extract or degrade hormones and other mediators (Chapters 11, 13).

11. Secrete cytokines during immune responses (Chapter 18).

12. Influence vascular smooth-muscle proliferation in the disease atherosclerosis.

Arterial Blood Pressure

What are the factors determining the pressure within an elastic container, such as a balloon filled with water? The pressure inside the balloon depends on (1) the volume of water, and (2) how easily the balloon walls can stretch. If the walls are very stretchable, large quantities of water can be added with only a small rise in pressure. Conversely, the addition of a small quantity of water causes a large pressure rise in a balloon that is difficult to stretch. The term used to denote how easily a structure stretches is **compliance:**

$$\text{Compliance} = \Delta \text{ volume}/\Delta \text{ pressure}$$

The higher the compliance of a structure, the more easily it can be stretched.

These principles apply to an analysis of arterial blood pressure. The contraction of the ventricles ejects blood into the pulmonary and systemic arteries during systole. If a precisely equal quantity of blood were to flow simultaneously out of the arteries, the total volume of blood in the arteries would remain constant, and arterial pressure would not change. Such is not the case, however. As shown in **Figure 12–30**, a volume of blood equal to only about one-third the stroke volume leaves the arteries during systole. The rest of the stroke volume remains in the arteries during systole, distending them and raising the arterial pressure. When ventricular contraction ends, the stretched arterial walls recoil passively, like a stretched rubber band being released, and blood continues to be driven into the arterioles during diastole. As blood leaves the arteries, the arterial volume and, therefore, the arterial pressure slowly fall, but the next ventricular contraction occurs while there is still adequate blood in the arteries to stretch them partially. Therefore, the arterial pressure does not fall to zero.

The aortic pressure pattern shown in **Figure 12–31a** is typical of the pressure changes that occur in all the large systemic arteries. The maximum arterial pressure reached during

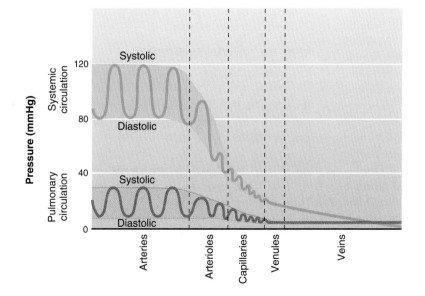

Figure 12–29

Pressures in the systemic and pulmonary vessels.

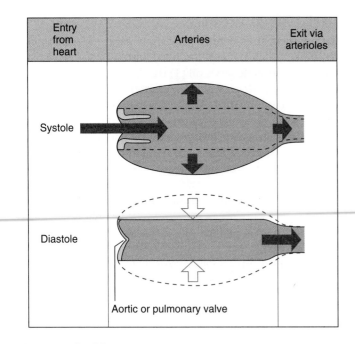

Figure 12–30

Movement of blood into and out of the arteries during the cardiac cycle. The lengths of the arrows denote relative quantities flowing into and out of the arteries and remaining in the arteries.

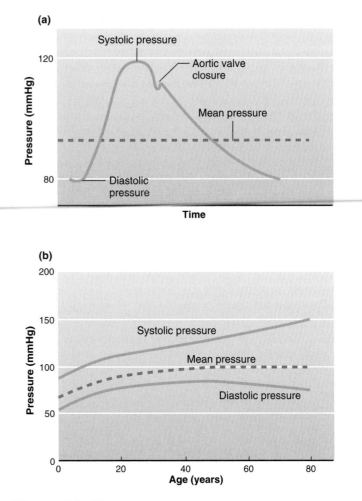

Figure 12–31

(a) Typical arterial pressure fluctuations during the cardiac cycle.
(b) Changes in arterial pressure with age in the U.S. population.

Adapted from National Institutes of Health Publication #04-5230, August 2004.

Figure 12–31 physiological *inquiry* 🅿️ℹ️

- At an elevated heart rate, the amount of time spent in diastole is reduced more than the amount of time spent in systole. How would you estimate the mean arterial blood pressure at a heart rate elevated to the point at which the times spent in systole and diastole are roughly equal?

Answer can be found at end of chapter.

peak ventricular ejection is called **systolic pressure (SP).** The minimum arterial pressure occurs just before ventricular ejection begins and is called **diastolic pressure (DP).** Arterial pressure is generally recorded as systolic/diastolic—that is, 120/80 mmHg in our example (see **Figure 12–31b** for average values at different ages in the population of the United States).

The difference between systolic pressure and diastolic pressure (120 − 80 = 40 mmHg in the example) is called the **pulse pressure.** It can be felt as a pulsation or throb in the arteries of the wrist or neck with each heartbeat. During diastole, nothing is felt over the artery, but the rapid rise in pressure at the next systole pushes out the artery wall, and it is this expansion of the vessel that produces the detectable throb.

The most important factors determining the magnitude of the pulse pressure are (1) stroke volume, (2) speed of ejection of the stroke volume, and (3) arterial compliance. Specifically, the pulse pressure a ventricular ejection produces is greater if the volume of blood ejected increases, if the speed at which it is ejected increases, or if the arteries are less compliant. This last phenomenon occurs in *arteriosclerosis,* the stiffening of the arteries that progresses with age and accounts for the increasing pulse pressure seen so often in older people (see Figure 12–31b).

It is evident from Figure 12–31a that arterial pressure is continuously changing throughout the cardiac cycle. The average, or **mean, arterial pressure (MAP)** in the cycle is not merely the value halfway between systolic pressure and diastolic pressure, because diastole lasts longer than systole. The true mean arterial pressure can be obtained by complex methods, but at a typical resting heart rate, it is approximately

equal to the diastolic pressure plus one-third of the pulse pressure (SP − DP), largely because diastole lasts about twice as long as systole:

$$\text{MAP} = \text{DP} + \tfrac{1}{3}(\text{SP} - \text{DP})$$

Thus, in our example:

$$\text{MAP} = 80 + \tfrac{1}{3}(40) = 93 \text{ mmHg}$$

The MAP is the most important of the pressures described because it is the pressure driving blood into the tissues averaged over the entire cardiac cycle. We can say mean "arterial" pressure without specifying which artery we are referring to because the aorta and other large arteries have such large diameters that they offer only negligible resistance to flow, and the mean pressures are therefore similar everywhere in the large arteries.

One additional important point should be made: We have stated that arterial compliance is an important determinant of pulse pressure, but for complex reasons, compliance does not significantly influence the mean arterial pressure. Thus, for example, a person with a low arterial compliance (due to arteriosclerosis) but an otherwise normal cardiovascular system will have a large pulse pressure but a normal mean arterial pressure. The determinants of mean arterial pressure are described in Section D. The method for measuring blood pressure is described next.

Measurement of Systemic Arterial Pressure

Both systolic and diastolic blood pressure are readily measured in human beings with the use of a device called a sphygmomanometer. An inflatable cuff containing a pressure gauge is wrapped around the upper arm, and a stethoscope is placed in a spot on the arm just below the cuff where the brachial artery lies.

The cuff is then inflated with air to a pressure greater than systolic blood pressure (**Figure 12–32**). The high pressure in the cuff is transmitted through the tissue of the arm and completely compresses the artery under the cuff, thereby preventing blood flow through the artery. The air in the cuff is then slowly released, causing the pressure in the cuff and on the artery to drop. When cuff pressure has fallen to a value just below the systolic pressure, the artery opens slightly and allows blood flow for a brief time at the peak of systole. During this interval, the blood flow through the partially compressed artery occurs at a very high velocity because of the small opening and the large pressure difference across the opening. The high-velocity blood flow is turbulent and, therefore, produces vibrations (called **Korotkoff's sounds**) that can be heard through the stethoscope. Thus, the pressure at which sounds are first heard as the cuff pressure decreases is identified as the systolic blood pressure.

As the pressure in the cuff decreases further, the duration of blood flow through the artery in each cycle becomes longer. When the cuff pressure reaches the diastolic blood pressure, all sound stops because flow is now continuous and nonturbulent through the open artery. Thus, diastolic pressure is identified as the cuff pressure at which sounds disappear.

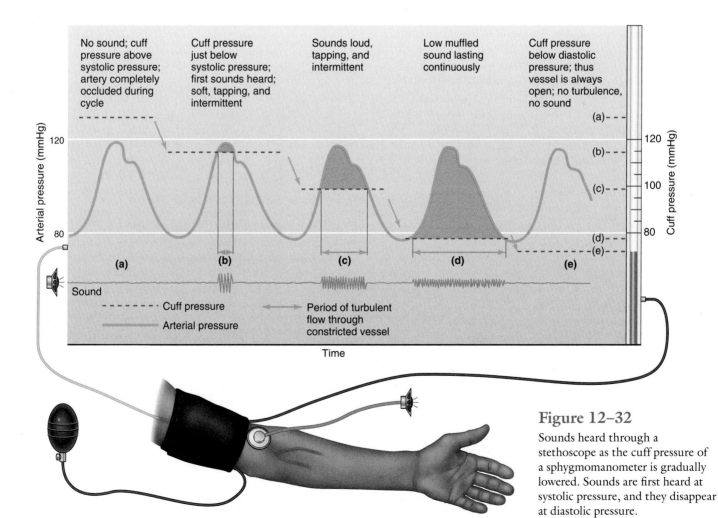

Figure 12–32

Sounds heard through a stethoscope as the cuff pressure of a sphygmomanometer is gradually lowered. Sounds are first heard at systolic pressure, and they disappear at diastolic pressure.

It should be clear from this description that the sounds heard during measurement of blood pressure are not the same as the heart sounds described earlier, which are due to closing of cardiac valves.

Arterioles

The arterioles play two major roles: (1) the arterioles in individual organs are responsible for determining the relative blood flows to those organs at any given mean arterial pressure, and (2) the arterioles, all together, are a major factor in determining mean arterial pressure itself. The first function will be described in this section, and the second in Section D.

Figure 12–33 illustrates the major principles of blood-flow distribution in terms of a simple model, a fluid-filled tank with a series of compressible outflow tubes. What determines the rate of flow through each exit tube? As stated in Section A of this chapter,

$$F = \Delta P/R$$

Because the driving pressure (the height of the fluid column in the tank) is identical for each tube, differences in flow are completely determined by differences in the resistance to flow offered by each tube. The lengths of the tubes are approximately the same, and the viscosity of the fluid is constant; therefore, differences in resistance are due solely to differences in the radii of the tubes. Obviously, the widest tubes have the greatest

flows. If the radius of each outflow tube can be independently altered, we can obtain various combinations of flows.

This analysis can now be applied to the cardiovascular system. The tank is analogous to the major arteries, which serve as a pressure reservoir, but are so large that they contribute little resistance to flow. Therefore, all the large arteries of the body can be considered a single pressure reservoir.

The arteries branch within each organ into progressively smaller arteries, which then branch into arterioles. The smallest arteries are narrow enough to offer significant resistance to flow, but the still narrower arterioles are the major sites of resistance in the vascular tree and are therefore analogous to the outflow tubes in the model. This explains the large decrease in mean pressure—from about 90 mmHg to 35 mmHg—as blood flows through the arterioles (see Figure 12–29). Pulse pressure also diminishes to the point that flow beyond the arterioles—that is, through capillaries, venules, and veins—is much less pulsatile.

Like the model's outflow tubes (see Figure 12–33), the arteriolar radii in individual organs are subject to independent adjustment. The blood flow (F) through any organ is represented by the following equation:

$$F_{organ} = (\text{MAP} - \text{venous pressure})/\text{Resistance}_{organ}$$

Venous pressure is normally close to zero, so we may write:

$$F_{organ} = \text{MAP}/\text{Resistance}_{organ}$$

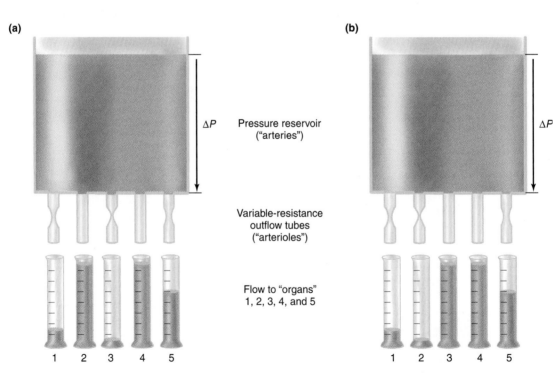

(a)

(b)

ΔP Pressure reservoir ("arteries")

Variable-resistance outflow tubes ("arterioles")

Flow to "organs" 1, 2, 3, 4, and 5

1 2 3 4 5

Figure 12–33

Physical model of the relationship between arterial pressure, arteriolar radius in different organs, and blood-flow distribution. In (a), blood flow is high through tube 2 and low through tube 3, whereas just the opposite is true for (b). This shift in blood flow was achieved by constricting tube 2 and dilating tube 3.

Chapter 12

Because the MAP is identical throughout the body, differences in flows between organs depend entirely on the relative resistances of their respective arterioles. Arterioles contain smooth muscle, which can either relax and cause the vessel radius to increase **(vasodilation)** or contract and decrease the vessel radius **(vasoconstriction).** Thus the pattern of blood-flow distribution depends upon the degree of arteriolar smooth muscle contraction within each organ and tissue. Look back at Figure 12–3, which illustrates the distribution of blood flows at rest; these are due to differing resistances in the various locations. Such distribution can change greatly—as during exercise, for example—by changing the various resistances.

How can resistance be changed? Arteriolar smooth muscle possesses a large degree of spontaneous activity (that is, contraction independent of any neural, hormonal, or paracrine input). This spontaneous contractile activity is called **intrinsic tone** (also called basal tone). It sets a baseline level of contraction that can be increased or decreased by external signals, such as neurotransmitters. These signals act by inducing changes in the muscle cells' cytosolic calcium concentration (see Chapter 9 for a description of excitation-contraction coupling in smooth muscle). An increase in contractile force above the vessel's intrinsic tone causes vasoconstriction, whereas a decrease in contractile force causes vasodilation. The mechanisms controlling vasoconstriction and vasodilation in arterioles fall into two general categories: (1) local controls, and (2) extrinsic (or reflex) controls.

Local Controls

The term **local controls** denotes mechanisms independent of nerves or hormones by which organs and tissues alter their own arteriolar resistances, thereby self-regulating their blood flows. This includes changes caused by autocrine/paracrine agents. This self-regulation includes the phenomena of active hyperemia, flow autoregulation, reactive hyperemia, and local response to injury.

Active Hyperemia

Most organs and tissues manifest an increased blood flow **(hyperemia)** when their metabolic activity is increased (**Figure 12–34a**); this is termed **active hyperemia.** For exam-

ple, the blood flow to exercising skeletal muscle increases in direct proportion to the increased activity of the muscle. Active hyperemia is the direct result of arteriolar dilation in the more active organ or tissue.

The factors that cause arteriolar smooth muscle to relax in active hyperemia are local chemical changes in the extracellular fluid surrounding the arterioles. These result from the increased metabolic activity in the cells near the arterioles. The relative contributions of the different factors implicated vary, depending upon the organs involved and on the duration of the increased activity. Thus, we will list, but not attempt to quantify, the local chemical changes that occur in the extracellular fluid.

Perhaps the most obvious change that occurs when tissues become more active is a decrease in the local concentration of oxygen, which is used in the production of ATP by oxidative phosphorylation. A number of other chemical factors *increase* when metabolism exceeds blood flow, including:

1. carbon dioxide, an end product of oxidative metabolism;
2. hydrogen ions (decrease in pH), for example, from lactic acid;
3. adenosine, a breakdown product of ATP;
4. potassium ions, accumulated from repeated action potential repolarization;
5. eicosanoids, breakdown products of membrane phospholipids;
6. osmolarity, from the breakdown of high-molecular-weight substances;
7. **bradykinin,** a peptide generated locally from a circulating protein called **kininogen** by the action of an enzyme, **kallikrein,** secreted by active gland cells; and
8. **nitric oxide,** a gas released by endothelial cells that acts on the immediately adjacent vascular smooth muscle. Its action will be discussed further in an upcoming section.

Local changes in all these chemical factors have been shown to cause arteriolar dilation under controlled experimental conditions, and they all probably contribute to the

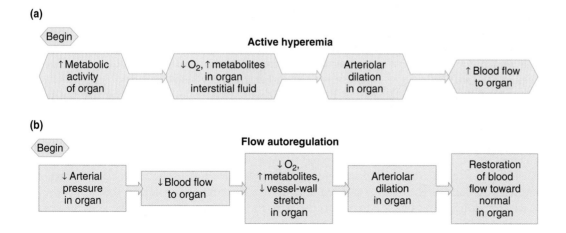

(a)

Begin → ↑Metabolic activity of organ → ↓O₂, ↑metabolites in organ interstitial fluid → Arteriolar dilation in organ → ↑Blood flow to organ

Active hyperemia

(b)

Begin → ↓Arterial pressure in organ → ↓Blood flow to organ → ↓O₂, ↑metabolites, ↓vessel-wall stretch in organ → Arteriolar dilation in organ → Restoration of blood flow toward normal in organ

Flow autoregulation

Figure 12–34

Local control of organ blood flow in response to (a) increases in metabolic activity, and (b) decreases in blood pressure. Decreases in metabolic activity or increases in blood pressure would produce changes opposite those shown here.

active-hyperemia response in one or more organs. It is likely, moreover, that additional important local factors remain to be discovered. All these chemical changes in the extracellular fluid act locally upon the arteriolar smooth muscle, causing it to relax. No nerves or hormones are involved.

It should not be too surprising that active hyperemia is most highly developed in skeletal muscle, cardiac muscle, and glands—tissues that show the widest range of normal metabolic activities in the body. It is highly efficient, therefore, that their supply of blood be primarily determined locally.

Flow Autoregulation

During active hyperemia, increased metabolic activity of the tissue or organ is the initial event leading to local vasodilation. However, locally mediated changes in arteriolar resistance can also occur when a tissue or organ suffers a change in its blood supply resulting from a change in blood pressure (**Figure 12–34b**). The change in resistance is in the direction of maintaining blood flow nearly constant in the face of the pressure change and is therefore termed **flow autoregulation.** For example, when arterial pressure in an organ is reduced, say, because of a partial blockage in the artery supplying the organ, local controls cause arteriolar vasodilation, which tends to maintain relatively constant flow.

What is the mechanism of flow autoregulation? One mechanism comprises the same metabolic factors described for active hyperemia. When an arterial pressure reduction lowers blood flow to an organ, the supply of oxygen to the organ diminishes, and the local extracellular oxygen concentration decreases. Simultaneously, the extracellular concentrations of carbon dioxide, hydrogen ions, and metabolites all increase because the blood cannot remove them as fast as they are produced. Also, eicosanoid synthesis is increased by unclear stimuli. Thus, the local metabolic changes occurring during decreased blood supply at constant metabolic activity are similar to those that occur during increased metabolic activity. This is because both situations reflect an initial imbalance between blood supply and level of cellular metabolic activity. Note, then, that the vasodilations of active hyperemia and of flow autoregulation in response to low arterial pressure do not differ in their major mechanisms, which involve local metabolic factors, but in the event—altered metabolism or altered blood pressure—that brings these mechanisms into play.

Flow autoregulation is not limited to circumstances in which arterial pressure decreases. The opposite events occur when, for various reasons, arterial pressure increases: The initial increase in flow due to the increase in pressure removes the local vasodilator chemical factors faster than they are produced and also increases the local concentration of oxygen. This causes the arterioles to constrict, thereby maintaining a relatively constant local flow in the face of the increased pressure.

Although our description has emphasized the role of local chemical factors in flow autoregulation, another mechanism also participates in this phenomenon in certain tissues and organs. Some arteriolar smooth muscle responds directly to increased stretch, caused by increased arterial pressure, by contracting to a greater extent. Conversely, decreased stretch, due to decreased arterial pressure, causes this vascular smooth muscle to decrease its tone. These direct responses of arteriolar smooth muscle to stretch are termed **myogenic responses.** They are caused by changes in calcium movement into the smooth muscle cells through stretch-sensitive calcium channels in the plasma membrane.

Reactive Hyperemia

When an organ or tissue has had its blood supply completely occluded, a profound transient increase in its blood flow occurs as soon as the occlusion is released. This phenomenon, known as **reactive hyperemia,** is essentially an extreme form of flow autoregulation. During the period of no blood flow, the arterioles in the affected organ or tissue dilate, owing to the local factors described previously. Blood flow, therefore, increases greatly through these wide-open arterioles as soon as the occlusion to arterial inflow is removed. This effect can be demonstrated by wrapping a string tightly around the base of your finger for several minutes. When it is removed, your finger will turn bright red due to the increase in blood flow.

Response to Injury

Tissue injury causes a variety of substances to be released locally from cells or generated from plasma precursors. These substances make arteriolar smooth muscle relax and cause vasodilation in an injured area. This phenomenon, a part of the general process known as inflammation, will be described in detail in Chapter 18.

Extrinsic Controls

Sympathetic Nerves

Most arterioles receive a rich supply of sympathetic postganglionic nerve fibers. These neurons release mainly norepinephrine, which binds to alpha-adrenergic receptors on the vascular smooth muscle to cause vasoconstriction.

In contrast, recall that the receptors for norepinephrine on heart muscle, including the conducting system, are mainly beta-adrenergic. This permits the pharmacologic use of beta-adrenergic antagonists to block the actions of norepinephrine on the heart but not the arterioles, and vice versa for alpha-adrenergic antagonists.

Control of the sympathetic nerves to arterioles can also be used to produce vasodilation. Because the sympathetic nerves are seldom completely quiescent but discharge at some intermediate rate that varies from organ to organ, they always are causing some degree of tonic constriction in addition to the vessels' intrinsic tone. Dilation can be achieved by decreasing the rate of sympathetic activity below this basal level.

The skin offers an excellent example of the role the sympathetic nerves play. At room temperature, skin arterioles are already under the influence of a moderate rate of sympathetic discharge. An appropriate stimulus—cold, fear, or loss of blood, for example—causes reflex enhancement of this sympathetic discharge, and the arterioles constrict further. In contrast, an increased body temperature reflexively inhibits the sympathetic nerves to the skin, the arterioles dilate, and the skin flushes as you radiate body heat.

In contrast to active hyperemia and flow autoregulation, the primary functions of sympathetic nerves to blood vessels are concerned not with the coordination of local metabolic needs and blood flow but with reflexes that serve whole body "needs." The most common reflex employing these nerves is that which regulates arterial blood pressure by influencing arteriolar resistance throughout the body (discussed in detail in the next section). Other reflexes redistribute blood flow to achieve a specific function (as in the previous example, to increase heat loss from the skin).

Parasympathetic Nerves

With few exceptions, there is little or no important parasympathetic innervation of arterioles. In other words, the great majority of blood vessels receive sympathetic but not parasympathetic input. This contrasts with the pattern of dual autonomic innervation seen in most tissues.

Noncholinergic, Nonadrenergic Autonomic Neurons

As described in Chapter 6, there is a population of autonomic postganglionic neurons that are labeled noncholinergic, nonadrenergic neurons because they release neither acetylcholine nor norepinephrine. Instead they release nitric oxide, a vasodilator, and, possibly, other noncholinergic vasodilator substances. These neurons are particularly prominent in the enteric nervous system, which plays a significant role in the control of the gastrointestinal system's blood vessels (Chapter 15). These neurons also innervate arterioles in certain other locations, for example, in the penis, where they mediate erection.

Hormones

Epinephrine, like norepinephrine released from sympathetic nerves, can bind to alpha-adrenergic receptors on arteriolar smooth muscle and cause vasoconstriction. The story is more complex, however, because many arteriolar smooth muscle cells possess the beta-2 subtype of adrenergic receptors as well as alpha-adrenergic receptors, and the binding of epinephrine to beta-2 receptors causes the muscle cells to relax rather than contract (**Figure 12–35**).

In most vascular beds, the existence of beta-2 adrenergic receptors on vascular smooth muscle is of little if any importance because the alpha-adrenergic receptors greatly outnumber them. The arterioles in skeletal muscle are an important exception, however. Because they have a large number of beta-2 adrenergic receptors, circulating epinephrine usually causes vasodilation in this vascular bed.

Another hormone important for arteriolar control is **angiotensin II,** which constricts most arterioles. This peptide is part of the renin-angiotensin system. Another important hormone that causes arteriolar constriction is **vasopressin,** which is released into the blood by the posterior pituitary gland (Chapter 11). The functions of vasopressin and angiotensin II will be described more fully in Chapter 14.

Finally, the hormone secreted by the cardiac atria—**atrial natriuretic peptide**—is a potent vasodilator. Whether this hormone, whose actions on the kidneys are described in

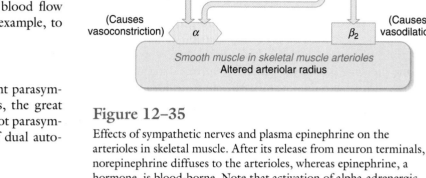

Figure 12–35

Effects of sympathetic nerves and plasma epinephrine on the arterioles in skeletal muscle. After its release from neuron terminals, norepinephrine diffuses to the arterioles, whereas epinephrine, a hormone, is blood-borne. Note that activation of alpha-adrenergic receptors and beta-2 adrenergic receptors produces opposing effects. For simplicity, norepinephrine is shown binding only to alpha-adrenergic receptors; it can also bind to beta-2 adrenergic receptors on the arterioles, but this occurs to a lesser extent.

Chapter 14, plays a widespread physiologic role in the control of arterioles is unsettled.

Endothelial Cells and Vascular Smooth Muscle

It should be clear from the previous sections that many substances can induce the contraction or relaxation of vascular smooth muscle. Many of these substances do so by acting directly on the arteriolar smooth muscle, but others act indirectly via the endothelial cells adjacent to the smooth muscle. Endothelial cells, in response to these latter substances as well as certain mechanical stimuli, secrete several paracrine agents that diffuse to the adjacent vascular smooth muscle and induce either relaxation or contraction, resulting in vasodilation or vasoconstriction, respectively.

One very important paracrine vasodilator released by endothelial cells is nitric oxide; note that we are dealing here with nitric oxide released from endothelial cells, not from nerve endings, as described earlier. (Before the identity of the vasodilator paracrine agent released by the endothelium was determined to be nitric oxide, it was called **endothelium-derived relaxing factor (EDRF),** and this name is still often used because substances other than nitric oxide may also fit this general definition.) Nitric oxide is released continuously in significant amounts by endothelial cells in the arterioles and contributes to arteriolar vasodilation in the basal state. In addition, its secretion rapidly and markedly increases in response to a large number of the chemical mediators involved in both reflex and local control of arterioles. For example, nitric oxide release is stimulated by bradykinin and histamine, substances produced locally during inflammation (Chapter 18).

Another vasodilator the endothelial cells release is the eicosanoid **prostacyclin (PGI$_2$).** Unlike the case for nitric oxide, there is little basal secretion of PGI_2, but secretion can increase markedly in response to various inputs. The roles of PGI_2 in the vascular responses to blood clotting are described in Section F of this chapter.

One of the important *vasoconstrictor* paracrine agents that the endothelial cells release in response to certain mechanical and chemical stimuli is **endothelin-1 (ET-1).** ET-1 is a member of the endothelin family of peptide paracrine agents secreted by a variety of cells in diverse tissues and organs, including the brain, kidneys, and lungs. Not only does ET-1 function as a paracrine agent, but under certain circumstances it can also achieve high enough concentrations in the blood to function as a hormone, causing widespread arteriolar vasoconstriction.

This discussion has so far focused only on arterioles. However, endothelial cells in arteries can also secrete various paracrine agents that influence the arteries' smooth muscle, and thus, their diameters and resistances to flow. The force the flowing blood exerts on the inner surface of the arterial wall (the endothelial cells) is called **shear stress;** it increases as the blood flow through the vessel increases. In response to this increased shear stress, arterial endothelium releases PGI_2, increased amounts of nitric oxide, and less ET-1. These agents cause the arterial vascular smooth muscle to relax and the artery to dilate. This **flow-induced arterial vasodilation** (which should be distinguished from *arteriolar* flow autoregulation) may be important in remodeling the arteries and in optimizing the blood supply to tissues under certain conditions.

Arteriolar Control in Specific Organs

Figure 12–36 summarizes the factors that determine arteriolar radius. The importance of local and reflex controls varies from organ to organ, and **Table 12–5** lists for reference the key features of arteriolar control in specific organs.

Capillaries

As mentioned at the beginning of Section A, at any given moment, approximately 5 percent of the total circulating blood is flowing through the capillaries. It is this 5 percent that is performing the ultimate function of the entire cardiovascular system—the exchange of nutrients, metabolic end products, and cell secretions. Some exchange also occurs in the venules, which can be viewed as extensions of capillaries.

The capillaries permeate almost every tissue of the body. Because most cells are no more than 0.1 mm (only a few cell widths) from a capillary, diffusion distances are very small, and exchange is highly efficient. There are an estimated 25,000 miles of capillaries in an adult, each individual capillary being only about 1 mm long with an inner diameter of 5 μm, just wide enough for an erythrocyte to squeeze through. (For comparison, a human hair is about 100 μm in diameter.)

The essential role of capillaries in tissue function has stimulated many questions concerning how capillaries develop and grow **(angiogenesis).** For example, what activates angiogenesis during wound healing and how do cancers stimulate growth of the new capillaries required for continued cancer growth? It is known that the vascular endothelial cells play a central role in the building of a new capillary network by cell locomotion and cell division. They are stimulated to do so by a variety of **angiogenic factors** (e.g., vascular endothelial growth factor [VEGF]) secreted locally by various tissue cells like fibroblasts, and by the endothelial cells themselves. Cancer cells also secrete angiogenic factors. The development of drugs to interfere with the secretion or action of these factors is a promising research area in anticancer therapy. For example, *angiostatin* is a substance that inhibits blood vessel growth and has been found to reduce the size of almost any tumor (or

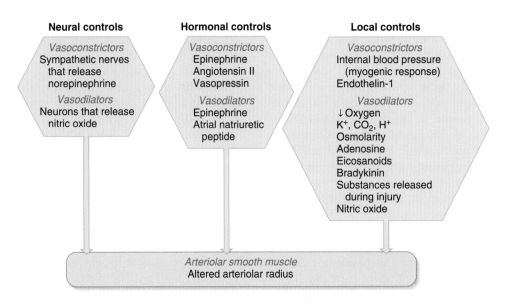

Figure 12–36

Major factors affecting arteriolar radius. Note that epinephrine can be a vasodilator or vasoconstrictor, depending on which adrenergic receptor subtype is present.

Table 12–5   Reference Summary of Arteriolar Control in Specific Organs

Heart

High intrinsic tone; oxygen extraction is very high at rest, so flow must increase when oxygen consumption increases if adequate oxygen supply is to be maintained.

Controlled mainly by local metabolic factors, particularly adenosine, and flow autoregulation; direct sympathetic influences are minor and normally overridden by local factors.

During systole, aortic semilunar cusps block the entrances to the coronary arteries, and vessels within the muscle wall are compressed; thus coronary flow occurs mainly during diastole.

Skeletal Muscle

Controlled by local metabolic factors during exercise.

Sympathetic nerves cause vasoconstriction (mediated by alpha-adrenergic receptors) in reflex response to decreased arterial pressure.

Epinephrine causes vasodilation via beta-2 adrenergic receptors when present in low concentration, and vasoconstriction, via alpha-adrenergic receptors, when present in high concentration.

GI Tract, Spleen, Pancreas, and Liver ("Splanchnic Organs")

Actually two capillary beds partially in series with each other; blood from the capillaries of the GI tract, spleen, and pancreas flows via the portal vein to the liver. In addition, the liver receives a separate arterial blood supply.

Sympathetic nerves cause vasoconstriction, mediated by alpha-adrenergic receptors, in reflex response to decreased arterial pressure and during stress. In addition, venous constriction causes displacement of a large volume of blood from the liver to the veins of the thorax.

Increased blood flow occurs following ingestion of a meal and is mediated by local metabolic factors, neurons, and hormones secreted by the GI tract.

Kidneys

Flow autoregulation is a major factor.

Sympathetic nerves cause vasoconstriction, mediated by alpha-adrenergic receptors, in reflex response to decreased arterial pressure and during stress. Angiotensin II is also a major vasoconstrictor. These reflexes help conserve sodium and water.

Brain

Excellent flow autoregulation.

Distribution of blood within the brain is controlled by local metabolic factors.

Vasodilation occurs in response to increased concentration of carbon dioxide in arterial blood.

Influenced relatively little by the autonomic nervous system.

Skin

Controlled mainly by sympathetic nerves, mediated by alpha-adrenergic receptors; reflex vasoconstriction occurs in response to decreased arterial pressure and cold, whereas vasodilation occurs in response to heat.

Substances released from sweat glands and noncholinergic, nonadrenergic neurons also cause vasodilation.

Venous plexus contains large volumes of blood, which contributes to skin color.

Lungs

Very low resistance compared to systemic circulation.

Controlled mainly by gravitational forces and passive physical forces within the lung.

Constriction, mediated by local factors, occurs in response to low oxygen concentration—just opposite that which occurs in the systemic circulation.

eliminate the tumor completely) in mice. Angiogenesis inhibitors are currently under study in people with cancer.

Anatomy of the Capillary Network

Capillary structure varies considerably from organ to organ, but the typical capillary (**Figure 12–37**) is a thin-walled tube of endothelial cells one layer thick resting on a basement membrane, without any surrounding smooth muscle or elastic tissue. Capillaries in several organs (e.g., the brain) have a second set of cells that surround the basement membrane and affect the ability of substances to penetrate the capillary wall.

The flat cells that constitute the endothelial tube are not attached tightly to each other but are separated by narrow, water-filled spaces termed **intercellular clefts.** The endothelial cells generally contain large numbers of endocytotic and exocytotic vesicles, and sometimes these fuse to form continuous **fused-vesicle channels** across the cell (**Figure 12–37a**).

Blood flow through capillaries depends very much on the state of the other vessels that constitute the microcirculation (**Figure 12–38**). For example, vasodilation of the arterioles supplying the capillaries causes increased capillary flow, whereas arteriolar vasoconstriction reduces capillary flow.

In addition, in some tissues and organs, blood does not enter capillaries directly from arterioles but from vessels called **metarterioles,** which connect arterioles to venules. Metarterioles, like arterioles, contain scattered smooth muscle cells. The site at which a capillary exits from a metarteriole is surrounded by a ring of smooth muscle, the **precapillary sphincter,** which relaxes or contracts in response to local metabolic factors. When contracted, the precapillary sphincter closes the entry to the capillary completely. The more active the tissue, the more precapillary sphincters are open at any moment and the more capillaries in the network are receiving blood. Precapillary sphincters may also exist where the capillaries exit from arterioles.

Velocity of Capillary Blood Flow

Figure 12–39a illustrates a simple mechanical model of a series of 1-cm-diameter balls being pushed down a single tube that branches into narrower tubes. Although each tributary tube has a smaller cross section than the wide tube, the sum of the tributary cross sections is three times greater than that of the wide tube. Let us assume that in the wide tube, each ball moves 3 cm/min. If the balls are 1 cm in diameter and they move two abreast, six balls leave the wide tube and enter the narrow tubes per minute, and six balls leave the narrow tubes per minute. At what speed does each ball move in the small tubes? The answer is 1 cm/min.

This example illustrates the following important principle: When a continuous stream moves through consecutive sets of tubes, the velocity of flow decreases as the sum of the cross-sectional areas of the tubes increases. This is precisely the case in the cardiovascular system (**Figure 12–39b**). The blood velocity is very great in the aorta, slows progressively in the arteries and arterioles, and then slows markedly as the blood passes through the huge cross-sectional area of the

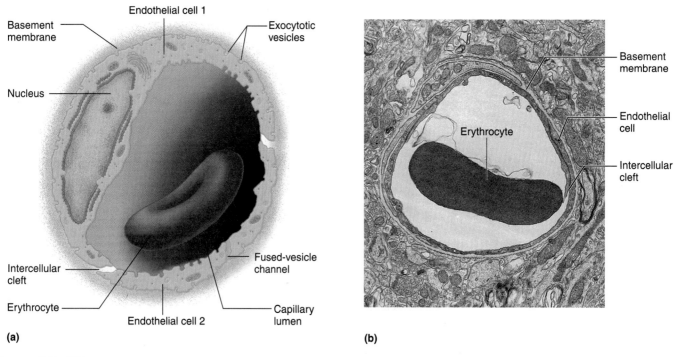

(a)   **(b)**

Figure 12–37

(a) Diagram of a capillary cross section. There are two endothelial cells in the figure, but the nucleus of only one is seen because the other is out of the plane of section. The fused-vesicle channel is part of endothelial cell 2. (b) Electron micrograph of a capillary containing a single erythrocyte; no nuclei are shown in this section. The long dimension of the blood cell is approximately 7 μm.

Figure adapted from Lentz. EM courtesy of Dr. Michael Hart.

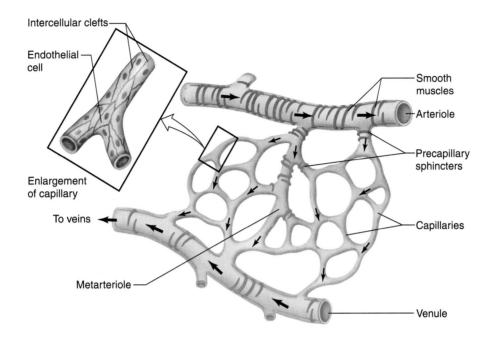

Figure 12–38

Diagram of microcirculation. Note the absence of smooth muscle in the capillaries.

Adapted from Chaffee and Lytle.

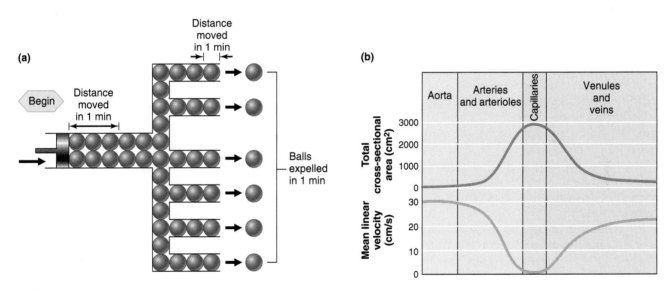

Figure 12–39

Relationship between total cross-sectional area and flow velocity. (a) The total cross-sectional area of the small tubes is three times greater than that of the large tube. Accordingly, flow velocity is one-third as great in the small tubes. (b) Cross-sectional area and velocity in the systemic circulation.

capillaries. Slow forward flow through the capillaries maximizes the time available for substances to exchange between the blood and interstitial fluid. The velocity of flow then progressively increases in the venules and veins because the cross-sectional area decreases. To reemphasize, blood velocity is not dependent on proximity to the heart, but rather on total cross-sectional area of the vessel type.

The cross-sectional area of the capillaries accounts for another important feature of capillaries: Because each capillary is very narrow, it offers considerable resistance to flow,

but the huge total number of capillaries provides such a large cross-sectional area that the total resistance of *all* the capillaries is much lower than that of the arterioles.

Diffusion Across the Capillary Wall: Exchanges of Nutrients and Metabolic End Products

The extremely slow forward movement of blood through the capillaries maximizes the time for substance exchange across the capillary wall. Three basic mechanisms allow substances to move between the interstitial fluid and the plasma: diffusion, vesicle transport, and bulk flow. Mediated transport constitutes a fourth mechanism in the capillaries of the brain. Diffusion and vesicle transport are described in this section, and bulk flow in the next.

In all capillaries, excluding those in the brain, diffusion is the only important means by which net movement of nutrients, oxygen, and metabolic end products occurs across the capillary walls. As described in the next section, there is some movement of these substances by bulk flow, but the amount is negligible.

Chapter 4 described the factors determining diffusion rates. Lipid-soluble substances, including oxygen and carbon dioxide, easily diffuse through the plasma membranes of the capillary endothelial cells. In contrast, ions and polar molecules are poorly soluble in lipid and must pass through small water-filled channels in the endothelial lining.

The presence of water-filled channels in the capillary walls causes the permeability of ions and small polar molecules to be quite high, although still much lower than that of lipid-soluble molecules. One location where these channels exist is in the intercellular clefts—that is, the narrow water-filled spaces between adjacent cells. The fused-vesicle channels that penetrate the endothelial cells provide another set of water-filled channels.

The water-filled channels allow only very small amounts of protein to diffuse through them. Very small amounts of protein may also cross the endothelial cells by vesicle transport—endocytosis of plasma at the luminal border and exocytosis of the endocytotic vesicle at the interstitial side.

Variations in the size of the water-filled channels account for great differences in the "leakiness" of capillaries in different organs. At one extreme are the "tight" capillaries of the brain, which have no intercellular clefts, only tight junctions. Therefore, water-soluble substances, even those of low molecular weight, can gain access to or exit from the brain interstitial space only by carrier-mediated transport through the blood-brain barrier.

At the other end of the spectrum are liver capillaries, which have large intercellular clefts as well as large holes in the plasma membranes of the endothelial cells so that even protein molecules can readily pass across them. This is important because two of the major functions of the liver are the synthesis of plasma proteins and the metabolism of substances bound to plasma proteins.

The leakiness of capillaries in most organs and tissues lies between these extremes of brain and liver capillaries.

What sequence of events is involved in the transfers of nutrients and metabolic end products between capillary blood and cells? Nutrients diffuse first from the plasma across the capillary wall into the interstitial fluid, where they gain entry to cells. Conversely, metabolic end products from the tissues move across the cells' plasma membranes into interstitial fluid, where they diffuse across the capillary endothelium into the plasma.

Transcapillary diffusion gradients for oxygen and nutrients occur as a result of cellular utilization of the substance. Those for metabolic end products arise as a result of cellular production of the substance. Consider two examples: glucose and carbon dioxide in muscle (**Figure 12–40**). Glucose is continuously transported from interstitial fluid into the muscle cell by carrier-mediated transport mechanisms. The removal of glucose from interstitial fluid lowers the interstitial fluid glucose concentration below the glucose concentration in capillary plasma and creates the gradient for glucose diffusion from the capillary into the interstitial fluid.

Simultaneously, carbon dioxide, which is continuously produced by muscle cells, diffuses into the interstitial fluid. This causes the carbon dioxide concentration in interstitial fluid to be greater than that in capillary plasma, producing a gradient for carbon dioxide diffusion from the interstitial fluid into the capillary.

Note that in both examples, metabolism—either utilization or production—of the substance is the event that ultimately establishes the transcapillary diffusion gradients.

If a tissue is to increase its metabolic rate, it must obtain more nutrients from the blood and must eliminate more metabolic end products. One mechanism for achieving that is active hyperemia. The second important mechanism is increased diffusion gradients between plasma and tissue: Increased cellular

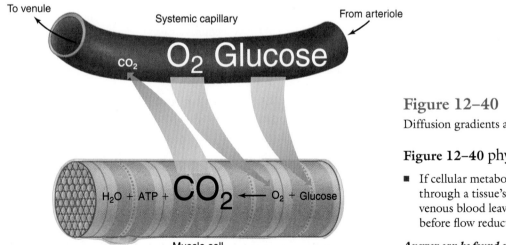

Figure 12–40

Diffusion gradients at a systemic capillary.

Figure 12–40 physiological *inquiry* pi

- If cellular metabolism was not changed, but the blood flow through a tissue's capillaries was reduced, how would the venous blood leaving that tissue differ compared to that before flow reduction?

Answer can be found at end of chapter.

Chapter 12

utilization of oxygen and nutrients lowers their tissue concentrations, whereas increased production of carbon dioxide and other end products raises their tissue concentrations. In both cases, the substance's transcapillary concentration difference increases, which also increases the rate of diffusion.

Bulk Flow Across the Capillary Wall: Distribution of the Extracellular Fluid

At the same time that the diffusional exchange of nutrients, oxygen, and metabolic end products is occurring across the capillaries, another, completely distinct process is also taking place across the capillary—the bulk flow of protein-free plasma. The function of this process is not exchange of nutrients and metabolic end products, but rather distribution of the extracellular fluid (**Figure 12–41**). Recall that extracellular fluid comprises the plasma and interstitial fluid. Normally, there is almost four times more interstitial fluid than plasma—11 L versus 3 L in a 70-kg person. This distribution is not fixed, however, and the interstitial fluid functions as a reservoir that can supply fluid to or receive fluid from the plasma.

As described in the previous section, the capillary wall is highly permeable to water and to almost all plasma solutes, except plasma proteins. Therefore, in the presence of a hydrostatic pressure difference across it, the capillary wall behaves like a porous filter, permitting protein-free plasma to move by bulk flow from capillary plasma to interstitial fluid through the water-filled channels. (This is technically termed *ultrafiltration* but we will refer to it simply as *filtration*.) *The concentrations of all the plasma solutes except protein are virtually the same in the filtering fluid as in plasma.*

The magnitude of the bulk flow is determined, in part, by the difference between the capillary blood pressure and the interstitial-fluid hydrostatic pressure. Normally, the former is much higher than the latter. Therefore, a considerable hydrostatic-pressure difference exists to filter protein-free plasma out of the capillaries into the interstitial fluid, with the protein remaining behind in the plasma.

Why doesn't all the plasma filter out into the interstitial space? The explanation is that the hydrostatic pressure difference favoring filtration is offset by an osmotic force opposing filtration. To understand this, we must review the principle of osmosis.

In Chapter 4, we described how a net movement of water occurs across a semipermeable membrane from a solution of high water concentration to a solution of low water concentration—that is, from a region with a low concentration of solute to which the membrane is impermeable (nonpenetrating solute) to a region of high nonpermeating solute concentration. Moreover, this osmotic flow of water "drags" along with it solutes that can penetrate the membrane. Thus, a difference in water concentration secondary to different concentrations of nonpenetrating solute on the two sides of a membrane can result in the movement of a solution containing both water and penetrating solutes in a manner similar to the bulk flow a hydrostatic pressure difference produces. Units of pressure (mmHg) are used in expressing this osmotic flow across a membrane just as for flow driven by a hydrostatic pressure difference.

This analysis can now be applied to osmotically induced flow across capillaries. The plasma within the capillary and the interstitial fluid outside it contain large quantities of low-molecular-weight penetrating solutes (also termed **crystalloids**); for example, sodium, chloride, and potassium. Because the capillary is highly permeable to all these crystalloids, their concentrations in the two solutions are essentially identical. Thus, the presence of the crystalloids causes no significant difference in water concentration. In contrast, the plasma proteins (also termed **colloids**), being essentially nonpenetrating, have a very low concentration in the interstitial fluid. The difference in protein concentration between the plasma and the interstitial fluid means that the water concentration of the plasma is slightly lower (by about 0.5 percent) than that of interstitial fluid, inducing an osmotic flow of water from the interstitial compartment into the capillary. Because the crystalloids in the interstitial fluid move along with the water, osmotic flow of fluid, like flow driven by a hydrostatic pressure difference, does not alter crystalloid concentrations in either plasma or interstitial fluid.

A key word in this last sentence is *concentrations*. The amount of water (the volume) and the amount of crystalloids in the two locations do change. Thus, an increased filtration of fluid from plasma to interstitial fluid increases the volume of the interstitial fluid and decreases the volume of the plasma, even though no changes in crystalloid concentrations occur.

In summary, opposing forces act to move fluid across the capillary wall (**Figure 12–42a**): (1) the difference between capillary blood hydrostatic pressure and interstitial fluid hydrostatic pressure favors filtration out of the capillary; and (2) the water-concentration difference between plasma and interstitial fluid, which results from differences in protein concentration, favors the **absorption** of interstitial fluid into the capillary. Therefore, the **net filtration pressure (NFP)** depends directly upon the algebraic sum of four variables: capillary hydrostatic pressure, P_c (favoring fluid movement out of the capillary); interstitial hydrostatic pressure, P_{IF} (favoring fluid movement into the capillary); the osmotic force due to plasma protein concentration, π_c (favoring fluid movement into the capillary); and

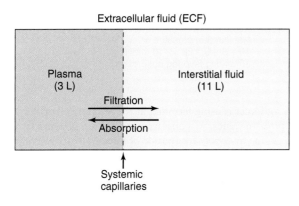

Extracellular fluid (ECF)

Plasma (3 L) | Interstitial fluid (11 L)

Filtration

Absorption

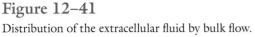

Systemic capillaries

Figure 12–41

Distribution of the extracellular fluid by bulk flow.

the osmotic force due to interstitial fluid protein concentration, π_{IF} (favoring fluid movement out of the capillary). Thus:

$$NFP = P_c + \pi_{IF} - P_{IF} - \pi_c$$

Note that we have arbitrarily assigned a positive value to the forces directed out of the capillary and negative values to the inward-directed forces. The four factors that determine net filtration pressure are termed the **Starling forces** (because Starling, the same physiologist who helped elucidate the Frank-Starling mechanism of the heart, was the first to develop the ideas).

We may now consider this movement quantitatively in the systemic circulation (**Figure 12–42b**). Much of the arterial blood pressure has already dissipated as the blood flows through the arterioles, so that hydrostatic pressure tending to push fluid out of the arterial end of a typical capillary is only about 35 mmHg. The interstitial fluid protein concentration at this end of the capillary would produce a flow of fluid out of the capillary equivalent to a hydrostatic pressure of 3 mmHg. Because the interstitial fluid hydrostatic pressure is virtually zero, the only inward-directed pressure at this end of the capil-

lary is the osmotic pressure due to plasma proteins, with a value of 28 mmHg. Thus, at the arterial end of the capillary, the net outward pressure exceeds the inward pressure by 10 mmHg, so bulk filtration of fluid will occur.

The only substantial difference in the Starling forces at the venous end of the capillary is that the hydrostatic blood pressure (P_c) has decreased from 35 to approximately 15 mmHg due to resistance as blood flowed along the capillary wall. The other three forces are essentially the same as at the arterial end, so the net inward pressure is 10 mmHg greater than the outward pressure, and bulk absorption of fluid will occur. Thus, net movement of fluid from the plasma into the interstitial space at the arterial end of capillaries tends to be balanced by fluid flow in the opposite direction at the venous end of the capillaries. For the aggregate of capillaries in the body, however, there is a small net filtration amounting to approximately 4 L/day (this number does not include the capillaries in the kidneys). The fate of this fluid will be described in the section on the lymphatic system.

In our example, we have assumed a typical capillary hydrostatic pressure varying from 35 down to 15 mmHg. In

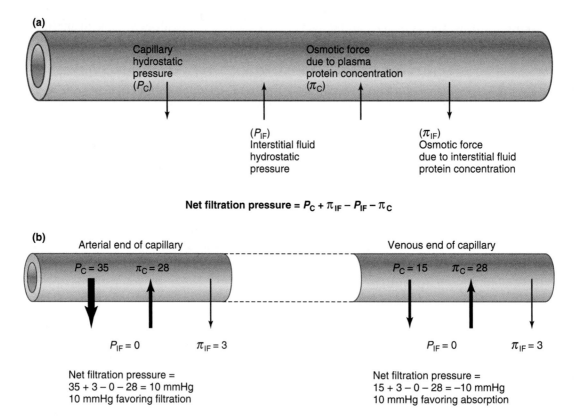

Figure 12–42

(a) The four factors determining fluid movement across capillaries. (b) Quantitation of forces causing filtration at the arterial end of the capillary and absorption at the venous end. Outward forces are arbitrarily assigned positive values, so a positive net filtration pressure favors filtration, while a negative pressure indicates net absorption of fluid will occur. Arrows in (b) denote magnitude of forces. No arrow is shown for interstitial fluid hydrostatic pressure (P_{IF}) in (b) because it is approximately zero.

Figure 12–42 physiological *inquiry* ⓟ𝑖

■ If an accident victim loses 1 L of blood, why would an intravenous injection of a liter of plasma be more effective for replacing the lost volume than injecting a liter of an equally concentrated crystalloid solution?

Answer can be found at end of chapter.

Chapter 12

reality, capillary hydrostatic pressures vary in different regions of the body and, as will be described in a later section, are strongly influenced by whether the person is lying down, sitting, or standing. Moreover, capillary hydrostatic pressure in any given region is subject to physiological regulation, mediated mainly by changes in the resistance of the arterioles in that region. As **Figure 12–43** shows, dilating the arterioles in a particular tissue raises capillary hydrostatic pressure in that region because less pressure is lost overcoming resistance between the arteries and the capillaries. Because of the increased capillary hydrostatic pressure, filtration is increased, and more protein-free fluid is lost to the interstitial fluid. In contrast, marked arteriolar constriction produces decreased capillary hydrostatic pressure and favors net movement of interstitial fluid into the vascular compartment. Indeed, the arterioles supplying a group of capillaries may be so dilated or so constricted that the capillaries manifest only filtration or only absorption, respectively, along their entire length.

We have presented the story of capillary filtration entirely in terms of the Starling forces, but one other factor is involved— the **capillary filtration coefficient.** This is a measure of how much fluid will filter per mmHg net filtration pressure. We previously ignored this factor because in most capillaries, it is not under physiological control. A major exception, however, occurs in the capillaries of the kidneys. As we will see in Chapter 14, certain of the kidney capillaries filter huge quantities of protein-free fluid because they have a very large capillary filtration coefficient that can be altered physiologically.

We must state again that capillary filtration and absorption play no significant role in the exchange of nutrients and metabolic end products between capillary and tissues. The reason is that the total quantity of a substance, such as glucose or carbon dioxide, moving into or out of a capillary as a result of net bulk flow is extremely small in comparison with the quantities moving by net diffusion.

Finally, this analysis of capillary fluid dynamics has considered only the systemic circulation. Precisely the same Starling forces apply to the capillaries in the pulmonary circulation, but the relative values of the four variables differ. In particular, because the pulmonary circulation is a low-resistance, low-pressure circuit, the normal pulmonary capillary hydrostatic pressure—the major force favoring movement of fluid out of the pulmonary capillaries into the interstitium—averages only about 7 mmHg. This is offset by a greater accumulation of proteins in lung interstitial fluid than is found in other tissues. Overall, Starling's forces in the lung slightly favor filtration as in other tissues, but extensive and active lymphatic drainage prevents the accumulation of extracellular fluid in the interstitial spaces and airways. The importance of this will be discussed further in the context of heart failure in Section E of this chapter.

Veins

Blood flows from capillaries into venules and then into veins. Some exchange of materials occurs between the interstitial fluid and the venules, just as in capillaries. Indeed, permeability to macromolecules is often greater for venules than for capillaries, particularly in damaged areas.

The veins are the last set of tubes through which blood flows on its way back to the heart. In the systemic circulation, the force driving this venous return is the pressure difference between the peripheral veins and the right atrium. The pressure in the first portion of the peripheral veins is generally quite low—only 10 to 15 mmHg—because most of the pressure imparted to the blood by the heart is dissipated by resistance as blood flows through the arterioles, capillaries, and venules. The right atrial pressure is normally close to 0 mmHg. Therefore, the total driving pressure for flow from the **peripheral veins** to the right atrium is only 10 to 15 mmHg. (The peripheral veins include all veins not contained within the chest cavity.) This pressure difference is adequate because of the low resistance to flow offered by the veins, which have large diameters. Thus, a major function of the veins is to act as low-resistance conduits for blood flow from the tissues to the heart. The peripheral veins of the arms and legs contain valves that permit flow only toward the heart.

In addition to their function as low-resistance conduits, the veins perform a second important function: Their diameters are reflexly altered in response to changes in blood volume, thereby maintaining peripheral venous pressure and venous return to the heart. In a previous section, we emphasized that the rate of venous return to the heart is a major determinant of end-diastolic ventricular volume and thereby stroke volume. Thus, we now see that peripheral venous pressure is an important determinant of stroke volume. We next describe how venous pressure is determined.

Determinants of Venous Pressure

The factors determining pressure in any elastic tube are the volume of fluid within it and the compliance of its walls. Consequently, total blood volume is one important determinant of venous pressure because, as we will see, at any given moment most blood is in the veins. Also, the walls of veins are thinner and much more compliant than those of arteries. Thus, veins can accommodate large volumes of blood with a

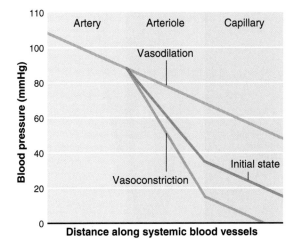

Figure 12–43

Effects of arteriolar vasodilation or vasoconstriction on capillary blood pressure in a single organ (under conditions of constant arterial pressure).

relatively small increase in internal pressure. Approximately 60 percent of the total blood volume is present in the systemic veins at any given moment (**Figure 12–44**), but the venous pressure averages less than 10 mmHg. (In contrast, the systemic arteries contain less than 15 percent of the blood, at a pressure of nearly 100 mmHg.)

The walls of the veins contain smooth muscle innervated by sympathetic neurons. Stimulation of these neurons releases norepinephrine, which causes contraction of the venous smooth muscle, decreasing the diameter and compliance of the vessels and raising the pressure within them. Increased venous pressure then drives more blood out of the veins into the right side of the heart. Note the different effect of venous constriction compared to that of arterioles: When arterioles constrict, it *reduces* forward flow through the systemic circuit, whereas constriction of veins *increases* forward flow. Although the sympathetic nerves are the most important input, venous smooth muscle, like arteriolar smooth muscle, also responds to hormonal and paracrine vasodilators and vasoconstrictors.

Two other mechanisms, in addition to contraction of venous smooth muscle, can increase venous pressure and facilitate venous return. These mechanisms are the **skeletal muscle pump** and the **respiratory pump.** During skeletal muscle contraction, the veins running through the muscle are partially compressed, which reduces their diameter and forces more blood back to the heart. Now we can describe a major function of the peripheral-vein valves: When the skeletal muscle pump raises venous pressure locally, the valves permit blood flow only toward the heart and prevent flow back toward the tissues (**Figure 12–45**).

The respiratory pump is somewhat more difficult to visualize. As Chapter 13 will describe, at the base of the chest cavity (thorax) is a large muscle called the diaphragm, which separates the thorax from the abdomen. During inspiration of air, the diaphragm descends, pushing on the abdominal contents and increasing abdominal pressure. This pressure increase is transmitted passively to the intraabdominal veins. Simultaneously, the pressure in the thorax decreases, thereby decreasing the pressure in the intrathoracic veins and right atrium. The net effect of the pressure changes in the abdomen and thorax is to increase the pressure difference between the peripheral veins and the heart. Thus, venous return is enhanced during inspiration (expiration would reverse this effect if not for the venous valves). The larger the inspiration, the greater the effect. Thus, breathing deeply and frequently, as in exercise, helps blood flow toward the heart.

You might get the (incorrect) impression from these descriptions that venous return and cardiac output are

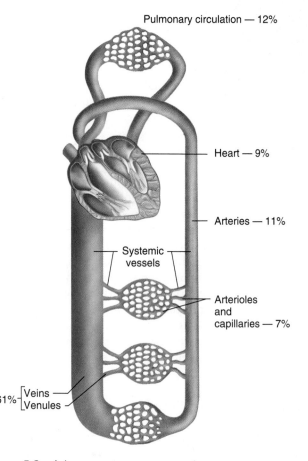

Figure 12–44

Distribution of the total blood volume in different parts of the cardiovascular system.

Adapted from Guyton.

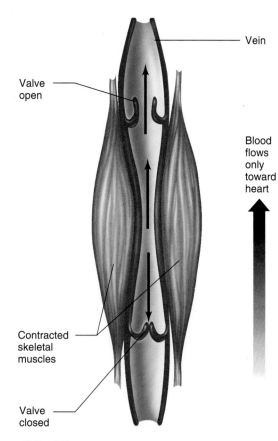

Figure 12–45

The skeletal muscle pump. During muscle contraction, venous diameter decreases, and venous pressure rises. The increase in pressure forces the flow only toward the heart because backward pressure forces the valves in the veins to close.

independent entities. However, any change in venous return almost immediately causes equivalent changes in cardiac output, largely through the Frank-Starling mechanism. *Venous return and cardiac output therefore must be identical except for very brief periods of time.*

In summary (**Figure 12–46**), venous smooth muscle contraction, the skeletal muscle pump, and the respiratory pump all work to facilitate venous return and thereby enhance cardiac output by the same amount.

The Lymphatic System

The **lymphatic system** is a network of small organs (lymph nodes) and tubes (**lymphatic vessels** or simply "lymphatics") through which **lymph**—a fluid derived from interstitial fluid—flows. The lymphatic system is not technically part of the cardiovascular system, but it is described in this chapter because its vessels provide a route for the movement of interstitial fluid to the cardiovascular system (**Figure 12–47a**).

Present in the interstitium of virtually all organs and tissues are numerous **lymphatic capillaries** that are completely

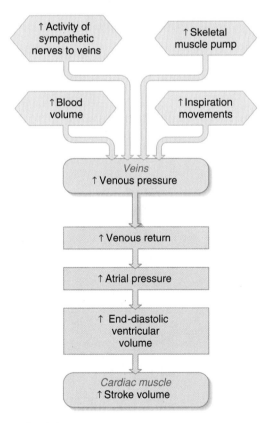

Figure 12–46

Major factors determining peripheral venous pressure, venous return, and stroke volume. Reversing the arrows in the boxes would indicate how these factors can decrease. The effects of increased inspiration on end-diastolic ventricular volume are actually quite complex, but for the sake of simplicity, they are shown here only as increasing venous pressure.

distinct from blood vessel capillaries. Like the latter, they are tubes made of only a single layer of endothelial cells resting on a basement membrane, but they have large water-filled channels that are permeable to all interstitial fluid constituents, including protein. The lymphatic capillaries are the first of the lymphatic vessels, for unlike the blood vessel capillaries, no tubes flow into them.

Small amounts of interstitial fluid continuously enter the lymphatic capillaries by bulk flow (the precise mechanisms that control this movement remain unclear). Now known as lymph, the fluid flows from the lymphatic capillaries into the next set of lymphatic vessels, which converge to form larger and larger lymphatic vessels. At various points in the body—in particular the neck, armpits, groin, and around the intestines—the lymph flows through lymph nodes (**Figure 12–47b**), the function of which is described in Chapter 18. Ultimately, the entire network ends in two large lymphatic ducts that drain into the veins near the junction of the jugular and subclavian veins in the upper chest. Valves at these junctions permit only one-way flow from lymphatic ducts into the veins. Thus, the lymphatic vessels carry interstitial fluid to the cardiovascular system.

The movement of interstitial fluid from the lymphatics to the cardiovascular system is very important because, as noted earlier, the amount of fluid filtered out of all the blood vessel capillaries (except those in the kidneys) exceeds that absorbed by approximately 4 L each day. This 4 L is returned to the blood via the lymphatic system. In the process, the small amounts of protein that leak out of blood vessel capillaries into the interstitial fluid are also returned to the cardiovascular system.

Failure of the lymphatic system due, for example, to occlusion by infectious organisms (as in the disease *elephantiasis*) allows the accumulation of excessive interstitial fluid. The result can be massive swelling of the involved area (**Figure 12–48**). Surgical removal of lymph nodes and vessels during the treatment of breast cancer can similarly allow interstitial fluid to pool in affected tissues. The accumulation of large amounts of interstitial fluid from whatever cause is termed *edema.*

In addition to draining excess interstitial fluid, the lymphatic system provides the pathway by which fat absorbed from the gastrointestinal tract reaches the blood (Chapter 15). The lymphatics also, unfortunately, are often the route by which cancer cells spread from their area of origin to other parts of the body (which is why cancer treatment sometimes includes the removal of lymph nodes).

Mechanism of Lymph Flow

In large part, the lymphatic vessels beyond the lymphatic capillaries propel the lymph within them by their own contractions. The smooth muscle in the wall of the lymphatics exerts a pumplike action by inherent rhythmical contractions. Because the lymphatic vessels have valves similar to those in veins, these contractions produce a one-way flow toward the point at which the lymphatics enter the circulatory system. The lymphatic vessel smooth muscle is responsive to stretch, so when no interstitial fluid accumulates, and therefore no lymph enters the lymphatics, the smooth muscle is inactive.

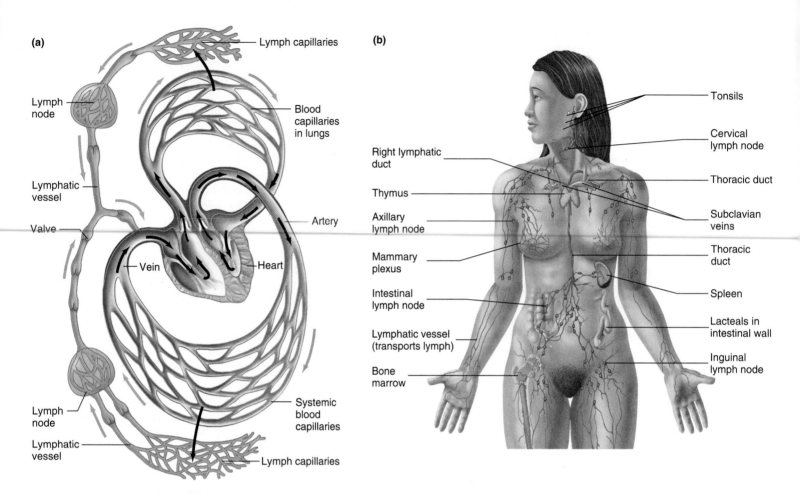

(a)

Lymph capillaries

Lymph node

Lymphatic vessel

Valve

Vein

Lymph node

Lymphatic vessel

Lymph capillaries

Blood capillaries in lungs

Artery

Heart

Systemic blood capillaries

(b)

Right lymphatic duct

Thymus

Axillary lymph node

Mammary plexus

Intestinal lymph node

Lymphatic vessel (transports lymph)

Bone marrow

Tonsils

Cervical lymph node

Thoracic duct

Subclavian veins

Thoracic duct

Spleen

Lacteals in intestinal wall

Inguinal lymph node

Figure 12–47

The lymphatic system (green) in relation to the cardiovascular system (blue and red). (a) The lymphatic system is a one-way system from interstitial fluid to the cardiovascular system. (b) Prior to re-entering the blood at the subclavian veins, lymph flows through lymph nodes in the neck, armpits, groin, and around the intestines.

Figure 12–47 physiological *inquiry* pi

■ How might periodic ingestion of extra fluids be expected to increase the flow of lymph?

Answer can be found at end of chapter.

As lymph formation increases, however, like when there is increased fluid filtration out of blood vessel capillaries, the increased fluid entering the lymphatics stretches the walls and triggers rhythmical contractions of the smooth muscle. This constitutes a negative feedback mechanism for adjusting the rate of lymph flow to the rate of lymph formation and thereby preventing edema.

In addition, the smooth muscle of the lymphatic vessels is innervated by sympathetic neurons, and excitation of these neurons in various physiological states such as exercise may contribute to increased lymph flow. Forces external to the lymphatic vessels also enhance lymph flow. These include the same external forces we described for veins—the skeletal muscle pump and respiratory pump.

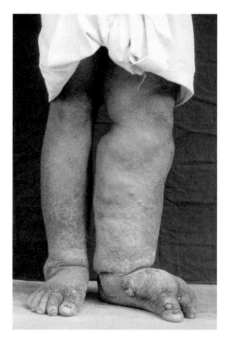

Figure 12–48

Elephantiasis is a disease resulting when mosquito-borne filaria worms block the return of lymph to the vascular system.

Causes of Edema

In addition to the blockage of lymph return discussed previously, other disease states produce edema by affecting one or more of the Starling forces.

Heart failure is a condition in which elevated venous pressure backs blood up into the capillaries, and the elevated hydrostatic pressure (P_c) causes filtration to occur faster than the lymphatics can remove interstitial fluid. The resulting edema can occur in either systemic or pulmonary capillary beds, as we will discuss in an upcoming section.

A more common experience is the swelling that occurs with injury—for example, when you sprain an ankle. Histamine, along with other chemical factors released locally in response to injury, dilate arterioles and thus elevate capillary pressure and filtration (review Figures 12–42 and 12–43). In addition, the chemicals released within injured tissue cause endothelial cells to distort, increasing the size of intercellular clefts and allowing plasma proteins to escape from the bloodstream more readily. This increases the protein osmotic force in the interstitial fluid (π_{IF}), adding to the tendency for filtration and edema to occur.

Finally, an abnormal decrease in plasma protein concentration also can result in edema. This condition reduces the main absorptive force at capillaries (π_c), thus allowing an increase in net filtration. Plasma protein concentration can be reduced by liver disease (decreased plasma protein production) or by kidney disease (loss of protein in the urine). In addition, as with liver disease, protein malnutrition (**kwashiorkor**) compromises the manufacture of plasma proteins. The resulting edema is particularly marked in the interstitial spaces within the abdominal cavity, producing the swollen-belly appearance commonly observed in people with insufficient protein in their diets.■

■ SECTION C SUMMARY

Arteries

I. The arteries function as low-resistance conduits and as pressure reservoirs for maintaining blood flow to the tissues during ventricular relaxation.

II. The difference between maximal arterial pressure (systolic pressure) and minimal arterial pressure (diastolic pressure) during a cardiac cycle is the pulse pressure.

III. Mean arterial pressure can be estimated as diastolic pressure plus one-third pulse pressure.

Arterioles

I. Arterioles, the dominant site of resistance to flow in the vascular system, play major roles in determining mean arterial pressure and in distributing flows to the various organs and tissues.

II. Arteriolar resistance is determined by local factors and by reflex neural and hormonal input.

   a. Local factors that change with the degree of metabolic activity cause the arteriolar vasodilation and increased flow of active hyperemia.

   b. Flow autoregulation, a change in resistance that maintains a constant flow in the face of changing arterial blood pressure, is due to local metabolic factors and to arteriolar myogenic responses to stretch.

   c. The sympathetic nerves, which innervate most arterioles, cause vasoconstriction via alpha-adrenergic receptors. In certain cases, noncholinergic, nonadrenergic neurons that release nitric oxide or other noncholinergic vasodilators also innervate blood vessels.

   d. Epinephrine causes vasoconstriction or vasodilation, depending on the proportion of alpha- and beta-2 adrenergic receptors in the organ.

   e. Angiotensin II and vasopressin cause vasoconstriction.

   f. Some chemical inputs act by stimulating endothelial cells to release vasodilator or vasoconstrictor paracrine agents, which then act on adjacent smooth muscle. These paracrine agents include the vasodilators nitric oxide (endothelium-derived relaxing factor) and prostacyclin, and the vasoconstrictor endothelin-1.

III. Table 12–5 summarizes arteriolar control in specific organs.

Capillaries

I. Capillaries are the site where nutrients and waste products are exchanged between blood and tissues.

II. Blood flows through the capillaries more slowly than through any other part of the vascular system because of the huge cross-sectional area of the capillaries.

III. Capillary blood flow is determined by the resistance of the arterioles supplying the capillaries and by the number of open precapillary sphincters.

IV. Diffusion is the mechanism that exchanges nutrients and metabolic end products between capillary plasma and interstitial fluid.

   a. Lipid-soluble substances move across the entire endothelial wall, whereas ions and polar molecules move through water-filled intercellular clefts or fused-vesicle channels.

   b. Plasma proteins move across most capillaries only very slowly, either by diffusion through water-filled channels or by vesicle transport.

   c. The diffusion gradient for a substance across capillaries arises as a result of cell utilization or production of the substance. Increased metabolism increases the diffusion gradient and increases the rate of diffusion.

V. Bulk flow of protein-free plasma or interstitial fluid across capillaries determines the distribution of extracellular fluid between these two fluid compartments.

   a. Filtration from plasma to interstitial fluid is favored by the hydrostatic pressure difference between the capillary and the interstitial fluid. Absorption from interstitial fluid to plasma is favored by the protein concentration difference between the plasma and the interstitial fluid.

   b. Filtration and absorption do not change the concentrations of crystalloids in the plasma and interstitial fluid because these substances move together with water.

   c. There is normally a small excess of filtration over absorption, which returns fluids to the bloodstream via lymphatic vessels.

Veins

  I. Veins serve as low-resistance conduits for venous return.

 II. Veins are very compliant and contain most of the blood in the vascular system.

   a. Sympathetically mediated vasoconstriction reflexly alters venous diameters so as to maintain venous pressure and venous return.

   b. The skeletal muscle pump and respiratory pump increase venous pressure locally and enhance venous return. Venous valves permit the pressure to produce flow only toward the heart.

The Lymphatic System

  I. The lymphatic system provides a one-way route for movement of interstitial fluid to the cardiovascular system.

 II. Lymph returns the excess fluid filtered from the blood vessel capillaries, as well as the protein that leaks out of the blood vessel capillaries.

 III. Lymph flow is driven mainly by contraction of smooth muscle in the lymphatic vessels, but also by the skeletal muscle pump and the respiratory pump.

Additional Clinical Examples

  I. Disease states that alter the Starling forces can result in edema (e.g., heart failure, tissue injury, liver disease, kidney disease, and protein malnutrition).

■ SECTION C KEY TERMS

absorption 397	kallikrein 389
active hyperemia 389	kininogen 389
angiogenesis 392	Korotkoff's sounds 387
angiogenic factors 392	local control 389
angiotensin II 391	lymph 401
atrial natriuretic peptide 391	lymphatic capillary 401
bradykinin 389	lymphatic system 401
capillary filtration	lymphatic vessel 401
coefficient 399	mean arterial pressure
colloid 397	(MAP) 386
compliance 385	metarteriole 394
crystalloid 397	myogenic response 390
diastolic pressure (DP) 386	net filtration pressure
endothelin-1 (ET-1) 392	(NFP) 397
endothelium-derived relaxing	nitric oxide 389
factor (EDRF) 391	peripheral veins 399
flow autoregulation 390	precapillary sphincter 394
flow-induced arterial	prostacyclin (PGI$_2$) 391
vasodilation 392	pulse pressure 386
fused-vesicle channel 394	reactive hyperemia 390
hyperemia 389	respiratory pump 400
intercellular cleft 394	shear stress 392
intrinsic tone 389	skeletal muscle pump 400

Starling force 398	vasodilation 389
systolic pressure (SP) 386	vasopressin 391
vasoconstriction 389	

■ SECTION C CLINICAL TERMS

angiostatin 392	elephantiasis 401
arteriosclerosis 386	kwashiorkor 403
edema 401	

■ SECTION C REVIEW QUESTIONS

1. Draw the pressure changes along the systemic and pulmonary vascular systems during the cardiac cycle.
2. What are the two main functions of the arteries?
3. What are normal values for systolic, diastolic, and mean arterial pressures? How is mean arterial pressure estimated?
4. What are two major factors that determine pulse pressure?
5. What denotes systolic and diastolic pressure in the measurement of arterial pressure with a sphygmomanometer?
6. What are the major sites of resistance in the systemic vascular system?
7. Name two functions of arterioles.
8. Write the formula relating flow through an organ to mean arterial pressure and to the resistance to flow that organ offers.
9. List the chemical factors thought to mediate active hyperemia.
10. Name the mechanism other than chemical factors that contributes to flow autoregulation.
11. What is the only autonomic innervation of most arterioles? What are the major adrenergic receptors influenced by these nerves? How can control of sympathetic nerves to arterioles achieve vasodilation?
12. Name four hormones that cause vasodilation or vasoconstriction of arterioles, and specify their effects.
13. Describe the role of endothelial paracrine agents in mediating arteriolar vasoconstriction and vasodilation, and give three examples.
14. Draw a flow diagram summarizing the factors affecting arteriolar radius.
15. What are the relative velocities of flow through the various segments of the vascular system?
16. Contrast diffusion and bulk flow. Which mechanism exchanges nutrients, oxygen, and metabolic end products across the capillary wall?
17. What is the only solute that has significant concentration differences across the capillary wall? How does this difference influence water concentration?
18. What four variables determine the net filtration pressure across the capillary wall? Give representative values for each of them in the systemic capillaries.
19. How do changes in local arteriolar resistance influence local capillary pressure?
20. What is the relationship between cardiac output and venous return in the steady state? What is the force driving venous return?
21. Contrast the compliances and blood volumes of the veins and arteries.
22. What three factors influence venous pressure?
23. Approximately how much fluid do the lymphatics return to the blood each day?
24. Describe the forces that cause lymph flow.

SECTION D Integration of Cardiovascular Function: Regulation of Systemic Arterial Pressure

In Chapter 1 we described the fundamental components of all reflex control systems: (1) an internal environmental variable maintained at a relatively stable level, (2) receptors sensitive to changes in this variable, (3) afferent pathways from the receptors, (4) an integrating center that receives and integrates the afferent inputs, (5) efferent pathways from the integrating center, and (6) effectors the efferent pathways "direct" to alter their activities. The control and integration of cardiovascular function will be described in these terms.

The major cardiovascular variable being regulated is the mean arterial pressure in the systemic circulation. This should not be surprising because this pressure is the driving force for blood flow through all the organs except the lungs. Maintaining it is therefore a prerequisite for ensuring adequate blood flow to these organs.

The mean systemic arterial pressure is the arithmetic product of two factors: (1) the cardiac output and (2) the **total peripheral resistance (TPR),** which is the sum of the resistances to flow offered by all the systemic blood vessels.

$$
\begin{array}{ccc}
\text{Mean systemic} & \text{Cardiac} & \text{Total peripheral} \\
\text{arterial pressure} = & \text{output} \times & \text{resistance} \\
\text{(MAP)} & \text{(CO)} & \text{(TPR)}
\end{array}
$$

These two factors, cardiac output and total peripheral resistance, set the mean systemic arterial pressure because they determine the average volume of blood in the systemic arteries over time, and it is this blood volume that causes the pressure. This relationship cannot be emphasized too strongly: *All changes in mean arterial pressure must be the result of changes in cardiac output and/or total peripheral resistance.* Keep in mind that mean arterial pressure will change only if the arithmetic product of cardiac output and total peripheral resistance changes. For example, if cardiac output doubles and total peripheral resistance decreases 50 percent, mean arterial pressure will not change because the product of cardiac output and total peripheral resistance has not changed. Because cardiac output is the volume of blood pumped into the arteries per unit time, it is fairly intuitive that it should be one of the two determinants of mean arterial volume and pressure. The contribution of total peripheral resistance to mean arterial pressure is less obvious, but it can be illustrated with the model introduced previously in Figure 12–33.

As shown in **Figure 12–49**, a pump pushes fluid into a container at the rate of 1 L/min. At steady state, fluid also leaves the container via outflow tubes at a total rate of 1 L/min. Therefore, the height of the fluid column (ΔP), which is the driving pressure for outflow, remains stable. We then disturb the steady state by dilating outflow tube 1, thereby increasing its radius, reducing its resistance, and increasing its flow. The total outflow for the system immediately becomes greater than 1 L/min, and more fluid leaves the reservoir than enters from the pump. Therefore, the volume and, thus, the height of the fluid column begin to decrease until a new steady state between

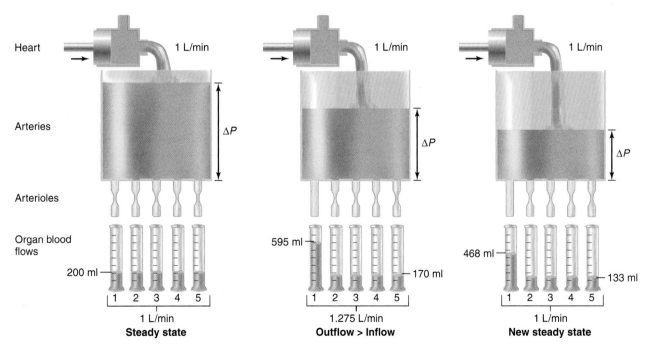

Figure 12–49

Dependence of arterial blood pressure upon total arteriolar resistance. Dilating one arteriolar bed affects arterial pressure and organ blood flow if no compensatory adjustments occur. The middle panel indicates a transient state before the new steady state occurs.

inflow and outflow is reached. In other words, at any given pump input, a change in total outflow resistance must produce changes in the volume and, thus, the height (pressure) in the reservoir.

This analysis can be applied to the cardiovascular system by again equating the pump with the heart, the reservoir with the arteries, and the outflow tubes with various arteriolar beds. As described earlier, small arteries and capillaries offer some resistance to flow, but the major site of resistance in the systemic blood vessels is the arterioles. Moreover, changes in total resistance are normally due to changes in the resistance of arterioles. Therefore, in our discussions we equate total peripheral resistance with total arteriolar resistance.

A physiological analogy to opening outflow tube 1 is exercise: during exercise, the skeletal muscle arterioles dilate, thereby decreasing resistance. If the cardiac output and the arteriolar diameters of all other vascular beds were to remain unchanged, the increased runoff through the skeletal muscle arterioles would cause a decrease in systemic arterial pressure.

We must reemphasize that it is the total arteriolar resistance that influences systemic arterial blood pressure. The distribution of resistances among organs is irrelevant in this regard. **Figure 12–50** illustrates this point. On the right, outflow tube 1 has been opened, as in the previous example, while tubes 2 to 4 have been simultaneously constricted. The increased resistance in tubes 2 to 4 compensates for the decreased resistance in tube 1; therefore total resistance remains unchanged, and the reservoir pressure is unchanged. Total outflow remains 1 L/min, although the distribution of flows is such that flow through tube 1 increases, flow through tubes 2 to 4 decreases, and flow through tube 5 is unchanged.

Applied to the systemic circulation, this process is analogous to altering the distribution of systemic vascular resistances. When the skeletal muscle arterioles (tube 1) dilate during exercise, the total resistance of the systemic circulation can still be maintained if arterioles constrict in other organs, such as the kidneys and gastrointestinal organs (tubes 2 to 4). In contrast, the brain arterioles (tube 5) remain unchanged, ensuring constant brain blood supply.

This type of resistance juggling can maintain total resistance only within limits, however. Obviously, if tube 1 opens very wide, even complete closure of the other tubes cannot prevent total outflow resistance from falling. In that situation, cardiac output must be increased to maintain pressure in the arteries. We will see that this is actually the case during exercise.

We have thus far explained in an intuitive way why cardiac output (CO) and total peripheral resistance (TPR) are the two variables that determine mean systemic arterial pressure. This intuitive approach, however, does not explain specifically why MAP is the arithmetic product of CO and TPR. This relationship can be derived formally from the basic equation relating flow, pressure, and resistance:

$$F = \Delta P / R$$

Rearranging terms algebraically, we have

$$\Delta P = F \times R$$

Because the systemic vascular system is a continuous series of tubes, this equation holds for the entire system—that is, from the arteries to the right atrium. Therefore, the ΔP term is mean systemic arterial pressure (MAP) minus the pressure in the right atrium, F is the cardiac output (CO), and R is the total peripheral resistance (TPR).

$$\text{MAP} - \text{Right atrial pressure} = \text{CO} \times \text{TPR}$$

Because the pressure in the right atrium is very close to 0 mmHg, we can drop this term and we are left with the equation presented earlier:

$$\text{MAP} = \text{CO} \times \text{TPR}$$

This equation is the fundamental equation of cardiovascular physiology. An analogous equation can also be applied to the pulmonary circulation:

$$\text{Mean pulmonary arterial pressure} = \text{CO} \times \text{Total pulmonary vascular resistance}$$

These equations provide a way to integrate almost all the information presented in this chapter. For example, we can now explain why mean pulmonary arterial pressure is much lower than mean systemic arterial pressure. The cardiac output through the pulmonary and systemic arteries is of course the same. Therefore, the pressures can differ only if the resistances differ. Thus, we can deduce that the pulmonary vessels offer much less resistance to flow than do the systemic vessels. In other words, the total pulmonary vascular resistance is lower than the total peripheral resistance.

Figure 12–51 presents the grand scheme of factors that determine mean systemic arterial pressure. None of this

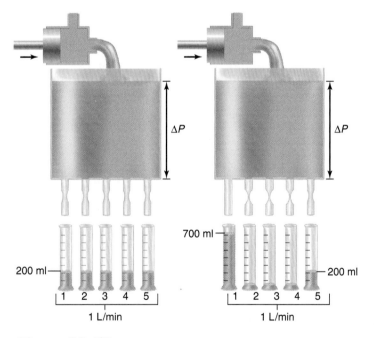

Figure 12–50

Compensation for dilation in one bed by constriction in others. When outflow tube 1 is opened, outflow tubes 2 to 4 are simultaneously tightened so that total outflow resistance, total runoff rate, and reservoir pressure all remain constant.

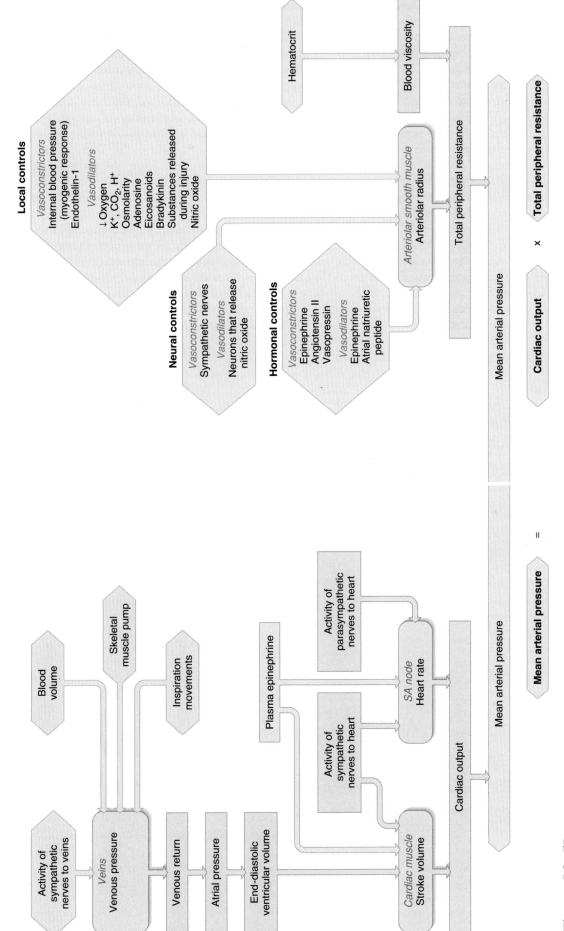

Figure 12–51

Summary of factors that determine systemic arterial pressure, a composite of Figures 12–28, 12–36, and 12–46, with the addition of the effect of hematocrit on resistance.

information is new—all of it was presented in previous figures. A change in only a single variable will produce a change in mean systemic arterial pressure by altering either cardiac output or total peripheral resistance. For example, **Figure 12–52** illustrates how the decrease in blood volume during hemorrhage leads to a decrease in mean arterial pressure.

Conversely, any deviation in arterial pressure, such as that occurring during hemorrhage, will elicit homeostatic reflexes so that cardiac output and/or total peripheral resistance will change in the direction required to minimize the initial change in arterial pressure.

In the short term—seconds to hours—these homeostatic adjustments to mean arterial pressure are brought about by reflexes called the baroreceptor reflexes. They utilize mainly changes in the activity of the autonomic nerves supplying the heart and blood vessels, as well as changes in the secretion of the hormones (epinephrine, angiotensin II, and vasopressin) that influence these structures. Over longer time spans, the baroreceptor reflexes become less important, and factors controlling blood volume play a dominant role in determining blood pressure. The next two sections describe these phenomena.

Baroreceptor Reflexes

Arterial Baroreceptors

It is only logical that the reflexes that homeostatically regulate arterial pressure originate primarily with arterial receptors that respond to changes in pressure. High in the neck, each of the two major vessels supplying the head, the common carotid arteries, divides into two smaller arteries (**Figure 12–53**). At this division, the wall of the artery is thinner than usual and contains a large number of branching, vine-like nerve endings. This portion of the artery is called the **carotid sinus** (the term *sinus* denotes a recess, space, or dilated channel). Its nerve endings are highly sensitive to stretch or distortion. The degree of wall stretching is directly related to the pressure within the artery. Thus, the carotid sinuses serve as pressure receptors, or **baroreceptors.** An area functionally similar to the carotid sinuses is found in the arch of the aorta and is termed the **aortic arch baroreceptor.** The two carotid sinuses and the aortic arch baroreceptor constitute the **arterial baroreceptors.** Afferent neurons travel from them to the brainstem and provide input to the neurons of cardiovascular control centers there.

Action potentials recorded in single afferent fibers from the carotid sinus demonstrate the pattern of baroreceptor response (**Figure 12–54**). In this experiment, the pressure in the carotid sinus is artificially controlled so that the pressure is steady, and not pulsatile (i.e., not varying as usual between

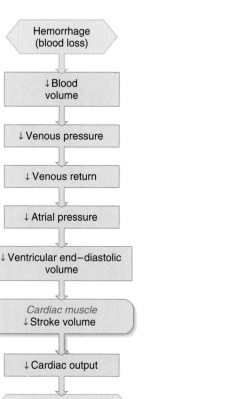

Figure 12–52

Sequence of events by which a decrease in blood volume leads to a decrease in mean arterial pressure.

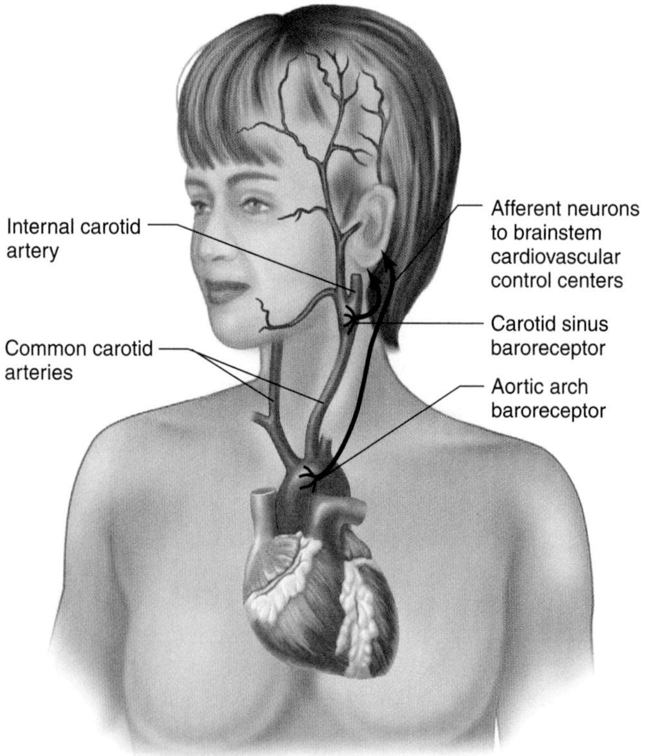

Figure 12–53

Locations of arterial baroreceptors.

Figure 12–53 physiological *inquiry* (pi)

■ When you first stand up after getting out of bed, how does the pressure detected by these baroreceptors change?

Answer can be found at end of chapter.

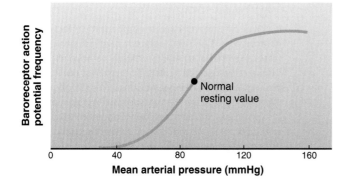

Figure 12–54

Effect of changing mean arterial pressure (MAP) on the firing of action potentials by afferent neurons from the carotid sinus. This experiment is done by pumping blood in a nonpulsatile manner through an isolated carotid sinus so as to be able to set the pressure inside it at any value desired.

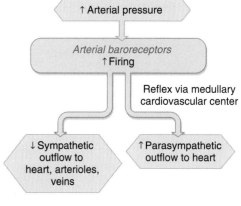

Figure 12–55

Neural components of the arterial baroreceptor reflex. If the initial change were a decrease in arterial pressure, all the arrows in the boxes would be reversed.

systolic and diastolic pressure). At a particular steady pressure, for example, 100 mmHg, there is a certain rate of discharge by the neuron. This rate can be increased by raising the arterial pressure, or it can be decreased by lowering the pressure. Thus, the rate of discharge of the carotid sinus is directly proportional to the mean arterial pressure.

If the experiment is repeated using the same mean pressures as before but allowing pressure pulsations, it is found that at any given mean pressure, the larger the pulse pressure, the faster the rate of firing by the carotid sinus. This responsiveness to pulse pressure adds a further element of information to blood pressure regulation, because small changes in factors such as blood volume may cause changes in arterial pulse pressure with little or no change in mean arterial pressure.

The Medullary Cardiovascular Center

The primary integrating center for the baroreceptor reflexes is a diffuse network of highly interconnected neurons called the **medullary cardiovascular center,** located in the medulla oblongata. The neurons in this center receive input from the various baroreceptors. This input determines the action potential frequency from the cardiovascular center along neural pathways that terminate upon the cell bodies and dendrites of the vagus (parasympathetic) neurons to the heart and the sympathetic neurons to the heart, arterioles, and veins. When the arterial baroreceptors increase their rate of discharge, the result is a decrease in sympathetic outflow to the heart, arterioles, and veins, and an increase in parasympathetic outflow to the heart (**Figure 12–55**). A decrease in baroreceptor firing rate results in the opposite pattern.

As parts of the baroreceptor reflexes, angiotensin II generation and vasopressin secretion are also altered to help restore blood pressure. Decreased arterial pressure elicits increased plasma concentrations of both these hormones, which raise arterial pressure by constricting arterioles. Chapter 14 will further describe the roles of angiotensin II and vaso-

pressin in terms of their effects on salt and water balance via the kidneys.

Operation of the Arterial Baroreceptor Reflex

Our description of the arterial baroreceptor reflex is now complete. If arterial pressure decreases, as during a hemorrhage (**Figure 12–56**), the discharge rate of the arterial baroreceptors also decreases. Fewer impulses travel up the afferent nerves to the medullary cardiovascular center, and this induces (1) increased heart rate because of increased sympathetic activity to the heart and decreased parasympathetic activity, (2) increased ventricular contractility because of increased sympathetic activity to the ventricular myocardium, (3) arteriolar constriction because of increased sympathetic activity to the arterioles (and increased plasma concentrations of angiotensin II and vasopressin), and (4) increased venous constriction because of increased sympathetic activity to the veins. The net result is an increased cardiac output (increased heart rate and stroke volume), increased total peripheral resistance (arteriolar constriction), and return of blood pressure toward normal.

Conversely, an increase in arterial blood pressure for any reason causes increased firing of the arterial baroreceptors, which reflexly induces a compensatory decrease in cardiac output and total peripheral resistance.

Having emphasized the great importance of the arterial baroreceptor reflex, we must now add an equally important qualification. The baroreceptor reflex functions primarily as a short-term regulator of arterial blood pressure. It is activated instantly by any blood pressure change and functions to restore blood pressure rapidly toward normal. Yet, if arterial pressure deviates from its normal set point for more than a few days, the arterial baroreceptors adapt to this new pressure; that is, they have a decreased frequency of action potential firing at any given pressure. Thus, in patients who have chronically elevated blood pressure, the arterial baroreceptors continue to oppose minute-to-minute changes in blood pressure, but at the higher set point.

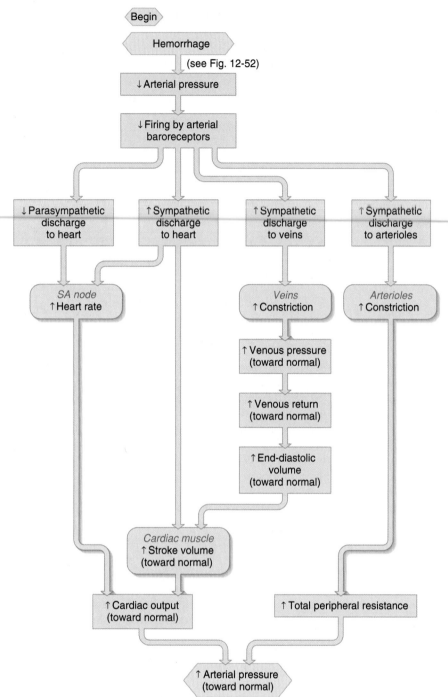

Figure 12–56

Figure 12–56

Arterial baroreceptor reflex compensation for hemorrhage. The compensatory mechanisms do not restore arterial pressure completely to normal. The increases designated "toward normal" are relative to prehemorrhage values; for example, the stroke volume is increased reflexly "toward normal" relative to the low point caused by the hemorrhage (i.e., before the reflex occurs), but it does not reach the level it had prior to the hemorrhage. For simplicity, the fact that plasma angiotensin II and vasopressin are also reflexly increased and help constrict arterioles is not shown.

Other Baroreceptors

The large systemic veins, the pulmonary vessels, and the walls of the heart also contain baroreceptors, most of which function in a manner analogous to the arterial baroreceptors. By keeping brain cardiovascular control centers constantly informed about changes in the systemic venous, pulmonary, atrial, and ventricular pressures, these other baroreceptors provide a further degree of regulatory sensitivity. In essence, they contribute a feedforward component of arterial pressure control. For example, a slight decrease in ventricular pressure reflexly increases the activity of the sympathetic nervous system even before the change lowers cardiac output and arterial pressure far enough for the arterial baroreceptors to detect it.

Blood Volume and Long-Term Regulation of Arterial Pressure

The fact that the arterial baroreceptors (and other baroreceptors as well) adapt to prolonged changes in pressure means that the baroreceptor reflexes cannot set long-term arterial pressure. The major mechanism for long-term regulation occurs through the blood volume. As described earlier, blood volume is a major determinant of arterial pressure because it influences, in turn, venous pressure, venous return, end-diastolic volume, stroke volume, and cardiac output. Thus, increased blood volume increases arterial pressure. But the opposite causal chain also exists—an increased arterial pressure reduces blood volume (more specifically, the plasma component of the blood volume) by increasing the excretion of salt and water by the kidneys, as will be described in Chapter 14.

Figure 12–57 illustrates how these two causal chains constitute negative feedback loops that determine both blood volume and arterial pressure. An increase in blood pressure, whatever the reason, causes a decrease in blood volume, which tends to bring the blood pressure back down. An increase in the blood volume, whatever the reason, raises the blood pressure, which tends to bring the blood volume back down. The important point is this: Because arterial pressure influences blood volume but blood volume also influences arterial pressure, blood pressure can stabilize, in the long run, only at a value at which blood volume is also stable. Consequently, changes in steady-state blood volume are the single most important long-term determinant of blood pressure.

Other Cardiovascular Reflexes and Responses

Stimuli acting upon receptors other than baroreceptors can initiate reflexes that cause changes in arterial pressure. For example, the following stimuli all cause an increase in blood pressure: decreased arterial oxygen concentration; increased arterial carbon dioxide concentration; decreased blood flow to the brain; and pain originating in the skin. In contrast, pain originating in the viscera or joints may cause *decreases* in arterial pressure.

Many physiological states such as eating and sexual activity are also associated with changes in blood pressure. For example, attending a stressful business meeting may raise mean blood pressure by as much as 20 mmHg, walking increases it 10 mmHg, and sleeping lowers it 10 mmHg. Mood also has a significant effect on blood pressure, which tends to be lower when people report that they are happy than when they are angry or anxious.

These changes are triggered by input from receptors or higher brain centers to the medullary cardiovascular center or, in some cases, to pathways distinct from these centers. For example, the fibers of certain neurons whose cell bodies are in the cerebral cortex and hypothalamus synapse directly on the sympathetic neurons in the spinal cord, bypassing the medullary center altogether.

There is a considerable degree of flexibility and integration in the control of blood pressure. For example, in an experimental animal, electrical stimulation of a discrete area of the hypothalamus elicits all the usually observed neurally mediated cardiovascular responses to an acute emotional situation. Stimulation of other brain sites elicits cardiovascular changes appropriate to the maintenance of body temperature, feeding, or sleeping. It seems such outputs are "preprogrammed." The complete pattern can be triggered by a natural stimulus that initiates the flow of information to the appropriate brain control center.

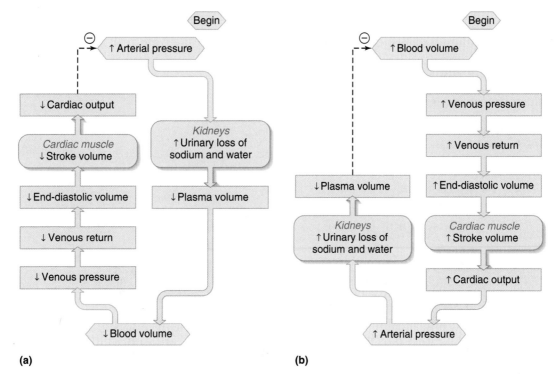

Figure 12–57

Causal reciprocal relationships between arterial pressure and blood volume. (a) An increase in arterial pressure due, for example, to an increased cardiac output induces a decrease in blood volume by promoting fluid excretion by the kidneys. This tends to restore arterial pressure to its original value. (b) An increase in blood volume due, for example, to increased fluid ingestion induces an increase in arterial pressure, which tends to restore blood volume to its original value by promoting fluid excretion by the kidneys. Because of these relationships, blood volume is a major determinant of arterial pressure.

Elevated Intracranial Pressure

A number of different circumstances can cause elevated pressure in the brain, including the presence of a rapidly growing cancerous tumor or a traumatic head injury that triggers internal hemorrhage or edema. What distinguishes these situations from similar problems elsewhere in the body is the fact that the enclosed bony cranium does not allow physical swelling toward the outside; consequently, pressure is directed inward. This inward pressure exerts a collapsing force on intracranial vasculature, and the reduction in radius greatly increases the resistance to blood flow (recall that resistance increases as the fourth power of a decrease in radius). Blood flow is thus reduced below the level needed to satisfy metabolic requirements, and brain oxygen levels

fall, while carbon dioxide and other metabolic wastes increase. Accumulated metabolites in the brain interstitial fluid powerfully stimulate sympathetic neurons controlling systemic arterioles, resulting in a large increase in TPR and, consequently, a major elevation in MAP (MAP = CO × TPR). In principle, this elevated systemic pressure is adaptive, in that it can overcome the collapsing pressures and force blood to flow through the brain once again. In practice, removal of the tumor or accumulated fluid is the only way to restore brain blood flow at a normal mean arterial pressure. This association of a rise in intracranial pressure with dramatically increased arterial blood pressure is known as **Cushing's phenomenon** (not to be confused with Cushing's syndrome or disease, which are described in Chapter 11).∎

■ SECTION D SUMMARY

I. Mean arterial pressure, the primary regulated variable in the cardiovascular system, equals the product of cardiac output and total peripheral resistance.

II. The factors that determine cardiac output and total peripheral resistance are summarized in Figure 12–51.

Baroreceptor Reflexes

I. The primary baroreceptors are the arterial baroreceptors, including the two carotid sinuses and the aortic arch. Nonarterial baroreceptors are located in the systemic veins, pulmonary vessels, and walls of the heart.

II. The firing rates of the arterial baroreceptors are proportional to mean arterial pressure and to pulse pressure.

III. An increase in firing of the arterial baroreceptors due to an increase in pressure causes, by way of the medullary cardiovascular center, an increase in parasympathetic outflow to the heart and a decrease in sympathetic outflow to the heart, arterioles, and veins. The result is a decrease in cardiac output and total peripheral resistance and, thus, a decrease in mean arterial pressure. The opposite occurs when the initial change is a decrease in arterial pressure.

Blood Volume and Long-Term Regulation of Arterial Pressure

I. The baroreceptor reflexes are short-term regulators of arterial pressure but adapt to a maintained change in pressure.

II. The most important long-term regulator of arterial pressure is the blood volume.

Other Cardiovascular Reflexes and Responses

I. Blood pressure can be influenced by many factors other than baroreceptors, including arterial blood gas concentrations, pain, emotions and sexual activity.

Additional Clinical Examples

I. Cushing's phenomenon is a clinical condition in which traumatic head injury leads to increased intracranial pressure, decreased brain blood flow, and a large increase in arterial blood pressure.

■ SECTION D KEY TERMS

aortic arch baroreceptor 408	medullary cardiovascular
arterial baroreceptors 408	center 409
baroreceptors 408	total peripheral resistance
carotid sinus 408	(TPR) 405

■ SECTION D CLINICAL TERMS

Cushing's phenomenon 412

■ SECTION D REVIEW QUESTIONS

1. Write the equation relating mean arterial pressure to cardiac output and total peripheral resistance.

2. What variable accounts for the fact that mean pulmonary arterial pressure is lower than mean systemic arterial pressure?

3. Draw a flow diagram illustrating the factors that determine mean arterial pressure.

4. Identify the receptors, afferent pathways, integrating center, efferent pathways, and effectors in the arterial baroreceptor reflex.

5. When the arterial baroreceptors decrease or increase their rate of firing, what changes in autonomic outflow and cardiovascular function occur?

6. Describe the role of blood volume in the long-term regulation of arterial pressure.

Hemorrhage and Other Causes of Hypotension

The term **hypotension** means a low blood pressure, regardless of cause. One general cause of hypotension is a loss of blood volume, as for example in a hemorrhage, which produces hypotension by the sequence of events shown previously in Figure 12–52. The most serious consequences of hypotension are reduced blood flow to the brain and cardiac muscle.

The immediate counteracting response to hemorrhage is the arterial baroreceptor reflex, as summarized in Figure 12–56.

Figure 12–58, which shows how five variables change over time when blood volume decreases, adds a further degree of clarification to Figure 12–56. The values of factors changed as a direct result of the hemorrhage—stroke volume, cardiac output, and mean arterial pressure—are restored by the baroreceptor reflex toward, but not all the way to, normal. In contrast, values not altered directly by the hemorrhage but only by the reflex response to hemorrhage—heart rate and total peripheral resistance—increase above their pre-hemorrhage values. The increased peripheral resistance results from increases in sympathetic outflow to the arterioles in many vascular beds but not those of the heart and brain. Thus, skin blood flow may decrease considerably because of arteriolar vasoconstriction—this is why the skin becomes cold and pale. Kidney and intestinal blood flow also decrease because the usual functions of these organs are less immediately essential for life.

A second important type of compensatory mechanism (one not shown in Figure 12–56) involves the movement of interstitial fluid into capillaries. This occurs because both the drop in blood pressure and the increase in arteriolar constriction decrease capillary hydrostatic pressure, thereby favoring the absorption of interstitial fluid (**Figure 12–59**). Thus, the initial event—blood loss and decreased blood volume—is in large part compensated for by the movement of interstitial fluid into the vascular system. This mechanism, referred to as **autotransfusion,** can restore the blood volume to virtually normal levels within 12 to 24 hours after a moderate hemorrhage (**Table 12–6**). At this time, the entire restoration of blood volume is due to expansion of the plasma volume and, thus, the hematocrit actually decreases.

The early compensatory mechanisms for hemorrhage (the baroreceptor reflexes and interstitial fluid absorption) are highly efficient, so that losses of as much as 1.5 L of blood—approximately 30 percent of total blood volume—can be sustained with only slight reductions of mean arterial pressure or cardiac output.

We must emphasize that absorption of interstitial fluid only *redistributes* the extracellular fluid. Ultimate *replacement* of

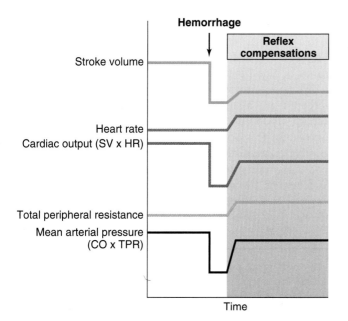

Figure 12–58

Five simultaneous graphs showing the time course of cardiovascular effects of hemorrhage. Note that the entire decrease in arterial pressure immediately following hemorrhage is secondary to the decrease in stroke volume and, thus, cardiac output. This figure emphasizes the relative proportions of the "increase" and "decrease" arrows of Figure 12–56. All variables shown are increased relative to the state immediately following the hemorrhage, but not necessarily to the state prior to the hemorrhage.

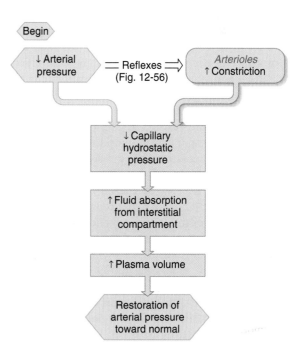

Figure 12–59

The autotransfusion mechanism compensates for blood loss by causing interstitial fluid to move into the capillaries.

Table 12-6	Fluid Shifts After Hemorrhage		
	Normal	Immediately After Hemorrhage	18 h After Hemorrhage
Total blood volume, ml	5000	4000	4900
Erythrocyte volume, ml	2300	1840	1840
Plasma volume, ml	2700	2160	3060

Table 12–6 physiological *inquiry* (pi)

- Calculate the hematocrit before and 18 hours after the hemorrhage, and explain the changes that are observed.

Answer can be found at end of chapter.

the fluid lost involves the control of fluid ingestion and kidney function. These slower-acting compensations are mediated by hormones, including renin, angiotensin and aldosterone, and are described in Chapter 14. Replacement of the lost erythrocytes requires the hormone **erythropoietin** to stimulate **erythropoiesis** (maturation of immature red blood cells); this is described in detail in the last section of this chapter. These replacement processes require days to weeks in contrast to the rapidly occurring reflex compensations illustrated in Figure 12–59.

Hemorrhage is a striking example of hypotension due to a decrease in blood volume. There is a second way, however, that hypotension can occur due to volume depletion that does not result from loss of whole blood. In such cases the loss is of salts—particularly sodium (along with chloride or bicarbonate)—and water. Such fluid loss may occur through the skin, as in severe sweating or burns. It may occur via the gastrointestinal tract, as through diarrhea or vomiting, or by unusually large urinary losses. Regardless of the route, the loss of fluid decreases circulating blood volume and produces symptoms and compensatory cardiovascular changes similar to those seen in hemorrhage.

Hypotension may also be caused by events other than blood or fluid loss. One major cause is a depression of cardiac pumping ability (for example, during a heart attack).

Another cause is strong emotion, which in rare cases can cause hypotension and fainting. Somehow, the higher brain centers involved with emotions inhibit sympathetic activity to the cardiovascular system and enhance parasympathetic activity to the heart, resulting in a markedly decreased arterial pressure and brain blood flow. This whole process, known as *vasovagal syncope,* is usually transient. It should be noted that the fainting that sometimes occurs in a person donating blood is usually due to hypotension brought on by emotion, not the blood loss, because losing 0.5 L of blood will not generally cause serious hypotension.

Massive release of endogenous substances that relax arteriolar smooth muscle may also cause hypotension by reducing total peripheral resistance. An important example is the hypotension that occurs during severe allergic responses (Chapter 18).

Shock

The term **shock** denotes any situation in which a decrease in blood flow to the organs and tissues damages them. Arterial pressure is usually, but not always, low in shock, and the classification of shock is quite similar to what we have already seen for hypotension. **Hypovolemic shock** is caused by a decrease in blood volume secondary to hemorrhage or loss of fluid other than blood. **Low-resistance shock** is due to a decrease in total peripheral resistance secondary to excessive release of vasodilators, as in allergy and infection. **Cardiogenic shock** is due to an extreme decrease in cardiac output from any of a variety of factors (for example, during a heart attack).

The cardiovascular system, especially the heart, suffers damage if shock is prolonged. As the heart deteriorates, cardiac output further declines and shock becomes progressively worse. Ultimately, shock may become irreversible even though blood transfusions and other appropriate therapy may temporarily restore blood pressure. See Chapter 19 for a detailed description of a clinical case involving low-resistance shock.

The Upright Posture

A decrease in the effective circulating blood volume occurs in the circulatory system when going from a lying, horizontal position to a standing, vertical one. Why this is so requires an understanding of the action of gravity upon the long, continuous columns of blood in the vessels between the heart and the feet.

The pressures we have given in previous sections of this chapter are for an individual in the horizontal position, in which all blood vessels are at approximately the same level as the heart. In this position, the weight of the blood produces negligible pressure. In contrast, when a person is vertical, the intravascular pressure everywhere becomes equal to the pressure generated by cardiac contraction plus an additional pressure equal to the weight of a column of blood from the heart to the point of measurement. In an average adult, for example, the weight of a column of blood extending from the heart to the feet amounts to 80 mmHg. In a foot capillary, therefore, the pressure increases from 25 (the average capillary pressure resulting from cardiac contraction) to 105 mmHg, the extra 80 mmHg being due to the weight of the column of blood.

This increase in pressure due to gravity influences the effective circulating blood volume in several ways. First, the increased hydrostatic pressure that occurs in the legs (as well as the buttocks and pelvic area) when a person is standing pushes outward on the highly distensible vein walls, causing marked distension. The result is pooling of blood in the veins; that is, much of the blood emerging from the capillaries simply goes into expanding the veins rather than returning to the heart. Simultaneously, the increase in capillary pressure caused by the gravitational force produces increased filtration of fluid out of the capillaries into the interstitial space. This is why our feet swell during prolonged standing. The combined effects of venous pooling and increased capillary filtration reduce the effective circulating blood volume very similarly to the effects caused by a mild hemorrhage. This explains why a person may sometimes feel faint upon standing up suddenly. This feeling is normally very transient, however, because the decrease in arterial pressure immediately causes reflex baroreceptor-mediated compensatory adjustments virtually identical to those shown in Figure 12–56 for hemorrhage.

The effects of gravity can be offset by contraction of the skeletal muscles in the legs. Even gentle contractions of the leg muscles without movement produce intermittent, complete emptying of the leg veins so that uninterrupted columns of venous blood from the heart to the feet no longer exist (**Figure 12–60**). The result is a decrease in both venous distension and pooling plus a significant reduction in capillary hydrostatic pressure and fluid filtration out of the capillaries. This phenomenon is illustrated by the fact that soldiers may faint while standing at attention for long periods of time because of minimal leg muscle contractions. Here, fainting may be considered adaptive because the venous and capillary pressure changes induced by gravity are eliminated once the person is horizontal. The pooled venous blood is mobilized, and the filtered fluid is absorbed back into the capillaries. Thus, the wrong thing to do for a person who has fainted is to hold him or her upright.

Exercise

During exercise, cardiac output may increase from a resting value of 5 L/min to a maximal value of 35 L/min in trained athletes. **Figure 12–61** illustrates the distribution of this cardiac output during strenuous exercise. As expected, most of the increase in cardiac output goes to the exercising muscles, but there are also increases in flow to skin if it beomes necessary to dissipate heat, and to the heart, required for the additional work performed by the heart in pumping the increased cardiac output. The increases in flow through these three vascular beds are the result of arteriolar vasodilation in them. In both skeletal and cardiac muscle, local metabolic factors mediate the vasodilation, whereas the vasodilation in skin is achieved mainly by a decrease in the firing of the sympathetic neurons to the skin. At the same time that arteriolar vasodilation is occurring

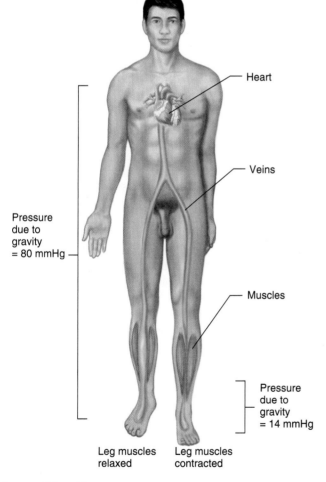

Figure 12–60

Role of contraction of the leg skeletal muscles in reducing capillary pressure and filtration in the upright position. The skeletal muscle contraction compresses the veins, causing intermittent emptying so that the columns of blood are interrupted.

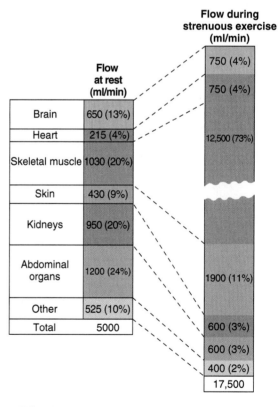

Figure 12–61

Distribution of the systemic cardiac output at rest and during strenuous exercise. The values at rest were previously presented in Figure 12–3.

Adapted from Chapman and Mitchell.

in these three beds, arteriolar vasoconstriction—manifested as decreased blood flow in Figure 12–61—is occurring in the kidneys and gastrointestinal organs, secondary to increased activity of the sympathetic neurons supplying them.

Vasodilation of arterioles in skeletal muscle, cardiac muscle, and skin causes a decrease in total peripheral resistance to blood flow. This decrease is partially offset by vasoconstriction of arterioles in other organs. Such resistance "juggling," however, is quite incapable of compensating for the huge dilation of the muscle arterioles, and the net result is a decrease in total peripheral resistance.

What happens to arterial blood pressure during exercise? As always, the mean arterial pressure is simply the arithmetic product of cardiac output and total peripheral resistance. During most forms of exercise (**Figure 12–62** illustrates the case for mild exercise), the cardiac output tends to increase somewhat more than the total peripheral resistance decreases, so that mean arterial pressure usually increases a small amount. Pulse pressure, in contrast, significantly increases because of an increase in both stroke volume and the speed at which the stroke volume is ejected. It should be noted that by "exercise," we are referring to cyclic contraction and relaxation of muscles occurring over a period of time, like jogging. A single, intense isometric contraction of muscles has a very different effect on blood pressure, and will be described shortly.

The increase in cardiac output during exercise is due to a large increase in heart rate and a small increase in stroke volume. The increase in heart rate is caused by a combination of decreased parasympathetic activity to the SA node and increased sympathetic activity. The increased stroke volume is due mainly to an increased ventricular contractility, manifested by an increased ejection fraction and mediated by the sympathetic nerves to the ventricular myocardium.

Note, however, in Figure 12–62 that there is a small increase (about 10 percent) in end-diastolic ventricular volume. Because of this increased filling, the Frank-Starling mechanism also contributes to the increased stroke volume, although not to the same degree as the increased contractility does. The increased contractility also accounts for the greater speed at which the stroke volume is ejected, as noted in the previous discussion of pulse pressure.

We have focused our attention on factors that act directly upon the heart to alter cardiac output during exercise. By themselves, however, these factors are insufficient to account for the elevated cardiac output. The fact is that cardiac output can be increased to high levels only if the peripheral processes favoring venous return to the heart are simultaneously activated to the same degree. Otherwise, the shortened filling time resulting from the high heart rate would lower end-diastolic volume, and thus stroke volume, by the Frank-Starling mechanism.

Factors promoting venous return during exercise are: (1) increased activity of the skeletal muscle pump, (2) increased depth and frequency of inspiration (the respiratory pump), (3) sympathetically mediated increase in venous tone, and (4) greater ease of blood flow from arteries to veins through the dilated skeletal muscle arterioles.

What control mechanisms elicit the cardiovascular changes in exercise? As described previously, vasodilation of arterioles in skeletal and cardiac muscle once exercise is underway represents active hyperemia as a result of local metabolic factors within the muscle. But what drives the enhanced sympathetic outflow to most other arterioles, the heart, and the veins, and the decreased parasympathetic outflow to the heart? The control of this autonomic outflow during exercise offers an excellent example of what we earlier referred to as a preprogrammed pattern, modified by continuous afferent input. One or more discrete control centers in the brain are activated during exercise by output from the cerebral cortex, and descending pathways from these centers to the appropriate autonomic preganglionic neurons elicit the firing pattern typical of exercise. These centers become active, and changes to cardiac and vascular function occur even before exercise begins. Thus, this constitutes a feedforward system.

Once exercise is underway, local chemical changes in the muscle can develop, particularly during intense exercise, because of imperfect matching between blood flow and metabolic demands. These changes activate chemoreceptors in the muscle. Afferent input from these receptors goes to the medullary cardiovascular center and facilitates the output reaching the autonomic neurons from higher brain centers (**Figure 12–63**). The result is a further increase in heart rate,

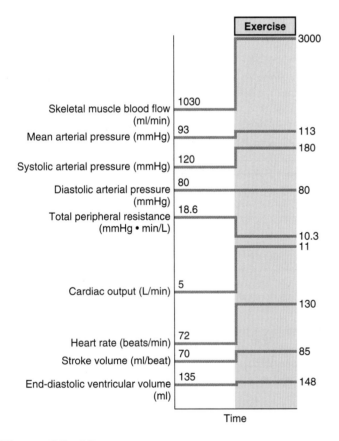

Figure 12–62

Summary of cardiovascular changes during mild upright exercise like jogging. The person was sitting quietly prior to the exercise. Total peripheral resistance was calculated from mean arterial pressure and cardiac output.

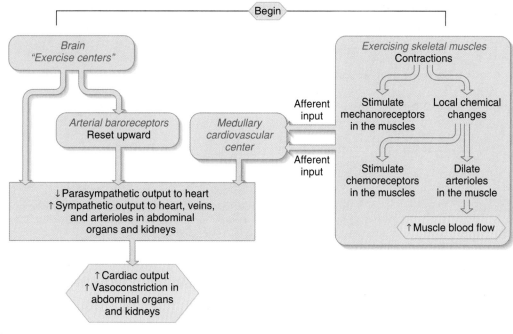

Figure 12–63

Control of the cardiovascular system during exercise. The primary outflow to the sympathetic and parasympathetic neurons is via pathways from "exercise centers" in the brain. Afferent input from mechanoreceptors and chemoreceptors in the exercising muscles and from reset arterial baroreceptors also influences the autonomic neurons by way of the medullary cardiovascular center.

myocardial contractility, and vascular resistance in the non-active organs. Such a system permits a fine degree of matching between cardiac pumping and total oxygen and nutrients required by the exercising muscles. Mechanoreceptors in the exercising muscles are also stimulated and provide input to the medullary cardiovascular center.

Finally, the arterial baroreceptors also play a role in the altered autonomic outflow. Knowing that the mean and pulsatile pressures rise during exercise, you might logically assume that the arterial baroreceptors will respond to these elevated pressures and signal for increased parasympathetic and decreased sympathetic outflow, a pattern designed to counter the rise in arterial pressure. In reality, however, exactly the opposite occurs; the arterial baroreceptors play an important role in *elevating* the arterial pressure over that existing at rest. The reason is that one neural component of the central command output goes to the arterial baroreceptors and "resets" them upward as exercise begins. This resetting causes the baroreceptors to respond as though arterial pressure had decreased, and their output (decreased action potential frequency) signals for decreased parasympathetic and increased sympathetic outflow.

Table 12–7 summarizes the changes that occur during moderate exercise—that is, exercise (like jogging, swimming, or fast walking) that involves large muscle groups for an extended period of time.

In closing, let's return to the other major category of exercise, which involves maintained high-force, slow shortening-velocity contractions, as in weightlifting. Here, too, cardiac output and arterial blood pressure increase, and the

arterioles in the exercising muscles undergo vasodilation due to local metabolic factors. However, there is a crucial difference: During maintained contractions, once the contracting muscles exceed 10 to 15 percent of their maximal force, the blood flow to the muscle is greatly reduced because the muscles are physically compressing the blood vessels that run through them. In other words, the arteriolar vasodilation is completely overcome by the physical compression of the blood vessels. Thus, the cardiovascular changes are ineffective in causing increased blood flow to the muscles, and these contractions can be maintained only briefly before fatigue sets in. Moreover, because of the compression of blood vessels, total peripheral resistance may go up considerably (instead of down, as in endurance exercise), contributing to a large rise in mean arterial pressure during the contraction. Frequent exposure of the heart to only this type of exercise can cause maladaptive changes in the left ventricle, including wall hypertrophy and diminished chamber volume.

Maximal Oxygen Consumption and Training

As the intensity of any endurance exercise increases, oxygen consumption also increases in exact proportion until reaching a point when it fails to rise despite a further increment in workload. This is known as **maximal oxygen consumption ($\dot{V}_{O_2}$max).** After $\dot{V}_{O_2}$max has been reached, work can be increased and sustained only very briefly by anaerobic metabolism in the exercising muscles.

Theoretically, $\dot{V}_{O_2}$max could be limited by (1) the cardiac output, (2) the respiratory system's ability to deliver oxygen to the blood, or (3) the exercising muscles' ability to use oxygen. In fact, in typical, healthy people (except for a few very highly

Table 12–7 Cardiovascular Changes During Moderate Exercise

Variable	Change	Explanation
Cardiac output	Increases	Heart rate and stroke volume both increase, the former to a much greater extent.
Heart rate	Increases	Sympathetic nerve activity to the SA node increases, and parasympathetic nerve activity decreases.
Stroke volume	Increases	Contractility increases due to increased sympathetic nerve activity to the ventricular myocardium; increased ventricular end-diastolic volume also contributes to increased stroke volume by the Frank-Starling mechanism.
Total peripheral resistance	Decreases	Resistance in heart and skeletal muscles decreases more than resistance in other vascular beds increases.
Mean arterial pressure	Increases	Cardiac output increases more than total peripheral resistance decreases.
Pulse pressure	Increases	Stroke volume and velocity of ejection of the stroke volume increase.
End-diastolic volume	Increases	Filling time is decreased by the high heart rate, but the factors favoring venous return—venoconstriction, skeletal muscle pump, and increased inspiratory movements—more than compensate for it.
Blood flow to heart and skeletal muscle	Increases	Active hyperemia occurs in both vascular beds, mediated by local metabolic factors.
Blood flow to skin	Increases	Sympathetic nerves to skin vessels are inhibited reflexly by the increase in body temperature.
Blood flow to viscera	Decreases	Sympathetic nerves to the blood vessels in the abdominal organs and kidneys are stimulated.
Blood flow to brain	Increases slightly	Autoregulation of brain arterioles maintains constant flow despite the increased mean arterial pressure.

trained athletes), cardiac output is the factor that determines $\dot{V}_{O_2}$max. With increasing workload (**Figure 12–64**), heart rate increases progressively until it reaches a maximum. Stroke volume increases much less and tends to level off at 75 percent of $\dot{V}_{O_2}$max (it actually starts to go back down in elderly people). The major factors responsible for limiting the rise in stroke volume and, thus, cardiac output are (1) the very rapid heart rate, which decreases diastolic filling time, and (2) inability of the peripheral factors favoring venous return (skeletal muscle pump, respiratory pump, venous vasoconstriction, arteriolar vasodilation) to increase ventricular filling further during the very short time available.

A person's $\dot{V}_{O_2}$max is not fixed at any given value but can be altered by his or her habitual level of physical activity. For example, prolonged bed rest may decrease $\dot{V}_{O_2}$max by 15 to 25 percent, whereas intense, long-term physical training may increase it by a similar amount. To be effective, the training must be endurance-type exercise and must reach certain minimal levels of duration, frequency, and intensity. For example, jogging 20 to 30 min three times weekly at 5 to 8 mi/h definitely produces a significant training effect in most people.

At rest, compared to values prior to training, the trained individual has an increased stroke volume and decreased heart rate with no change in cardiac output (see Figure 12–64). At $\dot{V}_{O_2}$max, cardiac output is increased compared to pretraining values; this is due entirely to an increased maximal stroke volume since training does not alter maximal heart rate (see Figure 12–64). The increase in stroke volume is due to a combination of (1) effects on the heart (remodeling of the ventricular walls produces moderate hypertrophy and an increase in chamber size), and (2) peripheral effects, including increased blood volume and increases in the number of blood vessels in skeletal muscle, which permit increased muscle blood flow and venous return.

Training also increases the concentrations of oxidative enzymes and mitochondria in the exercised muscles. These changes increase the speed and efficiency of metabolic reactions in the muscles and permit 200 to 300 percent increases in exercise endurance, but they do not increase $\dot{V}_{O_2}$max because they were not limiting it in the untrained individuals.

Aging is associated with significant changes in the heart's performance during exercise. Most striking is a decrease in the

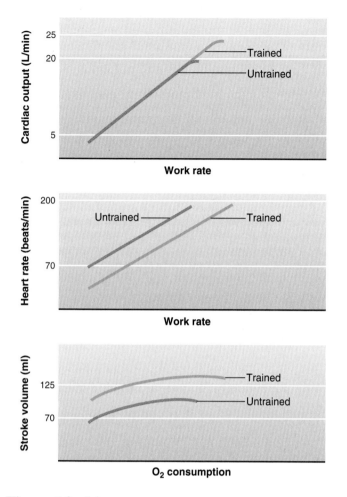

Figure 12–64

Changes in cardiac output, heart rate, and stroke volume with increasing workload in untrained and trained individuals.

maximum heart rate (and thus cardiac output) achievable. This results in particular from increased stiffness of the heart, and thus a decrease in the ability to rapidly fill during diastole.

Hypertension

Hypertension is defined as a chronically increased systemic arterial pressure. Although the clinical definition of hypertension is a blood pressure above 140/90 mmHg, new guidelines suggest that interventions to lower blood pressure should be instituted at systolic pressures of 130 to 139 mmHg, and diastolic pressures of 85 to 89 mmHg.

Theoretically, hypertension could result from an increase in cardiac output or in total peripheral resistance, or both. In reality, however, the major abnormality in most cases of well-established hypertension is increased total peripheral resistance caused by abnormally reduced arteriolar radius.

What causes the arteriolar constriction? In only a small fraction of cases is the cause known. For example, diseases that damage a kidney or decrease its blood supply are often associ-

ated with **renal hypertension.** The cause of the hypertension is increased release of renin from the kidneys, with subsequent increased generation of the potent vasoconstrictor angiotensin II. However, for more than 95 percent of the individuals with hypertension, the cause of the arteriolar constriction is unknown. Hypertension of unknown cause is called **primary hypertension** (formerly "essential hypertension").

Many hypotheses have been proposed to explain the increased arteriolar constriction of primary hypertension. At present, much evidence suggests that excessive sodium retention is a contributing factor in genetically predisposed ("salt-sensitive") persons. Many people with hypertension show a drop in blood pressure after being on low-sodium diets or receiving drugs, called **diuretics,** that cause increased sodium and water loss in the urine. Low dietary intake of calcium has also been implicated as a possible contributor to primary hypertension. Obesity and a sedentary lifestyle are definite risk factors, and weight reduction and exercise are frequently effective in causing some reduction of blood pressure in people with hypertension. Cigarette smoking, too, is a definite risk factor.

Hypertension causes a variety of problems. One of the organs most affected is the heart. Because the left ventricle in a hypertensive person must chronically pump against an increased arterial pressure (afterload), it develops an adaptive increase in muscle mass called **left ventricular hypertrophy.** In the early phases of the disease, this hypertrophy helps maintain the heart's function as a pump. With time, however, changes in the organization and properties of myocardial cells occur, and these result in diminished contractile function and heart failure. The presence of hypertension also enhances the possible development of atherosclerosis and heart attacks, kidney damage, and **stroke**—the rupture of a cerebral blood vessel, causing localized brain damage. Long-term data on the link between blood pressure and health show that for every 20 mmHg increase in systolic pressure and every 10 mmHg increase in diastolic pressure, the risk of heart disease and stroke doubles.

The major categories of drugs used to treat hypertension are summarized in **Table 12–8.** These drugs all act in ways that reduce cardiac output and/or total peripheral resistance. You will note in subsequent sections of this chapter that these same drugs are also used to treat heart failure and in both the prevention and treatment of heart attacks. One reason for this overlap is that these three diseases are causally interrelated. For example, as noted in this section, hypertension is a major risk factor for the development of heart failure and heart attacks. But the drugs often have multiple cardiovascular effects, which may play different roles in the treatment of the different diseases.

Heart Failure

Heart failure (also called **congestive heart failure**) is a collection of signs and symptoms that occur when the heart fails to pump an adequate cardiac output. This may happen for many reasons; two examples are pumping against a chronically elevated arterial pressure in hypertension, and undergoing structural damage due to decreased coronary blood flow. It has become

Table 12–8   Drugs Used to Treat Hypertension

1. **Diuretics:** These drugs increase urinary excretion of sodium and water (Chapter 14). They tend to decrease cardiac output with little or no change in total peripheral resistance.

2. **Beta-adrenergic receptor blockers:** These drugs exert their antihypertensive effects mainly by reducing cardiac output.

3. **Calcium channel blockers:** These drugs reduce the entry of calcium into vascular smooth muscle cells, causing them to contract less strongly and lowering total peripheral resistance. (Surprisingly, it has been found that despite their effectiveness in lowering blood pressure, at least some of these drugs may significantly increase the risk of a heart attack. Consequently, their use as therapy for hypertension is now under intensive review.)

4. **Angiotensin-converting enzyme (ACE) inhibitors:** As Chapter 14 will describe, the final step in the formation of angiotensin II, a vasoconstrictor, is mediated by an enzyme called angiotensin-converting enzyme. Drugs that block this enzyme therefore reduce the concentration of angiotensin II in plasma, which causes arteriolar vasodilation, lowering total peripheral resistance. The same effect can be achieved with drugs that block the receptors for angiotensin II. A reduction in plasma angiotensin II or blockage of its receptors is also protective against the development of heart wall changes that lead to heart failure.

5. Drugs that antagonize one or more components of the sympathetic nervous system: The major effect of these drugs is to reduce sympathetic mediated stimulation of arteriolar smooth muscle and thereby reduce total peripheral resistance. Examples include drugs that inhibit the brain centers that mediate the sympathetic outflow to arterioles, and drugs that block alpha-adrenergic receptors on the arterioles.

standard practice to separate people with heart failure into two categories: (1) those with diastolic dysfunction (problems with ventricular filling) and (2) those with systolic dysfunction (problems with ventricular ejection). Many people with heart failure, however, exhibit elements of both categories.

In **diastolic dysfunction,** the wall of the ventricle has reduced compliance. That is, it is abnormally stiff and therefore has a reduced ability to fill adequately at normal diastolic filling pressures. The result is a reduced end-diastolic volume (even though the end-diastolic pressure in the stiff ventricle may be quite high) and, therefore, a reduced stroke volume by the Frank-Starling mechanism. Note that in pure diastolic dysfunction, ventricular compliance is decreased, but ventricular contractility is normal.

Several situations may lead to decreased ventricular compliance, but by far the most common is the existence of systemic hypertension. As noted in the previous section, the left ventricle, pumping chronically against an elevated arterial pressure, hypertrophies. The structural and biochemical changes associated with this hypertrophy make the ventricle stiff and less able to expand.

In contrast to diastolic dysfunction, **systolic dysfunction** results from myocardial damage due, for example, to a heart attack (discussed next). This type of dysfunction is characterized by a decrease in cardiac contractility—a lower stroke volume at any given end-diastolic volume. This is manifested as a decrease in ejection fraction and, as illustrated in **Figure 12–65,** a downward shift of the ventricular function curve. The affected ventricle does not hypertrophy, but note that the end-diastolic volume increases.

The reduced cardiac output of heart failure, regardless of whether it is due to diastolic or systolic dysfunction, triggers the arterial baroreceptor reflexes. In this situation these

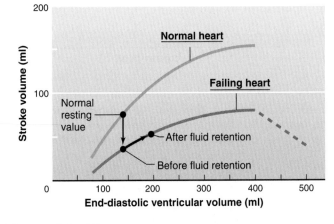

Figure 12–65

Relationship between end-diastolic ventricular volume and stroke volume in a normal heart and one with heart failure due to systolic dysfunction (decreased contractility). The normal curve was shown previously in Figure 12–24. With decreased contractility, the ventricular function curve is displaced downward; that is, there is a lower stroke volume at any given end-diastolic volume. Fluid retention causes an increase in end-diastolic volume and restores stroke volume toward normal by the Frank-Starling mechanism. Note that this compensation occurs even though contractility—the basic defect—has not been altered by the fluid retention.

reflexes are elicited more than usual because, for unknown reasons, the afferent baroreceptors are less sensitive. In other words, the baroreceptors discharge less rapidly than normal at any given mean or pulsatile arterial pressure, and the brain interprets this decreased discharge as a larger-than-usual fall in pressure. The results of the reflexes are (1) heart rate is increased through increased sympathetic and decreased parasympathetic discharge to the heart, and (2) total peripheral resistance is increased by increased sympathetic discharge to systemic arterioles as well as by increased plasma concentrations of the two major hormonal vasoconstrictors—angiotensin II and vasopressin.

The reflex increases in heart rate and total peripheral resistance are initially beneficial in restoring cardiac output and arterial pressure, just as if the changes in these parameters had been triggered by hemorrhage.

Maintained chronically throughout the period of cardiac failure, the baroreceptor reflexes also bring about fluid retention and an expansion—often massive—of the extracellular volume. This is because, as Chapter 14 describes, the neuroendocrine efferent components of the reflexes cause the kidneys to reduce their excretion of sodium and water. The retained fluid then causes expansion of the extracellular volume. Because the plasma volume is part of the extracellular fluid volume, plasma volume also increases. This in turn increases venous pressure, venous return, and end-diastolic ventricular volume, which tends to restore stroke volume toward normal by the Frank-Starling mechanism (see Figure 12–65). Thus, fluid retention is also, at least initially, an adaptive response to decreased cardiac output.

However, problems emerge as the fluid retention progresses. For one thing, when a ventricle with systolic dysfunction (as opposed to a normal ventricle) becomes very distended with blood, its force of contraction actually decreases and the situation worsens. Second, the fluid retention, with its accompanying elevation in venous pressure, causes edema—accumulation of interstitial fluid. Why does an increased venous pressure cause edema? The capillaries drain via venules into the veins; so when venous pressure increases, the capillary pressure also increases and causes increased filtration of fluid out of the capillaries into the interstitial fluid.

Thus, most of the fluid retained by the kidneys ends up as extra interstitial fluid rather than extra plasma. Swelling of the legs and feet is particularly prominent.

Most important in this regard, failure of the *left* ventricle—whether due to diastolic or systolic dysfunction—leads to **pulmonary edema,** the accumulation of fluid in the interstitial spaces of the lung or in the air spaces themselves, which impairs gas exchange. The reason for such accumulation is that the left ventricle fails to pump blood to the same extent as the right ventricle, so the volume of blood in all the pulmonary vessels increases. The resulting engorgement of pulmonary capillaries raises the capillary pressure above its normally very low value, causing filtration to occur at a rate faster than the lymphatics can remove the fluid. This situation usually worsens at night. During the day, because of the patient's upright posture, fluid accumulates in the legs; then the fluid is slowly absorbed back into the capillaries when the patient lies down at night, thus expanding the plasma volume and precipitating an attack of pulmonary edema.

Another component of the reflex response to heart failure that is at first beneficial but ultimately becomes maladaptive is the increase in total peripheral resistance, mediated by the sympathetic nerves to arterioles and by angiotensin II and vasopressin. By chronically maintaining the arterial blood pressure the failing heart must pump against, this increased resistance makes the failing heart work much harder.

One obvious treatment for heart failure is to correct, if possible, the precipitating cause (for example, hypertension). **Table 12–9** lists the types of drugs most often used for treatment. Finally, although cardiac transplantation is often the treatment of choice, the paucity of donor hearts, the high costs, and the challenges of postsurgical care render it a feasible option for only a very small number of patients. Considerable research has also been directed toward the development of artificial hearts, though success has been limited to date.

Table 12–9	**Types of Drugs Used to Treat Heart Failure**

1. ***Diuretics:*** Drugs that increase urinary excretion of sodium and water (Chapter 14). These drugs eliminate the excessive fluid accumulation contributing to edema and/or worsening myocardial function.

2. ***Cardiac inotropic drugs:*** Drugs such as ***digitalis*** that increase ventricular contractility by increasing cytosolic calcium concentration in the myocardial cell. The use of these drugs is currently controversial, however, because although they clearly improve the symptoms of heart failure, they do not prolong life and, in some studies, seem to have shortened it.

3. ***Vasodilator drugs:*** Drugs that lower total peripheral resistance and thus the arterial blood pressure (afterload) the failing heart must pump against. Some inhibit a component of the sympathetic nervous pathway to the arterioles, whereas others [angiotensin-converting enzyme (ACE) inhibitors] block the formation of angiotensin II (see Chapter 14). In addition, the ACE inhibitors prevent or reverse the maladaptive remodeling of the myocardium that is mediated by the elevated plasma concentration of angiotensin II in heart failure.

4. ***Beta-adrenergic receptor blockers:*** Drugs that block the major adrenergic receptors in the myocardium. The mechanism by which this action improves heart failure is unknown (indeed, you might have predicted that such an action, by blocking sympathetically induced increases in cardiac contractility, would be counterproductive). One hypothesis is that excess sympathetic stimulation of the heart reflexly produced by the decreased cardiac output of heart failure may cause an excessive elevation of cytosolic calcium concentration, which would lead to cell apoptosis and necrosis; beta-adrenergic receptor blockers would prevent this.

Cardiovascular Physiology

Coronary Artery Disease and Heart Attacks

We have seen that the myocardium does not extract oxygen and nutrients from the blood within the atria and ventricles, but depends upon its own blood supply via the coronary arteries. In *coronary artery disease,* changes in one or more of the coronary arteries cause insufficient blood flow (*ischemia*) to the heart. The result may be myocardial damage in the affected region and, if severe enough, death of that portion of the heart—a *myocardial infarction,* or *heart attack.* Many patients with coronary artery disease experience recurrent transient episodes of inadequate coronary blood flow, usually during exertion or emotional tension, before ultimately suffering a heart attack. The chest pain associated with these episodes is called *angina pectoris* (or, more commonly, angina).

The symptoms of myocardial infarction include prolonged chest pain, often radiating to the left arm, nausea, vomiting, sweating, weakness, and shortness of breath. Diagnosis is made by ECG changes typical of infarction and by measurement of certain proteins in plasma. These proteins are present in cardiac muscle and leak out into the blood when the muscle is damaged; the most commonly checked are the enzyme creatine kinase (CK), particularly the myocardial specific isoform (CK-MB), and cardiac troponin.

Approximately 1.1 million Americans have a new or recurrent heart attack each year, and over 40 percent of them die from it. Sudden cardiac deaths during myocardial infarction are due mainly to *ventricular fibrillation,* an abnormality in impulse conduction triggered by the damaged myocardial cells. This conduction pattern results in completely uncoordinated ventricular contractions that are ineffective in producing flow. (Note that ventricular fibrillation is fatal, whereas atrial fibrillation, as described earlier in this chapter, generally causes only minor cardiac problems.) A small fraction of individuals with ventricular fibrillation can be saved if emergency resuscitation procedures are applied immediately after the attack. This treatment is *cardiopulmonary resuscitation (CPR),* a repeated series of mouth-to-mouth respirations and chest compressions that circulate a small amount of oxygenated blood to the brain, heart, and other vital organs when the heart has stopped. CPR is then followed by definitive treatment, including *defibrillation,* a procedure in which electrical current is passed through the heart to try to halt the abnormal electrical activity causing the fibrillation. *Automatic electronic defibrillators (AEDs)* are now commonly found in public places. These devices make it relatively simple to render timely aid to victims of ventricular fibrillation.

The major cause of coronary artery disease is the presence of atherosclerosis in these vessels (**Figure 12–66**). *Atherosclerosis* is a disease of arteries characterized by a thickening of the portion of the arterial vessel wall closest to the lumen with plaques made up of (1) large numbers of abnormal smooth muscle cells, macrophages (derived from blood monocytes), and lymphocytes; (2) deposits of cholesterol and other fatty substances both in these cells and extracellularly; and (3) dense layers of connective tissue matrix. Such atherosclerotic plaques are one cause of aging-related arteriosclerosis.

Atherosclerosis reduces coronary blood flow through several mechanisms. The extra muscle cells and various deposits in the wall bulge into the lumen of the vessel and increase resistance to flow. Also, dysfunctional endothelial cells in the atherosclerotic area release excess vasoconstrictors (e.g., endothelin-1) and lower than normal amounts of vasodilators (nitric oxide and prostacyclin). These processes are progressive, sometimes leading ultimately to complete occlusion. Total occlusion is usually caused, however, by the formation of a blood clot (*coronary thrombosis*) in the narrowed atherosclerotic artery, and this triggers the heart attack.

The processes that lead to atherosclerosis are complex and still not completely understood. It is likely that the damage is initiated by agents that injure the endothelium and underlying smooth muscle, leading to an inflammatory and proliferative response that may well be protective at first but ultimately becomes excessive.

Cigarette smoking, high plasma concentrations of cholesterol and the amino acid homocysteine, hypertension, diabetes, obesity, a sedentary lifestyle, and stress can all increase the incidence and severity of the atherosclerotic process. These are all termed, therefore, "risk factors" for coronary artery disease, and prevention of this disease focuses on eliminating or minimizing risk factors through lifestyle changes and/or medications. In a sense, menopause can also be considered a risk factor for coronary artery disease because the incidence of heart attacks in women is very low until after menopause.

A few words about exercise are warranted here because of several potential confusions. While it is true that a sudden burst of strenuous physical activity can sometimes trigger a heart attack, the risk is greatly reduced in individuals who perform regular physical activity. The overall risk of heart attack at any time can be reduced as much as 35 to 55 percent by maintaining an active rather than sedentary lifestyle. In general, the more you exercise the better the protective effect, but any exercise is better than none. For example, even moderately paced walking three to four times a week confers significant benefit.

Regular exercise is protective against heart attacks for a variety of reasons. Among other things, it induces: (1) decreased myocardial oxygen demand due to decreases in resting heart rate and blood pressure; (2) increased diameter of coronary arteries; (3) decreased severity of hypertension and diabetes, two major risk factors for atherosclerosis; (4) decreased total plasma cholesterol concentration with simultaneous increase in the plasma concentration of a "good" cholesterol-carrying lipoprotein (HDL, discussed in Chapter 16); and (5) decreased tendency of blood to clot and improved ability of the body to dissolve blood clots.

Nutrition can also play a role in protecting against heart attacks. Reduction in the intake of saturated fat (a type abundant in red meat) and regular consumption of fruits, vegetables, whole grains, and fish may help by reducing the concentration of "bad" cholesterol (LDLs, discussed in Chapter 16) in the blood. This form of cholesterol contributes to the buildup of atherosclerotic plaques in blood vessels. Supplements like *folic acid* (a B vitamin; also called folate or folacin) may also be protective, in this case because folic acid helps reduce the blood concentration of the amino acid *homocysteine,* one of the risk factors for heart attacks. Homocysteine, cysteine with an

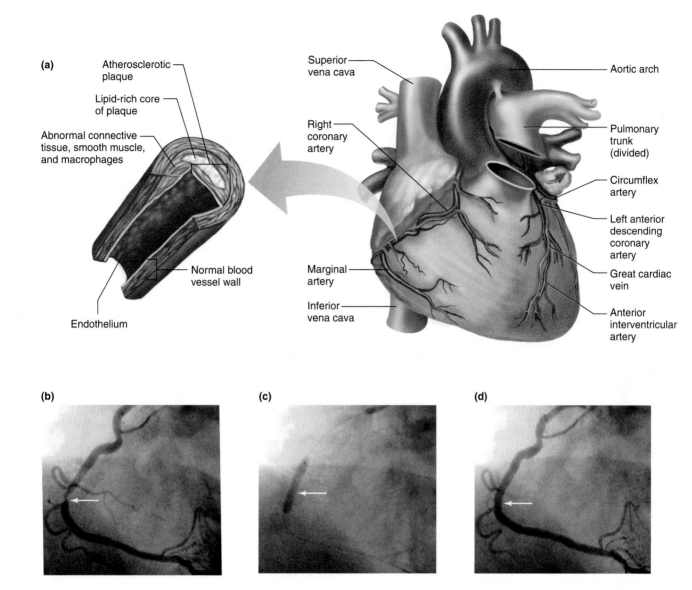

Figure 12-66

Coronary artery disease and its treatment. (a) Anterior view of the heart showing the major coronary vessels. Inset demonstrates narrowing due to atherosclerotic plaque. (b) Dye-contrast x-ray angiography performed by injecting radio-opaque dye shows a significant occlusion of the right coronary artery (arrow). (c) A guide wire is used to position and inflate a dye-filled balloon in the narrow region, and a wire-mesh stent is inserted. (d) Blood flows freely through the formerly narrowed region after the procedure.

(b), (c), and (d) courtesy of Matthew R. Wolff, M.D.

extra CH_2, is an intermediary in the metabolism of methionine and cysteine. In increased amounts, it exerts several pro-atherosclerotic effects, including damaging the endothelium of blood vessels. Folic acid is involved in the metabolism of homocysteine in a reaction that lowers the plasma concentration of this amino acid.

Finally, there is the question of alcohol and coronary artery disease. In many studies, moderate alcohol intake, red wine in particular, has been shown to reduce the risk of dying from a heart attack. However, alcohol also increases the chances of an early death from a variety of other diseases (cancer and cirrhosis of the liver, for example) and accidents. Because of

these complex health effects, it is now recommended that if people drink, they should have no more than one drink a day.

A number of different drugs can be used for the prevention and treatment of angina and coronary artery disease. For example, vasodilator drugs such as *nitroglycerin* (which is a vasodilator because it is converted in the body to nitric oxide) help by dilating the coronary arteries and the systemic arterioles and veins. The arteriolar effect lowers total peripheral resistance, thereby lowering arterial blood pressure and the work the heart must do to eject blood. The venous dilation, by lowering venous pressure, reduces venous return and thereby the stretch of the ventricle and its oxygen requirement during subsequent contraction. In addition, drugs that block β-adrenergic receptors are used to lower the arterial pressure in people with hypertension. They reduce myocardial work and cardiac output by inhibiting the effect of sympathetic nerves on heart rate and contractility. Drugs that prevent or reverse clotting within hours of its occurrence are also extremely important in the treatment (and prevention) of heart attacks. Use of these drugs, including aspirin, will be described in Section F of this chapter. Finally, a variety of drugs now in common use lower plasma cholesterol by influencing one or more metabolic pathways for cholesterol (Chapter 16). For example, one group of drugs, sometimes referred to as "statins," interferes with a critical enzyme involved in the liver's synthesis of cholesterol.

There are several surgical treatments for coronary artery disease after cardiac angiography (described earlier in this chapter) identifies the area of narrowing or occlusion. **Coronary balloon angioplasty** involves threading a catheter with a balloon at its tip into the occluded artery and then expanding the balloon (see Figure 12–66c). This procedure enlarges the lumen by stretching the vessel and breaking up abnormal tissue deposits. It is generally accompanied by permanent placing of **coronary stents** in the narrowed or occluded coronary vessel (see Figure 12–66d). Stents are tubes made of a stainless steel lattice that provide a scaffold within a vessel to open it and keep it open. Researchers are currently testing stents made of a hardened, biodegradeable polymer that is absorbed after six months to one year. Another surgical treatment is **coronary bypass,** in which a new vessel is attached across an area of occluded coronary artery. The new vessel is often a vein taken from elsewhere in the patient's body.

Atherosclerosis does not attack only the coronary vessels. Many arteries of the body are subject to this same occluding process, and wherever the atherosclerosis becomes severe, the resulting symptoms reflect the decrease in blood flow to the specific area. For example, occlusion of a cerebral artery due to atherosclerosis and its associated blood clotting can cause a stroke. (Recall that rupture of a cerebral vessel, as in hypertension, is another cause of stroke.) People with atherosclerotic cerebral vessels may also suffer reversible neurologic deficits, known as **transient ischemic attacks (TIAs),** lasting minutes to hours, without actually experiencing a stroke at the time.

Finally, note that both myocardial infarcts and strokes due to occlusion may result when a fragment of blood clot or fatty deposit breaks off and then lodges elsewhere, completely blocking a smaller vessel. The fragment is called an **embolus,** and the process is **embolism.** See Chapter 19 for a case study that highlights the dangers of an embolism.

■ SECTION E SUMMARY

Hemorrhage and Other Causes of Hypotension

I. The physiological responses to hemorrhage are summarized in Figures 12–52, 12–56, 12–58, and 12–59.

II. Hypotension can be caused by loss of body fluids, by cardiac malfunction, by strong emotion, and by liberation of vasodilator chemicals.

III. Shock is any situation in which blood flow to the tissues is low enough to cause damage to them.

The Upright Posture

I. In the upright posture, gravity acting upon unbroken columns of blood reduces venous return by increasing vascular pressures in the veins and capillaries in the limbs.

  a. The increased venous pressure distends the veins, causing venous pooling, and the increased capillary pressure causes increased filtration out of the capillaries.

  b. These effects are minimized by contraction of the skeletal muscles in the legs.

Exercise

I. The cardiovascular changes that occur in endurance-type exercise are illustrated in Figures 12–61 and 12–62.

II. The changes are due to active hyperemia in the exercising skeletal muscles and heart, to increased sympathetic outflow to the heart, arterioles, and veins, and to decreased parasympathetic outflow to the heart.

III. The increase in cardiac output depends not only on the autonomic influences on the heart, but on factors that help increase venous return.

IV. Training can increase a person's maximal oxygen consumption by increasing maximal stroke volume and thus cardiac output.

Hypertension

I. Hypertension is usually due to increased total peripheral resistance resulting from increased arteriolar vasoconstriction.

II. More than 95 percent of cases of hypertension are called *primary hypertension,* meaning the cause of the increased arteriolar vasoconstriction is unknown.

Heart Failure

I. Heart failure can occur as a result of diastolic or systolic dysfunction; in both cases, cardiac output becomes inadequate.

II. This leads to fluid retention by the kidneys and formation of edema because of increased capillary pressure.

III. Pulmonary edema can occur when the left ventricle fails.

Coronary Artery Disease and Heart Attacks

I. Insufficient coronary blood flow can cause damage to the heart.

II. Sudden death from a heart attack is usually due to ventricular fibrillation.

III. The major cause of reduced coronary blood flow is atherosclerosis, an occlusive disease of the arteries.

IV. People may suffer intermittent attacks of angina pectoris without actually suffering a heart attack at the time of the pain.

V. Atherosclerosis can also cause strokes and symptoms of inadequate blood flow in other areas.

VI. Coronary artery disease incidence is reduced by exercise, good nutrition, and avoiding smoking.

VII. Treatments for coronary artery disease include drugs that dilate blood vessels, reduce blood pressure, and prevent blood clotting. Balloon angioplasty and coronary bypass are surgical treatments.

■ SECTION E KEY TERMS

autotransfusion 413
erythropoiesis 414
erythropoietin 414
maximal oxygen consumption ($\dot{V}_{O_2}$max) 417

■ SECTION E REVIEW QUESTIONS

1. Draw a flow diagram illustrating the reflex compensation for hemorrhage.
2. What happens to plasma volume and interstitial fluid volume following a hemorrhage?
3. What causes hypotension during a severe allergic response?
4. How does gravity influence effective blood volume?
5. Describe the role of the skeletal muscle pump in decreasing capillary filtration.
6. List the directional changes that occur during exercise for all relevant cardiovascular variables. What are the specific efferent mechanisms that bring about these changes?
7. What factors enhance venous return during exercise?
8. Diagram the control of autonomic outflow during exercise.
9. What is the limiting cardiovascular factor in endurance exercise?
10. What changes in cardiac function occur at rest and during exercise as a result of endurance training?
11. What is the abnormality in most cases of established hypertension?
12. State how fluid retention can help restore stroke volume in heart failure.
13. How does heart failure lead to edema in the pulmonary and systemic vascular beds?
14. Name the major risk factors for atherosclerosis.
15. Describe changes in lifestyle that may help prevent coronary artery disease.
16. List some ways that coronary artery disease can be treated.

SECTION F   Blood and Hemostasis

Blood was defined earlier as a mixture of cellular components suspended in a fluid called plasma. In this section we will take a more detailed look at blood cells and plasma, and then discuss the complex mechanisms that prevent blood loss following injury.

Plasma

Plasma consists of a large number of organic and inorganic substances dissolved in water. Appendix F lists substances dissolved in plasma, and their typical concentrations.

The **plasma proteins** constitute most of the plasma solutes, by weight. Their role in exerting an osmotic pressure that favors the absorption of extracellular fluid into capillaries was described earlier. They can be classified into three broad groups: the **albumins,** the **globulins,** and **fibrinogen.** The first two have many overlapping functions, which are discussed in relevant sections throughout the book. Fibrinogen functions in clotting, discussed in detail in the latter part of this section. **Serum** is plasma with fibrinogen and other proteins involved in clotting removed as a result of clotting. The albumins are the most abundant of the three plasma protein groups and are synthesized by the liver. Cells normally do not take up plasma proteins; cells use plasma amino acids, not plasma proteins, to make their own proteins. Thus,

plasma proteins must be viewed differently from most of the other organic constituents of plasma, which use the plasma as a medium for transport to and from cells. In contrast, most plasma proteins perform their functions in the plasma itself or in the interstitial fluid.

In addition to the organic solutes, including proteins, nutrients, metabolic waste products, and hormones, plasma contains a variety of mineral electrolytes. These ions contribute much less to the *weight* of plasma than do the proteins, but in most cases they have much higher *molar concentrations*. This is because molarity is a measure not of weight, but of number of molecules per unit volume. Thus, there are many more ions than protein molecules, but the protein molecules are so large that a very small number of them greatly outweighs the much larger number of ions.

The Blood Cells

Erythrocytes

The major functions of erythrocytes are to carry oxygen taken in by the lungs and carbon dioxide produced by the cells. Erythrocytes contain large amounts of the protein **hemoglobin** with which oxygen and, to a lesser extent, carbon dioxide reversibly combine. Oxygen binds to iron atoms (Fe^{2+}) in the hemoglobin molecules. The average concentration of hemoglobin is 14 g/100 ml blood in women and 16 g/100 ml in men. Chapter 13 further describes the structure and functions of hemoglobin in the section discussing the transport of oxygen and carbon dioxide.

Erythrocytes have the shape of a biconcave disk—that is, a disk thicker at the edges than in the middle, like a doughnut with a center depression on each side instead of a hole (**Figure 12–67**). This shape and their small size (7 μm in diameter) impart to the erythrocytes a high surface area-to-volume ratio, so that oxygen and carbon dioxide can diffuse rapidly to and from the interior of the cell. The erythrocyte plasma membrane contains specific polysaccharides and proteins that differ from person to person, and these confer upon the blood its type, or group. Blood types are described in Chapter 18, in the context of the immune responses that occur in transfusion reactions.

The site of erythrocyte production is the soft interior of bones called **bone marrow,** specifically the red bone marrow. With differentiation, the erythrocyte precursors produce hemoglobin, but then they ultimately lose their nuclei and organelles—their machinery for protein synthesis. Young erythrocytes in the bone marrow still contain a few ribosomes, which produce a weblike (reticular) appearance when treated with special stains, an appearance that gives these young erythrocytes the name **reticulocyte.** Normally, only mature erythrocytes, which have lost these ribosomes, leave the bone marrow and enter the general circulation. In the presence of unusually rapid erythrocyte production, however, many reticulocytes do enter the blood, a phenomenon of clinical diagnostic usefulness.

Because erythrocytes lack nuclei and most organelles, they can neither reproduce themselves nor maintain their normal structure for very long. The average life span of an erythrocyte is approximately 120 days, which means that almost one percent of the body's erythrocytes are destroyed and must be replaced every day. This amounts to 250 billion cells per day! Destruction of damaged or dying erythrocytes normally occurs in the spleen and the liver. As we will later describe, most of the iron released in the process is conserved. The major breakdown product of hemoglobin is **bilirubin,** which is returned to the circulation and gives plasma its characteristic yellowish color (Chapter 15 will describe the fate of this substance).

The production of erythrocytes requires the usual nutrients needed to synthesize any cell: amino acids, lipids, and carbohydrates. In addition, both iron and certain growth factors, including the vitamins folic acid and vitamin B_{12}, are essential.

Iron

As noted previously, **iron** is the element to which oxygen binds on a hemoglobin molecule within an erythrocyte. Small amounts of iron are lost from the body via the urine, feces, sweat, and cells sloughed from the skin. Women lose an additional amount via menstrual blood. In order to remain in iron balance, the amount of iron lost from the body must be replaced by ingestion of iron-containing foods. Particularly rich sources of iron are meat, liver, shellfish, egg yolk, beans, nuts, and cereals. A significant upset of iron balance can result either in *iron deficiency,* leading to inadequate hemoglobin production, or in an excess of iron in the body, with serious toxic effects (*hemochromatosis*).

The homeostatic control of iron balance resides primarily in the intestinal epithelium, which actively absorbs iron from ingested foods. Normally, only a small fraction of ingested iron is absorbed. However, this fraction is increased or decreased in a negative feedback manner, depending upon the state of the body's iron balance—the more iron in the body, the less ingested iron is absorbed (the mechanism will be described in Chapter 15).

The body has a considerable store of iron, mainly in the liver, bound up in a protein called **ferritin.** Ferritin serves as a buffer against iron deficiency. About 50 percent of the total body iron is in hemoglobin, 25 percent is in other heme-containing proteins (mainly the cytochromes) in the cells of the body, and 25 percent is in liver ferritin. Moreover, the recycling of iron is very efficient (**Figure 12–68**). As old erythrocytes are destroyed in the spleen (and liver), their iron is released into the plasma and bound to an iron-transport plasma protein called **transferrin.** Transferrin delivers almost all of this iron to the bone marrow to be incorporated into new erythrocytes. Recirculation of erythrocyte iron is very important because it involves 20 times more iron per day than the body absorbs and excretes. On a much lesser scale, nonerythrocyte cells, some of which are continuously dying and being replaced, release iron from their cytochromes into the plasma and take up iron from it, transferrin serving as a carrier.

Folic Acid and Vitamin B_{12}

Folic acid, a vitamin found in large amounts in leafy plants, yeast, and liver, is required for synthesis of the nucleotide base thymine. It is, therefore, essential for the formation of DNA and thus for normal cell division. When this vitamin is not present in adequate amounts, impairment of cell division

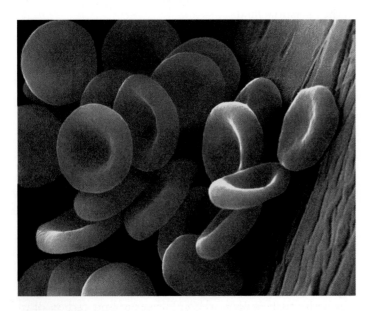

Figure 12–67
Electron micrograph of erythrocytes.

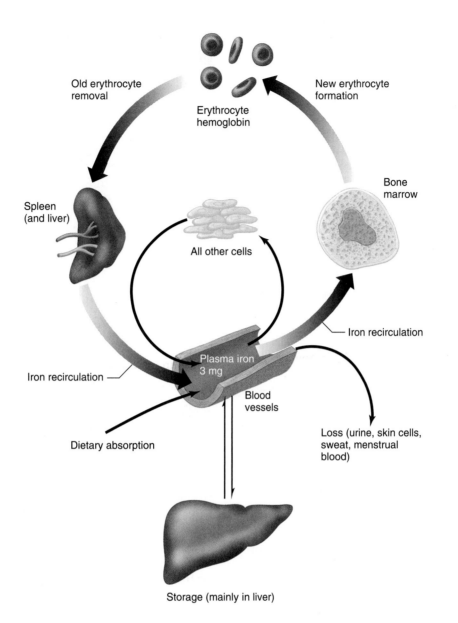

Old erythrocyte removal

New erythrocyte formation

Erythrocyte hemoglobin

Spleen (and liver)

Bone marrow

All other cells

Iron recirculation

Iron recirculation

Plasma iron 3 mg

Blood vessels

Dietary absorption

Loss (urine, skin cells, sweat, menstrual blood)

Storage (mainly in liver)

Figure 12–68

Summary of iron balance. The thickness of the arrows corresponds approximately to the amount of iron involved. In the steady state, the rate of gastrointestinal iron absorption equals the rate of iron loss via urine, skin, and menstrual flow.

Adapted from Crosby.

occurs throughout the body but is most striking in rapidly proliferating cells, including erythrocyte precursors. Thus, fewer erythrocytes are produced when folic acid is deficient.

The production of normal erythrocyte numbers also requires extremely small quantities (one-millionth of a gram per day) of a cobalt-containing molecule, **vitamin B_{12}** (also called cobalamin), because this vitamin is required for the action of folic acid. Vitamin B_{12} is found only in animal products, and strictly vegetarian diets are deficient in it. Also, the absorption of vitamin B_{12} from the gastrointestinal tract requires a protein called **intrinsic factor,** which is secreted by the stomach. Lack of this protein, therefore, causes vitamin B_{12} deficiency, and the resulting erythrocyte deficiency is known as *pernicious anemia.*

Regulation of Erythrocyte Production

In a typical person, the total volume of circulating erythrocytes remains remarkably constant because of reflexes that regulate the bone marrow's production of these cells. In the previous section, we stated that iron, folic acid, and vitamin B_{12} must be present for normal erythrocyte production, or **erythropoiesis.** However, none of these substances constitutes the signal that regulates the production rate.

The direct control of erythropoiesis is exerted primarily by a hormone called **erythropoietin,** which is secreted into the blood mainly by a particular group of hormone-secreting connective tissue cells in the kidneys (the liver also secretes this hormone, but to a much lesser extent). Erythropoietin acts on the bone marrow to stimulate the proliferation of erythrocyte progenitor cells and their differentiation into mature erythrocytes.

Erythropoietin is normally secreted in small amounts that stimulate the bone marrow to produce erythrocytes at a rate adequate to replace the usual loss. The erythropoietin secretion rate is increased markedly above basal values when there is a decreased oxygen delivery to the kidneys. Situations in which this occurs include insufficient pumping of blood by the heart, lung disease, anemia (a decrease in number of erythrocytes or in hemoglobin concentration), and exposure to high altitude. As a result of the increase in erythropoietin secretion, plasma erythropoietin concentration, erythrocyte production, and the oxygen-carrying capacity of the blood all increase. Therefore, oxygen delivery to the tissues returns toward normal (**Figure 12–69**).

Testosterone, the male sex hormone, also stimulates the release of erythropoietin. This accounts in part for the higher hematocrit in men than in women.

Anemia

As just described, **anemia** is defined as a decrease in the ability of the blood to carry oxygen due to (1) a decrease in the total number of erythrocytes, each having a normal quantity of hemoglobin, (2) a diminished concentration of hemoglobin per erythrocyte, or (3) a combination of both. Anemia has a wide variety of causes, summarized in **Table 12–10**.

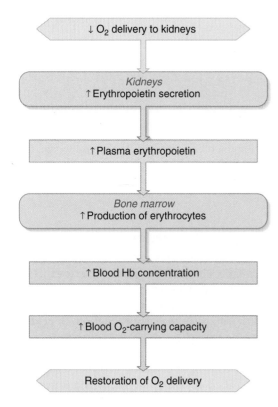

Figure 12–69

Reflex by which decreased oxygen delivery to the kidneys increases erythrocyte production via increased erythropoietin secretion.

Table 12–10	Major Causes of Anemia

1. Dietary deficiencies of iron (***iron-deficiency anemia***), vitamin B_{12}, or folic acid

2. Bone marrow failure due to toxic drugs or cancer

3. Blood loss from the body (***hemorrhage***) leading to iron deficiency

4. Inadequate secretion of erythropoietin in kidney disease

5. Excessive destruction of erythrocytes (e.g., sickle-cell anemia)

Sickle-cell anemia is due to a genetic mutation that alters one amino acid in the hemoglobin chain. At the low oxygen concentrations existing in many capillaries, the abnormal hemoglobin molecules interact with each other to form fiberlike structures that distort the erythrocyte membrane and cause the cell to form sickle shapes or other bizarre forms (a "sickle" is a crescent-shaped cutting blade). This causes both the blockage of capillaries, with consequent tissue damage and pain, and the destruction of the deformed erythrocytes, with consequent anemia. Sickle-cell anemia is an example of a disease that is manifested fully only in people homozygous for the mutated gene (that is, they have two copies of the mutated gene, one from each parent). In heterozygotes (one mutated copy and one normal gene), people who are said to have sickle-cell "trait," the normal gene codes for normal hemoglobin and the mutated gene for the abnormal hemoglobin. The erythrocytes in this case contain both types of hemoglobin, but symptoms are observed only when the oxygen concentration is unusually low, as at high altitude. The persistence of the sickle-cell mutation in humans is due to the fact that heterozygotes are more resistant to **malaria,** a blood infection caused by a protozoan parasite that is spread by mosquitoes in tropical regions. The mechanisms of this resistance are still being investigated.

Finally, there also exist conditions in which the problem is just the opposite of anemia, namely, more erythrocytes than normal; this is termed **polycythemia.** An example, to be described in Chapter 13, is the polycythemia that occurs in high-altitude dwellers. In this case, the increased number of erythrocytes is an adaptive response because it increases the oxygen-carrying capacity of blood exposed to low oxygen levels. As discussed earlier, however, raising the hematocrit increases the viscosity of blood, and thus the existence of polycythemia makes the flow of blood through blood vessels more difficult and puts a strain on the heart.

Leukocytes

If appropriate dyes are added to a drop of blood, which is then examined under a microscope, the various classes of leukocytes (**Table 12–11**) are clearly visible (**Figure 12–70**). They

Table 12–11	Numbers and Distributions of Erythrocytes, Leukocytes, and Platelets in Normal Human Blood

Total erythrocytes = 5,000,000 per mm^3 of blood

Total leukocytes = 7000 per mm^3 of blood

  Percent of total leukocytes:

    Polymorphonuclear granulocytes

      Neutrophils 50–70%

      Eosinophils 1–4%

      Basophils 0.1–0.3%

    Monocytes 2–8%

    Lymphocytes 20–40%

Total platelets = 250,000 per mm^3 of blood

are classified according to their structure and affinity for the various dyes.

The name **polymorphonuclear granulocytes** refers to the three classes of leukocytes that have multilobed nuclei and abundant membrane-surrounded granules. The granules of one group take up the red dye eosin, thus giving the cells their name, **eosinophils.** Cells of a second class have an affinity for a blue dye termed a "basic" dye and are called **basophils.** The granules of the third class have little affinity for either dye and are therefore called **neutrophils.** Neutrophils are by far the most abundant kind of leukocytes.

A fourth class of leukocyte is the **monocyte,** which is somewhat larger than the granulocyte and has a single oval or horseshoe-shaped nucleus and relatively few cytoplasmic granules. The final class of leukocytes is the **lymphocyte,** which contains little cytoplasm and, like the monocyte, a single relatively large nucleus.

Like the erythrocytes, all classes of leukocytes are produced in the bone marrow. In addition, monocytes and many lymphocytes undergo further development and cell division in tissues outside the bone marrow. These events and the specific leukocyte functions in the body's defenses are discussed in Chapter 18.

Platelets

The circulating platelets are colorless, non-nucleated cell fragments that contain numerous granules and are much smaller than erythrocytes. Platelets are produced when cytoplasmic portions of large bone marrow cells, termed **megakaryocytes,** pinch off and enter the circulation. Platelet functions in blood clotting are described later in this section.

Regulation of Blood Cell Production

In children, the marrow of most bones produces blood cells. By adulthood, however, only the bones of the chest, the base of the skull, and the upper portions of the limbs remain active. The bone marrow in an adult weighs almost as much as the liver, and it produces cells at an enormous rate.

All blood cells are descended from a single population of bone marrow cells called **pluripotent hematopoietic stem cells,** which are undifferentiated cells capable of giving rise to precursors (progenitors) of any of the different blood cells. When a pluripotent stem cell divides, its two daughter cells either remain pluripotent stem cells or become committed to a particular developmental pathway; what governs this "decision" is not known. The first branching yields either lymphoid stem cells, which give rise to the lymphocytes, or myeloid stem cells, the progenitors of all the other varieties (**Figure 12–71**). At some point, the proliferating offspring of the myeloid stem cells become committed to differentiate along only one path—for example, into erythrocytes.

Proliferation and differentiation of the various progenitor cells is stimulated, at multiple points, by a large number of protein hormones and paracrine agents collectively termed **hematopoietic growth factors (HGFs).** Thus, erythropoietin, the hormone described earlier, is an HGF. Others are listed for reference in **Table 12–12.** (Nomenclature can be confusing in this area because the HGFs belong to a still larger general family of messengers called cytokines, which are described in Chapter 18.)

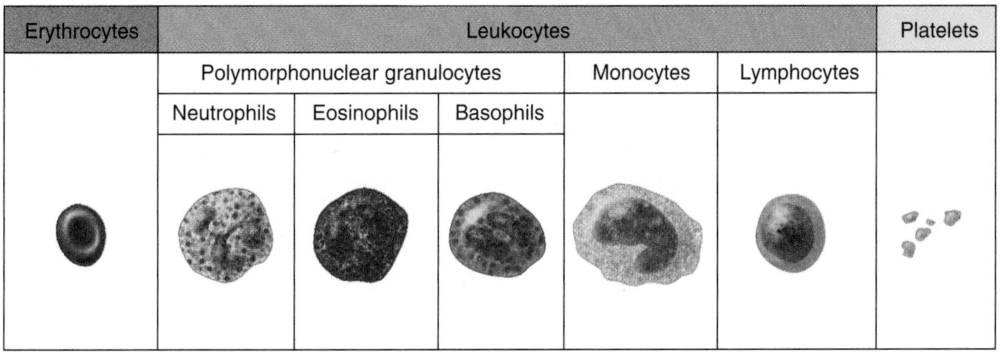

Erythrocytes	Leukocytes					Platelets
	Polymorphonuclear granulocytes			Monocytes	Lymphocytes	
	Neutrophils	Eosinophils	Basophils			

Figure 12–70

Classes of blood cells.

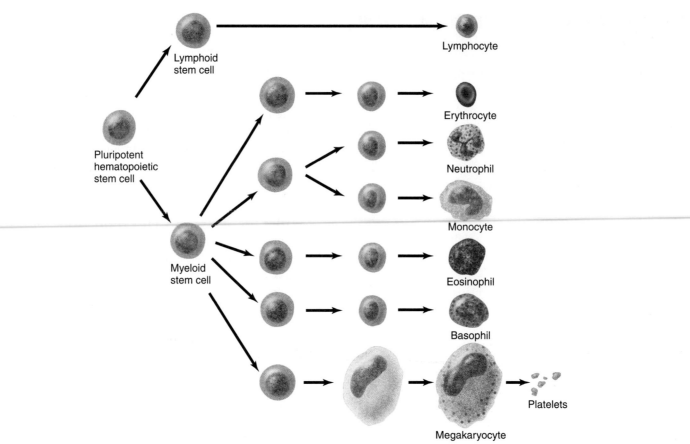

Figure 12–71

Production of blood cells by the bone marrow. For simplicity, no attempt has been made to differentiate the appearance of the various precursors.

Adapted form Golde and Gasson.

Table 12–12	Reference Table of Major Hematopoietic Growth Factors (HGFs)
Name	**Stimulates Progenitor Cells Leading To:**
Erythropoietin	Erythrocytes
Colony-stimulating factors (CSFs) (example: granulocyte CSF)	Granulocytes and monocytes
Interleukins (example: interleukin 3)	Various leukocytes
Thrombopoietin	Platelets (from megakaryocytes)
Stem cell factor	Many types of blood cells

The physiology of the HGFs is very complex because (1) there are so many types, (2) any given HGF is often produced by a variety of cell types throughout the body, and (3) HGFs often exert other actions in addition to stimulating blood cell production. There are, moreover, many interactions of the HGFs on particular bone marrow cells and processes. For example, although erythropoietin is the major stimulator of erythropoiesis, at least 10 other HGFs cooperate in the process. Finally, in several cases the HGFs not only stimulate differentiation and proliferation of progenitor cells, but they inhibit the usual programmed death (apoptosis) of these cells.

The administration of specific HGFs is proving to be of considerable clinical importance. Examples are the use of erythropoietin in persons having a deficiency of this hormone due to kidney disease, and the use of granulocyte colony-stimulating factor (G-CSF) to stimulate granulocyte production in individuals whose bone marrow has been damaged by anticancer drugs.

Hemostasis: The Prevention of Blood Loss

The stoppage of bleeding is known as **hemostasis** (don't confuse this word with homeostasis). Physiological hemostatic mechanisms are most effective in dealing with injuries in small vessels—arterioles, capillaries, and venules, which are the most common source of bleeding in everyday life. In contrast, the body usually cannot control bleeding from a medium or large artery. Venous bleeding leads to less rapid blood loss because veins have low blood pressure. Indeed, the drop in pressure induced by raising the bleeding part above the heart level may stop hemorrhage from a vein. In addition, if the venous bleeding is internal, the accumulation of blood in the tissues may increase interstitial pressure enough to eliminate the pressure gradient required for continued blood loss. Accumulation of blood in the tissues can occur as a result of bleeding from any vessel type and is known as a *hematoma.*

When a blood vessel is severed or otherwise injured, its immediate inherent response is to constrict (the mechanism is unclear). This short-lived response slows the flow of blood in the affected area. In addition, this constriction presses the opposed endothelial surfaces of the vessel together, and this contact induces a stickiness capable of keeping them "glued" together.

Permanent closure of the vessel by constriction and contact stickiness occurs only in the very smallest vessels of the microcirculation, however, and the staunching of bleeding ultimately depends upon two other interdependent processes that occur in rapid succession: (1) formation of a platelet plug; and (2) blood coagulation (clotting). The blood platelets are involved in both processes.

Formation of a Platelet Plug

The involvement of platelets in hemostasis requires their adhesion to a surface. Injury to a vessel disrupts the endothelium and exposes the underlying connective tissue collagen fibers. Platelets adhere to collagen, largely via an intermediary called **von Willebrand factor (vWF),** a plasma protein secreted by endothelial cells and platelets. This protein binds to exposed collagen molecules, changes its conformation, and becomes able to bind platelets. Thus, vWF forms a bridge between the damaged vessel wall and the platelets.

Binding of platelets to collagen triggers the platelets to release the contents of their secretory vesicles, which contain a variety of chemical agents. Many of these agents, including adenosine diphosphate (ADP) and serotonin, then act locally to induce multiple changes in the metabolism, shape, and surface proteins of the platelets, a process called **platelet activation.** Some of these changes cause new platelets to adhere to the old ones, a positive feedback phenomenon termed **platelet aggregation,** which rapidly creates a **platelet plug** inside the vessel.

Chemical agents in the platelets' secretory vesicles are not the only stimulators of platelet activation and aggregation. Adhesion of the platelets rapidly induces them to synthesize **thromboxane A₂,** a member of the eicosanoid family, from arachidonic acid in the platelet plasma membrane. Thromboxane A₂ is released into the extracellular fluid

and acts locally to further stimulate platelet aggregation and release of their secretory vesicle contents (**Figure 12–72**).

Fibrinogen, a plasma protein whose essential role in blood clotting is described in the next section, also plays a crucial role in the platelet aggregation produced by the factors previously described. It does so by forming the bridges between aggregating platelets. The receptors (binding sites) for fibrinogen on the platelet plasma membrane become exposed and activated during platelet activation.

The platelet plug can completely seal small breaks in blood vessel walls. Its effectiveness is further enhanced by another property of platelets—contraction. Platelets contain a very high concentration of actin and myosin (Chapter 9), which are stimulated to contract in aggregated platelets. This causes compression and strengthening of the platelet plug. (When they occur in a test tube, this contraction and compression are termed *clot retraction.*)

While the plug is being built up and compacted, the vascular smooth muscle in the damaged vessel is being stimulated to contract (see Figure 12–72), thereby decreasing the blood flow to the area and the pressure within the damaged vessel. This vasoconstriction is the result of platelet activity, for it is mediated by thromboxane A₂ and by several chemicals contained in the platelet's secretory vesicles.

Once started, why does the platelet plug not continuously expand, spreading away from the damaged endothelium

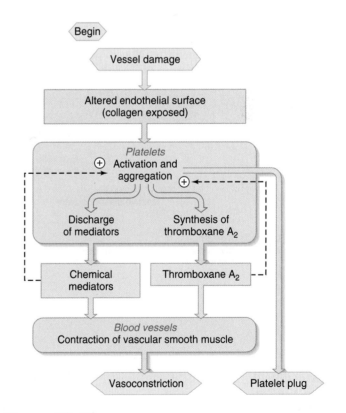

Figure 12–72

Sequence of events leading to formation of a platelet plug and vasoconstriction following damage to a blood vessel wall. Note the two positive feedbacks in the pathways.

along intact endothelium in both directions? One important reason involves the ability of the adjacent undamaged endothelial cells to synthesize and release the eicosanoid known as **prostacyclin** (also termed prostaglandin I_2, **PGI_2**), which is a profound inhibitor of platelet aggregation. Thus, whereas platelets possess the enzymes that produce thromboxane A_2 from arachidonic acid, normal endothelial cells contain a different enzyme that converts intermediates formed from arachidonic acid not to thromboxane A_2, but to prostacyclin (**Figure 12–73**). In addition to prostacyclin, the adjacent endothelial cells also release **nitric oxide,** which is not only a vasodilator (Section C) but also an inhibitor of platelet adhesion, activation, and aggregation.

To reiterate, the platelet plug is built up very rapidly and is the primary mechanism used to seal breaks in vessel walls.

In the following section, we will see that platelets are also essential for the next, more slowly occurring hemostatic event: blood coagulation.

Blood Coagulation: Clot Formation

Blood coagulation, or **clotting,** is the transformation of blood into a solid gel called a **clot** or **thrombus** and consisting mainly of a protein polymer known as **fibrin.** Clotting occurs locally around the original platelet plug and is the dominant hemostatic defense. Its function is to support and reinforce the platelet plug and to solidify blood that remains in the wound channel.

Figure 12–74 summarizes, in very simplified form, the events leading to clotting. These events, like platelet aggregation, are initiated when injury to a vessel disrupts the endothelium

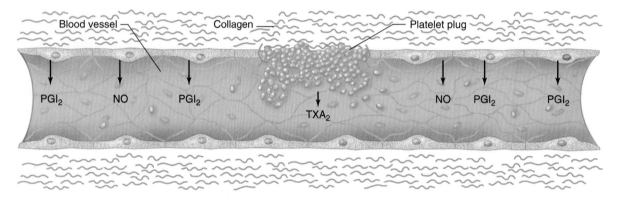

Figure 12–73

Prostacyclin (PGI_2) and nitric oxide (NO), both produced by endothelial cells, inhibit platelet aggregation and therefore prevent the spread of platelet aggregation from a damaged site. TXA_2 = thromboxane A_2.

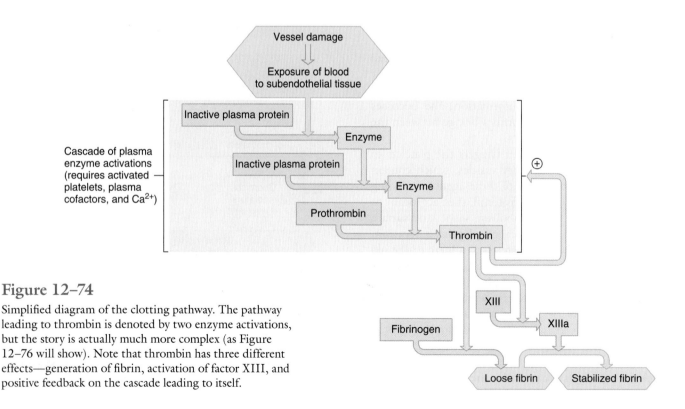

Figure 12–74

Simplified diagram of the clotting pathway. The pathway leading to thrombin is denoted by two enzyme activations, but the story is actually much more complex (as Figure 12–76 will show). Note that thrombin has three different effects—generation of fibrin, activation of factor XIII, and positive feedback on the cascade leading to itself.

and permits the blood to contact the underlying tissue. This contact initiates a locally occurring cascade of chemical activations. At each step of the cascade, an inactive plasma protein, or "factor," is converted (activated) to a proteolytic enzyme, which then catalyzes the generation of the next enzyme in the sequence. Each of these activations results from the splitting of a small peptide fragment from the inactive plasma protein precursor, thus exposing the active site of the enzyme. However, several of the plasma protein factors, following their activation, function not as enzymes but rather as cofactors for the enzymes.

For simplicity, Figure 12–74 gives no specifics about the cascade until the key point at which the plasma protein **prothrombin** is converted to the enzyme **thrombin.** Thrombin then catalyzes a reaction in which several polypeptides are split from molecules of the large, rod-shaped plasma protein fibrinogen. The fibrinogen remnants then bind to each other to form fibrin. The fibrin, initially a loose mesh of interlacing strands, is rapidly stabilized and strengthened by the enzymatically mediated formation of covalent cross-linkages. This chemical linking is catalyzed by an enzyme known as factor XIIIa, which is formed from plasma protein factor XIII in a reaction also catalyzed by thrombin.

Thus, thrombin catalyzes not only the formation of loose fibrin but also the activation of factor XIII, which stabilizes the fibrin network. But thrombin does even more than this—it exerts a profound positive feedback effect on its own formation. It does so by activating several proteins in the cascade and also by activating platelets. Therefore, once thrombin formation has begun, reactions leading to much more thrombin generation are activated by this initial thrombin. We will make use of this crucial fact later when we describe the specifics of the cascade leading to thrombin.

In the process of clotting, many erythrocytes and other cells are trapped in the fibrin meshwork (**Figure 12–75**), but the essential component of the clot is fibrin, and clotting can occur in the absence of all cellular elements except platelets.

Activated platelets are essential because several of the cascade reactions take place on the surface of the platelets. As noted earlier, platelet activation occurs early in the hemostatic response as a result of platelet adhesion to collagen, but in addition, thrombin is an important stimulator of platelet activation. The activation causes the platelets to display specific plasma membrane receptors that bind several of the clotting factors, and this permits the reactions to take place on the surface of the platelets. The activated platelets also display particular phospholipids, called **platelet factor (PF),** which function as a cofactor in the steps mediated by the bound clotting factors.

In addition to protein factors, plasma calcium is required at various steps in the clotting cascade. However, calcium concentration in the plasma can never get low enough to cause clotting defects because death would occur from muscle paralysis or cardiac arrhythmias before such low concentrations were reached.

Now we present the specifics of the early portions of the clotting cascade—those leading from vessel damage to the prothrombin-thrombin reaction. These early reactions consist of two seemingly parallel pathways that merge at the step just before the prothrombin-thrombin reaction. Under physiological conditions, however, the two pathways are not parallel but are actually activated sequentially, with thrombin serving as the link between them. It will be clearer, however, if we first discuss the two pathways as though they were separate and then deal with their actual interaction. The pathways are called (1) the **intrinsic pathway,** so named because everything necessary for it is in the blood; and (2) the **extrinsic pathway,** so named because a cellular element outside the blood is needed. **Figure 12–76** will be an essential reference for this entire discussion. Also, **Table 12–13** is a reference list of the names of and synonyms for the substances in these pathways.

The first plasma protein in the intrinsic pathway (upper left of Figure 12–76) is called factor XII. It can become activated to factor XIIa when it contacts certain types of surfaces,

Figure 12–75
Electron micrograph of erythrocytes enmeshed in fibrin.

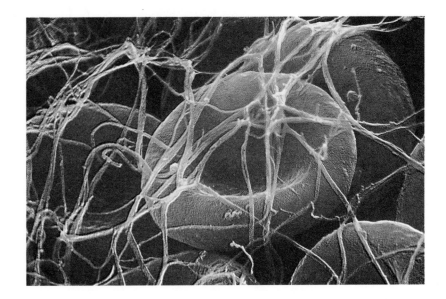

including the collagen fibers underlying damaged endothelium. The contact activation of factor XII to XIIa is a complex process that requires the participation of several other plasma proteins not shown in Figure 12–76. (Contact activation also explains why blood coagulates when it is taken from the body and put in a glass tube. This has nothing whatever to do with exposure to air, but happens because the glass surface acts like collagen and induces the same activation of factor XII and aggregation of platelets as a damaged vessel surface. A silicone coating delays clotting by reducing the activating effects of the glass surface.)

Factor XIIa then catalyzes the activation of factor XI to factor XIa, which activates factor IX to factor IXa. This last factor then activates factor X to factor Xa, which is the enzyme that converts prothrombin to thrombin. Note in Figure 12–76 that another plasma protein—factor VIIIa—serves as a cofactor (not an enzyme) in the factor IXa-mediated activation of

Table 12–13	Official Designations for Clotting Factors, Along with Synonyms More Commonly Used

Factor I (fibrinogen)

Factor Ia (fibrin)

Factor II (prothrombin)

Factor IIa (thrombin)

Factor III (tissue factor, tissue thromboplastin)

Factor IV (Ca^{2+})

Factors V, VII, VIII, IX, X, XI, XII, and XIII are the inactive forms of these factors; the active forms add an "a" (e.g., factor XIIa). There is no factor VI.

Platelet factor (PF)

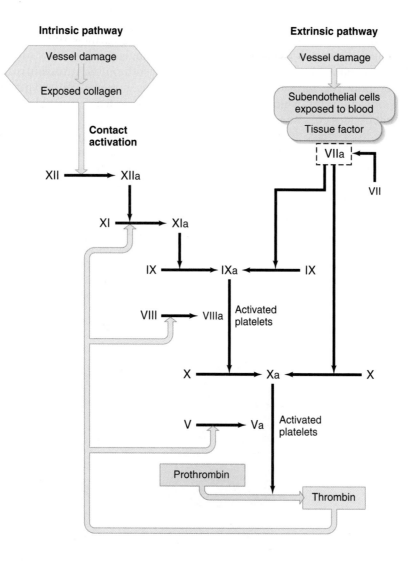

Figure 12–76

Two clotting pathways—intrinsic and extrinsic—merge and can lead to the generation of thrombin. Under most physiological conditions, however, factor XII and the contact activation step that begin the intrinsic pathway probably play little, if any, role in clotting. Rather, clotting is initiated solely by the extrinsic pathway, as described in the text. You might think that factors IX and X were accidentally transposed in the intrinsic pathway, but this is not the case; the order of activation really is XI, IX, and X. For the sake of clarity, the roles calcium plays in clotting are not shown.

factor X. The importance of factor VIII in clotting is emphasized by the fact that the disease **hemophilia,** characterized by excessive bleeding, is usually due to a genetic absence of this factor. (In a smaller number of cases, hemophilia is due to an absence of factor IX.)

Now we turn to the extrinsic pathway for initiating the clotting cascade (upper right of Figure 12–76). This pathway begins with a protein called **tissue factor,** which is not a plasma protein. It is located instead on the outer plasma membrane of various tissue cells, including fibroblasts and other cells in the walls of blood vessels below the endothelium. The blood is exposed to these subendothelial cells when vessel damage disrupts the endothelial lining. Tissue factor on these cells then binds a plasma protein, factor VII, which becomes activated to factor VIIa. The complex of tissue factor and factor VIIa on the plasma membrane of the tissue cell then catalyzes the activation of factor X. In addition, it catalyzes the activation of factor IX, which can then help activate even more factor X by way of the intrinsic pathway.

In summary, clotting can theoretically be initiated either by the activation of factor XII or by the generation of the tissue factor-factor VIIa complex. The two paths merge at factor Xa, which then catalyzes the conversion of prothrombin to thrombin, which catalyzes the formation of fibrin. As shown in Figure 12–76, thrombin also contributes to the activation of: (1) factors XI and VIII in the intrinsic pathway; and (2) factor V, with factor Va then serving as a cofactor for factor Xa. Not shown in the figure is the fact that thrombin also activates platelets.

As stated earlier, under physiological conditions, the two pathways just described actually are activated sequentially. To understand how this works, turn again to Figure 12–76; hold your hand over the first part of the intrinsic pathway so that you can eliminate the contact activation of factor XII and then begin the next paragraph's description at the top of the extrinsic pathway in the figure.

(1) The extrinsic pathway, with its tissue factor, is the usual way of initiating clotting in the body, and factor XII—the beginning of the full intrinsic pathway—normally plays little if any role (in contrast to its initiation of clotting in test tubes or within the body in several unusual situations). Thus, thrombin is initially generated only by the extrinsic pathway. The amount of thrombin is too small, however, to produce adequate, sustained coagulation. (2) It *is* large enough, though, to trigger thrombin's positive feedback effects on the intrinsic pathway—activation of factors XI and VIII and of platelets. (3) This is all that is needed to trigger the intrinsic pathway independently of factor XII. This pathway then generates the large amounts of thrombin required for adequate coagulation. Thus, the extrinsic pathway, via its initial generation of small amounts of thrombin, provides the means for recruiting the more potent intrinsic pathway without the participation of factor XII. In essence, thrombin eliminates the need for factor XII. Moreover, thrombin not only recruits the intrinsic pathway, but facilitates the prothrombin-thrombin step itself by activating factor V and platelets.

Finally, note that the liver plays several important indirect roles in clotting (**Figure 12–77**), and as a result persons with liver disease often have serious bleeding problems. First, the liver is the site of production for many of the plasma clotting factors. Second, the liver produces bile salts (Chapter 15), and these are important for normal intestinal absorption of the lipid-soluble substance **vitamin K.** The liver requires this vitamin to produce prothrombin and several other clotting factors.

Anticlotting Systems

Earlier we described how the release of prostacyclin and nitric oxide by endothelial cells inhibits platelet aggregation. Because this aggregation is an essential precursor for clotting, these agents reduce the magnitude and extent of clotting. In addition, however, the body has mechanisms for limiting clot formation itself and for dissolving a clot after it has formed.

Factors That Oppose Clot Formation

There are at least three different mechanisms that oppose clot formation, once underway, thereby helping to limit this process and prevent it from spreading excessively. Defects in any of these natural anticoagulant mechanisms are associated with abnormally high risk of clotting, a condition called **hypercoagulability** (see Chapter 19 for a case discussion of a patient with this condition).

The first anticoagulant mechanism acts during the initiation phase of clotting and utilizes the plasma protein called **tissue factor pathway inhibitor (TFPI),** which is secreted mainly by endothelial cells. This substance binds to tissue factor-factor VIIa complexes and inhibits the ability of these complexes to generate factor Xa. This anticoagulant mechanism is the reason that the extrinsic pathway by itself can generate only small amounts of thrombin.

The second anticoagulant mechanism is triggered by thrombin. As illustrated in **Figure 12–78**, thrombin can bind to an endothelial cell receptor known as **thrombomodulin.** This binding eliminates all of thrombin's clot-producing effects and causes the bound thrombin to bind a particular

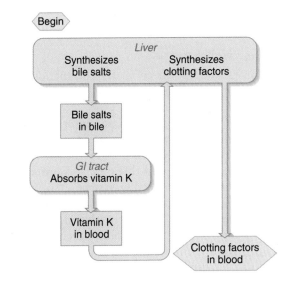

Figure 12–77

Roles of the liver in clotting.

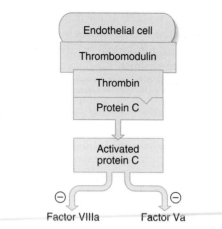

Figure 12–78

Thrombin indirectly inactivates factors VIIIa and Va via protein C. To activate protein C, thrombin must first bind to a thrombin receptor, thrombomodulin, on endothelial cells; this binding also eliminates thrombin's procoagulant effects. The ⊖ symbol indicates inactivation of factors Va and VIIIa.

Table 12–14	Actions of Thrombin
Procoagulant	Cleaves fibrinogen to fibrin
	Activates clotting factors XI, VIII, V, and XIII
	Stimulates platelet activation
Anticoagulant	Activates protein C, which inactivates clotting factors VIIIa and Va

plasma protein, **protein C** (distinguish this from protein kinase C, Chapter 5). The binding to thrombin activates protein C, which, in combination with yet another plasma protein, then inactivates factors VIIIa and Va. Thus, we saw earlier that thrombin directly activates factors VIII and V, and now we see that it indirectly inactivates them via protein C. **Table 12–14** summarizes the effects—both stimulatory and inhibitory—of thrombin on the clotting pathways.

A third naturally occurring anticoagulant mechanism is a plasma protein called **antithrombin III,** which inactivates thrombin and several other clotting factors. To do so, circulating antithrombin III must itself be activated, and this occurs when it binds to **heparin,** a substance present on the surface of endothelial cells. Antithrombin III prevents the spread of a clot by rapidly inactivating clotting factors that are carried away from the immediate site of the clot by the flowing blood.

The Fibrinolytic System

TFPI, protein C, and antithrombin III all function to *limit* clot formation. The system to be described now, however, dissolves a clot *after* it is formed.

A fibrin clot is not designed to last forever. It is a temporary fix until permanent repair of the vessel occurs. The **fibrinolytic** (or thrombolytic) **system** is the principal effector of clot removal. The physiology of this system (**Figure 12–79**) is analogous to that of the clotting system: It constitutes a plasma proenzyme, **plasminogen,** which can be activated to the active enzyme **plasmin** by protein **plasminogen activators.** Once formed, plasmin digests fibrin, thereby dissolving the clot.

The fibrinolytic system is proving to be every bit as complicated as the clotting system, with multiple types of plasminogen activators and pathways for generating them, as well as several inhibitors of these plasminogen activators. In describ-

ing how this system can be set into motion, we restrict our discussion to one example—the particular plasminogen activator known as **tissue plasminogen activator (t-PA),** which is secreted by endothelial cells. During clotting, both plasminogen and t-PA bind to fibrin and become incorporated throughout the clot. The binding of t-PA to fibrin is crucial because t-PA is a very weak enzyme in the absence of fibrin. The presence of fibrin profoundly increases the ability of t-PA to catalyze the generation of plasmin from plasminogen. Thus, fibrin is an important initiator of the fibrinolytic process that leads to its own dissolution.

The secretion of t-PA is the last of the various anticlotting functions exerted by endothelial cells that we have mentioned in this chapter. They are summarized in **Table 12–15.**

Anticlotting Drugs

Various drugs are used clinically to prevent or reverse clotting, and a brief description of their actions serves as a review of key clotting mechanisms. One of the most common uses of these drugs is in the prevention and treatment of myocardial infarction (heart attack), which, as described in Section E, is often the result of damage to endothelial cells. Such damage not only triggers clotting but interferes with the endothelial cells' normal *anticlotting* functions. For example, atherosclerosis interferes with the ability of endothelial cells to secrete nitric oxide.

Aspirin inhibits the cyclooxygenase enzyme in the eicosanoid pathways that generate prostaglandins and thromboxanes (Chapter 5). Because thromboxane A_2, produced by the platelets, is important for platelet aggregation, aspirin

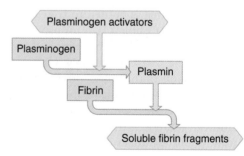

Figure 12–79

Basic fibrinolytic system. There are many different plasminogen activators and many different pathways for initiating their activity.

Table 12–15 Anticlotting Roles of Endothelial Cells

Action	Result
Normally provide an intact barrier between the blood and subendothelial connective tissue	Platelet aggregation and the formation of tissue factor-factor VIIa complexes are not triggered.
Synthesize and release PGI_2 and nitric oxide	These inhibit platelet activation and aggregation.
Secrete tissue factor pathway inhibitor	This inhibits the ability of tissue factor-factor VIIa complexes to generate factor Xa.
Bind thrombin (via thrombomodulin), which then activates protein C	Active protein C inactivates clotting factors VIIIa and Va.
Display heparin molecules on the surfaces of their plasma membranes	Heparin binds antithrombin III, and this molecule then inactivates thrombin and several other clotting factors.
Secrete tissue plasminogen activator	Tissue plasminogen activator catalyzes the formation of plasmin, which dissolves clots.

reduces both platelet aggregation and the ensuing coagulation. Importantly, low doses of aspirin cause a steady-state decrease in *platelet* cyclooxygenase (COX) activity but not *endothelial-cell* cyclooxygenase, and so the formation of prostacyclin—the prostaglandin that opposes platelet aggregation—is not impaired. (There is a reason for this difference between the responses of platelet and endothelial-cell cyclooxygenase to drugs: Platelets, once formed and released from megakaryocytes, have lost their ability to synthesize proteins, so that when their COX is blocked—the effect on any given COX molecule is irreversible—thromboxane A_2 synthesis is gone for that platelet's lifetime. In contrast, the endothelial cells produce new COX molecules to replace the ones blocked by the drug.) Aspirin appears to be highly effective at preventing heart attacks. In addition, the administration of aspirin following a heart attack significantly reduces the incidence of sudden death and a recurrent heart attack.

A variety of new drugs that interfere with platelet function by mechanisms different from those of aspirin also have great promise in the treatment or prevention of heart attacks. In particular, certain drugs block the binding of fibrinogen to platelets and thus interfere with platelet aggregation.

Drugs that interfere with the action of vitamin K, which is required for the synthesis of clotting factors by the liver, are collectively termed **oral anticoagulants.**

Heparin, the naturally occurring endothelial cell cofactor for antithrombin III, can also be administered as a drug, which then binds to endothelial cells. In addition to its role in facilitating the action of antithrombin III, heparin also inhibits platelet function.

In contrast to aspirin, the fibrinogen blockers, the oral anticoagulants, and heparin, all of which prevent clotting, the fifth type of drug—plasminogen activators—dissolves a clot after it is formed. The use of such drugs is termed **thrombolytic therapy.** Intravenous administration of **recombinant**

t-PA or a proteolytic drug called **streptokinase** within three hours after myocardial infarction significantly reduces myocardial damage and mortality. Recombinant t-PA has also been effective in reducing brain damage following a stroke caused by blood vessel occlusion. In addition, exciting new clinical studies suggest that a plasminogen activator found in vampire bat saliva may be even more effective than t-PA at protecting the brain after an ischemic stroke. Its name includes the genus and species of the animal—**Desmodus rotundus salivary plasminogen activator (DSPA).**

■ SECTION F SUMMARY

Plasma

I. Plasma is the liquid component of blood; it contains proteins, (albumins, globulins, and fibrinogen), nutrients, metabolic end products, hormones, and inorganic electrolytes.

II. Plasma proteins, synthesized by the liver, play many roles within the bloodstream, such as exerting osmotic pressure for absorption of interstitial fluid and participating in the clotting reaction.

The Blood Cells

I. The blood cells, which are suspended in plasma, include erythrocytes, leukocytes, and platelets.

II. Erythrocytes, which make up more than 99 percent of blood cells, contain hemoglobin, an oxygen-binding protein. Oxygen binds to the iron in hemoglobin.

   a. Erythrocytes are produced in the bone marrow and destroyed in the spleen and liver.

   b. Iron, folic acid, and vitamin B_{12} are essential for erythrocyte formation.

   c. The hormone erythropoietin, which is produced by the kidneys in response to low oxygen supply, stimulates erythrocyte differentiation and production by the bone marrow.

III. The leukocytes include three classes of polymorphonuclear granulocytes (neutrophils, eosinophils, and basophils), monocytes, and lymphocytes.

IV. Platelets are cell fragments essential for blood clotting.

V. Blood cells are descended from stem cells in the bone marrow. Hematopoietic growth factors control their production.

Hemostasis: The Prevention of Blood Loss

I. The initial response to blood vessel damage is vasoconstriction and the sticking together of the opposed endothelial surfaces.

II. The next events are formation of a platelet plug followed by blood coagulation (clotting).

III. Platelets adhere to exposed collagen in a damaged vessel and release the contents of their secretory vesicles.
   a. These substances help cause platelet activation and aggregation.
   b. This process is also enhanced by von Willebrand factor, secreted by the endothelial cells, and by thromboxane A$_2$, produced by the platelets.
   c. Fibrin forms the bridges between aggregating platelets.
   d. Contractile elements in the platelets compress and strengthen the plug.

IV. The platelet plug does not spread along normal endothelium because the latter secretes prostacyclin and nitric oxide, both of which inhibit platelet aggregation.

V. Blood is transformed into a solid gel when, at the site of vessel damage, plasma fibrinogen is converted into fibrin molecules, which then bind to each other to form a mesh.

VI. This reaction is catalyzed by the enzyme thrombin, which also activates factor XIII, a plasma protein that stabilizes the fibrin meshwork.

VII. The formation of thrombin from the plasma protein prothrombin is the end result of a cascade of reactions in which an inactive plasma protein is activated and then enzymatically activates the next protein in the series.
   a. Thrombin exerts a positive feedback stimulation of the cascade by activating platelets and several clotting factors.
   b. Activated platelets, which display platelet factor and binding sites for several activated plasma factors, are essential for the cascade.

VIII. In the body, the cascade usually begins via the extrinsic clotting pathway when tissue factor forms a complex with factor VIIa. This complex activates factor X, which then catalyzes the conversion of small amounts of prothrombin to thrombin. This thrombin then recruits the intrinsic pathway by activating factor XI and factor VIII, as well as platelets, and this pathway generates large amounts of thrombin.

IX. The liver requires vitamin K for the normal production of prothrombin and other clotting factors.

X. Clotting is limited by three events: (1) Tissue factor pathway inhibitor inhibits the tissue factor–factor VIIa complex; (2) protein C, activated by thrombin, inactivates factors VIIIa and Va; and (3) antithrombin III inactivates thrombin and several other clotting factors.

XI. Clots are dissolved by the fibrinolytic system.
   a. A plasma proenzyme, plasminogen, is activated by plasminogen activators to plasmin, which digests fibrin.
   b. Tissue plasminogen activator is secreted by endothelial cells and is activated by fibrin in a clot.

■ SECTION F KEY TERMS

albumin 425	plasma protein 425
antithrombin III 436	plasmin 436
basophil 429	plasminogen 436
bilirubin 426	plasminogen activator 436
blood coagulation 432	platelet activation 431
bone marrow 126	platelet aggregation 431
clot 432	platelet factor (PF) 433
clotting 432	platelet plug 431
eosinophil 429	pluripotent hematopoietic stem cell 429
erythropoiesis 427	
erythropoietin 427	polymorphonuclear granulocyte 429
extrinsic pathway 433	
ferritin 426	prostacyclin (PGI$_2$) 432
fibrin 432	protein C 436
fibrinogen 425	prothrombin 433
fibrinolytic system 436	reticulocyte 426
folic acid 426	serum 425
globulin 425	thrombin 433
hematopoietic growth factor (HGF) 429	thrombomodulin 435
	thromboxane A$_2$ 431
hemoglobin 425	thrombus 432
hemostasis 431	tissue factor 435
heparin 436	tissue factor pathway inhibitor (TFPI) 435
intrinsic factor 427	
intrinsic pathway 433	tissue plasminogen activator (t-PA) 436
iron 426	
lymphocyte 429	transferrin 426
megakaryocyte 429	vitamin B$_{12}$ 427
monocyte 429	vitamin K 435
neutrophil 429	von Willebrand factor (vWF) 431
nitric oxide 432	

■ SECTION F CLINICAL TERMS

anemia 428	iron deficiency 426
aspirin 436	iron-deficiency anemia 428
Desmodus rotundus salivary plasminogen activator (DSPA) 437	malaria 428
	oral anticoagulants 437
	pernicious anemia 427
hematoma 431	polycythemia 428
hemochromatosis 426	recombinant t-PA 437
hemophilia 435	sickle-cell anemia 428
hemorrhage 428	streptokinase 437
hypercoagulability 435	thrombolytic therapy 437

■ SECTION F REVIEW QUESTIONS

1. Give average values for total blood volume, erythrocyte volume, plasma volume, and hematocrit.
2. Which is the most abundant class of plasma protein?

3. Which solute is found in the highest concentration in plasma?
4. Summarize the production, life span, and destruction of erythrocytes.
5. What are the routes of iron gain, loss, and distribution, and how is iron recycled when erythrocytes are destroyed?
6. Describe the control of erythropoietin secretion and the effect of this hormone.
7. State the relative proportions of erythrocytes and leukocytes in blood.
8. Diagram the derivation of the different blood cell lines.
9. Describe the sequence of events leading to platelet activation and aggregation, and the formation of a platelet plug. What helps keep this process localized?
10. Diagram the clotting pathway beginning with prothrombin.
11. What is the role of platelets in clotting?
12. List all the procoagulant effects of thrombin.
13. How is the clotting cascade initiated? How does the extrinsic pathway recruit the intrinsic pathway?
14. Describe the roles of the liver and vitamin K in clotting.
15. List three ways in which clotting is limited.
16. Diagram the fibrinolytic system.
17. How does fibrin help initiate the fibrinolytic system?

Chapter 12 Test Questions

(Answers appear in Appendix A.)

1. Which of the following contains blood with the lowest oxygen content?
   a. aorta
   b. left atrium
   c. right ventricle
   d. pulmonary veins
   e. systemic arterioles

2. If other factors are equal, which of the vessels below would have the lowest resistance?
   a. length = 1 cm, radius = 1 cm
   b. length = 4 cm, radius = 1 cm
   c. length = 8 cm, radius = 1 cm
   d. length = 1 cm, radius = 2 cm
   e. length = 0.5 cm, radius = 2 cm

3. Which of the following correctly ranks pressures during isovolumetric contraction of a normal cardiac cycle?
   a. left ventricular > aortic > left atrial
   b. aortic > left atrial > left ventricular
   c. left atrial > aortic > left ventricular
   d. aortic > left ventricular > left atrial
   e. left ventricular > left atrial > aortic

4. Which of the following is *not* characteristic of the body's capillaries?
   a. large total surface area
   b. small individual diameter
   c. thin walls
   d. high blood velocity
   e. highly branched

5. Which of the following would *not* result in tissue edema?
   a. an increase in the concentration of plasma proteins
   b. an increase in the pore size of systemic capillaries
   c. an increase in venous pressure
   d. blockage of lymph vessels
   e. a decrease in the protein concentration of the plasma

6. Which statement comparing the systemic and pulmonary circuits is *true*?
   a. The blood flow is greater through the systemic.
   b. The blood flow is greater through the pulmonary.
   c. The absolute pressure is higher in the pulmonary.
   d. The blood flow is the same in both.
   e. The pressure gradient is the same in both.

7. What is mainly responsible for the delay between the atrial and ventricular contractions?
   a. the shallow slope of AV node pacemaker potentials
   b. slow action potential conduction velocity of AV node cells
   c. slow action potential conduction velocity along atrial muscle cell membranes
   d. slow action potential conduction in the Purkinje network of the ventricles
   e. greater parasympathetic nerve firing to the ventricles than to the atria

8. Which of the pressures below is closest to the mean arterial blood pressure in a person whose systolic blood pressure is 135 mmHg and pulse pressure is 50 mmHg?
   a. 110 mmHg
   b. 78 mmHg
   c. 102 mmHg
   d. 152 mmHg
   e. 85 mmHg

9. Which of the following would help restore homeostasis in the first few moments after a person's mean arterial pressure became elevated?
   a. a decrease in baroreceptor action potential frequency
   b. a decrease in action potential frequency along parasympathetic neurons to the heart
   c. an increase in action potential along sympathetic neurons to the heart
   d. a decrease in action potential frequency along sympathetic neurons to arterioles
   e. an increase in total peripheral resistance

10. Which is *false* about L-type calcium channels in cardiac ventricular muscle cells?
    a. They are open during the plateau of the action potential.
    b. They allow calcium entry that triggers sarcoplasmic reticulum calcium release.
    c. They are found in the T-tubule membrane.
    d. They open in response to depolarization of the membrane.
    e. They contribute to the pacemaker potential.

11. Which correctly pairs an ECG phase with the cardiac event responsible?
    a. P-wave: Depolarization of the ventricles
    b. P-wave: Depolarization of the AV node
    c. QRS-wave: Depolarization of the ventricles
    d. QRS-wave: Repolarization of the ventricles
    e. T-wave: Repolarization of the atria

12. When a person engages in strenuous, prolonged exercise
    a. blood flow to the kidneys is reduced.
    b. cardiac output is reduced.
    c. total peripheral resistance increases.
    d. systolic arterial blood pressure is reduced.
    e. blood flow to the brain is reduced.

13. Hematocrit is increased
    a. when a person has a vitamin B_{12} deficiency.
    b. by an increase in secretion of erythropoietin.
    c. when the number of white blood cells is increased.
    d. following a hemorrhage.
    e. in response to excess oxygen delivery to the kidneys.

14. The principal site of erythrocyte production is
    a. the liver.
    b. the kidneys.
    c. the bone marrow.
    d. the spleen.
    e. the lymph nodes.

15. Which is *not* part of the cascade leading to formation of a blood clot?
    a. contact between the blood and collagen found outside the blood vessels
    b. prothrombin converted to thrombin
    c. formation of a stabilized fibrin mesh
    d. activated platelets
    e. secretion of tissue plasminogen activator (t PA) by endothelial cells

Chapter 12 Quantitative and Thought Questions

(Answers appear in Appendix A.)

1. A person is found to have a hematocrit of 35 percent. Can you conclude that there is a decreased volume of erythrocytes in the blood?

2. Which would cause a greater increase in resistance to flow, a doubling of blood viscosity or a halving of tube radius?

3. If all plasma membrane calcium channels in contractile cardiac muscle cells were blocked with a drug, what would happen to the muscle's action potentials and contraction?

4. A person with a heart rate of 40 has no P waves but normal QRS complexes on the ECG. What is the explanation?

5. A person has a left ventricular systolic pressure of 180 mmHg and an aortic systolic pressure of 110 mmHg. What is the explanation?

6. A person has a left atrial pressure of 20 mmHg and a left ventricular pressure of 5 mmHg during ventricular filling. What is the explanation?

7. A patient is taking a drug that blocks beta-adrenergic receptors. What changes in cardiac function will the drug cause?

8. What is the mean arterial pressure in a person whose systolic and diastolic pressures are, respectively, 160 and 100 mmHg?

9. A person is given a drug that doubles the blood flow to her kidneys but does not change the mean arterial pressure. What must the drug be doing?

10. A blood vessel removed from an experimental animal dilates when exposed to acetylcholine. After the endothelium is scraped from the lumen of the vessel, it no longer dilates in response to this mediator. Explain.

11. A person is accumulating edema throughout the body. Average capillary pressure is 25 mmHg, and lymphatic function is normal. What is the most likely cause of the edema?

12. A person's cardiac output is 7 L/min and mean arterial pressure is 140 mmHg. What is the person's total peripheral resistance?

13. The following data are obtained for an experimental animal before and after administration of a drug. Before: Heart rate = 80 beats/min, stroke volume = 80 ml/beat. After: Heart rate = 100 beats/min, and stroke volume = 64 ml/beat. Total peripheral resistance remains unchanged. What has the drug done to mean arterial pressure?

14. When the nerves from all the arterial baroreceptors are cut in an experimental animal, what happens to mean arterial pressure?

15. What happens to the hematocrit within several hours after a hemorrhage?

Chapter 12 Answers to Physiological Inquiries

Figure 12–1 The hematocrit would be 33 percent because the red blood cell volume is the difference between total blood volume and plasma volume (4.5 – 3.0 = 1.5 L), and hematocrit is determined by the fraction of whole blood that is red blood cells (1.5 L /4.5 L = 0.33, or 33 percent).

Figure 12–5 No, the flow on side B would be less. The summed wall area causing friction would be the same in both sides. The formula for circumference of a circle is $2\pi r$, so the wall circumference in side A would be $2 \times 3.14 \times 2 = 12.56$, and for the two tubes on side B it would be $(2 \times 3.14 \times 1) + (2 \times 3.14 \times 1) = 12.56$. However, the total cross section through which flow occurs would be larger in side A than in side B. The formula for cross-sectional area of a circle is πr^2, so the area of side A would be $3.14 \times 2^2 = 12.56$, whereas the summed area of the tubes in side B would be $(3.14 \times 1^2) + (3.14 \times 1^2) = 6.28$. Thus, even with two outflow tubes on side B, there would be more flow through side A.

Figure 12–8 A: If this diagram included a systemic portal vessel, the order of structures in the lower box would be: aorta → arteries → arterioles → capillaries → venules→ portal vessel → capillaries → venules → veins → vena cava. Examples of portal vessels include the hepatic portal vein, which carries blood from the intestines to the liver (Chapter 15), and the hypothalamo-pituitary portal vessels (Figure 11–12).

Figure 12–13 Purkinje cell action potentials have a depolarizing pacemaker potential, like node cells (though the slope is much more gradual), and a rapid upstroke and broad plateau, like cardiac muscle cells.

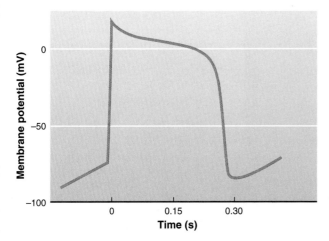

Figure 12–16 A reduction in current through voltage-gated K⁺ channels delays the repolarization of ventricular muscle cell action potentials. Thus, the T wave (ventricular repolarization) of the ECG wave is delayed relative to the QRS waves (ventricular depolarization). This fact gives the name to the condition, "Long Q-T Syndrome."

Figure 12–21 The patient most likely has a damaged semilunar valve that is stenotic and insufficient. A "whistling" murmur generally results from blood moving forward through a stenotic valve, whereas a lower-pitched "gurgling" murmur occurs when blood leaks backward through a valve that does not close properly. Systole and ejection occur between the two normal heart sounds, whereas diastole and filling occur after the second heart sound. Thus, a whistle between the heart sounds indicates a stenotic semilunar valve, and the gurgle following the second heart sound would arise from an insufficient semilunar valve. It is most likely that a single valve is both stenotic and insufficient in this case. Diagnosis could be confirmed either by determining where on the chest wall the sounds were loudest or by diagnostic imaging techniques.

Figure 12–25 Ejection fraction (EF) = Stroke volume (SV)/end-diastolic volume (EDV); end-systolic volume (ESV) = EDV – SV. Based on the graph, under control conditions the SV is 75 mL and during sympathetic stimulation it is 110 ml. Thus, Control ESV = 140 – 75 = 65 mL and EF = 75/140 = 53.6 percent; sympathetic ESV = 140 – 110 = 30 mL and EF = 110/140 = 78.6 percent.

Figure 12–31 At resting heart rate, the time spent in diastole is twice as long as that spent in systole (i.e., 1/3 of the total cycle is spent near systolic pressures) and the mean pressure is

approximately 1/3 of the distance from diastolic pressure to systolic pressure. At a heart rate in which equal time is spent in systole and diastole, the mean arterial blood pressure would be approximately halfway between those two pressures.

Figure 12–40 Venous blood leaving that tissue would be lower in oxygen and nutrients (like glucose) and higher in metabolic wastes (like carbon dioxide).

Figure 12–42 Injecting a liter of crystalloid to replace the lost blood would initially restore the volume (and thus the capillary hydrostatic pressure), but it would dilute the plasma proteins remaining in the bloodstream. As a result, the main force opposing capillary filtration (π_c) would be reduced, causing an increase in net filtration of fluid from the capillaries into the interstitial fluid space. A plasma injection, however, restores the plasma volume as well as the plasma proteins. Thus, the Starling forces remain in balance, and more of the injected volume remains within the vasculature.

Figure 12–47 Ingestion of fluids supports the net filtration of fluid at capillaries by transiently elevating vascular pressure (and thus P_c) and reducing the concentration of plasma proteins (and thus π_c). Although reflex mechanisms described in the next section and in Chapter 14 minimize and eventually reverse changes in blood pressure and plasma osmolarity, you could expect a transient rise in interstitial fluid formation and lymph flow after ingesting extra fluids.

Figure 12–53 There is a transient reduction in pressure at the baroreceptors when you first stand up. This occurs because gravity has a significant impact on blood flow. While lying down, the effect of gravity is minimal because baroreceptors and the rest of the vasculature are basically level with the heart. Upon standing, gravity resists the return of blood from below the heart (where the majority of the vascular volume exists). This transiently reduces cardiac output and, thus, blood pressure. Section E of this chapter provides a detailed description of this phenomenon, and explains how the body compensates for the effects of gravity.

Table 12–6 The hematocrit is the fraction of the total blood volume that is made up of erythrocytes. Thus, the normal hematocrit in this case was 2300/5000 × 100 = 46 percent. Immediately after the hemorrhage it was 1840/4000 × 100 = 46 percent, and 18 hours later it was 1840/1490 × 100 = 37 percent. The hemorrhage itself did not change hematocrit because erythrocytes and plasma were lost in equal proportions. However, over the next 18 hours there was a net shift of interstitial fluid into the blood plasma, due to a reduction in P_c. Because this occurs faster than does the production of new red blood cells, this "autotransfusion" resulted in a dilution of the remaining erythrocytes in the bloodstream. In the days and weeks that follow, increased erythropoietin will stimulate the replacement of the lost erythrocytes, and the lost ECF fluid volume will be replaced by ingestion and decreased urine output.

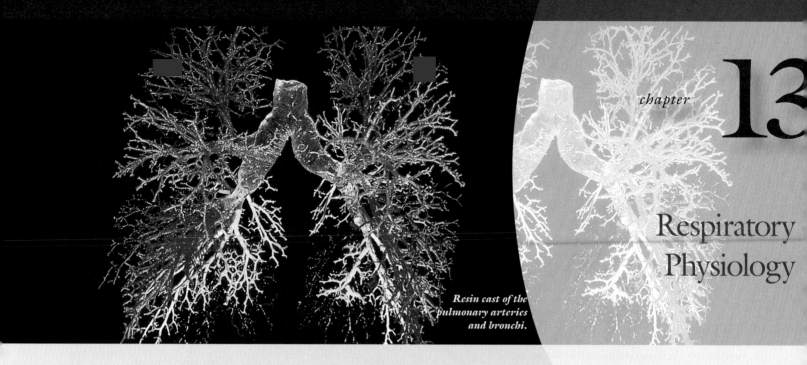

Resin cast of the
pulmonary arteries
and bronchi.

chapter

13

Respiratory Physiology

respiratory physiology can have two quite different meanings: (1) utilization of oxygen in the metabolism of organic molecules by cells, often termed internal or cellular respiration, as described in Chapter 3, and (2) the exchange of oxygen and carbon dioxide between an organism and the external environment, often called pulmonary physiology. The adjective **pulmonary** refers to the lungs. The second meaning is the subject of this chapter.

Human cells obtain most of their energy from chemical reactions involving oxygen. In addition, cells must be able to eliminate carbon dioxide, the major end product of oxidative metabolism. A unicellular organism can exchange oxygen and carbon dioxide directly with the external environment, but this is obviously impossible for most cells of a complex organism like a human being. Therefore, the evolution of large animals required the development of specialized

structures for the entire animal to exchange oxygen and carbon dioxide with the external environment. In humans and other mammals, the **respiratory system** includes the oral and nasal cavities, the lungs, the series of tubes leading to the lungs, and the chest structures responsible for moving air into and out of the lungs during breathing.

In addition to mediating gas exchange with the environment, the respiratory system serves other functions, as listed in **Table 13–1**.

Organization of the Respiratory System

There are two lungs, the right and left, each divided into lobes. The lungs consist mainly of tiny air-containing sacs called **alveoli** (singular, **alveolus**), which number approximately 300 million in an adult. The alveoli are the sites of gas exchange with the blood. The **airways** are the tubes that air flows through from the external environment to the alveoli and back.

　　Inspiration (inhalation) is the movement of air from the external environment through the airways into the alveoli during breathing. **Expiration** (exhalation) is movement in the opposite direction. An inspiration and an expiration constitute a **respiratory cycle.** During the entire respiratory cycle, the right ventricle of the heart pumps blood through the pulmonary arteries and arterioles and into the capillaries surrounding each alveolus. In a normal adult at rest, approximately 4 L of fresh air enters and leaves the alveoli per minute, while 5 L of blood, virtually the entire cardiac output, flows through the pulmonary capillaries. During heavy exercise, the air flow can increase twentyfold, and the blood flow five- to sixfold.

The Airways and Blood Vessels

During inspiration, air passes through either the nose or mouth into the **pharynx,** a passage common to both air and food (**Figure 13–1**). The pharynx branches into two tubes: the esophagus, through which food passes to the stomach, and the **larynx,** which is part of the airways. The larynx houses the **vocal cords,** two folds of elastic tissue stretched horizontally across its lumen. The flow of air past the vocal cords causes them to vibrate, producing sounds. The nose, mouth, pharynx, and larynx are collectively termed the **upper airways.**

　　The larynx opens into a long tube, the **trachea,** which in turn branches into two **bronchi** (singular, **bronchus**), one of which enters each lung. Within the lungs, there are more than 20 generations of branchings, each resulting in narrower, shorter, and more numerous tubes; their names are summarized in **Figure 13–2**. The walls of the trachea and bronchi contain rings of cartilage, which give them their cylindrical shape and support them. The first airway branches that

Table 13–1	Functions of the Respiratory System
1. Provides oxygen.	
2. Eliminates carbon dioxide.	
3. Regulates the blood's hydrogen ion concentration (pH) in coordination with the kidneys.	
4. Forms speech sounds (phonation).	
5. Defends against microbes.	
6. Influences arterial concentrations of chemical messengers by removing some from pulmonary capillary blood and producing and adding others to this blood.	
7. Traps and dissolves blood clots arising from systemic veins such as those in the legs.	

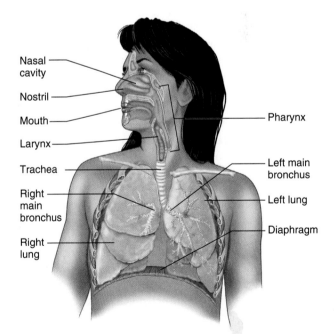

Figure 13–1

Organization of the respiratory system. The ribs have been removed in front, and the lungs are shown in a way that makes visible the major airways within them.

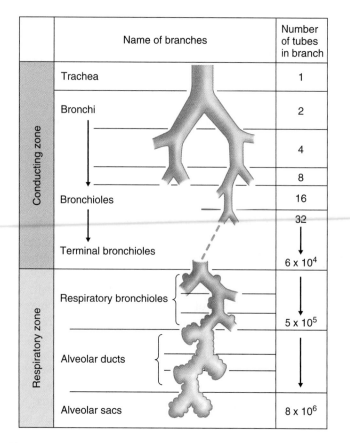

Name of branches		Number of tubes in branch
Trachea		1
Bronchi		2
		4
		8
Bronchioles		16
		32
Terminal bronchioles		6×10^4
Respiratory bronchioles		5×10^5
Alveolar ducts		
Alveolar sacs		8×10^6

Figure 13–2

Airway branching. Asymmetries in branching patterns between the right and left bronchial trees are not depicted. The diameters of the airways and alveoli are not drawn to scale.

no longer contain cartilage are termed **bronchioles.** Alveoli first begin to appear attached to the walls of the **respiratory bronchioles.** The number of alveoli increases in the alveolar ducts (see Figure 13–2), and the airways then end in grapelike clusters consisting entirely of alveoli (**Figure 13–3**). The airways, like blood vessels, are surrounded by smooth muscle, which contracts or relaxes to alter airway radius.

The airways beyond the larynx can be divided into two zones. The **conducting zone** extends from the top of the trachea to the beginning of the respiratory bronchioles. This zone contains no alveoli and has no gas exchange with the blood (**Table 13–2**). The **respiratory zone** extends from the respiratory bronchioles down. This zone contains alveoli and is the region where gases exchange with the blood.

The oral and nasal cavities trap airborne particles in nasal hairs and mucus. The epithelial surfaces of the airways, to the end of the respiratory bronchioles, contain cilia that constantly beat upward toward the pharynx. They also contain glands and individual epithelial cells that secrete mucus. Particulate matter, such as dust contained in the inspired air, sticks to the mucus, which is continuously and slowly moved by the cilia to the pharynx and then swallowed. This so-called mucous escalator is important in keeping the lungs clear of particulate matter and the many bacteria that enter the body on dust particles. Ciliary activity can be inhibited by many noxious agents including chronic cigarette smok-

ing. This is why smokers often cough up mucus that the cilia would normally have cleared.

The airway epithelium also secretes a watery fluid upon which the mucus can ride freely. The production of this fluid is impaired in the disease **cystic fibrosis,** the most common lethal genetic disease among Caucasians, in which the mucous layer becomes thick and dehydrated, obstructing the airways. The impaired secretion is due to a defect in the chloride channels involved in the secretory process.

Constriction of bronchioles in response to irritation helps to prevent particulate matter and irritants from entering the sites of gas exchange. Another protective mechanism against infection is provided by cells called macrophages that are present in the airways and alveoli. These cells engulf and destroy inhaled particles and bacteria that have reached the alveoli. Macrophages, like cilia, are injured by cigarette smoke and air pollutants.

The pulmonary blood vessels generally accompany the airways and also undergo numerous branchings. The smallest of these vessels branch into networks of capillaries that richly supply the alveoli (see Figure 13–3). The pulmonary circulation has a very low resistance compared to the systemic circulation, and for this reason the pressures within all pulmonary blood vessels are low.

Site of Gas Exchange: The Alveoli

The alveoli are tiny, hollow sacs whose open ends are continuous with the lumens of the airways (**Figure 13–4a**). Typically, a single alveolar wall separates the air in two adjacent alveoli. Most of the air-facing surfaces of the wall are lined by a continuous layer, one cell thick, of flat epithelial cells termed **type I alveolar cells.** Interspersed between these cells are thicker, specialized cells termed **type II alveolar cells** (**Figure 13–4b**) that produce a detergent-like substance called surfactant.

The alveolar walls contain capillaries and a very small interstitial space, which consists of interstitial fluid and a loose meshwork of connective tissue (see Figure 13–4b). In many places, the interstitial space is absent altogether, and the basement membranes of the alveolar-surface epithelium and the capillary-wall endothelium fuse. Thus the blood within an alveolar-wall capillary is separated from the air within the alveolus by an extremely thin barrier (0.2 µm, compared with the 7 µm diameter of an average red blood cell). The total surface area of alveoli in contact with capillaries is roughly the size of a tennis court. This extensive area and the thinness of the barrier permit the rapid exchange of large quantities of oxygen and carbon dioxide by diffusion.

In some of the alveolar walls, pores permit the flow of air between alveoli. This route can be very important when the airway leading to an alveolus is occluded by disease because some air can still enter the alveolus by way of the pores between it and adjacent alveoli.

Relation of the Lungs to the Thoracic (Chest) Wall

The lungs, like the heart, are situated in the **thorax,** the compartment of the body between the neck and abdomen. *Thorax* and *chest* are synonyms. The thorax is a closed compartment bounded at the neck by muscles and connective tissue and completely separated from the abdomen by a large, dome-

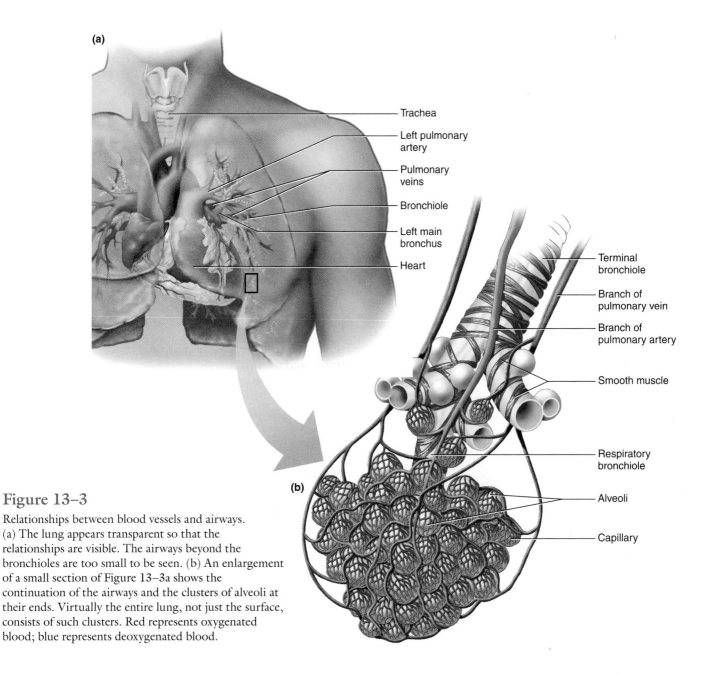

Figure 13–3

Relationships between blood vessels and airways.
(a) The lung appears transparent so that the
relationships are visible. The airways beyond the
bronchioles are too small to be seen. (b) An enlargement
of a small section of Figure 13–3a shows the
continuation of the airways and the clusters of alveoli at
their ends. Virtually the entire lung, not just the surface,
consists of such clusters. Red represents oxygenated
blood; blue represents deoxygenated blood.

Table 13–2	Functions of the Conducting Zone of the Airways

1. Provides a low-resistance pathway for air flow. Resistance
   is physiologically regulated by changes in contraction of
   airway smooth muscle and by physical forces acting upon the
   airways.

2. Defends against microbes, toxic chemicals, and other
   foreign matter. Cilia, mucus, and macrophages perform this
   function.

3. Warms and moistens the air.

4. Phonates (vocal cords).

shaped sheet of skeletal muscle called the **diaphragm.** The
wall of the thorax is formed by the spinal column, the ribs,
the breastbone (sternum), and several groups of muscles that
run between the ribs that are collectively called the **intercostal muscles.** The thoracic wall also contains large amounts of
connective tissue with elastic properties.

Each lung is surrounded by a completely closed sac, the
pleural sac, consisting of a thin sheet of cells called **pleura.** The
two pleural sacs are completely separate from each other. The
relationship between a lung and its pleural sac can be visualized
by imagining what happens when you push a fist into a fluid-
filled balloon. The arm shown in **Figure 13–5** represents the
major bronchus leading to the lung, the fist is the lung, and the
balloon is the pleural sac. The fist becomes coated by one surface of the balloon. In addition, the balloon is pushed back upon
itself so that its opposite surfaces lie close together but are separated by a thin layer of fluid. Unlike the hand and balloon, the
pleural surface coating the lung known as the **visceral pleura**

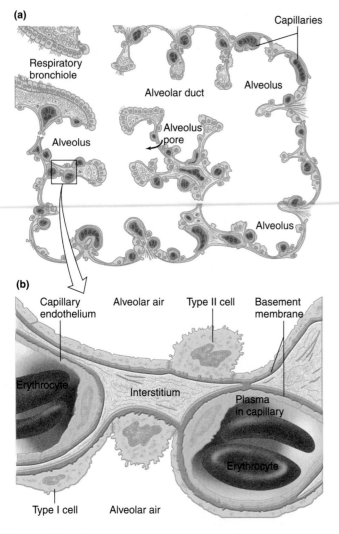

(a)

Respiratory bronchiole

Capillaries

Alveolar duct

Alveolus

Alveolus pore

Alveolus

Alveolus

(b)

Capillary endothelium

Alveolar air

Type II cell

Basement membrane

Erythrocyte

Interstitium

Plasma in capillary

Erythrocyte

Type I cell

Alveolar air

Figure 13–4

(a) Cross section through an area of the respiratory zone. There are 18 alveoli in this figure, only four of which are labeled. Two often share a common wall. (b) Schematic enlargement of a portion of an alveolar wall.

(a) From R.O. Greep and L. Weiss, *Histology,* 3d ed., McGraw-Hill New York, 1973. (b) Adapted from Gong and Drage.

is firmly attached to the lung by connective tissue. Similarly, the outer layer, called the **parietal pleura,** is attached to and lines the interior thoracic wall and diaphragm. The two layers of pleura in each sac are very close but not attached to each other. Rather, they are separated by an extremely thin layer of **intrapleural fluid,** the total volume of which is only a few milliliters. The intrapleural fluid totally surrounds the lungs and lubricates the pleural surfaces so that they can slide over each other during breathing. As we will see in the next section, changes in the hydrostatic pressure of the intrapleural fluid—the **intrapleural pressure (P_{ip})**—cause the lungs and thoracic wall to move in and out together during normal breathing.

A way to visualize the apposition of the two pleural surfaces is to put a drop of water between two glass microscope slides. The two slides can easily slide over each other but are very difficult to pull apart.

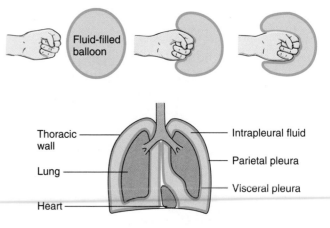

Fluid-filled balloon

Thoracic wall

Intrapleural fluid

Parietal pleura

Lung

Visceral pleura

Heart

Figure 13–5

Relationship of lungs, pleura, and thoracic wall, shown as analogous to pushing a fist into a fluid-filled balloon. Note that there is no communication between the right and left intrapleural fluids. For purposes of illustration, the volume of intrapleural fluid is greatly exaggerated. It normally consists of an extremely thin layer of fluid between the pleural membrane lining the inner surface of the thoracic wall (the parietal pleura) and the membrane lining the outer surface of the lungs (the visceral pleura).

Ventilation and Lung Mechanics

It is helpful to preview an inventory of steps involved in respiration (**Figure 13–6**) for orientation before examining the detailed descriptions of each step.

Ventilation is defined as the exchange of air between the atmosphere and alveoli. Like blood, air moves by *bulk flow,* from a region of high pressure to one of low pressure. Bulk flow can be described by the equation:

$$F = \Delta P/R \qquad (13–1)$$

Stated differently, flow (F) is proportional to the pressure difference (ΔP) between two points and inversely proportional to the resistance (R). For air flow into or out of the lungs, the relevant pressures are the gas pressure in the alveoli—the **alveolar pressure (P_{alv})**—and the gas pressure at the nose and mouth, normally **atmospheric pressure (P_{atm})** which is the pressure of the air surrounding the body:

$$F = (P_{alv} - P_{atm})/R \qquad (13–2)$$

A very important point must be made here: all pressures in the respiratory system, as in the cardiovascular system, are given *relative to atmospheric pressure,* which is 760 mmHg at sea level. For example, the alveolar pressure between breaths is said to be 0 mmHg, which means that it is the same as atmospheric pressure. From equation 13–2, when there is no air flow, $F = 0$; therefore, when there is no air flow, $P_{alv} - P_{atm} = 0$, and $P_{alv} = P_{atm}$.

During ventilation, air moves into and out of the lungs because the alveolar pressure is alternately less than and greater than atmospheric pressure (**Figure 13–7**). When P_{alv} is less than P_{atm}, the driving force for air flow is negative, indicating that air flow is inward (inspiration). When P_{alv} is greater than

P_{atm}, the driving force for air flow is positive (equation 13–2), indicating that air flow is outward (expiration). These alveolar pressure changes are caused, as we will see, by changes in the dimensions of the chest wall and lungs.

To understand how a change in lung dimensions causes a change in alveolar pressure, you need to learn one more basic concept—**Boyle's law,** which is represented by the equation $P_1V_1 = P_2V_2$ (**Figure 13–8**). At constant temperature, the relationship between the pressure (P) exerted by a fixed number of gas molecules and the volume (V) of their container is as follows: an increase in the volume of the container decreases the pressure of the gas, whereas a decrease in the container volume increases the pressure.

It is essential to recognize the correct causal sequences in ventilation. During inspiration and expiration, the volume of the "container"—the lungs—is made to change, and these changes then cause, by Boyle's law, the alveolar pressure changes that drive air flow into or out of the lungs. Our descriptions of ventilation must focus, therefore, on how the changes in lung dimensions are brought about.

There are no muscles attached to the lung surface to pull the lungs open or push them shut. Rather, the lungs are passive

① Ventilation: Exchange of air between atmosphere and alveoli by *bulk flow*
② Exchange of O_2 and CO_2 between alveolar air and blood in lung capillaries by *diffusion*
③ Transport of O_2 and CO_2 through pulmonary and systemic circulation by *bulk flow*
④ Exchange of O_2 and CO_2 between blood in tissue capillaries and cells in tissues by *diffusion*
⑤ Cellular utilization of O_2 and production of CO_2

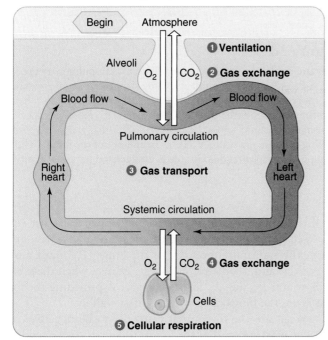

Figure 13–6
The steps of respiration.

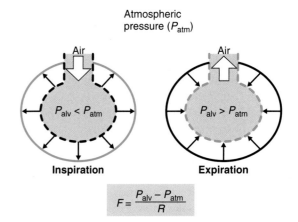

$$F = \frac{P_{alv} - P_{atm}}{R}$$

Figure 13–7
Relationships required for ventilation. When the alveolar pressure (P_{alv}) is less than atmospheric pressure (P_{atm}), air enters the lungs. Flow (F) is directly proportional to the pressure difference ($P_{alv} - P_{atm}$) and inversely proportional to airway resistance (R). Black lines show lung's position at beginning of inspiration or expiration, and blue lines at end.

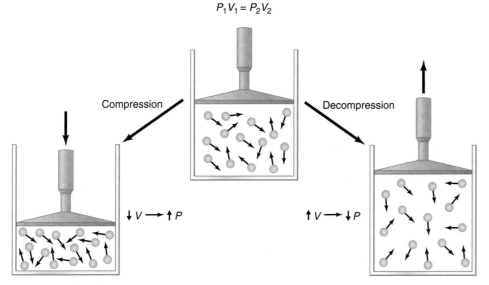

Figure 13–8
Boyle's law: The pressure exerted by a constant number of gas molecules (at a constant temperature) is inversely proportional to the volume of the container. As the container is compressed, the pressure in the container increases. When the container is decompressed, the pressure inside decreases.

elastic structures—like balloons—and their volume, therefore, depends on two factors. The first is the difference in pressure between the inside and outside of the lung, termed the **transpulmonary pressure (P_{tp})**. The second is how stretchable the lungs are, which determines how much they expand for a given change in P_{tp}. The rest of this section and the next three sections focus on transpulmonary pressure; stretchability will be discussed later in the section on lung compliance.

The pressure inside the lungs is the air pressure inside the alveoli (P_{alv}), and the pressure outside the lungs is the pressure of the intrapleural fluid surrounding the lungs (P_{ip}). Thus,

$$\text{Transpulmonary pressure} = P_{alv} - P_{ip} \qquad (13\text{-}3)$$
$$P_{tp} = P_{alv} - P_{ip}$$

Compare this equation to equation 13–2 (the equation that describes air flow into or out of the lungs), as it will be essential to distinguish these equations from each other (**Figure 13–9**).

Transpulmonary pressure is the **transmural pressure** that governs the static properties of the lungs. *Transmural* means "across a wall" and, by convention, is represented by the pressure in the inside of the structure (P_i) minus the pressure outside the structure (P_O). Inflation of a balloon-like structure like the lungs requires an increase in the transmural pressure such that P_i increases relative to P_O.

Table 13–3 and Figure 13–9 show the major transmural pressures of the respiratory system. The transmural pressure acting on the lungs (P_{tp}) is $P_{alv} - P_{ip}$ and on the chest wall (P_{cw}) is $P_{ip} - P_{atm}$. The muscles of the chest wall and the diaphragm contract and cause the chest wall to expand during inspiration. As the chest wall expands, P_{ip} decreases according to Boyle's law. P_{tp} becomes more positive as a result and the lungs expand. As this occurs, P_{alv} becomes more negative compared to P_{atm} (again due to Boyle's law), and air flows inward (inspiration, equation 13–2). Therefore, the transmural pressure of the lungs (P_{tp}) is increased to fill it with air by actively decreasing the pressure surrounding the lungs (P_{ip}) relative to the pressure inside the lungs (P_{alv}). When the respiratory muscles relax, elastic recoil of the lungs drives passive expiration back to the starting point.

How Is a Stable Balance Achieved Between Breaths?

Figure 13–10 illustrates the transmural pressures of the respiratory system at rest—that is, at the end of an unforced expiration when the respiratory muscles are relaxed and there is no

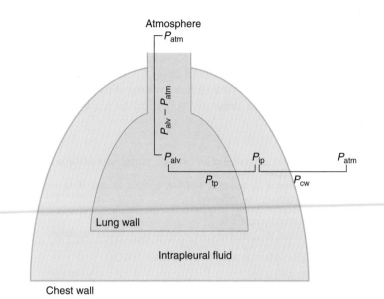

Figure 13–9

Pressure differences involved in ventilation. Transpulmonary pressure ($P_{tp} = P_{alv} - P_{ip}$) is a determinant of lung size. Intrapleural pressure (P_{ip}) at rest is a balance between the tendency of the lung to collapse and the tendency of the chest wall to expand. P_{cw} represents the transmural pressure across the chest wall ($P_{ip} - P_{atm}$). $P_{alv} - P_{atm}$ is the driving pressure gradient for airflow in and out of the lungs. (The volume of intrapleural fluid is greatly exaggerated for visual clarity.)

air flow. By definition, if there is no air flow, P_{alv} must equal P_{atm} (see equation 13–2). Because the lungs always have air in them, the transmural pressure of the lungs (P_{tp}) must always be positive and, therefore, $P_{alv} > P_{ip}$. At rest, when there is no air flow and $P_{alv} = 0$, P_{ip} must be negative, providing the force that keeps the lungs open and the chest wall in.

What are the forces that cause P_{ip} to be negative? The first, the **elastic recoil** of the lungs, is defined as the tendency of an elastic structure to oppose stretching or distortion. Even at rest, the lung contains air, and its natural tendency is to collapse because of elastic recoil. The lungs are held open by the positive P_{tp}, which, at rest, exactly opposes elastic recoil. The chest wall also has elastic recoil, and, at rest, its natural tendency is to expand.

At rest, all of these transmural pressures balance each other out. It is clear that the subatmospheric (negative) intrapleural pressure (P_{ip}) is the essential factor keeping the lungs

Table 13–3	Two Important Transmural Pressures of the Respiratory System		
Transmural Pressure	$P_i - P_O$	**Value at Rest**	**Explanatory Notes**
Transpulmonary (P_{tp})	$P_{alv} - P_{ip}$	0 – [–4] = 4 mmHg	Pressure difference holding lungs open (opposes inward elastic recoil of the lung)
Chest wall (P_{cw})	$P_{ip} - P_{atm}$	–4 – 0 = –4 mmHg	Pressure difference holding chest wall in (opposes outward elastic recoil of the chest wall)

P_i is pressure inside the structure, and P_O is pressure surrounding the structure.

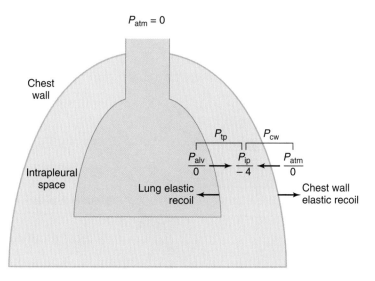

$P_{atm} = 0$

Chest wall

Intrapleural space

P_{tp} P_{cw}

P_{alv} P_{ip} P_{atm}
0   −4   0

Lung elastic recoil

Chest wall elastic recoil

Figure 13–10

Alveolar (P_{alv}), intrapleural (P_{ip}), transpulmonary (P_{tp}), and trans-chest-wall (P_{cw}) pressures (mmHg) at the end of an unforced expiration—that is, between breaths when there is no air flow. The transpulmonary pressure ($P_{alv} − P_{ip}$) exactly opposes the elastic recoil of the lung, and the lung volume remains stable. Similarly, trans-chest wall pressure ($P_{ip} − P_{atm}$) is balanced by the outward elastic recoil of the chest wall. Notice that transmural pressures are the pressure inside the wall minus the pressure outside the wall. (The volume of intrapleural fluid is greatly exaggerated for clarity.)

partially expanded between breaths. An extremely important question is: What is the reason for a subatmospheric (negative) P_{ip}?

As the lungs tend to collapse and the thoracic wall tends to expand, they move ever so slightly away from each other. This causes an infinitesimal enlargement of the fluid-filled intrapleural space between them. But fluid cannot expand the way air can, so even this tiny enlargement of the intrapleural space—so small that the pleural surfaces still remain in contact with each other—decreases the intrapleural pressure below atmospheric pressure. In this way, the elastic recoil of both the lungs and chest wall creates the subatmospheric intrapleural pressure that keeps them from moving apart more than a very tiny amount. Again, imagine trying to pull apart two glass slides that have a drop of water between them. The fluid pressure generated between the slides will be lower than atmospheric pressure.

The importance of the transpulmonary pressure in achieving this stable balance can be seen when, during surgery or trauma, the chest wall is pierced without damaging the lung. Atmospheric air rushes through the wound into the intrapleural space, a phenomenon called **pneumothorax,** and the intrapleural pressure goes from −4 mmHg to 0 mmHg. The transpulmonary pressure acting to hold the lung open is thus eliminated, and the lung collapses. At the same time, the chest wall moves outward because its elastic recoil is also no longer opposed. The thoracic cavity is divided into right and left sides by a structure called the **mediastinum,** so the pneumothorax is often unilateral (**Figure 13–11**).

Inspiration

Figures 13–12 and **13–13** summarize the events that occur during normal inspiration at rest. Inspiration is initiated by the neurally induced contraction of the diaphragm and the

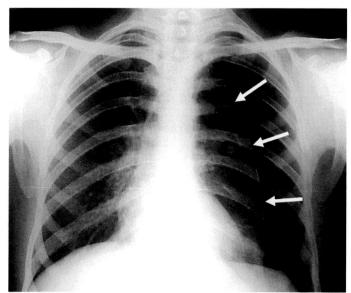

Figure 13–11

Pneumothorax due to trauma (motor vehicle collision). Arrows show outline of collapsed left lung. Right lung (shown on left) is not affected. Notice free air between collapsed left lung and chest wall.

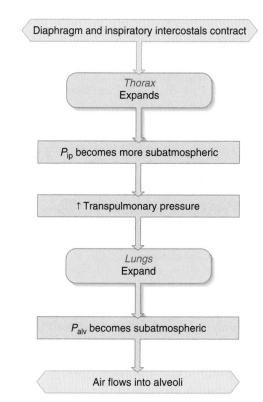

Diaphragm and inspiratory intercostals contract

↓

Thorax Expands

↓

P_{ip} becomes more subatmospheric

↓

↑ Transpulmonary pressure

↓

Lungs Expand

↓

P_{alv} becomes subatmospheric

↓

Air flows into alveoli

Figure 13–12

Sequence of events during inspiration. Figure 13–13 illustrates these events quantitatively.

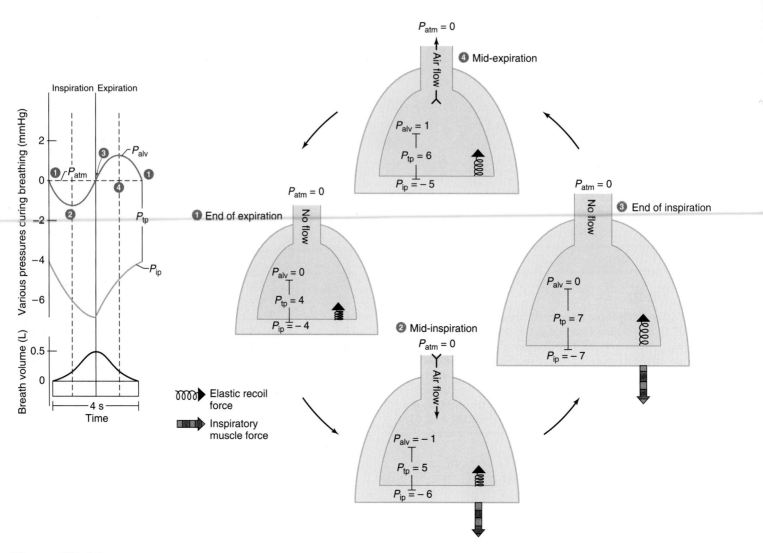

Figure 13–13

Summary of alveolar (P_{alv}), intrapleural (P_{ip}), and transpulmonary (P_{tp}) pressure changes and air flow during a typical respiratory cycle. At the end of expiration ❶, P_{alv} is equal to P_{atm} and there is no air flow. At mid-inspiration ❷, the chest wall is expanding, lowering P_{ip} and making P_{tp} more positive. This expands the lung making P_{alv} negative and results in an inward air flow. At end-inspiration ❸, the chest wall is no longer expanding but has yet to start passive contraction. Because lung size is not changing and the glottis is open to the atmosphere, P_{alv} is equal to P_{atm} and there is no air flow. As the respiratory muscles relax, the lungs and chest wall start to passively collapse due to elastic recoil. At mid-expiration ❹, the lung is collapsing, thus compressing alveolar gas. As a result, P_{alv} is positive relative to P_{atm}, and air flow is outward. The cycle starts over again at the end of expiration. Notice that throughout a typical respiratory cycle with a normal tidal volume, P_{ip} is negative relative to P_{atm}. In the graph on the left, the difference between P_{alv} and P_{ip} at any point along the curves is equivalent to P_{tp}.

Figure 13–13 physiological *inquiry* (p*i*)

■ How do the changes in P_{tp} between each step (❶–❹) explain whether the volume of the lung is increasing or decreasing?

Answer can be found at end of chapter.

inspiratory intercostal muscles located between the ribs. The adjective *inspiratory* here is a functional term, not an anatomical one. It denotes the several groups of intercostal muscles that contract during inspiration. The diaphragm is the most important inspiratory muscle that acts during normal quiet breathing. When activation of the **phrenic nerves** to the diaphragm causes it to contract, its dome moves downward into the abdomen, enlarging the thorax (**Figure 13–14**). Simultaneously,

activation of the intercostal nerves to the inspiratory intercostal muscles causes them to contract, leading to an upward and outward movement of the ribs and a further increase in thoracic size.

The crucial point is that contraction of the inspiratory muscles, by *actively* increasing the size of the thorax, upsets the stability set up by purely elastic forces between breaths. As the thorax enlarges, the thoracic wall moves ever so slightly

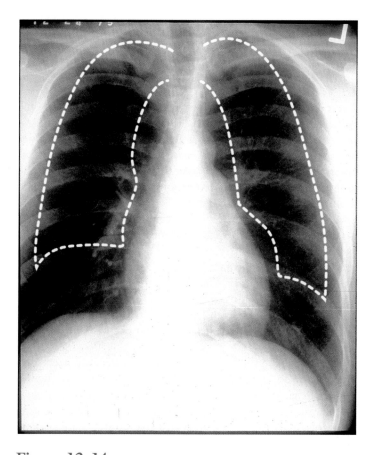

Figure 13–14

X-ray of chest at full inspiration. The dashed white line is an outline of the lungs in full expiration.

farther away from the lung surface. The intrapleural fluid pressure therefore becomes even more subatmospheric than it was between breaths. This decrease in intrapleural pressure *increases* the transpulmonary pressure. Therefore, the force acting to expand the lungs—the transpulmonary pressure—is now greater than the elastic recoil exerted by the lungs at this moment, and so the lungs expand further. Note in Figure 13–13 that, by the end of inspiration, equilibrium *across the lungs* is once again established because the more inflated lungs exert a greater elastic recoil, which equals the increased transpulmonary pressure. In other words, lung volume is stable whenever transpulmonary pressure is balanced by the elastic recoil of the lungs (that is, at the end of both inspiration and expiration when there is no air flow).

Thus, when contraction of the inspiratory muscles actively increases the thoracic dimensions, the lungs are passively forced to enlarge. The enlargement of the lungs causes an increase in the sizes of the alveoli throughout the lungs. By Boyle's law, the pressure within the alveoli decreases to less than atmospheric (see Figure 13–13). This produces the difference in pressure ($P_{alv} < P_{atm}$) that causes a bulk flow of air from the atmosphere through the airways into the alveoli. By the end of the inspiration, the pressure in the alveoli again equals atmospheric pressure because of this additional air, and air flow ceases.

Expiration

Figure 13–13 and **Figure 13–15** summarize the sequence of events that occur during expiration. At the end of inspiration, the nerves to the diaphragm and inspiratory intercostal muscles decrease their firing, and so these muscles relax. The diaphragm and chest wall are no longer actively pulled outward by the muscle contractions, and so they start to recoil inward to their original smaller dimensions that existed between breaths. This immediately makes the intrapleural pressure less subatmospheric thereby *decreasing* the transpulmonary pressure. Therefore, the transpulmonary pressure acting to expand the lungs is now smaller than the elastic recoil exerted by the more expanded lungs, and the lungs passively recoil to their original dimensions (see Figure 13–14).

As the lungs become smaller, air in the alveoli becomes temporarily compressed so that, by Boyle's law, alveolar pressure exceeds atmospheric pressure (see Figure 13–13). Therefore, air flows from the alveoli through the airways out into the atmosphere. Thus, expiration at rest is passive, depending only upon the relaxation of the inspiratory muscles and the elastic recoil of the stretched lungs.

Under certain conditions, such as during exercise, expiration of larger volumes is achieved by contraction of a different

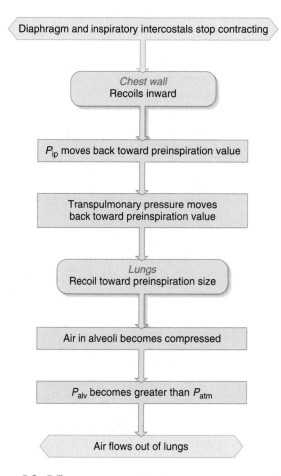

Figure 13–15

Sequence of events during expiration. Figure 13–13 illustrates these events quantitatively.

set of intercostal muscles and the abdominal muscles, which *actively* decreases thoracic dimensions. The *expiratory* intercostal muscles (again a functional term, not an anatomical one) insert on the ribs in such a way that their contraction pulls the chest wall downward and inward, thereby decreasing thoracic volume. Contraction of the abdominal muscles increases intra-abdominal pressure and forces the relaxed diaphragm up into the thorax.

Lung Compliance

To repeat, the degree of lung expansion at any instant is proportional to the transpulmonary pressure, $P_{alv} - P_{ip}$. But just how much any given change in transpulmonary pressure expands the lungs depends upon the stretchability, or compliance, of the lungs. **Lung compliance (C_L)** is defined as the magnitude of the change in lung volume (ΔV_L) produced by a given change in the transpulmonary pressure:

$$C_L = \Delta V_L / \Delta P_{tp} \qquad (13\text{–}4)$$

Thus, the greater the lung compliance, the easier it is to expand the lungs at any given change in transpulmonary pressure (**Figure 13–16**). Compliance can be considered the inverse of stiffness. A low lung compliance means that a greater-than-normal transpulmonary pressure must be developed across the lung to produce a given amount of lung expansion. In other words, when lung compliance is abnormally low (increased stiffness), intrapleural pressure must be made more subatmospheric than usual during inspiration to achieve lung expansion. This requires more vigorous contractions of the diaphragm and inspiratory intercostal muscles. Thus, the less compliant the lung, the more energy is required to produce a given amount of expansion. Persons with low lung compliance due to disease therefore tend to breathe shallowly and must breathe at a higher frequency to inspire an adequate volume of air.

Determinants of Lung Compliance

There are two major determinants of lung compliance. One is the stretchability of the lung tissues, particularly their elastic connective tissues. Thus, a thickening of the lung tissues decreases lung compliance. However, an equally important determinant of lung compliance is not the elasticity of the lung tissues, but the surface tension at the air-water interfaces within the alveoli.

The surface of the alveolar cells is moist, so the alveoli can be pictured as air-filled sacs lined with water. At an air-water interface, the attractive forces between the water molecules, known as **surface tension,** make the water lining like a stretched balloon that constantly tends to shrink and resists further stretching. Thus, expansion of the lung requires energy not only to stretch the connective tissue of the lung, but also to overcome the surface tension of the water layer lining the alveoli.

Indeed, the surface tension of pure water is so great that were the alveoli lined with pure water, lung expansion would require exhausting muscular effort, and the lungs would tend to collapse. It is extremely important, therefore, that the type II alveolar cells secrete the detergent-like substance mentioned earlier, known as **surfactant,** which markedly reduces the cohesive forces between water molecules on the alveolar surface. Therefore, surfactant lowers the surface tension, which increases lung compliance and makes it easier to expand the lungs (**Table 13–4**).

Surfactant is a mixture of both lipids and proteins, but its major component is a phospholipid that forms a monomolecular layer between the air and water at the alveolar surface. The amount of surfactant tends to decrease when breaths are small and constant. A deep breath, which people normally intersperse frequently in their breathing pattern, stretches the

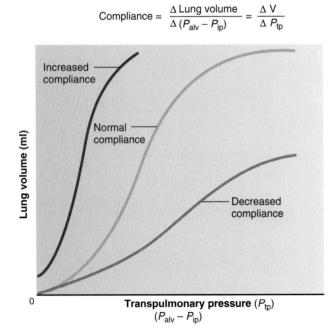

Compliance = $\dfrac{\Delta \text{ Lung volume}}{\Delta (P_{alv} - P_{ip})} = \dfrac{\Delta V}{\Delta P_{tp}}$

Increased compliance

Normal compliance

Decreased compliance

Lung volume (ml)

0

Transpulmonary pressure (P_{tp})
($P_{alv} - P_{ip}$)

Figure 13–16

A graphical representation of lung compliance. Changes in lung volume and transpulmonary pressure are measured as a subject takes progressively larger breaths. When compliance is lower than normal (the lung is stiffer), there is a lesser increase in lung volume for any given increase in transpulmonary pressure. When compliance is increased, as in emphysema, small decreases in P_{tp} allow the lung to collapse.

Figure 13–16 physiological *inquiry* (pi)

■ Premature infants with inadequate surfactant have decreased lung compliance (respiratory distress syndrome of the newborn). If surfactant is not available to administer for therapy, what can be done to inflate the lung?

Answer can be found at end of chapter.

Table 13-4	Some Important Facts About Pulmonary Surfactant

1. Pulmonary surfactant is a mixture of phospholipids and protein.

2. It is secreted by type II alveolar cells.

3. It lowers the surface tension of the water layer at the alveolar surface, which increases lung compliance, thereby making the lungs easier to expand.

4. Its surface tension is lower in smaller alveoli, thus stabilizing alveoli.

5. A deep breath increases its secretion by stretching the type II cells. Its concentration decreases when breaths are small.

6. Production in the fetal lung occurs in late gestation.

type II cells, which stimulates the secretion of surfactant. This is why patients who have had thoracic or abdominal surgery and are breathing shallowly because of the pain must be urged to take occasional deep breaths.

The **Law of Laplace** describes the relationship between pressure (P), surface tension (T), and the radius (r) of an alveolus shown in **Figure 13–17**:

$$P = 2T/r \qquad (13\text{–}5)$$

As the radius inside the alveolus decreases, the pressure increases. Now imagine two alveoli next to each other sharing an alveolar duct (see Figure 13–17). The radius of alveolus a (r_a) is greater than the radius of alveolus b (r_b). If surface tension (T) were equivalent between these two alveoli, alveolus b would have a higher pressure than alveolus a by the Law of Laplace. If P_b is higher than P_a, air would flow from alveolus b into alveolus a, and alveolus b would collapse. Therefore, small alveoli would be unstable and would collapse into large alveoli. Another important property of surfactant is that it stabilizes alveoli of different sizes by altering surface tension, depending on the surface area of the alveolus. As an alveolus gets smaller, the molecules of surfactant on its inside surface are less spread out, thus reducing surface tension. The reduction in surface tension helps to maintain a pressure in smaller alveoli equal to that in larger ones. This gives stability to alveoli of different sizes.

A striking example of what occurs when surfactant is deficient is the disease known as **respiratory distress syndrome of the newborn.** This is a leading cause of death in premature infants, in whom the surfactant-synthesizing cells may be too immature to function adequately. Respiratory movements in the fetus do not require surfactant because the lungs are filled with amniotic fluid, and the fetus receives oxygen from the maternal blood. Because of low lung compliance, the affected infant can inspire only by the most strenuous efforts, which may ultimately cause complete exhaustion, inability to breathe, lung collapse, and death. Therapy in such cases is assisted breathing with a mechanical ventilator and the administration of natural or synthetic surfactant given through the infant's trachea.

Airway Resistance

As previously stated, the volume of air that flows into or out of the alveoli per unit time is directly proportional to the pressure difference between the atmosphere and alveoli and is inversely proportional to the resistance to flow of the airways (see equation 13–2). The factors that determine airway resistance are analogous to those determining vascular resistance in the circulatory system: tube length, tube radius, and interactions between moving molecules (gas molecules, in this case). As in the circulatory system, the most important factor by far is the radius of the tube—airway resistance is inversely proportional to the fourth power of the airway radii.

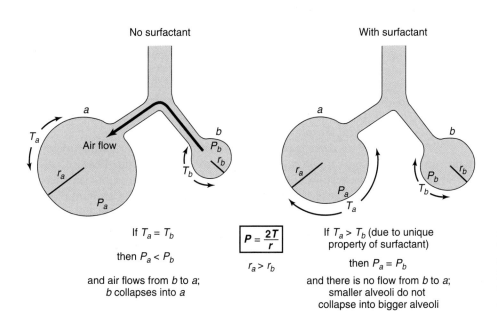

No surfactant

If $T_a = T_b$

then $P_a < P_b$

and air flows from b to a; b collapses into a

$$\boxed{P = \dfrac{2T}{r}}$$

$r_a > r_b$

With surfactant

If $T_a > T_b$ (due to unique property of surfactant)

then $P_a = P_b$

and there is no flow from b to a; smaller alveoli do not collapse into bigger alveoli

Figure 13–17

Stabilizing effect of surfactant. P is pressure inside the alveoli, T is a surface tension, and r is the radius of the alveolus. The Law of Laplace is described by the equation in the box.

Airway resistance to air flow is normally so small that very small pressure differences produce large volumes of air flow. As we have seen (see Figure 13–13), the average atmosphere-to-alveoli pressure difference during a normal breath when at rest is about 1 mmHg; yet approximately 500 ml of air is moved by this tiny difference.

Physical, neural, and chemical factors affect airway radii and therefore resistance. One important physical factor is the transpulmonary pressure, which exerts a distending force on the airways, just as on the alveoli. This is a major factor keeping the smaller airways—those without cartilage to support them—from collapsing. Because transpulmonary pressure increases during inspiration (see Figure 13–13), airway radius becomes larger and airway resistance lower as the lungs expand during inspiration. The opposite occurs during expiration.

A second physical factor holding the airways open is the elastic connective tissue fibers that link the outside of the airways to the surrounding alveolar tissue. These fibers are pulled upon as the lungs expand during inspiration, and in turn they help pull the airways open even more than between breaths. This is termed **lateral traction.** Thus, both the transpulmonary pressure and lateral traction act in the same direction, reducing airway resistance during inspiration.

Such physical factors also explain why the airways become narrower and airway resistance increases during a forced expiration. The increase in intrapleural pressure compresses the small conducting airways and decreases their radii. Therefore, because of increased airway resistance, there is a limit to how much one can increase the air flow rate during a forced expiration no matter how intense the effort. The harder one pushes, the greater the compression of the airways, further limiting expiratory air flow.

In addition to these physical factors, a variety of neuroendocrine and paracrine factors can influence airway smooth muscle and thereby airway resistance. For example, the hormone epinephrine relaxes airway smooth muscle by an effect on beta-adrenergic receptors, whereas the leukotrienes, members of the eicosanoid family, produced in the lungs during inflammation contract the muscle.

Why are we concerned with all the physical and chemical factors that *can* influence airway resistance when airway resistance is normally so low that it poses no impediment to air flow? The reason is that, under abnormal circumstances, changes in these factors may cause serious increases in airway resistance. Asthma and chronic obstructive pulmonary disease provide important examples, as we see next.

Asthma

Asthma is a disease characterized by intermittent episodes in which airway smooth muscle contracts strongly, markedly increasing airway resistance. The basic defect in asthma is chronic inflammation of the airways, the causes of which vary from person to person and include, among others, allergy, viral infections, and sensitivity to environmental factors. The underlying inflammation makes the airway smooth muscle hyperresponsive and causes it to contract strongly in response to such things as exercise (especially in cold, dry air), cigarette smoke, environmental pollutants, viruses, allergens, normally

released bronchoconstrictor chemicals, and a variety of other potential triggers. In fact, the incidence of asthma is increasing, possibly due in part to environmental pollution.

The first aim of therapy for asthma is to reduce the chronic inflammation and airway hyperresponsiveness with **anti-inflammatory drugs,** particularly inhaled glucocorticoids and leukotriene inhibitors. The second aim is to overcome acute excessive airway smooth muscle contraction with **bronchodilator drugs,** which relax the airways. The latter drugs work on the airways either by relaxing airway smooth muscle or by blocking the actions of bronchoconstrictors. For example, one class of bronchodilator drugs mimics the normal action of epinephrine on beta-adrenergic (beta-2) receptors. Another class of inhaled drugs block muscarinic cholinergic receptors, which have been implicated in bronchoconstriction.

Chronic Obstructive Pulmonary Disease

The term **chronic obstructive pulmonary disease** refers to emphysema, chronic bronchitis, or a combination of the two. These diseases, which cause severe difficulties not only in ventilation but in oxygenation of the blood, are among the major causes of disability and death in the United States. In contrast to asthma, increased smooth muscle contraction is *not* the cause of the airway obstruction in these diseases.

Emphysema is discussed later in this chapter; suffice it to say here that the cause of obstruction in this disease is destruction and collapse of the smaller airways.

Chronic bronchitis is characterized by excessive mucus production in the bronchi and chronic inflammatory changes in the small airways. The cause of obstruction is an accumulation of mucus in the airways and thickening of the inflamed airways. The same agents that cause emphysema—smoking, for example—also cause chronic bronchitis, which is why the two diseases frequently coexist.

Lung Volumes and Capacities

Normally the volume of air entering the lungs during a single inspiration, called the **tidal volume (V_t)**, is approximately equal to the volume leaving on the subsequent expiration. The tidal volume during normal quiet breathing is termed the resting tidal volume and is approximately 500 ml depending on body size. As illustrated in **Figure 13–18**, the maximal amount of air that can be increased above this value during deepest inspiration is termed the **inspiratory reserve volume (IRV)** and is about 3000 ml—i.e., six times greater than resting tidal volume.

After expiration of a resting tidal volume, the lungs still contain a very large volume of air. As described earlier, this is the resting position of the lungs and chest wall when there is no contraction of the respiratory muscles; this amount of air is termed the **functional residual capacity (FRC)** and averages about 2400 ml. Thus, the 500 ml of air inspired with each resting breath adds to and mixes with the much larger volume of air already in the lungs, and then 500 ml of the total is expired. Through maximal active contraction of the expiratory muscles, it is possible to expire much more of the air remaining after the resting tidal volume has been expired. This additional expired volume is termed the **expiratory reserve volume (ERV)** and is about 1200 ml. Even after a maximal active expiration, approximately 1200 ml of air still remains in the lungs; this is termed the **residual volume (RV).** Thus, the lungs are never completely emptied of air.

The **vital capacity (VC)** is the maximal volume of air a person can expire after a maximal inspiration. Under these conditions, the person is expiring both the resting tidal volume and inspiratory reserve volume just inspired, plus the expiratory reserve volume (see Figure 13–18). In other words, the vital capacity is the sum of these three volumes.

A variant on this method is the *forced expiratory volume in 1 s, (FEV₁),* in which the person takes a maximal inspiration and then exhales maximally as fast as possible. The important value is the fraction of the total "forced" vital capacity expired in 1 s. Normal individuals can expire approximately 80 percent of the vital capacity in one second.

Measurement of vital capacity and FEV are useful diagnostically and are known as *pulmonary function tests.* For example, people with *obstructive lung diseases* (increased airway resistance) typically have an FEV_1 that is less than 80 percent of the vital capacity because it is difficult for them to expire air rapidly through the narrowed airways. In contrast to obstructive lung diseases, *restrictive lung diseases* are characterized by normal airway resistance but impaired respiratory movements because of abnormalities in the lung tissue, the pleura, the chest wall, or the neuromuscular machinery. Restrictive lung diseases are thus characterized by a reduced vital capacity, but a normal ratio of FEV_1 to vital capacity.

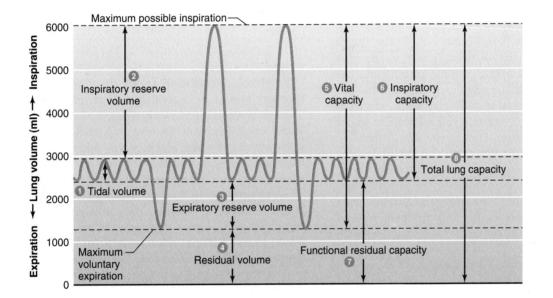

Figure 13–18

Lung volumes and capacities recorded on a spirometer, an apparatus for measuring inspired and expired volumes. When the subject inspires, the pen moves up; with expiration, it moves down. The capacities are the sums of two or more lung volumes. The lung volumes are the four distinct components of total lung capacity. Note that residual volume, total lung capacity, and functional residual capacity cannot be measured with a spirometer.

Respiratory Volumes and Capacities for an Average Young Adult Male		
Measurement	Typical Value	Definition
Respiratory Volumes		
❶ Tidal volume (TV)	500 ml	Amount of air inhaled or exhaled in one breath during relaxed, quiet breathing
❷ Inspiratory reserve volume (IRV)	3000 ml	Amount of air in excess of tidal inspiration that can be inhaled with maximum effort
❸ Expiratory reserve volume (ERV)	1200 ml	Amount of air in excess of tidal expiration that can be exhaled with maximum effort
❹ Residual volume (RV)	1200 ml	Amount of air remaining in the lungs after maximum expiration; keeps alveoli inflated between breaths and mixes with fresh air on next inspiration
Respiratory Capacities		
❺ Vital capacity (VC)	4700 ml	Amount of air that can be exhaled with maximum effort after maximum inspiration (ERV + TV + IRV); used to assess strength of thoracic muscles as well as pulmonary function
❻ Inspiratory capacity (IC)	3500 ml	Maximum amount of air that can be inhaled after a normal tidal expiration (TV + IRV)
❼ Functional residual capacity (FRC)	2400 ml	Amount of air remaining in the lungs after a normal tidal expiration (RV + ERV)
❽ Total lung capacity (TLC)	5900 ml	Maximum amount of air the lungs can contain (RV + VC)

Alveolar Ventilation

The total ventilation per minute, termed the **minute ventilation ($\dot{V}_E$),** is equal to the tidal volume multiplied by the respiratory rate:

$$\text{Minute ventilation} = \text{Tidal volume} \times \text{Respiratory rate}$$
$$\text{(ml/min)} \qquad \text{(ml/breath)} \qquad \text{(breaths/min)}$$

$$\dot{V}_E \quad = \quad V_t \quad \cdot \quad f \qquad \text{(13–6)}$$

For example, at rest, a normal person moves approximately 500 ml of air in and out of the lungs with each breath and takes 12 breaths each minute. The minute ventilation is therefore 500 ml/breath × 12 breaths/minute = 6000 ml of air per minute. However, because of dead space, not all this air is available for exchange with the blood, as we see next.

Dead Space

The conducting airways have a volume of about 150 ml. Exchanges of gases with the blood occur only in the alveoli and not in this 150 ml of the airways. Picture, then, what occurs during expiration of a tidal volume of 500 ml. The 500 ml of air is forced out of the alveoli and through the airways. Approximately 350 ml of this alveolar air is exhaled at the nose or mouth, but approximately 150 ml remains in the airways at the end of expiration. During the next inspiration (**Figure 13–19**), 500 ml of air flows into the alveoli, but the first 150 ml entering the alveoli is not atmospheric air but the 150 ml left behind in the airways from the last breath. Thus, only 350 ml of new atmospheric air enters the alveoli during the inspiration. The end result is that 150 ml of the 500 ml of atmospheric air entering the respiratory system during each inspiration never reaches the alveoli, but is merely moved in and out of the airways. Because these airways do not permit gas exchange with the blood, the space within them is termed the **anatomic dead space (V_D).**

Thus the volume of *fresh* air entering the alveoli during each inspiration equals the tidal volume *minus* the volume of air in the anatomic dead space. For the previous example:

Tidal volume (V_t) = 500 ml

Anatomic dead space (V_D) = 150 ml

Fresh air entering alveoli in one inspiration (V_A) =
500 ml – 150 ml = 350 ml

The total volume of fresh air entering the alveoli per minute is called the **alveolar ventilation ($\dot{V}_A$):**

$$\begin{array}{ccccc} \text{Alveolar} & \left(\text{Tidal} \right. & \text{Dead} & \text{Respiratory} \\ \text{ventilation} = & \text{volume} & - & \left. \text{space}\right) \times & \text{rate} \\ \text{(ml/min)} & \text{(ml/breath)} & \text{(ml/breath)} & \text{(breaths/min)} \end{array}$$

$$\dot{V}_A \quad = \quad (V_t \quad - \quad V_D) \quad \cdot \quad f \qquad \text{(13–7)}$$

Alveolar ventilation rather than minute ventilation, is the more important factor in the effectiveness of gas exchange. This generalization is demonstrated readily by the data in **Table 13–5**. In this experiment, subject A breathes rapidly and shallowly, B normally, and C slowly and deeply. Each subject has exactly the same minute ventilation; that is, each is moving the same amount of air in and out of the lungs per minute. Yet, when we subtract the anatomic dead-space ventilation from the minute ventilation, we find marked differences in alveolar ventilation. Subject A has no alveolar ventilation and would become unconscious in several minutes, whereas C has a considerably greater alveolar ventilation than B, who is breathing normally.

Another important generalization drawn from this example is that increased *depth* of breathing is far more effective in elevating alveolar ventilation than an equivalent increase in breathing *rate*. Conversely, a decrease in depth can lead to a critical reduction in alveolar ventilation. This is because a fixed volume of each tidal volume goes to the dead space. If the tidal volume decreases, the fraction of the tidal volume going to the dead space increases until, as in subject A, it may represent the entire tidal volume. On the other hand, any increase

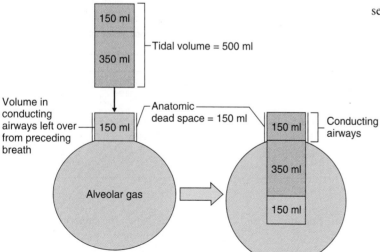

Figure 13–19

Effects of anatomic dead space on alveolar ventilation. Anatomic dead space is the volume of the conducting airways. Of a 500 ml tidal volume breath, 350 ml enters the airway involved in gas exchange. The remaining 150 ml remains in the conducting airways and does not participate in gas exchange.

	Table 13–5		Effect of Breathing Patterns on Alveolar Ventilation			
Subject	Tidal Volume (ml/breath)	×	Frequency (breaths/min) =	Minute Ventilation (ml/min)	Anatomic Dead-Space Ventilation (ml/min)	Alveolar Ventilation (ml/min)
A	150		40	6000	$150 \times 40 = 6000$	0
B	500		12	6000	$150 \times 12 = 1800$	4200
C	1000		6	6000	$150 \times 6 = 900$	5100

in tidal volume goes entirely toward increasing alveolar ventilation. These concepts have important physiological implications. Most situations that produce an increased ventilation, such as exercise, reflexly call forth a relatively greater increase in breathing depth than in breathing rate.

The anatomic dead space is not the only type of dead space. Some fresh inspired air is not used for gas exchange with the blood even though it reaches the alveoli because some alveoli may, for various reasons, have little or no blood supply. This volume of air is known as **alveolar dead space.** It is quite small in normal persons but may be very large in persons with several kinds of lung disease. As we shall see, local mechanisms that match air and blood flows minimize the alveolar dead space. The sum of the anatomic and alveolar dead spaces is known as the **physiologic dead space.** This is also known as wasted ventilation because it is air that is inspired but does not participate in gas exchange with blood flowing through the lungs.

Exchange of Gases in Alveoli and Tissues

We have now completed the discussion of the lung mechanics that produce alveolar ventilation, but this is only the first step in the respiratory process. Oxygen must move across the alveolar membranes into the pulmonary capillaries, be transported by the blood to the tissues, leave the tissue capillaries and enter the extracellular fluid, and finally cross plasma membranes to gain entry into cells. Carbon dioxide must follow a similar path, but in reverse.

In the steady state, the volume of oxygen that leaves the tissue capillaries and is consumed by the body cells per unit time is equal to the volume of oxygen added to the blood in the lungs during the same time period. Similarly, in the steady state, the rate at which carbon dioxide is produced by the body cells and enters the systemic blood is the same as the rate at which carbon dioxide leaves the blood in the lungs and is expired.

The amount of oxygen the cells consume and the amount of carbon dioxide they produce, however, are not necessarily identical. The balance depends primarily upon which nutrients are used for energy. The ratio of CO_2 produced to O_2 consumed is known as the **respiratory quotient (RQ).** On a mixed diet, the RQ is approximately 0.8; that is, 8 molecules of CO_2 are produced for every 10 molecules of O_2 consumed. The RQ is 1 for carbohydrate, 0.7 for fat, and 0.8 for protein.

Figure 13–20 presents typical exchange values during 1 min for a person at rest, assuming a cellular oxygen consumption of 250 ml/min, a carbon dioxide production of 200 ml/min, an alveolar ventilation of 4000 ml/min, and a cardiac output of 5000 ml/min.

Because only 21 percent of the atmospheric air is oxygen, the total oxygen entering the alveoli per min in our illustration is 21 percent of 4000 ml, or 840 ml/min. Of this inspired oxygen, 250 ml crosses the alveoli into the pulmonary capillaries, and the rest is subsequently exhaled. Note that blood entering the lungs already contains a large quantity of oxygen, to which the new 250 ml is added. The blood then flows from the lungs to the left side of the heart and is pumped by the left ventricle through the aorta, arteries, and arterioles into the tissue capillaries, where 250 ml of oxygen leaves the blood per minute for cells to take up and utilize. Thus, the quantities of oxygen added to the blood in the lungs and removed in the tissues are the same.

The story reads in reverse for carbon dioxide. A significant amount of carbon dioxide already exists in systemic arterial blood; to this is added an additional 200 ml per minute, the amount the cells produce, as blood flows through tissue capillaries. This 200 ml leaves the blood each minute as blood flows through the lungs and is expired.

Blood pumped by the heart carries oxygen and carbon dioxide between the lungs and tissues by bulk flow, but diffusion is responsible for the net movement of these molecules between the alveoli and blood, and between the blood and the cells of the body. Understanding the mechanisms involved in these diffusional exchanges depends upon some basic chemical and physical properties of gases, which we will now discuss.

Partial Pressures of Gases

Gas molecules undergo continuous random motion. These rapidly moving molecules collide and exert a pressure, the magnitude of which is increased by anything that increases the rate of movement. The pressure a gas exerts is proportional to temperature (because heat increases the speed at which molecules move) and the concentration of the gas—that is, the number of molecules per unit volume.

As **Dalton's law** states, in a mixture of gases, the pressure each gas exerts is independent of the pressure the others exert. This is because gas molecules are normally so far apart that they do not affect each other. Each gas in a mixture behaves as though no other gases are present, so the total

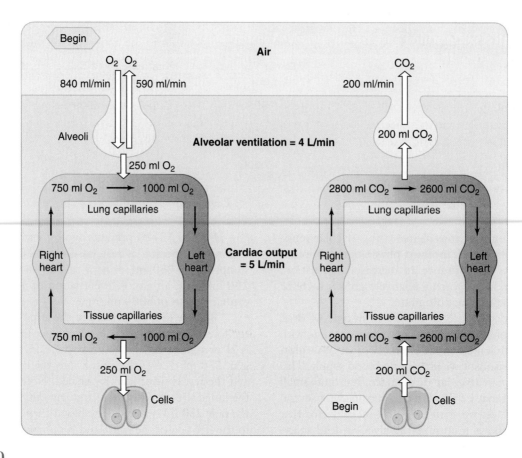

Figure 13–20

Summary of typical oxygen and carbon dioxide exchanges between atmosphere, lungs, blood, and tissues *during 1 min* in a resting individual. Note that the values in this figure for oxygen and carbon dioxide in blood are *not* the values per liter of blood, but rather the amounts transported *per minute* in the cardiac output (5 L in this example). The volume of oxygen in 1 L of arterial blood is 200 ml O_2/L of blood—that is, 1000 ml O_2/5 L of blood.

pressure of the mixture is simply the sum of the individual pressures. These individual pressures, termed **partial pressures,** are denoted by a P in front of the symbol for the gas. For example, the partial pressure of oxygen is expressed as P_{O_2}. The partial pressure of a gas is directly proportional to its concentration. Net diffusion of a gas will occur from a region where its partial pressure is high to a region where it is low.

Atmospheric air consists of approximately 79 percent nitrogen and approximately 21 percent oxygen, with very small quantities of water vapor, carbon dioxide, and inert gases. The sum of the partial pressures of all these gases is termed atmospheric pressure, or barometric pressure. It varies in different parts of the world as a result of local weather conditions and gravitational differences due to altitude, but at sea level it is 760 mmHg. Because the partial pressure of any gas in a mixture is the fractional concentration of that gas times the total pressure of all the gases, the P_{O_2} of atmospheric air is 0.21 × 760 mmHg = 160 mmHg at sea level.

Diffusion of Gases in Liquids

When a liquid is exposed to air containing a particular gas, molecules of the gas will enter the liquid and dissolve in it. **Henry's law** states that the amount of gas dissolved will be directly proportional to the partial pressure of the gas with which the liquid

is in equilibrium. A corollary is that, at equilibrium, the partial pressures of the gas molecules in the liquid and gaseous phases must be identical. Suppose, for example, that a closed container contains both water and gaseous oxygen. Oxygen molecules from the gas phase constantly bombard the surface of the water, some entering the water and dissolving. The number of molecules striking the surface is directly proportional to the P_{O_2} of the gas phase, so the number of molecules entering the water and dissolving in it is also directly proportional to the P_{O_2}. As long as the P_{O_2} in the gas phase is higher than the P_{O_2} in the liquid, there will be a net diffusion of oxygen into the liquid. Diffusion equilibrium will be reached only when the P_{O_2} in the liquid is equal to the P_{O_2} in the gas phase, and there will then be no further net diffusion between the two phases.

Conversely, if a liquid containing a dissolved gas at high partial pressure is exposed to a lower partial pressure of that same gas in a gas phase, a net diffusion of gas molecules will occur out of the liquid into the gas phase until the partial pressures in the two phases become equal.

The exchanges *between* gas and liquid phases described in the preceding two paragraphs are precisely the phenomena occurring in the lungs between alveolar air and pulmonary capillary blood. In addition, *within* a liquid, dissolved gas molecules also diffuse from a region of higher partial pressure to a region of lower partial pressure, an effect that underlies the exchange of gases between cells, extracellular fluid, and capillary blood throughout the body.

Why must the diffusion of gases into or within liquids be presented in terms of partial pressures rather than "concentrations," the values used to deal with the diffusion of all other solutes? The reason is that the concentration of a gas in a liquid is proportional not only to the partial pressure of the gas but also to the solubility of the gas in the liquid. The more soluble the gas, the greater its concentration will be at any given partial pressure. Thus, if a liquid is exposed to two different gases having the same partial pressures, at equilibrium the *partial pressures* of the two gases will be identical in the liquid, but the *concentrations* of the gases in the liquid will differ, depending upon their solubilities in that liquid.

With these basic gas properties as the foundation, we can now discuss the diffusion of oxygen and carbon dioxide across alveolar and capillary walls and plasma membranes. The partial pressures of these gases in air and in various sites of the body for a resting person at sea level appear in **Figure 13–21**. We start our discussion with the alveolar gas pressures because their values set those of systemic arterial blood. This fact cannot be emphasized too strongly: The alveolar P_{O_2} and P_{CO_2} determine the systemic arterial P_{O_2} and P_{CO_2}.

Alveolar Gas Pressures

Normal alveolar gas pressures are $P_{O_2} = 105$ mmHg and $P_{CO_2} = 40$ mmHg. (We do not deal with nitrogen, even though it is the most abundant gas in the alveoli, because nitrogen is biologically inert under normal conditions and does not undergo any net exchange in the alveoli.) Compare these values with the gas pressures in the air being breathed: $P_{O_2} = 160$ mmHg and $P_{CO_2} = 0.3$ mmHg, a value so low that we will simply treat it as zero. The alveolar P_{O_2} is lower than atmospheric P_{O_2} because some of the oxygen in the air entering the alveoli leaves them to enter the pulmonary capillaries. Alveolar P_{CO_2} is higher than atmospheric P_{CO_2} because carbon dioxide enters the alveoli from the pulmonary capillaries.

The factors that determine the precise value of alveolar P_{O_2} are (1) the P_{O_2} of atmospheric air, (2) the rate of alveolar ventilation, and (3) the rate of total-body oxygen consumption. Although equations exist for calculating the alveolar gas pressures from these variables, we will describe the interactions in a qualitative manner (**Table 13–6**). To start, we will assume that only one of the factors changes at a time.

First, a decrease in the P_{O_2} of the inspired air, such as would occur at high altitude, will decrease alveolar P_{O_2}. A decrease in alveolar ventilation will do the same thing (**Figure 13–22**) because less fresh air is entering the alveoli per unit time. Finally, an increase in the oxygen consumption in the cells will also lower alveolar P_{O_2} because a larger fraction of the oxygen in the entering fresh air will leave the alveoli to enter the blood for use by the tissues. (Recall that in the steady state, the volume of oxygen entering the blood in the lungs per unit time is always equal to the volume utilized by the tissues.) This discussion has been in terms of things that lower alveolar P_{O_2}; just reverse the direction of change of the three factors to see how to increase P_{O_2}.

The story for P_{CO_2} is analogous, again assuming that only one factor changes at a time. There is normally essentially no

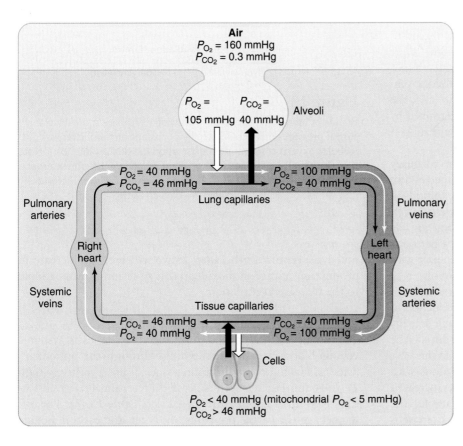

Air
$P_{O_2} = 160$ mmHg
$P_{CO_2} = 0.3$ mmHg

$P_{O_2} =$ 105 mmHg $P_{CO_2} =$ 40 mmHg   Alveoli

$P_{O_2} = 40$ mmHg
$P_{CO_2} = 46$ mmHg

$P_{O_2} = 100$ mmHg
$P_{CO_2} = 40$ mmHg

Lung capillaries

Pulmonary arteries

Pulmonary veins

Right heart

Left heart

Systemic veins

Systemic arteries

Tissue capillaries

$P_{CO_2} = 46$ mmHg
$P_{O_2} = 40$ mmHg

$P_{CO_2} = 40$ mmHg
$P_{O_2} = 100$ mmHg

Cells

$P_{O_2} < 40$ mmHg (mitochondrial $P_{O_2} < 5$ mmHg)
$P_{CO_2} > 46$ mmHg

Figure 13–21

Partial pressures of carbon dioxide and oxygen in inspired air at sea level and in various places in the body. The reason that the alveolar P_{O_2} and pulmonary vein P_{O_2} are not exactly the same is described later in the text. Note also that the P_{O_2} in the systemic arteries is shown as identical to that in the pulmonary veins; for reasons involving the anatomy of the blood flow to the lungs, the systemic arterial value is actually slightly less, but we have ignored this for the sake of clarity.

Table 13-6   Effects of Various Conditions on Alveolar Gas Pressures

Condition	Alveolar P_{O_2}	Alveolar P_{CO_2}
Breathing air with low P_{O_2}	Decreases	No change*
↑ Alveolar ventilation and unchanged metabolism	Increases	Decreases
↓ Alveolar ventilation and unchanged metabolism	Decreases	Increases
↑ Metabolism and unchanged alveolar ventilation	Decreases	Increases
↓ Metabolism and unchanged alveolar ventilation	Increases	Decreases
Proportional increases in metabolism and alveolar ventilation	No change	No change

*Breathing air with low P_{O_2} has no direct effect on alveolar P_{CO_2}. However, as described later in the text, people in this situation will reflexly increase their ventilation, and that will lower P_{CO_2}.

carbon dioxide in inspired air and so we can ignore that factor. A decreased alveolar ventilation will increase the alveolar P_{CO_2} (see Figure 13-22) because there is less inspired fresh air to dilute the carbon dioxide entering the alveoli from the blood. Increased production of carbon dioxide will also increase the alveolar P_{CO_2} because more carbon dioxide will be entering the alveoli from the blood per unit time. Recall that in the steady state, the volume of carbon dioxide entering the alveoli per unit time is always equal to the volume produced by the tissues. Just reverse the direction of the changes to decrease alveolar P_{CO_2}.

For simplicity, we assumed only one factor would change at a time, but if more than one factor changes, the effects will either add to or subtract from each other. For example, if oxygen consumption and alveolar ventilation both increase at the same time, their opposing effects on alveolar P_{O_2} will tend to cancel each other out, and alveolar P_{O_2} will not change.

This last example emphasizes that, at any particular atmospheric P_{O_2}, it is the *ratio* of oxygen consumption to alveolar ventilation that determines alveolar P_{O_2}—the higher the ratio, the lower the alveolar P_{O_2}. Similarly, alveolar P_{CO_2} is determined by the ratio of carbon dioxide production to alveolar ventilation—the higher the ratio, the higher the alveolar P_{CO_2}.

We can now define two terms that denote the adequacy of ventilation—that is, the relationship between metabolism and alveolar ventilation. Physiologists state these definitions in terms of carbon dioxide rather than oxygen. **Hypoventilation** exists when there is an increase in the ratio of carbon dioxide production to alveolar ventilation. In other words, a person is hypoventilating if the alveolar ventilation cannot keep pace with the carbon dioxide production. The result is that alveolar P_{CO_2} rises above the normal value. **Hyperventilation** exists when there is a decrease in the ratio of carbon dioxide production to alveolar ventilation—that is, when alveolar ventilation is actually too great for the amount of carbon dioxide being produced. The result is that alveolar P_{CO_2} decreases below the normal value.

Note that "hyperventilation" is not synonymous with "increased ventilation." Hyperventilation represents increased ventilation *relative to metabolism*. Thus, for example, the increased ventilation that occurs during moderate exercise is

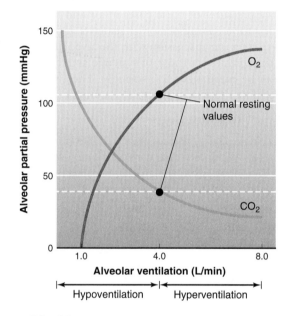

Figure 13-22

Effects of increasing or decreasing alveolar ventilation on alveolar partial pressures in a person having a constant metabolic rate (cellular oxygen consumption and carbon dioxide production). Note that alveolar P_{O_2} approaches zero when alveolar ventilation is about 1 L/min. At this point, all the oxygen entering the alveoli crosses into the blood, leaving virtually no oxygen in the alveoli.

not hyperventilation because, as we will see, the increase in production of carbon dioxide in this situation is proportional to the increased ventilation.

Gas Exchange Between Alveoli and Blood

The blood that enters the pulmonary capillaries is systemic venous blood pumped to the lungs through the pulmonary arteries. Having come from the tissues, it has a relatively high P_{CO_2} (46 mmHg in a normal person at rest) and a relatively low P_{O_2} (40 mmHg) (see Figure 13-21 and **Table 13-7**). The differences in the partial pressures of oxygen and carbon dioxide

Table 13–7	Normal Gas Pressure			
	Venous Blood	**Arterial Blood**	**Alveoli**	**Atmosphere**
P_{O_2}	40 mmHg	100 mmHg*	105 mmHg*	160 mmHg
P_{CO_2}	46 mmHg	40 mmHg	40 mmHg	0.3 mmHg

*The reason that the arterial P_{O_2} and alveolar P_{O_2} are not exactly the same is described later in this chapter.

on the two sides of the alveolar-capillary membrane result in the net diffusion of oxygen from alveoli to blood and of carbon dioxide from blood to alveoli. As this diffusion occurs, the P_{O_2} in the pulmonary capillary blood rises and the P_{CO_2} falls. The net diffusion of these gases ceases when the capillary partial pressures become equal to those in the alveoli.

In a normal person, the rates at which oxygen and carbon dioxide diffuse are so rapid and the blood flow through the capillaries so slow that complete equilibrium is reached well before the blood reaches the end of the capillaries (**Figure 13–23**).

Thus, the blood that leaves the pulmonary capillaries to return to the heart and be pumped into the systemic arteries has essentially the same P_{O_2} and P_{CO_2} as alveolar air. (They are not exactly the same, for reasons given later.) Accordingly, the factors described in the previous section—atmospheric P_{O_2}, cellular oxygen consumption and carbon dioxide production, and alveolar ventilation—determine the alveolar gas pressures, which then determine the systemic arterial gas pressures.

Given that diffusion between alveoli and pulmonary capillaries normally achieves complete equilibration, the more capillaries that participate in this process, the more total oxygen and carbon dioxide is exchanged. Many of the pulmonary capillaries at the apex (top) of each lung are normally closed at rest. During exercise, these capillaries open and receive blood, thereby enhancing gas exchange. The mechanism by which this occurs is a simple physical one; the pulmonary circulation at rest is at such a low blood pressure that the pressure in these apical capillaries is inadequate to keep them open, but the increased cardiac output of exercise raises pulmonary vascular pressures, which opens these capillaries.

The diffusion of gases between alveoli and capillaries may be impaired in a number of ways (see Figure 13–23), resulting in inadequate oxygen diffusion into the blood. For one thing, the total surface area of all of the alveoli in contact with pulmonary capillaries may be decreased. In lung infections or **pulmonary edema,** for example, some of the alveoli may become filled with fluid. Diffusion may also be impaired if the alveolar walls become severely thickened with connective tissue, as, for example, in the disease called **diffuse interstitial fibrosis.** Pure diffusion problems of these types are restricted to oxygen and usually do not affect the elimination of carbon dioxide, which is much more diffusible than oxygen.

Matching of Ventilation and Blood Flow in Alveoli

The major disease-induced cause of inadequate oxygen movement between alveoli and pulmonary capillary blood is not a problem with diffusion, but instead is due to the

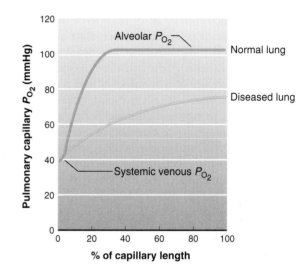

Figure 13–23

Equilibration of blood P_{O_2} with an alveolus with a P_{O_2} of 105 mmHg along the length of a pulmonary capillary. Note that an abnormal alveolar diffusion barrier (diseased alveolus) does not fully oxygenate the blood.

Figure 13–23 physiological *inquiry*

■ What is the effect of exercise on P_{O_2} at the end of a capillary in a normal region of the lung? In a region of the lung with diffusion limitation due to disease?

Answer can be found at end of chapter.

mismatching of the air supply and blood supply in individual alveoli.

The lungs are composed of approximately 300 million alveoli, each capable of receiving carbon dioxide from, and supplying oxygen to, the pulmonary capillary blood. To be most efficient, the correct proportion of alveolar air flow (ventilation) and capillary blood flow (perfusion) should be available to *each* alveolus. Any mismatching is termed **ventilation-perfusion inequality.**

The major effect of ventilation-perfusion inequality is to lower the P_{O_2} of systemic arterial blood. Indeed, largely because of gravitational effects on ventilation and perfusion, there is enough ventilation-perfusion inequality in normal people to lower the arterial P_{O_2} about 5 mmHg. One effect of upright posture is to increase the filling of blood vessels at the bottom of the lung due to gravity, which contributes to a difference in blood flow distribution in the lung. This is the major explanation of the fact, given earlier, that the P_{O_2} of blood in the pulmonary veins and systemic arteries is normally about 5 mmHg less than that of average alveolar air (see Table 13–7).

In disease states, regional changes in lung compliance, airway resistance, and vascular resistance can cause marked ventilation-perfusion inequalities. The extremes of this phenomenon are easy to visualize: (1) There may be ventilated alveoli with no blood supply at all (dead space or wasted ventilation) due to a blood clot, for example, or (2) there may be blood flowing through areas of lung that have no ventilation (this is termed a **shunt**) due to collapsed alveoli, for example. But the inequality need not be all-or-none to be significant.

Carbon dioxide elimination is also impaired by ventilation-perfusion inequality, but not nearly to the same degree as oxygen uptake. Although the reasons for this are complex, small increases in arterial CO_2 lead to increases in alveolar ventilation, which usually prevent increases in arterial P_{CO_2}. Nevertheless, severe ventilation-perfusion inequalities in disease states can lead to some elevation of arterial P_{CO_2}.

There are several local homeostatic responses within the lungs that minimize the mismatching of ventilation and blood flow and thereby maximize the efficiency of gas exchange (**Figure 13–24**). Probably the most important of these is a direct effect of low oxygen on pulmonary blood vessels. A decrease in ventilation within a group of alveoli—which might occur, for example, from a mucous plug blocking the small airways—leads to a decrease in alveolar P_{O_2} and the area around it, including the blood vessels. A decrease in P_{O_2} in these alveoli and nearby blood vessels leads to vasoconstriction, diverting blood flow away from the poorly ventilated area. This local adaptive effect, unique to the pulmonary arterial blood vessels, ensures that blood flow is directed away from diseased areas of the lung toward areas that are well-ventilated. Another factor to improve the match between ventilation and perfusion can occur if there is a local decrease in blood flow within a lung region due to, for example, a small blood clot in a pulmonary arteriole. A local decrease in blood flow brings less systemic CO_2 to that area, resulting in a local decrease in P_{CO_2}. This causes local bronchoconstriction, which diverts air flow away to areas of the lung with better perfusion.

The net adaptive effects of vasoconstriction and bronchoconstriction are to (1) supply less blood flow to poorly ventilated areas, thus diverting blood flow to well-ventilated areas, and (2) redirect air away from diseased or damaged alveoli and toward healthy alveoli. These factors greatly improve the efficiency of pulmonary gas exchange, but they are not perfect even in the healthy lung. There is always a small mismatch of ventilation and perfusion, which, as just described, leads to the normal alveolar-arterial O_2 gradient of about 5 mmHg.

Gas Exchange Between Tissues and Blood

As the systemic arterial blood enters capillaries throughout the body, it is separated from the interstitial fluid by only the thin capillary wall, which is highly permeable to both oxygen and carbon dioxide. The interstitial fluid, in turn, is separated from the intracellular fluid by the plasma membranes of the cells, which are also quite permeable to oxygen and carbon dioxide. Metabolic reactions occurring within cells are constantly consuming oxygen and producing carbon dioxide. Therefore, as shown in Figure 13–21, intracellular P_{O_2} is lower and P_{CO_2} higher than in blood. The lowest P_{O_2} of all—less than 5 mmHg—is in the mitochondria, the site of oxygen utilization. As a result, a net diffusion of oxygen occurs from blood into cells and, within the cells, into the mitochondria, and a net diffusion of carbon dioxide occurs from cells into blood. In this manner, as blood flows through systemic capillaries, its P_{O_2} decreases and its P_{CO_2} increases. This accounts for the systemic venous blood values shown in Figure 13–21 and Table 13–7.

In summary, the supply of new oxygen to the alveoli and the consumption of oxygen in the cells create P_{O_2} gradients that produce net diffusion of oxygen from alveoli to blood in the lungs and from blood to cells in the rest of the body. Conversely, the production of carbon dioxide by cells and its elimination from the alveoli via expiration create P_{CO_2} gradients that produce net diffusion of carbon dioxide from cells to blood in the rest of the body and from blood to alveoli in the lungs.

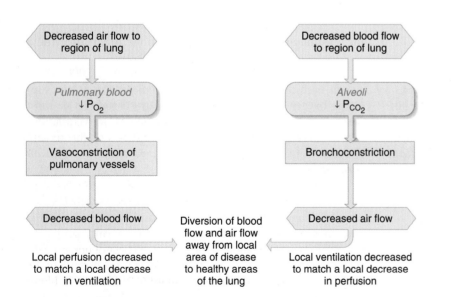

Figure 13–24

Local control of ventilation-perfusion matching.

Transport of Oxygen in Blood

Table 13–8 summarizes the oxygen content of systemic arterial blood, referred to simply as arterial blood. Each liter normally contains the number of oxygen molecules equivalent to 200 ml of pure gaseous oxygen at atmospheric pressure. The oxygen is present in two forms: (1) dissolved in the plasma and erythrocyte water and (2) reversibly combined with hemoglobin molecules in the erythrocytes.

As predicted by Henry's law, the amount of oxygen dissolved in blood is directly proportional to the P_{O_2} of the blood. Because the solubility of oxygen in water is relatively low, only 3 ml can be dissolved in 1 L of blood at the normal arterial P_{O_2} of 100 mmHg. The other 197 ml of oxygen in a liter of arterial blood, more than 98 percent of the oxygen content in the liter, is transported in the erythrocytes, reversibly combined with hemoglobin.

Each **hemoglobin** molecule is a protein made up of four subunits bound together. Each subunit consists of a molecular group known as **heme** and a polypeptide attached to the heme. The four polypeptides of a hemoglobin molecule are collectively called **globin.** Each of the four heme groups in a hemoglobin molecule (**Figure 13–25**) contains one atom of iron (Fe^{2+}), to which oxygen binds. Because each iron atom can bind one molecule of oxygen, a single hemoglobin molecule can bind four oxygen molecules (see Figure 2–21). However, for simplicity, the equation for the reaction between oxygen and hemoglobin is usually written in terms of a single polypeptide-heme chain of a hemoglobin molecule:

$$O_2 + Hb \rightleftharpoons HbO_2 \qquad (13\text{–}8)$$

Thus, this chain can exist in one of two forms—**deoxyhemoglobin (Hb)** and **oxyhemoglobin (HbO$_2$).** In a blood sample containing many hemoglobin molecules, the fraction of all the hemoglobin in the form of oxyhemoglobin is expressed as the **percent hemoglobin saturation:**

$$\text{Percent Hb saturation} = $$
$$\frac{O_2 \text{ bound to Hb}}{\text{Maximal capacity of Hb to bind } O_2} \times 100 \qquad (13\text{–}9)$$

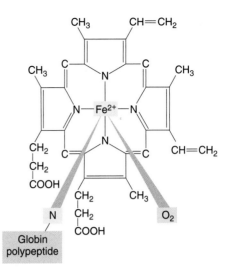

Figure 13–25

Heme. Oxygen binds to the iron atom (Fe^{2+}). Heme attaches to a polypeptide chain by a nitrogen atom to form one subunit of hemoglobin. Four of these subunits bind to each other to make a single hemoglobin molecule. See Figure 2–21 for the structure of hemoglobin.

For example, if the amount of oxygen bound to hemoglobin is 40 percent of the maximal capacity, the sample is said to be 40 percent saturated. The denominator in this equation is also termed the **oxygen-carrying capacity** of the blood.

What factors determine the percent hemoglobin saturation? By far the most important is the blood P_{O_2}. Before turning to this subject, however, it must be stressed that the *total amount* of oxygen carried by hemoglobin in the blood depends not only on the percent saturation of hemoglobin but also on how much hemoglobin is in each liter of blood. A significant decrease in hemoglobin in the blood is called *anemia.* For example, if a person's blood contained only half as much hemoglobin per liter as normal, then at any given percent saturation, the oxygen content of the blood would be only half as much.

What Is the Effect of P_{O_2} on Hemoglobin Saturation?

Based on equation 13–8 and the law of mass action, it is evident that increasing the blood P_{O_2} should increase the combination of oxygen with hemoglobin. The experimentally determined quantitative relationship between these variables is shown in **Figure 13–26**, which is called an **oxygen-hemoglobin dissociation curve.** (The term *dissociate* means "to separate," in this case, oxygen from hemoglobin; it could just as well have been called an oxygen-hemoglobin association curve.) The curve is sigmoid because, as stated earlier, each hemoglobin molecule contains four subunits. Each subunit can combine with one molecule of oxygen, and the reactions of the four subunits occur sequentially, with each combination facilitating the next one.

This combination of oxygen with hemoglobin is an example of cooperativity, as described in Chapter 3. The explanation in this case is as follows. The globin units of deoxyhemoglobin

Table 13–8	Oxygen Content of Systemic Arterial Blood at Sea Level

1 liter (L) arterial blood contains

3 ml	O_2 physically dissolved (1.5%)
197 ml	O_2 bound to hemoglobin (98.5%)
Total 200 ml	O_2

Cardiac output = 5 L/min

O_2 carried to tissues/min = 5 L/min × 200 ml O_2/L

                     = 1000 ml O_2/min

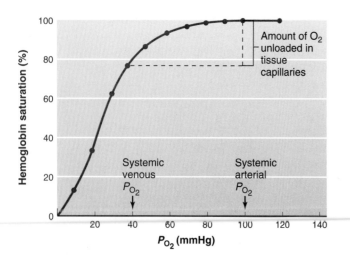

Figure 13–26

Oxygen-hemoglobin dissociation curve. This curve applies to blood at 37°C and a normal arterial hydrogen ion concentration. At any given blood hemoglobin concentration, the y-axis could also have plotted oxygen content, in milliliters of oxygen. At 100 percent saturation, the amount of hemoglobin in normal blood carries 200 ml of oxygen.

are tightly held by electrostatic bonds in a conformation with a relatively low affinity for oxygen. The binding of oxygen to a heme molecule breaks some of these bonds between the globin units, leading to a conformation change that leaves the remaining oxygen-binding sites more exposed. Thus, the binding of one oxygen molecule to deoxyhemoglobin increases the affinity of the remaining sites on the same hemoglobin molecule, and so on.

The shape of the oxygen-hemoglobin dissociation curve is extremely important in understanding oxygen exchange. The curve has a steep slope between 10 and 60 mmHg P_{O_2} and a relatively flat portion (or plateau) between 70 and 100 mmHg P_{O_2}. Thus, the extent to which oxygen combines with hemoglobin increases very rapidly as the P_{O_2} increases from 10 to 60 mmHg, so that at a P_{O_2} of 60 mmHg, 90 percent of the total hemoglobin is combined with oxygen. From this point on, a further increase in P_{O_2} produces only a small increase in oxygen binding.

This plateau at higher P_{O_2} values has a number of important implications. In many situations, including at high altitude and with pulmonary disease, a moderate reduction occurs in alveolar and therefore arterial P_{O_2}. Even if the P_{O_2} decreased from the normal value of 100 to 60 mmHg, the total quantity of oxygen carried by hemoglobin would decrease by only 10 percent because hemoglobin saturation is still close to 90 percent at a P_{O_2} of 60 mmHg. The plateau provides an excellent safety factor so that even a significant limitation of lung function still allows almost normal oxygen saturation of hemoglobin.

The plateau also explains why, in a healthy person at sea level, increasing the alveolar (and therefore the arterial) P_{O_2} either by hyperventilating or by breathing 100 percent oxygen adds very little additional oxygen to the blood. A small additional amount dissolves, but because hemoglobin is already almost completely saturated with oxygen at the normal arte-

rial P_{O_2} of 100 mmHg, it simply cannot pick up any more oxygen when the P_{O_2} is elevated beyond this point. This applies only to healthy people at sea level. If a person initially has a low arterial P_{O_2} because of lung disease or high altitude, then there would be a great deal of deoxyhemoglobin initially present in the arterial blood. Therefore, raising the alveolar and thereby the arterial P_{O_2} would result in significantly more oxygen transport.

The steep portion of the curve from 60 down to 20 mmHg is ideal for unloading oxygen in the tissues. That is, for a small decrease in P_{O_2} due to diffusion of oxygen from the blood to the cells, a large quantity of oxygen can be unloaded in the peripheral tissue capillary. Also, small shifts in the position of the curve due to various factors can significantly increase oxygen unloading as we will see for hydrogen ions in Figure 13–29.

We now retrace our steps and reconsider the movement of oxygen across the various membranes, this time including hemoglobin in our analysis. It is essential to recognize that the oxygen bound to hemoglobin does *not* contribute directly to the P_{O_2} of the blood; only dissolved oxygen does so. Therefore, oxygen diffusion is governed only by the dissolved portion, a fact that permitted us to ignore hemoglobin in discussing transmembrane partial pressure gradients. However, the presence of hemoglobin plays a critical role in determining the *total amount* of oxygen that will diffuse, as illustrated by a simple example (**Figure 13–27**).

Two solutions separated by a semipermeable membrane contain equal quantities of oxygen. The gas pressures in both solutions are equal, and no net diffusion of oxygen occurs. Addition of hemoglobin to compartment B disturbs this equilibrium because much of the oxygen combines with hemoglobin. Despite the fact that the total *quantity* of oxygen in

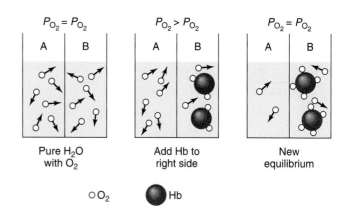

Figure 13–27

Effect of added hemoglobin on oxygen distribution between two compartments containing a fixed number of oxygen molecules and separated by a semipermeable membrane. At the new equilibrium, the P_{O_2} values are again equal to each other but lower than before the hemoglobin was added. However, the total oxygen, in other words, the oxygen dissolved plus that combined with hemoglobin, is now much higher on the right side of the membrane.

Adapted from Comroe.

compartment B is still the same, the number of *dissolved* oxygen molecules has decreased. Therefore, the P_{O_2} of compartment B is less than that of A, and so there is a net diffusion of oxygen from A to B. At the new equilibrium, the oxygen pressures are once again equal, but almost all the oxygen is in compartment B and has combined with hemoglobin.

Let us now apply this analysis to capillaries of the lungs and tissues (**Figure 13–28**). The plasma and erythrocytes entering the lungs have a P_{O_2} of 40 mmHg. As we can see from Figure 13–26, hemoglobin saturation at this P_{O_2} is 75 percent. The alveolar P_{O_2}—105 mmHg—is higher than the blood P_{O_2} and so oxygen diffuses from the alveoli into the plasma. This increases plasma P_{O_2} and induces diffusion of oxygen into the erythrocytes, elevating erythrocyte P_{O_2} and causing increased combination of oxygen and hemoglobin. Most of the oxygen diffusing into the blood from the alveoli does not remain dissolved but combines with hemoglobin. Therefore, the blood P_{O_2} normally remains less than the alveolar P_{O_2} until hemoglobin is virtually 100 percent saturated. Thus, the diffusion gradient favoring oxygen movement into the blood is maintained despite the very large transfer of oxygen.

In the tissue capillaries, the procedure is reversed. Because the mitochondria of the cells all over the body are utilizing oxygen, the cellular P_{O_2} is less than the P_{O_2} of the surrounding interstitial fluid. Therefore, oxygen is continuously diffusing into the cells. This causes the interstitial fluid P_{O_2} to always be less than the P_{O_2} of the blood flowing through the tissue capillaries, so net diffusion of oxygen occurs from the plasma within the capillary into the interstitial fluid. As a result, plasma P_{O_2} becomes lower than erythrocyte P_{O_2}, and oxygen diffuses out of the erythrocyte into the plasma. The lower-ing of erythrocyte P_{O_2} causes the dissociation of oxygen from hemoglobin, thereby liberating oxygen, which then diffuses out of the erythrocyte. The net result is a transfer, purely by diffusion, of large quantities of oxygen from hemoglobin to plasma to interstitial fluid to the mitochondria of tissue cells.

In most tissues under resting conditions, hemoglobin is still 75 percent saturated as the blood leaves the tissue capillaries. This fact underlies an important mechanism by which cells can obtain more oxygen whenever they increase their activity. For example, an exercising muscle consumes more oxygen, thereby lowering its tissue P_{O_2}. This increases the blood-to-tissue P_{O_2} gradient. As a result, the rate of oxygen diffusion from blood to cells increases. In turn, the resulting reduction in erythrocyte P_{O_2} causes additional dissociation of hemoglobin and oxygen. In this manner, the extraction of oxygen from blood in an exercising muscle is almost maximal and is much greater than the usual 25 percent. In addition, an increased blood flow to the muscles, called active hyperemia (Chapter 12), also contributes greatly to the increased oxygen supply.

Effect of Carbon Monoxide on Oxygen Carriage

Carbon monoxide is a colorless, odorless gas that is a product of the incomplete combustion of hydrocarbons, such as gasoline. It is a common cause of sickness and death due to poisoning, both intentional and accidental. Its most striking pathophysiological characteristic is its extremely high affinity—210 times that of oxygen—for the oxygen-binding sites in hemoglobin. For this reason, it reduces the amount of oxygen that combines with hemoglobin in pulmonary capillaries by competing for these sites. It also exerts a second deleterious

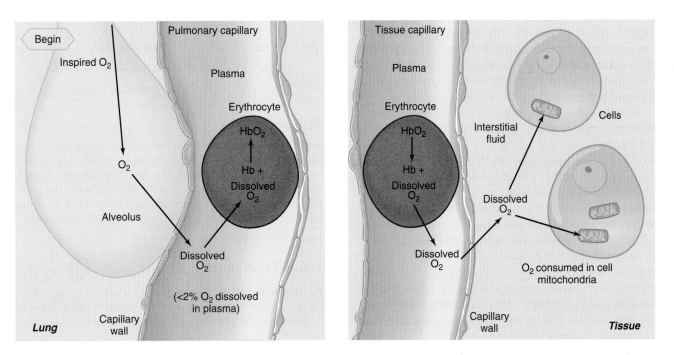

Figure 13–28

Oxygen movement in the lungs and tissues. Movement of inspired air into the alveoli is by bulk flow; all movements across membranes are by diffusion.

Respiratory Physiology

effect: it alters the hemoglobin molecule so as to shift the oxygen-hemoglobin dissociation curve to the left, thus decreasing the unloading of oxygen from hemoglobin in the tissues. As we will see later, the situation is worsened by the fact that persons suffering from carbon monoxide poisoning do not show any reflex increase in their ventilation.

Effects of Blood P_{CO_2}, H⁺ Concentration, Temperature, and DPG Concentration on Hemoglobin Saturation

At any given P_{O_2}, a variety of other factors influence the degree of hemoglobin saturation. These include blood P_{CO_2}, H⁺ concentration, temperature, and the concentration of a substance produced by erythrocytes called **2,3-diphosphoglycerate (DPG)** (also known as bisphosphoglycerate, BPG). As illustrated in **Figure 13–29**, an increase in any of these factors causes the dissociation curve to shift to the right. This means that at any given P_{O_2}, hemoglobin has less affinity for oxygen. In contrast, a decrease in any of these factors causes the dissociation curve to shift to the left, such that at any given P_{O_2}, hemoglobin has a greater affinity for oxygen.

The effects of increased P_{CO_2}, H⁺ concentration, and temperature are continuously exerted on the blood in tissue capillaries, because each of these factors is higher in tissue-capillary blood than in arterial blood. The P_{CO_2} is increased because of the carbon dioxide entering the blood from the tissues. For reasons to be described later, the H⁺ concentration is elevated because of the elevated P_{CO_2} and the release of metabolically produced acids such as lactic acid. The temperature is increased because of the heat produced by tissue metabolism. Hemoglobin exposed to this elevated blood P_{CO_2}, H⁺ concentration, and temperature as it passes through the tissue capillaries has a decreased affinity for oxygen. Thus, hemoglobin gives up even more oxygen than it would have if the decreased tissue-capillary P_{O_2} had been the only operating factor.

The more metabolically active a tissue is, the greater its P_{CO_2}, H⁺ concentration, and temperature will be. At any given P_{O_2}, this causes hemoglobin to release more oxygen during passage through the tissue's capillaries and provides the more active cells with additional oxygen. Here, then, is another local mechanism that increases oxygen delivery to tissues with increased metabolic activity.

What is the mechanism by which these factors influence the affinity of hemoglobin for oxygen? Carbon dioxide and hydrogen ions do so by combining with the globin portion of hemoglobin and altering the conformation of the hemoglobin molecule. Thus, these effects are a form of allosteric modulation (Chapter 3). An elevated temperature also decreases hemoglobin's affinity for oxygen by altering its molecular configuration.

Erythrocytes contain large quantities of DPG, which is present in only trace amounts in other cells. DPG, which is produced by the erythrocytes during glycolysis, reversibly binds with hemoglobin, allosterically causing it to have a lower affinity for oxygen (see Figure 13–29). The net result is that whenever DPG levels increase, there is enhanced unloading of oxygen from hemoglobin as blood flows through the tissues. Such an increase in DPG concentration is triggered by a variety of

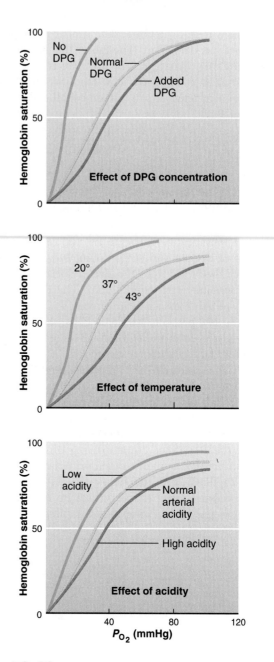

Figure 13–29

Effects of DPG concentration, temperature, and acidity on the relationship between P_{O_2} and hemoglobin saturation. The temperature of normal blood, of course, never diverges from 37°C as much as shown in the figure, but the principle is still the same when the changes are within the physiological range. High acidity and low acidity can be caused by high P_{CO_2} and low P_{CO_2}, respectively.

Adapted from Comroe.

Figure 13–29 physiological *inquiry* ⓟⁱ

- Researchers are developing blood substitutes to meet the demand for emergency transfusions. What would be the effect of artificial blood in which binding of O_2 is not altered by acidity?

Answer can be found at end of chapter.

conditions associated with inadequate oxygen supply to the tissues and helps to maintain oxygen delivery. For example, the increase in DPG is important during exposure to high altitude when the P_{O_2} of the blood is decreased because DPG increases the unloading of oxygen in the tissue capillaries.

Transport of Carbon Dioxide in Blood

In a resting person, metabolism generates about 200 ml of carbon dioxide per minute. When arterial blood flows through tissue capillaries, this volume of carbon dioxide diffuses from the tissues into the blood (**Figure 13–30a**). Carbon dioxide is much more soluble in water than is oxygen, so blood carries more dissolved carbon dioxide than dissolved oxygen. Even so, only 10 percent of the carbon dioxide entering the blood dissolves in the plasma and erythrocytes. In order to transport all of the CO_2 produced in the tissues to the lung, CO_2 in the blood must be carried in other forms.

Another 30 percent of the carbon dioxide molecules entering the blood react reversibly with the amino groups of hemoglobin to form **carbamino hemoglobin.** For simplicity, this reaction with hemoglobin is written as:

$$CO_2 + Hb \rightleftharpoons HbCO_2 \qquad (13\text{–}10)$$

This reaction is aided by the fact that deoxyhemoglobin, formed as blood flows through the tissue capillaries, has a greater affinity for carbon dioxide than does oxyhemoglobin.

The remaining 60 percent of the carbon dioxide molecules entering the blood in the tissues is converted to bicarbonate ions:

$$CO_2 + H_2O \underset{\substack{\text{carbonic} \\ \text{acid}}}{\overset{\substack{\text{carbonic} \\ \text{anhydrase}}}{\rightleftharpoons}} H_2CO_3 \rightleftharpoons \underset{\text{bicarbonate}}{HCO_3^-} + H^+ \qquad (13\text{–}11)$$

The first reaction in equation 13–11 is rate-limiting and is very slow unless catalyzed by the enzyme **carbonic anhydrase.** This enzyme is present in the erythrocytes but not in the plasma; therefore, this reaction occurs mainly in the erythrocytes. In contrast, carbonic acid dissociates very rapidly into a bicarbonate ion and a hydrogen ion without any enzyme assistance. Once formed, most of the bicarbonate moves out of the erythrocytes into the plasma via a transporter that exchanges one bicarbonate for one chloride ion (this is called the "chloride shift").

The reactions shown in equation 13–11 also explain why, as mentioned earlier, the H^+ concentration in tissue capillary blood and systemic venous blood is higher than that in arterial blood and increases as metabolic activity increases. The fate of these hydrogen ions will be discussed in the next section.

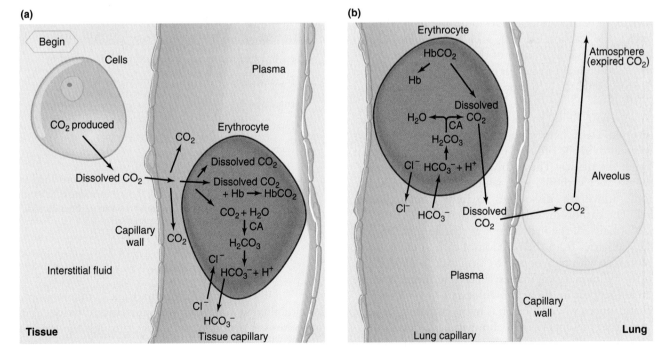

Figure 13–30

Summary of CO_2 movement. Expiration of CO_2 is by bulk flow, whereas all movements of CO_2 across membranes are by diffusion. Arrows reflect relative proportions of the fates of the CO_2. About two-thirds of the CO_2 entering the blood in the tissues ultimately is converted to HCO_3^- in the erythrocytes because carbonic anhydrase (CA) is located there, but most of the HCO_3^- then moves out of the erythrocytes into the plasma in exchange for chloride ions (the "chloride shift"). See Figure 13–31 for the fate of the hydrogen ions generated in the erythrocytes.

Because carbon dioxide undergoes these various fates in blood, it is customary to add up the amounts of dissolved carbon dioxide, bicarbonate, and carbon dioxide in carbamino hemoglobin to arrive at the **total blood carbon dioxide,** which is measured as a component of routine blood chemistry testing.

Just the opposite events occur as systemic venous blood flows through the lung capillaries (**Figure 13–30b**). Because the blood P_{CO_2} is higher than alveolar P_{CO_2}, a net diffusion of CO_2 from blood into alveoli occurs. This loss of CO_2 from the blood lowers the blood P_{CO_2} and drives reactions 13–10 and 13–11 to the left. HCO_3^- and H^+ combine to produce H_2CO_3, which then dissociates to CO_2 and H_2O. Similarly, $HbCO_2$ generates Hb and free CO_2. Normally, as fast as CO_2 is generated from HCO_3^- and H^+ and from $HbCO_2$, it diffuses into the alveoli. In this manner, all the CO_2 delivered into the blood in the tissues is now delivered into the alveoli, from where it is eliminated during expiration.

Transport of Hydrogen Ions Between Tissues and Lungs

As blood flows through the tissues, a fraction of oxyhemoglobin loses its oxygen to become deoxyhemoglobin, while simultaneously a large quantity of carbon dioxide enters the blood and undergoes the reactions that generate bicarbonate and hydrogen ions. What happens to these hydrogen ions?

Deoxyhemoglobin has a much greater affinity for H^+ than does oxyhemoglobin, so it binds (buffers) most of the hydrogen ions (**Figure 13–31**). Indeed, deoxyhemoglobin can be abbreviated HbH rather than Hb to denote its binding of H^+. In effect, the reaction is $HbO_2 + H^+ \rightleftharpoons HbH + O_2$. In this manner, only a small number of the hydrogen ions generated in the blood remain free. This explains why venous blood (pH = 7.36) is only slightly more acidic than arterial blood (pH = 7.40).

As the venous blood passes through the lungs, this reaction is reversed. Deoxyhemoglobin becomes converted to oxyhemoglobin and, in the process, releases the hydrogen ions it picked up in the tissues. The hydrogen ions react with bicarbonate to produce carbonic acid, which, under the influence of carbonic anhydrase, dissociates to form carbon dioxide and water. The carbon dioxide diffuses into the alveoli to be expired. Normally all the hydrogen ions that are generated in the tissue capillaries from the reaction of carbon dioxide and water recombine with bicarbonate to form carbon dioxide and water in the pulmonary capillaries. Therefore, none of these hydrogen ions appear in the *arterial* blood.

What happens when a person is hypoventilating or has a lung disease that prevents normal elimination of carbon dioxide? Not only would arterial P_{CO_2} increase as a result, but so would arterial H^+ concentration. Increased arterial H^+ concentration due to carbon dioxide retention is termed ***respiratory acidosis.*** Conversely, hyperventilation would lower the arterial values of both P_{CO_2} and H^+ concentration, producing ***respiratory alkalosis.***

The factors that influence the binding of CO_2 and O_2 by hemoglobin are summarized in **Table 13–9**.

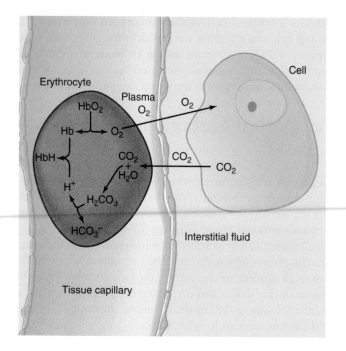

Figure 13–31

Binding of hydrogen ions by hemoglobin as blood flows through tissue capillaries. This reaction is facilitated because deoxyhemoglobin, formed as oxygen dissociates from hemoglobin, has a greater affinity for hydrogen ions than does oxyhemoglobin. For this reason, Hb and HbH are both abbreviations for deoxyhemoglobin.

Table 13–9	Effects of Various Factors on Hemoglobin

The affinity of hemoglobin for oxygen is decreased by:

1. Increased hydrogen ion concentration

2. Increased P_{CO_2}

3. Increased temperature

4. Increased DPG concentration

The affinity of hemoglobin for both hydrogen ions and carbon dioxide is decreased by increased P_{O_2}; that is, deoxyhemoglobin has a greater affinity for hydrogen ions and carbon dioxide than does oxyhemoglobin.

Another aspect of the remarkable hemoglobin molecule is its ability to bind and transport **nitric oxide.** A present hypothesis is that as blood passes through the lungs, hemoglobin picks up and binds not only oxygen but nitric oxide that is synthesized there, carries it to the peripheral tissues, and releases it along with oxygen. Simultaneously, via a different binding site, hemoglobin picks up and catabolizes nitric oxide produced in the peripheral tissues. Theoretically this cycle could play an important role in determining the peripheral concentration of nitric oxide and, thereby, the overall effect of this vasodilator agent. For example, by supplying net nitric oxide to the periphery, the process could cause additional vasodilation by systemic blood vessels. This would have effects on both local blood flow and systemic arterial blood pressure.

Finally, the fetus has a unique form of hemoglobin called **fetal hemoglobin,** which has a higher affinity for oxygen. This allows an increase in oxygen uptake in the placenta. Therefore, although fetal arterial P_{O_2} is lower than that in the air-breathing newborn, fetal hemoglobin allows adequate oxygen supply to the developing organs.

Control of Respiration

Neural Generation of Rhythmical Breathing

The diaphragm and intercostal muscles are skeletal muscles and therefore do not contract unless the nerves stimulate them to do so. Thus, breathing depends entirely upon cyclical respiratory muscle excitation of the diaphragm and the intercostal muscles by their motor nerves. Destruction of these nerves results in paralysis of the respiratory muscles and death, unless some form of artificial respiration can be instituted.

Inspiration is initiated by a burst of action potentials in the spinal motor nerves to inspiratory muscles like the diaphragm. Then the action potentials cease, the inspiratory muscles relax, and expiration occurs as the elastic lungs recoil. In situations such as exercise when the contraction of expiratory muscles facilitates expiration, the nerves to these muscles, which were not active during inspiration, begin firing during expiration.

By what mechanism are nerve impulses to the respiratory muscles alternately increased and decreased? Control of this neural activity resides primarily in neurons in the medulla oblongata, the same area of the brain that contains the major cardiovascular control centers. (For the rest of this chapter we will refer to the medulla oblongata simply as the medulla.) There are two main anatomical components of the **medullary respiratory center** (**Figure 13–32**). The neurons of the **dorsal respiratory group (DRG)** primarily fire during inspiration and have input to the spinal motor neurons that activate respiratory muscles involved in inspiration—the diaphragm and inspiratory intercostal muscles. The primary inspiratory muscle at rest is the diaphragm, which is innervated by the phrenic nerves. The **ventral respiratory group (VRG)** is the other main complex of neurons in the medullary respiratory center. The **respiratory rhythm generator** is located in the **Pre-Bötzinger Complex** of neurons in the upper part of the VRG. This rhythm generator appears to be composed of pacemaker cells as well as a complex neural network which, acting together, set the basal respiratory rate.

The lower part of the VRG contains nerves that fire both during inspiration and expiration. The inspiratory neurons of the

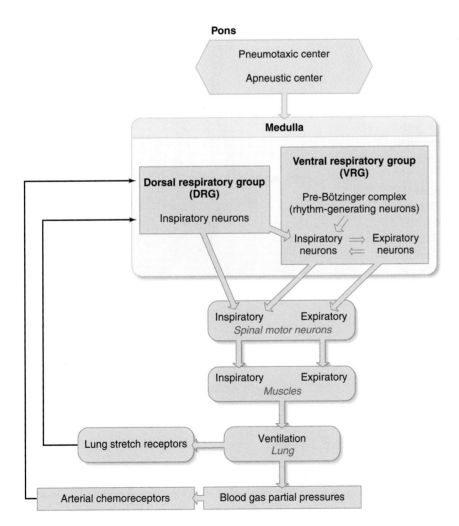

Figure 13–32

Brainstem respiratory control centers responsible for respiratory rhythm generation, activation of inspiratory and expiratory nerves and muscles, and for monitoring lung inflation via stretch receptors and alveolar ventilation via changes in arterial blood gas partial pressures.

VRG receive input from the inspiratory neurons of the DRG as well as the respiratory rhythm generator and, in turn, have input to the inspiratory motor neurons. The lower part of the VRG also contains expiratory neurons that appear to be most important when large increases in ventilation are required (for example, during strenuous exercise). During active expiration, motor nerves activated by the expiratory output from the VRG cause the expiratory muscles to contract. This helps to move air out of the lungs rather than depending only on the passive expiration that occurs during quiet breathing.

During quiet breathing, the respiratory rhythm generator activates inspiratory neurons in the VRG that depolarize the inspiratory spinal motor nerves, causing the inspiratory muscles to contract. When the inspiratory motor nerves stop firing, the inspiratory muscles relax, allowing passive expiration. There are complex interactions within the VRG; for example, inspiratory and expiratory neurons in the medulla show reciprocal inhibition. Thus, during increases in breathing, the inspiratory and expiratory motor nerves and muscles are not activated at the same time, but, rather, alternate in function.

The medullary inspiratory neurons receive a rich synaptic input from neurons in various areas of the pons, the part of the brainstem just above the medulla. This input fine-tunes the output of the medullary inspiratory neurons and may help terminate inspiration by inhibiting them. It is likely that an area of the lower pons called the **apneustic center** is the major source of this output, whereas an area of the upper pons called the **pneumotaxic center** modulates the activity of the apneustic center. The pneumotaxic center, also known as the **pontine respiratory group,** helps to smooth the transition between inspiration and expiration. The respiratory nerves in the medulla and pons also receive synaptic input from higher centers of the brain such that the pattern of respiration is controlled voluntarily during speaking, diving, and even with emotions and pain.

Another cutoff signal for inspiration comes from **pulmonary stretch receptors,** which lie in the airway smooth muscle layer and are activated by a large lung inflation. Action potentials in the afferent nerve fibers from the stretch receptors travel to the brain and inhibit the activity of the medullary inspiratory neurons. This is called the **Hering-Breuer reflex.** Thus, feedback from the lungs helps to terminate inspiration by inhibiting inspiratory nerves in the DRG. However, this reflex plays a role in setting respiratory rhythm only under conditions of very large tidal volumes, as in strenuous exercise. The arterial chemoreceptors described next also have important input to the respiratory control centers such that the rate and depth of respiration can be increased when the levels of arterial oxygen decrease, or when arterial carbon dioxide or hydrogen ion concentration increases.

A final point about the medullary inspiratory neurons are that they are quite sensitive to inhibition by drugs such as barbiturates and morphine. Death from an overdose of these drugs is often due directly to a cessation of ventilation.

Control of Ventilation by P_{O_2}, P_{CO_2}, and H^+ Concentration

Respiratory rate and tidal volume are not fixed but can be increased or decreased over a wide range. For simplicity, we will describe the control of ventilation without discussing whether rate or depth makes the greater contribution to the change.

There are many inputs to the medullary inspiratory neurons, but the most important for the automatic control of ventilation at rest come from peripheral (arterial) chemoreceptors and central chemoreceptors.

The **peripheral chemoreceptors,** located high in the neck at the bifurcation of the common carotid arteries and in the thorax on the arch of the aorta (**Figure 13–33**), are called the **carotid bodies** and **aortic bodies,** respectively. In both locations they are quite close to, but distinct from, the arterial baroreceptors and are in intimate contact with the arterial blood. The carotid bodies in particular are strategically located to monitor oxygen supply to the brain. The peripheral chemoreceptors are composed of specialized receptor cells stimulated mainly by a decrease in the arterial P_{O_2} and an increase in the arterial H^+ concentration (**Table 13–10**). These cells communicate synaptically with neuron terminals from which afferent nerve fibers pass to the brainstem. There they provide excitatory synaptic input to the medullary inspiratory neurons. The carotid body input is the predominant peripheral chemoreceptor involved in the control of respiration.

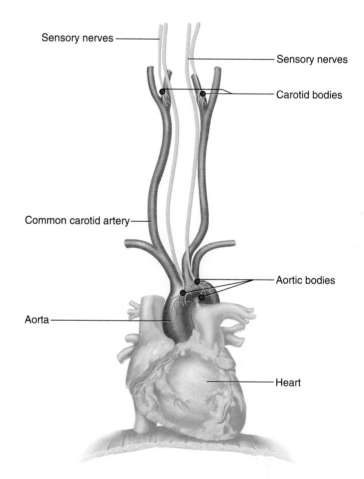

Sensory nerves

Sensory nerves

Carotid bodies

Common carotid artery

Aortic bodies

Aorta

Heart

Figure 13–33

Location of the carotid and aortic bodies. Note that each carotid body is quite close to a carotid sinus, the major arterial baroreceptor. Both right and left common carotid bifurcations contain a carotid sinus and a carotid body.

Table 13–10	Major Stimuli for the Central and Peripheral Chemoreceptors

Peripheral chemoreceptors—carotid bodies and aortic bodies—respond to changes in the arterial blood. They are stimulated by:

1. Decreased P_{O_2} (hypoxia)

2. Increased hydrogen ion concentration (metabolic acidosis)

3. Increased P_{CO_2} (respiratory acidosis)

Central chemoreceptors—located in the medulla oblongata—respond to changes in the *brain extracellular fluid*. They are stimulated by increased P_{CO_2} via associated changes in hydrogen ion concentration (see equation 13–11).

The **central chemoreceptors** are located in the medulla and, like the peripheral chemoreceptors, provide excitatory synaptic input to the medullary inspiratory neurons. They are stimulated by an increase in the H^+ concentration of the brain's extracellular fluid. As we will see later, such changes result mainly from changes in blood P_{CO_2}.

Control by P_{O_2}

Figure 13–34 illustrates an experiment in which healthy subjects breathe low P_{O_2} gas mixtures for several minutes. The experiment is performed in a way that keeps arterial P_{CO_2} constant so that the pure effects of changing only P_{O_2} can be studied. Little increase in ventilation is observed until the oxygen content of the inspired air is reduced enough to lower arterial P_{O_2} to 60 mmHg. Beyond this point, any further reduction in arterial P_{O_2} causes a marked reflex increase in ventilation.

This reflex is mediated by the peripheral chemoreceptors (**Figure 13–35**). The low arterial P_{O_2} increases the rate at which the receptors discharge, resulting in an increased number of action potentials traveling up the afferent nerve fibers and stimulating the medullary inspiratory neurons. The resulting increase in ventilation provides more oxygen to the alveoli and minimizes the decrease in alveolar and arterial P_{O_2} produced by the low P_{O_2} gas mixture.

It may seem surprising that we are insensitive to smaller reductions of arterial P_{O_2}, but look again at the oxygen-hemoglobin dissociation curve (see Figure 13–26). Total oxygen transport by the blood is not really reduced very much until the arterial P_{O_2} falls below about 60 mmHg. Therefore, increased ventilation would not result in much more oxygen being added to the blood until that point is reached.

To reiterate, the peripheral chemoreceptors respond to decreases in arterial P_{O_2}, as occurs in lung disease or exposure to high altitude. However, the peripheral chemoreceptors are *not* stimulated in situations in which modest reductions take place in the oxygen *content* of the blood but no change occurs in arterial P_{O_2}. As stated earlier, anemia is a decrease in the amount of hemoglobin present in the blood without a decrease in arterial P_{O_2}, because the concentration of dissolved oxygen in the blood is normal.

This same analysis holds true when oxygen content is reduced moderately by the presence of carbon monoxide, which, as described earlier, reduces the amount of oxygen combined with hemoglobin by competing for these sites. Because carbon monoxide does not affect the amount of oxygen that can dissolve in blood, the arterial P_{O_2} is unaltered, and no increase in peripheral chemoreceptor output occurs.

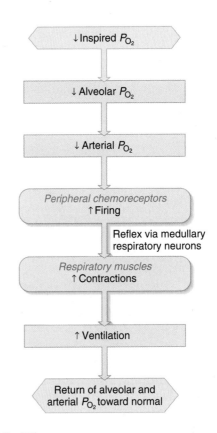

Figure 13–35

Sequence of events by which a low arterial P_{O_2} causes hyperventilation, which maintains alveolar (and, hence, arterial) P_{O_2} at a value higher than would exist if the ventilation had remained unchanged.

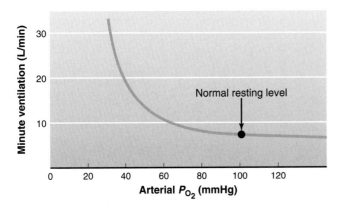

Figure 13–34

The effect on ventilation of breathing different oxygen mixtures. The arterial P_{CO_2} was maintained at 40 mmHg throughout the experiment.

Control by P_{CO_2}

Figure 13–36 illustrates an experiment in which subjects breathe air with variable quantities of carbon dioxide added. The presence of carbon dioxide in the inspired air causes an elevation of alveolar P_{CO_2} and thereby an elevation of arterial P_{CO_2}. Note that even a very small increase in arterial P_{CO_2} causes a marked reflex increase in ventilation. Experiments like this have documented that the reflex mechanisms controlling ventilation prevent small increases in arterial P_{CO_2} to a much greater degree than they prevent equivalent decreases in arterial P_{O_2}.

Of course we don't usually breathe bags of gas containing carbon dioxide. What is the physiological role of this reflex? If a defect in the respiratory system, such as emphysema, causes the body to retain carbon dioxide, the increase in arterial P_{CO_2} stimulates ventilation. This promotes the elimination of the carbon dioxide. Conversely, if arterial P_{CO_2} decreases below normal levels for whatever reason, this removes some of the stimulus for ventilation. This reduces ventilation and allows metabolically produced carbon dioxide to accumulate, thereby returning the P_{CO_2} to normal. In this manner, the arterial P_{CO_2} is stabilized near the normal value of 40 mmHg.

The ability of changes in arterial P_{CO_2} to control ventilation reflexly is largely due to associated changes in H^+ concentration (see equation 13–11). As summarized in **Figure 13–37**, both the peripheral and central chemoreceptors initiate the pathways that mediate these reflexes. The peripheral chemoreceptors are stimulated by the increased arterial H^+ concentration resulting from the increased P_{CO_2}. At the same time, because carbon dioxide diffuses rapidly across the membranes separating capillary blood and brain tissue, the increase in arterial P_{CO_2} causes a rapid increase in brain extracellular fluid P_{CO_2}. This increased P_{CO_2} increases *brain extracellular-fluid*

H^+ concentration, which stimulates the central chemoreceptors. Inputs from both the peripheral and central chemoreceptors stimulate the medullary inspiratory neurons to increase ventilation. The end result is a return of arterial and brain extracellular fluid P_{CO_2} and H^+ concentration toward normal.

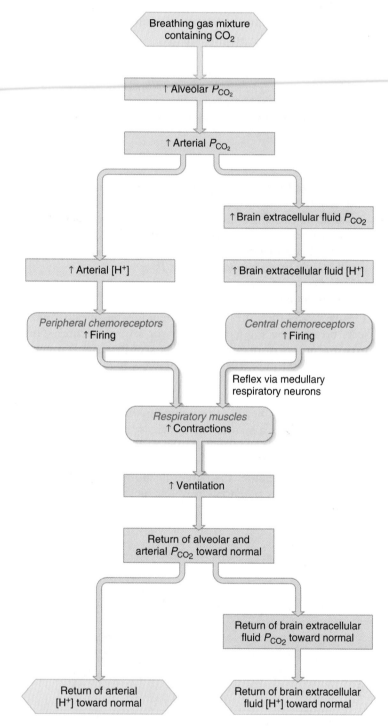

Figure 13–37

Pathways by which increased arterial P_{CO_2} stimulates ventilation. Note that the peripheral chemoreceptors are stimulated by an *increase* in H^+ concentration, whereas they are also stimulated by a *decrease* in P_{O_2} (see Figure 13–35).

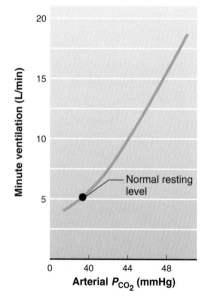

Figure 13–36

Effects on respiration of increasing arterial P_{CO_2} achieved by adding carbon dioxide to inspired air.

Of the two sets of receptors involved in this reflex response to elevated P_{CO_2}, the central chemoreceptors are the more important, accounting for about 70 percent of the increased ventilation.

It should also be noted that the effects of increased P_{CO_2} and decreased P_{O_2} not only exist as independent inputs to the medulla but potentiate each other's effects. The acute ventilatory response to combined low P_{O_2} and high P_{CO_2} is considerably greater than the sum of the individual responses.

Throughout this section, we have described the stimulatory effects of carbon dioxide on ventilation via reflex input to the medulla, but very high levels of carbon dioxide actually *inhibit* ventilation and may be lethal. This is because such concentrations of carbon dioxide act directly on the medulla to inhibit the respiratory neurons by an anesthesia-like effect. Other symptoms caused by very high blood P_{CO_2} include severe headaches, restlessness, and dulling or loss of consciousness.

Control by Changes in Arterial H⁺ Concentration That Are Not Due to Changes in Carbon Dioxide

We have seen that retention or excessive elimination of carbon dioxide causes respiratory acidosis and respiratory alkalosis, respectively. There are, however, many normal and pathological situations in which a change in arterial H⁺ concentration is due to some cause other than a primary change in P_{CO_2}. This is termed **metabolic acidosis** when H⁺ concentration is increased and **metabolic alkalosis** when it is decreased. In such cases, the peripheral chemoreceptors play the major role in altering ventilation.

For example, the addition of lactic acid to the blood, as in strenuous exercise, causes hyperventilation almost entirely by stimulation of the peripheral chemoreceptors (**Figures 13–38** and **13–39**). The central chemoreceptors are only minimally stimulated in this case because brain H⁺ concentration is increased to only a small extent, at least early on, by the hydrogen ions generated from the lactic acid. This is because

hydrogen ions penetrate the blood-brain barrier very slowly. In contrast, as described earlier, carbon dioxide penetrates the blood-brain barrier easily and changes brain H⁺ concentration.

The converse of the previous situation is also true: When arterial H⁺ concentration is lowered by any means other than by a reduction in P_{CO_2} (for example, by the loss of hydrogen ions from the stomach when vomiting), ventilation is reflexly depressed because of decreased peripheral chemoreceptor output.

The adaptive value such reflexes have in regulating arterial H⁺ concentration is shown in Figure 13–39. The increased ventilation induced by a metabolic acidosis reduces arterial P_{CO_2}, which lowers arterial H⁺ concentration back toward normal. Similarly, hypoventilation induced by a metabolic alkalosis results in an elevated arterial P_{CO_2} and a restoration of H⁺ concentration toward normal.

Notice that when a change in arterial H⁺ concentration due to some acid unrelated to carbon dioxide influences ventilation via the peripheral chemoreceptors, P_{CO_2} is displaced from normal. This is a reflex that regulates arterial H⁺ concentration at the expense of changes in arterial P_{CO_2}. Maintenance

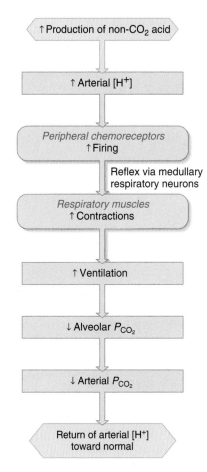

Figure 13–39

Reflexly induced hyperventilation minimizes the change in arterial hydrogen ion concentration when acids are produced in excess in the body. Note that under such conditions, arterial P_{CO_2} is reflexly reduced below its normal value.

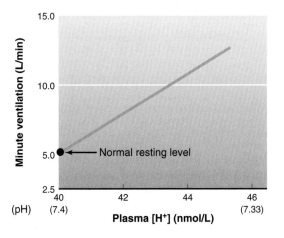

Figure 13–38

Changes in ventilation in response to an elevation of plasma hydrogen ion concentration produced by the administration of lactic acid.

Adapted from Lambertsen.

of normal arterial H^+ is necessary because most enzymes of the body function best at physiological pH.

Figure 13–40 summarizes the control of ventilation by P_{O_2}, P_{CO_2}, and H^+ concentration.

Control of Ventilation During Exercise

During exercise, the alveolar ventilation may increase as much as 20-fold. On the basis of our three variables—P_{O_2}, P_{CO_2}, and H^+ concentration—it might seem easy to explain the mechanism that induces this increased ventilation. Such is not the case, however, and the major stimuli to ventilation during exercise, at least moderate exercise, remain unclear.

Increased P_{CO_2} as the Stimulus?

It would seem logical that, as the exercising muscles produce more carbon dioxide, blood P_{CO_2} would increase. This is true, however, only for systemic *venous* blood but not for systemic *arterial* blood. Why doesn't arterial P_{CO_2} increase during exercise? Recall two facts from the section on alveolar gas pressures: (1) arterial P_{CO_2} is determined by alveolar P_{CO_2}, and (2) alveolar P_{CO_2} is determined by the ratio of carbon dioxide production to alveolar ventilation. During moderate exercise, the alveolar ventilation increases in exact proportion to the increased carbon dioxide production, so alveolar and there-

fore arterial P_{CO_2} do not change. In fact, in very strenuous exercise the alveolar ventilation increases relatively more than carbon dioxide production. In other words, during strenuous exercise, a person may hyperventilate, and thus alveolar and systemic arterial P_{CO_2} may actually decrease (**Figure 13–41**)!

Decreased P_{O_2} as the Stimulus?

The story is similar for oxygen. Although systemic *venous* P_{O_2} decreases during exercise due to an increase in oxygen consumption in the tissues, alveolar P_{O_2} and, therefore, systemic *arterial* P_{O_2} usually remain unchanged (see Figure 13–41). This is because cellular oxygen consumption and alveolar ventilation increase in exact proportion to each other, at least during moderate exercise.

This is a good place to recall an important point made in Chapter 12. In healthy individuals, ventilation is not the limiting factor in endurance exercise—cardiac output is. Ventilation can, as we have just seen, increase enough to maintain arterial P_{O_2}.

Increased H^+ Concentration as the Stimulus?

Because the arterial P_{CO_2} does not change during moderate exercise and decreases during strenuous exercise, there is no accumulation of excess H^+ resulting from carbon dioxide accumulation. However, during strenuous exercise, there *is* an increase in

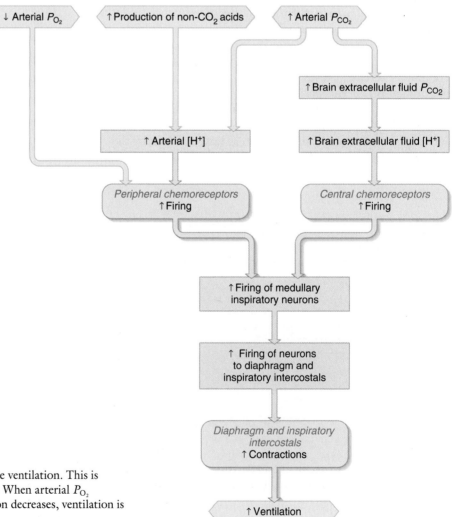

Figure 13–40

Summary of the major chemical inputs that stimulate ventilation. This is a combination of Figures 13–35, 13–37, and 13–39. When arterial P_{O_2} increases or when P_{CO_2} or hydrogen ion concentration decreases, ventilation is reflexly decreased.

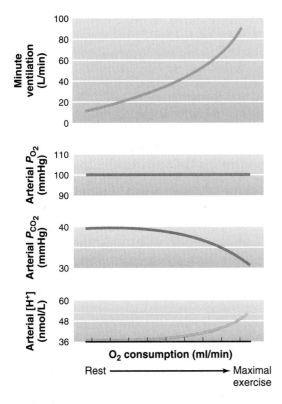

Figure 13–41

The effect of exercise on ventilation, arterial gas pressures, and hydrogen ion concentration. All these variables remain constant during moderate exercise; any change occurs only during strenuous exercise, when the person is actually hyperventilating (decrease in P_{CO_2}).

Adapted from Comroe.

arterial H^+ concentration (see Figure 13–41), due to the generation and release of lactic acid into the blood. This change in H^+ concentration is responsible, in part, for stimulating the hyperventilation accompanying strenuous exercise.

Other Factors

A variety of other factors play some role in stimulating ventilation during exercise. These include (1) reflex input from mechanoreceptors in joints and muscles; (2) an increase in body temperature; (3) inputs to the respiratory neurons via branches from axons descending from the brain to motor neurons supplying the exercising muscles; (4) an increase in the plasma epinephrine concentration; (5) an increase in the plasma potassium concentration due to movement of potassium out of the exercising muscles; and (6) a conditioned (learned) response mediated by neural input to the respiratory centers. The operation of this last factor can be seen in **Figure 13–42**. There is an abrupt increase—within seconds—in ventilation at the onset of exercise and an equally abrupt decrease at the end; these changes occur too rapidly to be explained by alteration of chemical constituents of the blood or by altered body temperature.

Figure 13–43 summarizes various factors that influence ventilation during exercise. The possibility that oscillatory changes in arterial P_{O_2}, P_{CO_2}, or H^+ concentration occur, despite unchanged average levels of these variables, and play some role has been proposed, but this remains unproven.

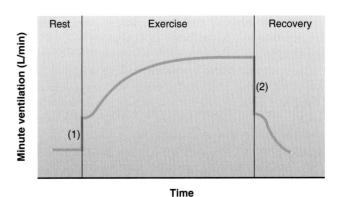

Figure 13–42

Ventilation changes during exercise. Note (1) the abrupt increase at the onset of exercise and (2) the equally abrupt but larger decrease at the end of exercise.

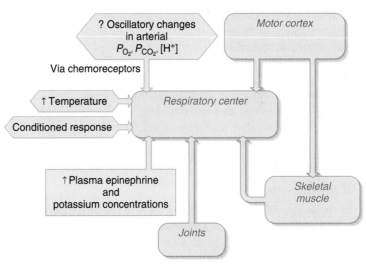

Figure 13–43

Summary of factors that stimulate ventilation during exercise. Note: ? indicates a theoretical input.

Figure 13–43 physiological *inquiry*

- The existence of chemoreceptors in the pulmonary artery has been suggested. Hypothesize a function for peripheral chemoreceptors located on and sensing the P_{O_2} and P_{CO_2} of the blood in the pulmonary artery.

Answer can be found at end of chapter.

Other Ventilatory Responses

Protective Reflexes

A group of responses protect the respiratory system from irritant materials. Most familiar are the cough and the sneeze reflexes, which originate in receptors located between airway epithelial cells. The receptors for the sneeze reflex are in the nose or pharynx, and those for cough are in the larynx, trachea, and bronchi. When the receptors initiating a cough are stimulated, the medullary respiratory neurons reflexly cause a deep inspiration and a violent expiration. In this manner, particles and secretions are moved from smaller to larger airways, and aspiration of materials into the lungs is also prevented.

Alcohol inhibits the cough reflex, which may partially explain the susceptibility of alcoholics to choking and pneumonia.

Another example of a protective reflex is the immediate cessation of respiration that is often triggered when noxious agents are inhaled. Chronic smoking may cause a loss of this reflex.

Voluntary Control of Breathing

Although we have discussed in detail the involuntary nature of most respiratory reflexes, the voluntary control of respiratory movements is important. Voluntary control is accomplished by descending pathways from the cerebral cortex to the motor neurons of the respiratory muscles. This voluntary control of respiration cannot be maintained when the involuntary stimuli, such as an elevated P_{CO_2} or H^+ concentration, become intense. An example is the inability to hold your breath for very long.

The opposite of breath-holding—deliberate hyperventilation—lowers alveolar and arterial P_{CO_2} and increases P_{O_2}. Unfortunately, swimmers sometimes voluntarily hyperventilate immediately before underwater swimming to be able to hold their breath longer. We say "unfortunately" because the low P_{CO_2} may still permit breath-holding at a time when the exertion is lowering the arterial P_{O_2} to levels that can cause unconsciousness and lead to drowning.

Besides the obvious forms of voluntary control, respiration must also be controlled during such complex actions as speaking, singing, and swallowing.

Reflexes from J Receptors

In the lungs, either in the capillary walls or the interstitium, are a group of receptors called **J receptors.** They are normally dormant but are stimulated by an increase in lung interstitial pressure caused by the collection of fluid in the interstitium. Such an increase occurs during the vascular congestion caused by either occlusion of a pulmonary vessel (**pulmonary embolus**) or left ventricular heart failure (Chapter 12), as well as by strong exercise in healthy people. The main reflex effects are rapid breathing (tachypnea) and a dry cough. In addition, neural input from J receptors gives rise to sensations of pressure in the chest and **dyspnea**—the feeling that breathing is labored or difficult.

Hypoxia

Hypoxia is defined as a deficiency of oxygen at the tissue level. There are many potential causes of hypoxia, but they can be classified into four general categories: (1) **hypoxic hypoxia** (also termed **hypoxemia**), in which the arterial P_{O_2} is reduced; (2) **anemic** or **carbon monoxide hypoxia,** in which the arterial P_{O_2} is normal but the total oxygen *content* of the blood is reduced because of inadequate numbers of erythrocytes, deficient or abnormal hemoglobin, or competition for the hemoglobin molecule by carbon monoxide; (3) **ischemic hypoxia** (also called hypoperfusion hypoxia), in which blood flow to the tissues is too low; and (4) **histotoxic hypoxia,** in which the quantity of oxygen reaching the tissues is normal, but the cell is unable to utilize the oxygen because a toxic agent—cyanide, for example—has interfered with the cell's metabolic machinery.

The primary causes of hypoxic hypoxia in disease are listed in **Table 13–11.** Exposure to the reduced P_{O_2} of high altitude also causes hypoxic hypoxia but is, of course, not a disease. The brief summaries in Table 13–11 provide a review of many of the key aspects of respiratory physiology and pathophysiology described in this chapter.

This table also emphasizes that some of the diseases that produce hypoxia also produce carbon dioxide retention and an increased arterial P_{CO_2} (**hypercapnea**). In such cases, treating only the oxygen deficit by administering oxygen may be inadequate because it does nothing about the hypercapnea. Indeed, such therapy may be dangerous. The primary respiratory drive in such patients is the hypoxia, because for several reasons the reflex ventilatory response to an increased P_{CO_2} may be lost in chronic situations. The administration of pure oxygen may cause such patients to stop breathing.

Why Do Ventilation-Perfusion Abnormalities Affect O_2 More than CO_2?

As described in Table 13–11, ventilation-perfusion inequalities often cause hypoxemia without associated increases in P_{CO_2}. The explanation for this resides in the fundamental difference between the transport of O_2 and CO_2 in the blood. Recall that the shape of the oxygen dissociation curve is sigmoidal (see Figure 13–26). An increase in P_{O_2} above 100 mmHg does not

Table 13–11	Causes of a Decreased Arterial P_{O_2} (Hypoxic Hypoxia) in Disease

1. **Hypoventilation** may be caused (a) by a defect anywhere along the respiratory control pathway, from the medulla through the respiratory muscles, (b) by severe thoracic cage abnormalities, and (c) by major obstruction of the upper airway. The hypoxemia of hypoventilation is always accompanied by an increased arterial P_{CO_2}.

2. **Diffusion impairment** results from thickening of the alveolar membranes or a decrease in their surface area. In turn, it causes blood P_{O_2} and alveolar P_{O_2} to fail to equilibrate. Often it is apparent only during exercise. Arterial P_{CO_2} is either normal, because carbon dioxide diffuses more readily than oxygen, or reduced, if the hypoxemia reflexly stimulates ventilation.

3. A **shunt** is (a) an anatomic abnormality of the cardiovascular system that causes mixed venous blood to bypass ventilated alveoli in passing from the right side of the heart to the left side, or (b) an intrapulmonary defect in which mixed venous blood perfuses unventilated alveoli (ventilation-perfusion = 0). Arterial P_{CO_2} generally does not rise because the effect of the shunt on arterial P_{CO_2} is counterbalanced by the increased ventilation reflexly stimulated by the hypoxemia.

4. **Ventilation-perfusion inequality** is by far the most common cause of hypoxemia. It occurs in chronic obstructive lung diseases and many other lung diseases. Arterial P_{CO_2} may be normal or increased, depending upon how much ventilation is reflexly stimulated.

add much oxygen to hemoglobin that is almost 100 percent saturated. If poorly ventilated, diseased alveoli are perfused, they contribute blood with low oxygen to the pulmonary vein and thus to the general circulation. If increases in ventilation ensue in order to compensate for this, the increase in P_{O_2} in the healthy part of the lung does not add much oxygen to the blood from that region because of the minimal increase in oxygen saturation. As blood from these different areas of the lung mix in the pulmonary vein, the net result is still deoxygenated blood (hypoxemia).

The situation for CO_2, however, is very different. The CO_2 content curve is relatively linear because CO_2 is transported in the blood mainly as highly soluble bicarbonate, which does not reach saturating levels at physiological concentrations. Therefore, although poorly ventilated areas of the lungs do cause increases in the CO_2 content of the blood entering the pulmonary vein, a compensatory increase in ventilation *lowers* CO_2 content below normal in the blood from the well-ventilated areas of the lung. The net result, as blood mixes in the pulmonary vein in this case, is essentially normal arterial CO_2 content and P_{CO_2}. Thus, clinically significant ventilation-perfusion mismatching can lead to low arterial P_{O_2} with normal P_{CO_2}.

Emphysema

The pathophysiology of emphysema, a major cause of hypoxia, offers an excellent review of many basic principles of respiratory physiology. *Emphysema* is characterized by the destruction of the alveolar walls leading to an increase in compliance. Furthermore, there is atrophy and collapse of the lower airways—those from the terminal bronchioles on down. The lungs actually self-destruct, attacked by proteolytic enzymes secreted by leukocytes in response to a variety of factors. Cigarette smoking is by far the most important of these factors; it stimulates the release of the proteolytic enzymes and destroys other enzymes that normally protect the lung against them.

As a result of alveolar-wall loss, adjacent alveoli fuse to form fewer but larger alveoli, and there is a loss of the pulmonary capillaries that were originally in the walls. The merging of alveoli, often into huge balloon-like structures, reduces the *total* surface area available for diffusion, and this impairs gas exchange. Moreover, because the destructive changes are not uniform throughout the lungs, some areas may receive large amounts of air and little blood, while others show just the opposite pattern. The result is marked ventilation-perfusion inequality.

In addition to problems in gas exchange, emphysema is associated with a marked increase in airway resistance, which greatly increases the work of breathing and, if severe enough, may cause hypoventilation. This is why emphysema is classified, as noted earlier in this chapter, as a "chronic *obstructive* pulmonary disease." The airway obstruction in emphysema is caused by the collapse of the lower airways. To understand this, recall that two physical factors passively holding the airways open are the transpulmonary pressure and the lateral traction of connective-tissue fibers attached to the airway exteriors. Both of these factors are diminished in emphysema because of the destruction of the lung elastic tissues, so the airways collapse.

In summary, patients with emphysema suffer from increased airway resistance, decreased total area available for diffusion, and ventilation-perfusion inequality. The result, particularly of the ventilation-perfusion inequality, is always

some degree of hypoxia. As explained above, an increase in arterial P_{CO_2} will not occur until the disease becomes extensive and prevents increases in alveolar ventilation.

Acclimatization to High Altitude

Atmospheric pressure progressively decreases as altitude increases. Thus, at the top of Mt. Everest (approximately 29,000 ft, or 9000 m), the atmospheric pressure is 253 mmHg, compared to 760 mmHg at sea level. The air is still 21 percent oxygen, which means that the inspired P_{O_2} is 53 mmHg (0.21×253 mmHg). Therefore, the alveolar and arterial P_{O_2} must decrease as persons ascend unless they breathe pure oxygen. The highest villages permanently inhabited by people are in the Andes at 19,000 ft (5700 m). These villagers work quite normally, and the only major precaution they take is that the women descend to lower altitudes during late pregnancy.

The effects of oxygen deprivation vary from one individual to another, but most people who ascend rapidly to altitudes above 10,000 ft experience some degree of **mountain sickness** (also called **altitude sickness**). This disorder consists of breathlessness, headache, nausea, vomiting, insomnia, fatigue, and impairment of mental processes. Much more serious is the appearance, in some individuals, of life-threatening pulmonary edema, which is the leakage of fluid from the pulmonary capillaries into the alveolar walls and eventually the airspaces themselves. Brain edema can also occur. Supplemental oxygen and diuretic therapy are used to treat mountain sickness.

Over the course of several days, the symptoms of mountain sickness usually disappear, although maximal physical capacity remains reduced. Acclimatization to high altitude is achieved by the compensatory mechanisms listed in **Table 13–12**.

Table 13–12	Acclimatization to the Hypoxia of High Altitude

1. The peripheral chemoreceptors stimulate ventilation.

2. Erythropoietin, a hormone secreted by the kidneys, stimulates erythrocyte synthesis, resulting in increased erythrocyte and hemoglobin concentration in blood.

3. DPG increases and shifts the hemoglobin dissociation curve to the right, facilitating oxygen unloading in the tissues. However, this DPG change is not always adaptive and may be maladaptive. For example, at very high altitudes, a right shift in the curve impairs oxygen *loading* in the lungs, an effect that outweighs any benefit from facilitation of *unloading* in the tissues.

4. Increases in capillary density (due to hypoxia-induced expression of the genes that code for angiogenic factors), number of mitochondria, and muscle myoglobin occur, all of which increase oxygen transfer.

5. The peripheral chemoreceptors stimulate an increased loss of sodium and water in the urine. This reduces plasma volume, resulting in a concentration of the erythrocytes and hemoglobin in the blood.

Finally, note that the responses to high altitude are essentially the same as the responses to hypoxia from any other cause. Thus, a person with severe hypoxia from lung disease may show many of the same changes—increased hematocrit, for example—as a high-altitude sojourner.

Nonrespiratory Functions of the Lungs

The lungs perform a variety of functions in addition to their roles in gas exchange and regulation of H^+ concentration. Most notable are the influences they have on the arterial concentrations of a large number of biologically active substances. Many substances (neurotransmitters and paracrine agents, for example) released locally into interstitial fluid may diffuse into capillaries and thus make their way into the systemic venous system. The lungs partially or completely remove some of these substances from the blood and thereby prevent them from reaching other locations in the body via the arteries. The cells that perform this function are the endothelial cells lining the pulmonary capillaries.

In contrast, the lungs may also produce new substances and add them to the blood. Some of these substances play local regulatory roles within the lungs, but if produced in large enough quantity, they may diffuse into the pulmonary capillaries and be carried to the rest of the body. For example, inflammatory responses (Chapter 18) in the lung may lead, via excessive release of potent chemicals such as histamine, to alterations of systemic blood pressure or flow. In at least one case, the lungs contribute a hormone, angiotensin II, to the blood (Chapter 14).

Finally, the lungs also act as a sieve that traps small blood clots generated in the systemic circulation, thereby preventing them from reaching the systemic arterial blood where they could occlude blood vessels in other organs.

ADDITIONAL CLINICAL EXAMPLES

This chapter has already described respiratory distress syndrome of the newborn, asthma, chronic obstructive pulmonary disease (COPD), and emphysema. *Acute respiratory distress syndrome* (**ARDS**) and *sleep apnea* are fascinating disorders of lung function and breathing that illustrate important characteristics of respiratory physiology.

Acute Respiratory Distress Syndrome (ARDS)

ARDS occurs mostly in adults, hence its alternate name, adult respiratory distress syndrome. As opposed to respiratory distress syndrome of the newborn, a deficiency in alveolar surfactant is the result of, not the cause of, ARDS. The hallmark of ARDS is the leaking of protein from the pulmonary blood vessels, which leads to an increase in the movement of liquid into the lung. ARDS can be caused by a wide variety of conditions, including pulmonary and systemic infections, aspiration of stomach contents, and severe trauma. The protein and fluid leakage into the lung invariably leads to a decrease in oxygen diffusion in the lung, leading to a decrease in oxygen in the blood (hypoxemia). Patients with ARDS are usually treated with supportive therapy, including mechanical pulmonary ventilation and supplemental oxygen therapy. Mortality from ARDS is very high (>50 percent), although this appears to be declining due to improved recognition and treatment.

Sleep Apnea

Sleep apnea is characterized by periodic cessation of respiration during sleep, which results in hypoxia and hypercapnia (asphyxia). In severe cases, this may occur more than 20 times an hour. There are two general types of sleep apnea. *Central sleep apnea* is primarily due to a decrease in neural output from the respiratory center in the medulla to the phrenic motor nerve output to the diaphragm. *Obstructive sleep apnea* is caused by increased airway resistance because of narrowing or collapse of the upper airways (primarily the pharynx) during inspiration (**Figure 13–44**). Obstructive sleep apnea may occur in as much as 4 percent of the adult population and with a greater frequency in the elderly and in men. Significant snoring may be an early sign of the eventual development of obstructive sleep apnea. Obesity is clearly a contributing factor because the excess fat in the neck decreases the diameter of the upper airways. A decrease in the activity of the upper airway dilating muscles, particularly during REM sleep, also contributes to airway collapse. Finally, anatomical narrowing and increased compliance of the upper airways contributes to periodic inspiratory obstruction during sleep.

There are many serious consequences of untreated sleep apnea, including hypertension of the pulmonary arteries (pulmonary hypertension) and added strain on the right ventricle of the heart. This can lead to heart failure and abnormal heart rhythm, leading to increased mortality. The periodic arousal that occurs during these apneic episodes results in serious disruption of normal sleep patterns and can lead to sleepiness during the day (*daytime somnolence*). Increased catecholamine release during these frequent arousals can also contribute to the development of high blood pressure.

There are a variety of treatments for obstructive sleep apnea. Surgery such as laser-assisted widening of the soft palate and uvula can sometimes be of benefit. Weight loss is often quite helpful. However, the mainstay of therapy is *continuous positive airway pressure* (**CPAP**) (**Figure 13–45**). The patient wears a small mask over the nose during sleep, which is attached to a positive pressure-generating device. By increasing airway pressure greater than P_{atm}, the collapse of the upper airways during inspiration is prevented. Although the CPAP nasal mask may seem obtrusive, many patients sleep much better with it, and many of the symptoms resolve with this treatment. ■

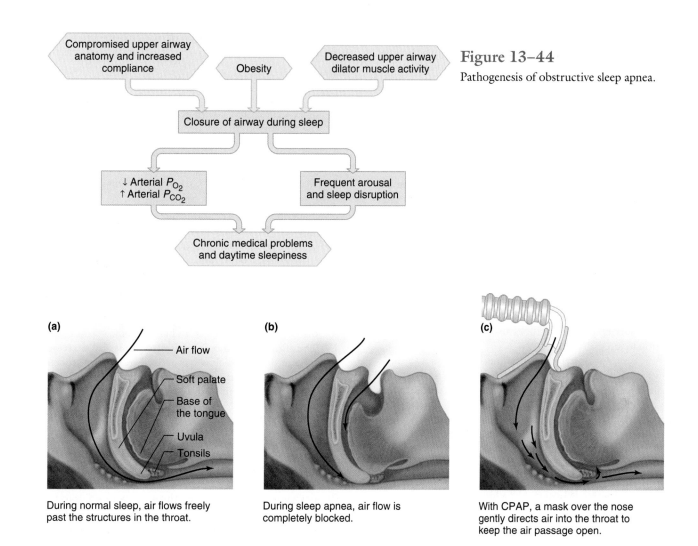

Figure 13–44

Pathogenesis of obstructive sleep apnea.

(a)
- Air flow
- Soft palate
- Base of the tongue
- Uvula
- Tonsils

During normal sleep, air flows freely past the structures in the throat.

(b)

During sleep apnea, air flow is completely blocked.

(c)

With CPAP, a mask over the nose gently directs air into the throat to keep the air passage open.

Figure 13–45

The pathophysiology and a standard treatment of obstructive sleep apnea. (a) Normal sleep with air flowing freely past the structures of the throat during an inspiration. (b) In obstructive sleep apnea (particularly with the patient sleeping in the supine position), the soft palate, uvula, and tongue occlude the airway, greatly increasing the resistance to air flow. (c) Continuous positive airway pressure (CPAP) is applied with a nasal mask preventing airway collapse.

■ **SUMMARY**

Organization of the Respiratory System

I. The respiratory system comprises the lungs, the airways leading to them, and the chest structures responsible for moving air into and out of them.
   a. The conducting zone of the airways consists of the trachea, bronchi, and terminal bronchioles.
   b. The respiratory zone of the airways consists of the alveoli, which are the sites of gas exchange, and those airways to which alveoli are attached.
   c. The alveoli are lined by type I cells and some type II cells, which produce surfactant.
   d. The lungs and interior of the thorax are covered by pleura; between the two pleural layers is an extremely thin layer of intrapleural fluid.
II. The lungs are elastic structures whose volume depends upon the pressure difference across the lungs—the transpulmonary pressure—and how stretchable the lungs are.

III. The steps involved in respiration are summarized in Figure 13–6. In the steady state, the net volumes of oxygen and carbon dioxide exchanged in the lungs per unit time are equal to the net volumes exchanged in the tissues.

Ventilation and Lung Mechanics

I. Bulk flow of air between the atmosphere and alveoli is proportional to the difference between the alveolar and atmospheric pressures and inversely proportional to the airway resistance: $F = (P_{alv} - P_{atm})/R$.
II. Between breaths at the end of an unforced expiration $P_{atm} = P_{alv}$, no air is flowing, and the dimensions of the lungs and thoracic cage are stable as the result of opposing elastic forces. The lungs are stretched and are attempting to recoil, whereas the chest wall is compressed and attempting to move outward. This creates a subatmospheric intrapleural pressure and hence a transpulmonary pressure that opposes the forces of elastic recoil.
III. During inspiration, the contractions of the diaphragm and inspiratory intercostal muscles increase the volume of the thoracic cage.

a. This makes intrapleural pressure more subatmospheric, increases transpulmonary pressure, and causes the lungs to expand to a greater degree than they do between breaths.

b. This expansion initially makes alveolar pressure subatmospheric, which creates the pressure difference between the atmosphere and alveoli to drive air flow into the lungs.

IV. During expiration, the inspiratory muscles cease contracting, allowing the elastic recoil of the lungs to return them to their original between-breath size.

a. This initially compresses the alveolar air, raising alveolar pressure above atmospheric pressure and driving air out of the lungs.

b. In forced expirations, the contraction of expiratory intercostal muscles and abdominal muscles actively decreases chest dimensions.

V. Lung compliance is determined by the elastic connective tissues of the lungs and the surface tension of the fluid lining the alveoli. The latter is greatly reduced, and compliance increased, by surfactant, produced by the type II cells of the alveoli. Surfactant also stabilizes alveoli by increasing surface tension in large alveoli.

VI. Airway resistance determines how much air flows into the lungs at any given pressure difference between atmosphere and alveoli. The major determinant of airway resistance is the radii of the airways.

VII. The vital capacity is the sum of resting tidal volume, inspiratory reserve volume, and expiratory reserve volume. The volume expired during the first second of a forced vital capacity (FVC) measurement is the FEV_1 and normally averages 80 percent of FVC.

VIII. Minute ventilation is the product of tidal volume and respiratory rate. Alveolar ventilation = (tidal volume − anatomic dead space) × respiratory rate.

Exchange of Gases in Alveoli and Tissues

I. Exchange of gases in lungs and tissues is by diffusion, as a result of differences in partial pressures. Gases diffuse from a region of higher partial pressure to one of lower partial pressure. Oxygen consumption is approximately 250 ml per minute whereas carbon dioxide production is approximately 200 ml per minute.

II. Normal alveolar gas pressure for oxygen is 105 mmHg and for carbon dioxide is 40 mmHg.

a. At any given inspired P_{O_2}, the ratio of oxygen consumption to alveolar ventilation determines alveolar P_{O_2}—the higher the ratio, the lower the alveolar P_{O_2}.

b. The higher the ratio of carbon dioxide production to alveolar ventilation, the higher the alveolar P_{CO_2}.

III. The average value at rest for systemic venous P_{O_2} is 40 mmHg and for P_{CO_2} is 46 mmHg.

IV. As systemic venous blood flows through the pulmonary capillaries, there is net diffusion of oxygen from alveoli to blood and of carbon dioxide from blood to alveoli. By the end of each pulmonary capillary, the blood gas pressures have become equal to those in the alveoli.

V. Inadequate gas exchange between alveoli and pulmonary capillaries may occur when the alveolar capillary surface area is decreased, when the alveolar walls thicken, or when there are ventilation-perfusion inequalities.

VI. Significant ventilation-perfusion inequalities cause the systemic arterial P_{O_2} to be reduced. An important mechanism for opposing mismatching is that a low local P_{O_2} causes local vasoconstriction, diverting blood away from poorly ventilated areas.

VII. In the tissues, net diffusion of oxygen occurs from blood to cells, and net diffusion of carbon dioxide from cells to blood.

Transport of Oxygen in Blood

I. Each liter of systemic arterial blood normally contains 200 ml of oxygen, more than 98 percent bound to hemoglobin and the rest dissolved.

II. The major determinant of the degree to which hemoglobin is saturated with oxygen is blood P_{O_2}.

a. Hemoglobin is almost 100 percent saturated at the normal systemic arterial P_{O_2} of 100 mmHg. The fact that saturation is already more than 90 percent at a P_{O_2} of 60 mmHg permits relatively normal uptake of oxygen by the blood even when alveolar P_{O_2} is moderately reduced.

b. Hemoglobin is 75 percent saturated at the normal systemic venous P_{O_2} of 40 mmHg. Thus, only 25 percent of the oxygen has dissociated from hemoglobin and entered the tissues.

III. The affinity of hemoglobin for oxygen is decreased by an increase in P_{CO_2}, hydrogen ion concentration, and temperature. All these conditions exist in the tissues and facilitate the dissociation of oxygen from hemoglobin.

IV. The affinity of hemoglobin for oxygen is also decreased by binding DPG, which is synthesized by the erythrocytes. DPG increases in situations associated with inadequate oxygen supply and helps maintain oxygen release in the tissues.

Transport of Carbon Dioxide in Blood

I. When carbon dioxide molecules diffuse from the tissues into the blood, 10 percent remains dissolved in plasma and erythrocytes, 30 percent combines in the erythrocytes with deoxyhemoglobin to form carbamino compounds, and 60 percent combines in the erythrocytes with water to form carbonic acid, which then dissociates to yield bicarbonate and hydrogen ions. Most of the bicarbonate then moves out of the erythrocytes into the plasma in exchange for chloride ions.

II. As venous blood flows through lung capillaries, blood P_{CO_2} decreases because of the diffusion of carbon dioxide out of the blood into the alveoli, and the reactions are reversed.

Transport of Hydrogen Ions Between Tissues and Lungs

I. Most of the hydrogen ions generated in the erythrocytes from carbonic acid during blood passage through tissue capillaries bind to deoxyhemoglobin because deoxyhemoglobin, formed as oxygen unloads from oxyhemoglobin, has a high affinity for hydrogen ions.

II. As the blood flows through the lung capillaries, hydrogen ions bound to deoxyhemoglobin are released and combine with bicarbonate to yield carbon dioxide and water.

Control of Respiration

I. Breathing depends upon cyclical inspiratory muscle excitation by the nerves to the diaphragm and intercostal muscles. This neural activity is triggered by the medullary inspiratory neurons.

II. The medullary respiratory center is composed of the dorsal respiratory group, which contains inspiratory neurons, and the ventral respiratory group, where the respiratory rhythm generator is located.

III. The most important inputs to the medullary inspiratory neurons for the involuntary control of ventilation are from the peripheral chemoreceptors—the carotid and aortic bodies—and the central chemoreceptors.

Chapter 13

IV. Ventilation is reflexly stimulated, via the peripheral chemoreceptors, by a decrease in arterial P_{O_2}, but only when the decrease is large.

V. Ventilation is reflexly stimulated, via both the peripheral and central chemoreceptors, when the arterial P_{CO_2} goes up even a slight amount. The stimulus for this reflex is not the increased P_{CO_2} itself, but the concomitant increased hydrogen ion concentration in arterial blood and brain extracellular fluid.

VI. Ventilation is also stimulated, mainly via the peripheral chemoreceptors, by an increase in arterial hydrogen ion concentration resulting from causes other than an increase in P_{CO_2}. The result of this reflex is to restore hydrogen ion concentration toward normal by lowering P_{CO_2}.

VII. Ventilation is reflexly inhibited by an increase in arterial P_{O_2} and by a decrease in arterial P_{CO_2} or hydrogen ion concentration.

VIII. During moderate exercise, ventilation increases in exact proportion to metabolism, but the signals causing this are not known. During very strenuous exercise, ventilation increases more than metabolism.

   a. The proportional increases in ventilation and metabolism during moderate exercise cause the arterial P_{O_2}, P_{CO_2}, and hydrogen ion concentration to remain unchanged.

   b. Arterial hydrogen ion concentration increases during very strenuous exercise because of increased lactic acid production. This accounts for some of the hyperventilation that occurs.

IX. Ventilation is also controlled by reflexes originating in airway receptors and by conscious intent.

Hypoxia

I. The causes of hypoxic hypoxia are listed in Table 13–11.

II. During exposure to hypoxia, as at high altitude, oxygen supply to the tissues is maintained by the five responses listed in Table 13–12.

Nonrespiratory Functions of the Lungs

I. The lungs influence arterial blood concentrations of biologically active substances by removing some from systemic venous blood and adding others to systemic arterial blood.

II. The lungs also act as sieves that trap and dissolve small clots formed in the systemic tissues.

Additional Clinical Examples

I. Acute respiratory distress syndrome is caused by a leakage of protein and fluid into the lung, which causes a decrease in oxygen diffusion.

II. Sleep apnea is a periodic cessation of breathing during sleep and is caused by decreased neural drive to breathe (central sleep apnea) or inspiratory occlusion of the upper airway (obstructive sleep apnea).

■ KEY TERMS

airway 443
alveolar dead space 457
alveolar pressure (P_{alv}) 446
alveolar ventilation ($\dot{V}_A$) 456
alveolus 443
anatomic dead space (V_D) 456
aortic body 470
apneustic center 470
atmospheric pressure (P_{atm}) 446
Boyle's law 447
bronchus 443
bronchiole 444
carbamino hemoglobin 467
carbonic anhydrase 467
carotid body 470
central chemoreceptor 471

conducting zone 444
Dalton's law 457
deoxyhemoglobin (Hb) 463
diaphragm 445
2,3-diphosphoglycerate (DPG) 466
dorsal respiratory group (DRG) 469
elastic recoil 448
expiration 443
expiratory reserve volume (ERV) 455
fetal hemoglobin 469
functional residual capacity (FRC) 454
globin 463
heme 463
hemoglobin 463
Henry's law 458
Hering-Breuer reflex 470
inspiration 443
inspiratory reserve volume (IRV) 454
intercostal muscle 445
intrapleural fluid 446
intrapleural pressure (P_{ip}) 446
J receptor 476
larynx 443
lateral traction 454
Law of Laplace 453
lung compliance (C_L) 452
mediastinum 449
medullary respiratory center 469
minute ventilation ($\dot{V}_E$) 456
nitric oxide 468
oxygen-carrying capacity 463
oxygen-hemoglobin dissociation curve 463
oxyhemoglobin (HbO$_2$) 463
parietal pleura 446
partial pressure 458

percent hemoglobin saturation 463
peripheral chemoreceptor 470
pharynx 443
phrenic nerves 450
physiologic dead space 457
pleura 445
pleural sac 445
pneumotaxic center 470
pontine respiratory group 470
Pre-Bötzinger complex 469
pulmonary 442
pulmonary stretch receptor 470
residual volume (RV) 455
respiratory bronchiole 444
respiratory cycle 443
respiratory physiology (two definitions) 442
respiratory quotient (RQ) 457
respiratory rhythm generator 469
respiratory system 443
respiratory zone 444
surface tension 452
surfactant 452
thorax 444
tidal volume (V_t) 454
total blood carbon dioxide 468
trachea 443
transmural pressure 448
transpulmonary pressure (P_{tp}) 448
type I alveolar cell 444
type II alveolar cell 444
upper airways 443
ventilation 446
ventral respiratory group (VRG) 469
visceral pleura 445
vital capacity (VC) 455
vocal cord 443

■ CLINICAL TERMS

acute respiratory distress syndrome (ARDS) 478
anemia 463
anemic hypoxia 476
anti-inflammatory drug 454
asthma 454
bronchodilator drug 454
carbon monoxide 465
carbon monoxide hypoxia 476
central sleep apnea 478
chronic bronchitis 478
chronic obstructive pulmonary disease 454
continuous positive airway pressure (CPAP) 478
cystic fibrosis 444
daytime somnolence 478

diffuse interstitial fibrosis 461
diffusion impairment 476
dyspnea 476
emphysema 477
forced expiratory volume in 1 s (FEV$_1$) 455
histotoxic hypoxia 476
hypercapnea 476
hyperventilation 460
hypoventilation 460
hypoxemia 476
hypoxia 476
hypoxic hypoxia 476
ischemic hypoxia 476
metabolic acidosis 473
metabolic alkalosis 473

■ REVIEW QUESTIONS

1. List the functions of the respiratory system.
2. At rest, how many liters of air and blood flow through the lungs per minute?
3. Describe four functions of the conducting portion of the airways.
4. Which respiration steps occur by diffusion and which by bulk flow?
5. What are normal values for intrapleural pressure, alveolar pressure, and transpulmonary pressure at the end of an unforced expiration?
6. Between breaths at the end of an unforced expiration, in what directions do the lungs and chest wall tend to move? What prevents them from doing so?
7. State typical values for oxygen consumption, carbon dioxide production, and cardiac output at rest. How much oxygen (in milliliters per liter) is present in systemic venous and systemic arterial blood?
8. Write the equation relating air flow into or out of the lungs to alveolar pressure, atmospheric pressure, and airway resistance.
9. Describe the sequence of events that cause air to move into the lungs during inspiration and out of the lungs during expiration. Diagram the changes in intrapleural pressure and alveolar pressure.
10. What factors determine lung compliance? Which is most important?
11. How does surfactant increase lung compliance? How does surfactant stabilize alveoli by preventing small alveoli from emptying into large alveoli?
12. How is airway resistance influenced by airway radii?
13. List the physical factors that alter airway resistance.
14. Contrast the causes of increased airway resistance in asthma, emphysema, and chronic bronchitis.
15. What distinguishes lung capacities, as a group, from lung volumes?
16. State the formula relating minute ventilation, tidal volume, and respiratory rate. Give representative values for each at rest.
17. State the formula for calculating alveolar ventilation. What is an average value for alveolar ventilation?
18. The partial pressure of a gas is dependent upon what two factors?
19. State the alveolar partial pressures for oxygen and carbon dioxide in a normal person at rest.
20. What factors determine alveolar partial pressures?
21. What is the mechanism of gas exchange between alveoli and pulmonary capillaries? In a normal person at rest, what are the gas pressures at the end of the pulmonary capillaries, relative to those in the alveoli?

22. Why does thickening of alveolar membranes impair oxygen movement but have little effect on carbon dioxide exchange?
23. What is the major result of ventilation-perfusion inequalities throughout the lungs? Describe homeostatic responses that minimize mismatching.
24. What generates the diffusion gradients for oxygen and carbon dioxide in the tissues?
25. In what two forms is oxygen carried in the blood? What are the normal quantities (in milliliters per liter) for each form in arterial blood?
26. Describe the structure of hemoglobin.
27. Draw an oxygen-hemoglobin dissociation curve. Put in the points that represent systemic venous and systemic arterial blood (ignore the rightward shift of the curve in systemic venous blood). What is the adaptive importance of the plateau? Of the steep portion?
28. Would breathing pure oxygen cause a large increase in oxygen transport by the blood in a normal person? In a person with a low alveolar P_{O_2}?
29. Describe the effects of increased P_{CO_2}, H^+ concentration, and temperature on the oxygen-hemoglobin dissociation curve. How are these effects adaptive for oxygen unloading in the tissues?
30. Describe the effects of increased DPG on the oxygen-hemoglobin dissociation curve. What is the adaptive importance of the effect of DPG on the curve?
31. Draw figures showing the reactions carbon dioxide undergoes entering the blood in the tissue capillaries and leaving the blood in the alveoli. What fractions are contributed by dissolved carbon dioxide, bicarbonate, and carbamino-hemoglobin?
32. What happens to most of the hydrogen ions formed in the erythrocytes from carbonic acid? What happens to blood H^+ concentration as blood flows through tissue capillaries?
33. What are the effects of P_{O_2} on carbamino-hemoglobin formation and H^+ binding by hemoglobin?
34. Describe the area of the brain in which automatic control of rhythmical respirations resides.
35. Describe the function of the pulmonary stretch receptors.
36. What changes stimulate the peripheral chemoreceptors? The central chemoreceptors?
37. Why doesn't moderate anemia or carbon monoxide exposure stimulate the peripheral chemoreceptors?
38. Is respiratory control more sensitive to small changes in arterial P_{O_2} or in arterial P_{CO_2}?
39. Describe the pathways by which increased arterial P_{CO_2} stimulates ventilation. What pathway is more important?
40. Describe the pathway by which a change in arterial H^+ concentration independent of altered carbon dioxide influences ventilation. What is the adaptive value of this reflex?
41. What happens to arterial P_{O_2}, P_{CO_2}, and H^+ concentration during moderate and strenuous exercise? List other factors that may stimulate ventilation during exercise.
42. List four general causes of hypoxic hypoxia.
43. Explain how ventilation-perfusion mismatch due to regional lung disease can cause hypoxic hypoxia but not hypercapnia.
44. Describe two general ways in which the lungs can alter the concentrations of substances other than oxygen, carbon dioxide, and H^+ in the arterial blood.
45. List two types of sleep apnea. Why does nasal CPAP prevent obstructive sleep apnea?

Chapter 13 Test Questions

(Answers appear in Appendix A.)

1. If $P_{atm} = 0$ mmHg and $P_{alv} = -2$ mmHg, then
   a. transpulmonary pressure (P_{tp}) is 2 mmHg.
   b. it is the end of the normal inspiration and there is no air flow.
   c. it is at the end of the normal expiration and there is no air flow.
   d. transpulmonary pressure (P_{tp}) is –2 mmHg.
   e. air is flowing into the lung.

2. Transpulmonary pressure (P_{tp}) increases by 2 mmHg during a normal inspiration. In subject A, 500 ml of air is inspired. In subject B, 250 ml of air is inspired for the same change in P_{tp}. Which is true?
   a. The compliance of the lung of subject B is less than that of subject A.
   b. The airway resistance of subject A is greater than that of subject B.
   c. The surface tension in the lung of subject B is less than that in subject A.
   d. The lung of subject A is deficient in surfactant.
   e. The compliance cannot be estimated from the data provided.

3. If alveolar ventilation is 4200 ml/min, respiratory frequency is 12 breaths per minute, and tidal volume is 500 ml, what is the anatomic dead-space ventilation?
   a. 1800 ml/min
   b. 6000 ml/min
   c. 350 ml/min
   d. 1200 ml/min
   e. It cannot be determined from the data provided.

4. Which of the following will increase alveolar P_{O_2}?
   a. Increase in metabolism and no change in alveolar ventilation
   b. Breathing air with 15 percent oxygen at sea level
   c. Increase in alveolar ventilation matched by an increase in metabolism
   d. Increased alveolar ventilation with no change in metabolism
   e. Carbon monoxide poisoning

5. Which of the following will cause the largest increase in systemic arterial oxygen saturation in the blood?
   a. An increase in red cell concentration (hematocrit) of 20 percent
   b. Breathing 100 percent O_2 in a normal subject at sea level
   c. An increase in arterial P_{O_2} from 40 to 60 mmHg
   d. Hyperventilation in a healthy subject at sea level
   e. Breathing a gas with 5 percent CO_2, 21 percent O_2, and 74 percent N_2 at sea level

6. In arterial blood with a P_{O_2} of 60 mmHg, which of the following situations will result in the lowest blood oxygen saturation?
   a. decreased DPG with normal body temperature and blood pH
   b. elevated body temperature, acidosis, and increased DPG
   c. decreased body temperature, alkalosis, and increased DPG
   d. normal body temperature with alkalosis
   e. elevated body temperature with alkalosis

7. Which of the following is *not* true about asthma?
   a. The basic defect is chronic airway inflammation.
   b. It is always caused by an allergy.
   c. The airway smooth muscle is hyperresponsive.
   d. It can be treated with inhaled steroid therapy.
   e. It can be treated with bronchodilator therapy.

8. Which of the following is true?
   a. Peripheral chemoreceptors increase firing with low arterial P_{O_2} but are not sensitive to an increase in arterial P_{CO_2}.
   b. The primary stimulus to the central chemoreceptors is low arterial P_{O_2}.
   c. Peripheral chemoreceptors increase firing during a metabolic alkalosis.
   d. The increase in ventilation during exercise is due to a decrease in arterial P_{O_2}.
   e. Peripheral and central chemoreceptors both increase firing when arterial P_{CO_2} increases.

9. Ventilation-perfusion inequalities lead to hypoxemia because
   a. the relationship between P_{CO_2} and the content of CO_2 in blood is sigmoidal.
   b. a decrease in ventilation-perfusion matching in a lung region causes pulmonary arteriolar vasodilation in that region.
   c. increases in ventilation cannot fully restore O_2 content in areas with low ventilation-perfusion matching.
   d. increases in ventilation cannot normalize P_{CO_2}.
   e. pulmonary blood vessels are not sensitive to changes in P_{O_2}.

10. After the expiration of a normal tidal volume, a subject breathes in as much air as possible. The volume of air inspired is the
    a. inspiratory reserve volume.
    b. vital capacity.
    c. inspiratory capacity.
    d. total lung capacity.
    e. functional residual capacity.

Chapter 13 Quantitative and Thought Questions

(Answers appear in Appendix A.)

1. At the end of a normal expiration, a person's lung volume is 2 L, his alveolar pressure is 0 mmHg, and his intrapleural pressure is –4 mmHg. He then inhales 800 ml, and at the end of inspiration the alveolar pressure is 0 mmHg and the intrapleural pressure is –8 mmHg. Calculate this person's lung compliance.

2. A patient is unable to produce surfactant. To inhale a normal tidal volume, will her intrapleural pressure have to be more or less subatmospheric during inspiration, relative to a healthy person?

3. A 70-kg adult patient is artificially ventilated by a machine during surgery at a rate of 20 breaths/min and a tidal volume of 250 ml/breath. Assuming a normal anatomic dead space of 150 ml, is this patient receiving an adequate alveolar ventilation?

4. Why must a person floating on the surface of the water and breathing through a snorkel increase his tidal volume and/or breathing frequency if alveolar ventilation is to remain normal?

5. A healthy person breathing room air voluntarily increases alveolar ventilation twofold and continues to do so until reaching new steady-state alveolar gas pressures for oxygen and carbon dioxide. Are the new values higher or lower than normal?

6. A person breathing room air has an alveolar P_{O_2} of 105 mmHg and an arterial P_{O_2} of 80 mmHg. Could hypoventilation due to, say, respiratory muscle weakness produce these values?

7. A person's alveolar membranes have become thickened enough to moderately decrease the rate at which gases diffuse across them at any given partial pressure differences. Will this person necessarily have a low arterial P_{O_2} at rest? During exercise?

8. A person is breathing 100 percent oxygen. How much will the oxygen content (in milliliters per liter of blood) of the arterial blood increase compared to when the person is breathing room air?

9. Which of the following have higher values in systemic venous blood than in systemic arterial blood: plasma P_{CO_2}, erythrocyte P_{CO_2}, plasma bicarbonate concentration, erythrocyte bicarbonate concentration, plasma hydrogen ion concentration, erythrocyte hydrogen ion concentration, erythrocyte carbamino concentration?

10. If the spinal cord were severed where it joins the brainstem, what would happen to respiration?

11. Which inspired gas mixture leads to the largest increase in minute ventilation?
    a. 10% O_2 / 5% CO_2
    b. 100% O_2 / 5% CO_2
    c. 21% O_2 / 5% CO_2
    d. 10% O_2 / 0% CO_2
    e. 0.1% CO / 5% CO_2

12. Patients with severe uncontrolled diabetes mellitus produce large quantities of certain organic acids. Can you predict the ventilation pattern in these patients and whether their arterial P_{O_2} and P_{CO_2} would increase or decrease?

13. Why does an inspired O_2 of 100 percent increase arterial P_{O_2} much more in a patient with ventilation-perfusion mismatch than in a patient with pure anatomical shunt?

Chapter 13 Answers to Physiological Inquiries

Figure 13–13

P_{alv}	P_{ip}	P_{tp} $(P_{alv} - P_{ip})$	Change in Lung Volume
❶ 0	−4	4	P_{tp} is increasing → lung volume↑
❷ −1	−6	5	P_{tp} is increasing → lung volume↑
❸ 0	−7	7	P_{tp} is decreasing → lung volume↓
❹ 1	−5	6	P_{tp} is decreasing → lung volume↓
❶ 0	−4	4	

Note: The actual volume increase or decrease in ml is determined by the compliance of the lung (see Figure 13–16).

Figure 13–16 Anything that increases P_{tp} during inspiration will, theoretically, increase lung volume. This can be done with positive airway pressure generated by mechanical ventilation, which will increase P_{alv}. This approach can work but also increases the risk of pneumothorax by inducing air leaks from the lung into the intrapleural space.

Figure 13–23 The increase in cardiac output with exercise greatly increases pulmonary blood flow and decreases the amount of time erythrocytes are exposed to increased oxygen from the alveoli. In a normal region of the lung, there is a large safety factor such that a large increase in blood flow still allows normal oxygen uptake. However, even small increases in the rate of capillary blood flow in a diseased portion of the lung will decrease oxygen uptake due to a loss of this safety factor.

Figure 13–29 Less O_2 will be unloaded in peripheral tissue as the blood is exposed to increased P_{CO_2} and decreased pH because the oxygen dissociation curve will not shift to the right as it does in real blood. Also, less O_2 will be loaded in the lungs as P_{CO_2} diffuses from blood into the alveoli because the oxygen dissociation curve will not shift to the left as it normally would with removal of CO_2 and decreased acidity.

Figure 13–43 These receptors might facilitate the increase in alveolar ventilation that occurs during exercise because pulmonary artery P_{O_2} will decrease and pulmonary artery P_{CO_2} will increase.

The Kidneys and Regulation of Water and Inorganic Ions

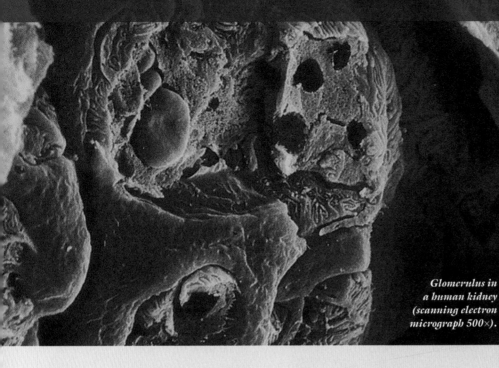

Glomerulus in a human kidney (scanning electron micrograph 500×).

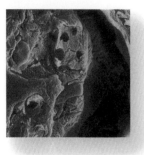

this chapter deals with the homeostatic regulation of the water and inorganic ion composition of the internal environment. The kidneys play the central role in these processes.

Regulation of the total-body balance of any substance can be studied in terms of the balance concept described in Chapter 1. Theoretically, a substance can appear in the body either as a result of ingestion or as a product of metabolism. On the loss side of the balance, a substance can be excreted from the body or can be metabolized. Therefore, if the quantity of any substance in the body is to be maintained at a nearly constant level over a period of time, the total amounts ingested and produced must equal the total amounts excreted and metabolized.

Reflexes that alter excretion via the urine constitute the major mechanisms that regulate the body balances of water and many of the inorganic ions that determine the properties of the extracellular fluid. The extracellular concentrations of these ions appeared in Table 4–1. We will first describe how the kidneys work in general and then apply this information to how they process specific substances like sodium, water, and potassium, and participate in reflexes that regulate these substances.

SECTION A   Basic Principles of Renal Physiology

Renal Functions

The adjective **renal** means "pertaining to the kidneys." The kidneys process the plasma portion of blood by removing substances from it and, in a few cases, by adding substances to it.

Table 14–1	Functions of the Kidneys

1. Regulation of water, inorganic ion balance, and acid-base balance (in cooperation with the lungs; Chapter 13)

2. Removal of metabolic waste products from the blood and their excretion in the urine

3. Removal of foreign chemicals from the blood and their excretion in the urine

4. Gluconeogenesis

5. Production of hormones/enzymes:
   a. Erythropoietin, which controls erythrocyte production (Chapter 12)
   b. Renin, an enzyme that controls the formation of angiotensin and influences blood pressure and sodium balance (this chapter)
   c. 1,25-dihydroxyvitamin D, which influences calcium balance (Chapter 11)

In so doing, they perform a variety of functions, as summarized in **Table 14–1**.

First, the kidneys play a central role in regulating the water concentration, inorganic ion composition, acid-base balance, and the fluid volume of the internal environment (e.g., blood volume). They do so by excreting just enough water and inorganic ions to keep the amounts of these substances in the body relatively constant. For example, if you increase your consumption of salt (sodium chloride), your kidneys will increase the amount of the salt excreted to match the intake. Alternatively, if there is not enough salt in the body, the kidneys will excrete very little salt.

Second, the kidneys excrete metabolic waste products into the urine as fast as they are produced. This keeps waste products, which can be toxic, from accumulating in the body. These metabolic wastes include **urea** from the catabolism of protein, **uric acid** from nucleic acids, **creatinine** from muscle creatine, the end products of hemoglobin breakdown (which give urine much of its color), and many others.

A third function of the kidneys is the urinary excretion of some foreign chemicals, such as drugs, pesticides, and food additives, and their metabolites.

A fourth function is **gluconeogenesis.** During prolonged fasting, the kidneys synthesize glucose from amino acids and other precursors and release it into the blood.

Finally, the kidneys act as secretory glands, releasing at least two hormones: erythropoietin (described in Chapter 12), and 1,25-dihydroxyvitamin D (described in Chapter 11). The kidneys also secrete an enzyme, renin, that is important in the control of blood pressure and sodium balance (described later in this chapter).

Structure of the Kidneys and Urinary System

The two kidneys lie in the back of the abdominal wall but not actually in the abdominal cavity. They are retroperitoneal, meaning they are just behind the peritoneum, the lining of this cavity. The urine flows from the kidneys through the **ureters** into the **bladder,** and then is eliminated via the **urethra** (**Figure 14–1**).

Each kidney contains approximately 1 million similar subunits called **nephrons.** Each nephron consists of (1) an initial filtering component called the **renal corpuscle,** and (2) a **tubule** that extends from the renal corpuscle (**Figure 14–2**). The renal corpuscle forms a filtrate from blood that is free of cells and proteins. This filtrate then leaves the renal corpuscle and enters the tubule. As it flows through the tubule, substances are added to or removed from it. Ultimately, the fluid remaining at the end of each nephron combines in the collecting ducts and exits the kidneys as urine.

Let us look first at the anatomy of the renal corpuscles—the filters. Each renal corpuscle contains a compact tuft of interconnected capillary loops called the **glomerulus** (plural, *glomeruli*), or **glomerular capillaries** (Figure 14–2 and **Figure 14–3**). Each glomerulus is supplied with blood by an arteriole called an **afferent arteriole.** The glomerulus protrudes into a fluid-filled capsule called **Bowman's capsule.** The combination of a glomerulus and a Bowman's capsule constitutes a renal corpuscle. As blood flows through the glomerulus, about 20 percent of the plasma filters into Bowman's capsule. The remaining blood then leaves the glomerulus by the **efferent arteriole.**

One way of visualizing the relationships within the renal corpuscle is to imagine a loosely clenched fist—the glomerulus—punched into a balloon—the Bowman's capsule. The part of Bowman's capsule in contact with the glomerulus becomes pushed inward but does not make contact with the opposite side of the capsule. Accordingly, a fluid-filled space called the **Bowman's space** exists within the capsule. Protein-free fluid filters from the glomerulus into this space.

Blood in the glomerulus is separated from the fluid in Bowman's space by a filtration barrier consisting of three layers (**Figures 14–3b** and **14–3c**). These include (1) the single-celled capillary endothelium, (2) a noncellular proteinaceous layer of basement membrane (also termed basal lamina) between the endothelium and the next layer, which is (3) the single-celled epithelial lining of Bowman's capsule. The epithelial cells in this region, called **podocytes,** are quite different from the simple flattened cells that line the rest of Bowman's capsule (the part of the "balloon" not in contact with the "fist"). They have an octopus-like structure in that they possess a large number of extensions, or foot processes. Fluid filters first across the endothelial cells, then through the basement membrane, and finally between the foot processes of the podocytes.

In addition to the capillary endothelial cells and the podocytes, there is a third cell type, **mesangial cells,** which are modified smooth muscle cells that surround the glomerular capillary loops but are not part of the filtration pathway. Their function will be described later.

The renal tubule is continuous with a Bowman's capsule. It is a very narrow, hollow cylinder made up of a single layer of epithelial cells (resting on a basement membrane). The epithelial cells differ in structure and function along the length of the tubule, and at least eight distinct segments are now recognized (see Figure 14–2). It is customary, however, to group two or more contiguous tubular segments when discussing function, and we will follow this practice. Thus, the segment of the tubule that drains Bowman's capsule is the **proximal tubule,** comprising the proximal convoluted tubule and the proximal straight tubule shown in Figure 14–2. The next portion of the tubule is the **loop of Henle,** which is a sharp, hairpin-like loop consisting of a **descending limb** coming from the proximal tubule and an **ascending limb** leading to the next tubular segment, the **distal convoluted tubule.** Fluid flows from the distal convoluted tubule into the **collecting duct system,** comprised of the **cortical collecting duct** and then the **medullary collecting duct.** The reasons for the terms *cortical* and *medullary* will be apparent shortly.

From Bowman's capsule to the collecting duct system, each nephron is completely separate from the others. This separation ends when multiple cortical collecting ducts merge.

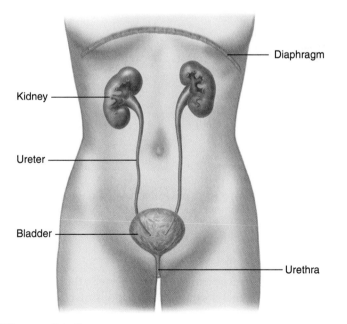

Figure 14–1

Urinary system in a woman. In the male, the urethra passes through the penis (Chapter 17). The diaphragm is shown for orientation.

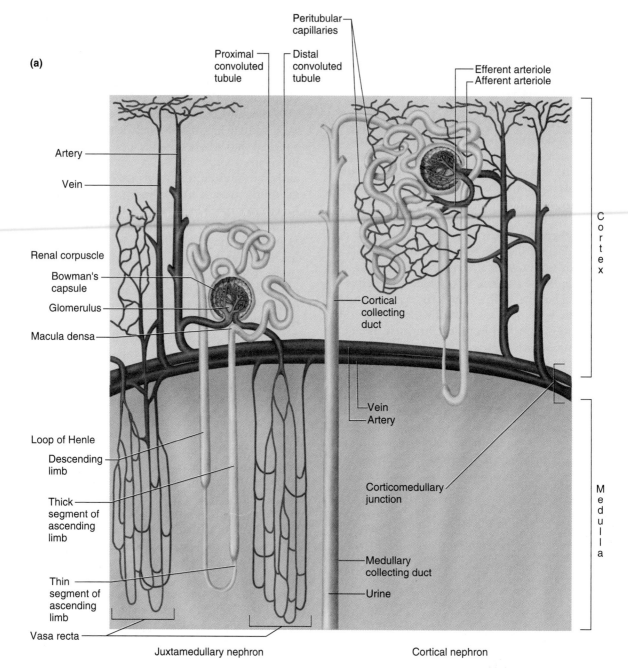

(a)

Peritubular capillaries

Proximal convoluted tubule

Distal convoluted tubule

Efferent arteriole
Afferent arteriole

Artery

Vein

Renal corpuscle

Bowman's capsule

Glomerulus

Macula densa

Cortical collecting duct

Loop of Henle

Descending limb

Thick segment of ascending limb

Thin segment of ascending limb

Vasa recta

Vein
Artery

Corticomedullary junction

Medullary collecting duct

Urine

Cortex

Medulla

Juxtamedullary nephron

Cortical nephron

Figure 14–2

Basic structure of a nephron. (a) Anatomical organization. The macula densa is not a distinct segment, but a plaque of cells in the ascending loop of Henle where the loop passes between the arterioles supplying its renal corpuscle of origin. The outer area of the kidney is called the cortex and the inner the medulla. Two types of nephrons are shown—the juxtamedullary have long loops of Henle that penetrate deeply into the medulla, while the cortical nephrons have short (or no) loops of Henle. Note that the efferent arterioles of juxtamedullary nephrons give rise to long looping vasa recta, while efferent arterioles of cortical nephrons give rise to peritubular capillaries. (b) *See next page.*

The result of additional mergings from this point on is that the urine drains into the kidney's central cavity, the **renal pelvis,** via several hundred large medullary collecting ducts. The renal pelvis is continuous with the ureter draining that kidney (**Figure 14–4**).

There are important regional differences in the kidney (see Figures 14–2 and 14–4). The outer portion is the **renal cortex,** and the inner portion the **renal medulla.** The cortex contains all the renal corpuscles. The loops of Henle extend from the cortex for varying distances down into the medulla. The medullary collecting ducts pass through the medulla on their way to the renal pelvis.

All along its length, each tubule is surrounded by capillaries, called the **peritubular capillaries.** Note that we have

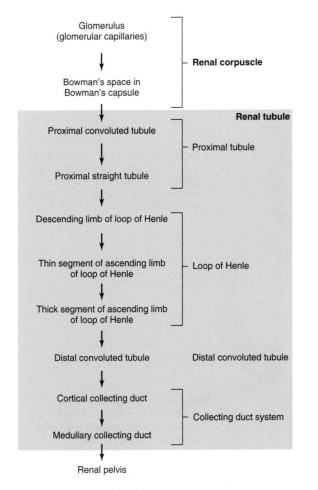

Glomerulus
(glomerular capillaries)

↓

Bowman's space in
Bowman's capsule

— Renal corpuscle

↓

Renal tubule

Proximal convoluted tubule

↓
— Proximal tubule

Proximal straight tubule

↓

Descending limb of loop of Henle

↓

Thin segment of ascending limb
of loop of Henle

— Loop of Henle

↓

Thick segment of ascending limb
of loop of Henle

↓

Distal convoluted tubule

Distal convoluted tubule

↓

Cortical collecting duct

↓
— Collecting duct system

Medullary collecting duct

↓

Renal pelvis

Figure 14–2 *continued*

Basic structure of a nephron. (b) Consecutive segments of the nephron. All segments in the yellow area are parts of the renal tubule; the terms to the right of the brackets are commonly used for several consecutive segments.

now mentioned two sets of capillaries in the kidneys—the glomerular capillaries (glomeruli) and the peritubular capillaries. Within each nephron, the two sets of capillaries are connected to each other by an efferent arteriole, the vessel by which blood leaves the glomerulus (see Figures 14–2 and 14–3a). Thus, the renal circulation is very unusual in that it includes *two* sets of arterioles and *two* sets of capillaries. After supplying the tubules with blood, the peritubular capillaries then join to form the veins by which blood leaves the kidney.

There are two general types of nephrons (see Figure 14–2a). About 15 percent of the nephrons are **juxtamedullary,** which means that the renal corpuscle lies in the part of the cortex closest to the cortical-medullary junction. The Henle's loops of these nephrons plunge deep into the medulla and, as we will see, are responsible for generating an osmotic gradient in the medulla responsible for the reabsorption of water. In close proximity to the juxtamedullary nephrons are long capillaries known

as the **vasa recta,** which also loop deeply into the medulla and then return to the cortical-medullary junction. The majority of nephrons are **cortical,** meaning their renal corpuscles are located in the outer cortex and their Henle's loops do not penetrate deep into the medulla. In fact, some cortical nephrons do not have a Henle's loop at all; they are involved in reabsorption and secretion but do not contribute to the hypertonic medullary interstitium described later in the chapter.

One additional anatomical detail involving both the tubule and the arterioles is important. Near its end, the ascending limb of each loop of Henle passes between the afferent and efferent arterioles of that loop's own nephron (see Figure 14–2). At this point there is a patch of cells in the wall of the ascending limb as it becomes the distal convoluted tubule called the **macula densa,** and the wall of the afferent arteriole contains secretory cells known as **juxtaglomerular (JG) cells.** The combination of macula densa and juxtaglomerular cells is known as the **juxtaglomerular apparatus (JGA)** (**Figures 14–3a** and **14–5**). The juxtaglomerular cells secrete renin into the blood.

Basic Renal Processes

As we have said, urine formation begins with the filtration of plasma from the glomerular capillaries into Bowman's space. This process is termed **glomerular filtration,** and the filtrate is called the **glomerular filtrate.** It is cell-free and, except for proteins, contains all the substances in virtually the same concentrations as in plasma. This type of filtrate is also termed an ultrafiltrate.

During its passage through the tubules, the filtrate's composition is altered by movements of substances from the tubules to the peritubular capillaries and vice versa (**Figure 14–6**). When the direction of movement is from tubular lumen to peritubular capillary plasma, the process is called **tubular reabsorption,** or simply reabsorption. Movement in the opposite direction—that is, from peritubular plasma to tubular lumen—is called **tubular secretion,** or simply secretion. Tubular secretion is also used to denote the movement of a solute from the cell interior to the lumen in the cases in which the kidney tubular cells themselves generate the substance.

To summarize: A substance can gain entry to the tubule and be excreted in the urine by glomerular filtration or tubular secretion or both. Once in the tubule, however, the substance need not be excreted but can be reabsorbed. Thus, the amount of any substance excreted in the urine is equal to the amount filtered plus the amount secreted, minus the amount reabsorbed.

$$\frac{\text{Amount}}{\text{excreted}} = \frac{\text{Amount}}{\text{filtered}} + \frac{\text{Amount}}{\text{secreted}} - \frac{\text{Amount}}{\text{reabsorbed}}$$

We must stress that not all these processes—filtration, secretion, and reabsorption—apply to all substances. For example, important solutes like glucose are completely reabsorbed, whereas toxins are secreted and not reabsorbed.

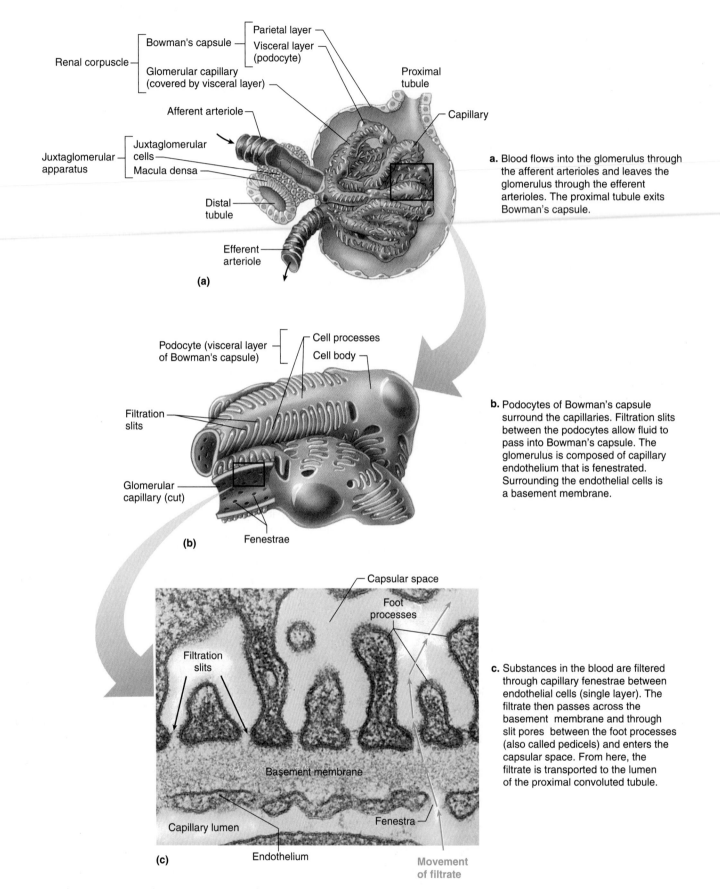

(a)

a. Blood flows into the glomerulus through the afferent arterioles and leaves the glomerulus through the efferent arterioles. The proximal tubule exits Bowman's capsule.

(b)

b. Podocytes of Bowman's capsule surround the capillaries. Filtration slits between the podocytes allow fluid to pass into Bowman's capsule. The glomerulus is composed of capillary endothelium that is fenestrated. Surrounding the endothelial cells is a basement membrane.

(c)

c. Substances in the blood are filtered through capillary fenestrae between endothelial cells (single layer). The filtrate then passes across the basement membrane and through slit pores between the foot processes (also called pedicels) and enters the capsular space. From here, the filtrate is transported to the lumen of the proximal convoluted tubule.

Figure 14–3

The renal corpuscle. (a) Anatomy of the renal corpuscle; (b) Inset view of podocytes and capillaries; (c) Transmission electron micrograph of a glomerular capillary and the glomerular membranes.

Daniel Friend from William Bloom and Don Fawcett, Textbook of Histology, 10th ed., W.B. Saunders Co. (Reproduced by permission of Edward Arnold)

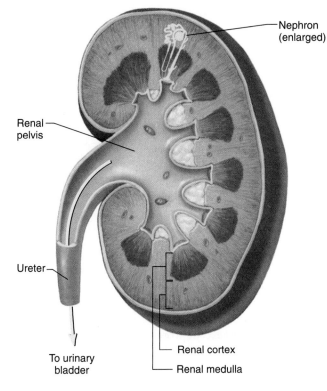

Renal pelvis

Ureter

To urinary bladder

Nephron (enlarged)

Renal cortex

Renal medulla

Figure 14–4

Section of a human kidney. For clarity, the juxtamedullary nephron illustrated to show nephron orientation is not to scale—its outline would not be clearly visible without a microscope. The outer kidney, which contains all the renal corpuscles, is the cortex, and the inner kidney is the medulla. Note that in the medulla, the loops of Henle and the collecting ducts run parallel to each other. The medullary collecting ducts drain into the renal pelvis.

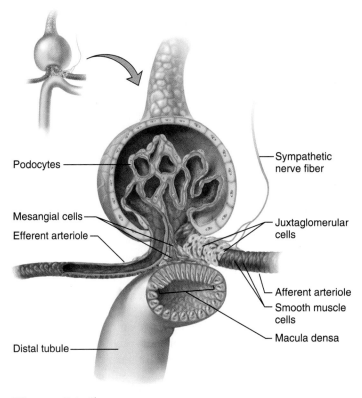

Podocytes

Mesangial cells

Efferent arteriole

Distal tubule

Sympathetic nerve fiber

Juxtaglomerular cells

Afferent arteriole

Smooth muscle cells

Macula densa

Figure 14–5

The juxtaglomerular apparatus.

To emphasize general principles, **Figure 14–7** illustrates the renal handling of three hypothetical substances. Approximately 20 percent of the plasma that enters the glomerular capillaries is filtered into Bowman's space. This filtrate, which contains X, Y, and Z in the same concentrations as in the capillary plasma, enters the proximal tubule and begins to flow through the rest of the tubule. Simultaneously, the remaining 80 percent of the plasma, containing X, Y, and Z, leaves the glomerular capillaries via the efferent arteriole and enters the peritubular capillaries.

Assume that the tubule can secrete 100 percent of the peritubular capillary substance X into the tubular lumen, but cannot reabsorb X. Therefore, by the combination of filtration and tubular secretion, the plasma that originally entered the renal artery is cleared of all of its substance X, which leaves the body via the urine.

By contrast, assume that the tubule can reabsorb, but not secrete, Y and Z. The amount of Y reabsorption is small, so that much of the filtered material is not reabsorbed and escapes from the body. For Z, however, the reabsorptive mechanism is so powerful that all the filtered Z is reabsorbed back into the plasma. Therefore, no Z is lost from the body.

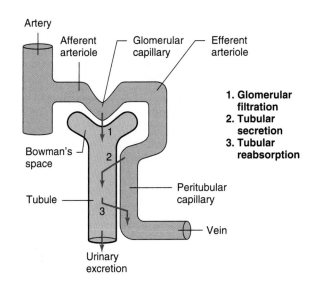

Artery

Afferent arteriole

Glomerular capillary

Efferent arteriole

Bowman's space

Tubule

Peritubular capillary

Vein

Urinary excretion

1. Glomerular filtration
2. Tubular secretion
3. Tubular reabsorption

Figure 14–6

The three basic components of renal function. This figure is to illustrate only the *directions* of reabsorption and secretion, not specific sites or order of occurrence. Depending on the particular substance, reabsorption and secretion can occur at various sites along the tubule.

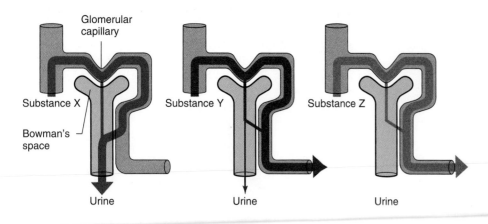

Figure 14–7

Renal handling of three hypothetical filtered substances X, Y, and Z. X is filtered and secreted but not reabsorbed. Y is filtered, and a fraction is then reabsorbed. Z is filtered and completely reabsorbed. The thickness of each line in this hypothetical example suggests the magnitude of the process.

Hence, for Z, the processes of filtration and reabsorption have canceled each other out, and the net result is as though Z had never entered the kidney.

A specific combination of filtration, tubular reabsorption, and tubular secretion applies to each substance in the plasma. The critical point is that, for many substances, the rates at which the processes proceed are subject to physiological control. By triggering changes in the rates of filtration, reabsorption, or secretion whenever the amount of a substance in the body is higher or lower than the normal limits, homeostatic mechanisms can regulate the substance's bodily balance. For example, consider what happens when a normally hydrated person drinks a lot of water. Within 1–2 h, all the excess water has been excreted in the urine, partly as a result of an increase in filtration but mainly as a result of decreased tubular reabsorption of water. In this example, the kidneys are the effector organs of a homeostatic process that maintains total body water within very narrow limits.

Although glomerular filtration, tubular reabsorption, and tubular secretion are the three basic renal processes, a fourth process—metabolism by the tubular cells—is also important for some substances. In some cases, the renal tubular cells remove substances from blood or glomerular filtrate and metabolize them, resulting in their disappearance from the body. In other cases, the cells *produce* substances and add them either to the blood or tubular fluid; the most important of these, as we will see, are ammonium ions, hydrogen ions, and bicarbonate ions.

In summary, one can study the normal renal processing of any given substance by asking a series of questions:

1. To what degree is the substance filterable at the renal corpuscle?
2. Is it reabsorbed?
3. Is it secreted?
4. What factors regulate the quantities filtered, reabsorbed, or secreted?
5. What are the pathways for altering renal excretion of the substance to maintain stable body balance?

Glomerular Filtration

As stated previously, the glomerular filtrate—that is, the fluid in Bowman's space—normally contains no cells but contains all plasma substances except proteins in virtually the same concentrations as in plasma. This is because glomerular filtration is a bulk-flow process in which water and all low-molecular-weight substances (including smaller peptides) move together. Most plasma proteins—the albumins and globulins—are excluded almost entirely from the filtrate. One reason for their exclusion is that the renal corpuscles restrict the movement of such high-molecular-weight substances. A second reason is that the filtration pathways in the corpuscular membranes are negatively charged, so they oppose the movement of these plasma proteins, most of which are negatively charged.

The only exceptions to the generalization that all nonprotein plasma substances have the same concentrations in the glomerular filtrate as in the plasma are certain low-molecular-weight substances that would otherwise be filterable but are bound to plasma proteins and therefore not filtered. For example, half the plasma calcium and virtually all of the plasma fatty acids are bound to plasma protein and so are not filtered.

Forces Involved in Filtration

Filtration across capillaries is determined by opposing Starling forces (described in Chapter 12). To review, they are the hydrostatic pressure difference across the capillary wall that favors filtration, and the protein concentration difference across the wall that creates an osmotic force that opposes filtration.

This also applies to the glomerular capillaries, as summarized in **Figure 14–8**. The blood pressure in the glomerular capillaries—the glomerular capillary hydrostatic pressure (P_{GC})—is a force favoring filtration. The fluid in Bowman's space exerts a hydrostatic pressure (P_{BS}) that opposes this filtration. Another opposing force is the osmotic force (π_{GC}) that results from the presence of protein in the glomerular capillary plasma. Recall that there is virtually no protein in the filtrate in Bowman's space because of the unique structure of the areas of filtration in the glomerulus, so the osmotic force in Bowman's space (π_{BS}) is zero. The unequal distribution of

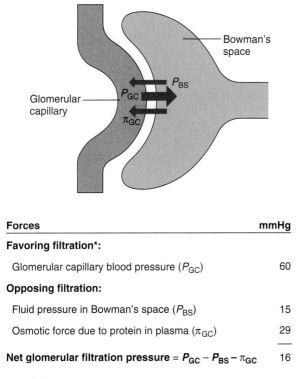

Forces	mmHg
Favoring filtration*:	
Glomerular capillary blood pressure (P_{GC})	60
Opposing filtration:	
Fluid pressure in Bowman's space (P_{BS})	15
Osmotic force due to protein in plasma (π_{GC})	29
Net glomerular filtration pressure = $P_{GC} - P_{BS} - \pi_{GC}$	16

Figure 14–8

Forces involved in glomerular filtration. The symbol π denotes the osmotic force due to the presence of protein in glomerular capillary plasma.

*The concentration of protein in Bowman's space is so low that π_{BS}, a force that would favor filtration, is considered zero.

Figure 14–8 physiological *inquiry* 🅿️

- What would be the effect of an increase in plasma albumin (the most abundant plasma protein) on glomerular filtration rate (GFR)?

Answer can be found at end of chapter.

protein causes the water concentration of the plasma to be slightly less than that of the fluid in Bowman's space, and this difference in water concentration favors fluid movement by bulk flow from Bowman's space into the glomerular capillaries—that is, it opposes glomerular filtration.

Note that in Figure 14–8, the value given for this osmotic force—29 mmHg—is slightly larger than the value—28 mmHg—for the osmotic force given in Chapter 12 for plasma in all arteries and nonrenal capillaries. The reason is that, unlike the situation elsewhere in the body, enough water filters out of the glomerular capillaries that the protein left behind in the plasma becomes more concentrated than in arterial plasma. In other capillaries, in contrast, little water filters out and the capillary protein concentration remains essentially unchanged from its value in arterial plasma. In other words, unlike the situation in other capillaries, the plasma protein concentration and, thus, the osmotic force, increases from the

beginning to the end of the glomerular capillaries. The value given in Figure 14–8 for the osmotic force is the average value along the length of the capillaries.

To summarize, the **net glomerular filtration pressure** is the sum of three relevant forces:

$$\text{Net glomerular filtration pressure} = P_{GC} - P_{BS} - \pi_{GC}$$

Normally the net filtration pressure is always positive because the glomerular capillary hydrostatic pressure (P_{GC}) is larger than the sum of the hydrostatic pressure in Bowman's space (P_{BS}) and the osmotic force opposing filtration (π_{GC}). The net glomerular filtration pressure initiates urine formation by forcing an essentially protein-free filtrate of plasma out of the glomerulus and into Bowman's space and then down the tubule into the renal pelvis.

Rate of Glomerular Filtration

The volume of fluid filtered from the glomeruli into Bowman's space per unit time is known as the **glomerular filtration rate (GFR).** GFR is determined not only by the net filtration pressure but also by the permeability of the corpuscular membranes and the surface area available for filtration. In other words, at any given net filtration pressure, the GFR will be directly proportional to the membrane permeability and the surface area. The glomerular capillaries are much more permeable to fluid than most other capillaries. Therefore, the net glomerular filtration pressure causes massive filtration of fluid into Bowman's space. In a 70-kg person, the GFR averages 180 L/day (125 ml/min)! This is much higher than the combined net filtration of 4 L/day of fluid across all the other capillaries in the body, as described in Chapter 12.

When we recall that the total volume of plasma in the cardiovascular system is approximately 3 L, it follows that the kidneys filter the entire plasma volume about 60 times a day. This opportunity to process such huge volumes of plasma enables the kidneys to regulate the constituents of the internal environment rapidly and to excrete large quantities of waste products.

GFR is not a fixed value but is subject to physiological regulation. This is achieved mainly by neural and hormonal input to the afferent and efferent arterioles, which causes changes in net glomerular filtration pressure (**Figure 14–9**). The glomerular capillaries are unique in that they are situated between two sets of arterioles—the afferent and efferent arterioles. Constriction of the afferent arterioles decreases hydrostatic pressure in the glomerular capillaries (P_{GC}). This is similar to arteriolar constriction in other organs and is due to a greater loss of pressure between arteries and capillaries (**Figure 14–9a**).

In contrast, efferent arteriolar constriction alone has the opposite effect on P_{GC} in that it *increases* it (**Figure 14–9b**). This occurs because the efferent arteriole lies beyond the glomerulus, so that efferent arteriolar constriction tends to "dam back" the blood in the glomerular capillaries, raising P_{GC}. Dilation of the efferent arteriole (**Figure 14–9c**)

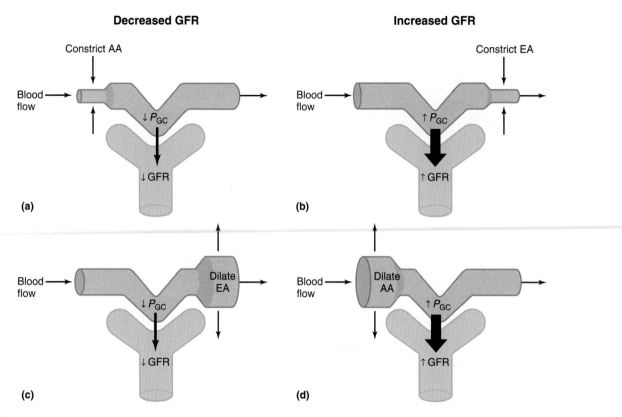

Figure 14–9

Control of GFR by constriction or dilation of afferent (AA) or efferent (EA) arterioles. Constriction of the afferent arteriole (a) or dilation of the efferent arteriole (c) reduces P_{GC}, thus decreasing GFR. Constriction of the efferent arteriole (b) or dilation of the afferent arteriole (d) increases P_{GC}, thus increasing GFR.

decreases P_{GC} and thus GFR, whereas dilation of the afferent arteriole increases P_{GC} and thus GFR (**Figure 14–9d**). Finally, simultaneous constriction or dilation of both sets of arterioles tends to leave P_{GC} unchanged because of the opposing effects.

In addition to the neuroendocrine input to the arterioles, there is also neural and humoral input to the mesangial cells that surround the glomerular capillaries. Contraction of these cells reduces the surface area of the capillaries, which causes a decrease in GFR at any given net filtration pressure.

It is possible to measure the total amount of any nonprotein or nonprotein-bound substance filtered into Bowman's space by multiplying the GFR by the plasma concentration of the substance. This amount is called the **filtered load** of the substance. For example, if the GFR is 180 L/day and plasma glucose concentration is 1 g/L, then the filtered load of glucose is 180 L/day × 1 g/L = 180 g/day.

Once the filtered load of the substance is known, it can be compared to the amount of the substance excreted. This indicates whether the substance undergoes net tubular reabsorption or net secretion. Whenever the quantity of a substance excreted in the urine is less than the filtered load, tubular reabsorption must have occurred. Conversely, if the amount excreted in the urine is greater than the filtered load, tubular secretion must have occurred.

Tubular Reabsorption

Table 14–2 summarizes data for a few plasma components that undergo filtration and reabsorption. It gives an idea of the magnitude and importance of reabsorptive mechanisms. The values in this table are typical for a healthy person on an average diet. There are at least three important conclusions we can draw from this table: (1) The filtered loads are enormous, generally larger than the amounts of the substances in the body. For example, the body contains about 40 L of water, but the volume of water filtered each day is 180 L. (2) Reabsorption of waste products is relatively incomplete (as in the case of urea), so that large fractions of their filtered loads are excreted in the urine. (3) Reabsorption of most useful plasma components, such as water, inorganic ions, and organic nutrients, is relatively complete so that the amounts excreted in the urine are very small fractions of their filtered loads.

An important distinction should be made between reabsorptive processes that can be controlled physiologically and those that cannot. The reabsorption rates of most organic nutrients, such as glucose, are always very high and are not physiologically regulated. Thus, the filtered loads of these substances are normally completely reabsorbed, with none appearing in the urine. For these substances, like substance Z in Figure 14–7, it is as though the kidneys do not exist because

Table 14–2	Average Values for Several Components that Undergo Filtration and Reabsorption		
Substance	Amount Filtered Per Day	Amount Excreted Per Day	Percent Reabsorbed
Water, L	180	1.8	99
Sodium, g	630	3.2	99.5
Glucose, g	180	0	100
Urea, g	54	30	44

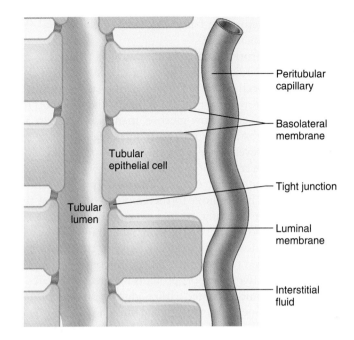

Figure 14–10

Diagrammatic representation of tubular epithelium.

the kidneys do not eliminate these substances from the body at all. Therefore, the kidneys do not regulate the plasma concentrations of these organic nutrients. Rather, the kidneys merely maintain whatever plasma concentrations already exist.

Recall that a major function of the kidneys is to eliminate soluble waste products. To do this, the blood is filtered in the glomeruli. One consequence of this is that substances necessary for normal body functions are filtered from the plasma into the tubular fluid. To prevent the loss of these important nonwaste products, the kidney has powerful mechanisms to reclaim useful substances from tubular fluid while allowing waste products to be excreted.

The reabsorptive rates for water and many ions, although also very high, are under physiological control. For example, if water intake is decreased, the kidney can increase water reabsorption to minimize water loss.

In contrast to glomerular filtration, the crucial steps in tubular reabsorption—those that achieve movement of a substance from tubular lumen to interstitial fluid—do *not* occur by bulk flow because there are inadequate pressure differences across the tubule and inadequate permeability of the tubular membranes. Instead, two other processes are involved: (1) The reabsorption of some substances from the tubular lumen is by diffusion, often across the tight junctions connecting the tubular epithelial cells (**Figure 14–10**). (2) The reabsorption of all other substances involves mediated transport, which requires the participation of transport proteins in the plasma membranes of tubular cells.

The final step in reabsorption is the movement of substances from the interstitial fluid into peritubular capillaries that occurs by a combination of diffusion and bulk flow. We will assume that this final process occurs automatically once the substance reaches the interstitial fluid.

Reabsorption by Diffusion

The reabsorption of urea by the proximal tubule provides an example of passive reabsorption by diffusion. An analysis of urea concentrations in the proximal tubule will help clarify

the mechanism. Because the corpuscular membranes are freely filterable to urea, the urea concentration in the fluid within Bowman's space is the same as that in the peritubular capillary plasma and the interstitial fluid surrounding the tubule. Then, as the filtered fluid flows through the proximal tubule, water reabsorption occurs (by mechanisms to be described later). This removal of water increases the concentration of urea in the tubular fluid so it is higher than in the interstitial fluid and peritubular capillaries. Therefore, urea diffuses down this concentration gradient from tubular lumen to peritubular capillary. Urea reabsorption is thus dependent upon the reabsorption of water. Reabsorption by diffusion in this manner occurs for a variety of lipid-soluble organic substances, both naturally occurring and foreign (e.g., the pesticide DDT).

Reabsorption by Mediated Transport

Figure 14–10 demonstrates that a substance reabsorbed by mediated transport must first cross the **luminal membrane** separating the tubular lumen from the cell interior. Then, the substance diffuses through the cytosol of the cell, and finally crosses the **basolateral membrane,** which begins at the tight junctions and constitutes the plasma membrane of the sides and base of the cell. This route is termed *transcellular* epithelial transport.

A substance need not be actively transported across *both* the luminal and basolateral membranes in order to be actively transported across the overall epithelium, thus moving from lumen to interstitial fluid against its electrochemical gradient. For example, sodium moves "downhill" (passively) into the cell across the luminal membrane either by diffusion or by facilitated diffusion and then is actively transported "uphill" out of the cell across the basolateral membrane via Na^+/K^+-ATPases in this membrane.

The reabsorption of many substances is coupled to the reabsorption of sodium. The cotransported substance moves uphill into the cell via a secondary active cotransporter as sodium moves downhill into the cell via this same cotransporter. This is precisely how glucose, many amino acids, and other organic substances undergo tubular reabsorption. The reabsorption of several inorganic ions is also coupled in a variety of ways to the reabsorption of sodium.

Many of the mediated transport reabsorptive systems in the renal tubule have a limit to the amounts of material they can transport per unit time (**transport maximum, T_m**). This is because the binding sites on the membrane transport proteins become saturated when the concentration of the transported substance increases to a certain level. An important example is the secondary active-transport proteins for glucose, located in the proximal tubule. As noted earlier, glucose does not usually appear in the urine because all of the filtered glucose is reabsorbed. This is illustrated in **Figure 14–11**, which shows the relationship between plasma glucose concentrations and the filtered load, reabsorption, and excretion of glucose. Plasma glucose concentration in a healthy person normally does not exceed 150 mg/100mL even after the person eats a sugary

meal. Notice that this level of plasma glucose is below the threshold at which glucose starts to appear in urine (***glucosuria***). Also notice that the T_m for the entire kidney is higher than the threshold for glucosuria. This is because the nephrons have a range of T_m values that, when averaged, give a T_m for the entire kidney, as shown in Figure 14–11. When plasma glucose concentration exceeds the transport maximum for a significant number of nephrons, glucose starts to appear in urine. In people with significant hyperglycemia (for example, in poorly controlled ***diabetes mellitus***), the plasma glucose concentration often exceeds the threshold value of 200 mg/100 mL, so that the filtered load exceeds the ability of the nephrons to reabsorb glucose. In other words, although the capacity of the kidney to reabsorb glucose can be normal in diabetes mellitus, the tubules cannot reabsorb the large increase in the filtered load of glucose. As you will learn in Chapter 16, the high filtered load of glucose can also lead to significant disruption of normal renal function (***diabetic nephropathy***).

The pattern described for glucose is also true for a large number of other organic nutrients. For example, most amino acids and water-soluble vitamins are filtered in large amounts each day, but almost all of these filtered molecules are reabsorbed by the proximal tubule. If the plasma concentration becomes high enough, however, reabsorption of the filtered load will not be as complete, and the substance will appear in larger amounts in the urine. Thus, people who ingest very large quantities of vitamin C have increased plasma concentrations of vitamin C. Eventually, the filtered load may exceed the tubular reabsorptive T_m for this substance, and any additional ingested vitamin C is excreted in the urine.

Tubular Secretion

Tubular secretion moves substances from peritubular capillaries into the tubular lumen. Like glomerular filtration, it constitutes a pathway from the blood into the tubule. Like reabsorption, secretion can occur by diffusion or by transcellular mediated transport. The most important substances secreted by the tubules are hydrogen ions and potassium. However, a large number of normally occurring organic anions, such as choline and creatinine, are also secreted; so are many foreign chemicals such as penicillin. Active secretion of a substance requires active transport either from the blood side (the interstitial fluid) into the cell (across the basolateral membrane) or out of the cell into the lumen (across the luminal membrane). As in reabsorption, tubular secretion is usually coupled to the reabsorption of sodium. Secretion from the interstitial space into the tubular fluid, which draws substances from the peritubular capillaries, is a mechanism to increase the ability of the kidney to dispose of substances at a higher rate rather than depending only on the filtered load.

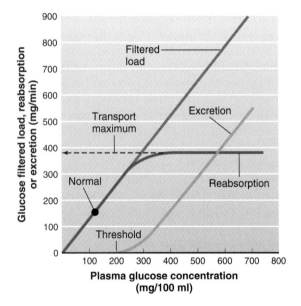

Figure 14–11

The relationship between plasma glucose concentration and the rate of glucose filtered (filtered load), reabsorbed, or excreted. The dotted line shows the transport maximum, which is the maximum rate at which glucose can be reabsorbed. Notice that as plasma glucose exceeds its threshold, glucose begins to appear in the urine.

Figure 14–11 physiological *inquiry* pi

- How would you calculate the filtered load and excretion rate of glucose?

Answer can be found at end of chapter.

Metabolism by the Tubules

We noted earlier that, during fasting, the cells of the renal tubules synthesize glucose and add it to the blood. They can also catabolize certain organic substances, such as peptides, taken up from either the tubular lumen or peritubular capillaries. Catabolism eliminates these substances from the body just as if they had been excreted into the urine.

Regulation of Membrane Channels and Transporters

Tubular reabsorption and/or secretion of many substances is under physiological control. For most of these substances, control is achieved by regulating the activity or concentrations of the membrane channel and transporter proteins involved in their transport. This regulation is achieved by hormones and paracrine/autocrine factors.

The recent explosion of information concerning the structure, function, and regulation of renal tubular-cell ion channels and transporters has made it possible to explain the underlying defects in some genetic diseases. For example, a genetic mutation can lead to an abnormality in the Na^+/glucose cotransporter that mediates active reabsorption of glucose in the proximal tubule. This can lead to the appearance of glucose in the urine (*familial renal glucosuria*). Contrast this condition to diabetes mellitus, in which the ability to reabsorb glucose is usually normal, but the filtered load of glucose exceeds the threshold for the tubules to reabsorb glucose (see Figure 14–11).

"Division of Labor" in the Tubules

To excrete waste products adequately, the GFR must be very large. This means that the filtered volume of water and the filtered loads of all the nonwaste plasma solutes are also very large. *The primary role of the proximal tubule is to reabsorb most of this filtered water and these solutes.* Furthermore, with potassium as the one major exception, the proximal tubule is the major site of solute secretion. Henle's loop also reabsorbs relatively large quantities of the major ions and, to a lesser extent, water.

Extensive reabsorption by the proximal tubule and Henle's loop ensures that the masses of solutes and the volume of water entering the tubular segments beyond Henle's loop are relatively small. These distal segments then do the fine-tuning for most substances, determining the final amounts excreted in the urine by adjusting their rates of reabsorption and, in a few cases, secretion. It should not be surprising, therefore, that most homeostatic controls act upon the more distal segments of the tubule.

The Concept of Renal Clearance

A useful way of quantifying renal function is in terms of clearance. The renal **clearance** of any substance is the volume of plasma from which that substance is completely removed ("cleared") by the kidneys per unit time. Every substance has its own distinct clearance value, but the units are always in volume of plasma per unit of time. The basic clearance formula for any substance S is

$$\text{Clearance of } S = \frac{\text{Mass of } S \text{ excreted per unit time}}{\text{Plasma concentration of } S}$$

Thus, the clearance of a substance is a measure of the volume of plasma completely cleared of the substance per unit time. This accounts for the mass of the substance excreted in the urine.

Because the mass of S excreted per unit time is equal to the urine concentration of S multiplied by the urine volume during that time, the formula for the clearance of S becomes

$$C_S = \frac{U_S V}{P_S}$$

where

C_S = clearance of S
U_S = urine concentration of S
V = urine volume per unit time
P_S = plasma concentration of S

Let us take the particularly important example of a polysaccharide named **inulin** (*not* insulin). This substance is an important research tool because its clearance is equal to the glomerular filtration rate. It is not normally found in the body, but we will administer it intravenously to a person at a rate sufficient to maintain a constant plasma concentration of 4 mg/L. Urine collected over a one-hour period has a volume of 0.1 L and an inulin concentration of 300 mg/L. Thus, inulin excretion equals 0.1 L/h × 300 mg/L, or 30 mg/h. How much plasma had to be completely cleared of its inulin to supply this 30 mg/h? We simply divide 30 mg/h by the plasma concentration, 4 mg/L, to obtain the volume cleared: 7.5 L/h. In other words, we are calculating the inulin clearance (C_{In}) from the measured urine volume per unit time (V), urine inulin concentration (U_{In}), and plasma inulin concentration (P_{In}):

$$C_{In} = \frac{U_{In} V}{P_{In}}$$

$$C_{In} = \frac{300 \text{ mg/L} \times 0.1 \text{ L/h}}{4 \text{ mg/L}}$$

$$C_{In} = 7.5 \text{ L/h}$$

Now for the crucial points. It is known that inulin is readily filtered at the renal corpuscle but is not reabsorbed, secreted, or metabolized by the tubule. Therefore, the mass of inulin excreted in our example—30 mg/h—must be equal to the mass filtered over that same time period (**Figure 14–12**). Thus, the clearance of inulin (C_{In}) must equal the volume of plasma originally filtered (GFR): C_{In} is equal to GFR.

The clearance of any substance handled by the kidneys in the same way as inulin—filtered, but not reabsorbed, secreted, or metabolized—would equal the GFR. Unfortunately, there are no substances normally present in the plasma that perfectly meet these criteria. For clinical purposes, the **creatinine clearance (C_{Cr})** is commonly used to approximate the GFR as follows. The waste product creatinine produced by muscle is filtered at the renal corpuscle and does not undergo reabsorption. It does undergo a small amount of secretion, however, so that some peritubular plasma is cleared of its creatinine by secretion. Therefore, C_{Cr} slightly overestimates the GFR but is close enough to be highly useful in most clinical situations.

This leads to an important generalization. When the clearance of any substance is greater than the GFR, that substance

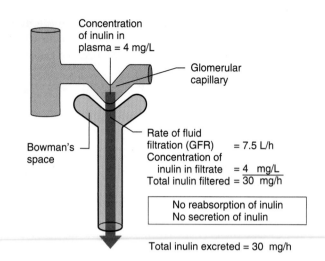

Figure 14–12

Example of renal handling of inulin, a substance that is filtered by the renal corpuscles but is neither reabsorbed nor secreted by the tubule. Therefore, the mass of inulin excreted per unit time is equal to the mass filtered during the same time period. As explained in the text, the clearance of inulin is equal to the glomerular filtration rate.

must undergo tubular secretion. Look back at our hypothetical substance X (see Figure 14–7): X is filtered, and all the X that escapes filtration is secreted; no X is reabsorbed. Consequently, all the plasma that enters the kidney per unit time is cleared of its X. Therefore, the clearance of X is a measure of **renal plasma flow.** A substance that is handled like X is the organic anion para-amino-hippurate (PAH), which is used for this purpose experimentally. (Like inulin, it must be administered intravenously.)

A similar logic leads to another important generalization. When the clearance of a filterable substance is less than the GFR, that substance must undergo reabsorption.

Micturition

Urine flow through the ureters to the bladder is propelled by contractions of the ureter wall smooth muscle. The urine is stored in the bladder and intermittently ejected during urination, or **micturition.**

The bladder is a balloon-like chamber with walls of smooth muscle collectively termed the **detrusor muscle.** The contraction of the detrusor muscle squeezes on the urine in the bladder lumen to produce urination. That part of the detrusor muscle at the base (or "neck") of the bladder where the urethra begins functions as the **internal urethral sphincter.** Just below the internal urethral sphincter, a ring of skeletal muscle surrounds the urethra. This is the **external urethral sphincter,** the contraction of which can prevent urination even when the detrusor muscle contracts strongly.

What factors influence these bladder structures (**Figure 14–13**)? (1) The detrusor muscle is innervated by parasympathetic neurons, which cause muscular contraction. Because of the arrangement of the smooth muscle fibers, when the detrusor muscle is relaxed, the internal urethral sphincter is closed. When the detrusor muscle contracts, changes in its shape pull open the internal urethral sphincter. (2) In addition, the internal sphincter receives sympathetic innervation, which causes contraction of the sphincter. (3) The external urethral sphincter, being skeletal muscle, is innervated by somatic motor neurons that cause contraction.

While the bladder is filling, there is little parasympathetic input to the detrusor muscle but strong sympathetic input to the internal urethral sphincter and strong input by the somatic motor neurons to the external urethral sphincter. Therefore, the detrusor muscle is relaxed, and the sphincters are closed.

What happens during micturition? As the bladder fills with urine, the pressure within it increases, which stimulates stretch receptors in the bladder wall. The afferent fibers from these receptors enter the spinal cord and stimulate the parasympathetic neurons, which then cause the detrusor muscle to

Bladder	Muscle	Innervation		
		Type	During filling	During micturition
	Detrusor (smooth muscle)	Parasympathetic (causes contraction)	Inhibited	Stimulated
	Internal urethral sphincter (smooth muscle)	Sympathetic (causes contraction)	Stimulated	Inhibited
	External urethral sphincter (skeletal muscle)	Somatic motor (causes contraction)	Stimulated	Inhibited

Figure 14–13

Control of the bladder.

contract. As noted previously, this contraction facilitates the opening of the internal urethral sphincter. Simultaneously, the afferent input from the stretch receptors reflexly inhibits the sympathetic neurons to the internal urethral sphincter, which further contributes to its opening. In addition, the afferent input also reflexly inhibits the somatic motor neurons to the external urethral sphincter, causing it to relax. Both sphincters are now open, and the contraction of the detrusor muscle can produce urination.

We have thus far described micturition as a local spinal reflex, but descending pathways from the brain can also profoundly influence this reflex, determining the ability to prevent or initiate micturition voluntarily. Loss of these descending pathways as a result of spinal cord damage eliminates one's ability to voluntarily control micturition. Prevention of micturition, learned during childhood, operates in the following way. As the bladder distends, the input from the bladder stretch receptors causes, via ascending pathways to the brain, a sense of bladder fullness and the urge to urinate. But in response to this, urination can be voluntarily prevented by activating descending pathways that stimulate both the sympathetic nerves to the internal urethral sphincter and the somatic motor nerves to the external urethral sphincter. In contrast, urination can be voluntarily initiated via the descending pathways to the appropriate neurons.

ADDITIONAL CLINICAL EXAMPLES

Incontinence

Incontinence is the involuntary release of urine, which can be a disturbing problem both socially and hygienically. The most common types are ***stress incontinence*** (due to sneezing, coughing, or exercise) and ***urge incontinence*** (associated with the desire to urinate). Incontinence is more common in women and may occur one to two times per week in more than 25 percent of women older than 60. It is very common in older women in nursing homes and assisted living facilities.

In women, stress incontinence is usually due to a loss of urethral support provided by the anterior vagina (see Figure 17–13a). Medications (such as estrogen replacement therapy to improve vaginal tone) can often relieve stress incontinence. Severe cases may require surgery to improve vaginal support of the bladder and urethra. The cause of urge incontinence is often unknown in individual patients. However, any irritation to the bladder or urethra (e.g., with a bacterial infection) can cause urge incontinence. Urge incontinence can be treated with drugs such as tolterodine or oxybutin, which antagonize the effects of the parasympathetic nerves on the detrusor muscle. Because these drugs are anticholinergic, they can have side effects such as blurred vision, constipation, and increased heart rate.■

■ SECTION A SUMMARY

Renal Function

I. The kidneys regulate the water and ionic composition of the body, excrete waste products, excrete foreign chemicals, produce glucose during prolonged fasting, and release factors and hormones into the blood (renin, 1,25-dihydroxyvitamin D, and erythropoietin). The first three functions are accomplished by continuous processing of the plasma.

Structure of the Kidneys and the Urinary System

I. Each nephron in the kidneys consists of a renal corpuscle and a tubule.
  a. Each renal corpuscle comprises a capillary tuft, termed a glomerulus, and a Bowman's capsule that the tuft protrudes into.
  b. The tubule extends from Bowman's capsule and is subdivided into the proximal tubule, loop of Henle, distal convoluted tubule, and collecting-duct system. At the level of the collecting ducts, multiple tubules join and empty into the renal pelvis, from which urine flows through the ureters to the bladder.
  c. Each glomerulus is supplied by an afferent arteriole, and an efferent arteriole leaves the glomerulus to branch into peritubular capillaries, which supply the tubule.

Basic Renal Processes

I. The three basic renal processes are glomerular filtration, tubular reabsorption, and tubular secretion. In addition, the kidneys synthesize and/or catabolize certain substances. The excretion of a substance is equal to the amount filtered plus the amount secreted, minus the amount reabsorbed.
II. Urine formation begins with glomerular filtration—approximately 180 L/day—of essentially protein-free plasma into Bowman's space.
  a. Glomerular filtrate contains all plasma substances other than proteins (and substances bound to proteins) in virtually the same concentrations as in plasma.
  b. Glomerular filtration is driven by the hydrostatic pressure in the glomerular capillaries and is opposed by both the hydrostatic pressure in Bowman's space and the osmotic force due to the proteins in the glomerular capillary plasma.
III. As the filtrate moves through the tubules, certain substances are reabsorbed either by diffusion or by mediated transport.
  a. Substances to which the tubular epithelium is permeable are reabsorbed by diffusion because water reabsorption creates tubule-interstitium concentration gradients for them.
  b. Active reabsorption of a substance requires the participation of transporters in the luminal or basolateral membrane.
  c. Tubular reabsorption rates are very high for nutrients, ions, and water, but lower for waste products.

d. Many of the mediated transport systems exhibit transport maximums. When the filtered load of a substance exceeds the transport maximum, large amounts may appear in the urine.

IV. Tubular secretion, like glomerular filtration, is a pathway for the entrance of a substance into the tubule.

The Concept of Renal Clearance

I. The clearance of any substance can be calculated by dividing the mass of the substance excreted per unit time by the plasma concentration of the substance.

II. GFR can be measured by means of the inulin clearance and estimated by means of the creatinine clearance.

Micturition

I. In the basic micturition reflex, bladder distension stimulates stretch receptors that trigger spinal reflexes; these reflexes lead to contraction of the detrusor muscle, mediated by parasympathetic neurons, and relaxation of both the internal and the external urethral sphincters, mediated by inhibition of the neurons to these muscles.

II. Voluntary control is exerted via descending pathways to the parasympathetic nerves supplying the detrusor muscle, the sympathetic nerves supplying the internal urethral sphincter, and the motor nerves supplying the external urethral sphincter.

Additional Clinical Examples

I. Incontinence is the involuntary release of urine that occurs most commonly in the elderly (particularly women).

■ SECTION A KEY TERMS

■ SECTION A CLINICAL TERMS

■ SECTION A REVIEW QUESTIONS

1. What are the functions of the kidneys?
2. What three hormones/factors do the kidneys secrete into the blood?
3. Fluid flows in sequence through what structures from the glomerulus to the bladder? Blood flows through what structures from the renal artery to the renal vein?
4. What are the three basic renal processes that lead to the formation of urine?
5. How does the composition of the glomerular filtrate compare with that of plasma?
6. Describe the forces that determine the magnitude of the GFR. What is a normal value of GFR?
7. Contrast the mechanisms of reabsorption for glucose and urea. Which one shows a T_m?
8. Diagram the sequence of events leading to micturition.

SECTION B

Regulation of Ion and Water Balance

Total-Body Balance of Sodium and Water

Table 14–3 summarizes total-body water balance. These are average values, that are subject to considerable normal variation. There are two sources of body water gain: (1) water produced from the oxidation of organic nutrients, and (2) water ingested in liquids and food (a rare steak is approximately 70 percent water). There are four sites that lose water to the external environment: skin, respiratory airways, gastrointestinal tract, and urinary tract. Menstrual flow constitutes a fifth potential source of water loss in women.

The loss of water by evaporation from the skin and the lining of the respiratory passageways is a continuous process. It is called **insensible water loss** because the person is unaware of its occurrence. Additional water can be made available for

Table 14–3	Average Daily Water Gain and Loss in Adults	
Intake		
In liquids		1200 ml
In food		1000 ml
Metabolically produced		350 ml
Total		**2550 ml**
Output		
Insensible loss (skin and lungs)		900 ml
Sweat		50 ml
In feces		100 ml
Urine		1500 ml
Total		**2550 ml**

evaporation from the skin by the production of sweat. Normal gastrointestinal loss of water in feces is generally quite small, but it can be severe in diarrhea. Gastrointestinal water loss can also be significant in vomiting.

Table 14–4 is a summary of total-body balance for sodium chloride. The excretion of sodium and chloride via the skin and gastrointestinal tract is normally small but increases markedly during severe sweating, vomiting, or diarrhea. Hemorrhage can also result in the loss of large quantities of both salt and water.

Under normal conditions, as Tables 14–3 and 14–4 show, salt and water losses equal salt and water gains, and no net change in body salt and water occurs. This matching of losses and gains is primarily the result of the regulation of urinary loss, which can be varied over an extremely wide range. For example, urinary water excretion can vary from approximately 0.4 L/day to 25 L/day, depending upon whether one is lost in the desert or participating in a beer-drinking contest. Similarly, some individuals ingest 20 to 25 g of sodium chloride per day, whereas a person on a low-salt diet may ingest only 0.05 g. Normal kidneys can readily alter the excretion of salt over this range to balance loss with gain.

Table 14–4	Daily Sodium Chloride Intake and Loss
Intake	
Food	10.50 g
Output	
Sweat	0.25 g
Feces	0.25 g
Urine	10.00 g
Total	**10.50 g**

Basic Renal Processes for Sodium and Water

Both sodium and water freely filter from the glomerular capillaries into Bowman's space because they have low molecular weights and circulate in the plasma in the free form (unbound to protein). They both undergo considerable reabsorption—normally more than 99 percent (see Table 14–2)—but no secretion. Most renal energy utilization goes to accomplish this enormous reabsorptive task. The bulk of sodium and water reabsorption (about two-thirds) occurs in the proximal tubule, but the major hormonal control of reabsorption is exerted on the distal convoluted tubules and collecting ducts.

The mechanisms of sodium and water reabsorption can be summarized in two generalizations: (1) sodium reabsorption is an active process occurring in all tubular segments except the descending limb of the loop of Henle; and (2) water reabsorption is by diffusion and is dependent upon sodium reabsorption.

Primary Active Sodium Reabsorption

The essential feature underlying sodium reabsorption throughout the tubule is the primary active transport of sodium out of the cells and into the interstitial fluid, as illustrated for the proximal tubule and cortical collecting duct in **Figure 14–14**. This transport is achieved by Na^+/K^+-ATPase pumps in the basolateral membrane of the cells. The active transport of sodium out of the cell keeps the intracellular concentration of sodium low compared to the tubular lumen, so sodium moves "downhill" out of the lumen into the tubular epithelial cells.

The mechanism of the downhill sodium movement across the luminal membrane into the cell varies from segment to segment of the tubule depending on which channels and/or transport proteins are present in their luminal membranes. For example, the luminal entry step in the proximal tubule cell occurs by cotransport with a variety of organic molecules such as glucose, or by countertransport with hydrogen ions. In the latter case, hydrogen ions move out of the cell to the lumen as sodium moves into the cell (**Figure 14–14a**). Thus, in the proximal tubule, sodium reabsorption drives the reabsorption of the cotransported substances and secretion of hydrogen ions. The luminal entry step for sodium in the cortical collecting duct occurs primarily by diffusion through sodium channels (**Figure 14–14b**).

The movement of sodium downhill from lumen into cell across the *luminal membrane* varies from one segment of the tubule to another. By contrast, the *basolateral membrane* step is the same in all sodium-reabsorbing tubular segments—the primary active transport of sodium out of the cell is via Na^+/K^+-ATPase pumps in this membrane. It is this transport process that lowers intracellular sodium concentration and so makes possible the downhill luminal entry step.

Coupling of Water Reabsorption to Sodium Reabsorption

As sodium and chloride are reabsorbed, water follows passively by osmosis (Chapter 4). **Figure 14–15** summarizes this coupling of solute and water reabsorption. (1) Sodium is transported from

(a)

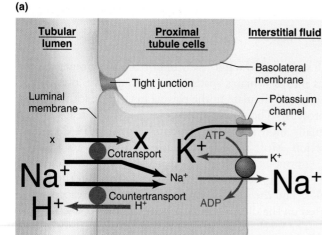

(b)

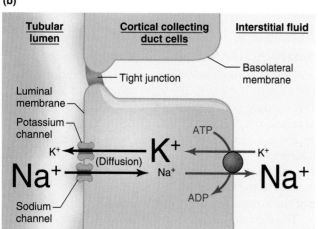

Figure 14–14

Mechanism of sodium reabsorption in the (a) proximal tubule and (b) cortical collecting duct. Figure 14–15 shows the movement of the reabsorbed sodium from the interstitial fluid into the peritubular capillaries. The sizes of the letters denote high and low concentrations. x represents organic molecules such as glucose and amino acids that are cotransported with Na^+. The fate of the potassium ions that the Na^+/K^+-ATPase pumps transport is discussed in the later section dealing with renal potassium handling.

Figure 14–14 physiological *inquiry* (p*i*)

- What would be the effect of a drug that blocks the sodium channels in the cortical collecting duct?

Answer can be found at end of chapter.

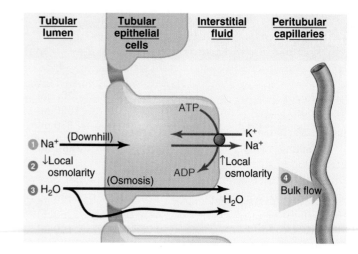

Figure 14–15

Coupling of water and sodium reabsorption. See text for explanation of circled numbers. The reabsorption of solutes other than sodium—for example, glucose, amino acids, and bicarbonate—also contributes to the difference in osmolarity between lumen and interstitial fluid, but the reabsorption of all these substances ultimately depends on direct or indirect cotransport and countertransport with sodium (see Figure 14–14a). Therefore, they are not shown in this figure.

adjacent to the cell (i.e., the local water concentration increases). At the same time, the appearance of solute in the interstitial fluid just outside the cell increases the local osmolarity (i.e., the local water concentration decreases). (3) The difference in water concentration between lumen and interstitial fluid causes net diffusion of water from the lumen across the tubular cells' plasma membranes and/or tight junctions into the interstitial fluid. (4) From there, water, sodium, and everything else dissolved in the interstitial fluid move together by bulk flow into peritubular capillaries as the final step in reabsorption.

Water movement across the tubular epithelium can only occur if the epithelium is permeable to water. No matter how large its concentration gradient, water cannot cross an epithelium impermeable to it. Water permeability varies from tubular segment to segment and depends largely on the presence of water channels, called **aquaporins,** in the plasma membranes. The water permeability of the proximal tubule is always very high, so this segment reabsorbs water molecules almost as rapidly as sodium ions. As a result, the proximal tubule reabsorbs large amounts sodium and water in the same proportions.

We will describe the water permeability of the next tubular segments—the loop of Henle and distal convoluted tubule—later. Now for the really crucial point: the water permeability of the last portions of the tubules, the cortical and medullary collecting ducts, can vary greatly due to physiological control. These are the only tubular segments in which water permeability is under such control.

The major determinant of this controlled permeability, and therefore, of passive water reabsorption in the collecting ducts, is a peptide hormone secreted by the posterior pituitary

the tubular lumen to the interstitial fluid across the epithelial cells. Other solutes, such as glucose, amino acids, and bicarbonate, whose reabsorption depends on sodium transport, also contribute to osmosis. (2) The removal of solutes from the tubular lumen lowers the local osmolarity of the tubular fluid

gland and known as **vasopressin,** or **antidiuretic hormone, (ADH).** Vasopressin stimulates the insertion into the luminal membrane of a particular group of aquaporin water channels made by the collecting duct cells. Thus, in the presence of a high plasma concentration of vasopressin, the water permeability of the collecting ducts increases dramatically. Therefore, water reabsorption is maximal, and the final urine volume is small—less than 1 percent of the filtered water.

Without vasopressin, the water permeability of the collecting ducts is extremely low, and very little water is reabsorbed from these sites. Therefore, a large volume of water remains behind in the tubule to be excreted in the urine. This increased urine excretion resulting from low vasopressin is termed **water diuresis. Diuresis** simply means a large urine flow from any cause. In a subsequent section, we will describe the control of vasopressin secretion.

The disease *diabetes insipidus,* which is distinct from the other kind of diabetes (diabetes mellitus, or "sugar diabetes"), illustrates what happens when the vasopressin system malfunctions. Diabetes insipidus is caused by the failure of the posterior pituitary to release vasopressin (*central diabetes insipidus*), or the inability of the kidney to respond to vasopressin (*nephrogenic diabetes insipidus*). Regardless of the type of diabetes insipidus, the permeability to water of the collecting ducts is low even if the patient is dehydrated. A constant water diuresis is present that can be as much as 25 L/day.

Note that in water diuresis, there is an increased urine flow, but not an increased solute excretion. In all other cases of diuresis, termed **osmotic diuresis,** the increased urine flow is the result of a primary increase in solute excretion. For example, failure of normal sodium reabsorption causes both increased sodium excretion and increased water excretion, because, as we have seen, water reabsorption is dependent on solute reabsorption. Another example of osmotic diuresis occurs in people with uncontrolled *diabetes mellitus:* In this case, the glucose that escapes reabsorption because of the huge filtered load retains water in the lumen, causing it to be excreted along with the glucose.

To summarize, any loss of solute in the urine must be accompanied by water loss (osmotic diuresis), but the reverse is not true. That is, water diuresis is not necessarily accompanied by equivalent solute loss.

Urine Concentration: The Countercurrent Multiplier System

Before reading this section you should review, by looking up in the glossary, several terms presented in Chapter 4— **hypoosmotic, isoosmotic,** and **hyperosmotic.**

In the section just concluded, we described how the kidneys produce a small volume of urine when the plasma concentration of vasopressin is high. Under these conditions, the urine is concentrated (hyperosmotic) relative to plasma. This section describes the mechanisms by which this hyperosmolarity is achieved.

The ability of the kidneys to produce hyperosmotic urine is a major determinant of the ability to survive with limited water intake. The human kidney can produce a maximal urinary concentration of 1400 mOsmol/L, almost five times the osmolarity of plasma, which is typically in the range of 285 to 300 mOsmol/L (rounded off to 300 mOsmol/L for convenience). The typical daily excretion of urea, sulfate, phosphate, other waste products, and ions amounts to approximately 600 mOsmol. Therefore, the minimal volume of urine water in which this mass of solute can be dissolved equals

$$\frac{600 \text{ mOsmol/day}}{1400 \text{ mOsmol/L}} = 0.444 \text{ L/day}$$

This volume of urine is known as the **obligatory water loss.** The loss of this minimal volume of urine contributes to dehydration when water intake is zero.

Urinary concentration takes place as tubular fluid flows through the *medullary* collecting ducts. The interstitial fluid surrounding these ducts is very hyperosmotic. In the presence of vasopressin, water diffuses out of the ducts into the interstitial fluid of the medulla and then enters the blood vessels of the medulla to be carried away.

The key question is: how does the medullary interstitial fluid become hyperosmotic? The answer involves several interrelated factors: 1. The countercurrent anatomy of the loop of Henle of juxtamedullary nephrons; 2. Reabsorption of NaCl in the ascending limbs of those loops of Henle; 3. Impermeability of those ascending limbs to water; 4. Trapping of urea in the medulla; and 5. Hairpin loops of vasa recta to minimize washout of the hyperosmotic medulla. Recall that Henle's loop forms a hairpin-like loop between the proximal tubule and the distal convoluted tubule (see Figure 14–2). The fluid entering the loop from the proximal tubule flows down the descending limb, turns the corner, and then flows up the ascending limb. The opposing flows in the two limbs is termed a countercurrent flow, and the entire loop functions as a **countercurrent multiplier system** to create a hyperosmotic medullary interstitial fluid.

Because the proximal tubule always reabsorbs sodium and water in the same proportions, the fluid entering the descending limb of the loop from the proximal tubule has the same osmolarity as plasma—300 mOsmol/L. For the moment, let's skip the descending limb because the events in it can only be understood in the context of what the *ascending* limb is doing. Along the entire length of the ascending limb, sodium and chloride are reabsorbed from the lumen into the medullary interstitial fluid (**Figure 14–16a**). In the upper (thick) portion of the ascending limb, this reabsorption is achieved by transporters that actively cotransport sodium and chloride. Such transporters are not present in the lower (thin) portion of the ascending limb, so the reabsorption there is a passive process. For simplicity in the explanation of the countercurrent multiplier, we shall treat the entire ascending limb as a homogeneous structure that actively reabsorbs sodium and chloride.

Very importantly, *the ascending limb is relatively impermeable to water,* so that little water follows the salt. The net result is that the interstitial fluid of the medulla becomes hyperosmotic compared to the fluid in the ascending limb because solute is reabsorbed without water.

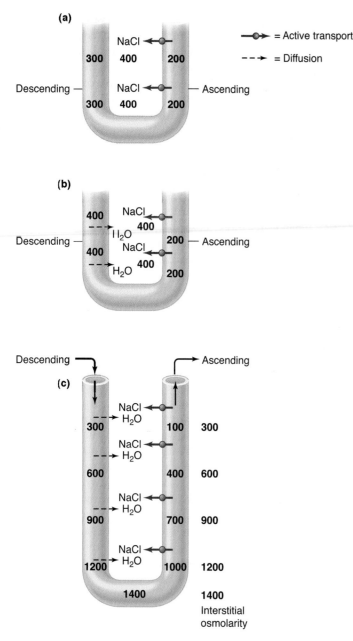

Figure 14–16

Generating a hyperosmolar medullary renal interstitium.(a) NaCl active transport in ascending limbs (impermeable to H_2O); (b) passive reabsorption of H_2O in descending limb; (c) add flow of fluid in lumen resulting in multiplication of osmolarity.

Now back to the descending limb. This segment, in contrast to the ascending limb, does not reabsorb sodium chloride and is highly permeable to water (**Figure 14–16b**). Therefore, a net diffusion of water occurs out of the descending limb into the more concentrated interstitial fluid until the osmolarities inside this limb and in the interstitial fluid are again equal. The interstitial hyperosmolarity is maintained during this equilibration because the ascending limb continues to pump

sodium chloride to maintain the concentration difference between it and the interstitial fluid.

Therefore, because of the diffusion of water, the osmolarities of the descending limb and interstitial fluid become equal, and both are higher—by 200 mOsmol/L in our example—than that of the ascending limb. This is the essence of the system: the loop countercurrent multiplier causes the interstitial fluid of the medulla to become concentrated. It is this hyperosmolarity that will draw water out of the collecting ducts and concentrate the urine. However, one more crucial feature—the "multiplication"—must be considered.

So far we have been analyzing this system as though the flow through the loop of Henle stops while the ion pumping and water diffusion are occurring. Now, let us see what happens when we allow flow through the entire length of the descending and ascending limbs of the loop of Henle (**Figure 14–16c**). The osmolarity difference—200 mOsmol/L—that exists at each horizontal level is "multiplied" as the fluid goes deeper into the medulla. By the time the fluid reaches the bend in the loop, the osmolarity of the tubular fluid and interstitium has been multiplied to a very high osmolarity that can be as high as 1400 mOsmol/L. Keep in mind that the active sodium chloride transport mechanism in the ascending limb (coupled with low water permeability in this segment) is the essential component of the system. Without it, the countercurrent flow would have no effect on loop and medullary interstitial osmolarity, which would simply remain 300 mOsmol/L throughout.

Now we have a concentrated medullary interstitial fluid, but we must still follow the fluid within the tubules from the loop of Henle through the distal convoluted tubule and into the collecting duct system, using **Figure 14–17** as our guide. Furthermore, urea reabsorption and trapping (described in detail later) contributes to the maximal medullary interstitial osmolarity. The countercurrent multiplier system concentrates the descending-loop fluid, but then lowers the osmolarity in the ascending loop so that the fluid entering the distal convoluted tubule is actually more dilute (hypoosmotic)—100 mOsmol/L in Figure 14–17—than the plasma. The fluid becomes even more dilute during its passage through the distal convoluted tubule because this tubular segment, like the ascending loop, actively transports sodium and chloride out of the tubule but is relatively impermeable to water. This hypoosmotic fluid then enters the cortical collecting duct. Because of the significant volume reabsorption, the flow of fluid at the end of the ascending limb is much less than the flow that entered the descending limb.

As noted earlier, vasopressin increases tubular permeability to water in both the cortical and medullary collecting ducts. In contrast, vasopressin does not influence water reabsorption in the parts of the tubule prior to the collecting ducts. Thus, regardless of the plasma concentration of this hormone, the fluid entering the cortical collecting duct is hypoosmotic. From there on, however, vasopressin is crucial. In the presence of high levels of vasopressin, water reabsorption occurs by diffusion from the hypoosmotic fluid in the cortical collecting duct until the fluid in this

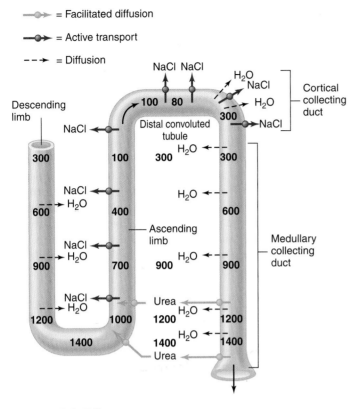

= Facilitated diffusion

= Active transport

= Diffusion

Figure 14–17

Simplified depiction of the generation of an interstitial fluid osmolarity gradient by the renal countercurrent multiplier system and its role in the formation of hyperosmotic urine in the presence of vasopressin. Notice that the hyperosmotic medulla depends on NaCl reabsorption and urea trapping (described in Figure 14–19).

Figure 14–17 physiological *inquiry* 🅿️*i*

■ Certain types of lung tumors secrete one or more hormones. What would happen to plasma and urine osmolarity and urine volume in a patient with a lung tumor that secretes vasopressin?

Answer can be found at end of chapter.

segment becomes isoosmotic to the interstitial fluid and peritubular plasma of the cortex—that is, until it is once again at 300 mOsmol/L.

The isoosmotic tubular fluid then enters and flows through the *medullary* collecting ducts. In the presence of high plasma concentrations of vasopressin, water diffuses out of the ducts into the medullary interstitial fluid as a result of the high osmolarity that the loop countercurrent multiplier system and urea trapping establish there. This water then enters the medullary capillaries and is carried out of the kidneys by the venous blood. Water reabsorption occurs all along the lengths of the medullary collecting ducts so that, in the presence of vasopressin, the fluid at the end of these ducts has essentially the same osmolarity as the interstitial fluid surrounding the bend in the loops—that is, at the bottom of the medulla. By this means, the final urine is hyperosmotic. By

retaining as much water as possible, the kidneys minimize the rate at which dehydration occurs during water deprivation.

In contrast, when plasma vasopressin concentration is low, both the cortical and medullary collecting ducts are relatively impermeable to water. As a result, a large volume of hypoosmotic urine is excreted, thereby eliminating an excess of water in the body.

The Medullary Circulation

A major question arises with the countercurrent system as described previously: Why doesn't the blood flowing through medullary capillaries eliminate the countercurrent gradient set up by the loops of Henle? One would think that as plasma with the usual osmolarity of 300 mOsm/L enters the highly concentrated environment of the medulla, there would be massive net diffusion of sodium and chloride into the capillaries and water out of them, and thus the interstitial gradient would be "washed away." However, the blood vessels in the medulla (vasa recta) form hairpin loops that run parallel to the loops of Henle and medullary collecting ducts. As shown in **Figure 14–18**, blood enters the top of the vessel loop at an osmolarity of 300 mOsm/L, and as the blood flows down the loop deeper and deeper into the medulla, sodium and chloride do indeed diffuse into, and water out of, the vessel. However, after the bend in the loop is reached, the blood then flows up the ascending vessel loop, where the process

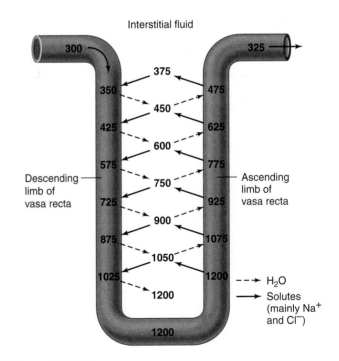

Figure 14–18

Function of the vasa recta to maintain the hypertonic interstitial renal medulla. All movements of water and solutes are by diffusion. Not shown is the simultaneously occurring uptake of interstitial fluid by bulk flow.

is almost completely reversed. Thus, the hairpin-loop structure of the vasa recta minimizes excessive loss of solute from the interstitium by *diffusion*. At the same time, both the salt and water being reabsorbed from the loops of Henle and collecting ducts are carried away in equivalent amounts by *bulk flow*, as determined by the usual capillary Starling forces. This maintains the steady-state countercurrent gradient set up by the loops of Henle. Because of NaCl and water reabsorbed from the loop of Henle and collecting ducts, the amount of blood flow leaving the vasa recta is at least two-fold higher than the blood flow entering the vasa recta. Finally the total blood flow going through all of the vasa recta is a small percentage of the total renal blood flow. This helps to minimize the washout of the hypertonic interstitium of the medulla.

The Recycling of Urea Helps to Establish a Hypertonic Medullary Interstitium

As was just described, the countercurrent multiplier establishes a hypertonic medullary interstitium that the vasa recta help to preserve. We already learned how the reabsorption of water in the proximal tubule mediates the reabsorption of urea by diffusion. As urea passes through the remainder of the nephron, it is reabsorbed, secreted into the tubule, and then reabsorbed again (**Figure 14–19**). This traps urea, an osmotically active molecule, in the medullary interstitium, thus increasing its osmolarity. In fact, as shown in Figure 14–17, urea contributes to the total osmolarity of the renal medulla.

Urea is freely filtered in the glomerulus. Approximately 50 percent of the filtered urea is reabsorbed in the proxi-

mal tubule, and the remaining 50 percent enters the loop of Henle. In the thin descending and ascending limbs of the loop of Henle, urea that has accumulated in the medullary interstitium is secreted back into the tubular lumen by facilitated diffusion. Therefore, virtually all of the urea that was originally filtered in the glomerulus is present in the fluid that enters the distal tubule. Some of the original urea is reabsorbed from the distal tubule and cortical collecting duct. Thereafter, about half of the urea is reabsorbed from the *medullary* collecting duct, whereas only 5 percent diffuses into the vasa recta. The remaining amount is secreted back into the loop of Henle. Fifteen percent of the urea originally filtered remains in the collecting duct and is excreted in the urine. This recycling of urea through the medullary interstitium and minimal uptake by the vasa recta traps urea there and contributes to the high osmolarity shown in Figure 14–17.

Renal Sodium Regulation

In healthy individuals, urinary sodium excretion increases when there is an excess of sodium in the body and decreases when there is a sodium deficit. These homeostatic responses are so precise that total-body sodium normally varies by only a few percent despite a wide range of sodium intakes and the occasional occurrence of large losses via the skin and gastrointestinal tract.

As we have seen, sodium is freely filterable from the glomerular capillaries into Bowman's space and is actively reabsorbed, but not secreted. Therefore:

Sodium excreted = Sodium filtered – Sodium reabsorbed

The body can adjust sodium excretion by changing both processes on the right of the equation. Thus, for example, when total-body sodium decreases for any reason, sodium excretion decreases below normal levels because sodium reabsorption increases.

The first issue in understanding the responses controlling sodium reabsorption is to determine what inputs initiate them; that is, what variables are receptors actually sensing? Surprisingly, there are no important receptors capable of detecting the total amount of sodium in the body. Rather, the responses that regulate urinary sodium excretion are initiated mainly by various cardiovascular baroreceptors, such as the carotid sinus, and by sensors in the kidney that monitor the filtered load of sodium.

As described in Chapter 12, baroreceptors respond to pressure changes within the cardiovascular system and initiate reflexes that rapidly regulate these pressures by acting on the heart, arterioles, and veins. The new information in this chapter is that *regulation of cardiovascular pressures by baroreceptors also simultaneously achieves regulation of total-body sodium.*

Sodium is the major extracellular solute constituting, along with associated anions, approximately 90 percent of these solutes. Thus, changes in total-body sodium result in similar changes in extracellular volume. Because extracellu-

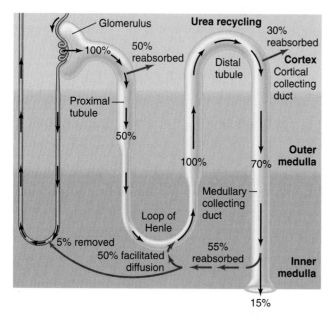

Figure 14–19

Urea recycling. The recycling of urea "traps" urea in the inner medulla, which increases osmolarity and helps to establish and maintain hypertonicity.

lar volume comprises plasma volume and interstitial volume, plasma volume is also directly related to total-body sodium. We saw in Chapter 12 that plasma volume is an important determinant of the blood pressures in the veins, cardiac chambers, and arteries. Thus, the chain linking total-body sodium to cardiovascular pressures is completed: low total-body sodium leads to low plasma volume, which leads to low cardiovascular pressures. These low pressures, via baroreceptors, initiate reflexes that influence the renal arterioles and tubules so as to lower GFR and increase sodium reabsorption. These latter events decrease sodium excretion, thereby retaining sodium in the body and preventing further decreases in plasma volume and cardiovascular pressures. Increases in total-body sodium have the reverse reflex effects.

To summarize, the amount of sodium in the body determines the extracellular fluid volume, the plasma volume component of which helps determine cardiovascular pressures, which initiate the responses that control sodium excretion.

Control of GFR

Figure 14–20 summarizes the major mechanisms by which an example of increased sodium loss elicits a decrease in GFR. The main direct cause of the reduced GFR is a reduced net glomerular filtration pressure. This occurs both as a consequence of a lowered arterial pressure in the kidneys and, more importantly, as a result of reflexes acting on the renal arterioles. Note that these reflexes are the basic baroreceptor reflexes described in Chapter 12—a decrease in cardiovascular pressures causes neurally mediated reflex vasoconstriction in many areas of the body. As we will see later, the hormones angiotensin II and vasopressin also participate in this renal vasoconstrictor response.

Conversely, an increase in GFR is usually elicited by neuroendocrine inputs when an increased total-body sodium level increases plasma volume. This increased GFR contributes to the increased renal sodium loss that returns extracellular volume to normal.

Control of Sodium Reabsorption

For the long-term regulation of sodium excretion, the control of sodium reabsorption is more important than the control of GFR. The major factor determining the rate of tubular sodium reabsorption is the hormone aldosterone.

Aldosterone and the Renin-Angiotensin System

The adrenal cortex produces a steroid hormone, **aldosterone,** which stimulates sodium reabsorption by the distal convoluted tubule and the cortical collecting ducts. An action affecting these late portions of the tubule is just what one would expect for a fine-tuning input because most of the filtered sodium has been reabsorbed by the time the filtrate reaches the distal parts of the nephron. When aldosterone is completely absent, approximately 2 percent of the filtered sodium (equivalent to 35 g of sodium chloride per day) is not reabsorbed, but excreted. In contrast, when the plasma concentration of aldosterone is high, essentially all the sodium reaching the distal tubule and cortical collecting ducts is reabsorbed. Normally,

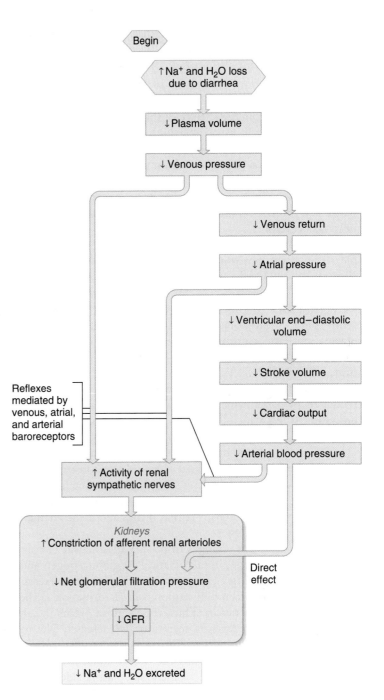

Figure 14–20

Direct and neurally mediated reflex pathways by which the GFR and thus sodium and water excretion decrease when plasma volume decreases.

the plasma concentration of aldosterone and the amount of sodium excreted lie somewhere between these extremes.

As opposed to vasopressin, which is a peptide and acts quickly, aldosterone is a steroid and acts more slowly because it induces changes in gene expression and protein synthesis. In the case of the nephron, the proteins participate in sodium transport. Look again at Figure 14–14b. Aldosterone induces the synthesis of all the channels and pumps shown in the cortical collecting duct. By this same mechanism, aldosterone also stimulates sodium absorption from the lumens of both the large intestine and the ducts carrying fluid from the sweat glands and salivary glands. In this manner, less sodium is lost in the feces and from the surface of the skin in sweat.

When a person eats a diet high in sodium, aldosterone secretion is low, whereas it is high when the person ingests a low-sodium diet or becomes sodium-depleted for some other reason. What controls the secretion of aldosterone under these circumstances? The answer is the hormone angiotensin II, which acts directly on the adrenal cortex to stimulate the secretion of aldosterone.

Angiotensin II is a component of the hormonal complex termed the **renin-angiotensin system,** summarized in **Figure 14–21. Renin** (pronounced REE-nin) is an enzyme secreted by the juxtaglomerular cells of the juxtaglomerular apparatuses in the kidneys. Once in the bloodstream, renin splits a small polypeptide, **angiotensin I,** from a large plasma protein, **angiotensinogen,** which is produced by the liver. Angiotensin I, a biologically inactive peptide, then undergoes further cleavage to form the active agent of the renin-angiotensin system, angiotensin II. This conversion is mediated by an enzyme known as **angiotensin-converting enzyme (ACE),** which is found in very high concentration on the luminal surface of capillary endothelial cells. Angiotensin II exerts many effects, but the most important are the stimulation of the secretion of aldosterone and the constriction of arterioles (described in Chapter 12). Plasma angiotensin II is high during salt depletion and low when salt intake is high. It is this change in angiotensin II that brings about the changes in aldosterone secretion.

What causes the changes in plasma angiotensin II concentration with changes in salt balance? Angiotensinogen and angiotensin-converting enzyme are usually present in excess, so the rate-limiting factor in angiotensin II formation is the plasma renin concentration. Thus, the chain of events in salt depletion is increased renin secretion → increased plasma renin concentration → increased plasma angiotensin I concentration → increased plasma angiotensin II concentration → increased aldosterone release → increased plasma aldosterone concentration.

What are the mechanisms by which sodium depletion causes an increase in renin secretion (**Figure 14–22**)? There are at least three distinct inputs to the juxtaglomerular cells: (1) the renal sympathetic nerves, (2) intrarenal baroreceptors, and (3) the macula densa.

The renal sympathetic nerves directly innervate the juxtaglomerular cells, and an increase in the activity of these nerves stimulates renin secretion. This makes sense because these nerves are reflexly activated via baroreceptors whenever a reduction in body sodium (and, therefore, plasma volume) lowers cardiovascular pressures (see Figure 14–20).

The other two inputs for controlling renin release—intrarenal baroreceptors and the macula densa—are contained within the kidneys and require no external neuroendocrine input (although such input can influence them). As noted earlier, the juxtaglomerular cells are located in the walls of the afferent arterioles. They are sensitive to the pressure within these arterioles, and so function as **intrarenal baroreceptors.** When blood pressure in the kidneys decreases, as occurs when plasma volume is decreased, these cells are stretched less and, therefore, secrete more renin (see Figure 14–22). Thus, the juxtaglomerular cells respond simultaneously to the combined effects of sympathetic input, triggered by baroreceptors external to the kidneys, and to their own pressure sensitivity.

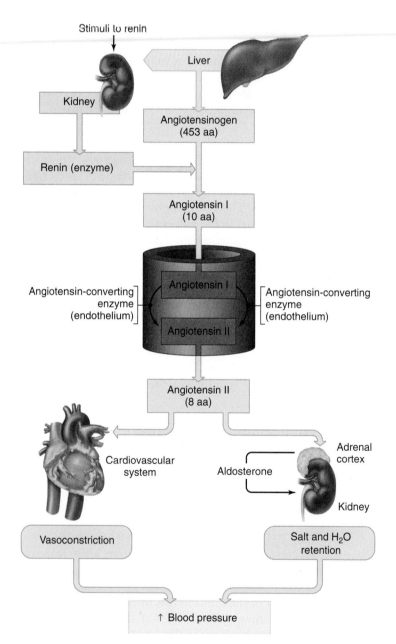

Figure 14–21

Summary of the renin-angiotensin system and the stimulation of aldosterone secretion by angiotensin II. Angiotensin-converting enzyme is located on the surface of capillary endothelial cells. The plasma concentration of renin is the rate-limiting factor in the renin-angiotensin system; that is, it is the major determinant of the plasma concentration of angiotensin II.

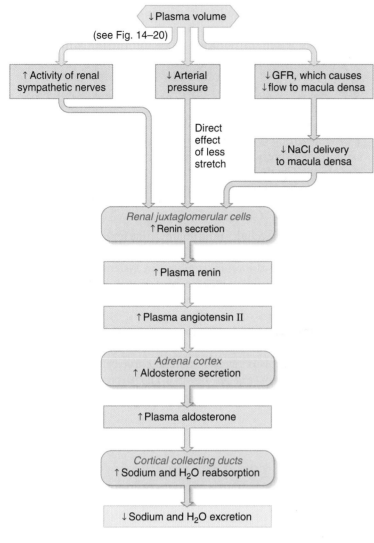

Figure 14–22

Pathways by which decreased plasma volume leads, via the renin-angiotensin system and aldosterone, to increased sodium reabsorption by the cortical collecting ducts and hence to decreased sodium excretion.

Figure 14–22 physiological *inquiry* (pi)

■ What would be the effect of denervation (removal of sympathetic neural input) of the kidneys on sodium and water excretion?

Answer can be found at end of chapter.

The other internal input to the juxtaglomerular cells is via the macula densa, which, as noted earlier, is located near the ends of the ascending loops of Henle (see Figure 14–2). The macula densa senses the sodium concentration in the tubular fluid flowing past it. A decreased salt concentration causes increased renin secretion. Therefore, in an indirect way, this mechanism is sensitive to changes in sodium intake. If salt intake is low, less sodium is filtered and less appears at the macula densa. Conversely, a high salt intake will cause a very low rate of release of renin. In addition, macula densa sodium

concentrations tend to decrease when arterial pressure is decreased due to decreased GFR and, therefore, tubular flow rate. This input therefore also results in increased renin release at the same time that the sympathetic nerves and intrarenal baroreceptors are doing so (see Figure 14–22).

The importance of this system is highlighted by the considerable redundancy in the control of renin secretion. Furthermore, as illustrated in Figure 14–22, the various mechanisms can all be participating at the same time.

By helping to regulate sodium balance and thereby plasma volume, the renin-angiotensin system contributes to the control of arterial blood pressure. However, this is not the only way in which it influences arterial pressure. Recall from Chapter 12 that angiotensin II is a potent constrictor of arterioles all over the body and that this effect on peripheral resistance increases arterial pressure.

Drugs have been developed to manipulate the angiotensin II and aldosterone components of the system. ACE inhibitors such as lisinopril reduce angiotensin II production from angiotensin I by inhibiting angiotensin-converting enzyme. Angiotensin II receptor blockers such as losartan prevent angiotensin II from binding to its receptor on target tissue (e.g., vascular smooth muscle and the adrenal cortex). Finally, there are drugs such as eplerenone that block the binding of aldosterone to its receptor in the kidney. Although these classes of drugs have different mechanisms of action, they are all effective in the treatment of hypertension. This highlights that many forms of hypertension can be attributed to the failure of the kidney to adequately excrete sodium and water.

Atrial Natriuretic Peptide

Another controller is **atrial natriuretic peptide (ANP),** also known as atrial natriuretic factor (ANF) or hormone (ANH). Cells in the cardiac atria synthesize and secrete ANP. ANP acts on several tubular segments to inhibit sodium reabsorption. It can also act on the renal blood vessels to increase GFR, which further contributes to increased sodium excretion. ANP also directly inhibits aldosterone secretion, which leads to an increase in sodium excretion. As would be predicted, the secretion of ANP increases when there is an excess of sodium in the body, but the stimulus for this increased secretion is not alterations in sodium concentration. Rather, using the same logic (only in reverse) that applies to the control of renin and aldosterone secretion, ANP secretion increases because of the expansion of plasma volume that accompanies an increase in body sodium. The specific stimulus is increased atrial distension (**Figure 14–23**).

Interaction of Blood Pressure and Renal Function

An important input controlling sodium reabsorption is arterial blood pressure. We have previously described how the arterial blood pressure constitutes a signal for important reflexes (involving the renin-angiotensin system and aldosterone) that influence sodium reabsorption. Now we are emphasizing that arterial pressure also acts locally on the tubules themselves. Specifically, an *increase* in arterial pressure *inhibits* sodium reabsorption and thereby increases sodium excretion in a process

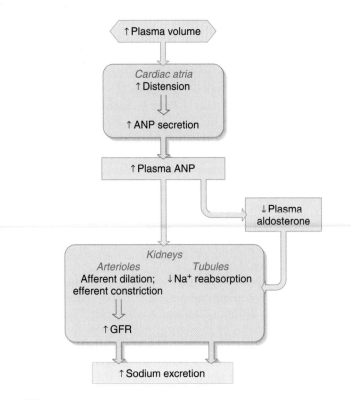

Figure 14–23

Atrial natriuretic peptide (ANP) increases sodium excretion.

previous section tend to have the same effects on water excretion as on sodium excretion. As is true for sodium, however, the rate of water reabsorption from the tubules is the most important factor determining how much water is excreted. As we have seen, this is determined by vasopressin, and so total-body water is regulated mainly by reflexes that alter the secretion of this hormone.

As described in Chapter 11, vasopressin is produced by a discrete group of hypothalamic neurons whose axons terminate in the posterior pituitary, which releases vasopressin into the blood. The most important of the inputs to these neurons come from baroreceptors and osmoreceptors.

Baroreceptor Control of Vasopressin Secretion

A decreased extracellular fluid volume, due for example, to diarrhea or hemorrhage, elicits an increase in aldosterone release via activation of the renin-angiotensin system. However, the decreased extracellular volume also triggers an increase in vasopressin secretion. This increased vasopressin increases the water permeability of the collecting ducts. More water is passively reabsorbed and less is excreted, so water is retained to help stabilize the extracellular volume.

This reflex is initiated by several baroreceptors in the cardiovascular system (**Figure 14–24**). The baroreceptors decrease their rate of firing when cardiovascular pressures decrease, as

termed **pressure natriuresis** ("natriuresis" means increased urinary sodium loss). The actual transduction mechanism of this direct effect is unknown.

Thus, an increased blood pressure reduces sodium reabsorption by two mechanisms: It reduces the activity of the renin-angiotensin-aldosterone system, and it also acts locally on the tubules. Conversely, a decreased blood pressure decreases sodium excretion both by stimulating the renin-angiotensin-aldosterone system and by acting on the tubules to enhance sodium reabsorption.

Now is a good time to look back at Figure 12–57, which describes the strong causal, reciprocal relationship between arterial blood pressure and blood volume, the result of which is that blood volume is perhaps the major long-term determinant of blood pressure. The direct effect of blood pressure on sodium excretion is, as Figure 12–57 shows, one of the major links in these relationships. An important hypothesis is that most people who develop hypertension do so because their kidneys, for some reason, do not excrete enough sodium in response to a normal arterial pressure. Consequently, at this normal pressure, some dietary sodium is retained, which causes the pressure to rise enough to produce adequate sodium excretion to balance sodium intake, although at an increased body sodium content.

Renal Water Regulation

Water excretion is the difference between the volume of water filtered (the GFR) and the volume reabsorbed. Thus, the changes in GFR initiated by baroreceptor afferent input described in the

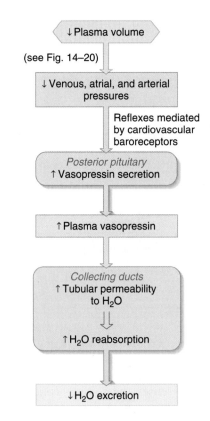

Figure 14–24

Baroreceptor pathway by which vasopressin secretion increases when plasma volume decreases. The opposite events (culminating in a decrease in vasopressin secretion) occur when plasma volume increases.

occurs when blood volume decreases. Therefore, the baroreceptors transmit fewer impulses via afferent neurons and ascending pathways to the hypothalamus, and the result is increased vasopressin secretion. Conversely, increased cardiovascular pressures cause more firing by the baroreceptors, resulting in a decrease in vasopressin secretion. The mechanism of this inverse relationship is an inhibitory neurotransmitter released by nerves in the afferent pathway.

In addition to its effect on water excretion, vasopressin, like angiotensin II, causes widespread arteriolar constriction. This helps restore arterial blood pressure toward normal (Chapter 12).

The baroreceptor reflex for vasopressin, as just described, has a relatively high threshold—that is, there must be a sizable reduction in cardiovascular pressures to trigger it. Therefore, this reflex, compared to the osmoreceptor reflex described next, generally plays a lesser role under most physiological circumstances, but it can become very important in pathological states such as hemorrhage.

Osmoreceptor Control of Vasopressin Secretion

We have seen how changes in extracellular volume simultaneously elicit reflex changes in the excretion of *both* sodium and water. This is adaptive because the situations causing extracellular volume alterations are very often associated with loss or gain of both sodium and water in proportional amounts. In contrast, changes in total-body water with no corresponding change in total-body sodium are compensated for by altering water excretion *without altering sodium excretion*.

A crucial point in understanding how such reflexes are initiated is realizing that changes in water alone, in contrast to sodium, have relatively little effect on extracellular volume. The reason is that water, unlike sodium, distributes throughout all the body fluid compartments, with about two-thirds entering the intracellular compartment rather than simply staying in the extracellular compartment, as sodium does. Therefore, cardiovascular pressures and baroreceptors are only slightly affected by pure water gains or losses. In contrast, the major effect of water loss or gain out of proportion to sodium loss or gain is a change in the osmolarity of the body fluids. This is a key point because, under conditions due predominantly to water gain or loss, the receptors that initiate the reflexes controlling vasopressin secretion are **osmoreceptors** in the hypothalamus. These receptors are responsive to changes in osmolarity.

As an example, imagine that you drink 2 L of water. The excess water lowers the body fluid osmolarity, which results in an inhibition of vasopressin secretion via the hypothalamic osmoreceptors (**Figure 14–25**). As a result, the water permeability of the collecting ducts decreases dramatically, water is not reabsorbed from these segments, and a large volume of hypoosmotic urine is excreted. In this manner, the excess water is eliminated.

At the other end of the spectrum, when the osmolarity of the body fluids increases because of water deprivation, vasopressin secretion is reflexly increased via the osmoreceptors, water reabsorption by the collecting ducts increases, and a very small volume of highly concentrated urine is excreted. By

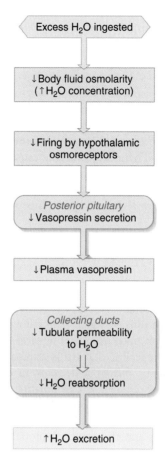

Figure 14–25

Osmoreceptor pathway that decreases vasopressin secretion and increases water excretion when excess water is ingested. The opposite events (an increase in vasopressin secretion) occur when osmolarity increases, as during water deprivation.

retaining relatively more water than solute, the kidneys help reduce the body fluid osmolarity back toward normal.

To summarize, regulation of body fluid osmolarity requires separation of water excretion from sodium excretion. That is, it requires the kidneys to excrete a urine that, relative to plasma, either contains more water than sodium and other solutes (water diuresis) or less water than solute (concentrated urine). This is made possible by two physiological factors: (1) osmoreceptors and (2) vasopressin-dependent water reabsorption without sodium reabsorption in the collecting ducts.

We have now described two afferent pathways controlling the vasopressin-secreting hypothalamic cells, one from baroreceptors and one from osmoreceptors. To add to the complexity, the hypothalamic cells receive synaptic input from many other brain areas, so that vasopressin secretion, and therefore urine volume and concentration, can be altered by pain, fear, and a variety of drugs. For example, ethanol inhibits vasopressin release, and this may account for the increased urine volume produced following the ingestion of alcohol, a urine volume well in excess of the volume of the beverage consumed.

A Summary Example: The Response to Sweating

Figure 14–26 shows the factors that control renal sodium and water excretion in response to severe sweating. You may notice the salty taste of sweat on your upper lip when you exercise. Sweat does contain sodium and chloride in addition to water, but is actually hypoosmotic compared to the body fluids from which it is derived. Therefore, sweating causes both a decrease in extracellular volume and an increase in body fluid osmolarity. The renal retention of water and sodium minimizes the deviations from normal caused by the loss of water and salt in the sweat.

Thirst and Salt Appetite

Deficits of salt and water must eventually be compensated for by ingestion of these substances, because the kidneys cannot create new sodium ions or water. The kidneys can only minimize their excretion until ingestion replaces the losses.

The subjective feeling of thirst is stimulated by an increase in plasma osmolarity and by a decrease in extracellular fluid volume (**Figure 14–27**). Plasma osmolarity is the single most important stimulus under normal physiological conditions. These are precisely the same two changes that stimulate vasopressin production, and the osmoreceptors and baroreceptors that control vasopressin secretion are the same as those for thirst. The brain centers that receive input from these receptors and that mediate thirst are located in the hypothalamus, very close to those areas that synthesize vasopressin.

Another influencing factor may be angiotensin II, which stimulates thirst by a direct effect on the brain, at least in experimental animals. Thus, the renin-angiotensin system may help regulate not only sodium balance but water balance as well, and constitutes one of the pathways by which thirst is stimulated when extracellular volume is decreased.

There are still other pathways controlling thirst. For example, dryness of the mouth and throat causes thirst, which is relieved by merely moistening them. Some kind of "metering" of water intake by other parts of the gastrointestinal tract also occurs. For example, a thirsty person given access to water stops drinking after replacing the lost water. This occurs well before most of the water has been absorbed from the gastrointestinal tract and has a chance to eliminate the stimulatory inputs to the systemic baroreceptors and osmoreceptors. This is probably mediated by afferent sensory nerves from the mouth, throat, and gastrointestinal tract, and probably prevents overhydration.

Salt appetite is an important part of sodium homeostasis and consists of two components: "hedonistic" appetite and "regulatory" appetite. Most mammals "like" salt and eat it whenever they can, regardless of whether they are salt-deficient. Human beings have a strong hedonistic appetite for salt, as manifested by almost universally large intakes of salt whenever it is cheap and readily available. For example, the average American consumes 10–15 g/day despite the fact that human beings can survive quite normally on less than 0.5 g/day. However, humans have relatively little regulatory salt appetite, at least until a bodily salt deficit becomes extremely large.

Potassium Regulation

Potassium is the most abundant intracellular ion. Although only 2 percent of total-body potassium is in the extracellular fluid, the potassium concentration in this fluid is extremely important for the function of excitable tissues, notably nerve

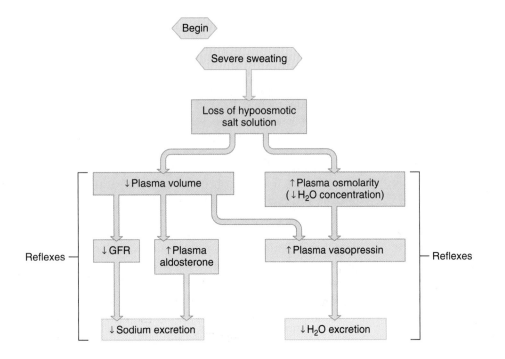

Figure 14–26

Pathways by which sodium and water excretion decrease in response to severe sweating. This figure is an amalgamation of Figures 14–20, 14–22, 14–24, and the reverse of 14–25.

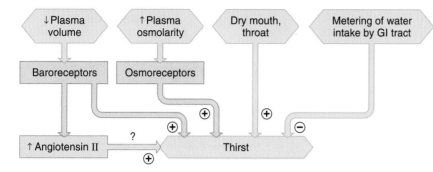

Figure 14–27

Inputs controlling thirst. The osmoreceptor input is the single most important stimulus under most physiological conditions. Psychological factors and conditioned responses are not shown. The question mark (?) indicates that evidence for the effects of angiotensin II on thirst comes primarily from experimental animals.

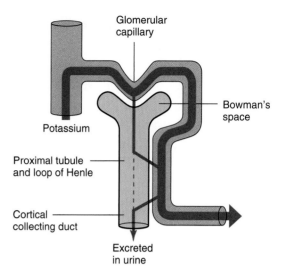

Figure 14–28

Simplified model of the basic renal processing of potassium.

and muscle. Recall from Chapter 6 that the resting membrane potentials of these tissues are directly related to the relative intracellular and extracellular potassium concentrations. Consequently, either increases (***hyperkalemia***) or decreases (***hypokalemia***) in extracellular potassium concentration can cause abnormal rhythms of the heart (***arrhythmias***) and abnormalities of skeletal muscle contraction.

A healthy person remains in potassium balance in the steady state by daily excreting an amount of potassium in the urine equal to the amount ingested minus the amounts eliminated in feces and sweat. Like sodium, potassium losses via sweat and the gastrointestinal tract are normally quite small, although vomiting or diarrhea can cause large quantities to be lost. The control of urinary potassium excretion is the major mechanism regulating body potassium.

Renal Regulation of Potassium

Potassium is freely filterable in the renal corpuscle. Normally, the tubules reabsorb most of this filtered potassium so that very little of the filtered potassium appears in the urine. However, the cortical collecting ducts can secrete potassium, and changes in potassium excretion are due mainly to changes in potassium secretion by this tubular segment (**Figure 14–28**).

During potassium depletion, when the homeostatic response is to minimize potassium loss, there is no potassium secretion by the cortical collecting ducts. Only the small amount of filtered potassium that escapes tubular reabsorption is excreted. With normal fluctuations in potassium intake, a variable amount of potassium is added to the small amount filtered and not reabsorbed. This maintains total-body potassium balance.

Figure 14–14b illustrates the mechanism of potassium secretion by the cortical collecting ducts. In this tubular segment, the K^+ pumped into the cell across the basolateral membrane by Na^+/K^+-ATPases diffuses into the tubular lumen through K^+ channels in the luminal membrane. Thus, the

secretion of potassium by the cortical collecting duct is associated with the reabsorption of sodium by this tubular segment. Potassium secretion does not occur in other sodium-reabsorbing tubular segments because there are few potassium channels in the luminal membranes of their cells. Rather, in these segments, the potassium pumped into the cell by Na^+/K^+-ATPase simply diffuses back across the basolateral membrane through potassium channels located there (see Figure 14–14a).

What factors influence potassium secretion by the cortical collecting ducts to achieve homeostasis of bodily potassium? The single most important factor is as follows: When a high-potassium diet is ingested (**Figure 14–29**), plasma potassium concentration increases, though very slightly, and this drives enhanced basolateral uptake via the Na^+/K^+-ATPase pumps. Thus, there is an enhanced potassium secretion. Conversely, a low-potassium diet or a negative potassium balance, such as results from diarrhea, lowers basolateral potassium uptake. This reduces potassium secretion and excretion, thereby helping to reestablish potassium balance.

A second important factor linking potassium secretion to potassium balance is the hormone aldosterone (see Figure 14–29). Besides stimulating tubular sodium reabsorption by the cortical collecting ducts, aldosterone simultaneously enhances tubular potassium secretion by this tubular segment.

The homeostatic mechanism by which an excess or deficit of potassium controls aldosterone production (see Figure 14–29) is different from the mechanism described earlier involving the renin-angiotensin system. The aldosterone-secreting cells of the adrenal cortex are sensitive to the potassium concentration of the extracellular fluid. Thus, an increased intake of

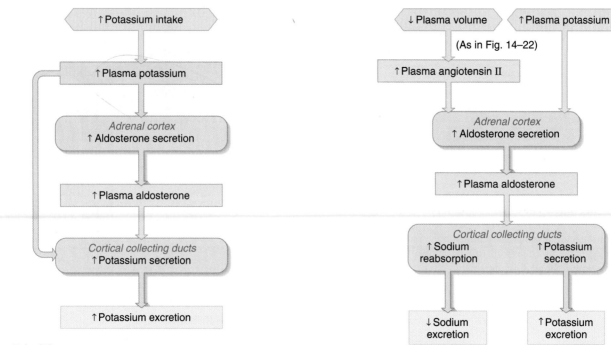

Figure 14–29

Pathways by which an increased potassium intake induces greater potassium excretion.

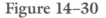

Figure 14–30

Summary of the control of aldosterone and its effects on sodium reabsorption and potassium secretion.

potassium leads to an increased extracellular potassium concentration, which in turn directly stimulates the adrenal cortex to produce aldosterone. The increased plasma aldosterone concentration increases potassium secretion and thereby eliminates the excess potassium from the body.

Conversely, a lowered extracellular potassium concentration decreases aldosterone production and thereby reduces potassium secretion. Less potassium than usual is excreted in the urine, thereby helping to restore the normal extracellular concentration.

Figure 14–30 summarizes the control and major renal tubular effects of aldosterone. The fact that a single hormone regulates both sodium and potassium excretion raises the question of potential conflicts between homeostasis of the two ions. For example, if a person was sodium-deficient and therefore secreting large amounts of aldosterone, the potassium-secreting effects of this hormone would tend to cause some potassium loss even though potassium balance was normal to start with. Usually, such conflicts cause only minor imbalances because there are a variety of other counteracting controls of sodium and potassium excretion.

Renal Regulation of Calcium and Phosphate

Calcium and phosphate balance are controlled primarily by parathyroid hormone and 1,25(OH)₂D, as described in detail in Chapter 11. Approximately 60 percent of plasma calcium is available for filtration in the kidney. The remaining plasma

calcium is protein-bound or complexed with anions. Because calcium is so important in the function of virtually every cell in the body, the kidney has powerful mechanisms to reabsorb calcium from the tubular fluid. More than 60 percent of calcium reabsorption is not under hormonal control and occurs in the proximal tubule. The hormonal control of calcium reabsorption occurs mainly in the distal convoluted tubule and early in the cortical collecting duct. When plasma calcium is low, the secretion of parathyroid hormone (PTH) from the parathyroid glands increases. PTH stimulates the opening of calcium channels in these parts of the nephron, thereby increasing calcium reabsorption. As discussed in Chapter 11, another important action of PTH in the kidney is to increase the activity of the 1-hydroxylase enzyme, thus activating 25(OH)-D to 1,25(OH)₂D, which then goes on to increase calcium and phosphate absorption in the gastrointestinal tract.

About half of the plasma phosphate is ionized and is filterable. Like calcium, most of the phosphate that is filtered is reabsorbed in the proximal tubule. Unlike calcium, phosphate reabsorption is decreased by PTH, thereby increasing the excretion of phosphate. Thus, when plasma calcium is low, and PTH and calcium reabsorption are increased as a result, phosphate excretion is increased.

Summary—Division of Labor

Table 14–5 summarizes the division of labor of renal function along the renal tubule. So far, we have discussed all of these processes except the transport of acids and bases, which Section C of this chapter will cover.

Table 14–5	Summary of "Division of Labor" in the Renal Tubules	
Tubular Segment	**Major Functions**	**Controlling Factors**
Glomerulus/Bowman's capsule	Forms ultrafiltrate of plasma	Starling forces (P_{GC}, P_{BS}, π_{GC})
Proximal tubule	Bulk reabsorption of solutes and water Secretion of solutes (except potassium) and organic acids and bases	Active transport of solutes with passive water reabsorption Parathyroid hormone inhibits phosphate reabsorption
Loop of Henle	Establishes medullary osmotic gradient (juxtamedullary nephrons) Secretion of urea	
Descending limb	Bulk reabsorption of water	Passive water reabsorption
Ascending limb	Reabsorption of NaCl	Active transport
Distal tubule and cortical collecting duct	Fine-tuning of the reabsorption/secretion of small quantity of solute remaining	Aldosterone stimulates sodium reabsorption and potassium excretion Parathyroid hormone stimulates calcium reabsorption
Cortical and medullary collecting duct	Fine-tuning of water reabsorption Reabsorption of urea	Vasopressin increases passive reabsorption of water

ADDITIONAL CLINICAL EXAMPLES

Hyperaldosteronism

Hyperaldosteronism encompasses a variety of chronic diseases involving an excess of the adrenal hormone aldosterone. The most common form is *primary aldosteronism,* which is often due to a noncancerous growth *(adenoma)* of the zona glomerulosa of the adrenal gland. These tumors release aldosterone in the absence of stimulation by angiotensin II (they are autonomous). Historically, this was called *Conn's syndrome.* In this condition, plasma aldosterone reaches very high levels. This stimulates sodium reabsorption and potassium excretion in the distal portions of the nephron.

As a result, patients develop hypertension that is difficult to control through standard medications, along with increased fluid volume, and low plasma potassium (hypokalemia). Because of the increased blood pressure and sodium delivery to the macula densa due to increased filtered load of sodium, renin release is greatly inhibited. Therefore, this is a renin-independent cause of hypertension and is one of the most common causes of endocrine hypertension. The usual treatment of Conn's syndrome is removal of the adrenal gland containing the tumor. The remaining adrenal gland then increases in size in compensation.■

■ SECTION B SUMMARY

Total-Body Balance of Sodium and Water

I. The body gains water via ingestion and internal production, and it loses water via urine, the gastrointestinal tract, and evaporation from the skin and respiratory tract (as insensible loss and sweat).

II. The body gains sodium and chloride by ingestion and loses them via the skin (in sweat), the gastrointestinal tract, and urine.

III. For both water and sodium, the major homeostatic control point for maintaining stable balance is renal excretion.

Basic Renal Processes for Sodium and Water

I. Sodium is freely filterable at the glomerulus, and its reabsorption is a primary active process dependent upon Na^+/K^+-ATPase pumps in the basolateral membranes of the tubular epithelium. Sodium is not secreted.

II. Sodium entry into the cell from the tubular lumen is always passive. Depending on the tubular segment, it is either through channels or by cotransport or countertransport with other substances.

III. Sodium reabsorption creates an osmotic difference across the tubule, which drives water reabsorption, largely through water channels (aquaporins).

IV. Water reabsorption is independent of the posterior pituitary hormone vasopressin until it reaches the collecting duct system, where vasopressin increases water permeability. A large volume of dilute urine is produced when plasma vasopressin concentration, and hence water reabsorption by the collecting ducts, is low.

V. A small volume of concentrated urine is produced by the renal countercurrent multiplier system when plasma vasopressin concentration is high.

a. The active transport of sodium chloride by the ascending loop of Henle causes increased osmolarity of the

The Kidneys and Regulation of Water and Inorganic Ions

interstitial fluid of the medulla but a dilution of the luminal fluid.

b. Vasopressin increases the permeability of the cortical collecting ducts to water, so water is reabsorbed by this segment until the luminal fluid is isoosmotic to plasma in the cortical peritubular capillaries.

c. The luminal fluid then enters and flows through the medullary collecting ducts, and the concentrated medullary interstitium causes water to move out of these ducts, made highly permeable to water by vasopressin. The result is concentration of the collecting duct fluid and the urine.

d. The hairpin-loop structure of the vasa recta prevents the countercurrent gradient from being washed away.

Renal Sodium Regulation

I. Sodium excretion is the difference between the amount of sodium filtered and the amount reabsorbed.

II. GFR, and hence the filtered load of sodium, is controlled by baroreceptor reflexes. Decreased vascular pressures cause decreased baroreceptor firing and hence increased sympathetic outflow to the renal arterioles, resulting in vasoconstriction and decreased GFR. These changes are generally relatively small under most physiological conditions.

III. The major control of tubular sodium reabsorption is the adrenal cortical hormone aldosterone, which stimulates sodium reabsorption in the cortical collecting ducts.

IV. The renin-angiotensin system is one of the two major controllers of aldosterone secretion. When extracellular volume decreases, renin secretion is stimulated by three inputs: (1) stimulation of the renal sympathetic nerves to the juxtaglomerular cells by extrarenal baroreceptor reflexes; (2) pressure decreases sensed by the juxtaglomerular cells, themselves acting as intrarenal baroreceptors; and (3) a signal generated by low sodium or chloride concentration in the lumen of the macula densa.

V. Many other factors influence sodium reabsorption. One of these, atrial natriuretic peptide, is secreted by cells in the atria in response to atrial distension; it inhibits sodium reabsorption and it also increases GFR.

VI. Arterial pressure acts locally on the renal tubules to influence sodium reabsorption, an increased pressure causing decreased reabsorption and hence increased excretion.

Renal Water Regulation

I. Water excretion is the difference between the amount of water filtered and the amount reabsorbed.

II. GFR regulation via the baroreceptor reflexes plays some role in regulating water excretion, but the major control is via vasopressin-mediated control of water reabsorption.

III. Vasopressin secretion by the posterior pituitary is controlled by cardiovascular baroreceptors and by osmoreceptors in the hypothalamus.

a. A low extracellular volume stimulates vasopressin secretion via the baroreceptor reflexes, and a high extracellular volume inhibits it.

b. Via the osmoreceptors, a high body fluid osmolarity stimulates vasopressin secretion, and a low osmolarity inhibits it.

Thirst and Salt Appetite

I. Thirst is stimulated by a variety of inputs, including baro-receptors, osmoreceptors, and possibly angiotensin II.

II. Salt appetite is not of major regulatory importance in human beings.

Potassium Regulation

I. A person remains in potassium balance by excreting an amount of potassium in the urine equal to the amount ingested minus the amounts lost in feces and sweat.

II. Potassium is freely filterable at the renal corpuscle and undergoes both reabsorption and secretion, the latter occurring in the cortical collecting ducts and serving as the major controlled variable determining potassium excretion.

III. When body potassium increases, extracellular potassium concentration also increases. This increase acts directly on the cortical collecting ducts to increase potassium secretion and also stimulates aldosterone secretion. The increased plasma aldosterone then also stimulates potassium secretion.

Renal Regulation of Calcium and Phosphate

I. About half of the plasma calcium and phosphate is ionized and filterable.

II. Most calcium and phosphate reabsorption occurs in the proximal tubule.

III. PTH increases calcium absorption in the distal convoluted tubule and early cortical collecting duct. PTH decreases phosphate reabsorption in the proximal tubule.

Additional Clinical Examples

I. The most common cause of hyperaldosteronism (too much aldosterone in the blood) is a noncancerous adrenal tumor (adenoma) that secretes aldosterone in the absence of stimulation from angiotensin II. This is sometimes called Conn's syndrome.

II. The excess aldosterone causes increased renal sodium reabsorption and fluid retention, and is a common cause of endocrine hypertension.

■ **SECTION B KEY TERMS**

aldosterone 507	hyperosmotic 503
angiotensin I 508	hypoosmotic 503
angiotensin II 508	insensible water loss 500
angiotensin-converting enzyme (ACE) 508	intrarenal baroreceptors 508
	isoosmotic 503
angiotensinogen 508	obligatory water loss 503
antidiuretic hormone (ADH) 503	osmoreceptor 511
	osmotic diuresis 503
aquaporin 502	pressure natriuresis 510
atrial natriuretic peptide (ANP) 509	renin 508
	renin-angiotensin system 508
countercurrent multiplier system 503	salt appetite 512
	vasopressin 503
diuresis 503	water diuresis 503

■ **SECTION B CLINICAL TERMS**

adenoma 515	hypokalemia 513
arrhythmia 513	nephrogenic diabetes
central diabetes insipidus 503	insipidus 503
diabetes insipidus 503	primary aldosteronism
hyperaldosteronism 515	(Conn's syndrome) 515
hyperkalemia 513	

■ SECTION B REVIEW QUESTIONS

1. What are the sources of water gain and loss in the body? What are the sources of sodium gain and loss?
2. Describe the distribution of water and sodium between the intracellular and extracellular fluids.
3. What is the relationship between body sodium and extracellular fluid volume?
4. What is the mechanism of sodium reabsorption, and how is the reabsorption of other solutes coupled to it?
5. What is the mechanism of water reabsorption, and how is it coupled to sodium reabsorption?
6. What is the effect of vasopressin on the renal tubules, and what are the sites affected?
7. Describe the characteristics of the two limbs of the loop of Henle with regard to their transport of sodium, chloride, and water.
8. Diagram the osmolarities in the two limbs of the loop of Henle, distal convoluted tubule, cortical collecting duct, cortical interstitium, medullary collecting duct, and medullary interstitium in the presence of vasopressin. What happens to the cortical and medullary collecting-duct values in the absence of vasopressin?
9. What two processes determine how much sodium is excreted per unit time?
10. Diagram the sequence of events in which a decrease in blood pressure leads to a decreased GFR.
11. List the sequence of events leading from increased renin secretion to increased aldosterone secretion.
12. What are the three inputs controlling renin secretion?
13. Diagram the sequence of events leading from decreased cardiovascular pressures or from an increased plasma osmolarity to an increased secretion of vasopressin.
14. What are the stimuli for thirst?
15. Which of the basic renal processes apply to potassium? Which of them is the controlled process, and which tubular segment performs it?
16. Diagram the steps leading from increased plasma potassium to increased potassium excretion.
17. What are the two major controls of aldosterone secretion, and what are this hormone's major actions?
18. Contrast the control of calcium and phosphate excretion by PTH.

SECTION C

Hydrogen Ion Regulation

Metabolic reactions are highly sensitive to the hydrogen ion concentration of the fluid in which they occur. This sensitivity is due to the influence that hydrogen ions have on enzyme function, which changes the shapes of proteins. Not surprisingly, then, the hydrogen ion concentration of the extracellular fluid is tightly regulated. At this point the reader might want to review the section on hydrogen ions, acidity, and pH in Chapter 2.

This regulation can be viewed in the same way as the balance of any other ion—that is, as matching gains and losses. When loss exceeds gain, the arterial plasma hydrogen ion concentration decreases and pH exceeds 7.4. This is termed **alkalosis.** When gain exceeds loss, the arterial plasma hydrogen ion concentration increases and the pH is less than 7.4. This is termed an **acidosis.**

Sources of Hydrogen Ion Gain or Loss

Table 14–6 summarizes the major routes for gains and losses of hydrogen ions. As described in Chapter 13, a huge quantity of CO_2—about 20,000 mmol—is generated daily as the result of oxidative metabolism. These CO_2 molecules participate in the generation of hydrogen ions during the passage of blood through peripheral tissues via the following reactions:

$$CO_2 + H_2O \underset{\text{carbonic anhydrase}}{\rightleftharpoons} H_2CO_3 \rightleftharpoons HCO_3^- + H^+ \qquad (14\text{-}1)$$

This source does not normally constitute a net gain of hydrogen ions. This is because the hydrogen ions generated via these reactions are reincorporated into water when the reactions

Table 14–6	Sources of Hydrogen Ion Gain or Loss

Gain

1. Generation of hydrogen ions from CO_2
2. Production of nonvolatile acids from the metabolism of proteins and other organic molecules
3. Gain of hydrogen ions due to loss of bicarbonate in diarrhea or other nongastric GI fluids
4. Gain of hydrogen ions due to loss of bicarbonate in the urine

Loss

1. Utilization of hydrogen ions in the metabolism of various organic anions
2. Loss of hydrogen ions in vomitus
3. Loss of hydrogen ions in the urine
4. Hyperventilation

are reversed during the passage of blood through the lungs (Chapter 13). Net retention of CO_2 does occur in hypoventilation or respiratory disease and causes a net gain of hydrogen ions. Conversely, net loss of CO_2 occurs in hyperventilation, and this causes net elimination of hydrogen ions.

The body also produces both organic and inorganic acids from sources other than CO_2. These are collectively termed **nonvolatile acids.** They include phosphoric acid and sulfuric acid, generated mainly by the catabolism of proteins, as well as lactic acid and several other organic acids. Dissociation of all of these acids yields anions and hydrogen ions. But simultaneously, the metabolism of a variety of organic anions utilizes hydrogen ions and produces bicarbonate. Thus, the metabolism of "nonvolatile" solutes both generates and utilizes hydrogen ions. With the high-protein diet typical in the United States, the generation of nonvolatile acids predominates in most people, and there is an average net production of 40 to 80 mmol of hydrogen ions per day.

A third potential source of the net gain or loss of hydrogen ions in the body occurs when gastrointestinal secretions leave the body. Vomitus contains a high concentration of hydrogen ions and so constitutes a source of net loss. In contrast, the other gastrointestinal secretions are alkaline. They contain very little hydrogen ion, but their concentration of bicarbonate is higher than in plasma. Loss of these fluids, as in diarrhea, in essence constitutes a *gain* of hydrogen ions. Given the mass action relationship shown in equation 14–1, *when a bicarbonate ion is lost from the body, it is the same as if the body had gained a hydrogen ion.* This is because loss of the bicarbonate causes the reactions shown in equation 14–1 to be driven to the right, thereby generating a hydrogen ion within the body. Similarly, when the body gains a bicarbonate ion, it is the same as if the body had lost a hydrogen ion, as the reactions of equation 14–1 are driven to the left.

Finally, the kidneys constitute the fourth source of net hydrogen ion gain or loss. That is, the kidneys can either remove hydrogen ions from the plasma or add them.

Buffering of Hydrogen Ions in the Body

Any substance that can reversibly bind hydrogen ions is called a **buffer.** Most hydrogen ions are buffered by extracellular and intracellular buffers. The normal extracellular fluid pH of 7.4 corresponds to a hydrogen ion concentration of only 0.00004 mmol/L (40 nanomol/L). Without buffering, the daily turnover of the 40 to 80 mmol of H^+ produced from nonvolatile acids generated in the body from metabolism would cause huge changes in body fluid hydrogen ion concentration.

The general form of buffering reactions is:

$$\text{Buffer} + H^+ \rightleftharpoons \text{HBuffer} \qquad (14\text{–}2)$$

HBuffer is a weak acid in that it can dissociate to Buffer plus H^+ or it can exist as the undissociated molecule (HBuffer). When H^+ concentration increases for any reason, the reaction is forced to the right, and more H^+ is bound by Buffer to form HBuffer. For example, when H^+ concentration is increased because of increased production of lactic acid, some of the hydrogen ions combine with the body's buffers, so the hydrogen ion concentration does not increase as much as it otherwise would have. Conversely, when H^+ concentration decreases because of the loss of hydrogen ions or the addition of alkali, equation 14–2 proceeds to the left and H^+ is released from HBuffer. In this manner, buffers stabilize H^+ concentration against changes in either direction.

The major extracellular buffer is the CO_2/HCO_3^- system summarized in equation 14–1. This system also plays some role in buffering within cells, but the major intracellular buffers are phosphates and proteins. An example of an intracellular protein buffer is hemoglobin, as described in Chapter 13.

Note that buffering does not eliminate hydrogen ions from the body or add them to the body; it only keeps them "locked up" until balance can be restored. How balance is achieved is the subject of the rest of our description of hydrogen ion regulation.

Integration of Homeostatic Controls

The kidneys are ultimately responsible for balancing hydrogen ion gains and losses so as to maintain a relatively constant plasma hydrogen ion concentration. Thus, the kidneys normally excrete the excess hydrogen ions from nonvolatile acids generated from metabolism—that is, all acids other than carbonic acid. Moreover, if there is an additional net gain of hydrogen ions due to increased production of these nonvolatile acids, to hypoventilation or respiratory malfunction, or to loss of alkaline gastrointestinal secretions, the kidneys increase the elimination of hydrogen ions from the body to restore balance. Alternatively, if there is a net loss of hydrogen ions from the body due to hyperventilation or vomiting, the kidneys replenish these hydrogen ions.

Although the kidneys are the ultimate hydrogen ion balancers, the respiratory system also plays a very important homeostatic role. We have pointed out that hypoventilation, respiratory malfunction, and hyperventilation can cause a hydrogen ion imbalance. Now we emphasize that when a hydrogen ion imbalance is due to a nonrespiratory cause, then ventilation is reflexly altered so as to help compensate for the imbalance. We described this phenomenon in Chapter 13 (see Figure 13–38). An elevated arterial hydrogen ion concentration stimulates ventilation, which causes reduced arterial P_{CO_2}, and thus, by mass action, reduces hydrogen ion concentration. Alternatively, a decreased plasma hydrogen ion concentration inhibits ventilation, thereby increasing arterial P_{CO_2} and the hydrogen ion concentration.

Thus, the respiratory system and kidneys work together. The respiratory response to altered plasma hydrogen ion concentration is very rapid (minutes) and keeps this concentration from changing too much until the more slowly responding kidneys (hours to days) can actually eliminate the imbalance. If the respiratory system is the actual cause of the hydrogen ion imbalance, then the kidneys are the sole homeostatic responder. Conversely, malfunctioning kidneys can create a hydrogen ion imbalance by eliminating too little or too much hydrogen ion from the body, and then the respiratory response is the only one in control.

Renal Mechanisms

The kidneys eliminate or replenish hydrogen ions from the body by altering plasma bicarbonate concentration. The key to understanding how altering plasma bicarbonate concentration eliminates or replenishes hydrogen ions was stated earlier. That is, the excretion of a bicarbonate in the urine increases the plasma hydrogen ion concentration just as if a hydrogen ion had been added to the plasma. Similarly, the addition of a bicarbonate to the plasma lowers the plasma hydrogen ion concentration just as if a hydrogen ion had been removed from the plasma.

Thus, when the plasma hydrogen ion concentration decreases (alkalosis) for whatever reason, the kidneys' homeostatic response is to excrete large quantities of bicarbonate. This increases plasma hydrogen ion concentration toward normal. In contrast, when plasma hydrogen ion concentration increases (acidosis), the kidneys do not excrete bicarbonate in the urine. Rather, kidney tubular cells produce *new* bicarbonate and add it to the plasma. This lowers the plasma hydrogen ion concentration toward normal.

Bicarbonate Handling

Bicarbonate is completely filterable at the renal corpuscles and undergoes significant tubular reabsorption in the proximal tubule, ascending loop of Henle, and cortical collecting ducts. Bicarbonate can also be secreted in the collecting ducts. Therefore:

HCO_3^- excretion =
  HCO_3^- filtered + HCO_3^- secreted − HCO_3^- reabsorbed

For simplicity, we will ignore the secretion of bicarbonate because it is always much less than tubular reabsorption, and we will treat bicarbonate excretion as the difference between filtration and reabsorption.

Bicarbonate reabsorption is an active process, but it is not accomplished in the conventional manner of simply having an active pump for bicarbonate ions at the luminal or basolateral membrane of the tubular cells. Instead, bicarbonate reabsorption depends on the tubular secretion of hydrogen ions, which combine in the lumen with filtered bicarbonates.

Figure 14–31 illustrates the sequence of events. Begin this figure inside the cell with the combination of CO_2 and H_2O to form H_2CO_3, a reaction catalyzed by the enzyme carbonic anhydrase. The H_2CO_3 immediately dissociates to yield H^+ and bicarbonate (HCO_3^-). The HCO_3^- diffuses down its concentration gradient across the basolateral membrane into interstitial fluid and then into the blood. Simultaneously, the H^+ is secreted into the lumen. Depending on the tubular segment, this secretion is achieved by some combination of primary H^+-ATPase pumps, primary H^+/K^+-ATPase pumps, and Na^+/H^+ countertransporters.

The secreted H^+, however, is not excreted. Instead, it combines in the lumen with a filtered HCO_3^- and generates CO_2 and H_2O, both of which can diffuse into the cell and be available for another cycle of hydrogen ion generation. The overall result is that the bicarbonate filtered from the plasma at the renal corpuscle has disappeared, but its place in the plasma has been taken by the bicarbonate that was produced inside the cell. In this manner, no net change in plasma bicarbonate concentration has occurred. It may seem inaccurate to refer to this process as bicarbonate "reabsorption" because the bicar-

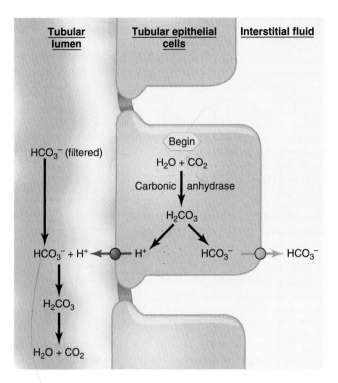

Figure 14–31

Reabsorption of bicarbonate. Begin looking at this figure inside the cell, with the combination of CO_2 and H_2O to form H_2CO_3. As shown in the figure, active H^+-ATPase pumps are involved in the movement of H^+ out of the cell across the luminal membrane; in several tubular segments, this transport step is also mediated by Na^+/H^+ countertransporters and/or H^+/K^+-ATPase pumps.

bonate that appears in the peritubular plasma is not the same bicarbonate ion that was filtered. Yet, the overall result is the same as if the filtered bicarbonate had been reabsorbed in the conventional manner like a sodium or potassium ion.

Except in response to alkalosis, discussed in the next section, the kidneys normally reabsorb all filtered bicarbonate, thereby preventing the loss of bicarbonate in the urine.

Addition of New Bicarbonate to the Plasma

An essential concept shown in Figure 14–31 is that as long as there are still significant amounts of filtered bicarbonate ions in the lumen, almost all secreted hydrogen ions will combine with them. But what happens to any secreted hydrogen ions once almost all the bicarbonate has been reabsorbed and is no longer available in the lumen to combine with the hydrogen ions?

The answer, illustrated in **Figure 14–32**, is that the extra secreted hydrogen ions combine in the lumen with a filtered nonbicarbonate buffer, usually HPO_4^{2-}. (Other filtered buffers can also participate, but HPO_4^{2-} is the most important.) The hydrogen ion is then excreted in the urine as part of an $H_2PO_4^-$ ion. Now for the critical point: note in Figure 14–32 that, under these conditions, the bicarbonate generated within the tubular cell by the carbonic anhydrase reaction and entering the plasma constitutes a *net gain* of bicarbonate by the plasma, not merely a replacement for a filtered bicarbonate. Thus, when a secreted hydrogen ion combines in the lumen

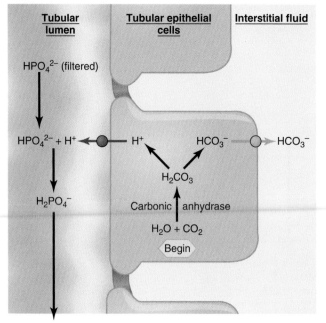

Figure 14–32

Renal contribution of new HCO_3^- to the plasma as achieved by tubular secretion of H^+. The process of intracellular H^+ and HCO_3^- generation, with H^+ moving into the lumen and HCO_3^- into the plasma, is identical to that shown in Figure 14–31. Once in the lumen, however, the H^+ combines with filtered phosphate (HPO_4^{2-}) rather than filtered HCO_3^- and is excreted as $H_2PO_4^-$. As described in the legend for Figure 14–31, the transport of hydrogen ions into the lumen is accomplished not only by H^+-ATPase pumps but, in several tubular segments, by Na^+/H^+ countertransporters and/or H^+/K^+-ATPase pumps as well.

with a buffer other than bicarbonate, the overall effect is not merely one of bicarbonate conservation, as in Figure 14–31, but rather of addition to the plasma of a *new* bicarbonate. This raises the bicarbonate concentration of the plasma and alkalinizes it.

To repeat, significant numbers of hydrogen ions combine with filtered nonbicarbonate buffers like HPO_4^{2-} only after the filtered bicarbonate has virtually all been reabsorbed. The main reason is that there is such a large load of filtered bicarbonate—25 times more than the load of filtered nonbicarbonate buffers—competing for the secreted hydrogen ions.

There is a second mechanism by which the tubules contribute new bicarbonate to the plasma that involves not hydrogen ion secretion, but rather the renal production and secretion of ammonium ions (NH_4^+) (**Figure 14–33**). Tubular cells, mainly those of the proximal tubule, take up glutamine from both the glomerular filtrate and peritubular plasma and metabolize it. In the process, both NH_4^+ and bicarbonate are formed inside the cells. The NH_4^+ is actively secreted via Na^+/NH_4^+ countertransport into the lumen and excreted, while the bicarbonate moves into the peritubular capillaries and constitutes new plasma bicarbonate.

A comparison of Figures 14–32 and 14–33 demonstrates that the overall result—renal contribution of new bicarbonate

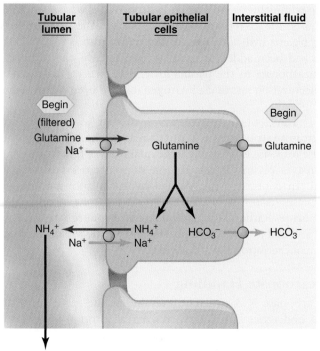

Figure 14–33

Renal contribution of new HCO_3^- to the plasma as achieved by renal metabolism of glutamine and excretion of ammonium (NH_4^+). Compare this figure to Figure 14–32. This process occurs mainly in the proximal tubule.

to the plasma—is the same regardless of whether it is achieved by: (1) H^+ secretion and excretion on nonbicarbonate buffers such as phosphate (see Figure 14–32); or (2) by glutamine metabolism with NH_4^+ excretion (see Figure 14–33). It is convenient, therefore, to view the latter as representing H^+ excretion "bound" to NH_3, just as the former case constitutes H^+ excretion bound to nonbicarbonate buffers. Thus, the amount of H^+ excreted in the urine in these two forms is a measure of the amount of new bicarbonate added to the plasma by the kidneys. Indeed, "urinary H^+ excretion" and "renal contribution of new bicarbonate to the plasma" are really two sides of the same coin.

The kidneys normally contribute enough new bicarbonate to the blood by excreting hydrogen ions to compensate for the hydrogen ions from nonvolatile acids generated in the body.

Renal Responses to Acidosis and Alkalosis

We can now apply this material to the renal responses to the presence of an acidosis or alkalosis. These are summarized in **Table 14–7**.

Clearly, these homeostatic responses require that the rates of hydrogen ion secretion, glutamine metabolism, and ammonium excretion be subject to physiological control by changes in blood hydrogen ion concentration. The specific pathways and mechanisms that bring about these rate changes are very complex, however, and are not presented here.

Table 14–7	Renal Responses to Acidosis and Alkalosis

Responses to Acidosis

1. Sufficient hydrogen ions are secreted to reabsorb all the filtered bicarbonate.

2. Still more hydrogen ions are secreted, and this contributes new bicarbonate to the plasma as these hydrogen ions are excreted bound to nonbicarbonate urinary buffers such as HPO_4^{2-}.

3. Tubular glutamine metabolism and ammonium excretion are enhanced, which also contributes new bicarbonate to the plasma.

Net result: More new bicarbonate ions than usual are added to the blood, and plasma bicarbonate is increased, thereby compensating for the acidosis. The urine is highly acidic (lowest attainable pH = 4.4).

Responses to Alkalosis

1. Rate of hydrogen ion secretion is inadequate to reabsorb all the filtered bicarbonate, so significant amounts of bicarbonate are excreted in the urine, and there is little or no excretion of hydrogen ions on nonbicarbonate urinary buffers.

2. Tubular glutamine metabolism and ammonium excretion are decreased so that little or no new bicarbonate is contributed to the plasma from this source.

Net result: Plasma bicarbonate concentration is decreased, thereby compensating for the alkalosis. The urine is alkaline (pH > 7.4).

Classification of Acidosis and Alkalosis

To repeat, acidosis refers to any situation in which the hydrogen ion concentration of arterial plasma is elevated whereas alkalosis denotes a reduction. All such situations fit into two distinct categories (**Table 14–8**): (1) *respiratory acidosis or alkalosis;* (2) *metabolic acidosis or alkalosis.*

As its name implies, respiratory acidosis results from altered alveolar ventilation. Respiratory acidosis occurs when the respiratory system fails to eliminate carbon dioxide as fast as it is produced. Respiratory alkalosis occurs when the respiratory system eliminates carbon dioxide faster than it is produced. As described earlier, the imbalance of arterial hydrogen ion concentrations in such cases is completely explainable in terms of mass action. Thus, the hallmark of a respiratory acidosis is an elevation in both arterial P_{CO_2} and hydrogen ion concentration, whereas that of respiratory alkalosis is a reduction in both.

Metabolic acidosis or alkalosis includes all situations other than those in which the primary problem is respiratory. Some common causes of metabolic acidosis are excessive production of lactic acid (during severe exercise or hypoxia) or of ketone bodies (in uncontrolled diabetes mellitus or fasting, as described in Chapter 16). Metabolic acidosis can also result from excessive loss of bicarbonate, as in diarrhea. Another cause of metabolic alkalosis is persistent vomiting, with its associated loss of hydrogen ions as HCl from the stomach.

What is the arterial P_{CO_2} in metabolic acidosis or alkalosis? By definition, metabolic acidosis and alkalosis must be due to something other than excess retention or loss of carbon dioxide, so you might have predicted that arterial P_{CO_2} would

Table 14–8	Changes in the Arterial Concentrations of Hydrogen Ions, Bicarbonate, and Carbon Dioxide in Acid-Base Disorders

Primary Disorder	H⁺	HCO₃⁻	CO₂	Cause of HCO₃⁻ Change	Cause of CO₂ Change
Respiratory acidosis	↑	↑	↑	Renal compensation	Primary abnormality
Respiratory alkalosis	↓	↓	↓	Renal compensation	Primary abnormality
Metabolic acidosis	↑	↓	↓	Primary abnormality	Reflex ventilatory compensation
Metabolic alkalosis	↓	↑	↑	Primary abnormality	Reflex ventilatory compensation

Table 14–8 physiological *inquiry* (pi)

■ A patient has an arterial P_{O_2} of 50 mmHg, an arterial P_{CO_2} of 60 mmHg, and an arterial pH of 7.36. Classify the acid-base disturbance and hypothesize a cause.

Answer can be found at end of chapter.

be unchanged, but this is not the case. As emphasized earlier in this chapter, the elevated hydrogen ion concentration associated with metabolic acidosis *reflexly* stimulates ventilation and lowers arterial P_{CO_2}. By mass action, this helps restore the hydrogen ion concentration toward normal. Conversely, a person with metabolic alkalosis will reflexly have ventilation inhibited. The result is an increase in arterial P_{CO_2} and, by mass action, an associated restoration of hydrogen ion concentration toward normal.

To reiterate, the plasma P_{CO_2} changes in metabolic acidosis and alkalosis are not the *cause* of the acidosis or alkalosis, but the *result* of compensatory reflex responses to nonrespiratory abnormalities. Thus, in metabolic as opposed to respiratory conditions, the arterial plasma P_{CO_2} and hydrogen ion concentration move in opposite directions, as summarized in Table 14–8.

■ SECTION C SUMMARY

Sources of Hydrogen Ion Gain or Loss

I. Total-body balance of hydrogen ions is the result of both metabolic production of these ions and of net gains or losses via the respiratory system, gastrointestinal tract, and urine (Table 14–6).

II. A stable balance is achieved by regulation of urinary losses.

Buffering of Hydrogen Ions in the Body

I. Buffering is a means of minimizing changes in hydrogen ion concentration by combining these ions reversibly with anions such as bicarbonate and intracellular proteins.

II. The major extracellular buffering system is the CO_2/HCO_3^- system, and the major intracellular buffers are proteins and phosphates.

Integration of Homeostatic Controls

I. The kidneys and the respiratory system are the homeostatic regulators of plasma hydrogen ion concentration.

II. The kidneys are the organs that achieve body hydrogen ion balance.

III. A decrease in arterial plasma hydrogen ion concentration causes reflex hypoventilation, which raises arterial P_{CO_2} and, hence, raises plasma hydrogen ion concentration toward normal. An increase in plasma hydrogen ion concentration causes reflex hyperventilation, which lowers arterial P_{CO_2} and, hence, lowers hydrogen ion concentration toward normal.

Renal Mechanisms

I. The kidneys maintain a stable plasma hydrogen ion concentration by regulating plasma bicarbonate concentration. They can either excrete bicarbonate or contribute new bicarbonate to the blood.

II. Bicarbonate is reabsorbed when hydrogen ions, generated in the tubular cells by a process catalyzed by carbonic anhydrase, are secreted into the lumen and combine with filtered bicarbonate. The secreted hydrogen ions are not excreted in this situation.

III. In contrast, when the secreted hydrogen ions combine in the lumen with filtered phosphate or other nonbicarbonate buffer, they are excreted, and the kidneys have contributed new bicarbonate to the blood.

IV. The kidneys also contribute new bicarbonate to the blood when they produce and excrete ammonium.

Classification of Acidosis and Alkalosis

I. Acid-base disorders are categorized as respiratory or metabolic.

  a. Respiratory acidosis is due to retention of carbon dioxide, and respiratory alkalosis to excessive elimination of carbon dioxide.

  b. All other causes of acidosis or alkalosis are termed *metabolic* and reflect gain or loss, respectively, of hydrogen ions from a source other than carbon dioxide.

■ SECTION C KEY TERMS

buffer 518	nonvolatile acid 517

■ SECTION C CLINICAL TERMS

acidosis 517	metabolic alkalosis 521
alkalosis 517	respiratory acidosis 521
metabolic acidosis 521	respiratory alkalosis 521

■ SECTION C REVIEW QUESTIONS

1. What are the sources of gain and loss of hydrogen ions in the body?
2. List the body's major buffer systems.
3. Describe the role of the respiratory system in the regulation of hydrogen ion concentration.
4. How does the tubular secretion of hydrogen ions occur, and how does it achieve bicarbonate reabsorption?
5. How does hydrogen ion secretion contribute to the renal addition of new bicarbonate to the blood? What determines whether a secreted hydrogen ion will achieve these results or will instead cause bicarbonate reabsorption?
6. How does the metabolism of glutamine by the tubular cells contribute new bicarbonate to the blood and ammonium to the urine?
7. What two quantities make up "hydrogen ion excretion?" Why can this term be equated with "contribution of new bicarbonate to the plasma?"
8. How do the kidneys respond to the presence of an acidosis or alkalosis?
9. Classify the four types of acid-base disorders according to plasma hydrogen ion concentration, bicarbonate concentration, and P_{CO_2}.

Diuretics and Kidney Disease

Diuretics

Drugs used clinically to increase the volume of urine excreted are known as **diuretics**. Most of these agents act on the tubules to inhibit the reabsorption of sodium, along with chloride and/or bicarbonate, resulting in increased excretion of these ions. Because water reabsorption is dependent upon sodium reabsorption, water reabsorption is also reduced, resulting in increased water excretion.

A large variety of clinically useful diuretics are available and are classified according to the specific mechanisms by which they inhibit sodium reabsorption. For example, **loop diuretics**, such as furosemide, act on the ascending limb of the loop of Henle to inhibit the transport protein that mediates the first step in sodium reabsorption in this segment—cotransport of sodium and chloride (and potassium) into the cell across the luminal membrane.

Except for one category of diuretics, called **potassium-sparing diuretics**, all diuretics not only increase sodium excretion but also cause increased potassium excretion, which is often an unwanted side effect. The potassium-sparing diuretics inhibit sodium reabsorption in the cortical collecting duct, and they simultaneously inhibit potassium secretion there. Potassium-sparing diuretics either block the action of aldosterone (e.g., spironolactone or eplerenone) or block the epithelial sodium channel in the cortical collecting duct (e.g., triamterine or amiloride). This explains why they do not cause increased potassium excretion. Osmotic diuretics such as mannitol are filtered but not reabsorbed, thus retaining water in the urine. This is the same reason that uncontrolled diabetes mellitus and its associated glucosuria can cause excessive water loss and dehydration (see Figure 16–12).

Diuretics are among the most commonly used medications. For one thing, they are used to treat diseases characterized by renal retention of salt and water. As emphasized earlier in this chapter, the regulation of blood pressure normally produces stability of total-body sodium mass and extracellular volume because of the close correlation between these variables. In contrast, in several types of disease, this correlation is disrupted and the reflexes that maintain blood pressure can cause renal retention of sodium. Sodium excretion may decrease to almost nothing despite continued sodium ingestion, leading to abnormal expansion of the extracellular fluid **(edema)**. Diuretics are used to prevent or reverse this renal retention of sodium and water.

The most common example of this phenomenon is **congestive heart failure** (Chapter 12). A person with a failing heart manifests a decreased GFR and increased aldosterone secretion, both of which contribute to the virtual absence of sodium in the urine. The net result is extracellular volume expansion and edema. The sodium-retaining responses are triggered by the lower cardiac output (a result of cardiac failure) and the decrease in arterial blood pressure that results directly from this decrease in cardiac output.

Another disease in which diuretics are often used is hypertension (Chapter 12). The decrease in body sodium and water resulting from the diuretic-induced excretion of these substances brings about arteriolar dilation and a lowering of the blood pressure. The precise mechanism by which decreased body sodium causes arteriolar dilation is not known.

Kidney Disease

Many diseases affect the kidneys. Infections, allergies, congenital defects, kidney stones (accumulation of mineral deposits in nephron tubules), tumors, and toxic chemicals are some possible sources of kidney damage. Obstruction of the urethra or a ureter may cause injury from the buildup of pressure and may predispose the kidneys to bacterial infection. A common cause of renal failure is poorly controlled diabetes mellitus. The increase in blood glucose interferes with normal renal filtration and tubular function (Chapter 16).

One frequent sign of kidney disease is the appearance of protein in the urine. In normal kidneys, there is a very tiny amount of protein in the glomerular filtrate because the corpuscular membranes are not completely impermeable to proteins, particularly those with lower molecular weights. However, the cells of the proximal tubule completely remove this filtered protein from the tubular lumen, and no protein appears in the final urine. In contrast, diseased renal corpuscles may become much more permeable to protein, and diseased proximal tubules may lose their ability to remove filtered protein from the tubular lumen. The result is that protein appears in the urine.

Although many diseases of the kidney are self-limited and produce no permanent damage, others worsen if untreated. The symptoms of profound renal malfunction are relatively independent of the damaging agent and are collectively known as **uremia**, literally, "urine in the blood."

The severity of uremia depends upon how well the impaired kidneys can preserve the constancy of the internal environment. Assuming that the person continues to ingest a normal diet containing the usual quantities of nutrients and electrolytes, what problems arise? The key fact to keep in mind is that the kidney destruction markedly reduces the number of functioning nephrons. Accordingly, the many substances, particularly potentially toxic waste products, that gain entry to the tubule by filtration build up in the blood. In addition, the excretion of potassium is impaired because there are too few nephrons capable of normal tubular secretion of this ion. The person may also develop acidosis because the reduced number of nephrons fail to add enough new bicarbonate to the blood to compensate for the daily metabolic production of nonvolatile acids.

The remarkable fact is how large the safety factor is in renal function. In general, the kidneys are still able to perform their regulatory function quite well as long as 10 percent of the nephrons are functioning. This is because these remaining nephrons undergo alterations in function—filtration, reabsorption, and secretion—to compensate for the missing nephrons. For example, each remaining nephron increases its rate of potassium secretion so that the total amount of potassium the kidneys excrete is maintained at normal levels. The limits of regulation are restricted, however. To use potassium as our example again, if someone with severe renal disease were to go on a diet high in potassium, the remaining nephrons might not be able to secrete enough potassium to prevent potassium retention.

Other problems arise in uremia because of abnormal secretion of the hormones the kidneys produce. Thus, decreased secretion of erythropoietin results in anemia (Chapter 12). Decreased ability to form $1,25\text{-}(OH)_2D$ results in deficient absorption of calcium from the gastrointestinal tract, with a resulting decrease in plasma calcium, increase in PTH, and inadequate bone calcification (secondary hyperparathyroidism). Erythropoietin and $1,25\text{-}(OH)_2D$ (calcitriol) can be administered to patients with uremia.

In the case of the secreted enzyme renin, there is rarely too little secretion, but rather, too much secretion by the juxtaglomerular cells of the damaged kidneys. The main reasons for the increase in renin are decreased sodium delivery to the macula densa and decreased perfusion of affected nephrons (intrarenal baroreceptor). The result is increased plasma angiotensin II concentration and the development of **renal hypertension.** ACE inhibitors and angiotensin II receptor blockers can be used to lower blood pressure and improve sodium and water balance.

Hemodialysis, Peritoneal Dialysis, and Transplantation

Failing kidneys may reach a point when they can no longer excrete water and ions at rates that maintain body balances of these substances, nor can they excrete waste products as fast as they are produced. Dietary alterations can help minimize but not eliminate these problems. For example, lowering potassium intake reduces the amount of potassium to be excreted. The clinical techniques used to perform the kidneys' excretory functions are hemodialysis and peritoneal dialysis. The general term *dialysis* means to separate substances using a permeable membrane.

The artificial kidney is an apparatus that utilizes a process termed **hemodialysis** to remove wastes and excess substances from the blood (**Figure 14–34**). During hemodialysis, blood is

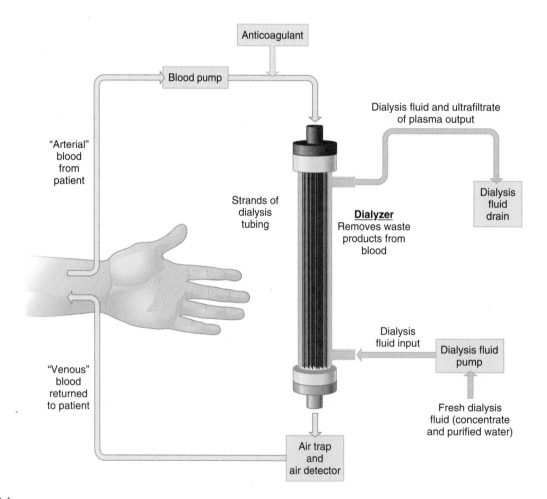

Figure 14–34

Simplified diagram of hemodialysis. Note that blood and dialysis fluid flow in opposite directions through the dialyzer (countercurrent). The blood flow can be 400 ml/min, and the dialysis fluid flow rate can be 1000 ml/min! During a three- to four-hour dialysis session, approximately 72 to 96 L of blood and 3000 to 4000 L of dialysis fluid passes through the dialyzer. The dialyzer is composed of many strands of very thin dialysis tubing. Blood flows inside each tube, and dialysis fluid bathes the outside of the dialysis tubing. This provides a large surface area for diffusion of waste products out of the blood and into the dialysis fluid.

pumped from one of the patient's arteries through tubing that is surrounded by special dialysis fluid. The tubing then conducts the blood back into the patient by way of a vein. The dialysis tubing is generally made of cellophane that is highly permeable to most solutes but relatively impermeable to protein and completely impermeable to blood cells—characteristics quite similar to those of capillaries. The dialysis fluid is a salt solution with ionic concentrations similar to or lower than those in normal plasma, and it contains no creatinine, urea, or other substances to be completely removed from the plasma. As blood flows through the tubing, the concentrations of nonprotein plasma solutes tend to reach diffusion equilibrium with those of the solutes in the bath fluid. For example, if the plasma potassium concentration of the patient is above normal, potassium diffuses out of the blood across the cellophane tubing and into the dialysis fluid. Similarly, waste products and excesses of other substances also diffuse into the dialysis fluid and thus are eliminated from the body.

Patients with acute reversible renal failure may require hemodialysis for only days or weeks. Patients with chronic irreversible renal failure require treatment for the rest of their lives, however, unless they receive a kidney transplant. Such patients undergo hemodialysis several times a week.

Another way of removing excess substances from the blood is *peritoneal dialysis*, which uses the lining of the patient's own abdominal cavity (peritoneum) as a dialysis membrane. Fluid is injected, via a needle inserted through the abdominal wall, into this cavity and allowed to remain there for hours, during which solutes diffuse into the fluid from the person's blood. The dialysis fluid is then removed by reinserting the needle and is replaced with new fluid. This procedure can be performed several times daily by a patient who is simultaneously doing normal activities.

The long-term treatment of choice for most patients with permanent renal failure is kidney transplantation. Rejection of the transplanted kidney by the recipient's body is a potential problem, but great strides have been made in reducing the frequency of rejection (Chapter 18). Many people who might benefit from a transplant, however, do not receive one. Currently, the major source of kidneys for transplantation is recently deceased persons. Improved public understanding should lead to many more individuals giving permission in advance to have their kidneys and other organs used following their death. Recently, donation from a living, related donor has become more common. Because of the large safety factor, the donor can function quite normally with one kidney.

■ SECTION D SUMMARY

Diuretics

I. Most diuretics inhibit reabsorption of sodium and water, thereby enhancing the excretion of these substances. Different diuretics act on different nephron segments.

Kidney Disease

I. Many of the symptoms of uremia—general renal malfunction—are due to retention of substances because of reduced GFR and, in the case of potassium and hydrogen ion, reduced secretion. Other symptoms are due to inadequate secretion of erythropoietin and 1,25-dihydroxyvitamin D, and too much secretion of renin.

II. Either hemodialysis or peritoneal dialysis can be used chronically to eliminate water, ions, and waste products retained during uremia.

■ SECTION D CLINICAL TERMS

congestive heart failure 523   peritoneal dialysis 525
diuretics 523   potassium-sparing diuretic 523
edema 523   renal hypertension 524
hemodialysis 524   uremia 523
loop diuretic 523

■ SECTION D REVIEW QUESTIONS

1. List the different types of diuretics and briefly summarize their mechanisms of action.
2. List several diseases that diuretics can be used to treat.
3. What substances are found in the urine of patients with kidney disease that are usually not found in the urine of healthy people?
4. What is uremia?
5. What is the cause of renal hypertension?
6. Briefly summarize how renal and peritoneal dialysis work.

Chapter 14 Test Questions

(Answers appear in Appendix A.)

1. Which of the following will lead to an increase in glomerular fluid filtration in the kidney?
   a. An increase in the protein concentration in the plasma.
   b. An increase in the fluid pressure in Bowman's space.
   c. An increase in the glomerular capillary blood pressure.
   d. A decrease in the glomerular capillary blood pressure.
   e. Constriction of the afferent arteriole.

2. Which of the following is true about renal clearance?
   a. It is the amount of a substance excreted per unit time.
   b. A substance with clearance >GFR undergoes only filtration.
   c. A substance with clearance >GFR undergoes filtration and secretion.
   d. It can be calculated knowing only the filtered load of a substance and the rate of urine production.
   e. Creatinine clearance approximates renal plasma flow.

3. Which of the following will *not* lead to a diuresis?
   a. excessive sweating
   b. central diabetes insipidus
   c. nephrogenic diabetes insipidus
   d. excessive water intake
   e. uncontrolled diabetes mellitus

4. Which of the following contributes directly to the generation of a hypertonic medullary interstitium in the kidney?
   a. active sodium transport in the descending limb of Henle's loop
   b. active water reabsorption in the ascending limb of Henle's loop
   c. active sodium reabsorption in the distal convoluted tubule
   d. water reabsorption in the cortical collecting duct
   e. secretion of urea into Henle's loop

5. An increase in renin is caused by
   a. a decrease in sodium intake.
   b. a decrease in renal sympathetic nerve activity.
   c. an increase in blood pressure in the renal artery.
   d. an aldosterone-secreting adrenal tumor.
   e. essential hypertension.

6. An increase in parathyroid hormone will
   a. increase plasma 25(OH) D.
   b. decrease plasma 1,25-(OH)$_2$D.
   c. decrease calcium excretion.
   d. increase phosphate reabsorption.
   e. increase calcium reabsorption in the proximal tubule.

7. Which of the following is a component of the renal response to metabolic acidosis?
   a. reabsorption of hydrogen ions
   b. secretion of bicarbonate into the tubular lumen
   c. secretion of ammonium into the tubular lumen

   d. secretion of glutamine into the interstitial fluid
   e. carbonic anhydrase-mediated production of HPO$_4^{2-}$

8. Which of the following is consistent with respiratory alkalosis?
   a. an increase in alveolar ventilation during mild exercise
   b. hyperventilation
   c. an increase in plasma bicarbonate
   d. an increase in arterial CO$_2$
   e. urine pH < 5.0

9. Which is true about the difference between cortical and juxtamedullary nephrons?
   a. Most nephrons are juxtamedullary.
   b. The efferent arterioles of cortical nephrons give rise to most of the vasa recta
   c. The afferent arterioles of the juxtamedullary nephrons give rise to most of the vasa recta.
   d. All cortical nephrons have a loop of Henle.
   e. Juxtamedullary nephrons generate a hyperosmotic medullary interstitium.

10. Which of the following is consistent with untreated chronic renal failure?
    a. proteinuria
    b. hypokalemia
    c. increased plasma 1,25-(OH)$_2$D
    d. increased plasma erythropoeitin
    e. increased plasma bicarbonate

Chapter 14 Quantitative and Thought Questions

(Answers appear in Appendix A.)

1. Substance T is present in the urine. Does this prove that it is filterable at the glomerulus?

2. Substance V is not normally present in the urine. Does this prove that it is neither filtered nor secreted?

3. The concentration of glucose in plasma is 100 mg/100 ml, and the GFR is 125 ml/min. How much glucose is filtered per minute?

4. A person is excreting abnormally large amounts of a particular amino acid. Just from the theoretical description of T_m-limited reabsorptive mechanisms in the text, list several possible causes.

5. The concentration of urea in urine is always much higher than the concentration in plasma. Does this mean that urea is secreted?

6. If a person takes a drug that blocks the reabsorption of sodium, what will happen to the reabsorption of water, urea, chloride, glucose, and amino acids and to the secretion of hydrogen ions?

7. Compare the changes in GFR and renin secretion occurring in response to a moderate hemorrhage in two individuals—one

taking a drug that blocks the sympathetic nerves to the kidneys and the other not taking such a drug.

8. If a person is taking a drug that completely inhibits angiotensin-converting enzyme, what will happen to aldosterone secretion when the person goes on a low-sodium diet?

9. In the steady state, is the amount of sodium chloride excreted daily in the urine of a normal person ingesting 12 g of sodium chloride per day: (a) 12 g/day or (b) less than 12 g/day? Explain.

10. A young woman who has suffered a head injury seems to have recovered but is thirsty all the time. What do you think might be the cause?

11. A patient has a tumor in the adrenal cortex that continuously secretes large amounts of aldosterone. What is this condition called, and what effects does this have on the total amount of sodium and potassium in her body?

12. A person is taking a drug that inhibits the tubular secretion of hydrogen ions. What effect does this drug have on the body's balance of sodium, water, and hydrogen ions?

Chapter 14 Answers to Physiological Inquiries

Figure 14–8 GFR will decrease because the increase in plasma osmotic force from albumin will oppose filtration.

Figure 14–11 Filtered load = GFR × Plasma glucose concentration
Excretion rate = Urine glucose concentration × Urine flow rate

Figure 14–14b It would decrease sodium reabsorption from the tubular fluid. This will result in an increase in urinary sodium excretion. The osmotic force of sodium will carry water with it, thus increasing urine output. Examples of such diuretics are triamterene and amiloride.

Figure 14–17 The increased vasopressin would cause maximal water reabsorption. Urine volume would be low (antidiuresis) and urine osmolarity would remain high. The continuous water reabsorption would cause a decrease in plasma sodium concentration (hyponatremia) due to dilution of sodium. Consequently, the plasma would have very low osmolarity. The decreased plasma osmolarity would not inhibit vasopressin secretion from the tumor because it is not controlled by the hypothalamic osmoreceptors. This is called the *syndrome of inappropriate antidiuretic hormone (SIADH)* and is one of several possible causes of hyponatremia in humans.

Figure 14–22 Under normal conditions, the redundant control of renin release, as indicated in this figure, as well as the participation of vasopressin (see Figure 14–24), would allow the maintenance of normal sodium and water balance even with denervated kidneys. However, during severe decreases in plasma volume, like in dehydration, the denervated kidney may not produce sufficient renin to maximally decrease sodium excretion.

Table 14–8 The patient has respiratory acidosis with renal compensation (hypercapnia with a normalization of arterial pH). The patient is hypoxic, which, with normal lung function, usually leads to hyperventilation and respiratory alkalosis. Therefore, the patient is likely to have chronic lung disease resulting in hypoxemia and retention of carbon dioxide (hypercapnia). We know it is chronic because the kidneys have had time to compensate for the acidosis by increasing the bicarbonate added to the blood thus restoring arterial pH almost to normal (see Figures 14–31 to 14–33).

chapter 5

The Digestion and Absorption of Food

Radiograph of abdomen with radioopaque contrast (barium enema).

the **gastrointestinal (GI) system** (**Figure 15–1**) includes the **gastrointestinal tract,** consisting of the mouth, pharynx, esophagus, stomach, small intestine, and large intestine, and the accessory organs, consisting of the salivary glands, liver, gallbladder, and pancreas. The accessory organs are not part of the tract but secrete substances into it via connecting ducts. The GI tract is also known as the alimentary canal or the digestive tract. The overall function of the gastrointestinal system is to process ingested foods into molecular forms that are then transferred, along with salts and water, to the body's internal environment, where the circulatory system can distribute them to cells. The gastrointestinal system is under the local neural control of the enteric nervous system and of the central nervous system.

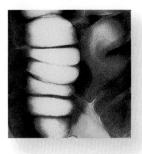

The adult gastrointestinal tract is a tube approximately 9 m (30 feet) in length, running through the body from mouth to anus. The lumen of the tract, like the hole in a doughnut, is continuous with the external environment, which means that its contents are technically outside the body. This fact is relevant to understanding some of the tract's properties. For example, the large intestine is inhabited by billions of bacteria, most of which are harmless and even beneficial in this location. However, if the same bacteria enter the internal environment, as may happen, for example, in the case of a ruptured appendix, they may cause a severe infection.

Most food enters the gastrointestinal tract as large particles containing macromolecules, such as proteins and polysaccharides, which are unable to cross the intestinal epithelium. Before ingested food can be absorbed, therefore, it must be dissolved and broken down into small molecules. This dissolving and breaking-down process is called **digestion** and is accomplished by the action of hydrochloric acid in the stomach, bile from the liver, and a variety of digestive enzymes released by the system's exocrine glands. Each of these substances is released into the lumen of the GI tract through the process of **secretion.**

The molecules produced by digestion then move from the lumen of the gastrointestinal tract across a layer of epithelial cells and enter the blood or lymph. This process is called **absorption.**

While digestion, secretion, and absorption are taking place, contractions of smooth muscles in the gastrointestinal tract wall serve two functions. They mix the luminal contents with the various secretions, and they move the contents through the tract from mouth to anus. These contractions are referred to as the **motility** of the gastrointestinal tract.

The functions of the gastrointestinal system can be described in terms of these four processes—digestion, secretion, absorption, and motility (**Figure 15–2**)—and the mechanisms controlling them.

The gastrointestinal system is designed to maximize absorption, and within fairly wide limits, will absorb as much of any particular substance as is ingested. With a few important exceptions (to be described later), therefore, the gastrointestinal system does not regulate the amount of nutrients absorbed or their concentrations in the internal environment. The regulation of the plasma concentration of the absorbed nutrients is primarily the function of the kidneys (Chapter 14) and a number of endocrine glands (Chapters 11 and 16).

Small amounts of certain metabolic end products are excreted via the gastrointestinal tract, primarily by way of the bile. The lungs and kidneys are responsible for the elimination of most of the body's waste products. The material known as **feces** leaves the system at the end of the gastrointestinal tract. Feces consists almost entirely of bacteria and ingested material that was neither digested nor absorbed—that is, material that was never actually part of the internal environment.

Overview: Functions of the Gastrointestinal Organs

Figure 15–3 presents an overview of the secretions and functions of the gastrointestinal organs. The gastrointestinal tract begins at the **mouth,** where digestion starts with chewing, which breaks up large pieces of food into smaller particles we can swallow. **Saliva,** secreted by three pairs of **salivary glands** (see Figure 15–1) located in the head, drains into the mouth through a series of short ducts. Saliva, which contains mucus, moistens and lubricates the food particles before swallowing. It also contains the enzyme **amylase,** which partially digests polysaccharides (complex sugars). A third function of saliva is to dissolve some of the food molecules. Only in the dissolved state can these molecules react with chemoreceptors in the mouth, giving rise to the sensation of taste (Chapter 7). Finally, saliva has antibacterial properties. See **Table 15–1** for the major functions of saliva.

The next segments of the tract, the **pharynx** and **esophagus,** do not contribute to digestion but provide the pathway for ingested materials to reach the stomach. The muscles in the walls of these segments control swallowing.

The **stomach** is a saclike organ located between the esophagus and the small intestine. Its functions are to store, dissolve, and partially digest the macromolecules in food and to regulate the rate at which the contents of the stomach empty into the small intestine. The glands lining the stomach wall secrete a strong acid, **hydrochloric acid,** and several protein-digesting enzymes collectively known as **pepsin.** Actually, a precursor of pepsin known as pepsinogen is secreted and converted to pepsin in the lumen of the stomach.

The primary function of hydrochloric acid is to dissolve the particulate matter in food. The acidic environment

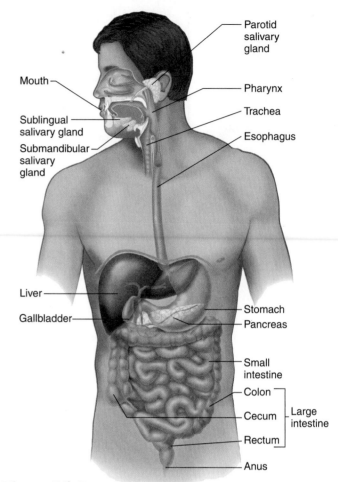

Figure 15–1

Anatomy of the gastrointestinal system. The liver overlies the gallbladder and a portion of the stomach, and the stomach overlies part of the pancreas.

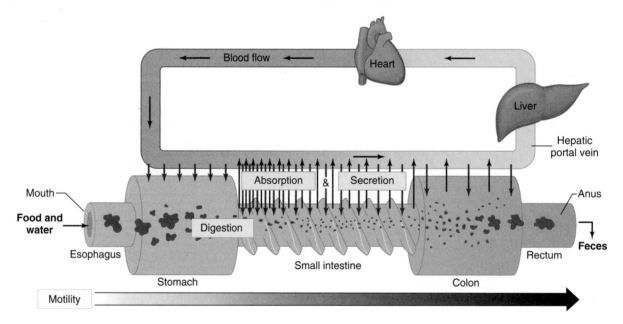

Figure 15–2

Four processes the gastrointestinal tract carries out: digestion, secretion, absorption, and motility. Outward (black) arrows indicate absorption of the products of digestion, water, minerals, and vitamins into the blood, most of which occurs in the small intestine. Inward-pointing (red) arrows represent the secretion of enzymes and bile salts into the GI tract. The wavy configuration of the small intestine represents muscular contractions (motility) throughout the tract.

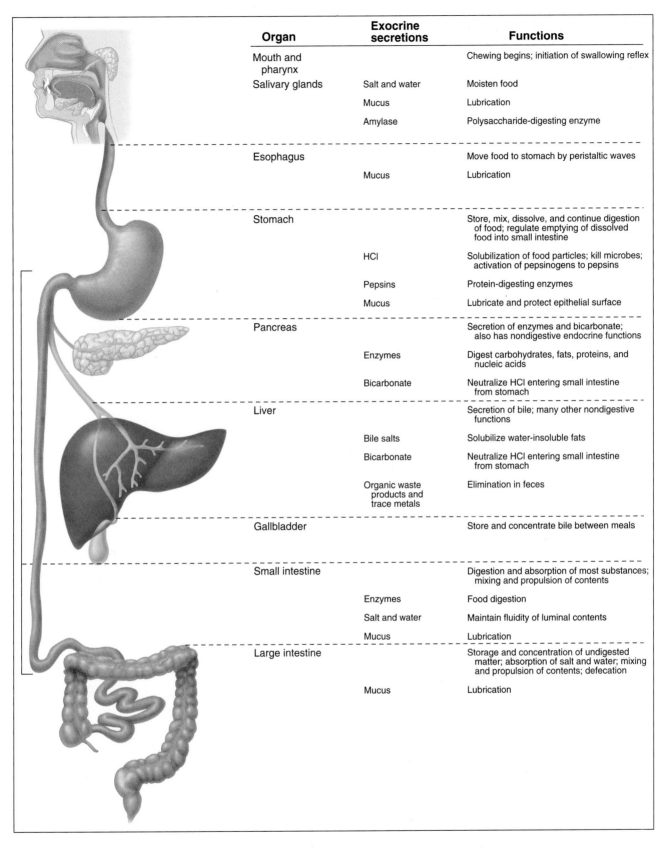

Organ	Exocrine secretions	Functions
Mouth and pharynx		Chewing begins; initiation of swallowing reflex
Salivary glands	Salt and water	Moisten food
	Mucus	Lubrication
	Amylase	Polysaccharide-digesting enzyme
Esophagus		Move food to stomach by peristaltic waves
	Mucus	Lubrication
Stomach		Store, mix, dissolve, and continue digestion of food; regulate emptying of dissolved food into small intestine
	HCl	Solubilization of food particles; kill microbes; activation of pepsinogens to pepsins
	Pepsins	Protein-digesting enzymes
	Mucus	Lubricate and protect epithelial surface
Pancreas		Secretion of enzymes and bicarbonate; also has nondigestive endocrine functions
	Enzymes	Digest carbohydrates, fats, proteins, and nucleic acids
	Bicarbonate	Neutralize HCl entering small intestine from stomach
Liver		Secretion of bile; many other nondigestive functions
	Bile salts	Solubilize water-insoluble fats
	Bicarbonate	Neutralize HCl entering small intestine from stomach
	Organic waste products and trace metals	Elimination in feces
Gallbladder		Store and concentrate bile between meals
Small intestine		Digestion and absorption of most substances; mixing and propulsion of contents
	Enzymes	Food digestion
	Salt and water	Maintain fluidity of luminal contents
	Mucus	Lubrication
Large intestine		Storage and concentration of undigested matter; absorption of salt and water; mixing and propulsion of contents; defecation
	Mucus	Lubrication

Figure 15–3

Functions of the gastrointestinal organs.

The Digestion and Absorption of Food

Table 15–1	Major Functions of Saliva
1. Moistens and lubricates food	
2. Digestion of polysaccharides by amylase	
3. Dissolves food	
4. Antibacterial actions	

in the **gastric** (adjective for "stomach") lumen alters the ionization of polar molecules, especially proteins, disrupting the extracellular network of connective tissue proteins that form the structural framework of the tissues in food. The proteins and polysaccharides released by hydrochloric acid's dissolving action are partially digested in the stomach by pepsin and amylase, the latter contributed by the salivary glands. Fat is a major food component that is not dissolved by acid.

Hydrochloric acid also kills most of the bacteria that enter along with food. This process is not completely effective, and some bacteria survive to colonize and multiply in the gastrointestinal tract, particularly the large intestine.

The digestive actions of the stomach reduce food particles to a solution known as **chyme,** which contains molecular fragments of proteins and polysaccharides, droplets of fat, and salt, water, and various other small molecules ingested in the food. Virtually none of these molecules, except water, can cross the epithelium of the gastric wall, and thus little absorption of organic nutrients occurs in the stomach.

Most absorption and the final stages of digestion occur in the next section of the tract, the **small intestine,** a tube about 2.4 cm in diameter and 3 m in length, that leads from the stomach to the large intestine. (The small intestine is almost twice as long if removed from the abdomen because the muscular wall loses its tone.) Hydrolytic enzymes in the small intestine break down molecules of intact or partially digested carbohydrates, fats, and proteins into monosaccharides, fatty acids, and amino acids. Some of these enzymes are on the luminal surface of the intestinal lining cells, whereas others are secreted by the pancreas and enter the intestinal lumen. The products of digestion are absorbed across the epithelial cells and enter the blood and/or lymph. Vitamins, minerals, and water, which do not require enzymatic digestion, are also absorbed in the small intestine.

The small intestine is divided into three segments: an initial short segment, the **duodenum,** is followed by the **jejunum** and then by the longest segment, the **ileum.** Normally, most of the chyme entering from the stomach is digested and absorbed in the first quarter of the small intestine, in the duodenum and jejunum.

Two major organs—the pancreas and liver—secrete substances that flow via ducts into the duodenum. The **pancreas,** an elongated gland located behind the stomach, has both endocrine (Chapter 16) and exocrine functions, but only the latter are directly involved in gastrointestinal function and are described in this chapter. The exocrine portion of the pancreas secretes digestive enzymes and a fluid rich in bicarbonate ions. The high acidity of the chyme coming from the stomach would inactivate the pancreatic enzymes in the small intestine if the acid were not neutralized by the bicarbonate ions in the pancreatic fluid.

The **liver,** a large organ located in the upper right portion of the abdomen, has a variety of functions, which are described in various chapters. This is a convenient place to provide, in **Table 15–2,** a comprehensive reference list of these **hepatic** (meaning "pertaining to the liver") functions and the chapters in which they are described. The breadth of hepatic function is highlighted by the devastating and often lethal effects of liver failure. We will be concerned in this chapter only with the liver's exocrine functions that are directly related to the secretion of **bile.**

Bile contains bicarbonate ions, cholesterol, phospholipids, bile pigments, a number of organic wastes and—most important—a group of substances collectively termed **bile salts.** The bicarbonate ions, like those from the pancreas, help neutralize acid from the stomach, whereas the bile salts, as we shall see, solubilize dietary fat. These fats would otherwise be insoluble in water, and their solubilization increases the rates at which they are digested and absorbed.

Bile is secreted by the liver into small ducts that join to form the common hepatic duct. Between meals, secreted bile is stored in the **gallbladder,** a small sac underneath the liver that branches from the common hepatic duct. The gallbladder concentrates the organic molecules in bile by absorbing salts and water. During a meal, the smooth muscles in the gallbladder wall contract, causing a concentrated bile solution to be injected into the duodenum via the **common bile duct** (**Figure 15–4**), an extension of the common hepatic duct. The gallbladder can be surgically removed without impairing bile secretion by the liver or its flow into the intestinal tract. In fact, many animals that secrete bile do not have a gallbladder.

In the small intestine, monosaccharides and amino acids are absorbed by specific transporter-mediated processes in the plasma membranes of the intestinal epithelial cells, whereas fatty acids enter these cells by diffusion. Most mineral ions are actively absorbed by transporters, and water diffuses passively down osmotic gradients.

The motility of the small intestine, brought about by the smooth muscles in its walls, (1) mixes the luminal contents with the various secretions, (2) brings the contents into contact with the epithelial surface where absorption takes place, and (3) slowly advances the luminal material toward the large intestine. Because most substances are absorbed in the small intestine, only small volumes of water, salts, and undigested material pass on to the **large intestine.** The large intestine temporarily stores the undigested material (some of which is metabolized by bacteria) and concentrates it by absorbing salts and water. Contractions of the **rectum,** the final segment of the large intestine, and relaxation of associated sphincter muscles expel the feces in a process called **defecation.**

The average American adult consumes about 500–800 g of food and 1200 ml of water per day, but this is only a frac-

Table 15–2 Summary of Liver Functions

A. Exocrine (digestive) functions (Chapter 15)

1. Synthesizes and secretes bile salts, which are necessary for adequate digestion and absorption of fats.
2. Secretes into the bile a bicarbonate-rich solution, that helps neutralize acid in the duodenum.

B. Endocrine functions

1. In response to growth hormone, secretes insulin-like growth factor I (IGF-I), which promotes growth by stimulating cell division in various tissues, including bone (Chapter 11).
2. Contributes to the activation of vitamin D (Chapter 11).
3. Forms triiodothyronine (T_3) from thyroxine (T_4) (Chapter 11).
4. Secretes angiotensinogen, which renin acts upon to form angiotensin I (Chapter 14).
5. Metabolizes hormones (Chapter 11).
6. Secretes cytokines involved in immune defenses (Chapter 18).

C. Clotting functions

1. Produces many of the plasma clotting factors, including prothrombin and fibrinogen (Chapter 12).
2. Produces bile salts, which are essential for the gastrointestinal absorption of vitamin K, which is, in turn, needed for the production of the clotting factors (Chapter 12).

D. Plasma proteins

1. Synthesizes and secretes plasma albumin (Chapter 12), acute phase proteins (Chapter 18), binding proteins for various hormones (Chapter 11) and trace elements (Chapter 12), lipoproteins (Chapter 16), and other proteins mentioned elsewhere in this table.

E. Organic metabolism (Chapter 16)

1. Converts plasma glucose into glycogen and triglycerides during absorptive period.
2. Converts plasma amino acids to fatty acids, which can be incorporated into triglycerides during absorptive period.
3. Synthesizes triglycerides and secretes them as lipoproteins during absorptive period.
4. Produces glucose from glycogen (glycogenolysis) and other sources (gluconeogenesis) during postabsorptive period and releases the glucose into the blood.
5. Converts fatty acids into ketones during fasting.
6. Produces urea, the major end product of amino acid (protein) catabolism, and releases it into the blood.

F. Cholesterol metabolism (Chapter 16)

1. Synthesizes cholesterol and releases it into the blood.
2. Secretes plasma cholesterol into the bile.
3. Converts plasma cholesterol into bile salts.

G. Excretory and degradative functions

1. Secretes bilirubin and other bile pigments into the bile (Chapter 15).
2. Excretes, via the bile, many endogenous and foreign organic molecules as well as trace metals (Chapter 18).
3. Biotransforms many endogenous and foreign organic molecules (Chapter 18).
4. Destroys old erythrocytes (Chapter 12).

---

tion of the material entering the lumen of the gastrointestinal tract. An additional 7000 ml of fluid from salivary glands, gastric glands, pancreas, liver, and intestinal glands are secreted into the tract each day (**Figure 15–5**). Of the approximately 8 L of fluid entering the tract, 99 percent is absorbed; only about 100 ml are normally lost in the feces. This small amount of fluid loss represents only 4 percent of the total fluids lost by the body each day. Most fluid loss is via the kidneys and respiratory system. Almost all the salts in the secreted fluids are also reabsorbed into the blood. Moreover, the secreted digestive enzymes are themselves digested, and the resulting amino acids are absorbed into the blood.

Finally, a critical component is the role of the central nervous system in the control of gastrointestinal functions. The

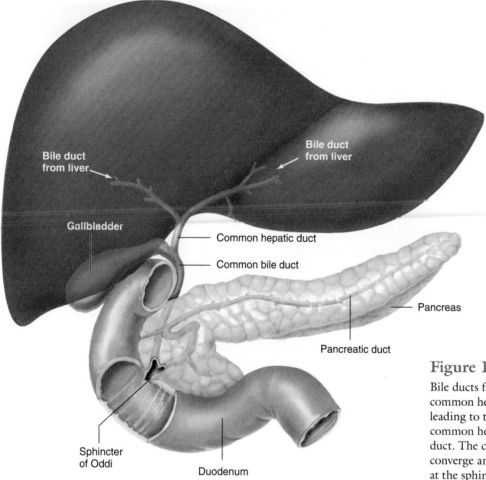

Figure 15–4

Bile ducts from the liver converge to form the common hepatic duct, from which branches the duct leading to the gallbladder. Beyond this branch, the common hepatic duct becomes the common bile duct. The common bile duct and the pancreatic duct converge and empty their contents into the duodenum at the sphincter of Oddi.

CNS receives information from the GI tract (afferent input) and has a vital influence on GI function (efferent output).

This completes our overview of the gastrointestinal system. Because its major tasks are digestion and absorption, we begin our more detailed description with these processes. Subsequent sections of the chapter will then describe, organ by organ, regulation of the secretions and motility that produce the optimal conditions for digestion and absorption. A prerequisite for this physiology, however, is an understanding of the structure of the gastrointestinal tract wall.

Structure of the Gastrointestinal Tract Wall

From the mid-esophagus to the anus, the wall of the gastrointestinal tract has the general structure illustrated in **Figure 15–6**. Most of the luminal (inside) surface is highly convoluted, a feature that greatly increases the surface area available for absorption. From the stomach on, this surface is covered by a single layer of epithelial cells linked together along the edges of their luminal surfaces by tight junctions.

Included in this epithelial layer are exocrine cells that secrete mucus into the lumen of the tract and endocrine cells that release hormones into the blood. Invaginations of the epithelium into the underlying tissue form exocrine glands that secrete acid, enzymes, water, and ions, as well as mucus, into the lumen.

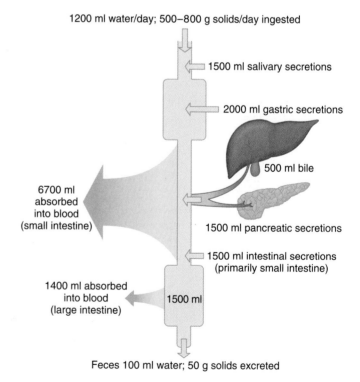

Figure 15–5

Average amounts of solids and fluid ingested, secreted, absorbed, and excreted from the gastrointestinal tract daily.

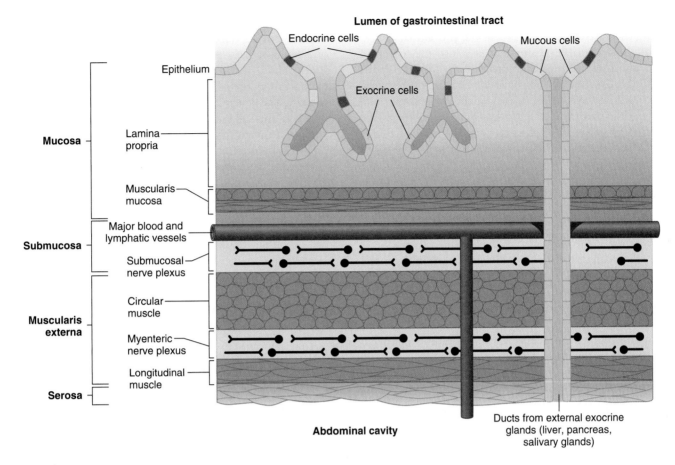

Figure 15–6

Structure of the gastrointestinal wall in longitudinal section. Not shown are the smaller blood vessels and lymphatics, neural connections between the two nerve plexuses, and neural terminations on muscles, glands, and epithelium.

Just below the epithelium is the lamina propria, which is a layer of connective tissue through which pass small blood vessels, nerve fibers, and lymphatic vessels. (These structures do not appear in Figure 15–6 but are in **Figure 15–7**.) The lamina propria is separated from underlying tissues by the mucularis mucosa, which is a thin layer of smooth muscle. The combination of these three layers—the epithelium, lamina propria, and muscularis mucosa—is called the **mucosa** (see Figure 15–6).

Beneath the mucosa is the **submucosa,** which is a second connective tissue layer. This layer contains a network of nerve cells, termed the **submucosal plexus,** and blood and lymphatic vessels whose branches penetrate into both the overlying mucosa and the underlying layers of smooth muscle called the **muscularis externa.** Contractions of these muscles provide the forces for moving and mixing the gastrointestinal contents. The muscularis externa has two layers: (1) a relatively thick inner layer of **circular muscle,** whose fibers are oriented in a circular pattern around the tube so that contraction produces a narrowing of the lumen, and (2) a thinner outer layer of **longitudinal muscle,** whose contraction shortens the tube. Between these two muscle layers is a second network of nerve cells known as the **myenteric plexus.**

Finally, surrounding the outer surface of the tube is a thin layer of connective tissue called the **serosa.** Thin sheets

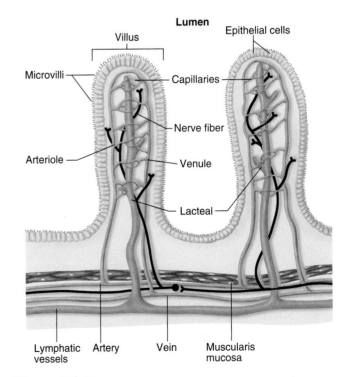

Figure 15–7

Structure of villi in small intestine.

The Digestion and Absorption of Food

of connective tissue connect the serosa to the abdominal wall, supporting the gastrointestinal tract in the abdominal cavity.

Extending from the luminal surface into the lumen of the small intestine are fingerlike projections known as **villi** (see Figure 15–7). The surface of each villus is covered with a layer of epithelial cells whose surface membranes form small projections called **microvilli** (also known collectively as the **brush border**) (**Figure 15–8**). The combination of folded mucosa, villi, and microvilli increases the small intestine's surface area about 600-fold over that of a flat-surfaced tube having the same length and diameter. The human small intestine's total surface area is about 250 to 300 m^2, roughly the area of a tennis court.

Epithelial surfaces in the gastrointestinal tract are continuously being replaced by new epithelial cells. In the small intestine, new cells arise by cell division from cells at the base of the villi. These cells differentiate as they migrate to the top of the villus, replacing older cells that die and are discharged into the intestinal lumen. These dead cells release into the lumen their intracellular enzymes, which then contribute to the digestive process. About 17 billion epithelial cells are replaced each day, and the entire epithelium of the small intestine is replaced approximately every five days. It is because of this rapid cell turnover that the lining of the intestinal tract is so susceptible to damage by agents that inhibit cell division, like anticancer drugs and by radiation therapy.

The center of each intestinal villus is occupied both by a single, blind-ended lymphatic vessel termed a **lacteal** and by a capillary network (see Figure 15–7). As we will see, most of the fat absorbed in the small intestine enters the lacteals. Material absorbed by the lacteals reaches the general circulation by eventually emptying from the lymphatic system into the thoracic duct.

Other absorbed nutrients enter the blood capillaries. The venous drainage from the small intestine, as well as from the large intestine, pancreas, and portions of the stomach, does not empty directly into the vena cava but passes first, via the **hepatic portal vein,** to the liver. There it flows through a second capillary network before leaving the liver to return to the heart. Thus, material absorbed into the intestinal capillaries, in contrast to the lacteals, can be processed by the liver before entering the general circulation. This is important because the liver contains enzymes that can metabolize (detoxify) harmful compounds that may have been ingested, thereby preventing them from entering the circulation. It also explains why certain drugs (e.g., testosterone) are given by injection or skin patch (because oral administration may injure the liver).

The gastrointestinal tract also has a variety of immune functions, allowing it to produce antibodies and fight infectious organisms that are not destroyed by the acidity of the stomach. For example, the small intestine has **Peyer's patches** and immune cells that secrete inflammatory mediators (e.g., cytokines; Chapter 18), that alter motility. These mediators may play a role in causing inflammation in certain autoimmune disorders such as *Crohn's disease* and *colitis* (*inflammatory bowel disease*), which are described at the end of this chapter.

Digestion and Absorption

Carbohydrate

The average daily intake of carbohydrates is about 250 to 300 g per day in a typical American diet. This represents about half the average daily intake of calories. About two-thirds of this carbohydrate is the plant polysaccharide starch, and most of the remainder consists of the disaccharides sucrose (table sugar) and lactose (milk sugar) (**Table 15–3**). Only small amounts of monosaccharides are normally present in the diet. Cellulose and certain other complex polysaccharides found in vegetable matter—referred to as **fiber**—are not broken down

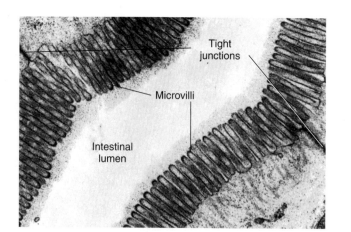

Figure 15–8

Microvilli on the surface of intestinal epithelial cells.

From D. W. Fawcett, *J. Histochem. Cytochem,* 13: 75–91 (1965). Courtesy of Susumo Ito.

Table 15–3	Carbohydrates in Food	
Class	**Examples**	**Made of**
Polysaccharides	Starch	Glucose
	Cellulose	Glucose
	Glycogen	Glucose
Disaccharides	Sucrose	Glucose-fructose
	Lactose	Glucose-galactose
	Maltose	Glucose-glucose
Monosaccharides	Glucose	
	Fructose	
	Galactose	

by the enzymes in the small intestine and pass on to the large intestine, where they are partially metabolized by bacteria.

The digestion of starch by salivary amylase begins in the mouth and briefly continues in the upper part of the stomach before gastric acid destroys the amylase. Starch digestion is completed in the small intestine by pancreatic amylase. The products of both amylases are the disaccharide maltose and a mixture of short, branched chains of glucose molecules. These products, along with ingested sucrose and lactose, are broken down into monosaccharides—glucose, galactose, and fructose—by enzymes located on the luminal membranes of the small intestine epithelial cells (brush border). These monosaccharides are then transported across the intestinal epithelium into the blood. Fructose enters the epithelial cells by facilitated diffusion, whereas glucose and galactose undergo secondary active transport coupled to sodium. These monosaccharides then leave the epithelial cells and enter the blood by way of facilitated diffusion transporters in the basolateral membranes of the epithelial cells. Most ingested carbohydrates are digested and absorbed within the first 20 percent of the small intestine.

Protein

A normal adult requires only 40 to 50 g of protein per day to supply essential amino acids and replace the nitrogen contained in amino acids that is converted to urea. A typical American diet contains about 70 to 90 g of protein per day. This represents about one-sixth of the average daily caloric intake. In addition, a large amount of protein, in the form of enzymes and mucus, is secreted into the gastrointestinal tract or enters it via the disintegration of epithelial cells. Regardless of source, most of the protein in the lumen is broken down into amino acids and absorbed by the small intestine.

Proteins are broken down to peptide fragments in the stomach by pepsin, and in the small intestine by **trypsin** and **chymotrypsin,** the major proteases secreted by the pancreas. These fragments are further digested to free amino acids by **carboxypeptidase** from the pancreas and **aminopeptidase,** located on the luminal membranes of the small intestine epithelial cells. These last two enzymes split off amino acids from the carboxyl and amino ends of peptide chains, respectively. At least 20 different peptidases are located on the luminal membrane of the epithelial cells, with various specificities for the peptide bonds they attack.

The free amino acids then enter the epithelial cells by secondary active transport coupled to sodium. There are multiple transporters with different specificities for the 20 types of amino acids. Short chains of two or three amino acids are also absorbed by a secondary active transport coupled to the hydrogen ion gradient. This contrasts with carbohydrate absorption, in which molecules larger than monosaccharides are not absorbed. Thus, as was the case for carbohydrates, luminal absorption of amino acids is a process that requires energy (ATP). Within the epithelial cell, these di- and tripeptides are hydrolyzed to amino acids, which then leave the cell and enter the blood through a facilitated diffusion carrier in the basolateral membranes. As with carbohydrates, protein digestion and absorption are largely completed in the upper portion of the small intestine.

Very small amounts of intact proteins are able to cross the intestinal epithelium and gain access to the interstitial fluid. They do so by a combination of endocytosis and exocytosis. The absorptive capacity for intact proteins is much greater in infants than in adults, and antibodies (proteins involved in the immunological defense system of the body) secreted into the mother's milk can be absorbed by the infant, providing some passive immunity until the infant begins to produce its own antibodies.

Fat

The average daily intake of fat is 70 to 100 g per day in a typical American diet. This represents about one-third of the average daily caloric intake. Most is in the form of triglycerides. Fat digestion occurs almost entirely in the small intestine. The major digestive enzyme in this process is pancreatic **lipase,** which catalyzes the splitting of bonds linking fatty acids to the first and third carbon atoms of glycerol, producing two free fatty acids and a monoglyceride as products:

$$\text{Triglyceride} \xrightarrow{\text{lipase}} \text{Monoglyceride} + 2 \text{ Fatty acids}$$

The fats in the ingested foods are insoluble in water and aggregate into large lipid droplets in the upper portion of the stomach. This is like a mixture of oil and vinegar after shaking. Because pancreatic lipase is a water-soluble enzyme, its digestive action in the small intestine can take place only at the *surface* of a lipid droplet. Therefore, if most of the ingested fat remained in large lipid droplets, the rate of lipid digestion would be very slow. The rate of digestion is, however, substantially increased by division of the large lipid droplets into a number of much smaller droplets, each about 1 mm in diameter, thereby increasing their surface area and accessibility to lipase action. This process is known as **emulsification,** and the resulting suspension of small lipid droplets is an emulsion.

The emulsification of fat requires (1) mechanical disruption of the large fat droplets into smaller droplets, and (2) an emulsifying agent, which acts to prevent the smaller droplets from reaggregating back into large droplets. The mechanical disruption is provided by contractile activity, occurring in the lower portion of the stomach and in the small intestine, which grinds and mixes the luminal contents. Phospholipids in food along with phospholipids and bile salts secreted in the bile provide the emulsifying agents.

Phospholipids are amphipathic molecules (Chapter 2) consisting of two nonpolar fatty acid chains attached to glycerol, with a charged phosphate group located on glycerol's third carbon. Bile salts are formed from cholesterol in the liver and are also amphipathic (**Figure 15–9**). The nonpolar portions of the phospholipids and bile salts associate with the nonpolar interior of the lipid droplets, leaving the polar portions exposed at the water surface. There they repel other lipid droplets that are similarly coated with these emulsifying agents, thereby preventing their reaggregation into larger fat droplets (**Figure 15–10**).

The coating of the lipid droplets with these emulsifying agents, however, impairs the accessibility of the water-soluble lipase to its lipid substrate. To overcome this problem, the pancreas secretes a protein known as **colipase,** which is amphipathic and lodges on the lipid droplet surface. Colipase binds the lipase enzyme, holding it on the surface of the lipid droplet.

Although emulsification speeds up digestion, absorption of the water-insoluble products of the lipase reaction would still be very slow if it were not for a second action of the bile salts, the formation of **micelles,** which are similar in structure to emulsion droplets but much smaller—4 to 7 nm in diameter. Micelles consist of bile salts, fatty acids, monoglycerides, and phospholipids all clustered together with the polar ends of each molecule oriented toward the micelle's surface and the nonpolar portions forming the micelle's core (**Figure 15–11**). Also included in the core of the micelle are small amounts of fat-soluble vitamins and cholesterol.

How do micelles increase absorption? Although fatty acids and monoglycerides have an extremely low solubility in water, a few molecules do exist in solution and are free to diffuse across the lipid portion of the luminal plasma membranes of the epithelial cells lining the small intestine. Micelles, containing the products of fat digestion, are in equilibrium with the small concentration of fat digestion products that are free in solution. Thus, micelles are continuously breaking down and reforming. When a micelle breaks down, its contents are released into the solution and they become available to diffuse across the intestinal lining. As the concentrations of free lipids fall, because of their diffusion into epithelial cells, more lipids are released into the free phase as micelles break down (see Figure 15–11). Thus, the micelles provide a means of keeping most of the insoluble fat digestion products in small, soluble aggregates, while at the same time replenishing the small amount of products in solu-

tion and, therefore, free to diffuse into the intestinal epithelium. Note that it is not the micelle that is absorbed, but rather the individual lipid molecules released from the micelle.

Although fatty acids and monoglycerides enter epithelial cells from the intestinal lumen, it is triglycerides that are released on the other side of the cell into the interstitial fluid. In other words, during their passage through the epithelial cells, fatty acids and monoglycerides are resynthesized into triglycerides. This occurs in the smooth endoplasmic reticulum, where the enzymes for triglyceride synthesis are located. This process lowers the concentration of cytosolic free fatty acids and monoglycerides and thus maintains a diffusion gradient for these molecules into the cell. Within this organelle, the resynthesized fat aggregates into small droplets coated with amphipathic proteins that perform an emulsifying function similar to that of bile salts.

The exit of these fat droplets from the cell follows the same pathway as a secreted protein. Vesicles containing the droplet pinch off the endoplasmic reticulum, are processed through the Golgi apparatus, and eventually fuse with the plasma membrane, releasing the fat droplet into the interstitial fluid. These 1 micron-diameter, extracellular fat droplets are known as **chylomicrons.** Chylomicrons contain not only triglycerides but other lipids (including phospholipids, cholesterol, and fat-soluble vitamins) that have been absorbed by the same process that led to fatty acid and monoglyceride movement into the epithelial cells of the small intestine.

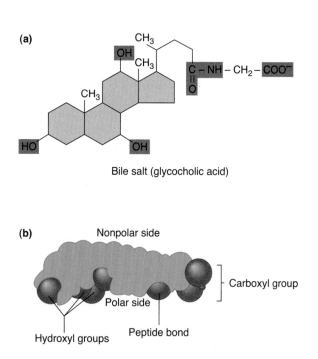

Figure 15–9

Structure of bile salts. (a) Chemical formula of glycocholic acid, one of several bile salts secreted by the liver (polar groups in color). Note the similarity to the structure of steroids (see Figure 11–3). (b) Three-dimensional structure of a bile salt, showing its polar and nonpolar surfaces.

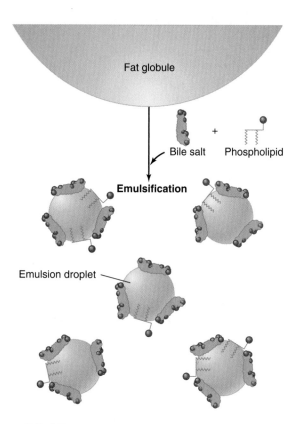

Figure 15–10

Emulsification of fat by bile salts and phospholipids. Note that the nonpolar sides (green) of bile salts and phospholipids are oriented toward fat, while the polar sides (red) of these compounds are oriented outward.

The chylomicrons released from the epithelial cells pass into lacteals—lymphatic capillaries in the intestinal villi—rather than into the blood capillaries. The chylomicrons cannot enter the blood capillaries because the basement membrane (an extracellular glycoprotein layer) at the outer surface of the capillary provides a barrier to the diffusion of large chylomicrons. In contrast, the lacteals do not have basement membranes and have large, slit pores between their endothelial cells that allow the chylomicrons to pass into the lymph. The lymph from the small intestine, as from everywhere else in the body, eventually empties into systemic veins via the thoracic duct. In Chapter 16 we describe how the lipids in the circulating blood chylomicrons are made available to the cells of the body.

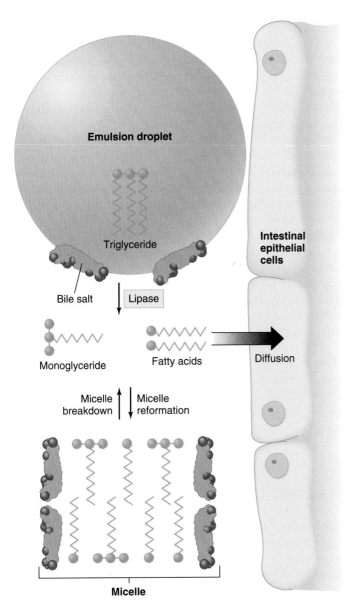

Figure 15–11

The products of fat digestion by lipase are held in solution in the micellar state, combined with bile salts and phospholipids. For simplicity, the phospholipids and colipase (see text) are not shown, and the size of the micelle is greatly exaggerated. Note that micelles and free fatty acids are in equilibrium so that as fatty acids are absorbed, more can be released from the micelles.

The Digestion and Absorption of Food

Figure 15–12 summarizes the pathway fat takes in moving from the intestinal lumen into the lymphatic system.

Vitamins

The fat-soluble vitamins—A, D, E, and K—follow the pathway for fat absorption described in the previous section. They are solubilized in micelles; thus, any interference with the secretion of bile or the action of bile salts in the intestine decreases the absorption of the fat-soluble vitamins (***malabsorption***). Malabsorption syndromes can lead to deficiency of fat-soluble vitamins. For example, ***nontropical sprue***—a loss of intestinal surface area due to a sensitivity to the wheat protein **gluten**—can lead to vitamin D malabsorption, which ultimately results in a decrease in calcium absorption in the GI tract (Chapter 11).

With one exception, water-soluble vitamins are absorbed by diffusion or mediated transport. The exception, vitamin B_{12} (cyanocobalamin), is a very large, charged molecule. To

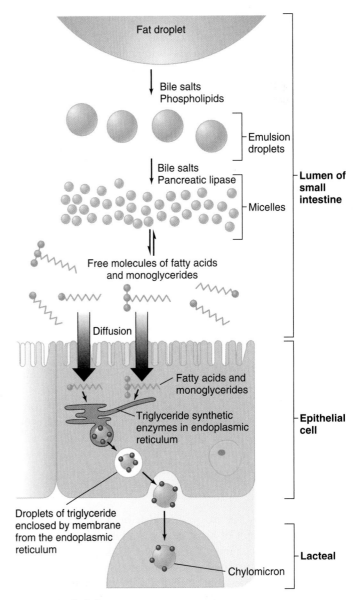

Figure 15–12

Summary of fat absorption across the epithelial cells of the small intestine.

be absorbed, vitamin B_{12} must first bind to a protein, known as **intrinsic factor,** secreted by the acid-secreting cells in the stomach. Intrinsic factor with bound vitamin B_{12} then binds to specific sites on the epithelial cells in the lower portion of the ileum, where vitamin B_{12} is absorbed by endocytosis. As described in Chapter 12, vitamin B_{12} is required for erythrocyte formation, and deficiencies result in *pernicious anemia.* This form of anemia may occur when the stomach either has been removed (for example, to treat ulcers or gastric cancer) or fails to secrete intrinsic factor (often due to autoimmune destruction of parietal cells). Because the absorption of vitamin B_{12} occurs in the lower part of the ileum, removal or dysfunction of this segment due to disease can also result in pernicious anemia. Although normal individuals can absorb oral vitamin B_{12}, it is not very effective in patients with pernicious anemia because of the absence of intrinsic factor. Therefore, the treatment of pernicious anemia usually requires injections of vitamin B_{12} .

Water and Minerals

Water is the most abundant substance in chyme. Approximately 8000 ml of ingested and secreted water enter the small intestine each day, but only 1500 ml pass on to the large intestine because 80 percent of the fluid is absorbed in the small intestine. Small amounts of water are absorbed in the stomach, but the stomach has a much smaller surface area available for diffusion and lacks the solute-absorbing mechanisms that create the osmotic gradients necessary for net water absorption. The epithelial membranes of the small intestine are very permeable to water, and net water diffusion occurs across the epithelium whenever a water concentration difference is established by the active absorption of solutes. The mechanisms coupling solute and water absorption by epithelial cells were described in Chapter 4.

Sodium ions account for much of the actively transported solute because they constitute the most abundant solute in chyme. Sodium absorption is a primary active process, using the Na^+/K^+-ATPase pumps as described in Chapter 4 and similar to that for renal tubular sodium and water reabsorption (Chapter 14). Chloride and bicarbonate ions are absorbed with the sodium ions and contribute another large fraction of the absorbed solute.

Other minerals present in smaller concentrations, such as potassium, magnesium, and calcium, are also absorbed, as are trace elements such as iron, zinc, and iodide. Consideration of the transport processes associated with each of these is beyond the scope of this book, and we shall briefly consider as an example the absorption of only one—iron. Calcium absorption and its regulation were described in Chapter 11.

Iron

Iron is necessary for normal health because it is the O_2-binding component of hemoglobin, and it is also a key component of many enzymes. Only about 10 percent of ingested iron is absorbed into the blood each day. Iron ions are actively transported into intestinal epithelial cells, where most of them are incorporated into ferritin, the protein-iron complex that functions as an intracellular iron store (Chapter 12). The absorbed iron that does not bind to ferritin is released on the blood side, where it circulates throughout the body bound to the plasma protein transferrin. Most of the iron bound to ferritin in the epithelial cells is released back into the intestinal lumen when the cells at the tips of the villi disintegrate, and the iron is then excreted in the feces.

Iron absorption depends on the body's iron content. When body stores are ample, the increased concentration of free iron in the plasma and intestinal epithelial cells leads to an increased transcription of the gene encoding the ferritin protein and thus an increased synthesis of ferritin. This results in the increased binding of iron in the intestinal epithelial cells and a reduction in the amount of iron released into the blood. When body stores of iron decrease (for example, after a loss of blood), the production of intestinal ferritin decreases. This leads to a decrease in the amount of iron bound to ferritin, thereby increasing the unbound iron released into the blood.

Once iron has entered the blood, the body has very little means of excreting it, so it accumulates in tissues. Although the control mechanisms for iron absorption tend to maintain the iron content of the body at a fairly constant level, a very large ingestion of iron can overwhelm them, leading to an increased deposition of iron in tissues and producing toxic effects such as skin pigmentation, diabetes mellitus, liver and heart failure, and decreased testicular function. This condition is termed *hemochromatosis.* Some people have genetically defective control mechanisms and therefore develop hemochromatosis even when iron ingestion is normal. They can be treated with frequent blood withdrawal (*phlebotomy*), which removes iron contained in red blood cells (hemoglobin) from the body.

Iron absorption also depends on the types of food ingested because it binds to many negatively charged ions in food, which can retard its absorption. For example, iron in ingested liver is much more absorbable than iron in egg yolk because the latter contains phosphates that bind the iron to form an insoluble and unabsorbable complex.

The absorption of iron is typical of that of most trace metals in several respects: (1) cellular storage proteins and plasma carrier proteins are involved, and (2) the control of absorption, rather than urinary excretion, is the major mechanism for the homeostatic control of the body's content of the trace metal.

How Are Gastrointestinal Processes Regulated?

Unlike control systems that regulate variables in the internal environment, the control mechanisms of the gastrointestinal system regulate conditions in the lumen of the tract. With few exceptions, like those just discussed for iron and other trace metals, these control mechanisms are governed by the volume and composition of the luminal contents rather than by the nutritional state of the body.

Basic Principles

Gastrointestinal reflexes are initiated by a relatively small number of luminal stimuli: (1) distension of the wall by the volume of the luminal contents; (2) chyme osmolarity (total solute concentration); (3) chyme acidity; and (4) chyme concentrations of specific digestion products like monosaccharides, fatty acids, peptides, and amino acids. These stimuli act on mechanoreceptors, osmoreceptors, and chemoreceptors located in the wall of the tract which trigger reflexes that influence the effectors—the muscle layers in the wall of the tract and the exocrine glands that secrete substances into its lumen.

Neural Regulation

The gastrointestinal tract has its own local nervous system, known as the **enteric nervous system,** in the form of two nerve networks, the myenteric plexus and the submucosal plexus (see Figure 15–6). These neurons either synapse with other neurons in the plexus or end near smooth muscles, glands, and epithelial cells. Many axons leave the myenteric plexus and synapse with neurons in the submucosal plexus, and vice versa, so that neural activity in one plexus influences the activity in the other. Moreover, stimulation at one point in the plexus can lead to impulses that are conducted both up and down the tract. For example, stimuli in the upper part of the small intestine may affect smooth muscle and gland activity in the stomach as well as in the lower part of the intestinal tract. In general, the myenteric plexus influences smooth muscle activity whereas the submucosal plexus influences secretory activity.

The enteric nervous system contains adrenergic and cholinergic neurons as well as neurons that release other neurotransmitters, such as nitric oxide, several neuropeptides, and ATP.

Many of the effectors mentioned earlier—muscle cells and exocrine glands—are supplied by neurons that are part of the enteric nervous system. This permits neural reflexes that are completely within the tract—that is, independent of the CNS. In addition, nerve fibers from both the sympathetic and parasympathetic branches of the autonomic nervous system enter the intestinal tract and synapse with neurons in both plexuses. Via these pathways, the CNS can influence the motility and secretory activity of the gastrointestinal tract.

Thus, two types of neural reflex arcs exist (**Figure 15–13**): (1) **short reflexes** from receptors through the nerve plexuses to effector cells; and (2) **long reflexes** from receptors in the tract to the CNS by way of afferent nerves, and back to the nerve plexuses and effector cells by way of autonomic nerve fibers.

Finally, it should be noted that not all neural reflexes are initiated by signals *within* the tract. Hunger, the sight or smell of food, and the emotional state of an individual can have significant effects on the gastrointestinal tract, effects that are mediated by the CNS via autonomic neurons.

Hormonal Regulation

The hormones that control the gastrointestinal system are secreted mainly by endocrine cells scattered throughout the epithelium of the stomach and small intestine. That is, these cells are not clustered into discrete organs like the thyroid or adrenal glands. One surface of each endocrine cell is exposed to the lumen of the gastrointestinal tract. At this surface, various chemical substances in the chyme stimulate the cell to release its hormones from the opposite side of the cell into the

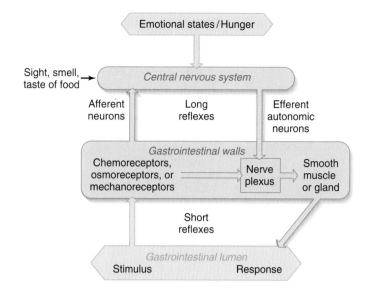

Figure 15–13

Long and short neural reflex pathways activated by stimuli in the gastrointestinal tract. The long reflexes utilize neurons that link the central nervous system to the gastrointestinal tract. Chemoreceptors are stimulated by chemicals, osmoreceptors are sensitive to changes in osmolarity (salt concentration), and mechanoreceptors respond to distention of the gastrointestinal wall.

blood. The gastrointestinal hormones reach their target cells primarily via the circulation.

The four best-understood gastrointestinal hormones are **secretin, cholecystokinin (CCK), gastrin,** and **glucose-dependent insulinotropic peptide (GIP).** They, as well as several candidate hormones, also exist in the CNS and in gastrointestinal plexus neurons, where they function as neurotransmitters or neuromodulators.

Table 15–4, which summarizes the characteristics of these GI hormones, not only serves as a reference for future discussions but also illustrates the following generalizations: (1) each hormone participates in a feedback control system that regulates some aspect of the GI luminal environment, and (2) most GI hormones affect more than one type of target cell.

These two generalizations can be illustrated by CCK. The presence of fatty acids and amino acids in the small intestine triggers CCK secretion from cells in the small intestine into the blood. Circulating CCK then stimulates the pancreas to increase the secretion of digestive enzymes. CCK also causes the gallbladder to contract, delivering to the intestine the bile salts required for micelle formation. As fat and amino acids are absorbed, the stimuli (fatty acids and amino acids in the lumen) for CCK release are removed.

In many cases, a single effector cell contains receptors for more than one hormone, as well as receptors for neurotransmitters and paracrine agents. The result is a variety of inputs that can affect the cell's response. One such event is the phenomenon known as **potentiation,** which is exemplified by the interaction between secretin and CCK. Secretin strongly

Table 15–4	Properties of Gastrointestinal Hormones			
	Gastrin	**CCK**	**Secretin**	**GIP**
Chemical class	Peptide	Peptide	Peptide	Peptide
Site of production	Antrum of stomach	Small intestine	Small intestine	Small intestine
Stimuli for hormone release	Amino acids, peptides in stomach; parasympathetic nerves	Amino acids, fatty acids in small intestine	Acid in small intestine	Glucose, fat in small intestine
Factors inhibiting hormone release	Acid in stomach; somatostatin			
		Target Organ Responses		
Stomach Acid secretion Motility Growth	Stimulates Stimulates Stimulates	Inhibits Inhibits	Inhibits Inhibits	
Pancreas Bicarbonate secretion Enzyme secretion Insulin secretion Growth of exocrine pancreas	Stimulates	Potentiates secretin's actions Stimulates Stimulates	Stimulates Potentiates CCK's actions Stimulates	Stimulates
Liver (bile ducts) Bicarbonate secretion		Potentiates secretin's actions	Stimulates	
Gallbladder Contraction		Stimulates		
Sphincter of Oddi		Relaxes		
Small intestine Motility Growth	Stimulates ileum Stimulates			
Large intestine	Stimulates mass movement			

Table 15–4 physiological *inquiry* (pi)

- Gastrinomas are tumors of the GI tract that secrete gastrin, leading to very high plasma concentrations. What might be some of the effects of a gastrinoma?

Answer can be found at end of chapter.

stimulates pancreatic bicarbonate secretion, whereas CCK is a weak stimulus of bicarbonate secretion. Both hormones together, however, stimulate pancreatic bicarbonate secretion more strongly than would be predicted by the sum of their individual stimulatory effects. This is because CCK amplifies the response to secretin. One of the consequences of potentiation is that small changes in the plasma concentration of one gastrointestinal hormone can have large effects on the actions of other gastrointestinal hormones.

Many other peptide hormones that have recently been described are released by or influence gastrointestinal function. **Leptin** is released from fat cells and influences food intake and metabolic rate. **Ghrelin** is released from the stomach during fasting. The details of these GI hormones will be described in Chapter 16. Briefly, leptin induces satiety, while ghrelin stimulates hunger.

In addition to their stimulation (or in some cases inhibition) of effector cell functions, the gastrointestinal hormones also have trophic (growth-promoting) effects on various tissues, including the gastric and intestinal mucosa and the exocrine portions of the pancreas. Certain gastrointestinal hormones (like GIP) also influence the endocrine pancreas, thereby amplifying the insulin response to eating and absorbing a meal (Chapter 16).

Phases of Gastrointestinal Control

The neural and hormonal control of the gastrointestinal system is, in large part, divisible into three phases—cephalic, gastric, and intestinal—according to where the stimulus is perceived.

The **cephalic phase** is initiated when receptors in the head (*cephalic*) are stimulated by sight, smell, taste, and chewing. Various emotional states can also initiate this phase. The efferent pathways for these reflexes are primarily mediated by parasympathetic fibers carried in the vagus nerves. These fibers activate neurons in the gastrointestinal nerve plexuses, which in turn affect secretory and contractile activity.

Four types of stimuli in the stomach initiate the reflexes that constitute the **gastric phase** of regulation: distension, acidity, amino acids, and peptides formed during the digestion of ingested protein. The responses to these stimuli are mediated by short and long neural reflexes and by release of the hormone gastrin.

Finally, the **intestinal phase** is initiated by stimuli in the intestinal tract: distension, acidity, osmolarity, and various digestive products. The intestinal phase is mediated by both short and long neural reflexes and by the gastrointestinal hormones secretin, CCK, and GIP, all of which are secreted by endocrine cells in the small intestine.

We reemphasize that each of these phases is named for the site at which the various stimuli initiate the reflex and not for the sites of effector activity. Each phase is characterized by efferent output to virtually all organs in the gastrointestinal tract. Also, these phases do not occur in temporal sequence except at the very beginning of a meal. Rather, during ingestion and the much longer absorptive period, reflexes characteristic of all three phases may be occurring simultaneously.

Keeping in mind the neural and hormonal mechanisms available for regulating gastrointestinal activity, we can now examine the specific contractile and secretory processes that occur in each segment of the gastrointestinal system.

Mouth, Pharynx, and Esophagus

Chewing

Chewing is controlled by the somatic nerves to the skeletal muscles of the mouth and jaw. In addition to the voluntary control of these muscles, rhythmical chewing motions are reflexly activated by the pressure of food against the gums, hard palate at the roof of the mouth, and tongue. Activation of these mechanoreceptors leads to reflexive inhibition of the muscles holding the jaw closed. The resulting relaxation of the jaw reduces the pressure on the various mechanoreceptors, leading to a new cycle of contraction and relaxation.

Although chewing prolongs the subjective pleasure of taste, it does not appreciably alter the rate at which food is digested and absorbed. On the other hand, attempting to swallow a large particle of food can lead to choking if the particle lodges over the trachea, blocking the entry of air into the lungs.

Saliva

The secretion of saliva is controlled by both sympathetic and parasympathetic neurons. Unlike their antagonistic activity in most organs, both systems stimulate salivary secretion, with the parasympathetics producing the greater response. There is no hormonal regulation of salivary secretion. In the absence of ingested material, a low rate of salivary secretion keeps the mouth moist. The smell or sight of food induces a cephalic phase of salivary secretion. This reflex can be conditioned to other cues, a phenomenon made famous by Pavlov. Salivary secretion can increase markedly in response to a meal. This reflex response is initiated by chemoreceptors (acidic fruit juices are a particularly strong stimulus) and pressure receptors in the walls of the mouth and on the tongue.

Increased saliva secretion is accomplished by a large increase in blood flow to the salivary glands, which is mediated primarily by an increase in parasympathetic neural activity. The volume of saliva secreted per gram of tissue is the largest secretion of any of the body's exocrine glands.

Sjögren's syndrome is a fascinating immune disorder in which many different exocrine glands are rendered nonfunctional by the infiltration of white blood cells and immune complexes. The loss of salivary gland function frequently occurs and can be treated by taking frequent sips of water and with oral fluoride treatment to prevent tooth decay.

Swallowing

Swallowing is a complex reflex initiated when pressure receptors in the walls of the pharynx are stimulated by food or drink forced into the rear of the mouth by the tongue. These receptors send afferent impulses to the **swallowing center** in the medulla oblongata of the brainstem. This center then elicits swallowing via efferent fibers to the muscles in the pharynx and esophagus as well as to the respiratory muscles.

As the ingested material moves into the pharynx, the soft palate elevates and lodges against the back wall of the pharynx, preventing food from entering the nasal cavity (**Figure 15–14b**). Impulses from the swallowing center inhibit respiration, raise the larynx, and close the **glottis** (the area around the vocal cords and the space between them), keeping food from moving into the trachea. As the tongue forces the food farther back into the pharynx, the food tilts a flap of tissue, the **epiglottis,** backward to cover the closed glottis (**Figure 15–14c**), thereby preventing food from entering the trachea (**aspiration**).

The next stage of swallowing occurs in the esophagus, the foot-long tube that passes through the thoracic cavity, penetrates the diaphragm (which separates the thoracic cavity from the abdominal cavity), and joins the stomach a few centimeters below the diaphragm. Skeletal muscles surround the upper third of the esophagus, smooth muscles the lower two-thirds.

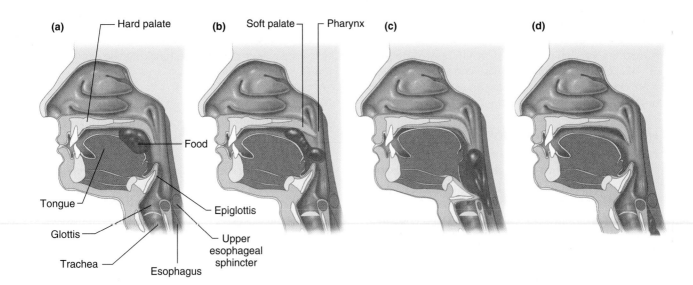

Figure 15–14

Movements of food through the pharynx and upper esophagus during swallowing. (a) The tongue pushes the food bolus to the back of the pharynx. (b) The soft palate elevates to prevent food from entering the nasal passages. (c) The epiglottis covers the glottis to prevent food or liquid from entering the trachea (aspiration), and the upper esophageal sphincter relaxes. (d) Food descends into the esophagus.

Figure 15–14b physiological *inquiry* 🅟🅘

- What are some of the consequences of aspiration?

Answer can be found at end of chapter.

As described in Chapter 13, the pressure in the thoracic cavity is negative relative to atmospheric pressure, and this subatmospheric pressure is transmitted across the thin wall of the intrathoracic portion of the esophagus to the lumen. In contrast, the luminal pressure in the pharynx at the opening to the esophagus is equal to atmospheric pressure, and the pressure at the opposite end of the esophagus in the stomach is slightly greater than atmospheric pressure. Therefore, pressure differences could tend to force both air (from above) and gastric contents (from below) into the esophagus. This does not occur, however, because both ends of the esophagus are normally closed by the contraction of **sphincter** muscles. A ring of skeletal muscles surrounds the esophagus just below the pharynx and forms the **upper esophageal sphincter,** whereas the smooth muscles in the last portion of the esophagus form the **lower esophageal sphincter** (**Figure 15–15**).

The esophageal phase of swallowing begins with relaxation of the upper esophageal sphincter. Immediately after the food has passed, the sphincter closes, the glottis opens, and breathing resumes. Once in the esophagus, the food moves toward the stomach by a progressive wave of muscle contractions that proceeds along the esophagus, compressing the lumen and forcing the food ahead. Such waves of contraction in the muscle layers surrounding a tube are known as **peristaltic waves.** One esophageal peristaltic wave takes about nine seconds to reach the stomach. Swallowing can occur even when a person is upside down or in zero gravity (outer space)

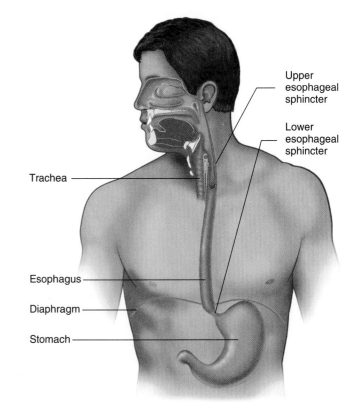

Figure 15–15

Location of upper and lower esophageal sphincters.

because it is not primarily gravity but the peristaltic wave that moves the food to the stomach.

The lower esophageal sphincter opens and remains relaxed throughout the period of swallowing, allowing the arriving food to enter the stomach. After the food passes, the sphincter closes, resealing the junction between the esophagus and the stomach.

Swallowing is an example of a reflex in which multiple responses occur in a sequence determined by the pattern of synaptic connections between neurons in a brain coordinating center. Both skeletal and smooth muscles are involved, so the swallowing center must direct efferent activity in both somatic nerves (to skeletal muscle) and autonomic nerves (to smooth muscle). Simultaneously, afferent fibers from receptors in the esophageal wall send information to the swallowing center; this can alter the efferent activity. For example, if a large food particle does not reach the stomach during the initial peristaltic wave, the maintained distension of the esophagus by the particle activates receptors that initiate reflexes, causing repeated waves of peristaltic activity (**secondary peristalsis**). This is usually not accompanied by the initial pharyngeal events of swallowing.

The ability of the lower esophageal sphincter to maintain a barrier between the stomach and the esophagus when swallowing is not taking place is aided by the fact that the last portion of the esophagus lies below the diaphragm and is subject to the same abdominal pressures as the stomach. In other words, if the pressure in the abdominal cavity increases, for example, during cycles of respiration or contraction of the abdominal muscles, the pressures on both the gastric contents and the terminal segment of the esophagus are raised together. This prevents the formation of a pressure gradient between the stomach and esophagus that could force the stomach's contents into the esophagus.

During pregnancy, the growth of the fetus not only increases the pressure on the abdominal contents but also pushes the terminal segment of the esophagus through the diaphragm into the thoracic cavity. The sphincter is therefore no longer assisted by changes in abdominal pressure. Consequently, during the last half of pregnancy, increased abdominal pressure tends to force some of the gastric contents up into the esophagus. The hydrochloric acid from the stomach irritates the esophageal walls, producing pain known as *heartburn* (because the pain appears to be located in the area of the heart). Heartburn often subsides in the last weeks of pregnancy prior to delivery, as the uterus descends lower into the pelvis, decreasing the pressure on the stomach.

Heartburn also occurs in the absence of pregnancy. Some people have less efficient lower esophageal sphincters, resulting in repeated episodes of gastric contents refluxing into the esophagus (*gastroesophageal reflux*), heartburn, and in extreme cases, ulceration, scarring, obstruction, or perforation of the lower esophagus. Heartburn can also occur after a large meal, which can raise the pressure in the stomach enough to force acid into the esophagus. Gastroesophageal reflux can also cause coughing and irritation of the larynx in the absence of any esophageal symptoms, and it has even been implicated in the onset of asthmatic symptoms in susceptible individuals.

The lower esophageal sphincter not only undergoes brief periods of relaxation during a swallow but also in the absence of a swallow. During these periods of relaxation, small amounts of the acid contents from the stomach normally reflux into the esophagus. The acid in the esophagus triggers a secondary peristaltic wave and also stimulates increased salivary secretion, which helps to neutralize the acid and clear it from the esophagus.

Stomach

The epithelial layer lining the stomach invaginates into the mucosa, forming many tubular glands. Glands in the thin-walled upper portions of the **body** of the stomach (**Figure 15–16**) secrete mucus, hydrochloric acid, and the enzyme precursor **pepsinogen.** The uppermost part of the body of the stomach is called the **fundus.** The lower portion of the stomach, the **antrum,** has a much thicker layer of smooth muscle. The glands in this region secrete little acid but contain the endocrine cells that secrete the hormone gastrin.

The cells at the opening of the glands secrete mucus (**Figure 15–17**). Lining the walls of the glands are **parietal cells** (also known as oxyntic cells), which secrete acid and intrinsic factor, and **chief cells,** which secrete pepsinogen. Thus, each of the three major exocrine secretions of the stomach—mucus, acid, and pepsinogen—is secreted by a different cell type. The gastric glands in the antrum also contain **enteroendocrine cells,** which secrete gastrin. In addition, **enterochromaffin-like (ECL) cells,** which release the paracrine agent histamine, and endocrine (D) cells that secrete the peptide messenger **somatostatin,** are scattered throughout the tubular glands.

HCl Secretion

The stomach secretes about 2 L of hydrochloric acid per day. The concentration of hydrogen ions in the lumen of the stomach may reach >150 mM, which is 1 to 3 million times higher than the concentration in the blood.

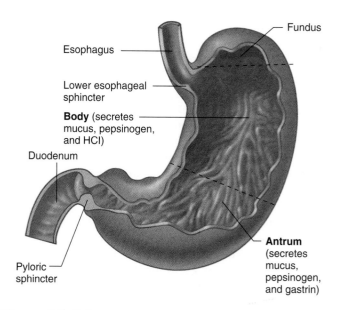

Figure 15–16

The two regions of the stomach: body and antrum. The fundus is the uppermost portion of the body of the stomach.

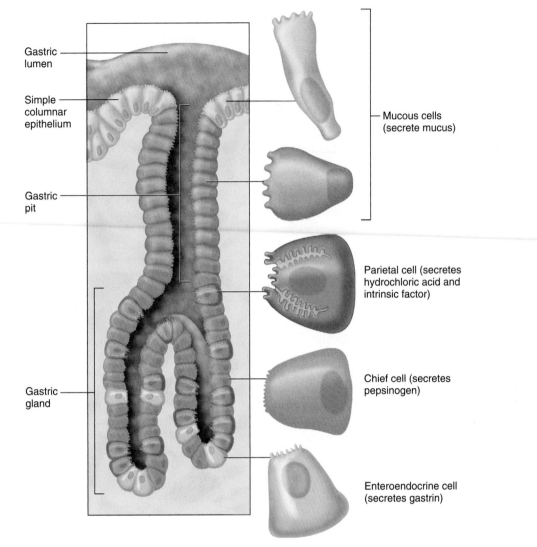

Gastric lumen

Simple columnar epithelium

Gastric pit

Gastric gland

Mucous cells (secrete mucus)

Parietal cell (secretes hydrochloric acid and intrinsic factor)

Chief cell (secretes pepsinogen)

Enteroendocrine cell (secretes gastrin)

Figure 15–17

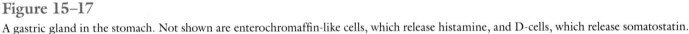

A gastric gland in the stomach. Not shown are enterochromaffin-like cells, which release histamine, and D-cells, which release somatostatin.

Primary H$^+$/K$^+$-ATPases in the luminal membrane of the parietal cells pump hydrogen ions into the lumen of the stomach (**Figure 15–18**). This primary active transporter also pumps potassium into the cell, which then leaks back into the lumen through potassium channels. Excessive vomiting can lead to potassium depletion due to this leak, and to metabolic alkalosis due to loss of hydrogen ions. As hydrogen ions are secreted into the lumen, bicarbonate ions are being secreted on the opposite side of the cell into the blood, in exchange for chloride ions.

Increased acid secretion, stimulated by factors described in the next paragraph, results from the transfer of H$^+$/K$^+$-ATPase proteins from the membranes of intracellular vesicles to the plasma membrane by fusion of these vesicles with the membrane, thus increasing the number of pump proteins in the plasma membrane. This process is analogous to that described in Chapter 14 for the transfer of water channels to the plasma membrane of kidney collecting-duct cells in response to ADH.

Four chemical messengers regulate the insertion of H$^+$/K$^+$-ATPases into the plasma membrane and therefore acid secretion: gastrin (a gastric hormone), acetylcholine (ACh, a neurotransmitter), histamine, and somatostatin (two paracrine agents). Parietal cell membranes contain receptors for all four of these molecules (**Figure 15–19**). Somatostatin inhibits acid secretion, whereas the other three stimulate secretion. Histamine is particularly important in stimulating acid secretion because it markedly potentiates the response to the other two stimuli, gastrin and ACh. As we will discuss later when considering ulcers, this potentiating effect of histamine is the reason that drugs that block histamine receptors in the stomach suppress acid secretion.

Not only do these chemical messengers act directly on the parietal cells, they also influence each other's secretion. During a meal, the rate of acid secretion increases markedly as stimuli arising from the cephalic, gastric, and intestinal phases alter the release of the four chemical messengers described in the previous paragraph. During the cephalic phase, increased

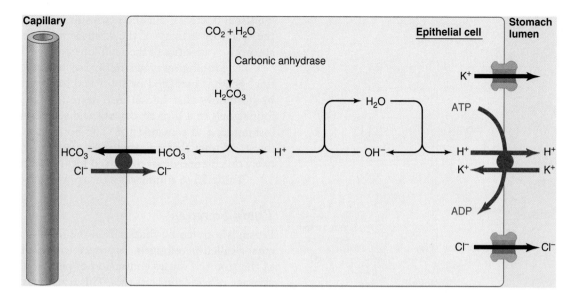

Figure 15–18

Secretion of hydrochloric acid by parietal cells. The hydrogen ions secreted into the lumen by primary active transport are derived from the breakdown of water molecules, leaving hydroxyl ions (OH^-) behind. These hydroxyl ions are neutralized by combination with other hydrogen ions generated by the reaction between carbon dioxide and water, a reaction catalyzed by the enzyme carbonic anhydrase, which is present in high concentrations in parietal cells. The bicarbonate ions formed by this reaction move out of the parietal cell on the blood side in exchange for chloride ions.

Figure 15–18 physiological *inquiry* 🅟🅘

- Why doesn't the high concentration of hydrogen ions in the stomach lumen destroy the lining of the stomach wall?

Answer can be found at end of chapter.

activity of the parasympathetic nerves to the stomach's enteric nervous system results in the release of ACh from the plexus neurons, gastrin from the gastrin-releasing cells, and histamine from ECL cells (**Figure 15–20**).

Once food has reached the stomach, the gastric phase stimuli—distension from the volume of ingested material and the presence of peptides and amino acids released by the digestion of luminal proteins—produce a further increase in acid secretion. These stimuli use some of the same neural pathways used during the cephalic phase. Nerve endings in the mucosa of the stomach respond to these luminal stimuli and send action potentials to the enteric nervous system, which in turn, can relay signals to the gastrin-releasing cells, histamine-releasing cells, and parietal cells. In addition, peptides and amino acids can act directly on the gastrin-releasing endocrine cells to promote gastrin secretion.

The concentration of acid in the gastric lumen is itself an important determinant of the rate of acid secretion for the following reason. Hydrogen ions (acid) stimulate the release of somatostatin from endocrine cells in the gastric wall. Somatostatin then acts on the parietal cells to inhibit acid secretion; it also inhibits the release of gastrin and histamine. The net result is a negative feedback control of acid secretion. As the contents of the gastric lumen become more acidic, the stimuli that promote acid secretion decrease.

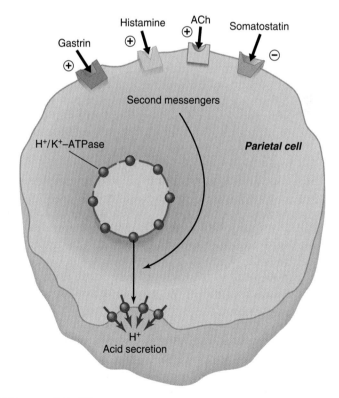

Figure 15–19

The four inputs to parietal cells that regulate acid secretion by generating second messengers. These second messengers control the transfer of the H^+/K^+-ATPase pumps in cytoplasmic vesicle membranes to the plasma membrane.

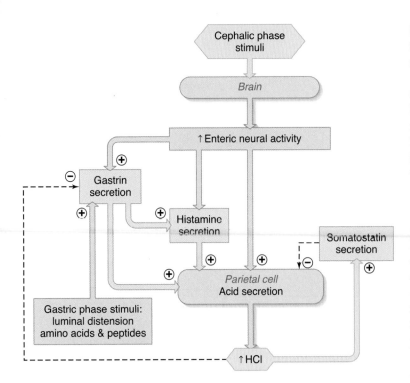

Figure 15–20

Cephalic and gastric phases controlling acid secretion by the stomach. The dashed line and ⊖ indicate that an increase in acidity inhibits the secretion of gastrin and that somatostatin inhibits the release of HCl. HCl inhibition of gastrin and somatostatin inhibition of HCl are negative feedback loops limiting overproduction of HCl.

Increasing the protein content of a meal increases acid secretion. This occurs for two reasons. First, protein ingestion increases the concentration of peptides in the lumen of the stomach. These peptides, as we have seen, stimulate acid secretion. The second reason is more complicated and reflects the effects of proteins on luminal acidity. Before food enters the stomach, the H^+ concentration in the lumen is high because there are few buffers present to bind any secreted hydrogen ions; therefore, the rate of acid secretion is low because high acidity inhibits acid secretion. The protein in food is an excellent buffer; however, as protein enters the stomach, the H^+ concentration decreases as H^+ binds to proteins. This decrease in acidity removes the inhibition of acid secretion. The more protein in a meal, the greater the buffering of acid, and the more acid secreted.

We now come to the intestinal phase that controls acid secretion—the phase in which stimuli in the early portion of the small intestine influence acid secretion by the stomach. High acidity in the duodenum triggers reflexes that inhibit gastric acid secretion. This inhibition is beneficial because the digestive activity of enzymes and bile salts in the small intestine is strongly inhibited by acidic solutions. This reflex limits gastric acid production when the H^+ concentration in the duodenum increases due to the entry of chyme from the stomach.

Acid, distension, hypertonic solutions, solutions containing amino acids, and fatty acids in the small intestine reflexly inhibit gastric acid secretion. Thus, the extent to which acid secretion is inhibited during the intestinal phase varies, depending upon the volume and composition of the intestinal contents, but the net result is the same—balancing the secretory activity of the stomach with the digestive and absorptive capacities of the small intestine.

The inhibition of gastric acid secretion during the intestinal phase is mediated by short and long neural reflexes and by hormones that inhibit acid secretion by influencing the four signals that directly control acid secretion: ACh, gastrin, histamine, and somatostatin. The hormones released by the intestinal tract that reflexly inhibit gastric activity are collectively called **enterogastrones** and include secretin and CCK.

Table 15–5 summarizes the control of acid secretion.

Pepsin Secretion

Pepsin is secreted by chief cells in the form of an inactive precursor called pepsinogen. Exposure to low pH in the lumen of the stomach causes conversion of pepsinogen to pepsin by cleavage of acid-labile linkages. This reaction is faster when pH is lower; in fact, it is almost instantaneous when pH<2. Once formed, pepsin itself can act on pepsinogen to produce more pepsin (**Figure 15–21**).

The synthesis and secretion of pepsinogen, followed by its intraluminal activation to pepsin, provides an example of a process that occurs with many other secreted proteolytic enzymes in the gastrointestinal tract. Because these enzymes are synthesized in inactive forms, collectively referred to as **zymogens,** any substrates that these enzymes might be able to act upon inside the cell producing them are protected from digestion, thus preventing damage to the cells.

Pepsin is active only in the presence of a high H^+ concentration (low pH). It is irreversibly inactivated when it enters the small intestine, where the bicarbonate ions secreted into the small intestine neutralize the hydrogen ions.

The primary pathway for stimulating pepsinogen secretion is input to the chief cells from the enteric nervous system. During the cephalic, gastric, and intestinal phases, most of the factors that stimulate or inhibit acid secretion exert the same effect on pepsinogen secretion. Thus, pepsinogen secretion parallels acid secretion.

Pepsin is not essential for protein digestion because in its absence, as occurs in some pathological conditions, protein can be completely digested by enzymes in the small intestine. However, pepsin accelerates protein digestion and normally accounts for about 20 percent of total protein digestion. It is also important in the digestion of collagen contained in the connective tissue matrix of meat.

Gastric Motility

An empty stomach has a volume of only about 50 ml, and the diameter of its lumen is only slightly larger than that of the small intestine. When a meal is swallowed, however, the smooth muscles in the fundus and body relax before the arrival of food, allowing the stomach's volume to increase to as much as 1.5 L with little increase in pressure. This **receptive relaxation** is mediated by the parasympathetic nerves to the stomach's enteric nerve plexuses, with coordination provided by afferent input from the stomach via the vagus nerve and by the swallowing center in the brain. Nitric oxide and serotonin released by enteric neurons mediate this relaxation.

Table 15-5   Control of HCL Secretion During a Meal

Stimuli	Pathways	Result
Cephalic phase 　Sight 　Smell 　Taste 　Chewing	Parasympathetic nerves to enteric nervous system	↑HCl secretion
Gastric contents (gastric phase) 　Distension 　↑Peptides 　↓H⁺ concentration	Long and short neural reflexes and direct stimulation of gastrin secretion	↑HCl secretion
Intestinal contents (intestinal phase) 　Distension 　↑H⁺ concentration 　↑Osmolarity 　↑Nutrient concentrations	Long and short neural reflexes; secretin, CCK, and other duodenal hormones	↓HCl secretion

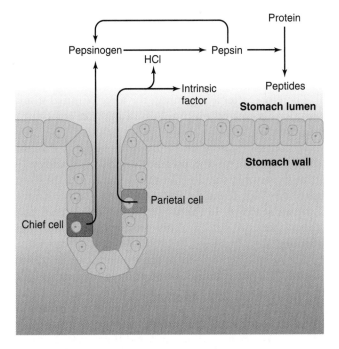

Figure 15–21

Conversion of pepsinogen to pepsin in the lumen of the stomach. An increase in HCl acidifies the stomach contents. High acidity (low pH) cleaves pepsin from pepsinogen. The pepsin thus formed also catalyzes its own production by acting on additional molecules of pepsinogen. The parietal cells also secrete intrinsic factor, which is needed to absorb vitamin B_{12} in the small intestine.

As in the esophagus, the stomach produces peristaltic waves in response to the arriving food. Each wave begins in the body of the stomach and produces only a ripple as it proceeds toward the antrum, a contraction too weak to produce much mixing of the luminal contents with acid and pepsin.

As the wave approaches the larger mass of wall muscle surrounding the antrum, however, it produces a more powerful contraction, which both mixes the luminal contents and *closes* the **pyloric sphincter,** a ring of smooth muscle and connective tissue between the antrum and the duodenum (**Figure 15–22**). The pyloric sphincter muscles contract upon arrival of a peristaltic wave. As a consequence of the sphincter closing, only a small amount of chyme is expelled into the duodenum with each wave. Most of the antral contents are forced backward toward the body of the stomach, thereby contributing to the mixing activity in the antrum. Recall that the lower esophageal sphincter prevents this retrograde movement of stomach contents from entering the esophagus.

What is responsible for producing gastric peristaltic waves? Their rhythm (three per minute) is generated by pacemaker cells in the longitudinal smooth muscle layer. These smooth muscle cells undergo spontaneous depolarization-repolarization cycles (slow waves) known as the **basic electrical rhythm** of the stomach. These slow waves are conducted through gap junctions along the stomach's longitudinal muscle layer and also induce similar slow waves in the overlying circular muscle layer. In the absence of neural or hormonal input, however, these depolarizations are too small to cause significant contractions. Excitatory neurotransmitters and hormones act upon the smooth muscle to further depolarize the membrane, thereby bringing it closer to threshold. Action potentials may be generated at the peak of the slow wave cycle if threshold is reached (**Figure 15–23**) and may thus cause larger contractions. The number of spikes fired with each wave determines the strength of the muscle contraction.

Thus, whereas the frequency of contraction is determined by the intrinsic basic electrical rhythm and remains essentially constant, the force of contraction, and therefore the amount of gastric emptying per contraction, are determined reflexly by neural and hormonal input to the antral smooth muscle.

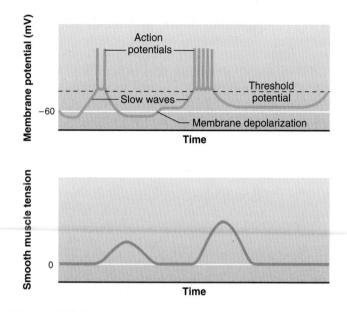

Figure 15–23

Slow wave oscillations in the membrane potential of gastric smooth muscle fibers trigger bursts of action potentials when threshold potential is reached at the wave peak. Membrane depolarization brings the slow wave closer to threshold, increasing the action potential frequency and thus the force of smooth muscle contraction.

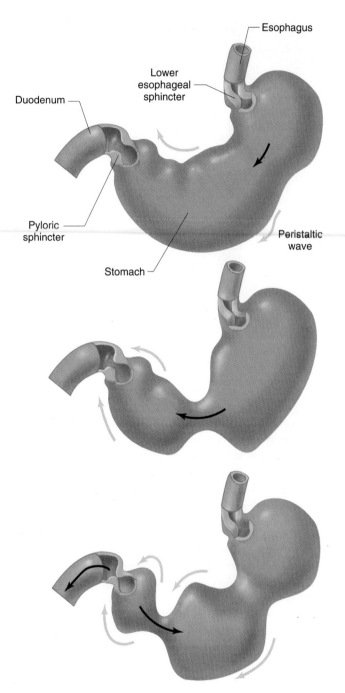

Figure 15–22

Peristaltic waves passing over the stomach force a small amount of luminal material into the duodenum. Black arrows indicate movement of luminal material; purple arrows indicate movement of the peristaltic wave in the stomach wall.

The initiation of these reflexes depends upon the contents of both the stomach and small intestine. All the factors previously discussed that regulate acid secretion (see Table 15–5) can also alter gastric motility. For example, gastrin, in sufficiently high concentrations, increases the force of antral smooth muscle contractions. Distension of the stomach also increases the force of antral contractions through long and short reflexes triggered by mechanoreceptors in the stomach

wall. Therefore, after a large meal, the force of initial stomach contractions is greater, which results in a greater emptying per contraction.

In contrast, gastric emptying is inhibited by distension of the duodenum, or the presence of fat, high-acidity (low pH), or hypertonic solutions in the lumen of the duodenum (**Figure 15–24**). These are the same factors that inhibit acid and pepsin secretion in the stomach. Fat is the most potent of these chemical stimuli. This prevents overfilling of the duodenum.

Autonomic nerve fibers to the stomach can be activated by the CNS independently of the reflexes originating in the stomach and duodenum and can influence gastric motility. An increase in parasympathetic activity increases gastric motility, whereas an increase in sympathetic activity decreases motility. Via these pathways, pain and emotions can alter motility; however, different people show different gastrointestinal responses to apparently similar emotional states.

As we have seen, a hypertonic solution in the duodenum is one of the stimuli inhibiting gastric emptying. This reflex prevents the fluid in the duodenum from becoming too hypertonic. It does so by slowing the rate of entry of chyme and thereby the delivery of large molecules that can rapidly be broken down into many small molecules by enzymes in the small intestine. A patient whose stomach has been removed because of disease (e.g., cancer) must eat a number of small meals. A large meal, in the absence of the controlled emptying by the stomach, could rapidly enter the intestine, producing a hypertonic solution. This hypertonic solution could cause enough water to flow (by osmosis) into the intestine from

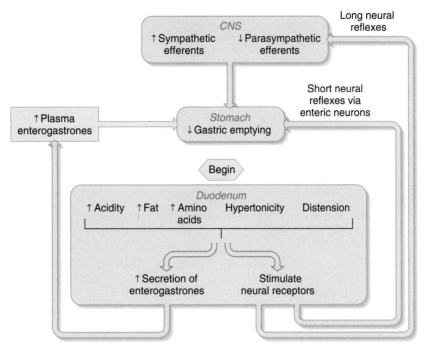

Figure 15–24

Intestinal-phase pathways inhibiting gastric emptying.

the blood to lower the blood volume and produce circulatory complications. The large distension of the intestine by the entering fluid can also trigger vomiting in these patients. All of these symptoms produced by the rapid entry of large quantities of ingested material into the small intestine are known as the ***dumping syndrome.***

Once the contents of the stomach have emptied over a period of several hours, the peristaltic waves cease and the empty stomach is mostly quiescent. During this time, however, there are brief intervals of peristaltic activity that we will describe along with the events controlling intestinal motility.

Pancreatic Secretions

The exocrine portion of the pancreas secretes bicarbonate ions and a number of digestive enzymes into ducts that converge into the pancreatic duct, which joins the common bile duct from the liver just before it enters the duodenum (see Figure 15–4). The enzymes are secreted by gland cells at the pancreatic end of the duct system, whereas bicarbonate ions are secreted by the epithelial cells lining the ducts (**Figure 15–25**).

The mechanism of bicarbonate secretion is analogous to that of hydrochloric acid secretion by the stomach (see Figure 15–18), except that the directions of hydrogen ion and bicarbonate ion movement are reversed. Hydrogen ions, derived from a carbonic anhydrase-catalyzed reaction between carbon dioxide and water, are actively transported out of the duct cells by an H^+/K^+-ATPase pump and released into the blood. The bicarbonate ions are secreted into the duct lumen.

The enzymes the pancreas secretes digest fat, polysaccharides, proteins, and nucleic acids to fatty acids, sugars, amino acids, and nucleotides, respectively. A partial list of these

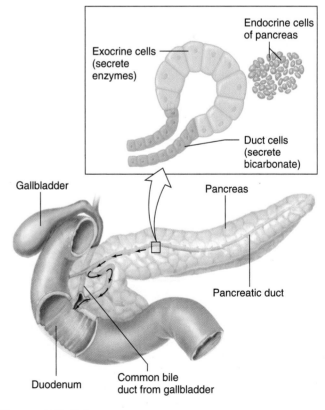

Figure 15–25

Structure of the pancreas.

Table 15–6	Pancreatic Enzymes	
Enzyme	**Substrate**	**Action**
Trypsin, chymotrypsin, elastase	Proteins	Break peptide bonds in proteins to form peptide fragments
Carboxypeptidase	Proteins	Splits off terminal amino acid from carboxyl end of protein
Lipase	Fats	Splits off two fatty acids from triglycerides, forming free fatty acids and monoglycerides
Amylase	Polysaccharides	Splits polysaccharides into glucose and maltose
Ribonuclease, deoxyribonuclease	Nucleic acids	Split nucleic acids into free mononucleotides

enzymes and their activities appears in **Table 15–6.** The proteolytic enzymes are secreted in inactive forms (zymogens), as described for pepsinogen in the stomach, and then activated in the duodenum by other enzymes. Like pepsinogen, the secretion of zymogens protects pancreatic cells from autodigestion. A key step in this activation is mediated by **enterokinase,** which is embedded in the luminal plasma membranes of the intestinal epithelial cells. It is a proteolytic enzyme that splits off a peptide from pancreatic **trypsinogen,** forming the active enzyme trypsin. Trypsin is also a proteolytic enzyme, and once activated, it activates the other pancreatic zymogens by splitting off peptide fragments (**Figure 15–26**). This activating function is in addition to the role of trypsin in digesting ingested protein.

The nonproteolytic enzymes secreted by the pancreas (e.g., amylase and lipase) are released in fully active form.

Pancreatic secretion increases during a meal, mainly as a result of stimulation by the hormones secretin and CCK (see Table 15–4). Secretin is the primary stimulant for bicarbonate secretion, whereas CCK mainly stimulates enzyme secretion.

Because the function of pancreatic bicarbonate is to neutralize acid entering the duodenum from the stomach, it is appropriate that the major stimulus for secretin release is increased acidity in the duodenum (**Figure 15–27**). In analogous fashion, CCK stimulates the secretion of digestive enzymes, including those for fat and protein digestion, so it is appropriate that the stimuli for its release are fatty acids and amino acids in the duodenum (**Figure 15–28**).

Luminal acid and fatty acids also act on afferent nerve endings in the intestinal wall, initiating reflexes that act on the pancreas to increase both enzyme and bicarbonate secre-

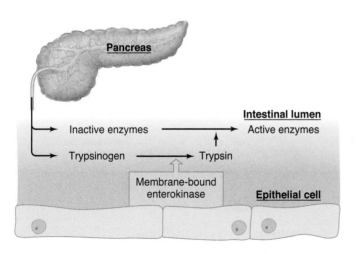

Figure 15–26

Activation of pancreatic enzyme precursors in the small intestine.

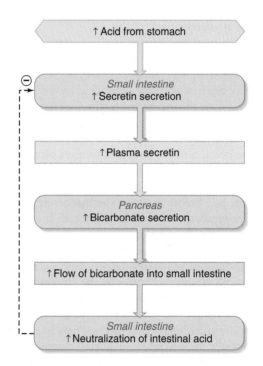

Figure 15–27

Hormonal regulation of pancreatic bicarbonate secretion. Dashed line and ⊖ indicates that neutralization of intestinal acid (↑pH) turns off secretin secretion (negative feedback).

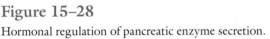

↑Intestinal fatty acids and amino acids

Small intestine
↑CCK secretion

↑Plasma CCK

Pancreas
↑Enzyme secretion

↑Flow of enzymes into small intestine

Small intestine
↑Digestion of fats and protein

Figure 15–28

Hormonal regulation of pancreatic enzyme secretion.

tion. Thus, the organic nutrients in the small intestine initiate, via hormonal and neural reflexes, the secretions involved in their own digestion.

Although most of the pancreatic exocrine secretions are controlled by stimuli arising from the intestinal phase of digestion, cephalic and gastric stimuli also play a role by way of the parasympathetic nerves to the pancreas. Thus, the taste of food or the distension of the stomach by food will lead to increased pancreatic secretion.

Bile Secretion and Liver Function

As stated earlier, bile is secreted by liver cells into a number of small ducts, the **bile canaliculi** (**Figure 15–29**), which converge to form the common hepatic duct (see Figure 15–4). Bile contains six major ingredients: (1) bile salts; (2) lecithin (a phospholipid); (3) bicarbonate ions and other salts; (4) cholesterol; (5) bile pigments and small amounts of other metabolic end products; and (6) trace metals. Bile salts and lecithin are synthesized in the liver and, as we have seen, help solubilize fat in the small intestine. Bicarbonate ions neutralize acid in the duodenum, and the last three ingredients represent substances extracted from the blood by the liver and excreted via the bile.

From the standpoint of gastrointestinal function, the most important components of bile are the bile salts. During the digestion of a fatty meal, most of the bile salts entering the intestinal tract via the bile are absorbed by specific sodium-coupled transporters in the ileum (the last segment of the small intestine). The absorbed bile salts are returned via the portal vein to the liver, where they are once again secreted into

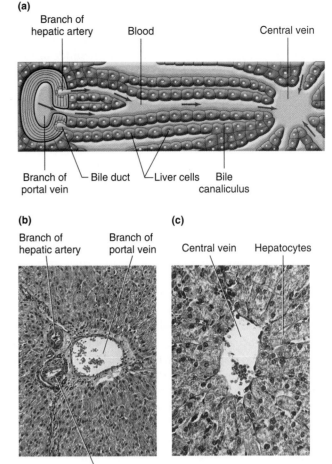

Figure 15–29

(a) A small section of the liver showing the location of bile canaliculi and ducts with respect to blood and liver cells. Bile (green) is formed by uptake by liver cells (hepatocyte) of bile salts and secretion into bile canaliculi. (b) Photomicrograph of the liver showing a portal space with its characteristic small artery and bile duct surrounded by connective tissue. H&E stain, medium magnification. (c) A central (centrolobular) vein. H&E stain, medium magnification.

(a) Adapted from Kappas and Alvares; (b, c) Courtesy of M.F. Santos from Junqueira and Cameiro Basic Histology Image Library (McGraw-Hill).

the bile. Uptake of bile salts from portal blood into **hepatocytes** (liver cells) is driven by secondary active transport coupled to sodium. This recycling pathway from the liver to the intestine and back to the liver is known as the **enterohepatic circulation** (**Figure 15–30**). A small amount (5 percent) of the bile salts escapes this recycling and is lost in the feces, but the liver synthesizes new bile salts from cholesterol to replace them. During the digestion of a meal, the entire bile salt content of the body may be recycled several times via the enterohepatic circulation.

In addition to synthesizing bile salts from cholesterol, the liver also secretes cholesterol extracted from the blood into the bile. Bile secretion, followed by excretion of cholesterol in the feces, is one of the mechanisms for maintaining cholesterol homeostasis in the blood (Chapter 16) and is also the process by which some cholesterol-lowering drugs work. Dietary fiber also sequesters bile and thereby lowers plasma cholesterol. Cholesterol is insoluble in water, and its solubility in bile is achieved by its incorporation into micelles (whereas in blood, cholesterol is incorporated into lipoproteins). Gallstones, consisting of precipitated cholesterol, will be discussed at the end of this chapter.

Bile pigments are substances formed from the heme portion of hemoglobin when old or damaged erythrocytes are digested in the spleen and liver. The predominant bile pigment is **bilirubin,** which is extracted from the blood by liver cells and actively secreted into the bile. Bilirubin is yellow and contributes to the color of bile. During their passage through the intestinal tract, some of the bile pigments are absorbed into the blood and are eventually excreted in the urine, giving urine its yellow color. After entering the intestinal tract, some bilirubin is modified by bacterial enzymes to form the brown pigments that give feces their characteristic color.

The components of bile are secreted by two different cell types. The bile salts, cholesterol, lecithin, and bile pigments are secreted by hepatocytes, whereas most of the bicarbonate-rich salt solution is secreted by the epithelial cells lining the bile ducts. Secretion of the salt solution by the bile ducts, just like the secretion by the pancreas, is stimulated by secretin in response to the presence of acid in the duodenum.

Bile salt secretion is controlled by the concentration of bile salts in the blood, unlike the pancreas, whose secretions are controlled by intestinal hormones. The greater the plasma concentration of bile salts, the greater their secretion into the bile canaliculi. Absorption of bile salts from the intestine during the digestion of a meal leads to their increased plasma concentration and thus to an increased rate of bile salt secretion by the liver. Although bile secretion is greatest during and just after a meal, the liver is always secreting some bile. Surrounding the common bile duct at the point where it enters the duodenum is a ring of smooth muscle known as the **sphincter of Oddi.** When this sphincter is closed, the dilute bile secreted by the liver is shunted into the gallbladder. Here, the organic components of bile become concentrated as NaCl and water are absorbed into the blood.

Shortly after the beginning of a fatty meal, the sphincter of Oddi relaxes and the gallbladder contracts, discharging concentrated bile into the duodenum. The signal for gallbladder contraction and sphincter relaxation is the intestinal hormone CCK—appropriately so, because as we have seen, the presence of fat in the duodenum is a major stimulus for this hormone's release. It is from this ability to cause contraction of the gallbladder that cholecystokinin received its name: *chole,* bile; *cysto,* bladder; *kinin,* to move. **Figure 15–31** summarizes the factors controlling the entry of bile into the small intestine.

Small Intestine

Secretion

Approximately 1500 ml of fluid is secreted by the walls of the small intestine from the blood into the lumen each day. One of the causes of water movement (secretion) into the lumen is

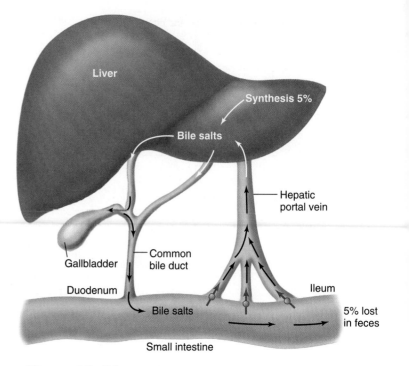

Figure 15–30

Enterohepatic circulation of bile salts. Bile salts are secreted into bile (green) and enter the duodenum through the common bile duct. Bile salts are reabsorbed from the intestinal lumen into hepatic portal blood (red arrows). The liver (hepatocytes) reclaims bile salts from hepatic portal blood.

Figure 15–30 physiological *inquiry* (p*i*)

■ In addition to the hepatic portal vein, can you name another portal vein system and explain the meaning of the term "portal"?

Answer can be found at end of chapter.

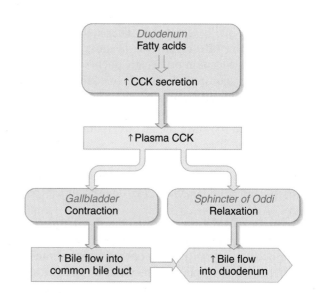

Figure 15–31

Regulation of bile entry into the small intestine.

that the intestinal epithelium at the base of the villi secretes a number of mineral ions, notably sodium, chloride, and bicarbonate ions into the lumen, and water follows by osmosis. These secretions, along with mucus, lubricate the surface of the intestinal tract and help protect the epithelial cells from excessive damage by the digestive enzymes in the lumen. Some damage to these cells still occurs, however, and the intestinal epithelium has one of the highest cell renewal rates of any tissue in the body.

Chloride is the primary ion determining the magnitude of fluid secretion. Various hormonal and paracrine signals—as well as certain bacterial toxins—can increase the opening frequency of these channels and thus increase fluid secretion.

As stated earlier, water movement into the lumen also occurs when the chyme entering the small intestine from the stomach is hypertonic because of a high concentration of solutes in the meal, and because digestion breaks down large molecules into many more small molecules. This hypertonicity causes the osmotic movement of water from the isotonic plasma into the intestinal lumen.

Absorption

Normally, virtually all of the fluid secreted by the small intestine is absorbed back into the blood. In addition, a much larger volume of fluid, which includes salivary, gastric, hepatic, and pancreatic secretions, as well as ingested water, is simultaneously absorbed from the intestinal lumen into the blood. Thus, overall there is a large net absorption of water from the small intestine. Absorption is achieved by the transport of ions, primarily sodium, from the intestinal lumen into the blood, with water following by osmosis.

Motility

In contrast to the peristaltic waves that sweep over the stomach, the most common motion in the small intestine during digestion of a meal is a stationary contraction and relaxation of intestinal segments, with little apparent net movement toward the large intestine (**Figure 15–32**). Each contracting segment is only a few centimeters long, and the contraction lasts a few seconds. The chyme in the lumen of a contracting segment is forced both up and down the intestine. This rhythmical contraction and relaxation of the intestine, known as **segmentation,** produces a continuous division and subdivision of the intestinal contents, thoroughly mixing the chyme in the lumen and bringing it into contact with the intestinal wall.

These segmenting movements are initiated by electrical activity generated by pacemaker cells in or associated with the circular smooth muscle layer. As with the slow waves in the stomach, this intestinal basic electrical rhythm produces oscillations in the smooth muscle membrane potential. If threshold is reached, action potentials are triggered that increase muscle contraction. The frequency of segmentation is set by the frequency of the intestinal basic electrical rhythm, but unlike the stomach, which normally has a single rhythm (three per minute), the intestinal rhythm varies along the length of the intestine, each successive region having a slightly lower frequency than the one above. For example, segmentation in the

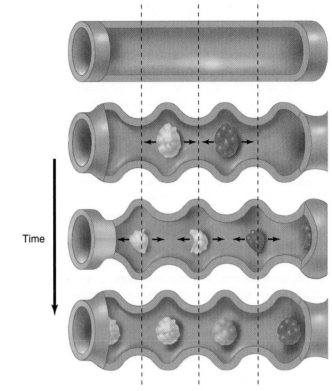

Site of first contraction

Time

Figure 15–32

Segmentation contractions in a portion of the small intestine in which segments of the intestines contract and relax in a rhythmic pattern but do *not* undergo peristalsis. This is the rhythm encountered during a meal. Dotted lines are reference points to show the site of the first contraction in time (starting at the top). As contractions occur at the next site, the chyme is divided and pushed back and forth, mixing the luminal contents.

duodenum occurs at a frequency of about 12 contractions/min, whereas in the last portion of the ileum the rate is only 9 contractions/min. Segmentation produces, therefore, a slow migration of the intestinal contents toward the large intestine because more chyme is forced downward, on the average, than upward.

The intensity of segmentation can be altered by hormones, the enteric nervous system, and autonomic nerves; parasympathetic activity increases the force of contraction, and sympathetic stimulation decreases it. Thus, cephalic phase stimuli, including emotional states, can alter intestinal motility. As is true for the stomach, these inputs produce changes in the force of smooth muscle contraction but do not significantly change the frequencies of the basic electrical rhythms.

After most of a meal has been absorbed, the segmenting contractions cease and are replaced by a pattern of peristaltic activity known as the **migrating myoelectric complex (MMC).** Beginning in the lower portion of the stomach, repeated waves of peristaltic activity travel about 2 feet along the small intestine and then die out. The next MMC starts slightly farther down the small intestine so that peristaltic activity slowly migrates down the small intestine, taking about two hours to reach the large intestine. By the time the MMC reaches the end of the ileum, new waves are beginning in the stomach, and the process repeats.

The MMC moves any undigested material still remaining in the small intestine into the large intestine and also prevents bacteria from remaining in the small intestine long enough to grow and multiply excessively. In diseases characterized by an aberrant MMC, bacterial overgrowth in the small intestine can become a major problem. Upon the arrival of a meal in the stomach, the MMC rapidly ceases in the intestine and is replaced by segmentation.

A rise in the plasma concentration of the intestinal hormone **motilin** is thought to initiate the MMC. Feeding inhibits the release of motilin; motilin stimulates MMCs via both the enteric and autonomic nervous systems.

The contractile activity in various regions of the small intestine can be altered by reflexes initiated at different points along the gastrointestinal tract. For example, segmentation intensity in the ileum increases during periods of gastric emptying, known as the **gastroileal reflex.** Large distensions of the intestine, injury to the intestinal wall, and various bacterial infections in the intestine lead to a complete cessation of motility, the **intestino-intestinal reflex.**

As much as 500 ml of air may be swallowed during a meal. Most of this air travels no farther than the esophagus, from which it is eventually expelled by belching. Some of the air reaches the stomach, however, and is passed on to the intestines, where its percolation through the chyme as the intestinal contents mix produces gurgling sounds that can be quite loud.

Large Intestine

The large intestine is a tube 6.5 cm (2.5 inches) in diameter and about 1.5 m (5 feet) long. Its first portion, the **cecum,** forms a blind-ended pouch from which extends the **appendix,** a small, fingerlike projection that may participate in immune function but is not essential (**Figure 15–33**). The **colon** consists of three relatively straight segments—the ascending, transverse, and descending portions. The terminal portion of the descending colon is S-shaped, forming the sigmoid colon, which empties into a relatively straight segment of the large intestine, the rectum, which ends at the anus.

Although the large intestine has a greater diameter than the small intestine, its epithelial surface area is far smaller, because the large intestine is much shorter than the small intestine, its surface is not convoluted, and its mucosa lacks villi. The secretions of the large intestine are scanty, lack digestive enzymes, and consist mostly of mucus and fluid containing bicarbonate and potassium ions. The primary function of the large intestine is to store and concentrate fecal material before defecation.

Chyme enters the cecum through the **ileocecal sphincter.** This sphincter is normally closed, but after a meal, when the gastroileal reflex increases ileal contractions, it relaxes each time the terminal portion of the ileum contracts, allowing chyme to enter the large intestine. Distension of the large intestine, on the other hand, produces a reflex contraction of

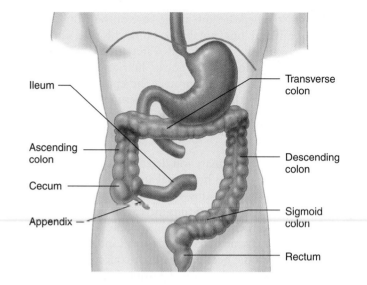

Figure 15–33

The large intestine begins with the cecum.

the sphincter, preventing fecal material from moving back into the small intestine.

About 1500 ml of chyme enters the large intestine from the small intestine each day. This material is derived largely from the secretions of the lower small intestine because most of the ingested food is absorbed before reaching the large intestine. Fluid absorption by the large intestine normally accounts for only a small fraction of the fluid absorbed by the gastrointestinal tract each day.

The primary absorptive process in the large intestine is the active transport of sodium from lumen to blood, with the accompanying osmotic absorption of water. If fecal material remains in the large intestine for a long time, almost all the water is absorbed, leaving behind hard fecal pellets. There is normally a net movement of potassium from blood into the large intestine lumen, probably due to an active mechanism stimulated by cAMP. Severe depletion of total-body potassium can result when large volumes of fluid are excreted in the feces. There is also a net movement of bicarbonate ions into the lumen coupled to chloride absorption from the lumen, and loss of this bicarbonate (a base) in patients with prolonged diarrhea can cause a metabolic acidosis (Chapter 14).

The large intestine also absorbs some of the products formed by the bacteria colonizing this region. Undigested polysaccharides (fiber) are metabolized to short-chain fatty acids by bacteria in the large intestine and absorbed by passive diffusion. The bicarbonate the large intestine secretes helps to neutralize the increased acidity resulting from the formation of these fatty acids. These bacteria also produce small amounts of vitamins, especially vitamin K, for absorption into the blood. Although this source of vitamins generally provides only a small part of the normal daily requirement, it may make a significant contribution when dietary vitamin intake is low. An individual who depends on absorption of vitamins formed by bacteria in the large intestine may become vitamin deficient if treated with antibiotics that inhibit other species of bacteria as well as the disease-causing bacteria.

Other bacterial products include gas (**flatus**), which is a mixture of nitrogen and carbon dioxide, with small amounts of the gases hydrogen, methane, and hydrogen sulfide. Bacterial fermentation of undigested polysaccharides produces these gases in the colon (except for nitrogen, which is derived from swallowed air), at the rate of about 400 to 700 ml/day. Certain foods (beans, for example) contain large amounts of carbohydrates that cannot be digested by intestinal enzymes but are readily metabolized by bacteria in the large intestine, producing large amounts of gas.

Motility and Defecation

Contractions of the circular smooth muscle in the large intestine produce a segmentation motion with a rhythm considerably slower (one every 30 min) than that in the small intestine. Because of the slow propulsion of the large intestine contents, material entering the large intestine from the small intestine remains for about 18 to 24 h. This provides time for bacteria to grow and multiply. Three to four times a day, generally following a meal, a wave of intense contraction known as a **mass movement** spreads rapidly over the transverse segment of the large intestine toward the rectum. This usually coincides with the gastroileal reflex. Unlike a peristaltic wave, in which the smooth muscle at each point relaxes after the wave of contraction has passed, the smooth muscle of the large intestine remains contracted for some time after a mass movement. The large intestine is innervated by parasympathetic and sympathetic nerves. Parasympathetic input increases segmental contractions, whereas sympathetic input decreases colonic contractions.

The **anus,** the exit from the rectum, is normally closed by the **internal anal sphincter,** composed of smooth muscle, and the **external anal sphincter,** composed of skeletal muscle under voluntary control. The sudden distension of the walls of the rectum produced by the mass movement of fecal material into it initiates the neurally mediated **defecation reflex.**

The conscious urge to defecate, mediated by mechanoreceptors, accompanies distension of the rectum. The reflex response consists of a contraction of the rectum, relaxation of the internal anal sphincter, but *contraction* of the external anal sphincter (initially), and increased peristaltic activity in the sigmoid colon. Eventually, a pressure is reached in the rectum that triggers reflex *relaxation* of the external anal sphincter, allowing the feces to be expelled.

Brain centers can, however, via descending pathways to somatic nerves to the external anal sphincter, override the reflex signals that eventually would relax the sphincter, thereby keeping the external sphincter closed and allowing a person to delay defecation. In this case, the prolonged distension of the rectum initiates a reverse peristalsis, driving the rectal contents back into the sigmoid colon. The urge to defecate then subsides until the next mass movement again propels more feces into the rectum, increasing its volume and again initiating the defecation reflex. Voluntary control of the external anal sphincter is learned during childhood. Spinal cord damage can lead to a loss of voluntary control over defecation.

Defecation is normally assisted by a deep breath, followed by closure of the glottis and contraction of the abdominal and thoracic muscles, producing an increase in abdominal pressure that is transmitted to the contents of the large intestine and rectum. This maneuver (termed the Valsalva maneuver) also causes a rise in intrathoracic pressure, which leads to a transient rise in blood pressure followed by a fall in pressure as the venous return to the heart is decreased. The cardiovascular changes resulting from excessive strain during defecation may precipitate a stroke or heart attack, especially in constipated elderly people with cardiovascular disease.

Pathophysiology of the Gastrointestinal Tract

Because the end result of gastrointestinal function is the absorption of nutrients, salts, and water, most malfunctions of this organ system affect either the nutritional state of the body or its salt and water content. The following are a few examples of disordered gastrointestinal function.

Ulcers

Considering the high concentration of acid and pepsin secreted by the stomach, it is natural to wonder why the stomach does not digest itself. Several of the factors that protect the walls of the stomach from being digested include: (1) The surface of the mucosa is lined with cells that secrete a slightly alkaline mucus that forms a thin layer over the luminal surface. Both the protein content of mucus and its alkalinity neutralize hydrogen ions in the immediate area of the epithelium. Thus, mucus forms a chemical barrier between the highly acidic contents of the lumen and the cell surface. (2) The tight junctions between the epithelial cells lining the stomach restrict the diffusion of hydrogen ions into the underlying tissues. (3) Damaged epithelial cells are replaced every few days by new cells arising by the division of cells within the gastric pits.

At times, these protective mechanisms can prove inadequate, and erosion (***ulcers***) of the gastric surface can develop. Ulcers can occur not only in the stomach but also in the lower part of the esophagus and in the duodenum. Indeed, duodenal ulcers are about 10 times more frequent than gastric ulcers, affecting about 10 percent of the U.S. population. Damage to blood vessels in the tissues underlying the ulcer may cause bleeding into the gastrointestinal lumen (**Figure 15–34**). On occasion, the ulcer may penetrate the entire wall, resulting in leakage of the luminal contents into the abdominal cavity. A device used to diagnose gastric and duodenal ulcers is the ***endoscope*** (see Figure 15–34). This uses either fiberoptic or video technology to directly visualize the gastric and duodenal mucosa. Furthermore, the endoscopist can apply local treatments and take samples of tissue (***biopsy***) during upper endoscopy. Similar devices can be used to visualize the colon (flexible ***sigmoidoscopy*** or ***colonoscopy***) and the bronchi (***bronchoscopy***).

Ulcer formation involves breaking the mucosal barrier and exposing the underlying tissue to the corrosive action of acid and pepsin, but it is not always clear what produces the initial damage to the barrier. Although acid is essential for

(a)

Endoscope

Trachea

From power source

To video monitor

Diaphragm

Duodenum

Pyloric sphincter

Upper esophageal sphincter

Lower esophageal sphincter

Stomach

Endoscope

(b)

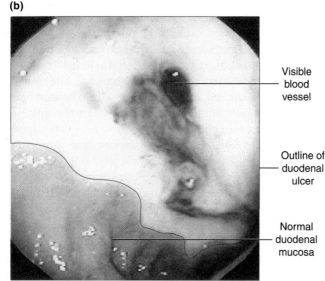

Visible blood vessel

Outline of duodenal ulcer

Normal duodenal mucosa

Figure 15–34

(a) Video endoscopy of the upper GI tract: The physician passes the endoscope through the mouth (or nose) down the esophagus, stomach, and into the duodenum. A light source at the tip of the endoscope illuminates the mucosa. The tip also has a miniature video chip, which transmits images up the endoscope to a video recorder (image shown in Figure 15–34b). Local treatments can be applied and small tissue samples (biopsies) can be taken with the endoscope. Earlier versions of this device used fiberoptic technology. (b) Image of duodenal ulcer taken through an endoscope (shown in Figure 15–34a). The entire ulcer is about 2 cm across (the width of the image).

(b) Courtesy of Fernando Carballo, M.D.

ulcer formation, it is not necessarily the primary factor, and many patients with ulcers have normal or even subnormal rates of acid secretion.

Many factors, including genetic susceptibility, drugs, alcohol, bile salts, and an excessive secretion of acid and pepsin, may contribute to ulcer formation. The major factor, however, is the presence of a bacterium, **_Helicobacter pylori,_** that is present in the stomachs of a majority of patients with ulcers or **_gastritis_** (inflammation of the stomach walls). Suppression of these bacteria with antibiotics usually helps heal the damaged mucosa.

Once an ulcer has formed, the inhibition of acid secretion can remove the constant irritation and allow the ulcer to heal. Two classes of drugs are potent inhibitors of acid secretion. One class of inhibitors acts by blocking a specific class of histamine receptors (H_2) found on parietal cells, which stimulate acid secretion. An example of an H_2 receptor antagonist is cimetidine. The second class of drugs directly inhibits the H^+/K^+-ATPase pump in parietal cells. Examples of these so-called proton pump inhibitors are omeprazole and lansoprazole. Although both classes of drugs are effective in healing ulcers, the ulcers tend to recur if the *Helicobacter pylori* bacteria are not removed.

Despite popular notions, the role of stress in producing ulcers remains unclear. Once the ulcer has been formed, however, emotional stress can aggravate it by increasing acid secretion.

Vomiting

Vomiting is the forceful expulsion of the contents of the stomach and upper intestinal tract through the mouth. Like swallowing, vomiting is a complex reflex coordinated by a region in the brainstem medulla oblongata, in this case known as the **vomiting center.** Neural input to this center from receptors in many different regions of the body can initiate the vomiting reflex. For example, excessive distension of the stomach or small intestine, various substances acting upon chemoreceptors in the intestinal wall or in the brain, increased pressure within the skull, rotating movements of the head (motion sickness), intense pain, and tactile stimuli applied to the back of the throat can all initiate vomiting. The **area postrema** in the brain, which is outside the blood-brain barrier, is sensitive to toxins in the blood and can initiate vomiting. There are many chemicals (*emetics*) that can stimulate vomiting via receptors in the stomach, duodenum, or brain. A powerful household emetic is *syrup of ipecac,* which stimulates vomiting by a dual action on the enteric mucosa and the brain.

What is the adaptive value of this reflex? Obviously, the removal of ingested toxic substances before they can be absorbed is beneficial. Moreover, the nausea that usually accompanies vomiting may have the adaptive value of conditioning the individual to avoid the future ingestion of foods containing such toxic substances. Why other types of stimuli, such as those producing motion sickness, have become linked to the vomiting center is not clear.

Vomiting is usually preceded by increased salivation, sweating, increased heart rate, pallor, and nausea. The events leading to vomiting begin with a deep breath, closure of the glottis, and elevation of the soft palate. The abdominal muscles then contract, raising the abdominal pressure, which is transmitted to the stomach's contents. The lower esophageal sphincter relaxes, and the high abdominal pressure forces the contents of the stomach into the esophagus. This initial sequence of events, which can occur repeatedly without expulsion via the mouth, is known as *retching.* Vomiting occurs when the abdominal contractions become so strong that the increased intrathoracic pressure forces the contents of the esophagus through the upper esophageal sphincter.

Vomiting is also accompanied by strong contractions in the upper portion of the small intestine—contractions that tend to force some of the intestinal contents back into the stomach for expulsion. Thus, some bile may be present in the vomitus.

Excessive vomiting can lead to large losses of the water and salts that normally would be absorbed in the small intestine. This can result in severe dehydration, upset the body's salt balance, and produce circulatory problems due to a decrease in plasma volume. The loss of acid from vomiting results in a metabolic alkalosis (Chapter 14). A variety of antiemetic drugs can suppress vomiting.

Gallstones

As described earlier, bile contains not only bile salts but also cholesterol and phospholipids, which are water-insoluble and are maintained in soluble form in the bile as micelles. When the concentration of cholesterol in the bile becomes high in relation to the concentrations of phospholipid and bile salts, cholesterol crystallizes out of solution, forming *gallstones.* This can occur if the liver secretes excessive amounts of cholesterol or if the cholesterol becomes overly concentrated in the gallbladder as a result of salt and water absorption. Although cholesterol gallstones are the most frequently encountered gallstones in the Western world, the precipitation of bile pigments can also occasionally be responsible for gallstone formation.

Why some individuals develop gallstones and others do not is still unclear. Women, for example, have about twice the incidence of gallstone formation as men, and Native Americans have a very high incidence compared with other ethnic groups in the United States.

If a gallstone is small, it may pass through the common bile duct into the intestine with no complications. A larger stone may become lodged in the opening of the gallbladder, causing painful contractile spasms of the smooth muscle. A more serious complication arises when a gallstone lodges in the common bile duct, thereby preventing bile from entering the intestine. A large decrease in bile can decrease fat digestion and absorption. Furthermore, impaired absorption of the fat-soluble vitamins A, D, K, and E can occur, leading to, for example, clotting problems (vitamin K deficiency) and calcium malabsorption (due to vitamin D deficiency). The fat that is not absorbed enters the large intestine and eventually appears in the feces (a condition known as *steatorrhea*). Furthermore, bacteria in the large intestine convert some of this fat into fatty acid derivatives that alter salt and water movements, leading to a net flow of fluid into the large intestine. The result is diarrhea and fluid loss.

Because the duct from the pancreas joins the common bile duct just before it enters the duodenum, a gallstone that becomes lodged at this point prevents both bile and pancreatic secretions from entering the intestine. This results in failure both to neutralize acid and to adequately digest most organic nutrients, not just fat. The end result is severe nutritional deficiencies.

The buildup of very high pressure in a blocked common bile duct is transmitted back to the liver and interferes with the further secretion of bile. As a result, bilirubin, which is normally secreted into the bile by uptake from the blood in the liver, accumulates in the blood and diffuses into tissues, producing a yellowish coloration of the skin and eyes known as *jaundice.*

It should be emphasized, however, that bile duct obstruction is not the only cause of jaundice. Bilirubin accumulation in the blood can occur if hepatocytes are damaged by liver disease and therefore fail to secrete bilirubin into the bile. It can also occur if the level of bilirubin in the blood exceeds the capacity of the normal liver to secrete it, as in diseases that result in an increased breakdown of red blood cells—*hemolytic jaundice.* At birth, the liver's capacity to secrete bilirubin is not fully developed. During the first few days of life this may result in jaundice, which normally clears spontaneously. Excessive accumulation of bilirubin during the neonatal period, as occurs, for example, with hemolytic disease of the newborn (Chapter 18), carries a risk of bilirubin-induced neurological damage during a critical phase in the development of the nervous system.

Although surgery may be necessary to remove an inflamed gallbladder (*cholecystectomy*) or stones from an obstructed duct, newer techniques use drugs to dissolve gallstones. Patients who have had a cholecystectomy still make bile and transport it to the small intestine via the bile duct. Therefore, fat digestion and absorption can be maintained, but bile secretion and fat intake in the diet are no longer coupled. Thus, large, fatty meals are difficult to digest because of the absence of a large pool of bile normally released from the gallbladder in response to CCK. A diet low in fat content is usually advisable.

Lactose Intolerance

Lactose is the major carbohydrate in milk. It cannot be absorbed directly but must first be digested into its components—glucose and galactose—which are readily absorbed by active transport. Lactose is digested by the enzyme **lactase,** which is embedded in the luminal plasma membranes of intestinal epithelial cells. Lactase is usually present at birth and allows the nursing infant to digest the lactose in breast milk. Most Asians and many people with ancestors from certain areas of Africa undergo a decline in lactase production at about two years of age. This leads to *lactose intolerance*—the inability to completely digest lactose such that its concentration increases in the small intestine. Most people of Northern European descent maintain the ability to digest lactose as adults.

Because the absorption of water requires prior absorption of solute to provide an osmotic gradient, the unabsorbed lactose in persons with lactose intolerance prevents some of the water from being absorbed. This lactose-containing fluid is passed on to the large intestine, where bacteria digest the lactose. They then metabolize the released monosaccharides, producing large quantities of gas (which distends the colon, producing pain) and short-chain fatty acids, which cause fluid movement into the lumen of the large intestine, producing diarrhea. The response to ingestion of milk or dairy products by adults whose lactase levels have diminished varies from mild discomfort to severely dehydrating diarrhea, according to the volume of milk and dairy products ingested and the amount of lactase present in the intestine. The person can avoid these symptoms by either drinking milk in which the lactose has been predigested or taking pills containing lactase along with the milk.

Inflammatory Bowel Disease

The general term inflammatory bowel disease (IBD) comprises two related diseases—Crohn's disease and ulcerative colitis, both of which have already been mentioned earlier in this chapter. Both diseases involve chronic inflammation of the bowel. Crohn's disease can occur anywhere along the GI tract from the mouth to the anus, although it is most common near the end of the ileum. Colitis is confined to the colon. The incidence of IBD in the United States is 7 to 11 per 100,000 people and is most common in Caucasians, particularly Ashkenazi Jews. The most common ages of onset for IBD are in the late teens to early twenties and then again in people older than 60.

Although the precise cause or causes of IBD are not certain, it seems that it occurs as a combination of environmental and genetic factors. There appears to be a genetic predisposition for an abnormal response of the bowel mucosa to infection and the presence of normal luminal bacteria. Therefore, IBD appears to result from inappropriate immune and tissue repair responses to essentially normal microorganisms in the intestinal lumen.

Ulcerative colitis is caused by disruption of the normal mucosa with the presence of bleeding, edema, and ulcerations (losses of tissue due to inflammation). In its most extreme, the bowel wall can get so thin and the loss of tissue so great that holes (perforations) that go all the way through the bowel wall can occur. Active Crohn's disease shows inflammation and thickening of the bowel wall such that the lumen can become narrowed to the point where it may even become blocked, or obstructed, which can be very painful.

The main symptoms of ulcerative colitis are diarrhea, rectal bleeding, and abdominal cramps. The part of the small intestine at the end of the ileum is the most common site of Crohn's disease, so the first symptoms felt by patients with this disease are often pain in the lower right abdomen and diarrhea. Because it is often accompanied by fever due to the exaggerated immune response, the initial symptoms can be mistaken for acute appendicitis (Chapter 19). Because of its obstructive nature, the abdominal pain in Crohn's disease is often relieved temporarily by defecation.

The current initial treatment of IBD is the use of 5-aminosalicylate drugs, such as **sulfasalazine,** which appear to have both antibacterial and anti-inflammatory effects. In more severe cases, the use of glucocorticoids as anti-inflammatory agents can be very useful, although their overuse has significant risks such as loss of bone mass. It is often helpful to make adjustments in the diet to allow the inflamed bowel time to heal. Finally, new drug therapy using immunosuppressive medicines such as **tacrolimus** and **cyclosporine** show promise. When IBD becomes severe, it is sometimes necessary to perform surgery to remove the diseased bowel.

Constipation and Diarrhea

Many people have a mistaken belief that, unless they have a bowel movement every day, the absorption of "toxic" substances from fecal material in the large intestine will somehow poison them. Attempts to identify such toxic agents in the blood following prolonged periods of fecal retention have been unsuccessful, and there appears to be no physiological necessity for having bowel movements at frequent intervals. Whatever maintains a person in a comfortable state is physiologically adequate, whether this means a bowel movement after every meal, once a day, or only once a week.

On the other hand, some symptoms—headache, loss of appetite, nausea, and abdominal distension—may arise when defecation has not occurred for several days or even weeks, depending on the individual. These symptoms of **constipation** are caused not by toxins, but by distension of the rectum. The longer that fecal material remains in the large intestine, the more water is absorbed and the harder and drier the feces become, making defecation more difficult and sometimes painful.

Decreased motility of the large intestine is the primary factor causing constipation. This often occurs in the elderly, or it may result from damage to the colon's enteric nervous system or from emotional stress.

One of the factors increasing motility in the large intestine, and thus opposing the development of constipation, is distension. As noted earlier, dietary fiber (cellulose and other complex polysaccharides) is not digested by the enzymes in the small intestine and is passed on to the large intestine, where its bulk produces distension and thereby increases motility. Bran, most fruits, and vegetables are examples of foods that have a relatively high fiber content.

Laxatives, agents that increase the frequency or ease of defecation, act through a variety of mechanisms. Fiber provides a natural laxative. Some laxatives, such as mineral oil, simply lubricate the feces, making defecation easier and less painful. Others contain magnesium and aluminum salts, which are poorly absorbed and therefore lead to water retention in the intestinal tract. Still others, such as castor oil, stimulate the motility of the colon and inhibit ion transport across the wall, thus affecting water absorption.

Excessive use of laxatives in an attempt to maintain a preconceived notion of regularity leads to a decreased responsiveness of the large intestine to normal defecation-promoting signals. In such cases, a long period without defecation may occur following cessation of laxative intake, appearing to confirm the necessity of taking laxatives to promote regularity.

Diarrhea is characterized by large, frequent, watery stools. Diarrhea can result from decreased fluid absorption, increased fluid secretion, or both. The increased motility that accompanies diarrhea probably does not cause most cases of diarrhea (by decreasing the time available for fluid absorption), but rather results from the distension produced by increased luminal fluid.

A number of bacterial, protozoan, and viral diseases of the intestinal tract cause secretory diarrhea. **Cholera,** which is endemic in many parts of the world, is caused by a bacterium that releases a toxin that stimulates the production of cyclic AMP in the secretory cells at the base of the intestinal villi. This leads to an increased frequency in the opening of the chloride channels in the luminal membrane and hence increased secretion of chloride ions. An accompanying osmotic flow of water into the intestinal lumen occurs, resulting in massive diarrhea that can be life-threatening due to dehydration and decreased blood volume that leads to circulatory shock. The salt and water lost by this severe form of diarrhea can be balanced by ingesting a simple solution containing salt and glucose. The active absorption of these solutes is accompanied by absorption of water, which replaces the fluid lost by diarrhea. **Traveler's diarrhea,** produced by several species of bacteria, produces a secretory diarrhea by the same mechanism as the cholera bacterium, but is less severe.

In addition to decreased blood volume due to salt and water loss, other consequences of severe diarrhea are potassium depletion and metabolic acidosis (Chapter 14) resulting from the excessive fecal loss of potassium and bicarbonate ions, respectively.

■ **SUMMARY**

I. The gastrointestinal system transfers digested organic nutrients, minerals, and water from the external environment to the internal environment. The four processes used to accomplish this function are (a) digestion, (b) secretion, (c) absorption, and (d) motility.
 a. The system is designed to maximize the absorption of most nutrients, not to regulate the amount absorbed.
 b. The system does not play a major role in the removal of waste products from the internal environment.

Overview: Functions of the Gastrointestinal Organs

I. Figure 15–3 summarizes the names and functions of the gastrointestinal organs.
II. Each day the gastrointestinal tract secretes about six times more fluid into the lumen than is ingested. Only 1 percent of this fluid is excreted in the feces.

Structure of the Gastrointestinal Tract Wall

I. Figure 15–6 diagrams the structure of the wall of the gastrointestinal tract.
 a. The area available for absorption in the small intestine is greatly increased by the folding of the intestinal wall and by the presence of villi and microvilli on the surface of the epithelial cells.
 b. The epithelial cells lining the intestinal tract are continuously replaced by new cells arising from cell division at the base of the villi.
 c. The venous blood from the small intestine, containing absorbed nutrients other than fat, passes to the liver via the hepatic portal vein before returning to the heart. Fat is absorbed into the lymphatic vessels (lacteals) in each villus.

Digestion and Absorption

I. Starch is digested by amylases secreted by the salivary glands and pancreas. The resulting products, as well as ingested disaccharides, are digested to monosaccharides by enzymes in the luminal membranes of epithelial cells in the small intestine.
 a. Most monosaccharides are then absorbed by secondary active transport.
 b. Some polysaccharides, such as cellulose, cannot be digested and pass to the large intestine, where bacteria metabolize them.
II. Proteins are broken down into small peptides and amino acids, which are absorbed by secondary active transport in the small intestine.
 a. The breakdown of proteins to peptides is catalyzed by pepsin in the stomach and by the pancreatic enzymes trypsin and chymotrypsin in the small intestine.
 b. Peptides are broken down into amino acids by pancreatic carboxypeptidase and intestinal aminopeptidase.

c. Small peptides consisting of two to three amino acids can be actively absorbed into epithelial cells and then broken down to amino acids, which are released into the blood.

III. The digestion and absorption of fat by the small intestine requires mechanisms that solubilize the fat and its digestion products.

   a. Large fat globules leaving the stomach are emulsified in the small intestine by bile salts and phospholipids secreted by the liver.

   b. Lipase from the pancreas digests fat at the surface of the emulsion droplets, forming fatty acids and monoglycerides.

   c. These water-insoluble products of lipase action, when combined with bile salts, form micelles, which are in equilibrium with the free molecules.

   d. Free fatty acids and monoglycerides diffuse across the luminal membranes of epithelial cells, where they are enzymatically recombined to form triglycerides, which are released as chylomicrons from the blood side of the cell by exocytosis.

   e. The released chylomicrons enter lacteals in the intestinal villi and pass, by way of the lymphatic system and the thoracic duct, to the venous blood returning to the heart.

IV. Fat-soluble vitamins are absorbed by the same pathway used for fat absorption. Most water-soluble vitamins are absorbed in the small intestine by diffusion or mediated transport. Vitamin B_{12} is absorbed in the ileum by endocytosis after combining with intrinsic factor secreted into the lumen by parietal cells in the stomach.

V. Water is absorbed from the small intestine by osmosis following the active absorption of solutes, primarily sodium chloride.

How are Gastrointestinal Processes Regulated?

I. Most gastrointestinal reflexes are initiated by luminal stimuli: (a) distension, (b) osmolarity, (c) acidity, and (d) digestion products.

   a. Neural reflexes are mediated by short reflexes in the enteric nervous system and by long reflexes involving afferent and efferent neurons to and from the CNS.

   b. Endocrine cells scattered throughout the epithelium of the stomach secrete gastrin, and cells in the small intestine secrete secretin, CCK, and GIP. Table 15–4 lists the properties of these hormones.

   c. The three phases of gastrointestinal regulation—cephalic, gastric, and intestinal—are named for the location of the stimulus that initiates the response.

II. Chewing breaks up food into particles suitable for swallowing, but it is not essential for the eventual digestion and absorption of food.

III. Salivary secretion is stimulated by food in the mouth acting reflexly via chemoreceptors and pressure receptors and by sensory stimuli (e.g., sight or smell of food). Both sympathetic and parasympathetic stimulation increase salivary secretion.

IV. Food moved into the pharynx by the tongue initiates swallowing, which is coordinated by the swallowing center in the brainstem medulla oblongata.

   a. Food is prevented from entering the trachea by inhibition of respiration and by closure of the glottis.

   b. The upper esophageal sphincter relaxes as food is moved into the esophagus, and then the sphincter closes.

   c. Food is moved through the esophagus toward the stomach by peristaltic waves. The lower esophageal sphincter remains open throughout swallowing.

   d. If food does not reach the stomach with the first peristaltic wave, distension of the esophagus initiates secondary peristalsis.

V. Table 15–5 summarizes the factors controlling acid secretion by parietal cells in the stomach.

VI. Pepsinogen, secreted by the gastric chief cells in response to most of the same reflexes that control acid secretion, is converted to the active proteolytic enzyme pepsin in the stomach's lumen, primarily by acid.

VII. Peristaltic waves sweeping over the stomach become stronger in the antrum, where most mixing occurs. With each wave, only a small portion of the stomach's contents are expelled into the small intestine through the pyloric sphincter.

   a. Cycles of membrane depolarization, the basic electrical rhythm generated by gastric smooth muscle, determine gastric peristaltic wave frequency. Contraction strength can be altered by neural and hormonal changes in membrane potential, which is imposed on the basic electrical rhythm.

   b. Distension of the stomach increases the force of contractions and the rate of emptying. Distension of the small intestine, and fat, acid, or hypertonic solutions in the intestinal lumen inhibit gastric contractions.

VIII. The exocrine portion of the pancreas secretes digestive enzymes and bicarbonate ions, all of which reach the duodenum through the pancreatic duct.

   a. The bicarbonate ions neutralize acid entering the small intestine from the stomach.

   b. Most of the proteolytic enzymes, including trypsin, are secreted by the pancreas in inactive forms. Trypsin is activated by enterokinase located on the membranes of the small intestine cells; trypsin then activates other inactive pancreatic enzymes.

   c. The hormone secretin, released from the small intestine in response to increased luminal acidity, stimulates pancreatic bicarbonate secretion. The small intestine releases CCK in response to the products of fat and protein digestion. CCK then stimulates pancreatic enzyme secretion.

   d. Parasympathetic stimulation increases pancreatic secretion.

IX. The liver secretes bile, the major ingredients of which are bile salts, cholesterol, lecithin, bicarbonate ions, bile pigments, and trace metals.

   a. Bile salts undergo continuous enterohepatic recirculation during a meal. The liver synthesizes new bile salts to replace those lost in the feces.

   b. The greater the bile salt concentration in the hepatic portal blood, the greater the rate of bile secretion.

   c. Bilirubin, the major bile pigment, is a breakdown product of hemoglobin and is absorbed from the blood by the liver and secreted into the bile.

   d. Secretin stimulates bicarbonate secretion by the cells lining the bile ducts in the liver.

   e. Bile is concentrated in the gallbladder by the absorption of NaCl and water.

   f. Following a meal, the release of CCK from the small intestine causes the gallbladder to contract and the sphincter of Oddi to relax, thereby injecting concentrated bile into the intestine.

X. In the small intestine, the digestion of polysaccharides and proteins increases the osmolarity of the luminal contents, producing water flow into the lumen.

XI. Sodium, chloride, bicarbonate, and water are secreted by the small intestine. However, most of these secreted substances, as well as those entering the small intestine from other sources, are absorbed back into the blood.

XII. Intestinal motility is coordinated by the enteric nervous system and modified by long and short reflexes and hormones.

   a. During and shortly after a meal, the intestinal contents are mixed by segmenting movements of the intestinal wall.

   b. After most of the food has been digested and absorbed, the migrating myoelectric complex (MMC), which moves the undigested material into the large intestine by a migrating segment of peristaltic waves, replaces segmentation.

XIII. The primary function of the large intestine is to store and concentrate fecal matter before defecation.

   a. Water is absorbed from the large intestine secondary to the active absorption of sodium, leading to the concentration of fecal matter.

   b. Flatus is produced by bacterial fermentation of undigested polysaccharides.

   c. Three to four times a day, mass movements in the colon move its contents into the rectum.

   d. Distension of the rectum initiates defecation, which is assisted by a forced expiration against a closed glottis.

   e. Defecation can be voluntarily controlled through somatic nerves to the skeletal muscles of the external anal sphincter.

Pathophysiology of the Gastrointestinal Tract

I. The factors that normally prevent breakdown of the mucosal barrier and formation of ulcers are (1) secretion of an alkaline mucus, (2) tight junctions between epithelial cells, and (3) rapid replacement of epithelial cells.

   a. The bacterium *Helicobacter pylori* is a major cause of damage to the mucosal barrier, leading to ulcers.

   b. Drugs that block histamine receptors or inhibit the H^+/K^+-ATPase pump inhibit acid secretion and promote ulcer healing.

II. Vomiting is coordinated by the vomiting center in the brainstem medulla oblongata. Contractions of abdominal muscles force the contents of the stomach into the esophagus (retching); if the contractions are strong enough, they force the contents of the esophagus through the upper esophageal sphincter into the mouth (vomiting).

III. Precipitation of cholesterol or, less often, bile pigments in the gallbladder forms gallstones, which can block the exit of the gallbladder or common bile duct. In the latter case, the failure of bile salts to reach the intestine causes decreased fat digestion and absorption, and the accumulation of bile pigments in the blood and tissues causes jaundice.

IV. Lactase activity, which is present at birth, undergoes a genetically determined decrease during childhood in many individuals. In the absence of lactase, lactose cannot be digested, and its presence in the small intestine can cause diarrhea and increased flatus production when milk products are ingested.

V. Inflammatory bowel disease (IBD) is comprised of Crohn's disease and ulcerative colitis. They result from a genetic predisposition to having an inappropriate immune response to infection.

VI. Constipation is primarily the result of decreased colonic motility. The symptoms of constipation are produced by overdistension of the rectum, not by the absorption of toxic bacterial products.

VII. Diarrhea can be caused by decreased fluid absorption, increased fluid secretion, or both.

■ KEY TERMS

■ CLINICAL TERMS

■ REVIEW QUESTIONS

1. List the four processes that accomplish the functions of the gastrointestinal system.
2. List the primary functions performed by each of the organs in the gastrointestinal system.
3. Approximately how much fluid is secreted into the gastrointestinal tract each day compared with the amount of food and drink ingested? How much of this appears in the feces?
4. What structures are responsible for the large surface area of the small intestine?
5. Where does the venous blood go after leaving the small intestine?
6. Identify the enzymes involved in carbohydrate digestion and the mechanism of carbohydrate absorption in the small intestine.
7. List three ways in which proteins or their digestion products can be absorbed from the small intestine.
8. Describe the process of fat emulsification.
9. What is the role of micelles in fat absorption?

10. Describe the movement of fat digestion products from the intestinal lumen to a lacteal.
11. How does the absorption of fat-soluble vitamins differ from that of water-soluble vitamins?
12. Specify two conditions that may lead to failure to absorb vitamin B_{12}.
13. How are salts and water absorbed in the small intestine?
14. Describe the role of ferritin in the absorption of iron.
15. List the four types of stimuli that initiate most gastrointestinal reflexes.
16. Describe the location of the enteric nervous system and its role in both short and long reflexes.
17. Name the four best understood gastrointestinal hormones and state their major functions.
18. Describe the neural reflexes leading to increased salivary secretion.
19. Describe the sequence of events that occur during swallowing.
20. List the cephalic, gastric, and intestinal phase stimuli that stimulate or inhibit acid secretion by the stomach.
21. Describe the function of gastrin and the factors controlling its secretion.
22. By what mechanism is pepsinogen converted to pepsin in the stomach?
23. Describe the factors that control gastric emptying.
24. Describe the mechanisms controlling pancreatic secretion of bicarbonate and enzymes.
25. How are pancreatic proteolytic enzymes activated in the small intestine?
26. List the major constituents of bile.
27. Describe the recycling of bile salts by the enterohepatic circulation.
28. What determines the rate of bile secretion by the liver?
29. Describe the effects of secretin and CCK on the bile ducts and gallbladder.
30. What causes water to move from the blood to the lumen of the duodenum following gastric emptying?
31. Describe the type of intestinal motility found during and shortly after a meal and the type found several hours after a meal.
32. Describe the production of flatus by the large intestine.
33. Describe the factors that initiate and control defecation.
34. Why is the stomach's wall normally not digested by the acid and digestive enzymes in the lumen?
35. Describe the process of vomiting.
36. What are the consequences of blocking the common bile duct with a gallstone?
37. What are the consequences of the failure to digest lactose in the small intestine?
38. Contrast the factors that cause constipation with those that produce diarrhea.

Chapter 15 Test Questions

(Answers appear in Appendix A.)

Questions 1–4: Match the following gastrointestinal hormones to their descriptions.

   a. gastrin   c. secretin
   b. CCK   d. GIP

1. It is stimulated by the presence of acid in the small intestine and stimulates bicarbonate release from the pancreas and bile ducts.
2. It is stimulated by glucose and fat in the small intestine and increases insulin and amplifies the insulin responses to glucose.

3. It is inhibited by acid in the stomach and stimulates acid secretion from the stomach.
4. It is stimulated by amino acids and fatty acids in the small intestine and stimulates pancreatic enzyme secretion.

5. Which of the following is true about pepsin?
   a. Most pepsin is released directly from chief cells.
   b. Pepsin is most active at high pH.
   c. Pepsin is essential for protein digestion.
   d. Pepsin accelerates protein digestion.
   e. Pepsin accelerates fat digestion.

6. Micelles increase the absorption of fat by
   a. binding the lipase enzyme and holding it on the surface of the lipid emulsion droplet.
   b. keeping the insoluble products of fat digestion in small aggregates.
   c. promoting direct absorption across the intestinal epithelium.
   d. metabolizing triglyceride to monoglyceride.
   e. facilitating absorption into the lacteals.

7. Which of the following inhibits gastric HCl secretion during a meal?
   a. stimulation of the parasympathetic nerves to the enteric nervous system
   b. the sight and smell of food
   c. distension of the duodenum
   d. presence of peptides in the stomach
   e. distension of the stomach

8. Which component of bile is not primarily secreted by hepatocytes?
   a. bicarbonate
   b. bile salts
   c. cholesterol
   d. lecithin
   e. bilirubin

9. Which of the following is true about segmentation in the small intestine?
   a. It is a type of peristalsis.
   b. It moves chyme only from the duodenum to the ileum.
   c. Its frequency is the same in each intestinal segment.
   d. It is unaffected by cephalic phase stimuli.
   e. It produces a slow migration of chyme to the large intestine.

10. Which of the following is the primary absorptive process in the large intestine?
    a. active transport of sodium from the lumen to the blood
    b. absorption of water
    c. active transport of potassium from the lumen to the blood
    d. active absorption of bicarbonate into the blood
    e. active secretion of chloride from the blood

Chapter 15 Quantitative and Thought Questions

(Answers appear in Appendix A.)

1. If the salivary glands were unable to secrete amylase, what effect would this have on starch digestion?

2. Whole milk or a fatty snack consumed before the ingestion of alcohol decreases the rate of intoxication. By what mechanism may fat be acting to produce this effect?

3. A patient brought to a hospital after a period of prolonged vomiting has an elevated heart rate, decreased blood pressure, and below-normal blood potassium and acidity. Explain these symptoms in terms of the consequences of excessive vomiting.

4. Can fat be digested and absorbed in the absence of bile salts? Explain.

5. How might damage to the lower portion of the spinal cord affect defecation?

6. One of the older but no longer used procedures in the treatment of ulcers is abdominal vagotomy, surgical cutting of the vagus (parasympathetic) nerves to the stomach. By what mechanism might this procedure help ulcers to heal and decrease the incidence of new ulcers?

Chapter 15 Answers to Physiological Inquiries

Table 15–4 The most common finding is an abnormally high production of gastric (hydrochloric) acid due to gastrin stimulation of the parietal cell of the stomach (see Figure 15–20). This high acidity can cause injury to the duodenum because the pancreas cannot produce sufficient quantities of bicarbonate to neutralize it (see Figure 15–27). The low pH in the duodenum can also inactivate pancreatic enzymes (see Figure 15–28) which can ultimately lead to diarrhea due to unabsorbed nutrients, and to increased fat in the stool. The spectrum of findings in a patient with a gastrinoma is called the Zollinger-Ellison syndrome.

Figure 15–14b Aspiration of food during swallowing can lead to occlusion (blockage) of the airways, which can result in a disruption of oxygen delivery and carbon dioxide removal from the pulmonary system. Aspiration of stomach contents can lead to severe lung damage primarily due to the low pH of the material (see description of acute respiratory distress syndrome in Chapter 13).

Figure 15–18 Mucus secreted by the cells in the gastric pit (see Figure 15–17) creates a protective coating and traps bicarbonate. This gastric mucosal barrier protects the stomach from the luminal acidity.

Figure 15–30 A portal vein carries blood from one capillary bed to another capillary bed (rather than from capillaries to venules as described in Chapter 12). The hypothalamo-pituitary portal veins carry hypophysiotropic hormones from the capillaries of the median eminence to the anterior pituitary gland where they stimulate or inhibit the release of pituitary hormones (Chapter 11).

chapter 16

Regulation of Organic Metabolism and Energy Balance

Genetically obese and normal mouse.

Chapter 3 introduced the concepts of energy and of organic metabolism at the level of the individual cell. This chapter deals with two topics that are concerned in one way or another with those same concepts, but for the entire body. First, this chapter describes how the metabolic pathways for carbohydrate, fat, and protein are controlled so as to provide continuous sources of energy to the various tissues and organs at all times. Next, the determinants of total-body energy balance and the regulation of body temperature are described.

SECTION A Control and Integration of Carbohydrate, Protein, and Fat Metabolism

Events of the Absorptive and Postabsorptive States

The regular availability of food is a very recent event in the history of humankind, and indeed is still not universal. Thus, it should not be surprising that mechanisms have evolved for survival during alternating periods of plenty and fasting. There are two functional states or periods the body undergoes in providing energy for cellular activities: the **absorptive state,** during which ingested nutrients enter the blood from the gastrointestinal tract, and the **postabsorptive state,** during which the gastrointestinal tract is empty of nutrients and the body's own stores must supply energy. Because an average meal requires approximately four hours for complete absorption, our usual three-meal-a-day pattern places us in the postabsorptive state during the late morning, again in the late afternoon, and during most of the night. We will refer to going more than 24 h without eating as fasting.

During the absorptive period, some of the ingested nutrients supply the energy needs of the body, and the remainder are added to the body's energy stores, to be called upon during the next postabsorptive period. Total-body energy stores are adequate for the average person to withstand a fast of many weeks, provided water is available.

Figures 16–1 and **16–2** summarize the major pathways to be described in this chapter. Although they may appear formidable at first glance, they should present little difficulty after we have described the component parts. You should refer to these figures frequently during the following discussion.

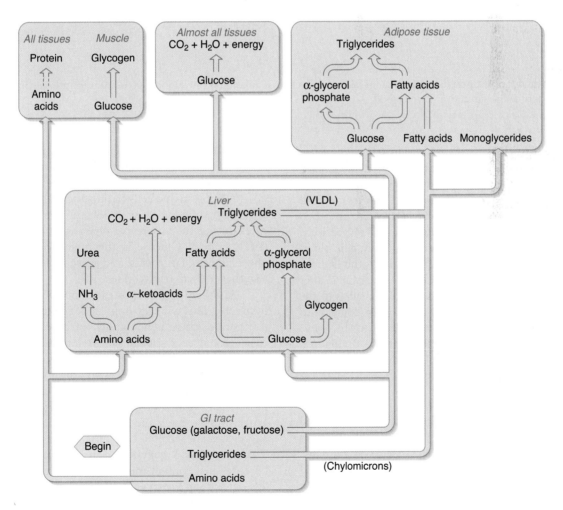

Figure 16–1

Major metabolic pathways of the absorptive state. The arrow from amino acids to protein is dashed to denote the fact that excess amino acids are not stored as protein (see text). All arrows between boxes denote transport of the substance via the blood. VLDL = very-low-density lipoproteins. Energy = ATP.

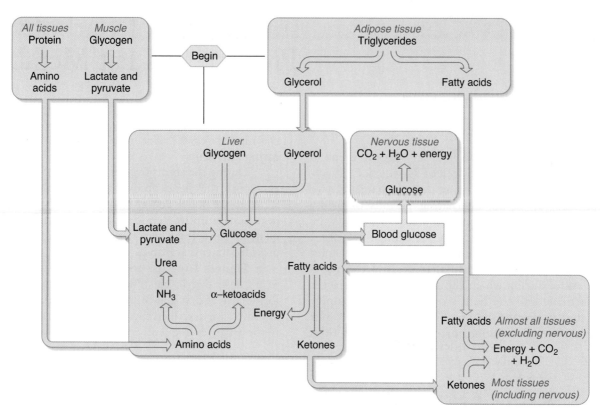

Figure 16–2

Major metabolic pathways of the postabsorptive state. The central focus is regulation of the blood glucose concentration. All arrows between boxes denote transport of the substance via the blood.

Absorptive State

An average meal contains all three of the major nutrients—carbohydrate, protein, and fat—with carbohydrate constituting most of the meal's energy content (calories). Recall from Chapter 15 that carbohydrate and protein are absorbed primarily as monosaccharides and amino acids, respectively, into the blood supplying the gastrointestinal tract. The blood leaves the gastrointestinal tract to go directly to the liver by way of the hepatic portal vein. This allows the liver to alter the nutrient composition of the blood before it returns to the heart to be pumped to the rest of the body. In contrast to carbohydrate and amino acids, fat is absorbed into the *lymph* as triglycerides in chylomicrons. The lymph then drains into the systemic venous system. Thus, the liver cannot first modify absorbed fat before it reaches other tissues.

Absorbed Carbohydrate

Some of the carbohydrate absorbed from the gastrointestinal tract is galactose and fructose. Because these sugars are either converted to glucose by the liver or enter essentially the same metabolic pathways as glucose, we will simply refer to absorbed carbohydrate as glucose.

Glucose is the body's major energy source during the absorptive state. Much of the absorbed glucose enters cells and is catabolized to carbon dioxide and water, providing energy for ATP formation. Skeletal muscle makes up the majority of body mass, so it is the major consumer of glucose, even at rest. Skeletal muscle not only catabolizes glucose during the absorptive phase, but also converts some of the glucose to the polysaccharide glycogen, which is then stored in the muscle for future use.

Adipose tissue cells (adipocytes) also catabolize glucose for energy, but the most important fate of glucose in adipocytes during the absorptive phase is its transformation to fat (triglycerides). Glucose is the precursor of both α-glycerol phosphate and fatty acids, and these molecules are then linked together to form triglycerides.

Another large fraction of the absorbed glucose enters the liver cells. This is a very important point: During the absorptive period, there is net *uptake* of glucose by the liver. It is either stored as glycogen, as in skeletal muscle, or transformed to α-glycerol phosphate and fatty acids, which are then used to synthesize triglycerides, as in adipose tissue. Some of the fat synthesized from glucose in the liver is stored there, but most is packaged, along with specific proteins, into molecular aggregates of lipids and proteins called **lipoproteins.** These aggregates are secreted by the liver cells and enter the blood. They are called **very-low-density lipoproteins (VLDL)** because they contain much more fat than protein, and fat is less dense than protein. The synthesis of VLDLs by liver cells occurs by processes similar to those for the synthesis of chylomicrons by intestinal mucosal cells, as Chapter 15 described.

Because of their large size, VLDL complexes do not readily penetrate capillary walls, once in the bloodstream. Instead, their triglycerides are hydrolyzed mainly to monoglycerides (glycerol linked to one fatty acid chain) and fatty acids by the enzyme **lipoprotein lipase.** This enzyme is located on the blood-facing surface of capillary endothelial cells, especially those in adipose tissue. In adipose tissue capillaries, the fatty acids generated diffuse across the capillary wall and into the adipocytes. There they combine with α-glycerol phosphate, supplied by glucose metabolites, to form triglycerides once again. Thus, most of the fatty acids in the VLDL triglycerides originally synthesized from glucose by the *liver* end up being stored in triglyceride in *adipose tissue.* The monoglycerides formed in the blood by the action of lipoprotein lipase in adipose tissue capillaries circulate to the liver, where they are metabolized.

To summarize, the major fates of glucose during the absorptive phase are utilization for energy, storage as glycogen in liver and skeletal muscle, and storage as fat in adipose tissue.

Absorbed Triglycerides

As described in Chapter 15, absorbed chylomicrons enter the lymph, which flows into the systemic circulation. The biochemical processing of these chylomicron triglycerides in plasma is quite similar to that just described for VLDLs produced by the liver. The fatty acids of plasma chylomicrons are released, mainly within adipose tissue capillaries, by the action of endothelial lipoprotein lipase. The released fatty acids then enter adipocytes and combine with α-glycerol phosphate, synthesized in the adipocytes from glucose metabolites, to form triglycerides.

The importance of glucose for triglyceride synthesis in adipocytes cannot be overemphasized. Adipocytes do not have the enzyme required for phosphorylation of glycerol, so α-glycerol phosphate can be formed in these cells only from glucose metabolites and not from glycerol or any other fat metabolites.

In contrast to α-glycerol phosphate, there are three major sources of the fatty acids found in adipose tissue triglyceride: (1) glucose that enters adipose tissue and is converted to fatty acids; (2) glucose that is converted in the liver to VLDL triglycerides, which are transported via the blood to the adipose tissue; and (3) ingested triglycerides transported to adipose tissue in chylomicrons. As we have seen, sources (2) and (3) require the action of lipoprotein lipase to release the fatty acids from the circulating triglycerides.

This description has emphasized the *storage* of ingested fat. For simplicity, Figure 16–1 does not include the fraction of the ingested fat that is not stored, but is oxidized during the absorptive state by various organs to provide energy. The relative amounts of carbohydrate and fat used for energy during the absorptive period depend largely on the content of the meal.

Absorbed Amino Acids

Some absorbed amino acids enter liver cells. They are used to synthesize a variety of proteins, including liver enzymes and plasma proteins, or they are converted to carbohydrate-like intermediates known as **α-ketoacids** by removal of the amino group. This process is called deamination. The amino groups are used to synthesize urea in the liver, which enters the blood and is excreted by the kidneys. The α-ketoacids can enter the

Krebs tricarboxylic acid cycle and be catabolized to provide energy for the liver cells. They can also be converted to fatty acids, thereby participating in fat synthesis by the liver.

Most ingested amino acids are not taken up by the liver cells, however, but instead enter other cells (see Figure 16–1), where they may be used to synthesize proteins. All cells require a constant supply of amino acids for protein synthesis and participate in protein metabolism.

Protein synthesis is represented by a dashed arrow in Figure 16–1 to call attention to an important fact: There is a net synthesis of protein during the absorptive period, but this basically just replaces the proteins catabolized during the postabsorptive period. In other words, excess amino acids are not stored as protein in the sense that glucose is stored as glycogen or that both glucose and fat are stored as fat. Rather, ingested amino acids in excess of those needed to maintain a stable rate of protein turnover are merely converted to carbohydrate or fat. Therefore, eating large amounts of protein does not in itself cause increases in body protein. Increased daily consumption of protein does, however, provide the amino acids needed to support the high rates of protein synthesis occurring in growing children or in adults who increase muscle mass by weight-lifting exercise.

Table 16–1 summarizes nutrient metabolism during the absorptive period.

Postabsorptive State

As the absorptive period ends, net synthesis of glycogen, fat, and protein ceases, and net catabolism of all these substances begins to occur. The overall significance of these events can be understood in terms of the essential problem during the postabsorptive period: No glucose is being absorbed from the intestinal tract, yet the plasma glucose concentration must be maintained because the brain normally utilizes only glucose for energy. If plasma glucose concentration decreases too much, alterations of neural activity can occur, ranging from subtle impairment of mental function to seizures, coma, and even death.

The events that maintain plasma glucose concentration fall into two categories: (1) reactions that provide sources of blood glucose, and (2) cellular utilization of fat for energy, thus "sparing" glucose.

Table 16–1	Summary of Nutrient Metabolism During the Absorptive Period
1. Energy is provided primarily by absorbed carbohydrate in a typical meal.	
2. There is net uptake of glucose by the liver.	
3. Some carbohydrate is stored as glycogen in liver and muscle, but most carbohydrates and fats in excess of that used for energy are stored as fat in adipose tissue.	
4. There is some synthesis of body proteins, but some of the amino acids in dietary protein are used for energy or converted to fat.	

Sources of Blood Glucose

The sources of blood glucose during the postabsorptive period are as follows (see Figure 16–2):

1. **Glycogenolysis,** the hydrolysis of glycogen stores to monomers of glucose-6-phosphate, occurs in the liver and skeletal muscle. In the liver, glucose-6-phosphate is enzymatically converted to glucose, which then enters the blood. Hepatic glycogenolysis begins within seconds of an appropriate stimulus, such as sympathetic nervous system activation. Thus, it is the first line of defense in maintaining plasma glucose concentration. The amount of glucose available from this source, however, can supply the body's needs for only several hours before hepatic glycogen is nearly depleted.

   Glycogenolysis also occurs in skeletal muscle, which contains approximately the same amount of glycogen as the liver. Unlike the liver, however, muscle lacks the enzyme necessary to form glucose from the glucose-6-phosphate formed during glycogenolysis. Instead, the glucose-6-phosphate undergoes glycolysis within muscle to yield ATP, pyruvate, and lactate. Some of the lactate enters the blood, circulates to the liver, and is converted into glucose, which can then leave the liver cells to enter the blood. Thus, muscle glycogen contributes to the blood glucose indirectly via the liver.

2. The catabolism of triglycerides in adipose tissue yields glycerol and fatty acids, a process termed **lipolysis.** The glycerol and fatty acids then enter the blood by diffusion. The glycerol reaching the liver is converted to glucose. Thus, an important source of glucose during the postabsorptive period is the glycerol released when adipose tissue triglyceride is broken down.

3. A few hours into the postabsorptive period, protein becomes another source of blood glucose. Large quantities of protein in muscle and other tissues can be catabolized without serious cellular malfunction. There are, of course, limits to this process, and continued protein loss during a prolonged fast ultimately means functional disintegration, sickness, and death. Before this point is reached, however, protein breakdown can supply large quantities of amino acids. These amino acids enter the blood and are taken up by the liver, where some can be converted via the α-ketoacid pathway to glucose. This glucose is then released into the blood.

Items 1 through 3 describe the synthesis by the liver of glucose from pyruvate, lactate, glycerol, and amino acids. Synthesis from any of these precursors is known as **gluconeogenesis**— that is, "formation of new glucose." During a 24-h fast, gluconeogenesis provides approximately 180 g of glucose. The kidneys can also perform gluconeogenesis, but mainly during a prolonged fast.

Glucose Sparing (Fat Utilization)

The 180 g of glucose per day produced by gluconeogenesis in the liver (and kidneys) during fasting supplies 720 kcal of energy. As described later in this chapter, normal total energy expenditure for an average adult is 1500 to 3000 kcal/day. Therefore, gluconeogenesis cannot supply all the energy needs of the body. An adjustment must therefore take place during the transition from the absorptive to the postabsorptive state: Most organs and tissues, other than those of the nervous system, markedly reduce their glucose catabolism and increase their fat utilization, the latter becoming the major energy source. This metabolic adjustment, termed **glucose sparing,** "spares" the glucose produced by the liver for the nervous system's use.

The essential step in this adjustment is lipolysis, the catabolism of adipose tissue triglyceride, which liberates glycerol and fatty acids into the blood. We described lipolysis in the previous section in terms of its importance in providing glycerol to the liver for conversion to glucose. Now, we focus on the liberated fatty acids, which circulate bound to plasma albumin. (Despite this binding to protein, they are known as free fatty acids [FFA] because they are "free" of their attachment to glycerol.) The circulating fatty acids are taken up and metabolized by almost all tissues, *excluding the nervous system.* They provide energy in two ways (Chapter 3): (1) They first undergo beta oxidation to yield hydrogen atoms (that go on to oxidative phosphorylation) and acetyl CoA; and (2) the acetyl CoA enters the Krebs cycle and is catabolized to carbon dioxide and water.

The liver is unique, however, in that most of the acetyl CoA it forms from fatty acids during the postabsorptive state does not enter the Krebs cycle but is processed into three compounds collectively called **ketones,** or ketone bodies. (Note that ketones are not the same as α-ketoacids, which, as we have seen, are metabolites of amino acids.) Ketones are released into the blood and provide an important energy source during prolonged fasting for the many tissues, *including* those of the nervous system, capable of oxidizing them via the Krebs cycle. One of the ketones is acetone, some of which is exhaled and accounts in part for the distinctive breath odor of individuals undergoing prolonged fasting.

The net result of fatty acid and ketone utilization during fasting is the provision of energy for the body while at the same time sparing glucose for the brain and nervous system. Moreover, as just emphasized, the brain can use ketones for an energy source, and it does so increasingly as ketones build up in the blood during the first few days of a fast. The survival value of this phenomenon is significant: When the brain reduces its glucose requirement by utilizing ketones, much less protein breakdown is required to supply amino acids for gluconeogenesis. Consequently, the protein stores will last longer, and the ability to withstand a long fast without serious tissue damage is enhanced.

Table 16–2 summarizes the events of the postabsorptive period. The combined effects of glycogenolysis, gluconeogenesis, and the switch to fat utilization are so efficient that, after several days of complete fasting, the plasma glucose concentration is reduced by only a few percent. After one month, it is decreased by only 25 percent (although in very thin persons, this happens much sooner).

Table 16–2	Summary of Nutrient Metabolism During the Postabsorptive Period

1. Glycogen, fat, and protein syntheses are curtailed, and net breakdown occurs.

2. Glucose is formed in the liver both from the glycogen stored there and by gluconeogenesis from blood-borne lactate, pyruvate, glycerol, and amino acids. The kidneys also perform gluconeogenesis during a prolonged fast.

3. The glucose produced in the liver (and kidneys) is released into the blood, but its utilization for energy is greatly reduced in muscle and other nonneural tissues.

4. Lipolysis releases adipose tissue fatty acids into the blood, and the oxidation of these fatty acids by most cells and of ketones produced from them by the liver provides most of the body's energy supply.

5. The brain continues to use glucose but also starts using ketones as they build up in the blood.

Endocrine and Neural Control of the Absorptive and Postabsorptive States

We now turn to the endocrine and neural factors that control and integrate these metabolic pathways. We will focus primarily on the following questions, summarized in **Figure 16–3**: (1) What controls net anabolism of protein, glycogen, and triglyceride in the absorptive phase, and net catabolism in the postabsorptive phase? (2) What induces the cells to utilize primarily glucose for energy during the absorptive phase, but fat during the postabsorptive phase? (3) What stimulates net glucose uptake by the liver during the absorptive phase, but gluconeogenesis and glucose release during the postabsorptive phase?

The most important controls of these transitions from feasting to fasting, and vice versa, are two pancreatic hormones—insulin and glucagon. Also playing a role are the hormones epinephrine and cortisol from the adrenal glands and the sympathetic nerves to liver and adipose tissue.

Insulin and glucagon are peptide hormones secreted by the **islets of Langerhans** (or, simply, pancreatic islets), clusters of endocrine cells in the pancreas. There are several distinct types of islet cells, each of which secretes a different

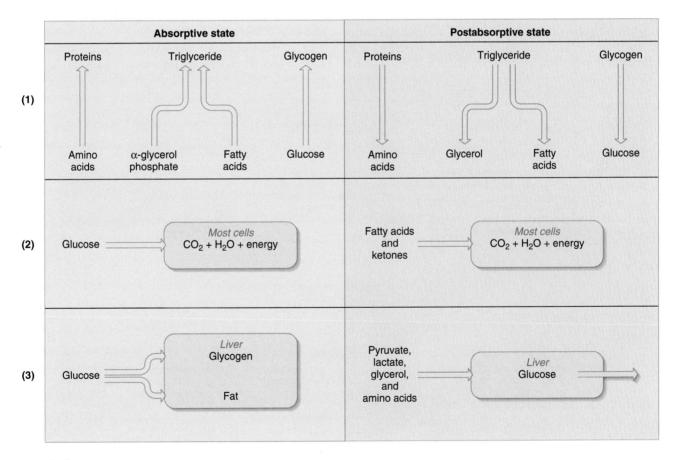

Figure 16–3

Summary of critical points in transition from the absorptive state to the postabsorptive state. The term *absorptive state* could be replaced with *actions of insulin,* and the term *postabsorptive state* with *results of decreased insulin.* The numbers at the left margin refer to discussion questions in the text.

hormone. The **beta cells** (or B cells) are the source of insulin, and the **alpha cells** (or A cells) of glucagon. There is at least one other molecule—somatostatin—secreted by still other islet cells called **delta cells** (or D cells). Pancreatic somatostatin is the same peptide chemically as the hypothalamic somatostatin, which controls growth hormone secretion from the anterior pituitary. The physiological functions of pancreatic somatostatin in humans are not fully established, but the peptide is known to be capable of inhibiting the secretion of both insulin and glucagon. Thus, it may act as a paracrine regulator of pancreatic secretion of these two hormones, preventing oversecretion of either.

Insulin

Insulin—sometimes called the "storage hormone"—is the most important controller of organic metabolism. Its secretion, and thus its plasma concentration, is increased during the absorptive state and decreased during the postabsorptive state.

The metabolic effects of insulin are exerted mainly on muscle cells (both cardiac and skeletal), adipose tissue cells, and liver cells. **Figure 16–4** summarizes the most important responses of these target cells. Compare the top portion of this figure to Figure 16–1 and to the left panel of Figure 16–3 and you will see that the responses to an increase in insulin are the same as the events of the absorptive-state pattern. Conversely, the effects of a reduction in plasma insulin are the same as the events of the postabsorptive pattern in Figure 16–2 and the right panel of Figure 16–3. The reason for these correspondences is that an increased plasma concentration of insulin is the major cause of the absorptive-state events, and a decreased plasma concentration of insulin is the major cause of the postabsorptive events.

Like all peptide hormones, insulin induces its effects by binding to specific receptors on the plasma membranes of its target cells. This binding triggers signal transduction pathways that influence the plasma membrane transport proteins and intracellular enzymes of the target cell. For example, in muscle cells and adipose tissue cells, an increased insulin concentration stimulates cytoplasmic vesicles that contain a particular type of glucose transporter (GLUT-4) in their membrane to fuse with the plasma membrane (**Figure 16–5**). The increased number of plasma membrane glucose transporters resulting from this fusion then causes a greater rate of glucose movement from the extracellular fluid into the cells by facilitated diffusion. Recall from Chapter 4 that glucose enters virtually all body cells by facilitated diffusion. There are multiple subtypes of glucose transporters that mediate this process, however, and the subtype GLUT-4, which is regulated by insulin, is found mainly in muscle and adipose tissue cells. Of great significance is that the cells of the brain express a different subtype of GLUT, one that has very high affinity for glucose and whose activity is not insulin-dependent. This ensures that even if plasma insulin levels are very low, as in prolonged fasting, cells of the brain can continue to transport glucose from the blood and maintain CNS function.

A description of the many enzymes whose activities and/or concentrations are influenced by insulin is beyond the

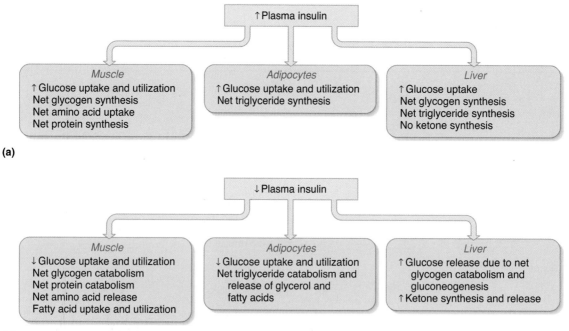

(a)

(b)

Figure 16–4

Summary of overall target-cell responses to (a) an increase or (b) a decrease in the plasma concentration of insulin. The responses in (a) are virtually identical to the absorptive state events of Figure 16–1 and the left panel of Figure 16–3; the responses in (b) are virtually identical to the postabsorptive state events of Figure 16–2 and the right panel of Figure 16–3.

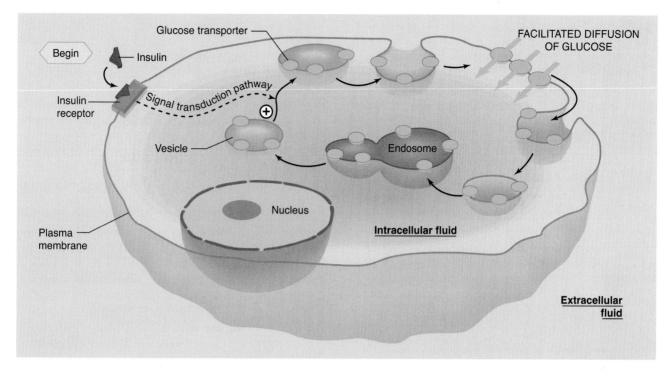

Figure 16–5

Stimulation by insulin of the translocation of glucose transporters from cytoplasmic vesicles to the plasma membrane in muscle cells and adipose tissue cells. Note that these transporters are constantly recycled by endocytosis from the plasma membrane back through endosomes into vesicles. As long as insulin levels are elevated, the entire cycle continues, and the number of transporters in the plasma membrane stays high. This is how insulin decreases the plasma level of glucose. In contrast, when insulin levels decrease, the cycle is broken, the vesicles accumulate in the cytoplasm, and the number of transporters in the plasma membrane decreases. Thus, without insulin the plasma glucose level would increase, because glucose transport from plasma to cells would be reduced.

Figure 16–5 physiological *inquiry* ⓟⓘ

- What advantage is there to having insulin-dependent glucose transporters pre-packaged in a cell?

Answer can be found at end of chapter.

scope of this book, but the overall pattern is shown in **Figure 16–6** for reference and to illustrate several principles. The essential information to understand about insulin's actions is the target cells' ultimate responses—that is, the material summarized in Figure 16–4. Figure 16–6 shows some of the specific biochemical reactions that underlie these responses.

A major principle illustrated by Figure 16–6 is that, in each of its target cells, insulin brings about its ultimate responses by multiple actions. Take, for example, its effects on muscle cells. In these cells, insulin favors glycogen formation and storage by (1) increasing glucose transport into the cell, (2) stimulating the key enzyme (**glycogen synthase**) that catalyzes the rate-limiting step in glycogen synthesis, and (3) inhibiting the key enzyme (**glycogen phosphorylase**) that catalyzes glycogen catabolism. Thus, insulin favors glucose transformation to and storage as glycogen in muscle through three pathways. Similarly, for protein synthesis in muscle cells, insulin (1) increases the number of active plasma membrane transporters for amino acids, thereby increasing amino acid transport into the cells, (2) stimulates the ribosomal enzymes that mediate the synthesis of protein from these

amino acids, and (3) inhibits the enzymes that mediate protein catabolism.

Control of Insulin Secretion

The major controlling factor for insulin secretion is the plasma glucose concentration. An increase in plasma glucose concentration, as occurs after a meal, acts on the beta cells of the islets of Langerhans to stimulate insulin secretion, whereas a decrease in plasma glucose removes the stimulus for insulin secretion. The feedback nature of this system is shown in **Figure 16–7**: Following a meal, the increase in plasma glucose concentration stimulates insulin secretion. The insulin stimulates the entry of glucose into muscle and adipose tissue, as well as net uptake, rather than net output, of glucose by the liver. These effects eventually reduce the blood concentration of glucose to its premeal level, thereby removing the stimulus for insulin secretion and causing it to return to its previous level.

In addition to plasma glucose concentration, there are many other factors that control insulin secretion (**Figure 16–8**). For example, elevated amino acid concentrations stimulate insulin secretion. This is another negative feedback control: Amino

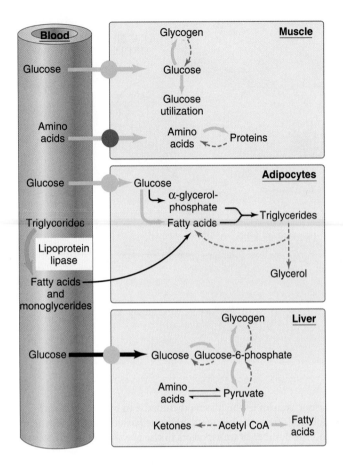

Figure 16–6

Illustration of the key biochemical events that underlie the responses of target cells to insulin as summarized in Figure 16–4. Each green arrow denotes a process stimulated by insulin, whereas a dashed red arrow denotes inhibition by insulin. Except for the effects on the transport proteins for glucose and amino acids, all other effects are exerted on insulin-sensitive enzymes. The bowed arrows denote pathways whose reversibility is mediated by different enzymes (Chapter 3); such enzymes are commonly the ones influenced by insulin and other hormones. The black arrows are processes that are not *directly* affected by insulin, but are enhanced in the presence of increased insulin as the result of mass action.

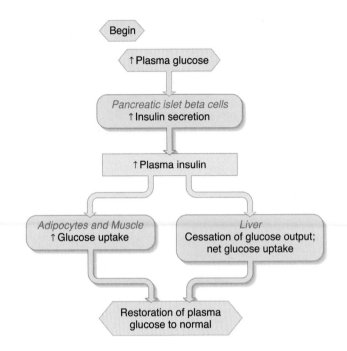

Figure 16–7

Nature of plasma glucose control over insulin secretion. As glucose levels increase in plasma (e.g., after a meal containing carbohydrate), insulin secretion is rapidly stimulated. The increase in insulin stimulates glucose transport from extracellular fluid into cells, thus decreasing plasma glucose concentrations. Insulin also acts to inhibit hepatic glucose output.

Figure 16–7 physiological *inquiry* (pi)

■ Notice that the brain is not listed as being insulin-sensitive. Why is that advantageous?

Answer can be found at end of chapter.

acid concentrations increase in the blood after ingestion of a protein-containing meal, and the increased plasma insulin stimulates the uptake of these amino acids by muscle and other cells, thereby lowering their concentrations.

There are also important hormonal controls over insulin secretion. For example, a hormone—glucose-dependent insulinotropic peptide (GIP)—secreted by endocrine cells in the gastrointestinal tract in response to eating stimulates the release of insulin. This response provides a feedforward component to glucose regulation during the ingestion of a meal. Thus, insulin secretion rises earlier than it would have if plasma glucose were the only controller, thereby minimizing the peak in plasma glucose concentration. This mechanism minimizes the likelihood of large increases in plasma glucose after a meal.

Finally, the autonomic neurons to the islets of Langerhans also influence insulin secretion. Activation of the parasympathetic neurons, which occurs during the ingestion of a meal, stimulates the secretion of insulin and constitutes a second type of feedforward regulation. In contrast, activation of the sympathetic neurons to the islets or an increase in the plasma concentration of epinephrine (the hormone secreted by the adrenal medulla) inhibits insulin secretion. The significance of this relationship for the body's response to low plasma glucose (**hypoglycemia**), stress, and exercise—all situations in which sympathetic activity is increased—will be described later in this chapter.

In summary, insulin plays the primary role in controlling the metabolic adjustments required for feasting or fasting. Other hormonal and neural factors, however, also play significant roles. They all oppose the action of insulin in one way or another and are known as **glucose-counterregulatory controls.** As described next, the most important of these are glucagon, epinephrine, sympathetic nerves, cortisol, and growth hormone.

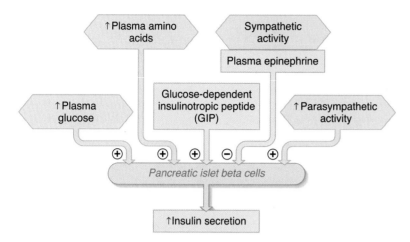

Figure 16–8

Major controls of insulin secretion. The ⊕ and ⊖ symbols represent stimulatory and inhibitory actions, respectively. GIP is a gastrointestinal hormone that acts as a feedforward signal to the pancreas.

Glucagon

As noted earlier, **glucagon** is the peptide hormone produced by the alpha cells of the pancreatic islets. The major physiological effects of glucagon occur within the liver and oppose those of insulin (**Figure 16–9**). Thus, glucagon (1) increases glycogen breakdown, (2) increases gluconeogenesis, and (3) increases the synthesis of ketones. The overall results are to increase the plasma concentrations of glucose and ketones, which are important for the postabsorptive period, and to prevent hypoglycemia. The effects, if any, of glucagon on adipocyte function in humans are still unresolved.

From a knowledge of these effects, you might predict that glucagon secretion should increase during the postabsorptive period and periods of prolonged fasting; this is indeed the case. The major stimulus for glucagon secretion at these times is hypoglycemia. The adaptive value of such a reflex is clear: A decreasing plasma glucose concentration induces an increase in the secretion of glucagon into the blood, which, by its effects on metabolism, serves to restore normal blood glucose concentration by glycogenolysis and gluconeogenesis. At the same time, glucagon supplies ketones for utilization by the brain. Conversely, an increased plasma glucose concentration inhibits glucagon's secretion, thereby helping to return the plasma glucose concentration toward normal. Thus, during the postabsorptive state, plasma insulin concentration is *low* and plasma glucagon concentration is *high,* and this combined change accounts almost entirely for the transition from the absorptive to the postabsorptive state. Said in a different way, this shift is best explained by a rise in the glucagon: insulin ratio in the plasma.

The secretion of glucagon, like that of insulin, is controlled not only by the plasma concentration of glucose and other nutrients but also by neural and hormonal inputs to the islets. For example, the sympathetic nerves to the islets stimulate glucagon secretion—just the opposite of their effect on insulin secretion.

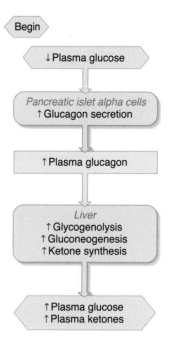

Figure 16–9

Nature of plasma glucose control over glucagon secretion.

Epinephrine and Sympathetic Nerves to Liver and Adipose Tissue

As noted earlier, epinephrine and the sympathetic nerves to the pancreatic islets inhibit insulin secretion and stimulate glucagon secretion. In addition, epinephrine also affects nutrient metabolism directly (**Figure 16–10**). Its major direct effects include stimulation of (1) glycogenolysis in both the liver and skeletal muscle, (2) gluconeogenesis in the liver, and (3) lipolysis in adipocytes. Activation of the sympathetic nerves to the liver and adipose tissue elicits essentially the same responses from these organs as does circulating epinephrine.

In adipocytes, epinephrine stimulates the activity of an enzyme called **hormone-sensitive lipase (HSL).** Once activated, HSL causes the breakdown of triglycerides to free fatty acids and glycerol. Both are then released into the blood, where they serve directly as a fuel source (fatty acids) or as a gluconeogenic precursor (glycerol). Not surprisingly, insulin inhibits the activity of HSL during the absorptive state.

Thus, enhanced sympathetic nervous system activity exerts effects on organic metabolism—specifically, increased plasma concentrations of glucose, glycerol, and fatty acids—that are opposite those of insulin.

As might be predicted from these effects, hypoglycemia leads to increases in both epinephrine secretion and sympathetic nerve activity to the liver and adipose tissue. This is the same stimulus that leads to increased glucagon secretion, although the receptors and pathways are totally different. When the plasma glucose concentration decreases, glucose receptors in the central nervous system (and, possibly, the liver) initiate the reflexes that lead to increased activity in the sympathetic pathways to the adrenal medulla, liver, and adipose tissue. The adaptive value of the response is the same as that for

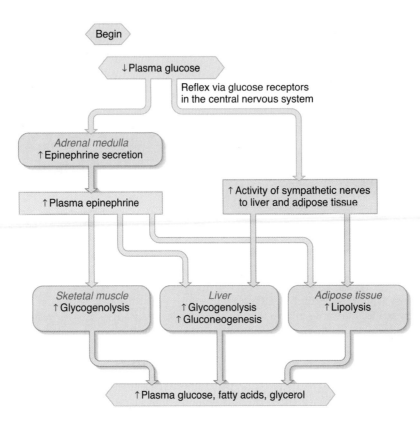

Figure 16–10

Participation of the sympathetic nervous system in the response to a low plasma glucose concentration (hypoglycemia). Glycogenolysis in skeletal muscle contributes to restoring plasma glucose by releasing lactate and pyruvate, which are converted to glucose in the liver. Recall also from Figure 16–8 and the text that the sympathetic nervous system inhibits insulin and stimulates glucagon secretion, which further contributes to the increased plasma fuel sources.

the glucagon response to hypoglycemia: Blood glucose returns toward normal, and fatty acids are supplied for cell utilization.

Cortisol

Cortisol, the major glucocorticoid produced by the adrenal cortex, plays an essential permissive role in the adjustments to fasting. We have described how fasting is associated with the stimulation of both gluconeogenesis and lipolysis; however, neither of these critical metabolic transformations occurs to the usual degree in a person deficient in cortisol. In other words, the plasma cortisol level does not need to increase much during fasting, but the presence of cortisol in the blood maintains the concentrations of the key liver and adipose tissue enzymes required for gluconeogenesis and lipolysis (e.g., HSL). Therefore, in response to fasting, people with a cortisol deficiency develop hypoglycemia serious enough to interfere with cellular function. Moreover, cortisol can play more than a permissive role when its plasma concentration does increase, as it does during stress. At high concentrations, cortisol elicits many metabolic events ordinarily associated with fasting (**Table 16–3**). In fact, cortisol actually reduces the sensitivity of muscle and adipose cells to insulin, which helps to maintain plasma glucose levels during fasting, thereby providing a regular source of energy for the brain. Clearly, here is another hormone, in addition to glucagon and epinephrine, that can exert actions opposite those of insulin. Indeed, people with very high plasma levels of cortisol, due either to abnormally high secretion or to glucocorticoid administration for medical reasons, can develop symptoms similar to those seen in individuals, such as certain diabetics, whose cells do not respond properly to insulin.

Growth Hormone

The primary physiological effects of growth hormone are to stimulate both growth and protein synthesis. Compared to these effects, those it exerts on carbohydrate and lipid metabolism are minor. Nonetheless, as is true for cortisol, either severe deficiency or marked excess of growth hormone does produce significant abnormalities in lipid and carbohydrate metabolism. Growth hormone's effects on these nutrients, in contrast to those on protein metabolism, are similar to those of cortisol and opposite those of insulin. Growth hormone (1) renders adipocytes more responsive to lipolytic stimuli, (2) increases gluconeogenesis by the liver, and (3) reduces the ability of insulin to cause glucose uptake by muscle and adipose tissue. These three effects are often termed growth hormone's "anti-insulin effects." Because of these effects, some of the symptoms seen in people with acromegaly (excess growth hormone production; Chapter 11) are similar to those seen in people with insulin resistance.

Table 16–3 — Effects of Cortisol on Organic Metabolism

1. Basal concentrations are permissive for stimulation of gluconeogenesis and lipolysis in the postabsorptive state.

2. Increased plasma concentrations cause:

   a. Increased protein catabolism

   b. Increased gluconeogenesis

   c. Decreased glucose uptake by muscle cells and adipose tissue cells

   d. Increased triglyceride breakdown

Net result: Increased plasma concentrations of amino acids, glucose, and free fatty acids

Summary of Hormonal Controls

To a great extent, insulin may be viewed as the "hormone of plenty." Its secretion and plasma concentration are increased during the absorptive period and decreased during postabsorption. These changes are adequate to cause most of the metabolic changes associated with these periods. In addition, opposed in various ways to insulin's effects are the actions of the major glucose-counterregulatory controls—glucagon, epinephrine and the sympathetic nerves to the liver and adipose tissue, cortisol, and growth hormone (**Table 16–4**). Glucagon and the sympathetic nervous system definitely play roles in preventing hypoglycemia. The rates of secretion of cortisol and growth hormone are not usually coupled to the absorptive-postabsorptive pattern. Nevertheless, their presence in the blood at basal concentrations is necessary for normal adjustment of lipid and carbohydrate metabolism to the postabsorptive period, and excessive amounts of either hormone cause abnormally elevated plasma glucose concentrations.

Energy Homeostasis in Exercise and Stress

During exercise, large quantities of fuels must be mobilized to provide the energy required for muscle contraction. These include plasma glucose and fatty acids as well as the muscle's own glycogen.

The plasma glucose used during exercise is supplied by the liver, both by breakdown of its glycogen stores and by gluconeogenesis. Glycerol is made available to the liver by a marked increase in adipose tissue lipolysis, with a resultant release of glycerol and fatty acids into the blood, the fatty acids serving as a fuel source for the exercising muscle.

What happens to plasma glucose concentration during exercise? It changes very little in short-term, mild-to-moderate exercise and may even increase slightly with strenuous, short-term activity. However, during prolonged exercise (**Figure 16–11**)—more than about 90 min—plasma glucose concentration does decrease, but usually by less than 25 percent. Clearly, glucose output by the liver increases approximately in proportion to increased glucose utilization during exercise, at least until the later stages of prolonged exercise when it begins to lag somewhat.

The metabolic profile seen in an exercising person—increases in hepatic glucose production, triglyceride breakdown, and fatty acid utilization—is similar to that seen in a fasting person, and the endocrine controls are also the same. Exercise is characterized by a decrease in insulin secretion and an increase in glucagon secretion (see Figure 16–11), and the changes in the plasma concentrations of these two hormones are the major controls during exercise. In addition, activity of the sympathetic nervous system increases (including secretion of epinephrine), and cortisol and growth hormone secretion both increase as well.

What triggers increased glucagon secretion and decreased insulin secretion during exercise? One signal, at least during *prolonged* exercise, is the modest decrease in plasma glucose that occurs (see Figure 16–11). This is the same signal that controls

Table 16–4 — Summary of Glucose-Counterregulatory Controls*

	Glucagon	Epinephrine	Cortisol	Growth Hormone
Glycogenolysis	✓	✓		
Gluconeogenesis	✓	✓	✓	✓
Lipolysis		✓	✓	✓
Inhibition of glucose uptake by muscle cells and adipose tissue cells			✓	✓

*A ✓ indicates that the hormone stimulates the process; no ✓ indicates that the hormone has no major physiological effect on the process. Epinephrine stimulates glycogenolysis in both liver and skeletal muscle, whereas glucagon does so only in liver.

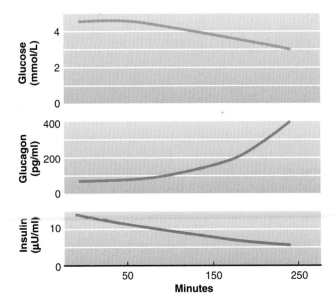

Figure 16–11

Plasma concentrations of glucose, glucagon, and insulin during prolonged (240 min) moderate exercise at a fixed intensity.

Adapted from Felig and Wahren.

the secretion of these hormones in fasting. Other inputs at all intensities of exercise include increased circulating epinephrine and increased activity of the sympathetic neurons supplying the pancreatic islets. Thus, the increased sympathetic nervous system activity characteristic of exercise not only contributes directly to fuel mobilization by acting on the liver and adipose tissue, but contributes indirectly by inhibiting the secretion of insulin and stimulating that of glucagon. This sympathetic output is not triggered by changes in plasma glucose concentration but is mediated by the central nervous system as part of the neural response to exercise.

One component of the response to exercise is quite different from the response to fasting: In exercise, glucose uptake and utilization by the muscles are increased, whereas during fasting they are markedly reduced. How is it that, dur-

ing exercise, the movement of glucose via facilitated diffusion into muscle can remain high in the presence of reduced plasma insulin and increased plasma concentrations of cortisol and growth hormone, all of which decrease glucose uptake by skeletal muscle? By an as-yet-unidentified mechanism, muscle contraction causes migration of an intracellular store of glucose transporters to the plasma membrane. Thus, exercising muscles require more glucose than do muscles at rest, but less insulin to induce glucose transport into muscle cells.

Exercise and the postabsorptive state are not the only situations characterized by the endocrine profile of decreased insulin and increased glucagon, sympathetic activity, cortisol, and growth hormone. This profile also occurs in response to a variety of nonspecific stresses, both physical and emotional. The adaptive value of these endocrine responses to stress is that the resulting metabolic shifts prepare the body for exercise ("fight or flight") in the face of real or threatened injury. In addition, the amino acids liberated by the catabolism of body protein stores because of decreased insulin and increased cortisol not only provide energy via gluconeogenesis but also constitute a potential source of amino acids for tissue repair should injury occur.

Chronic, intense exercise can also be stressful for the human body. In such cases, certain nonessential functions shut down so that nutrients can be directed primarily to muscle. One of these nonessential functions is reproduction. Thus, adolescents engaged in rigorous daily training regimens, such as gymnastics, may show delayed puberty. Similarly, women who perform chronic, intense exercise may become temporarily infertile, a condition known as ***exercise-induced amenorrhea*** (the lack of regular menstrual cycles; Chapter 17). This condition is seen in a variety of occupations that combine weight loss and strenuous exercise, such as may occur in professional ballerinas. Whether exercise-induced infertility occurs in men is uncertain, but most evidence suggests it does not.

It should be clear from this discussion that the maintenance of plasma glucose and other nutrients within a homeostatic range is vitally important for proper functioning of the tissues and organs in the body. When the regulation of these substances is abnormal, the consequences may be severe, as we see next.

ADDITIONAL CLINICAL EXAMPLES

Diabetes Mellitus

The name *diabetes,* meaning "syphon" or "running through," denotes the increased urinary volume excreted by people suffering from this disease. *Mellitus,* meaning "sweet," distinguishes this urine from the large quantities of nonsweet ("insipid") urine produced by persons suffering from vasopressin deficiency. As described in Chapter 14, the latter disorder is known as diabetes insipidus, and the unmodified word *diabetes* is often used as a synonym for ***diabetes mellitus,*** a disease that affects over 15 million people in the United States and that is increasing at an alarming rate.

Diabetes can be due to a deficiency of insulin or to a decreased responsiveness to insulin. Thus, diabetes is not

one but two diseases with different causes. Classification of these diseases rests on how much insulin a person's pancreas is secreting. In ***type 1 diabetes mellitus*** (***T1DM;*** formerly called insulin-dependent diabetes mellitus), insulin is completely or almost completely absent from the islets of Langerhans and the plasma. Therefore, therapy with insulin is essential. In ***type 2 diabetes mellitus*** (***T2DM,*** formerly called non-insulin-dependent diabetes mellitus), insulin is usually present in plasma at nearly normal or even above-normal levels, but cellular sensitivity to insulin is lower than normal (***insulin resistance***). Therefore, therapy may involve some combination of insulin or drugs that increase cellular sensitivity to insulin.

T1DM is less common, affecting approximately 5 percent of diabetic patients in the United States. T1DM is due to the total or near-total autoimmune destruction of the pancreatic beta cells by the body's own white blood cells. The triggering events for this autoimmune response are not yet fully established. Treatment of T1DM always involves the administration of insulin (by injection, because oral administration of insulin would not be effective due to the actions of gastrointestinal enzymes). Recent therapies for T1DM make use of alternative routes of insulin administration, such as a recently approved inhaled form of insulin. Transplantation of a whole pancreas or normal islet cells into a person with T1DM is also effective in reducing insulin requirements.

Because of insulin deficiency, *untreated* patients with T1DM always have elevated glucose concentrations in their blood. The increase in plasma glucose occurs because (1) glucose fails to enter insulin's target cells normally, and (2) the liver continuously makes glucose by glycogenolysis and gluco-neogenesis, and secretes the glucose into the blood. Recall also that insulin normally suppresses lipolysis and ketone formation. Thus, another result of the insulin deficiency is pronounced lipolysis with subsequent elevation of plasma glycerol and fatty acids. Many of the fatty acids are then converted by the liver into ketones, which are released into the blood.

If extreme, these metabolic changes culminate in the acute life-threatening emergency called ***diabetic ketoacidosis*** (**Figure 16–12**). Some of the problems are due to the effects that extremely elevated plasma glucose concentration produces on renal function. Chapter 14 pointed out that a typical person does not excrete glucose because all glucose filtered at the renal corpuscle is reabsorbed by the tubules. However, the elevated plasma glucose of diabetes increases the filtered load of glucose beyond the maximum tubular reabsorptive capacity, and therefore large amounts of glucose are excreted. For the same reasons, large amounts of ketones may also appear in the urine. These urinary losses deplete the body of nutrients and lead to weight loss. Far worse, however, is the fact that these unreabsorbed solutes cause an osmotic diuresis—increased urinary excretion of sodium and water, which can lead, by the sequence of events shown in Figure 16–12, to hypotension, brain damage, and death. It should be noted, however, that apart from this extreme example, diabetics are more often prone to hypertension, not hypotension (due to several causes, including vascular and kidney damage, and the obesity often associated with T2DM).

The other serious abnormality in diabetic ketoacidosis is the increased plasma hydrogen ion concentration caused by the accumulation of ketones. As described in Chapter 3, ketones are four-carbon breakdown products of fatty acids. Two ketones, known as hydroxybutyric acid and acetoacetic acid, are acidic at the pH of blood. This increased hydrogen ion concentration causes brain dysfunction that can contribute to coma and death.

Diabetic ketoacidosis is seen primarily in patients with *untreated* T1DM—that is, those with almost total inability to secrete insulin. However, at least 90 percent of diabetics are in the T2DM category and rarely develop metabolic derangements severe enough to develop diabetic ketoacidosis.

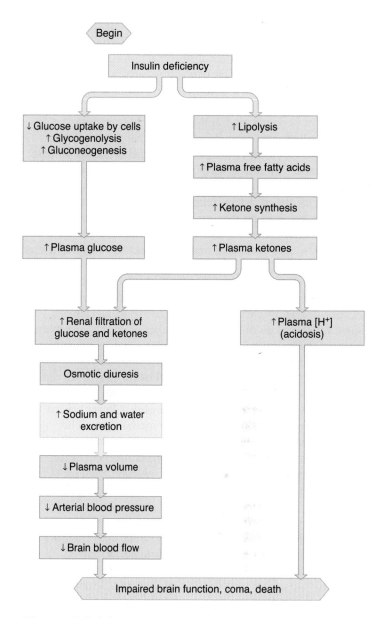

Figure 16–12

Diabetic ketoacidosis: Events caused by severe untreated insulin deficiency in type 1 diabetes mellitus.

T2DM is a syndrome mainly of overweight adults, typically starting in middle life. However, T2DM is *not* an age-dependent syndrome. As the incidence of childhood obesity has soared in the United States, so too has the incidence of T2DM in children and adolescents. Given the earlier mention of progressive weight loss in T1DM as a symptom of diabetes, why is it that most people with T2DM are overweight? One reason is that people with T2DM, in contrast to those with T1DM, do not excrete enough glucose in the urine to cause weight loss.

Several factors combine to cause T2DM. One major problem is target cell hyporesponsiveness to insulin, termed insulin resistance. Obesity accounts for much of the insulin resistance in T2DM. Obesity in any person—diabetic or not—induces some degree of insulin resistance, particularly in muscle and adipose tissue cells. One hypothesis is that the excess adipose tissue overproduces a messenger that causes downregulation of insulin-responsive glucose transporters or in some other way blocks insulin's actions. Another hypothesis is that excess fat deposition outside adipose tissue (for example, in muscle) causes a decrease in insulin sensitivity.

Most people with T2DM not only have insulin resistance but also have a defect in the ability of their beta cells to secrete insulin in response to a rise in plasma glucose concentration. In other words, although insulin resistance is the primary factor inducing hyperglycemia in T2DM, an as-yet-unidentified defect in beta cell function prevents these cells from responding maximally to the hyperglycemia.

The most effective therapy for obese persons with T2DM is weight reduction because obesity is a major cause of insulin resistance. An exercise program is also very important because insulin sensitivity is increased by frequent endurance-type exercise, independent of changes in body weight. This occurs, at least in part, because exercise training causes a substantial increase in the total number of plasma membrane glucose transporters in skeletal muscles.

If plasma glucose concentration is not adequately controlled by a program of weight reduction, exercise, and dietary modification (specifically, low-fat diets), then the person may be given orally active drugs that lower plasma glucose concentration by a variety of mechanisms. The *sulfonylureas* lower plasma glucose by acting on the beta cells to stimulate insulin secretion. Other drugs increase insulin sensitivity or decrease hepatic gluconeogenesis. Finally, in some cases the use of high doses of insulin itself is warranted.

Unfortunately, people with either form of diabetes mellitus tend to develop a variety of chronic abnormalities, including atherosclerosis, hypertension, kidney failure, small-vessel and nerve disease, susceptibility to infection, and blindness. Elevated plasma glucose contributes to most of these abnormalities either by causing the intracellular accumulation of certain glucose metabolites that exert harmful effects on cells when present in high concentrations, or by linking glucose to proteins, thereby altering their function.

This discussion of diabetes has focused on insulin, but it is now clear that the hormones that elevate plasma glucose concentration may contribute to the severity of the disease. Glucagon is quite important in this regard. Most diabetics, particularly those with T1DM, have inappropriately high plasma glucagon concentrations, which contribute to the metabolic abnormalities typical of diabetes. One reason these individuals have high plasma levels of glucagon is that insulin normally inhibits glucagon secretion, and the low insulin of T1DM releases glucagon secretion from this inhibition.

Finally, as we have seen, all the systems that increase plasma glucose concentration are activated during stress, which explains why stress worsens diabetic symptoms. Because diabetic ketoacidosis itself constitutes a severe stress, a positive feedback cycle is triggered: a lack of insulin induces ketoacidosis, which elicits activation of the glucose-counterregulatory systems, which worsens the ketoacidosis.

Hypoglycemia

Hypoglycemia is broadly defined as an abnormally low plasma glucose concentration. Plasma glucose concentration can decrease to very low values, usually during the postabsorptive state, in persons with several types of disorders. *Fasting hypoglycemia* and the relatively uncommon disorders responsible for it can be understood in terms of the regulation of blood glucose concentration. They include (1) an excess of insulin due to an insulin-producing tumor, drugs that stimulate insulin secretion, or taking too much insulin (if the person is diabetic); and (2) a defect in one or more glucose-counterregulatory controls, for example, inadequate glycogenolysis and/or gluconeogenesis due to liver disease, glucagon deficiency, or cortisol deficiency.

Fasting hypoglycemia causes many symptoms. Some—increased heart rate, trembling, nervousness, sweating, and anxiety—are accounted for by activation of the sympathetic nervous system caused reflexly by the hypoglycemia. Other symptoms, such as headache, confusion, dizziness, uncoordination, and slurred speech, are direct consequences of too little glucose reaching the brain. More serious brain effects, including convulsions and coma, can occur if the plasma glucose concentration falls low enough.

In contrast, low plasma glucose concentration has not been shown to routinely produce either acute or chronic symptoms of fatigue, lethargy, loss of libido, depression, or many other symptoms for which popular opinion frequently holds it responsible. Most of the symptoms commonly ascribed to hypoglycemia have other causes.

Increased Plasma Cholesterol

Earlier, we described the flow of lipids to and from adipose tissue in the form of fatty acids and triglycerides complexed with proteins. One very important lipid—**cholesterol**—was not mentioned earlier because it, unlike the fatty acids and triglycerides, does not serve as a metabolic fuel. Instead, cholesterol is a precursor for plasma membranes, bile salts, steroid hormones, and other specialized molecules. Thus, cholesterol has many important functions in the body. Unfortunately, it can also cause problems. Specifically, high plasma concentrations of cholesterol enhance the development of *atherosclerosis,* the arterial thickening that leads to heart attacks, strokes, and other forms of cardiovascular damage (Chapter 12).

Figure 16–13 illustrates a schema for cholesterol balance. The two sources of cholesterol are dietary cholesterol and cholesterol synthesized within the body. Dietary cholesterol comes from animal sources, egg yolk being by far the richest in this lipid (a single egg contains about 250 mg of cholesterol). Not all ingested cholesterol is absorbed into the blood, however—much of it simply passes through the length of the gastrointestinal tract and is excreted in the feces.

What about cholesterol synthesis within the body? Almost all cells can synthesize some of the cholesterol required for their own plasma membranes, but most cannot do so in adequate amounts and depend upon receiving cholesterol from the blood. This is also true of the endocrine

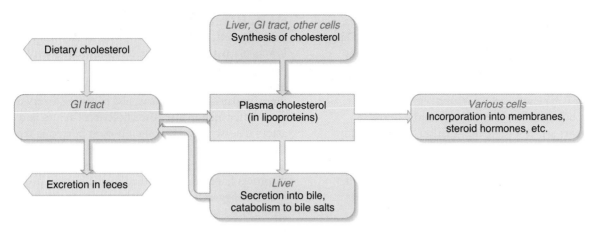

Figure 16–13

Cholesterol balance. Most of the cholesterol that is converted to bile salts, stored in the gallbladder, and secreted into the intestine gets recycled back to the liver. Changes in dietary cholesterol can modify plasma cholesterol concentration, but not usually dramatically. Cholesterol synthesis by the liver is up-regulated when dietary cholesterol is reduced, and vice versa.

cells that produce steroid hormones from cholesterol. Thus, most cells *remove* cholesterol from the blood. In contrast, the liver and cells lining the gastrointestinal tract can produce large amounts of cholesterol, most of which *enters* the blood.

Now for the other side of cholesterol balance—the pathways, all involving the liver, for net cholesterol loss from the body. First, some plasma cholesterol is picked up by liver cells and secreted into the bile, which carries it to the intestinal tract. Here it is treated much like ingested cholesterol, some being absorbed back into the blood and the remainder being excreted in the feces. Second, much of the cholesterol picked up by the liver cells is metabolized into bile salts (Chapter 15). After their production by the liver, these bile salts, like secreted cholesterol, flow through the bile duct into the small intestine. (As described in Chapter 15, many of these bile salts are then reclaimed by absorption back into the blood across the wall of the lower small intestine.)

The liver is clearly the major organ that controls cholesterol homeostasis, for it can add newly synthesized cholesterol to the blood and it can remove cholesterol from the blood, secreting it into the bile or metabolizing it to bile salts. The homeostatic control mechanisms that keep plasma cholesterol concentrations within a normal range operate on all of these hepatic processes, but the single most important response involves cholesterol production. The liver's synthesis of cholesterol is inhibited whenever dietary—and therefore plasma—cholesterol is increased. This is because cholesterol inhibits the enzyme (called HMG-CoA reductase) critical for cholesterol synthesis by the liver.

Thus, as soon as the plasma cholesterol level starts increasing because of increased cholesterol ingestion, hepatic synthesis of cholesterol is inhibited, and the plasma concentration of cholesterol remains close to its original value. Conversely, when dietary cholesterol is reduced and plasma cholesterol begins to fall, hepatic synthesis is stimulated (released from inhibition). This increased production opposes any further decrease in plasma cholesterol. The sensitivity of this negative feedback control

of cholesterol synthesis differs greatly from person to person, but it is the major reason why, for most people, it is difficult to change plasma cholesterol very much in either direction by altering only dietary cholesterol.

Thus far, the relative constancy of plasma cholesterol has been emphasized. There are, however, environmental and physiological factors that can significantly alter plasma cholesterol concentrations. Perhaps the most important of these factors are the quantity and type of dietary fatty acids. Ingesting saturated fatty acids, the dominant fatty acids of animal fat (particularly high in red meats, most cheeses, and whole milk), increases plasma cholesterol. In contrast, eating either polyunsaturated fatty acids (the predominant plant fatty acids) or monounsaturated fatty acids such as those in olive or peanut oil, decreases plasma cholesterol. The various fatty acids exert their effects on plasma cholesterol by altering cholesterol synthesis, excretion, and metabolism to bile salts.

A variety of drugs now in common use are also capable of lowering plasma cholesterol by influencing one or more of the metabolic pathways for cholesterol—for example, inhibiting the critical enzyme for hepatic cholesterol synthesis—or by interfering with intestinal absorption of bile salts.

The story is more complicated than this, however, because not all plasma cholesterol has the same function or significance for disease. Like most other lipids, cholesterol circulates in the plasma as part of various lipoprotein complexes. These include chylomicrons (Chapter 15), VLDL (this chapter), **low-density lipoproteins (LDL),** and **high-density lipoproteins (HDL).** LDL are the main cholesterol carriers, and they *deliver* cholesterol to cells throughout the body. LDL bind to plasma membrane receptors specific for a protein component of the LDL, and the LDL are taken up by the cells by endocytosis. In contrast to LDL, HDL *remove* excess cholesterol from blood and tissue, including the cholesterol-loaded cells of atherosclerotic plaques (Chapter 12). They then deliver this cholesterol to the liver, which secretes it into the bile or converts it to bile salts. HDL also delivers cholesterol to steroid-producing endocrine cells.

Uptake of the HDL by the liver and these endocrine cells is facilitated by the presence in their plasma membranes of large numbers of receptors specific for HDL, which bind to the receptors and then are taken into the cells.

LDL cholesterol is often designated "bad" cholesterol because high plasma levels are associated with increased deposition of cholesterol in arterial walls and a higher incidence of heart attacks. (The designation "bad" should not obscure the fact that LDL is essential for supplying cells with the cholesterol they require to synthesize cell membranes and, in the case of the gonads and adrenal glands, steroid hormones.) Using the same criteria, HDL cholesterol has been designated "good" cholesterol.

The best single indicator of the likelihood of developing atherosclerotic heart disease is, therefore, not total plasma cholesterol but rather the ratio of plasma LDL cholesterol to plasma HDL cholesterol—the lower the ratio, the lower the risk. Cigarette smoking, a known risk factor for heart attacks, lowers plasma HDL, whereas weight reduction (in overweight persons) and regular exercise usually increase it. Estrogen not only lowers LDL but raises HDL, which explains, in part, why premenopausal women have less coronary artery disease than men. After menopause, the cholesterol values and coronary artery disease rates in women not on hormone replacement therapy (Chapter 17) become similar to those in men.

Finally, a great variety of cholesterol metabolism disorders have been identified in the human population. In *familial hypercholesterolemia,* for example, LDL receptors are reduced in number or are nonfunctional. Consequently, LDL accumulates in the blood to very high levels. If untreated, this disease may result in atherosclerosis and heart disease at unusually young ages. ■

■ SECTION A SUMMARY

Events of the Absorptive and Postabsorptive States

I. During absorption, energy is provided primarily by absorbed carbohydrate. Net synthesis of glycogen, triglyceride, and protein occurs.
   a. Some absorbed carbohydrate not used for energy is converted to glycogen, mainly in the liver and skeletal muscle, but most is converted, in liver and adipocytes, to α-glycerol phosphate and fatty acids, which then combine to form triglycerides. The liver releases its triglycerides in very-low-density lipoproteins, the fatty acids of which are picked up by adipocytes.
   b. The fatty acids of some absorbed triglycerides are used for energy, but most are rebuilt into fat in adipose tissue.
   c. Some absorbed amino acids are converted to proteins, but excess amino acids are converted to carbohydrate and fat.
   d. There is a net upake of glucose by the liver.

II. In the postabsorptive state, blood glucose level is maintained by a combination of glucose production by the liver and a switch from glucose utilization to fatty acid and ketone utilization by most tissues.
   a. Synthesis of glycogen, fat, and protein is curtailed, and net breakdown of these molecules occurs.
   b. The liver forms glucose by glycogenolysis of its own glycogen and by gluconeogenesis from lactate and pyruvate (from the breakdown of muscle glycogen), glycerol (from adipose tissue lipolysis), and amino acids (from protein catabolism).
   c. Glycolysis is decreased, and most of the body's energy supply comes from the oxidation of fatty acids released by adipose tissue lipolysis and of ketones produced from fatty acids by the liver.
   d. The brain continues to use glucose but also starts using ketones as they build up in the blood.

Endocrine and Neural Control of the Absorptive and Postabsorptive States

I. The major hormones secreted by the pancreatic islets of Langerhans are insulin by the beta cells and glucagon by the alpha cells.

II. Insulin is the most important hormone controlling metabolism.
   a. In muscle, it stimulates glucose uptake, glycolysis, and net synthesis of glycogen and protein. In adipose tissue, it stimulates glucose uptake and net synthesis of triglyceride. In liver, it inhibits gluconeogenesis and glucose release and stimulates the net synthesis of glycogen and triglycerides.
   b. The major stimulus for insulin secretion is an increased plasma glucose concentration, but secretion is also influenced by many other factors summarized in Figure 16–8.

III. Glucagon, epinephrine, cortisol, and growth hormone all exert effects on carbohydrate and lipid metabolism that are opposite, in one way or another, to those of insulin. They raise plasma concentrations of glucose, glycerol, and fatty acids.
   a. Glucagon's physiological actions are on the liver, where it stimulates glycogenolysis, gluconeogenesis, and ketone synthesis.
   b. The major stimulus for glucagon secretion is hypoglycemia, but secretion is also stimulated by other inputs, including the sympathetic nerves to the islets.
   c. Epinephrine released from the adrenal medulla in response to hypoglycemia stimulates glycogenolysis in the liver and muscle, gluconeogenesis in the liver, and lipolysis in adipocytes. The sympathetic nerves to liver and adipose tissue exert effects similar to those of epinephrine.
   d. Cortisol is permissive for gluconeogenesis and lipolysis; in higher concentrations, it stimulates gluconeogenesis and blocks glucose uptake. These last two effects are also exerted by growth hormone.

Energy Homeostasis in Exercise and Stress

I. During exercise, the muscles use as their energy sources plasma glucose, plasma fatty acids, and their own glycogen.
   a. Glucose is provided by the liver, and fatty acids are provided by adipose tissue lipolysis.
   b. The changes in plasma insulin, glucagon, and epinephrine are similar to those that occur during the postabsorptive period and are mediated mainly by the sympathetic nervous system.

II. Stress causes hormonal changes similar to those caused by exercise.

Additional Clinical Examples

I. T1DM is due to insulin deficiency and can lead to diabetic ketoacidosis if untreated.

II. T2DM is usually associated with obesity and is caused by a combination of insulin resistance and a defect in beta cell responsiveness to elevated plasma glucose concentration. Plasma insulin concentration is usually normal or elevated.

III. Hypoglycemia is defined as abnormally low glucose levels in the blood. It may arise in the fed or fasting state. Symptoms of hypoglycemia are similar to those of sympathetic nervous system activation. However, severe hypoglycemia can lead to brain dysfunction and even death if untreated.

IV. Plasma cholesterol is a precursor for the synthesis of plasma membranes, bile salts, and steroid hormones.

V. Cholesterol synthesis by the liver is controlled so as to homeostatically regulate plasma cholesterol concentration; it varies inversely with ingested cholesterol.

VI. The liver also secretes cholesterol into the bile and converts it to bile salts.

VII. Plasma cholesterol is carried mainly by low-density lipoproteins, which deliver it to cells; high-density lipoproteins carry cholesterol from cells to the liver and steroid-producing cells. The LDL/HDL ratio correlates with the incidence of coronary heart disease.

■ SECTION A KEY TERMS

absorptive state   567
alpha cell   572
α-ketoacid   569
beta cell   572
cholesterol   580
delta cell   572
glucagon   575
gluconeogenesis   570
glucose-counterregulatory
    control   574
glucose sparing   570
glycogenolysis   570
glycogen phosphorylase   573
glycogen synthase   573
high-density lipoprotein
    (HDL)   581

hormone-sensitive lipase
    (HSL)   575
hypoglycemia   574
insulin   572
islets of Langerhans   571
ketone   570
lipolysis   570
lipoprotein   568
lipoprotein lipase   569
low-density lipoprotein
    (LDL)   581
postabsorptive state   567
very-low-density lipoprotein
    (VLDL)   568

■ SECTION A CLINICAL TERMS

atherosclerosis   580
diabetes mellitus   578
diabetic ketoacidosis   579
exercise-induced
    amenorrhea   578
familial hypercholesterolemia   582
fasting hypoglycemia   580

insulin resistance   578
sulfonylureas   580
type 1 diabetes mellitus
    (T1DM)   578
type 2 diabetes mellitus
    (T2DM)   578

■ SECTION A REVIEW QUESTIONS

1. Using a diagram, summarize the events of the absorptive period.
2. In what two organs does major glycogen storage occur?
3. How do the liver and adipose tissue metabolize glucose during the absorptive period?
4. How does adipose tissue metabolize absorbed triglyceride, and what are the three major sources of the fatty acids in adipose tissue triglyceride?
5. What happens to most of the absorbed amino acids when a high-protein meal is ingested?
6. Using a diagram, summarize the events of the postabsorptive period; include the four sources of blood glucose and the pathways leading to ketone formation.
7. Distinguish between the roles of glycerol and free fatty acids during fasting.
8. List the overall responses of muscle, adipose tissue, and liver to insulin. What effects occur when plasma insulin concentration decreases?
9. Describe five inputs controlling insulin secretion and the physiological significance of each.
10. List the effects of glucagon on the liver and their consequences.
11. Discuss two inputs controlling glucagon secretion and the physiological significance of each.
12. List the metabolic effects of epinephrine and the sympathetic nerves to the liver and adipose tissue, and state the net results of each.
13. Describe the permissive effects of cortisol and the effects that occur when plasma cortisol concentration increases.
14. List the effects of growth hormone on carbohydrate and lipid metabolism.
15. Which hormones stimulate gluconeogenesis? Glycogenolysis in the liver? Glycogenolysis in skeletal muscle? Lipolysis? Blockade of glucose uptake?
16. Describe how plasma glucose, insulin, glucagon, and epinephrine levels change during exercise and stress. What causes the changes in the concentrations of the hormones?
17. Describe the metabolic disorders of severe T1DM.
18. How does obesity contribute to T2DM?
19. Hypersecretion of which hormones can induce a diabetic state?
20. Using a diagram, describe the sources of cholesterol gain and loss. Include three roles the liver plays in cholesterol metabolism, and describe the controls over these processes.
21. What are the effects of saturated and unsaturated fatty acids on plasma cholesterol?
22. What is the significance of the ratio of LDL cholesterol to HDL cholesterol?

SECTION B   Regulation of Total-Body Energy Balance and Temperature

Basic Concepts of Energy Expenditure

The breakdown of organic molecules liberates the energy locked in their molecular bonds. Cells use this energy to perform the various forms of biological work, such as muscle contraction, active transport, and molecular synthesis. The first law of thermodynamics states that energy can be neither created nor destroyed, but can be converted from one form to another. Thus, internal energy liberated (ΔE) during breakdown of an organic molecule can either appear as heat (H) or be used to perform work (W).

$$\Delta E = H + W$$

During metabolism, about 60 percent of the energy released from organic molecules appears immediately as heat, and the rest is used for work. The energy used for work must first be incorporated into molecules of ATP. The subsequent breakdown of ATP serves as the immediate energy source for the work. The body is incapable of converting heat to work, but the heat released in its chemical reactions helps to maintain body temperature.

Biological work can be divided into two general categories: (1) **external work**—the movement of external objects by contracting skeletal muscles; and (2) **internal work**—all other forms of work, including skeletal muscle activity not used in moving external objects. As just stated, much of the energy liberated from nutrient catabolism appears immediately as heat. What may not be obvious is that internal work, too, is ultimately transformed to heat except during periods of growth. For example, internal work is performed during cardiac contraction, but this energy appears ultimately as heat generated by the friction of blood flow through the blood vessels.

Thus, the total energy liberated when cells catabolize organic nutrients may be transformed into body heat, can be used to do external work, or can be stored in the body in the form of organic molecules. The **total energy expenditure** of the body is therefore given by the equation

Total energy expenditure = Internal heat produced
+ External work performed + Energy stored

Metabolic Rate

The basic metric unit of energy is the joule. When quantifying the energy of metabolism, however, another unit is used, called the **calorie** (equal to 4.184 joules). One calorie is the amount of heat required to raise the temperature of one gram of water from 14.5°C to 15.5°C. Because the amount of energy stored in food is quite high relative to a calorie, a more convenient expression of energy in this context is the **kilocalorie (kcal),** which is equal to 1000 calories. (In the field of nutrition, the terms "Calorie" with a capital C and "kilocalorie" are synonyms; they are 1000 "calories," with a small c.) Total energy expenditure per unit time is called the **metabolic rate.**

Because many factors cause the metabolic rate to vary (**Table 16–5**), the most common method for evaluating it specifies certain standardized conditions and measures what is known as the **basal metabolic rate (BMR).** In the basal condition, the subject is at mental and physical rest in a room at a comfortable temperature and has not eaten for at least 12 hours (i.e., he or she is in a postabsorptive state). These conditions are arbitrarily designated "basal," even though the metabolic rate during sleep may be lower than the BMR. The BMR is often called the "metabolic cost of living," and most of the energy involved is expended by the heart, muscle, liver, kidneys, and brain. For the following discussion, the term *BMR* can be applied to metabolic rate only when the specified conditions are met. Thus, a person who has recently eaten or is exercising has a metabolic rate, but not a basal metabolic rate. The next sections describe several of the important determinants of BMR and metabolic rate.

Thyroid Hormones

The thyroid hormones are the single most important determinant of BMR regardless of size, age, or gender. TH increases the oxygen consumption and heat production of most body tissues, a notable exception being the brain. This ability to increase BMR is termed a **calorigenic effect.**

Long-term excessive TH, as in people with hyperthyroidism, induces a host of effects secondary to the calorigenic effect. For example, the increased metabolic demands mark-

Table 16–5	Some Factors Affecting the Metabolic Rate

Sleep (↓ during sleep)

Age (↓ with ↑ age)

Gender (women less than men at any given size)

Fasting (BMR decreases, which conserves energy stores)

Height, weight, and body surface area

Growth

Pregnancy, menstruation, lactation

Infection or other disease

Body temperature

Recent ingestion of food

Muscular activity

Emotional stress

Environmental temperature

Circulating levels of various hormones, especially epinephrine, thyroid hormone, and leptin

(The presence of, or an increase in any of these factors causes an increase in metabolic rate.)

edly increase hunger and food intake. The greater intake often remains inadequate to meet metabolic needs. The resulting net catabolism of protein and fat stores leads to loss of body weight. Of importance is the fact that the more metabolically active a particular cell is, the greater its requirements for vitamin cofactors. Therefore, even with increased dietary intake, the onset of hyperthyroidism may result in symptoms of vitamin deficiency. Also, the greater heat production activates heat-dissipating mechanisms, such as skin vasodilation and sweating, and the person feels intolerant to warm environments. In contrast, the hypothyroid person may experience cold intolerance.

The calorigenic effect of TH is only one of a wide variety of effects these hormones exert. The major functions of the thyroid hormones have all been described in this chapter and in Chapter 11 and are listed for reference in **Table 16–6**.

As described in Chapter 11, secretion of the thyroid hormones is stimulated by the anterior pituitary hormone thyroid-stimulating hormone (TSH), itself stimulated by the hypophysiotropic hormone, thyrotropin-releasing hormone (TRH). The thyroid hormones, in turn, exert a negative feedback effect on the hypothalamo-pituitary system. What is unusual about this hormonal system is that there is no known stimulus that activates the negative feedback elimination of the stimulus (as, for example, the way changes in plasma glucose influence insulin secretion). It is thought that the thyroid hormones set a background for the various parameters, such as BMR, that they influence. Starvation, however, is associated with decreased TH production, which tends to slow the consumption rate of endogenous fuel stores.

Epinephrine

Epinephrine is another hormone that exerts a calorigenic effect. This effect may be related to its stimulation of glycogen and triglyceride catabolism, as ATP hydrolysis and energy liberation occur during both the breakdown and subsequent resynthesis of these molecules. Thus, when epinephrine secretion by the adrenal medulla is stimulated, the metabolic rate rises. This accounts for part of the greater heat production associated with emotional stress.

Food-Induced Thermogenesis

The ingestion of food rapidly increases the metabolic rate by 10 to 20 percent for a few hours after eating. This effect is known as **food-induced thermogenesis.** Ingested protein produces the greatest effect, while carbohydrate and fat produce less. Most of the increased heat production is caused by the processing of the absorbed nutrients by the liver, not by the energy expended by the gastrointestinal tract in digestion and absorption. Because of the contribution of food-induced thermogenesis, BMR tests must be performed in the postabsorptive state.

Food-induced thermogenesis is the rapid increase in energy expenditure in response to ingestion of a meal. As we will see, *prolonged* alterations in food intake (either increased or decreased total calories) also have significant effects on metabolic rate.

Muscle Activity

The factor that can most increase metabolic rate is altered skeletal muscle activity. Even minimal increases in muscle contraction significantly increase metabolic rate, and strenuous exercise may raise energy expenditure more than 15-fold (**Table 16–7**). Thus, depending on the degree of physical activity, total energy expenditure may vary for a healthy young adult from a value of approximately 1500 kcal/24 h to more than 7000 kcal/24 h (for a lumberjack). Changes in muscle activity also account in part for the changes in metabolic rate that occur during sleep (decreased muscle contraction) and during exposure to a low environmental temperature (increased muscle contraction due to shivering).

Regulation of Total-Body Energy Stores

Under normal conditions for body weight to remain stable, the total energy expenditure (metabolic rate) of the body must equal the total energy intake. We have already identified the ultimate forms of energy expenditure: internal heat production, external work, and net molecular synthesis (energy storage). The source of input is the energy contained in ingested food. Therefore:

Energy from food intake =
    Internal heat produced + External work + Energy stored

Table 16–6	Major Functions of the Thyroid Hormones (TH)

1. Required for normal maturation of the nervous system in the fetus and infant
   Deficiency: Mental retardation (cretinism)

2. Required for normal bodily growth because they facilitate the secretion of and response to growth hormone
   Deficiency: Deficient growth in children

3. Required for normal alertness and reflexes at all ages
   Deficiency: Mentally and physically slow and lethargic; delayed reflexes
   Excess: Restless, irritable, anxious, wakeful; hyper-reflexic

4. Major determinant of the rate at which the body produces heat during the basal metabolic state
   Deficiency: Low BMR, cold intolerance; decreased food appetite
   Excess: High BMR, heat intolerance; increased food appetite, increased catabolism of nutrients

5. Facilitates the activity of the sympathetic nervous system by stimulating the synthesis of one class of receptors (beta receptors) for epinephrine and norepinephrine
   Excess: Symptoms similar to those observed with activation of the sympathetic nervous system (e.g., increased heart rate)

Table 16-7	Energy Expenditure During Different Types of Activity for a 70-kg (154-lb) Person	
Form of Activity		**Energy kcal/h**
Lying still, awake		77
Sitting at rest		100
Typewriting rapidly		140
Dressing or undressing		150
Walking on level ground at 4.3 km/h (2.6mi/h)		200
Bicycling on level ground at 9 km/h (5.3 mi/h)		304
Walking on 3 percent grade at 4.3 km/h (2.6 mi/h)		357
Sawing wood or shoveling snow		480
Jogging at 9 km/h (5.3 mi/h)		570
Rowing at 20 strokes/min		828

This equation includes no term for loss of fuel from the body via excretion of nutrients because normally only negligible losses occur via the urine, feces, and sloughed hair and skin. In certain diseases, however, the most important being diabetes mellitus, urinary losses of organic molecules may be quite large and would have to be included in the equation.

Rearranging the equation to focus on energy storage gives:

Energy stored =
Energy from food intake − (Internal heat produced + External work)

Thus, whenever energy intake differs from the sum of internal heat produced and external work, changes in energy storage occur; that is, the total-body energy content increases or decreases. Normally, energy storage is mainly in the form of fat in adipose tissue.

It is worth emphasizing at this point that "body weight" and "total-body energy content" are not synonymous. Body weight is determined not only by the amount of fat, carbohydrate, and protein in the body, but also by the amounts of water, bone, and other minerals. For example, an individual can lose body weight quickly as the result of sweating or an unusual increase in urinary output, or can gain large amounts of weight as a result of water retention, as occurs, for example, during heart failure. Moreover, even focusing only on the nutrients, a constant body weight does not mean that total-body energy content is constant. The reason is that 1 g of fat contains 9 kcal, whereas 1 g of either carbohydrate or protein contains 4 kcal. Thus, for example, aging is usually associated with a gain of fat and a loss of protein; the result is that even though the person's body weight may stay constant, the total-body energy content has increased. Apart from these qualifications, however, in the remainder of this chapter changes in body weight are equated with changes in total-body energy content and, more specifically, changes in body fat stores.

Body weight in adults is usually regulated around a stable set point. Theoretically this regulation can be achieved by reflexly adjusting caloric intake and/or energy expenditure in response to changes in body weight. It had long been assumed that regulation of caloric intake was the only important adjustment, and the next section will describe this process. However, it is now clear that energy expenditure can also be adjusted in response to changes in body weight.

A typical demonstration of this process in human beings is as follows: Total daily energy expenditure was measured in nonobese subjects at their usual body weight and again after they either lost 10 percent of their body weight by underfeeding or gained 10 percent by overfeeding. At their new body weight, the overfed subjects manifested a large (15 percent) increase in both resting and nonresting energy expenditure, and the underfed subjects showed a similar decrease. These changes in energy expenditure were much greater than could be accounted for simply by the altered metabolic mass of the body or having to move a larger or smaller body.

The generalization that emerges from this and other similar studies is that a dietary-induced change in total-body energy stores triggers, in negative feedback fashion, an alteration in energy expenditure that opposes the gain or loss of energy stores. This phenomenon helps explain why some dieters lose about 5 to 10 pounds of fat fairly easily and then become stuck at a plateau. It also helps explain why some very thin people have difficulty trying to gain much weight. Another unsettled question is whether such "metabolic resistance" to changes in body weight persists indefinitely or is only a transient response to rapid changes in body weight.

Control of Food Intake

The control of food intake can be analyzed in the same way as any other biological control system. As the previous section emphasized, the variable being maintained in this system is total-body energy content or, more specifically, total fat stores. An essential component of such a control system is a hormone—**leptin**—synthesized by adipose tissue cells themselves, and released from the cells in proportion to the amount of fat in the adipose tissue. This hormone acts on the hypothalamus to cause a reduction in food intake, in part by inhibiting the release of **neuropeptide Y,** a hypothalamic neurotransmitter that stimulates eating. Leptin also stimulates the metabolic rate and, therefore, plays an important role in the changes in energy expenditure that occur in response to overfeeding or underfeeding, as described in the previous section. Thus, as illustrated in **Figure 16–14**, leptin functions in a negative feedback system to maintain a stable total-body energy content by "telling" the brain how much fat is being stored.

Leptin appears to exert many other effects on the hypothalamus and anterior pituitary. For example, during long-

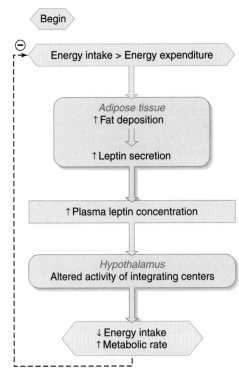

Figure 16–14

Postulated role of leptin in the control of total-body energy stores. Note that the direction of the arrows within the boxes would be reversed if energy (food) intake were less than energy expenditure.

Figure 16–14 physiological *inquiry*

- Under what circumstances might the appetite-suppressing action of leptin be counterproductive?

Answer can be found at end of chapter.

term fasting, there is a marked decrease in the secretion of the sex steroids and thyroid hormones, and an increase in the secretion of adrenal glucocorticoids. These adaptations make sense when we consider that reproduction is energetically costly, that thyroid hormones increase energy usage, and that adrenal steroids stimulate production of substrates for gluconeogenesis that reach the liver via the circulation. In experimental animals, these effects are almost completely eliminated by administering leptin. This suggests that leptin normally exerts a modulatory effect on the pathways that control the secretion of these hormones (the possible role of leptin in puberty is described in Chapter 17).

It should be emphasized that leptin is important for *long-term* matching of caloric intake to energy expenditure. In addition, it is thought that various other signals act on the hypothalamus (and other brain areas) over short periods of time to regulate individual meal length and frequency (**Figure 16–15**). These **satiety signals** (factors that reduce appetite) cause the person to cease feeling hungry and set the time period before hunger returns. For example, the rate of insulin-dependent glucose utilization by certain areas of the hypothalamus increases during eating, and this probably con-

stitutes a satiety signal. Insulin, which increases during food absorption, also acts as a direct satiety signal. The increase in metabolic rate induced by eating tends to raise body temperature slightly, which acts as yet another satiety signal. Finally, some satiety signals are initiated by the presence of food within the gastrointestinal tract. These include neural signals triggered by stimulation of both stretch receptors and chemoreceptors in the stomach and duodenum, as well as by several of the hormones (cholecystokinin, for example) released from the stomach and duodenum during eating.

Food intake is also strongly influenced by the reinforcement, both positive and negative, of such things as smell, taste, and texture. In addition, the behavioral concepts of reinforcement, drive, and motivation, described in Chapter 8, must be incorporated into any comprehensive model of food-intake control.

Another significant factor that may alter food intake is stress. However, it is difficult to make generalizations about the effects of stress on food intake in humans. Some people respond to stress by refraining from eating, whereas others may overeat.

Although we have focused on leptin and other factors as satiety signals, it is important to realize that a primary function of leptin is to increase metabolic rate. If a person is subjected to starvation, his or her adipocytes begin to shrink, as catabolic hormones mobilize triglycerides from fat cells. This decrease in size causes a proportional reduction in leptin secretion from the shrinking cells. The decrease in leptin concentration removes the signal that normally inhibits appetite and speeds up metabolism. The result is that a loss of fat mass leads to a decrease in leptin, and thus a decrease in BMR and an increase in appetite. This may be the true evolutionary significance of leptin, namely that its disappearance from the blood results in a decreased BMR, thus prolonging life during periods of starvation.

In addition to leptin, another recently discovered hormone appears to be an important regulator of appetite. **Ghrelin** (pronounced GREH-lin) is a 28-amino-acid peptide synthesized and released primarily from endocrine cells in the fundus of the stomach. Ghrelin is also produced in smaller amounts from other gastrointestinal and non-gastrointestinal tissues.

Ghrelin has several major functions. One is to increase growth hormone release from the pituitary—this is the derivation of the word *ghrelin*. The major function of ghrelin pertinent to this chapter is to increase hunger by stimulating NPY and other neuropeptides in the feeding centers in the arcuate nuclei of the hypothalamus. Ghrelin also decreases the breakdown of fat and increases gastric motility and acid production. It makes sense, then, that the major stimuli to ghrelin are fasting (ghrelin levels increase just before a meal) and a low-calorie diet.

Ghrelin, therefore, participates in several feedback loops. Fasting or a low-calorie diet leads to an increase in ghrelin. This stimulates hunger and, if food is available, food intake. The food intake subsequently reduces ghrelin, possibly through stomach distention, caloric absorption, or some other mechanism. The increase in ghrelin before a meal increases gastric motility and acid secretion, preparing the stomach for the impending meal.

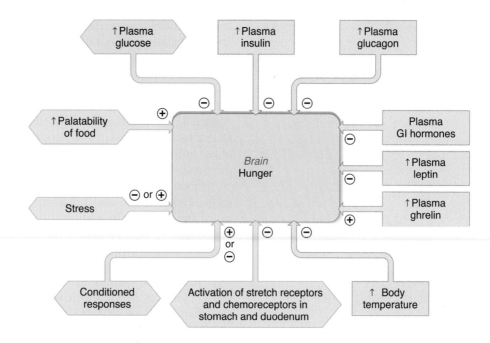

Figure 16–15

Short-term inputs controlling appetite and, consequently, food intake. The minus signs denote hunger suppression, and the plus signs denote hunger stimulation.

Figure 16–15 physiological _inquiry_ (pi)

■ As shown, stretch receptors in the gut after a meal can suppress hunger. Would drinking a large glass of water before a meal be an effective means of dieting?

Answer can be found at end of chapter.

With the epidemic of obesity in the United States and other countries, it is hoped that drugs can be developed that inhibit the ghrelin-sensitive neurons in the hypothalamus, limiting hunger and, hopefully, allowing weight loss. Interestingly, stomach bypass surgery for the morbidly obese decreases ghrelin—this may be one mechanism that makes this surgery successful in facilitating the initial weight loss and maintaining it later.

Overweight and Obesity

The definition of **overweight** is functional, a state in which an increased amount of fat in the body results in a significant impairment of health from a variety of diseases, notably hypertension, atherosclerosis, heart disease, diabetes, and sleep apnea. **Obesity** denotes a particularly large accumulation of fat—that is, extreme overweight. The difficulty has been establishing just how much fat constitutes "overweight"—that is, in determining at what point fat accumulation begins to constitute a health risk. This is evaluated by epidemiological studies that correlate disease rates with some measure of the amount of fat in the body. Currently, the preferred simple method for assessing the latter is not the body weight but the **body mass index (BMI),** which is calculated by dividing the weight (in kilograms) by the square of the height (in meters). For example, a 70-kg person with a height of 180 cm would have a BMI of 21.6 ($70/1.8^2$).

Current National Institutes of Health guidelines categorize BMIs of greater than 25 as overweight (i.e., as having some increased health risk because of excess fat) and those greater than 30 as obese, with a markedly increased health risk. According to these criteria, more than half of U.S. women and men age 20 and older are now considered to be overweight and one-quarter or more to be clinically obese! Even more troubling is that the incidence of childhood obesity is increasing in the United States and other countries. These guidelines, however, are controversial. First, the epidemiological studies do not always agree as to where along the continuum of BMIs between 25 and 30 health risks begin to occur. Second, even granting increased risk above a BMI of 25, the studies do not always account for confounding factors associated with being overweight or even obese, particularly a sedentary lifestyle. Instead, the increased health risk may be at least partly due to lack of physical activity, not body fat per se.

To add to the complexity, there is growing evidence that not just total fat but where the fat is located has important consequences. Specifically, people with mostly abdominal fat ("apples") are at greater risk for developing serious conditions such as diabetes and cardiovascular diseases than people whose fat is mainly in the lower body ("pears")—on the buttocks and thighs. There is currently no agreement as to the explanation of this phenomenon, but there are important differences in the physiology of adipose tissue cells in these regions. For example, adipose tissue cells in the abdomen are much more adept at breaking down fat stores and releasing the products into the blood.

What is known about the underlying causes of obesity? Identical twins who have been separated soon after birth and raised in different households manifest strikingly similar body weights and incidences of obesity as adults. Twin studies thus indicate that genetic factors play an important role in obesity. It has been postulated that natural selection favored the evolution in our ancestors of so-called **"thrifty genes,"** which boosted the ability to store fat from each feast in order to sustain people through the next fast. Given today's relative abundance of high-fat foods in many countries, such an adaptation is now a liability.

Despite the importance of genetic factors, psychological, cultural, and social factors can also play a significant role. For example, the increasing incidence of obesity in the United States and other industrialized nations during the past 30 years cannot be explained by changes in our genes.

Much recent research has focused on possible abnormalities in the leptin system as a cause of obesity. In one strain of mice (shown in the chapter-opening photo), the gene that codes for leptin is mutated so that adipose tissue cells produce an abnormal, inactive leptin, resulting in hereditary obesity. The same is *not* true, however, for the vast majority of obese people. The leptin secreted by these people is normally active, and leptin concentrations in the blood are elevated, not reduced. This observation indicates that leptin secretion is not at fault in these people. Thus, such people may be leptin-resistant in much the same way that people with type 2 diabetes are insulin-resistant. Moreover, there are multiple genes that interact with one another and with environmental factors to influence a person's susceptibility to weight gain.

The methods and goals of treating obesity are now undergoing extensive rethinking. An increase in body fat must be due to an excess of energy intake over energy expenditure, and low-calorie diets have long been the mainstay of therapy. However, it is now clear that such diets alone have limited effectiveness in obese people; over 90 percent regain all or most of the lost weight within five years. Another important reason for the ineffectiveness of such diets is that, as described earlier, the person's metabolic rate decreases as leptin levels decrease, sometimes falling low enough to prevent further weight loss on as little as 1000 calories a day. Because of this, many obese people continue to gain weight or remain in stable energy balance on a caloric intake equal to or less than the amount consumed by people of normal weight. These persons must either have less physical activity than normal or have lower basal metabolic rates. Finally, at least half of obese people—those who are more than 20 percent overweight—who try to diet down to desirable weights suffer medically, physically, and psychologically. This is what would be expected if the body were "trying" to maintain body weight (more specifically fat stores) at the higher set point.

Such studies, taken together, indicate that crash diets are not an effective long-term method for controlling weight. Instead caloric intake should be set at a level that can be maintained for the rest of one's life. Such an intake in an overweight person should lead to a slow, steady weight loss of no more than one pound per week until the body weight stabilizes at a new, lower level. Most important, any program of weight loss should include increased physical activity. The exercise itself uses calories, but more importantly, it partially offsets the tendency, described earlier, for the metabolic rate to decrease during long-term caloric restriction and weight loss. Also, the combination of exercise and caloric restriction may cause the person to lose more fat and less protein than with caloric restriction alone, although a recent study suggests this may not always be true. To restate the information of the previous two sentences in terms of control systems, exercise seems to lower the set point around which the body regulates total-body fat stores.

Let us calculate how rapidly a person can expect to lose weight on a reducing diet (assuming, for simplicity, no change in energy expenditure). Suppose a person whose steady-state metabolic rate per 24 h is 2000 kcal goes on a 1000 kcal/day diet. How much of the person's own body fat will be required to supply this additional 1000 kcal/day? Because fat contains 9 kcal/g:

$$\frac{1000 \text{ kcal/day}}{9 \text{ kcal/g}} = 111 \text{ g/day, or } 777 \text{ g/week}$$

Approximately another 77 g of water is lost from the adipose tissue along with this fat (adipose tissue is 10 percent water), so that the grand total for one week's loss equals 854 g, or 1.8 pounds. Thus, even on this severe diet, the person can reasonably expect to lose approximately this amount of weight per week, assuming no decrease in metabolic rate occurs. Actually, the amount of weight lost during the first week will probably be considerably greater because a large amount of water may be lost early in the diet, particularly when the diet contains little carbohydrate. This early loss is not really elimination of excess fat but often underlies the extravagant claims made for fad diets.

Eating Disorders: Anorexia Nervosa and Bulimia Nervosa

Two of the major eating disorders are found primarily in adolescent girls and young women. The typical person with *anorexia nervosa* becomes pathologically obsessed with her weight and body image. She may decrease her food intake so severely that she may die of starvation. It is not known whether the cause of anorexia nervosa is primarily psychological or biological. There are many other abnormalities associated with it—cessation of menstrual periods, low blood pressure, low body temperature, and altered secretion of many hormones, including increased levels of ghrelin. It is likely that these are simply the results of starvation, although it is possible that some represent signs, along with the eating disturbances, of primary hypothalamic malfunction.

Bulimia nervosa, usually simply called bulimia, is a disease characterized by recurrent episodes of binge eating. It is usually associated with regular self-induced vomiting, use of laxatives or diuretics, as well as strict dieting, fasting, or vigorous exercise to lose weight or to prevent weight gain. Like individuals with anorexia nervosa, those with bulimia manifest a persistent overconcern with body weight, although they generally remain within 10 percent of their ideal weight. This disease, too, is accompanied by a variety of physiological abnormalities, but it is unknown in some cases whether they are causal or secondary.

In addition to anorexia and bulimia, rare lesions or tumors within the hypothalamic centers that normally regulate appetite can result in over- or underfeeding.

What Should We Eat?

In the last few years, more and more dietary factors have been associated with the cause or prevention of many diseases, including not only coronary artery disease but hypertension, cancer, birth defects, osteoporosis, and a variety of other chronic diseases. These associations come mainly from animal studies, epidemiologic studies on people, and basic research concerning potential mechanisms. The problem is that the findings are often difficult to interpret and may be conflicting. To synthesize all this material in the form of simple, clear recommendations to the general public is a monumental task, and all such attempts have been subjected to intense criticism. One of the most commonly used sets, issued by the National Research Council, is presented in **Table 16-8**.

Table 16-8	Summary of National Research Council Dietary Recommendations

1. Reduce fat intake to 30 percent or less of total calories. Reduce saturated fatty acid intake to less than 10 percent of calories and intake of cholesterol to less than 300 mg daily.

2. Every day eat five or more servings of a combination of vegetables and fruits, especially green and yellow vegetables and citrus fruits. Also, increase starches and other complex carbohydrates by eating six or more daily servings of a combination of breads, cereals, and legumes.

3. Maintain protein intake at moderate levels.

4. Balance food intake and physical activity to maintain appropriate body weight.

5. Alcohol consumption is not recommended. For those who drink alcoholic beverages, limit consumption to the equivalent of one ounce of pure alcohol in a single day.

6. Limit total daily intake of sodium to 2.4 g or less.

7. Maintain adequate calcium intake.

8. Avoid taking dietary supplements in excess of the RDA (Recommended Dietary Allowance) in any one day.

9. Maintain an optimal intake of fluoride, particularly during the years of primary and secondary tooth formation and growth.

Regulation of Body Temperature

In the preceding discussion, it was emphasized that energy expenditure is linked to our ability to maintain a stable, warm body temperature. In this section, we discuss the mechanisms by which the body gains or loses heat in a variety of healthy or pathological settings.

Birds and mammals, including people, are capable of maintaining their body temperatures within very narrow limits despite wide fluctuations in ambient temperature and are termed **homeothermic.** The relatively stable and high body temperature frees biochemical reactions from fluctuating with the external temperature. However, the maintenance of a warm body temperature (approximately 37°C in healthy persons) imposes a requirement for precise regulatory mechanisms because further, large elevations of temperature cause nerve malfunction and protein denaturation. Some people suffer convulsions at a body temperature of 41°C (106°F), and 43°C is considered to be the absolute limit for survival.

Several important generalizations about normal human body temperature should be stressed at the outset: (1) Oral temperature averages about 0.5°C less than rectal, which is generally used as an estimate of internal temperature (also known as **core body temperature**). Thus, not all parts of the body have the same temperature. (2) Internal temperature varies several degrees in response to activity patterns and changes in external temperature. (3) There is a characteristic circadian fluctuation of about 1°C (**Figure 16-16**), with temperature being lowest during the night and highest during the day. (4) An added variation in women is a higher temperature during the second half of the menstrual cycle due to the effects of progesterone.

Temperature regulation can be studied by our usual balance methods. The total heat content gained or lost by the

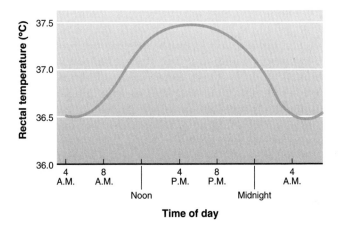

Figure 16-16

Circadian changes in core (measured as rectal) body temperature in a typical person. This figure does not take into account daily minor swings in temperature due to such things as exercise, eating, and menstrual cycle, nor are the absolute values on the *y*-axis representative of all individuals.

Adapted from Scales et al.

body is determined by the net difference between heat gain (from the environment and produced in the body) and heat loss. Maintaining a stable body temperature means that, in the steady state, heat production must equal heat loss. Some of the basic principles of heat gain were described earlier in this chapter in the section on metabolic rate, and those governing heat loss are described next. Then we will present the reflexes that act upon these processes to regulate body temperature.

Mechanisms of Heat Loss or Gain

The surface of the body can lose heat to the external environment by radiation, conduction, convection, and by the evaporation of water (**Figure 16–17**). Before defining each of these processes, however, it must be emphasized that radiation, conduction, and convection can under certain circumstances lead to heat *gain* instead of loss.

Radiation is the process by which the surfaces of all objects constantly emit heat in the form of electromagnetic waves. The rate of emission is determined by the temperature of the radiating surface. Thus, if the body surface is warmer than the various surfaces in the environment, net heat is lost from the body, the rate being directly dependent upon the

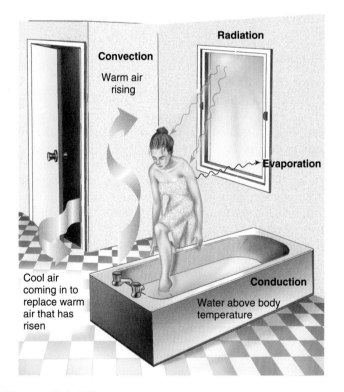

Figure 16–17

Mechanisms of heat transfer.

Figure 16–17 physiological *inquiry*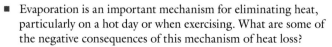

- Evaporation is an important mechanism for eliminating heat, particularly on a hot day or when exercising. What are some of the negative consequences of this mechanism of heat loss?

Answer can be found at end of chapter.

temperature difference between the surfaces. Conversely, the body gains heat by absorbing electromagnetic energy emitted by the sun.

Conduction is the loss or gain of heat by transfer of thermal energy during collisions between adjacent molecules. In essence, heat is "conducted" from molecule to molecule. The body surface loses or gains heat by conduction through direct contact with cooler or warmer substances, including the air or water. Not all substances, however, conduct heat equally. Water is a better conductor of heat than is air and, thus, more heat is lost from the body in water than in air of similar temperature.

Convection is the process whereby conductive heat loss or gain is aided by movement of the air or water next to the body. For example, air next to the body is heated by conduction. Because warm air is less dense than cool air, the heated air around the body surface rises, thus carrying away the heat just taken from the body. The air that moves away is replaced by cooler air, which in turn follows the same pattern. Convection is always occurring because warm air is less dense and therefore rises, but it can be greatly facilitated by external forces such as wind or fans. Thus, convection aids conductive heat exchange by continuously maintaining a supply of cool air. Therefore, in the rest of this chapter, the term *conduction* will also imply convection.

Evaporation of water from the skin and membranes lining the respiratory tract is the other major process causing loss of body heat. A very large amount of energy—600 kcal/L—is required to transform water from the liquid to the gaseous state. Thus, whenever water vaporizes from the body's surface, the heat required to drive the process is conducted from the surface, thereby cooling it.

Temperature-Regulating Reflexes

Temperature regulation offers a classic example of a biological control system. The balance between heat production and heat loss is continuously being disturbed, either by changes in metabolic rate (exercise being the most powerful influence) or by changes in the external environment (e.g., air temperature) that alter heat loss or gain. The resulting changes in body temperature are detected by thermoreceptors. These receptors initiate reflexes that change the output of various effectors so that heat production and/or loss are changed and body temperature is restored toward normal.

Figure 16–18 summarizes the components of these reflexes. There are two categories of thermoreceptors, one in the skin (**peripheral thermoreceptors**) and the other (**central thermoreceptors**) in deep body structures, including abdominal organs and thermoreceptive neurons in the hypothalamus. Because it is the core body temperature, not the skin temperature, that is kept constant, the central thermoreceptors provide the essential negative feedback component of the reflexes. The peripheral thermoreceptors provide feedforward information, as described in Chapter 1, and also account for the ability to identify a hot or cold area of the skin.

The hypothalamus serves as the primary overall integrator of the reflexes, but other brain centers also exert some control over specific components of the reflexes. Output from the

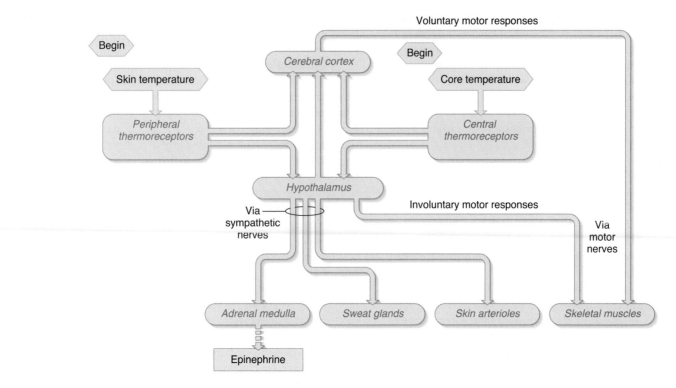

Figure 16–18

Summary of temperature-regulating mechanisms beginning with peripheral thermoreceptors and central thermoreceptors. The dashed arrow from the adrenal medulla indicates that this hormonal pathway is of minor importance in adult human beings. The solid arrows denote neural pathways. The hypothalamus influences sympathetic nerves via descending pathways.

hypothalamus and the other brain areas to the effectors is via: (1) sympathetic nerves to the sweat glands, skin arterioles, and the adrenal medulla; and (2) motor neurons to the skeletal muscles.

Control of Heat Production

Changes in muscle activity constitute the major control of heat production for temperature regulation. The first muscle changes in response to a decrease in core body temperature are a gradual and general increase in skeletal muscle contraction. This may lead to shivering, which consists of oscillating, rhythmical muscle contractions and relaxations occurring at a rapid rate. During shivering, the efferent motor nerves to the skeletal muscles are influenced by descending pathways under the primary control of the hypothalamus. Because almost no external work is performed by shivering, virtually all the energy liberated by the metabolic machinery appears as internal heat, a process known as **shivering thermogenesis.** People also use their muscles for voluntary heat-producing activities such as foot stamping and hand rubbing. The opposite muscle reactions occur in response to heat. Basal muscle contraction is reflexly decreased, and voluntary movement is also diminished. These attempts to reduce heat production are limited, however, both because basal muscle contraction is quite low to start with and because any increased core temperature produced by the heat acts *directly* on cells to increase metabolic rate. In other words, an increase in cellular temperature directly accelerates the rate at which all of its chemical reactions occur. This is due to the increased thermal motion of dissolved molecules, making it more likely that they will encounter each other. The result is that ATP is expended at a higher rate because ATP is needed for many of a cell's chemical reactions. This, in turn, results in a compensatory increase in ATP production from cellular fuel stores, which also generates heat as a by-product of fuel metabolism (Chapter 3). Thus, increasing cellular temperature can itself result in the production of additional heat through increased metabolism.

Muscle contraction is not the only process controlled in temperature-regulating reflexes. In most experimental animals, chronic cold exposure induces an increase in metabolic rate (heat production) that is not due to increased muscle activity and is termed **nonshivering thermogenesis.** Its causes are an increased adrenal secretion of epinephrine and increased sympathetic activity to adipose tissue, with some contribution by thyroid hormone as well. However, nonshivering thermogenesis is quite minimal, if present at all, in adult human beings, and there is no increased secretion of thyroid hormone in response to cold. Nonshivering thermogenesis does occur in infants, however, whose shivering mechanism is not yet fully developed. Newborn infants possess in addition to normal (white) adipocytes a type of adipose tissue called brown fat, or **brown adipose tissue.** This type of adipose tissue is responsive to thyroid hormone, epinephrine, and the sympathetic nervous system, and it contains large amounts of a class of proteins called uncoupling proteins. These proteins uncouple oxidation from phosphorylation (Chapter 3) and in effect, make fuel metabolism less efficient (less ATP is generated). The major product of this inefficient metabolism is heat, which then contributes to maintaining body temperature in infants. Although very small amounts of brown adipose tissue are present in adults, its physiological role in adults, if any, is unknown.

Control of Heat Loss by Radiation and Conduction

For purposes of temperature control, the body may be thought of as a central core surrounded by a shell consisting of skin and subcutaneous tissue; we will refer to this complex outer shell simply as skin. The temperature of the central core is regulated at approximately 37°C, but the temperature of the outer surface of the skin changes markedly.

If the skin and its underlying tissue were a perfect insulator, no heat would ever be lost from the core. The temperature of the outer skin surface would equal the environmental temperature, and net conduction would be zero. The skin is not a perfect insulator, however, so the temperature of its outer surface generally is somewhere between that of the external environment and that of the core. Instead of acting as an insulator, the skin functions as a variable regulator of heat exchange. Its effectiveness in this capacity is subject to physiological control by a change in blood flow. The more blood reaching the skin from the core, the more closely the skin's temperature approaches that of the core. In effect, the blood vessels can carry heat to the skin surface to be lost to the external environment. These vessels are controlled largely by vasoconstrictor sympathetic nerves, which are reflexly stimulated in response to cold and inhibited in response to heat. There is also a population of sympathetic neurons to the skin whose neurotransmitters cause active vasodilation. Certain areas of skin participate much more than others in all these vasomotor responses, and so skin temperatures vary with location.

Finally, there are three *behavioral* mechanisms for altering heat loss by radiation and conduction: changes in surface area, changes in clothing, and choice of surroundings. Curling up into a ball, hunching the shoulders, and similar maneuvers in response to cold reduce the surface area exposed to the environment, thereby decreasing heat loss by radiation and conduction. In human beings, clothing is also an important component of temperature regulation, substituting for the insulating effects of feathers in birds and fur in other mammals. The outer surface of the clothes forms the true "exterior" of the body surface. The skin loses heat directly to the air space trapped by the clothes, which in turn pick up heat from the inner air layer and transfer it to the external environment. The insulating ability of clothing is determined primarily by the thickness of the trapped air layer.

Clothing is important not only at low temperatures, but also at very high temperatures. When the environmental temperature is greater than body temperature, conduction favors heat *gain* rather than heat loss. Heat gain also occurs by radiation during exposure to the sun. People therefore insulate themselves in such situations by wearing clothes. The clothing, however, must be loose to allow adequate movement of air to permit evaporation. Wearing loose-fitting clothes is

actually far more cooling than going nude in a hot environment and during direct exposure to the sun.

The third behavioral mechanism for altering heat loss is to seek out warmer or colder surroundings, as for example by moving from a shady spot into the sunlight. Raising or lowering the thermostat of a house or turning on an air conditioner also fits this category.

Control of Heat Loss by Evaporation

Even in the absence of sweating, there is loss of water by diffusion through the skin, which is not completely waterproof. A similar amount is lost from the respiratory lining during expiration. These two losses are known as **insensible water loss** and amount to approximately 600 ml/day in human beings. Evaporation of this water can account for a significant fraction of total heat loss. In contrast to this passive water loss, sweating requires the active secretion of fluid by **sweat glands** and its extrusion into ducts that carry it to the skin surface.

Production of sweat is stimulated by sympathetic nerves to the glands. (These nerves release acetylcholine rather than the usual sympathetic neurotransmitter norepinephrine.) Sweat is a dilute solution containing sodium chloride as its major solute. Sweating rates of over 4 L/h have been reported; the evaporation of 4 L of water would eliminate almost 2400 kcal of heat from the body!

Sweat must evaporate in order to exert its cooling effect. The most important factor determining evaporation rate is the water vapor concentration of the air—that is, the relative humidity. The discomfort suffered on humid days is due to the failure of evaporation; the sweat glands continue to secrete, but the sweat simply remains on the skin or drips off.

Integration of Effector Mechanisms

Table 16–9 summarizes the effector mechanisms regulating temperature, none of which is an all-or-none response but a graded, progressive increase or decrease in activity. By altering heat loss, changes in skin blood flow alone can regulate body temperature over a range of environmental temperatures (approximately 25 to 30°C or 75 to 86°F for a nude individual) known as the **thermoneutral zone.** At temperatures lower than this, even maximal vasoconstriction cannot prevent heat loss from exceeding heat production, and the body must increase its heat production to maintain temperature. At environmental temperatures above the thermoneutral zone, even maximal vasodilation cannot eliminate heat as fast as it is produced, and another heat-loss mechanism—sweating—therefore comes strongly into play. At environmental temperatures above that of the body, heat is actually added to the body by radiation and conduction. Under such conditions, evaporation is the sole mechanism for heat loss. A person's ability to tolerate such temperatures is determined by the humidity and by his/her maximal sweating rate. For example, when the air is completely dry, a person can tolerate an environmental temperature of 130°C (225°F) for 20 min or longer, whereas very moist air at 46°C (115°F) is bearable for only a few minutes.

Table 16–9 Summary of Effector Mechanisms in Temperature Regulation

Desired Effect	Mechanism
	Stimulated by Cold
Decrease heat loss	1. Vasoconstriction of skin vessels
	2. Reduction of surface area (curling up, etc.)
	3. Behavioral response (put on warmer clothes, raise thermostat setting, etc.)
Increase heat production	1. Increased muscle tone
	2. Shivering and increased voluntary activity
	3. Increased secretion of epinephrine (minimal in adults)
	4. Increased food appetite
	Stimulated by Heat
Increase heat loss	1. Vasodilation of skin vessels
	2. Sweating
	3. Behavioral response (put on cooler clothes, turn on fan, etc.)
Decrease heat production	1. Decreased muscle tone and voluntary activity
	2. Decreased secretion of epinephrine (minimal in adults)
	3. Decreased food appetite

Temperature Acclimatization

Changes in sweating onset, volume, and composition determine people's chronic adaptation to high temperatures. A person newly arrived in a hot environment has poor ability to do work; body temperature rises and severe weakness may occur. After several days, there is a great improvement in work tolerance, with much less increase in body temperature, and the person is said to have acclimatized to the heat. Body temperature does not rise as much because sweating begins sooner and the volume of sweat produced is greater.

There is also an important change in the composition of the sweat, namely, a marked reduction in its sodium concentration. This adaptation, which minimizes the loss of sodium from the body via sweat, is due to increased secretion of the adrenal mineralocorticoid hormone aldosterone. The sweat gland secretory cells produce a solution with a sodium concentration similar to that of plasma, but some of the sodium is absorbed back into the blood as the secretion flows along the sweat gland ducts toward the skin surface. Aldosterone stimulates this absorption in a manner identical to its stimulation of sodium reabsorption in the renal tubules.

Cold acclimatization has been much less studied than heat acclimatization because of the difficulty of subjecting people to total-body cold stress over long enough periods to produce acclimatization. Moreover, people who live in cold climates generally dress very warmly and so would not develop acclimatization to the cold.

ADDITIONAL CLINICAL EXAMPLES

In addition to obesity, anorexia, bulimia, and temperature acclimatization, several other features of energy usage and thermoregulation are of great clinical interest.

Fever and Hyperthermia

Fever is an elevation of body temperature due to a resetting of the "thermostat" in the hypothalamus. A person with a fever still regulates body temperature in response to heat or cold but at a higher set point. The most common cause of fever is infection, but physical trauma and stress can also induce fever.

The onset of fever during infection is often gradual, but it is most striking when it occurs rapidly in the form of a chill. The temperature setting of the brain thermostat is suddenly raised. Because of this, the person feels cold, even though his or her actual body temperature may be normal. Thus, the typical actions that are used to increase body temperature, such as vasoconstriction and shivering, occur. The person also curls up and puts on more blankets. This combination of decreased heat loss and increased heat production serves to drive body temperature up to the new set point, where it stabilizes. It will continue to be regulated at this new value until the thermostat is reset to normal and the fever "breaks."

The person then feels hot, throws off the covers, and manifests profound vasodilation and sweating.

What is the basis for the thermostat resetting? Chemical messengers collectively termed **endogenous pyrogen (EP)** are released from macrophages (as well as other cell types) in the presence of infection or other fever-producing stimuli. The next steps vary depending on the precise stimulus for the release of EP. As illustrated in **Figure 16–19,** in some cases EP probably circulates in the blood to act upon the thermoreceptors in the hypothalamus (and perhaps other brain areas), altering their input to the integrating centers. In other cases, EP may be produced by macrophage-like cells in the liver and stimulate neural receptors there that give rise to afferent neural input to the hypothalamic thermoreceptors. In both cases, the immediate cause of the resetting is a local synthesis and release of prostaglandins within the hypothalamus. *Aspirin* reduces fever by inhibiting this prostaglandin synthesis.

The term *EP* was coined at a time when the identity of the chemical messenger(s) was not known. At least one peptide, **interleukin 1 (IL-1),** is now known to function as an EP, but other peptides—for example, **interleukin 6 (IL-6)**—play

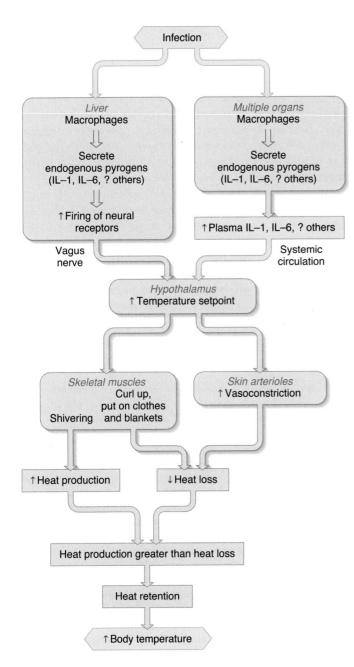

Figure 16–19

Pathway by which infection causes fever (IL-1 = interleukin 1, IL-6 = interleukin 6). The effector responses serve to *raise* body temperature during an infection.

a role, too. In addition to their effects on temperature, IL-1 and the other peptides have many other effects (described in Chapter 18) that enhance resistance to infection and promote the healing of damaged tissue.

The story is even more complicated, however, because in response to a rising temperature the hypothalamus and other tissues release messengers that prevent excessive fever or contribute to the resetting of body temperature when the fever-causing stimulus is eliminated. Such messengers are termed **endogenous cryogens.** One known endogenous cryogen is vasopressin, which functions in this regard as a neurotransmitter rather than as a hormone.

One would expect fever, which is such a consistent feature of infection, to play some important protective role. Most evidence suggests that this is the case. For example, increased body temperature stimulates a large number of the body's defensive responses to infection. The likelihood that fever is a beneficial response raises important questions about the use of aspirin and other drugs to suppress fever during infection. It must be emphasized that these questions apply to the usual modest fevers. There is no question that an extremely high fever can be harmful, particularly in its effects on the central nervous system, and must be vigorously opposed with drugs and other forms of therapy.

To reiterate, fever is an increased body temperature caused by an elevation of the thermal set point. When body temperature is elevated for any other reason beyond a narrow normal range, it is termed *hyperthermia.* The most common cause of hyperthermia in a typical person is exercise; the rise in body temperature above set point is due to retention of some of the internal heat generated by the exercising muscles.

As shown in **Figure 16–20**, heat production rises immediately during the initial stage of exercise and exceeds heat loss, causing heat storage in the body and a rise in the core temperature. This rise in core temperature triggers reflexes, via the central thermoreceptors, that cause increased heat loss. As skin blood flow and sweating increase, the discrepancy between heat production and heat loss starts to diminish but does not disappear. Therefore, core temperature continues to rise. Ultimately, core temperature will be high enough to drive (via the central thermoreceptors) the heat-loss reflexes at a rate such that heat loss once again equals heat production. At this point, core temperature stabilizes at this elevated value despite continued exercise.

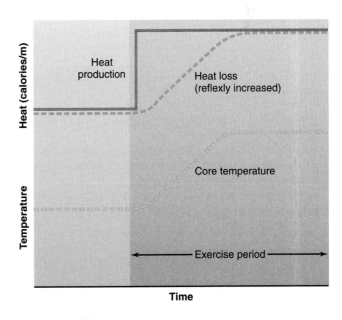

Figure 16–20

Thermal changes during exercise. Heat loss is reflexly increased, and when it once again equals heat production, core temperature stabilizes.

Heat Exhaustion and Heat Stroke

Heat exhaustion is a state of collapse, often taking the form of fainting, due to hypotension brought on by (1) depletion of plasma volume secondary to sweating, and (2) extreme dilation of skin blood vessels. Recall from Chapter 12 that blood pressure, cardiac output, and total peripheral resistance are related according to the equation MAP = CO × TPR. Thus, decreases in both cardiac output (due to the decreased plasma volume) and peripheral resistance (due to the vasodilation) contribute to the hypotension. Heat exhaustion occurs as a direct consequence of the activity of heat-loss mechanisms. Because these mechanisms have been so active, the body temperature is only modestly elevated. In a sense, heat exhaustion is a safety valve that, by forcing a cessation of work in a hot environment when heat-loss mechanisms are overtaxed, prevents the larger rise in body temperature that would cause the far more serious condition of heat stroke.

In contrast to heat exhaustion, *heat stroke* represents a complete breakdown in heat-regulating systems so that body temperature keeps going up and up. It is an extremely dangerous situation characterized by collapse, delirium, seizures, or prolonged unconsciousness—all due to greatly elevated body temperature. It almost always occurs in association with exposure to or overexertion in hot and humid environments. In some individuals, particularly the elderly, heat stroke may appear with no apparent prior period of severe sweating, but in most cases, it comes on as the end stage of prolonged untreated heat exhaustion. Exactly what triggers the transition to heat stroke is not clear, although impaired circulation to the brain due to dehydration is one factor. The striking finding, however, is that even in the face of a rapidly rising body temperature, the person fails to sweat. Heat stroke is a positive-feedback situation in which the rising body temperature directly stimulates metabolism—that is, heat production—which further raises body temperature. For both heat exhaustion and heat stroke, the remedy is external cooling, fluid replacement, and cessation of activity. ■

Basic Concepts of Energy Expenditure

I. The energy liberated during a chemical reaction appears either as heat or work.

II. Total energy expenditure = heat produced + external work done + energy stored.

III. Metabolic rate is influenced by the many factors summarized in Table 16–5.

IV. Metabolic rate is increased by the thyroid hormones and epinephrine. The other functions of the thyroid hormones are summarized in Table 16–6.

Regulation of Total-Body Energy Stores

I. Energy storage as fat can be positive when the metabolic rate is less than, or negative when the metabolic rate is greater than the energy content of ingested food.
  a. Energy storage is regulated mainly by reflex adjustment of food intake.
  b. In addition, the metabolic rate increases or decreases to some extent when food intake is chronically increased or decreased, respectively.

II. Food intake is controlled by leptin, which is secreted by adipose tissue cells, and a variety of satiety factors, as summarized in Figure 16–15.

III. Being overweight or obese, the result of an imbalance between food intake and metabolic rate, increases the risk of many diseases.

Regulation of Body Temperature

I. Core body temperature shows a circadian rhythm, with temperature highest during the day and lowest at night.

II. The body exchanges heat with the external environment by radiation, conduction, convection, and evaporation of water from the body surface.

III. The hypothalamus and other brain areas contain the integrating centers for temperature-regulating reflexes, and both peripheral and central thermoreceptors participate in these reflexes.

IV. Body temperature is regulated by altering heat production and/or heat loss so as to change total-body heat content.
  a. Heat production is altered by increasing muscle tone, shivering, and voluntary activity.
  b. Heat loss by radiation, conduction, and convection depends on the temperature difference between the skin surface and the environment.
  c. In response to cold, skin temperature is decreased by decreasing skin blood flow through reflex stimulation of the sympathetic nerves to the skin. In response to heat, skin temperature is increased by inhibiting these nerves.
  d. Behavioral responses such as putting on more clothes also influence heat loss.
  e. Evaporation of water occurs all the time as insensible loss from the skin and respiratory lining. Additional water for evaporation is supplied by sweat, stimulated by the sympathetic nerves to the sweat glands.
  f. Increased heat production is essential for temperature regulation at environmental temperatures below the thermoneutral zone, and sweating is essential at temperatures above this zone.

V. Temperature acclimatization to heat is achieved by an earlier onset of sweating, an increased volume of sweat, and a decreased sodium concentration of the sweat.

Additional Clinical Examples

I. Fever is due to a resetting of the temperature set point so that heat production is increased and heat loss is decreased in order to raise body temperature to the new set point and keep it there. The stimulus is endogenous pyrogen, in the form of interleukin 1 and other peptides as well.

II. The hyperthermia of exercise is due to the increased heat produced by the muscles, and it is partially offset by skin vasodilation.

III. Extreme increases in body temperature can result in heat exhaustion or heat stroke. In heat exhaustion, blood pressure decreases due to vasodilation. In heat stroke, the normal thermoregulatory mechanisms fail, and thus heat stroke can be fatal.

1. State the formula relating total energy expenditure, heat produced, external work, and energy storage.
2. What two hormones alter the basal metabolic rate?
3. State the equation for total-body energy balance. Describe the three possible states of balance with regard to energy storage.
4. What happens to the basal metabolic rate after a person has either lost or gained weight?
5. List five satiety signals.
6. List three beneficial effects of exercise in a weight-loss program.
7. Compare and contrast the four mechanisms for heat loss.
8. Describe the control of skin blood vessels during exposure to cold or heat.
9. With a diagram, summarize the reflex responses to heat or cold. What are the dominant mechanisms for temperature regulation in the thermoneutral zone and in temperatures below and above this range?
10. What changes are exhibited by a heat-acclimatized person?
11. Summarize the sequence of events leading to a fever and contrast this to the sequence leading to hyperthermia during exercise.

Chapter 16 Test Questions

(Answers appear in Appendix A.)

1. Which is *incorrect*?
   a. Fatty acids can be converted into glucose in the liver.
   b. Glucose can be converted into fatty acids in adipose cells.
   c. Certain amino acids can be converted into glucose by the liver.
   d. Triglycerides are absorbed from the GI tract in the form of chylomicrons.
   e. The absorptive state is characterized by ingested nutrients entering the blood from the GI tract.

2. During the postabsorptive state, epinephrine stimulates breakdown of adipose triglycerides by
   a. inhibiting lipoprotein lipase.
   b. stimulating hormone-sensitive lipase.
   c. increasing production of glycogen.
   d. inhibiting hormone-sensitive lipase.
   e. promoting increased adipose ketone production.

3. Which is true of strenuous, prolonged exercise?
   a. It results in an increase in plasma glucagon levels.
   b. It results in an increase in plasma insulin levels.
   c. Plasma glucose levels do not change.
   d. Skeletal muscle uptake of glucose is inhibited.
   e. Plasma levels of cortisol and growth hormone both decrease.

4. Untreated type 1 diabetes mellitus is characterized by
   a. decreased sensitivity of adipose and skeletal muscle cells to insulin.
   b. higher-than-normal plasma levels of insulin.
   c. loss of body fluid due to increased urine production.

   d. age-dependent onset (only occurs in adults).
   e. obesity.

5. Which is *not* a function of insulin?
   a. to stimulate amino acid transport across cell membranes
   b. to inhibit hepatic glucose output
   c. to inhibit glucagon secretion
   d. to stimulate lipolysis in adipocytes
   e. to stimulate glycogen synthase in skeletal muscle

6. The calorigenic effect of thyroid hormones
   a. refers to the ability of thyroid hormones to increase the body's oxygen consumption.
   b. helps maintain body temperature.
   c. helps explain why hyperthyroidism is sometimes associated with symptoms of vitamin deficiencies.
   d. is the most important determinant of basal metabolic rate.
   e. All of the above are true.

7. Which of the following mechanisms of heat exchange results from local air currents?
   a. radiation
   b. convection
   c. conduction
   d. evaporation
   e. none of the above

True/False
For questions 8–15, answer true or false.

8. Nonshivering thermogenesis occurs outside the thermoneutral zone.

9. Skin and core temperature are both kept constant in homeotherms.

10. Leptin inhibits, and ghrelin stimulates appetite.

11. Actively contracting skeletal muscles have an increased need for insulin.

12. Body mass index is calculated as height in meters divided by weight in kilograms.

13. In conduction, heat moves from a surface of higher temperature to one of lower temperature.

14. Skin blood vessels constrict in response to elevated core body temperature.

15. Evaporative cooling is most efficient in dry weather.

Chapter 16 Quantitative and Thought Questions

(Answers appear in Appendix A.)

1. What happens to the triglyceride concentrations in the plasma and in adipose tissue after administration of a drug that blocks the action of lipoprotein lipase?

2. A resting, unstressed person has increased plasma concentrations of free fatty acids, glycerol, amino acids, and ketones. What situations might be responsible and what additional plasma measurement would distinguish among them?

3. A healthy volunteer is given an injection of insulin. The plasma concentrations of which hormones increase as a result?

4. If the sympathetic preganglionic fibers to the adrenal medulla were cut in an animal, would this eliminate the sympathetically mediated component of increased gluconeogenesis and lipolysis during exercise? Explain.

5. A patient with T1DM suffers a broken leg. Would you advise this person to increase or decrease his insulin dosage?

6. A person has a defect in the ability of her small intestine to absorb bile salts. What effect will this have on her plasma cholesterol concentration?

7. A well-trained athlete is found to have a moderately elevated plasma cholesterol concentration. What additional measurements would you advise this person to have done?

8. What are the sources of heat loss for a person immersed up to the neck in a 40°C bath?

9. Lizards can regulate their body temperatures only through behavioral means. Can you predict what they do when they are infected with bacteria?

Chapter 16 Answers to Physiological Inquiries

Figure 16–5 Having the transporters already synthesized and packaged into intracellular vesicle membranes means that glucose transport can be tightly and quickly coupled with changes in glucose concentrations in the blood. This protects the body against the harmful effects of excess blood glucose levels, and also prevents urinary loss of glucose by keeping the rate of glucose filtration below the maximum rate at which the kidney can reabsorb it. This tight coupling could not occur if the transporters needed to be synthesized each time a cell was stimulated by insulin.

Figure 16–7 The brain is absolutely necessary for immediate survival, and can maintain glucose uptake from the plasma in the fasted state when insulin levels are very low.

Figure 16–14 The body's normal response to leptin is to reduce appetite and increase metabolic rate. This would not be adaptive during times when it is important to increase body energy (fat) stores. An example of such a situation is pregnancy, when gaining weight in the form of increased fat mass is important for providing energy to the growing fetus. In nature, another example is the requirement of hibernating animals to store large amounts of fat prior to hibernation. In these cases, the effects of leptin are decreased or ignored by the brain.

Figure 16–15 In the short term, drinking water before a meal may reduce appetite by stretching the stomach, and this may contribute to eating a smaller meal. However, as described in Chapter 15, water is quickly absorbed by the GI tract and provides no calories, and thus hunger will soon return once the meal is over.

Figure 16–17 The amount of fluid in the body decreases as water evaporates from the surface of the skin. This fluid must be replaced by drinking or the body will become dehydrated. In addition, sweat is salty (as you may have noticed by the salt residue remaining on hats or clothing once the sweat has dried). This means that the body's salt content also needs to be restored. This is a good example of how maintaining homeostasis for one variable (body temperature) may result in disruption of homeostasis for other variables (water and salt balance).

chapter 17

Reproduction

Scanning electron micrograph of a single sperm cell on the surface of an egg.

SECTION A General Terminology and Concepts; Sex Determination and Differentiation

The primary reproductive organs are known as the **gonads:** the **testes** (singular, *testis*) in the male and the **ovaries** (singular, *ovary*) in the female. In both sexes, the gonads serve dual functions: The first of these is **gametogenesis,** which is the production of the reproductive cells, or **gametes.** These are **spermatozoa** (singular, *spermatozoan,* usually shortened to **sperm**) in males and **ova** (singular, *ovum*) in females. Secondly, the gonads secrete steroid hormones, often termed **sex hormones** or **gonadal steroids.** The major sex hormones are **testosterone** in the male and **estradiol** and **progesterone** in the female.

As described in Chapter 11, testosterone belongs to a group of steroid hormones that have similar masculinizing actions and are collectively called **androgens.** In the male, most of the circulating testosterone is synthesized in the testes. Other circulating androgens are produced by the adrenal cortex, but they are much less potent than testosterone and are unable to maintain male reproductive function if testosterone secretion is inadequate. Some adrenal androgens, like dehydroepiandrosterone (DHEA), are sold as dietary supplements and touted as miracle drugs. Supposedly, DHEA can stop or reverse the aging process and the diseases associated with it, cure depression, strengthen the immune system, and improve athletic performance. Despite the claims, there are few long-term studies demonstrating the benefits or risks of this hormone. DHEA is itself a weak androgen but can be converted in the body to testosterone in both men and women.

Estrogens are a class of steroid hormones secreted in large amounts by the ovaries and placenta. There are actually three major estrogens: *Estradiol* is the predominant estrogen in the plasma, is produced by the ovary and placenta, and is often used synonymously with the generic term estrogen. **Estrone** is also produced by the ovary and placenta. **Estriol** is usually found only in pregnant women in whom it is produced by the placenta. In all cases, estrogens are produced from androgens by the enzyme **aromatase.** Because levels of the different estrogens vary widely depending on the circumstances, and because they have similar actions in the female, we will refer to them throughout this chapter as *estrogen.*

Estrogens are not unique to females, nor are androgens to males. Plasma estrogen in males is derived from the release of small amounts by the testes and from the conversion of androgens to estrogen by the aromatase enzyme in some nongonadal tissues (notably adipose tissue). Conversely, in females, small amounts of androgens are secreted by the ovaries and larger amounts by the adrenal cortex. Some of these androgens are then converted to estrogen in nongonadal tissues, just as in men, and contribute to the plasma estrogen.

Progesterone in females is a major secretory product of the ovary at specific times of the menstrual cycle as well as from the placenta during pregnancy. Progesterone is also an intermediate in the synthetic pathways for adrenal steroids, estrogen, and androgens.

As described in Chapters 5 and 11, all steroid hormones act in the same general way. They bind to intracellular receptors, and the hormone-receptor complex then binds to DNA in the nucleus to alter the rate of formation of particular mRNAs. The result is a change in the rates of synthesis of the proteins coded for by the genes being transcribed. The resulting change in the concentrations of these proteins in the target cells accounts for the responses to the hormone.

The duct systems through which the sperm or eggs are transported and the glands lining or emptying into the ducts are termed the **accessory reproductive organs.** In the female, the breasts are also included in this category. The **secondary sexual characteristics** comprise the many external differences between males and females. Examples are hair distribution, body shape, and average adult height. The secondary sexual characteristics are not directly involved in reproduction.

Reproductive function is largely controlled by a chain of hormones (**Figure 17–1**). The first hormone in the chain is **gonadotropin-releasing hormone (GnRH).** As described in Chapter 11, GnRH is one of the hypophysiotropic hormones involved in the control of anterior pituitary gland function. It is secreted by neuroendocrine cells in the hypothalamus, and it reaches the anterior pituitary via the hypothalamo-pituitary portal blood vessels. The brain is, therefore, the primary regulator of reproduction.

The cell bodies of the GnRH neurons receive input from throughout the brain as well as from hormones in the blood. Secretion of GnRH is triggered by action potentials in GnRH-producing hypothalamic neuroendocrine cells. These action potentials occur periodically in brief bursts, with virtually no secretion in between. The pulsatile pattern of GnRH secretion is important because the cells of the anterior pituitary that secrete the gonadotropins lose sensitivity to GnRH if the concentration of this hormone remains constantly elevated.

In the anterior pituitary, GnRH stimulates the release of the pituitary **gonadotropins—follicle-stimulating hormone (FSH)** and **luteinizing hormone (LH)** (see Figure 17–1). These two glycoproteins were named for their effects in the female, but their molecular structures are the same in both sexes. The two hormones act upon the gonads, the result being (1) the maturation of sperm or ova and (2) stimulation of sex hormone secretion. In turn, the sex hormones exert many effects on all portions of the reproductive system, including the gonads from which they come and other parts of the body as well. In addition, the gonadal steroids exert feedback effects on the secretion of GnRH, FSH, and LH. Gonadal protein hormones such as **inhibin** also exert feedback effects on the anterior pituitary.

Each link in this hormonal chain is essential. A decrease in function of the hypothalamus or the anterior pituitary can result in failure of gonadal steroid secretion and gametogenesis just as if the gonads themselves were diseased.

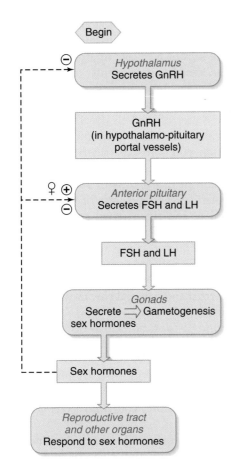

Figure 17–1

General pattern of reproduction control in both males and females. GnRH, like all hypothalamic hypophysiotropic hormones, reaches the anterior pituitary via the hypothalamo-pituitary portal vessels. The arrow within the box marked "gonads" denotes the fact that the sex hormones act locally, as paracrine agents, to influence the gametes. ⊖ indicates negative feedback inhibition. ♀ ⊕ indicates estrogen stimulation of FSH and LH in the middle of the menstrual cycle in women (positive feedback).

As a result of changes in the amount and pattern of hormone secretions, reproductive function changes markedly during a person's lifetime and may be divided into the stages summarized in **Table 17–1**.

General Principles of Gametogenesis

At any point in gametogenesis, the developing gametes are termed **germ cells.** These cells undergo either mitosis or the type of cell division known as meiosis (described later). Because the general principles of gametogenesis are essentially the same in males and females, they are introduced in this section, with features specific to the male or female described later.

The first stage in gametogenesis is proliferation of the primordial germ cells by mitosis. With the exception of the gametes, the DNA of each nucleated human cell is contained in 23 pairs of chromosomes, giving a total of 46. The two corresponding chromosomes in each pair are said to be homologous to each other, with one coming from each parent. In

Table 17–1	Stages in the Control of Reproductive Function

1. During the initial stage, which begins during fetal life and ends in the first year of life (infancy), GnRH, the gonadotropins, and gonadal sex hormones are secreted at relatively high levels.

2. From infancy to puberty, the secretion rates of these hormones are very low, and reproductive function is quiescent.

3. Beginning at puberty, hormonal secretion rates increase markedly, showing large cyclical variations in women during the menstrual cycle. This ushers in the period of active reproduction.

4. Finally, reproductive function diminishes later in life, largely because the gonads become less responsive to the gonadotropins. The ability to reproduce ceases entirely in women.

mitosis, the 46 chromosomes of the dividing cell are replicated. The cell then divides into two new cells called daughter cells. Each of the two daughter cells resulting from the division receives a full set of 46 chromosomes identical to those of the original cell. Thus, each daughter cell receives identical genetic information during mitosis.

In this manner, mitosis of primordial germ cells, each containing 46 chromosomes, provides a supply of identical germ cells for the next stages. The timing of mitosis in germ cells differs greatly in females and males. In the female, mitosis of germ cells in the ovary occurs primarily during fetal development. In the male, some mitosis occurs in the embryonic testes to generate the population of germ cells present at birth, but mitosis really begins in earnest at puberty and usually continues throughout life.

The second stage of gametogenesis is **meiosis,** in which each resulting gamete receives only 23 chromosomes from a 46-chromosome germ cell, one chromosome from each homologous pair. Because a sperm and an ovulated egg each has only 23 chromosomes, their union at fertilization results once again in a cell with a full complement of 46 chromosomes.

The process of meiosis is depicted in **Figure 17–2** (letters are keyed to the text). Meiosis consists of two cell divisions in succession. The events preceding the first meiotic division are identical to those preceding a *mitotic* division. During the interphase period that precedes a mitotic division, chromosomal DNA is replicated. Thus, after DNA replication, an interphase cell has 46 chromosomes, but each chromosome consists of two identical strands of DNA, termed sister chromatids, which are joined together by a centromere (a).

As the first meiotic division begins, homologous chromosomes, each consisting of two identical sister chromatids, come together and line up adjacent to each other. Thus, (b) 23 pairs of homologous chromosomes (called **bivalents**) are formed. (c) The sister chromatids of each chromosome condense into thick, rodlike structures. Then, (d) within each homologous

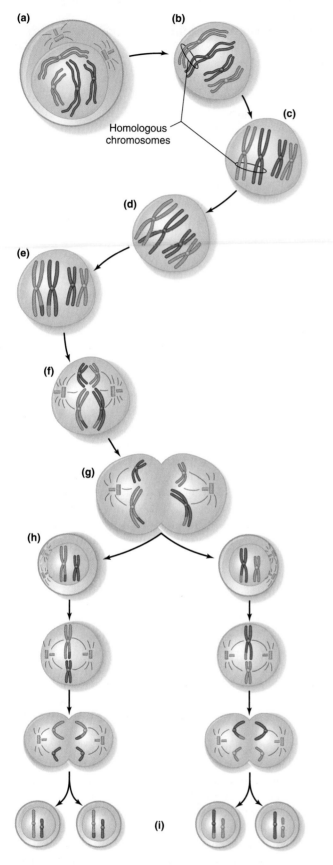

Homologous chromosomes

pair, corresponding segments of homologous chromosomes align closely. This allows two nonsister chromatids to undergo an exchange of sites of breakage in a process called **crossing-over** (e). Thus, crossing-over results in the recombination of genes on homologous chromosomes.

Following crossing-over, the homologous chromosomes line up in the center of the cell (f). The orientation of each pair on the equator is random, meaning that sometimes the maternal portion points to a particular pole of the cell, and sometimes the paternal portion does so. The cell then divides (the first division of meiosis), with the maternal chromatids of any particular pair going to one of the two cells resulting from the division, and the paternal chromatids going to the other (g). Because of the random orientation of the homologous pairs at the equator, it is extremely unlikely that all 23 maternal chromatids will end up in one cell and all 23 paternal chromatids in the other. Over 8 million (2^{23}) different combinations of maternal and paternal chromosomes can result during this first meiotic division.

The second division of meiosis occurs without any further replication of DNA. The sister chromatids—both of which were originally either maternal or paternal—of each chromosome separate and move apart into the new daughter cells (h to i). The daughter cells resulting from the second meiotic division, therefore, contain 23 one-chromatid chromosomes (i).

To summarize, meiosis produces daughter cells having only 23 chromosomes, and two events during the first meiotic division contribute to the enormous genetic variability of the daughter cells: (1) crossing-over and (2) the random distribution of maternal and paternal chromatid pairs between the two daughter cells.

Sex Determination

Genetic inheritance sets the gender of the individual, or **sex determination,** which is established at the moment of fertilization. Gender is determined by genetic inheritance of two chromosomes called the **sex chromosomes.** The larger of the sex chromosomes is called the **X chromosome** and the smaller, the **Y chromosome.** Males possess one X and one Y, whereas females have two X chromosomes. Thus, the genetic difference between male and female (**genotype**) is simply the difference in one chromosome.

The ovum can contribute only an X chromosome, whereas half of the sperm produced during meiosis are X and half are Y. When the sperm and the egg join, 50 percent should have XX and 50 percent XY. Interestingly, however, sex ratios at birth are not exactly 1:1; rather, there tends to be a slight preponderance of male births, possibly due to functional differences in sperm carrying the X versus Y chromosome.

Figure 17–2

Stages of meiosis in a generalized germ cell. For simplicity, the initial cell (a), which is in interphase, is given only four chromosomes rather than 46, the number in a human cell. Also, cytoplasm is shown only in (a), (h), and (i). Chromosomes from one parent are purple, and those from the other parent are blue. The letters are keyed to descriptions in the text. From (h) to (i), the size of the cells can vary quite dramatically in ova development.

An easy method exists for determining whether a person's cells contain two X chromosomes, the typical female pattern. When two X chromosomes are present, only one functions, and the nonfunctional X chromosome condenses to form a nuclear mass termed the **sex chromatin (Barr body),** which is readily observable with a light microscope. Scrapings from the cheek mucosa or white blood cells are convenient sources of cells to be examined. The single X chromosome in male cells rarely condenses to form sex chromatin.

A more exacting technique for determining sex chromosome composition employs tissue culture visualization of all the chromosomes—a **karyotype.** This technique is used to identify a group of genetic sex abnormalities characterized by such unusual chromosomal combinations such as XXX, XXY, and XO (only one sex chromosome). The end result of such combinations is usually the failure of normal anatomical and functional sexual development.

Sex Differentiation

The multiple processes involved in the development of the reproductive system in the fetus are collectively called **sex differentiation.** It is not surprising that people with atypical chromosomal combinations can manifest atypical sexual development. However, careful study has also revealed individuals with normal chromosomal combinations but abnormal sexual appearance and function (**phenotype**). In these people, sex differentiation has been atypical, and their phenotype may not correspond with their genotype—that is, the presence of XX or XY chromosomes.

It will be important to bear in mind during the following description one essential generalization: the genes directly determine only whether the individual will have testes or ovaries. The rest of sex differentiation depends upon the presence or absence of substances produced by the genetically determined gonads, in particular, the testes.

Differentiation of the Gonads

The male and female gonads derive embryologically from the same site—an area called the urogenital ridge. Until the sixth week of uterine life, primordial gonads are undifferentiated. In the genetic male, the testes begin to develop during the seventh week. A gene on the Y chromosome (the **SRY gene,** for **s**ex-determining **r**egion of the **Y** chromosome) is expressed at this time in the urogenital ridge cells and triggers this development. In the absence of a Y chromosome and, consequently, the SRY gene, testes do not develop. Instead, ovaries begin to develop in the same area at about 11 weeks.

By what mechanism does the SRY gene induce the formation of the testes? This gene codes for a protein, SRY, which sets into motion a sequence of gene activations ultimately leading to the formation of testes from the various embryonic cells in the urogenital ridge.

There is an unusual and important fact concerning the behavior of the X and Y chromosomes during meiosis. As described earlier in this chapter, during meiosis, homologous chromosomes come together, line up point for point, and then exchange fragments with each other, resulting in an exchange of genes (recombination) on these chromosomes. Such crossing-over involving the Y and X genes, however, could allow the SRY gene—the male-determining gene—to get into the female's genome. To prevent this, by mechanisms still not understood, the Y and X chromosomes do not undergo recombination (except at the very tips, where the SRY gene is not located and the Y chromosome has the same genes as the X).

Differentiation of Internal and External Genitalia

The internal duct system and external genitalia of the fetus are capable of developing into either sexual phenotype. Before the functioning of the fetal gonads, the primitive reproductive tract includes a double genital duct system comprised of the **Wolffian ducts** and **Müllerian ducts,** and a common opening for the genital ducts and urinary system to the outside. Normally, most of the reproductive tract develops from only one of these duct systems. In the male, the Wolffian ducts persist and the Müllerian ducts regress, whereas in the female, the opposite happens. The external genitalia in the two genders and the outer part of the vagina do not develop from these duct systems, however, but from other structures at the body surface.

Which of the two duct systems and types of external genitalia develops depends on the presence or absence of fetal testes. These testes secrete (1) testosterone from the Leydig cells and (2) protein hormone called **Müllerian-inhibiting substance (MIS)** from the Sertoli cells (**Figure 17–3a**). SRY protein induces the expression of the gene for MIS; MIS then causes the degeneration of the Müllerian duct system. Simultaneously, testosterone causes the Wolffian ducts to differentiate into the epididymis, vas deferens, ejaculatory duct, and seminal vesicles. Externally and somewhat later, under the influence primarily of **dihydrotestosterone (DHT)** produced from testosterone in target tissue, a penis forms, and the tissue near it fuses to form the scrotum. The testes will ultimately descend into the scrotum, stimulated to do so by testosterone. Failure of the testes to descend is called **cryptorchidism** and is common in infants with decreased androgen secretion. Because sperm production requires about 2°C lower temperature than normal core body temperature, sperm production is usually decreased in cryptorchidism. Treatments include hormone therapy and surgical approaches to move the testes into the scrotum.

In contrast, the female fetus, not having testes (because of the absence of the SRY gene), does not secrete testosterone and MIS. In the absence of MIS, the Müllerian system does not degenerate but rather develops into fallopian tubes and a uterus. In the absence of testosterone, the Wolffian ducts degenerate, and a vagina and female external genitalia develop from the structures at the body surface (**Figure 17–3b**). Ovaries, though present in the female fetus, do not play a role in these developmental processes. In other words, female development will occur automatically unless stopped from doing so by the presence of factors released from functioning testes.

There are various conditions in which normal sex differentiation does not occur. For example, in **androgen insensitivity syndrome** (also called **testicular feminization**), the genotype is XY and testes are present, but the phenotype (external

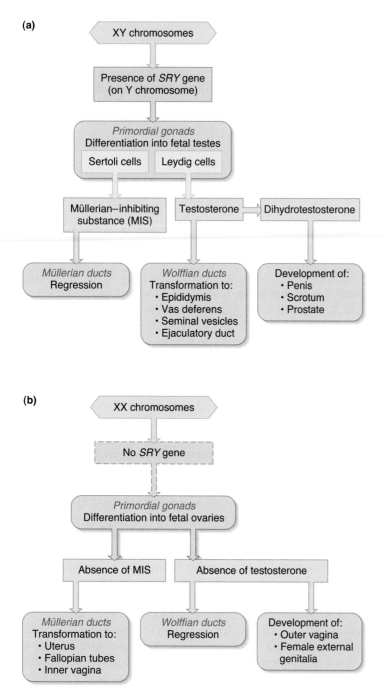

(a)

XY chromosomes

↓

Presence of *SRY* gene
(on Y chromosome)

↓

Primordial gonads
Differentiation into fetal testes

Sertoli cells   Leydig cells

↓   ↓

Müllerian–inhibiting
substance (MIS)   Testosterone → Dihydrotestosterone

↓   ↓   ↓

Müllerian ducts
Regression

Wolffian ducts
Transformation to:
• Epididymis
• Vas deferens
• Seminal vesicles
• Ejaculatory duct

Development of:
• Penis
• Scrotum
• Prostate

(b)

XX chromosomes

↓

No *SRY* gene

↓

Primordial gonads
Differentiation into fetal ovaries

↓   ↓

Absence of MIS   Absence of testosterone

↓   ↓

Müllerian ducts
Transformation to:
• Uterus
• Fallopian tubes
• Inner vagina

Wolffian ducts
Regression

Development of:
• Outer vagina
• Female external
genitalia

Figure 17–3

Sex differentiation. (a)Male. (b) Female. The *SRY* gene codes for the SRY protein. Conversion of testosterone to dihydrotestosterone occurs primarily in target tissue.

Figure 17–3a physiological *inquiry*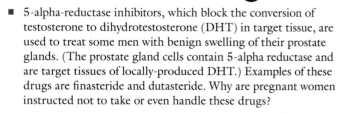

■ 5-alpha-reductase inhibitors, which block the conversion of testosterone to dihydrotestosterone (DHT) in target tissue, are used to treat some men with benign swelling of their prostate glands. (The prostate gland cells contain 5-alpha reductase and are target tissues of locally-produced DHT.) Examples of these drugs are finasteride and dutasteride. Why are pregnant women instructed not to take or even handle these drugs?

Answer can be found at end of chapter.

genitalia and vagina) is female. It is caused by a mutation in the androgen receptor gene. Under the influence of SRY, the fetal testes differentiate as usual, and they secrete both MIS and testosterone. MIS causes the Müllerian ducts to regress, but the inability of the Wolffian ducts to respond to testosterone also causes them to regress, and so no duct system develops. The tissues that develop into external genitalia are also unresponsive to androgen, so female external genitalia and a vagina develop. The testes do not descend, and they are usually removed when the diagnosis is made. The syndrome is usually not detected until menstrual cycles fail to begin at puberty.

Sexual Differentiation of the Central Nervous System and Homosexuality

With regard to sexual behavior, differences in the brain may form during development. For example, genetic female monkeys given testosterone during their late fetal life manifest evidence of masculine sex behavior, such as mounting, as adults.

In this regard, a potentially important difference in human brain anatomy has been reported: The size of a particular nucleus (neuronal cluster) in the hypothalamus is significantly larger in men. A subsequent study showed that the nucleus is also larger in heterosexual men compared to homosexual men, although there is considerable variability and the findings should not be overinterpreted. A similar sexually dimorphic area exists in experimental animals and is known to be involved in male-type sexual behavior and is influenced during development by testosterone.

Another approach to evaluating the genetics and hormone dependency of sexual behavior and gender preference is the use of twin and family studies. The pooled data from many studies show that 57 percent of identical twin brothers of homosexual men were also homosexual, compared to 24 percent of fraternal twins and 13 percent of nontwin brothers. The numbers for homosexual women are similar. This suggests a genetic component to sexual orientation.

■ **SECTION A SUMMARY**

The gonads have a dual function—gametogenesis and the secretion of sex hormones.

General Principles of Gametogenesis

I. The first stage of gametogenesis is mitosis of primordial germ cells.
II. This is followed by meiosis, which is a sequence of two cell divisions resulting in each gamete receiving 23 chromosomes.
III. Crossing-over and random distribution of maternal and paternal chromatids to the daughter cells during meiosis cause genetic variability in the gametes.

Sex Determination

I. Gender is determined by the two sex chromosomes: Males are XY, and females are XX.

Sex Differentiation

I. A gene on the Y chromosome is responsible for the development of testes. In the absence of a Y chromosome, testes do not develop and ovaries do instead.

II. When functioning male gonads are present, they secrete testosterone and MIS, so a male reproductive tract and external genitalia develop. In the absence of testes, the female system develops.

■ SECTION A REVIEW QUESTIONS

1. Describe the stages of gametogenesis and how meiosis results in genetic variability.
2. State the genetic difference between males and females and a method for identifying genetic sex.
3. Describe the sequence of events, the timing, and the control of the development of the gonads and the internal and external genitalia.

SECTION B Male Reproductive Physiology

Anatomy

The male reproductive system includes the two testes, the system of ducts that store and transport sperm to the exterior, the glands that empty into these ducts, and the penis. The duct system, glands, and penis constitute the male accessory reproductive organs.

The testes are suspended outside the abdomen in the **scrotum,** which is an outpouching of the abdominal wall and is divided internally into two sacs, one for each testis. During fetal development, the testes are located in the abdomen, but during the seventh month of **gestation,** they usually descend into the scrotum. This descent is essential for normal sperm production during adulthood, since sperm formation requires a temperature ~2°C lower than normal internal body temperature. Cooling is achieved by air circulating around the scrotum and by a heat-exchange mechanism in the blood vessels supplying the testes. In contrast to spermatogenesis, testosterone secretion can usually occur normally at internal body temperature, and so failure of testes descent usually does not impair testosterone secretion.

The sites of **spermatogenesis** (sperm formation) in the testes are the many tiny, convoluted **seminiferous tubules** (**Figure 17–4**). The combined length of these tubes is 250 m (the length of over 2.5 football fields). Each seminiferous tubule is bounded by a basement membrane. In the center of each tubule is a fluid-filled lumen containing spermatozoa. The tubular wall is composed of developing germ cells and another cell type, to be described later, called Sertoli cells.

The **Leydig cells,** or interstitial cells, which lie in small connective tissue spaces between the tubules, are the cells that synthesize and release testosterone. Thus, the sperm-producing and testosterone-producing functions of the testes are carried out by different structures—the seminiferous tubules and Leydig cells, respectively.

The seminiferous tubules from different areas of a testis converge to form a network of interconnected tubes, the **rete testis** (**Figure 17–5**). Small ducts called efferent ductules leave the rete testis, pierce the fibrous covering of the testis, and empty into a single duct within a structure called the **epididymis** (plural, *epididymides*). The epididymis is loosely attached to the outside of the testis. The duct of the epididymis is so convoluted that, when straightened out at dissection, it measures 6 m. The epididymis draining each testis leads to a **vas deferens,** a large, thick-walled tube lined with smooth muscle. The vas deferens and the blood vessels and nerves supplying the testis are bound together in the **spermatic cord,** which passes to the testis through a slitlike passage, the inguinal canal, in the abdominal wall.

After entering the abdomen, the two vas deferens—one from each testis—continue to the back of the urinary bladder base (**Figure 17–6**). The ducts from two large glands, the **seminal vesicles,** which lie behind the bladder, join the two vas deferens to form the two **ejaculatory ducts.** The ejaculatory ducts then enter the **prostate gland** and join the urethra, coming from the bladder. The prostate gland is a single donut-shaped gland below the bladder and surrounding the upper part of the urethra, into which it secretes fluid through hundreds of tiny openings in the side of the

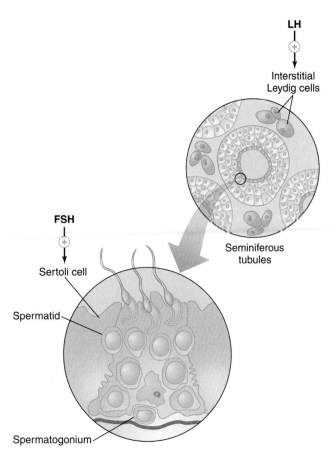

Figure 17–4

Cross-section of an area of testis. The Sertoli cells (stimulated by FSH to increase spermatogenesis and produce inhibin) are in the seminiferous tubules, the sites of sperm production. The tubules are separated from each other by interstitial space (colored light blue) that contains Leydig cells (stimulated by LH to produce testosterone) and blood vessels.

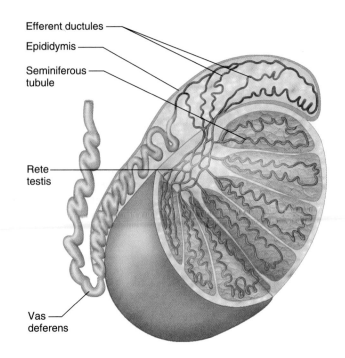

Figure 17–5

Section of a testis. The upper portion of the testis has been removed to show its interior.

Spermatogenesis

The various stages of spermatogenesis are summarized in **Figure 17–7.** The undifferentiated germ cells, which are termed **spermatogonia** (singular, *spermatogonium*) begin to divide mitotically at puberty. The daughter cells of this first division then divide again and again for a specified number of division cycles so that a clone of spermatogonia is produced from each stem cell spermatogonium. Some differentiation occurs in addition to cell division. The cells that result from the final mitotic division and differentiation in the series are called **primary spermatocytes,** and these are the cells that will undergo the first meiotic division of spermatogenesis.

It should be emphasized that if all the cells in the clone produced by each stem cell spermatogonium followed this pathway, the spermatogonia would disappear—that is, they would all be converted to primary spermatocytes. This does not occur because, at an early point, one of the cells of each clone "drops out" of the mitosis-differentiation cycle to remain a stem cell spermatogonium that will later enter into its own full sequence of divisions. One cell of the clone it produces will do likewise, and so on. Thus, the supply of undifferentiated spermatogonia does not decrease.

Each primary spermatocyte increases markedly in size and undergoes the first meiotic division (see Figure 17–7) to form two **secondary spermatocytes,** each of which contains 23 two-chromatid chromosomes. Each secondary spermatocyte undergoes the second meiotic division (see Figure 17–2h to i) to form **spermatids.** Thus, each primary spermatocyte, containing 46 two-chromatid chromosomes, produces four spermatids, each containing 23 one-chromatid chromosomes.

urethra. The urethra emerges from the prostate gland and enters the penis. The paired **bulbourethral glands,** lying below the prostate, drain into the urethra just after it leaves the prostate.

The prostate gland and seminal vesicles secrete most of the fluid in which ejaculated sperm are suspended. This fluid, plus the sperm cells, constitute **semen,** the sperm contributing a small percentage of the total volume. The glandular secretions contain a large number of different chemical substances, including (1) nutrients, (2) buffers for protecting the sperm against the acidic vaginal secretions, (3) chemicals (particularly from the seminal vesicles) that increase sperm motility, and (4) prostaglandins. The function of the prostaglandins, which are produced by the seminal vesicles, is still not clear. The bulbourethral glands contribute a small volume of lubricating mucoid secretions.

In addition to providing a route for sperm from the seminiferous tubules to the exterior, several of the duct system segments perform additional functions to be described in the section on sperm transport.

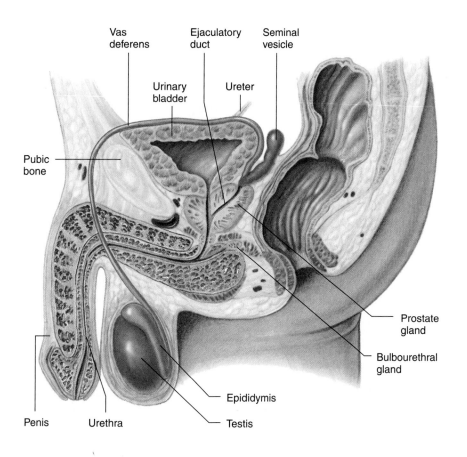

Figure 17–6

Anatomic organization of the male reproductive tract. This figure shows the testis, epididymis, vas deferens, ejaculatory duct, seminal vesicle, and bulbourethral gland on only one side of the body, but they are all paired structures. The urinary bladder and a ureter are shown for orientation but are not part of the reproductive tract. Once the ejaculatory ducts join the urethra in the prostate, the urinary and reproductive tracts have merged.

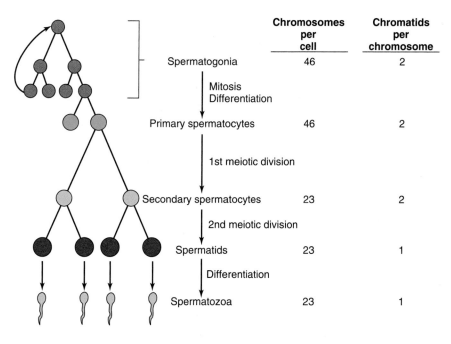

	Chromosomes per cell	Chromatids per chromosome
Spermatogonia	46	2
Primary spermatocytes	46	2
Secondary spermatocytes	23	2
Spermatids	23	1
Spermatozoa	23	1

Figure 17–7

Summary of spermatogenesis, which begins at puberty. Each spermatogonium yields, by mitosis, a clone of spermatogonia; for simplicity, the figure shows only two such cycles, with a third mitotic cycle generating two primary spermatocytes. The arrow from one of the spermatogonia back to a stem cell spermatogonium denotes the fact that one cell of the clone does not go on to generate primary spermatocytes, but reverts to an undifferentiated spermatogonium that gives rise to a new clone. Note that each primary spermatocyte produces four spermatozoa.

The final phase of spermatogenesis is the differentiation of the spermatids into spermatozoa (sperm). This process involves extensive cell remodeling, including elongation, but no further cell divisions. The head of a sperm (**Figure 17–8**) consists almost entirely of the nucleus, which contains the genetic information (DNA). The tip of the nucleus is covered by the **acrosome,** a protein-filled vesicle containing several enzymes that play an important role in fertilization. Most of the tail is a flagellum—a group of contractile filaments that produce whiplike movements capable of propelling the sperm at a velocity of 1 to 4 mm per min. Mitochondria form the midpiece of the sperm and provide the energy for movement.

The entire process of spermatogenesis, from primary spermatocyte to sperm, takes approximately 64 days. The typical human male manufactures approximately 30 million sperm per day.

Thus far, spermatogenesis has been described without regard to its orientation within the seminiferous tubules or the participation of **Sertoli cells,** the second type of cell in the seminiferous tubules, with which the developing germ cells are closely associated. Each seminiferous tubule is bounded by a basement membrane. Each Sertoli cell extends from the basement membrane all the way to the lumen in the center of the tubule and is joined to adjacent Sertoli cells by means of tight junctions (**Figure 17–9**). Thus, the Sertoli cells form an unbroken ring around the outer circumference of the seminiferous tubule. The tight junctions divide the tubule into two compartments—a basal compartment between the basement membrane and the tight junctions, and a central compartment, beginning at the tight junctions and including the lumen.

The ring of interconnected Sertoli cells forms the **Sertoli cell barrier** (blood-testes barrier), which prevents the movement of many chemicals from the blood into the lumen of the seminiferous tubule and helps retain luminal fluid. This ensures proper conditions for germ-cell development and differentiation in the tubules. The arrangement of Sertoli cells also permits different stages of spermatogenesis to take place in different compartments and, therefore, in different environments.

Mitotic cell divisions and differentiation of spermatogonia to yield primary spermatocytes take place entirely in the basal compartment. The primary spermatocytes then move through the tight junctions of the Sertoli cells (which open in front of them while at the same time forming new tight junctions behind them) to gain entry into the central compartment. In this central compartment, the meiotic divisions of spermatogenesis occur, and the spermatids differentiate into sperm while contained in recesses formed by invaginations of the Sertoli cell plasma membranes. When sperm formation is complete, the cytoplasm of the Sertoli cell around the sperm retracts, and the sperm are released into the lumen to be bathed by the luminal fluid.

Sertoli cells serve as the route by which nutrients reach developing germ cells, and they also secrete most of the fluid found in the tubule lumen. This fluid has a highly characteristic ionic composition. It also contains **androgen-binding protein (ABP),** which binds the total testosterone secreted by the Leydig cells and crosses the Sertoli cell barrier to enter the tubule. This protein maintains a high concentration of total testosterone in the lumen of the tubule. The dissociation of free testosterone from ABP continuously bathes the developing spermatocytes and Sertoli cells in testosterone.

Sertoli cells do more than influence the ionic and nutritional environment of the germ cells. In response to FSH from the anterior pituitary and to local testosterone produced in the Leydig cell, Sertoli cells secrete a variety of chemical messengers. These function as paracrine agents to stimulate proliferation and differentiation of the germ cells.

In addition, the Sertoli cells secrete the protein hormone inhibin and paracrine agents that affect Leydig cell function. The many functions of Sertoli cells, several of which remain to be described later in this chapter, are summarized in **Table 17–2**.

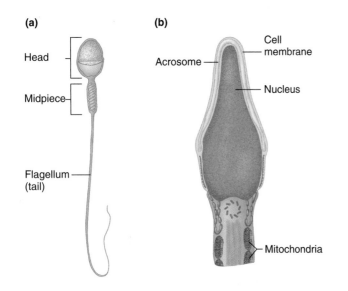

Figure 17–8

(a) Diagram of a human mature sperm. (b) A close-up of the head drawn from a different angle. The acrosome contains enzymes required for fertilization of the ovum.

Table 17–2	Functions of Sertoli Cells
1.	Provide Sertoli cell barrier to chemicals in the plasma
2.	Nourish developing sperm
3.	Secrete luminal fluid, including androgen-binding protein
4.	Respond to stimulation by testosterone and FSH to secrete paracrine agents that stimulate sperm proliferation and differentiation
5.	Secrete the protein hormone inhibin, which inhibits FSH secretion from the pituitary
6.	Secrete paracrine agents that influence the function of Leydig cells
7.	Phagocytize defective sperm
8.	Secrete, during embryonic life, Müllerian-inhibiting substance (MIS), which causes the primordial female duct system to regress

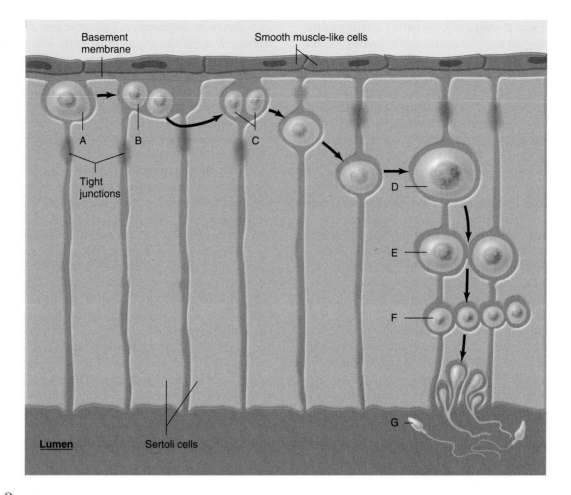

Figure 17–9

Relation of the Sertoli cells and germ cells. The Sertoli cells form a ring (barrier) around the entire tubule. For convenience of presentation, the various stages of spermatogenesis are shown as though the germ cells move down a line of adjacent Sertoli cells; in reality, all stages beginning with any given spermatogonium take place between the same two Sertoli cells. Spermatogonia (A and B) are found only in the basal compartment (between the tight junctions of the Sertoli cells and the basement membrane of the tubule). After several mitotic cycles (A to B), the spermatogonia (B) give rise to primary spermatocytes (C). Each of the latter crosses a tight junction, enlarges (D), and divides into two secondary spermatocytes (E), which divide into spermatids (F), which in turn differentiate into spermatozoa (G). This last step involves loss of cytoplasm by the spermatids.

Adapted from Tung.

Transport of Sperm

From the seminiferous tubules, the sperm pass through the rete testis and efferent ducts into the epididymis and from there to the vas deferens. The vas deferens and the portion of the epididymis closest to it serve as a storage reservoir for sperm until ejaculation.

Movement of the sperm as far as the epididymis results from the pressure that the Sertoli cells create by continuously secreting fluid into the seminiferous tubules. The sperm themselves are normally nonmotile at this time.

During passage through the epididymis, the concentration of the sperm increases dramatically due to fluid absorption from the lumen of the epididymis. Therefore, as the sperm pass from the end of the epididymis into the vas deferens, they are a densely packed mass whose transport is no longer facilitated by fluid movement. Instead, peristaltic con-

tractions of the smooth muscle in the epididymis and vas deferens cause the sperm to move.

The absence of a large quantity of fluid accounts for the fact that **vasectomy,** the surgical tying-off and removal of a segment of each vas deferens, does not cause the accumulation of much fluid behind the tie-off point. The sperm, which are still produced after vasectomy, do build up, however, and eventually dissolve, with their chemical components absorbed into the bloodstream. Vasectomy does not affect testosterone secretion because it does not alter the function of the Leydig cells. The next step in sperm transport is ejaculation.

Erection

The penis consists almost entirely of three cylindrical, vascular compartments running its entire length. Normally the small arteries supplying the vascular compartments are constricted so that the compartments contain little blood and the penis is flaccid. During sexual excitation, the small arteries dilate, blood flow increases, the three vascular compartments become engorged with blood at high pressure, and the penis becomes rigid (**erection**). The vascular dilation is initiated by neural input to the small arteries of the penis. As the vascular compartments expand, the veins emptying them are passively compressed, furthur increasing the local pressure, thus contributing to the engorgement while blood flow remains elevated. This entire process occurs rapidly with complete erection sometimes taking only 5 to 10 s.

What are the neural inputs to the small arteries of the penis? At rest, the dominant input is from sympathetic neurons that release norepinephrine, which causes the arterial smooth muscle to contract. During erection this sympathetic input is inhibited. Much more important is the activation of nonadrenergic, noncholinergic autonomic neurons to the arteries (**Figure 17–10**). These neurons and associated endothelial cells release **nitric oxide,** which relaxes the arterial smooth muscle.

Which receptors and afferent pathways initiate these reflexes? The primary stimulus comes from mechano-receptors in the genital region, particularly in the head of the penis. The afferent fibers carrying the impulses synapse in the lower spinal cord on interneurons that control the efferent outflow.

It must be stressed, however, that higher brain centers, via descending pathways, may also exert profound stimulatory or inhibitory effects upon the autonomic neurons to the small arteries of the penis. Thus, mechanical stimuli from areas other than the penis, as well as thoughts, emotions, sights, and odors, can induce erection in the complete absence of penile stimulation (or prevent erection even though stimulation is present).

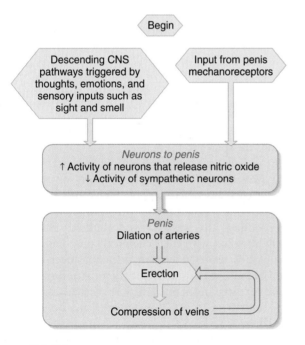

Figure 17–10

Reflex pathways for erection. Nitric oxide, a vasodilator, is the most important neurotransmitter to the arteries in this reflex.

Erectile dysfunction (also termed impotence) is the consistent inability to achieve or sustain an erection of sufficient rigidity for sexual intercourse, and is a common problem. Although it can be mild to moderate in degree, complete erectile dysfunction is present in as many as 10 percent of adult American males between the ages of 40 and 70. During this period of life, its rate almost doubles. The organic causes are multiple and include damage to or malfunction of the efferent nerves or descending pathways, endocrine disorders, various therapeutic and "recreational" drugs (e.g., alcohol), and certain diseases, particularly diabetes mellitus. Erectile dysfunction can also be due to psychological factors (such as depression), which are mediated by the brain and the descending pathways.

There are now a group of orally active **phosphodiesterase type 5** (PDE5) **inhibitors** including sildenafil (*Viagra®*), vardenafil (*Levitra®*), and tadalafil (*Cialis®*) that greatly improve the ability of many men with erectile dysfunction to achieve and maintain an erection. The most important event leading to erection is the dilation of penile arteries by nitric oxide, released from autonomic neurons. Nitric oxide stimulates the enzyme guanylyl cyclase, which catalyzes the formation of cyclic GMP, as described in Chapter 5. This second messenger then continues the signal transduction pathway leading to the relaxation of the arterial smooth muscle. The sequence of events is terminated by an enzyme-dependent breakdown of cGMP. PDE5 inhibitors block the action of this enzyme and thereby permit a higher concentration of cGMP to exist.

Ejaculation

The discharge of semen from the penis, called **ejaculation,** is primarily a spinal reflex mediated by afferent pathways from penile mechanoreceptors. When the level of stimulation is high enough, a patterned sequence of discharge of the efferent neurons ensues. This sequence can be divided into two phases: (1) the smooth muscles of the epididymis, vas deferens, ejaculatory ducts, prostate, and seminal vesicles contract as a result of sympathetic nerve stimulation, emptying the sperm and glandular secretions into the urethra (**emission**); and (2) the semen, with an average volume of 3 ml and containing 300 million sperm, is then expelled from the urethra by a series of rapid contractions of the urethral smooth muscle as well as the skeletal muscle at the base of the penis. During ejaculation, the sphincter at the base of the urinary bladder is closed so that sperm cannot enter the bladder, nor can urine be expelled from it. Note that erection involves inhibition of sympathetic nerves (to the small arteries of the penis), while ejaculation involves stimulation of sympathetic nerves (to the smooth muscles of the duct system).

The rhythmical muscular contractions that occur during ejaculation are associated with intense pleasure and many systemic physiological changes, collectively termed an **orgasm.** Marked skeletal muscle contractions occur throughout the body, and there is a transient increase in heart rate and blood pressure.

Once ejaculation has occurred, there is a latent period during which a second erection is not possible. The latent period is quite variable but may last from minutes to hours.

As is true of erectile dysfunction, **premature ejaculation** or failure to ejaculate can be the result of influence by higher brain centers.

Hormonal Control of Male Reproductive Functions

Control of the Testes

Figure 17–11 summarizes the control of the testes. In a normal adult man, the GnRH-secreting neuroendocrine cells in the hypothalamus fire a brief burst of action potentials approximately every 90 min, secreting GnRH at these times. The GnRH reaching the anterior pituitary via the hypothalamo-pituitary portal vessels during each periodic pulse triggers the release of both LH and FSH from the same cell type, although not necessarily in equal amounts. Thus, systemic plasma concentrations of FSH and LH also show pulsatility—rapid increases followed by slow decreases over the next 90 min or so as the hormones are slowly removed from the plasma.

There is a clear separation of the actions of FSH and LH within the testes (see Figure 17–11). FSH acts primarily on the Sertoli cells to stimulate the secretion of paracrine agents needed for spermatogenesis. LH, by contrast, acts primarily on the Leydig cells to stimulate testosterone secretion. In addition to its many important systemic effects as a hormone, the testosterone secreted by the Leydig cells also acts locally, as a paracrine agent, by moving from the interstitial spaces into the seminiferous tubules. Testosterone enters Sertoli cells, where it facilitates spermatogenesis. Thus, despite the absence of a *direct* effect on cells in the seminiferous tubules, LH exerts an essential *indirect* effect because the testosterone secretion stimulated by LH is required for spermatogenesis.

The last components of the hypothalamo-pituitary control of male reproduction that remain to be discussed are the negative feedback effects exerted by testicular hormones. Even though FSH and LH are produced by the same cell type, their secretion rates can be altered to different degrees by negative feedback inputs.

Testosterone inhibits LH secretion in two ways (see Figure 17–11): (1) it acts on the hypothalamus to decrease the amplitude of GnRH bursts, which results in a decrease in the secretion of gonadotropins; and (2) it acts directly on the anterior pituitary to decrease the LH response to any given amount of GnRH.

How do the testes reduce FSH secretion? The major inhibitory signal, exerted directly on the anterior pituitary, is the protein hormone inhibin secreted by the Sertoli cells (see Figure 17–11). This is a logical completion of a negative feedback loop such that FSH stimulates Sertoli cells to increase both spermatogenesis and inhibin production, and inhibin decreases FSH release.

Despite all these complexities, the total amounts of GnRH, LH, FSH, testosterone, and inhibin secreted and of sperm produced are relatively constant from day to day in the adult male. This is completely different from the large cyclical variations of activity so characteristic of the female reproductive processes.

Testosterone

In addition to its essential paracrine action within the testes on spermatogenesis and its negative feedback effects on the hypothalamus and anterior pituitary, testosterone exerts many other effects, as summarized in **Table 17–3**.

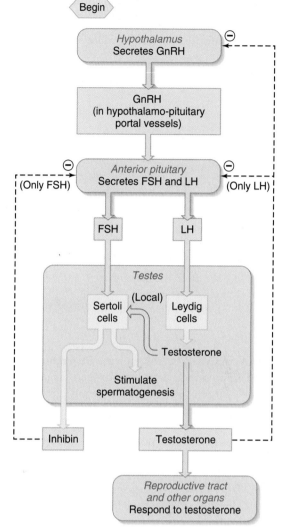

Figure 17–11

Summary of hormonal control of male reproductive function. Note that FSH acts only on the Sertoli cells, whereas LH acts primarily on the Leydig cells. The secretion of FSH is inhibited mainly by inhibin, a protein hormone secreted by the Sertoli cells, and the secretion of LH is inhibited mainly by testosterone, the steroid hormone secreted by the Leydig cells. Testosterone, acting locally on Sertoli cells, stimulates spermatogenesis, while FSH stimulates inhibin release from Sertoli cells.

Figure 17–11 physiological *inquiry*

- Men with decreased anterior pituitary gland function often have decreased sperm production as well as low testosterone levels. Would you expect the administration of testosterone alone to restore sperm production to normal?

Answer can be found at end of chapter.

In Chapter 11 we mentioned that some hormones must undergo transformation in their target cells in order to be effective. This is true of testosterone in many (but not all) of its target cells. In some cells, like in the adult prostate, after its entry into the cytoplasm, testosterone is converted to dihydrotestosterone (DHT), which is more potent than testosterone. This

Table 17–3	Effects of Testosterone in the Male

1. Required for initiation and maintenance of spermatogenesis (acts via Sertoli cells)

2. Decreases GnRH secretion via an action on the hypothalamus

3. Inhibits LH secretion via a direct action on the anterior pituitary

4. Induces differentiation of male accessory reproductive organs and maintains their function

5. Induces male secondary sex characteristics; opposes action of estrogen on breast growth

6. Stimulates protein anabolism, bone growth, and cessation of bone growth

7. Required for sex drive and may enhance aggressive behavior

8. Stimulates erythropoietin secretion by the kidneys

conversion is catalyzed by the enzyme **5α-reductase,** which is expressed in a wide variety of androgen target tissues. In certain other target cells (e.g., the brain), testosterone is transformed not to dihydrotestosterone but to estradiol, which is the active hormone in these cells. The enzyme aromatase catalyzes this conversion. Note in this latter case that the "male" sex hormone must be converted to the "female" sex hormone to be active in the male. The fact that, depending on the target cells, testosterone may act as testosterone, dihydrotestosterone, or estradiol has important pathophysiological implications because some men lack 5α-reductase or aromatase in some tissues. Therefore, they will exhibit certain signs of testosterone deficiency but not others. For example, an XY fetus with 5α-reductase deficiency will have normal differentiation of male reproductive duct structures (an effect of testosterone per se) but will not have normal development of external male genitalia, which requires DHT.

Therapy for **prostate cancer** makes use of these facts: prostate cancer cells are stimulated by dihydrotestosterone, so the cancer can be treated with inhibitors of 5α-reductase. Furthermore, **male pattern baldness** may also be treated with 5α-reductase inhibitors because DHT tends to promote hair loss from the scalp.

Accessory Reproductive Organs

The fetal differentiation, and later growth and function of the entire male duct system, glands, and penis all depend upon testosterone. Following **castration** (removal of the gonads) in the adult male, all the accessory reproductive organs decrease in size, the glands markedly reduce their secretion rates, and the smooth-muscle activity of the ducts is diminished. Sex drive (**libido**), erection, and ejaculation are usually impaired. These defects disappear upon the administration of testosterone. This also applies to **hypogonadism,** which is described at the end of this section.

Puberty

Puberty is the period during which the reproductive organs mature and reproduction becomes possible. In males, this usually occurs between 12 and 16 years of age. Surprisingly, some of the first signs of puberty are due not to gonadal steroids but to increased secretion of adrenal androgens, probably under the stimulation of adrenocorticotropic hormone (ACTH). These androgens cause the very early development of pubic and axillary (armpit) hair, as well as the early stages of the pubertal growth spurt in concert with growth hormone and insulin-like growth factor I (Chapter 11). All other developments in puberty, however, reflect increased activity of the hypothalamic-anterior pituitary-gonadal axis.

Increased GnRH secretion at puberty causes increased secretion of pituitary gonadotropins, which stimulate the seminiferous tubules and testosterone secretion. Testosterone, in addition to its critical role in spermatogenesis, induces the pubertal changes that occur in the accessory reproductive organs, secondary sex characteristics, and sex drive. The mechanism of the brain change that results in increased GnRH secretion at puberty remains unknown. One important event is that the brain becomes less sensitive to the negative feedback effects of gonadal hormones at the time of puberty.

Secondary Sex Characteristics and Growth

Virtually all the male secondary sex characteristics are dependent on testosterone and its metabolite, DHT. For example, a male castrated before puberty has minimal facial, underarm, or pubic hair. Other androgen-dependent secondary sexual characteristics are deepening of the voice resulting from the growth of the larynx, thick secretion of the skin oil glands (often causing acne), and the masculine pattern of fat distribution. Androgens also stimulate bone growth, mostly through the stimulation of growth hormone secretion. Ultimately, however, androgens terminate bone growth by causing closure of the bones' epiphyseal plates. Androgens are "anabolic steroids" in that they exert a direct stimulatory effect on protein synthesis in muscle. Finally, androgens stimulate the secretion of the hormone erythropoietin by the kidneys; this is a major reason why men have a higher hematocrit than women.

Behavior

Androgens are essential in males for the development of sex drive at puberty, and they play an important role in maintaining sex drive (libido) in the adult male. A controversial question is whether androgens influence other human behaviors in addition to sexual behavior. It has proven very difficult to answer this question with respect to human beings, but there is little doubt that androgen-dependent behavioral differences based on gender do exist in other mammals. For example, aggression is clearly greater in male animals and is androgen-dependent.

Andropause

Changes in the male reproductive system with aging are less drastic than those in women (described later in this chapter). Once testosterone and pituitary gonadotropin secretions are initiated at puberty, they continue, at least to some extent, throughout adult life. There is a steady decrease, however, in testosterone secretion, beginning at about 40 years of age, which apparently reflects slow deterioration of testicular function and failure of the gonads to respond to the pituitary gonadotropins. Along with the decreasing testosterone levels, libido decreases, and sperm become less motile. Despite these events, many men continue to be fertile in their seventies and eighties.

With aging, some men manifest increased emotional problems, such as depression, and this is sometimes referred to as the **andropause** (or **male climacteric**). It is not clear, however, what role hormone changes play in this phenomenon.

ADDITIONAL CLINICAL EXAMPLES

We have already discussed erectile dysfunction and prostate cancer. Several additional conditions highlight the pathophysiology of male reproduction.

Hypogonadism

A decrease in testosterone release from the testes can be caused by a wide variety of disorders. In general, they can be classified into testicular failure (primary hypogonadism) or a failure to supply the testes with appropriate gonadotropic stimulus (secondary hypogonadism). The loss of normal testicular androgen production before puberty can lead to a failure to develop secondary sex characteristics (deepening of the voice, pubic and axillary hair, and increased libido) as well as a failure to develop normal sperm production.

A relatively common genetic cause of hypogonadism is **Klinefelter's syndrome.** The most common cause of this disorder, occurring in 1 in 500 male births, is an extra X chromosome (XXY) caused by meiotic nondisjunction. The classic form is caused by the failure of the two sex chromosomes to separate during the first meiotic division in gametogenesis (from Steps g to h in Figure 17–2). The extra X chromosome can come from either the egg or the sperm. That is, if nondisjunction occurs in the ovary leading to an XX ovum, an XXY genotype will result if fertilized by a Y sperm. If nondisjunction occurs in the testes leading to an XY sperm, an XXY genotype will result if that sperm fertilizes a normal (single X) ovum.

The XXY genotype leads to a male child of normal appearance before puberty. However, after puberty, the testes remain small and firm. Because of abnormal Leydig cell function, testosterone levels are low. The lack of local testosterone production in addition to the abnormal development of the seminiferous tubules leads to decreased sperm production. Normal secondary sex characteristics do not appear, and breast size increases (**gynecomastia**) (**Figure 17–12**). These men have relatively high gonadotropin levels (LH and FSH) due to loss of androgen and inhibin negative feedback. Men with Klinefelter's syndrome can be treated with androgen replacement therapy to increase libido and decrease breast size. This will not restore normal spermatogenesis, however.

Hypogonadism in men can also be caused by a decrease in LH and FSH secretion (secondary hypogonadism). Although there are many causes of this loss of function of the pituitary cells that secrete LH and FSH, **hyperprolactinemia** (increased prolactin in the blood) is one of the most

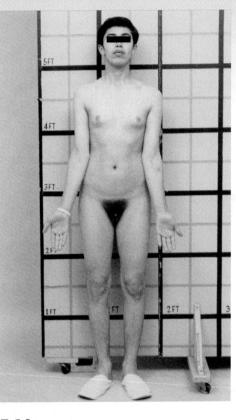

Figure 17–12

Klinefelter's syndrome in a 20-year-old man. Note relatively increased lower/upper body segment ratio, gynecomastia, small penis, and sparse body hair with a female pubic hair pattern.

Courtesy of Glenn D. Braunstein, MD.

common. Although prolactin has minimal effects in men under normal conditions, the pituitary gland still has cells that secrete prolactin (lactotrophs). Pituitary tumors arising from prolactin-secreting cells can develop and secrete very high levels of prolactin. One of the effects of high prolactin is to inhibit LH and FSH secretion. (This occurs in men and women.) Sometimes, these tumors can get so big that they can press on the optic chiasm and cause loss of vision. One potential treatment takes advantage of the fact that the primary controller of prolactin release is inhibition by hypothalamic dopamine. Dopamine agonists can be given to reduce the size and activity of prolactin-secreting tumors. Once these tumors are reduced in size, they can be removed by surgery, although the tumors often recur. ■

Anatomy

I. The male gonads, the testes, produce sperm in the seminiferous tubules and secrete testosterone from the Leydig cells.

Spermatogenesis

I. The meiotic divisions of spermatogenesis result in sperm containing 23 chromosomes, compared to the original 46 of the spermatogonia.

II. The developing germ cells are intimately associated with the Sertoli cells, which perform many functions, as summarized in Table 17–2.

Transport of Sperm

I. From the seminiferous tubules, the sperm pass into the epididymis, where they are concentrated and become mature.

II. The epididymis and vas deferens store the sperm, and the seminal vesicles and prostate secrete most of the semen.

III. Erection of the penis occurs because of vascular engorgement accomplished by relaxation of the small arteries and passive occlusion of the veins.

IV. Ejaculation includes emission—emptying of semen into the urethra—followed by expulsion of the semen from the urethra.

Hormonal Control of Male Reproductive Functions

I. Pulses of hypothalamic GnRH stimulate the anterior pituitary to secrete FSH and LH, which then act on the testes: FSH on the Sertoli cells to stimulate spermatogenesis and inhibin secretion, and LH on the Leydig cells to stimulate testosterone secretion.

II. Testosterone, acting locally on the Sertoli cells, is essential for maintaining spermatogenesis.

III. Testosterone exerts a negative feedback inhibition on both the hypothalamus and the anterior pituitary to reduce mainly LH secretion. Inhibin exerts a negative feedback inhibition on FSH secretion.

IV. Testosterone maintains the accessory reproductive organs and male secondary sex characteristics and stimulates the growth of muscle and bone. In many of its target cells, it must first undergo transformation to dihydrotestosterone or to estrogen.

Puberty

I. A change in brain function at the onset of puberty results in increases in the hypothalamic-anterior pituitary-gonadal axis (because of increases in GnRH).

II. The first signs of puberty are the appearance of pubic and axillary hair.

Andropause

I. The andropause is a decrease in testosterone with aging (but usually not a complete cessation of androgen production).

Additional Clinical Examples

I. Male hypogonadism is a decrease in testicular function. Klinefelter's syndrome (usually XXY genotype) is a common cause of male hypogonadism.

II. Prolactin-secreting pituitary tumors lead to hyperprolactinemia. This causes a decrease in LH and FSH, which results in hypogonadism.

acrosome 608	prostate gland 605
androgen-binding protein (ABP) 608	puberty 612
	rete testis 605
bulbourethral gland 606	scrotum 605
ejaculation 610	secondary spermatocyte 606
ejaculatory duct 605	semen 606
emission 610	seminal vesicle 605
epididymis 605	seminiferous tubule 605
erection 610	Sertoli cell 608
5α-reductase 612	Sertoli cell barrier 608
gestation 605	spermatic cord 605
Leydig cell 605	spermatid 606
libido 612	spermatogenesis 605
nitric oxide 610	spermatogonium 606
orgasm 610	vas deferens 605
primary spermatocyte 606	

andropause (male climacteric) 613	Klinefelter's syndrome 613
	male pattern baldness 612
castration 612	phosphodiesterase type 5 inhibitor 610
erectile dysfunction 610	
gynecomastia 613	premature ejaculation 610
hyperprolactinemia 613	prostate cancer 612
hypogonadism 612	vasectomy 609

1. Describe the sequence of events leading from spermatogonia to sperm.
2. List the functions of the Sertoli cells.
3. Describe the path sperm take from the seminiferous tubules to the urethra.
4. Describe the roles of the prostate gland, seminal vesicles, and bulbourethral glands in the formation of semen.
5. Describe the neural control of erection and ejaculation.
6. Diagram the hormonal chain controlling the testes. Contrast the effects of FSH and LH.
7. What are the feedback controls from the testes to the hypothalamus and pituitary?
8. Define puberty in the male. When does it usually occur?
9. List the effects of androgens on accessory reproductive organs, secondary sex characteristics, growth, protein metabolism, and behavior.
10. Describe the conversion of testosterone to DHT and estrogen.
11. How does hyperprolactinemia cause hypogonadism?

Unlike the continuous sperm production of the male, the maturation of the female gamete, the egg, followed by its release from the ovary—**ovulation**—is cyclical. The structure and function of the female reproductive system (e.g., the uterus) are synchronized with these ovarian cycles. In human beings, these cycles are called **menstrual cycles.** The length of a menstrual cycle varies considerably from woman to woman and even in any particular woman, but averages about 28 days. The first day of menstrual flow (**menstruation**) is designated as day 1.

Menstruation is the result of events occurring in the uterus. However, the uterine events of the menstrual cycle are due to cyclical changes in hormone secretion by the ovaries. The ovaries are also the sites for the maturation of gametes. One oocyte normally becomes fully mature and is ovulated around the middle of each menstrual cycle.

It is the interactions among the ovaries, hypothalamus, and anterior pituitary gland that produce the cyclical changes in the ovaries that result in (1) maturation of a gamete each cycle, and (2) hormone secretions that cause cyclical changes in all of the female reproductive organs (particularly the uterus). The uterine changes prepare this organ to receive and nourish the developing conceptus, and only when there is no pregnancy does menstruation occur. After describing the full menstrual cycle in the absence of pregnancy, we describe the events of pregnancy, delivery, and lactation.

Anatomy

The female reproductive system includes the two ovaries and the female reproductive tract—two **fallopian tubes** (or oviducts), the uterus, the cervix, and the vagina. These structures are termed the **female internal genitalia** (**Figure 17–13**). Unlike in the male, the urinary and reproductive duct systems of the female are entirely separate from each other.

The ovaries are almond-sized organs in the upper pelvic cavity, one on each side of the uterus. The ends of the fallopian tubes are not directly attached to the ovaries but open into the abdominal cavity close to them. The opening of each fallopian tube is funnel-shaped and surrounded by long, fingerlike projections (the **fimbriae**) lined with ciliated epithelium. The other ends of the fallopian tubes are attached to the uterus and empty directly into its cavity. The **uterus** is a hollow, thick-walled, muscular organ lying between the urinary bladder and rectum. It is the source of menstrual flow, and it is where the fetus develops during pregnancy. The lower portion of the uterus is the **cervix.** A small opening in the cervix leads to the **vagina,** the canal leading from the uterus to the outside.

The **female external genitalia** (**Figure 17–14**) include the mons pubis, labia majora, labia minora, clitoris, vestibule of the vagina, and vestibular glands. The term **vulva** is another name for all these structures. The mons pubis is the rounded fatty prominence over the junction of the pubic bones. The labia majora, the female homolog of the scrotum, are two prominent skin folds that form the outer lips of the vulva. (The terms *homolog* and *analogous* mean that the two structures are derived embryologically from the same source and/or have similar functions.) The labia minora are small skin folds lying between the labia majora. They surround the urethral and vaginal openings, and the area thus enclosed is the vestibule, into which glands empty. The vaginal opening lies behind the opening of the urethra. Partially overlying the vaginal opening is a thin fold of mucous membrane, the **hymen.** The **clitoris,** the female homolog of the penis, is an erectile structure located at the front of the vulva.

Ovarian Functions

As noted at the beginning of this chapter, the ovary, like the testis, serves several functions: (1) **oogenesis,** the production of gametes during the fetal period, (2) maturation of the oocyte, (3) expulsion of the mature oocyte (ovulation), and (4) secretion of the female sex steroid hormones, (estrogen and progesterone), as well as the peptide hormone inhibin. Before ovulation, the maturation of the oocyte and endocrine functions of the ovaries take place in a single structure, the follicle. After ovulation, the follicle, now without an egg, differentiates into a corpus luteum, which only has an endocrine function. For comparison, recall that in the testes, the production of gametes and the secretion of sex steroids take place in different compartments—in the seminiferous tubules and in the Leydig cells, respectively.

Oogenesis

The female germ cells, like those of the male, have different names at different stages of development. However, the term **egg** will be used to refer to the germ cells at any stage.

At birth, the ovaries contain an estimated total of 2 to 4 million eggs, and no new ones appear after birth. Thus, in marked contrast to the male, the newborn female already has all the germ cells she will ever have. Only a few, perhaps 400, will be ovulated during a woman's lifetime. All the others degenerate at some point in their development so that few, if any, remain by the time a woman reaches approximately 50 years of age. One result of this developmental pattern is that the eggs ovulated near age 50 are 35 to 40 years older than those ovulated just after puberty. It is possible that certain chromosomal defects more common among children born to older women are the result of aging changes in the egg.

During early *in utero* development, the primitive germ cells, or **oogonia** (singular, *oogonium*), a term analogous to spermatogonia in the male, undergo numerous mitotic divisions (**Figure 17–15**). Around the seventh month after conception, the fetal oogonia cease dividing. Current thinking is that from this point on, no new germ cells are generated.

During fetal life, all the oogonia develop into **primary oocytes** (analogous to primary spermatocytes), which then begin a first meiotic division by replicating their DNA. They do not, however, complete the division in the fetus. Accordingly, all the eggs present at birth are primary oocytes

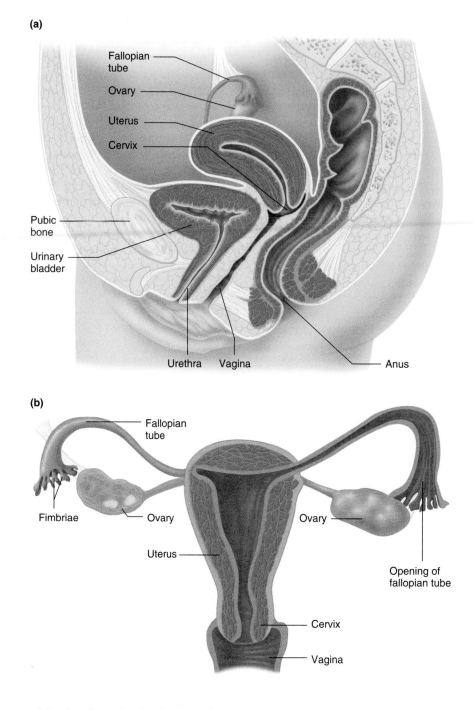

Figure 17–13

Female reproductive system. (a) Section through a female pelvis. (b) Diagram cut away on the right to show the continuity between the organs of the reproductive duct system—fallopian tubes, uterus, and vagina.

containing 46 chromosomes, each with two sister chromatids. The cells are said to be in a state of **meiotic arrest.**

This state continues until puberty and the onset of renewed activity in the ovaries. Indeed, only those primary oocytes destined for ovulation will ever complete the first meiotic division, for it occurs just before the egg is ovulated. This division is analogous to the division of the primary spermatocyte, and each daughter cell receives 23 chromosomes, each with two chromatids. In this division, however, one of the

two daughter cells, the **secondary oocyte,** retains virtually all the cytoplasm. The other, called the first polar body, is very small and nonfunctional. Thus, the primary oocyte, which is already as large as the egg will be, passes on to the secondary oocyte just half of its chromosomes but almost all of its nutrient-rich cytoplasm.

The second meiotic division occurs in a fallopian tube *after ovulation,* but only if the secondary oocyte is fertilized— that is, penetrated by a sperm. As a result of this second mei-

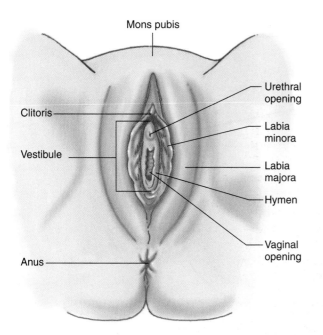

Figure 17–14

Female external genitalia.

otic division, the daughter cells each receive 23 chromosomes, each with a single chromatid. Once again, one daughter cell, now called an ovum, retains nearly all the cytoplasm. The other daughter cell, the second polar body, is very small and nonfunctional. The net result of oogenesis is that each primary oocyte can produce only one ovum (see Figure 17–15). In contrast, each primary spermatocyte produces four viable spermatozoa.

Follicle Growth

Throughout their life in the ovaries, the eggs exist in structures known as **follicles.** Follicles begin as **primordial follicles,** which consist of one primary oocyte surrounded by a single layer of cells called **granulosa cells.** Further development from the primordial follicle stage (**Figure 17–16**) is characterized by an increase in the size of the oocyte, a proliferation of the granulosa cells into multiple layers, and the separation of the oocyte from the inner granulosa cells by a thick layer of material, the **zona pellucida.** The granulosa cells secrete estrogen, small amounts of progesterone just before ovulation, and the peptide hormone inhibin.

Despite the presence of a zona pellucida, the inner layer of granulosa cells remains closely associated with the oocyte by means of cytoplasmic processes that traverse the zona pellucida and form gap junctions with the oocyte. Through these gap junctions, nutrients and chemical messengers are passed to the oocyte. For example, the granulosa cells produce one or more factors that act on the primary oocytes to maintain them in meiotic arrest.

As the follicle grows by mitosis of granulosa cells, connective tissue cells surrounding the granulosa cells differentiate and form layers known as the **theca,** which play an important role in estrogen secretion by the granulosa cells. Shortly after this, the primary oocyte reaches full size (115 μm in diameter), and a fluid-filled space, the **antrum,** begins to form in the midst of the granulosa cells as a result of fluid they secrete.

The progression of some primordial follicles to the preantral and early antral stages (see Figure 17–16) occurs throughout infancy and childhood, and then during the entire menstrual cycle. Therefore, although most of the follicles in

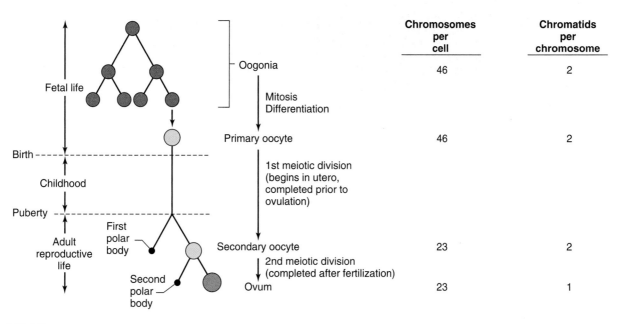

	Chromosomes per cell	Chromatids per chromosome
Oogonia	46	2
Primary oocyte	46	2
Secondary oocyte	23	2
Ovum	23	1

Figure 17–15

Summary of oogenesis. Compare with the male pattern in Figure 17–7. The secondary oocyte is ovulated and does not complete its meiotic division unless it is penetrated (fertilized) by a sperm. Thus, it is a semantic oddity that the oocyte is not termed an egg or ovum until after fertilization occurs. Note that each primary oocyte yields only one secondary oocyte, which can yield only one ovum.

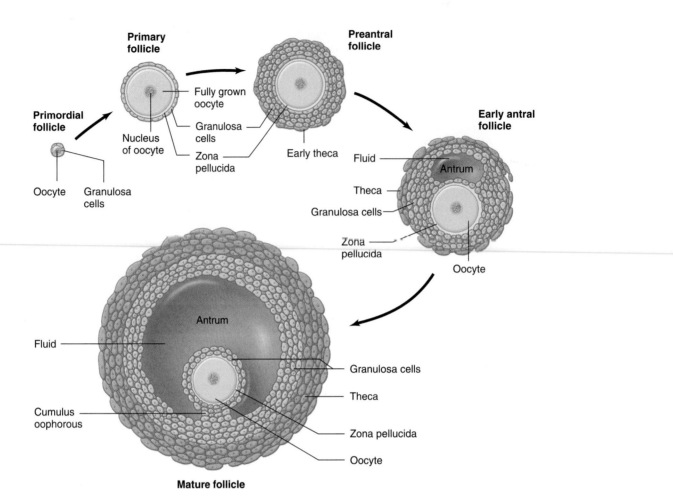

Figure 17–16

Development of a human oocyte and ovarian follicle. The fully mature follicle is 1.5 cm in diameter. Blood vessels are not shown.

Adapted from Erickson et al.

the ovaries are still primordial, a relatively constant number of preantral and early antral follicles are also always present. At the beginning of each menstrual cycle, 10 to 25 of these preantral and early antral follicles begin to develop into larger antral follicles. About one week into the cycle, a further selection process occurs: only one of the larger antral follicles, the **dominant follicle,** continues to develop. The exact process by which a follicle is selected for dominance is not known, but it is likely related to the amount of estrogen produced locally within the follicle. (This is probably why hyperstimulation of infertile women with gonadotropin injections can result in the development of many follicles.) The nondominant follicles (in both ovaries) that had begun to enlarge undergo a degenerative process called **atresia** which is an example of programmed cell death, or apoptosis. The eggs in the degenerating follicles also die.

Atresia is not limited to just antral follicles, however, for follicles can undergo atresia at any stage of development. Indeed, this process is already occurring in the female fetus, so that the 2 to 4 million follicles and eggs present at birth represent only a small fraction of those present earlier in gestation. Atresia then continues all through prepubertal life so that only

200,000 to 400,000 follicles remain when active reproductive life begins. Of these, all but about 400 will undergo atresia during a woman's reproductive life. Therefore, 99.99 percent of the ovarian follicles present at birth will undergo atresia.

The dominant follicle enlarges mainly as a result of an increase in fluid, causing the antrum to expand. As this occurs, the granulosa cell layers surrounding the egg form a mound that projects into the antrum and is called the **cumulus ooph-orous** (see Figure 17–16). As the time of ovulation approaches, the egg (a primary oocyte) emerges from meiotic arrest and completes its first meiotic division to become a secondary oocyte. The cumulus separates from the follicle wall so that it and the oocyte float free in the antral fluid. The mature follicle (also called a **graafian follicle**) becomes so large (diameter about 1.5 cm) that it balloons out on the surface of the ovary.

Ovulation occurs when the thin walls of the follicle and ovary rupture at the site where they are joined because of enzymatic digestion. The secondary oocyte, surrounded by its tightly adhering zona pellucida and granulosa cells, as well as the cumulus, is carried out of the ovary and onto the ovarian surface by the antral fluid. All this happens on approximately day 14 of the menstrual cycle.

Occasionally, two or more follicles reach maturity, and more than one egg may be ovulated. This is the most common cause of multiple births. In such cases the siblings are **fraternal,** not identical, because the eggs carry different sets of genes. We will describe later how identical twins form.

Formation of the Corpus Luteum

After the mature follicle discharges its antral fluid and egg, it collapses around the antrum and undergoes a rapid transformation. The granulosa cells enlarge greatly, and the entire glandlike structure formed is called the **corpus luteum,** which secretes estrogen, progesterone, and inhibin. If the discharged egg, now in a fallopian tube, is not fertilized, the corpus luteum reaches its maximum development within approximately 10 days. It then rapidly degenerates by apoptosis. As we will see, it is the loss of corpus luteum function that leads to menstruation and the beginning of a new menstrual cycle.

In terms of ovarian function, therefore, the menstrual cycle may be divided into two phases approximately equal in length and separated by ovulation (**Figure 17–17**): (1) the **follicular phase,** during which a mature follicle and secondary oocyte develop; and (2) the **luteal phase,** beginning after ovulation and lasting until the death of the corpus luteum.

Sites of Synthesis of Ovarian Hormones

The sites of ovarian hormone syntheses can be summarized as follows: estrogen is synthesized and released into the blood during the follicular phase mainly by the granulosa cells. After ovulation, estrogen is synthesized and released by the corpus luteum. Progesterone, the other major ovarian steroid hormone, is synthesized and released in very small amounts by the granulosa and theca cells just before ovulation, but its major source is the corpus luteum. Inhibin, a peptide hormone, is secreted by both the granulosa cells and the corpus luteum.

Control of Ovarian Function

The major factors controlling ovarian function are analogous to the controls described for testicular function. They constitute a hormonal system made up of GnRH, the anterior pituitary gonadotropins FSH and LH, and gonadal sex hormones—estrogen and progesterone.

As in the male, the entire sequence of controls depends upon the pulsatile secretion of GnRH from hypothalamic neuroendocrine cells. In the female, however, the frequency and amplitude of these pulses during a 24-hour period change over the course of the menstrual cycle. So does the responsiveness both of the anterior pituitary to GnRH and of the ovaries to FSH and LH.

Let us look first at the patterns of hormone concentrations in systemic plasma during a normal menstrual cycle (**Figure 17–18**). (GnRH is not shown because its important concentration is not in systemic plasma, but rather in the plasma within the hypothalamo-pituitary portal vessels.) In Figure 17–18, the lines are plots of average daily concentrations; that is, the increases and decreases during a single day stemming from episodic secretion have been averaged. For now, ignore both the legend and circled numbers in this figure because we are concerned here only with hormonal patterns and not the explanations of these patterns.

FSH increases in the early part of the follicular phase and then steadily decreases throughout the remainder of the cycle except for a small midcycle peak. LH is constant during most of the follicular phase but then shows a very large midcycle increase—the **LH surge**—peaking approximately 18 h *before* ovulation. This is followed by a rapid decrease and then a further slow decline during the luteal phase.

After remaining fairly low and stable for the first week, estrogen increases rapidly during the second week as the dominant ovarian follicle grows and secretes more estrogen. Estrogen then starts decreasing shortly before LH has peaked.

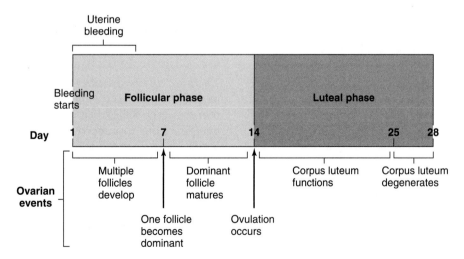

Figure 17–17

Summary of ovarian events during a menstrual cycle (if fertilization does not occur). The first day of the cycle is named for a uterine event—the onset of bleeding—even though ovarian events are used to denote the cycle phases.

Figure 17–18

Summary of systemic plasma hormone concentrations and ovarian events during the menstrual cycle. The events marked by the circled numbers are described later in the text and are listed here to provide a summary. The arrows in this legend denote causality. ❶ FSH and LH secretion increase (because plasma estrogen concentration is low and exerting little negative feedback). → ❷ Multiple antral follicles begin to enlarge and secrete estrogen. → ❸ Plasma estrogen concentration begins to rise. ❹ One follicle becomes dominant (probably by chance) and secretes very large amounts of estrogen. → ❺ Plasma estrogen level increases markedly. → ❻ FSH secretion and plasma FSH concentration decrease, causing atresia of nondominant follicles, but then ❼ increasing plasma estrogen exerts a "positive" feedback on gonadotropin secretion. → ❽ An LH surge is triggered. → ❾ The egg completes its first meiotic division and cytoplasmic maturation while the follicle secretes less estrogen accompanied by some progesterone, ❿ ovulation occurs, and ⓫ the corpus luteum forms and begins to secrete large amounts of both estrogen and progesterone. → ⓬ Plasma estrogen and progesterone increase. → ⓭ FSH and LH secretion are inhibited and their plasma concentrations decrease. ⓮ The corpus luteum begins to degenerate (cause unknown) and decrease its hormone secretion. → ⓯ Plasma estrogen and progesterone concentrations decrease. → ⓰ FSH and LH secretion begin to increase, and a new cycle begins (back to ❶).

Figure 17–18 physiological *inquiry* p*i*

■ 1. Why do blood FSH levels increase at the end of the luteal phase ⓰? 2. What naturally occurring event could rescue the corpus luteum and prevent its degeneration starting in the middle of the luteal phase?

Answers can be found at end of chapter.

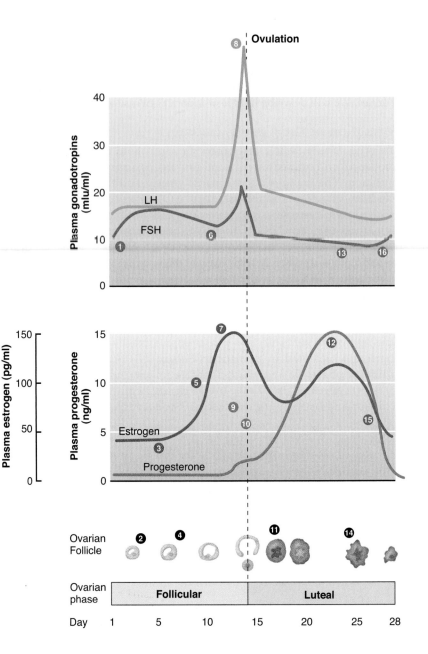

This is followed by a second increase, due to secretion by the corpus luteum, and finally, a rapid decrease during the last days of the cycle. Very small amounts of progesterone are released by the ovaries during the follicular phase until just before ovulation. Very soon after ovulation, the developing corpus luteum begins to release large amounts of progesterone, and from this point the progesterone pattern is similar to that for estrogen.

Not shown in Figure 17–18 is the plasma concentration of inhibin. Its pattern is similar to that of estrogen: it increases during the late follicular phase, remains high during the luteal phase, and then decreases as the corpus luteum degenerates.

The following discussion will explain how these hormonal changes are interrelated to produce a self-cycling pattern. The numbers in Figure 17–18 are keyed to the text. The feedback effects of the ovarian hormones to be described in the text are summarized for reference in **Table 17–4**.

Follicle Development and Estrogen Synthesis During the Early and Middle Follicular Phases

There are always a number of preantral and early antral follicles in the ovary between puberty and menopause. Further development of the follicle beyond these stages requires stimulation by FSH. Prior to puberty, the plasma concentration of FSH is too low to induce such development. This changes during puberty, and menstrual cycles commence. The increase in FSH secretion that occurs as one cycle ends and the next begins (numbers ⓰ to ❶ in Figure 17–18) provides this stimulation, and a group of preantral and early antral follicles enlarge ❷. The increase in FSH at the end of the cycle (⓰ to ❶) is due to decreased progesterone, estrogen, and inhibin (removal of negative feedback).

During the next week or so, there is a division of labor between the actions of FSH and LH on the follicles: FSH acts on the granulosa cells, and LH acts on the theca cells.

Table 17–4	Summary of Major Feedback Effects of Estrogen, Progesterone, and Inhibin

1. **Estrogen,** in **low plasma concentrations,** causes the anterior pituitary to secrete less FSH and LH in response to GnRH and also may inhibit the hypothalamic neurons that secrete GnRH.

   Result: Negative feedback inhibition of FSH and LH secretion during the early and middle follicular phase.

2. **Inhibin** acts on the pituitary to inhibit the secretion of FSH.

   Result: Negative feedback inhibition of FSH secretion throughout the cycle.

3. **Estrogen, when increasing dramatically,** causes anterior pituitary cells to secrete more LH and FSH in response to GnRH. Estrogen can also stimulate the hypothalamic neurons that secrete GnRH.

   Result: Positive feedback stimulation of the LH surge, which triggers ovulation.

4. High plasma concentrations of **progesterone,** in the presence of estrogen, inhibit the hypothalamic neurons that secrete GnRH.

   Result: Negative feedback inhibition of FSH and LH secretion and prevention of LH surges during the luteal phase and pregnancy.

The reason is that, at this point in the cycle, granulosa cells have FSH receptors but no LH receptors, and theca cells have just the reverse. FSH stimulates the granulosa cells to multiply and produce estrogen, and it also stimulates enlargement of the antrum. Some of the estrogen produced diffuses into the blood and maintains a relatively stable plasma concentration ❸. Estrogen also functions as a paracrine/autocrine agent within the follicle, where, along with FSH and growth factors, it stimulates the proliferation of granulosa cells, which further increases estrogen production.

The granulosa cells, however, require help to produce estrogen because they are deficient in the enzymes required to produce the androgen precursors of estrogen (Chapter 11). The granulosa cells are aided by the theca cells. As shown in **Figure 17–19**, LH acts upon the theca cells, stimulating them not only to proliferate but also to synthesize androgens. The androgens diffuse into the granulosa cells and are converted to estrogen by aromatase. Thus, the secretion of estrogen by the granulosa cells requires the interplay of both types of follicle cells and both pituitary gonadotropins.

At this point, it is worthwhile to emphasize the similarities that the two types of follicle cells bear to cells of the testes during this period of the cycle: the granulosa cell is similar to the Sertoli cell in that it controls the microenvironment in which the germ cell develops and matures, and it is stimulated by both FSH and the major gonadal sex hormone. The theca cell is similar to the Leydig cell in that it produces mainly androgens and is stimulated to do so by LH.

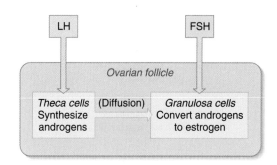

Figure 17–19

Control of estrogen synthesis during the early and middle follicular phases. (The major androgen secreted by the theca cells is androstenedione.) Androgen diffusing from theca to granulosa cell passes through the basement membrane (not shown).

By the beginning of the second week, one follicle has become dominant (number ❹ in Figure 17–18), and the other developing follicles degenerate. The reason for this is that, as shown in Figure 17–18, the plasma concentration of FSH, a crucial factor necessary for the survival of the follicle cells, begins to decrease, and there is no longer enough FSH to prevent atresia. But why, then, does the dominant follicle survive? There are several reasons why this follicle, having gained a head start, is able to keep going. First, its granulosa cells have achieved a greater sensitivity to FSH because of increased numbers of FSH receptors. Second, its granulosa cells now begin to be stimulated not only by FSH but by LH as well. We emphasized in the previous section that, during the first week or so of the follicular phase, LH acts only on the theca cells. As the dominant follicle matures, this situation changes, and LH receptors, induced by FSH, also begin to appear in large numbers on the granulosa cells. The increase in local estrogen within the follicle results from these factors.

The dominant follicle now starts to secrete enough estrogen that the plasma concentration of this steroid begins to increase ❺. We can now also explain why plasma FSH starts to decrease at this time. Estrogen, at these still relatively low concentrations, is exerting a *negative feedback* inhibition on the secretion of gonadotropins (Table 17–4 and **Figure 17–20**). A major site of estrogen action is the anterior pituitary, where it reduces the amount of FSH and LH secreted in response to any given amount of GnRH. Estrogen probably also acts on the hypothalamus to decrease the amplitude of GnRH pulses and, hence, the total amount of GnRH secreted over any time period.

Therefore, as expected from this negative feedback, the plasma concentration of FSH (and LH, to a lesser extent) begins to decrease as a result of the increasing level of estrogen as the follicular phase continues (❻ in Figure 17–18). One reason that FSH decreases more than LH is that the granulosa cells also secrete inhibin, which, as in the male, inhibits mainly the secretion of FSH (see Figure 17–20).

LH Surge and Ovulation

The inhibitory effect of estrogen on gonadotropin secretion occurs when plasma estrogen concentration is relatively low, as during the early and middle follicular phases. In contrast,

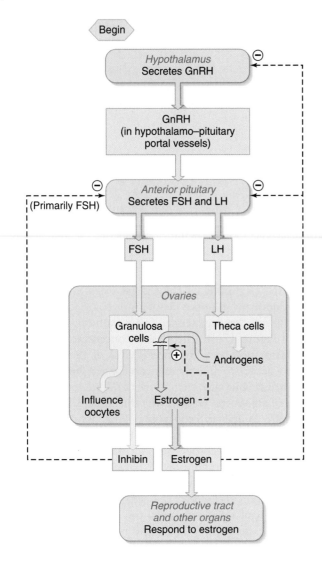

Figure 17–20

Summary of hormonal control of ovarian function during the early and middle follicular phases. Compare with the analogous pattern of the male (see Figure 17–11). Inhibin is a protein hormone that inhibits FSH secretion. The wavy broken arrows in the granulosa cells denote the conversion of androgens to estrogen in these cells, as shown in Figure 17–19. The dotted line within the ovaries indicates that estrogen increases granulosa cell function (local positive feedback).

Figure 17–20 physiological *inquiry*

- A 30-year-old woman has failed to have menstrual cycles for the past few months; her pregnancy test is negative. Her plasma FSH and LH levels are elevated whereas her plasma estrogen levels are low. What is the likely cause of her failure to menstruate?

Answer can be found at end of chapter.

increasing plasma concentrations of estrogen for one to two days, as occurs during the estrogen peak of the late follicular phase (❼ in Figure 17–18), act upon the pituitary to enhance the sensitivity of LH-releasing mechanisms to GnRH (Table 17–4 and **Figure 17–21**) and may also stimulate GnRH release from the hypothalamus. These are the positive-feedback effects of estrogen.

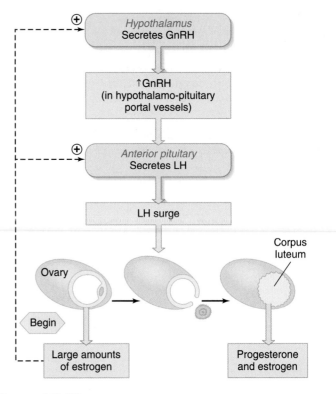

Figure 17–21

In the late follicular phase, the dominant follicle secretes large amounts of estrogen, which act on the anterior pituitary and, possibly, the hypothalamus to cause an LH surge. The increased plasma LH then triggers both ovulation and formation of the corpus luteum. These actions of LH are mediated via the granulosa cells.

The net result is that rapidly increasing estrogen leads to the LH surge (❽ in Figure 17–18). As shown in Figure 17–18 ❾, an increase in FSH and progesterone also occurs at the time of the LH surge, but it is not known whether this has a physiological role in the regulation of the cycle.

The midcycle surge of LH is the primary event that induces ovulation. The high plasma concentration of LH acts upon the granulosa cells to cause the events, presented in **Table 17–5**, that culminate in ovulation ❿, as indicated by the dashed vertical line in Figure 17–18.

The function of the granulosa cells in mediating the effects of the LH surge is the last in the series of these cells' functions described in this chapter. They are all summarized in **Table 17–6**.

The Luteal Phase

The LH surge not only induces ovulation by the mature follicle but also stimulates the reactions that transform the remaining granulosa and theca cells of that follicle into a corpus luteum (⓫ in Figure 17–18). A low but adequate LH concentration maintains the function of the corpus luteum for about 14 days.

During its short life in the nonpregnant woman, the corpus luteum secretes large quantities of progesterone and estrogen ⓬, as well as inhibin. In the presence of estrogen, the high plasma concentration of progesterone causes a decrease in the secretion of the gonadotropins by the pituitary. It probably does

Table 17–5 — Effects of the LH Surge on Ovarian Function

1. The primary oocyte completes its first meiotic division and undergoes cytoplasmic changes that prepare the ovum for implantation should fertilization occur. These LH effects on the oocyte are mediated by messengers released from the granulosa cells in response to LH.

2. Antrum size (fluid volume) and blood flow to the follicle increase markedly.

3. The granulosa cells begin releasing progesterone and decreasing the release of estrogen, which accounts for the midcycle decrease in plasma estrogen concentration and the small rise in plasma progesterone just before ovulation.

4. Enzymes and prostaglandins, synthesized by the granulosa cells, break down the follicular-ovarian membranes. These weakened membranes rupture, allowing the oocyte and its surrounding granulosa cells to be carried out onto the surface of the ovary.

5. The remaining granulosa cells of the ruptured follicle (along with the theca cells of that follicle) are transformed into the corpus luteum, which begins to release progesterone and estrogen.

Table 17–6 — Functions of Granulosa Cells

1. Nourish oocyte

2. Secrete chemical messengers that influence the oocyte and the theca cells

3. Secrete antral fluid

4. Are the site of action for estrogen and FSH in the control of follicle development during early and middle follicular phases

5. Express aromatase, which converts androgen (from theca cells) to estrogen

6. Secrete inhibin, which inhibits FSH secretion via an action on the pituitary

7. Are the site of action for LH induction of changes in the oocyte and follicle culminating in ovulation and formation of the corpus luteum

this by acting on the hypothalamus to *suppress* the secretion of GnRH. (The progesterone also prevents any LH surges during the first half of the luteal phase despite the high concentrations of estrogen at this time.) The increase in plasma inhibin concentration in the luteal phase also contributes to the suppression of FSH secretion. Consequently, during the luteal phase of the cycle, plasma concentrations of the gonadotropins are very low ⓭. The feedback suppression of gonadotropins in the luteal phase is summarized in **Figure 17–22**.

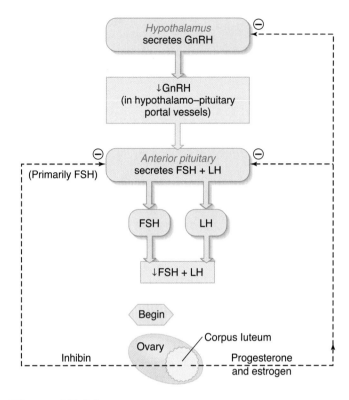

Figure 17–22

Suppression of FSH and LH during luteal phase. If implantation of a developing conceptus does not occur and hCG does not appear in the blood, the corpus luteum dies, progesterone and estrogen decrease, menstruation occurs, and the next menstrual cycle begins.

The corpus luteum has a finite life in the absence of an increase in gonadotropin secretion. Thus, it degenerates within two weeks if pregnancy does not occur ⓮.

With degeneration of the corpus luteum, plasma progesterone and estrogen concentrations decrease ⓯. The secretion of FSH and LH (and probably GnRH, as well) increase (⓰ and ❶) as a result of being freed from the inhibiting effects of high concentrations of ovarian hormones. The cycle then begins anew.

This completes the description of the control of ovarian function. It should be emphasized that, although the hypothalamus and anterior pituitary are essential controllers, events within the *ovary* are the real sources of timing for the cycle. When the ovary secretes enough estrogen, the LH surge is induced, which in turn causes ovulation. When the corpus luteum degenerates, the decrease in hormone secretion allows the gonadotropin levels to increase enough to promote the growth of another group of follicles. Thus, ovarian events, via hormonal feedbacks, control the hypothalamus and anterior pituitary.

Uterine Changes in the Menstrual Cycle

The phases of the menstrual cycle can also be described in terms of uterine events (**Figure 17–23**). Day 1 is, as noted earlier, the first day of menstrual flow, and the entire duration

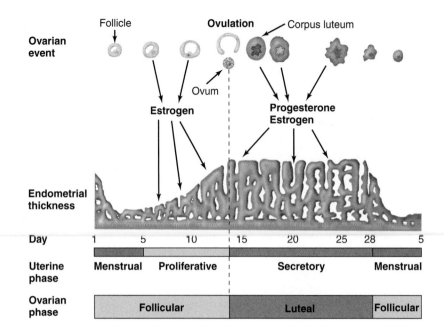

Figure 17–23

Relationships between ovarian and uterine changes during the menstrual cycle. Refer to Figure 17–18 for specific hormonal changes.

of menstruation is known as the **menstrual phase** (generally about 3 to 5 days in a typical 28-day cycle). During this time, the epithelial lining of the uterus—the **endometrium**—degenerates, resulting in the menstrual flow. The menstrual flow then ceases, and the endometrium begins to thicken as it regenerates under the influence of estrogen. This period of growth, the **proliferative phase,** lasts for the 10 days or so between cessation of menstruation and the occurrence of ovulation. Soon after ovulation, the endometrium increases secretory activity under the influence of progesterone and estrogen. Thus, the part of the menstrual cycle between ovulation and the onset of the next menstruation is called the **secretory phase.**

As shown in Figure 17–23, the ovarian follicular phase includes the uterine menstrual and proliferative phases, whereas the ovarian luteal phase is the same as the uterine secretory phase.

The uterine changes during a menstrual cycle are caused by changes in the plasma concentrations of estrogen and progesterone (see Figure 17–23). During the proliferative phase, an increasing plasma estrogen level stimulates growth of both the endometrium and the underlying uterine smooth muscle (**myometrium**). In addition, it induces the synthesis of receptors for progesterone in endometrial cells. Then, following ovulation and formation of the corpus luteum (during the secretory phase), progesterone acts upon this estrogen-primed endometrium to convert it to an actively secreting tissue. The endometrial glands become coiled and filled with glycogen, the blood vessels become more numerous, and enzymes accumulate in the glands and connective tissue. These changes are essential to make the endometrium a hospitable environment for implantation and nourishment of the developing embryo.

Progesterone also inhibits myometrial contractions, in large part by opposing the stimulatory actions of estrogen and

locally generated prostaglandins. This is very important to ensure that a fertilized egg, once it arrives in the uterus, will not be swept out by uterine contractions before it can implant in the wall. Uterine quiescence is maintained by progesterone throughout pregnancy and is essential to prevent premature delivery.

Estrogen and progesterone also have important effects on the secretion of mucus by the cervix. Under the influence of estrogen alone, this mucus is abundant, clear, and watery. All of these characteristics are most pronounced at the time of ovulation and allow sperm deposited in the vagina to move easily through the mucus on their way to the uterus and fallopian tubes. In contrast, progesterone, present in significant concentrations only after ovulation, causes the mucus to become thick and sticky—in essence a "plug" that prevents bacteria from entering the uterus from the vagina. The antibacterial blockage protects the uterus and the fetus if conception has occurred.

The decrease in plasma progesterone and estrogen levels that results from degeneration of the corpus luteum deprives the highly developed endometrium of its hormonal support and causes menstruation. The first event is profound constriction of the uterine blood vessels, which leads to a diminished supply of oxygen and nutrients to the endometrial cells. Disintegration starts in the entire lining, except for a thin, underlying layer that will regenerate the endometrium in the next cycle. Also, the uterine smooth muscle begins to undergo rhythmical contractions.

Both the vasoconstriction and uterine contractions are mediated by prostaglandins produced by the endometrium in response to the decrease in plasma estrogen and progesterone. The major cause of menstrual cramps, ***dysmenorrhea,*** is overproduction of these prostaglandins, leading to excessive uterine contractions. The prostaglandins also affect smooth muscle elsewhere in the body, which accounts for some of the systemic symptoms that sometimes accompany the cramps, such as nausea, vomiting, and headache.

After the initial period of vascular constriction, the endometrial arterioles dilate, resulting in hemorrhage through the weakened capillary walls. The menstrual flow consists of this blood mixed with endometrial debris. Typical blood loss per menstrual period is about 50 to 150 ml.

The major events of the menstrual cycle are summarized in **Table 17–7**. This table, in essence, combines the information in Figures 17–18 and 17–23.

Other Effects of Estrogen and Progesterone

Estrogen has other effects in addition to its paracrine function within the ovaries, its effects on the anterior pituitary and the hypothalamus, and its uterine actions. They are summarized in **Table 17–8**.

Progesterone also exerts a variety of effects (Table 17–8). Because plasma progesterone is markedly elevated only after ovulation has occurred, several of these effects can be used to indicate whether ovulation has taken place. First, progesterone inhibits proliferation of the cells lining the vagina. Second, there is a small increase (approximately 0.5°C) in body temperature that usually occurs after ovulation and persists throughout the luteal phase; this change is probably due to an action of progesterone on temperature regulatory centers in the brain.

Note that in its myometrial and vaginal effects, as well as several others listed in Table 17–8, progesterone exerts an "antiestrogen effect," probably by decreasing the number of estrogen receptors. In contrast, the synthesis of progesterone receptors is stimulated by estrogen in many tissues (for example, the endometrium), and so responsiveness to progesterone usually requires the presence of estrogen (**estrogen priming**).

Like all steroid hormones, both estrogen and progesterone act in the cell nucleus, and their biochemical mechanism of action is at the level of gene transcription.

In closing this section, brief mention should be made of the transient physical and emotional symptoms that appear in many women prior to the onset of menstrual flow and disappear within a few days after the start of menstruation. The symptoms—which may include painful or swollen breasts, headache, backache, depression, anxiety, irritability, and other physical, emotional, and behavioral changes—are often attributed to estrogen or progesterone excess. The plasma concentrations of these hormones, however, are usually normal in women having these symptoms, and their cause is not actually known. In order of increasing severity of symptoms, the overall problem is categorized as *premenstrual tension, premenstrual syndrome (PMS),* or *premenstrual dysphoric disorder (PMDD),* the last-named being so severe as to be temporarily disabling. These symptoms appear to result

Table 17–7	Summary of the Menstrual Cycle
Day(s)	**Major Events**
1–5	Estrogen and progesterone are low because the previous corpus luteum is regressing. *Therefore:* (a) Endometrial lining sloughs. (b) Secretion of FSH and LH is released from inhibition, and their plasma concentrations increase. *Therefore:* Several growing follicles are stimulated to mature.
7	A single follicle (usually) becomes dominant.
7–12	Plasma estrogen increases because of secretion by the dominant follicle. *Therefore:* Endometrium is stimulated to proliferate.
7–12	LH and FSH decrease due to estrogen and inhibin negative feedback. *Therefore:* Degeneration (atresia) of nondominant follicles occurs.
12–13	LH surge is induced by increasing plasma estrogen. *Therefore:* (a) Oocyte is induced to complete its first meiotic division and undergo cytoplasmic maturation. (b) Follicle is stimulated to secrete digestive enzymes and prostaglandins.
14	Ovulation is mediated by follicular enzymes and prostaglandins.
15–25	Corpus luteum forms and, under the influence of low but adequate levels of LH, secretes estrogen and progesterone, increasing plasma concentrations of these hormones. *Therefore:* (a) Secretory endometrium develops. (b) Secretion of FSH and LH is inhibited, lowering their plasma concentrations. *Therefore:* No new follicles develop.
25–28	Corpus luteum degenerates (if implantation of the conceptus does not occur). *Therefore:* Plasma estrogen and progesterone concentrations decrease. *Therefore:* Endometrium begins to slough at conclusion of day 28, and a new cycle begins.

Table 17–8	Some Effects of Female Sex Steroids

Estrogen

1. Stimulates growth of ovary and follicles (local effects)

2. Stimulates growth of smooth muscle and proliferation of epithelial linings of reproductive tract. In addition:
   a. Fallopian tubes: Increases contractions and ciliary activity.
   b. Uterus: Increases myometrial contractions and responsiveness to oxytocin. Stimulates secretion of abundant, watery cervical mucus. Prepares endometrium for progesterone's actions by inducing progesterone receptors.
   c. Vagina: Increases layering of epithelial cells.

3. Stimulates external genitalia growth, particularly during puberty

4. Stimulates breast growth, particularly ducts and fat deposition during puberty

5. Stimulates female body configuration development during puberty: narrow shoulders, broad hips, female fat distribution (deposition on hips and breasts)

6. Stimulates fluid secretion from lipid (sebum)-producing skin glands (sebaceous glands). (This "anti-acne" effect opposes the acne-producing effects of androgen.)

7. Stimulates bone growth and ultimate cessation of bone growth (closure of epiphyseal plates); protects against osteoporosis; does not have an anabolic effect on skeletal muscle

8. Vascular effects (deficiency produces "hot flashes")

9. Has feedback effects on hypothalamus and anterior pituitary (see Table 17–4)

10. Stimulates prolactin secretion but inhibits prolactin's milk-inducing action on the breasts

11. Protects against atherosclerosis by effects on plasma cholesterol (Chapter 16), blood vessels, and blood clotting (Chapter 12)

Progesterone

1. Converts the estrogen-primed endometrium to an actively secreting tissue suitable for implantation of an embryo

2. Induces thick, sticky cervical mucus

3. Decreases contractions of fallopian tubes and myometrium

4. Decreases proliferation of vaginal epithelial cells

5. Stimulates breast growth, particularly glandular tissue

6. Inhibits milk-inducing effects of prolactin

7. Has feedback effects on hypothalamus and anterior pituitary (see Table 17–4)

8. Increases body temperature

from a complex interplay between the sex steroids and brain neurotransmitters.

Androgens in Women

Androgens are present in the blood of women as a result of production by the adrenal glands and ovaries. These androgens play several important roles in the female, including stimulation of the growth of pubic hair, axillary hair, and, possibly, skeletal muscle, and maintenance of sex drive. Excess androgens may cause *virilism*: the female fat distribution disappears, a beard appears along with the male body hair distribution, the voice lowers in pitch, the skeletal muscle mass enlarges, the clitoris enlarges, and the breasts diminish in size.

Puberty

Puberty in females is a process similar to that in males (described earlier in this chapter). It usually starts earlier in girls (10 to 12 years old) than in boys. In the female, GnRH, the pituitary gonadotropins, and estrogen are all secreted at very low levels during childhood. For this reason, there is no follicle maturation beyond the early antral stage, and menstrual cycles do not occur. The female accessory sex organs remain small and nonfunctional, and there are minimal secondary sex characteristics. The onset of puberty is caused, in large part, by an alteration in brain function that increases the secretion of GnRH. This hypophysiotropic hormone in turn stimulates the secretion of pituitary gonadotropins, which stimulate follicle development and estrogen secretion. Estrogen, in addition to its critical role in follicle development, induces the changes in the accessory sex organs and secondary sex characteristics associated with puberty. **Menarche,** the first menstruation, is a late event of puberty (averaging 12.3 years of age in the United States).

As in males, the mechanism of the brain change that results in increased GnRH secretion in girls at puberty remains unknown. The brain may become less sensitive to the negative feedback effects of gonadal hormones at the time of puberty. Also, the adipose-tissue hormone leptin (Chapter 16) is known to stimulate the secretion of GnRH and may play a role in puberty. This may explain why the onset of puberty tends to correlate with the attainment of a certain level of energy stores (fat) in the girl's body. The effect of leptin on GnRH could also explain why girls who exercise extensively and are extremely thin (decreased leptin) often fail to initiate menstrual cycles at the expected age (delayed menarche). The onset of puberty is not abrupt but develops over several years, as evidenced by slowly rising plasma concentrations of the gonadotropins and testosterone or estrogen.

Female Sexual Response

The female response to sexual intercourse is characterized by marked increases in blood flow and muscular contraction in many areas of the body. For example, increasing sexual excitement is associated with vascular engorgement of the breasts and erection of the nipples, resulting from contraction of smooth muscle fibers in them. The clitoris, which has a rich supply of sensory nerve endings, increases in diameter and length as a result of increased blood flow. During intercourse, the blood flow to the vagina increases and the vaginal epithelium is lubricated by mucus.

Orgasm in the female, as in the male, is accompanied by pleasurable feelings and many physical events. There is a sudden increase in skeletal muscle activity involving almost all parts of the body; the heart rate and blood pressure increase, and there is a transient rhythmical contraction of the vagina and uterus. Orgasm seems to play a minimal role in ensuring fertilization because conception can occur in the absence of an orgasm.

Sexual desire in women is probably more dependent upon androgens, secreted by the adrenal glands and ovaries, than estrogen. Sex drive is also maintained beyond menopause, a time when estrogen levels become very low. New studies have suggested that low-dose androgen therapy may be useful for the treatment of decreased libido in women. These effects are mediated by a direct effect of androgen and by conversion of androgens to estrogen by aromatase in the brain.

Pregnancy

For pregnancy to occur, the introduction of sperm must occur between five days before and one day after ovulation. This is because the sperm, following their ejaculation into the vagina, remain capable of fertilizing an egg for up to four to six days, and the ovulated egg remains viable for only 24 to 48 h.

Egg Transport

At ovulation, the egg is extruded onto the surface of the ovary. Recall that the fimbriae at the ends of the fallopian tubes are lined with ciliated epithelium. At ovulation, the smooth muscle of the fimbriae causes them to pass over the ovary while the cilia beat in waves toward the interior of the duct. These ciliary motions sweep the egg into the fallopian tube as it emerges onto the ovarian surface.

Within the fallopian tube, egg movement, driven almost entirely by fallopian-tube cilia, is so slow that the egg takes about four days to reach the uterus. Thus, if fertilization is to occur, it must do so in the fallopian tube because of the short viability of the unfertilized egg.

Intercourse, Sperm Transport, and Capacitation

Ejaculation, described earlier in this chapter, results in deposition of semen into the vagina during intercourse. The act of intercourse itself provides some impetus for the transport of sperm out of the vagina to the cervix because of the fluid pressure of the ejaculate. Passage into the cervical mucus by the swimming sperm is dependent on the estrogen-induced changes in consistency of the mucus described earlier. Sperm can enter the uterus within minutes of ejaculation. Furthermore, the sperm can survive for up to a day or two within the cervical mucus, from which they can be released to enter the uterus. Transport of the sperm through the length of the uterus and into the fallopian tubes occurs via the sperm's own propulsions and uterine contractions.

The mortality rate of sperm during the trip is huge. One reason for this is that the vaginal environment is acidic, a protection against yeast and bacterial infections. Another is the length and energy requirements of the trip. Of the several hundred million sperm deposited in the vagina in an ejaculation, only about 100 to 200 reach the fallopian tube. This is one of the major reasons there must be so many sperm in the ejaculate for fertilization to occur.

Sperm are not able to fertilize the egg until they have resided in the female tract for several hours and been acted upon by secretions of the tract. This process, called **capacitation,** causes: (1) the previously regular wavelike beats of the sperm's tail to be replaced by a more whiplike action that propels the sperm forward in strong surges, and (2) the sperm's plasma membrane to become altered so that it will be capable of fusing with the surface membrane of the egg.

Fertilization

Fertilization begins with the fusion of a sperm and egg, usually within a few hours after ovulation. The egg usually must be fertilized within 24 to 48 hours of ovulation. Many sperm, after moving between the cumulus of granulosa cells still surrounding the egg, bind to the zona pellucida. The zona pellucida proteins function as receptors for sperm surface proteins. The sperm head has many of these proteins and so becomes bound simultaneously to many sperm receptors on the zona pellucida.

This binding triggers what is termed the **acrosome reaction** in the bound sperm: the plasma membrane of the sperm head is altered so that the underlying membrane-bound acrosomal enzymes are now exposed to the outside—that is, to the zona pellucida. The enzymes digest a path through the zona pellucida as the sperm, using its tail, advances through this coating. The first sperm to penetrate the entire zona pellucida and reach the egg's plasma membrane fuses with this membrane. The head of the sperm then slowly passes into the egg's cytoplasm.

Viability of the newly fertilized egg, now called a **zygote,** depends upon preventing the entry of additional sperm. A specific mechanism mediates this **block to polyspermy.** The initial fusion of the sperm and egg plasma membranes triggers a reaction that changes membrane potential, preventing additional sperm from binding. Subsequently, during the **cortical reaction,** cytosolic secretory vesicles located around the egg's periphery release their contents, by exocytosis, into the narrow space between the egg plasma membrane and the zona pellucida. Some of these molecules are enzymes that enter the zona pellucida and cause both inactivation of its sperm-binding sites and hardening of the entire zona pellucida. This prevents additional sperm from binding to the zona pellucida and those sperm already advancing through it from continuing.

The fertilized egg completes its second meiotic division over the next few hours, and the one daughter cell with practically no cytoplasm—the second polar body—is extruded and disintegrates. The two sets of chromosomes—23 from the egg and 23 from the sperm, which are surrounded by distinct membranes and are known as pronuclei—migrate to the center of the cell. During this period of a few hours, the DNA of the chromosomes in both pronuclei is replicated, the pronuclear membranes break down, the cell is ready to undergo a mitotic division, and fertilization is complete. Fertilization also triggers activation of the egg enzymes required for the ensuing cell divisions and embryogenesis. The major events of fertilization are summarized in **Figure 17–24**. If fertilization had not occurred, the egg would have slowly disintegrated and been phagocytized by cells lining the uterus.

Rarely, a fertilized egg remains in a fallopian tube and embeds itself in the tube wall. Even more rarely, a fertilized egg may move backwards out of the fallopian tube into the abdominal cavity, where implantation can occur. Both kinds of *ectopic pregnancies* cannot succeed, and surgery is necessary to end the pregnancy (unless there is a spontaneous abortion) because of the risk of maternal hemorrhage.

Early Development, Implantation, and Placentation

The events from the LH surge to the rescue of the corpus luteum are summarized in **Table 17–9**. The **conceptus**—a collective term for everything ultimately derived from the original zygote (fertilized egg) throughout the pregnancy—remains in the fallopian tube for three to four days. The major reason is that estrogen maintains the contraction of the smooth muscle near where the fallopian tube enters the wall of the uterus. As plasma progesterone levels rise, this smooth muscle relaxes and allows the conceptus to pass. During its stay in the fallopian tube, the conceptus undergoes a number of mitotic cell divisions, a process known as **cleavage.** These divisions, however, are unusual in that no cell growth occurs before each division.

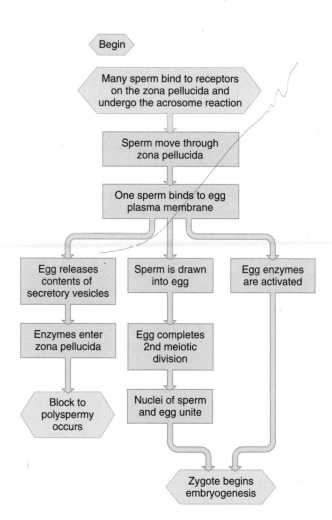

Figure 17–24

Events leading to fertilization, block to polyspermy, and the beginning of embryogenesis.

Table 17–9	Summary of Events from Ovulation and Fertilization to Implantation	
Days After LH Peak	**Event**	**Location**
1	Ovulation	Ovary
2	Fertilization	Fallopian tube
2–4	Cell division to ~32 cells	Fallopian tube
5	Blastocyst enters the uterine cavity	Uterus
6–7	Implantation	Uterus
9–10	Human chorionic gonadotropin (hCG) from implanted blastocyst (trophoblast cells) rescues corpus luteum (see Figure 17–29)	Trophoblast → maternal ovary

Thus, the 16- to 32-cell conceptus that reaches the uterus is essentially the same size as the original fertilized egg.

Each of these cells is **totipotent**—that is, has the capacity to develop into an entire individual. Therefore, identical (monozygotic) twins result when, at some point during cleavage, the dividing cells become completely separated into two independently growing cell masses. In contrast, fraternal (dizygotic) twins result when two eggs are ovulated and fertilized.

After reaching the uterus, the conceptus floats free in the intrauterine fluid, from which it receives nutrients, for approximately three days, all the while undergoing further cell divisions. Soon the conceptus reaches the stage known as a **blastocyst,** by which point the cells have lost their totipotentiality and have begun to differentiate. The blastocyst consists of an outer layer of cells called the **trophoblast, an inner cell mass,** and a central fluid-filled cavity (**Figure 17–25**). During subsequent development, the inner cell mass will give rise to the developing human—called an **embryo** during the first two months and a **fetus** after that—and some of the membranes associated with it. The trophoblast will surround the embryo and fetus throughout development and be involved in its nutrition as well as in the secretion of several important hormones.

The period during which the zygote develops into a blastocyst corresponds with days 14 to 21 of the typical menstrual cycle. During this period, the uterine lining is being prepared by progesterone (secreted by the corpus luteum) to receive the blastocyst. By approximately the twenty-first day of the cycle (that is, seven days after ovulation), **implantation**—the embedding of the blastocyst into the endometrium—begins (see Figure 17–25). The trophoblast cells are quite sticky, particularly in the region overlying the inner cell mass, and it is this portion of the blastocyst that adheres to the endometrium and initiates implantation.

The initial contact between blastocyst and endometrium induces rapid proliferation of the trophoblast, the cells of which penetrate between endometrial cells. Proteolytic enzymes secreted by the trophoblast allow the blastocyst to bury itself in the endometrial layer. The endometrium, too, is undergoing changes at the site of contact. Implantation requires communication—via several paracrine agents—between the blastocyst and the cells of the endometrium. Implantation is soon completed (**Figure 17–26**), and the nutrient-rich endometrial cells provide the metabolic fuel and raw materials required for early growth of the embryo.

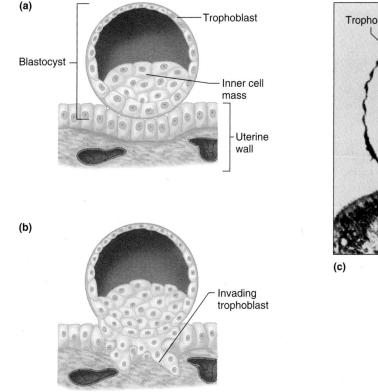

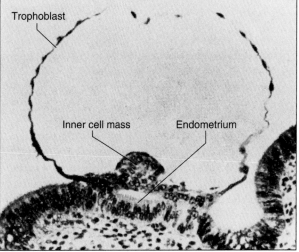

Figure 17–25

Contact (a) and implantation (b) of the blastocyst into the uterine wall at about 6–7 days after the previous LH peak. The trophoblast cells secrete hGC into the maternal circulation, which rescues the corpus luteum and maintains pregnancy. The trophoblast eventually develops into a component of the placenta. (c) Monkey blastocyst in contact with uterine wall (150×).

(c) Source: Section through a macaque blastocyst, after Heuser and Streeter. Reproduced with permission from R. O'Rahilly and F. Müller, *Human Embryology and Teratology,* Wiley-Liss, New York, 3rd Edition, 2001.

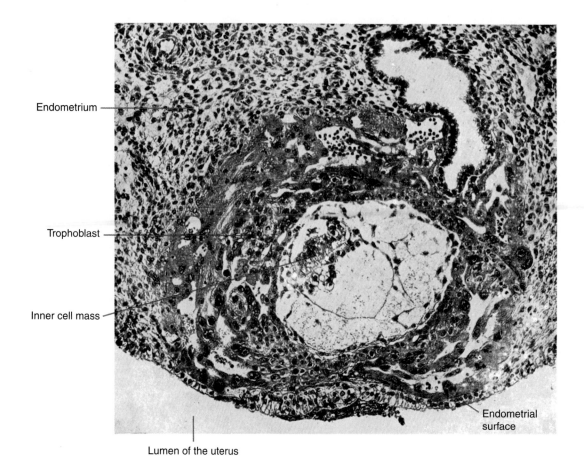

Endometrium

Trophoblast

Inner cell mass

Endometrial surface

Lumen of the uterus

Figure 17–26

Eleven-day human embryo completely embedded in the uterine lining.

From A. T. Hertig and J. Rock, *Carnegie Contrib. Embryol.* 29:127 (1941).

This simple nutritive system, however, is only adequate to provide for the embryo during the first few weeks, when it is very small. The structure that takes over this function is the **placenta,** a combination of interlocking fetal and maternal tissues that serves as the organ of exchange between mother and fetus for the remainder of the pregnancy.

The embryonic portion of the placenta is supplied by the outermost layers of trophoblast cells, the **chorion,** and the maternal portion by the endometrium underlying the chorion. Fingerlike projections of the trophoblast cells, called **chorionic villi,** extend from the chorion into the endometrium (**Figure 17–27**). The villi contain a rich network of capillaries that are part of the embryo's circulatory system. The endometrium around the villi is altered by enzymes and paracrine agents secreted from the cells of the invading villi so that each villus becomes completely surrounded by a pool, or **sinus,** of maternal blood supplied by maternal arterioles.

The maternal blood enters these placental sinuses via the uterine artery; the blood flows through the sinuses and then exits via the uterine veins. Simultaneously, blood flows from the fetus into the capillaries of the chorionic villi via the **umbilical arteries** and out of the capillaries back to the fetus via the **umbilical vein.** All of these umbilical vessels are contained in the **umbilical cord,** a long, ropelike structure that connects the fetus to the placenta.

Five weeks after implantation, the placenta has become well established, the fetal heart has begun to pump blood, and the entire mechanism for nutrition of the embryo and, subsequently, fetus and excretion of its waste products is in operation. A layer of epithelial cells in the villi and of endothelial cells in the fetal capillaries separate the maternal and fetal blood. Waste products move from blood in the fetal capillaries across these layers into the maternal blood, and nutrients, hormones, and growth factors move in the opposite direction. Some substances, such as oxygen and carbon dioxide, move by diffusion. Others, such as glucose, use transport proteins in the plasma membranes of the epithelial cells. Still other substances (e.g., several amino acids and hormones) are produced by the trophoblast layers of the placenta itself and added to the fetal and maternal blood. Note that there is an exchange of materials between the two bloodstreams but no actual mixing of the fetal and maternal blood. Umbilical veins carry oxygen and nutrient-rich blood from the placenta to the fetus, whereas umbilical arteries carry blood with waste products and a low oxygen content to the placenta.

Meanwhile, a space called the **amniotic cavity** has formed between the inner cell mass and the chorion (**Figure 17–28**).

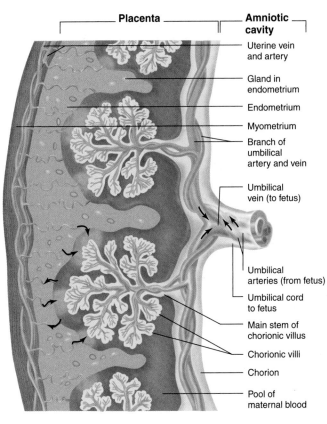

Figure 17-27

Interrelations of fetal and maternal tissues in the formation of the placenta. See Figure 17-28 for the orientation of the placenta.

From B. M. Carlson, *Patten's Foundations of Embryology,* 5th ed., McGraw-Hill, New York, 1988.

The epithelial layer lining the cavity is derived from the inner cell mass and is called the **amnion,** or **amniotic sac.** It eventually fuses with the inner surface of the chorion so that only a single combined membrane surrounds the fetus. The fluid in the amniotic cavity, the **amniotic fluid,** resembles the fetal extracellular fluid, and it buffers mechanical disturbances and temperature variations.

The fetus, floating in the amniotic cavity and attached by the umbilical cord to the placenta, develops into a viable infant during the next eight months. Note in Figure 17-28 that eventually only the amniotic sac separates the fetus from the uterine lumen.

Amniotic fluid can be sampled by *amniocentesis* as early as the sixteenth week of pregnancy. This is done by inserting a needle into the amniotic cavity. Some genetic diseases can be diagnosed by the finding of certain chemicals either in the fluid or in sloughed fetal cells suspended in the fluid. The chromosomes of these fetal cells can also be examined for diagnosis of certain disorders as well as to determine the sex of the fetus. Another technique for fetal diagnosis is *chorionic villus sampling.* This technique, which can be performed as early as 9 to 12 weeks of pregnancy, involves obtaining tissue from a chorionic villus of the placenta. This technique, however, carries a higher risk of inducing the loss of the fetus

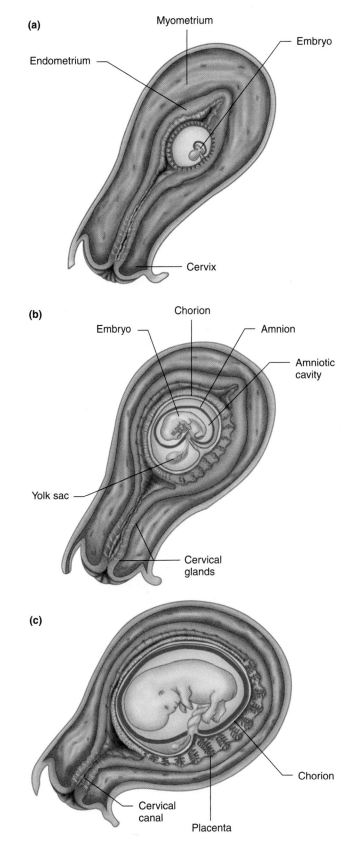

Figure 17-28

The uterus at (a) 3, (b) 5, and (c) 8 weeks after fertilization. Embryos and their membranes are drawn to actual size. Uterus is within actual size range. The yolk sac is formed from the trophoblast. It has no nutritional function in humans but is important in embryonic development.

From B. M. Carlson, *Patten's Foundations of Embryology,* 5th ed., McGraw-Hill, New York, 1988.

(*miscarriage*) than does amniocentesis. A third technique for fetal diagnosis is ultrasound, which provides a "picture" of the fetus without the use of x-rays. A fourth technique for screening for fetal abnormalities involves obtaining only *maternal* blood and analyzing it for several normally occurring proteins whose concentrations change in the presence of these abnormalities. For example, particular changes in the concentrations of two hormones produced during pregnancy—human chorionic gonadotropin and estriol—and alpha-fetoprotein (a major fetal plasma protein that crosses the placenta into the maternal blood) can identify many cases of **Down syndrome,** a genetic form of mental retardation associated with distinct facial and body features.

Maternal nutrition is crucial for the fetus. Malnutrition early in pregnancy can cause specific abnormalities that are **congenital;** that is, existing at birth. Malnutrition retards fetal growth and results in infants with higher-than-normal death rates, reduced growth after birth, and an increased incidence of learning disabilities and other medical problems. Specific nutrients, not just total calories, are also very important. For example, there is an increased incidence of neural defects in the offspring of mothers who are deficient in the B-vitamin folate (also called folic acid and folacin).

The developing embryo and fetus are also subject to considerable influences by a host of non-nutrient factors such as noise, radiation, chemicals, and viruses, to which the mother may be exposed. For example, drugs taken by the mother can reach the fetus via transport across the placenta and can impair fetal growth and development. In this regard, it must be emphasized that aspirin, alcohol, and the chemicals in cigarette smoke are very potent agents, as are illicit drugs such as cocaine. Any agent that can cause birth defects in the fetus is known as a **teratogen.**

Because half of the fetal genes—those from the father—differ from those of the mother, the fetus is in essence a foreign transplant in the mother. The integrity of the fetal-maternal blood barrier also protects the fetus from immunological attack by the mother.

Hormonal and Other Changes During Pregnancy

Throughout pregnancy, plasma concentrations of estrogen and progesterone continually increase (**Figure 17–29**). Estrogen stimulates the growth of the uterine muscle mass, which will eventually supply the contractile force needed to deliver the fetus. Progesterone inhibits uterine contractility so that the fetus is not expelled prematurely. During approximately the first two months of pregnancy, almost all the estrogen and progesterone are supplied by the corpus luteum.

Recall that if pregnancy had not occurred, the corpus luteum would have degenerated within two weeks after its formation. The persistence of the corpus luteum during pregnancy is due to a hormone called **human chorionic gonadotropin (hCG),** which the trophoblast cells start to secrete around the time they start their endometrial invasion. hCG gains entry to the maternal circulation, and the detection of this hormone in the mother's plasma and/or urine is used as

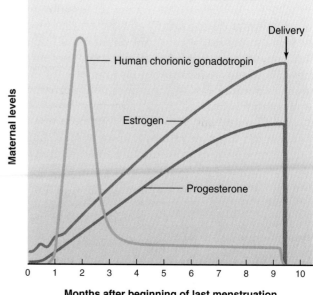

Months after beginning of last menstruation

Figure 17–29

Maternal levels of estrogen, progesterone, and human chorionic gonadotropin during pregnancy. Curves depicting hormone concentrations are not drawn to scale.

Figure 17–29 physiological *inquiry*

■ Why do progesterone and estrogen levels continue to increase during pregnancy even though human chorionic gonadotropin (hCG) decreases?

Answer can be found at end of chapter.

a test for pregnancy. This glycoprotein is very similar to LH, and it not only prevents the corpus luteum from degenerating but strongly stimulates its steroid secretion. Thus, the signal that preserves the corpus luteum comes from the conceptus, not the mother's tissues.

The secretion of hCG reaches a peak 60 to 80 days after the last menstruation (see Figure 17–29). It then decreases just as rapidly, so that by the end of the third month it has reached a low level that remains relatively constant for the duration of the pregnancy. Associated with this decrease in hCG secretion, the placenta begins to secrete large quantities of estrogen and progesterone. The very marked increases in plasma concentrations of estrogen and progesterone during the last six months of pregnancy are due entirely to their secretion by the trophoblast cells of the placenta, and the corpus luteum regresses after three months.

An important aspect of placental steroid secretion is that the placenta has the enzymes required for the synthesis of progesterone but not those needed for the formation of androgens, which are the precursors of estrogen. The placenta is supplied with androgens via the maternal ovaries and adrenal glands and by the *fetal* adrenal glands and liver. The placenta converts the androgens into estrogen by expressing the aromatase enzyme.

The secretion of GnRH and, therefore, of LH and FSH is powerfully inhibited by high concentrations of progesterone in the presence of estrogen. Both of these gonadal steroids are secreted in high concentrations by the corpus luteum and then by the placenta throughout pregnancy, so the secretion of the pituitary gonadotropins remains extremely low. As a consequence, there are no ovarian or menstrual cycles during pregnancy.

The trophoblast cells of the placenta produce not only hCG and steroids, but also inhibin and many other hormones that can influence the mother. One unique hormone that is secreted in very large amounts has effects similar to those of both prolactin and growth hormone. This protein hormone, **human placental lactogen,** mobilizes fats from adipose tissue and stimulates glucose production in the liver (growth-hormone-like) in the mother. It also stimulates breast development (prolactin-like) in preparation for lactation. Some of the many other physiological changes, hormonal and nonhormonal, in the mother during pregnancy are summarized in **Table 17–10**.

Approximately 5 to 10 percent of pregnant women retain too much fluid (edema) and have protein in the urine and hypertension. These are the symptoms of *preeclampsia;* when convulsions also occur, the condition is termed *eclampsia.* These two syndromes are collectively called *toxemia of pregnancy.* This

Table 17–10	Maternal Responses to Pregnancy
	Response
Placenta	Secretion of estrogen, progesterone, human chorionic gonadotropin, inhibin, human placental lactogen, and other hormones
Anterior pituitary	Increased secretion of prolactin Secretes very little FSH and LH
Adrenal cortex	Increased secretion of aldosterone and cortisol
Posterior pituitary	Increased secretion of vasopressin
Parathyroids	Increased secretion of parathyroid hormone
Kidneys	Increased secretion of renin, erythropoietin, and 1,25-dihydroxyvitamin D Retention of salt and water. *Cause:* Increased aldosterone, vasopressin, and estrogen
Breasts	Enlarge and develop mature glandular structure *Cause:* Estrogen, progesterone, prolactin, and human placental lactogen
Blood volume	Increased *Cause:* Total erythrocyte volume is increased by erythropoietin, and plasma volume by salt and water retention. However, plasma volume usually increases more than red cells, thereby leading to small decreases in hematocrit.
Bone turnover	Increased *Cause:* Increased parathyroid hormone and 1,25-dihydroxyvitamin D
Body weight	Increased by average of 12.5 kg, 60 percent of which is water
Circulation	Cardiac output increases, total peripheral resistance decreases (vasodilation in uterus, skin, breasts, GI tract, and kidneys), and mean arterial pressure stays constant
Respiration	Hyperventilation occurs (arterial P_{CO_2} decreases) due to the effects of increased progesterone
Organic metabolism	Metabolic rate increases Plasma glucose, gluconeogenesis, and fatty acid mobilization all increase. *Cause:* Hyporesponsiveness to insulin due to insulin antagonism by human placental lactogen and cortisol
Appetite and thirst	Increased (particularly after the first trimester)
Nutritional RDAs*	Increased

*RDA—Recommended daily allowance

can result in decreased growth rate and death of the fetus. The factors responsible for eclampsia are unknown, but the evidence strongly implicates abnormal vasoconstriction of the maternal blood vessels and inadequate invasion of the endometrium by trophoblast cells, resulting in poor blood perfusion of the placenta.

Pregnancy Sickness

Some women suffer from **pregnancy sickness** (popularly called morning sickness), which is characterized by nausea and vomiting during the first three months (first trimester) of pregnancy. The exact cause is unknown, but high concentrations of estrogen and other substances may be responsible.

Parturition

A normal human pregnancy lasts approximately 40 weeks, counting from the first day of the last menstrual cycle, or approximately 38 weeks from the day of ovulation and conception. Safe survival of premature infants is now possible at about the twenty-fourth week of pregnancy, but treatment of these infants often requires heroic efforts and often with significant deficits in the infant.

During the last few weeks of pregnancy, a variety of events occur in the uterus, culminating in the birth (delivery) of the infant, followed by the placenta. All of these events, including delivery, are termed **parturition.** Throughout most of pregnancy, the smooth muscle cells of the myometrium are relatively disconnected from each other, and the uterus is sealed at its outlet by the firm, inflexible collagen fibers that constitute the cervix. These features are maintained mainly by progesterone. During the last few weeks of pregnancy, as a result of ever-increasing levels of estrogen, the smooth muscle cells synthesize *connexin,* proteins that form gap junctions between the cells, which allows the myometrium to undergo coordinated contractions. Simultaneously, the cervix becomes soft and flexible due to an enzymatically mediated breakup of its collagen fibers. The synthesis of the enzymes is mediated by a variety of messengers, including estrogen and placental prostaglandins, the synthesis of which is stimulated by estrogen. The peptide hormone **relaxin** secreted by the ovaries is also involved. Estrogen has yet another important effect on the myometrium during this period: It induces the synthesis of receptors for the posterior pituitary hormone oxytocin, which is a powerful stimulator of uterine smooth muscle contraction.

Delivery is produced by strong rhythmical contractions of the myometrium. Actually, weak and infrequent uterine contractions begin at approximately 30 weeks and gradually increase in both strength and frequency. During the last month, the entire uterine contents shift downward so that the near-term fetus is brought into contact with the cervix. In over 90 percent of births, the baby's head is downward and acts as the wedge to dilate the cervical canal when labor begins (**Figure 17–30**). Occasionally, a baby is oriented with some other part of the body downward (***breech presentation***). This can require the surgical delivery of the fetus, placenta, and associated membranes through an abdominal and uterine incision (***cesarean section***).

At the onset of labor and delivery or before, the amniotic sac ruptures, and the amniotic fluid flows through the vagina. When labor begins in earnest, the uterine contractions become strong and occur at approximately 10- to 15-min intervals. The contractions begin in the upper portion of the uterus and sweep downward.

As the contractions increase in intensity and frequency, the cervix is gradually forced open (dilation) to a maximum diameter of approximately 10 cm (4 in). Until this point, the contractions have not moved the fetus out of the uterus. Now the contractions move the fetus through the cervix and vagina. At this time the mother, by bearing down to increase abdominal pressure, adds to the effect of uterine contractions to deliver the baby. The umbilical vessels and placenta are still functioning, so that the baby is not yet on its own, but within minutes of delivery both the umbilical vessels and the placental vessels completely constrict, stopping blood flow to the placenta. The entire placenta becomes separated from the underlying uterine wall, and a wave of uterine contractions delivers the placenta as the **afterbirth.**

Usually, parturition proceeds automatically from beginning to end and requires no significant medical intervention. In a small percentage of cases, however, the position of the baby or some maternal complication can interfere with normal delivery (e.g., breech presentation). The head-first position of the fetus is important for several reasons: (1) If the baby is not oriented head first, another portion of its body is in contact with the cervix and is generally a far less effective wedge. (2) Because of the head's large diameter compared with the rest of the body, if the body were to go through the cervical canal first, the canal might obstruct the passage of the head, leading to problems when the partially delivered baby tries to breathe. (3) If the umbilical cord becomes caught between the canal wall and the baby's head or chest, mechanical compression of the umbilical vessels can result. Despite these potential problems, however, many babies who are not oriented head first are born without significant difficulties.

What mechanisms control the events of parturition?

1. The autonomic neurons to the uterus are of little importance because anesthetizing them does not interfere with delivery.
2. The smooth muscle cells of the myometrium have inherent rhythmicity and are capable of autonomous contractions, which are facilitated as the muscle is stretched by the growing fetus.
3. The pregnant uterus near term and during labor secretes several prostaglandins (PGE_2 and $PGF_{2\alpha}$) that are potent stimulators of uterine smooth muscle contraction.
4. **Oxytocin,** one of the hormones released from the posterior pituitary, is an extremely potent uterine muscle stimulant. It not only acts directly on uterine smooth muscle but also stimulates it to synthesize the prostaglandins. Oxytocin is reflexly secreted from the posterior pituitary as a result of neural input to the hypothalamus, originating from receptors in the uterus, particularly the cervix. Also, as noted previously, the

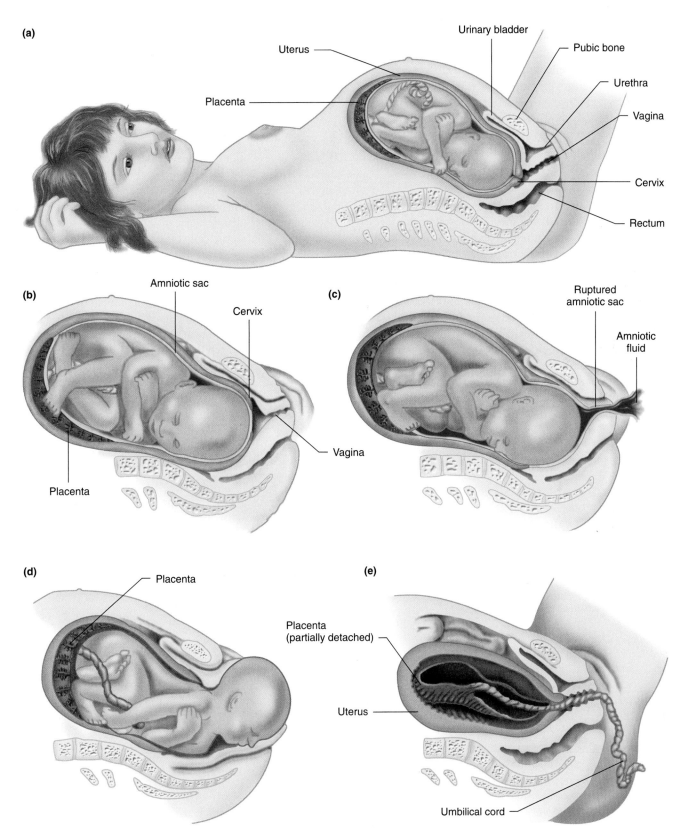

(a)
Uterus
Placenta
Urinary bladder
Pubic bone
Urethra
Vagina
Cervix
Rectum

(b)
Amniotic sac
Cervix
Placenta
Vagina

(c)
Ruptured amniotic sac
Amniotic fluid

(d)
Placenta

(e)
Placenta (partially detached)
Uterus
Umbilical cord

Figure 17–30

Stages of parturition. (a) Parturition has not yet begun. (b) The cervix is dilating. (c) The cervix is completely dilated, and the fetus's head is entering the cervical canal; the amniotic sac has ruptured and the amniotic fluid escapes. (d) The fetus is moving through the vagina. (e) The placenta is coming loose from the uterine wall in preparation for its expulsion.

number of oxytocin receptors in the uterus increases during the last few weeks of pregnancy. Thus, the contractile response to any given plasma concentration of oxytocin is greatly increased at parturition.

5. Throughout pregnancy, progesterone exerts an essential powerful inhibitory effect upon uterine contractions by decreasing the sensitivity of the myometrium to estrogen, oxytocin, and prostaglandins. Unlike the situation in many other species, however, the rate of progesterone secretion does not decrease before or during parturition in women (until after delivery of the placenta, the source of the progesterone); therefore, progesterone withdrawal does not play a role in parturition.

These mechanisms are shown in a unified pattern in **Figure 17–31**. Once started, the uterine contractions exert a positive feedback effect upon themselves via both local facilitation of inherent uterine contractions and reflex stimulation of oxytocin secretion. But precisely what the relative importance of all these factors is in *initiating* parturition remains unclear. One hypothesis is that the feto-placental unit, rather than the mother, is the source of the initiating signals to start parturition. That is, the fetus begins to outstrip the ability of the placenta to supply oxygen and nutrients, and to remove waste products. This leads to the fetal production of hormonal signals like ACTH. Another theory is that a "placental clock" acting via placental production of CRH signals the fetal production of ACTH. Either way, ACTH-mediated changes in

fetal adrenal steroid production seem to be an important signal to the mother to begin parturition.

The actions of prostaglandins on parturition are the last in a series of prostaglandin effects on the female reproductive system. They are summarized in **Table 17–11**.

Lactation

The secretion of milk by the breasts, or **mammary glands,** is termed **lactation.** The breasts contain ducts that branch all through the tissue and converge at the nipples (**Figure 17–32**). These ducts start in saclike glands called **alveoli** (the same term is used to denote the lung air sacs). The breast alveoli, which are the sites of milk secretion, look like bunches of grapes with stems terminating in the ducts. The alveoli and the ducts immediately adjacent to them are surrounded by specialized contractile cells called **myoepithelial cells.**

Before puberty, the breasts are small with little internal glandular structure. With the onset of puberty in females, the increased estrogen causes a marked enhancement of duct growth and branching but relatively little development of the alveoli, and much of the breast enlargement at this time is due to fat deposition. Progesterone secretion also commences at puberty during the luteal phase of each cycle, and this hormone contributes to breast growth by stimulating the growth of alveoli.

During each menstrual cycle, the breasts undergo fluctuations in association with the changing blood concentrations of estrogen and progesterone. These changes are small

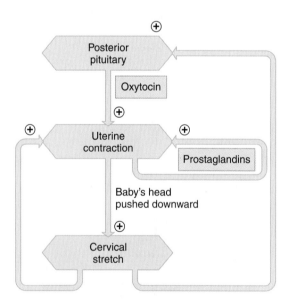

Figure 17–31

Factors stimulating uterine contractions during parturition. Note the positive feedback nature of several of the inputs.

Figure 17–31 physiological *inquiry* 🅟

- If a full-term fetus is oriented feet-first in the uterus, parturition may not proceed in a timely manner. Why?

Answer can be found at end of chapter.

Table 17–11	Some Effects of Prostaglandins* on the Female Reproductive System	
Site of Production	**Action of Prostaglandins**	**Result**
Late antral follicle	Stimulate production of digestive enzymes	Rupture of follicle
Corpus luteum	May interfere with hormone secretion and function	Death of corpus luteum
Uterus	Constrict blood vessels in endometrium	Onset of menstruation
	Cause changes in endometrial blood vessels and cells early in pregnancy	Facilitates implantation
	Increase contraction of myometrium	Helps to initiate both menstruation and parturition
	Cause cervical ripening	Facilitates cervical dilation during parturition

*The term *prostaglandins* is used loosely here, as is customary in reproductive physiology, to include all the eicosanoids.

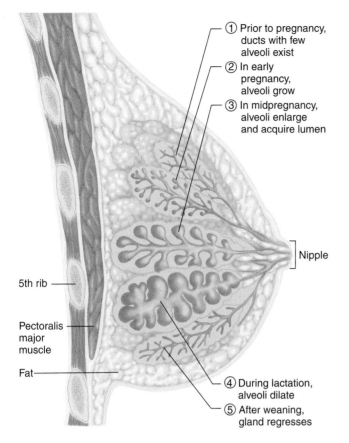

Figure 17–32

Anatomy of the breast. The numbers refer to the sequential changes that occur over time.

Adapted from Elias et al.

① Prior to pregnancy, ducts with few alveoli exist

② In early pregnancy, alveoli grow

③ In midpregnancy, alveoli enlarge and acquire lumen

Nipple

5th rib

Pectoralis major muscle

Fat

④ During lactation, alveoli dilate

⑤ After weaning, gland regresses

compared with the marked breast enlargement that occurs during pregnancy as a result of the stimulatory effects of high plasma concentrations of estrogen, progesterone, prolactin, and human placental lactogen. Except for prolactin, which is secreted by the maternal anterior pituitary, these hormones are secreted by the placenta. Under the influence of these hormones, both the ductal and the alveolar structures become fully developed.

As described in Chapter 11, other factors influence the anterior pituitary cells that secrete prolactin. They are inhibited by **dopamine,** which is secreted by the hypothalamus. They are probably stimulated by at least one **prolactin-releasing factor (PRF),** also secreted by the hypothalamus (the chemical identity of PRF is still uncertain). The dopamine and PRF secreted by the hypothalamus are hypophysiotropic hormones that reach the anterior pituitary by way of the hypothalamo-pituitary portal vessels.

Under the dominant inhibitory influence of dopamine, prolactin secretion is low before puberty. It then increases considerably at puberty in girls but not in boys, stimulated by the increased plasma estrogen concentration that occurs at this time. During pregnancy, there is a marked further increase in prolactin secretion due to stimulation by estrogen.

Prolactin is the major hormone stimulating the production of milk. However, despite the fact that prolactin is elevated and the breasts are markedly enlarged and fully developed as pregnancy progresses, there is usually no secretion of milk. This is because estrogen and progesterone, in large concentrations, prevent milk production by inhibiting this action of prolactin on the breasts. Thus, although estrogen causes an increase in the secretion of prolactin and acts with prolactin in promoting breast growth and differentiation, it, along with progesterone, is antagonistic to prolactin's ability to induce milk production. Delivery removes the source—the placenta—of the large amounts of estrogen and progesterone and, thereby, releases milk production from inhibition.

The decrease in estrogen following parturition also causes *basal* prolactin secretion to decrease from its peak late-pregnancy levels. After several months, prolactin returns toward prepregnancy levels even if the mother continues to nurse. Superimposed upon this basal level, however, are large secretory bursts of prolactin during each nursing period. The episodic pulses of prolactin are signals to the breasts to maintain milk production. These pulses usually cease several days after the mother completely stops nursing her infant, but continue as long as nursing continues.

The reflexes mediating the surges of prolactin (**Figure 17–33**) are initiated by afferent input to the hypothalamus from nipple receptors stimulated by suckling. This input's major effect is to inhibit the hypothalamic neurons that release dopamine.

One other reflex process is essential for nursing. Milk is secreted into the lumen of the alveoli, but the infant cannot suck the milk out of the breast. It must first be moved into the ducts, from which it can be sucked. This movement is called the **milk ejection reflex** (also called milk letdown) and is accomplished by contraction of the myoepithelial cells surrounding the alveoli. The contraction is under the control of oxytocin, which is reflexly released from posterior pituitary neurons in response to suckling (see Figure 17–33). Higher brain centers can also exert an important influence over oxytocin release: a nursing mother may actually leak milk when she hears her baby cry or even thinks about nursing.

Suckling also inhibits the hypothalamo-pituitary-ovarian axis at a variety of steps, with a resultant block of ovulation. This is probably due to increased prolactin and direct effects on the hypothalamic GnRH release. If suckling is continued at high frequency, ovulation can be delayed for months to years. This "natural" birth control may help to space out pregnancies. When supplements are added to the baby's diet and the frequency of suckling is decreased, however, most women will resume ovulation even though they continue to nurse. However, ovulation may resume even without a decrease in nursing. Failure to use adequate birth control may result in an unplanned pregnancy in nursing women.

Initially after delivery, the breasts secrete a watery, protein-rich fluid called **colostrum.** After about 24 to 48 hours, the secretion of milk itself begins. Milk contains four major nutrients: water, protein, fat, and the carbohydrate lactose (milk sugar).

Colostrum and milk also contain antibodies and other messengers of the immune system, all of which are important

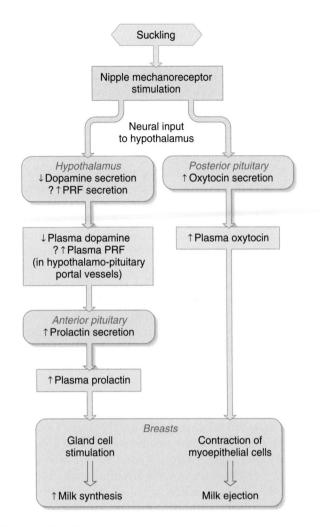

Figure 17–33

Major controls of the secretion of prolactin and oxytocin during nursing. The importance of PRF (prolactin-releasing factors) is not known (indicated by ?).

for the protection of the newborn, as well as for longer-term activation of the child's own immune system. Milk also contains many growth factors and hormones thought to help in tissue development and maturation, as well as a large number of neuropeptides and endogenous opioids that may subtly shape the infant's brain and behavior. Some of these substances are synthesized by the breasts themselves, not just transported from blood to milk. The reasons the milk proteins can gain entry to the newborn's blood are that (1) the low gastric acidity of the newborn does not denature them, and (2) the newborn's intestinal epithelium is more permeable to proteins than is the adult epithelium.

Unfortunately, infectious agents, including the virus that causes AIDS, can be transmitted through breast milk, as can some drugs. For example, the concentration of alcohol in breast milk is approximately the same as in maternal plasma.

Breast-feeding for at least the first 6 to 12 months is strongly advocated. In less-developed countries, where alternative formulas are often either contaminated or nutritionally inadequate because of improper dilution or inadequate refrigeration, breast-feeding significantly reduces infant sickness and mortality. In the United States, effects on infant survival are not usually apparent, but breast-feeding reduces the severity of gastrointestinal infections, has positive effects on mother-infant interaction, is economical, and has long-term health benefits. Cow's milk has many but not all of the constituents of mother's milk, but often in very different concentrations, and it is very difficult to reproduce mother's milk in a commercial formula.

Contraception

Physiologically, pregnancy is said to begin not at fertilization but after implantation is complete, approximately one week *after* fertilization. Birth control procedures that work prior to implantation are called **contraceptives** (**Table 17–12**). Procedures that cause the death of the embryo or fetus after implantation are called **abortifacients.**

Some forms of contraception, such as vasectomy, tubal ligation, vaginal diaphragms, vaginal caps, spermicides, and condoms, prevent sperm from reaching the egg. (In addition, condoms significantly reduce the risk of **sexually transmitted diseases [STDs]** such as AIDS, syphilis, gonorrhea, chlamydia, and herpes.)

Oral contraceptives are based on the fact that estrogen and progesterone can inhibit pituitary gonadotropin release, thereby preventing ovulation. One type of oral contraceptive is a combination of a synthetic estrogen and a progesterone-like substance (a progestogen or progestin). Another type is the so-called mini-pill, which contains only the progesterone-like substance. In actuality, the oral contraceptives, particularly the minipill, do not always prevent ovulation, but they are still effective because they have other contraceptive effects. For example, progestogens affect the composition of the cervical mucus, reducing the ability of the sperm to pass through the cervix, and they also inhibit the estrogen-induced proliferation of the endometrium, making it inhospitable for implantation.

Another method of delivering a contraceptive progestogen is via tiny capsules that are implanted beneath the skin and last for five years. Yet another method is the intramuscular injection of a different progestogen substance (e.g., **Depo-Provera**® or **Lunelle**®) every 1 to 3 months. Alternate methods of providing highly efficacious hormonal contraception include skin patches and vaginal rings.

The **intrauterine device (IUD)** works beyond the point of fertilization but before implantation has begun or is complete. The presence of one of these small objects in the uterus somehow interferes with the endometrial preparation for acceptance of the blastocyst.

In addition to the methods used before intercourse (precoital contraception), there are a variety of drugs used within 72 h *after* intercourse (postcoital or emergency contraception). These most commonly interfere with ovulation, transport of the conceptus to the uterus, or implantation. One approach is a high dose of estrogen, or two large doses (12 h apart) of a combined estrogen-progestin oral contraceptive. Another approach has used the drug **RU 486 (mifepristone),** which has antiprogesterone activity because it binds competitively

Table 17–12　Some Forms of Contraception

Method	First Year Failure Rate*	Physiological Mechanism of Effectiveness
Barrier Methods 　Condoms (♂ and ♀) 　Diaphragm/cervical cap (♀)	12%	Prevent sperm from entering uterus
Spermicides (♀)	20%	Kill sperm in the vagina (after insemination)
Sterilization 　Vasectomy (♂) 　Tubal ligation (♀)	<0.5%	Prevents sperm from becoming part of seminal fluid Prevents sperm from reaching egg
Intrauterine device (IUD) (♀)	3%	Prevents implantation of blastocyst
Estrogens and/or Progestins 　Oral contraceptive pill (♀) 　Emergency oral contraception (♀) 　Injectable or implantable progestins (♀) 　Transdermal (skin patch) (♀) 　Vaginal ring (♀)	 3% 1% <0.5% <1% <1%	Prevent ovulation by suppressing LH surge (negative feedback); thicken cervical mucus (prevents sperm from entering uterus); alter endometrium to prevent implantation of blastocyst

*From Hall, J. E., Infertility and Fertility Control, *Harrison's Principles of Internal Medicine,* McGraw-Hill, 2004. Rosen, M., and Cedars, M. I. Female Reproductive Endocrinology and Infertility. *Basic and Clinical Endocrinology,* 7th ed., McGraw-Hill, 2004; ACOG Practice Bulletin. Emergency Contraception. *Obstet. Gynecol.* 2005, 106:1443–52. See also www.fda.gov. Failure rate assumes proper and consistent use.

Notes

Spermicides are often used in combination with diaphragm/cervical cap and condoms.

Only condoms are effective in preventing sexually transmitted diseases.

Rhythm method (abstinence around time of ovulation) and coitus interruptus (withdrawal) are not listed because they are not reliable.

Only total abstinence is 100% effective in preventing pregnancy.

to progesterone receptors in the uterus but does not activate them. Antagonism of progesterone's effects causes the endometrium to erode and the contractions of the fallopian tubes and myometrium to increase. RU 486 can also be used later in pregnancy as an abortifacient.

The ***rhythm method*** uses abstention from sexual intercourse near the time of ovulation. Unfortunately, it is difficult to time ovulation precisely, even with laboratory techniques. For example, the small rise in body temperature or change in vaginal epithelium, all of which are indicators of ovulation, occur only *after* ovulation. This combined with the marked variability of the time of ovulation in many women—from day 5 to day 15 of the cycle—explains why the rhythm method has a high failure rate.

There are still no effective chemical agents for male contraception. One potential approach is to decrease gonadotropin levels, which would decrease spermatogenesis. Testosterone would then have to be given to maintain libido.

Infertility

Approximately 12 percent of men and women of reproductive age in the United States are infertile. The numbers of infertile men and women are approximately equal until about age 30, after which infertility becomes more prevalent in women. In many cases, infertility can be successfully treated with drugs, artificial insemination, or corrective surgery.

When the cause of infertility cannot be treated, it can sometimes be circumvented in women by the technique of ***in vitro fertilization.*** First, the woman is injected with drugs that stimulate multiple egg production. Immediately before ovulation, at least one egg is then removed from the ovary via a needle inserted into the ovary through the top of the vagina or the lower abdominal wall. The egg is placed in a dish for several days with sperm. After the fertilized egg has developed into a cluster of two to eight cells, it is then transferred to the woman's uterus. The success rate of this procedure, when one egg is transferred, is only about 15 to 20 percent.

Menopause

Around the age of 50, on the average, menstrual cycles become less regular. Ultimately they cease entirely, and this cessation is known as **menopause.** The phase of life beginning with menstrual irregularity is termed **perimenopause.** It involves many physical and emotional changes accompanying the cessation of reproductive function.

Menopause and the irregular function leading to it are caused primarily by ovarian failure. The ovaries lose their ability to respond to the gonadotropins, mainly because most, if not all, ovarian follicles and eggs have disappeared by this time through atresia. The hypothalamus and anterior pituitary continue to function relatively normally as demonstrated by the fact that the gonadotropins are secreted in greater amounts. The main reason for this is that the decreased plasma estrogen and inhibin do not exert as much negative feedback on gonadotropin secretion.

A small amount of estrogen usually persists in plasma beyond menopause, mainly from the peripheral conversion of adrenal androgens to estrogen by aromatase, but the level is inadequate to maintain estrogen-dependent tissues. The breasts and genital organs gradually atrophy to a large degree. Thinning and dryness of the vaginal epithelium can cause sexual intercourse to be painful. Because estrogen is a potent bone-protective hormone, significant decreases in bone mass may occur (**osteoporosis**). This results in an increased risk of bone fractures in postmenopausal women. The **hot flashes** so typical of menopause are periodic sudden feelings of warmth, dilation of the skin arterioles, and marked sweating. In addition, the incidence of cardiovascular disease increases after menopause. Women have much less coronary artery disease than men until after menopause, when the incidence becomes similar in both sexes, a pattern that is due to the protective effects of estrogen: estrogen exerts beneficial actions on plasma cholesterol and also exerts multiple direct protective actions on vessel walls. Recent studies, however, have questioned the long-term protective effects of estrogen in the prevention of heart disease.

Many of the symptoms associated with menopause as well as the development of osteoporosis can be reduced by the administration of estrogen. Recent studies also indicate that estrogen use may reduce the risk of developing Alzheimer's disease and may also be useful in the treatment of this disease.

The desirability of administering estrogen to postmenopausal women is controversial, however, because estrogen administration increases the risk of developing uterine endometrial cancer and breast cancer. The increased risk of endometrial cancer can be reduced by the administration of a progestogen along with estrogen, but the progestogen does not influence the risk of breast cancer. The progestogen only slightly lessens estrogen's protective effect against coronary artery disease.

Relevant to the question of hormone replacement therapy (as well as to the hormonal treatment of breast and uterine cancer) is the development of drugs (for example, **tamoxifen**) that exert some proestrogenic and some antiestrogenic effects. These drugs are collectively termed **selective estrogen receptor modulators (SERMs)** because they activate estrogen receptors in certain tissues but not in others; moreover, in these latter tissues SERMs act as estrogen antagonists. Obviously, the ideal would be to have a SERM that has the pro-estrogenic effects of protecting against osteoporosis, heart attacks, and Alzheimer's disease, but opposes the development of breast and uterine cancers. What makes this type of SERM possible? One important contributor is that there are two distinct forms of estrogen receptors, which are affected differentially by different SERMs.

ADDITIONAL CLINICAL EXAMPLES

Dysmenorrhea, pregnancy sickness, infertility, and postmenopausal osteoporosis have already been discussed. There are several other clinical conditions that further illustrate important physiological principles related to reproduction.

Amenorrhea

Amenorrhea is defined as a failure to have a normal menstruation for approximately six months. The two general types are primary amenorrhea (the failure to have a normal menarche) and secondary amenorrhea (the loss of previously normal menstrual cycles).

There are many reasons why a girl may fail to initiate normal menarche. It is generally believed that a phenotypic female who does not have the onset of menses by age 16 should be evaluated. One cause of amenorrhea, the androgen insensitivity syndrome (testicular feminization) was discussed in Section A of this chapter. Other possible causes of primary amenorrhea are ovarian enzyme defects and lack of normal pituitary function.

The most common cause of secondary amenorrhea is pregnancy. Menstrual cycles cease after normal implantation of the blastocyst due primarily to hCG-induced increases in estrogen and progesterone production from the corpus luteum, which suppress maternal pituitary gonadotropin secretion. Another cause of secondary amenorrhea is **hyperprolactinemia,** which occurs for similar reasons in women as in men (discussed in the previous section). Tumors of the lactotrophs of the pituitary hypersecrete prolactin, which suppresses LH and FSH secretion. Thus, menstrual cycles cannot continue since gonadotropin levels are low. This is often accompanied by **galactorrhea**—inappropriate milk production—because prolactin stimulates the mammary gland. Hyperprolactinemia can be treated with dopamine agonists because prolactin is primarily under the inhibitory control of hypothalamic dopamine.

Excessive exercise and **anorexia nervosa** (self-imposed starvation) can cause primary or secondary amenorrhea. There are a variety of theories for why this is so. One unifying theory is that the brain can sense a loss of body fat, possibly via decreased leptin levels, and that this leads the hypothalamus to cease GnRH pulses. From a teleological view, this makes sense because pregnant women must supply a large caloric input to the developing fetus and a lack of body fat would indicate inadequate caloric capacity. The prepubertal appearance of female gymnasts who have minimal body fat indicates hypogonadism and probably amenorrhea, which can persist for many years after menarche would normally take place.

Congenital Adrenal Hyperplasia

Congenital adrenal hyperplasia is caused by adrenal androgen overproduction in the fetus. It is almost always caused by a partial defect in the ability of the fetal adrenal gland to synthesize cortisol due to a mutation in an enzyme in the cortisol synthetic pathway (**Figure 17–34**). The

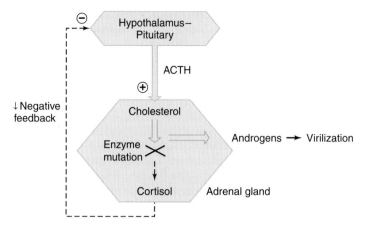

Figure 17–34

Mechanism of virilization in female fetuses with congenital adrenal hyperplasia. An enzyme defect (usually partial) in the steroidogenic pathway leads to decreased production of cortisol and a shift of precursors into the adrenal androgen pathway. Because cortisol negative feedback is decreased, ACTH release from the fetal pituitary increases. Although cortisol can eventually be normalized, it is at the expense of ACTH-stimulated adrenal hypertrophy and excess fetal adrenal androgen production.

decrease in cortisol in the fetal blood leads to an increase in plasma ACTH, due to a loss of glucocorticoid negative feedback. The increase in fetal ACTH stimulates the fetal adrenal cortex to try to make more cortisol. Remember, the adrenal cortex can synthesize androgens from the same precursor as cortisol. ACTH stimulation results in an increase in androgen production because the precursors cannot be efficiently converted to cortisol. This increase in fetal androgen production results in **virilization** of an XX fetus (masculinized external genitalia). If untreated in the fetus, the XX baby is usually born with ***ambiguous genitalia***—it

is not obvious whether the baby is a phenotypic boy or girl. These babies require treatment with cortisol replacement.

We have already discussed amniocentesis and chorionic villus sampling in the previous section. These techniques can be used to obtain a sample of fetal cells. If a woman has previously had a baby with congenital adrenal hyperplasia, her subsequent pregnancies can be screened. If an XX fetus is found to have one of the several possible mutations in the enzymes of the pathway that synthesizes cortisol, the mother can be given low doses of synthetic glucocorticoids, which cross the placenta and suppress fetal ACTH release. This lowers fetal adrenal androgen production and allows normal development of the external genitalia.

Precocious Puberty

Precocious puberty is defined as the premature appearance of secondary sex characteristics. The age of the normal onset of puberty is controversial, although it is generally thought that pubertal onset before the age of 6 to 7 in girls and 8 to 9 in boys should be investigated. Precocious puberty is usually caused by an increase in gonadal steroid production. This leads to an early onset of the puberty growth spurt, maturation of the skeleton, breast development (in girls), and enlargement of the genitalia in boys. Therefore, these children are usually taller at an early age. However, because gonadal steroids also stop the pubertal growth spurt by inducing epiphyseal closure, final adult height is usually less than predicted.

Although there are a variety of causes for the premature increase in gonadal steroids, *true* (or complete) precocious puberty is caused by the premature activation of GnRH and LH/FSH secretion. This is often caused by tumors or infections in the area of the central nervous system that control GnRH release. Treatments that decrease LH and FSH release are important to allow normal development. ■

■ SECTION C SUMMARY

Anatomy

I. The female internal genitalia are the ovaries, fallopian tubes, uterus, cervix, and vagina.

II. The female external genitalia include the mons pubis, labia, clitoris, and vestibule of the vagina. These are also called the vulva.

Ovarian Functions

I. The female gonads, the ovaries, produce eggs and secrete estrogen, progesterone, and inhibin.

II. The two meiotic divisions of oogenesis result in each ovum having 23 chromosomes, in contrast to the 46 of the original oogonia.

III. The follicle consists of the egg, inner layers of granulosa cells surrounding the egg, and outer layers of theca cells.

IV. At the beginning of each menstrual cycle, a group of preantral and early antral follicles continues to develop, but soon only the dominant follicle continues its development to full maturity and ovulation.

V. Following ovulation, the remaining cells of that follicle differentiate into the corpus luteum, which lasts about 10 to 14 days if pregnancy does not occur.

VI. The menstrual cycle can be divided, according to ovarian events, into a follicular phase and a luteal phase, which last approximately 14 days each and are separated by ovulation.

Control of Ovarian Function

I. The menstrual cycle results from a finely tuned interplay of hormones secreted by the ovaries, the anterior pituitary, and the hypothalamus.

II. During the early and middle follicular phases, FSH stimulates the granulosa cells to proliferate and secrete estrogen, and LH stimulates the theca cells to proliferate and produce the androgens that the granulosa cells use to make estrogen.

a. During this time, estrogen exerts negative feedback on the anterior pituitary to inhibit the secretion of the gonadotropins. It probably also inhibits the secretion of GnRH by the hypothalamus.

b. Inhibin preferentially inhibits FSH secretion.

III. During the late follicular phase, plasma estrogen increases to elicit a surge of LH, which then causes, via the granulosa cells, completion of the egg's first meiotic division and cytoplasmic maturation, ovulation, and formation of the corpus luteum.

IV. During the luteal phase, under the influence of small amounts of LH, the corpus luteum secretes progesterone and estrogen. Regression of the corpus luteum results in a cessation of the secretion of these hormones.

V. Secretion of GnRH and the gonadotropins is inhibited during the luteal phase by the combination of progesterone, estrogen, and inhibin.

Uterine Changes in the Menstrual Cycle

I. The ovarian follicular phase is equivalent to the uterine menstrual and proliferative phases, the first day of menstruation being the first day of the cycle. The ovarian luteal phase is equivalent to the uterine secretory phase.

  a. Menstruation occurs when the plasma estrogen and progesterone levels decrease as a result of regression of the corpus luteum.

  b. During the proliferative phase, estrogen stimulates growth of the endometrium and myometrium and causes the cervical mucus to be readily penetrable by sperm.

  c. During the secretory phase, progesterone converts the estrogen-primed endometrium to a secretory tissue and makes the cervical mucus relatively impenetrable to sperm. It also inhibits uterine contractions.

Other Effects of Estrogen and Progesterone

I. The many effects of estrogen and progesterone are summarized in Table 17–8.

Androgens in Women

I. Androgens are produced in women and have several functions including growth of pubic and axillary hair.

II. Excess androgen can cause virilism.

Puberty

I. At puberty, the hypothalamic-anterior-pituitary-gonadal axis becomes active as a result of a change in brain function that permits increased secretion of GnRH.

II. The first signs of puberty are changes in nipples and the appearance of pubic or axillary hair.

Female Sexual Response

I. Sexual intercourse results in increases in blood flow and muscular contractions throughout the body.

II. Androgens appear to be important in libido in women.

Pregnancy

I. After ovulation, the egg is swept into the fallopian tube, where a sperm, having undergone capacitation and the acrosome reaction, fertilizes it.

II. Following fertilization, the egg undergoes its second meiotic division, and the nuclei of the egg and sperm fuse. Reactions in the ovum block penetration by other sperm and trigger cell division and embryogenesis.

III. The conceptus undergoes cleavage, eventually becoming a blastocyst, which implants in the endometrium on approximately day 7 after ovulation.

  a. The trophoblast gives rise to the fetal part of the placenta, whereas the inner cell mass develops into the embryo proper.

  b. Although they do not mix, fetal blood and maternal blood both flow through the placenta, exchanging gases, nutrients, hormones, waste products, and other substances.

  c. The fetus is surrounded by amniotic fluid in the amniotic sac.

IV. The progesterone and estrogen required to maintain the uterus during pregnancy come from the corpus luteum for the first two months of pregnancy, their secretion stimulated by human chorionic gonadotropin produced by the trophoblast.

V. During the last seven months of pregnancy, the corpus luteum regresses, and the placenta itself produces large amounts of progesterone and estrogen.

VI. The high levels of progesterone, in the presence of estrogen, inhibit the secretion of GnRH and thereby that of the gonadotropins, so that menstrual cycles do not occur.

VII. Delivery occurs by rhythmical contractions of the uterus, which first dilate the cervix and then move the infant, followed by the placenta, through the vagina. The contractions are stimulated in part by oxytocin, released from the posterior pituitary in a reflex triggered by uterine mechanoreceptors, and by uterine prostaglandins.

VIII. The breasts develop markedly during pregnancy as a result of the combined influences of estrogen, progesterone, prolactin, and human placental lactogen.

  a. Prolactin secretion is stimulated during pregnancy by estrogen acting on the anterior pituitary, but milk is not synthesized because high concentrations of estrogen and progesterone inhibit the milk-producing action of prolactin on the breasts.

  b. As a result of the suckling reflex, large bursts of prolactin and oxytocin are released during nursing. The prolactin stimulates milk production and the oxytocin causes milk ejection.

Menopause

I. Around the age of 50, a woman's menstrual cycles become less regular and ultimately disappear—menopause.

  a. The cause of menopause is a decrease in the number of ovarian follicles and their hyporesponsiveness to the gonadotropins.

  b. The symptoms of menopause are largely due to the marked decrease in plasma estrogen concentration.

Additional Clinical Examples

I. Amenorrhea is either the absence of menarche (primary) or the loss of normal menstrual cycles (secondary).

II. Congenital adrenal hyperplasia is caused by an adrenal enzyme defect and a shift of substrate to the androgen pathway. Androgen production can lead to virilization of an XX fetus.

III. Precocious puberty is the early onset of puberty often caused by LH and FSH overproduction.

■ S E C T I O N   C   R E V I E W   Q U E S T I O N S

1. Draw the female reproductive tract.
2. Describe the various stages from oogonium to mature ovum.
3. Describe the progression from a primordial follicle to a dominant follicle.
4. Name three hormones produced by the ovaries and name the cells that produce them.
5. Diagram the changes in plasma concentrations of estrogen, progesterone, LH, and FSH during the menstrual cycle.
6. What are the analogies between the granulosa cells and the Sertoli cells and between the theca cells and the Leydig cells?
7. List the effects of FSH and LH on the follicle.
8. Describe the effects of estrogen and inhibin on gonadotropin secretion during the early, middle, and late follicular phases.
9. List the effects of the LH surge on the egg and the follicle.
10. What are the effects of the sex steroids and inhibin on gonadotropin secretion during the luteal phase?
11. Describe the hormonal control of the corpus luteum and the changes that occur in the corpus luteum in a nonpregnant cycle and in a cycle when pregnancy occurs.
12. What happens to the sex steroids and the gonadotropins as the corpus luteum degenerates?
13. Compare the phases of the menstrual cycle according to uterine and ovarian events.
14. Describe the effects of estrogen and progesterone on the endometrium, cervical mucus, and myometrium.
15. Describe the uterine events associated with menstruation.
16. List the effects of estrogen on the accessory sex organs and secondary sex characteristics.
17. List the effects of progesterone on the breasts, cervical mucus, vaginal epithelium, and body temperature.
18. What are the sources and effects of androgens in women?
19. How does the egg get from the ovary to a fallopian tube?
20. Where does fertilization normally occur?
21. Describe the events that occur during fertilization.
22. How many days after ovulation does implantation occur, and in what stage is the conceptus at that time?
23. Describe the structure of the placenta and the pathways for exchange between maternal and fetal blood.

24. State the sources of estrogen and progesterone during different stages of pregnancy. What is the dominant estrogen of pregnancy, and how is it produced?
25. What is the state of gonadotropin secretion during pregnancy, and what is the cause?
26. What anatomical feature permits coordinated contractions of the myometrium?
27. Describe the mechanisms and messengers that contribute to parturition.
28. List the effects of prostaglandins on the female reproductive system.
29. Describe the development of the breasts after puberty and during pregnancy, and list the major hormones responsible.
30. Describe the effects of estrogen on the secretion and actions of prolactin during pregnancy.
31. Diagram the suckling reflex for prolactin release.
32. Diagram the milk ejection reflex.
33. List two main types of amenorrhea and give examples of each.
34. What is the state of estrogen and gonadotropin secretion before puberty and after menopause?
35. List the hormonal and anatomical changes that occur after menopause.
36. Why would glucocorticoids given to a pregnant woman help to treat her fetus with congenital adrenal hyperplasia?

Chapter 17 Test Questions

(Answers appear in Appendix A.)

1. Development of normal female internal and external genitalia requires
   a. Müllerian-inhibiting substance.
   b. expression of the *SRY* gene.
   c. insensitivity to circulating testosterone.
   d. complete absence of testosterone.
   e. absence of a Y chromosome.

2. Which is *not* characteristic of a normal postpubertal male?
   a. Inhibin from the Sertoli cells decreases FSH secretion.
   b. Testosterone has paracrine effects on the Sertoli cells.
   c. Testosterone stimulates GnRH from the hypothalamus.
   d. Testosterone inhibits LH secretion.
   e. GnRH from the hypothalamus is released in pulses.

Matching Questions (3–7)
Use each answer only once.
   a. Day 1 of menstrual cycle
   b. Day 7 of menstrual cycle
   c. Day 13 of menstrual cycle
   d. Day 23 of menstrual cycle
   e. Day 26 of menstrual cycle

3. Progesterone from the corpus luteum peaks.
4. Estrogen positive feedback is peaking.
5. One follicle becomes dominant.
6. Estrogen and progesterone are both decreasing.
7. Increase in FSH stimulates antral follicles to begin to secrete estrogen.

8. The Leydig cell is primarily characterized by
   a. aromatization of testosterone.
   b. secretion of inhibin.
   c. secretion of testosterone.
   d. expression of receptors only to FSH.
   e. transformation into the corpus luteum.

9. During the third trimester of pregnancy, the placenta is *not* the primary source of which hormone in maternal blood?
   a. estrogen   d. inhibin
   b. prolactin   e. hCG
   c. progesterone

10. Menopause is characterized primarily by
   a. primary ovarian failure.
   b. loss of estrogen secretion from the ovary due to a decrease in LH.
   c. loss of estrogen secretion from the ovary due to a decrease in FSH.
   d. a decrease in FSH and LH due to increased inhibin.
   e. a decrease in FSH and LH due to a decrease in GnRH pulses.

Chapter 17 Quantitative and Thought Questions

(Answers appear in Appendix A.)

1. What symptom will be common to a person whose Leydig cells have been destroyed and to a person whose Sertoli cells have been destroyed? What symptom will not be common?

2. A male athlete taking large amounts of an androgenic steroid becomes sterile (unable to produce sperm capable of causing fertilization). Explain.

3. A man who is sterile is found to have the following: no evidence of demasculinization, an increased blood concentration of FSH, and a normal plasma concentration of LH. What is the most likely basis of his sterility?

4. If you were a scientist trying to develop a male contraceptive acting on the anterior pituitary, would you try to block the secretion of FSH or LH? Explain the reason for your choice.

5. A 30-year-old man has very small muscles, a sparse beard, and a high-pitched voice. His plasma concentration of LH is elevated. Explain the likely cause of all these findings.

6. There are disorders of the adrenal cortex in which excessive amounts of androgens are produced. If this occurs in a woman, what will happen to her menstrual cycles?

7. Women with inadequate secretion of GnRH are often treated for their sterility with drugs that mimic the action of this hormone. Can you suggest a possible reason that such treatment is often associated with multiple births?

8. Which of the following would be a signal that ovulation is soon to occur: the cervical mucus becoming thick and sticky, an increase in body temperature, or a marked rise in plasma LH?

9. The absence of what phenomenon would interfere with the ability of sperm obtained by masturbation to fertilize an egg in a test tube?

10. If a woman seven months pregnant is found to have a marked decrease in plasma estrogen but a normal plasma progesterone for that time of pregnancy, what would you conclude?

11. What types of drugs might you work on if you were trying to develop one to stop premature labor?

12. If a genetic male failed to produce MIS during fetal life, what would the result be?

13. Could the symptoms of menopause be treated by injections of FSH and LH?

Chapter 17 Answers to Physiological Inquiries

Figure 17–3a These drugs would be absorbed by the pregnant woman and cross the placenta to enter the fetal circulation. These drugs would block production of dihydrotestosterone in target tissues with 5-alpha reductase activity, thereby interfering with the development of normal sexual differentiation of the penis, scrotum, and prostate in the male fetus.

Figure 17–11 Testosterone alone usually does not restore spermatogenesis to normal. FSH is necessary to stimulate spermatogenesis from the Sertoli cell independently of local testosterone production. Furthermore, giving testosterone as a drug is usually not sufficient to replace the local production of testosterone in the testes necessary to maintain spermatogenesis. Therefore, gonadotropins with a mixture of activity for receptors to LH (to stimulate local testosterone production) and FSH (to stimulate the Sertoli cells) usually must be given to restore spermatogenesis.

Figure 17–18 1. FSH increases because the corpus luteum is degenerating. The loss of the negative feedback by progesterone and estrogen from the corpus luteum relieves the pituitary of this inhibitory effect and allows FSH to increase, thus stimulating a group of follicles for the next menstrual cycle. 2. If conception occurs and the developing blastocyst implants (pregnancy), the trophoblast cells of the implanted blastocyst release a gonadotropin—human chorionic gonadotropin (hCG)—into the maternal blood, thus rescuing the corpus luteum in very early pregnancy. Production of progesterone from the corpus luteum of pregnancy prevents menses and the loss of the implanted embryo. The measurement of hCG in maternal blood or urine is the basis of the pregnancy test.

Figure 17–20 The elevated pituitary gonadotropins suggest a lack of estrogen and inhibin negative feedback, pointing to premature ovarian failure as a diagnosis. One cause of premature ovarian failure is autoimmune ovarian destruction. Like Graves' disease and Addison's disease (see Chapters 11 and 19), premature ovarian failure is a form of endocrine autoimmunity.

Figure 17–29 hCG stimulates progesterone and estrogen from the corpus luteum early in pregnancy. The placenta takes over this function during the second trimester of pregnancy such that most of the maternal estrogen and progesterone later in pregnancy is from the placenta. Placental production of these steroids does not require gonadotropin stimulation.

Figure 17–31 The feet may not provide sufficient cervical stretch to maintain the positive feedback stimulation of oxytocin and uterine contraction.

Defense Mechanisms of the Body

Human immunodeficiency viruses budding from a T cell.

immunology is the study of the physiological defenses by which the body (the host) recognizes itself from nonself (foreign matter). In the process, it destroys or neutralizes (renders harmless) foreign matter, both living and nonliving. In distinguishing self from nonself, immune defenses (1) protect against infection by **microbes**—viruses, bacteria, fungi, and parasites; (2) isolate or remove nonmicrobial foreign substances; and (3) destroy cancer cells that arise in the body, a function known as **immune surveillance.**

Immune defenses, or immunity, can be classified into two categories: nonspecific and specific, which interact with each other. **Nonspecific immune defenses** protect against foreign substances or cells without having to recognize their specific identities. The mechanisms of protection used by these defenses are not unique to the particular foreign substance or cell. **Specific immune defenses** (also called acquired immunity) depend upon

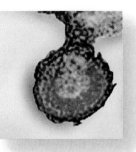

specific recognition, by lymphocytes, of the substance or cell to be attacked. This is followed by an attack unique for that substance or cell. Nonspecific and specific immune responses function in synchrony. For example, components of nonspecific immunity provide instructions that enable the specific responses to select appropriate targets to attack and to have strategies for their elimination.

Before introducing the cells that participate in immune defenses, let us first describe the microbes we will be most concerned with in this chapter—bacteria and viruses. These are the dominant infectious organisms in the United States and other industrialized nations. On a global basis, however, infections with parasitic eukaryotic organisms are responsible for a huge amount of illness and death. For example, several hundred million people now have malaria, a disease caused by infection of the *Plasmodium* species.

Bacteria are unicellular organisms that have an outer coating (the cell wall) in addition to a plasma membrane, but no intracellular membrane-bound organelles. Bacteria can damage tissues at the sites of bacterial replication, or they can release toxins that enter the blood and disrupt physiological functions in other parts of the body.

Viruses are essentially nucleic acids surrounded by a protein coat. Unlike bacteria, viruses lack both the enzyme machinery for metabolism and the ribosomes essential for protein synthesis. Thus, they cannot multiply by themselves but must exist inside other cells and use the biochemical apparatus of those cells. The viral nucleic acid directs the host cell to synthesize the proteins required for viral replication, with the required nucleotides and energy sources also supplied by the host cell. The effect of viral habitation and replication within a cell depends on the type of virus. After entering a cell, some viruses (the common cold virus, for example) multiply rapidly, kill the cell, and then move on to other cells. Other viruses, such as the one that causes genital herpes, can lie dormant in infected cells before suddenly undergoing the rapid replication that causes cell damage. Finally, certain viruses can transform their host cells into cancer cells.

Cells Mediating Immune Defenses

The cells that carry out immune defenses collectively make up the **immune system,** but they do not constitute a "system" in the sense of anatomically connected organs like the gastrointestinal and urinary systems. Instead, they are a diverse collection of cells found both in the blood and lymph and in tissues throughout the body. Because of the large number of cells and the far larger number of chemical messengers that participate in immune defenses, a miniglossary defining the cells and messengers discussed in this chapter is given toward the end of this chapter in Table 18–12.

The most numerous of the immune system cells are the various types of white blood cells collectively known as **leukocytes.** These include the neutrophils, basophils, eosinophils, monocytes, and lymphocytes. (The first three types are also grouped under the general term *polymorphonuclear granulocyte* because their cell nuclei have unusual, irregular appearances.) The anatomy, production, and blood concentrations of these cells were described in Chapter 12 and should be reviewed at this time. Unlike erythrocytes, leukocytes use the blood mainly for transportation and leave the circulatory system to enter the tissues where they function.

Plasma cells are not really a distinct cell line but differentiate from a particular set of lymphocytes (the B lymphocytes) during immune responses. Despite their name, plasma cells are primarily found in the tissues in which they differentiate from their parent lymphocytes. The major function of plasma cells is to synthesize and secrete antibodies.

Macrophages are found in virtually all organs and tissues, their structures varying somewhat from location to location. They are derived from monocytes that pass through the walls of blood vessels to enter the tissues and transform into macrophages. In keeping with one of their major functions, the engulfing of particles and microbes by endocytosis, macrophages are strategically placed where they will encounter their targets. For example, they are found in large numbers in the various epithelia in contact with the external environment, such as the skin and internal surfaces of respiratory and digestive system tubes. In several organs, they line the vessels through which blood or lymph flows.

There are several cell populations that are not macrophages and are not descended from monocytes, but exert various macrophage functions. These are collectively termed **macrophage-like** (or **dendritic**) **cells.** They are found scattered in almost all tissues.

Mast cells are found throughout connective tissues, particularly beneath the epithelial surfaces of the body. They are derived from the differentiation of a unique set of bone marrow cells that have entered the blood and then left the blood vessels to enter connective tissue, where they differentiate and undergo cell division. Thus, mature mast cells, unlike basophils, with which they share many characteristics, are not normally found in the blood. The most striking anatomical feature of mast cells is their very large number of cytosolic vesicles, which secrete locally acting chemical messengers such as histamine.

The sites of production and functions of all these cells are briefly listed in **Table 18–1** for reference and will be described in subsequent sections. Suffice it for now to emphasize two points. **Lymphocytes** serve as recognition cells in specific immune defenses and are essential for all aspects of these responses. Neutrophils, monocytes, macrophages, and dendritic cells have a variety of activities, but particularly important is their ability to secrete inflammatory mediators and to function as **phagocytes.** A phagocyte denotes any cell capable of **phagocytosis,** the form of endocytosis whereby the phagocytic cell engulfs and usually destroys particulate matter.

Cytokines

The cells of the immune system secrete a multitude (more than 100, to date) of protein messengers that regulate host cell division (mitosis) and function in both nonspecific and specific immune defenses. The collective term for these messengers, each of which has its own unique name, is **cytokines.** They are not produced by distinct specialized glands but rather by a variety of individual cells. The great majority of their actions occur at the site at which they are secreted, the cytokine acting as an autocrine/paracrine agent. In some cases, however, the cytokine circulates in the blood to exert hormonal effects on distant organs and tissues involved in host defenses.

Cytokines link the components of the immune system together. They are the chemical communication network that allows different immune system cells to "talk" to one another. This is called *cross-talk,* and it is essential for the precise timing of the functions of the immune system. Most cytokines are secreted by more than one type of immune system cell and by nonimmune cells as well (for example, by endothelial cells and fibroblasts). This often produces cascades of cytokine secretion, in which one cytokine stimulates the release of another, and so on. Any given cytokine may exert actions on an extremely broad range of target cells. For example, the cytokine interleukin 2 influences the function of virtually every cell of the immune system. There is great redundancy in cytokine action; that is, different cytokines can have very similar effects. Cytokines are also involved in many nonimmunological processes, such as bone formation and uterine function.

This chapter will be limited to a discussion of a few of the important cytokines and their most important functions, which are summarized in **Table 18–2.**

Nonspecific Immune Defenses

Nonspecific immune defenses protect against foreign cells or matter without having to recognize their specific identities. These defenses recognize some *general* property marking the invader as foreign. The most common such identity tags are often found in particular classes of carbohydrates or lipids that are in microbial cell walls. Plasma-membrane receptors on certain immune cells, as well as a variety of circulating proteins (particularly a family called complement) can bind to these carbohydrates and lipids at crucial steps in nonspecific responses. This use of a system based on carbohydrate and lipid for detecting the presence of foreign cells is a key feature that distinguishes nonspecific defenses from specific ones, which recognize foreign cells mainly by specific proteins the foreign cells produce.

The nonspecific immune defenses include defenses at the body surfaces, the response to injury known as inflammation, and a family of antiviral proteins called interferons.

Defenses at Body Surfaces

The body's first lines of defense against microbes are the barriers offered by surfaces exposed to the external environment. Very few microorganisms can penetrate the intact skin, and the various skin glands, salivary glands, and the lacrymal (tear) glands all secrete antimicrobial chemicals. These may include antibodies, enzymes such as lysozyme, which destroy bacterial cell walls, and an iron-binding protein called lactoferrin. This protein prevents bacteria from obtaining the iron they need to function properly.

The mucus secreted by the epithelial linings of the respiratory and upper gastrointestinal tracts also contains antimicrobial chemicals, but more important, mucus is sticky. Particles that adhere to it are prevented from entering the blood. They are either swept by ciliary action up into the pharynx and then swallowed, as occurs in the upper respiratory tract, or are phagocytized by macrophages in the various linings.

Other specialized surface defenses are the hairs at the entrance to the nose, the cough and sneeze reflexes, and the acid secretion of the stomach, which kills microbes.

Inflammation

Inflammation is the local response to infection or injury. The functions of inflammation are to destroy or inactivate foreign invaders and to set the stage for tissue repair. The key mediators are the cells that function as phagocytes. As noted earlier, the most important phagocytes are neutrophils, macrophages, and dendritic cells.

In this section, inflammation is described as it occurs in the nonspecific defenses induced by the invasion of microbes. Most of the same responses can be elicited by a variety of other injuries—cold, heat, and trauma, for example. Moreover, we will see later that inflammation accompanies many *specific* immune defenses in which the inflammation becomes amplified.

Table 18–1 Cells Mediating Immune Defenses

Name	Site Produced	Functions
Leukocytes (white blood cells)		
Neutrophils	Bone marrow	1. Phagocytosis 2. Release chemicals involved in inflammation (vasodilators, chemotaxins, etc.)
Basophils	Bone marrow	Carry out functions in blood similar to those of mast cells in tissues (see below)
Eosinophils	Bone marrow	1. Destroy multicellular parasites 2. Participate in immediate hypersensitivity reactions
Monocytes	Bone marrow	1. Carry out functions in blood similar to those of macrophages in tissues (see below) 2. Enter tissues and transform into macrophages
Lymphocytes	Mature in bone marrow (B cells and NK cells) and thymus (T cells); activated in peripheral lymphoid organs	Serve as recognition cells in specific immune responses and are essential for all aspects of these responses
B cells		1. Initiate antibody-mediated immune responses by binding specific antigens to the B cell's plasma membrane receptors, which are immunoglobulins 2. During activation are transformed into plasma cells, which secrete antibodies 3. Present antigen to helper T cells
Cytotoxic T cells (CD8 cells)		Bind to antigens on plasma membrane of target cells (virus-infected cells, cancer cells, and tissue transplants) and directly destroy the cells
Helper T cells (CD4 cells)		Secrete cytokines that help to activate B cells, cytotoxic T cells, NK cells, and macrophages
NK cells		1. Bind directly and nonspecifically to virus-infected cells and cancer cells and kill them 2. Function as killer cells in antibody-dependent cellular cytotoxicity (ADCC)
Plasma cells	Peripheral lymphoid organs; differentiate from B cells during immune responses	Secrete antibodies
Macrophages	Bone marrow; reside in almost all tissues and organs; differentiate from monocytes	1. Phagocytosis 2. Extracellular killing via secretion of toxic chemicals 3. Process and present antigens to helper T cells 4. Secrete cytokines involved in inflammation, activation and differentiation of helper T cells, and systemic responses to infection or injury (the acute phase response)
Dendritic (macrophage-like) cells	Almost all tissues and organs; microglia in the central nervous system	Same as macrophages
Mast cells	Bone marrow; reside in almost all tissues and organs; differentiate from bone marrow cells	Release histamine and other chemicals involved in inflammation

Table 18–2 Features of Selected* Cytokines

Cytokine	Source	Target Cells	Major Functions
Interleukin 1, tumor necrosis factor, and interleukin 6	Antigen-presenting cells such as macrophages	Helper T cells; certain brain cells; numerous systemic cells	Stimulate IL-2 secretion and IL-2 receptor expression; induce fever; stimulate systemic responses to inflammation, infection, and injury
Interleukin 2	Most immune cells	Helper T cells; cytotoxic T cells; NK cells; B cells	Stimulate proliferation Promote conversion to plasma cells
Interferons	Most cell types	Most cell types	Stimulate cells to produce antiviral proteins (nonspecific response)
Interferon gamma	NK cells and activated helper T cells	NK cells and macrophages	Stimulate proliferation and secretion of cytotoxic compounds
Chemokines	Damaged cells, including endothelial cells	Neutrophils and other leukocytes	Facilitate accumulation of leukocytes at sites of injury and inflammation
Colony-stimulating factors	Macrophages	Bone marrow	Stimulate proliferation of neutrophils and monocytes

*Note: This list is not meant to be exhaustive. There are >100 known cytokines.

The sequence of local events in a typical nonspecific inflammatory response to a bacterial infection—one caused, for example, by a cut with a bacteria-covered knife—is summarized in **Table 18–3**. The familiar signs of tissue injury and inflammation are local redness, swelling, heat, and pain.

The events of inflammation that underlie these signs are induced and regulated by a large number of chemical mediators, some of which are summarized for reference in **Table 18–4** (not all of these will be described in this chapter). Note in this table that some of these mediators are cytokines. Any given event of inflammation, such as vasodilation, may be induced by multiple mediators. Moreover, any given mediator may induce more than one event. Based on their origins, the mediators fall into two general categories: (1) peptides (kinins, for example) generated in the infected area by enzymatic actions on proteins that circulate in the plasma; and (2) substances secreted into the extracellular fluid from cells that either already exist in the infected area (injured cells or mast cells, for example) or enter it during inflammation (neutrophils, for example).

Let us now go step by step through the process summarized in Table 18–3, assuming that the bacterial infection in our example is localized to the tissue just beneath the skin. If the invading bacteria enter the blood or lymph, then similar inflammatory responses would take place in any other tissue or organ the blood-borne or lymph-borne microorganisms reach.

Vasodilation and Increased Permeability to Protein
A variety of chemical mediators dilate most of the microcirculation vessels in an infected and/or damaged area. The mediators also cause the local capillaries and venules to become permeable to proteins by inducing their endothelial cells to

Table 18–3 Sequence of Events in a Nonspecific Local Inflammatory Response to Bacteria

1. Entry of bacteria into tissue; injury to tissues causes release of chemicals to initiate the following events

2. Vasodilation of the microcirculation in the infected area, leading to increased blood flow

3. Large increase in protein permeability of the capillaries and venules in the infected area, with resulting diffusion of protein and filtration of fluid into the interstitial fluid

4. Chemotaxis: Movement of leukocytes from the venules into the interstitial fluid of the infected area

5. Destruction of bacteria in the tissue either through phagocytosis or by other mechanisms

6. Tissue repair

contract, opening spaces between them through which the proteins can move.

The adaptive value of these vascular changes is twofold: (1) the increased blood flow to the inflamed area (which accounts for the redness and heat) increases the delivery of proteins and leukocytes; and (2) the increased permeability to protein ensures that the plasma proteins that participate in inflammation—many of which are normally restrained by the intact endothelium—can gain entry to the interstitial fluid.

Table 18–4	Some Important Local Inflammatory Mediators
Mediator	**Source**
Kinins	Generated from enzymatic action on plasma proteins
Complement	Generated from enzymatic action on plasma proteins
Products of blood clotting	Generated from enzymatic action on plasma proteins
Histamine	Secreted by mast cells and injured cells
Eicosanoids	Secreted by many cell types
Platelet-activating factor	Secreted by many cell types
Cytokines, including chemokines	Secreted by injured cells, monocytes, macrophages, neutrophils, lymphocytes, and several nonimmune cell types, including endothelial cells and fibroblasts
Lysosomal enzymes, nitric oxide, and other oxygen-derived substances	Secreted by injured cells, neutrophils, and macrophages

As described in Chapter 12, the vasodilation and increased permeability to protein, however, cause net filtration of plasma into the interstitial fluid and the development of edema. This accounts for the swelling in an inflamed area, which is simply a consequence of the changes in the microcirculation and has no known adaptive value of its own.

Chemotaxis

With the onset of inflammation, circulating neutrophils begin to move out of the blood across the endothelium of capillaries and venules to enter the inflamed area. This multistage process is known as **chemotaxis.** It involves a variety of protein and carbohydrate **adhesion molecules** on both the endothelial cell and the neutrophil. It is regulated by messenger molecules released by cells in the injured area, including the endothelial cells. These messengers are collectively called **chemoattractants** (also called **chemotaxins** or chemotactic factors).

In the first stage, the neutrophil is loosely tethered to the endothelial cells by certain adhesion molecules. This event, known as **margination,** occurs as the neutrophil rolls along the vessel surface. In essence, this initial reversible event exposes the neutrophil to chemoattractants being released in the injured area. These chemoattractants act on the neutrophil to induce the rapid appearance of another class of adhesion molecules in its plasma membrane—molecules that bind tightly to their matching molecules in the endothelial cells. Thus, the neutrophils collect along the site of injury, rather than being washed away with the flowing blood.

In the next stage, known as **diapedesis,** a narrow projection of the neutrophil is inserted into the space between two endothelial cells, and the entire neutrophil squeezes through the endothelial wall and into the interstitial fluid. In this way, huge numbers of neutrophils migrate into the inflamed area.

Once in the interstitial fluid, neutrophils migrate toward the site of tissue damage (chemotaxis). This occurs because damaged cells release chemoattractants. Thus, neutrophils tend to move toward the microbes that entered into an injured area.

Movement of leukocytes from the blood into the damaged area is not limited to neutrophils. Monocytes follow later, and once in the tissue they undergo anatomical and functional changes that transform them to macrophages. As we will see later, in specific immune defenses lymphocytes undergo chemotaxis, as can basophils and eosinophils under certain conditions.

An important aspect of the multistep chemotaxis process is that it provides selectivity and flexibility for the migration of the various leukocyte types. Multiple adhesion molecules that are relatively distinct for the different leukocytes are controlled by different sets of chemoattractants. Particularly important in this regard are those cytokines that function as chemoattractants for distinct subsets of leukocytes. For example, one type of cytokine stimulates the chemotaxis of neutrophils, whereas another stimulates that of eosinophils. Thus, subsets of leukocytes can be stimulated to enter particular tissues at designated times during an inflammatory response, depending on the type of invader and the cytokine response it induces. The various cytokines that have chemoattractant actions are collectively referred to as **chemokines.**

Killing by Phagocytes

Once neutrophils and other leukocytes arrive at the site of an injury, they begin the process of destroying invading microbes by phagocytosis. The initial step in phagocytosis is contact between the surfaces of the phagocyte and microbe (**Figure 18–1**). One of the major triggers for phagocytosis during this contact is the interaction of phagocyte receptors with certain

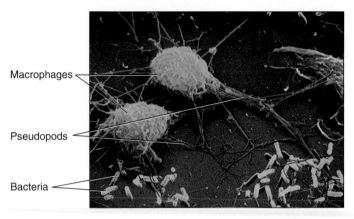

Figure 18–1

Macrophages contacting bacteria and preparing to engulf them.

Macrophages

Pseudopods

Bacteria

carbohydrates or lipids in the microbial cell walls. Contact is not always sufficient to trigger engulfment, however, particularly with bacteria that are surrounded by a thick, gelatinous capsule. Instead, chemical factors produced by the body can bind the phagocyte tightly to the microbe and thereby enhance phagocytosis. Any substance that does this is known as an **opsonin,** from the Greek word that means "to prepare for eating."

As the phagocyte engulfs the microbe (**Figure 18–2**), the internal, microbe-containing sac formed in this step is called a **phagosome.** A layer of plasma membrane separates the

microbe from the cytosol of the phagocyte. The phagosome membrane then makes contact with one of the phagocyte's lysosomes, which is filled with a variety of hydrolytic enzymes. The membranes of the phagosome and lysosome fuse, and the combined vesicles are now called a **phagolysosome.** Inside the phagolysosome, the lysosomal enzymes break down the microbe's macromolecules. In addition, other enzymes in the phagolysosome membrane produce **nitric oxide** as well as **hydrogen peroxide** and other oxygen derivatives, all of which are extremely destructive to the microbe's macromolecules.

Such intracellular destruction is not the only way phagocytes can kill microbes. The phagocytes also release antimicrobial substances into the extracellular fluid, where these chemicals can destroy the microbes without prior phagocytosis.

Some of these substances (for example, nitric oxide) secreted into the extracellular fluid (**Figure 18–3**) also function as inflammatory mediators. Thus, when phagocytes enter the area and encounter microbes, positive feedback mechanisms cause inflammatory mediators, including chemokines, to be released that bring in more phagocytes.

Complement

The family of plasma proteins known as **complement** provides another means for extracellular killing of microbes without prior phagocytosis. Certain complement proteins are always circulating in the blood in an inactive state. Upon activation of a complement protein in response to infection or damage, a cascade occurs so that this active protein activates a second complement protein, which activates a third, and so on. In this

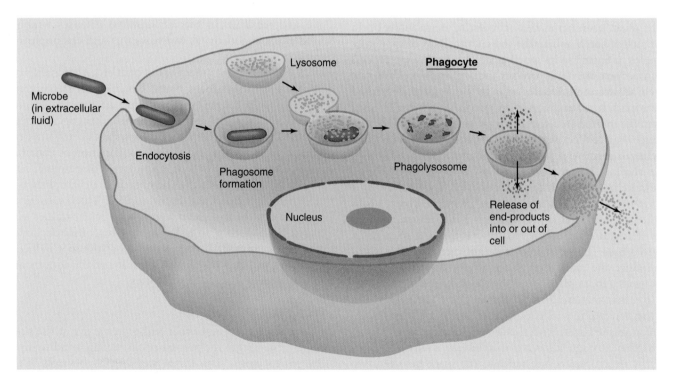

Figure 18–2

Phagocytosis and intracellular destruction of a microbe. After destruction has taken place in the phagolysosome, the end-products are released to the outside of the cell by exocytosis or used by the cell for its own metabolism.

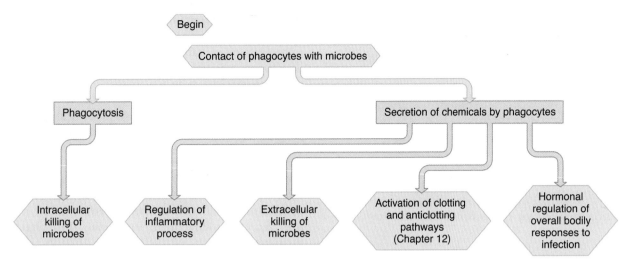

Figure 18–3

Role of phagocytes in nonspecific immune defenses. Hormonal regulation of overall bodily responses to infection, partly addressed in Chapter 11, will also be discussed later in this chapter.

way, multiple active complement proteins are generated in the extracellular fluid of the infected area from inactive complement molecules that have entered from the blood. Because this system consists of at least 30 distinct proteins, it is extremely complex, and we will identify the roles of only a few of the individual complement proteins.

Five of the active proteins generated in the complement cascade form a multiunit protein called the **membrane attack complex (MAC).** The MAC embeds itself in the microbial plasma membrane and forms porelike channels in the membrane, making it leaky. Water and salts enter the microbe, which disrupts the intracellular ionic environment and kills the microbe.

In addition to supplying a means for direct killing of microbes, the complement system serves other important functions in inflammation (**Figure 18–4**). Some of the activated complement molecules along the cascade cause, either directly or indirectly (by stimulating the release of other inflammatory mediators), vasodilation, increased microves-

sel permeability to protein, and chemotaxis. Also, one of the complement molecules—**C3b**—acts as an opsonin to attach the phagocyte to the microbe (**Figure 18–5**).

As we will see later, antibodies, a class of proteins secreted by lymphocytes, are required to activate the very first complement protein (C1) in the full sequence known as the classical complement pathway. However, lymphocytes are not involved in *nonspecific* inflammation, our present topic. How, then, is the complement sequence initiated during nonspecific inflammation? The answer is that there is an **alternate complement pathway,** one that is not antibody-dependent and that bypasses C1. The alternate pathway is initiated as the result of interactions between carbohydrates on the surface of the microbes and inactive complement molecules beyond C1. These interactions lead to the formation of active C3b, the opsonin described in the previous paragraph, and the activation of the subsequent complement molecules in the pathway. However, not all microbes have a surface conducive to initiating the alternate pathway.

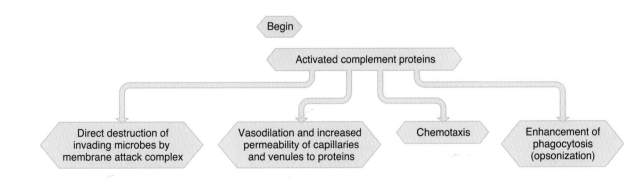

Figure 18–4

Functions of complement proteins. The effects on blood vessels and chemotaxis are exerted both directly by complement molecules and indirectly via other inflammatory mediators (for example, histamine) that are released by the complement molecules.

Other Opsonins in Nonspecific Defenses

In addition to complement C3b, other plasma proteins can bind nonspecifically to carbohydrates or lipids in the cell wall of microbes and facilitate opsonization. Many of these, for example **C-reactive protein,** are produced by the liver and are always found at some concentration in the plasma. Their production and plasma concentrations, however, are greatly increased during inflammation.

Tissue Repair

The final stage of inflammation is tissue repair. Depending upon the tissue involved, multiplication of organ-specific cells by cell division may or may not occur during this stage. For

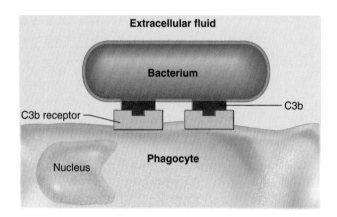

Figure 18–5

Function of complement C3b as an opsonin. One portion of C3b binds nonspecifically to carbohydrates on the surface of the bacterium, whereas another portion binds to specific receptor sites for C3b on the plasma membrane of the phagocyte. The structures are not drawn to scale.

example, liver cells multiply but skeletal muscle cells do not. In any case, fibroblasts (a type of connective tissue cell) that reside in the area divide rapidly and begin to secrete large quantities of collagen, while blood vessel cells proliferate in the process of angiogenesis. All of these events are brought about by chemical mediators, particularly a group of locally produced growth factors. Finally, remodeling occurs as the healing process winds down. The final repair may be imperfect, leaving a scar.

Interferons

Interferons are a family of cytokines that nonspecifically inhibit viral replication inside host cells. In response to infection by a virus, most cell types produce interferons and secrete them into the extracellular fluid. Interferons then bind to plasma membrane receptors on the secreting cell and on other cells, whether they are infected or not (**Figure 18–6**). This binding triggers the synthesis of dozens of different antiviral proteins by the cell. If the cell is already infected or eventually becomes infected, these proteins interfere with the ability of the viruses to replicate.

Interferons are not specific. Many kinds of viruses induce interferon synthesis, and interferons in turn can inhibit the multiplication of many kinds of viruses.

Specific Immune Defenses

Because of the complexity of specific immune defenses, it is useful to present a brief orientation before describing in more detail the various components of the response.

Overview

Lymphocytes are the essential cells in specific immune defenses. Unlike nonspecific defense mechanisms, lymphocytes must recognize the specific foreign matter to be attacked. Any for-

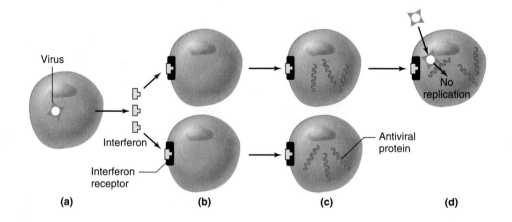

Figure 18–6

Role of interferon in preventing viral replication. (a) Most cell types, when infected with viruses, secrete interferon, which enters the interstitial fluid and binds to interferon receptors on the secreting cells themselves (autocrine function), and (b) adjacent cells (paracrine function). In addition, some interferon enters the blood and binds to interferon receptors on far-removed cells (endocrine function). The binding of interferon to its receptors induces the synthesis of proteins (c) that inhibit viral replication should viruses enter the cell (d).

Figure 18–6 physiological *inquiry* (p*i*)

■ Are there other examples besides immune secretions in which a single substance may act as both an endocrine and paracrine agent?

Answer can be found at end of chapter.

eign molecule that can trigger a specific immune response against itself or the cell bearing it is called an **antigen.** (Some immunologists prefer the more technically correct term *immunogen*.) Most antigens are either proteins or very large polysaccharides. The term *antigen* is a functional term; that is, any molecule, regardless of its location or function, that can induce a specific immune response against itself is by definition an antigen. Thus, it is any molecule or cell that the host does not recognize as self. Antigens include the protein coats of viruses, specific proteins on foreign cells, cancer cells, or transplanted cells, and toxins. It is the ability of lymphocytes to distinguish one antigen from another that confers specificity upon the immune responses in which they participate.

A typical specific immune response can be divided into three stages: (1) the encounter and recognition of an antigen by lymphocytes, (2) lymphocyte activation, and (3) the attack launched by the activated lymphocytes and their secretions.

Stage 1: During its development, each lymphocyte synthesizes and inserts into its plasma membrane a single type of receptor that can bind to a specific antigen. If, at a later time, the lymphocyte ever encounters that antigen, the antigen becomes bound to the receptors. This binding is the physicochemical meaning of the word *recognize* in immunology. As a result, the ability of lymphocytes to distinguish one antigen from another is determined by the nature of their plasma membrane receptors. *Each lymphocyte is specific for just one type of antigen.* The progeny of this specific antigen-stimulated lymphocyte are called **clones.** It is estimated that in a typical person the lymphocyte population expresses more than 100 million distinct antigen receptors, each with the potential to form a different clone.

Stage 2: The binding of antigen to receptor must occur for **lymphocyte activation.** Upon binding to an antigen, the lymphocyte undergoes a cell division, and the two resulting daughter cells then also divide (even though only one of them still has the antigen attached to it) and so on. In other words, the original binding of antigen by a single lymphocyte specific for that antigen triggers multiple cycles of cell divisions. As a result, many lymphocytes form that are identical to the one that started the cycles and can recognize the antigen; this is called **clonal expansion.** After activation, some lymphocytes will function as effector lymphocytes to carry out the attack response. Others will be set aside as **memory cells,** poised to recognize the antigen if it returns in the future.

Stage 3: The activated effector lymphocytes launch an attack against all antigens of the kind that initiated the immune response. Theoretically, it takes only one or two antigen molecules to *initiate* the specific immune response that will then result in an attack on all of the other antigens of that specific kind in the body. One group of lymphocytes, activated B

cells, differentiate into plasma cells, which secrete antibodies into the blood. These antibodies then recruit and guide other molecules and cells to perform the actual attack. Another type of lymphocyte, activated cytotoxic T cells, directly attack and kill the cells bearing the antigens. Once the attack is successfully completed, the great majority of the B cells, plasma cells, helper T cells, and cytotoxic T cells that participated in it die by apoptosis. The timely death of these effector cells is a homeostatic response that prevents the immune defense from becoming excessive and possibly destroying its own tissues. However, memory cells persist even after the immune response has been successfully completed.

Lymphoid Organs and Lymphocyte Origins

Our first task is to describe the organs and tissues in which lymphocytes originate and come to reside. Then the various types of lymphocytes alluded to in the overview and summarized in Table 18–1 will be described.

Lymphoid Organs

Like all leukocytes, lymphocytes circulate in the blood. At any moment, the great majority of lymphocytes are not actually in the blood, however, but in a group of organs and tissues collectively called the **lymphoid organs.** These are subdivided into primary and secondary lymphoid organs.

The **primary lymphoid organs** are the bone marrow and thymus. These organs supply the secondary lymphoid organs with mature lymphocytes—that is, lymphocytes already programmed to perform their functions when activated by antigen. The bone marrow and thymus are not normally sites in which lymphocytes undergo activation during an immune response.

The **secondary lymphoid organs** include the lymph nodes, spleen, tonsils, and lymphocyte accumulations in the linings of the intestinal, respiratory, genital, and urinary tracts. It is in the secondary lymphoid organs that lymphocytes are activated to participate in specific immune responses.

We have stated that the bone marrow and thymus supply mature lymphocytes to the secondary lymphoid organs. Most of the lymphocytes in the secondary organs are not, however, the same cells that originated in the primary lymphoid organs. The explanation of this seeming paradox is that, once in the secondary organ, a mature lymphocyte coming from the bone marrow or thymus can undergo cell division to produce additional identical lymphocytes, which in turn undergo cell division and so on. In other words, all lymphocytes are *descended* from ancestors that matured in the bone marrow or thymus but may not themselves have arisen in those organs. Recall that all the progeny cells derived by cell division from a single lymphocyte constitute a lymphocyte clone.

A distinction must be made between the lymphoid organs and the lymphatic system, described in Chapter 12. The latter is a network of lymphatic vessels and the lymph nodes found along these vessels. Of all the lymphoid organs, only the lymph nodes also belong to the lymphatic system.

There are no anatomical links, other than via the cardiovascular system, between the various lymphoid organs. Let us look briefly at these organs, with the exception of the bone marrow, which was described in Chapter 12.

The **thymus** lies in the upper part of the chest. Its size varies with age, being relatively large at birth and continuing to grow until puberty, when it gradually atrophies and is replaced by fatty tissue. Before its atrophy, the thymus consists mainly of mature lymphocytes that will eventually migrate via the blood to the secondary lymphoid organs. It also contains endocrine cells that secrete a group of hormones, collectively called **thymopoietin,** that exert a still poorly understood regulatory effect on lymphocytes of thymic origin.

Recall from Chapter 12 that the fluid flowing along the lymphatic vessels is called lymph, which is interstitial fluid that has entered the lymphatic capillaries and is routed to the large lymphatic vessels that drain into systemic veins. During this trip, the lymph flows through **lymph nodes** scattered along the vessels. Lymph, therefore, is the route by which lymphocytes in the lymph nodes encounter the antigens that activate them. Each node is a honeycomb of lymph-filled sinuses (**Figure 18–7**) with large clusters of lymphocytes (the lymphatic nodules) between the sinuses. They also contain many macrophages and dendritic cells.

The **spleen** is the largest of the secondary lymphoid organs and lies in the left part of the abdominal cavity between the stomach and the diaphragm. The spleen is to the circulating blood what the lymph nodes are to the lymph. Blood percolates through the vascular meshwork of the spleen's interior, where large collections of lymphocytes, macrophages, and dendritic cells are found. The macrophages of the spleen, in addition to interacting with lymphocytes, also phagocytize aging or dead erythrocytes.

The **tonsils** and **adenoids** are a group of small, rounded lymphoid organs in the pharynx. They are filled with lymphocytes, macrophages, and dendritic cells, and they have openings ("crypts") to the surface of the pharynx. Their lymphocytes respond to microbes that arrive by way of ingested food as well as through inspired air. Similarly, the lymphocytes in the linings of the various tracts exposed to the external environment respond to infectious agents that penetrate the linings from the lumen of the tract.

At any moment in time, some lymphocytes are on their way from the bone marrow or thymus to the secondary lymphoid organs. The vast majority, though, are cells that are participating in lymphocyte traffic *between* the secondary lymphoid organs, blood, lymph, and all the tissues of the body. Lymphocytes from all the secondary lymphoid organs constantly enter the lymphatic vessels that drain them (all lymphoid organs, not just lymph nodes, are drained by lymphatic vessels) and are carried to the blood. Simultaneously, some blood lymphocytes are pushing through the endothelium of venules all over the body to enter the interstitial fluid. From there, they

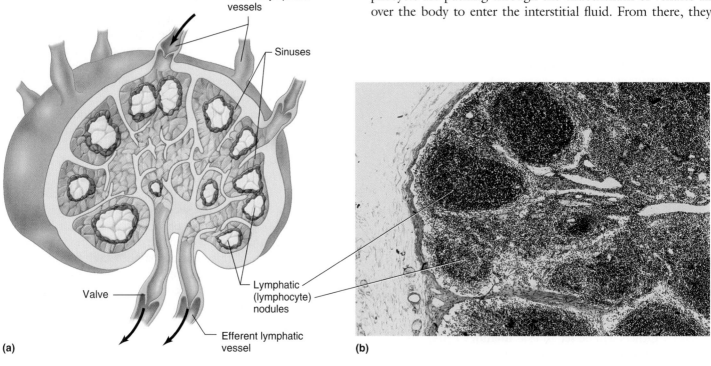

(a)

(b)

Figure 18–7

Anatomy of a lymph node as seen in (a) a sketch and in (b) a section viewed by light-microscopy.

Figure 18–7 physiological *inquiry*

- The nonspecific immune response includes vasodilation of the microcirculation and an increase in protein permeability of the capillaries (see Table 18–3). How might these changes enhance the specific immune response during an infection?

Answer can be found at end of chapter.

move into lymphatic capillaries and along the lymphatic vessels to lymph nodes. They may then leave the lymphatic vessels to take up residence in the node.

This recirculation is going on all the time, not just during an infection, although the migration of lymphocytes into an inflamed area is greatly increased by the chemotaxis process (Table 18–3). Lymphocyte trafficking greatly increases the likelihood that any given lymphocyte will encounter the antigen it is specifically programmed to recognize. (In contrast to the lymphocytes, polymorphonuclear granulocytes and monocytes do not recirculate; once they leave the bloodstream to enter a tissue they remain there or die.)

Lymphocyte Origins

The multiple populations and subpopulations of lymphocytes are summarized in Table 18–1. **B lymphocytes,** or simply **B cells,** mature in the bone marrow and then are carried by the blood to the secondary lymphoid organs (**Figure 18–8**). This process of maturation and migration continues throughout a person's life. All generations of lymphocytes that subsequently

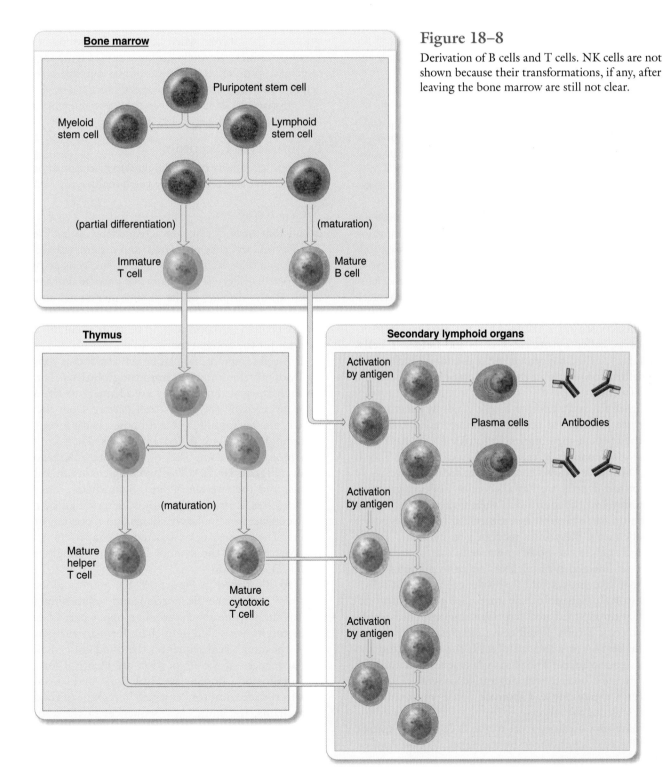

Figure 18–8

Derivation of B cells and T cells. NK cells are not shown because their transformations, if any, after leaving the bone marrow are still not clear.

arise from these cells by cell division in the secondary lymphoid organs will be identical to the parent cells; that is, they will be B-cell clones.

In contrast to the B cells, other lymphocytes leave the bone marrow in an immature state during fetal and early neonatal life. They are carried to the thymus and mature there before moving to the secondary lymphoid organs. These cells are called **T lymphocytes** or **T cells.** Like B cells, T cells also undergo cell division in secondary lymphoid organs, the progeny being identical to the original T cells and thus part of that T-cell clone.

In addition to the B and T cells, there is another distinct population of lymphocytes, **natural killer (NK) cells.** These cells arise in the bone marrow, but their precursors and life history are still unclear. As we will see, NK cells, unlike B and T cells, are not specific to a given antigen.

Functions of B Cells and T Cells

Upon activation, B cells differentiate into plasma cells, which secrete **antibodies,** proteins that travel all over the body to reach antigens identical to those that stimulated their production. In the body fluids outside of cells, the antibodies combine with these antigens and guide an attack that eliminates the antigens or the cells bearing them.

Antibody-mediated responses are also called *humoral* responses, the adjective humoral denoting communication "by way of soluble chemical messengers" (in this case, antibodies in the blood). Antibody-mediated responses have an extremely wide diversity of targets and are the major defense against bacteria, viruses, and other microbes in the extracellular fluid, and against toxic molecules (toxins).

T cells constitute a family that has two major functional subsets, called **cytotoxic T cells** and **helper T cells.** A third subset, called suppressor or regulatory T cells, have been hypothesized to inhibit the function of both B cells and cytotoxic T cells.

Another way to categorize T cells is not by function but rather by the presence of certain proteins, called CD4 and CD8, in their plasma membranes. Cytotoxic T cells have CD8 and so are also commonly called CD8+ cells; helper T cells express CD4 and so are also commonly called CD4+ cells.

Cytotoxic T cells are "attack" cells. Following activation, they travel to the location of their target, bind to them via antigen on these targets, and directly kill their targets via secreted chemicals. Responses mediated by cytotoxic T cells are directed against the body's own cells that have become cancerous or infected with viruses (or certain bacteria and parasites that, like viruses, take up residence inside host cells).

It is worth emphasizing the important geographical difference in antibody-mediated responses and responses mediated by cytotoxic T cells. The B cells (and plasma cells derived from them) remain in whatever location the recognition and activation steps occurred. The plasma cells send their antibodies forth, via the blood, to seek out antigens identical to those that triggered the response. Cytotoxic T cells must enter the blood and seek out the targets.

We have now assigned roles to the B cells and cytotoxic T cells. What role is performed by the helper T cells? As their name implies, these cells do not themselves function as attack cells but rather assist in the activation and function of both B cells and cytotoxic T cells. Helper T cells go through the usual first two stages of the immune response. First, they combine with antigen and then undergo activation. Once activated, they secrete cytokines that act on B cells and cytotoxic T cells that have also bound antigen. This is a very important point. With only a few exceptions, B cells and cytotoxic T cells cannot function adequately unless they are stimulated by cytokines from helper T cells.

Helper T cells will be considered as though they were a homogeneous cell population, but in fact, there are different subtypes of helper T cells, distinguished by the different cytokines they secrete when activated. By means of these different cytokines, they help different sets of lymphocytes, macrophages, and NK cells. Some of the cytokines secreted by helper T cells also act as inflammatory mediators.

Figure 18–9 summarizes the basic interactions among B, cytotoxic T, and helper T cells.

Lymphocyte Receptors

To repeat, the ability of lymphocytes to distinguish one antigen from another is determined by the lymphocytes' receptors.

B-Cell Receptors

Recall that once B cells are activated by antigen and helper T-cell cytokines, they proliferate and differentiate into plasma cells, which secrete antibodies. The plasma cells derived from a particular B cell can secrete only one particular antibody. Each B cell always displays on its plasma membrane copies of the particular antibody its plasma cell progeny can produce. This surface protein (glycoprotein, to be more accurate) acts as the receptor for the antigen specific to it.

B-cell receptors and plasma cell antibodies constitute the family of proteins known as **immunoglobulins.** The receptors themselves, even though they are identical to the antibodies to be secreted by the plasma cell derived from the activated B cell, are technically not antibodies because only *secreted* immunoglobulins are called antibodies. Each immunoglobulin molecule is composed of four interlinked polypeptide chains (**Figure 18–10**). The two long chains are called heavy chains, and the two short ones, light chains. There are five major classes of immunoglobulins, determined by the amino acid sequences in the heavy chains and a portion of the light chains. The classes are designated by the letters A, D, E, G, and M following the symbol Ig for immunoglobulin; thus IgA, IgD, and so on.

As illustrated in Figure 18–10, immunoglobulins have a "stem" called the **Fc** portion and comprising the lower half of the two heavy chains. The upper part of each heavy chain and its associated light chain form an **antigen binding site**—the amino acid sequences that bind antigen. The amino acid sequences of the Fc portion are identical for all immunoglobulins of a single class (IgA, IgD, and so on). In contrast to the identical (or "constant") regions of the heavy and light chains, the amino acid sequences of the antigen binding sites vary from immunoglobulin to immunoglobulin in a given class, and are thus known as variable ends. Each of the

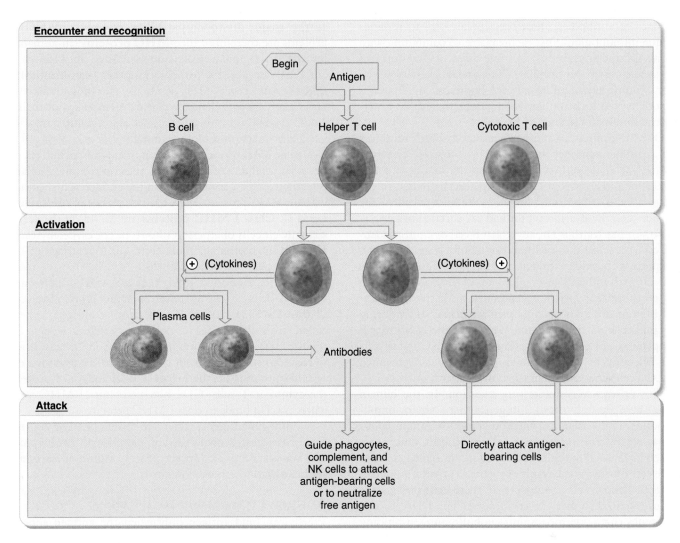

Figure 18–9

Summary of the roles of B, cytotoxic T, and helper T cells in immune responses. Events of the attack phase are described in later sections. The ⊕ symbol denotes a stimulatory effect (activation) of cytokines.

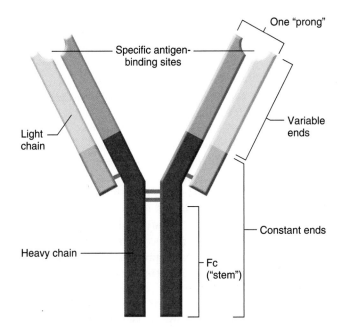

Figure 18–10

Immunoglobulin structure. The Fc portions and an extended region of the heavy chains are the same for all immunoglobulins of a particular class. A small portion of the light chains are also the same for a given immunoglobulin class. Collectively, these portions of the heavy and light chains are called "constant ends." Each "prong" contains a variable amino acid sequence, which represents the single antigen binding site. The links between chains represent disulfide bonds.

five classes of antibodies, therefore, contains up to millions of unique immunoglobulins, each capable of combining with only one specific antigen (or, in some cases, several antigens whose structures are very similar). The interaction between an antigen binding site of an immunoglobulin and an antigen is analogous to the lock-and-key interactions that apply generally to the binding of ligands by proteins.

One more point should be mentioned: B-cell receptors can bind antigen whether the antigen is a molecule dissolved in the extracellular fluid or is present on the surface of a foreign cell, such as a microbe, floating free in the fluids. In the latter case, the B cell becomes linked to the foreign cell via the bonds between the B-cell receptor and the surface antigen.

To summarize thus far, any given B cell or clone of identical B cells possesses unique immunoglobulin receptors—that is, receptors with unique antigen binding sites. Thus, the body arms itself with millions of small clones of different B cells in order to ensure that specific receptors exist for the vast number of different antigens the organism *might* encounter during its lifetime. The particular immunoglobulin that any given B cell displays as a receptor on its plasma membrane (and that its plasma cell progeny will secrete as antibodies) is determined during the cell's maturation in the bone marrow.

This raises a very interesting question: In the human genome there are only about 200 genes that code for immunoglobulins. How, then, can the body produce immunoglobulins having millions of different antigen binding sites, given that each immunoglobulin requires coding by a distinct gene? This diversity arises as the result of a genetic process unique to developing lymphocytes because only these cells possess the enzymes required to catalyze the process. The DNA in each of the genes that code for immunoglobulin antigen binding sites is cut into small segments, randomly rearranged along the gene, and then rejoined to form new DNA molecules. This cutting and rejoining varies from B cell to B cell, thus resulting in great diversity of the genes coding for the immunoglobulins of all the B cells taken together.

T-Cell Receptors

T-cell receptors for antigens are two-chained proteins that, like immunoglobulins, have specific regions that differ from one T-cell clone to another. However, T-cell receptors remain embedded in the T-cell membrane and are not secreted like immunoglobulins. As in B-cell development, multiple DNA rearrangements occur during T-cell maturation, leading to millions of distinct T-cell clones—distinct in that the cells of any given clone possess receptors of a single specificity. For T cells, this maturation occurs during their residence in the thymus.

In addition to their general structural differences, the B- and T-cell receptors differ in a much more important way: *The T-cell receptor cannot combine with antigen unless the antigen is first complexed with certain of the body's own plasma membrane proteins.* The T-cell receptor then combines with the entire complex of antigen and body (self) protein.

The self plasma membrane proteins that must be complexed with the antigen in order for T-cell recognition to occur constitute a group of proteins coded for by genes found on a single chromosome (chromosome 6) and known collectively as the **major histocompatibility complex (MHC).** The proteins are therefore called **MHC proteins** (in humans, also known as the human leukocyte-associated antigens, or HLA antigens). Because no two persons other than identical twins have the same sets of MHC genes, no two individuals have the same MHC proteins on the plasma membranes of their cells. MHC proteins are, in essence, cellular "identity tags"—that is, genetic markers of biological self.

The MHC proteins are often called "restriction elements" because the ability of a T cell's receptor to recognize an antigen is restricted to situations in which the antigen is first complexed with an MHC protein. There are two classes of MHC proteins: I and II. **Class I MHC proteins** are found on the surface of virtually all cells of a person's body except erythrocytes. **Class II MHC proteins** are found only on the surface of macrophages, B cells, and dendritic cells.

Now for another important point: The different subsets of T cells do not all have the same MHC requirements (**Table 18–5**). Cytotoxic T cells require antigen to be associated with class I MHC proteins, whereas helper T cells require class II MHC proteins. One reason for this difference in requirements stems from the presence, as described earlier, of CD4 proteins on the helper T cells and CD8 proteins on the cytotoxic T cells; CD4 binds to class II MHC proteins, whereas CD8 binds to class I MHC proteins.

How do antigens, which are foreign, end up on the surface of the body's own cells complexed with MHC proteins? The answer is provided by the process known as **antigen presentation,** to which we now turn.

Antigen Presentation to T Cells

T cells can bind antigen only when the antigen appears on the plasma membrane of a host cell complexed with the cell's MHC proteins. Cells bearing these complexes, therefore, function as **antigen-presenting cells (APCs).**

Presentation to Helper T Cells

Helper T cells require class II MHC proteins to function. Only macrophages, B cells, and dendritic cells express class II MHC proteins and therefore can function as APCs for helper T cells.

Table 18–5	MHC Restriction of the Lymphocyte Receptors
Cell Type	**MHC Restriction**
B	Do not interact with MHC proteins
Helper T	Class II, found only on macrophages, dendritic cells, and B cells
Cytotoxic T	Class I, found on all nucleated cells of the body
NK	Interaction with MHC proteins not required for activation

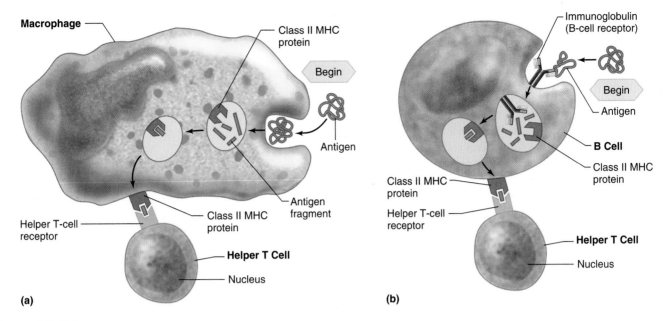

Figure 18–11

Sequence of events by which antigen is processed and presented to a helper T cell by (a) a macrophage or (b) a B cell. In both cases, begin the figure with the antigen in the extracellular fluid.

Adapted from Gray, Sette, and Buus.

The function of the macrophage (or dendritic cell) as an APC for helper T cells is easier to visualize (**Figure 18–11a**) because the macrophage forms a link between nonspecific and specific immune defenses. After a microbe or noncellular antigen has been phagocytized by a macrophage in a *nonspecific* response, it is partially broken down into smaller peptide fragments by the macrophage's proteolytic enzymes. The resulting digested fragments then bind (within endosomes) to class II MHC proteins synthesized by the macrophage. The fragments actually fit into a deep groove in the center of the MHC proteins. The fragment-MHC complex is then transported to the cell surface, where it is displayed in the plasma membrane. It is to this entire complex on the cell surface of the macrophage (or dendritic cell) that a specific helper T cell binds.

Note that it is not the intact antigen but rather the peptide fragments, called antigenic determinants or **epitopes,** of the antigen that are complexed to the MHC proteins and presented to the T cell. Despite this, it is customary to refer to "antigen" presentation rather than "epitope" presentation.

How B cells process antigen and present it to helper T cells is essentially the same as just described for macrophages (**Figure 18–11b**). It must be emphasized that the ability of B cells to present antigen to helper T cells is a *second* function of B cells in response to antigenic stimulation, the other being the differentiation of the B cells into antibody-secreting plasma cells.

The binding between helper T-cell receptor and an antigen bound to class II MHC proteins on an APC is the essential *antigen-specific* event in helper T-cell activation. However, this binding by itself will not result in T-cell activation. In addition, *nonspecific* interactions occur between other (nonantigenic) pairs of proteins on the surfaces of the attached helper T cell and APC, and these provide a necessary **costimulus** for T-cell activation (**Figure 18–12**).

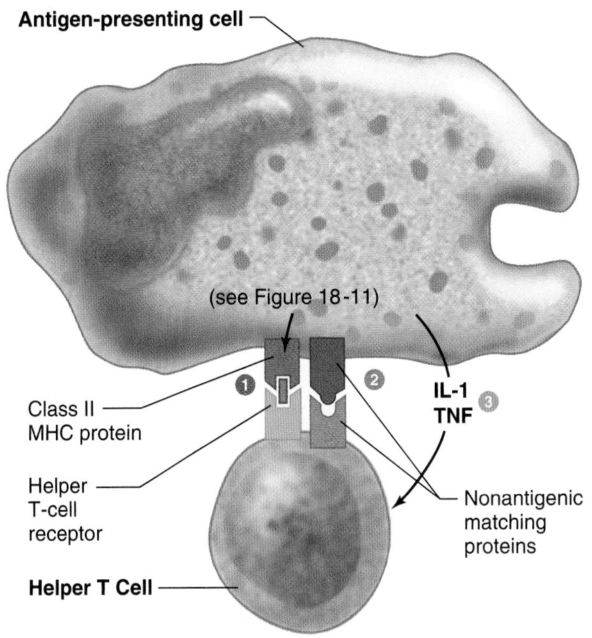

Figure 18–12

Three events are required for the activation of helper T cells: (1) presentation of the antigen bound to a class II MHC protein on an antigen-presenting cell (APC); (2) the binding of matching nonantigenic proteins in the plasma membranes of the APC and the helper T cell (costimulus); and (3) secretion by the APC of the cytokines interleukin 1 (IL-1), tumor necrosis factor (TNF), and other cytokines, which act on the helper T cell.

Finally, the antigenic binding of the APC to the T cell, along with the costimulus, causes the APC to secrete large amounts of the cytokines **interleukin 1 (IL-1)** and **tumor necrosis factor (TNF),** which act as paracrine agents on the attached helper T cell to provide yet another important stimulus for activation.

Thus, the APC participates in the activation of a helper T cell in three ways: (1) antigen presentation, (2) provision of a costimulus in the form of a matching nonantigenic plasma membrane protein, and (3) secretion of IL-1, TNF, and other cytokines (see Figure 18–12).

The activated helper T cell itself now secretes various cytokines that have both autocrine effects on the helper T cell and paracrine effects on adjacent B cells and any nearby cytotoxic T cells, NK cells, and still other cell types. These processes will be described in later sections.

Presentation to Cytotoxic T Cells

Because class I MHC proteins are synthesized by virtually all nucleated cells, any such cell can act as an APC for a cytotoxic T cell. This distinction helps explain the major function of cytotoxic T cells—destruction of *any* of the body's own cells that have become cancerous or infected with viruses. The key point is that the antigens that complex with class I MHC proteins arise *within* body cells. They are endogenous antigens, synthesized by a body cell.

How do such antigens arise? In the case of viruses, once a virus has taken up residence inside a host cell, the viral nucleic acid causes the host cell to manufacture viral proteins that are foreign to the cell. A cancerous cell has had one or more of its genes altered by chemicals, radiation, or other factors. The altered genes, called **oncogenes,** code for proteins that are not normally found in the body. Such proteins act as antigens.

In both virus-infected cells and cancerous cells, some of the endogenously produced antigenic proteins are hydrolyzed by cytosolic enzymes (in proteasomes) into peptide fragments, which are transported into the endoplasmic reticulum. There they are complexed with the host cell's class I MHC proteins and then shuttled by exocytosis to the plasma membrane surface, where a cytotoxic T cell specific for the complex can bind to it (**Figure 18–13**).

NK Cells

As noted earlier, NK (natural killer) cells constitute a distinct class of lymphocytes. They have several functional similarities to those of cytotoxic T cells. For example, their major targets are virus-infected cells and cancer cells, and they attack and kill these target cells directly, after binding to them. However, unlike cytotoxic T cells, NK cells are not antigen specific; that is, each NK cell can attack virus-infected cells or cancer cells without recognizing a specific antigen. They have neither T-cell receptors nor the immunoglobulin receptors of B cells, and the exact nature of the NK-cell surface receptors that permits the cells to identify their targets is unknown (except in one case presented later). MHC proteins are not involved in the activation of NK cells.

Why, then, do we deal with them in the context of *specific* immune responses? Because, as will be described subsequently, their participation in an immune response is greatly enhanced either by certain antibodies or by cytokines secreted by helper T cells activated during specific immune responses.

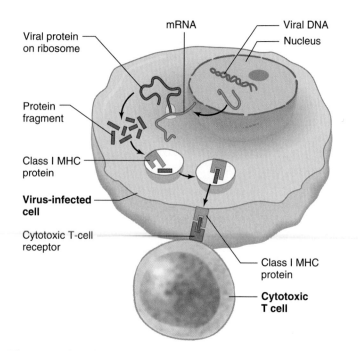

Figure 18–13

Processing and presentation of viral antigen to a cytotoxic T cell by an infected cell. Begin this figure with the viral DNA in the cell's nucleus. The viral DNA induces the infected cell to produce viral protein, which is then hydrolyzed (by proteasomes). The fragments are complexed to the cell's class I MHC proteins in the endoplasmic reticulum, and these complexes are then shuttled to the plasma membrane.

Adapted from Gray, Sette, and Buus.

Development of Immune Tolerance

Our basic framework for understanding specific immune responses requires consideration of one more crucial question: How does the body develop what is called **immune tolerance**—lack of immune responsiveness to self? This may seem a strange question given the definition of an antigen as a foreign molecule that can generate an immune response. How is it, though, that the body "knows" that its own molecules, particularly proteins, are not foreign but are self molecules?

Recall that the huge diversity of lymphocyte receptors is ultimately the result of multiple random DNA cutting/recombination processes. It is virtually certain, therefore, that in each person, clones of lymphocytes would have emerged with receptors that could bind to that person's own proteins. The existence and functioning of such lymphocytes would be disastrous because such binding would launch an immune attack against the cells expressing these proteins. There are at least two mechanisms, called *clonal deletion* and *clonal inactivation*, that explain why normally there are no active lymphocytes that respond to self components.

First, during fetal and early postnatal life, T cells are exposed to a wide mix of self proteins in the thymus. Those T cells with receptors capable of binding self proteins are destroyed by apoptosis (programmed cell death). This process is called **clonal deletion.** The second process, **clonal inactivation,** occurs not in the thymus but in the periphery and causes potentially self-reacting T cells to become nonresponsive.

What are the mechanisms of clonal deletion and inactivation during fetal and early postnatal life? Recall that full activation of a helper T cell requires not only an antigen-specific stimulus but a nonspecific costimulus (interaction between complementary nonantigenic proteins on the APC and the T cell). If this costimulus is *not* provided, the helper T cell not only fails to become activated by antigen, but dies or becomes inactivated forever. This is the case during early life, although what accounts for the costimulus not being delivered is unclear. B cells can also undergo clonal deletion and inactivation.

This completes the framework for understanding specific immune defenses. The next two sections utilize this framework in presenting typical responses from beginning to end, highlighting the interactions between lymphocytes and describing the attack mechanisms used by the various pathways.

Antibody-Mediated Immune Responses: Defenses Against Bacteria, Extracellular Viruses, and Toxins

A classical antibody-mediated response is one that results in the destruction of bacteria. The sequence of events, which is quite similar to the response to a virus in the extracellular fluid, is summarized in **Table 18–6** and **Figure 18–14**.

Antigen Recognition and Lymphocyte Activation

This process starts the same way as for nonspecific responses, with the bacteria penetrating one of the body's linings and entering the interstitial fluid. The bacteria then enter the lymphatic system and/or bloodstream and are carried to the lymph nodes and/or the spleen, respectively. There a B cell, using its immunoglobulin receptor, recognizes the bacterial surface antigen and binds the bacterium.

In a few cases (notably bacteria with cell-wall polysaccharide capsules), this binding is all that is needed to trigger B-cell activation. For the great majority of antigens, however, antigen binding is not enough, and signals in the form of cytokines released into the interstitial fluid by helper T cells near the antigen-bound B cells are also needed.

For helper T cells to react against bacteria by secreting cytokines, they must bind to a complex of antigen and class II MHC protein on an APC. Let us assume that in this case the APC is a macrophage that has phagocytized one of the bacteria, hydrolyzed its proteins into peptide fragments, complexed them with class II MHC proteins, and displayed the complexes on its surface. A helper T cell specific for the complex then binds to it, beginning the activation of the helper T cell. Moreover, the macrophage helps this activation process in two other ways: It provides a costimulus via nonantigenic plasma membrane proteins, and it secretes IL-1 and TNF.

IL-1 and TNF stimulate the helper T cell to secrete another cytokine named **interleukin 2 (IL-2)** and to express the receptor for IL-2. IL-2, acting as an autocrine agent, then provides a proliferative stimulus to the activated helper T cell (see Figure 18–14). The cell divides, beginning the mitotic cycles that lead to the formation of a clone of activated helper T cells, and these cells then release not only IL-2 but other cytokines as well.

Certain of these cytokines provide the additional signals required to activate nearby antigen-bound B cells to proliferate and differentiate into plasma cells, which then secrete antibodies.

Thus, as shown in Figure 18–14, a series of protein messengers interconnects the various cell types, the helper T cells serving as the central coordinator. The macrophage releases IL-1 and TNF, which act on the helper T cell to stimulate the release of IL-2, which stimulates the helper T cell to multiply. The activated progeny then release still other cytokines that help activate antigen-bound B cells.

As stated earlier, however, some of the B-cell progeny do not differentiate into plasma cells but instead into memory cells, whose characteristics permit them to respond more rapidly

Table 18–6	Summary of Events in Antibody-Mediated Immunity Against Bacteria

1. In secondary lymphoid organs, bacterial antigen binds to specific receptors on the plasma membranes of B cells.

2. Simultaneously, antigen-presenting cells (APCs), for example, macrophages, (a) present to helper T cells processed antigen complexed to class II MHC proteins on the APCs, (b) provide a costimulus in the form of another membrane protein, and (c) secrete IL-1, TNF, and other cytokines, which act on the helper T cells.

3. In response, the helper T cells secrete IL-2, which stimulates the helper T cells themselves to proliferate and secrete IL-2 and other cytokines. These activate antigen-bound B cells to proliferate and differentiate into plasma cells. Some of the B cells differentiate into memory cells rather than plasma cells.

4. The plasma cells secrete antibodies specific for the antigen that initiated the response, and the antibodies circulate all over the body via the blood.

5. These antibodies combine with antigen on the surface of the bacteria anywhere in the body.

6. Presence of antibody bound to antigen facilitates phagocytosis of the bacteria by neutrophils and macrophages. It also activates the complement system, which further enhances phagocytosis and can directly kill the bacteria by the membrane attack complex. It may also induce antibody-dependent cellular cytotoxicity mediated by NK cells that bind to the antibody's Fc portion.

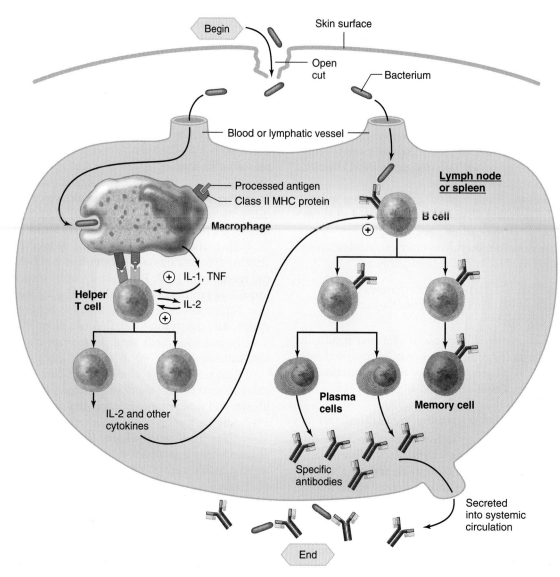

Figure 18–14

Summary of events by which a bacterial infection leads to antibody synthesis in peripheral lymphoid organs. The secreted antibodies travel by the blood to the site of infection, where they bind to bacteria of the type that induced the response. The attack triggered by antibody's binding to bacteria is described in the text. (As illustrated in Figure 18–11b, an antigen-bound B cell, rather than a macrophage, as shown in this figure, can function as the antigen-presenting cell to the helper T cell. Also for clarity, the intracellular processing of the antigen by the macrophage [see Figure 18–11a] is not shown in this figure.)

and vigorously should the antigen reappear at a future time (see Figure 18–14).

The example we have been using employed a macrophage as the APC to helper T cells, but B cells can also serve in this role (see Figure 18–11). The binding of the helper T cell to the antigen-bound B cell ensures maximal stimulation of the B cell by the cytokines secreted by that helper T cell and any of its progeny that remain nearby.

Antibody Secretion

After their differentiation from B cells, plasma cells produce thousands of antibody molecules per second before they die in a day or so. We mentioned earlier that there are five major classes of antibodies. The most abundant are the **IgG** antibodies, commonly called **gamma globulin,** and **IgM** antibodies.

These two groups together provide the bulk of specific immunity against bacteria and viruses in the extracellular fluid. **IgE** antibodies participate in defenses against multicellular parasites and also mediate allergic responses. **IgA** antibodies are secreted by plasma cells in the linings of the gastrointestinal, respiratory, and genitourinary tracts; these antibodies generally act locally in the linings or on their surfaces. They are also secreted by the mammary glands and therefore are the major antibodies in milk. The functions of **IgD** are still unclear.

In the kind of infection described in this chapter, the B and plasma cells, sitting on the nodes near the infected tissues, recognize antigen and are activated to make antibodies. The antibodies (mostly IgG and IgM) circulate through the lymph and blood to return to the infected site. At sites of infection, the antibodies leave the blood (recall that nonspecific inflammation had already made capillaries and venules leaky at these sites) and combine with the type of bacterial surface antigen that initiated the immune response (see Figure 18–14). These antibodies then direct the attack (see following discussion) against the bacteria to which they are now bound.

Thus, immunoglobulins play two distinct roles in immune responses: (1) during the initial recognition step, those on the surface of B cells bind to antigen brought to them; and (2) those secreted by the plasma cells (antibodies) bind to bacteria bearing the same antigens, "marking" them as the targets to be attacked.

The Attack: Effects of Antibodies

The antibodies bound to antigen on the microbial surface do not directly kill the microbe but instead link up the microbe physically to the actual killing mechanisms—phagocytes (neutrophils and macrophages), complement, or NK cells. This linkage not only triggers the attack mechanism but ensures that the killing effects are restricted to the microbe. Linkage to specific antibodies normally protects adjacent normal structures from the toxic effects of the chemicals employed by the killing mechanisms.

Direct Enhancement of Phagocytosis Antibodies can act directly as opsonins. The mechanism is analogous to that for complement C3b (see Figure 18–5) in that the antibody links the phagocyte to the antigen. As shown in **Figure 18–15**, the phagocyte has membrane receptors that bind to the Fc portion of an antibody. This linkage promotes attachment of the antigen to the phagocyte and the triggering of phagocytosis of the bacterium.

Activation of the Complement System As described earlier in this chapter, the plasma complement system is activated in *nonspecific* inflammatory responses via the alternate complement pathway. In contrast, in *specific* immune responses, the presence of antibody of the IgG or IgM class bound to antigen activates the **classical complement pathway.** The first molecule in this pathway, C1, binds to the Fc portion of an antibody that has combined with antigen (**Figure 18–16**). This results in activation of the enzymatic portions of C1, thereby initiating the entire classical pathway. The end product of this cascade, the membrane attack complex (MAC), can kill the cells the antibody is bound to by making their membranes leaky. In addition, as we saw in

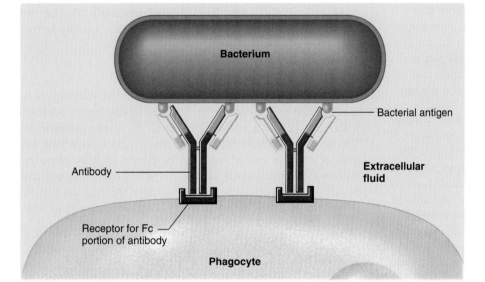

Figure 18–15

Direct enhancement of phagocytosis by antibody. The antibody links the phagocyte to the bacterium. Compare this mechanism of opsonization to that mediated by complement C3b (see Figure 18–5).

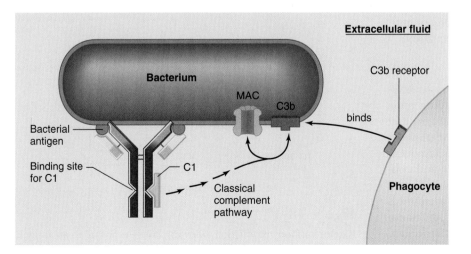

Figure 18–16

Activation of classical complement pathway by binding of antibody to bacterial antigen. C1 is activated by its binding to the Fc portion of the antibody. The membrane attack complex (MAC) is then generated, along with C3b, which acts as an opsonin by binding the bacteria to a phagocyte.

Figure 18–5, another activated complement molecule (C3b) functions as an opsonin to enhance phagocytosis of the microbe by neutrophils and macrophages (see Figure 18–16). Thus, antibodies enhance phagocytosis both directly (see Figure 18–15) and via activation of complement C3b.

It is important to note that C1 binds not to the unique antigen binding sites in the antibody's prongs, but rather to complement binding sites in the Fc portion. Because the latter are the same in virtually all antibodies of the IgG and IgM classes, the complement molecule will bind to *any* antigen-bound antibodies belonging to these classes. In other words, there is only one set of complement molecules, and once activated, they do essentially the same thing regardless of the specific identity of the invader.

Antibody-Dependent Cellular Cytotoxicity We have seen that both a particular complement molecule (C1) and a phagocyte can bind nonspecifically to the Fc portion of an antibody bound to antigen. NK cells can also do this (just substitute an NK cell for the phagocyte in Figure 18–15). Thus, antibodies can link target cells to NK cells, which then kill the targets directly by secreting toxic chemicals. This is called **antibody-dependent cellular cytotoxicity (ADCC)** because the killing (cytotoxicity) is carried out by cells (NK cells), but the process depends upon the presence of antibody. Note that it is the antibodies that confer specificity upon ADCC, just as they do on antibody-dependent phagocytosis and complement activation. This mechanism for bringing NK cells into play is the one exception, mentioned earlier, to the generalization that the mechanism by which NK cells identify their targets is unclear.

Direct Neutralization of Bacterial Toxins and Viruses Toxins secreted by bacteria into the extracellular fluid can act as antigens to induce antibody production. The antibodies then combine with the free toxins, thus preventing interaction of the toxin with susceptible cells. Because each antibody has two binding sites for antigen, clump-like chains of antibody-antigen complexes form, and these clumps are then phagocytized.

A similar binding process occurs as part of the major antibody-mediated mechanism for eliminating viruses in the extracellular fluid. Certain of the viral surface proteins serve as antigens, and the antibodies produced against them combine with them, preventing attachment of the virus to plasma membranes of potential host cells. This prevents the virus from entering a cell. As with bacterial toxins, chains of antibody-virus complexes are formed and can be phagocytized.

Active and Passive Humoral Immunity

The response of the antibody-producing machinery to invasion by a foreign antigen varies enormously, depending upon whether the machinery has previously been exposed to that antigen. Antibody production occurs slowly over several weeks following the first contact with an antigen, but any subsequent infection by the same invader elicits an immediate and considerable outpouring of additional specific antibodies (**Figure 18–17**). This response, which is mediated by the memory B cells described earlier, confers a greatly enhanced resistance toward subsequent infection with that particular microorganism. Resistance built up as a result of the body's contact with

microorganisms and their toxins or other antigenic components is known as **active immunity.**

Until the twentieth century, the only way to develop active immunity was to suffer an infection, but now the injection of microbial derivatives in vaccines is used. A ***vaccine*** may consist of small quantities of living or dead microbes, small quantities of toxins, or harmless antigenic molecules derived from the microorganism or its toxin. The general principle is always the same: Exposure of the body to the agent results in an active immune response along with the induction of the memory cells required for rapid, effective response to possible future infection by that particular organism.

A second kind of immunity, known as **passive immunity,** is simply the direct transfer of antibodies from one person to another, the recipient thereby receiving preformed antibodies. Such transfers occur between mother and fetus because IgG can move across the placenta. Also, breast-fed children receive IgA antibodies in the mother's milk. These are important sources of protection for the infant during the first months of life, when the antibody-synthesizing capacity is relatively poor.

The same principle is used clinically when specific antibodies (produced by genetic engineering) or pooled gamma globulin is given to patients exposed to or suffering from certain infections such as hepatitis. Because antibodies are proteins with a limited life span, the protection afforded by this transfer of antibodies is relatively short-lived, usually lasting only a few weeks or months.

Summary

It is now possible to summarize the interplay between nonspecific and specific immune defenses in resisting a bacterial infection. When a particular bacterium is encountered for the first time, *nonspecific* defense mechanisms resist its entry and, if entry is gained, attempt to eliminate it by phagocytosis and nonphagocytic killing in the inflammatory process. Simultaneously, bacterial antigens induce the relevant specific B-cell clones to differentiate into plasma cells capable of anti-

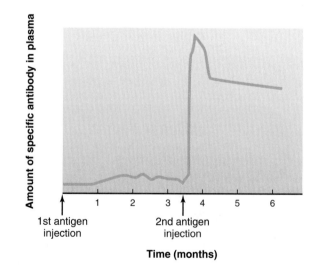

Figure 18–17

Rate of antibody production following initial exposure to an antigen and subsequent exposure to the same antigen.

Chapter 18

body production. If the nonspecific defenses are rapidly successful, these slowly developing *specific* immune responses may never play an important role. If the nonspecific responses are only partly successful, the infection may persist long enough for significant amounts of antibody to be produced. The presence of antibody leads to both enhanced phagocytosis and direct destruction of the foreign cells, as well as to neutralization of any toxins the bacteria secrete. All subsequent encounters with that type of bacterium will activate the specific responses much sooner and with greater intensity. That is, the person may have active immunity against that bacteria.

The defenses against viruses in the extracellular fluid are similar, resulting in destruction or neutralization of the virus.

Defenses Against Virus-Infected Cells and Cancer Cells

The previous section described how antibody-mediated immune responses constitute the major long-term defense against exogenous antigens—bacteria, viruses, and individual foreign molecules that enter the body and are encountered by the immune system in the extracellular fluid. This section now details how the body's own cells that have become infected by viruses (or other intracellular microbes) or transformed into cancer cells are destroyed.

What is the value of destroying virus-infected host cells? Such destruction results in release of the viruses into the extracellular fluid, where they can be directly neutralized by circulating antibody, as just described. Generally, only a few host cells are sacrificed in this way, but once viruses have had a chance to replicate and spread from cell to cell, so many virus-infected host cells may be killed by the body's own defenses that organ malfunction may occur.

Role of Cytotoxic T Cells

Figure 18–18 summarizes a typical cytotoxic T-cell response triggered by viral infection of body cells. The response triggered by a cancer cell would be similar. As described earlier, a virus-infected or cancer cell produces foreign proteins, "endogenous antigens," which are processed and presented on the plasma membrane of the cell complexed with class I MHC proteins. Cytotoxic T cells specific for the particular antigen can bind to the complex, but just as with B cells, binding to antigen alone does not cause activation of the cytotoxic T cell. Cytokines from adjacent activated helper T cells are also needed.

What role do the helper T cells play in these cases? Figure 18–14 illustrates the most likely mechanism. Macrophages phagocytize free extracellular viruses (or, in the case of cancer, antigens released from the surface of the cancerous cells)

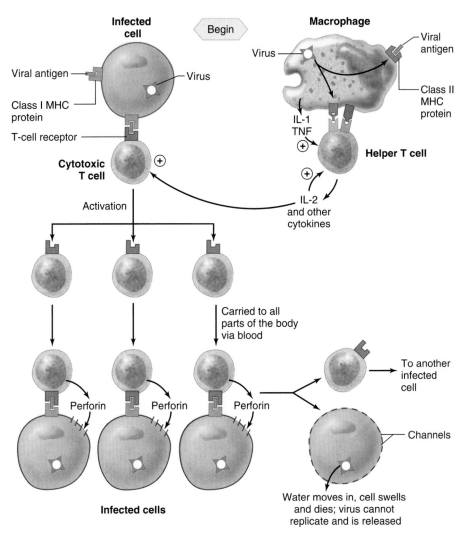

Figure 18–18

Summary of events in the killing of virus-infected cells by cytotoxic T cells. The released viruses can then be phagocytized. The sequence would be similar if the inducing cell were a cancer cell rather than a virus-infected cell.

and then process and present antigen, in association with class II MHC proteins, to the helper T cells. In addition, the macrophages provide a costimulus and also secrete IL-1 and TNF. The activated helper T cell releases IL-2 and other cytokines. IL-2 then acts as an autocrine agent to stimulate proliferation of the helper T cell.

The IL-2 also acts as a paracrine agent on the cytotoxic T cell bound to the surface of the virus-infected or cancer cell, stimulating this attack cell to proliferate. Other cytokines secreted by the activated helper T cell perform the same functions. Why is proliferation important if a cytotoxic T cell has already found and bound to its target? The answer is that there is rarely just one virus-infected cell or one cancer cell. By expanding the clone of cytotoxic T cells capable of recognizing the particular antigen, proliferating attack cells increase the likelihood that the other virus-infected or cancer cells will be encountered by the specific type of cytotoxic T cell.

There are several mechanisms of target cell killing by activated cytotoxic T cells, but one of the most important is as follows (see Figure 18–18): The cytotoxic T cell releases, by exocytosis, the contents of its secretory vesicles into the extracellular space between itself and the target cell to which it is bound. These vesicles contain a protein, **perforin** (also called pore-forming protein), which is similar in structure to the proteins of the complement system's membrane attack complex. Perforin inserts into the target cell's membrane and forms channels through the membrane. In this manner, it causes the attacked cell to become leaky and die. The fact that perforin is released directly into the space between the tightly attached cytotoxic T cell and the target ensures that innocent host bystander cells will not be killed because perforin is not at all specific.

Some cytotoxic T cells generated during proliferation following an initial antigenic stimulation do not complete their full activation at this time but remain as memory cells. Thus, active immunity exists for cytotoxic T cells just as for B cells.

Role of NK Cells and Activated Macrophages

Although cytotoxic T cells are very important attack cells against virus-infected and cancer cells, they are not the only ones. NK cells and activated macrophages also destroy such cells by secreting toxic chemicals.

In the section on antibody-dependent cellular cytotoxicity (ADCC), we pointed out that NK cells can be linked to target cells by antibodies, and this certainly constitutes one potential method of bringing them into play against virus-infected or cancer cells. In most cases, however, strong antibody responses are not triggered by virus-infected or cancer cells, and the NK cell must bind *directly* to its target, without the help of antibodies. As noted earlier, NK cells do not have antigen specificity; rather, they nonspecifically bind to any virus-infected and cancer cell.

The major signals for NK cells to proliferate and secrete their toxic chemicals are IL-2 and a member of the interferon family—**interferon-gamma**—secreted by the helper T cells that have been activated specifically by the targets (**Figure 18–19**). (Whereas essentially all body cells can produce the other interferons, as described earlier, only activated helper T cells and NK cells can produce interferon-gamma.)

Thus, the attack by the NK cells is nonspecific, but a specific immune response on the part of the helper T cells is required to bring the NK cells into play. Moreover, there is a positive feedback mechanism at work here because activated NK cells can themselves secrete interferon-gamma (see Figure 18–19).

IL-2 and interferon-gamma act not only on NK cells but on macrophages in the vicinity to enhance their ability to kill cancer cells and cells infected with viruses and other microbes. Macrophages stimulated by IL-2 and interferon-gamma are called **activated macrophages** (see Figure 18–19). They secrete large amounts of many chemicals that are capable of killing cells by a variety of mechanisms.

Table 18–7 summarizes the multiple defenses against viruses described in this chapter.

Systemic Manifestations of Infection

There are many *systemic* responses to infection—that is, responses of organs and tissues distant from the site of infection or immune response. These systemic responses are collectively known as the **acute phase response** (**Figure 18–20**). It is natural to think of them as part of the disease, but the fact is

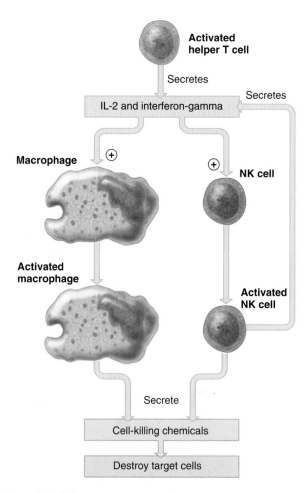

Figure 18–19

Role of IL-2 and interferon-gamma, secreted by activated helper T cells, in stimulating the killing ability of NK cells and macrophages.

Table 18–7 Summary of Host Responses to Viruses

	Main Cells Involved	Comment on Action
Nonspecific defenses	Body surface linings	Provide physical barrier; antiviral chemicals
Anatomical barriers	Tissue macrophages	Provide phagocytosis of extracellular virus
Inflammation	Most cell types after viruses enter	Interferon nonspecifically prevents viral replication inside host cells
Interferon	them	
Specific defenses	Plasma cells (derived from B cells)	Antibodies neutralize virus and thus prevent viral entry into cell
Antibody-mediated	that secrete antibodies	Antibodies activate complement, which leads to enhanced phagocytosis of extracellular virus
		Antibodies recruit NK cells via antibody-mediated cellular cytotoxicity
Helper	Helper T cells	Secrete interleukins; keep NK cells, macrophages, cytotoxic T cells, and helper T cells active; also help convert B cells to plasma cells.
Direct cell killing	Cytotoxic T cells, NK cells, and activated macrophages	Destroy host cell via secreted chemicals and thus induce release of virus into extracellular fluid where it can be phagocytized
		Activity stimulated by IL-2 and interferon-gamma

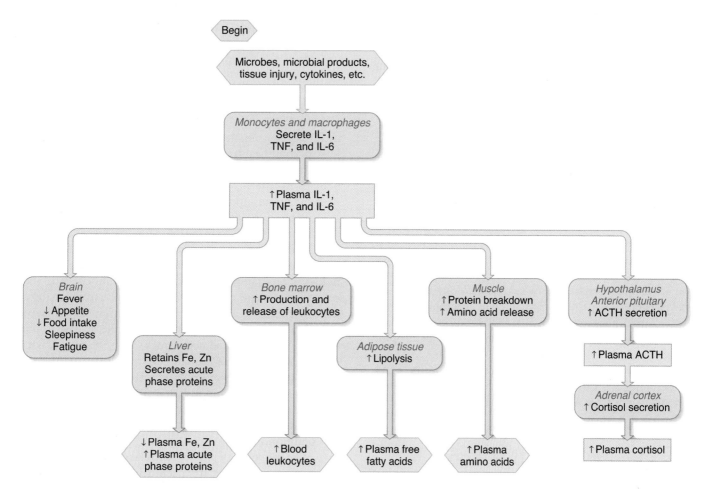

Figure 18–20

Systemic responses to infection or injury (the acute phase response). Other cytokines probably also participate. This figure does not include all the components of the acute phase response; for example, IL-1 and several other cytokines also stimulate the secretion of insulin and glucagon. The effect of cortisol on the immune response is inhibitory; cortisol provides a negative feedback action to prevent excessive immune activity (Chapter 11).

that most of them actually represent the body's own adaptive responses to the infection.

The single most common and striking systemic sign of infection is fever, the mechanism of which is described in Chapter 16. Present evidence suggests that fever is often beneficial because an increase in body temperature enhances many of the protective responses described in this chapter.

Decreases in the plasma concentrations of iron and zinc occur in response to infection and are due to changes in the uptake and/or release of these trace elements by liver, spleen, and other tissues. The decrease in plasma iron concentration has adaptive value because bacteria require a high concentration of iron to multiply. The role of the decrease in zinc is not known.

Another adaptive response to infection is the secretion by the liver of a group of proteins known collectively as **acute phase proteins.** These proteins exert many effects on the inflammatory process that serve to minimize the extent of local tissue damage. In addition, they are important for tissue repair and for clearance of cell debris and the toxins released from microbes. An example of an acute phase protein is C-reactive protein, which functions as a nonspecific opsonin to enhance phagocytosis.

Another response to infection, increased production and release of neutrophils and monocytes by the bone marrow, is of obvious value. Also occurring is a release of amino acids from muscle; the amino acids provide the building blocks for the synthesis of proteins required to fight the infection and for tissue repair. Increased release of fatty acids from adipose tissue also occurs, providing a source of energy. The secretion of many hormones, notably cortisol, is increased in the acute phase response, exerting negative feedback actions on immune function.

All of these systemic responses to infection and many others are elicited by one or more of the cytokines released from activated macrophages and other cells (see Figure 18–20). In particular IL-1, TNF, and another cytokine—**interleukin 6 (IL-6),** all of which serve local roles in immune responses, also serve as hormones to elicit distant responses such as fever.

Several other cytokines are also known to participate in the acute phase response. For example, colony-stimulating factors (Chapter 12), which are secreted by macrophages, lymphocytes, endothelial cells, and fibroblasts, provide a major stimulus to the bone marrow to produce more neutrophils and monocytes.

The participation of macrophages in the acute phase response completes our discussion of these cells, the various functions of which are summarized in **Table 18–8.**

Factors That Alter the Body's Resistance to Infection

There are many factors that determine the body's capacity to resist infection; a few important examples are presented here.

Protein-calorie malnutrition is, worldwide, the single greatest contributor to decreased resistance to infection. Because inadequate amino acids are available to synthesize essential proteins, immune function is impaired. Deficits of specific nutrients other than protein can also lower resistance to infection.

Table 18–8	Role of Macrophages in Immune Responses

1. In nonspecific inflammation, macrophages phagocytize particulate matter, including microbes. They also secrete antimicrobial chemicals and protein messengers (cytokines) that function as local inflammatory mediators. The inflammatory cytokines include IL-1 and TNF.

2. Macrophages process and present antigen to cytotoxic T cells and helper T cells.

3. The secreted IL-1 and TNF (see number 1 above) stimulate helper T cells to secrete IL-2 and to express the receptor for IL-2.

4. During specific immune responses, macrophages perform the same killing and inflammation-inducing functions as in (1) but are more efficient because antibodies act as opsonins and because the cells are transformed into activated macrophages by IL-2 and interferon-gamma, both secreted by helper T cells.

5. The secreted IL-1, TNF, and IL-6 mediate many of the systemic responses to infection or injury.

A preexisting disease, infectious or noninfectious, can also predispose the body to infection. People with diabetes mellitus, for example, are more likely to develop infections, at least partially explainable on the basis of defective leukocyte function. Moreover, any injury to a tissue lowers its resistance, perhaps by altering the chemical environment or interfering with the blood supply.

Both stress and a person's state of mind can either enhance or reduce resistance to infection (and cancer). There are multiple mechanisms that constitute the links in these "mind-body" interactions, as revealed by the field called psychoneuroimmunology. For example, lymphoid tissue is innervated, and the cells that mediate immune defenses have receptors for many neurotransmitters and hormones. Conversely, as we have seen, some of the cytokines the immune cells release have important effects on the brain and endocrine system. Moreover, lymphocytes secrete several of the same hormones produced by endocrine glands. Thus the immune system can alter neural and endocrine function, and in turn neural and endocrine activity can modify immune function. For example, it has been shown that the production of antibodies can be altered by psychological conditioning.

The influence of physical exercise on the body's resistance to infection and cancer has been debated for decades. Present evidence indicates that the intensity, duration, chronicity, and psychological stress of the exercise all have important influences, both negative and positive, on a host of immune functions (for example, the level of circulating NK cells). Most experts in the field believe that, despite all these complexities, modest exercise and physical conditioning have net beneficial effects on the immune system and on host resistance.

Another factor associated with decreased immune function is sleep deprivation. For example, loss of a single night's sleep has been observed to reduce the activity of blood NK cells. The mechanism of this response is uncertain but the results have been replicated by numerous investigators.

Resistance to infection will be impaired if one of the basic resistance mechanisms itself is deficient, as, for example, in people who have a genetic deficiency that impairs their ability to produce antibodies. These people experience frequent and sometimes life-threatening infections that can be prevented by regular replacement injections of gamma globulin. Another genetic defect is *combined immunodeficiency,* which is an absence of both B and T cells. If untreated, infants with this disorder usually die within their first year of life from overwhelming infections. Combined immunodeficiency can be cured by a bone marrow transplant, which supplies both B cells and cells that will migrate to the thymus and become T cells.

An environmentally induced decrease in the production of leukocytes is also an important cause of lowered resistance. This can occur, for example, in patients given drugs specifically to inhibit the rejection of tissue or organ transplants (see the section on graft rejection that follows).

In terms of the numbers of people involved, the most important example of the lack of a basic resistance mechanism is the disease called *acquired immune deficiency syndrome* *(AIDS).*

Acquired Immune Deficiency Syndrome (AIDS)

AIDS is caused by *human immunodeficiency virus (HIV),* which incapacitates the immune system. HIV belongs to the retrovirus family, whose nucleic acid core is RNA rather than DNA. Retroviruses possess an enzyme called reverse transcriptase, which, once the virus is inside a host cell, transcribes the virus's RNA into DNA, which is then integrated into the host cell's chromosomes. Replication of the virus inside the cell causes the death of the cell.

Unfortunately, the cells that HIV preferentially (but not exclusively) enters are helper T cells. HIV infects these cells because the CD4 protein on the plasma membrane of helper T cells acts as a receptor for one of the HIV's surface proteins (gp120). Thus, the helper T cell binds the virus, making it possible for the virus to enter the cell. Very importantly, this binding of the HIV gp120 protein to CD4 is not sufficient to grant the HIV entry into the helper T cell. In addition, another surface protein on the helper T cell, one that serves normally as a receptor for certain chemokines, must serve as a coreceptor for the gp120. It has been found that persons who have a mutation in this chemokine receptor are highly resistant to infection with HIV. Much research is now focused on the possible therapeutic use of chemicals that can interact with and block this coreceptor.

Once in the helper T cell, the replicating HIV directly kills the helper T cell and also indirectly causes its death via the body's usual immune attack. The attack is mediated in this case mainly by cytotoxic T cells attacking the virus-infected cells. In addition, by still poorly understood mechanisms, HIV causes the death of many *uninfected* helper T cells by apoptosis. Without adequate numbers of helper T cells, neither B cells nor cytotoxic T cells can function normally. Thus, the AIDS patient dies from infections and cancers that the immune system would ordinarily readily handle.

AIDS was first described in 1981, and it has since reached epidemic proportions worldwide. The great majority of persons now infected with HIV have no symptoms of AIDS. It is important to distinguish between the presence of the symptomatic disease—AIDS—and asymptomatic infection with HIV. (The latter is diagnosed by the presence of anti-HIV antibodies or HIV RNA in the blood.) It is thought, however, that most infected persons will eventually develop AIDS, although at highly varying rates.

The path from HIV infection to AIDS commonly takes about 10 years in untreated persons. Typically, during the first five years the rapidly replicating viruses continually kill large numbers of helper T cells in lymphoid tissues, but these are replaced by new cells. Therefore, the number of helper T cells stays normal (about 1000 cells/mm^3 of blood), and the person is asymptomatic. During the next five years, this balance is lost; the number of helper T cells, as measured in blood, decreases to about half the normal level, but many people still remain asymptomatic. As the helper T cell count continues to decrease, however, the symptoms of AIDS begin—infections with bacteria, viruses, fungi, and parasites. These are accompanied by systemic symptoms of weight loss, lethargy, and fever—all caused by high levels of the cytokines that induce the acute phase response. Certain unusual cancers (such as *Kaposi's sarcoma*) also occur with relatively high frequency. In untreated persons, death usually ensues within two years after the onset of AIDS symptoms.

The major routes of transmission of HIV are through (1) transfer of contaminated blood or blood products from one person to another; (2) unprotected sexual intercourse with an infected partner; (3) transmission from an infected mother to her fetus across the placenta during pregnancy and delivery; or (4) transfer via breast milk during nursing.

There are two components to the therapeutic management of HIV-infected persons: one directed against the virus itself to delay progression of the disease, and one to prevent or treat the opportunistic infections and cancers that ultimately cause death. The present recommended treatment for HIV infection itself is a simultaneous battery of at least four drugs. Two of these inhibit the action of the HIV enzyme (reverse transcriptase) that converts the viral RNA into the host's DNA, a third drug inhibits the HIV enzyme (α-protease) that cleaves a large protein into smaller units required for the assembly of new HIV, and a fourth drug blocks fusion of the virus with the T cell. The use of this complex and expensive regimen (called *HAART,* for highly active anti-retroviral therapy) greatly reduces the replication of HIV in the body and ideally should be introduced very early in the course of HIV infection, not just after the appearance of AIDS.

The ultimate hope for prevention of AIDS is the development of a vaccine. For a variety of reasons related to the nature of the virus (it generates large numbers of distinct subspecies) and the fact that it infects helper T cells, which are crucial for immune responses, vaccine development is not an easy task.

Antibiotics

The most important of the drugs employed in helping the body to resist microbes, mainly bacteria, are antibiotics. An **antibiotic** is any molecule or substance that kills bacteria. Antibiotics may be produced by one strain of bacteria to defend against other strains. Since the mid-twentieth century, commercial manufacture of antibiotics such as **penicillin** has revolutionized our ability to treat disease.

Antibiotics exert a wide variety of effects, including inhibition of bacterial cell-wall synthesis, protein synthesis, and DNA replication. Fortunately, a number of the reactions involved in the synthesis of protein by bacteria, and the proteins themselves, are sufficiently different from those in human cells that certain antibiotics can inhibit them without interfering with the body's own protein synthesis. For example, the antibiotic **erythromycin** blocks the movement of ribosomes along bacterial messenger RNA.

Antibiotics, however, must not be used indiscriminately. For one thing, they may exert allergic reactions, and they may exert toxic effects on the *body's* cells. A second reason for judicious use is the escalating and very serious problem of drug resistance. Most large bacterial populations contain a few mutants that are resistant to the drug, and these few may be capable of multiplying into large populations resistant to the effects of that particular antibiotic. Alternatively, the antibiotic can induce the expression of a latent gene that confers resistance. Finally, resistance can be transferred from one resistant microbe directly to another previously nonresistant microbe by means of DNA passed between them. (One example of how drug resistance can spread by these phenomena is that many bacterial strains that were once highly susceptible to penicillin now produce an enzyme that cleaves the penicillin molecule.) A third reason for the judicious use of antibiotics is that these agents may actually contribute to a new infection by eliminating certain species of relatively harmless bacteria that ordinarily prevent the growth of more dangerous ones. One site in which this may occur is the large intestine, where the loss of harmless bacteria may account for the symptoms of cramps and diarrhea that occur in some individuals taking certain types of antibiotics.

Harmful Immune Responses

Until now, we have focused on the mechanisms of immune responses and their protective effects. The following section discusses how immune responses can sometimes actually be harmful or unwanted.

Graft Rejection

The major obstacle to successful transplantation of tissues and organs is that the immune system recognizes the transplants, called grafts, as foreign and launches an attack against them. This is called **graft rejection.** Although B cells and macrophages play some role, cytotoxic T cells and helper T cells are mainly responsible for graft rejection.

Except in the case of identical twins, the class I MHC proteins on the cells of a graft differ from the recipient's. So do the class II molecules present on the macrophages in the graft (recall that virtually all organs and tissues have macrophages). Consequently, the MHC proteins of both classes are recognized as foreign by the recipient's T cells, and the cells bearing these proteins are destroyed by the recipient's cytotoxic T cells with the aid of helper T cells.

Some of the tools aimed at reducing graft rejection are radiation and drugs that kill actively dividing lymphocytes and thereby decrease the recipient's T-cell population. A very effective drug, however, is **cyclosporin,** which does not kill lymphocytes but rather blocks the production of IL-2 and other cytokines by helper T cells. This eliminates a critical signal for proliferation of both the helper T cells themselves and the cytotoxic T cells. Synthetic adrenal corticosteroids in large doses are also used to reduce the rejection.

There are several problems with the use of drugs like cyclosporin and potent synthetic adrenal corticosteroids: (1) immunosuppression with them is nonspecific, so patients taking them are at increased risk for infections and cancer; (2) they exert other toxic side effects; and (3) they must be used continuously to inhibit rejection. An important new kind of therapy, one that may be able to avoid these problems, is under study. Recall that immune tolerance for self proteins is achieved by clonal deletion and/or inactivation, and that the mechanism for this is absence of a nonantigenic costimulus at the time the antigen is first encountered. The hope is that, at the time of graft surgery, treatment with drugs that block the complementary proteins constituting the costimulus may induce a permanent state of immune tolerance toward the graft.

The Fetus as a Graft

During pregnancy, the fetal trophoblast cells of the placenta lie in direct contact with maternal immune cells. Because half of the fetal genes are paternal, all proteins coded for by these genes are foreign to the mother. Why does the mother's immune system refrain from attacking the trophoblast cells, which express such proteins, and thus fail to reject the placenta? This issue is far from solved, but one critical mechanism (there are certainly others) is as follows: Trophoblast cells, unlike virtually all other nucleated cells, do not express the usual class I MHC proteins. Instead, they express a unique class I MHC protein that maternal immune cells do not recognize as foreign.

Transfusion Reactions

Transfusion reactions, the illness caused when erythrocytes are destroyed during blood transfusion, are a special example of tissue rejection, one that illustrates the fact that antibodies rather than cytotoxic T cells can sometimes be the major factor in rejection. Erythrocytes do not have MHC proteins, but they do have plasma membrane proteins and carbohydrates (the latter linked to the membrane by lipids) that can function as antigens when exposed to another person's blood. There are more than 400 erythrocyte antigens, but the ABO system of carbohydrates is the most important for transfusion reactions.

Some people have the gene that results in synthesis of the A antigen, some have the gene for the B antigen, some have both genes, and some have neither gene. (Genes cannot code for the carbohydrates that function as antigens; rather, they code for the particular enzymes that catalyze the formation of the carbohydrates.) The erythrocytes of those with neither gene are said to have O-type erythrocytes. Consequently, the possible blood types are A, B, AB, and O (**Table 18–9**).

Table 18–9 Human ABO Blood Groups

Blood Group	Percent*	Antigen on RBC	Genetic Possibilities Homozygous	Genetic Possibilities Heterozygous	Antibody in Blood
A	42	A	AA	AO	Anti-B
B	10	B	BB	BO	Anti-A
AB	3	A and B	—	AB	Neither anti-A nor anti-B
O	45	Neither A nor B	OO	—	Both anti-A and anti-B

*In the United States

Type A individuals always have anti-B antibodies in their plasma. Similarly, type B individuals have plasma anti-A antibodies. Type AB individuals have neither anti-A nor anti-B antibody, and type O individuals have both. These anti-erythrocyte antibodies are called natural antibodies. How they arise naturally—that is, without exposure to the appropriate antigen-bearing erythrocytes—is not clear.

With this information as background, we can predict what happens if a type A person is given type B blood. There are two incompatibilities: (1) the recipient's anti-B antibodies cause the transfused cells to be attacked, and (2) the anti-A antibodies in the transfused plasma cause the recipient's cells to be attacked. The latter is generally of little consequence, however, because the transfused antibodies become so diluted in the recipient's plasma that they are ineffective in inducing a response. It is the destruction of the transfused cells by the recipient's antibodies that produces the problem.

Similar analyses show that the following situations would result in an attack on the transfused erythrocytes: a type B person given either A or AB blood; a type A person given either type B or AB blood; a type O person given A, B, or AB blood. Type O people are, therefore, sometimes called universal donors, whereas type AB people are universal recipients. These terms are misleading, however, because besides antigens of the ABO system, there are a host of other erythrocyte antigens and plasma antibodies against them. Therefore, except in a dire emergency, the blood of donor and recipient must be tested for incompatibilities directly by the procedure called *cross-matching.* The recipient's serum is combined on a glass slide with the prospective donor's erythrocytes (a "major" cross-match), and the mixture is observed for rupture (hemolysis) or clumping (agglutination) of the erythrocytes; this indicates a mismatch. In addition, the recipient's erythrocytes can be combined with the prospective donor's serum (a "minor" cross-match), looking again for mismatches.

Another group of erythrocyte membrane antigens of medical importance is the Rh system of proteins. There are more than 40 such antigens, but the one most likely to cause a problem is called Rh_o, known commonly as the **Rh factor** because it was first studied in rhesus monkeys. Human erythrocytes either have the antigen (Rh-positive) or lack it (Rh-negative). About 85 percent of the U.S. population is Rh-positive.

Antibodies in the Rh system, unlike the "natural antibodies" of the ABO system, follow the classical immunity pattern in that no one has anti-Rh antibodies unless exposed to Rh-positive cells from another person. This can occur if an Rh-negative person is subjected to multiple transfusions with Rh-positive blood, but its major occurrence involves the mother-fetus relationship. During pregnancy, some of the fetal erythrocytes may cross the placental barriers into the maternal circulation. If the mother is Rh-negative and the fetus is Rh-positive, this can induce the mother to synthesize anti-Rh antibodies. This occurs mainly during separation of the placenta at delivery. Thus, a first Rh-positive pregnancy rarely offers any danger to the fetus because delivery occurs before the mother makes the antibodies. In future pregnancies, however, these antibodies will already be present in the mother and can cross the placenta to attack and hemolyze the erythrocytes of an Rh-positive fetus. This condition, which can cause an anemia severe enough to cause the death of the fetus in utero or of the newborn, is called **hemolytic disease of the newborn.** The risk increases with each Rh-positive pregnancy as the mother becomes more and more sensitized.

Fortunately, this disease can be prevented by giving an Rh-negative mother human gamma globulin against Rh-positive erythrocytes within 72 h after she has delivered an Rh-positive infant. These antibodies bind to the antigenic sites on any Rh-positive erythrocytes that might have entered the mother's blood during delivery and prevent them from inducing antibody synthesis by the mother. The administered antibodies are eventually catabolized.

You may be wondering whether ABO incompatibilities are also a cause of hemolytic disease of the newborn. For example, a woman with type O blood has antibodies to both the A and B antigens. If her fetus is type A or B, this theoretically should cause a problem. Fortunately, it usually does not, partly because the A and B antigens are not strongly expressed in fetal erythrocytes and partly because the antibodies, unlike the anti-Rh antibodies, are of the IgM type, which do not readily cross the placenta.

Allergy (Hypersensitivity)

Allergy, or *hypersensitivity,* refers to diseases in which immune responses to environmental antigens cause inflammation and damage to the body itself. Antigens that cause allergy are called *allergens;* common examples include those in ragweed pollen and poison ivy. Most allergens themselves are relatively or completely harmless—it is the immune responses to them that cause the damage. In essence, then, allergy is immunity gone wrong, for the response is inappropriate to the stimulus.

A word about terminology is useful here: There are three major types of hypersensitivity, as categorized by the different immunologic effector pathways involved in the inflammatory response. The term *allergy* is sometimes used popularly to denote only one of these types, that mediated by IgE antibodies. We will follow the common practice, however, of using the term *allergy* in its broader sense as synonymous with *hypersensitivity.*

To develop a particular allergy, a genetically predisposed person must first be exposed to the allergen. This initial exposure causes "sensitization." It is the subsequent exposures that elicit the damaging immune responses we recognize as the disease. The diversity of allergic responses reflects the different immunological effector pathways elicited. The classification of allergic diseases is based on these mechanisms (**Table 18–10**).

In one type of allergy, the inflammatory response is independent of antibodies. It is due to pronounced secretion of cytokines by helper T cells activated by antigen in the area. These cytokines themselves act as inflammatory mediators and also activate macrophages to secrete their potent mediators. Because it takes several days to develop, this type of allergy is known as ***delayed hypersensitivity.*** The skin rash that appears after contact with poison ivy is an example.

In contrast to this are the various types of antibody-mediated allergic responses. One important type is called ***immune-complex hypersensitivity.*** It occurs when so many antibodies (of either the IgG or IgM types) combine with free antigens that large numbers of antigen-antibody complexes precipitate out on the surface of endothelial cells or are trapped in capillary walls, particularly those of the renal corpuscles. These immune complexes activate complement, which then induces an inflammatory response that damages the tissues immediately surrounding the complexes.

The more common type of antibody-mediated allergic responses, however, are those called ***immediate hypersensitivity,*** because they are usually very rapid in onset. They are also called ***IgE-mediated hypersensitivity*** because they involve IgE antibodies.

Immediate Hypersensitivity In immediate hypersensitivity, initial exposure to the antigen leads to some antibody synthesis and, more important, to the production of memory B cells that mediate active immunity. Upon re-exposure, the antigen elicits a more powerful antibody response. So far, none of this is unusual, but the difference is that the particular antigens that elicit immediate hypersensitivity reactions stimulate, in genetically susceptible persons, the production of type IgE antibodies. Production of IgE requires the participation of a particular subset of helper T cells that are activated by the allergens presented by B cells. These activated helper T cells then release cytokines that preferentially stimulate differentiation of the B cells into IgE-producing plasma cells.

Upon their release from plasma cells, IgE antibodies circulate throughout the body and become attached, via binding sites on their Fc portions, to connective-tissue mast cells (**Figure 18–21**). When the same antigen type subsequently enters the body and combines with the IgE bound to the mast cell, this triggers the mast cell to secrete many inflammatory mediators, including **histamine,** various eicosanoids, and chemokines. All of these mediators then initiate a local inflammatory response. (The entire sequence of events just described for mast cells can also occur with basophils in the circulation.)

Thus, the symptoms of IgE-mediated allergy reflect the various effects of these inflammatory mediators and the body site in which the antigen-IgE-mast cell combination occurs. For example, when a previously sensitized person inhales ragweed pollen, the antigen combines with IgE on mast cells in the respiratory passages. The mediators released cause increased secretion of mucus, increased blood flow, swelling of the epithelial lining, and contraction of the smooth muscle surrounding the airways. Thus, there follow the symptoms of congestion, runny nose, sneezing, and difficulty in breathing that characterize hay fever. Immediate hypersensitivities to penicillin and insect venoms sometimes occur, and these are usually correlated with IgE production.

Allergic symptoms are usually localized to the site of antigen entry. If very large amounts of the chemicals released by the mast cells (or blood basophils) enter the circulation, however, systemic symptoms may result and cause severe hypotension and bronchiolar constriction. This sequence of events, called ***anaphylaxis,*** can cause death due to circulatory and respiratory failure. It can be elicited in some sensitized people by the antigen in a single bee sting.

The very rapid components of immediate hypersensitivity often proceed to a ***late-phase reaction*** lasting many hours or days, during which large numbers of leukocytes, particularly eosinophils, migrate into the inflamed area. The chemoattractants involved are cytokines released by mast cells and helper T cells activated by the allergen. The eosinophils, once in the area, secrete mediators that prolong the inflammation and sensitize the tissues, so that less allergen is needed the next time to evoke a response.

Given the inappropriateness of most immediate hypersensitivity responses, how did such a system evolve? The normal physiological function of the IgE-mast cell-eosinophil pathways is to repel invasion by multicellular parasites that cannot be phagocytized. The mediators released by the mast cells stimulate the inflammatory response against the parasites, and the eosinophils serve as the major killer cells against them by secreting several toxins. How this system also came to be inducible by harmless agents is not clear.

Table 18–10	Major Types of Hypersensitivity

1. **Delayed hypersensitivity**—Mediated by helper T cells and macrophages; independent of antibodies

2. **Immune-complex hypersensitivity**—Mediated by antigen-antibody complexes deposited in tissue

3. **Immediate hypersensitivity**—Mediated by IgE antibodies, mast cells, and eosinophils

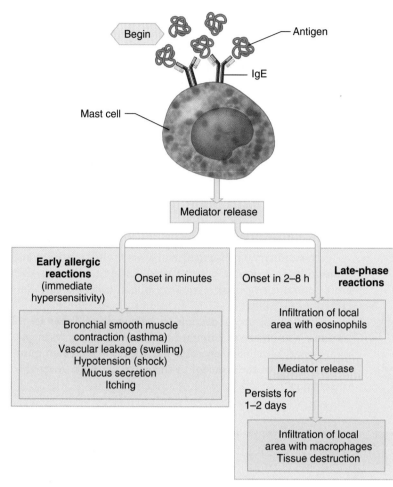

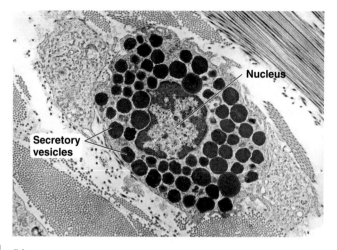

(b)

(a)

Figure 18–21

Immediate hypersensitivity allergic response. (a) Sequence of events. (b) Colorized electron micrograph of a mast cell, showing numerous secretory vesicles.

Autoimmune Disease

While allergy is due to an inappropriate response to an environmental antigen, *autoimmune disease* is due to an inappropriate immune attack triggered by the body's own proteins acting as antigens. The immune attack, mediated by autoantibodies and self-reactive T cells, is directed specifically against the body's own cells that contain these proteins.

We explained earlier how the body is normally in a state of immune tolerance toward its own cells. Unfortunately, there are situations in which this tolerance breaks down and the body does in fact launch antibody- or killer cell-mediated attacks against its own cells and tissues. A growing number of human diseases are being recognized as autoimmune in origin. Examples are *multiple sclerosis,* in which myelin is attacked; *myasthenia gravis,* in which the nicotinic receptors for acetylcholine on skeletal muscle cells are the target; *rheumatoid arthritis,* in which connective tissues in joints are damaged; and *Type 1 diabetes mellitus,* in which the insulin-producing cells of the pancreas are destroyed. Some possible causes for the body's failure to recognize its own cells are summarized in **Table 18–11.**

Excessive Inflammatory Responses

Recall that complement, other inflammatory mediators, and the toxic chemicals secreted by neutrophils and macrophages are not specific with regard to their targets. Consequently, sometimes during an inflammatory response directed against

microbes there can be so much generation or release of these substances that adjacent normal tissues may be damaged. These substances can also cause potentially lethal systemic responses. For example, macrophages release very large amounts of IL-1 and TNF, both of which are powerful inflammatory mediators (in addition to their other effects) in response to an infection with certain types of bacteria. These cytokines can cause profound vasodilation throughout the body, precipitating a type of hypotension called *septic shock.* This is often accompanied by dangerously high fevers. In other words, it is not the bacteria themselves that cause septic shock but rather the cytokines released in response to the bacteria.

Another important example of damage produced by excessive inflammation in response to microbes is the dementia that occurs in AIDS. HIV does not itself attack neurons, but it does infect microglia. Such invasion causes the microglia, which function as macrophage-like cells, to produce very high levels of inflammatory cytokines and other molecules that are toxic to neurons. (Microglia are also implicated in noninfectious brain disorders, like *Alzheimer's disease,* that are characterized by inflammation.)

Excessive chronic inflammation can also occur in the absence of microbial infection. Thus, various major diseases, including *asthma,* rheumatoid arthritis, and *inflammatory bowel disease,* are categorized as *chronic inflammatory diseases.* The causes of these diseases, and the interplay between

Table 18–11	Some Possible Causes of Autoimmune Attack

1. There may be failure of clonal deletion in the thymus or of clonal inactivation in the periphery. This is particularly true for "sequestered antigens," such as certain proteins that are unavailable to the immune system during critical early-life periods.

2. Normal body proteins may be altered by combination with drugs or environmental chemicals. This leads to an attack on the cells bearing the now "foreign" protein.

3. In immune attacks on virus-infected bodily cells, so many cells may be destroyed that disease results.

4. Genetic mutations in the body's cells may yield new proteins that serve as antigens.

5. The body may encounter microbes whose antigens are so close in structure to certain of the body's own proteins that the antibodies or cytotoxic T cells produced against these microbial antigens also attack cells bearing the self proteins.

6. Proteins normally never encountered by lymphocytes may become exposed as a result of some other disease.

genetic and environmental factors, are still poorly understood. Some, like rheumatoid arthritis, are mainly autoimmune in nature, but all appear to be associated with positive feedback increases in the production of cytokines and other inflammatory mediators.

Yet another example of excessive inflammation in a non-infectious state is the development of atherosclerotic plaques in blood vessels (Chapter 12). It is likely that, in response to endothelial cell dysfunction, the vessel wall releases inflammatory cytokines (IL-1, for example) that promote all stages of atherosclerosis—excessive clotting, chemotaxis of various leukocytes (as well as smooth muscle cells), and so on. The endothelial-cell dysfunction is caused by initially subtle vessel wall injury by lipoproteins and other factors, including elevated blood pressure and homocysteine (Chapter 12).

In summary, the various mediators of inflammation and immunity are a double-edged sword: In usual amounts they are essential for normal resistance, but in excessive amounts they can cause illness.

This completes the section on immunology. **Table 18–12** presents a summary of immune mechanisms in the form of a miniglossary of cells and chemical mediators involved in immune responses. All of the material in this table has been covered in this chapter.

Table 18–12	A Miniglossary of Cells and Chemical Mediators Involved in Immune Functions

Cells

Activated macrophages Macrophages whose killing ability has been enhanced by cytokines, particularly IL-2 and interferon-gamma.

Antigen-presenting cells (APC) Cells that present antigen, complexed with MHC proteins, on their surface to T cells.

B cells Lymphocytes that, upon activation, proliferate and differentiate into antibody-secreting plasma cells; provide major defense against bacteria, viruses in the extracellular fluid, and toxins; and can function as antigen-presenting cells to helper T cells.

Cytotoxic T cells The class of T lymphocytes that, upon activation by specific antigen, directly attack the cells bearing that type of antigen; are major killers of virus-infected cells and cancer cells; and bind antigen associated with class I MHC proteins.

Eosinophils Leukocytes involved in destruction of parasites and in immediate hypersensitivity responses.

Helper T cells The class of T cells that, via secreted cytokines, play a stimulatory role in the activation of B cells and cytotoxic T cells; also can activate NK cells and macrophages; and bind antigen associated with class II MHC proteins.

Lymphocytes The type of leukocyte responsible for specific immune defenses; categorized mainly as B cells, T cells, and NK cells.

Macrophages Cell type that (1) functions as a phagocyte, (2) processes and presents antigen to helper T cells, and (3) secretes cytokines involved in inflammation, activation of lymphocytes, and the systemic acute phase response to infection or injury.

Dendritic (macrophage-like) cells Several cell types that exert functions similar to those of macrophages.

Mast cells Tissue cells that bind IgE and release inflammatory mediators in response to parasites and immediate hypersensitivity reactions.

Memory cells B cells and cytotoxic T cells that differentiate during an initial immune response and respond rapidly during a subsequent exposure to the same antigen.

Monocytes A type of leukocyte; leaves the bloodstream and is transformed into a macrophage; has functions similar to those of macrophages.

Natural killer (NK) cells Class of lymphocytes that bind to cells bearing foreign antigens without specific recognition and kill them directly; major targets are virus-infected cells and cancer cells; participate in antibody-dependent cellular cytotoxicity (ADCC).

Neutrophils Leukocytes that function as phagocytes and also release chemicals involved in inflammation.

Plasma cells Cells that differentiate from activated B lymphocytes and secrete antibodies.

T cells Lymphocytes derived from precursors that differentiated in the thymus; see cytotoxic T cells and helper T cells.

Chemical Mediators

Acute phase proteins Group of proteins secreted by the liver during systemic response to injury or infection; stimulus for their secretion is IL-1, IL-6, and other cytokines.

Antibodies Immunoglobulins secreted by plasma cells; combine with the type of antigen that stimulated their production and direct an attack against the antigen or a cell bearing it.

C1 The first protein in the classical complement pathway.

Chemoattractants A general name given to any chemical mediator that stimulates chemotaxis of neutrophils or other leukocytes.

Chemokines Any cytokine that functions as a chemoattractant.

Chemotaxin A synonym for chemoattractant.

Complement A group of plasma proteins that, upon activation, kill microbes directly and facilitate the various steps of the inflammatory process, including phagocytosis; the classical complement pathway is triggered by antigen-antibody complexes, whereas the alternate pathway can operate independently of antibody.

C-reactive protein One of several proteins that function as nonspecific opsonins; production by the liver is increased during the acute phase response.

Cytokines General term for protein messengers that regulate immune responses; secreted by macrophages, monocytes, lymphocytes, neutrophils, and several nonimmune cell types; function both locally and as hormones.

Eicosanoids General term for products of arachidonic acid metabolism (prostaglandins, thromboxanes, leukotrienes); function as important inflammatory mediators.

Histamine An inflammatory mediator secreted mainly by mast cells; acts on microcirculation to cause vasodilation and increased permeability to protein.

IgA The class of antibodies secreted by the lining of the body's various "tracts."

IgD A class of antibodies whose function is unknown.

IgE The class of antibodies that mediate immediate hypersensitivity and resistance to parasites.

IgG The most abundant class of plasma antibodies.

IgM A class of antibodies that is produced first in all immune responses. Along with IgG, it provides the bulk of specific humoral immunity against bacteria and viruses.

Immunoglobulin (Ig) Proteins that function as B-cell receptors and antibodies; the five major classes are IgA, IgD, IgE, IgG, and IgM.

Interferon Group of cytokines that nonspecifically inhibits viral replication; interferon-gamma also stimulates the killing ability of NK cells and macrophages.

Interferon-gamma (See *Interferon*)

Interleukin 1 (IL-1) Cytokine secreted by macrophages (and other cells) that activates helper T cells, exerts many inflammatory effects, and mediates many of the systemic acute phase responses, including fever.

Interleukin 2 (IL-2) Cytokine secreted by activated helper T cells that causes helper T cells, cytotoxic T cells, and NK cells to proliferate, and causes activation of macrophages.

Interleukin 6 (IL-6) Cytokine secreted by macrophages (and other cells) that exerts multiple effects on immune system cells, inflammation, and the acute phase response.

Kinins Peptides that split from kininogens in inflamed areas and facilitate the vascular changes associated with inflammation; they also activate neuronal pain receptors.

Leukotrienes A class of eicosanoids that are generated by the lipoxygenase pathway and function as inflammatory mediators.

Membrane attack complex (MAC) Group of complement proteins that form channels in the surface of a microbe, making it leaky and killing it.

Natural antibodies Antibodies to the erythrocyte antigens (of the A or B type).

Opsonin General name given to any chemical mediator that promotes phagocytosis.

Perforin Protein secreted by cytotoxic T cells and NK cells that forms channels in the plasma membrane of the target cell, making it leaky and killing it; its structure and function are similar to that of the MAC in the complement system.

Tumor necrosis factor (TNF) Cytokine secreted by macrophages (and other cells) that has many of the same actions as IL-1.

Systemic Lupus Erythematosus

Among the many immunological diseases discussed in this chapter were those that fell under the general category of autoimmune disease. Because these are so common, and their consequences can be so severe, it is worth exploring them in greater detail. One well-known autoimmune disease with particularly serious symptoms is known as *systemic lupus erythematosus (SLE)*. SLE affects roughly two to three people per 10,000 population. As with most autoimmune diseases, the majority (90 percent) of SLE sufferers are female. Although the disease can be manifested at any age, it most commonly appears in women of childbearing age. SLE is distinguished from another form of lupus, known as drug-induced lupus erythematosus. In the latter, certain individuals being treated with drugs such as *hydralazine* (a vasodilator effective in reducing high blood pressure) have been known to develop lupus-like symptoms. True SLE, however, is a disease of the immune system.

The two major immune dysfunctions in SLE are hyperactivity of T and B cells, with overexpression of "self" antibodies, and decreased negative regulation of the immune response. In some other autoimmune diseases, one or a small number of antigens appear to be the root of the disorder, and these are often localized to one or a few organs. In SLE, however, the situation is much more grave. This is because there are many self proteins that become antigenic, and many of these are in the nuclei of all cells. Thus, the chief (but not only) site of autoantibody production in SLE is directed against proteins associated with cellular DNA and RNA. Because all cells share the same DNA and RNA, there are few—if any—parts of the body that are not susceptible to immune attack in SLE. Exactly what initiates the immune response in SLE is unclear. However, it is known that most people with this disease are photosensitive; that is, their skin cells are readily damaged by ultraviolet light from the sun. When these cells die, their nuclear contents become exposed to phagocytes and other components of the immune system. As a result, symptoms of SLE tend to flare up when a person with the disease is exposed to excessive sunlight.

SLE has a strong genetic component, as evidenced by the fact that approximately 40 to 50 percent of identical twins share the disease when one is afflicted. Moreover, there is an increased frequency of five specific class II MHC variants in people with SLE, as well as deficient or abnormal complement proteins. Still, there are almost certainly environmental triggers that elicit the disease in genetically susceptible people (because, as stated, in half the cases where one twin has SLE, the other does not). There is no conclusive evidence that infections due to viral invasion are a trigger for the development of SLE. Other than sunlight, other triggers that appear to be associated with the appearance of SLE are certain chemicals and foods, such as alfalfa sprouts.

SLE can be mild or severe, intermittent or chronic. In all cases, though, the effects are widespread. Typically, connective tissue damage is extensive, with repeated inflammatory reactions in joints, muscle, and skin. The outer covering of the heart (pericardium) may become inflamed, gastrointestinal activity may be affected, and retinal damage is sometimes observed. Even the brain is not spared, as cognitive dysfunction and even seizures may arise in severe cases. The skin often develops inflamed patches, notably on the face along the cheeks and bridge of the nose (forming the so-called "*butterfly rash*" seen in some patients with SLE). Perhaps the greatest danger occurs when immune complexes and immunoglobulins accumulate in the glomeruli of the nephrons of the kidney. This often leads to *nephritis* (inflammation of the nephrons) and results in damaged, leaky glomeruli. The appearance of protein in the urine, therefore, is a clinical finding associated with SLE. Finally, certain proteins on the plasma membranes of red blood cells and platelets may also become antigenic in SLE. When the immune system attacks these structures, the result is lysis of red blood cells and destruction and loss of platelets (*thrombocytopenia*). Loss of red blood cells in this manner contributes to the condition known as *hemolytic anemia,* a not uncommon manifestation of SLE.

In addition to the production of self antibodies in large numbers, there also appears to be a failure of the immune system to regulate itself. Thus, the immune attacks, once begun, do not stop after a few days but instead continue indefinitely. Some investigators believe this may be related to a deficiency or inactivity of suppressor T cells, but this has not been proven.

The treatments for SLE depend on its severity and the overall physical condition of the patient. In mild flare-ups, aspirin and other nonsteroidal anti-inflammatory drugs may be sufficient to control pain and inflammation, together with changes in lifestyle to avoid potential triggers. In more advanced cases, immunosuppression with high doses of synthetic adrenal corticosteroids is employed. ■

■ SUMMARY

Cells Mediating Immune Defenses

I. Immune defenses may be nonspecific, so that the identity of the target is not recognized, or they may be specific, so that it is recognized.

II. The cells of the immune system are leukocytes (neutrophils, eosinophils, basophils, monocytes, and lymphocytes), plasma cells, macrophages, dendritic or macrophage-like cells, and mast cells. The leukocytes use the blood for transportation but function mainly in the tissues.

III. Cells of the immune system (as well as some other cells) secrete protein messengers that regulate immune responses and are collectively called cytokines.

Nonspecific Immune Defenses

I. External barriers to infection are the skin, the linings of the respiratory, gastrointestinal, and genitourinary tracts, the cilia of these linings, and antimicrobial chemicals in glandular secretions.

II. Inflammation, the local response to injury or infection, includes vasodilation, increased vascular permeability to protein, phagocyte chemotaxis, destruction of the invader via phagocytosis or extracellular killing, and tissue repair.

  a. The mediators controlling these processes, summarized in Table 18–4, are either released from cells in the area or generated extracellularly from plasma proteins.

  b. The main cells that function as phagocytes are the neutrophils, monocytes, macrophages, and dendritic cells. These cells also secrete many inflammatory mediators.

  c. One group of inflammatory mediators—the complement family of plasma proteins activated during nonspecific inflammation by the alternate complement pathway—not only stimulates many of the steps of inflammation but mediates extracellular killing via the membrane attack complex.

  d. The final response to infection or tissue damage is tissue repair.

III. Interferon stimulates the production of intracellular proteins that nonspecifically inhibit viral replication.

Specific Immune Defenses

I. Lymphocytes mediate specific immune responses.

II. Specific immune responses occur in three stages.

  a. A lymphocyte programmed to recognize a specific antigen encounters it and binds to it via plasma membrane receptors specific for the antigen.

  b. The lymphocyte undergoes activation—a cycle of cell divisions and differentiation.

  c. The multiple active lymphocytes produced in this manner launch an attack all over the body against the specific antigens that stimulated their production.

III. The lymphoid organs are categorized as primary (bone marrow and thymus) or secondary (lymph nodes, spleen, tonsils, and lymphocyte collections in the linings of the body's tracts).

  a. The primary lymphoid organs are the sites of maturation of lymphocytes that will then be carried to the secondary lymphoid organs, which are the major sites of lymphocyte cell division and specific immune responses.

  b. Lymphocytes undergo a continuous recirculation among the secondary lymphoid organs, lymph, blood, and all the body's organs and tissues.

IV. The three broad populations of lymphocytes are B, T, and NK cells.

  a. B cells mature in the bone marrow and are carried to the secondary lymphoid organs, where additional B cells arise by cell division.

  b. T cells leave the bone marrow in an immature state, are carried to the thymus, and undergo maturation there. These cells then travel to the secondary lymphoid organs and new T cells arise from them by cell division.

  c. NK cells originate in the bone marrow.

V. B cells and T cells have different functions.

  a. B cells, upon activation, differentiate into plasma cells, which secrete antibodies. Antibody-mediated responses constitute the major defense against bacteria, viruses, and toxins in the extracellular fluid.

  b. Cytotoxic T cells directly attack and kill virus-infected cells and cancer cells, without the participation of antibodies.

  c. Helper T cells stimulate B cells and cytotoxic T cells via the cytokines they secrete. With few exceptions, this help is essential for activation of the B cells and cytotoxic T cells.

VI. B-cell plasma-membrane receptors are copies of the specific antibody (immunoglobulin) that the cell is capable of producing.

  a. Any given B cell or clone of B cells produces antibodies that have a unique antigen binding site.

  b. Antibodies are composed of four interlocking polypeptide chains; the variable regions of the antibodies are the sites that bind antigen.

VII. T-cell surface plasma-membrane receptors are not immunoglobulins, but they do have specific antigen binding sites that differ from one T-cell clone to another.

  a. The T-cell receptor binds antigen only when the antigen is complexed to one of the body's own plasma membrane MHC proteins.

  b. Class I MHC proteins are found on all nucleated cells of the body, whereas class II MHC proteins are found only on macrophages, B cells, and dendritic cells. Cytotoxic T cells require antigen to be complexed to class I proteins, whereas helper T cells require class II proteins.

VIII. Antigen presentation is required for T cell activation.

  a. Only macrophages, B cells, and dendritic cells function as antigen-presenting cells (APCs) for helper T cells. The antigen is internalized by the APC and hydrolyzed to peptide fragments, which are complexed with class II MHC proteins. This complex is then shuttled to the plasma membrane of the APC, which also delivers a nonspecific costimulus to the T cell and secretes interleukin 1 (IL-1) and tumor necrosis factor (TNF).

  b. A virus-infected cell or cancer cell can function as an APC for cytotoxic T cells. The viral antigen or cancer-associated antigen is synthesized by the cell itself and hydrolyzed to peptide fragments, which are complexed to class I MHC proteins. The complex is then shuttled to the plasma membrane of the cell.

IX. NK cells have the same targets as cytotoxic T cells, but they are not antigen-specific; most of their mechanisms of target identification are not understood.

X. Immune tolerance is the result of clonal deletion and clonal inactivation.

XI. In antibody-mediated responses, the membrane receptors of a B cell bind antigen, and at the same time a helper T cell also binds antigen in association with a class II MHC protein on a macrophage or other APC.

  a. The helper T cell, activated by the antigen, by a nonantigenic protein costimulus, and by IL-1 and TNF secreted by the APC, secretes IL-2, which then causes the helper T cell to proliferate into a clone of cells that secrete additional cytokines.

  b. These cytokines then stimulate the antigen-bound B cell to proliferate and differentiate into plasma cells, which secrete antibodies. Some of the activated B cells become memory cells, which are responsible for active immunity.

  c. There are five major classes of secreted antibodies: IgG, IgM, IgA, IgD, and IgE. The first two are the major antibodies against bacterial and viral infection.

  d. The secreted antibodies are carried throughout the body by the blood and combine with antigen. The antigen-antibody complex enhances the inflammatory response, in large part by activating the complement system. Complement proteins mediate many steps of inflammation, act as opsonins, and directly kill antibody-bound cells via the membrane attack complex.

   e. Antibodies of the IgG class also act directly as opsonins and link target cells to NK cells, which directly kill the target cells.

   f. Antibodies also neutralize toxins and extracellular viruses.

XII. Virus-infected cells and cancer cells are killed by cytotoxic T cells, NK cells, and activated macrophages.

   a. A cytotoxic T cell binds, via its membrane receptor, to cells bearing a viral antigen or cancer-associated antigen in association with a class I MHC protein.

   b. Activation of the cytotoxic T cell also requires cytokines secreted by helper T cells, themselves activated by antigen presented by a macrophage. The cytotoxic T cell then releases perforin, which kills the attached target cell by making it leaky.

   c. NK cells and macrophages are also stimulated by helper T-cell cytokines, particularly IL-2 and interferon-gamma, to attack and kill virus-infected or cancer cells.

Systemic Manifestations of Infection

   I. The acute phase response is summarized in Figure 18–20.
   II. The major mediators of this response are IL-1, TNF, and IL-6.

Factors that Alter the Body's Resistance to Infection

   I. The body's capacity to resist infection is influenced by nutritional status, the presence of other diseases, psychological factors, and the intactness of the immune system.
   II. AIDS is caused by a retrovirus that destroys helper T cells and therefore reduces the ability to resist infection and cancer.
   III. Antibiotics interfere with the synthesis of macromolecules by bacteria.

Harmful Immune Responses

   I. Rejection of tissue transplants is initiated by MHC proteins on the transplanted cells and is mediated mainly by cytotoxic T cells.
   II. Transfusion reactions are mediated by antibodies.
   a. Transfused erythrocytes will be destroyed if the recipient has natural antibodies against the antigens (type A or type B) on the cells.
   b. Antibodies against Rh-positive erythrocytes can be produced following the exposure of an Rh-negative person to such cells.
   III. Allergies (hypersensitivity reactions) caused by allergens are of several types.
   a. In delayed hypersensitivity, the inflammation is due to the interplay of helper T cell cytokines and macrophages. Immune-complex hypersensitivity is due to complement activation by antigen-antibody complexes.
   b. In immediate hypersensitivity, antigen binds to IgE antibodies, which are themselves bound to mast cells. The mast cells then release inflammatory mediators such as histamine that produce the symptoms of allergy. The late phase of immediate hypersensitivity is mediated by eosinophils.
   IV. Autoimmune attacks are directed against the body's own proteins, acting as antigens. Reasons for the failure of immune tolerance are summarized in Table 18–11.
   V. Normal tissues can be damaged by excessive inflammatory responses to microbes.

Additional Clinical Examples

   I. Systemic lupus erythematosus (SLE) is an autoimmune disease in which several antigens on different types of cells are mistakenly recognized as "nonself."
   II. The major clinical problems in SLE are nephritis, anemia, and thrombocytopenia.

■ KEY TERMS

■ REVIEW QUESTIONS

1. What are the major cells of the immune system and their general functions?
2. Describe the major anatomical and biochemical barriers to infection.
3. Name the three cell types that function as phagocytes.
4. List the sequence of events in an inflammatory response and describe each step.
5. Name the sources of the major inflammatory mediators.
6. What triggers the alternate pathway for complement activation? What roles does complement play in inflammation and cell killing?
7. Describe the antiviral role of interferon.
8. Name the lymphoid organs. Contrast the functions of the bone marrow and thymus with those of the secondary lymphoid organs.
9. Name the various populations and subpopulations of lymphocytes and discuss their roles in specific immune responses.
10. Contrast the major targets of antibody-mediated responses and responses mediated by cytotoxic T cells and NK cells.
11. How do the Fc and combining-site portions of antibodies differ?
12. What are the differences between B-cell receptors and T-cell receptors? Between cytotoxic T-cell receptors and helper T-cell receptors?
13. Compare and contrast antigen presentation to helper T cells and cytotoxic T cells.
14. Compare and contrast cytotoxic T cells and NK cells.
15. What two processes contribute to immune tolerance?
16. Diagram the sequence of events in an antibody-mediated response, including the role of helper T cells, interleukin 1, and interleukin 2.
17. Contrast the general functions of the different antibody classes.
18. How is complement activation triggered in the classical complement pathway, and how does complement "know" what cells to attack?
19. Name two ways in which the presence of antibodies enhances phagocytosis.
20. How do NK cells "know" which cells to attack in ADCC?
21. Diagram the sequence of events by which a virus-infected cell is attacked and destroyed by cytotoxic T cells. Include the roles of cytotoxic T cells, helper T cells, interleukin 1, and interleukin 2.
22. Contrast the extracellular and intracellular phases of immune responses to viruses, discussing the role of interferon.
23. List the systemic responses to infection or injury and the mediators responsible for them.
24. What factors influence the body's resistance to infection?
25. What is the major defect in AIDS, and what causes it?
26. What is the major cell type involved in graft rejection?
27. Diagram the sequences of events in immediate hypersensitivity.

Chapter 18 Test Questions

(Answers appear in Appendix A.)

1. Which of the following is an opsinin?
   a. IL-2
   b. C1 protein
   c. C3b protein
   d. C-reactive protein
   e. membrane attack complex

2. Which are important in nonspecific immune defenses?
   a. interferons
   b. clonal inactivation
   c. lymphocyte activation

   d. secretion of antibodies from plasma cells
   e. class 1 MHC proteins

3. A second exposure to a given foreign antigen elicits a rapid and pronounced immune response because
   a. passive immunity occurs after the first exposure.
   b. some B cells differentiate into memory B cells after the first exposure.
   c. there are a greater number of antigen-presenting cells available due to the earlier exposure.
   d. the array of class II MHC proteins expressed by antigen-presenting cells is permanently altered by the first exposure.
   e. Both a and b are correct.

4. Which statement is incorrect?
   a. The most abundant immunoglobulins are IgG and IgM antibodies.
   b. IgG antibodies are involved in specific immune responses against bacteria and viruses in the extracellular fluid.
   c. IgM antibodies are primarily involved in immune defense mechanisms found in the surface or lining of the gastrointestinal, respiratory, and genitourinary tracts.
   d. All antibodies of a given class have an Fc portion that is identical in amino acid sequence.
   e. Antibodies can exist at the surface of a B cell, or be circulating freely in the blood.

True/False

5. Antibiotics are useful for treating illnesses caused by viruses.
6. Chronic inflammatory diseases may occur even in the absence of any infection.
7. All T cells are derived from lymphocytes, but not all lymphocytes are T cells.
8. Edema, which occurs during inflammation, has important adaptive value in helping defend against infection or injury.
9. Bone marrow and the thymus gland are examples of secondary lymphoid organs.

Chapter 18 Quantitative and Thought Questions

(Answers appear in Appendix A.)

1. If an individual failed to develop a thymus because of a genetic defect, what would happen to the immune responses mediated by antibodies and those mediated by cytotoxic T cells?

2. What abnormalities would a person with a neutrophil deficiency display? A person with a monocyte deficiency?

3. An experimental animal is given a drug that blocks phagocytosis. Will this drug prevent the animal's immune system from killing foreign cells via the complement system?

4. If the Fc portion of a patient's antibodies is abnormal, what effects could this have on antibody-mediated responses?

5. Would you predict that patients with AIDS would develop fever in response to an infection? Explain.

Chapter 18 Answers to Physiological Inquiries

Figure 18–6 Many molecules in the body act this way. For example, somatostatin acts locally in the stomach to control acid production (paracrine) and is secreted into the hypothalamo-pituitary portal veins to control growth hormone secretion (endocrine). Testosterone acts locally within the testes (paracrine) and reaches other targets through the blood (endocrine).

Figure 18–7 Vasodilation and increased protein permeability of the microcirculation both contribute to an increase in the rate of filtration of fluid from the plasma into the interstitial space. Because lymph vessels are the main route by which fluid and protein are returned from the interstitial space to the circulatory system (see Figure 12–47), these changes will lead to increased flow of lymph. As that fluid flows through the lymph nodes, lymphocytes are exposed to antigens from the invading pathogen, thus activating the specific immune response.

Medical Physiology: Integration Using Clinical Cases

Rushing a patient to the emergency room

Physiology is one of the pillars of the health-related professions, including nursing, occupational health, physical therapy, dentistry, and medicine. In fact, the term *pathophysiology*—the changes in function associated with disease—highlights the intertwining of physiology and medicine. You need a thorough understanding of physiological principles to properly diagnose and treat diseases and disorders. We are aware that many users of this textbook may not be planning a career in the health professions. However, teachers of physiology can attest to the fact that the use of clinical examples is an extremely effective approach to highlight and reinforce the understanding of the functions and interactions of the organ systems of the body.

This chapter uses clinical cases to allow you to continue to explore the material you learned from this book and, at the same time, review some of the basic principles of physiology. More importantly, this chapter illustrates the concept of *integrative physiology*. In real life, virtually every significant clinical case involves many different organ systems. The true art of medicine is the ability of clinicians to recall these basic principles and put them together in the evaluation of the patient. Each case has a section called "Physiological Integration" to highlight this fact. As you read these sections, you should consider the relationships among disease, integrative physiology, and homeostasis, the last of which has been a theme throughout this textbook.

The cases are organized so that they get more complicated and challenging as the chapter proceeds. For this reason, we encourage you to read the cases in sequence. Furthermore, some of the conditions described in this chapter are not explicitly described in the book and may be new to you. Interspersed at key points in the chapter are several places where you will be asked to "Reflect and Review." In some cases, specific answers to these questions are not provided in the case itself. We encourage you to answer these questions as the case unfolds by, if necessary, referring back to the appropriate section of the book. Furthermore, we have annotated each case with figure and table numbers to facilitate review of material covered in previous chapters. We hope that this will motivate you to synthesize and integrate information from throughout the book and perhaps even go beyond what you have learned. In fact, you may enjoy consulting other sources to answer some of the more challenging questions or learn more about specific aspects of each case that interest you.

We hope you enjoy this chapter and we encourage the readers of the book to send us comments and suggestions for additional cases.

CASE 19–1

A Woman with Palpitations and Heat Intolerance

Case Presentation

A 23-year-old woman visits her family physician with complaints of a 12-month history of increasing nervousness, irritability, and **palpitations** (a noticeable increase in the force of her heartbeat). Furthermore, she feels very warm in a room when everyone else feels comfortable. She has lost 30 pounds of body weight over this period despite having a voracious appetite and increased food intake.

Reflect and Review #1
■ Describe the control of body temperature (see Figure 16–18). What may have contributed to the woman's feelings of excessive warmth?

Two years ago, she was jogging about 20 miles per week. However, she had not done any running for the past year because she "didn't feel up to it" and complained of some weakness in her legs. She said she often felt irritable and had mood swings. Her menstrual periods have been less frequent over the past year. Her previous medical history was normal for a person her age.

Reflect and Review #2
■ What could cause a change in the frequency of menstrual periods (see Figure 17–18 and Table 17–7)?
■ Is there anything that could affect menstrual periods that might also account for the feelings of excessive warmth?

Physical Examination

The patient is a 5' 7" (170 cm) tall, 110 pound (50 kg) woman. Her systolic/diastolic blood pressure is 140/60 mmHg (normal for a young, healthy woman is about 110/70 mmHg). Her resting pulse rate is 100 beats/minute. Before she became ill, her resting heart rate was about 60–70 beats/minute. Her respiratory rate is 17 breaths per minute (normal for her was approximately 12–14 breaths per minute). Her skin is warm and moist. Her eyes are bulging out (**proptosis** or **exopthalmos**) (**Figure 19–1**). Finally, when she is asked to gaze to the far right, her right eye does not move as far as does her left eye and she says she has double vision (**diplopia**).

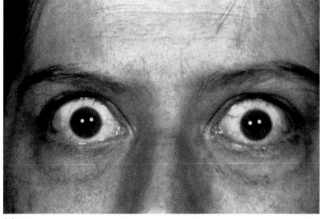

(a)

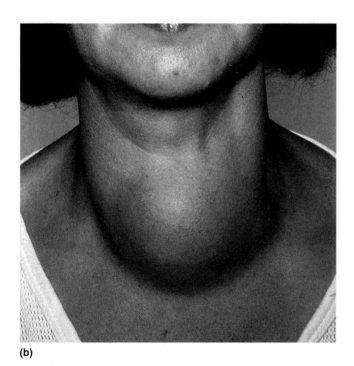

(b)

Figure 19–1

Patient showing (a) proptosis and (b) goiter.

Reflect and Review #3

■ Briefly describe the control of systemic blood pressure, heart rate, and respiratory rate (see Figures 12–23 and 12–51, and Figure 13–32). What might be causing her hypertension, *tachycardia* (increased heart rate), and *tachypnea* (increased respiratory rate)?

■ Describe the muscles that control eye movement (see Figure 7–32). Based on Figure 7–32, what eye muscles do you think are affected in this woman?

Upon further examination, the physician notes an enlargement of a structure in the front, lower part of her neck (see Figure 19–1). It is smooth (no bumps or nodules felt) and painless. When the patient swallows, this enlarged structure moves up and down. When a stethoscope is placed over this structure, the physician can hear a swishing sound (called a *bruit* [pronounced BREW-ee]) with each heartbeat.

Reflect and Review #4

■ What structure might be responsible for the swelling in the patient's lower neck (see Figures 11–21 and 11–31, and Figure 15–15)? What are the major functions of this structure?

Her patellar tendon (knee jerk) reflexes are hyperactive. When she holds her hands out straight, she exhibits fine tremors (shaking).

Reflect and Review #5

■ What are the neural pathways involved in the knee jerk reflex (see Figure 10–6)? Could the enlarged structure in her neck account for the abnormal reflexes observed?

Laboratory Tests

The family physician considers the history and physical exam and decides to order some blood tests. The results are shown in **Table 19–1**.

Reflect and Review #6

■ Describe the feedback control loops of the hormones whose values were abnormal (see Figure 11–23 and Figure 17–20). Which, if any, of these hormones might account for the symptoms in this patient?

Table 19–1	Laboratory Results for Patient	
Blood Measurements*	**Result**	**Normal Range**
Sodium	136 mEq/L	(135–145 mEq/L)
Potassium	5.0 mEq/L	(3.8–5.2 mEq/L)
Chloride	102 mEq/L	(95–105 mEq/L)
pH	7.39	(7.38–7.45)
Calcium (total)	9.6 mg/dL	(9.0–10.5 mg/dL)
Parathyroid hormone	15 pg/mL	(10–75 pg/mL)
Glucose (fasting)	80 mg/dL	(70–110 mg/dL)
Prolactin	10.4 ng/mL	(1.4–24.2 ng/mL)
Estrogen (midcycle)	100 pg/mL	(150–750 pg/mL)
Total T_4	20 μg/dL	(5–11 μg/dL)
Free T_4	2.8 ng/dL	(0.8–1.6 ng/dL)
Thyroid-stimulating hormone (TSH)	0.01 μU/mL	(0.3–4.0 μU/mL)

T_4, thyroxine

*In actuality, these measurements are performed in blood serum.

- Why is the serum glucose sample obtained in the fasted state (see Figure 16–8)? Does the serum glucose concentration rule out diabetes mellitus as a factor in this patient's illness?

Diagnosis

The most likely explanation for the findings is an increase in thyroid hormone in the patient's blood. When elevated thyroid hormone causes significant symptoms, it is part of a condition called **hyperthyroidism,** which, when severe enough, can lead to **thyrotoxicosis.** The enlarged organ in the neck is likely the thyroid gland, although large thyroid glands (**goiters**) also occur in hypothyroidism (see Figure 11–24 for an extreme example). In order to interpret the thyroid function tests shown in Table 19–1, first review the control of thyroid hormone synthesis and release (see Figure 11–23).

There are two circulating thyroid hormones—thyroxine (T_4) and triiodothyronine (T_3). Whereas T_4 is the principal hormone released by the thyroid gland, T_3 is actually more potent and is actively produced in target tissues by the removal of one iodine molecule from T_4. The release of T_4 by the thyroid gland is normally controlled by thyroid-stimulating hormone (TSH) secreted by the anterior pituitary gland. Binding of TSH to its G-protein-coupled plasma membrane receptor acti-

vates adenylyl cyclase and cAMP formation, which then stimulates cAMP-dependent protein kinase (see Figure 5–6). Like most anterior pituitary tropic/trophic hormones, TSH not only stimulates the activity of the thyroid gland, it also stimulates its growth. As with most other pituitary-target hormone systems, the target gland hormone (T_4) inhibits the release of the anterior pituitary hormone controlling it (in this case, TSH) via negative feedback (see Figures 11–20 and 11–23).

There are several reasons why the thyroid gland in this patient could be producing too much thyroid hormone, leading to thyrotoxicosis. The most common condition to focus on here is called **Graves' disease.** In this condition, the thyroid gland is stimulated by antibodies that activate the receptor for TSH on the follicular cell of the thyroid (**Figure 19–2**). Therefore, these TSH receptor-stimulating antibodies mimic the action of TSH, but are distinct from authentic TSH from the anterior pituitary gland. These **thyroid-stimulating immunoglobulins (TSIs)** are characteristic of an autoimmune disorder in which the patient makes antibodies that bind to one or more proteins expressed in his or her own tissues (see Table 18–11). TSIs are produced by B-lymphocytes that, in addition to residing in lymph nodes, can actually infiltrate the thyroid gland in Graves' disease. In Chapter 9, you learned about a disease called myasthenia gravis, in which autoantibodies bind to and destroy the nicotinic acetylcholine recep-

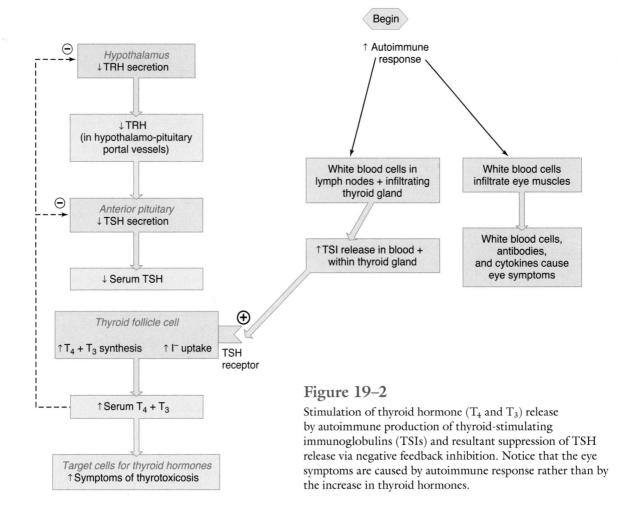

Figure 19–2

Stimulation of thyroid hormone (T_4 and T_3) release by autoimmune production of thyroid-stimulating immunoglobulins (TSIs) and resultant suppression of TSH release via negative feedback inhibition. Notice that the eye symptoms are caused by autoimmune response rather than by the increase in thyroid hormones.

tor in the neuromuscular junction. This is typical of antibody-antigen reactions, in which antigens are removed from the body (Chapter 18). In Graves' disease, however, the autoantibodies are highly unusual in that they not only recognize and bind to the TSH receptor on thyroid follicular cells, but this binding *stimulates* rather than destroys the receptor. Therefore, TSIs stimulate the thyroid gland to synthesize and secrete excess T_4 independently of TSH. The increase in T_4 would be predicted to suppress the secretion of TSH from the anterior pituitary gland by negative feedback, which is consistent with the low serum TSH level measured in the patient's blood. The elevated serum T_4 probably also suppressed the synthesis and release of thyrotropin-releasing hormone (TRH) from the hypothalamus via negative feedback. (Serum TRH levels are not determined in such situations because TRH is secreted directly into the hypothalamo-pituitary portal circulation. The actual amount of TRH from the hypothalamus that reaches the systemic circulation is too small for its measurement in blood in a peripheral vein to be useful.)

The total and free (not bound to plasma proteins) T_4 concentrations in the blood of this patient are elevated, confirming the diagnosis of hyperthyroidism. Measurement of free T_4 is helpful because most of the circulating thyroid hormone in the blood is bound to plasma proteins, so measuring the serum T_4 that is not bound to plasma proteins proves that there is an increase in the amount of biologically active T_4. The suppressed TSH confirms that the T_4 is elevated independently of stimulation from the anterior pituitary gland. This suppression of serum TSH, as in our patient, is one of the hallmarks of Graves' disease and is often used as a screening test for many disorders of the thyroid gland. Although TSI concentrations can be measured in the serum in some patients, it is often not necessary because the diagnosis of Graves' disease is the most likely, considering that it represents the majority of patients with hyperthyroidism.

Although Graves' disease was by far the most likely diagnosis, the physician ordered some additional tests to rule out other possible causes of the symptoms. Serum electrolytes were measured because they are important in the generation and maintenance of membrane potentials (Chapter 6) and their abnormalities can lead to weakness and palpitations. Serum calcium and parathyroid hormone were measured because weakness is a common finding in primary hyperparathyroidism (Chapter 11). A normal fasting plasma glucose and pH indicated that diabetes mellitus was probably not the cause of the patient's weakness and fatigue. A normal prolactin level indicated that she does not have hyperprolactinemia, which can cause abnormalities in the menstrual cycle (see Additional Clinical Examples in Chapter 17).

Physiological Integration

Thyroid disease is relatively common, particularly in women, in whom up to 10 percent will develop hyper- or hypothyroidism by the age of 60 to 65. Thyroid hormone has a wide range of effects throughout the body and, therefore, an understanding of all the organ systems is extremely useful in understanding the symptoms of thyroid disease.

One of the main effects of thyroid hormone is calorigenic—it increases the basal metabolic rate (BMR). This increase in metabolic rate is caused by activation of intracellular thyroid hormone receptors (see Figure 5–4) that are expressed in cells throughout the body. This leads to increased expression of Na^+/K^+-ATPases as well as the synthesis of other cellular proteins involved in oxygen consumption and metabolic rate in many tissues (see Figure 3–45). The resultant increase in heat production by our patient explains the warmness and moistness of her skin and her heat intolerance. It also explains why, despite eating more, she is losing weight because she is burning more fuel than she is ingesting.

The nervousness, irritability, and emotional swings are likely due to effects of thyroid hormone on the central nervous system, although the exact cellular mechanism of this is not well understood. The symptoms also appear to be due to an increased sensitivity within the central nervous system to circulating catecholamines. The muscle weakness is probably due to a thyroid hormone-induced increase in muscle protein turnover, local metabolic changes, and loss of muscle mass. Despite this, there appears to be an increase in the speed of muscle contraction and relaxation, contributing to the hyperactive reflexes observed in our patient.

Her thyroid gland is enlarged because TSIs are mimicking the actions of TSH to stimulate the thyroid gland to grow. The enlarged thyroid with increased metabolic activity explains why a bruit was heard over the thyroid gland. The thyroid gland has a high blood flow per gram of tissue even in normal individuals. The increase in thyroid function in Graves' disease leads to a marked increase in blood flow to the thyroid—so much so that it is audible with a stethoscope during systole in some patients.

Her increased systolic blood pressure and heart rate can be explained in several ways. First, there are direct effects of thyroid hormones on the heart, such as an increase in transcription of the myosin genes. Second, as described in Chapter 11, Section C, thyroid hormone has permissive effects to potentiate the effects of catecholamines on the cardiovascular system. Finally, the small decrease in diastolic pressure may result from arteriolar vasodilation and reduced total peripheral resistance in response to increased tissue temperature and metabolite concentrations (see Figure 12–51).

Increased thyroid hormone can directly inhibit the release of the pituitary gonadotropins FSH and LH. This can lead to a decrease in release of gonadal steroids from the ovaries, an irregular pattern or complete loss of menstrual periods, and a lack of ovulation. This explains the lower serum estrogen levels at the middle of the menstrual cycle in our patient.

The eye findings are among the most striking in many patients with Graves' disease (see Figure 19–1). The proptosis (bulging out of the eye) is likely due to the autoimmune component of the disease, rather than to a direct effect of thyroid hormone. Supporting this idea is that proptosis can occur before the development of hyperthyroidism, and excessive thyroid hormone therapy for hypothyroidism usually does not cause proptosis. Furthermore, proptosis is caused by infiltration of white blood cells into the extra-ocular muscles behind the eye, rather than being caused by TSIs. These cells release chemicals that result in inflammation (see Figure 19–1 and Figure 7–32), causing the muscles to swell and forcing the eyeball forward. Sometimes, particular muscles of the eye are more affected than others, which explains the double vision of our patient when she gazes to one side.

Therapy

The most important component of the treatment is to lower the thyroid hormone levels. There are three general approaches to accomplish this. Removal of the thyroid gland is the most obvious, but currently the least frequently used approach. Removing a large, hyperactive thyroid gland has definite surgical risks and is usually not performed unless absolutely necessary. The drugs **methimazole** and **propylthiouracil** can be used because they block the synthesis of thyroid hormone by reducing the oxidation and organification of iodide (see Figure 11–22). Although these drugs are effective in some patients, they sometimes lose effectiveness and do not provide a definitive cure.

In the United States, a common approach is a more permanent, nonsurgical treatment. This involves the partial destruction of the thyroid gland with a high dose of orally administered **radioactive iodide.** Remember that iodide is a critical component of thyroid hormone (see Figure 11–22),

and the thyroid gland has a mechanism to trap iodide by secondary active transport from the blood into the follicular cell. Radioactive iodide administered to the patient is trapped by the thyroid; the local emission of radioactive decay destroys most of the thyroid gland over time. However, the procedure does not work equally well in all people. In fact, sometimes patients have so much of their thyroid gland destroyed that they become hypothyroid. Such people must take T_4 pills for the rest of their lives to maintain thyroid hormones in the normal range.

In the short term, while waiting for the treatments to take effect, patients benefit from treatment with beta-adrenergic receptor blockers (see Table 12–9) to reduce the effects of increased sensitivity to circulating catecholamines. This often helps control the palpitations and increased heart rate, as well as some of the other symptoms such as nervousness and tremors. Because proptosis is not caused by the increase in T_4, its treatment can be accomplished, if necessary, with anti-inflammatory drugs, such as glucocorticoids, or surgery or radiation therapy of the eye muscles.

With adequate treatment, patients generally get better over time with most, if not all, of the symptoms resolving. Our patient was treated with radioactive iodide, and all of her symptoms slowly resolved over several months.

CASE 19–2

A Man with Chest Pain After a Long Airplane Flight

Case Presentation

A 50-year-old, obese man has just returned from vacationing in Hawaii. He took an 8-h flight during which he sat by the window and did not leave his seat. In the taxi on the way home from the airport, he starts to feel chest pain, and has shortness of breath, increased respiratory rate, and nausea. Thinking he is having a heart attack (**myocardial infarction**), he asks the taxi driver to take him to the hospital emergency room.

Physical Examination

An examination of the patient at the hospital indicates that he has dull, aching chest pain, is clearly upset and anxious, short of breath, and overweight. He is 68 in (173 cm) tall and weighs 300 lbs (136 kg). The emergency room physician immediately orders an electrocardiogram (ECG), primarily to rule out a heart attack. The ECG shows an increased heart rate (100 beats per min), but does not show changes consistent with a heart attack or with left heart failure.

Reflect and Review #7
- What are the main factors that control heart rate (see Figure 12–23)? Might any of them explain the elevated heart rate in our patient?
- How might damage to the heart be detected in an ECG? (See Figures 12–15 and 12–16 for a general discussion of ECG.)

A chest x-ray is performed in an attempt to determine the cause of the patient's chest pain and shortness of breath. The results indicate no abnormalities such as pneumonia or collapse of lung lobes (**atelectasis**).

Laboratory Tests

Based on the patient's history and symptoms, the physician obtains a sample of the patient's arterial blood in order to measure the levels of oxygen, carbon dioxide, bicarbonate, hy-

Table 19–2	Blood Gas, Bicarbonate, and Hemoglobin Results While Patient Breathes Room Air	
Blood Measurement	**Result**	**Normal Range**
Arterial P_{O_2}	60 mmHg	>90 mmHg
Arterial P_{CO_2}	30 mmHg	35–45 mmHg
Arterial pH	7.50	7.38–7.45
Bicarbonate	22 mmol/L	23–27 mmol/L
Hemoglobin	15 g/dL	14–18 g/dL

drogen ions (pH), and hemoglobin. The findings are shown in **Table 19–2**.

Reflect and Review #8

- What is the cause of the change in arterial pH in our patient (see Table 14–8)?

The results of these tests reveal that the patient has hypoxic hypoxia (hypoxemia), as indicated by the low arterial P_{O_2}, and has an acute respiratory alkalosis, as indicated by the low arterial P_{CO_2} and bicarbonate, and high arterial pH. The normal hemoglobin concentration indicates that the patient is not anemic.

Reflect and Review #9

- What are some possible causes of hypoxemia (see Table 13–11)?
- What are the two main types of alkalosis (see Table 14–8)?
- How do we know the alkalosis in our patient was acute (of recent, short-term origin) (see Table 14–8)?

The patient is given 100 percent oxygen to breathe through a mask over his mouth and nose. This results in an increase in arterial P_{O_2} to 205 mmHg, a small increase in arterial P_{CO_2} to 32 mmHg, and a small decrease in arterial pH to 7.48. The normal response to breathing 100 percent oxygen is an increase in arterial P_{O_2} to greater than 600 mmHg, and no change in arterial P_{CO_2} or pH.

Reflect and Review #10

- Explain why increasing arterial P_{O_2} with supplemental oxygen caused the observed changes in arterial P_{CO_2} and pH (see Figure 13–40).

Diagnosis

Because a heart attack has been ruled out, the physician suspects that the patient has at least one **pulmonary embolism.** An **embolism** (plural, emboli) is a blockage of blood flow through a blood vessel produced by an obstruction. It is often caused by a blood clot—or **thrombus**—in the pulmonary arteries/arterioles. These clots usually arise from larger clots in leg veins.

To confirm his diagnosis, the physician orders a **ventilation-perfusion scan.** In the ventilation scan, the patient inhales a small amount of radioactive gas. Special imaging devices are then used to detect the inhaled radiation and visualize which parts of the lung are adequately ventilated. Poorly ventilated areas of the lung will contain less radioactive gas. In the perfusion scan, a small amount of albumin, a naturally occurring plasma protein, tagged with a radioactive tracer, is injected into a vein. As the radioactive protein enters the pulmonary circulation, its distribution can be monitored using the same imaging device as just described. This procedure allows the physician to determine if parts of the lungs are receiving less than their normal share of blood flow because poorly perfused areas of the lungs will contain less radioactive albumin. The ventilation scan was normal, but the perfusion scan showed abnormalities. **Figure 19–3** shows the results of the perfusion scan, demonstrating dramatic decreases in perfusion in specif-

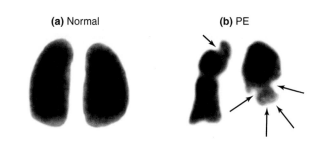

(a) Normal   **(b)** PE

Figure 19–3

Pulmonary embolism (PE) from a deep vein thrombosis shown on a lung perfusion scan (posterior) with radiolabeled albumin. (a) A normal perfusion scan. (b) Multiple perfusion defects are shown (arrows).

Source: From Access Medicine On-Line–Current Medical Dx & Tx. McGraw-Hill, 2006.

ic regions of the lung. These results supported the physician's diagnosis of several pulmonary emboli.

There are a variety of materials that can occlude pulmonary arterial blood vessels, including air, fat, foreign bodies, parasite eggs, and tumor cells. The most common embolus is a thrombus that can theoretically come from any large vein, but usually comes from the deep veins of the muscles in the calves (**deep vein thrombosis**). The fact that our patient sat on an 8-h flight without moving around greatly increased the chances for the formation of a deep vein thrombosis in the leg. This is because blood that is allowed to pool in the leg veins has a tendency to form clots (see Figure 12–45). After the abnormal lung perfusion scan, an ultrasound examination of the legs was performed to confirm whether clots were present in the leg veins. The results showed a large clot in the femoral and popliteal veins in the right leg.

Pulmonary embolism is a common and potentially fatal result of deep vein thrombosis. In fact, pulmonary embolism and deep vein thrombosis can be considered part of one syndrome. It may cause as many as 200,000 deaths each year in the United States. Most cases are not diagnosed until after death (on post-mortem examination) either because the symptoms are initially mild or because the syndrome is misdiagnosed. Most small clots that form in small veins in the calves of the lower legs remain fixed in place associated with the lining of the vein, and do not cause symptoms. However, if a clot enlarges and migrates into larger veins such as the femoral and popliteal veins, as in our patient, it can break off and migrate up the vena cava, through the right atrium and right ventricle, and into the pulmonary arterial circulation, where it can become lodged (see Figure 12–2 for an overview of the circulatory system). When this happens, blood flow is reduced or cut off to one or more large segments of the lung.

Reflect and Review #11

- Why will regional decreases in pulmonary blood flow lead to hypoxemia (see Figure 13–24 and Table 13–11)?

Fortunately, these clots are too large to pass through the pulmonary circulation into the systemic circulation. When clots do form on the arterial side of the circulation, they can occlude arterial blood vessels, thereby depriving vital organs of oxygen and nutrients and preventing the removal of toxic waste products. If this occurs in the cerebral arterial circulation, it can lead to a stroke. If this occurs in the coronary arteries, it can lead to a heart attack.

Physiological Integration

The presence of hypoxemia and hyperventilation (the cause of the acute respiratory alkalosis), and the history and symptoms, suggest that the patient is suffering from an acute decrease in pulmonary blood flow to some parts of the lung. This results in a clinically significant ventilation/perfusion inequality (see Table 13–11). The hyperventilation is only partly due to the mild hypoxemia, because the arterial P_{O_2} of 60 mmHg in our patient, although low, is just at the threshold oxygen level that stimulates the peripheral chemoreceptors (see Figures 13–33 and 13–34). Other causes of hyperventilation may be anxiety and pain, which may also explain the increased heart rate observed in the patient at the emergency room.

The ventilation/perfusion inequality means that the patient is ventilating areas of the lung to which blood is not flowing, leading to increased alveolar dead space (see Figure 13–19). The extra blood flows to other nearby lung regions leading to a local decrease in the ratio of ventilation to perfusion (physiologic shunt). This results in some blood leaving the lung without adequate oxygenation. Remember that disruption of the delicate balance between regional ventilation and perfusion throughout the lung results in a failure to fully oxygenate the blood leaving the lung. In addition, hypoxia within the pulmonary circulation leads to vasoconstriction of the arterioles in the lungs and an increase in pulmonary artery pressure (see Figure 13–24).

We know that the respiratory alkalosis was acute and not a long-standing problem because the arterial pH was still alkaline from the hyperventilation. This indicates that the kidneys did not have time to respond to the change in pH by increasing bicarbonate excretion in the urine (see Table 14–7).

Why did the pulmonary embolism cause a decrease in arterial P_{O_2} but did not increase and, in fact, decreased arterial P_{CO_2}? Remember from Chapter 13 that the relationship between partial pressure and content is sigmoid for oxygen but relatively linear for CO_2. Because of the plateau of O_2 content as P_{O_2} increases above 60 mmHg (see Figure 13–26), increasing alveolar O_2 in over-ventilated regions of the lung does not significantly increase O_2 content of the blood leaving that region. Therefore, although hyperventilation does increase O_2 in some alveoli, it does not compensate for the decrease in O_2 content in some pulmonary capillaries due to ventilation-perfusion inequalities. Increasing ventilation can decrease the CO_2 content of blood due to the linearity of the relationship between P_{CO_2} and CO_2 content of the blood. The overall net effect is acute respiratory alkalosis due to decreased arterial P_{CO_2}. Interestingly, the hypoxemia can be partially overcome if the patient breathes gas that is enriched in oxygen because, although ventilation and perfusion are not well matched, there is not complete shunting of blood in the lungs. Thus, increasing alveolar P_{O_2} can still increase oxygenation of some areas of the lung with ventilation-perfusion mismatching, at least

somewhat. The arterial P_{CO_2} may have increased a little and pH decreased a little on supplemental O_2 because the improved arterial P_{O_2} decreased peripheral chemoreceptor stimulation and the degree of hyperventilation lessened (see Figure 13–34).

Our patient's initial complaint was chest pain, which made him think he was having a heart attack. He was actually fortunate to have chest pain because it caused him to go to the emergency room, and that may have saved his life. Although the exact reasons for chest pain in pulmonary embolism are uncertain, one possibility is that the clots result in an acute increase in pulmonary artery pressure, which can result in pain.

Why did this man have a pulmonary embolism? There are several risk factors for the development of deep vein thrombosis that can result in pulmonary embolism. Prolonged sitting often causes a stagnant pooling of blood in the lower legs (see Figures 12–45 and 12–60). That is why it is highly recommended that we avoid sitting for extended periods of time. Even just sitting at a computer for hours is discouraged. Contraction of the leg skeletal muscles compresses the leg veins. This results in intermittent emptying of the veins, decreasing the chances for clot formation. Obesity also increased the risk of deep vein thrombosis in our patient by further increasing the pooling of blood in the leg veins, increasing the amount of certain clotting factors in the blood, and changing platelet function.

There are also a number of gene defects that can lead to an increased tendency to form clots, a condition called inherited *hypercoagulability*. The most common is resistance to activated protein C (see Figure 12–78), which can occur in up to 3 percent of healthy adults in the United States. In fact, our patient was tested and found to have resistance to activated protein C. Therefore, the combination of obesity, sitting for a prolonged period of time, and hypercoagulability is the likely cause of deep vein thrombosis and pulmonary embolism in our patient.

Therapy

As soon as the diagnosis of pulmonary embolism was made, our patient was immediately started on intravenous heparin and *recombinant tissue plasminogen activator* (*rec-tPA*). Heparin is an anticlotting factor that counteracts the hypercoagulability. Rec-tPA is a synthetic form of a naturally occurring molecule that helps dissolve clots. The ventilation-perfusion scan was repeated a few days later and lung blood flow was almost normal. Supplemental oxygen was reduced over this time and then stopped when blood gases normalized.

Considering that this patient now has a proven inherited cause of hypercoagulability, he has an increased probability to have another deep vein thrombosis and even pulmonary embolism in the near future. It is also possible that some of his family members have the same defect, for which they should be tested and adequately counseled. Our patient was sent home and continued to receive oral anticoagulants for 6 months (see description of anticlotting drugs in Chapter 12, Section F), and was actively followed by his primary care physician. He was encouraged to lose weight because obesity increases the risk of a deep vein thrombosis occurring again. Some physicians even advocate life-long anticoagulation therapy for a patient such as ours.

A Man with Abdominal Pain, Fever, and Circulatory Failure

Case Presentation

A 21-year-old healthy college student was canoeing deep in the Canadian wilderness when he felt the first twinge of abdominal pain. Thinking that he either ate some undercooked fish or strained a muscle while paddling, he stopped to rest for a day but the pain steadily intensified. He began to shiver and felt extremely cold even though it was a warm day. These symptoms worsened during the 36 h it took to paddle to the outpost camp and be airlifted to the nearest medical center.

Reflect and Review #12
■ Based on your knowledge of the homeostatic control of body temperature, why might this young man feel cold despite the signs that his body temperature is elevated (see Figures 16–18 and 16–19)?
■ What might be the cause of the abdominal pain (see Figure 7–18)?

Physical Examination

On arrival at the hospital, the young man is confused and lapsing in and out of consciousness. He is rushed to the emergency room for examination. His temperature is 39.2°C (normal range ~36.5°–37.5°C), heart rate is 140 beats/min (normal range 65–85), respiration rate is 34 breaths/min (normal ~12), and blood pressure is 84/44 mmHg (normal ~120/80). He is taking deep breaths but his lungs are clear when listened to with a stethoscope. His abdomen is rigid and extremely tender when gently pressed on, especially in the lower-right quadrant. Upon questioning, his friends state that he has not urinated in over 24 hours. Therefore, a hollow tube called a *catheter* is inserted through the urethra into the urinary bladder to collect his urine. However, an abnormally small amount of urine (10 ml) was collected (see Figures 14–20 through 14–24 for a review of the control of urine output).

Reflect and Review #13
■ What mechanisms link low systemic blood pressure in this patient to the low urine output (see Figure 14–20)?
■ What organs are located in the lower-right quadrant of the abdominal cavity (see Figures 15–1 and 15–33)?

Laboratory Tests

Additional measurements were then performed, and the results are shown in **Table 19–3**.

Reflect and Review #14
■ Explain the relationship between arterial P_{CO_2} and pH values. Why is his arterial bicarbonate so low (see Table 14–10)?

■ What functions do white blood cells serve? What might be the cause of their abnormal values in this patient (see Figure 12–71 and Table 18–1)?
■ What metabolic processes produce lactate (lactic acid)? Under what circumstances would that production be increased above normal (see Figure 3–41 and Figure 13–38)?
■ Why did creatinine in the blood increase (see Figure 14–12)?

Diagnosis

A catheter is placed into an arm vein so that an intravenous infusion of isotonic saline (NaCl) can be started. Antibiotics are added to the saline to fight the apparent infection. A *computed tomography (CT)* scan of the abdomen is performed, which reveals an inflamed appendix (**Figure 19–4**). The patient is admitted to the intensive care unit (ICU) for continued intravenous fluid replacement, physiological monitoring, and the insertion of additional catheters that can be used for the measurement of arterial and central venous blood pressures.

The patient is then taken to the operating room for abdominal exploration. Surgeons remove an inflamed appendix that is found to have a small hole (*perforation*) and shows signs of *necrosis* (dying or dead tissue).

Table 19–3	Initial Laboratory Results with the Patient Breathing Room Air	
Blood Measurement*	**Result**	**Normal Range**
White blood cells	25,000 per mm³	4300–10,800 per mm³
Arterial P_{O_2}	90 mmHg	90–100 mmHg
Arterial P_{CO_2}	28 mmHg	35–45 mmHg
Arterial pH	7.25	7.38–7.45
Arterial bicarbonate	13 mmol/L	23–27 mmol/L
Lactate	8 mmol/L	0.5–2.2 mmol/L
Glucose	5 mmol/L	4–6 mmol/L
Creatinine	2.2 mg/dL	0.8–1.4 mg/dL

*In actuality, these measurements are done in whole blood or blood serum.

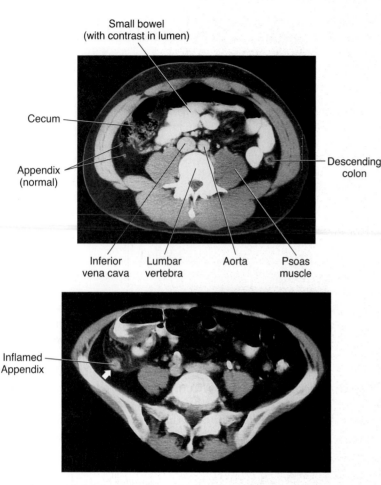

Small bowel
(with contrast in lumen)

Cecum

Appendix
(normal)

Descending
colon

Inferior
vena cava

Lumbar
vertebra

Aorta

Psoas
muscle

Inflamed
Appendix

Figure 19–4

Normal abdominal CT scan (top) identifying major structures. CT scan on the bottom shows an inflamed appendix (arrow).

Reflect and Review #15

■ Where is the appendix located (see Figure 15–33)?

A bacterial infection of the membranes surrounding the abdominal organs is found. This type of infection, called **peritonitis,** results in **pus** (yellow liquid made up of white blood cells, bacteria, and cellular debris) being produced. The pus is removed, the abdominal organs are thoroughly washed with saline and antibiotics, and the patient is returned to the ICU where arterial and central venous blood pressures and urine output are monitored.

Reflect and Review #16

■ What is the purpose of monitoring central venous blood pressure (see Figure 12–46)? Suggest other variables to monitor in this patient.

In the hours after surgery, the patient is maintained on mechanical ventilation. Gurgling breath sounds and falling arterial oxygen partial pressure indicate the presence of fluid in his lungs. Supplemental oxygen is provided to minimize the decrease in arterial oxygen by having the patient breathe a mixture of air enriched in oxygen. Widespread swelling of body tissues indicates that interstitial fluid volume is increasing, and

his blood pressure and urine output remain dangerously below normal. In addition to providing continued intravenous fluids and antibiotic therapy, the ICU staff infuses norepinephrine and vasopressin (vasoconstrictors), and methylprednisolone (a synthetic glucocorticoid given at pharmacological doses). For the next several days, the patient is critically ill while his condition is continuously monitored. Appropriate treatment adjustments are implemented as needed to attempt to normalize his blood volume, blood pressure, plasma lactate, plasma pH, and gas partial pressures in his blood.

This patient's condition began as acute **appendicitis,** but the delay in treatment allowed it to progress to the potentially lethal condition known as **septic shock.** Although **Escherichia coli** and other bacterial species are normally present in the large intestine and its associated appendix, blockage of the lumen of the appendix or the blood supply to the appendix can allow those normally harmless bacteria to multiply out of control. When this happens, the appendix becomes distended and the pressure inside the appendix increases significantly due to inflammation. Eventually, these factors can lead to ulceration of the mucosa of the appendix, followed by rupture of the organ. This releases bacteria into the peritoneal cavity. The bacteria then release toxins that diffuse into the blood vessels in the abdomen, leading to a dramatic cascade of events (**Figure 19–5**). When a bacterial infection is accompanied by a **systemic inflammatory response** (defined by symptoms

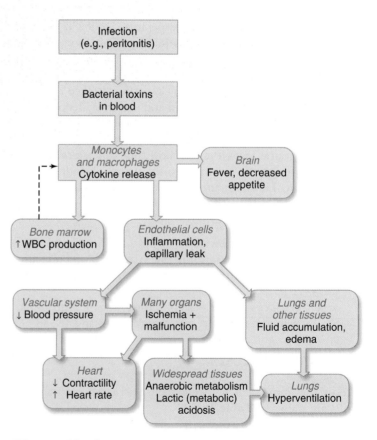

Figure 19–5

Cascade of some of the events from a serious infection to widespread organ failure in septic shock.

such as elevated body temperature, pulse rate, respiratory rate, and white blood cell count), the condition is referred to as *sepsis*. The most common sites of bacterial infections leading to sepsis are the lungs, abdomen (as in our patient), urinary tract, and sites where catheters penetrate the skin or blood vessels. If sepsis progresses to septic shock, patients also develop a significant decrease in blood pressure (a decrease in systolic pressure of greater than 40 mmHg or a mean arterial pressure less than 65 mmHg) that is not reversible by intravenous infusion of large volumes of isotonic saline solution. This type of circulatory failure is an example of ***low-resistance shock***, as described in Chapter 12.

Physiological Integration

Bacterial infections stimulate the body to mount a rapid and widespread defense reaction (see Figure 18–20). Monocytes and macrophages (two types of white blood cells) secrete a variety of signaling molecules known generally as cytokines (see Table 18–2), which include substances such as interleukins and tumor necrosis factor. Target tissues for cytokines include (1) the brain, where they mediate the onset of fever, a decrease in appetite, fatigue, and an increase in ACTH secretion; (2) the bone marrow, where they stimulate an increase in the rate of white blood cell production; and (3) endothelial cells throughout the vasculature, where they stimulate processes leading to inflammation and increased capillary leakiness. Many species of bacteria release toxins, which greatly accelerate and exaggerate cytokine release and effects, often resulting in a maladaptive or life-threatening over-reaction. The systemic inflammatory response has far-reaching effects on all body systems.

Such was the case of our patient by the time he finally reached the hospital. The set point for his body temperature was reset upward by circulating cytokines, resulting in ***fever***, and he shivered and felt chilled as his body attempted to warm itself toward the new, higher set point. The onslaught of cytokines and other inflammatory mediators (see Table 18–4) accelerated as his white blood cell count increased and bacterial toxins were released into his circulation. Excessive amounts of those chemicals caused widespread injury to the microvascular endothelium and led to leakage of fluid out of capillaries.

When capillaries become excessively leaky, bulk flow favors the exit of fluid from the circulation (see Figure 12–41). Plasma proteins escape into the interstitial fluid, creating a significant osmotic force that draws fluid out through capillary pores. This is due to Starling forces, which are described in Chapter 12 (see Figure 12–42). This loss of fluid causes a drastic reduction in circulating blood volume, to the point where even baroreceptor reflexes (see Chapter 12, Section D) are unable to maintain arterial blood pressure. Dramatic elevation of heart rate is evidence of activation of the baroreceptor reflexes via the cardiovascular control centers in the brain to raise blood pressure toward normal. Even relatively large intravenous fluid infusions fail to reverse this hypotension because much of the infused fluid simply escapes into the interstitial space. Accumulation of fluid in the interstitial space leads to the tissue edema observed in our patient, and leakiness of pulmonary capillaries eventually led to fluid in his lungs (***pulmonary edema***).

Lowered systemic arterial blood pressure makes it difficult to produce adequate blood flow through the tissues. When blood flow is inadequate to meet demands for oxygen and nutrients (***ischemia***), tissues, organs and organ systems malfunction. For example, our patient's inability to form urine resulted from low blood flow through his kidneys (see Figure 14–20). The increase in serum creatinine was evidence that glomerular filtration rate was decreased (see Figure 14–12).

A more general consequence of reduced oxygen availability is that cells must resort to anaerobic pathways to manufacture ATP, and lactic acid (lactate) is produced as a by-product (see Figure 3–41 and Figure 9–22). This led to the marked metabolic acidosis seen in our patient. His hyperventilation was driven by the peripheral and central chemoreceptors, in an attempt to compensate by removing CO_2-derived acid from the plasma (see Figure 13–37). Another mechanism designed to combat acidosis is the addition of new bicarbonate to the plasma and the excretion of H^+ via the kidney (Chapter 14, Section C), but the decrease in renal blood flow and glomerular filtration rate rendered this mechanism ineffective. His oxygen delivery to tissues was further compromised by the fluid buildup in his lungs. The added barrier to oxygen diffusion from lung alveoli into pulmonary capillaries (see Figure 13–28) reduced the oxygen partial pressure of his systemic arterial blood.

Therapy

Septic shock is an extremely challenging condition to treat, with mortality rates of 40–60 percent. One of the most important factors in determining patient survival is early recognition of the condition and onset of treatment. As soon as it has been determined that a patient is septic and is progressing toward septic shock, survival depends on rapid and continuous assessment of his or her physiological condition and timely therapeutic responses to changing conditions. Among the variables monitored, in addition to those listed in Table 19–3, are body temperature, heart rate, blood pressure, arterial and venous oxygen saturation, mean arterial and central venous blood pressures, urine output, and specific biochemical plasma indicators of the function of other organs, such as the liver. Using this information, physicians can take steps to improve cardiovascular and respiratory function while battling the infection that is the root cause of the condition.

Immediate interventions in the treatment of septic shock are aimed at restoring systemic oxygen delivery and thus relieving the widespread tissue hypoxia that is a hallmark of the condition. Mean arterial blood pressure is increased by infusions of isotonic saline and by treatment with vasoconstrictors such as norepinephrine and vasopressin (see Figure 12–51). The extra circulating fluid volume increases cardiac output by increasing venous pressure and cardiac filling (see Figure 12–46), whereas norepinephrine (the neurotransmitter normally released from postsynaptic sympathetic nerve endings) increases cardiac contractility and arteriolar vasoconstriction (see Figure 12–51). Maintaining mean arterial pressure between 65 and 90 mmHg is necessary to ensure adequate flow of blood through the tissues. Central venous pressure is monitored because it is a good index of venous return and the volume of fluid within the cardiovascular system (see Figure 12–57). The oxygen content of the blood is maintained by ventilating the lungs with supplemental oxygen to make sure that hemoglobin is saturated with oxygen (see Figure 13–26). It is also helpful to reduce the patient's demand for oxygen by paralyzing the respiratory muscles with drugs and providing mechanical ventilation, usually through a tube placed in the trachea attached to a positive-pressure pump. Otherwise, the increase in rate and depth of breathing that is typical of a patient in septic shock causes a marked increase in oxygen use by the respiratory muscles, and directs blood flow away from other organs already suffering from lack of oxygen.

The infection must be treated while also restoring cardiovascular function. Antibiotics that act on a wide variety of types of bacteria are administered as soon as possible after sepsis is diagnosed. The source of the infection is then located, accumulated pus and dead tissue are removed, and the surrounding tissue is thoroughly cleaned. Ideally, samples of blood and/or pus from the site of infection can be grown in culture, and within 48 h the specific bacterial species involved in the infection can be identified. The intravenous antibiotic therapy can then be altered to use drugs known to specifically target the invading species.

Recent clinical studies have suggested other therapeutic measures that can increase the survival rate of patients with septic shock. Pharmacological doses of glucocorticoid injections have also shown promise in some patients with septic shock. These hormones activate mechanisms throughout many tissues of the body that help the body cope with stress (see Table 11–3). Important among those effects are the inhibition of the inflammatory response and the enhancement of the sensitivity of vascular smooth muscle to adrenergic agents like norepinephrine.

Over a six-day period, the condition of our patient gradually improved. His blood pressure increased and stabilized, and the intravenous fluid and norepinephrine infusions were gradually reduced and then stopped. The edema in his lungs and tissues slowly subsided, he regained consciousness, and he was eventually able to maintain oxygen saturation in his arterial blood without mechanical ventilation. During his two-week hospital stay, the brain, liver, and kidney function returned to normal, and he had no apparent long-term organ damage from his ordeal. He has been extremely fortunate; approximately 500,000 cases of severe septic shock occur in the United States each year and less than half of those patients survive. His youth and relatively good initial physical condition were most likely instrumental in helping him beat the odds.

■ SUMMARY

Case 19–1: A Woman with Palpitations and Heat Intolerance

I. Case Presentation
   a. Her symptoms are nervousness, palpitations, feelings of warmth in a cool room, and significant weight loss despite eating a lot.

II. Physical Examination
   a. Her systolic blood pressure is increased and her diastolic pressure is decreased. Her resting heart rate is 100 beats per minute.

   b. She has an enlarged thyroid gland (goiter) and her eyes bulge out (proptosis).
   c. She has increased knee jerk reflexes and her hands are shaking.

III. Laboratory Tests
   a. She has increased thyroid hormone and decreased thyroid-stimulating hormone in the blood.

IV. Diagnosis
   a. She is diagnosed with hyperthyroidism (excess thyroid hormone activity).
   b. Hyperthyroidism is usually caused by Graves' disease—an autoimmune disease.

V. Physiological Integration
   a. Autoimmune production of thyroid-stimulating immunoglobulins (TSIs) stimulates the thyroid gland to produce too much thyroid hormone and to enlarge. The excess thyroid hormone suppresses the release of thyroid-stimulating hormone from the anterior pituitary gland.
   b. Infiltration of the muscles controlling eye movement by white blood cells leads to inflammation and proptosis.
   c. Increased thyroid hormone in the blood leads to an increase in sensitivity to catecholamines, resulting in an increase in systolic blood pressure and heart rate.
   d. Increased thyroid hormone leads to increased metabolic rate in a variety of tissues. This causes heat intolerance, hyperactive reflexes, and a small decrease in diastolic pressure.

VI. Therapy
   a. Three possible therapies include radioactive iodide administration to destroy much of the thyroid gland, drugs that block the synthesis of thyroid hormone, or surgical removal of the thyroid gland.

Case 19–2: A Man with Chest Pain After a Long Airplane Flight

I. Case Presentation
   a. A man has chest pain and shortness of breath after an 8-h flight.

II. Physical Examination
   a. He has an increased heart rate but his ECG does not show evidence of a heart attack.
   b. His chest x-ray is essentially normal.

III. Laboratory Tests
   a. He is hypoxemic and has an acute respiratory alkalosis.

IV. Diagnosis
   a. His ventilation-perfusion scan shows evidence of a pulmonary embolism (blockage of pulmonary blood flow).
   b. An ultrasound of his legs shows a deep vein thrombosis.
   c. A clot formed in his leg veins because he sat for a long period of time. In addition, there is evidence that he has a genetic disorder of coagulation. The clot migrated to the lung, causing a pulmonary embolus.

V. Physiological Integration
   a. Hypoxemia is caused by a dramatic disruption of the regional balance between ventilation and perfusion throughout the lung.
   b. Hyperventilation due to anxiety and pain, as well as hypoxemia, caused an acute respiratory alkalosis.

VI. Therapy
   a. Treatment focuses on anticoagulation with heparin (to prevent clotting) and recombinant tissue plasminogen activator (to dissolve the clots).
   b. Long-term anticoagulation therapy is recommended.

Case 19–3: A Man with Abdominal Pain, Fever, and Circulatory Failure

I. Case Presentation
    a. A young man has increasing abdominal pain over three days.

II. Physical Examination
    a. He has a fever, increased heart and respiratory rates, and low blood pressure.
    b. He has pain and rigidity localized to the lower-right quadrant of his abdomen.
    c. His urine output is low.

III. Laboratory Tests
    a. He has increased white blood cells suggesting an infection.
    b. He has a metabolic (lactic) acidosis with a respiratory compensation (low arterial P_{CO_2}).
    c. He has an elevated creatinine, which indicates a decrease in glomerular filtration rate.

IV. Diagnosis
    a. A computed tomography (CT) scan shows an inflamed appendix, suggesting a diagnosis of appendicitis. The low blood pressure suggests septic shock due to peritonitis (ruptured appendix).
    b. The diagnosis is confirmed in an abdominal exploration during which a perforated appendix is removed. The membranes near it are infected, proving peritonitis.

V. Physiological Integration
    a. Toxins from the bacteria have caused the low blood pressure because of vasodilation.
    b. The lower glomerular filtration rate is due to low blood pressure and decrease renal perfusion.

VI. Therapy
    a. Therapy consists of intravenous fluids before and after surgery to support cardiac output and blood pressure and vasoconstrictor drugs to maintain blood pressure.
    b. Antibiotic therapy is given to fight the peritoneal infection.

■ CLINICAL TERMS

appendicitis 692
atelectasis 688
bruit 685
catheter 691
computed tomography (CT) 691
deep vein thrombosis 689
diplopia 684
embolism 689
Escherichia coli 692
exopthalmos 684
fever 693
goiter 686
Graves' disease 686
hypercoagulability 690
hyperthyroidism 686
ischemia 693
low-resistance shock 693
methimazole 688
myocardial infarction 688
necrosis 691
palpitation 684
perforation 691
peritonitis 692
proptosis 684
propylthiouracil 688
pulmonary edema 693
pulmonary embolism 689
pus 692
radioactive iodide 688
recombinant tissue plasminogen activator (rec-tPA) 690
sepsis 693
septic shock 692
systemic inflammatory response 692
tachycardia 685
tachypnea 685
thrombus 689
thyroid-stimulating immunoglobulin (TSI) 686
thyrotoxicosis 686
ventilation-perfusion scan 689

Appendix A

Chapter 1

Test Questions

1-1 b The four basic cell types are epithelial, muscle, nervous, and connective.

1-2 a Steady-state requires energy input, but equilibrium does not.

1-3 c Muscles carry out the response (removing the hand from the stove).

1-4 c Circadian rhythms are typically entrained by the light/dark cycle, but in the absence of such cues, the rhythms "free run" with their own endogenous cycle length.

1-5 b Intracellular fluid volume is greater than the sum of plasma and interstitial fluid.

Quantitative and Thought Questions

1-1 No. There may in fact be a genetic difference, but there is another possibility: The altered skin blood flow in the cold could represent an *acclimatization* undergone by each Eskimo during his or her lifetime as a result of performing such work repeatedly.

1-2 This could occur in many ways. For example, suppose that an individual were to become dehydrated. What would happen to his or her plasma sodium *concentration*? Initially, the loss of fluid would result in an increased sodium concentration, even though the absolute amount of sodium may not have changed much. The increase in sodium concentration would trigger endocrine and renal responses designed to return the sodium concentration to normal. Another example occurs during mountain climbing. At high altitude, a person who is not acclimatized to low oxygen pressures will hyperventilate (i.e., greatly increase the rate and depth of breathing) to get more oxygen into his or her blood. One consequence of hyperventilation, though, is that more of the carbon dioxide in the body is exhaled. Carbon dioxide tends to produce hydrogen ions in the blood (Chapter 14). Thus, ascent to high altitude leads to alkaline blood, which must then be compensated for by renal, endocrine, and other responses.

Chapter 2

Test Questions

2-1 e The continued creation of new free radicals is a chain reaction, and contributes to the potentially damaging effects of a given free radical.

2-2 d

2-3 b This is a dehydration reaction. The reverse reaction would be hydrolysis.

2-4 b Uracil is found in RNA; thymine is found in DNA.

2-5 b

Chapter 3

Test Questions

3-1 a

3-2 b Transcription refers to the conversion of a gene's DNA into RNA; translation is the conversion of mRNA into protein.

3-3 a Allosteric modulation occurs at a site separate from the ligand-binding site. The resulting change in three dimensional structure of the protein may enhance or reduce the ability of the protein to bind its ligand.

3-4 b

3-5 c

3-6 d Catabolism refers to the breakdown of fatty acids into usable forms for the production of ATP.

Quantitative and Thought Questions

3-1 Nucleotide bases in DNA pair A to T and G to C. Given the base sequence of one DNA strand as:

A-G-T-G-C-A-A-G-T-C-T

a. The complementary strand of DNA would be:

T-C-A-C-G-T-T-C-A-G-A

b. The sequence in RNA transcribed from the first strand would be:

U-C-A-C-G-U-U-C-A-G-A

Recall that uracil U replaces thymine T in RNA.

3-2 The triplet code G-T-A in DNA will be transcribed into mRNA as C-A-U, and the anticodon in tRNA corresponding to C-A-U is G-U-A.

3-3 If the gene were only composed of the triplet exon code words, the gene would be 300 nucleotides in length because a triplet of three nucleotides codes for one amino acid. However, because of the presence of intron segments in most genes, which account for 75 to 90 percent of the nucleotides in a gene, the gene would be between 1200 and 3 nucleotides long; moreover, it would also contain termination codons. Thus, the exact size of a gene cannot be determined by knowing the number of amino acids in the protein the gene codes.

3-4 A drug could decrease acid secretion by (1) binding to the membrane sites that normally inhibit acid secretion, which would produce the same effect as the body's natural messengers that inhibit acid secretion;

(2) binding to a membrane protein that normally stimulates acid secretion but not itself triggering acid secretion, thereby preventing the body's natural messengers from binding (competition); or (3) having an allosteric effect on the binding sites, which would increase the affinity of the sites that normally bind inhibitor messengers or decrease the affinity of those sites that normally bind stimulatory messengers.

3-5 The reason for a lack of insulin effect could be either a decrease in the number of available binding sites insulin can bind to or a decrease in the affinity of the binding sites for insulin so that less insulin is bound. A third possibility, which does not involve insulin binding, would be a defect in the way the binding site triggers a cell response once it has bound insulin.

3-6 (a) Acid secretion could be increased to 40 mmol/h by (1) increasing the concentration of compound X from 2 pM to 8 pM, thereby increasing the number of binding sites occupied; or (2) increasing the affinity of the binding sites for compound X, thereby increasing the amount bound without changing the concentration of compound X. (b) Increasing the concentration of compound X from 18 to 28 pM will not increase acid secretion because, at 18 pM, all the binding sites are occupied (the system is saturated), and there are no further binding sites available.

3-7 The maximum rate at which the end product E can be formed is 5 molecules per second, the rate of the slowest (rate-limiting) reaction in the pathway.

3-8 Under normal conditions, the concentration of oxygen at the level of the mitochondria in cells, including muscle at rest, is sufficient to saturate the enzyme that combines oxygen with hydrogen to form water. The rate-limiting reactions in the electron transport chain depend on the available concentrations of ADP and P_i, which are combined to form ATP.

Thus, increasing the oxygen concentration above normal levels will not increase ATP production. If a muscle is contracting, it will break down ATP into ADP and P_i, which become the major rate-limiting substrates for increasing ATP production. With intense muscle activity, the level of oxygen may fall below saturating levels, limiting the rate of ATP production, and intensely active muscles must then use anaerobic glycolysis to provide additional ATP. Under these circumstances, increasing the oxygen concentration in the blood will increase the rate of ATP production. As discussed in Chapter 12, it is not the concentration of oxygen in the blood that is increased during exercise but the rate of blood flow to a muscle, resulting in greater quantities of oxygen delivery to the tissue.

3-9 During starvation, in the absence of ingested glucose, the body's stores of glycogen are rapidly depleted. Glucose, which is the major fuel used by the brain, must now be synthesized from other types of molecules. Most of this newly formed glucose comes from the breakdown of proteins to amino acids and their conversion to glucose. To a lesser extent, the glycerol portion of fat is converted to glucose. The fatty acid portion of fat cannot be converted to glucose.

3-10 Ammonia is formed in most cells during the oxidative deamination of amino acids and then travels to the liver via the blood. The liver detoxifies the ammonia by converting it to the nontoxic compound urea. Because the liver is the site in which ammonia is converted to urea, diseases that damage the liver can lead to an accumulation of ammonia in the blood, which is especially toxic to nerve cells. Note that it is not the liver that produces the ammonia.

Chapter 4

Test Questions

4-1 c Channels are proteins that span the membrane and are opened by ligands, voltage, or mechanical stimuli.

4-2 d Facilitated diffusion does not require ATP. Recall that secondary active transport *indirectly* requires ATP because ion pumps were needed to establish the electrochemical gradient for a particular ion (such as Na^+).

4-3 b After the initial movement of water out of the cells due to osmosis, the urea concentration quickly equilibrates across each cell's plasma membrane, removing any osmotic stimulus.

4-4 e Segregation of function on different surfaces of the cell, and the ability to secrete chemicals (e.g., from the pancreas), are two of the most important features of epithelial cells.

4-5 a Diffusion is slowed by the resistance of a membrane.

4-6 e Because ions are charged, both the chemical and the electrical gradients determine its rate and direction of diffusion.

Quantitative and Thought Questions

4-1 (a) During diffusion, the net flux always occurs from high to low concentration. Thus, it will be from 2 to 1 in A and from 1 to 2 in B. (b) At equilibrium, the concentrations of solute in the two compartments will be equal: 4 mM in case A and 31 mM in case B. (c) Both will reach diffusion equilibrium at the same rate because the difference in concentration across the membrane is the same in each case, 2 mM [(3 − 5) = −2, and (32 − 30) = 2]. The two one-way fluxes will be much larger in B than in A, but the net flux has the same magnitude in both cases, although it is oriented in opposite directions.

4-2 The ability of one amino acid to decrease the flux of a second amino acid across a cell membrane is an example of the competition of two molecules for the same binding site, as explained in Chapter 3. The binding site for alanine on the transport protein can also bind leucine. The higher the concentration of alanine, the greater the number of binding sites that it occupies, and the fewer available for binding leucine. Thus, less leucine will move into the cell.

4-3 The net transport will be out of the cell in the direction from the higher-affinity site on the intracellular surface to the lower-affinity site on the extracellular surface. More molecules will be bound to the transporter on the higher-affinity side of the membrane, and thus more will move out of the cell than into it, until the concentration in the extracellular fluid becomes large enough that the number of molecules bound to transporters at the extracellular surface is equal to the number bound at the intracellular surface.

4-4 Although ATP is not used directly in secondary active transport, it is necessary for the primary active transport of sodium out of cells. Because it is the sodium concentration gradient across the plasma membrane that provides the energy for most secondary active transport systems, a decrease in ATP production will decrease primary active sodium transport, leading to a decrease in the sodium concentration gradient and thus to a decrease in secondary active transport.

4-5 The solution with the greatest osmolarity will have the lowest water concentration. Recall that NaCl forms two ions in solution and $CaCl_2$ forms three. Thus, the osmolarities are:

A. $20 + 30 + (2 \times 150) + (3 \times 10) = 380$ mOsm
B. $10 + 100 + (2 \times 20) + (3 \times 50) = 300$ mOsm
C. $100 + 200 + (2 \times 10) + (3 \times 20) = 380$ mOsm
D. $30 + 10 + (2 \times 60) + (3 \times 100) = 460$ mOsm

Solution D has the lowest water concentration. Solution B is isoosmotic since it has the same osmolarity as intracellular fluid.

4-6 Initially the osmolarity of compartment 1 is $(2 \times 200) + 100 = 500$ mOsm and that of 2 is $(2 \times 100) + 300 = 500$ mOsm. The two solutions thus have the same osmolarity, and there is no difference in water concentration across the membrane. Because the membrane is permeable to urea, this substance will undergo net diffusion until it reaches the same concentration (200 mM) on the two sides of the membrane. In other words, in the steady state it will not affect the volumes of the compartments. In contrast, the higher initial NaCl concentration in compartment 1 than in compartment 2 will cause, by osmosis, the movement of water from compartment 2 to compartment 1 until the concentration of NaCl in both is 150 mM. Note that the same volume change would have occurred if there were no urea present in either compartment. It is only the concentration of nonpenetrating solutes (NaCl in this case) that determines the volume change, regardless of the concentration of any penetrating solutes that are present.

4-7 The osmolarities and nonpenetrating-solute concentrations are:

Solution	Osmolarity, mOsm	Nonpenetrating solute concentration, mOsm
A	$(2 \times 150) + 100 = 400$	$2 \times 150 = 300$
B	$(2 \times 100) + 150 = 350$	$2 \times 100 = 200$
C	$(2 \times 200) + 100 = 500$	$2 \times 200 = 400$
D	$(2 \times 100) + 50 = 250$	$2 \times 100 = 200$

Only the concentration of nonpenetrating solutes (NaCl in this case) will determine the change in cell volume. The intracellular concentration of nonpenetrating solute is 300 mOsm, so solution A will produce no change in cell volume. Solutions B and D will cause cells to swell because they have a lower concentration of nonpenetrating solute (higher water concentration) than the intracellular fluid. Solution C will cause cells to shrink because it has a higher concentration of nonpenetrating solute than the intracellular fluid.

4-8 Solution A is isotonic because it has the same concentration of nonpenetrating solutes as intracellular fluid (300 mOsm). Solution A is also hyperosmotic because its total osmolarity is greater than 300 mOsm, as is also true for solutions B and C. Solution B is hypotonic because its concentration of nonpenetrating solutes is less than 300 mOsm. Solution C is hypertonic because its concentration of nonpenetrating solutes is greater than 300 mOsm. Solution D is hypotonic (less than 300 mOsm of nonpenetrating solutes) and also hypoosmotic (having a total osmolarity of less than 300 mOsm).

4-9 Exocytosis is triggered by an increase in cytosolic calcium concentration. Calcium ions are actively transported out of cells, in part by secondary countertransport coupled to the downhill entry of sodium ions on the same transporter. If the intracellular concentration of sodium ions were increased, the sodium concentration gradient across the membrane would be decreased, and this would decrease the secondary active transport of calcium out of the cell. This would lead to an increase in cytosolic calcium concentration, which would trigger increased exocytosis.

Chapter 5

Test Questions

5-1 b
5-2 a
5-3 e
5-4 a Calmodulin is a calcium-binding protein that is inactive in the absence of Ca^{2+}.

5-5 d Lipid-soluble messengers cross the plasma membrane and act primarily on cytosolic and nuclear receptors.

5-6 b

Quantitative and Thought Questions

5-1 Patient A's drug very likely acts to block phospholipase A_2, whereas patient B's drug blocks lipoxygenase (see Figure 5–12).

5-2 The chronic loss of exposure of the heart's receptors to norepinephrine causes an up-regulation of this receptor type (i.e., more receptors in the heart for norepinephrine). The drug, being an agonist of norepinephrine (i.e., able to bind to norepinephrine's receptors and activate them) is now more effective because there are more receptors for it to combine with.

5-3 None. You are told that all six responses are mediated by the cAMP system, thus, blockage of any of the steps listed in the question would eliminate all six of the responses. This is because the cascade for all six responses is identical from the receptor through the formation of cAMP and activation of cAMP-dependent protein kinase. Thus, the drug must be acting at a point beyond this kinase (e.g., at the level of the phosphorylated protein mediating this response).

5-4 Not in most cells, because there are other physiological mechanisms by which signals impinging on the cell can increase cytosolic calcium concentration. These include (1) second-messenger-induced release of calcium from the endoplasmic reticulum and (2) voltage-sensitive calcium channels.

5-5 Intracellular second messengers do not disappear immediately upon removal of a first messenger. Instead, some second messengers (such as cAMP) may linger inside the cell for seconds, minutes, or even longer after the first messenger is gone.

Chapter 6

Test Questions

6-1 b Afferent neurons have peripheral axon terminals associated with sensory receptors, cell bodies in the dorsal root ganglion of the spinal cord, and central axon terminals that project into the spinal cord.

6-2 c Oligodendrocytes form myelin sheaths in the central nervous system.

6-3 d Insert the given chloride concentrations into the Nernst equation; remember to use –1 as the valence (z).

6-4 d A, B, and C all are correct. Using the Nernst equation to calculate the sodium equilibrium potential gives values of +31, +36, and +40 mV for a, b, and c. If the membrane potential was +42 mV, the outward electrical force on sodium would be greater than the inward concentration gradient, so sodium would move out of the cell in each of these cases.

6-5 e Neither sodium nor potassium is in equilibrium at the resting membrane potential, but the action of the Na^+/K^+-ATPase pump prevents the small but steady leak of both ions from dissipating the concentration gradients.

6-6 a Because Na^+ is further away from its electrochemical equilibrium than is K^+, there would be more sodium entry than potassium exit, causing local depolarization and local current flow that would decrease with distance from the site of the stimulus.

6-7 c Due to the persistent open state of the voltage-gated potassium channels, for a brief time at the end of an action potential the membrane is hyperpolarized. When the voltage-gated potassium channels eventually close, the potassium leak channels once again determine the resting membrane potential.

6-8 d The IPSP caused by neuron B would summate with (subtract from) the amplitude of the EPSP caused by neuron A's firing.

6-9 a Dopamine, like norepinephrine and epinephrine, is a catecholamine neurotransmitter manufactured by enzymatic modification of the amino acid tyrosine.

6-10 b Norepinephrine is the neurotransmitter released by postganglionic neurons onto smooth muscle cells.

Quantitative and Thought Questions

6-1 Little change in the resting membrane potential would occur when the pump first stops because the pump's *direct* contribution to charge separation is very small. With time, however, the membrane potential would depolarize progressively toward zero because the sodium and potassium concentration gradients, which depend on the Na^+/K^+-ATPase pumps and which give rise to the membrane potential, run down.

6-2 The resting potential would decrease (i.e., become less negative) because the concentration gradient causing net diffusion of this positively charged ion out of the cell would be smaller. The action potential would fire more easily (i.e., with smaller stimuli) because the resting potential would be closer to threshold. It would repolarize more slowly because repolarization depends on net potassium diffusion from the cell, and the concentration gradient driving this diffusion is lower. Also, the afterhyperpolarization would be smaller.

6-3 The hypothalamus was probably damaged. It plays a critical role in appetite, thirst, and sexual capacity.

6-4 The drug probably blocks cholinergic muscarinic receptors. These receptors on effector cells mediate the actions of parasympathetic nerves. Therefore, the drug would remove the slowing effect of these nerves on the heart, allowing the heart to speed up. Blocking their effect on the salivary glands would cause the dry mouth. We know that the drug is not blocking cholinergic nicotinic receptors because the skeletal muscles are not affected.

6-5 Because the membrane potential of the cells in question depolarizes (i.e., becomes less negative) when chloride channels are blocked, one can assume there was net chloride diffusion into the cells through these channels prior to treatment with the drug. Therefore, one can also predict that this passive inward movement was being exactly balanced by active transport of chloride out of the cells.

6-6 Without acetylcholinesterase, more acetylcholine would remain bound to the receptors, and all the actions normally caused by acetylcholine would be accentuated. Thus, there would be marked narrowing of the pupils, airway constriction, stomach cramping and diarrhea, sweating, salivation, slowing of the heart, and fall in blood pressure. On the other hand, in skeletal muscles, which must repolarize after excitation in order to be excited again, there would be weakness, fatigue, and finally inability to contract. In fact, lethal poisoning by high doses of cholinesterase inhibitors occurs because of paralysis of the muscles involved in respiration. Low doses of these compounds are used therapeutically.

6-7 These potassium channels, which open after a short delay following the initiation of an action potential, increase potassium diffusion out of the cell, hastening repolarization. They also account for the increased potassium permeability that causes the afterhyperpolarization. Therefore, the action potential would be broader (that is, longer in duration) and would return to resting level more slowly, and the afterhyperpolarization would be absent.

Chapter 7

Test Questions

7-1 a For example, photons of light are the adequate stimulus for photoreceptors of the eye, and sound is the adequate stimulus for hair cells of the ear.

7-2 b Receptor potentials generate only local currents in the receptor membrane that transduces the stimulus, but when they reach the first node of Ranvier, they depolarize the membrane to threshold, and there the voltage-gated sodium channels first initiate action potentials. Beyond that point, the receptor potential decreases with distance, whereas action potentials propagate all the way to the central axon terminals.

7-3 d Lateral inhibition increases the contrast between the region at the center of a stimulus and regions at the edges of the stimulus, which increases the acuity of stimulus localization.

7-4 a The occipital lobe of the cortex is the initial site of visual processing. (Review Figure 7–14.)

7-5 e Somatic sensations include those from the skin, muscles, bones, tendons, and joints, but not encoding of sound by cochlear hair cells.

7-6 b A myopic (nearsighted) person has an eyeball that is too long. When the ciliary muscles are relaxed and the lens is as flat as possible, parallel light rays from distant objects focus in front of the retina, while diverging rays from near objects are able to focus on the retina. (Recall that with normal vision, it takes ciliary muscle contraction and a rounded lens to focus on near objects.)

7-7 d When the right optic tract is destroyed, perception of images formed on the right half of the retina in both eyes is lost, so nothing is visible at the left side of a person's field of view (Review Figure 7–29.)

7-8 a Pressure waves traveling down the cochlea make the cochlear duct vibrate, moving the basilar membrane against the stationary tectorial membrane and bending the hair cells that bridge the gap between the two.

7-9 c With the sudden head rotation from left-to-right, inertia of the endolymph causes it to rotate from right-to-left with respect to the semicircular canal that lies in the horizontal plane. This fluid flow bends the cupula and embedded hair cells within the ampulla, which influences the firing of action potentials along the vestibular nerve.

7-10 d "Umami" is derived from the Japanese word meaning delicious, and the stimulation of these taste receptors by glutamate produces the perception of a rich, meaty flavor.

Quantitative and Thought Questions

7-1 (a) Use drugs to block transmission in the pathways that convey information about pain to the brain. For example, if substance P is the neurotransmitter at the central endings of the nociceptor afferent fibers, give a drug that blocks the substance P receptors. (b) Cut the dorsal root at the level of entry of the nociceptor fibers to prevent transmission of their action potentials into the central nervous system. (c) Give a drug that activates receptors in the descending pathways that block transmission of the incoming or ascending pain information. (d) Stimulate the neurons in these same descending pathways to increase their blocking activity (stimulation-produced analgesia or, possibly, acupuncture). (e) Cut the ascending pathways that transmit information from the nociceptor afferents. (f) Deal with emotions, attitudes, memories, and so on to decrease sensitivity to the pain. (g) Stimulate nonpain, low-threshold afferent fibers to block transmission through the pain pathways (TENS). (h) Block transmission in the afferent nerve with a local anesthetic such as Novocaine or Lidocaine.

7-2 Information regarding temperature is carried via the anterolateral system to the brain. Fibers of this system cross to the opposite side of the body in the spinal cord at the level of entry of the afferent fibers (see Figure 7–19a). Damage to the left side of the spinal cord or any part of the left side of the brain that contains fibers of the pathways for temperature would interfere with awareness of a heat stimulus on the right. Thus, damage to the somatosensory cortex of the left cerebral hemisphere (i.e., opposite the stimulus) would interfere with awareness of the stimulus. Injury to the spinal cord at the point at which fibers of the anterolateral system from the two halves of the spinal cord cross to the opposite side would interfere with the awareness of heat applied to either side of the body, as would the unlikely event that damage occurred to relevant areas of both sides of the brain.

7-3   Vision would be restricted to the rods; therefore, it would be normal at very low levels of illumination (when the cones would not be stimulated anyway), but at higher levels of illumination clear vision of fine details would be lost, and everything would appear in shades of gray, with no color vision. In very bright light, there would be no vision because of bleaching of the rods' rhodopsin.

7-4   (a) The individual lacks a functioning primary visual cortex. (b) The individual lacks a functioning visual association cortex.

Chapter 8

Test Questions

8-1 d
8-2 c
8-3 a
8-4 b
8-5 e   See Figure 8–6.
8-6 b   If by experience you discover that a persistent stimulus like the noise from a fan does not have relevance, there is a reduction in conscious attention directed toward that stimulus. This is an example of "habituation."
8-7 c   The mesolimbic dopamine pathway mediates the perception of reward that is associated with adaptive behaviors, including goal-directed behaviors related to preserving homeostasis, like eating and drinking.
8-8 d   Serotonin-specific reuptake inhibitors (SSRIs) are the most widely used antidepressant drugs, although other types of antidepressants additionally enhance signaling by norepinephrine.
8-9 a   Short-term, or "working" memories are transferred into new long-term memories in the process of consolidation, which requires a functional hippocampus. When the hippocampus is destroyed, previously formed long-term memories remain intact, but the ability to form new memories is lost.
8-10 c   Broca's area is located near the region of the left frontal lobe motor cortex that controls the face; when it is damaged, individuals have "expressive aphasia." This means that they comprehend language but are unable to articulate their own thoughts into words.

Quantitative and Thought Questions

8-1   Dopamine is depleted in the basal ganglia of people with Parkinson's disease, and they are therapeutically given dopamine agonists, usually L-dopa. This treatment raises dopamine levels in other parts of the brain, however, where the dopamine levels were previously normal. Schizophrenia is associated with increased brain dopamine levels, and symptoms of this disease appear when dopamine levels are high. The converse therapeutic problem can occur during the treatment of schizophrenics with dopamine-lowering drugs, which sometimes cause the symptoms of Parkinson's disease to appear.

8-2   Experiments on anesthetized animals often involve either stimulating a brain part to observe the effects of increased neuronal activity, or damaging ("lesioning") an area to observe resulting deficits. Such experiments on animals, which lack the complex language mechanisms humans have, cannot help with language studies. Diseases sometimes mimic these two experimental situations, and behavioral studies of the resulting language deficits in people with aphasia, coupled with study of their brains after death, have provided a wealth of information.

Chapter 9

Test Questions

9-1 a   A single skeletal muscle fiber, or cell, is composed of many myofibrils.
9-2 e   The dark stripe in a striated muscle that constitutes the A band results from the aligned thick filaments within myofibrils, so thick filament length is equal to A-band width.
9-3 b   As filaments slide during a shortening contraction, the I band becomes narrower, so the distance between the Z line and the thick filaments (at the end of the A band) must decrease.
9-4 d   DHP receptors act as voltage sensors in the T-tubule membrane and are physically linked to ryanodine receptors in the sarcoplasmic reticulum membrane. When an action potential depolarizes the T-tubule membrane, DHP receptors change conformation and trigger the opening of the ryanodine receptors. This allows calcium to flood from the interior of the sarcoplasmic reticulum into the cytosol.
9-5 c   In an isometric twitch, tension begins to rise as soon as excitation-contraction is complete and the first cross-bridges begin to attach. In an isotonic twitch, excitation-contraction coupling takes the same amount of time, but the fiber is delayed from shortening until after enough cross-bridges have attached to move the load.
9-6 b   In the first few seconds of exercise, mass-action favors transfer of the high-energy phosphate from creatine phosphate to ADP by the enzyme creatine kinase.

9-7 d Fast-oxidative-glycolytic fibers are an intermediate type that are designed to contract rapidly but to resist fatigue. They utilize both aerobic and anaerobic energy systems, and thus they are red fibers with high myoglobin (which facilitates production of ATP by oxidative phosphorylation), but they also have a moderate ability to generate ATP through glycolytic pathways. (Refer to Table 9–3.)

9-8 c In smooth muscle cells, dense bodies serve the same functional role as Z lines do in striated muscle cells—they serve as the anchoring point for the *thin* filaments.

9-9 b When myosin light-chain kinase transfers a phosphate group from ATP to the myosin light chains of the cross-bridges, binding and cycling of cross-bridges is activated.

9-10 d Stretching a sheet of single-unit smooth muscle cells opens mechanically gated ion channels, which causes a depolarization that propagates through gap junctions, followed by calcium entry and contraction. This does not occur in multiunit smooth muscle.

9-11 e The amount of calcium released during a typical resting heart beat exposes less than half of the thin filament cross-bridge binding sites. Autonomic neurotransmitters and hormones can increase or decrease the amount of calcium released to the cytosol during EC coupling.

Quantitative and Thought Questions

9-1 Under resting conditions, the myosin has already bound and hydrolyzed a molecule of ATP, resulting in an energized molecule of myosin ($M \cdot ADP \cdot P_i$). Because ATP is necessary to detach the myosin cross-bridge from actin at the end of cross-bridge movement, the absence of ATP will result in rigor mortis, in which case the cross-bridges become bound to actin but do not detach, leaving myosin bound to actin ($A \cdot M$).

9-2 No. The transverse tubules conduct the muscle action potential from the plasma membrane into the interior of the fiber, where it can trigger the release of calcium from the sarcoplasmic reticulum. If the transverse tubules were not attached to the plasma membrane, an action potential could not be conducted to the sarcoplasmic reticulum, and there would be no release of calcium to initiate contraction.

9-3 The length-tension relationship states that the maximum tension developed by a muscle decreases at lengths below L_0. During normal shortening, as the sarcomere length becomes shorter than the optimal length, the maximum tension that can be generated decreases. With a light load, the muscle will continue to shorten until its maximal tension just equals the load. No further shortening is possible because at shorter sarcomere lengths the tension would be less than the load. The heavier the load, the less the distance shortened before reaching the isometric state.

9-4 Maximum tension is produced when the fiber is (1) stimulated by an action potential frequency that is high enough to produce a maximal tetanic tension, and (2) at its optimum length L_0, where the thick and thin filaments have overlap sufficient to provide the greatest number of cross-bridges for tension production.

9-5 Moderate tension—for example, 50 percent of maximal tension—is accomplished by recruiting sufficient numbers of motor units to produce this degree of tension. If activity is maintained at this level for prolonged periods, some of the active fibers will begin to fatigue and their contribution to the total tension will decrease. The same level of total tension can be maintained, however, by recruiting new motor units as some of the original ones fatigue. At this point, for example, one might have 50 percent of the fibers active, 25 percent fatigued, and 25 percent still unrecruited. Eventually, when all the fibers have fatigued and there are no additional motor units to recruit, the whole muscle will fatigue.

9-6 The oxidative motor units, both fast and slow, will be affected first by a decrease in blood flow because they depend on blood flow to provide both the fuel—glucose and fatty acids—and the oxygen required to metabolize the fuel. The fast-glycolytic motor units will be affected more slowly because they rely predominantly on internal stores of glycogen, which is anaerobically metabolized by glycolysis.

9-7 Two factors lead to the recovery of muscle force. (1) Some new fibers can be formed by the fusion and development of undifferentiated satellite cells. This will replace some, but not all, of the fibers that were damaged. (2) Some of the restored force results from hypertrophy of the surviving fibers. Because of the loss of fibers in the accident, the remaining fibers must produce more force to move a given load. The remaining fibers undergo increased synthesis of actin and myosin, resulting in increases in fiber diameter and thus their force of contraction.

9-8 In the absence of extracellular calcium ions, skeletal muscle contracts normally in response to an action potential generated in its plasma membrane because the calcium required to trigger contraction comes entirely from the sarcoplasmic reticulum within the muscle fibers. If the motor neuron to the muscle is stimulated in a calcium-free medium, however, the muscle will not contract because the influx of calcium from the extracellular fluid into the motor nerve terminal is necessary to trigger the release of acetylcholine that in turn triggers an action potential in the muscle.

In a calcium-free solution, smooth muscles would not respond either to stimulation of the nerve or to the plasma membrane. Stimulating the nerve would have no effect because calcium entry into presynaptic terminals is necessary for neurotransmitter release. Stimulating the smooth muscle cell membrane would also not cause a response in the absence of calcium because in all of the various types of smooth muscle, calcium must enter from outside the cell to trigger contraction. In some cases, the external calcium directly initiates contraction, and in others it triggers the release of calcium from the sarcoplasmic reticulum (calcium-induced calcium release).

9-9 The simplest model to explain the experimental observations is as follows. Upon parasympathetic nerve stimulation, a neurotransmitter is released that binds to receptors on the membranes of smooth muscle cells and triggers contraction. The substance released, however, is not acetylcholine (ACh) for the following reason.

Action potentials in the parasympathetic nerves are essential for initiating nerve-induced contraction. When the nerves were prevented from generating action potentials by blockage of their voltage-gated sodium channels, there was no response to nerve stimulation. ACh is the neurotransmitter released from most, but not all, parasympathetic endings. When the muscarinic receptors for ACh were blocked, however, stimulation of the parasympathetic nerves still produced a contraction, providing evidence that some substance other than ACh is being released by the neurons and producing contraction.

9-10 Elevation of extracellular fluid calcium concentration would increase the amount of calcium entering the cytosol through L-type calcium channels. This would result in a greater depolarization of cardiac muscle cell membranes during action potentials. The strength of cardiac muscle contractions would also be increased because this larger calcium entry would trigger more calcium release through ryanodine receptor channels, and thus there would be a greater activation of cross-bridge cycling.

Chapter 10

Test Questions

10-1 b The basal nuclei, sensorimotor cortex, thalamus, brainstem, and cerebellum are all middle-level structures that create a motor program based on the intention to carry out a voluntary movement.

10-2 c When a given muscle is stretched, muscle-spindle stretch receptors send action potentials along afferent fibers that synapse directly on alpha motor neurons to extrafusal fibers to that muscle, causing it to contract back toward the prestretched length.

10-3 a Afferent action potentials from pain receptors in the injured left foot would stimulate the withdrawal reflex of the left leg (activation of flexor muscles and inhibition of extensors) and the opposite pattern in the right leg (the crossed-extensor reflex).

10-4 d Activating the gamma motor neurons would cause contraction of the ends of intrafusal muscle fibers,

stretching the muscle-spindle receptors, and the resulting action potentials would monosynaptically excite the alpha motor neurons innervating the extrafusal fibers of the stretch receptors.

10-5 c See Figure 10–10.

10-6 T Most descending corticospinal pathways cross the midline of the body in the medulla oblongata.

10-7 F Upper motor neuron disorders are typically characterized by hypertonia and spasticity.

10-8 F The reverse is actually true.

10-9 F In Parkinson's, a deficit of dopamine from neurons of the substantia nigra results in "resting tremors."

10-10 T *Clostridium botulinum* toxin specifically blocks the release of neurotransmitter from neurons that normally inhibit motor neurons. The resulting imbalance of excitatory and inhibitory inputs causes spastic contractions of muscles.

Quantitative and Thought Questions

10-1 None. The gamma motor neurons are important in preventing the muscle spindle stretch receptors from going slack, but when this reflex is tested, the intrafusal fibers are not flaccid. The test is performed with a bent knee, which stretches the extensor muscles in the thigh (and the intrafusal fibers within the stretch receptors). The stretch receptors are therefore responsive.

10-2 The efferent pathway of the reflex arc (the alpha motor neurons) would not be activated, the effector cells (the extrafusal muscle fibers) would not be activated, and there would be no reflex response.

10-3 The drawing must have excitatory synapses on the motor neurons of both ipsilateral extensor and ipsilateral flexor muscles.

10-4 A toxin that interferes with the inhibitory synapses on motor neurons would leave unbalanced the normal excitatory input to these neurons. Thus, the otherwise normal motor neurons would fire excessively, which would result in increased muscle contraction. This is exactly what happens in lockjaw as a result of the toxin produced by the tetanus bacillus.

10-5 In mild cases of tetanus, agonists (stimulators) of the inhibitory interneuron neurotransmitter gamma-aminobutyric acid (GABA) can shift the balance back toward the inhibition of alpha motor neurons. In more severe cases, paralysis can be induced by administering long-lasting drugs that block the nicotinic acetylcholine receptors at the neuromuscular junction.

Chapter 11

Test Questions

11-1 c
11-2 a
11-3 e
11-4 b

11-5 d

11-6 a At any given concentration of hormone, more A is bound to receptor than B.

11-7 d Goiter results from dysfunction of the thyroid gland.

11-8 e Recall that thyroid hormone potentiates the effects of epinephrine and the SNS.

11-9 b

11-10 e Recall that there exists a large store of iodinated thyroglobulin in thyroid follicles.

11-11 c Low plasma calcium decreases the filtered load of calcium. It also stimulates parathyroid hormone, which increases calcium reabsorption from the distal tubule. This helps to prevent the further loss of calcium in the urine.

11-12 d Parathyroid hormone is a potent stimulator of calcium resorption from bone.

11-13 T T_4 is the chief circulating form, but T_3 is more active.

11-14 F Acromegaly is associated with hyperglycemia and hypertension.

11-15 T

Quantitative and Thought Questions

11-1 Epinephrine falls to very low levels during rest and fails to increase during stress. The sympathetic preganglionics provide the only major control of the adrenal medulla.

11-2 The increased concentration of binding protein causes more TH to be bound, thereby lowering the plasma concentration of *free* TH. This causes less negative-feedback inhibition of TSH secretion by the anterior pituitary, and the increased TSH causes the thyroid to secrete more TH until the free concentration has returned to normal. The end result is an increased *total* plasma TH—most bound to the protein—but a normal free TH. There is no hyperthyroidism because it is only the free concentration that exerts effects on TH's target cells.

11-3 Destruction of the anterior pituitary or hypothalamus. These symptoms reflect the absence of, in order, growth hormone, the gonadotropins, and ACTH (the symptom is due to the resulting decrease in cortisol secretion). The problem is either primary hyposecretion of anterior pituitary hormones or secondary hyposecretion because the hypothalamus is not secreting hypophysiotropic hormones normally.

11-4 Vasopressin and oxytocin (the posterior pituitary hormones) secretion would decrease. The anterior pituitary hormones would not be affected because the influence of the hypothalamus on these hormones is exerted not by connecting nerves but via the hypophysiotropic hormones in the portal vascular system.

11-5 The secretion of GH increases. Somatostatin, coming from the hypothalamus, normally exerts an inhibitory effect on the secretion of this hormone.

11-6 Norepinephrine and many other neurotransmitters are released by neurons that terminate on the hypothalamic neurons that secrete the hypophysiotropic hormones. Therefore, manipulation of these neurotransmitters will alter secretion of the hypophysiotropic hormones and thereby the anterior pituitary hormones.

11-7 The high dose of the cortisol-like substance inhibits the secretion of ACTH by feedback inhibition of (1) hypothalamic corticotropin releasing hormone and (2) the response of the anterior pituitary to this hypophysiotropic hormone. The lack of ACTH causes the adrenal to atrophy and decrease its secretion of cortisol.

11-8 The hypothalamus. The low basal TSH indicates either that the pituitary is defective or that it is receiving inadequate stimulation (TRH) from the hypothalamus. If the thyroid itself were defective, basal TSH would be elevated because of less negative-feedback inhibition by TH. The TSH increase in response to TRH shows that the pituitary is capable of responding to a stimulus and so is unlikely to be defective. Therefore, the problem is that the hypothalamus is secreting too little TRH (in reality, this is very rare).

11-9 *In utero* malnutrition. Neither growth hormone nor the thyroid hormones influence *in utero* growth.

11-10 Androgens stimulate growth but also cause the ultimate cessation of growth by closing the epiphyseal plates. Therefore, there might be a rapid growth spurt in response to the androgens but a subsequent premature cessation of growth. Estrogens exert similar effects.

Chapter 12

Test Questions

12-1 c Blood in the right ventricle is relatively deoxygenated after returning from the tissues.

12-2 e Resistance decreases as the fourth power of an increase in radius, and in direct proportion to a decrease in vessel length.

12-3 d See Figure 12–19.

12-4 d The large total cross-section of capillaries results in very slow blood velocity.

12-5 a Increasing colloid osmotic pressure would decrease filtration of fluid from capillaries into the tissues.

12-6 d Pressures are higher in the systemic circuit, but because the cardiovascular system is a closed loop, the flow must be the same in both.

12-7 b The AV node is the only conduction point between atria and ventricles, and the slow propagation through it delays the beginning of ventricular contraction.

12-8 c The diastolic pressure in this example is 85; adding 1/3 of the pulse pressure gives a MAP of 101.7 mmHg.

12-9 d Reduced firing to arterioles would reduce total peripheral resistance and thus reduce mean arterial pressure toward normal.

12-10 e Ventricular muscle cells do not have a pacemaker potential, and the L-type calcium channel is not open during this phase of the action potential even in autorhythmic cells.

12-11 c

12-12 a Increased sympathetic nerve firing and norepinephrine release during exercise constricts vascular beds in the kidneys, GI tract, and other tissues to compensate for the large dilation of muscle vascular beds.

12-13 b Reduced oxygen delivery to the kidneys increases the secretion of erythropoietin, which stimulates bone marrow to increase production of erythrocytes.

12-14 c

12-15 e t-PA is part of the fibrinolytic system that dissolves clots.

Quantitative and Thought Questions

12-1 No. Decreased erythrocyte volume is certainly one possible explanation, but there is a second: The person might have a normal erythrocyte volume but an abnormally increased plasma volume. Convince yourself of this by writing the hematocrit equation as: erythrocyte volume/(erythrocyte volume + plasma volume).

12-2 A halving of tube radius. Resistance is directly proportional to blood viscosity but inversely proportional to the *fourth power* of tube radius.

12-3 The plateau of the action potential and the contraction would be absent. You might think that contraction would persist because most calcium in excitation-contraction coupling in the heart comes from the sarcoplasmic reticulum. However, the signal for the release of this calcium is the calcium entering across the plasma membrane.

12-4 The SA node is not functioning, and the ventricles are being driven by a pacemaker in the AV node or the bundle of His.

12-5 The person has a narrowed aortic valve. Normally, the resistance across the aortic valve is so small that there is only a tiny pressure difference between the left ventricle and the aorta during ventricular ejection. In the example given here, the large pressure difference indicates that resistance across the valve must be very high.

12-6 This question is analogous to question 12-5 in that the large pressure difference across a valve while the valve is open indicates an abnormally narrowed valve—in this case, the left AV valve.

12-7 Decreased heart rate and contractility. These are effects mediated by the sympathetic nerves on beta-adrenergic receptors in the heart.

12-8 120 mmHg. MAP = DP + 1/3 (SP – DP).

12-9 The drug must have caused the arterioles in the kidneys to dilate enough to reduce their resistance by 50 percent. Blood flow to an organ is determined by mean arterial pressure and the organ's resistance

to flow. Another important point can be deduced here: If mean arterial pressure has not changed even though renal resistance has dropped 50 percent, then either the resistance of some other organ or actual cardiac output has gone up.

12-10 The experiment suggests that acetylcholine causes vasodilation by releasing nitric oxide or some other vasodilator from endothelial cells.

12-11 A low plasma protein concentration. Capillary pressure is, if anything, lower than normal and so cannot be causing the edema. Another possibility is that capillary permeability to plasma proteins has increased, as occurs in burns.

12-12 20 mmHg/L per minute. TPR = MAP/CO.

12-13 Nothing. Cardiac output and TPR have remained unchanged, so their product, MAP, also remains unchanged. This question emphasizes that MAP depends on cardiac output but not on the combination of heart rate and stroke volume that produces the cardiac output.

12-14 It increases. There are a certain number of impulses traveling up the nerves from the arterial baroreceptors. When these nerves are cut, the number of impulses reaching the medullary cardiovascular center goes to zero, just as it would physiologically if the mean arterial pressure were to decrease markedly. Accordingly, the medullary cardiovascular center responds to the absent impulses by reflexly increasing arterial pressure.

12-15 It decreases. The hemorrhage causes no immediate change in hematocrit because erythrocytes and plasma are lost in the same proportion. As interstitial fluid starts entering the capillaries, however, it expands the plasma volume and decreases hematocrit. (This is too soon for any new erythrocytes to be synthesized.)

Chapter 13

Test Questions

13-1 e If alveolar pressure (P_{alv}) is negative with respect to atmospheric pressure (P_{atm}), the driving force for airflow is inward (from the atmosphere into the lung).

13-2 a For the same change in transpulmonary pressure, a less compliant (i.e., stiffer) lung will have a smaller change in lung volume.

13-3 a Total minute ventilation is comprised of dead space plus alveolar ventilation. Minute ventilation is respiratory frequency (12 breaths per minute) multiplied by tidal volume (500 ml/breath) = 6 ml/min. Subtract from that alveolar ventilation (4200 ml/min) and one gets 1800 ml/min.

13-4 d An increase in alveolar P_{O_2} results from an increase in alveolar ventilation (supply of oxygen) relative to metabolic rate (consumption of oxygen).

13-5 c The relationship between arterial P_{O_2} and arterial oxygen saturation is described by the oxygen-hemoglobin dissociation curve. The greatest increase in oxygen saturation for the same change in P_{O_2} occurs at the steepest part of the curve—between a P_{O_2} of 40 and 60 mmHg.

13-6 b Increases in blood temperature, decreases in blood pH, and increases in DPG shifts the oxygen-hemoglobin curve downward, leading to a lower oxygen saturation at the same P_{O_2}.

13-7 b There are forms of asthma that are not primarily due to the presence of allergens. Examples are exercise- or cold-air-induced asthma.

13-8 e Respiratory acidosis (increase in blood P_{CO_2} and decrease in pH) is a major stimulus to ventilation— this is mediated both by afferents from the peripheral chemoreceptors and by an increase in central chemoreceptor activity.

13-9 c Because of the shape of the oxygen-hemoglobin dissociation curve, small increases in P_{O_2} due to increases in ventilation cannot fully saturate hemoglobin. When the desaturated blood mixes with saturated blood, the average is still hypoxemic.

13-10 c Remember that a lung capacity is the sum of at least two volumes. Inspiratory capacity is the sum of tidal volume and inspiratory reserve volume.

Quantitative and Thought Questions

13-1 200 ml/mmHg.

Lung compliance = Δ lung volume/Δ $(P_{alv} - P_{ip})$
= 800 ml/[0 − (−8)] mmHg–
[0 − (−4)] mmHg
= 800 ml/4 mmHg = 200 ml/mmHg

13-2 More subatmospheric than normal. A decreased surfactant level causes the lungs to be less compliant (i.e., more difficult to expand). Therefore, a greater transpulmonary pressure $(P_{alv} - P_{ip})$ is required to expand them a given amount.

13-3 No.

Alveolar ventilation = (tidal volume − dead space) × breathing rate
= (250 ml − 150 ml) × 20 breaths/min
= 2 ml/min

whereas normal alveolar ventilation is approximately 4 ml/min.

13-4 The volume of the snorkel constitutes an additional dead space, so total pulmonary ventilation must be increased if alveolar ventilation is to remain constant.

13-5 The alveolar P_{O_2} will be higher than normal, and the alveolar P_{CO_2} will be lower. If you do not understand why, review the factors that determine the alveolar gas pressures.

13-6 No. Hypoventilation reduces arterial P_{O_2} but only because it reduces alveolar P_{O_2}. That is, in hypoventilation, *both* alveolar and arterial P_{O_2} are decreased to essentially the same degree. In this problem, alveolar P_{O_2} is normal, and so the person is not hypoventilating. The low arterial P_{O_2} must therefore represent a defect that causes a discrepancy between alveolar P_{O_2} and arterial P_{O_2}. Possibilities include impaired diffusion, a shunting of blood from the right side of the heart to the left through a hole in the heart wall, and a mismatch between air flow and blood flow in the alveoli.

13-7 Not at rest, if the defect is not too severe. Recall that equilibration of alveolar air and pulmonary capillary blood is normally so rapid that it occurs well before the end of the capillaries. Therefore, even though diffusion may be retarded, as in this problem, there may still be enough time for equilibration to be reached. In contrast, the time for equilibration is decreased during exercise, and failure to equilibrate is much more likely to occur, resulting in a lowered arterial P_{O_2}.

13-8 Only a few percent (specifically, from approximately 200 ml O_2/L blood to approximately 215 ml O_2/L blood). The reason the increase is so small is that almost all the oxygen in blood is carried bound to hemoglobin, and hemoglobin is almost 100 percent saturated at the arterial P_{O_2} achieved by breathing room air. The high arterial P_{O_2} achieved by breathing 100 percent oxygen does cause a directly proportional increase in the amount of oxygen *dissolved* in the blood (the additional 15 ml), but this still remains a small fraction of the total oxygen in the blood. Review the numbers given in the chapter.

13-9 All. The reasons are all given in the chapter.

13-10 It would cease. Respiration depends on descending input from the medulla to the nerves supplying the diaphragm and the inspiratory intercostal muscles.

13-11 (a) The combination of hypercapnia (elevated P_{CO_2} due to increased inspired CO_2) and hypoxia (due to decreased inspired O_2) greatly augments ventilation by stimulating central and peripheral chemoreceptors. Although CO decreases O_2 content, chemoreceptors are not stimulated and ventilation does not increase.

13-12 These patients have profound hyperventilation, with marked increases in both the depth and rate of ventilation. The stimulus, mainly via the peripheral chemoreceptors, is the marked increase in their arterial hydrogen ion concentration due to the acids produced. The hyperventilation causes an increase in their arterial P_{O_2} and a decrease in their arterial P_{CO_2}.

13-13 In pure anatomical shunt, blood passes through the lung without exposure to any alveolar air. Therefore, increases in alveolar P_{O_2} caused by increased inspired O_2 will not affect the P_{O_2} of the shunt blood. By contrast, there is still some blood flowing through a region of the lung with a ventilation-perfusion mismatch. Therefore, an increase in P_{O_2} in the alveoli can increase the P_{O_2} in this blood, which, when mixing with blood leaving other areas of the lung, can increase the blood in the pulmonary vein and hence the arterial circulation.

Chapter 14

Test Questions

14-1 c The main driving force favoring fluid filtration from the glomerular capillary to Bowman's space is glomerular capillary blood pressure (P_{GC}).

14-2 c In order for a substance to appear in the urine at a faster rate than its filtration rate, it must also be actively secreted into the tubular fluid.

14-3 a Excessive sweating will decrease blood volume. This will lead to compensatory mechanisms to preserve total body water, including a decrease in urine production (antidiuresis).

14-4 e Urea is trapped in the medullary interstitium and is an osmotically active particle. The resultant increase in tonicity helps to maintain the gradient for medullary passive water reabsorption.

14-5 a A decrease in sodium intake stimulates renin because of the decrease in sodium delivery to the macula densa. This is detected and results in an increase in renin release from the juxtaglomerular cells.

14-6 c Parathyroid hormone stimulates calcium reabsorption in the distal tubules of the nephron, thereby decreasing calcium excretion. Because parathyroid hormone is increased in hypocalcemic states, the resulting decrease in calcium excretion helps to restore blood calcium to normal.

14-7 c Secretion of ammonium into the renal tubule is one way to rid the body of excess hydrogen ion (metabolic acidosis).

14-8 b Increases in ventilation greater than metabolic rate "blow off" CO_2 and results in a decrease in arterial P_{CO_2}. Because of the buffering of bicarbonate, this increases arterial pH (respiratory alkalosis).

14-9 e Cortical nephrons either have short or absent Henle's loops. Only juxtamedullary nephrons have long Henle's loops, which plunge into the renal medulla and create a hyperosmotic interstitium via countercurrent multiplication.

14-10 a When the renal corpuscles become diseased, they greatly increase their permeability to protein. Furthermore, diseased proximal tubules cannot remove the filtered protein from the tubular lumen. This results in increased protein in the urine (proteinuria).

Quantitative and Thought Questions

14-1 No. These are possibilities, but there is another. Substance T may be secreted by the tubules.

14-2 No. It is a possibility, but there is another. Substance V may be filtered and/or secreted, but the substance V entering the lumen via these routes may be completely reabsorbed.

14-3 125 mg/min. The amount of any substance filtered per unit time is given by the product of the GFR and the filterable plasma concentration of the substance—in this case, 125 ml/min × 100 mg/ 100 ml = 125 mg/min.

14-4 The plasma concentration might be so high that the T_m for the amino acid is exceeded, so all the filtered amino acid is not reabsorbed. A second possibility is that there is a specific defect in the tubular transport for this amino acid. A third possibility is that some other amino acid is present in the plasma in high concentration and is competing for reabsorption.

14-5 No. Urea is filtered and then partially reabsorbed. The reason its concentration in the tubule is higher than in the plasma is that relatively more water is reabsorbed than urea. Therefore, the urea in the tubule becomes concentrated. Despite the fact that urea *concentration* in the urine is greater than in the plasma, the *amount excreted* is less than the filtered load (that is, net reabsorption has occurred).

14-6 They would all be decreased. The transport of all these substances is coupled, in one way or another, to that of sodium.

14-7 GFR would not go down as much, and renin secretion would not go up as much as in a person not receiving the drug. The sympathetic nerves are a major pathway for both responses during hemorrhage.

14-8 There would be little if any increase in aldosterone secretion. The major stimulus for increased aldosterone secretion is angiotensin II, but this substance is formed from angiotensin I by the action of angiotensin-converting enzyme, and so blockade of this enzyme would block the pathway.

14-9 (b) Urinary excretion in the steady state must be less than ingested sodium chloride by an amount equal to that lost in the sweat and feces. This is normally quite small, less than 1 g/day, so that urine excretion in this case equals approximately 11 g/day.

14-10 If the hypothalamus had been damaged, there might be inadequate secretion of ADH. This would cause loss of a large volume of urine, which would tend to dehydrate the person and make her thirsty. Of course, the area of the brain involved in thirst might have suffered damage.

14-11 This is primary aldosteronism or Conn's syndrome. Because aldosterone stimulates sodium reabsorption and potassium secretion, there will be total-body retention of sodium and loss of potassium. Interestingly, the person in this situation actually retains very little sodium because urinary sodium excretion returns to normal after a few days despite the continued presence of the high aldosterone. One explanation for this is that GFR and atrial natriuretic factor both increase as a result of the initial sodium retention.

14-12 Sodium and water balance would become negative because of increased excretion of these substances in the urine. The person would also develop a decreased plasma bicarbonate concentration and metabolic acidosis because of increased bicarbonate excretion. The effects on acid-base status are explained by the fact that hydrogen ion secretion—blocked by the drug—is needed both for bicarbonate reabsorption and for the excretion of hydrogen ion (contribution of new bicarbonate to the blood). The increased sodium excretion reflects the fact that much sodium reabsorption by the proximal tubule is achieved by Na^+/H^+ countertransport. By blocking hydrogen ion secretion, therefore, the drug also partially blocks sodium reabsorption. The increased water excretion occurs because the failure to reabsorb sodium and bicarbonate decreases water reabsorption (remember that water reabsorption is secondary to solute reabsorption), resulting in an osmotic diuresis.

Chapter 15

Test Questions

15-1 c When the stomach contents, which are very acidic, move into the small intestine, it stimulates the release of secretin, which circulates to the pancreas and stimulates the release of bicarbonate into the small intestine. This neutralizes the acid and protects the small intestine.

15-2 d GIP release is a feedforward mechanism to signal the islet cells in the pancreas that the products of food digestion are on their way to the blood. This results in an augmented insulin response to a meal.

15-3 a Gastrin is a major controller of acid secretion by the stomach. When the stomach becomes very acidic, gastrin release is inhibited, preventing continued acid production.

15-4 b Cholecystokinin is the primary signal from the small intestine to the pancreas to increase digestive enzyme release into the small intestine.

15-5 d The enzyme pepsin is produced from pepsinogen in the presence of acid. This zymogen accelerates protein digestion.

15-6 b Because fat is insoluble in an aqueous environment, micelles keep fat droplets from re-aggregating and small enough to be absorbed.

15-7 c Distention of the duodenum signals the stomach that the meal has moved on and continued acid secretion in the stomach is not necessary until the next meal.

15-8 a Bicarbonate in the bile is secreted by the epithelial cells lining the bile ducts.

15-9 e Although the primary movement of chyme in segmentation is back and forth, the overall, net movement of chyme is from the small intestine to the large intestine.

15-10 a The active transport of sodium in the large intestine is the driving force for the osmotic absorption of water.

Quantitative and Thought Questions

15-1 If the salivary glands fail to secrete amylase, the undigested starch that reaches the small intestine will still be digested by the amylase the pancreas secretes. Thus, starch digestion is not significantly affected by the absence of salivary amylase.

15-2 Alcohol can be absorbed across the stomach wall, but absorption is much more rapid from the small intestine with its larger surface area. Ingestion of foods containing fat releases enterogastrones from the small intestine, and these hormones inhibit gastric emptying and thus prolong the time alcohol spends in the stomach before reaching the small intestine. Milk, contrary to popular belief, does not "protect" the lining of the stomach from alcohol by coating it with a fatty layer. Rather, the fat content of milk decreases the rate of absorption of alcohol by decreasing the rate of gastric emptying.

15-3 Vomiting results in the loss of fluid and acid from the body. The fluid comes from the luminal contents of the stomach and duodenum, most of which was secreted by the gastric glands, pancreas, and liver and thus is derived from the blood. The cardiovascular symptoms of this patient are the result of the decrease in blood volume that accompanies vomiting.

The secretion of acid by the stomach produces an equal number of bicarbonate ions, which are released into the blood. Normally these bicarbonate ions are neutralized by hydrogen ions released into the blood by the pancreas when this organ secretes bicarbonate ions. Because gastric acid is lost during vomiting, the pancreas is not stimulated to secrete bicarbonate by the usual high-acidity signal from the duodenum, and no corresponding hydrogen ions are formed to neutralize the bicarbonate released into the blood by the stomach. As a result, the acidity of the blood decreases. Loss of potassium from the loss of stomach contents can also lead to hypokalemia.

15-4 Fat can be digested and absorbed in the absence of bile salts, but in greatly decreased amounts. Without adequate emulsification of fat by bile salts and phospholipids, only the fat at the surface of large lipid droplets is available to pancreatic lipase, and the rate of fat digestion is very slow. Without the formation of micelles with the aid of bile salts, the products of fat digestion become dissolved in the large lipid droplets, where they are not readily available for diffusion into the epithelial cells. In the absence of bile salts, only about 50 percent of the ingested fat is digested and absorbed. The undigested fat is passed on to the large intestine, where bacteria produce compounds that increase colonic motility and promote the secretion of fluid into the lumen of the large intestine, leading to diarrhea.

15-5 Damage to the lower portion of the spinal cord produces a loss of voluntary control over defecation due to disruption of the somatic nerves to the skeletal muscle of the external anal sphincter. Damage to the somatic nerves leaves the external sphincter in a continuously relaxed state. Under these conditions, defecation occurs whenever the rectum becomes distended and the defection reflex is initiated.

15-6 Vagotomy decreases the secretion of acid by the stomach. Impulses in the parasympathetic nerves directly stimulate acid secretion by the parietal cells and also cause the release of gastrin, which in turn stimulates acid secretion. Impulses in the vagus nerves are increased during both the cephalic and gastric phases of digestion. Vagotomy, by decreasing the amount of acid secreted, decreases irritation of existing ulcers, which promotes healing and decreases the probability of acid contributing to the production of new ulcers.

Chapter 16

Test Questions

16-1 a Glucose can be converted to fat, but fatty acids cannot be converted to glucose.

16-2 b HSL is an intracellular enzyme that acts on triglycerides.

16-3 a Glucagon acts to prevent hypoglycemia from occurring.

16-4 c If untreated, Type 1 DM causes an osmotic diuresis when the transport maximum for glucose is exceeded in the kidney.

16-5 d Insulin stimulates lipogenesis, not lipolysis.

16-6 e Recall that vitamin deficiencies occur even with normal dietary intake of vitamins, because the metabolic rate is increased in hyperthyroidism.

16-7 b

16-8 T

16-9 F Core temperature is generally kept fairly constant, but skin temperature can vary.

16-10 T

16-11 F As muscles begin contracting during exercise, they become partially insulin-independent.

16-12 F BMI equals body mass in kg divided by (height in meters)^2.

16-13 T

16-14 F Skin vessels dilate in such conditions in order to help dissipate heat by bringing warm blood close to the skin surface.

16-15 T

Quantitative and Thought Questions

16-1 The concentration in plasma would increase, and the amount stored in adipose tissue would decrease. Lipoprotein lipase cleaves plasma triglycerides, so its blockade would decrease the rate at which these molecules were cleared from plasma and would decrease the availability of the fatty acids in them for the synthesis of intracellular triglycerides. However, this would only reduce but not eliminate such synthesis, because the adipose tissue cells could still synthesize their own fatty acids from glucose.

16-2 The person might be an insulin-dependent diabetic or might be a healthy fasting person; plasma glucose would be increased in the first case but decreased in the second. Plasma insulin concentration would be useful because it would be decreased in both cases. The fact that the person was resting and unstressed

was specified because severe stress or strenuous exercise could also produce the plasma changes mentioned. Plasma glucose would increase during stress and decrease during strenuous exercise.

16-3 Glucagon, epinephrine, cortisol, and growth hormone. The insulin will produce hypoglycemia, which then induces reflex increases in the secretion of all these hormones.

16-4 It might reduce it but not eliminate it. The sympathetic effects on organic metabolism during exercise are mediated not only by circulating epinephrine but also by sympathetic nerves to the liver (glycogenolysis and gluconeogenesis), to adipose tissue (lipolysis), and to the pancreatic islets (inhibition of insulin secretion and stimulation of glucagon secretion).

16-5 Increase. The stress of the accident will initially elicit increased activity of all the glucose counterregulatory controls and will therefore necessitate more insulin to oppose these influences.

16-6 It will lower plasma cholesterol concentration. Bile salts are formed from cholesterol, and losses of these bile salts in the feces will be replaced by the synthesis of new ones from cholesterol. Chapter 15 describes how bile salts are normally absorbed from the small intestine so that very few of those secreted into the bile are normally lost from the body.

16-7 Plasma concentrations of HDL and LDL. It is the ratio of LDL cholesterol to HDL cholesterol that best correlates with the development of atherosclerosis (HDL cholesterol is "good" cholesterol). The answer to this question would have been the same regardless of whether the person was an athlete, but the question was phrased this way to emphasize that people who exercise generally have increased HDL cholesterol.

16-8 Heat loss from the head, mainly via convection and sweating, is the major route for loss under these conditions. The rest of the body is *gaining* heat by conduction, and sweating is of no value in the rest of the body because the water cannot evaporate. Heat is also lost via the expired air (insensible loss), and some people actually begin to pant under such conditions. The rapid, shallow breathing increases air flow and heat loss without causing hyperventilation.

16-9 They seek out warmer places, if available, so that their body temperature increases. That is, they use behavior to develop a fever. This is excellent evidence that the hyperthermia of infection is a fever (i.e., a set-point change).

Chapter 17

Test Questions

17-1 e Without the presence of the Y chromosome in the testes and the local production of SRY protein, the undifferentiated gonads are programmed to differentiate into ovaries.

17-2 c Only females exhibit gonadal steroid (estrogen) positive feedback on GnRH release.

17-3 d The luteal phase of the ovary, when progesterone production is maximal, occurs after ovulation but before the end of the menstrual cycle.

17-4 c Estrogen stimulates LH release (positive feedback) just before the LH surge and ovulation (usually on day 14).

17-5 b One follicle becomes dominant early in the menstrual cycle.

17-6 e The death of the corpus luteum (in the absence of pregnancy and hCG) results in a dramatic decrease in ovarian progesterone and estrogen production.

17-7 a The loss of ovarian steroid production with the death of the corpus luteum releases the pituitary from negative feedback and allows FSH to increase. This stimulates the maturation of a small number of follicles for the next menstrual cycle.

17-8 c The primary function of the Leydig cell is the production of testosterone in response to stimulation with LH.

17-9 b Prolactin is produced by the maternal pituitary. It is homologous to but not the same peptide as human placental lactogen, which is produced by the placenta.

17-10 a The primary event in menopause is the loss of ovarian function. The decrease in estrogen leads to an increase in pituitary gonadotropin release (loss of negative feedback).

Quantitative and Thought Questions

17-1 Sterility due to lack of spermatogenesis would be the common symptom. The Sertoli cells are essential for spermatogenesis, and so is testosterone produced by the Leydig cells. The person with Leydig cell destruction, but not the person with Sertoli cell destruction, would also have other symptoms of testosterone deficiency.

17-2 The androgens act on the hypothalamus and anterior pituitary to inhibit the secretion of the gonadotropins. Therefore, spermatogenesis is inhibited. Importantly, even if this man were given FSH, the sterility would probably remain because the lack of LH would cause deficient testosterone secretion, and *locally* produced testosterone is needed for spermatogenesis (i.e., the exogenous androgen cannot do this job).

17-3 Impaired function of the seminiferous tubules, notably of the Sertoli cells. The increased plasma FSH concentration is due to the lack of negative feedback inhibition of FSH secretion by inhibin, itself secreted by the Sertoli cells. The Leydig cells seem to be functioning normally in this person because the lack of demasculinization and the normal plasma LH indicate normal testosterone secretion.

17-4 FSH secretion. FSH acts on the Sertoli cells and LH acts on the Leydig cells, so sterility would result in either case, but the loss of LH would also cause undesirable elimination of testosterone and its effects.

17-5 These findings are all due to testosterone deficiency. You would also expect to find that the testes and penis were small if the deficiency occurred before puberty.

17-6 They will be eliminated or become very irregular. The androgens act on the hypothalamus to inhibit the secretion of GnRH and on the pituitary to inhibit the response to GnRH. The result is inadequate secretion of gonadotropins and therefore inadequate stimulation of the ovaries. In addition to the loss of regular menstrual cycles, the woman may suffer some degree of masculinization of the secondary sex characteristics because of the combined effects of androgen excess and estrogen deficiency.

17-7 Such treatment may cause so much secretion of FSH that multiple follicles become dominant and have their eggs ovulated during the LH surge.

17-8 An increased plasma LH. The other two are due to increased plasma progesterone and so do not occur until *after* ovulation and formation of the corpus luteum.

17-9 The absence of sperm capacitation. When test-tube fertilization is performed, special techniques are used to induce capacitation.

17-10 The fetus is in difficulty. The placenta produces progesterone entirely on its own, whereas estriol secretion requires participation of the fetus, specifically, the fetal adrenal cortex.

17-11 Prostaglandin antagonists, oxytocin antagonists, and drugs that lower cytosolic calcium concentration. You may not have thought of the last category because calcium is not mentioned in this context in the chapter, but as in all muscle, calcium is the immediate cause of contraction in the myometrium.

17-12 This person would have normal male external genitals and testes, although the testes may not have descended fully, but would also have some degree of development of uterine tubes, a uterus, and a vagina. These internal female structures would tend to develop because no MIS was present to cause degeneration of the Müllerian duct system.

17-13 No. These two hormones are already elevated in menopause, and the problem is that the ovaries are unable to respond to them with estrogen secretion. Thus, the treatment must be with estrogen itself.

Chapter 18
Test Questions

18-1 c
18-2 a
18-3 b This is known as active immunity.

18-4 c IgA antibodies act in this way.

18-5 F Antibiotics are bactericidal. They are sometimes given in viral diseases to eliminate or prevent secondary infections caused by bacteria, however.

18-6 T For example, rheumatoid arthritis and inflammatory bowel disease are not associated with infection.

18-7 T Some lymphocytes are B cells.

18-8 F Edema is a consequence of inflammation and has no known adaptive value.

18-9 F These are the primary lymphoid organs. An example of a secondary organ is a lymph node.

Quantitative and Thought Questions

18-1 Both would be impaired because T cells would not differentiate. The absence of cytotoxic T cells would eliminate responses mediated by these cells. The absence of helper T cells would impair antibody-mediated responses because most B cells require cytokines from helper T cells to become activated.

18-2 Neutrophil deficiency would impair nonspecific inflammatory responses to bacteria. Monocyte deficiency, by causing macrophage deficiency, would impair both nonspecific inflammation and specific immune responses.

18-3 The drug might reduce but would not eliminate the action of complement, because this system destroys cells directly (via the membrane attack complex) as well as by facilitating phagocytosis.

18-4 Antibodies would bind normally to antigen but might not be able to activate complement, act as opsonins, or recruit NK cells in ADCC. The reason for these defects is that the sites to which complement C1, phagocytes, and NK cells bind are all located in the Fc portion of antibodies.

18-5 They do develop fever, although often not to the same degree as normal. They can do so because IL-1 and other cytokines secreted by macrophages cause fever, whereas the defect in AIDS is failure of helper T cell function.

Appendix B

Effects on Cardiovascular System 415–19

Atrial pumping 416–17
Cardiac output (increases) 415–19, 424
   Distribution during exercise 405, 415–16
Control mechanisms 416–17
Coronary blood flow (increases) 367
Gastrointestinal blood flow (decreases) 415–16
Heart attacks (protective against) 422
Heart rate (increases) 416, 417
Lymph flow (increases) 401–402
Maximal oxygen consumption (increases) 417–19
Mean arterial pressure (increases) 416–17
Renal blood flow (decreases) 415–16
Skeletal muscle blood flow (increases) 389, 415, 417
Skin blood flow (increases) 415
Stroke volume (increases) 416, 418, 419
Summary 416
Venous return (increases) 416–17
   Role of respiratory pump 400–401, 417
   Role of skeletal muscle pump 400–401, 417

Effects on Organic Metabolism 577–78

Cortisol secretion (increases) 576
Diabetes mellitus (protects against) 578–80
Epinephrine secretion (increases) 575–76
Fuel homeostasis 577–78, 582
Fuel source 86, 273, 577–78
Glucagon secretion (increases) 575
Glucose mobilization from liver (increases) 577–78
Glucose uptake by muscle (increases) 577–78
Growth hormone secretion (increases) 578
Insulin secretion (decreases) 575
Metabolic rate (increases) 585
Plasma glucose changes 575
Plasma HDL (increases) 581
Plasma lactic acid (increases) 274, 473
Sympathetic nervous system activity (increases) 577, 578

Effects on Respiration 474–75

Alveolar gas pressures (no change in moderate
   exercise) 474–75
Capillary diffusion 462, 465
Control of respiration in exercise 473, 474–75
Oxygen debt 273
Pulmonary capillaries (dilate) 415–16
Ventilation (increases) 474–75
   Breathing depth (increases) 273, 476
   Expiration 451–52
   Respiratory rate (increases) 273, 470–74
   Role of Hering-Breuer reflex 470
   Stimuli 474–75

Effects on Skeletal Muscle 277–78

Adaptation to exercise 277–78
Arterioles (dilate) 405–408
Changes with aging 278
Fatigue 273–74
Glucose uptake and utilization (increase) 273
Hypertrophy 256
Local blood flow (increases) 389, 405, 415–16
Local metabolic rate (increases) 75
Local temperature (increases) 75
Nutrient utilization 577–78
Oxygen extraction from blood (increases) 273
Recruitment of motor units 277

Other Effects

Aging 278, 418–19
Body temperature (increases) 595
Central command fatigue 274
Gastrointestinal blood flow (decreases) 415–16
Immune function 670
Menstrual function 578, 640
Metabolic acidosis 473
Metabolic rate (increases) 584–85
Muscle fatigue 273–74
Osteoporosis (protects against) 345, 355, 640
Stress 577–78, 582
Weight loss 580, 589

Types of Exercise

Aerobic exercise 277–78
Endurance exercise 277, 417–19
Long-distance running 273, 277
Moderate exercise 277–78
Swimming 476
Weightlifting 277, 278, 417

Appendix C

Diseases and Disorders

Acquired immune deficiency syndrome (AIDS) 638, 671

Acromegaly 350–51

Acute respiratory distress syndrome (ARDS) 478

Addiction 187

Addison's disease 344

Adenoma 515

Adrenal insufficiency 344

Allergy (hypersensitivity) 343, 673–74

Altitude sickness 477

Alzheimer's disease 167, 246, 675

Amenorrhea 578, 640

Amnesia 246, 250

Anaphylaxis 674

Androgen insensitivity syndrome 603–604

Andropause 612–13, 614

Anemia 428, 463, 540, 678

Anemic hypoxia 476

Anorexia nervosa 589–90, 640

Anterograde amnesia 246

Aphasia 247–48

Appendicitis 692

Arrhythmias 513

Arteriosclerosis 386

Arthritis 343

Asthma 132f, 454, 675–76

Astigmatism 211

Atherosclerosis 344, 422, 580

Atrial fibrillation 377

Attention deficit hyperactivity disorder (ADHD) 239

Autism 308

Autoimmune disease 342, 675, 676t

Autoimmune thyroiditis 340

AV conduction disorders 371

Bipolar disorder 243

Botulism 166, 266

Brain death 237–38

Breech presentation 634

Bulimia nervosa 589–90

Carbon monoxide hypoxia 171, 468–69

Cardiogenic shock 414

Cataract 210

Central diabetes insipidus 503

Central sleep apnea 478

Cerebellar disease 307–308

Cholera 561

Chronic bronchitis 454

Chronic fatigue syndrome 344

Chronic inflammatory disease 675–76

Chronic obstructive pulmonary disease 454, 477

Colitis 536

Color blindness 227

Coma 237–38

Combined immunodeficiency 671

Concussion 250

Congenital adrenal hyperplasia 640–41

Congestive heart failure 419, 523

Conn's syndrome 515

Coronary artery disease 422–24

Coronary thrombosis 422

Cretinism 339

Crohn's disease 536, 560

Cryptorchidism 603

Cushing's disease 344–45

Cushing's phenomenon 412

Cushing's syndrome 344–45

Cystic fibrosis 444

Deep vein thrombosis 689, 690

Delayed hypersensitivity 674

Denervation atrophy 277

Depressive disorder (depression) 243–44

Diabetes insipidus 503

Diabetes mellitus 104, 328, 496, 578–80, 670

Diastolic dysfunction 420

Diffuse interstitial fibrosis 461

Diffusion impairment 476t

Disuse atrophy 277

Down syndrome 632

Duchenne muscular dystrophy 280, 281f

Dumping syndrome 551

Dwarfism 348

Dysmenorrhea 348

Eclampsia 633

Ectopic pacemakers 370

Ectopic pregnancy 628

Elephantiasis 401, 402f

Embolism 424, 689

Embolus 424

Emphysema 454, 472, 477

Epidural hematoma 250

Epilepsy 233, 249

Erectile dysfunction 610

Exercise-induced amenorrhea 578

Familial hypercholesterolemia 582

Familial renal glucosuria 497

Fasting hypoglycemia 580

Gallstones 559–60

Gastroesophageal reflux 545

Gigantism 350

Glaucoma 211

Graft rejection 672

Graves' disease 340–41, 686–87

Growth hormone insensitivity syndrome 348

Hashimoto's disease 340

Heart attack 422–24

Heart failure 382, 403, 419–21, 424

Heart murmurs 377–78

Heat exhaustion 596

Heat intolerance 341, 684–88, 694

Heat stroke 596

Hemochromatosis 426, 540

Hemolytic disease of the newborn 673

Hemophilia 435

Hemorrhage 413–14, 424, 428t

Histotoxic hypoxia 476

Thrombus 432, 689

Thyroiditis 687–88

Thyrotoxicosis 340, 686

Toxemia of pregnancy 633

Transfusion reaction 672–73

Transient ischemic attack (TIA) 424

Tuberculosis 344

Type 1 (T1DM) diabetes mellitus 578–80, 675

Type 2 (T2DM) diabetes mellitus 578–80

Ulcers 557–59

Upper motor neuron disorder 309

Urbach-Weithe disease 249

Uremia 523

Urge incontinence 499

Vasovagal syncope 414

Ventilation-perfusion inequality 461–62, 476–77

Ventricular fibrillation 422

Vertigo 224

Signs and Symptoms

Acidosis 517, 520–22

Akinesia 306

Aldosteronism 514

Alkalosis 517, 520–22

Altered states of consciousness 242–45

Ambiguous genitalia 641

Angina pectoris 382, 422

Atelectasis 688

Atrophy 277

Bradykinesia 306

Bruit 685

Butterfly rash 678

Catatonia 243

Clasp-knife phenomenon 309

Cold intolerance 340

Constipation 560–61

Cramp 309

Cross-tolerance 245

Daytime somnolence 478

Diabetic ketoacidosis 579, 580

Diabetic nephropathy 496

Diarrhea 561

Diplopia 684

Dyspnea 476

Edema 401, 403, 523

Exophthalmos 684

Farsightedness 211

Fever 81, 594–95, 691–94, 695

Flaccidity 310

Galactorrhea 640

Gastritis 558

Glucosuria 496

Goiter 338, 340f, 686

Gynecomastia 613

Heartburn 545

Hematoma 431

Hemolytic anemia 678

Hemolytic jaundice 559–60

Hot flashes 640

Hyperalgesia 206

Hypercalcemia 355–56

Hypercapnea 476

Hypercholesterolemia 424

Hyperkalemia 513

Hyperopic 211

Hyperparathyroidism 355–56

Hypertonia 309

Hypocalcemia 355–56

Hypocalcemic tetany 356

Hypoglycemia 580

Hypokalemia 513

Hypoparathyroidism 340–41

Hypotonia 310

Hypoxemia 689, 690

Hypoxia 476, 481

Immunosuppression 345

Insulin resistance 578

Intention tremor 307

Ischemia 422, 693

Jaundice 559

Late-phase reaction 674

Left ventricular hypertrophy 419

Masculinization 322

Morning sickness 634

Motion sickness 224

Muscle cramps 280

Myopic 210

Myxedema 340

Nearsightedness 210, 211f

Necrosis 691

Nystagmus 223

Palpitations 684–88, 694

Perforation 691

Plasmapheresis 280

Polycythemia 428

Prognathism 350

Proptosis 684

Pus 692

Referred pain 205

Retching 559

Rigidity 309

Spasm 309

Spasticity 309

Steatorrhea 559

Tachycardia 685

Tachypnea 685

Tetany 270

Thrombocytopenia 678

Tolerance 187, 245

Traveler's diarrhea 561

Uremia 523, 524

Virilism 641

Virilization 641

Withdrawal 244, 245t

Infectious or Causative Agents

Allergen 673–74

Anabolic steroids 349–50

Bacteria 647, 650t, 663–67, 693

Carbon monoxide 171, 468–69

Clostridium tetani 166, 311

Escherichia coli 692

Helicobacter pylori 558, 559

Human immunodeficiency virus (HIV) 671

LSD 169

Microbes 646

Nicotine 186, 245t

Pharmacological effects 325, 343

PTH-related peptide (PTHrp) 355

Sarin 167

Teratogen 632

Tetanus toxin 166

Tetrodotoxin 154

Thyroid-stimulating immunoglobulins (TSI) 686

Viruses 647, 663–68

Treatments, Diagnostics, and Therapeutic Drugs

Glossary

A

A band one of the transverse bands making up repeated striations of cardiac and skeletal muscle; region of aligned myosin-containing thick filaments

absolute refractory period time during which an excitable membrane cannot generate an action potential in response to any stimulus

absorption movement of materials across an epithelial layer from body cavity or compartment toward the blood

absorptive state period during which nutrients enter bloodstream from gastrointestinal tract

accessory reproductive organ duct through which sperm or egg is transported, or a gland emptying into such a duct (in the female, the breasts are usually included)

acclimatization (ah-climb-ah-tih-ZAY-shun) environmentally induced improvement in functioning of a physiological system with no change in genetic endowment

accommodation adjustment of eye for viewing various distances by changing shape of lens

A cell *see* alpha cell

acetylcholine (ACh) (ass-ih-teel-KOH-leen) a neurotransmitter released by pre- and postganglionic parasympathetic neurons, preganglionic sympathetic neurons, somatic neurons, and some CNS neurons

acetylcholinesterase (ass-ih-teel-koh-lin-ES-ter-ase) enzyme that breaks down acetylcholine into acetic acid and choline

acetyl coenzyme A (acetyl CoA) (ASS-ih-teel koh-EN-zime A, koh-A) metabolic intermediate that transfers acetyl groups to Krebs cycle and various synthetic pathways

acetyl group —COCH$_3$

acid molecule capable of releasing a hydrogen ion; solution having an H$^+$ concentration greater than that of pure water (that is, pH less than 7); *see also* strong acid, weak acid

acidic solution any solution with a pH less than 7.0

acidity concentration of free, unbound hydrogen ion in a solution; the higher the H$^+$ concentration, the greater the acidity

acidosis (ass-ih-DOH-sis) any situation in which arterial H$^+$ concentration is elevated above normal resting levels; *see also* metabolic acidosis, respiratory acidosis

acquired reflex behaviors that appear to be stereotypical and automatic, but which in fact result from considerable conscious effort to be learned (also called *learned reflex*)

acrosome (AK-roh-sohm) cytoplasmic vesicle containing digestive enzymes and located at head of a sperm

acrosome reaction process that occurs in the sperm after it binds to the zona pellucida of the egg, exposing acrosomal enzymes

actin (AK-tin) a protein that, with myosin, constitutes the contractile apparatus of muscle cells; it also is part of the cytoskeleton found in all cells.

action potential electrical signal propagated by nerve and muscle cells; an all-or-none depolarization of membrane polarity; has a threshold and refractory period and is conducted without decrement

activated macrophage macrophage whose killing ability has been enhanced by cytokines, particularly IL-2 and interferon-gamma

activation *see* lymphocyte activation

activation energy energy necessary to disrupt existing chemical bonds during a chemical reaction

active hyperemia (hy-per-EE-me-ah) increased blood flow through a tissue associated with increased metabolic activity

active immunity resistance to reinfection acquired by contact with microorganisms, their toxins, or other antigenic material; *compare* passive immunity

active site region of enzyme to which substrate binds

active transport energy-requiring system that uses transporters to move ions or molecules across a membrane against an electrochemical difference; *see also* primary active transport, secondary active transport

active zone region within a nerve terminal where neurotransmitter vesicles are clustered prior to secretion

activity *see* enzyme activity

acuity sharpness or keenness of perception

acute (ah-KUTE) lasting a relatively short time; *compare* chronic

acute phase protein group of proteins secreted by liver during systemic response to injury or infection

acute phase response responses of tissues and organs distant from site of infection or immune response

adaptation (evolution) a biological characteristic that favors survival in a particular environment; (neural) decrease in action potential frequency in a neuron despite constant stimulus

adenine one of the four bases making up DNA; also a breakdown product of ATP used as a neurotransmitter

adenoid lymphoid tissue; also known as pharyngeal tonsil

adenosine diphosphate (ADP) (ah-DEN-oh-seen dy-FOS-fate) two-phosphate product of ATP breakdown

adenosine monophosphate (AMP) one-phosphate derivative of ATP

adenosine triphosphate (ATP) major molecule that transfers energy from metabolism to cell functions during its breakdown to ADP and release of P$_i$

adenylyl cyclase (ad-DEN-ah-lil SY-klase) enzyme that catalyzes transformation of ATP to cyclic AMP

adequate stimulus the modality of stimulus to which a particular sensory receptor is most sensitive

adhesion molecule group of proteins secreted from endothelial cells that mediate attachment of leukocytes to endothelium during injury

adipocyte (ad-DIP-oh-site) cell specialized for triglyceride synthesis and storage; fat cell

adipose tissue (AD-ah-poze) tissue composed largely of fat-storing cells

adrenal cortex (ah-DREE-nal KOR-tex) endocrine gland that forms outer shell of each adrenal gland; secretes steroid hormones—mainly cortisol, aldosterone, and androgens; *compare* adrenal medulla

adrenal gland one of a pair of endocrine glands above each kidney; each gland consists of outer *adrenal cortex* and inner *adrenal medulla*

adrenal medulla (meh-DUL-ah) endocrine gland that forms inner core of each adrenal gland; secretes amine hormones, mainly epinephrine; *compare* adrenal cortex

adrenergic (ad-ren-ER-jik) pertaining to norepinephrine or epinephrine; compound that acts like norepinephrine or epinephrine

adrenocorticotropic hormone (ACTH) (ad-ren-oh-kor-tih-koh-TROH-pik) polypeptide hormone secreted by anterior pituitary; stimulates adrenal cortex to secrete cortisol; also called *corticotropin*

adsorptive endocytosis receptor-mediated invagination of plasma membranes by which both extracellular fluid and specific extracellular molecules are moved into cytosol

aerobic (air-OH-bik) in presence of oxygen

afferent (AF-er-ent) carrying toward

afferent arteriole vessel in kidney that carries blood from artery to renal corpuscle

afferent division (PNS) neurons in the peripheral nervous system that project to the central nervous system

afferent neuron neuron that carries information from sensory receptors at its peripheral endings to CNS; cell body lies outside CNS

afferent pathway component of reflex arc that transmits information from receptor to integrating center

affinity strength with which ligand binds to its binding site

afterbirth placenta and associated membranes expelled from uterus after delivery of infant

afterhyperpolarization decrease in membrane potential in neurons at the end of the action potential

afterload the work the heart does while ejecting blood; a function of the arterial blood pressure, as well as the diameter and thickness of the ventricles

agonist (AG-ah-nist) chemical messenger that binds to receptor and triggers cell's response; often refers to drug that mimics action of chemical normally in the body

airway tube through which air flows between external environment and lung alveoli

albumin (al-BU-min) most abundant plasma protein

aldosterone (al-doh-STEER-own or al-DOS-stir-own) mineralocorticoid steroid hormone secreted by adrenal cortex; regulates electrolyte balance

alkaline solution any solution having H+ concentration lower than that of pure water (that is, having a pH greater than 7)

alkalosis (alk-ah-LOH-sis) any situation in which arterial blood H+ concentration is reduced below normal resting levels; *see also* metabolic alkalosis, respiratory alkalosis

all-or-none pertaining to event that occurs maximally or not at all

allosteric modulation (al-low-STAIR-ik) control of protein binding site properties by modulator molecules that bind to regions of the protein other than the binding site altered by them

allosteric protein protein whose binding site characteristics are subject to allosteric modulation

alpha-adrenergic receptor one type of plasma-membrane receptor for epinephrine and norepinephrine; also called alpha adrenoceptor; *compare* beta-adrenergic receptor

alpha cell glucagon-secreting cell of pancreatic islets of Langerhans

alpha-glycerol phosphate three-carbon molecule that combines with fatty acids to form triglyceride; also called glycerol 3-phosphate

alpha helix coiled regions of proteins or DNA formed by hydrogen bonds

α-ketoacid (AL-fuh KEY-toh) molecule formed from amino acid metabolism and containing carbonyl (—CO—) and carboxyl (—COOH) groups

alpha motor neuron motor neuron that innervates extrafusal skeletal muscle fibers

alpha rhythm prominent 8- to 13-Hz oscillation on the electroencephalograms of awake, relaxed adults with their eyes closed

alternate complement pathway sequence for complement activation that bypasses first steps in classical pathway and is not antibody dependent

alveolar dead space (al-VEE-oh-lar) volume of fresh inspired air that reaches alveoli but does not undergo gas exchange with blood

alveolar pressure (P_{alv}) air pressure in pulmonary alveoli

alveolar ventilation ($\dot{V}_A$) volume of atmospheric air entering alveoli each minute

alveolus (al-VEE-oh-lus) (lungs) thin-walled, air-filled "outpocketing" from terminal air passageways in lungs; (glands) cell cluster at end of duct in secretory gland

amacrine cell (AM-ah-krin) a specialized type of neuron found in the retina of the eye that integrates information between local photoreceptor cells

amine hormone (ah-MEEN) hormone derived from amino acid tyrosine; includes thyroid hormones, epinephrine, norepinephrine, and dopamine

amino acid (ah-MEEN-oh) molecule containing amino group, carboxyl group, and side chain attached to a carbon atom; molecular subunit of protein

amino acid side chain the variable portions of amino acids; may contain acidic or basic charged regions, or may be hydrophobic

amino group —NH_2; ionizes to —NH_3^+

aminopeptidase (ah-meen-oh-PEP-tih-dase) one of a family of enzymes located in the intestinal epithelial membrane; breaks peptide bond at amino end of polypeptide

ammonia NH_3; produced during amino acid breakdown; converted in liver to urea; ionized form is ammonium

amnion another term for amniotic sac

amniotic cavity (am-nee-AHT-ik) fluid-filled space surrounding the developing fetus enclosed by amniotic sac

amniotic fluid liquid within amniotic cavity that has a composition similar to extracellular fluid

amniotic sac membrane surrounding fetus *in utero*

AMPA receptor receptor protein found in the membrane of some brain neurons, named for its binding to alpha-amino-3 hydroxy-5 methyl-4 isoxazole proprionic acid

amphipathic (am-fuh-PATH-ik) a molecule containing polar or ionized groups at one end and nonpolar groups at the other

ampulla structure in the wall of the semicircular canals containing hair cells that respond to head movement

amylase (AM-ih-lase) enzyme that partially breaks down polysaccharides

anabolism (an-NAB-oh-lizm) cellular synthesis of organic molecules

anaerobic (an-ih-ROH-bik) in the absence of oxygen

anatomic dead space space in respiratory tract airways where gas exchange does not occur with blood

androgen (AN-dro-jen) any hormone with testosterone-like actions

androgen-binding protein synthesized and secreted by Sertoli cell of the testes—binds to and increases local testosterone concentration in fluid in the seminiferous tubule

anemia (ah-NEE-me-ah) reduction in total blood hemoglobin

angiogenesis (an-gee-oh-JEN-ah-sis) the development and growth of capillaries; stimulated by angiogenic factors

angiogenic factor chemical signal that induces the development and growth of blood vessels

angiotensin I small polypeptide generated in plasma by renin's action on angiotensinogen

angiotensin II hormone formed by action of angiotensin-converting enzyme on angiotensin I; stimulates aldosterone secretion from adrenal cortex, vascular smooth-muscle contraction, and thirst

angiotensin-converting enzyme (ACE) enzyme on capillary endothelial cells that catalyzes removal of two amino acids from angiotensin I to form angiotensin II

angiotensinogen (an-gee-oh-ten-SIN-oh-gen) plasma protein precursor of angiotensin I; produced by liver

anion (AN-eye-on) negatively charged ion; *compare* cation

antagonist (muscle) muscle whose action opposes intended movement; (drug) molecule that competes with another for a receptor and binds to the receptor but does not trigger the cell's response

anterior toward or at the front

anterior pituitary anterior portion of pituitary gland; synthesizes, stores, and releases ACTH, GH, TSH, prolactin, FSH, and LH

anterograde (AN-ter-oh-grayd) movement of a substance or action potential in the forward direction from a neuron's dendrites and/or cell body, toward the axon terminal

anterolateral pathway ascending neural pathway running in the anterolateral column of the spinal cord white matter; conveys information about pain and temperature

antibody (AN-tih-bah-dee) immunoglobulin secreted by plasma cell; combines with type of antigen that stimulated its production; directs attack against antigen or cell bearing it

antibody-dependent cellular cytotoxicity (ADCC) killing of target cells by toxic chemicals secreted by NK cells; the target cells are linked to the NK cells by antibodies

antibody-mediated response humoral immune response mediated by circulating antibodies; major defense against microbes and toxins in the extracellular fluid

anticodon (an-tie-KOH-don) three-nucleotide sequence in tRNA able to base-pair with complementary codon in mRNA during protein synthesis

antidiuretic hormone (ADH) (an-ty-dy-yor-ET-ik) *see* vasopressin

antigen (AN-tih-jen) any foreign molecule that stimulates a specific immune response

antigen binding site one of the two variable "prongs" on an immunoglobulin capable of binding to a specific antigen

antigen presentation process by which an antigen-presenting cell, such as a macrophage, combines proteolytic fragments of a foreign antigen with host cell class II MHC proteins, which are transported to the host cell's surface

antigen-presenting cell (APC) cell that presents antigen, complexed with MHC proteins on its surface, to T cells

antithrombin III a plasma protein activated by heparin that limits clot formation by inactivating thrombin and other clotting factors

antrum (AN-trum) (gastric) lower portion of stomach (that is, region closest to pyloric sphincter); (ovarian) fluid-filled cavity in maturing ovarian follicle

anus lowest opening of the digestive tract through which fecal matter is extruded

aorta (a-OR-tah) largest artery in body; carries blood from left ventricle of heart to thorax and abdomen

aortic arch baroreceptor (a-OR-tik) *see* arterial baroreceptor

aortic body chemoreceptor chemoreceptor located near aortic arch; sensitive to arterial blood O_2 pressure and H^+ concentration

aortic valve valve between left ventricle of heart and aorta

apneustic center area in the lower pons in the brain with input to the medullary inspiratory neurons; helps to terminate inspiration

apoptosis (ay-pop-TOE-sis) programmed cell death that typically occurs during differentiation and development

appendix small fingerlike projection from cecum of large intestine

aquaporin (ah-qua-PORE-in) protein membrane channel through which water can diffuse

aqueous (AY-kwee-us) watery; prepared with water

aqueous humor fluid filling the anterior chamber of the eye

arachidonic acid (ah-rak-ah-DON-ik) polyunsaturated fatty acid precursor of eicosanoids

arachnoid mater (ah-RAK-noid) the middle of three membranes (meninges) covering the brain

area postrema a circumventricular organ outside the blood-brain barrier

aromatase enzyme that converts androgens to estrogens; located predominantly in the ovaries, the placenta, the brain, and adipose tissue

arrhythmia (ay-RYTH-me-ah) any variation from normal heartbeat rhythm

arterial baroreceptor nerve endings sensitive to stretch or distortion produced by arterial blood pressure changes; located in carotid sinus or aortic arch; also called carotid sinus and aortic arch baroreceptors

arteriole (are-TEER-ee-ole) blood vessel between artery and capillary, surrounded by smooth muscle; primary site of vascular resistance

arteriosclerosis (are-TEER-ee-oh-sklare-OH-sis) "hardening" of arterial walls that can have different causes, including deposition of collagenous fibers that occurs with aging

artery (ARE-ter-ee) thick-walled elastic vessel that carries blood away from heart to arterioles

ascending limb portion of Henle's loop of renal tubule leading to distal convoluted tubule

ascending pathway neural pathway that goes to the brain; also called sensory pathway

aspartate (ah-SPAR-tate) an excitatory neurotransmitter in CNS; ionized form of the amino acid aspartic acid

aspiration inhalation of liquid or a foreign body into the airways

association cortex *see* cortical association area

astrocyte a form of glial cell that regulates composition of extracellular fluid around neurons and forms part of the blood-brain barrier

atmospheric pressure (P_{atm}) air pressure surrounding the body (760 mmHg at sea level)

atom smallest unit of matter that has unique chemical characteristics; has no net charge; combines with other atoms to form chemical substances

atomic nucleus dense region, consisting of protons and neutrons, at center of atom

atomic number number of protons in nucleus of atom

atomic weight value that indicates an atom's mass relative to mass of other types of atoms based on the assignment of a value of 12 to carbon atom

ATP *see* adenosine triphosphate

ATPase (aa-tea-PEE-ase) enzyme that breaks down ATP to ADP and inorganic phosphate

atresia degeneration of nondominant follicles in the ovary

atrial natriuretic peptide (ANP) (nay-tree-yor-ET-ik) peptide hormone secreted by cardiac atrial cells in response to atrial distension; causes increased renal sodium excretion

atrioventricular (AV) node (ay-tree-oh-ven-TRIK-you-lar) region at base of right atrium near interventricular septum, containing specialized cardiac muscle cells through which electrical activity must pass to go from atria to ventricles

atrioventricular (AV) valve valve between atrium and ventricle of heart; AV valve on right side of heart is the *tricuspid valve,* and that on left side is the *mitral valve*

atrium (AY-tree-um) chamber of heart that receives blood from veins and passes it on to ventricle on same side of heart

atrophy (AT-roh-fee) wasting away; decrease in size

atropine (AT-roh-peen) a drug that specifically blocks the binding of acetylcholine to muscarinic acetylcholine receptors

auditory (AW-dih-tor-ee) pertaining to sense of hearing

auditory cortex region of cerebral cortex that receives nerve fibers from auditory pathways

autocrine agent (AW-toh-crin) chemical messenger secreted into extracellular fluid that acts upon cell that secreted it; *compare* paracrine agent

automaticity (aw-toh-mah-TISS-ih-tee) capable of spontaneous, rhythmical self-excitation

autonomic ganglion group of neuron cell bodies in the peripheral nervous system

autonomic nervous system (aw-toh-NAHM-ik) component of efferent division of peripheral nervous system that consists of sympathetic and parasympathetic subdivisions; innervates cardiac muscle, smooth muscle, and glands; *compare* somatic nervous system

autoreceptor a receptor on a cell affected by a chemical messenger released from the same cell

autoregulation (aw-toh-reg-you-LAY-shun) ability of an individual organ to control (self-regulate) its vascular resistance independent of neural and hormonal influence; *see also* flow autoregulation

autosome chromosome that contains genes for proteins governing most cell structures and functions; *compare* sex chromosome

autotransfusion shift of fluid from the interstitial space to the blood following a decrease in blood pressure

axo-axonic synapse presynaptic synapse where an axon stimulates the presynaptic terminal of another axon

axon (AX-ahn) extension from neuron cell body; propagates action potentials away from cell body; also called a nerve fiber

axon hillock part of the axon nearest the cell body where the action potential begins

axon terminal end of axon; forms synaptic or neuroeffector junction with postjunctional cell

axonal transport process involving intracellular filaments by which materials are moved from one end of axon to other

B

B cell (immune system) lymphocyte that, upon activation, proliferates and differentiates into antibody-secreting plasma cell; (endocrine cell) *see* beta cell

B lymphocyte *see* B cell

barometric pressure *see* atmospheric pressure

baroreceptor receptor sensitive to pressure and to rate of change in pressure; *see also* arterial baroreceptor, intrarenal baroreceptor

Barr body sex chromatin nuclear mass formed by the nonfunctional X chromosome in female cell nuclei

basal (BAY-sul) resting level

basal cell cell found within taste buds that can divide and differentiate to replace worn-out taste receptor cells

basal metabolic rate (BMR) metabolic rate when a person is at mental and physical rest but not sleeping, at comfortable temperature, and has fasted at least 12 h; also called metabolic cost of living

basal nuclei nuclei deep in cerebral hemispheres that code and relay information associated with control of body movements; specifically, caudate nucleus, globus pallidus, and putamen; also called *basal ganglia*

base (acid-base) any molecule that can combine with H^+; (nucleotide) molecular ring of carbon and nitrogen that, with a phosphate group and a sugar, constitutes a nucleotide

basement membrane thin layer of extracellular proteinaceous material upon which epithelial and endothelial cells sit

basic electrical rhythm spontaneous depolarization-repolarization cycles of pacemaker cells in longitudinal smooth muscle layer of stomach and intestines; coordinates repetitive muscular activity of GI tract

basilar membrane (BAS-ih-lar) membrane that separates cochlear duct and scala tympani in inner ear; supports organ of Corti

basolateral membrane (bay-so-LAH-ter-al) sides of epithelial cell other than luminal surface; also called serosal or blood side of cell

basophil (BAY-so-fill) polymorphonuclear granulocytic leukocyte whose granules stain with basic dyes; enters tissues and becomes mast cell

beta-adrenergic receptor (BAY-ta ad-ren-ER-jik) a type of plasma membrane receptor for epinephrine and norepinephrine; *compare* alpha-adrenergic receptor; also called *beta adrenoceptor*

beta cell insulin-secreting cell in pancreatic islets of Langerhans; also called *B cell*

beta-endorphin putative hormone released from the anterior pituitary gland, believed to play a role in adaptation to stress and pain relief; also acts as a neurotransmitter

beta-lipoprotein a protein formed from the proopiomelanocortin precursor in the anterior pituitary gland; further processing results in the putative hormone beta-endorphin

beta oxidation (ox ih-DAY-shun) series of reactions that generate hydrogen atoms (for oxidative phosphorylation) from breakdown of fatty acids to acetyl CoA

beta rhythm low, fast EEG oscillations in alert, awake adults who are paying attention to (or thinking hard about) something

beta sheet a form of secondary protein structure determined by the relative hydrophobicity of amino acid side chains

bicarbonate (by-KAR-bah-nate) HCO_3^-

bicuspid valve another term for the left atrioventricular valve, also called the mitral valve

bile fluid secreted by liver into bile canaliculi; contains bicarbonate, bile salts, cholesterol, lecithin, bile pigments, metabolic end products, and certain trace metals

bile canaliculi (kan-al-IK-you-lee) small ducts adjacent to liver cells into which bile is secreted

bile pigment colored substance, derived from breakdown of heme group of hemoglobin, secreted in bile

bile salt one of a family of steroid molecules produced from cholesterol and secreted in bile by the liver; promotes solubilization and digestion of fat in small intestine

bilirubin (bil-eh-RUE-bin) yellow substance resulting from heme breakdown; excreted in bile as a bile pigment

binding site region of protein to which a specific ligand binds

biogenic amine (by-oh-JEN-ik ah-MEEN) one of family of neurotransmitters having basic formula $R-NH_2$; includes dopamine, norepinephrine, epinephrine, serotonin, and histamine

biological clock neurons that drive body rhythms

bipolar cell type of nerve cell that has one input branch and one output branch

bivalent paired homologous chromosomes, each with two sister chromatids, that are produced during meiosis

bladder urinary bladder; thick-walled sac composed of smooth muscle; stores urine prior to urination

blastocyst (BLAS-toh-cyst) particular early embryonic stage consisting of ball of developing cells surrounding central cavity

block to polyspermy process that prevents more than one sperm cell from fertilizing an ovum

blood pressurized contents of the cardiovascular system composed of a liquid phase (plasma) and cellular phase (red and white blood cells, platelets)

blood-brain barrier group of anatomical barriers and transport systems in brain capillary endothelium that controls kinds of substances entering brain extracellular space from blood and their rates of entry

blood coagulation (koh-ag-you-LAY-shun) blood clotting

blood sugar glucose

blood type blood classification according to presence of A and/or B antigens or lack of them (O)

blood vessel tubular structures of various sizes that transport blood throughout the body

body (of stomach) middle portion of the stomach; secretes mucus, pepsinogen, and hydrochloric acid

body mass index (BMI) method for assessing degree of obesity; calculated as weight in kilograms divided by square of height in meters

bone age an x-ray determination of the degree of bone development; often used in assessing reasons for unusual stature in children

bone marrow highly vascular, cellular substance in central cavity of some bones; site of erythrocyte, leukocyte, and platelet synthesis

Bowman's capsule blind sac at beginning of tubular component of kidney nephron

Bowman's space fluid-filled space within Bowman's capsule into which protein-free fluid filters from the glomerulus

Boyle's law pressure of a fixed amount of gas in a container is inversely proportional to container's volume

bradykinin (braid-ee-KY-nin) protein formed by action of the enzyme kallikrein on precursor

brain self-stimulation phenomenon in which animals will press a bar to get electrical stimulation of certain parts of their brains

brainstem brain subdivision consisting of medulla oblongata, pons, and midbrain and located between spinal cord and forebrain

brainstem pathway descending motor pathway whose cells of origin are in the brainstem

Broca's area (BRO-kahz) region of left frontal lobe associated with speech production

bronchiole (BRON-key-ole) small airway distal to bronchus

bronchus (BRON-kus) large-diameter air passage that enters lung; located between trachea and bronchioles

brown adipose tissue type of adipose (fat) tissue found in newborns and in many mammals, with a higher heat-producing capacity than ordinary white fat; may be important in regulating body temperature in extreme conditions

brush border small projections (microvilli) of epithelial cells covering the villi of the small intestine; major absorptive surface of the small intestine

buffer weak acid or base that can exist in undissociated (H buffer) or dissociated (H^+ + buffer) form

buffering reversible hydrogen-ion binding by anions when H^+ concentration changes; tends to minimize changes in acidity of a solution when acid is added or removed

bulbourethral gland (bul-bo-you-REETH-ral) one of paired glands in male that secretes fluid components of semen into the urethra

bulk flow movement of fluids or gases from region of higher pressure to one of lower pressure

bundle branch pathway composed of cells that rapidly conduct electrical signals down the right and left sides of the interventricular septum; these pathways connect the bundle of His to the Purkinje network

bundle of His (HISS) nervelike structure composed of modified heart cells that carries electrical impulses from the atrioventricular node down the interventricular septum

C

calcitonin hormone from the thyroid gland that inhibits bone resorption

calmodulin (kal-MOD-you-lin) intracellular calcium-binding protein that mediates many of calcium's second-messenger functions

calmodulin-dependent protein kinase an intracellular enzyme that, when activated by calcium and the protein calmodulin, phosphorylates many protein substrates within cells; it is a component of many intracellular signaling mechanisms

calorie (cal) unit of heat-energy measurement; amount of heat needed to raise temperature of 1 g of water 1°C; *compare* kilocalorie

calorigenic effect (kah-lor-ih-JEN-ik) increase in metabolic rate caused by epinephrine or thyroid hormones

cAMP-dependent protein kinase (KY-nase) *see* cyclic AMP-dependent protein kinase

capacitation *see* sperm capacitation

capillary one of the smallest blood vessels

capillary filtration coefficient a measure of fluid flow across a capillary wall at a given hydrostatic pressure gradient; determined by porosity of the barrier

capsaicin (kap-SAY-sin) the molecule in chili peppers that causes sensations of heat and pain in the mouth and throat

carbamino hemoglobin (kar-bah-MEEN-oh HE-moe-glow-bin) compound resulting from combination of carbon dioxide and amino groups in hemoglobin

carbohydrate substance composed of carbon, hydrogen, and oxygen according to general formula $C_n(H_2O)_n$, where *n* is any whole number

carbonic acid (kar-BAHN-ik) H_2CO_3; an acid formed from H_2O and CO_2

carbonic anhydrase (an-HY-drase) enzyme that catalyzes the reaction $CO_2 + H_2O \rightleftharpoons H_2CO_3$

carbon monoxide CO; gas that reacts with hemoglobin; decreases blood oxygen-carrying capacity and shifts oxygen-hemoglobin dissociation curve to the left; also acts as an intracellular messenger in neurons

carboxyl group (kar-BOX-il) —COOH; ionizes to carboxyl ion (—COO^-)

carboxypeptidase (kar-box-ee-PEP-tih-dase) enzyme secreted into small intestine by exocrine pancreas as precursor, procarboxypeptidase; breaks peptide bond at carboxyl end of protein

cardiac (KAR-dee-ak) pertaining to the heart

cardiac cycle one contraction-relaxation sequence of heart

cardiac muscle heart muscle

cardiac output blood volume pumped by each ventricle per minute (not total output pumped by both ventricles)

cardiovascular center neuron cluster in brainstem medulla oblongata that serves as a major integrating center for reflexes affecting heart and blood vessels

cardiovascular system heart and blood vessels

carotid (kuh-RAH-tid) pertaining to two major arteries (carotid arteries) in neck that convey blood to head

carotid body chemoreceptor chemoreceptor near main branching of carotid artery; sensitive to blood O_2 pressure and H^+ concentration

carotid sinus region of internal carotid artery just above main carotid branching; location of carotid baroreceptors

carotid sinus baroreceptor *see* arterial baroreceptor

carrier *see* transporter

catabolism (kuh-TAB-oh-lizm) cellular breakdown of organic molecules

catalyst (KAT-ah-list) substance that accelerates chemical reactions but does not itself undergo any net chemical change during the reaction

catch-up growth a period of rapid growth during which a child attains his or her predicted height for a given age after a temporary period of slow growth due to illness or malnourishment

catecholamine (kat-eh-COLE-ah-meen) dopamine, epinephrine, or norepinephrine, all of which have similar chemical structures

cation (KAT-eye-on) ion having net positive charge; *compare* anion

caveolae (kav-ee-OH-lee) small invaginations of the plasma membrane that pinch off and form endocytotic vesicles that deliver their contents directly to the cytosol

cecum (SEE-come) dilated pouch at beginning of large intestine into which the ileum, colon, and appendix open

cell body in cells with long extensions, the part that contains the nucleus

cell differentiation *see* differentiation

cell organelle (or-guh-NEL) membrane-bound compartment, nonmembranous particle, or filament that performs specialized functions in cell

center of gravity point in a body at which body mass is in perfect balance; if the body were suspended from a string attached to this point, there would be no movement

central chemoreceptor receptor in brainstem medulla oblongata that responds to H^+ concentration changes of brain extracellular fluid

central command fatigue muscle fatigue due to failure of appropriate regions of cerebral cortex to excite motor neurons

central nervous system (CNS) brain plus spinal cord

central sulcus a deep infolding on each half of the brain that separates the parietal and central lobes

central thermoreceptor temperature receptor in hypothalamus, spinal cord, abdominal organ, or other internal location

centriole (SEN-tree-ole) small cytoplasmic body having nine fused sets of microtubules; participates in nuclear and cell division

centrosome region of cell cytoplasm in which microtubule formation and elongation occur, particularly during cell division

cephalic phase (seh-FAL-ik) (of gastrointestinal control) initiation of the neural and hormonal reflexes regulating gastrointestinal functions by stimulation of receptors in head, that is, cephalic receptors—sight, smell, taste, and chewing—as well as by emotional states

cerebellum (ser-ah-BEL-um) brain subdivision lying behind forebrain and above brainstem; deals with muscle movement control

cerebral cortex (SER-ah-brul or sah-REE-brul) cellular layer covering the cerebrum

cerebral hemisphere either left or right half of the cerebral cortex

cerebral ventricle one of four interconnected spaces in the brain; filled with cerebrospinal fluid

cerebrospinal fluid (CSF) (sah-ree-broh-SPY-nal) fluid that fills cerebral ventricles and the subarachnoid space surrounding brain and spinal cord

cerebrum (SER-ah-brum or sah-REE-brum) part of the brain that, with diencephalon, forms the forebrain

cervix (SIR-vix) lower portion of uterus; cervical opening connects uterine and vaginal lumens

cGMP-dependent protein kinase (KY-nase) *see* cyclic GMP-dependent protein kinase

channel small passage in plasma membrane formed by integral membrane proteins and through which certain small-diameter molecules and ions can diffuse; *see also* ligand-gated channel, mechanically gated channel, voltage-gated channel

channel gating process of opening and closing ion channels

chemical element specific type of atom

chemical equilibrium state when rates of forward and reverse components of a chemical reaction are equal, and no net change in reactant or product concentration occurs

chemical reaction breaking of some chemical bonds and formation of new ones, which changes one type of molecule to another; *see also* irreversible reaction, reversible reaction

chemical specificity *see* specificity

chemical synapse (SIN-apse) synapse at which neurotransmitters released by one neuron diffuse across an extracellular gap to influence a second neuron's activity

chemiosmotic hypothesis the proposed mechanism by which ATP is formed during oxidative phosphorylation; the hypothesis proposes that the movement of protons across mitochondrial inner membranes is coupled with ATP production

chemoattractant any mediator that causes chemotaxis; also called *chemotaxin*

chemokine any cytokine that functions as a chemoattractant

chemoreceptor afferent nerve ending (or cell associated with it) sensitive to concentrations of certain chemicals

chemotaxin (kee-moh-TAX-in) *see* chemoattractant

chemotaxis (kee-moh-TAX-iss) movement of cells, particularly phagocytes, in a specific direction in response to a chemical stimulus

chief cell gastric gland cell that secretes pepsinogen, precursor of pepsin

cholecalciferol (vitamin D₃) (kohl-ee-kal-SIF-er-ol) animal vitamin D

cholecystokinin (CCK) (koh-lee-sis-toh-KY-nin) peptide hormone secreted by duodenum that regulates gastric motility and secretion, gallbladder contraction, and pancreatic enzyme secretion; possible satiety signal

cholesterol particular steroid molecule; precursor of steroid hormones and bile salts and a component of plasma membranes

cholesterol esterase enzyme that removes a fatty acid from a molecule of esterified cholesterol; required for the production of free cholesterol in steroidogenic glands

cholinergic (koh-lin-ER-jik) pertaining to acetylcholine; a compound that acts like acetylcholine

chondrocyte (KON-droh-site) cell types that form new cartilage

chordae tendineae (KORE-day TEN-den-ay) strong, fibrous cords that connect papillary muscles to the edges of atrioventricular valves; they prevent backward flow of blood during ventricular systole

chorion outermost fetal membrane derived from trophoblast cells; becomes part of the placenta

chorionic villi fingerlike projections of the trophoblast cells extending from the chorion into the endometrium of the uterus

choroid (KORE-oyd) pigmented layer of eye that lies next to retina

choroid plexus highly vascular epithelial structure lining portions of cerebral ventricles; responsible for much of cerebrospinal fluid formation

chromatid (KROM-ah-tid) one of two identical strands of chromatin resulting from DNA duplication during mitosis or meiosis

chromatin (KROM-ih-tin) combination of DNA and nuclear proteins; principal component of chromosomes

chromophore retinal light-sensitive component of a photopigment

chromosome highly coiled, condensed form of chromatin formed in cell nucleus during mitosis and meiosis

chronic (KRON-ik) persisting over a long time; *compare* acute

chylomicron (kye-loh-MY-kron) small droplet consisting of lipids and protein released from intestinal epithelial cells into the lacteals during fat absorption

chyme (kyme) solution of partially digested food in stomach and intestinal lumens

chymotrypsin enzyme secreted by exocrine pancreas; breaks certain peptide bonds in proteins and polypeptides

cilia (SIL-ee-ah) hairlike projections from specialized epithelial cells that sweep back and forth in a synchronized way to propel material along epithelial surface

ciliary muscle involved in movement and shape of the lens during accommodation

circadian rhythm (sir-KAY-dee-an) occurring in an approximately 24 h cycle

circular muscle smooth-muscle layer in stomach and intestinal walls that has muscle fibers circumferentially oriented around these organs

circulatory system (SIRK-you-la-tor-ee) the heart and system of vessels that deliver blood to all parts of the body

citric acid cycle *see* Krebs cycle

classical complement pathway antibody-dependent system for activating complement; begins with complement molecule Cl

clathrin (clathrin-coated pits) a cytosolic protein that binds to regions of the plasma membrane and helps initiate receptor-mediated endocytosis; clathrin forms localized regions known as clathrin-coated pits that invaginate to become endocytotic vesicles

clearance volume of plasma from which a particular substance has been completely removed in a given time

cleavage mitotic cell division

clitoris (KLIT-or-iss) small body of erectile tissue in female external genitalia; homologous to penis

clonal deletion destruction by apoptosis in the thymus of those T cells that have receptors capable of binding to self proteins

clonal expansion lymphocyte cell divisions initiated by binding of an antigen to a lymphocyte cell membrane receptor

clonal inactivation process occurring in the periphery (that is, not in the thymus) that causes potentially self-reacting T cells to become nonresponsive

clone one of a set of genetically identical molecules, cells, or organisms

clot solid phase of blood, formed from platelets, trapped blood cells, and a polymer of the protein fibrin

clotting phase transition of blood from a liquid cell suspension into a solid, gel-like mass

coactivation pattern of nerve firing to a muscle, in which alpha and gamma motor neurons simultaneously activate contractile (extrinsic) fibers while increasing tension in the ends of the muscle spindle (instrinsic) fibers

coagulation (koh-ag-you-LAY-shun) blood clotting

cochlea (KOK-lee-ah) inner ear; fluid-filled spiral-shaped compartment that contains cochlear duct

cochlear duct (KOK-lee-er) fluid-filled membranous tube that extends length of inner ear, dividing it into compartments; contains organ of Corti

coding process by which neural signals from sensory receptors are converted into action potentials in the CNS

codon (KOH-don) three-base sequence in mRNA that determines the position of a specific amino acid during protein synthesis or that designates the end of the coded sequence of a protein

coenzyme (koh-EN-zime) organic cofactor; generally serves as a carrier that transfers atoms or small molecular fragments from one reaction to another; is not consumed in the reaction and can be reused

coenzyme A *see* acetyl coenzyme A

cofactor (KOH-fact-or) organic or inorganic substance that binds to a specific region of an enzyme and is necessary for the enzyme's activity

colipase protein secreted by pancreas that binds lipase, bringing it in contact with lipid droplets in the small intestine

collagen fiber (KOL-ah-jen) strong, fibrous protein that functions as extracellular structural element in connective tissue

collateral branch of a nerve axon

collecting duct system portion of renal tubules between distal convoluted tubules and renal pelvis; comprises *cortical collecting duct* and *medullary collecting duct*

colloid (KOL-oid) large molecule, mainly protein, to which capillaries are relatively impermeable; also, part of the inner structure of the thyroid gland

colon (KOH-lun) a portion of the large intestine, specifically the part extending from cecum to rectum

colony-stimulating factor (CSF) collective term for several hematopoietic growth factors that stimulate production of neutrophils and monocytes

colostrum watery, protein-rich liquid secreted by mother's breasts for first 24 to 48 hours after delivery of baby

commissure (KOM-ih-shur) bundle of nerve fibers linking right and left halves of the brain

common bile duct carries bile from gallbladder to small intestine

competition ability of similar molecules to combine with the same binding site or receptor

complement (KOM-plih-ment) one of a group of plasma proteins that, upon activation, kills microbes directly and facilitates the inflammatory process, including phagocytosis

compliance stretchability; *see also* lung compliance

concentration amount of material per unit volume of solution

concentration gradient gradation in concentration that occurs between two regions having different concentrations

concentric contraction muscle activity that involves shortening of muscle length

conceptus collective term for the fertilized egg and everything derived from it

conducting system network of cardiac muscle fibers specialized to conduct electrical activity between different areas of heart

conducting zone air passages that extend from top of trachea to beginning of respiratory bronchioles and have walls too thick for gas exchange between air and blood

conduction (heat) transfer of thermal energy during collisions of adjacent molecules

cone one of two retinal receptor types for photic energy; gives rise to color vision

conformation three-dimensional shape of a molecule

congenital existing at birth; usually referring to a birth defect

connective tissue one of the four major categories of tissues in the body; major component of extracellular matrices, cartilage, and bone

connective tissue cell cell specialized to form extracellular elements that connect, anchor, and support body structures

conscious experience things of which a person is aware; thoughts, feelings, perceptions, ideas, and reasoning during any state of consciousness

consciousness *see* conscious experience, state of consciousness

consolidation process by which short-term memories are converted into long-term memories

contractility (kon-trak-TIL-ity) force of heart contraction that is independent of sarcomere length

contraction operation of the force-generating process in a muscle

contraction time time between beginning of force development and peak twitch tension by the muscle

contralateral on the opposite side of the body

control system *see* homeostatic control system

convection (kon-VEK-shun) process by which a fluid or gas next to a warm body is heated by conduction, moves away, and is replaced by colder fluid or gas that in turn follows the same cycle

convergence (neuronal) many presynaptic neurons synapsing upon one postsynaptic neuron; (of eyes) turning of eyes inward (that is, toward nose) to view near objects

cooperativity interaction between functional binding sites in a multimeric protein

core body temperature temperature of inner body

cornea (KOR-nee-ah) transparent structure covering front of eye; forms part of eye's optical system and helps focus an object's image on retina

coronary pertaining to blood vessels of heart

coronary artery vessel delivering oxygenated blood to the muscular walls of the heart

coronary blood flow blood flow to heart muscle

corpus callosum (KOR-pus kal-LOH-sum) wide band of nerve fibers connecting the two cerebral hemispheres; a brain commissure

corpus luteum (LOO-tee-um) ovarian structure formed from the follicle after ovulation; secretes estrogen and progesterone

cortex (KOR-tex) outer layer of organ; *see also* adrenal cortex, cerebral cortex; *compare* medulla

cortical association area region of cerebral cortex that receives input from various sensory types, memory stores, and so on, and performs further perceptual processing

cortical collecting duct primary site of sodium reabsorption at the end of the nephron

cortical nephron functional unit of the kidney contained in the renal cortex and with a small (or no) loop of Henle

cortical reaction release of factors by the ovum that hardens the zona pellucida

corticobulbar pathway (kor-tih-koh-BUL-bar) descending pathway having its neuron cell bodies in cerebral cortex; its axons pass without synapsing to region of brainstem motor neurons

corticospinal pathway descending pathway having its neuron cell bodies in cerebral cortex; its axons pass without synapsing to region of spinal motor neurons; also called the pyramidal tract; *compare* brainstem pathway, corticobulbar pathway

corticosteroid (kor-tih-koh-STEER-oid) steroid produced by adrenal cortex or drug that resembles one

corticotropin-releasing hormone (CRH) (kor-tih-koh-TROH-pin) hypophysiotropic peptide hormone that stimulates ACTH (corticotropin) secretion by anterior pituitary

cortisol (KOR-tih-sol) main glucocorticoid steroid hormone secreted by adrenal cortex; regulates various aspects of organic metabolism

costimulus nonspecific interactions between proteins on the surface of antigen-presenting cells and helper T cells; required for T cell activation

cotransmitter chemical messenger released with a neurotransmitter from synapse or neuroeffector junction

cotransport form of secondary active transport in which net movement of actively transported substance and "downhill" movement of molecule supplying the energy are in the same direction

countercurrent multiplier system mechanism associated with loops of Henle that creates a region having high interstitial fluid osmolarity in renal medulla

countertransport form of secondary active transport in which net movement of actively transported molecule is in direction opposite "downhill" movement of molecule supplying the energy

covalent bond (koh-VAY-lent) chemical bond between two atoms in which each atom shares one of its electrons with the other

covalent modulation alteration of a protein's shape, and therefore its function, by the covalent binding of various chemical groups to it

cranial nerve one of 24 peripheral nerves (12 pairs) that join brainstem or forebrain with structures outside CNS

C-reactive protein an acute phase protein that functions as a nonspecific opsonin

creatine phosphate (CP) (KREE-ah-tin) molecule that transfers phosphate and energy to ADP to generate ATP

creatinine (kree-AT-ih-nin) waste product derived from muscle creatine

creatinine clearance (C_{cr}) plasma volume from which creatinine is removed by the kidneys per unit time; approximates glomerular filtration rate

cristae (mitochondrial) the inner membrane of mitochondria, which may assume sheetlike or tubular appearances; site containing cytochrome P450 enzymes involved in steroid hormone production

critical period time during development when a system is most readily influenced by factors, sometimes irreversibly

cross-bridge in muscle, myosin projection extending from thick filament and capable of exerting force on thin filament, causing the filaments to slide past each other

cross-bridge cycle sequence of events between binding of a cross-bridge to actin, its release, and reattachment during muscle contraction

crossed-extensor reflex increased activation of extensor muscles contralateral to limb flexion

crossing-over process in which segments of maternal and paternal chromosomes exchange with each other during chromosomal pairing in meiosis

crystalloid low molecular weight solute

C3b a complement molecule that attaches phagocytes to microbes

cumulus oophorous layers of granulosa cells that surround the egg within the dominant follicle

cupula a gelatinous mass within the semicircular canals that contains stereocilia and responds to head movement

current movement of electric charge; in biological systems, this is achieved by ion movement

cusp a flap or "leaflet" of a heart valve

cutaneous (cue-TAY-nee-us) pertaining to skin

cyclic AMP (cAMP) cyclic 3′,5′-adenosine monophosphate; cyclic nucleotide that serves as a second messenger for many "first" chemical messengers

cyclic AMP-dependent protein kinase (KY-nase) enzyme that is activated by cyclic AMP and then phosphorylates specific proteins, thereby altering their activity; also called *protein kinase A*

cyclic endoperoxide eicosanoid formed from arachidonic acid by cyclooxygenase

cyclic GMP (cGMP) cyclic 3′,5′-guanosine monophosphate; cyclic nucleotide that acts as second messenger in some cells

cyclic GMP-dependent protein kinase (KY-nase) enzyme that is activated by cyclic GMP and then phosphorylates specific proteins, thereby altering their activity; also called *protein kinase G*

cyclooxygenase (COX) (sy-klo-OX-ah-jen-ase) enzyme that acts on arachidonic acid and initiates production of cyclic endoperoxides, prostaglandins, and thromboxanes

cytochrome (SY-toh-krom) one of a series of enzymes that couples energy to ATP formation during oxidative phosphorylation

cytochrome P450 enzyme that mediates hydroxylation reactions in the biosynthesis of steroid hormones

cytokine (SY-toh-kine) general term for protein extracellular messengers that regulate immune responses; secreted by macrophages, monocytes, lymphocytes, neutrophils, and several nonimmune cell types

cytoplasm (SY-toh-plasm) region of cell interior outside the nucleus

cytosine (C) (SY-toh-seen) pyrimidine base in DNA and RNA

cytoskeleton cytoplasmic filamentous network associated with cell shape and movement

cytosol (SY-toh-sol) intracellular fluid that surrounds cell organelles and nucleus

cytotoxic T cell (SY-toh-TOX-ik) T lymphocyte that, upon activation by specific antigen, directly attacks a cell bearing that type of antigen and destroys it; major killer of virus-infected and cancer cells

D

Dalton's law pressure exerted by each gas in a mixture of gases is independent of the pressure exerted by the other gases

dark adaptation process by which photoreceptors in the retina adjust to darkness

daughter cell one of the two new cells formed when a cell divides

dead space volume of inspired air that cannot be exchanged with blood; *see also* alveolar dead space, anatomic dead space

deamination *see* oxidative deamination

declarative memory memories of facts and events

decremental decreasing in amplitude

decussation (dek-uh-SAY-shun) crossover of neuronal pathways from one side of the body to the other

defecation (def-ih-KAY-shun) expulsion of feces from rectum

defecation reflex urge to extrude feces caused by sudden distension of the walls of the rectum

dehydration type of chemical reaction in which two smaller molecules, such as amino acids, are joined to form a larger molecule; a single molecule of water is lost in the process

delta rhythm slow-wave, high-amplitude EEG waves associated with the deepest stages of slow-wave sleep

dendrite (DEN-drite) highly branched extension of neuron cell body; receives synaptic input from other neurons

dendritic cell a type of immune cell with macrophage-like properties

dense body cytoplasmic structure to which thin filaments of a smooth muscle fiber are anchored

deoxyhemoglobin (Hb, HbH) (dee-ox-see-HEE-moh-gloh-bin) hemoglobin not combined with oxygen; reduced hemoglobin

deoxyribonucleic acid (DNA) (dee-ox-see-ry-boh-noo-KLAY-ik) nucleic acid that stores and transmits genetic information; consists of double strand of nucleotide subunits that contain deoxyribose

deoxyribose a ribose molecule with a single hydroxyl group removed; a component of DNA

depolarize to change membrane potential value toward zero so that cell interior becomes less negative than resting level

descending limb (of Henle's loop) segment of renal tubule into which proximal tubule drains

descending pathway neural pathways that go from the brain down to the spinal cord

desmosome (DEZ-moh-some) junction that holds two cells together; consists of plasma membranes of adjacent cells linked by fibers, yet separated by a 20-nm extracellular space filled with a cementing substance

detrusor muscle (duh-TRUSS-or) the smooth muscle that forms the wall of the urinary bladder

developmental acclimatization potentially irreversible change in structure or function of one or more organ systems that occurs during early life and favors survival in specific environments

diacylglycerol (DAG) (dy-aa-syl-GLIS-er-ol) second messenger that activates protein kinase C, which then phosphorylates a large number of other proteins

diapedesis (dye-app-uh-DEE-suhs) passage of leukocytes out of the blood and into the surrounding tissue

diaphragm (DY-ah-fram) dome-shaped skeletal muscle sheet that separates the abdominal and thoracic cavities; principal muscle of respiration

diastole (dy-ASS-toh-lee) period of cardiac cycle when ventricles are relaxing

diastolic pressure (DP) (dy-ah-STAL-ik) minimum blood pressure during cardiac cycle

diencephalon (dy-en-SEF-ah-lon) core of anterior part of brain; lies beneath cerebral hemispheres and contains *thalamus* and *hypothalamus*

differentiation (dif-fer-en-she-AY-shun) process by which unspecialized cells acquire specialized structural and functional properties

diffusion (dif-FU-shun) movement of molecules from one location to another because of random thermal molecular motion; net diffusion always occurs from a region of higher concentration to a region of lower concentration

diffusion equilibrium state during which diffusion fluxes in opposite directions are equal; that is, the net flux equals zero

diffusion potential voltage difference created by net diffusion of ions

digestion process of breaking down large particles and high-molecular-weight substances into small molecules

dihydropyridine (DHP) receptor (die-hydro-PEER-a-deen) nonconducting calcium channels in the T-tubule membranes of skeletal muscle cells, which act as voltage sensors in excitation-contraction coupling

dihydrotestosterone (dy-hy-droh-tes-TOS-ter-own) steroid formed by enzyme-mediated alteration of testosterone; active form of testosterone in certain of its target cells

1,25-dihydroxyvitamin D (1,25-OH₂D) $(1,25\text{-OH}_2\text{D})$ (1-25-dy-hy-DROX-ee-vy-tah-min DEE) hormone that is formed by kidneys and is the active form of vitamin D

diiodotyrosine (DIT) a doubly iodinated tyrosine molecule that is an intermediate in the formation of thyroid hormones

2,3-diphosphoglycerate (DPG) (2-3-dy-fos-foh-GLISS-er-ate) substance produced by erythrocytes during glycolysis; binds reversibly to hemoglobin, causing it to release oxygen

disaccharide (dy-SAK-er-ide) carbohydrate molecule composed of two monosaccharides

disc layer of membranes in outer segment of photoreceptor; contains photopigments

dissociation separation from

distal (DIS-tal) farther from reference point; *compare* proximal

distal convoluted tubule portion of kidney tubule between loop of Henle and collecting duct system

disulfide bond R—S—S—R

diuresis (dy-uh-REE-sis) increased urine excretion

diuretic (dy-uh-RET-ik) substance that inhibits fluid reabsorption in renal tubule, thereby increasing urine excretion

diurnal (dy-URN-al) daily; occurring in a 24-h cycle

divergence (dy-VER-gence) (neuronal) one presynaptic neuron synapsing upon many postsynaptic neurons; (of eyes) turning of eyes outward to view distant objects

dl deciliter; 0.1 L

dominant follicle most mature developing follicle in the ovary from which the mature egg is ovulated

dopamine (DOPE-ah-meen) biogenic amine (catecholamine) neurotransmitter and hormone; precursor of epinephrine and norepinephrine; *see also* prolactin-inhibiting hormone

dorsal (DOR-sal) toward or at the back

dorsal column pathway ascending pathway for somatosensory information; runs through dorsal area of spinal white matter

dorsal horn and ventral horn regions of gray matter in the spinal cord containing cell bodies of interneurons and motor neurons

dorsal respiratory group neurons in the medullary respiratory center that fire during inspiration

dorsal root group of afferent nerve fibers that enters dorsal region of spinal cord

dorsal root ganglion group of sensory nerve cell bodies that have axons projecting to the dorsal horn of the spinal cord

down-regulation decrease in number of target-cell receptors for a given messenger in response to a chronic high concentration of that messenger; *compare* up-regulation

dual innervation (in-ner-VAY-shun) innervation of an organ or gland by both sympathetic and parasympathetic nerve fibers

duodenum (due-oh-DEE-num) first portion of small intestine (between stomach and jejunum)

dura mater thick, outermost membrane (meninges) covering the brain

dynein (DIE-neen) motor protein that uses the energy from ATP to transport attached cellular cargo molecules along microtubules

dynorphin (dye-NOR-fin) one of a group of endogenous opioid peptides that act as neuromodulators in the brain

E

eardrum *see* tympanic membrane

eccentric contraction muscle activity that is accompanied by lengthening of the muscle generally by an external load that exceeds muscle force

ECG lead combination of a reference electrode (designated negative) and a recording electrode (designated positive) that are placed on the surface of the body and provide a "view" of the electrical activity of the heart

edema (ed-DEE-mah) accumulation of excess fluid in interstitial space

EEG arousal transformation of EEG pattern from alpha to beta rhythm during increased levels of attention

effector (ee-FECK-tor) cell or cell collection whose change in activity constitutes the response in a control system

efferent (EF-er-ent) carrying away from

efferent arteriole renal vessel that conveys blood from glomerulus to peritubular capillaries

efferent division (PNS) neurons in the peripheral nervous system that project out of the central nervous system

efferent neuron neuron that carries information away from CNS

efferent pathway component of reflex arc that transmits information from integrating center to effector

egg female germ cell at any of its stages of development

eicosanoid (eye-KOH-sah-noid) general term for modified fatty acids that are products of arachidonic acid metabolism (cyclic endoperoxides, prostaglandins, thromboxanes, and leukotrienes); function as paracrine/autocrine agents

ejaculation (ee-jak-you-LAY-shun) discharge of semen from penis

ejaculatory duct (ee-JAK-you-lah-tory) continuation of vas deferens after it is joined by seminal vesicle duct; joins urethra in prostate gland

ejection fraction (EF) the ratio of stroke volume to end-diastolic volume; $EF = SV/EDV$

elastic recoil tendency of an elastic structure to oppose stretching or distortion

elastin fiber a protein with elastic or springlike properties; found in large arteries and in the airways

electrical force force that causes charged particles to move toward regions having an opposite charge and away from regions having a like charge

electrical potential (E) (or electric potential difference) *see* potential

electrical signal graded potential or action potential

electrical synapse (SIN-apse) synapse at which local currents resulting from electrical activity flow between two neurons through gap junctions joining them

electrocardiogram (ECG, EKG) (ee-lek-troh-KARD-ee-oh-gram) recording at skin surface of the electrical currents generated by cardiac muscle action potentials

electrochemical difference force determining direction and magnitude of net charge movement; combination of electrical and chemical gradients

electrochemical gradient the driving force across a plasma membrane that dictates whether an ion will move into or out of a cell; established by both the concentration difference and the electrical charge difference between the cytosolic and extracellular surfaces of the membrane

electrode (ee-LEK-trode) probe used to stimulate electrically, or record from, the body surface or a tissue

electroencephalogram (EEG) (eh-lek-troh-en-SEF-ah-loh-gram) recording of brain electrical activity from scalp

electrogenic pump (elec-troh-JEN-ik) active transport system that directly separates electrical charge, thereby producing a potential difference

electrolyte (ee-LEK-troh-lite) substance that dissociates into ions when in aqueous solution

electromagnetic radiation radiation composed of waves with electrical and magnetic components; includes gamma rays, x-rays, and ultraviolet, visible light, infrared, and radio waves

electron (ee-LEK-tron) subatomic particle that carries one unit of negative charge

electron transport chain a series of metal-containing proteins within mitochondria that participate in the flow of electrons from proteins to molecular oxygen; they are key components of the energy-producing processes in all cells

embryo (EM-bree-oh) organism during early stages of development; in human beings, the first two months of intrauterine life

emission (ee-MISH-un) movement of male genital duct contents into urethra prior to ejaculation

emotional behavior outward expression and display of inner emotions

emulsification (eh-mul-suh-fah-KAY-shun) division of large lipid droplets into very small droplets

end-diastolic volume (EDV) (dy-ah-STAH-lik) amount of blood in ventricle just prior to systole

endocrine gland (EN-doh-krin) group of epithelial cells that secrete into the extracellular space hormones that then diffuse into bloodstream; also called a *ductless gland*

endocrine system all the body's hormone-secreting glands

endocytosis (en-doh-sy-TOH-sis) process in which plasma membrane folds into the cell, forming small pockets that pinch off to produce intracellular, membrane-bound vesicles; *see also* phagocytosis

endogenous cryogen chemical messenger released by the hypothalamus and other tissues, which acts to reset the body's thermostat and reduce fever

endogenous opioid (en-DAHJ-en-us OH-pee-oid) an example of certain neuropeptides—endorphin, dynorphin, or enkephalin

endogenous pyrogen (EP) (en-DAHJ-en-us PY-roh-jen) any of the cytokines (including interleukin 1 and interleukin 6) that act physiologically in the brain to cause fever

endolymph extracellular fluid found in the cochlea and vestibular apparatus

endometrium (en-doh-MEE-tree-um) glandular epithelium lining uterine cavity

endoperoxide *see* cyclic endoperoxide

endoplasmic reticulum (en-doh-PLAS-mik reh-TIK-you-lum) cell organelle that consists of interconnected network of membrane-bound branched tubules and flattened sacs; two types are distinguished: *rough,* with ribosomes attached, and *smooth,* which is smooth-surfaced (does not contain ribosomes)

endosome (EN-doh-some) intracellular vesicles and tubular elements between Golgi apparatus and plasma membrane; sorts and distributes vesicles during endo- and exocytosis

endothelial cell *see* endothelium

endothelin-1 (en-doh-THEE-lin) one member of a family of peptides secreted by many tissues that can act as a paracrine or hormonal signal; one major action is vasoconstriction

endothelium (en-doh-THEE-lee-um) thin layer of cells that lines heart cavities and blood vessels

endothelium-derived relaxing factor (EDRF) nitric oxide and possibly other substances; secreted by vascular endothelium, it relaxes vascular smooth muscle and causes arteriolar dilation

end-plate potential (EPP) depolarization of motor end plate of skeletal muscle fiber in response to acetylcholine; initiates action potential in muscle plasma membrane

end-product inhibition inhibition of a metabolic pathway by final product's action upon allosteric site on an enzyme (usually the rate-limiting enzyme) in the pathway

end-systolic volume (ESV) (sis-TAH-lik) amount of blood remaining in ventricle after ejection

energy ability to produce change; measured by amount of work performed during a given change

enkephalin (en-KEF-ah-lin) peptide neurotransmitter at some synapses activated by opiate drugs; an endogenous opioid

enteric nervous system (en-TAIR-ik) neural network residing in and innervating walls of gastrointestinal tract

enterochromaffin-like (ECL) cell histamine-secreting cell of the stomach

enteroendocrine cell cell located in the gastric gland in the stomach; secretes gastrin

enterogastrones (en-ter-oh-GAS-trones) collective term for hormones released by intestinal tract; inhibit stomach activity

enterohepatic circulation (en-ter-oh-hih-PAT-ik) reabsorption of bile salts (and other substances) from intestines, passage to liver (via hepatic portal vein), and secretion back to intestines (via bile)

enterokinase (en-ter-oh-KY-nase) enzyme in luminal plasma membrane of intestinal epithelial cells; converts pancreatic trypsinogen to trypsin

entrainment (en-TRAIN-ment) adjusting biological rhythm to environmental cues

enzyme (EN-zime) protein catalyst that accelerates specific chemical reactions but does not itself undergo net chemical change during the reaction

enzyme activity rate at which enzyme converts reactant to product; may be measure of the properties of enzyme's active site as altered by allosteric or covalent modulation; affects rate of enzyme-mediated reaction

eosinophil (ee-oh-SIN-oh-fil) polymorphonuclear granulocytic leukocyte whose granules take up red dye eosin; involved in parasite destruction and allergic responses

ependymal cell (ep-END-ih-mel) type of glial cell that lines internal cavities of the brain and produces cerebrospinal fluid

epicardium (epp-ee-KAR-dee-um) layer of connective tissue closely affixed to outer surface of the heart

epididymis (ep-ih-DID-eh-mus) portion of male reproductive duct system located between seminiferous tubules and vas deferens

epiglottis (ep-ih-GLOT-iss) thin cartilage flap that folds down, covering trachea, during swallowing

epinephrine (E) (ep-ih-NEF-rin) amine hormone secreted by adrenal medulla and involved in regulation of organic metabolism; a biogenic amine (catecholamine) neurotransmitter; also called *adrenaline*

epiphyseal closure (ep-ih-FIZ-ee-al) conversion of epiphyseal growth plate to bone

epiphyseal growth plate actively proliferating cartilage near bone ends; region of bone growth

epiphysis (eh-PIF-ih-sis) end of long bone

epithelial cell (ep-ih-THEE-lee-al) cell at surface of body or hollow organ; specialized to secrete or absorb ions and organic molecules; with other epithelial cells, forms an *epithelium*

epithelial tissue one of the four major tissue types in the body, comprised of aggregates of epithelial cells

epithelial transport molecule movement from one extracellular compartment across epithelial cells into a second extracellular compartment

epithelium (ep-ih-THEE-lee-um) tissue that covers all body surfaces, lines all body cavities, and forms most glands

epitope (EP-ih-tope) antigenic portion of a molecule complexed to the MHC protein and presented to the T cell; also called an *antigenic determinant*

equilibrium (ee-quah-LIB-ree-um) no net change occurs in a system; requires no energy

equilibrium potential voltage gradient across a membrane that is equal in force but opposite in direction to concentration force affecting a given ion species

erection penis or clitoris becoming stiff due to vascular congestion

ergocalciferol (vitamin D$_2$) plant vitamin D

error signal steady-state difference between level of regulated variable in a control system and set point for that variable

erythrocyte (eh-RITH-roh-site) red blood cell

erythropoiesis (eh-rith-roh-poy-EE-sis) erythrocyte production

erythropoietin (eh-rith-roh-POY-ih-tin) peptide hormone secreted mainly by kidney cells; stimulates red blood cell production; one of the hematopoietic growth factors

esophagus (eh-SOF-uh-gus) portion of digestive tract that connects throat (pharynx) and stomach

essential amino acid amino acid that cannot be formed by the body at all (or at a rate adequate to meet metabolic requirements) and so must be obtained from diet

essential nutrient substance required for normal or optimal body function but synthesized by the body either not at all or in amounts inadequate to prevent disease

estradiol (es-tra-DY-ol) steroid hormone of estrogen family; major female sex hormone

estriol (ES-tree-ol) estrogen present in pregnancy; produced primarily by the placenta

estrogen (ES-troh-jen) group of steroid hormones that have effects similar to estradiol on female reproductive tract

estrogen priming increase in responsiveness to progesterone caused by prior exposure to estrogen (e.g., in the uterus)

estrone estrogen that is less prominent than estradiol

eukaryotic cell cell containing a membrane-enclosed nucleus with genetic material; plant and animal cells

eustachian tube (yoo-STAY-shee-an) duct connecting the middle ear with the nasopharynx

evaporation the loss of body water by perspiration, resulting in cooling

excitability ability to produce electric signals

excitable membrane membrane capable of producing action potentials

excitation-contraction coupling in muscle fibers, mechanism linking plasma membrane stimulation with cross-bridge force generation

excitatory amino acid amino acid that acts as an excitatory (depolarizing) neurotransmitter in the nervous system

excitatory postsynaptic potential (EPSP) (post-sin-NAP-tic) depolarizing graded potential in postsynaptic neuron in response to activation of excitatory synapse

excitatory synapse (SIN-apse) synapse that, when activated, increases likelihood that postsynaptic neuron will undergo action potentials or increases frequency of existing action potentials

excitotoxicity (eggs-SY-toe-tocks-ih-city) spreading damage to brain cells due to release of glutamate from ruptured neurons

excretion elimination of a substance from the body

exocrine gland (EX-oh-krin) cluster of epithelial cells specialized for secretion and having ducts that lead to an epithelial surface

exocytosis (ex-oh-sy-TOH-sis) process in which intracellular vesicle fuses with plasma membrane, the vesicle opens, and its contents are liberated into the extracellular fluid

exon (EX-on) DNA gene region containing code words for a part of the amino acid sequence of a protein

expiration (ex-pur-A-shun) movement of air out of lungs

expiratory reserve volume (ERV) (ex-PY-ruh-tor-ee) volume of air that can be exhaled by maximal contraction of expiratory muscles after normal resting expiration

extension straightening a joint

extensor muscle muscle whose activity straightens a joint

external anal sphincter ring of skeletal muscle around lower end of rectum

external auditory canal outer canal of the ear between the pinna and the tympanic membrane

external environment environment surrounding external surface of an organism

external genitalia (jen-ih-TAH-lee-ah) (female) mons pubis, labia majora and minora, clitoris, vestibule of the vagina, and vestibular glands; (male) penis and scrotum

external urethral sphincter ring of skeletal muscle that surrounds the urethra at base of bladder

external work movement of external objects by skeletal muscle contraction

extracellular fluid fluid outside cell; interstitial fluid and plasma

extracellular matrix (MAY-trix) a complex consisting of a mixture of proteins (and, in some cases, minerals) interspersed with extracellular fluid

extrafusal fiber primary muscle fiber in skeletal muscle, as opposed to modified (intrafusal) fiber in muscle spindle

extrapyramidal system *see* brainstem pathway

extrinsic (ex-TRIN-sik) coming from outside

extrinsic pathway formation of fibrin clots by pathway using tissue factor on cells in interstitium; once activated, it also recruits the intrinsic clotting pathway beyond factor XII

F

facilitated diffusion (fah-SIL-ih-tay-ted) system using a transporter to move molecules from high to low concentration across a membrane; energy not required

fallopian tube one of two tubes that carries egg from ovary to uterus

fast fiber skeletal muscle fiber that contains myosin having high ATPase activity

fast-glycolytic fiber type of skeletal muscle fiber that has high intrinsic contraction speed and abundant capacity for production of ATP by anaerobic glycolysis

fast-oxidative-glycolytic fiber type of skeletal muscle fiber that has high intrinsic contraction speed and abundant capacity for production of ATP by aerobic oxidative phosphorylation

fat mobilization increased breakdown of triglycerides and release of glycerol and fatty acids into blood

fat-soluble vitamin *see* vitamin

fatty acid carbon chain with carboxyl group at one end through which chain can be linked to glycerol to form triglyceride; *see also* polyunsaturated fatty acid, saturated fatty acid, unsaturated fatty acid

Fc "stem" part of antibody

feces (FEE-sees) material expelled from large intestine during defecation

feedback *see* negative feedback, positive feedback

feedforward aspect of some control systems that allows system to anticipate changes in a regulated variable

female external genitalia mons pubis, labia majora, labia minora, clitoris, outer vagina, and its glands

female internal genitalia (jen-ih-TALE-ee-ah) ovaries, uterine tubes, uterus, and vagina

ferritin (FAIR-ih-tin) iron-binding protein that stores iron in body

fertilization union of sperm and egg

fetal hemoglobin oxygen-carrying molecule with high oxygen affinity

fetus (FEE-tus) human being from third month of intrauterine life until birth

fever increased body temperature due to setting of "thermostat" of temperature-regulating mechanisms at higher-than-normal level

fiber *see* muscle fiber, nerve fiber

fibrin (FY-brin) protein polymer resulting from enzymatic cleavage of fibrinogen; can turn blood into gel (clot)

fibrinogen (fy-BRIN-oh-jen) plasma protein precursor of fibrin

fibrinolytic system (fye-brin-oh-LIT-ik) cascade of plasma enzymes that breaks down clots; also called thrombolytic system

fight-or-flight response activation of sympathetic nervous system during stress

filtered load amount of any substance filtered from renal glomerular capillaries into Bowman's capsule

filtration movement of essentially protein-free plasma out across capillary walls due to a pressure gradient across the wall

fimbria (FIM-bree-ah) opening of the fallopian tube; it has fingerlike projections lined with ciliated epithelium through which the ovulated egg passes into the fallopian tube

first messenger extracellular chemical messenger

5α-reductase intracellular enzyme that converts testosterone to dihydrotestosterone

flatus (FLAY-tus) intestinal gas expelled through anus

flavine adenine dinucleotide (FAD) coenzyme derived from the B vitamin riboflavin; transfers hydrogen from one substrate to another

flexion (FLEK-shun) bending a joint

flow autoregulation ability of individual arterioles to alter their resistance in response to changing blood pressure so that relatively constant blood flow is maintained

flow-induced arterial vasodilation mechanism for relaxing arteriolar smooth muscles that involves detection of shear stress by endothelial cells, which release paracrine inhibitors of contraction

fluid endocytosis invagination of a plasma membrane by which a cell can engulf extracellular fluid

fluid-mosaic model (moh-ZAY-ik) cell membrane structure consists of proteins embedded in bimolecular lipid that has the physical properties of a fluid, allowing membrane proteins to move laterally within it

flux amount of a substance crossing a surface in a unit of time; *see also* net flux

folic acid (FOH-lik) vitamin of B-complex group; essential for formation of nucleotide thiamine

follicle (FOL-ih-kel) egg and its encasing follicular, granulosa, and theca cells at all stages prior to ovulation; also called *ovarian follicle*

follicle-stimulating hormone (FSH) protein hormone secreted by anterior pituitary in males and females that acts on gonads; a gonadotropin

follicular phase (fuh-LIK-you-lar) that portion of menstrual cycle during which follicle and egg develop to maturity prior to ovulation

food-induced thermogenesis the creation of heat within the body following a meal, particularly one rich in protein; at least part of the heat is generated secondarily to the increased activity of the gastrointestinal tract

foot process large extension of sarcoplasmic reticulum calcium channels (ryanodine receptors), which connect them to the T-tubule membrane and mediate excitation-contraction coupling in skeletal muscle; also known as *junctional feet*

forebrain large, anterior brain subdivision consisting of right and left cerebral hemispheres (the cerebrum) and diencephalon

formed element solid phase of blood, including cells (erythrocytes and leukocytes) and cell fragments (platelets)

fovea centralis (FOH-vee-ah) area near center of retina where cones are most concentrated; gives rise to most acute vision

Frank-Starling mechanism the relationship between stroke volume and end-diastolic volume such that stroke volume increases as end-diastolic volume increases; also called *Starling's law of the heart*

fraternal twins dizygotic twins that occur when two eggs are fertilized

free radical atom that has an unpaired electron in its outermost orbital; molecule containing such an atom

free-running rhythm cyclical activity driven by biological clock in absence of environmental cues

frequency number of times an event occurs per unit time

frontal lobe region of anterior cerebral cortex where motor areas, Broca's speech center, and some association cortex are located

fructose (FRUK-tose) five-carbon sugar; present in sucrose (table sugar)

F-type sodium channel the "funny" sodium-conducting channel mainly responsible for the inward flow of positive current in autorhythmic cardiac cells

functional residual capacity lung volume after relaxed expiration

functional site binding site on allosteric protein that, when activated, carries out protein's physiological function; also called active site

fundus upper portion of the stomach; secretes mucus, pepsinogen, and hydrochloric acid

fused tetanus (TET-ah-nuss) skeletal muscle activation in which action potential frequency is sufficiently high to cause a smooth, sustained, maximal strength contraction

fused-vesicle channel endocytotic or exocytotic vesicles that have fused to form a continuous water-filled channel through capillary endothelial cell

G

GABA (gamma-aminobutyric acid) an amino acid neurotransmitter commonly occurring at inhibitory synapses in the central nervous system

gallbladder small sac under the liver; concentrates bile and stores it between meals; contraction of gallbladder ejects bile, which eventually flows into small intestine

gamete (GAM-eet) germ cell or reproductive cell; sperm in male and egg in female

gametogenesis (gah-mee-toh-JEN-ih-sis) gamete production

gamma globulin immunoglobulin G (IgG), most abundant class of plasma antibodies

gamma motor neuron small motor neuron that controls intrafusal muscle fibers in muscle spindles

ganglion (GANG-glee-on) (pl. ganglia) generally reserved for cluster of neuron cell bodies outside CNS

ganglion cell retinal neuron that is postsynaptic to bipolar cells; axons of ganglion cells form optic nerve

gap junction protein channels linking cytosol of adjacent cells; allows ions and small molecules to flow between cytosols of the connected cells

gastric (GAS-trik) pertaining to the stomach

gastric phase (of gastrointestinal control) initiation of neural and hormonal gastrointestinal reflexes by stimulation of stomach wall

gastrin (GAS-trin) peptide hormone secreted by antral region of stomach; stimulates gastric acid secretion

gastroileal reflex (gas-troh-IL-ee-al) increase in contractions of ileum during gastric emptying

gastrointestinal (GI) system (gas-troh-in-TES-tin-al) gastrointestinal tract plus salivary glands, liver, gallbladder, and pancreas

gastrointestinal tract mouth, pharynx, esophagus, stomach, and small and large intestines

gating opening or closing ion channels

gene unit of hereditary information; portion of DNA containing information required to determine a protein's amino acid sequence

gene cloning process of forming identical DNA sequences using genetic engineering techniques

genetic code three-nucleotide sequence in a gene; indicates the location of a particular amino acid in the protein specified by that gene

genome complete set of an organism's genes

genotype the set of alleles present in an individual; determines genetic sex (XX, female; XY, male)

germ cell cell that gives rise to male or female gametes (sperm and eggs)

gestation (jess-TAY-shun) length of time of intrauterine fetal development (usually about nine months in humans)

ghrelin (GREH-lin) hormone released from cells of the stomach; stimulates hunger

gland *see* endocrine gland, exocrine gland

glial cell (GLEE-al) nonneuronal cell in CNS; helps regulate extracellular environment of CNS; also called *neuroglia*

globin (GLOH-bin) collective term for the four polypeptide chains of the hemoglobin molecule

globulin (GLOB-you-lin) one of a family of proteins found in blood plasma

glomerular capillary very small blood vessel within the glomerulus of the kidney through which plasma is filtered

glomerular filtrate ultrafiltrate of plasma produced in the glomerulus that is usually free of cells and large proteins

glomerular filtration process by which components of plasma in the glomerular capillary are passed to the Bowman's space of the glomerulus—process is governed by net glomerular filtration pressure

glomerular filtration rate (GFR) volume of fluid filtered from renal glomerular capillaries into Bowman's capsule per unit time

glomerulus (gloh-MER-you-lus) tufts of glomerular capillaries at beginning of kidney nephron

glottis opening between vocal cords through which air passes, and surrounding area

glucagon (GLOO-kah-gahn) peptide hormone secreted by alpha cells of pancreatic islets of Langerhans; leads to rise in plasma glucose

glucocorticoid (gloo-koh-KOR-tih-koid) steroid hormone produced by adrenal cortex and having major effects on nutrient metabolism

gluconeogenesis (gloo-koh-nee-oh-JEN-ih-sis) formation of glucose by the liver or kidneys from pyruvate, lactate, glycerol, or amino acids

glucose major monosaccharide in the body; a six-carbon sugar, $C_6H_{12}O_6$; also called blood sugar

glucose-counterregulatory control neural or hormonal factors that oppose insulin's actions; glucagon, epinephrine, sympathetic nerves to liver and adipose tissue, cortisol, and growth hormone

glucose-dependent insulinotropic peptide (GIP) intestinal hormone; stimulates insulin secretion in response to glucose and fat in small intestine

glucose-6-phosphate (FOS-fate) first intermediate in glycolytic pathway

glucose sparing switch from glucose to fat utilization by most cells during postabsorptive state

glutamate (GLU-tah-mate) anion formed from the amino acid glutamic acid; a major excitatory CNS neurotransmitter

glutamine (GLOO-tah-meen) glutamate having an extra NH_3

gluten wheat protein

glycerol (GLISS-er-ol) three-carbon carbohydrate; forms backbone of triglyceride

glycine (GLY-seen) an amino acid; a neurotransmitter at some inhibitory synapses in CNS

glycocalyx (gly-koh-KAY-lix) fuzzy coating on extracellular surface of plasma membrane; consists of short, branched carbohydrate chains

glycogen (GLY-koh-jen) highly branched polysaccharide composed of glucose subunits; major carbohydrate storage form in body

glycogenolysis (gly-koh-jen-NOL-ih-sis) glycogen breakdown to glucose

glycogen phosphorylase intracellular enzyme required to begin the process of breaking down glycogen into glucose; inhibited by insulin

glycogen synthase intracellular enzyme required to synthesize glycogen; stimulated by insulin

glycolysis (gly-KOL-ih-sis) metabolic pathway that breaks down glucose to two molecules of pyruvate (aerobically) or two molecules of lactate (anaerobically)

glycolytic fiber skeletal muscle fiber that has a high concentration of glycolytic enzymes and large glycogen stores; white muscle fiber

glycoprotein protein containing covalently linked carbohydrates

Goldman-Hodgkin-Katz (GHK) equation calculation for electrochemical equilibrium when a membrane is permeable to more than one ion

Golgi apparatus (GOAL-gee) cell organelle consisting of flattened membranous sacs; usually near nucleus; processes newly synthesized proteins for secretion or distribution to other organelles

Golgi tendon organ tension-sensitive mechanoreceptor ending of afferent nerve fiber; wrapped around collagen bundles in tendon

gonad (GOH-nad) gamete-producing reproductive organ—testes in male and ovaries in female

gonadal steroid hormone synthesized in the testes (testosterone) and ovaries (estrogen and progesterone)

gonadotropic hormone (goh-nad-oh-TROH-pik) hormone secreted by anterior pituitary that controls gonadal function; FSH or LH; also called *gonadotropin*

gonadotropin glycoprotein hormone secreted by pituitary (LH, FSH) and placenta (hCG) that influence gonadal function

gonadotropin-releasing hormone (GnRH) hypophysiotropic hormone that stimulates LH and FSH secretion by anterior pituitary in males and females

G protein one protein from a family of regulatory proteins that reversibly bind guanosine nucleotides; plasma membrane G proteins interact with membrane ion channels or enzymes

graafian follicle (GRAF-ee-un) mature follicle just before ovulation

graded potential membrane potential change of variable amplitude and duration that is conducted decrementally; has no threshold or refractory period

gradient (GRAY-dee-ent) continuous increase or decrease of a variable over distance

gram atomic mass amount of element in grams equal to the numerical value of its atomic weight

granulosa cell (gran-you-LOH-sah) cell that contributes to the layers surrounding egg and antrum in ovarian follicle; secretes estrogen, inhibin, and other messengers that influence the egg

gray matter area of brain and spinal cord that appears gray in unstained specimens and consists mainly of cell bodies and unmyelinated portions of nerve fibers

growth cone tip of developing axon

growth factor one of a group of peptides that is highly effective in stimulating cell division and/or differentiation of certain cell types

growth hormone (GH) peptide hormone secreted by anterior pituitary; stimulates insulin-like growth factor I release; enhances body growth by stimulating protein synthesis

growth hormone-releasing hormone (GHRH) hypothalamic peptide hormone that stimulates growth hormone secretion by anterior pituitary

growth-inhibiting factor one of a group of peptides that modulates growth by inhibiting cell division in specific tissues

guanine (G) (GWAH-neen) purine base in DNA and RNA

guanosine triphosphate (GTP) (GWAH-noh-seen tri-FOS-fate) energy-transporting molecule similar to ATP except that it contains the base guanine rather than adenine

guanylyl cyclase (GUAN-ah-lil) enzyme that catalyzes transformation of GTP to cyclic GMP

gustation (gus-TAY-shun) the sense of taste

gyrus (JYE-rus) sinuous raised ridges on the outer surface of the cerebral cortex

H

habituation (hab-bit-you-A-shun) reversible decrease in response strength upon repeatedly administered stimulation

hair cell mechanoreceptor in organ of Corti and vestibular apparatus

heart muscular pump that generates blood pressure and flow in the cardiovascular system

heart rate number of heart contractions per minute

heart sound noise that results from vibrations due to closure of atrioventricular valves (first heart sound) or pulmonary and aortic valves (second heart sound)

heavy chain pair of large, coiled polypeptides that makes up the rod and globular head of a myosin molecule

helicotrema outer point in the cochlea where the scala vestibuli and scala tympani meet

helper T cell T cell that, via secreted cytokines, enhances the activation of B cells and cytotoxic T cells

hematocrit (heh-MAT-oh-krit) percentage of total blood volume occupied by blood cells

hematopoietic growth factor (HGF) (heh-MAT-oh-poi-ET-ik) group of protein hormones and paracrine agents that stimulate proliferation and differentiation of various types of blood cells

heme (heem) iron-containing organic molecule bound to each of the four polypeptide chains of hemoglobin or to cytochromes

hemodynamics the factors describing what determines the movement of blood, in particular, pressure, flow, and resistance

hemoglobin (HEE-moh-gloh-bin) protein composed of four polypeptide chains, each attached to a heme; located in erythrocytes and transports most blood oxygen

hemoglobin saturation percent of hemoglobin molecules combined with oxygen

hemorrhage (HEM-er-age) bleeding

hemostasis (hee-moh-STAY-sis) stopping blood loss from a damaged vessel

Henle's loop *see* loop of Henle

Henry's law amount of gas dissolved in a liquid is proportional to the partial pressure of gas with which the liquid is in equilibrium

heparin (HEP-ah-rin) anticlotting agent found on endothelial cell surfaces; binds antithrombin III to tissues; used as an anticoagulant drug

hepatic (hih-PAT-ik) pertaining to the liver

hepatic portal vein vein that conveys blood from capillaries in the intestines and portions of the stomach and pancreas to capillaries in the liver

hepatocyte parenchymal cell of the liver

Hering-Breuer reflex inflation of the lung stimulates afferent nerves, which inhibit the inspiratory nerves in the medulla and thereby help to terminate inspiration

hertz (Hz) (hurts) cycles per second; measure used for wave frequencies

heterozygous (het-er-oh-ZY-gus) condition of having maternal and paternal copies of a gene with slightly different nucleotide sequences (alleles); *compare* homozygous

hexose a six-carbon sugar, like glucose

high-density lipoprotein (HDL) lipid-protein aggregate having low proportion of lipid; promotes removal of cholesterol from cells

hippocampus (hip-oh-KAM-pus) portion of limbic system associated with learning and emotions

histamine (HISS-tah-meen) inflammatory chemical messenger secreted mainly by mast cells; monoamine neurotransmitter

histone class of proteins that participate in the packaging of DNA within the nucleus; strands of DNA form coils around the histones

homeostasis (home-ee-oh-STAY-sis) relatively stable condition of internal environment that results from regulatory system actions

homeostatic control system (home-ee-oh-STAT-ik) collection of interconnected components that keeps a physical or chemical variable of internal environment within a predetermined normal range of values

homeothermic (home-ee-oh-THERM-ik) capable of maintaining body temperature within very narrow limits

homologous (hoh-MAHL-ah-gus) corresponding in origin, structure, and position

homozygous (hoh-moh-ZY-gus) condition of having maternal and paternal copies of a gene with identical nucleotide sequences (alleles); *compare* heterozygous

horizontal cell a specialized type of neuron found in the retina of the eye that integrates information from local photoreceptor cells

hormone chemical messenger synthesized by specific endocrine cells in response to certain stimuli and secreted into the blood, which carries it to target cells

hormone-sensitive lipase (HSL) an enzyme present in adipose tissue that acts to break down triglycerides into glycerol and fatty acids, which then enter the circulation; it is inhibited by insulin and stimulated by catecholamines

human chorionic gonadotropin (hCG) (kor-ee-ON-ik go-NAD-oh-troh-pin) protein hormone secreted by trophoblastic cells of embryo; maintains secretory activity of corpus luteum during first 3 months of pregnancy

human placental lactogen (plah-SEN-tal LAK-toh-jen) hormone produced by placenta that has effects similar to those of growth hormone and prolactin

hydrochloric acid (hy-droh-KLOR-ik) HCl; strong acid secreted into stomach lumen by parietal cells

hydrogen bond weak chemical bond between two molecules or parts of the same molecule, in which negative region of one polarized substance is electrostatically attracted to a positively charged region of polarized hydrogen atom in the other

hydrogen ion (EYE-on) H^+; single proton; H^+ concentration of a solution determines its acidity

hydrogen peroxide H_2O_2; chemical produced by phagosome and highly destructive to macromolecules

hydrolysis (hy-DRAHL-ih-sis) breaking of chemical bond with addition of elements of water (—H and —OH) to the products formed; also called hydrolytic reaction

hydrophilic (hy-droh-FIL-ik) attracted to, and easily dissolved in, water

hydrophobic (hy-droh-FOH-bik) not attracted to, and insoluble in, water

hydrostatic pressure (hy-droh-STAT-ik) pressure exerted by fluid

hydroxyapatite crystals composed primarily of calcium and phosphate deposited in bone matrix (mineralization)

hydroxyl group (hy-DROX-il) —OH

hydroxyl radical a highly reactive oxygen derivative (free radical) that can be formed in small amounts during oxidative phosphorylation

hymen membrane that partially covers the opening to the vagina

hyper- increased

hypercalcemia increased plasma calcium

hypercapnea increased arterial P_{CO_2}

hyperemia (hy-per-EE-me-ah) increased blood flow; *see also* active hyperemia

hyperosmotic (hy-per-oz-MAH-tik) having total solute concentration greater than normal extracellular fluid

hyperpolarize to change membrane potential so cell interior becomes more negative than its resting state

hypertension chronically increased arterial blood pressure

hyperthermia increased body temperature above the set point

hypertonic (hy-per-TAH-nik) solutions containing a higher concentration of effectively membrane-impermeable solute particles than normal (isotonic) extracellular fluid

hypertrophy (hy-PER-troh-fee) enlargement of a tissue or organ due to increased cell size rather than increased cell number

hyperventilation increased ventilation adequate to reduce arterial P_{CO_2}

hypo- too little; below

hypocalcemia the condition of low blood (and interstitial) calcium concentration

hypoglycemia (hy-poh-gly-SEE-me-ah) low blood glucose (sugar) concentration

hypoosmotic (hy-poh-oz-MAH-tik) having total solute concentration less than that of normal extracellular fluid

hypophysiotropic hormone (hy-poh-fiz-ee-oh-TROH-pik) any hormone secreted by hypothalamus that controls secretion of an anterior pituitary hormone

hypotension low blood pressure

hypothalamic releasing hormone (hy-poh-thah-LAM-ik) *see* hypophysiotropic hormone

hypothalamo-pituitary portal vessels small veins that link the capillaries of the median eminence at the base of the hypothalamus to capillaries that bathe the cells of the anterior pituitary gland; neurohormones are secreted from the hypothalamus into these vessels

hypothalamus (hy-poh-THAL-ah-mus) brain region below thalamus; responsible for integration of many basic neural, endocrine, and behavioral functions, especially those concerned with regulation of internal environment

hypotonic (hy-poh-TAH-nik) solutions containing a lower concentration of effectively nonpenetrating solute particles than normal (isotonic) extracellular fluid

hypoventilation decrease in ventilation that causes an increase in arterial P_{CO_2}

hypoxemia low levels of oxygen in the blood

hypoxia low levels of oxygen in the air, blood, or tissues

H zone one of transverse bands making up striated pattern of cardiac and skeletal muscle; light region that bisects A band

I

I band one of transverse bands making up repeating striations of cardiac and skeletal muscle; located between A bands of adjacent sarcomeres and bisected by Z line

IgA class of antibodies secreted by, and acting locally in, lining of gastrointestinal, respiratory, and genitourinary tracts

IgD class of antibodies whose function is unknown

IgE class of antibodies that mediate immediate hypersensitivity and resistance to parasites

IgG gamma globulin; most abundant class of antibodies

IgM class of antibodies that, along with IgG, provide major specific humoral immunity against bacteria and viruses

ileocecal sphincter (il-ee-oh-SEE-kal) ring of smooth muscle separating small and large intestines (that is, ileum and cecum)

ileum (IL-ee-um) final, longest segment of small intestine

immune defense *see* nonspecific immune defense, specific immune defense

immune surveillance (sir-VAY-lence) recognition and destruction of cancer cells that arise in body

immune system widely dispersed cells and tissues that participate in the elimination of foreign cells, microbes, and toxins from the body

immune tolerance the lack of immune responses to self components

immunity physiological mechanisms that allow body to recognize materials as foreign or abnormal and to neutralize or eliminate them; *see also* active immunity, passive immunity

immunoglobulin (Ig) (im-mun-o-GLOB-you-lin) proteins that are antibodies and antibody-like receptors on B cells (five classes are IgG, IgA, IgD, IgM, and IgE)

immunology the study of the defenses by which the body destroys or neutralizes foreign cells, microbes, and toxins

implantation (im-plan-TAY-shun) event during which fertilized egg becomes embedded in uterine wall

inactivation gate portion of the voltage-gated sodium or potassium channel that closes the channel

incus one of three bones in the inner ear that transmit movements of the tympanic membrane to the inner ear

inferior vena cava (VEE-nah KAY-vah) large vein that carries blood from lower half of body to right atrium of heart

inflammation (in-flah-MAY-shun) local response to injury or infection characterized by swelling, pain, heat, and redness

infundibulum (in-fun-DIBB-yoo-lum) the stalk connecting the median eminence at the base of the hypothalamus with the pituitary gland

inhibin (in-HIB-in) protein hormone secreted by seminiferous-tubule Sertoli cells and ovarian granulosa cells; inhibits FSH secretion

inhibitory postsynaptic potential (IPSP) hyperpolarizing graded potential that arises in postsynaptic neuron in response to activation of inhibitory synaptic endings upon it

inhibitory synapse (SIN-apse) synapse that, when activated, decreases likelihood that postsynaptic neuron will fire an action potential (or decreases frequency of existing action potentials)

initial segment first portion of axon plus the part of the cell body where axon arises

initiation factor a protein required for ribosomal assembly and the establishment of an initiation complex that allows new protein synthesis to begin

inner cell mass portion of the blastocyst that becomes the embryo

inner ear cochlea; contains organ of Corti

inner emotion emotional feelings that are entirely within a person

inner segment portion of photoreceptor that contains cell organelles; synapses with bipolar cells of retina

innervate to supply with nerves

inorganic pertaining to substances that do not contain carbon; *compare* organic

inorganic phosphate (P_i) (FOS-fate) $H_2PO_4^-$, HPO_4^{2-}, or PO_4^{3-}

inositol trisphosphate (IP_3) (in-OS-ih-tol-tris-FOS-fate) second messenger that causes release of calcium from endoplasmic reticulum into cytosol

insensible water loss water loss of which a person is unaware—that is, loss by evaporation from skin (excluding sweat) and respiratory passage lining

inspiration air movement from atmosphere into lungs

inspiratory reserve volume (IRV) maximal air volume that can be inspired above resting tidal volume

insulin (IN-suh-lin) peptide hormone secreted by beta cells of pancreatic islets of Langerhans; has metabolic and growth-promoting effects; stimulates glucose and amino acid uptake by most cells and stimulates protein, fat, and glycogen synthesis

insulin-like growth factor I (IGF-I) insulin-like growth factor that mediates mitosis-stimulating effect of growth hormone on bone and other tissues and has feedback effect on pituitary

insulin-like growth factor II mitogenic hormone active during fetal life; postnatal role, if any, is unknown

integral membrane protein protein embedded in membrane lipid layer; may span entire membrane or be located at only one side

integrating center cells that receive one or more signals and send out appropriate response; also called an *integrator*

integrin (in-TEH-grin) transmembrane protein in plasma membrane; binds to specific proteins in extracellular matrix and on adjacent cells to help organize cells into tissues

intercalated disk (in-TER-kuh-lay-tid) structure connecting adjacent cardiac myocytes, having components for tensile strength (desmosomes) and low-resistance electrical pathways (gap junctions)

intercellular cleft a narrow, water-filled space between capillary endothelial cells

intercellular fluid fluid that lies between cells; also called interstitial fluid

intercostal muscle (in-ter-KOS-tal) skeletal muscle that lies between ribs and whose contraction causes rib cage movement during breathing

interferon (in-ter-FEER-on) family of proteins that nonspecifically inhibit viral replication inside host cells; interferon-gamma also stimulates the killing ability of macrophages and NK cells

interferon-gamma *see* interferon

interleukin (in-ter-LOO-kin) a family of cytokines with many effects on immune responses and host defenses

interleukin 1 (IL-1) cytokine secreted by macrophages and other cells that activates helper T cells, exerts many inflammatory effects, and mediates many of the systemic, acute-phase responses, including fever

interleukin 2 (IL-2) cytokine secreted by activated helper T cells that causes antigen-activated helper T, cytotoxic T, and NK cells to proliferate; also causes activation of macrophages

interleukin 6 (IL-6) cytokine secreted by macrophages and other cells that exerts multiple effects on immune system cells, inflammation, and the acute-phase response

intermediate filament actin-containing filament associated with desmosomes

internal anal sphincter smooth-muscle ring around lower end of rectum

internal environment extracellular fluid (interstitial fluid and plasma)

internal genitalia (jen-ih-TAY-lee-ah) *see* female internal genitalia

internal urethral sphincter (you-REE-thrul) part of smooth muscle of urinary bladder wall that opens and closes the bladder outlet

internal work energy-requiring activities in body; *see also* work; *compare* external work

interneuron neuron whose cell body and axon lie entirely in CNS

interstitial fluid extracellular fluid surrounding tissue cells; excludes plasma

interstitium (in-ter-STISH-um) interstitial space; fluid-filled space between tissue cells

interventricular septum the muscular wall separating the right and left ventricles of the heart

intestinal phase (of gastrointestinal control) initiation of neural and hormonal gastrointestinal reflexes by simulation of intestinal tract walls

intestino-intestinal reflex cessation of contractile activity in intestines in response to various stimuli in intestine

intracellular fluid fluid in cells; cytosol plus fluid in cell organelles, including nucleus

intrafusal fiber modified skeletal muscle fiber in muscle spindle

intrapleural fluid (in-trah-PLUR-al) thin fluid film in thoracic cavity between pleura lining the inner wall of thoracic cage and pleura covering lungs

intrapleural pressure (P_{ip}) pressure in pleural space; also called *intrathoracic pressure*

intrarenal baroreceptor pressure-sensitive juxtaglomerular cells of afferent arterioles, which respond to decreased renal arterial pressure by secreting more renin

intrathoracic pressure *see* intrapleural pressure

intrinsic (in-TRIN-sik) situated entirely within a part

intrinsic factor glycoprotein secreted by stomach epithelium and necessary for absorption of vitamin B_{12} in the ileum

intrinsic pathway intravascular sequence of fibrin clot formation initiated by factor XII or, more usually, by the initial thrombin generated by the extrinsic clotting pathway

intrinsic tone spontaneous low-level contraction of smooth muscle, independent of neural, hormonal, or paracrine input

intron (IN-trahn) regions of noncoding nucleotides in a gene

inulin polysaccharide that is filtered but not reabsorbed, secreted, or metabolized in the renal tubules; used to measure glomerular filtration rate

inversely proportional relationship in which, as one factor increases by a given amount, the other decreases by a given amount

iodide trapping active transport of iodide from plasma across the thyroid follicular cell membrane, followed by diffusion of iodide into the colloid of the follicle

iodine chemical found in certain foods and as an additive to table salt; concentrated by the thyroid gland, where it is incorporated into the structure of thyroid hormone

ion (EYE-on) atom or small molecule containing unequal number of electrons and protons and therefore carrying a net positive or negative electric charge

ionic bond (eye-ON-ik) strong electrical attraction between two oppositely charged ions

ionization (eye-on-ih-ZAY-shun) process of removing electrons from or adding them to an atom or small molecule to form an ion

ionotropic receptor (eye-ohn-uh-TROPE-ik) membrane protein through which ionic current is controlled by the binding of extracellular signaling molecules

ipsilateral (ip-sih-LAT-er-al) on the same side of the body

iris ringlike structure surrounding pupil of eye

irreversible reaction chemical reaction that releases large quantities of energy and results in almost all the reactant molecules being converted to product; *compare* reversible reaction

ischemia (iss-KEY-me-ah) reduced blood supply

islet of Langerhans (EYE-let of LAN-ger-hans) cluster of pancreatic endocrine cells; distinct islet cells secrete insulin, glucagon, somatostatin, and pancreatic polypeptide

isometric contraction (eye-soh-MET-rik) contraction of muscle under conditions in which it develops tension but does not change length

isoosmotic (eye-soh-oz-MAH-tik) having the same total solute concentration as extracellular fluid

isotonic (eye-soh-TAH-nik) containing the same number of effectively nonpenetrating solute particles as normal extracellular fluid; *see also* isotonic contraction

isotonic contraction contraction of muscle under conditions in which load on the muscle remains constant but muscle changes length

isotope an atom consisting of one or more additional neutrons than protons in its nucleus

isovolumetric ventricular contraction (eye-soh-vol-you-MET-rik) early phase of systole when atrioventricular and aortic valves are closed and ventricular size remains constant

isovolumetric ventricular relaxation early phase of diastole when atrioventricular and aortic valves are closed and ventricular size remains constant

J

JAK kinase cytoplasmic kinase bound to a receptor but not intrinsic to it

jejunum (jeh-JU-num) middle segment of small intestine

J receptor receptor in the lung capillary walls or interstitium that responds to increased lung interstitial pressure

junctional feet *see foot process*

juxtaglomerular apparatus (JGA) (jux-tah-gloh-MER-you-lar) renal structure consisting of macula densa and juxtaglomerular cells; site of renin secretion and sensors for renin secretion and control of glomerular filtration rate

juxtaglomerular (JG) cell renin-secreting cells in the afferent arterioles of the renal nephron in contact with the macula densa

juxtamedullary nephron functional unit of the kidney with glomeruli in the deep cortex and a long loop of Henle, which plunges into the medulla

K

kallikrein (KAL-ih-cryn) an enzyme produced by gland cells that catalyzes the conversion of the circulating protein kininogen into the signaling molecule bradykinin

karyotype chromosome characteristics of a cell, usually visualized with a microscope

keto acid a class of breakdown products formed from the deamination of amino acids

ketone (KEY-tone) product of fatty acid metabolism that accumulates in blood during starvation and in severe untreated diabetes mellitus; acetoacetic acid, acetone, or B-hydroxybutyric acid; also called *ketone body*

kilocalorie (kcal) (KIL-oh-kal-ah-ree) amount of heat required to change the temperature of 1 L water by 1°C; calorie used in nutrition; also called *Calorie* and *large calorie*

kinase (KY-nase) enzyme that transfers a phosphate (usually from ATP) to another molecule

kinesin (ky-NEE-sin) motor protein that uses the energy from ATP to transport attached cellular cargo along microtubules

kinesthesia (kin-ess-THEE-zee-ah) sense of movement derived from movement at a joint

kinin (KY-nin) peptide that splits from kininogen; facilitates vascular changes and activates pain receptors

kininogen (ky-NIN-oh-jen) plasma protein from which kinins are generated in an inflamed area

knee jerk reflex often used in clinical assessment of nerve and muscle function; striking the tendon just below the kneecap causes reflex contraction of anterior thigh muscles, which extends the knee

Korotkoff's sounds (Kor-OTT-koff) sounds caused by turbulent blood flow during determination of blood pressure with a pressurized cuff

Krebs cycle mitochondrial metabolic pathway that utilizes fragments derived from carbohydrate, protein, and fat breakdown and produces carbon dioxide, hydrogen (for oxidative phosphorylation), and small amounts of ATP; also called tricarboxylic acid cycle or citric acid cycle

L

labeled lines principle describing the idea that a unique anatomical pathway of neurons connects a given sensory receptor directly to the CNS neurons responsible for processing that modality and location on the body

labyrinth complicated bony structure that houses the cochlea and vestibular apparatus

lactase (LAK-tase) small intestine enzyme that breaks down lactose (milk sugar) into glucose and galactose

lactate ionized form of lactic acid

lactation (lak-TAY-shun) production and secretion of milk by mammary glands

lacteal (lak-TEEL) blind-ended lymph vessel in center of each intestinal villus

lactic acid (LAK-tik) three-carbon molecule formed by glycolytic pathway in absence of oxygen; dissociates to form lactate and hydrogen ions

lactose (LAK-tose) disaccharide composed of glucose and galactose; also called *milk sugar*

laminar flow (LAM-ih-ner) when a fluid (e.g., blood) flows smoothly through a tube in concentric layers, without any turbulence

large intestine part of the gastrointestinal tract between the small intestine and rectum; absorbs salts and water

larynx (LAR-inks) part of air passageway between pharynx and trachea; contains the vocal cords

latch state contractile state of some smooth muscles in which force can be maintained for prolonged periods with very little energy use; cross-bridge cycling slows to the point where thick and thin filaments are effectively "latched" together

latent period (LAY-tent) period lasting several milliseconds between action potential initiation in a muscle fiber and beginning of mechanical activity

lateral position farther from the midline

lateral inhibition method of refining sensory information in afferent neurons and ascending pathways whereby fibers inhibit each other, the most active fibers causing the greatest inhibition of adjacent fibers

lateral sac enlarged region at end of each sarcoplasmic reticulum segment; adjacent to transverse tubule

lateral traction force (in the lung) holding small airways open; exerted by elastic connective tissue linked to surrounding alveolar tissue

Law of Laplace (lah-PLAHS) transmural pressure difference = 2 × surface tension divided by the radius of a hollow ball (e.g., an alveolus)

law of mass action maxim that an increase in reactant concentration causes a chemical reaction to proceed in direction of product formation; the opposite occurs with decreased reactant concentration

L-dopa L-dihydroxyphenylalanine; precursor to dopamine formation

leak potassium channels potassium channels that are open when a membrane is at rest

learning acquisition and storage of information as a result of experience

lecithin (LESS-ih-thin) a phospholipid

lengthening contraction contraction as an external force pulls a muscle to a longer length despite opposing forces generated by the active cross-bridges

lens adjustable part of eye's optical system, which helps focus object's image on retina

leptin adipose-derived hormone that acts within the brain to decrease appetite and increase metabolism

leukocyte (LOO-koh-site) white blood cell

leukotrienes (LOO-koh-treens) type of eicosanoid that is generated by lipoxygenase pathway and functions as inflammatory mediator

levodopa (L-dopa) a drug used in the treatment of Parkinson's disease

Leydig cell (LY-dig) testosterone-secreting endocrine cell that lies between seminiferous tubules of testes; also called *interstitial cell*

LH surge large rise in luteinizing hormone secretion by anterior pituitary about day 14 of menstrual cycle

libido (luh-BEE-doh) sex drive

ligand (LY-gand) any molecule or ion that binds to protein surface by noncovalent bonds

ligand-gated channel membrane channel operated by the binding of specific molecules to channel proteins

light adaptation process by which photoreceptors in the retina adjust to sudden bright light

light chain pair of small polypeptides bound to each globular head of a myosin molecule; function is to *modulate* contraction

limbic system (LIM-bik) interconnected brain structures in cerebrum; involved with emotions and learning

lipase (LY-pase) enzyme that hydrolyzes triglyceride to monoglyceride and fatty acids; *see also* lipoprotein lipase

lipid (LIP-id) molecule composed primarily of carbon and hydrogen and characterized by insolubility in water

lipid bilayer a sheet consisting of two layers of amphipathic lipids; nonprotein part of a cell membrane

lipolysis (ly-POL-ih-sis) triglyceride breakdown

lipoprotein (lip-oh-PROH-teen) lipid aggregate partially coated by protein; involved in lipid transport in blood

lipoprotein lipase capillary endothelial enzyme that hydrolyzes triglyceride in lipoprotein to monoglyceride and fatty acids

lipoxygenase (ly-POX-ih-jen-ase) enzyme that acts on arachidonic acid and leads to leukotriene formation

liver large organ located in the upper right portion of the abdomen with exocrine, endocrine, and metabolic functions

load external force acting on muscle

local control mechanism existing within tissues that modulates activity independent of neural or hormonal input (e.g., blood flow into a local vascular bed)

local current movement of positive ions toward more negative membrane region and simultaneous movement of negative ions in opposite direction

local homeostatic response (home-ee-oh-STAT-ik) response acting in immediate vicinity of a stimulus, without nerves or hormones, and having net effect of counteracting stimulus

longitudinal muscle thin outer layer of the intestine; contraction shortens the length of the tube

long-loop negative feedback inhibition of anterior pituitary and/or hypothalamus by hormone secreted by third endocrine gland in a sequence

long reflex neural loop from afferents in the gastrointestinal tract to the central nervous system and back to nerve plexuses and effector cells via the autonomic nervous system; involved in the control of motility and secretory activity

long-term depression condition in which nerves show reduced responses to stimuli after an earlier stimulation

long-term memory information stored in the brain for prolonged periods

long-term potentiation (LTP) process by which certain synapses undergo long-lasting increase in effectiveness when heavily used

loop of Henle (HEN-lee) hairpinlike segment of kidney nephron with *descending* and *ascending limbs;* situated between proximal and distal tubules

low-density lipoprotein (LDL) (lip-oh-PROH-teen) protein-lipid aggregate that is major carrier of plasma cholesterol to cells

lower esophageal sphincter smooth muscle of last portion of esophagus; can close off esophageal opening into the stomach

lower motor neurons neurons that synapse directly onto muscle cells and stimulate their contraction

L-type calcium channel voltage-gated channel permitting calcium entry into heart cells during the action potential; L denotes the long-lasting open time that characterizes these channels

lumen (LOO-men) space in hollow tube or organ

luminal (LOO-min-ul) pertaining to lumen

luminal membrane portion of plasma membrane facing the lumen; also called *apical* or *mucosal membrane*

lung compliance (C_L) (come PLY-ance) change in lung volume caused by a given change in transpulmonary pressure; the greater the lung compliance, the more stretchable the lung wall

luteal phase (LOO-tee-al) last half of menstrual cycle following ovulation; corpus luteum is active ovarian structure

luteinizing hormone (LH) (LOO-tee-en-iz-ing) peptide gonadotropic hormone secreted by anterior pituitary; rapid increase in females at midmenstrual cycle initiates ovulation; stimulates Leydig cells in males

lymph (limf) fluid in lymphatic vessels

lymphatic capillary (lim-FAT-ik) smallest-diameter vessel type of the lymphatic system; site of entry of excess extracellular fluid

lymphatic system (lim-FAT-ik) network of vessels that conveys lymph from tissues to blood and to lymph nodes along these vessels

lymphatic vessel any vessel of the lymph system in which excess interstitial fluid is transported and returned to the circulation; along the way, the fluid (lymph) passes through lymph nodes

lymph node small organ, containing lymphocytes, located along lymph vessel; a site of lymphocyte cell division and initiation of specific immune responses

lymphocyte (LIMF-oh-site) type of leukocyte responsible for specific immune defenses; B cells, T cells, and NK cells

lymphocyte activation cell division and differentiation of lymphocytes following antigen binding

lymphoid organ (LIMF-oid) bone marrow, lymph node, spleen, thymus, tonsil, or aggregate of lymphoid follicles; *see also* primary lymphoid organ, secondary lymphoid organ

lysosome (LY-soh-some) membrane-bound cell organelle containing digestive enzymes in a highly acid solution that break down bacteria, large molecules that have entered the cell, and damaged components of the cell

M

macromolecule large molecule composed of up to thousands of atoms

macrophage (MAK-roh-fahje) cell that phagocytizes foreign matter, processes it, presents antigen to lymphocytes, and secretes cytokines (monokines) involved in inflammation, activation of lymphocytes, and systemic acute phase response to infection or injury; *see also* activated macrophage, macrophage-like cell

macrophage-like cell one of several cell types that exert functions similar to those of macrophages

macula densa (MAK-you-lah DEN-sah) specialized sensor cells of renal tubule at end of loop of Henle; component of juxtaglomerular apparatus

major histocompatibility complex (MHC) group of genes that code for major histocompatibility complex proteins, which are important for specific immune function

malabsorption a decrease in the absorption of minerals, vitamins, and nutrients in the gastrointestinal tract

malaria (mul-LARE-ee-uh) parasitic blood disease caused by a protozoan spread by the bites of infected mosquitoes

malleus one of three bones in the inner ear that transmit movements of the tympanic membrane to the inner ear

mammary gland milk-secreting gland in breast

margination initial step in leukocyte action in inflamed tissues, in which leukocytes adhere to the endothelial cell

mass fundamental property of an object equivalent to the amount of matter in the object

mass movement contraction of large segments of colon; propels fecal matter into rectum

mast cell tissue cell that releases histamine and other chemicals involved in inflammation

matrix (mitochondrial) the innermost mitochondrial compartment

maximal oxygen consumption ($\dot{V}_{O_2}$ max) peak rate of oxygen use as physical exertion is increased; increments in workload above this point must be fueled by anaerobic metabolism

maximal tubular capacity (T_m) *see* transport maximum

mean arterial pressure (MAP) average blood pressure during cardiac cycle; approximately diastolic pressure plus one-third pulse pressure

mechanically gated channel membrane ion channel that is opened or closed by deformation or stretch of the plasma membrane

mechanoreceptor (meh-KAN-oh-ree-sep-tor) sensory neuron specialized to respond to mechanical stimuli such as touch receptors in the skin and stretch receptors in muscle

median eminence (EM-ih-nence) region at base of hypothalamus containing capillary tufts into which hypophysiotropic hormones are secreted

mediastinum (mee-dee-uh-STY-num) membrane separating right and left thoracic compartments

mediate (MEE-dee-ate) bring about

mediated transport movement of molecules across membrane by binding to protein transporter; characterized by specificity, competition, and saturation; includes facilitated diffusion and active transport

medulla (meh-DUL-ah) innermost portion of an organ; *compare* cortex; *see* adrenal medulla, medulla oblongata

medulla oblongata (ob-long-GOT-ah) part of the brainstem closest to the spinal cord

medullary cardiovascular center *see* cardiovascular center

medullary collecting duct terminal component of the nephron in which vasopressin-sensitive passive water reabsorption occurs

medullary inspiratory center neurons in the medulla oblongata that set the pace for inspiration; their rate of firing is rhythmical but can be overridden by conscious control

megakaryocyte (meg-ah-KAR-ee-oh-site) large bone marrow cell that gives rise to platelets

meiosis (my-OH-sis) process of cell division leading to gamete (sperm or egg) formation; daughter cells receive only half the chromosomes present in original cell

meiotic arrest state of primary oocytes from fetal development until puberty, after which meiosis is completed

membrane attack complex (MAC) group of complement proteins that form channels in microbe surface and destroy microbe

membrane potential voltage difference between inside and outside of cell

memory *see* declarative memory, procedural memory, working memory

memory cell B cell or T cell that differentiates during an initial infection and responds rapidly during subsequent exposure to same antigen

memory encoding processes by which an experience is transformed to a memory of that experience

menarche (MEN-ark-ee) onset, at puberty, of menstrual cycling in women

meninges (men-IN-jees) protective membranes that cover brain and spinal cord

menopause (MEN-ah-paws) cessation of menstrual cycling in middle age

menstrual cycle (MEN-stroo-al) cyclical rise and fall in female reproductive hormones and processes, beginning with menstruation

menstrual phase time during menstrual cycle in which menstrual blood is present

menstruation (men-stroo-AY-shun) flow of menstrual fluid from uterus; also called *menstrual period*

menthol an alcohol derived from mint oil that activates ion channels found in temperature receptors that sense cool temperatures

mesangial cell modified smooth-muscle cell that surrounds renal glomerular capillary loops; they help to control glomerular filtration rate

mesolimbic dopamine pathway neural pathway through the limbic system that uses dopamine as its neurotransmitter and is involved in reward

messenger RNA (mRNA) ribonucleic acid that transfers genetic information for a protein's amino acid sequence from DNA to ribosome

metabolic acidosis (met-ah-BOL-ik ass-ih-DOH-sis) acidosis due to the buildup of acids other than carbonic acid (from carbon dioxide)

metabolic alkalosis (al-kah-LOH-sis) alkalosis resulting from the removal of hydrogen ions by mechanisms other than respiratory removal of carbon dioxide

metabolic end product final molecule produced by a metabolic reaction or series of reactions

metabolic pathway sequence of enzyme-mediated chemical reactions by which molecules are synthesized and broken down in cells

metabolic rate total body energy expenditure per unit time

Glossary

metabolism (meh-TAB-uhl-izm) chemical reactions that occur in a living organism

metabolite (meh-TAB-oh-lite) substance produced by metabolism

metabolize change by chemical reactions

metabotropic receptor (meh-tab-oh-TRO-pik) membrane receptor in neurons that initiates formation of second messengers when bound with ligand

metarteriole (MET-are-teer-ee-ole) blood vessel that directly connects arteriole and venule

methyl group —CH_3

MHC protein plasma-membrane protein coded for by a major histocompatibility complex; restricts T-cell receptor's ability to combine with antigen on cell; categorized as class I and class II

micelle (MY-sell) soluble cluster of amphipathic molecules in which molecules' polar regions line surface and nonpolar regions orient toward center; formed from fatty acids, monoglycerides, and bile salts during fat digestion in small intestine

microbe bacterium, virus, fungus, or other parasite

microcirculation blood circulation in arterioles, capillaries, and venules

microfilament rodlike cytoplasmic actin filament that forms major component of cytoskeleton

microglia a type of glial cell that acts as a macrophage

microtubule tubular cytoplasmic filament composed of the protein tubulin; provides internal support for cells and allows change in cell shape and organelle movement in cell

microvillus (my-kroh-VIL-us) small fingerlike projection from epithelial-cell surface; greatly increase surface area of cell; characteristic of epithelium lining small intestine and kidney nephrons

micturition (mik-chur-RISH-un) urination

midbrain the most rostral section of the brainstem

middle ear air-filled space in temporal bone; contains three ear bones that conduct sound waves from tympanic membrane to cochlea

migrating myoelectric complex (MMC) pattern of peristaltic waves that pass over small segments of intestine after absorption of meal

milk ejection reflex process by which milk is moved from mammary gland alveoli into ducts, from which it can be sucked; due to oxytocin

milliliter (ml) (MIL-ih-lee-ter) volume equal to 0.001 L

millimol (mmol) (MIL-ih-mole) amount equal to 0.001 mol

millivolt (mV) (MIL-ih-volt) electrical potential equal to 0.001 V

mineral inorganic substance (that is, without carbon); major minerals in body are calcium, phosphorus, potassium, sulfur, sodium, chloride, and magnesium

mineralization the process of calcifying bone collagen to form lamellar bone

mineralocorticoid (min-er-al-oh-KORT-ih-koid) steroid hormone produced by adrenal cortex; has major effect on sodium and potassium balance; major mineralocorticoid is aldosterone

minute ventilation total ventilation per minute; equals tidal volume times respiratory rate

mitochondrion (my-toh-KON-dree-un) rod-shaped or oval cytoplasmic organelle that produces most of cell's ATP; site of Krebs cycle and oxidative phosphorylation enzymes

mitogen (MY-tuh-jen) chemical that stimulates cell division

mitosis (my-TOH-sis) process in cell division in which DNA is duplicated and copies of each chromosome are passed to daughter cells as the nucleus divides

mitral valve (MY-tral) valve between left atrium and left ventricle of heart

M line transverse stripe occurring at the center of the A band in cardiac and skeletal muscle; location of energy-generating enzymes and proteins connecting adjacent thick filaments

modality (moh-DAL-ih-tee) type of sensory stimulus

modulation *see* allosteric modulation, covalent modulation

modulator molecule ligand that, by acting at an allosteric regulatory site, alters properties of other binding sites on a protein and thus regulates its functional activity

mol weight of a substance in grams equal to its molecular weight; 1 mol = 6×10^{23} molecules

molarity (moh-LAR-ih-tee) number of moles of solute per liter of solution

molecular weight sum of atomic weights of all atoms in molecule

molecule chemical substance formed by linking atoms together

monoamine oxidase (MAO) enzyme that breaks down catecholamines in nerve terminal and synapse

monocyte (MAH-noh-site) type of leukocyte; leaves bloodstream and is transformed into a macrophage

monoglyceride (mah-noh-GLISS-er-ide) glycerol linked to one fatty acid side chain

monoiodotyrosine a singly iodinated tyrosine molecule that is an intermediate in the synthesis of thyroid hormones

monosaccharide (mah-noh-SAK-er-ide) carbohydrate consisting of one sugar molecule, which generally contains five or six carbon atoms

monosynaptic reflex (mah-noh-sih-NAP-tik) reflex in which the afferent neuron directly activates motor neurons

monounsaturated fatty acid a fatty acid, such as oleic acid, in which one carbon-carbon double bond is formed within the hydrocarbon chain due to the removal of two hydrogen atoms

mood a long-term inner emotion that affects how individuals perceive their environment

motilin (moh-TIL-in) candidate intestinal hormone thought to control normal GI motor activity

motility, gastric movement of the gastrointestinal tract mediated by muscular contractions

motivation *see* primary motivated behavior

motor having to do with muscles and movement

motor control hierarchy brain areas having a role in skeletal muscle control are rank-ordered in three functional groups

motor control system CNS parts that contribute to control of skeletal muscle movements

motor cortex strip of cerebral cortex along posterior border of frontal lobe; gives rise to many axons descending in corticospinal and multineuronal pathways; also called *primary motor cortex*

motor end plate specialized region of muscle cell plasma membrane that lies directly under axon terminal of a motor neuron

motor neuron somatic efferent neuron, which innervates skeletal muscle

motor neuron pool all the motor neurons for a given muscle

motor program pattern of neural activity required to perform a certain movement

motor unit motor neuron plus the muscle fibers it innervates

mouth general term for the expanded uppermost portion of the digestive tract

mucosa (mu-KOH-sah) three layers of gastrointestinal tract wall nearest lumen—that is, *epithelium, lamina propria,* and *muscularis mucosa*

Müllerian duct (mul-AIR-ee-an) part of embryo that, in a female, develops into reproductive system ducts, but in a male, degenerates

Müllerian-inhibiting substance (MIS) protein secreted by fetal testes that causes Müllerian ducts to degenerate

multimeric protein a protein in which two or more proteins are associated via hydrogen bonds, hydrophobic attractions, and other forces, to yield a single, larger protein

multineuronal pathway pathway made up of chains of neurons functionally connected by synapses; also called *multisynaptic pathway*

multiunit smooth muscle smooth muscle that exhibits little, if any, propagation of electrical activity from fiber to fiber and whose contractile activity is closely coupled to its neural input

muscarinic receptor (mus-kur-IN-ik) acetylcholine receptor that responds to the mushroom poison muscarine; located on smooth muscle, cardiac muscle, some CNS neurons, and glands

muscle number of muscle fibers bound together by connective tissue

muscle cell specialized cell containing actin and myosin filaments and capable of generating force and movement

muscle fatigue decrease in muscle tension with prolonged activity

muscle fiber muscle cell

muscle spindle a receptor organ, made up of specialized muscle fibers, that detects stretch of skeletal muscles

muscle-spindle stretch receptor capsule-enclosed arrangement of afferent nerve fiber endings around specialized skeletal muscle fibers; sensitive to stretch

muscle tension force exerted by a contracting muscle on object

muscle thick filament *see* thick filament

muscle tissue one of the four major tissue types in the body, comprising smooth, cardiac, and skeletal muscle; can be under voluntary or involuntary control

muscle tone degree of resistance of muscle to passive stretch due to ongoing contractile activity; *see also* smooth muscle tone

muscularis externa two layers of muscle in the gastrointestinal tract consisting of circular and longitudinal muscle

mutagen (MUTE-uh-jen) factor in the environment that increases mutation rate

mutation (mu-TAY-shun) any change in base sequence of DNA that changes genetic information

myelin (MY-uh-lin) insulating material covering axons of many neurons; consists of layers of myelin-forming cell plasma membrane wrapped around axon

myenteric plexus (my-en-TER-ik PLEX-us) nerve cell network between circular and longitudinal muscle layers in esophagus, stomach, and intestinal walls

myo- (MY-oh) pertaining to muscle

myoblast (MY-oh-blast) embryological cell that gives rise to muscle fibers

myocardium (my-oh-KARD-ee-um) cardiac muscle, which forms heart walls

myoepithelial cell (my-oh-ep-ih-THEE-lee-al) specialized contractile cell in certain exocrine glands; contraction forces gland's secretion through ducts

myofibril (my-oh-FY-bril) bundle of thick and thin contractile filaments in cytoplasm of striated muscle; myofibrils exhibit a repeating sarcomere pattern along longitudinal axis of muscle

myogenic response (my-oh-JEN-ik) response originating in muscle

myoglobin (my-oh-GLOH-bin) muscle fiber protein that binds oxygen

myometrium (my-oh-MEE-tree-um) uterine smooth muscle

myosin (MY-oh-sin) contractile protein that forms thick filaments in muscle fibers

myosin ATPase enzymatic site on globular head of myosin that catalyzes ATP breakdown to ADP and P_i, releasing the chemical energy used to produce force of muscle contraction

myosin light-chain kinase smooth-muscle protein kinase; when activated by Ca-calmodulin, phosphorylates myosin light chain

myosin light-chain phosphatase enzyme that removes high-energy phosphate from myosin; important in the relaxation of smooth muscle cells

N

Na⁺/K⁺-ATPase pump primary active transport protein that splits ATP and releases energy used to transport sodium out of cell and potassium in

NAD⁺ *see* nicotinamide adenine dinucleotide

natural killer (NK) cell type of lymphocyte that binds to virus-infected and cancer cells without specific recognition and kills them directly; participates in antibody-dependent cellular cytotoxicity

necrosis death of a cell or population of cells within a tissue or organ, usually due to oxygen and nutrient deprivation

negative balance loss of substance from body exceeds gain, and total amount in body decreases, also used for physical parameters such as body temperature and energy; *compare* positive balance

negative feedback characteristic of control systems in which system's response opposes the original change in the system; *compare* positive feedback

nephron (NEF-ron) functional unit of kidney; has vascular and tubular components

Nernst equation calculation for electrochemical equilibrium across a membrane for any single ion

nerve group of many nerve fibers traveling together in peripheral nervous system

nerve cell cell in nervous system specialized to initiate, integrate, and conduct electrical signals; also called *neuron*

nerve fiber axon of a neuron

nerve tissue one of the four major tissue types in the body, responsible for coordinated control of muscle activity, reflexes, and conscious thought

net amount remaining after deductions have been made; final amount

net filtration pressure algebraic sum of inward- and outward-directed forces that determine the direction and magnitude of fluid flow across a capillary wall

net flux difference between two one-way fluxes

net glomerular filtration pressure sum of the relevant forces resulting in glomerular filtration; it is the hydrostatic pressure within the glomerular capillary (P_{GC}) minus the hydrostatic pressure in Bowman's space (P_{BS}) and minus the osmotic force in the glomerular capillary (π_{GC})

neuroeffector junction "synapse" between a neuron and muscle or gland cell

neuroglia *see* glial cell

neurohormone chemical messenger that is released by a neuron and travels in bloodstream to its target cell

neuromodulator chemical messenger that acts on neurons, usually by a second-messenger system, to alter response to a neurotransmitter

neuromuscular junction synapselike junction between an axon terminal of an efferent nerve fiber and a skeletal muscle fiber

neuron (NUR-ahn) *see* nerve cell

neuropeptide family of more than 50 neurotransmitters composed of 2 or more amino acids; often also functions as chemical messenger in nonneural tissues

neuropeptide Y a peptide found in the brain whose actions include control of reproduction, appetite, and metabolism

neurotransmitter chemical messenger used by neurons to communicate with each other or with effectors

neurotrophic factor (neur-oh-TRO-fic) protein that stimulates growth and differentiation of some neurons

neutral solution a solution that is neither basic nor acidic (pH 7.0)

neutron noncharged component of the nucleus of an atom

neutrophil (NOO-troh-fil) polymorphonuclear granulocytic leukocyte whose granules show preference for neither eosin nor basic dyes; functions as phagocyte and releases chemicals involved in inflammation

nicotinamide adenine dinucleotide (NAD⁺) coenzyme derived from the B vitamin niacin; transfers hydrogen from one substrate to another

nicotinic receptor (nik-oh-TIN-ik) acetylcholine receptor that responds to nicotine; primarily, receptors at motor end plate and on postganglionic autonomic neurons

nitric oxide a gas that functions as intercellular messenger, including neurotransmitters; is endothelium-derived relaxing factor; destroys intracellular microbes

n-methyl-d-aspartate (NMDA) receptor ionotropic glutamate receptor involved in learning and memory

nociceptor (NOH-sih-sep-tor) sensory receptor whose stimulation causes pain

node of Ranvier (RAHN-vee-ay) space between adjacent myelin-forming cells along myelinated axon where axonal plasma membrane is exposed to extracellular fluid; also called *neurofibril node*

nonpenetrating solute dissolved substance that does not passively diffuse across a plasma membrane

nonpolar pertaining to molecule or region of molecule containing predominantly chemical bonds in which electrons are shared equally between atoms; having few polar or ionized groups

nonshivering thermogenesis the creation of bodily heat by processes other than shivering; for example, certain hormones can stimulate metabolism in brown adipose tissue, resulting in heat production in infants (but this does not occur to any significant extent in adults)

nonspecific ascending pathway chain of synaptically connected neurons in CNS that are activated by sensory units of several different types; signals general information; *compare* specific ascending pathway

nonspecific immune defense response that nonselectively protects against foreign material without having to recognize its specific identity

nonsteroidal anti-inflammatory drug (NSAID) inhibitor of enzymes within the synthetic pathway leading from arachidonic acid to leukotrienes and prostaglandins

nonvolatile acid organic (e.g., lactic) or inorganic (e.g., phosphoric and sulfuric) acid not derived directly from carbon dioxide

noradrenergic referring to neurons that release norepinephrine as a neurotransmitter or membrane receptors that bind norepinephrine

norepinephrine (NE) (nor-ep-ih-NEF-rin) biogenic amine (catecholamine) neurotransmitter released at most sympathetic postganglionic endings, from adrenal medulla, and in many CNS regions

NREM sleep sleep state associated with large, slow EEG waves and considerable postural-muscle tone but not dreaming; also called *slow-wave sleep*

nuclear bag fiber specialized stretch receptor in skeletal muscle spindles that responds to both the magnitude of muscle stretch and the speed at which it is stretched

nuclear chain fiber specialized stretch receptor in skeletal muscle spindles that responds in direct proportion to the length of a muscle

nuclear envelope double membrane surrounding cell nucleus

nuclear pore opening in nuclear envelope through which molecular messengers pass between nucleus and cytoplasm

nucleic acid (noo-KLAY-ik) nucleotide polymer in which phosphate of one nucleotide is linked to the sugar of the adjacent one; stores and transmits genetic information; includes DNA and RNA

nucleolus (noo-KLEE-oh-lus) densely staining nuclear region containing portions of DNA that code for ribosomal proteins

nucleosome (NOO-clee-oh-some) nuclear complexes of several histones and their associated coils of DNA

nucleotide (NOO-klee-oh-tide) molecular subunit of nucleic acid; purine or pyrimidine base, sugar, and phosphate

nucleus (NOO-klee-us) (pl. nuclei) (cell) large membrane-bound organelle that contains cell's DNA; (neural) cluster of neuron cell bodies in CNS

O

obligatory water loss minimal amount of water required to excrete waste products

occipital lobe (ok-SIP-ih-tul) posterior region of cerebral cortex where primary visual cortex is located

odorant molecule received by the olfactory system that induces a sensation of smell

Ohm's law current (I) is directly proportional to voltage (E) and inversely proportional to resistance (R) such that $I = E/R$

olfaction (ol-FAK-shun) sense of smell

olfactory (ol-FAK-tor-ee) pertaining to sense of smell

olfactory bulb (ohl-FAK-tor-ee) anterior protuberance of the brain containing cells that process odor inputs

olfactory cortex region on the inferior and medial surface of the frontal lobe of the cerebral cortex where information about the sense of smell is processed

olfactory epithelium mucous membrane in upper part of nasal cavity containing receptors for sense of smell

oligodendrocyte (oh-lih-goh-DEN-droh-site) type of glial cell; responsible for myelin formation in CNS

oncogene (ON-koh-jeen) altered gene that can lead to cancer

oogenesis (oh-uh-JEN-ih-sis) gamete production in female

oogonium (oh-uh-GOH-nee-um) primitive germ cell that gives rise to primary oocyte

opioid (OH-pee-oid) *see* endogenous opioid

opponent color cell ganglion cells in the retina that are inhibited by input from one type of cone photoreceptor but activated by another type of cone photoreceptor

opsin (OP-sin) protein component of photopigment

opsonin (op-SOH-nin) any substance that binds a microbe to a phagocyte and promotes phagocytosis

opthalmoscope medical instrument with magnifying lenses and light, designed for viewing the interior of the eye through the pupil

optic chiasm (KYE-azm) place at base of brain at which optic nerves meet; some neurons cross here to other side of brain

optic disc region of the retina where neurons to the brain exit the eye; lack of photoreceptors here results in a "blind spot"

optic nerve bundle of neurons connecting the eye to the optic chiasm

optic tract bundle of neurons connecting the optic chiasm to the lateral geniculate nucleus of the thalamus

optimal length (L_0) sarcomere length at which muscle fiber develops maximal isometric tension

organ collection of tissues joined in structural unit to serve common function

organ of Corti (KOR-tee) structure in inner ear capable of transducing sound-wave energy into action potentials

organelle *see* cell organelle

organic pertaining to carbon-containing substances; *compare* inorganic

organ system organs that together serve an overall function

orgasm (OR-gazm) inner emotions and systemic physiological changes that mark apex of sexual intercourse, usually accompanied in the male by ejaculation

orienting response behavior in response to a novel stimulus; that is, the person stops what he or she is doing, looks around, listens intently, and turns toward stimulus

osmol (OZ-mole) 1 mole of solute ions and molecules

osmolarity (oz-moh-LAR-ih-tee) total solute concentration of a solution; measure of water concentration in that the higher the solution osmolarity, the lower the water concentration

osmoreceptor (OZ-moh-ree-sep-tor) receptor that responds to changes in osmolarity of surrounding fluid

osmosis (oz-MOH-sis) net diffusion of water across a selective barrier from region of higher water concentration (lower solute concentration) to region of lower water concentration (higher solute concentration)

osmotic diuresis increase in urine flow resulting from increased solute excretion (e.g., glucose in uncontrolled diabetes mellitus)

osmotic pressure (oz-MAH-tik) pressure that must be applied to a solution on one side of a membrane to prevent osmotic flow of water across the membrane from a compartment of pure water; a measure of the solution's osmolarity

osteoblast (OS-tee-oh-blast) cell type responsible for laying down protein matrix of bone; called osteocyte after calcified matrix has been set down

osteoclast (OS-tee-oh-clast) cell that breaks down previously formed bone

osteocyte cell transformed from osteoblast when surrounded by mineralized bone matrix

osteoid collagen matrix in bone that becomes mineralized

otolith (OH-toe-lith) calcium carbonate crystal embedded in the mucus covering of the auditory hair cell

outer segment light-sensitive portion of the photoreceptor containing photopigments

oval window membrane-covered opening between middle ear cavity and scala vestibuli of inner ear

ovarian follicle *see* follicle

ovary (OH-vah-ree) gonad in female

overshoot part of the action potential in which the membrane potential goes above zero

ovulation (ov-you-LAY-shun) release of egg, surrounded by its zona pellucida and granulosa cells, from ovary

ovum (pl. ova) gamete of female; egg

oxidative (OX-ih-day-tive) using oxygen

oxidative deamination (dee-am-ih-NAY-shun) reaction in which an amino group ($-NH_2$) from an amino acid is replaced by oxygen to form a keto acid

oxidative fiber muscle fiber that has numerous mitochondria and therefore a high capacity for oxidative phosphorylation; red muscle fiber

oxidative phosphorylation (fos-for-ih-LAY-shun) process by which energy derived from reaction between hydrogen and oxygen to form water is transferred to ATP during its formation

oxygen-carrying capacity maximum amount of oxygen the blood can carry; in general, proportional to the amount of hemoglobin per unit volume of blood

oxygen debt decrease in energy reserves during exercise that results in an increase in oxygen consumption and an increased production of ATP by oxidative phosphorylation following the exercise

oxygen-hemoglobin dissociation curve S-shaped (sigmoid) relationship between the gas pressure of oxygen (partial pressure of O_2) and amount of oxygen bound to hemoglobin per unit blood (hemoglobin saturation)

oxyhemoglobin (HbO_2) (ox-see-HEE-moh-gloh-bin) hemoglobin combined with oxygen

oxytocin (ox-see-TOH-sin) peptide hormone synthesized in hypothalamus and released from posterior pituitary; stimulates mammary glands to release milk and uterus to contract

P

pacemaker neurons that set rhythm of biological clocks independent of external cues; any nerve or muscle cell that has an inherent autorhythmicity and determines activity pattern of other cells

pacemaker potential spontaneous gradual depolarization to threshold of some nerve and muscle cells' plasma membrane

pancreas elongated gland behind the stomach with both exocrine (secretes digestive enzymes into the gastrointestinal tract) and endocrine (secretes insulin into the blood) functions

papillary muscle (PAP-ih-lair-ee) muscular projections from interior of ventricular chambers that connect to atrioventricular valves and prevent backward flow of blood during ventricular contraction

paracellular pathway the space between adjacent cells of an epithelium through which some molecules diffuse as they cross the epithelium

paracrine agent (PAR-ah-krin) chemical messenger that exerts its effects on cells near its secretion site; by convention, excludes neurotransmitters; *compare* autocrine agent

paradoxical sleep *see* REM sleep

parasympathetic division (par-ah-sim-pah-THET-ik) portion of autonomic nervous system whose preganglionic fibers leave CNS from brainstem and sacral portion of spinal cord; most of its postganglionic fibers release acetylcholine; *compare* sympathetic division

parathyroid gland one of four parathyroid-hormone secreting glands on thyroid gland surface

parathyroid hormone (PTH) peptide hormone secreted by parathyroid glands; regulates calcium and phosphate concentrations of extracellular fluid

parietal cell (pah-RY-ih-tal) gastric gland cell that secretes hydrochloric acid and intrinsic factor

parietal lobe region of cerebral cortex containing sensory cortex and some association cortex

parietal pleura (pah-RYE-it-al plu-rah) serous membranes covering the inside of the chest wall, the diaphragm, and the mediastinum

partial pressure (P) that part of total gas pressure due to molecules of one gas species; measure of concentration of a gas in a gas mixture

parturition events leading to and including delivery of infant

passive immunity resistance to infection resulting from direct transfer of antibodies or sensitized T cells from one person (or animal) to another; *compare* active immunity

pathway series of connected nerves that move a particular type of information from one part of the brain to another part

pentose any five carbon monosaccharide

pepsin (PEP-sin) family of several protein-digesting enzymes formed in the stomach; breaks protein down to peptide fragments

pepsinogen (pep-SIN-ah-jen) inactive precursor of pepsin; secreted by chief cells of gastric mucosa

peptide (PEP-tide) short polypeptide chain; by convention, having less than 50 amino acids

peptide bond polar covalent chemical bond joining the amino and carboxyl groups of two amino acids; forms protein backbone

peptide hormone any of a family of hormones, like insulin, composed of approximately two to 50 amino acids; generally soluble in acid, unlike larger protein hormones, which are insoluble

peptidergic neuron that releases peptides

percent hemoglobin saturation *see* hemoglobin saturation

perception understanding of objects and events of external world that we acquire from neural processing of sensory information

perforin protein secreted by cytotoxic T cells; forms channels in plasma membrane of target cell, which destroys it

perfusion blood flow

pericardium (per-ah-KAR-dee-um) connective-tissue sac surrounding heart

perimenopause beginning period leading to cessation of menstruation

peripheral chemoreceptor carotid or aortic body; responds to changes in arterial blood P_{O_2} and H^+ concentration

peripheral membrane protein hydrophilic proteins associated with cytoplasmic surface of cell membrane

peripheral nervous system nerve fibers extending from CNS

peripheral thermoreceptor cold or warm receptor in skin or certain mucous membranes

peripheral vein blood vessel outside the chest cavity that returns blood from capillaries toward the heart

peristaltic wave (per-ih-STAL-tik) progressive wave of smooth muscle contraction and relaxation that proceeds along wall of a tube, compressing the tube and causing its contents to move

peritoneum (per-ih-toh-NEE-um) membrane lining abdominal and pelvic cavities and covering organs there

peritubular capillary capillary closely associated with renal tubule

permeability coefficient (P) number that defines the proportionality between a flux and a concentration gradient and depends on the properties of the membrane and the diffusing molecule

permissiveness the facilitation of the action of one hormone by another; for example, the effects of epinephrine are exacerbated by thyroid hormone and by cortisol

peroxisome (per-OX-ih-some) cell organelle that destroys certain toxic products by oxidative reactions

Peyer's patches lymphatic tissue located in the lamina propria of the ileum of the small intestine

pH expression of a solution's acidity; negative logarithm to base 10 of H^+ concentration; pH decreases as acidity increases

phagocyte (FAH-go-site) any cell capable of phagocytosis

phagocytosis (fag-uh-sy-TOH-sis) engulfment of particles by a cell

phagolysosome an intracellular vesicle formed when a lysosome and a phagosome combine; the contents of the lysosome begin the process of destroying the contents of the phagosome

phagosome plasma membrane-bound, intracellular sac formed when a phagocyte engulfs a microbe

pharynx (FAIR-inks) throat; passage common to routes taken by food and air

phase shift a resetting of the internal clock due to altered environmental cues

phasic (FAYZ-ik) intermittent; *compare* tonic

phenotype (FEE-noh-type) gender based on physical appearance

phosphate group (FOS-fate) —PO_4^{2-}

phosphodiesterase (fos-foh-dy-ES-ter-ase) enzyme that catalyzes cyclic AMP breakdown to AMP

phospholipase A$_2$ (fos-fo-LY-pase A-two) enzyme that splits arachidonic acid from plasma membrane phospholipid

phospholipase C receptor-controlled plasma-membrane enzyme that catalyzes

phosphatidylinositol bisphosphate breakdown to inositol trisphosphate and diacylglycerol

phospholipid (fos-foh-LIP-id) lipid subclass similar to triglyceride except that a phosphate group (—PO_4^{2-}) and small nitrogen-containing molecule are attached to third hydroxyl group of glycerol; major component of cell membranes

phosphoprotein phosphatase (FOS-fah-tase) enzyme that removes phosphate from protein

phosphoric acid (fos-FOR-ik) acid generated during catabolism of phosphorus-containing compounds; dissociates to form inorganic phosphate and hydrogen ions

phosphorylation (fos for ah LAY-shun) addition of phosphate group to an organic molecule

photopigment light-sensitive molecule altered by absorption of photic energy of certain wavelengths; consists of opsin bound to a chromophore

photoreceptor sensory cell specialized to respond to light; contains pigments that make it sensitive to different light wavelengths

phrenic nerves main motor nerves innervating the diaphragm and providing the impulses to inspire

physiological dead space sum of the anatomic and alveolar dead spaces; it is the part of the respiratory tree in which gas exchange with blood does not occur

physiological genomics application of the understanding of how genes are regulated to the study of physiology; the link between genes and physiology

physiology (fiz-ee-OL-uh-jee) branch of biology dealing with the mechanisms by which living organisms function

pia mater (PEE-ah MAH-ter) innermost of three membranes (meninges) covering the brain

pigment epithelium dark, innermost layer of the retina; absorbs light that bypasses photopigments

pinocytosis (pin-oh-sy-TOH-sis) endocytosis when the vesicle encloses extracellular fluid or specific molecules in the extracellular fluid that have bound to proteins on the extracellular surface of the plasma membrane

pitch degree of how high or low a sound is perceived

pituitary gland (pih-TOO-ih-tar-ee) endocrine gland that lies in bony pocket below hypothalamus; constitutes anterior pituitary and posterior pituitary

pituitary gonadotropin *see* gonadotropic hormone

placenta (plah-SEN-tah) interlocking fetal and maternal tissues that serve as organ of molecular exchange between fetal and maternal circulations

plasma (PLAS-muh) liquid portion of blood; component of extracellular fluid

plasma cell cell that differentiates from activated B lymphocytes and secretes antibodies

plasma membrane membrane that forms outer surface of cell and separates cell's contents from extracellular fluid

plasma membrane effector protein plasma-membrane protein that serves as ion channel or enzyme in signal transduction sequence

plasma protein most are albumins, globulins, or fibrinogen

plasmin (PLAZ-min) proteolytic enzyme able to decompose fibrin and thereby to dissolve blood clots

plasminogen (plaz-MIN-oh-jen) inactive precursor of plasmin

plasminogen activator any plasma protein that activates proenzyme plasminogen

plasticity (plas-TISS-ih-tee) ability of neural tissue to change its responsiveness to stimulation because of its past history of activation

platelet (PLATE-let) cell fragment present in blood; plays several roles in blood clotting

platelet activation changes in the metabolism, shape, and surface proteins of platelets that begin the clotting process

platelet aggregation (ag-reh-GAY-shun) positive feedback process resulting in platelets sticking together

platelet factor (PF) phospholipid exposed in membranes of aggregated platelets; important in activation of several plasma factors in clot formation

platelet plug blockage of a vessel by activated, adherent platelets

pleura (PLUR-ah) thin cellular sheet attached to thoracic cage interior (*parietal pleura*) and, folding back upon itself, attached to lung surface (*visceral pleura*); forms two enclosed *pleural sacs* in thoracic cage

pleural sac membrane enclosing each lung

pluripotent hematopoietic stem cells (plur-ih-POH-tent) single population of bone marrow cells from which all blood cells are descended

pneumotaxic center (NOO-moh-tak-sik) area of the upper pons in the brain that modulates activity of the apneustic center

podocyte epithelial cells lining Bowman's capsule, whose foot processes form filtration slits

polar covalent bond covalent chemical bond in which two electrons are shared unequally between two atoms; atom to which the electrons are drawn becomes slightly negative, while other atom becomes slightly positive; also called *polar bond*

polarized (POH-luh-rized) having two electrical poles, one negative and one positive

polar molecule pertaining to molecule or region of molecule containing polar covalent bonds or ionized groups; part of molecule to which electrons are drawn becomes slightly negative, and region from which electrons are drawn becomes slightly positive; molecule is soluble in water

polymer (POL-ih-mer) large molecule formed by linking together smaller similar subunits

polymodal neuron sensory neuron that responds to more than one type of stimulus

polymorphonuclear granulocyte (pol-ee-morf-oh-NUK-lee-er GRAN-you-loh-site) subclass of leukocytes consisting of eosinophils, basophils, and neutrophils

polypeptide (pol-ee-PEP-tide) polymer consisting of amino acid subunits joined by peptide bonds; also called *peptide* or *protein*

polysaccharide (pol-ee-SAK-er-ide) large carbohydrate formed by linking monosaccharide subunits together

polysynaptic reflex (pol-ee-sih-NAP-tik) reflex employing one or more interneurons in its reflex arc

polyunsaturated fatty acid fatty acid that contains more than one double bond

pons large area of the brainstem containing many nerve axons

pontine respiratory group neurons in the pons that modulate respiratory rhythms

pool the readily available quantity of a substance in the body; often equals amounts in extracellular fluid

portal system a type of circulation characterized by two capillary beds connected by veins called portal veins

portal vein vessel through which blood from several abdominal organs flows to the liver

portal vessel any blood vessel that links two capillary networks

positive balance gain of substance exceeds loss, and amount of that substance in body increases; *compare* negative balance

positive feedback characteristic of control systems in which an initial disturbance sets off train of events that increases the disturbance even further; *compare* negative feedback

postabsorptive state (post-ab-SORP-tive) period during which nutrients are not being absorbed by gastrointestinal tract and energy must be supplied by body's endogenous stores

posterior toward or at the back

posterior pituitary portion of pituitary from which oxytocin and vasopressin are released

postganglionic neuron (post-gang-glee-ON-ik) autonomic-nervous-system neuron or nerve fiber whose cell body lies in a ganglion; conducts impulses away from ganglion toward periphery; *compare* preganglionic neuron

postsynaptic density area in the postsynaptic cell membrane that contains neurotransmitter receptors and structural proteins important for synapse function

postsynaptic neuron (post-sin-NAP-tik) neuron that conducts information away from a synapse

postsynaptic potential local potential that arises in postsynaptic neuron in response to activation of synapses upon it; *see also* excitatory postsynaptic potential, inhibitory postsynaptic potential

postural reflex reflex that maintains or restores upright, stable posture

potential (or potential difference) voltage difference between two points; *see also* action potential, graded potential

potential difference a difference in charge between two points

potentiation (poh-ten-she-AY-shun) presence of one agent enhances response to a second such that final response is greater than sum of the two individual responses

potocytosis (poh-toe-si-toe-sis) a type of receptor-mediated endocytosis in which vesicle contents are delivered directly to the cytosol

power stroke the step of a cross-bridge cycle involving physical rotation of the globular head

preattentive processing neural processes that occur to direct our attention to a particular aspect of the environment

pre-Botzinger complex neurons of the ventral respiratory group in the medulla that are the respiratory rhythm generator

precapillary sphincter (SFINK-ter) smooth-muscle ring around capillary where it exits from thoroughfare channel or arteriole

preganglionic neuron autonomic-nervous-system neuron or nerve fiber whose cell body lies in CNS and whose axon terminals lie in a ganglion; conducts action potentials from CNS to ganglion; *compare* postganglionic neuron

preinitiation complex a group of transcription factors and accessory proteins that associate with promoter regions of specific genes; the complex is required for gene transcription to commence

preload the amount of filling of ventricles just prior to contraction; the end-diastolic volume

premotor area region of the cerebral cortex found on the lateral sides of the brain in front of the primary motor cortex; involved in planning and enacting complex muscle movements

pressure natriuresis increase in sodium excretion induced by a local action within the renal tubules due to an increase in the arterial pressure within the kidney

presynaptic facilitation (pre-sin-NAP-tik) excitatory input to neurons through synapses at the nerve terminal

presynaptic inhibition inhibitory input to neurons through synapses at the nerve terminal

presynaptic neuron neuron that conducts action potentials toward a synapse

primary active transport active transport in which chemical energy is transferred directly from ATP to transporter protein

primary cortical receiving area region of cerebral cortex where specific ascending pathways end; somatosensory, visual, auditory, or taste cortex

primary lymphoid organ organs that supply secondary lymphoid organs with mature lymphocytes; bone marrow and thymus

primary motivated behavior behavior related directly to achieving homeostasis

primary motor cortex *see* motor cortex

primary oocyte (OH-uh-site) female germ cell that undergoes first meiotic division to form secondary oocyte and polar body

primary protein structure the amino acid sequence of a protein

primary response gene (PRG) gene influenced by transcription factors generated in response to first messengers

primary RNA transcript an RNA molecule transcribed from a gene before intron removal and splicing

primary spermatocyte (sper-MAT-uh-site) male germ cell derived from spermatogonia; undergoes meiotic division to form two secondary spermatocytes

primordial follicle (FAH-lik-el) an immature oocyte encased in a single layer of granulosa cells

procedural memory the memory of how to do things

process long extension from neuron cell body

product molecule formed in enzyme-catalyzed chemical reaction

progesterone (proh-JES-ter-own) steroid hormone secreted by corpus luteum and placenta; stimulates uterine gland secretion, inhibits uterine smooth-muscle contraction, and stimulates breast growth

program related sequence of neural activity preliminary to motor act

prohormone peptide precursor from which are cleaved one or more active peptide hormones

prokaryotic cell cell such as a bacterium that does not contain its genetic information within a membrane-enclosed nucleus

prolactin (pro-LAK-tin) peptide hormone secreted by anterior pituitary; stimulates milk secretion by mammary glands

prolactin-inhibiting hormone (PIH) dopamine, which serves as a hypophysiotropic hormone to inhibit prolactin secretion by anterior pituitary

prolactin-releasing factor putative hypothalamic factor that stimulates prolactin release

proliferative phase (pro-LIF-er-ah-tive) stage of menstrual cycle between menstruation and ovulation during which endometrium repairs itself and grows

promoter specific nucleotide sequence at beginning of gene that controls the initiation of gene transcription; determines which of the paired strands of DNA is transcribed into RNA

propagation (prop-ah-GAY-shun) conduction of nerve impulse

proprioception (PROH-pree-oh-cep-shun) sense of posture and position; sensory information dealing with the position of the body in space and its parts relative to one another

prostacyclin (PGI₂) eicosanoid that inhibits platelet aggregation in blood clotting; also called prostaglandin I_2 (PGI₂)

prostaglandin (pros-tah-GLAN-din) one class of a group of modified unsaturated fatty acids (eicosanoids) that function mainly as paracrine or autocrine agents

prostate gland (PROS-tate) large gland encircling urethra in the male; secretes seminal fluid into urethra

protease (PROH-tee-ase) an enzyme capable of breaking peptide bonds in a protein

proteasome a complex of proteins capable of denaturing (unfolding) other proteins and assisting in protein degradation

protein large polymer consisting of one or more sequences of amino acid subunits joined by peptide bonds

protein binding site *see* binding site

protein C plasma protein that inhibits clotting

protein kinase (KY-nase) any enzyme that phosphorylates other proteins by transferring to them a phosphate group from ATP

protein kinase C enzyme that phosphorylates certain intracellular proteins when activated by diacylglycerol

proteolysis the process whereby peptides and proteins are cleaved into smaller molecules, by the actions of specific enzymes (proteases)

proteolytic (proh-tee-oh-LIT-ik) breaks down protein

prothrombin (proh-THROM-bin) inactive precursor of thrombin; produced by liver and normally present in plasma

proton (PROH-tahn) positively charged subatomic particle

proximal (PROX-sih-mal) nearer; closer to reference point; *compare* distal

proximal tubule first tubular component of a nephron after Bowman's capsule; comprises *convoluted* and *straight segments*

puberty attainment of sexual maturity when conception becomes possible; as commonly used, refers to 3 to 5 years of sexual development that culminates in sexual maturity

pulmonary (PUL-mah-nar ee) pertaining to lungs

pulmonary artery large, branching vessel carrying oxygen-poor blood away from the heart toward the lungs

pulmonary circulation circulation through lungs; portion of cardiovascular system between pulmonary trunk, as it leaves the right ventricle, and pulmonary veins, as they enter the left atrium

pulmonary stretch receptor afferent nerve ending lying in airway smooth muscle and activated by lung inflation

pulmonary surfactant *see* surfactant

pulmonary trunk large artery that carries blood from right ventricle of heart to lungs

pulmonary valve valve between right ventricle of heart and pulmonary trunk

pulmonary vein large, converging vessel that returns oxygen-rich blood toward the heart from the lungs

pulse pressure difference between systolic and diastolic arterial blood pressures

pupil opening in iris of eye through which light passes to reach retina

purine (PURE-ene) double-ring, nitrogen-containing subunit of nucleotide; adenine or guanine

Purkinje fiber (purr-KIN-gee) specialized myocardial cell that constitutes part of conducting system of heart; conveys excitation from bundle branches to ventricular muscle

P wave component of electrocardiogram reflecting atrial depolarization

pyloric sphincter (py-LOR-ik) ring of smooth muscle between stomach and small intestine

pyramidal cell large neuron with characteristic pyramid-shaped cell body and apical dendrite

pyramidal system descending nervous system pathways that originate in the cerebral cortex, cross over the midline in the medulla, and control fine movements of the distal extremities

pyramidal tract *see* corticospinal pathway

pyrimidine (pi-RIM-ih-deen) single-ring, nitrogen-containing subunit of nucleotide; cytosine, thymine, or uracil

pyrogen *see* endogenous pyrogen

pyruvate (PY-roo-vayt) anion formed when pyruvic acid loses a hydrogen ion

pyruvic acid (py-ROO-vik) three-carbon intermediate in glycolysis that, in absence of oxygen, forms lactic acid or, in presence of oxygen, enters Krebs cycle

Q

QRS complex component of electrocardiogram corresponding to ventricular depolarization

quaternary protein structure formed when two or more proteins associate with each other by hydrogen bonds and other forces; the individual proteins are then termed subunits

R

radiation emission of heat from the surface of an object

rapid eye movement sleep *see* REM sleep

rapidly adapting receptor sensory receptor that fires for a brief period at the onset and/or offset of a stimulus

rate-limiting enzyme enzyme in metabolic pathway most easily saturated with substrate; determines rate of entire metabolic pathway

rate-limiting reaction slowest reaction in metabolic pathway; catalyzed by rate-limiting enzyme

reactant (ree-AK-tent) molecule that enters a chemical reaction; called the substrate in enzyme-catalyzed reactions

reaction *see* chemical reaction

reactive hyperemia (hy-per-EE-me-ah) transient increase in blood flow following release of occlusion of blood supply

receptive field (or neuron) area of body that, if stimulated, results in activity in that neuron

receptive relaxation relaxation of the smooth muscles of the stomach (fundus and body) when food is swallowed; mediated by parasympathetic nerves in the enteric nerve plexuses

receptor (in sensory system) specialized peripheral ending of afferent neuron, or separate cell intimately associated with it, that detects changes in some aspect of environment; (in intercellular chemical communication) specific binding site in plasma membrane or interior of target cell with which a chemical messenger combines to exert its effects

receptor activation change in receptor conformation caused by combination of messenger with receptor

receptor desensitization temporary inability of a receptor to respond to its ligand due to prior ligand binding

receptor-mediated endocytosis the specific uptake of ligands in the extracellular fluid by regions of the plasma membrane that invaginate and form intracellular vesicles

receptor potential graded potential that arises in afferent neuron ending, or a specialized cell intimately associated with it, in response to stimulation

receptor tyrosine kinase the major type of receptor protein that is itself an enzyme; these receptors are on plasma membranes and respond to many different water-soluble chemical messengers

reciprocal innervation inhibition of motor neurons activating muscles whose contraction would oppose an intended movement

recognition binding of antigen to receptor specific for that antigen on lymphocyte surface

recombinant DNA (re-KOM-bih-nent) DNA formed by joining portions of two DNA molecules previously fragmented by a restriction enzyme

recruitment activation of additional cells in response to increased stimulus strength; increasing the number of active motor units in a muscle

rectum short segment of large intestine between sigmoid colon and anus

red muscle fiber muscle fiber having high oxidative capacity and large amount of myoglobin

reflex (REE-flex) biological control system linking stimulus with response and mediated by a reflex arc

reflex arc neural or hormonal components that mediate a reflex; usually includes receptor, afferent pathway, integrating center, efferent pathway, and effector

reflex response final change due to action of stimulus upon reflex arc; also called *effector response*

refractory period (reh-FRAK-tor-ee) time during which an excitable membrane does not respond to a stimulus that normally causes response; *see also* absolute refractory period, relative refractory period

regulatory site site on protein that interacts with modulator molecule; alters functional site properties

relative refractory period time during which excitable membrane will produce action potential, but only to a stimulus of greater strength than the usual threshold strength

relaxation return of muscle to a low force-generating state, caused by detachment of cross-bridges

relaxin hormone secreted by the ovary before parturition

releasing hormone *see* hypophysiotropic hormone

REM sleep (rem) sleep state associated with small, rapid EEG oscillations, complete loss of tone in postural muscles, and dreaming; also called *rapid eye movement sleep, paradoxical sleep*

renal (REE-nal) pertaining to kidneys

renal corpuscle combination of glomerulus and Bowman's capsule

renal cortex outer portion of the kidney

renal medulla inner portion of the kidney

renal pelvis cavity at base of each kidney; receives urine from collecting duct system and empties it into ureter

renal plasma flow the total amount of plasma (blood minus red cell volume) that passes through both kidneys per unit time

renal tubule fluid-filled tube extending from the renal corpuscle; site of renal fluid and electrolyte exchange

renin (REE-nin) peptide secreted by kidneys; acts as an enzyme that catalyzes splitting off of angiotensin I from angiotensinogen in plasma

renin-angiotensin system hormonal system consisting of renin-stimulated angiotensin I production followed by conversion to angiotensin II by angiotensin-converting enzyme

replicate (REP-lih-kayt) duplicate

repolarize return transmembrane potential to its resting level

residual volume (RV) air volume remaining in lungs after maximal expiration

resistance (R) hindrance to movement through a particular substance, tube, or opening

respiration (cellular) oxygen utilization in metabolism of organic molecules; (respiratory system) oxygen and carbon dioxide exchange between organism and external environment

respiratory acidosis increased arterial H^+ concentration due to carbon dioxide retention

respiratory alkalosis decreased arterial H^+ concentration when carbon dioxide elimination from the lungs exceeds its production

respiratory bronchiole largest branch of the respiratory tree in which the units of gas exchange (alveoli) appear

respiratory cycle changes in the lung volumes from the beginning of an inspiration, including the expiration, to the beginning of the next inspiration

respiratory pump mechanism whereby reductions in intrathoracic pressure during the breathing cycle tend to favor the return of blood to the heart from peripheral veins

respiratory quotient (RQ) ratio of carbon dioxide produced to oxygen consumed during metabolism

respiratory rate number of breaths per minute

respiratory rhythm generator neural network in the brainstem that generates output to the phrenic nerve

respiratory zone portion of airways from beginning of respiratory bronchioles to alveoli; contains alveoli across which gas exchange occurs

resting membrane potential voltage difference between inside and outside of cell in absence of excitatory or inhibitory stimulation; also called *resting potential*

rest-or-digest homeostatic state characteristic of parasympathetic nervous system activation

restriction element *see* MHC protein

retching strong involuntary attempt to vomit but without stomach contents passing through upper esophageal sphincter

rete testes (REE-tee TES-teez) network of canals at the ends of the seminiferous tubules in the testes

reticular activating system extensive neuron network extending through brainstem core; receives and integrates information from many afferent pathways and from other CNS regions

reticulocyte (ruh-TIK-you-low-site) name given to immature red blood cells that have a weblike pattern in the cytosol due to the persistence of ribosomes

retina thin layer of neural tissue lining back of eyeball; contains receptors for vision

retinal (ret-in-AL) form of vitamin A that forms chromophore component of photopigment

retrograde movement of a substance or action potential backward along a neuron, from axon terminals toward the cell body and dendrites

reuptake active process that recaptures excess secreted neurotransmitter back into the presynaptic cell; can be inhibited with drugs

reversible reaction chemical reaction in which energy release is small enough for reverse reaction to occur readily; *compare* irreversible reaction

Rh factor group of erythrocyte plasma-membrane antigens that may (Rh^+) or may not (Rh^-) be present

rhodopsin (roh-DOP-sin) photopigment in rods

ribonucleic acid (RNA) (ry-boh-noo-KLAY-ik) single-stranded nucleic acid involved in transcription of genetic information and translation of that information into protein structure; contains the sugar ribose; *see also* messenger RNA, ribosomal RNA, transfer RNA

ribose the sugar backbone of RNA

ribosomal RNA (rRNA) (ry-boh-SOME-al) type of RNA used in ribosome assembly; becomes part of ribosome

ribosome (RY-boh-some) cytoplasmic particle that mediates linking together of amino acids to form proteins; attached to endoplasmic reticulum as bound ribosome, or suspended in cytoplasm as free ribosome

right and left bundle branch *see* bundle branch

rigor mortis (RIG-or MOR-tiss) stiffness of skeletal muscles after death due to failure of cross-bridges to dissociate from actin because of the loss of ATP

RNA polymerase (POL-ih-muh-rase) enzyme that forms RNA by joining together appropriate nucleotides after they have base-paired to DNA

rod one of two receptor types for photic energy; contains the photopigment rhodopsin

rough endoplasmic reticulum *see* endoplasmic reticulum

round window membrane-covered opening in the cochlea that responds to fluid movement in the scala tympani

ryanodine receptor calcium-release channel found in the lateral sacs of the sarcoplasmic reticulum in skeletal muscle cells

S

saccade (sah-KADE) short, jerking eyeball movement

saccule structure in the semicircular canals that responds to changes in linear movement of the head by mechanical forces on otoliths located on its surface

saliva watery solution of salts and proteins, including mucins and amylase, secreted by salivary glands

salivary gland one of three pairs of exocrine glands around the mouth that produce saliva

salt appetite desire for salt, consisting of hedonistic and regulatory components

saltatory conduction propagation of action potentials along a myelinated axon such that the action potentials jump from one node of Ranvier in the myelin sheath to the next

sarcomere (SAR-kuh-meer) repeating structural unit of myofibril; composed of thick and thin filaments; extends between two adjacent Z lines

sarcoplasmic reticulum (sar-koh-PLAZ-mik reh-TIK-you-lum) endoplasmic reticulum in muscle fiber; site of storage and release of calcium ions

satellite cell undifferentiated cell found within skeletal muscle tissue that can fuse and develop into new muscle fiber following muscle injury

satiety signal (sah-TY-ih-tee) input to food control centers that causes hunger to cease and sets time period before hunger returns

saturated fatty acid fatty acid whose carbon atoms are all linked by single covalent bonds

saturation occupation of all available binding sites by their ligand

scala tympani (SCALE-ah TIM-pah-nee) fluid-filled inner-ear compartment that receives sound waves from basilar membrane and transmits them to round window

scala vestibuli (ves-TIB-you-lee) fluid-filled inner-ear compartment that receives sound waves from oval window and transmits them to basilar membrane and cochlear duct

Schwann cell nonneural cell that forms myelin sheath in peripheral nervous system

sclera (SKLAIR-ah) the tough, outermost tissue layer of the eyeball

scrotum (SKROH-tum) sac that contains testes and epididymides

secondary active transport active transport in which energy released during transmembrane movement of one substance from higher to lower concentration is transferred to the simultaneous movement of another substance from lower to higher concentration

secondary lymphoid organ lymph node, spleen, tonsil, or lymphocyte accumulation in gastrointestinal, respiratory, urinary, or reproductive tract; site of stimulation of lymphocyte response

secondary oocyte daughter cell (23 chromosomes) retaining most cytoplasm resulting from first meiotic division in the ovary

secondary peristalsis (per-ih-STAL-sis) esophageal peristaltic waves not immediately preceded by pharyngeal phase of swallow

secondary protein structure the helical and beta-sheet structure of a protein

secondary sexual characteristic external difference between male and female not directly involved in reproduction

secondary spermatocyte a 23-chromosome cell resulting from the first meiotic division of the primary spermatocyte in the testes

second messenger intracellular substance that serves as relay from plasma membrane to intracellular biochemical machinery, where it alters some aspect of cell's function

secretin (SEEK-reh-tin) peptide hormone secreted by upper small intestine; stimulates pancreas to secrete bicarbonate into small intestine

secretion (sih-KREE-shun) elaboration and release of organic molecules, ions, and water by cells in response to specific stimuli

secretory phase (SEEK-rih-tor-ee) stage of menstrual cycle following ovulation during which secretory type of endometrium develops

secretory vesicle membrane-bound vesicle produced by Golgi apparatus; contains protein to be secreted by cell

segmentation (seg-men-TAY-shun) series of stationary rhythmical contractions and relaxations of rings of intestinal smooth muscle; mixes intestinal contents

selective attention paying attention to or focusing on a particular stimulus or event while ignoring other ongoing sources of information

semen (SEE-men) sperm-containing fluid of male ejaculate

semicircular canal passage in temporal bone; contains sense organs for equilibrium and movement

seminal vesicle one of pair of exocrine glands in males that secrete fluid into vas deferens

seminiferous tubule (sem-ih-NIF-er-ous) tubule in testis in which sperm production occurs; lined with Sertoli cells

semipermeable membrane (sem-ee-PER-me-ah-bul) membrane permeable to some substances (usually water) but not to others (some solutes)

sensation the mental perception of a stimulus

sensorimotor cortex (sen-sor-ee-MOH-tor) all areas of cerebral cortex that play a role in skeletal muscle control

sensory information information that originates in stimulated sensory receptors

sensory pathway a group of neuron chains, each chain consisting of three or more neurons connected end-to-end by synapses; carries action potentials to those parts of the brain involved in conscious recognition of sensory information

sensory receptor a cell or portion of a cell that contains structures or chemical molecules sensitive to changes in an energy form in the outside world or internal environment; in response to activation by this energy, the sensory receptor initiates action potentials in that cell or an adjacent one

sensory system part of nervous system that receives, conducts, or processes information that leads to perception of a stimulus

sensory transduction neural process of changing a sensory stimulus into a change in neuronal function

sensory unit afferent neuron plus receptors it innervates

serosa (sir-OH-sah) connective-tissue layer surrounding outer surface of stomach and intestines

serotonin (sair-oh-TONE-in) biogenic amine neurotransmitter; paracrine agent in blood platelets and digestive tract; also called *5-hydroxytryptamine,* or *5-HT*

Sertoli cell (sir-TOH-lee) cell intimately associated with developing germ cells in seminiferous tubule; creates blood-testis barrier, secretes fluid into seminiferous tubule, and mediates hormonal effects on tubule

Sertoli cell barrier barrier to the movement of chemicals from the blood into the lumen of the seminiferous tubules in the testes

serum (SEER-um) blood plasma from which fibrinogen and other clotting proteins have been removed as result of clotting

set point steady-state value maintained by homeostatic control system

sex chromatin (CHROM-ah-tin) nuclear mass not usually found in cells of males; condensed X chromosome

sex chromosome X or Y chromosome

sex determination genetic basis of individual's sex, XY determining male, and XX, female

sex differentiation development of male or female reproductive organs

sex hormone estrogen, progesterone, testosterone, or related hormones

shaft portion of bone between epiphyseal plates

shear stress force exerted perpendicular to a surface (e.g., force exerted on the walls of a vessel by fluid flowing past)

shivering thermogenesis neurally induced cycles of contraction and relaxation of skeletal muscle in response to decreased body temperature; little or no external work is performed, and thus the increased metabolism of muscle leads primarily to heat production

short-loop negative feedback influence of hypothalamus by an anterior pituitary hormone

short reflex local neural loop from gastrointestinal receptors to nerve plexuses

short-term memory storage of incoming neural information for seconds to minutes; may be converted into long-term memory

sigmoid colon (SIG-moid) S-shaped terminal portion of colon

signal sequence initial portion of newly synthesized protein (if protein is destined for secretion)

signal transduction pathway sequence of mechanisms that relay information from plasma-membrane receptor to cell's response mechanism; *see also* transduction

single-unit smooth muscle smooth muscle that responds to stimulation as single unit because gap junctions join muscle fibers, allowing electrical activity to pass from cell to cell

sinoatrial (SA) node (sy-noh-AY-tree-al) region in right atrium of heart containing specialized cardiac muscle cells that depolarize spontaneously faster than other cells in the conducting system; determines heart rate

sinus vascular channel for the passage of blood or lymph

skeletal muscle striated muscle attached to bone or skin and responsible for skeletal movements and facial expression; controlled by somatic nervous system

skeletal muscle pump pumping effect of contracting skeletal muscles on blood flow through underlying vessels

skeletomotor fiber *see* extrafusal fiber

sleep *see* NREM sleep, REM sleep

sliding-filament mechanism process of muscle contraction in which shortening occurs by thick and thin filaments sliding past each other

slow fiber muscle fiber whose myosin has low ATPase activity

slowly adapting receptor sensory receptor that fires repeatedly as long as a stimulus is ongoing

slow-oxidative fiber type of skeletal muscle fiber that has slow intrinsic contraction speed but fatigues very slowly due to abundant capacity for production of ATP by aerobic oxidative phosphorylation

slow wave slow, rhythmic oscillation of smooth-muscle membrane potentials toward and away from threshold, due to regular fluctuations in ionic permeability

slow-wave sleep *see* NREM sleep

small intestine longest portion of the gastrointestinal tract; between the stomach and large intestine

smooth endoplasmic reticulum *see* endoplasmic reticulum

smooth muscle nonstriated muscle that surrounds hollow organs and tubes; *see also* multiunit smooth muscle, single-unit smooth muscle

smooth muscle tone smooth-muscle tension due to low-level cross-bridge activity in absence of external stimuli

SNARE protein soluble N-ethylmaleimide-sensitive fusion protein attachment protein receptor

sodium inactivation turning off of increased sodium permeability at action potential peak

soft palate (PAL-et) nonbony region at back of roof of mouth

solute (SOL-yoot) substances dissolved in a liquid

solution liquid (solvent) containing dissolved substances (solutes)

solvent liquid in which substances are dissolved

soma cell body of neuron

somatic (soh-MAT-ik) pertaining to the body; related to body's framework or outer walls, including skin, skeletal muscle, tendons, and joints

somatic nervous system component of efferent division of peripheral nervous system; innervates skeletal muscle; *compare* autonomic nervous system

somatic receptor neural receptor in the framework or outer wall of the body that responds to mechanical stimulation of skin or hairs and underlying tissues, rotation or bending of joints, temperature changes, or painful stimuli

somatic sensation feelings/perceptions coming from muscle, skin, and bones

somatosensory cortex (suh-mat-uh-SEN-suh-ree) strip of cerebral cortex in parietal lobe in which nerve fibers transmitting somatic sensory information synapse

somatostatin (SS) (suh-mat-uh-STAT-in) hypophysiotropic hormone that inhibits growth hormone secretion by anterior pituitary; possible neurotransmitter; also found in stomach and pancreatic islets

somatotopic map a representation of the different regions of the body formed by neurons of the cerebral cortex

sound wave air disturbance due to variations between regions of high air molecule density (compression) and low density (rarefaction)

spatial summation (SPAY-shul) adding together effects of simultaneous inputs to different places on a neuron to produce potential change greater than that caused by single input

specific ascending pathway chain of synaptically connected neurons in CNS, all activated by sensory units of same type

specific immune defense response that depends upon recognition of specific foreign material for reaction to it

specificity selectivity; ability of binding site to react with only one, or a limited number of, types of molecules

sperm *see* spermatozoan

spermatic cord structure including the vas deferens and blood vessels and nerves supplying the testes

spermatid (SPER-mah-tid) immature sperm

spermatogenesis (sper-mah-toh-JEN-ih-sis) sperm formation

spermatogonium (sper-mah-toh-GOH-nee-um) undifferentiated germ cell that gives rise to primary spermatocyte

spermatozoan (spur-ma-toh-ZOH-in) male gamete; also called sperm

sperm capacitation (kah-pas-ih-TAY-shun) process by which sperm in female reproductive tract gains ability to fertilize egg

sphincter (SFINK-ter) smooth-muscle ring that surrounds a tube, closing tube as muscle contracts

sphincter of Oddi (OH-dye) smooth-muscle ring surrounding common bile duct at its entrance into duodenum

sphygmomanometer (sfig-moh-mah-NOM-eh-ter) device consisting of inflatable cuff and pressure gauge for measuring arterial blood pressure

spinal nerve one of 86 peripheral nerves (43 pairs) that join spinal cord

spinal reflex reflex whose afferent and efferent components are in spinal nerves; can occur in absence of brain control

spleen largest lymphoid organ; located between stomach and diaphragm

spliceosome protein and nuclear RNA complex that removes introns and links exons together during gene transcription

SRY gene gene on the Y chromosome that determines development of testes in genetic male

stable balance net loss of substance from body equals net gain, and amount of substance in body neither increases nor decreases; *compare* negative balance, positive balance

stapedius (stah-PEE-dee-us) skeletal muscle that attaches to the stapes and protects the auditory apparatus by dampening the movement of the ear ossicles during persistent, loud sounds

stapes one of three bones in the inner ear that transmit movements of the tympanic membrane to the inner ear

starch moderately branched plant polysaccharide composed of glucose subunits

Starling force factor that determines direction and magnitude of fluid movement across capillary wall

Starling's law of the heart *see* Frank-Starling mechanism

state of consciousness degree of mental alertness—that is, whether awake, drowsy, asleep, and so on

steady state no net change; continual energy input to system is required, however, to prevent net change; *compare* equilibrium

stem cell cell that in adult body divides continuously and forms supply of cells for differentiation

stereocilia (ster-ee-oh-SIL-ee-ah) nonmotile cilia containing actin filaments

steroid (STER-oid) lipid subclass; molecule consists of four interconnected carbon rings to which polar groups may be attached

steroid hormone any of a family of hormones, like progesterone, whose structure is derived from cholesterol

steroid hormone receptor superfamily class of intracellular receptor proteins that bind steroid hormones and other lipophilic molecules and induce changes in gene transcription

stimulus detectable change in internal or external environment

stomach expandable, saclike structure in the gastrointestinal tract between the esophagus and small intestine; site of initial digestion of proteins

"stop" signal three-nucleotide sequence in mRNA that signifies end of protein coding sequence

stress environmental change that must be adapted to if health and life are to be maintained; event that elicits increased cortisol secretion

stretch receptor afferent nerve ending that is depolarized by stretching; *see also* muscle-spindle stretch receptor

stretch reflex monosynaptic reflex, mediated by muscle-spindle stretch receptor, in which muscle stretch causes contraction of that muscle

striated muscle (STRY-ay-ted) muscle having transverse banding pattern due to repeating sarcomere structure; *see also* cardiac muscle, skeletal muscle

stroke volume blood volume ejected by a ventricle during one heartbeat

strong acid acid that ionizes completely to form hydrogen ions and corresponding anions when dissolved in water; *compare* weak acid

strychnine an alkaloid nervous system poison that blocks the action of the inhibitory neurotransmitter, glycine

subarachnoid space space between the arachnoid and pia mater meninges containing cerebrospinal fluid

subcortical nuclei groups of cells in brain below the cerebral cortex

submucosa layer of tissue beneath the gastrointestinal mucosa

submucosal plexus (sub-mu-KOH-zal PLEX-us) nerve-cell network in submucosa of esophageal, stomach, and intestinal walls

substance P neuropeptide neurotransmitter released by afferent neurons in pain pathway as well as other sites

substantia nigra (sub-STAN-sha NIE-gra) a subcortical nucleus containing dark-staining neurons that release dopamine and are important for suppressing extraneous muscle activity

substrate (SUB-strate) reactant in enzyme-mediated reaction

substrate-level phosphorylation (fos-for-ih-LAY-shun) direct transfer of phosphate group from metabolic intermediate to ADP to form ATP

subsynaptic membrane (sub-sih-NAP-tik) the part of postsynaptic neuron's plasma membrane under synapse

subthreshold potential (sub-THRESH-old) depolarization less than threshold potential

subthreshold stimulus stimulus capable of depolarizing membrane but not by enough to reach threshold

sucrose (SOO-krose) disaccharide composed of glucose and fructose; also called *table sugar*

sulcus (plural, *sulci*) a deep groove between gyri on the surface of the cerebral cortex

sulfate SO_4^{2-}

sulfhydryl group (sulf-HY-drul) —SH

sulfuric acid (sulf-YOR-ik) acid generated during catabolism of sulfur-containing compounds; dissociates to form sulfate and hydrogen ions

summation (sum-MAY-shun) increase in muscle tension or shortening in response to rapid, repetitive stimulation relative to single twitch

superior vena cava (VEE-nah KAY-vah) large vein that carries blood from upper half of body to right atrium of heart

superoxide anion potentially damaging, highly reactive oxygen molecule containing an extra electron

supersensitivity increased response to a ligand due to receptor up-regulation

supplementary motor cortex region of the cerebral cortex found on the medial side of brain hemispheres in front of the primary motor cortex; involved in planning and enacting complex muscle movements

suprachiasmatic nucleus group of cells in the hypothalamus involved in production of circadian rhythms

surface tension attractive forces between water molecules at an air-water interface resulting in net force that acts to reduce surface area

surfactant (sir-FAK-tent) detergent-like phospholipid-protein mixture produced by pulmonary type II alveolar cells; reduces surface tension of fluid film lining alveoli

swallowing center area of the medulla oblongata in the central nervous system that receives afferent neural input from the mouth and sends efferent output to the muscles of the pharynx, esophagus, and respiratory system, coordinating swallowing

sweat gland gland beneath the skin that is capable of secreting a salty fluid through ducts to the surface of the skin in response to heat-induced neural signals from the autonomic nervous system

sympathetic division portion of autonomic nervous system whose preganglionic fibers leave CNS at thoracic and lumbar portions of spinal cord; *compare* parasympathetic division

sympathetic trunk one of paired chains of interconnected sympathetic ganglia that lie on either side of vertebral column

sympathomimetic (sym-path-oh-mih-MET-ik) produces effects similar to those of sympathetic nervous system

synapse (SIN-apse) anatomically specialized junction between two neurons where electrical activity in one neuron influences excitability of second; *see also* chemical synapse, electrical synapse, excitatory synapse, inhibitory synapse

synaptic cleft narrow extracellular space separating pre- and postsynaptic neurons at chemical synapse

synaptic delay length of time it takes for electrical changes to move from the presynaptic to the postsynaptic membrane

synaptic potential *see* postsynaptic potential

synaptic vesicle cellular structure that holds and releases neurotransmitter at the synapse

synaptotagmin (sin-ap-toh-TAG-min) protein present in wall of synaptic vesicle that binds calcium and helps stimulate the process of exocytosis

synergistic muscle (sin-er-JIS-tik) muscle that exerts force to aid intended motion

systemic circulation (sis-TEM-ik) circulation from left ventricle through all organs except lungs and back to heart

systole (SIS-toh-lee) period of ventricular contraction

systolic pressure (SP) (sis-TAHL-ik) maximum arterial blood pressure during cardiac cycle

T

target cell cell influenced by a certain hormone

taste bud sense organ that contains chemoreceptors for taste

T cell lymphocyte derived from precursor that differentiated in thymus; *see also* cytotoxic T cell, helper T cell

tectorial membrane (tek-TOR-ee-al) structure in organ of Corti in contact with receptor cell hairs

temporal lobe region of cerebral cortex where primary auditory cortex and Wernicke's speech center are located

temporal summation membrane potential produced as two or more inputs, occurring at different times, are added together; potential change is greater than that caused by single input

tendon (TEN-don) collagen fiber bundle that connects skeletal muscle to bone and transmits muscle contraction force to the bone

tension force; *see also* muscle tension

tensor tympani skeletal muscle that attaches to the ear drum and protects the auditory apparatus from loud sounds by dampening the movement of the tympanum

tertiary protein structure the three-dimensional folded structure of a protein formed by hydrogen bonds, hydrophobic attractions, electrostatic interactions, and cysteine cross-bridges

testis (TES-tiss) (pl. testes) gonad in male

testosterone (test-TOS-ter-own) steroid hormone produced in interstitial cells of testes; major male sex hormone

tetanus (TET-ah-nus) maintained mechanical response of muscle to high-frequency stimulation; also the disease lockjaw

thalamus (THAL-ah-mus) subdivision of diencephalon; integrating center for sensory input on its way to cerebral cortex; also contains motor nuclei

theca (THEE-kah) cell layer that surrounds ovarian-follicle granulosa cells

thermogenesis (ther-moh-JEN-ih-sis) heat generation

thermoneutral zone temperature range over which changes in skin blood flow can regulate body temperature

thermoreceptor sensory receptor for temperature and temperature changes, particularly in low (cold receptor) or high (warm receptor) range

theta rhythm slow-frequency, high-amplitude waves of the EEG associated with early stages of slow-wave sleep

thick filament myosin filament in muscle cell

thin filament actin filament in muscle cell

thoracic cavity (thor-ASS-ik) chest cavity

thoracic wall chest wall

thorax (THOR-aks) closed body cavity between neck and diaphragm; contains lung, heart, thymus, large vessels, and esophagus; also called the chest

threshold potential (THRESH-old) membrane potential above which an excitable cell fires an action potential

threshold stimulus stimulus capable of depolarizing membrane just to threshold

thrifty gene gene postulated to have evolved in order to increase the body's ability to store fat

thrombin (THROM-bin) enzyme that catalyzes conversion of fibrinogen to fibrin; has multiple other actions in blood clotting

thrombolytic system *see* fibrinolytic system

thrombomodulin an endothelial receptor to which thrombin can bind, thereby eliminating thrombin's clot-producing effects and causing it to bind and activate protein C

thrombosis (throm-BOH-sis) clot formation in body

thromboxane an eicosanoid derived from arachidonic acid by the action of cyclooxygenase; among other functions, thromboxanes are involved in platelet aggregation

thrombus (THROM-bus) blood clot

thymine (T) (THIGH-meen) pyrimidine base in DNA but not RNA

thymopoietin (thigh-moh-POY-uh-tin) thymus-derived hormone that regulates lymphocyte development

thymus (THIGH-mus) lymphoid organ in upper part of chest; site of T-lymphocyte differentiation

thyroglobulin (thigh-roh-GLOB-you-lin) large protein precursor of thyroid hormones in colloid of follicles in thyroid gland; storage form of thyroid hormones

thyroid gland endocrine gland in neck; secretes thyroid hormones and calcitonin

thyroid hormone (TH) collective term for amine hormones released from thyroid gland—that is, thyroxine (T_4) and triiodothyronine (T_3)

thyroid peroxidase enzyme within the thyroid gland that mediates many of the steps of thyroid hormone synthesis

thyroid-stimulating hormone (TSH) glycoprotein hormone secreted by anterior pituitary; induces secretion of thyroid hormone; also called thyrotropin

thyrotropin-releasing hormone (TRH) hypophysiotropic hormone that stimulates thyrotropin and prolactin secretion by anterior pituitary

thyroxine (T_4) (thigh-ROCKS-in) tetraiodothyronine; iodine-containing amine hormone secreted by thyroid gland

tidal volume (V_t) air volume entering or leaving lungs with single breath during any state of respiratory activity

tight junction cell junction in which extracellular surfaces of the plasma membrane of two adjacent cells are joined together; extends around epithelial cell and restricts molecule diffusion through space between cells

time-averaged mean value obtained for a given physiological variable over an extended period of time; for example, average serum hormone levels can be estimated over 24 hours by collecting their metabolites in the total daily urine output

tip link small, extracellular fiber connecting adjacent stereocilia that activates ion channels when the cilia are bent

tissue aggregate of single type of specialized cell; also denotes general cellular fabric of a given organ

tissue factor protein involved in initiation of clotting via the extrinsic pathway; located on plasma membrane of subendothelial cells

tissue factor pathway inhibitor a plasma protein secreted by endothelial cells; one of several mechanisms for protecting against excessive blood coagulation

tissue plasminogen activator (t-PA) plasma protein produced by endothelial cells; after binding to fibrinogen, activates the proenzyme plasminogen

titin protein that extends from the Z line to the thick filaments and M line of skeletal muscle sarcomere

T lymphocyte see T cell

tone maintained functional activity; see also muscle tone

tonic (TAH-nik) continuous activity; compare phasic

tonsil one of several small lymphoid organs in pharynx

total blood carbon dioxide sum total of dissolved carbon dioxide, bicarbonate, and carbamino-CO_2

total energy expenditure sum of external work done plus heat produced plus energy stored by body

total peripheral resistance (TPR) total resistance to flow in systemic blood vessels from beginning of aorta to ends of venae cavae

totipotent cells of the conceptus that have the capacity to develop into a normal, mature fetus

trace element mineral present in body in extremely small quantities

trachea (TRAY-key-ah) single airway connecting larynx with bronchi; windpipe

tract large, myelinated nerve fiber bundle in CNS

transamination (trans-am-in-NAY-shun) reaction in which an amino acid amino group ($-NH_2$) is transferred to a ketoacid, the ketoacid thus becoming an amino acid

transcellular pathway crossing an epithelium by movement into an epithelial cell, diffusion through the cytosol of that cell, and exit across the opposite membrane

transcription formation of RNA containing, in linear sequence of its nucleotides, the genetic information of a specific gene; first stage of protein synthesis

transcription factor one of a class of proteins that act as gene switches, regulating the transcription of a particular gene by activating or repressing the initiation process

transducin (trans-DOO-sin) G protein in disc membranes of photoreceptor; initiates inactivation of cGMP

transduction process by which stimulus energy is transformed into a response

transepithelial transport see epithelial transport

transferrin (trans-FAIR-in) iron-binding protein that carries iron in plasma

transfer RNA (tRNA) type of RNA; different tRNAs combine with different amino acids and with codon on mRNA specific for that amino acid, thus arranging amino acids in sequence to form specific protein

translation during protein synthesis, assembly of amino acids in correct order according to genetic instructions in mRNA; occurs on ribosomes

transmembrane protein a protein that spans the plasma membrane and contains both hydrophilic and hydrophobic regions; often acts as a receptor or an ion channel

transmural pressure pressure difference exerted on the two sides of a wall

transporter integral membrane protein that mediates passage of molecule through membrane; also called carrier

transport maximum (T_m) upper limit to amount of material that carrier-mediated transport can move across the renal tubule

transpulmonary pressure (P_{tp}) difference in pressure between the inside and outside of the lung (alveolar pressure minus the intrapleural pressure)

transverse tubule (T-tubule) tubule extending from striated-muscle plasma membrane into the fiber, passing between opposed sarcoplasmic reticulum segments; conducts muscle action potential into muscle fiber

tricarboxylic acid cycle see Krebs cycle

tricuspid valve (try-CUS-pid) valve between right atrium and right ventricle of heart

triglyceride subclass of lipids composed of glycerol and three fatty acids; also called fat, neutral fat, or acylglycerol

triiodothyronine (T_3) (try-eye-oh-doh-THIGH-roh-neen) iodine-containing amine hormone secreted by thyroid gland

triplet code three-base sequence in DNA and RNA that specifies particular amino acid

trophic (TROF-ik) growth promoting

trophoblast (TROH-foh-blast) outer layer of blastocyst; gives rise to fetal portion of placental tissue

tropic hormone hormone that stimulates the secretion of another hormone

tropomyosin (troh-poh-MY-oh-sin) regulatory protein capable of reversibly converting binding sites on actin; associated with muscle thin filaments

troponin (troh-POH-nin) regulatory protein bound to actin and tropomyosin of striated muscle thin filaments; site of calcium binding that initiates contractile activity

trypsin (TRIP-sin) enzyme secreted into small intestine by exocrine pancreas as precursor trypsinogen; breaks certain peptide bonds in proteins and polypeptides

trypsinogen (trip-SIN-oh-jen) inactive precursor of trypsin; secreted by exocrine pancreas

T-tubule see transverse tubule

T-type calcium channel channel that carries inward calcium current that briefly supports diastolic depolarization of cardiac pacemaker cells (T: "transient")

tubular reabsorption transfer of materials from kidney tubule lumen to peritubular capillaries

tubular secretion transfer of materials from peritubular capillaries to kidney tubule lumen

tubulin (TOOB-you-lin) the major protein component of microtubules

tumor necrosis factor (TNF) (neh-KROH-sis) cytokine secreted by macrophages (and other cells); has many of the same functions as IL-1

T wave component of electrocardiogram corresponding to ventricular repolarization

twitch mechanical response of muscle to single action potential

tympanic membrane (tim-PAN-ik) membrane stretched across end of ear canal; also called eardrum

type I alveolar cell a flat epithelial cell that with others forms a continuous layer lining the air-facing surface of the pulmonary alveoli

type II alveolar cell pulmonary cell that produces surfactant

tyrosine (TY-roh-seen) amino acid; precursor of catecholamines and thyroid hormones

tyrosine kinase protein kinase that phosphorylates tyrosine portion of proteins; may be part of plasma membrane receptor

U

ubiquitin (you-BIK-wit-in) small intracellular peptide that attaches to proteins and directs them to proteasomes

ultrafiltrate (ul-tra-FIL-trate) protein-free fluid formed from plasma as it is forced through capillary walls by pressure gradient

umami (oo-MOM-ee) unique taste sensation roughly equivalent to "flavorfulness"

umbilical artery artery transporting blood from the fetus into the capillaries of the chorionic villi

umbilical cord (um-BIL-ih-kul) long, ropelike structure that connects the fetus to the placenta and contains umbilical arteries and vein

umbilical vein vein transporting blood from the chorionic villi capillaries back to the fetus

unfused tetanus stimulation of skeletal muscle at a low-to-moderate action potential frequency that results in oscillating, submaximal force

unsaturated fatty acid fatty acid containing one or more double bonds

upper airway part of the respiratory tree consisting of the nose, mouth, pharynx, and larynx

upper esophageal sphincter (ih-sof-ih-JEE-al SFINK-ter) skeletal muscle ring surrounding esophagus just below pharynx that, when contracted, closes entrance to esophagus

upper motor neuron neuron of the motor cortex and descending pathways involved in motor control; they are not technically "motor neurons" because they synapse on neurons, not muscle cells

up-regulation increase in number of target-cell receptors for given messenger in response to chronic low extracellular concentration of that messenger; *see also* supersensitivity; *compare* down-regulation

uracil (U) (YOU-rah-sil) pyrimidine base; present in RNA but not DNA

urea (you-REE-ah) major nitrogenous waste product of protein breakdown and amino acid catabolism

ureter (YOU-rih-ter) tube that connects kidney to bladder

urethra (you-REE-thrah) tube that connects bladder to outside of body

uric acid (YOU-rik) waste product derived from nucleic acid catabolism

urinary bladder *see* bladder

uterus (YOU-ter-us) hollow organ in pelvic region of females; houses fetus during pregnancy; also called *womb*

utricle structure in the semicircular canals that responds to changes in linear movement of the head by mechanical forces on otoliths located on its surface

V

vagina (vah-JY-nah) canal leading from uterus to outside of body; also called *birth canal*

vagus nerve (VAY-gus) cranial nerve X; major parasympathetic nerve

van der Waals forces (walls) weak attractive forces between nonpolar regions of molecules

varicosity (vair-ih-KOS-ih-tee) swollen region of axon; contains neurotransmitter-filled vesicles; analogous to presynaptic ending

vasa recta (VAY-zuh) blood vessels that form loops parallel to the loops of Henle in the renal medulla

vascular system closed system of blood vessels that includes all arteries, arterioles, capillaries, venules, and veins

vas deferens (vas DEF-er-enz) one of paired male reproductive ducts that connect epididymis of testis to urethra; also called *ductus deferens*

vasoconstriction (vays-oh-kon-STRIK-shun) decrease in blood vessel diameter due to vascular smooth muscle contraction

vasodilation (vays-oh-dy-LAY-shun) increase in blood vessel diameter due to vascular smooth muscle relaxation

vasopressin (vays-oh-PRES-sin) peptide hormone synthesized in hypothalamus and released from posterior pituitary; increases water permeability of kidneys' collecting ducts and causes vasoconstriction; also called *antidiuretic hormone (ADH)*

vasovagal syncope (vays-oh-VAY-gal SIN-koh-pay) transient fainting episode occurring when strong emotions decrease sympathetic activity and increase parasympathetic activity to the cardiovascular system; unconsciousness results from reduced blood flow to the brain secondary to a fall in arterial blood pressure

vault recently discovered cytoplasmic structures composed of protein and RNA; their function is uncertain but may involve cytoplasmic-nuclear transport and modulation of a cell's sensitivity to certain drugs

vein any vessel that returns blood to heart; *see also* portal vein

vena cava (VEE-nah KAY-vah) (pl. venae cavae) one of two large veins that returns systemic blood to heart; *see also* inferior vena cava, superior vena cava

venous return (VR) blood volume flowing *to* heart per unit time

ventilation air exchange between atmosphere and alveoli; alveolar air flow

ventral (VEN-tral) toward or at the front of body

ventral respiratory group region of the brainstem containing expiratory neurons important during exercise

ventral root one of two groups of efferent fibers that leave ventral side of spinal cord

ventricle (VEN-trih-kul) cavity, as in cerebral ventricle or heart ventricle; lower chamber of heart

ventricular ejection phase of the cardiac pump cycle during ventricle contraction when blood exits through the semilunar valves

ventricular filling phase of the cardiac pump cycle during which the ventricles are resting and blood enters through the atrioventricular valves

ventricular function curve relation of the increase in stroke volume as end-diastolic volume increases, all other factors being equal

venule (VEEN-ule) small vessel that carries blood from capillary network to vein

very-low-density lipoprotein (VLDL) (lip-oh-PROH-teen) lipid-protein aggregate having high proportion of fat

vesicle (VES-ih-kul) small, membrane-bound organelle within cells

vestibular apparatus *see* vestibular system

vestibular receptor hair cell in semicircular canal, utricle, or saccule

vestibular system sense organ in temporal bone of skull; consists of three semicircular canals, a utricle, and a saccule; also called *sense organ of balance, vestibular apparatus*

vestibulocochlear nerve (ves-tibb-yoo-loh-KOKE-lee-ar) eighth cranial nerve; transmits sensory information about sound and motion from the inner ear to the brain

villus (VIL-us) fingerlike projection from highly folded surface of small intestine; covered with single-layered epithelium

virus nucleic acid core surrounded by protein coat; lacks enzyme machinery for energy production and ribosomes for protein synthesis; thus cannot survive or reproduce except inside other cells, using their biochemical apparatus

viscera (VISS-er-ah) organs in thoracic and abdominal cavities

visceral pleura (VISS-er-al PLOO-rah) serous membranes covering the surface of the lung

viscosity (viss-KOS-ih-tee) measure of friction between adjacent layers of a flowing liquid; property of fluid that makes it resist flow

visible spectrum wavelengths of electromagnetic radiation capable of stimulating photoreceptors of the eye

visual field part of world being viewed at a given time

vital capacity (VC) maximal amount of air that can be expired, regardless of time required, following maximal inspiration

vitalism (VY-tal-ism) view that explanation of life processes requires a "life force" rather than physicochemical processes alone

vitamin organic molecule required in trace amounts for normal health and growth; not manufactured in the body and must be supplied by diet; classified as water-soluble (vitamins C and the B complex) and fat-soluble (vitamins A, D, E, and K)

vitamin B$_{12}$ an essential vitamin found in animal products that plays an important role in the production of red blood cells

vitamin D secosteroid absorbed in the diet or released from the skin under UV light. There are two forms: D$_2$ is from plants whereas D$_3$ is from animals

vitamin D$_2$ *see* ergocalciferol

vitamin D$_3$ *see* cholecalciferol

vitamin K a lipid-soluble substance absorbed from the diet and manufactured by bacteria of the large intestine; required for production of numerous factors involved in blood clotting

vitreous humor jellylike fluid filling the posterior chamber of the eye

vocal cord one of two elastic-tissue bands stretched across laryngeal opening and caused to vibrate when air moves past them, producing sounds

volt (V) unit of measurement of electrical potential between two points

voltage measure of potential of separated electrical charges to do work; measure of electrical force between two points

voltage-gated channel cell-membrane ion channel opened or closed by changes in membrane potential

voluntary movement consciously carried-out motions mediated by the somatic nervous system and skeletal muscle contraction

vomiting center neurons in brainstem medulla oblongata that coordinate vomiting reflex

von Willebrand factor (vWF) (von-VILL-ih-brand) plasma protein secreted by endothelial cells; facilitates adherence of platelets to damaged vessel wall

vulva (VUL-vah) female external genitalia; mons pubis, labia majora and minora, clitoris, vestibule of vagina, and vestibular glands

W

waste product product from a metabolic reaction or series of reactions that serves no function

water diuresis increase in urine flow due to increased fluid output (usually due to decreased secretion or action of vasopressin)

water-soluble vitamin *see* vitamin

wavelength distance between two successive wave peaks in oscillating medium

weak acid acid whose molecules do not completely ionize to form hydrogen ions when dissolved in water; *compare* strong acid

Wernicke's area brain area involved in language comprehension

white matter portion of CNS that appears white in unstained specimens and contains primarily myelinated nerve fibers

white muscle fiber muscle fiber lacking appreciable amounts of myoglobin

withdrawal reflex bending of those joints that withdraw an injured part away from a painful stimulus

Wolffian duct (WOLF-ee-an) part of embryonic duct system that, in male, remains and develops into reproductive system ducts, but in female, degenerates

work measure of energy required to produce physical displacement of matter; *see also* external work, internal work

working memory short-term memory storage process serving as initial depository of information

X

X chromosome *see* sex chromosome

Y

Y chromosome *see* sex chromosome

Z

Z line structure running across myofibril at each end of striated muscle sarcomere; anchors one end of thin filaments and titin

zona pellucida (ZOH-nah peh-LOO-sih-dah) thick, clear layer separating egg from surrounding granulosa cells

zonular fiber connecting the ciliary muscles with the lens of the eye

zygote (ZY-goat) a newly fertilized egg

zymogen (ZY-moh-jen) enzyme precursor requiring some change to become active

References

Berne, R. M., and N. M. Levy: *Cardiovascular Physiology,* 8th ed., Mosby, St. Louis, 2001.

Bloom, F. E., A. Lazerson, and L. Hofstadter: *Brain, Mind, and Behaviour,* Freeman, New York, 2001.

Boron, W. F., and E. L. Boulpaep: *Medical Physiology: A Cellular and Molecular Approach,* Saunders, Philadelphia, 2003.

Braunwald, E., et al.: *Harrison's Principles of Internal Medicine,* 15th ed., McGraw-Hill, New York, 2001.

Carlson, B. M.: *Patten's Foundations of Embryology,* 6th ed., McGraw-Hill, New York, 1996.

Chaffee, E. E., and I. M. Lytle: *Basic Physiology and Anatomy,* 4th ed., Lippincott, Philadelphia, 1980.

Chapman, C. B., and J. H. Mitchell: *Scientific American,* May 1965.

Comroe, J. H.: *Physiology of Respiration,* Year Book, Chicago, 1965.

Crosby, W. H.: *Hospital Practice,* February 1987.

Davis, H., and H. R. Silverman: *Hearing and Deafness,* Holt, Rinehart, and Winston, New York, 1970.

Dowling, J. E., and B. B. Boycott: *Proceedings of the Royal Society, London B,* **166:**80 (1966).

Elias, H., J. E. Pauly, and E. R. Burns: *Histology and Human Microanatomy,* 4th ed., Wiley, New York, 1978.

Erickson, G. F., D. A. Magoffin, C. A. Dyer, and C. Hofeditz: *Endocrine Reviews,* Summer 1985.

Felig, P., and J. Wahren: *New England Journal of Medicine,* **293:**1078 (1975).

Felig, P., and L. A. Frohman: *Endocrinology and Metabolism,* 4th ed., McGraw-Hill, New York, 2001.

Ganong, W. F.: *Review of Medical Physiology,* Appelton and Lange (McGraw-Hill), 2001.

Gardner, E.: *Fundamentals of Neurology,* 5th ed., Saunders, Philadelphia, 1968.

Golde, D. W., and J. C. Gasson: *Scientific American,* July 1988.

Goodman, H. M.: *Basic Medical Endocrinology,* Raven, New York, 1988.

Gray, H. M., A. Sette, and S. Buus (drawing by G. B. Kelvin): *Scientific American,* November 1989.

Greenspan, F. S., and D. G. Gardner: *Basic and Clinical Endocrinology,* 7th ed., McGraw-Hill, New York, 2004.

Hedge, G. A., H. D. Colby, and R. L. Goodman: *Clinical Endocrine Physiology,* Saunders, Philadelphia, 1987.

Hoffman, B. F., and P. E. Cranefield: *Electrophysiology of the Heart,* McGraw-Hill, New York, 1960.

Hubel, D. H., and T. N. Wiesel: *Journal of Physiology,* **154:**572 (1960).

Hudspeth, A. J.: Scientific American, January 1983.

Kandel, E. R., and J. H. Schwartz: *Principles of Neural Science,* 2nd ed., Elsevier/North-Holland, New York, 1985.

Kappas, A., and A. P. Alvares: *Scientific American,* June 1975.

Lambersten, C. J.: in P. Bard (ed.), *Medical Physiological Psychology,* 11th ed., Mosby, St. Louis, 1961.

Lehninger, A. L.: *Biochemistry,* Worth, New York, 1970.

Lentz, T. L.: *Cell Fine Structure,* Saunders, Philadelphia, 1971.

Little, R. C.: *Physiology of the Heart and Circulation,* 2nd ed., Year Book, Chicago, 1981.

Meyer-Franke, A., and B. Barres: *Current Biology,* **4:**847 (1994).

Moore-Ede, M. C., and F. M. Sulzman: *The Clocks That Time Us,* Harvard, Cambridge (MA), 1982.

Nauta, W. J. H., and M. Fiertag: *Fundamental Neuroanatomy,* Freeman, New York, 1986.

Olds, J.: *Scientific American,* October 1956.

Rasmujssen, A. T.: *Outlines of Neuroanatomy,* 2nd ed., Wm. C. Brown, Dubuque, IA, 1943.

Rushmer, R. F.: *Cardiovascular Dynamics,* 2nd ed., Saunders, Philadelphia, 1961.

Saladin, K. S.: *Anatomy and Physiology: The Unity of Form and Function,* 2nd ed., McGraw-Hill, New York, 2001.

Scales, W. E., A. J. Vander, M. B. Brown, and M. J. Kluger: *American Journal of Physiology,* **65:**1840 (1988).

Seeley, R. R., T. R. Stephens, and P. Tate: *Anatomy and Physiology,* 6th ed., McGraw-Hill, New York, 2003.

Silberberg, M. S.: *Chemistry. The Molecular Nature of Matter and Change,* 3rd ed., McGraw-Hill, New York, 2003.

Slone, R. M., et al.: *Thoracic Imaging,* McGraw-Hill, New York, 1999.

Snyder, S. H.: *Drugs and the Brain,* Freeman, New York, 1996.

The Stay Well Company: *Snoring and Sleep Apnea,* San Bruno, CA, 2001.

Tung, K.: *Hospital Practice,* June 1988.

Young, M. P.: *Proc. R. Soc. Lond. B,* **252:**13 (1993).

Van de Graaff, K. M.: *Human Anatomy,* 6th ed., McGraw-Hill, New York, 2002.

von Bekesy, G.: *Scientific American,* August 1957.

Walmsley, B., F. J. Alvarez, and R. E. W. Fyffe: *Trends in Neuroscience,* **21:**81 (1998).

Wang, C. C., and M. I. Grossman: *American Journal of Physiology,* **164:**527 (1951).

Wersall, J., L. Gleisner, and P. G. Lundquist: in A. V. S. de Reuck and J. Knight (eds.), *Myostatic, Kinesthetic, and Vestibular Mechanisms,* Ciba Foundation Symposium, Little, Brown, Boston, 1967.

Credits

Photos

Chapter 1
Opener: © Photodisc Red/Getty Images Royalty Free

Chapter 2
Opener: © Kenneth Eward/BioGrafx/Photo Researchers

Chapter 3
Opener: © Professors P. Motta & T. Naguro/ Science Photo Library/Photo Researchers; **p. 43 (top):** © Professors P. Motta & T. Naguro/ Science Photo Library/Photo Researchers; **3.1:** From: *Histology, A Text & Atlas,* by Johannes A. G. Rhodin, © 1974 by Johannes Rhodin. Used by permission of Oxford University Press, Inc.; **3.3:** From K. R. Porter in T. W. Goodwin and O. Lindberg (eds.) *Biological Structure and Function, Vol. 1,* Academic Press, Inc., New York, 1961; **3.6a:** From J. D. Robertson in Michael Locke (ed.), *Cell Membranes in Development,* Academic Press, Inc., New York, 1964; **3.10c:** Reproduced from *The Journal of Cell Biology,* 1963, 17:375-412 by copyright permission of The Rockefeller University Press; **3.11 (left):** Courtesy of Keith R. Porter; **3.12:** Electron micrograph from D. W. Fawcett, *The Cell, An Atlas of Fine Structure,* W. B. Saunders Company, Philadelphia, 1966; **3.13:** © Dr. Dennis Kunkel/Visuals Unlimited; **3.14:** Courtesy of Keith R. Porter

Chapter 4
Opener: © VVG/Science Photo Library/Photo Researchers

Chapter 5
Opener: © Pr. Lavery, Laboratory of Biochemical Theory/ Copyright © Eurelios/ Phototake — All rights reserved

Chapter 6
Opener: © Su-Chun Zhang, MD, PhD; **6.2c:** © C. Raines/Science VU/Visuals Unlimited

Chapter 7
Opener: © Kenn Kostuk/Peter Arnold; **7.22c:** © Lisa Klancher; **7.27:** Courtesy of Beckman Vision Center at UCSF School of Medicine/D. Coppenhagen, S. Mittman and M. Maglio; **7.38a:** From C.M. Hackney and D.N. Furness (1995) "Mechanotransduction in vertebrate hair cells: the structure and function of the stereociliary bundle" in *American Journal Physiology,* 268: (Cell Physiol. 37): CA-C13. Micrograph by C.M. Hackney and D. N. Furness, Keele University; **7.44c:** © Ed Reschke; **7.46:** © Annabella Bluesky/SPL/Photo Researchers

Chapter 8
Opener: © Richard T. Nowitz/Photo Researchers; **8.12 & 8.16:** Courtesy Marcus E. Raichle, MD, Washington University School of Medicine; **8.17:** © Sally Shaywitz, et al., 1995 NMR Research/Yale Medical School

Chapter 9
Opener: © Steve Gschmeissner/Photo Researchers; **p. 255 (top):** © Steve Gschmeissner/Photo Researchers; **9.1ab:** © Ed Reschke; **9.1c:** © McGraw-Hill Higher Education; **9.3a:** © Marion L. Greaser, University of Wisconsin; **9.4a:** From H.E. Huxley, *Journal of Molecular Biology,* 37:507-520, 1968; **9.14a:** © Don W. Fawcett/Photo Researchers; **9.24:** © Dr. Gladden Willis/Visuals Unlimited; **9.32 & 9.39a:** © Ed Reschke

Chapter 10
Opener: © Tim Pannell/CORBIS RF; **p. 297:** © Tim Pannell/CORBIS RF; **10.14a,b:** © Kevin Strang

Chapter 11
Opener: Copyright © ISM/Phototake; **p. 316:** Copyright © ISM/Phototake; **11.21b:** © Biophoto Associates/Photo Researchers; **11.24:** © CNRI/Phototake; **11.25a,b:** From P. Felig and L. Frohman, *Endocrinology & Metabolism,* 2001, fig. 11.44 (left & right) page 481; **11.29a-c:** From Gagel, McCutcheon, "Pituitary Gigantism" Vol. 340: 524-524, Feb. 18, 1999. Copyright © 1999 Massachusetts Medical Society. All rights reserved

Chapter 12
Opener: © Lunagrafix, Inc./Photo Researchers; **p. 360:** © Lunagrafix, Inc./Photo Researchers; **12.7b1 & b2:** © Dr. Wallace Alpine; **12.37b:** © Michael Noel Hart, M.D., Dept. of Pathology, University of Wisconsin, Madison; **12.48:** © R. Umesh Chandran, TDR, WHO/ SPL/Photo Researchers; **12.66a-c:** © Matthew R. Wolff, M.D., University of Wisconsin, Madison; **12.67:** © Dr. Dennis Kunkel/Visuals Unlimited; **12.75:** © NIBSC/SPL/Photo Researchers

Chapter 13
Opener: © SPL/Photo Researchers; **p. 443:** © SPL/Photo Researchers; **13.11:** © Zephyr/SPL/ Photo Researchers; **13.14:** © Southern Illinois University, School of Medicine, Springfield

Chapter 14
Opener: © Manfred Kage/Peter Arnold; **p. 486:** © Manfred Kage/Peter Arnold; **14.3c:** Copyright © Daniel Friend from Don W. Fawcett & Ronald P. Jensh, *Bloom & Fawcett's Concise Histology* 2nd edition, 2002. Reproduced by permission of Edward Arnold

Chapter 15
Opener: © SPL/Photo Researchers; **p. 529:** © SPL/Photo Researchers; **15.8:** From D.W. Fawcett, *J. Histochem. Cytochem,* 13:75-91, 1965. Courtesy of Susumo Ito; **15.29a,b:** © Dr. Marinilce F. Santos, Institute of Biomedical Sciences, University of Sao Paulo, Brazil; **15.34b:** Courtesy Fernando Carballo, M.D.

Chapter 16
Opener: Photo by John Sholtis, The Rockefeller University, New York, NY. © 1995 Amgen Inc.

Chapter 17
Opener: © David M. Phillips/Photo Researchers; **17.12:** © Glenn D. Braunstein, M.D., Cedars-Sinai Medical Center, Los Angeles, CA; **17.25c:** Section through a macaque blastocyst, after Heuser and Streeter. Reproduced with permission from R. O'Rahilly and F. Muller, *Human Embryology and Teratology,* Wiley-Liss, New York, 3rd Edition, 2001; **17.26:** From A.T. Hertiz and J. Rock, *Carnegie Contrib. Embryology,* 29:127(1941)

Chapter 18
Opener: © Superstock; **p. 647:** © Superstock; **18.1:** © Manfred Kage/Peter Arnold, Inc.; **18.7b:** © Dr. Gopal Murti/SPL/Photo Researchers; **18.21b:** © Dr. Fred Hossler/Visuals Unlimited

Chapter 19
Opener: © Image Source Limited/Phototake; **p. 684:** © Image Source Limited/Phototake; **19.1a,b:** © Custom Medical Stock; **19.3 (top & bottom):** From Lawrence M. Tierney, *Current Medical Diagnosis and Treatment,* 2006; **19.4 (top & bottom):** © Dr. David Schwartz, New York University School of Medicine from Schwartz DT, Reisdorff EJ, *Emergency Radiology,* McGraw-Hill 2000

Index

Dehydration, 24, 274
Dehydroepiandosterone (DHEA), 322, 350, 600
Delayed hypersensitivity, 674
Delta cells (D cells), 572
Delta rhythm, 234
Dendrites, 138
Dendritic cells, 647, 649t, 676t
Denervation atrophy, 277
Dense bodies, 284
Deoxyhemoglobin (Hb), 463, 468
Deoxyribonuclease, 552t
Deoxyribonucleic acid. *See* DNA
Deoxyribose, 36
Dependence, and substance abuse, 244–45
Depolarization, of graded potentials, 149, 151f
Depo-Provera, 638
Depression, 243–44
Descending limb, 487, 515t
Descending pathways, 199f, 297–98, 308, 312
Desmodus rotundus salivary plasminogen activator (DSPA), 437
Desmosomes, 48, 49f
Detrusor muscle, 498
Development. *See also* Age and aging; Children; Fetus; Growth
    central nervous system, 174
    cortisol, 342
    thyroid hormones, 339
Developmental acclimatization, 13
Diabetes insipidus, 503
Diabetes mellitus. *See also* Insulin
    definition and description of, 578–80
    facilitated diffusion, 104
    hyporesponsiveness, 328
    reabsorption of glucose by kidneys, 496
    resistance to infections, 670
Diabetes nephropathy, 496
Diabetic ketoacidosis, 579, 580
Diacylglycerol (DAG), 130, 134t
Diagnosis
    chest pain after airplane flight, 689–90
    palpitations and heat intolerance, 686–87
    of substance dependence, 245t
Diapedesis, 651
Diaphragm, 445
Diarrhea, 561
Diastole and diastolic pressure (DP)
    cardiac cycle, 373, 374f, 375, 377
    defined, 386
    early phase of, 377
Diastolic dysfunction, 420
Diencephalon, 174
Diet. *See also* Food; Nutrients and nutrition
    cholesterol levels, 580–82
    essential nutrients, 91–92, 93
    diabetes mellitus, 580
    fiber, 536–37, 561
    hydrogen ions and high-protein, 518
    iodine, 340
    iron, 426
    obesity and weight loss, 589, 590t
    pregnancy, 633t
    vitamins, 92
Diffuse interstitial fibrosis, 461
Diffusion. *See also* Diffusion impairment
    capillary wall, 395–97
    of gases in liquids, 458–59
    kidneys, 495
    magnitude and direction of, 97–99, 116

medullary circulation, 506
membranes, 99–102, 116
Diffusion equilibrium, 97
Diffusion impairment, 476t
Diffusion rate, 99
Digestion, 529, 536–40, 561–62
Digestive system, primary functions of, 4t
Digitalis, 421t
Dihydropyridine (DHP) receptor, 262
Dihydrotestosterone (DHT), 318t, 603, 611–12
Diiodotyrosine (DIT), 337
Dimethoxymethylamphetamine (DOM), 244f
Dimethyltryptamine (DMT), 244f
2,3-Diphosphoglycerate (DPG), 466–67, 477
Diplopia, 684
Direction, of diffusion, 97–99
Disaccharides, 28, 536
Discs, of photoreceptors, 212
Disease. *See also* Pathophysiology; *specific diseases*
    cardiovascular system, 413–25
    hypoxic hypoxia, 476t
    metabolic rate, 584t
    stress, 344–45
    synaptic mechanisms, 166
Distal convoluted tubule, 487, 515t
Disulfide bond, 34–35
Disuse atrophy, 277
Diuresis, 503
Diuretics, 419, 420t, 421t, 523
Divergence, of presynaptic cells, 159
Division of labor, in kidneys, 514, 515t
DNA (deoxyribonucleic acid). *See also* Genetics
    gametogenesis, 601–602
    protein synthesis, 55–57, 61t
    RNA composition compared to, 38t
    structure of, 35–36, 37f
Docking proteins, 65
Dominant follicle, 618
Dopamine
    anterior pituitary, 334
    catecholamines, 167
    functions of, 318t
    lactation, 637
    molecular structure of, 244f
    motivation and reward systems, 240, 241
    synthesis, 317
Dopaminergic neurons, 166
Dorsal column pathway, 206, 207f
Dorsal horns, 177
Dorsal respiratory group (DRG), 469
Dorsal root ganglia, 177
Dorsal roots, 177
Double helix, 56
Down-regulation, of receptors, 122t, 123, 324
Down syndrome, 632
Drugs. *See also* Pharmacological effects; *specific drugs*
    AIDS and HIV, 671
    anticlotting, 436–37
    antidepressants, 243
    chemical mediators, 241
    cholesterol levels, 581
    coronary artery disease, 424
    diabetes mellitus, 580
    heart failure, 421t
    hypertension, 419, 420t
    incontinence, 499

nicotine, 187
perception, 201
pregnancy and fetal development, 632
psychoactive substances, 244
renin-angiotensin system, 509
synaptic transmission, 165–66
ulcers, 558–59
Dual innervation, 182–83
Duchenne muscular dystrophy, 280, 281f
Dumping syndrome, 551
Duodenal ulcers, 557, 558f
Duodenum, 532, 550
Dura matter, 185
Dwarfism, 348
Dynamic constancy, 6
Dyneins, 139, 140f
Dynorphins, 170
Dysmenorrhea, 624
Dyspnea, 476
Dystrophin, 280

E

Ear, and hearing, 217–22
Early diastole, 377
Eating disorders, 589–90
Eccentric contraction, 267
ECG leads, 371, 372f
Eclampsia, 633
Ectopic pacemakers, 370
Ectopic pregnancies, 628
Edema, 401, 403, 523
EEG arousal, 234
Effector, 10
Effector mechanisms, and body temperature, 593, 594t
Efferent arteriole, 487, 494f
Efferent division, 178
Efferent neurons, 139
Efferent pathway, 10
Egg, 615, 627–28. *See also* Ova
Eicosanoids, 131–33, 651t, 677t
Ejaculation, 610
Ejaculatory ducts, 605, 607f
Ejection fraction (EF), 380
Elastase, 552t
Elastic recoil, 448
Elastin fibers, 3
Elderly. *See* Age and aging
Electrical forces, and ion channels, 100
Electrical potential, 144, 146
Electrical synapses, 159–60
Electricity, basic principles of, 144
Electrocardiogram (ECG), 371, 372f, 422
Electrochemical gradient, 100, 148
Electroconvulsive therapy (ECT), 243–44
Electroencephalogram (EEG), 233, 235f, 238
Electrogenic pump, 148
Electrolytes, 21
Electromagnetic spectrum, 208
Electron(s), 19
Electron micrographs, 43, 44f, 45
Electron transport chain, 83
Elements
    atomic composition of body, 19t, 20
    ionic forms of in body, 22t
Elephantiasis, 401, 402f
Embolism, 424, 689
Embolus, 424

Gonadotropin(s), 600
Gonadotropin-releasing hormone (GnRH)
  anterior pituitary, 334
  control of testes, 611
  follicle development, 621
  functions of, 318t
  mechanisms of action, 600
  pregnancy, 633
  puberty, 626, 627
G protein(s), 127, 130–31
G-protein-coupled receptors, 127
Graafian follicle, 618
Graded potential, 149–57
Graft rejection, 672
Gram atomic mass, 20
Granulocyte colony-stimulating factor
  (G-CSF), 430
Granulosa cells, 617, 621, 623t
Graves' disease, 340–41, 686–87
Gravity, and posture, 310, 414, 415
Gray matter, 174, 175f
Growth. *See also* Development
  bone, 346, 351
  environmental factors, 347, 351
  follicles, 617–19
  metabolic rate, 584t
  hormonal influences on, 347–50, 351
  secondary sex characteristics, 612
  thyroid hormones, 339
Growth cone, 142
Growth factors, 319t, 347
Growth hormone, 319t, 347–49, 576
Growth hormone insensitivity syndrome, 348
Growth hormone-releasing hormone (GHRH),
  318t, 334, 348
Growth-inhibiting factors, 347
Guanine, 36, 37f, 38
Guanosine diphosphate (GDP), 82
Guanosine triphosphate (GTP), 82
Guanylyl cyclase, 126, 213, 610
Gustation, 224
Gymnastics, 578
Gynecomastia, 613
Gyrus, 174, 175f

H

HAART (highly active anti-retroviral therapy), 671
Habituation, 238
Hair cells, 221, 223
Harvey, William, 361
Hashimoto's disease, 340
H⁺-ATPase, 106
Hbuffer, 518
Head trauma, 250
Health. *See also* Diseases; Pathophysiology
  cardiovascular system, 413–25
  essential nutrients, 91
Hearing, and sensory system, 217–22, 227, 228
Hearing aids, 222
Heart. *See also* Cardiac muscle; Heart attacks;
  Heart disease; Heart failure; Heart rate
  anatomy of, 365–67, 383
  arterioles, 393t
  autonomic nervous system, 184t
  cardiac cycle, 373–78, 383
  cardiac output, 378–82, 383
  heartbeat coordination, 367–73, 383
  hormones, 318t

  measurement of cardiac function, 382, 383
  overview of cardiovascular system, 364t
Heart attacks, 422–24
Heartbeat coordination, 367–73, 383
Heartburn, 545
Heart disease, 371
Heart failure, 382, 403, 419–21, 424. *See also*
  Congestive heart failure
Heart murmurs, 377–78
Heart rate
  cardiac output, 378
  exercise, 416, 418, 419f
  heart failure, 421
  sequence of excitation, 368
Heart sounds, 377–78
Heart valve defects, 378f
Heat acclimatization, 594
Heat exhaustion, 596
Heat intolerance, 341, 684–88, 694
Heat loss and gain, 591, 593, 594t
Heat production, 592–93, 594t
Heat stroke, 596
Heavy chains, 259
Hedonistic appetite, for salt, 512
Helicobacter pylori, 558, 559
Helicotrema, 220
Helper T cells (CD4 cells)
  antigen presentation, 660–62
  antigen recognition, 663
  functions of, 649t, 658, 659f
  HIV, 671
  role of in immune responses, 676t
  viral infection, 669t, 671
Hematocrit, 360, 363
Hematoma, 431
Hematopoietic growth factors (HGFs), 429, 430t
Heme, 463
Hemochromatosis, 426, 540
Hemodialysis, 524–25
Hemodynamics, 362
Hemoglobin
  erythrocytes, 425–26
  iron, 540
  structure of, 36f
  transport of oxygen, 463–66, 468t
Hemolytic anemia, 678
Hemolytic disease of the newborn, 673
Hemolytic jaundice, 559–60
Hemophilin, 435
Hemorrhage, 413–14, 424, 428t
Hemostasis, 431–38
Henry's law, 458, 463
Heparin, 436, 437, 690
Hepatic functions, 532. *See also* Liver
Hepatic portal vein, 536
Hepatitis, 666
Hepatocytes, 553
Hering-Breuer reflex, 470
Heroin, 186, 245t
Hexoses, 28
High-density lipoproteins (HDL), 581–82
High-frequency fatigue, 273
Hippocampus, 176f, 243
Histamine
  biogenic amines as neurotransmitters, 167t
  defined, 677t
  edema, 403
  immediate hypersensitivity, 674

  inflammation of lungs, 478
  local inflammatory mediators, 651t
Histones, 55
Histotoxic hypoxia, 476
H⁺/K⁺-ATPase, 106, 546
Homeostasis. *See also* Homeostatic control
  systems
  defined, 6
  as defining feature of physiology, 6–7
  endocrine control of calcium, 352–56
  of energy in exercise and stress, 577–78, 582
  intracellular chemical messengers, 11–12
  processes related to, 13–15
Homeostatic control systems. *See also*
  Homeostasis
  components of, 9–11
  feedback systems, 8
  feedforward regulation, 9
  general characteristics of, 7, 9t
  kidneys, 518, 522
  resetting of set points, 8–9
Homeothermic temperature, 590
Homocysteine, 422–23
Homosexuality, 604
Horizontal cells, 214
Hormone(s). *See also* Endocrine system;
  Estrogen; Follicle-stimulating hormone;
  Gonadotropin-releasing hormone; Hormone
  replacement therapy; Insulin; Luteinizing
  hormone; Testosterone
  arterioles, 391
  blood and transport of, 323–24, 328
  control of secretion, 326–27, 329
  defined, 316
  gastrointestinal regulation, 541–43
  growth, 347, 351
  homeostasis, 11
  male reproductive functions, 611–13, 614
  mechanisms of action, 324–25, 329
  metabolic rate, 584t
  metabolism and excretion, 324, 328
  neurotransmitters, 167
  ovarian, 619
  pregnancy, 632–34
  smooth muscles, 287–89
  structure and synthesis, 317–23, 328
Hormone-receptor binding, 325
Hormone replacement therapy, 640
Hormone-sensitive lipase (HSL), 575
Hot flashes, 640
Human chorionic gonadotropin (hCG), 319t, 632
Human immunodeficiency virus (HIV), 671
Human physiology, scope of, 2. *See also*
  Physiology
Human placental lactogen (hPL), 319t, 633
Humoral hypercalcemia of malignancy, 355
Hydralazine, 678
Hydrocephalus, 186
Hydrochloric acid, 26, 530, 532, 545–48
Hydrogen. *See also* Hydrogen ions
  atoms, 21
  hemoglobin saturation, 466–67
  transport between tissues and lungs, 468–69
Hydrogen bonds, 23–24, 36t, 38f
Hydrogen ions, 26, 470–75, 480, 517–22
Hydrogen peroxide, 84, 652
Hydrolysis, 24, 260, 267t, 286
Hydrophilic molecules, 24

S

Saccades, 216
Saccule, 222, 223
Sacral nerves, 179f
Saiety signals, 587
Saliva, 530, 532t, 543
Salivary glands, 184t, 530, 531t
Salt appetite, 512, 516
Saltatory conduction, 155
Sarcomere, 257, 258
Sarcoplasmic reticulum, 261, 262–63, 286
Sarin, 167
Satellite cells, 256
Saturated fatty acids, 29, 581
Saturation
    messenger-receptor interactions, 122
    protein binding sites, 67–68
Scala tympani, 220
Scala vestibuli, 219–20
Schizophrenia, 242–43
Schwann cells, 139
Sciatic nerve, 179f
Sclera, 209
Scrotum, 605
Seasonal affective depressive disorder (SADD), 244
Secondary active transport, 104, 106–108
Secondary adrenal insufficiency, 344
Secondary hyperparathyroidism, 356
Secondary hypersecretion, 328
Secondary hyposecretion, 327
Secondary lymphoid organs, 655
Secondary oocyte, 616
Secondary peristalsis, 545
Secondary protein structure, 33–34
Secondary sexual characteristics, 600, 612
Secondary spermatocytes, 606
Second messengers, 126, 131, 134t
Secretin, 318t, 541–42, 552
Secretion
    antibodies, 664–65
    defined, 529
    of hormones, 326–27
    of insulin, 573–74
    of proteins, 63–65
Secretory phase, of menstrual cycle, 624
Secretory vesicles, 52
Segmentation, 555
Selective attention, 238–39
Selective estrogen receptor modulators (SERMs), 355, 640
"Self" antibodies, 678
Semen, 606
Semicircular canals, 222–23
Seminal vesicles, 605, 607f
Seminiferous tubules, 605
Semipermeable membrane, 110
Sensation, 192
Sensitization, and allergy, 674
Sensorimotor cortex, 298f, 304
Sensory information, 192
Sensory neglect, 239
Sensory pathway, 198
Sensory receptors, 139, 192–94, 202
Sensory system
    association cortex and perceptual processing, 200–201, 202
    chemical senses, 224–27, 228–29

defined, 192
hearing, 217–22, 228
neural pathways, 198–200, 202
primary sensory coding, 194–98, 202
sensory receptors, 192–94, 202
somatic sensation, 203–207, 228
vestibular system, 222–27, 228
vision, 208–17, 228
Sensory transduction, 192
Sensory unit, 194
Sepsis, 693
Septa nuclei, 176f
Septal defect, 377
Septic shock, 675, 692, 693–94
Serosa, 535–36
Serotonin, 167t, 168
Serotonin reuptake blockers, 168
Serotonin-specific reuptake inhibitors (SSRIs), 243
Sertoli cell(s), 608, 609f
Sertoli cell barrier, 608
Serum, 425
Set point, 7, 8–9
Sex chromatin, 603
Sex chromosomes, 602
Sex determination, 602–603
Sex differentiation, 603–604
Sex hormones, 349–50, 600
Sexual behavior, 604
Sexually transmitted diseases (STDs), 638
Sexual response, 627, 642
Shaft, of bone, 346
Shear stress, 392
Shivering, 7, 592
Shivering thermogenesis, 592
Shock, 414
Shortening velocity, 277
Short-loop negative feedback, 336
Short reflexes, 541
Short-term memory, 246
Shunt, 462
Sickle-cell anemia, 35, 428
Sigmoidoscopy, 557
Signal sequence, 63–64
Signal transduction pathways, 123–35
Sildenafil, 610
Single-fiber contraction, 266–71, 281–82
Single-unit smooth muscle, 289
Sinoatrial (SA) node, 368, 369–71, 378–79, 381t
Sinus, 630
Sjögren's syndrome, 543
Skeletal muscle
    arterioles, 393t
    characteristics of cells, 292t
    energy metabolism, 272–74, 282
    molecular mechanisms of contraction, 258–66, 281
    single-fiber contraction, 266–71, 281–82
    structure of, 255–58, 281
    types of fibers, 274, 282
    whole-muscle contraction, 274–79, 282
Skeletal muscle pump, 400–401
Skin
    arterioles and sympathetic nerves, 390, 393t
    autonomic nervous system, 184t
    immune defenses, 648
    temperature regulation, 593
Skin receptors, 203f
Sleep. *See also* Sleep apnea

consciousness, 234–35
metabolic rate, 584t
Sleep apnea, 235, 478, 479f
Sleep deprivation, 235
Sleep-wake cycles, 236
Sliding-filament mechanism, of muscle contraction, 258–61
Slow fibers, 274
Slowly adapting receptors, 198, 199f
Slow-oxidative fibers, 274, 275f, 276t
Slow waves, 287
Small intestine, 531t, 532, 542t, 554–56
Smell, 226
Smoking, of tobacco
    chronic obstructive pulmonary disease, 454
    heart disease, 422
    hypertension, 419
    physiological effects of, 186, 187
    pregnancy, 632
    protective respiratory reflexes, 476
Smooth endoplasmic reticulum, 51
Smooth muscle
    characteristics of cells, 292t
    contraction, 285–90, 292–93, 389
    defined, 255
    endothelial cells, 391–92
    lymphatic vessels, 402
    myogenic responses, 390
    structure of, 284–85, 292
    types of, 289–90
Smooth muscle tone, 287
SNAREs, 161, 166
Sneeze reflex, 475–76
Sodium
    active transport, 115
    digestion and absorption, 540
    filtration and reabsorption in kidney, 495t
    homeostasis, 7, 15
    ions, 20
    NA^+/K^+-ATPase pump, 105, 106f, 107
    renal processes for water and, 501–506, 515–16
    renal regulation of, 506–10
    resting membrane potential, 145–48
    thirst and salt appetite, 512
    total-body balance of water and, 500–501, 515
Solutes, 24
Solutions, 24–27, 111. *See also* Concentration
Solvent, 24
Soma, 138
Somatic motor neurons, 182
Somatic nervous system, 178, 180f
Somatic receptors, 200
Somatic sensation, 203–207, 228
Somatosensory cortex, 200, 206, 207f, 304
Somatosensory system, 206
Somatostatin (SS), 318t, 334, 348, 545
Somatotopic map, 304
Sound, and hearing, 217–21
Sound waves, 217, 233
Spasms, 309
Spasticity, 309
Spatial summation, 163
Specific ascending pathways, 199, 200f
Specific immune defenses, 646–47, 654–68, 679–80
Specificity, of receptors, 121–22
Spectrum, and light, 208–209
Sperm, 600, 608f, 609–10, 614, 627